Kiessling / Puschmann / Schmieder / Schneider
Contact Lines for Electric Railways

Authors:

Dr.-Ing. Friedrich Kiessling (chapters 2, 7, 10, 13)
Amundsenstrasse 9b, 14469 Potsdam, Germany
fkiess@josebus.org

Former Scientific Advisor and Head of Rail Electrification Engineering Department at Siemens AG, Transportation Systems. About 50 years of professional experience, convenor and member of CENELEC and IEC standardisation bodies, convenor of TSI working group for the Energy subsystem. Author of numerous publications on traction power supply systems.

Dipl.-Ing. Rainer Puschmann (chapters 3, 4, 12, 15, 16, 17)
Unterer Kirchenweg 11, 91338 Igensdorf, Germany
rainer.puschmann@expert-consultancy.com

Chief Expert at Expert Consultancy, former Head of Design, After Sales Services and Engineering Segments at Siemens AG, Transportation Systems, and SPL Powerlines Group. Certified expert for design reviews and acceptances at Railway Federal Authority and Railway CERT. Author of many publications on overhead contact lines and traction power supply systems.

Dr.-Ing. Axel Schmieder (chapters 1, 3, 9, 11, 14)
Mozartstraße 33b, 91052 Erlangen, Germany
axel.schmieder@siemens.com

Siemens AG, Mobility Division, Rail Electrification, Principal Key Expert and Independent Auditor for contact lines, former activities at Deutsche Reichsbahn and Deutsche Bahn for Railway Electrification. About 40 years experience in this field. Convenor and member of IEC, CENELEC and DKE standardisation groups. Author of books and numerous publications on rail electrification.

Dr.-Ing. Egid Schneider (chapters 5, 6, 8)
Am Mühlberg 8, 91085 Weisendorf, Germany
tue.Schneider@t-online.de

Former Senior Expert at Siemens AG, Transportation Systems. Experienced in design and development of traction power supplies for 25 years. Convenor and member of several CENELEC and IEC standardization bodies. Author of numerous publications on traction power supply systems.

Cover photo:
Overhead contact line type Siemens Sicat H1.0 and ICE 3 train on the high-speed line Frankfurt am Main to Cologne. The photo is presented with kind permission of Deutsche Bahn.

Contact Lines
for Electric Railways

Planning
Design
Implementation
Maintenance

by Friedrich Kiessling, Rainer Puschmann,
Axel Schmieder and Egid Schneider

Third, completely revised
and enlarged edition, 2018

Publicis Publishing

The Deutsche Nationalbibliothek lists this publication in the Deutsche Nationalbibliografie;
detailed bibliographic data are available in the Internet at http://dnb.d-nb.de.

www.publicis-books.de

Print ISBN 978-3-89578-420-0
Print ISBN 978-3-89578-961-8

Third, completely revised and enlarged edition, 2018

Publisher: Publicis Publishing, Erlangen, Germany
© 2018 by Publicis Pixelpark Erlangen – eine Zweigniederlassung der Publicis Pixelpark GmbH

Printed in Germany

Foreword to the third English Edition

The book "Contact Lines for Electric Railways" has attained the status of a standard text book for contact line engineering worldwide. As it is some years since the second edition went out of print, Siemens AG, Mobility Division, Rail Electrification, and SPL Powerlines Group GmbH are pleased to present this completely revised English edition to the international audience. We are confident that this work will meet the requirements of engineers in railway operating entities, contracting and engineering companies, as well as supervising bodies. It has also been designed as a teaching and learning aid.

As noted in Werner Breitling's Foreword to the second English edition, contact lines are an essential component of electric traction supply systems. They are the interface between the fixed installations and moving energy consumers that are the electric traction units. Contact line design needs to be adapted to the technical and operational requirements of individual electric railway systems, including low-voltage DC systems or medium-voltage AC systems. It is a valuable resource from tramway applications running at relatively low speeds to high-speed trains in a plethora of climatic environments and topographic conditions, including in the open, in long tunnels under the sea or through mountains. The importance of the overhead contact line as a basis for ecologically beneficial transport of persons and goods continues to grow.

The structure and contents of this book reflects the authors objectives and considers all aspects of contact lines:

- an overview of different kinds of power supply systems feeding electrically driven trains
- an introduction to the development of energy transmission to railways from the beginnings up to today's high-speed traffic
- the electrical and mechanical aspects including theoretical considerations, electric and mechanical calculations and design of flexible overhead contact lines and rigid contact rails
- the theory of interaction between contact lines and pantographs, especially at high speeds and its verification by measurements
- necessary measures to protect passengers and the public from electrical dangers
- detailed descriptions and examples of components
- comprehensive presentation of contact line planning
- construction and maintenance of contact lines
- examples describe various possibilities for achieving required solutions
- relevant IEC and Cenelec standards that have replaced former national standards are listed in an Annex

The authors are active or former employees of Siemens AG, Rail Electrification, or of Powerlines Group GmbH, formed out of the former Siemens contact line activities in Austria and Germany. The authors describe the technology represented by both companies and the contents of the book describes international state-of-the-art technology. The presented examples demonstrate the potential of both companies. The authors conceived the book and prepared the draft under their own initiative. Siemens AG and Powerlines Group GmbH supported its drafting and printing. As the responsible executives of these entities, we are pleased that the revised third English edition has been completed. We hope this edition has similar success as its predecessors.

Erlangen/Wolkersdorf,
October 2017

Dr.-Ing. Elmar Zeiler
Head of Rail Electrification
Siemens AG, Mobility Division

Mag. Gerhard Ehringer
Chief Executive Officer
Powerlines Group GmbH

Preface to the first English edition

The first edition of "Fahrleitungen elektrischer Bahnen" (Contact Lines for Electric Railways) was published in German in 1997 by B.G. Teubner-Verlag Stuttgart. The first edition was out of print quickly, so a second, revised edition was published in 1999. The co-authors of this book, Professor Dr. sc Anatoli Ignatjewitsch Gukow and Dr. sc. Peter Schmidt, died unexpectedly in 1999 and 2000, respectively. Both had essential roles in the production of the German edition.

There were no comparable works available and the book enjoyed wide distribution and attracted great interest, even in non-German-speaking countries instigating the need for translations in other languages.

Prior to the first English edition, substantial parts of the book were revised and adapted to include international overhead contact line designs. The revisions were based on international standards as published by IEC and EN.

Advice and comments from readers were also incorporated. More attention was paid to 50 Hz railways and local public transportation systems. New calculation methods, up-to-date examples of completed electrification projects and recently developed overhead contact line components have also been included.

The aims of the book are explained in the preface to the first edition, which also appears in this edition. The world-wide spread of high-speed railway systems, the need to ensure inter operability and the expansion of local public traffic systems are intensifying the demands made on electric railways, the qualifications of staff involved and supporting documentation. So, this edition especially aims to describe the theoretical principles underlying overhead contact lines and to offer possible solutions for their application, whilst taking current international developments in this complex field into consideration. At the same time, the book is intended as a co-operative contribution with projects carried out in parts of the world where German is not spoken.

The authors would like to thank the Transportation Systems Electrification Department of Siemens AG and especially the heads of this department, Dr. Werner Kruckow and Peter Schraut, who supported the preparation of the English edition. Beat Furrer of Furrer & Frey AG, Bern, Switzerland sponsored the preparation of the manuscript. The authors also thank the publishing company for its excellent technical facilities. Thanks are extended to Gernot Hirsinger for preparing the translation as well as Bela Jozsa, Norm Grady, Terry Wilkinson, John Allan and Jan Liddicut from Melbourne/Australia, who edited the English version and ensured that the complicated subject matter was understandable to English speaking readers. The authors thank Dr. Wilhelm Baldauf of Deutsche Bahn AG and Dr. Egid Schneider at Siemens AG, who supported them with contributions to certain sections of this revised edition, and Michael Schwarz for desk top editing.

The authors hope this book will promote co-operation amongst colleagues working in this field in as many countries as possible, and that it will contributes to their mutual technical understanding. They look forward to readers' comments and their advice on the content and design of the book.

Erlangen, September 2001   *Friedrich Kießling, Rainer Puschmann, Axel Schmieder*

Preface to the second English edition

The first English edition of this book was based on the second German version printed in 1998 and published in 2001. This first English edition was well accepted by railway companies, manufactures, contractors, consultants and at universities and encouraged translations into Chinese in 2003 and Spanish in 2008. Meanwhile the English edition sold out so the publisher commissioned the authors to prepare a revised edition of the book to be published in 2009.

The authors of the first English edition invited Egid Schneider from Siemens AG to participate as co-author. He accepted the offer and revised the chapters dealing with electrical aspects of contact line systems. Like the first English edition, this book addresses the requirements of an international audience so the individual chapters of the first edition were revised and enlarged. The design of contact lines has been affected by the European Directives on Interoperability and the related Technical Specifications. These documents initiated the preparation and issuing of several new European Standards which have also been considered in this new edition.

Since 2001, new high-speed rail networks have been introduced in a number of countries and others enlarged. In Europe high-speed operation commenced in Austria, the Netherlands and Switzerland. The networks in France, Germany, Italy and Spain were extended. In China, high-speed lines designed for commercial speeds of 350 km/h were commissioned. In Switzerland, a new tunnel, through the Alps, designed for high speed operation was finished and put into service. For these applications, contact lines were designed anew and or adapted from previous designs. Examples of new lines, new materials and components that were introduced to the market are dealt with in the text.

To present a comprehensive book on the subject matter the authors needed and received assistance and support from professional partners. In this context the authors would like to thank

- Jörg Schneppendahl and Roland Edel, Siemens AG, Erlangen, Railway Electrification, Johannes Emmelheinz and Daniel Leckel, Siemens AG, Erlangen, Railway Maintenance for sponsoring the revision and publication of the book,
- Gerhard Seitfudem of Publicis Corporate Publishing, Erlangen, for the excellent processing of the book including four-colour printing,
- Bela Jozsa who was assisted by Terry Wilkinson, Norm Grady and Jan Liddicut of Melbourne/Australia, who checked the draft and ensured that the complicated subject matter would be understood by English speaking readers,
- Christian Courtois, Paris, Hans-Herbert Meyer, Bergheim, Albrecht Brodkorb and Wieland Burkert, Erlangen, for their contributions,
- those readers who commented on errors and proposed amendments,
- Michael Schwarz and Jörg Unglaub, Erlangen, who took care of the desk top editing.

The authors hope that the revised new edition meets the requirements of people world wide who are active in the contact line business world wide. The authors welcome comments on the contents and usefulness of this book.

Erlangen, March 2009

Friedrich Kiessling, Rainer Puschmann, Axel Schmieder, Egid Schneider

Preface to the third English edition

The second English edition of this book was printed in 2009. It was based on the first English edition published in 2001 and followed by Chinese and Spanish editions published in 2004 and 2008, respectively. The Spanish version drafted mainly by Tomas Vega will be followed by a Russian edition currently in preparation. All editions have been well received by the international electrification community. A third German edition was published in 2014 by Publicis Publishing, Erlangen as a Siemens technical book as were the two previous English editions. The second English edition has been sold out for some years, so the publisher and the electrification entities Siemens and SPL Powerlines asked the authors to write a new English edition for the international market. All the authors responsible for the 2009 edition were able and willing to contribute to this third English edition.

Like the previous editions the book meets the requirements of an international audience. Compared with the previous edition, there are 17 chapters instead of 14 with subjects including contact line protection, components, implemented installations, management and maintenance dealt with in separate chapters. Chapters 1 to 8 deal with basic aspects of contact line design, chapters 9 to 14 are devoted to design and planning and chapters 15 to 17 are dedicated to installation, worked examples, management and maintenance.

With the publication of directive 96/49/EC in 1996, the replacement of national standards by European Cenelec or international IEC standards commenced. Since then, standards relevant to contact lines have been in continuous development. Since the publication of our last edition, the contact line standard EN 50119 has been amended twice, in 2009 and 2013. Currently EN 50119 is under revision and a new version will be published in 2019. Based on EN 50119 a new version of IEC 60913 was published in 2013. Also the EN 50122 and IEC 62128 standard series dealing with protective provisions and safety during maintenance in fixed installations have been revised. Because of lead times involved, any book needs to be drafted well in advance of printing and it is not feasible to always consider the most recent changes in the standards. Therefore, this edition is based mainly on standards published in the first half of 2017.

Since the last edition, remarkable progress has been made in electrically powered railways. In China, several new high-speed lines have been added to the worlds largest high-speed network, whilst in Europe, in countries such as The Netherlands, Spain and Turkey the high-speed network was been greatly extended. In Switzerland, the Gotthard Base Tunnel was completed and commercial services begun. In Austria, in the Vienna area, railway traffic was restructured with a new Central station and a tunnel under the Vienna forest. In Denmark, the electrification of a 1 300 km long network commenced, with the first section being commissioned in 2017. In Germany, the high-speed Leipzig–Erfurt line was commissioned as an important link between Berlin and Munich. In the UK the Great Western and Midland Main lines are under construction.

More and more cities are extending existing tram lines, metro systems, city railways and trolley bus networks utilising new materials for the overhead contact lines.

For all these lines modified or new contact line designs were required. Information on these projects can be found in the book.

To be able to prepare and present a complex book on a technical subject such as this, the authors required personal and financial assistance from professional partners. In this regard the authors would like to thank:

- Dr.-Ing. Elmar Zeiler, Siemens AG, Mobility Division, Rail Electrification Segment, Erlangen, and Magister Gerhard Ehringer, Powerlines Group GmbH, Wolkersdorf, for sponsoring the preparation and printing of this third edition of the book
- Andreas Thon, Florian Beulcke, Siemens AG, Erlangen, and Ralf Hickethier, SPL Powerlines Germany GmbH, Forchheim, for their continuous interest in the progress of the drafting
- Dr. Gerhard Seitfudem, Publicis Publishing; Erlangen, for the excellent production of the book including four-colour printing
- all readers who commented on errors and proposed amendments
- Bela Jozsa as principal editor and his co-editors Jan Liddicut, Norm Grady and Terry Wilkinson, all from Australia, who checked the draft and ensured that the difficult subject would be understood by the English speaking audience
- Michael Schwarz, Erlangen, and Thomas Nickel, Forchheim, who took care of the desk top publishing

The authors hope that, as with previous editions, this book will fill the needs of professionals worldwide, practicing in the contact line industry. The authors also welcome any recommendations for improvements and comments on errors.

Potsdam, Igensdorf,
Erlangen, Weisendorf
October 2017

Friedrich Kießling, Rainer Puschmann,
Axel Schmieder, Egid Schneider

Contents

1 Power supply systems

1.0 Symbols and abbreviations

Symbol	Definition	Unit
4QC	four quadrant chopper	–
ACLR	Automatic overhead Contact Line Re-closing	–
ACLT	Automatic overhead Contact Line Testing	–
ACLRT	Automatic overhead Contact Line Residual voltage Testing	–
ADIF	Administrador de Infraestructuras Ferroviarias	–
AED	Automatic Earthing Device	–
AGP	feeder-related testing device used in 15 kV substations	–
ASD	Automatic Sychronisation Device	–
BC	Braking Chopper	–
BSS	Block-type Substation	–
CCS	Control and Communication System	–
CNCC	Central Network Control Centre	–
CRCS	Central Rotating Converter Station	–
CSCS	Central Static Converter Station	–
CP	Coupling Post	–
DB	German Railway	–
DCR	Decentral pulsating Rectifier	–
DCS	Decentral Converter Station	–
DMM	Digital Meter Monitoring	–
DRCS	Decentral Rotating Converter Station	–
DSCS	Decentral Static Converter Station	–
FM	Field Module	–
IGBT	Insulated Gate Bipolar Transistor	–
JPP	Joint Power Plant	–
LCU	Local Control Unit	–
L1 (2, 3)	phase conductor 1 (2, 3)	–
M	moment	Nm
MMDC	Modular Multilevel Direct Converter	–
MNCC	Main Network Control Centre	–
NCC	Network Command Centre	–
NF	Negative Feeder	–
ÖBB	Austrian National Railway	–
OBB	Operating Busbar	–
OCL	Overhead Contact Line	–
OCLD	Overhead Contact Line Disconnector	–
OS	Operating Signal	–
P	power	kW
PC	Potential Connection	–

Symbol	Definition	Unit
PCN	Potential Connection of Non-power equipment	–
PP	Power Plant	–
RC	Return Current	–
RCBB	Return Current Busbar	–
RCC	Return Conductor Cubicle	–
RCM	Remote Control Module	–
RF/BM	Rectifier with Battery Management	–
RSS	Rectifier Substation	–
S_K''	short circuit power of the three-phase network	MVA
S_l	traction power	MVA
SBB	Swiss National Railway	–
SC	Short Circuiter	–
SCD	Short-Circuiting Device	–
SCS	central Static Converter Station	–
SD	Switching Device	–
SG	Synchronous Generator	–
SG	Switch Gear	–
SM	Synchronous Motor	–
SMC	Switching Control Centre	–
SP	Switching Post	–
SS	substation	–
ST	railway station	–
SW	switching station	–
TBB	Testing Busbar	–
TEN	Trans European railway Network	–
TC	Telecommunication	–
TL/TR	Transmission Line and Transformer	–
U_i	inverse voltage	V
U_n	nominal voltage	V
$U_{\max 1}$	highest permanent voltage	V
$U_{\max 2}$	highest non-permanent voltage	V
$U_{\max 3}$	highest overvoltage	V
$U_{\min 1}$	lowest permanent voltage	V
$U_{\min 2}$	lowest non-permanent voltage	V
VPN	Virtual Private Network	–
Zes	DB's master network control centre	–
$\mathrm{d}i/\mathrm{d}t$	slew rate	A/s
f	frequency	Hz
n	number of revolutions	1/s, 1/min
n_u	multiple of nominal voltage	–
p	number of pole-pairs	–
u_U	voltage inbalance	%

1.1   Functions of traction power supply

Electric traction has the function of safely transporting people and/or goods with the aid of electrified traction lines. The objective of the *traction power supply* is to ensure uninterrupted, reliable and safe operation of the electric traction vehicles. Technically, the traction power supply comprises the total of the fixed installations of the electric traction system [1.1, 1.2, 1.3]. The traction power supply can be subdivided into *traction power generation*, *traction power transmission*, *traction power feeding*, *power collection* by electric traction vehicles and *return current circuit*. The supply of moving consumers through contact lines represents the significant difference between electric traction systems and the public grid. In most cases, the traction power is supplied from the public power supply networks to the traction substations along the railway lines.

At German Railway (DB) and four other European railways, the traction power is supplied by AC 110 kV or 132 kV 16,7 Hz transmission lines as part of a HV traction power supply grid. The *traction power distribution* is performed by the *traction power substations* and overhead contact lines.

This book is dedicated to *contact lines* which form the *traction power feeding* system. Contact line systems can be subdivided into:

 – overhead contact line installations
 – third rail installations
 – overhead conductor rail installations

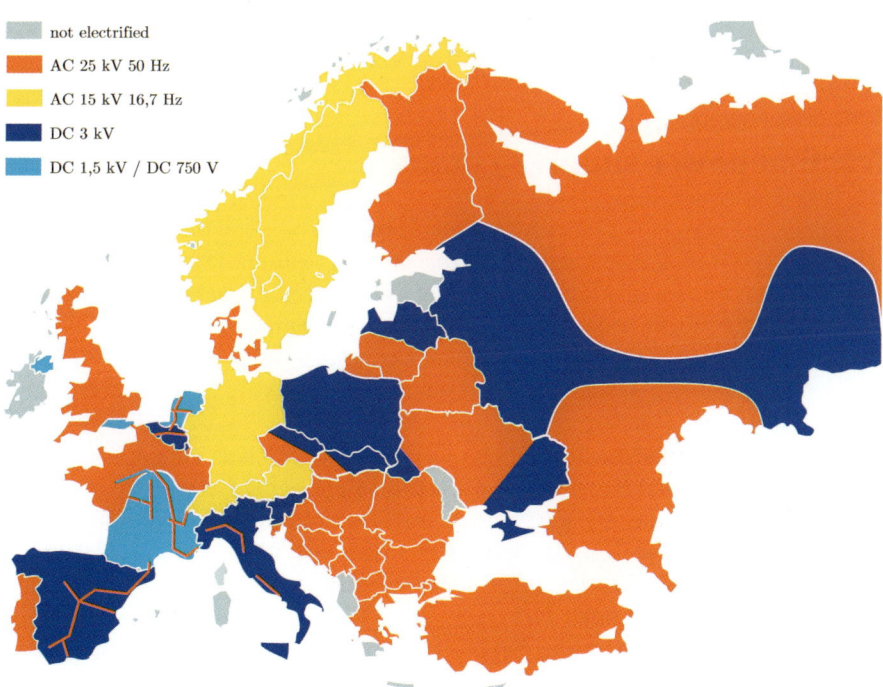

not electrified
AC 25 kV 50 Hz
AC 15 kV 16,7 Hz
DC 3 kV
DC 1,5 kV / DC 750 V

Figure 1.1: Traction power systems for mainline railways in Europe (situation 2014, see also [1.4]).

Table 1.1: Power supply systems of electrical operated main line and regional railways in line kilometer [1.11].

Continent		DC			AC			
		1,5 kV	3,0 kV	Other	15 kV 16,7 Hz	25 kV 50 Hz	Other	Total
Europa [1]	km	7 510	61 440	4 943	36 735	63 858	553	175 039
	%	4	35	3	21	37	0	100
Asia and	km	8 700	3 800	107	0	51 666	4 628	68 901
Australia	%	13	6	0	0	75	7	100
America	km	0	1 956	304	0	441	815	3 516
	%	0	56	9	0	13	23	100
Africa	km	62	7 148	0	0	3 716	861	11 787
	%	1	61	0	0	32	7	100
World	km	16 272	74 344	5 354	36 735	119 681	6 857	259 243
	%	6	29	2	14	46	3	100

[1] Asian part of Russia included

To comply with the requirements for reliable operation of electric traction, the following criteria are applicable, particularly with regard to contact lines:

- the provision of uninterrupted traction power at the pantographs of the traction vehicles
- the ability of the network to continuously absorb regenerated braking energy
- compliance with specified and standardized quality parameters for the voltages available at the pantographs of electric traction vehicles

In addition, special consideration needs to be given to *electrical loads on traction systems* as they differ from the loads on the public energy supply grid because they are not only heavily dependent on time but also continuously varying in location of consumption.

1.2   Traction power supply systems

The type of voltage and current is generally used to distinguish between the various types of electrical energy supply for electric traction. Originally, direct current was used for electric traction. The reason for this was the extremely favourable, hyperbolic traction/speed diagram of the series commutator motors used as drives in railway applications.

Globally, approximately one third of all electric traction systems still use *direct current*. The low voltage used existing direct current traction systems necessitates high currents to transmit the required traction power.

At the beginning of the twentieth century, efforts were made to combine the traction advantages of the series motor with the transforming capability of *alternating current*. At that time, the objective was a single-phase AC series motor drive, which was to be fed with single-phase AC at the frequency of the public grids, in Europe that was 50 Hz. Because of the state of technical development at that time several problems arose with this frequency, which could not be solved satisfactorily at that time:

- the heavy commutator wear of the 50 Hz single-phase series motor by a frequency-proportional induced voltage in the single brush winding
- the high and frequency-proportional, inductive interference in cables running in parallel to the electric traction system

Table 1.2: Length of power supply systems of European electrically operated main-line and regional railways [1.11].

Country	DC			AC			
	1,5 kV	3,0 kV	Other	15 kV 16,7 Hz	25 kV 50 Hz	Other	Total
Austria	0	0	0	3 526	0	84	3 610
Belarus	0	0	0	0	874	0	874
Belgium	0	2 647	0	0	303	0	2 950
Bosnia	0	0	0	0	590	0	590
Bulgaria	0	0	0	0	2 880	0	2 880
Croatia	0	133	0	0	1 066	0	1 199
Czech Republic	24	1 764	0	1	1 306	0	3 095
Denmark	172	0	0	0	458	0	630
Estonia	0	133	0	0	0	0	133
Finland	0	0	0	0	3 067	0	3 067
France	5 904	0	63	59	9 138	0	15 164
Georgia	37	0	1 562	0	0	0	1 599
Germany	0	0	505	19 231	0	0	19 736
Great Britain	19	0	2 014	0	3 345	0	5 378
Greece	0	0	0	0	764	0	764
Hungary	0	0	0	0	2 858	0	2 858
Ireland	49	0	0	0	0	0	49
Italy	181	12 453	32	0	407	0	13 073
Latvia	0	0	257	0	0	0	257
Lithuania	0	0	0	0	122	0	122
Luxembourg	0	19	0	0	243	0	262
Macedonia	0	0	0	0	223	0	223
Monaco	0	0	0	0	2	0	2
Montenegro	0	0	0	0	169	0	169
Netherlands	0	0	0	0	291	0	291
Norway	0	0	0	2 700	0	0	2 700
Poland	0	11 799	0	0	0	0	11 799
Portugal	25	0	0	0	1 411	0	1 436
Rumania	0	0	0	0	3 292	0	3 292
Russia [1]	0	18 800	0	0	21 500	0	40 300
Serbia	0	0	0	0	1 196	0	1 196
Slovenia	0	1 006	0	0	0	0	1 006
Slovakia	42	807	6	2	758	0	1 615
Spain	746	6 950	186	0	1 347	0	9 229
Sweden	65	0	0	7 531	0	0	7 596
Switzerland	246	0	317	3 685	0	469	4 718
Turkey	0	0	0	0	1 928	0	1 928
Ukraine	0	4 930	0	0	4 320	0	9 250

[1] Asian part of Russia included

- the high values of voltage asymmetry in the 50 Hz three-phase network supply caused by the traction power single-phase supply
- the public power networks were not strong enough to provide the power for the railway electrification

In Germany, development efforts led to *single-phase AC supply* with a frequency $50 \text{Hz}/3 = 16 \, 2/3 \, \text{Hz}$, where the electrical energy is generated and distributed as single phase in a separate railway high-voltage network. Three German administrations introduced this type of traction power during the years 1912/1913 [1.5, 1.6]. This power supply was independently introduced in Austria, Switzerland, Norway and Sweden and has proven to be powerful and

Table 1.3: Nominal voltages and permissible limits according to EN 50 163,
Tables 1 and A.1.

Type of power supply	U_{min2} V	U_{min1} V	U_n V	U_{max1} V	U_{max2} V	U_{max3} V
DC 600 V		400	600	720	800	–
DC 750 V		500	750	900	1 000	1 270
DC 1,5 kV		1 000	1 500	1 800	1 950	2 540
DC 3,0 kV		2 000	3 000	3 600	3 900	5 075
AC 15 kV 16,7 Hz	11 000	12 000	15 000	17 250	18 000	24 300
AC 25 kV 50 Hz	17 500	19 000	25 000	27 500	29 000	38 750

U_n nominal voltage
U_{min1} lowest permanent voltage
U_{min2} lowest non-permanent voltage, duration between U_{min1} and U_{min2} 2 min
U_{max1} highest permanent voltage
U_{max2} highest non-permanent voltage, duration U_{max1} and U_{max2} 5 min
U_{max3} highest overvoltage, duration not more than 20 ms

effective, also for the electrical power supply of high-speed and high-capacity traffic. In 2000 Austrian, German and Swiss railways changed their nominal frequency from 16 2/3 Hz to 16,7 Hz [1.7] to [1.9]. Since then, the relevant standards such as EN 50 163 and EN 50 188 have used the latter value as nominal frequency. Therefore, 16,7 Hz is used throughout this book to characterize the Central and Northern European power supply.

Initial experience with an *AC 50 Hz traction power supply* was gained at the Höllentalbahn in Germany starting in 1940 [1.10]. As a result of the enormous progress made since in the field of power electronics, AC 25 kV 50 Hz is currently preferred in countries now starting to electrify their railways or new high-speed lines, where DC power supply was used before. The same conclusion applies to countries with 60 Hz as frequency of public power supply.

The most common traction power systems are:
- direct current with 0,6 kV, 0,75 kV, 1,5 kV and 3 kV
- alternating current with 16,7 Hz 15 kV
- alternating current with 50 Hz or 60 Hz 25 kV

According to statistics published in [1.11] at the end of the year 2010, the length of electrified railway networks worldwide was approximately 260 000 km, the distribution of which is listed in Table 1.1. Table 1.2 and Figure 1.1 show power supply systems in several European countries. Additional information on electrification systems in Europe is given in [1.4].

Other traction power supply systems are DC 0,8 kV; DC 0,85 kV; DC 0,86 kV; DC 0,9 kV; DC 1,0 kV; DC 1,125 kV; DC 1,2 kV; DC 1,25 kV; DC 1,35 kV and DC 1,3 kV, having in total a portion of 2,1 % on electrified lines worldwide, and AC 11 kV 16,7 Hz; AC 6 kV 25 Hz; AC 6,6 kV 50 Hz; AC 20 kV 50 Hz; AC 50 kV 50 Hz; AC 20 kV 60 Hz and AC 25 kV 60 Hz,

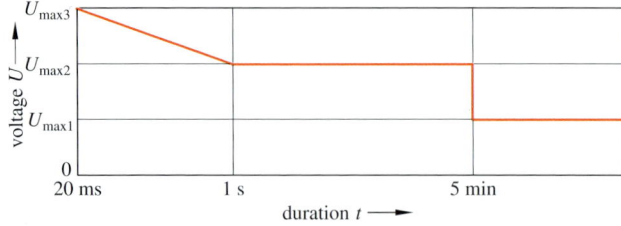

Figure 1.2: Maximum value of voltage U according to duration. Source: EN 50 163, Figure A.1

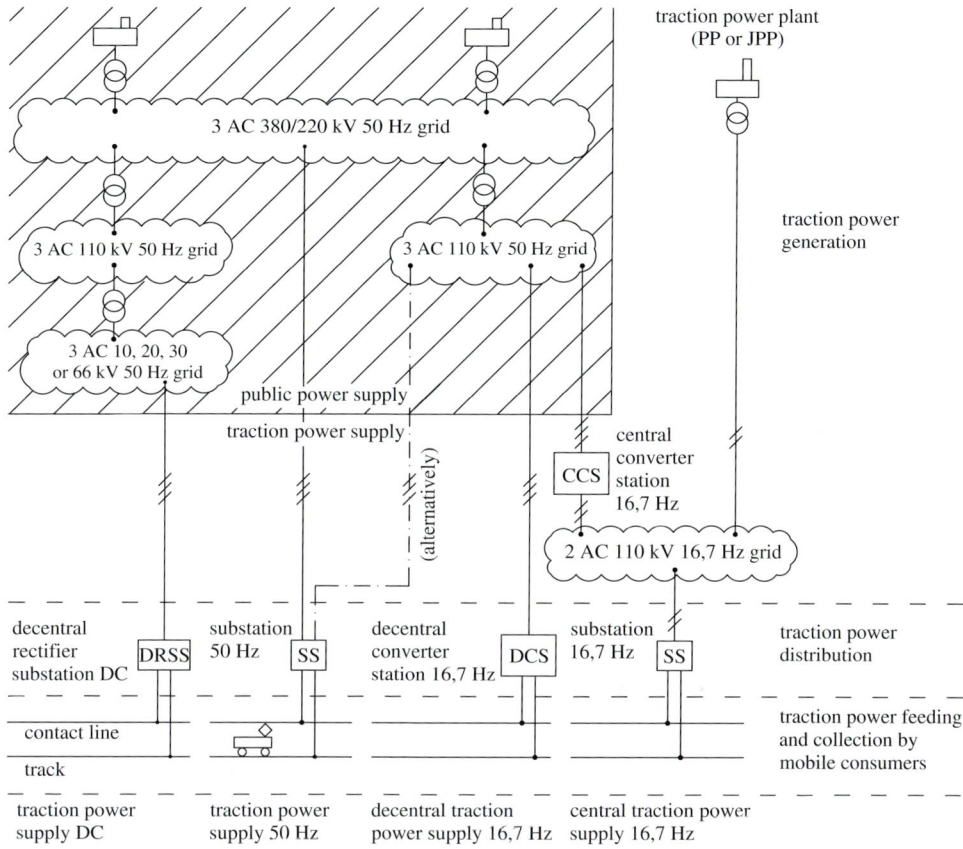

Figure 1.3: Structure of traction power supply systems [1.14].

having in total a portion of 2,6 % on electrified lines worldwide. Urban mass transit lines use mainly DC 0,6 kV, 0,75 kV or 1,5 kV.

In Table 1.3 the nominal voltages according to EN 50 163 are shown with their operational limits. Under normal operational conditions the voltages at the pantograph have to be between $U_{\min 1}$ and $U_{\max 2}$. The duration of the voltages between $U_{\min 1}$ and $U_{\min 2}$ should not exceed two minutes. In addition the duration of the voltages between $U_{\max 1}$ and $U_{\max 2}$ should not exceed five minutes.

Power supply selection and design of AC or DC railway require comprehensive studies and analysis based on line, vehicle and operational data. They comprise the electric rating of substations and contact lines with:

    – comparisons of systems and alternatives
    – economic design of systems
    – planning of power surplus for future extensions
    – calculation of energy consumption
    – determination of investment

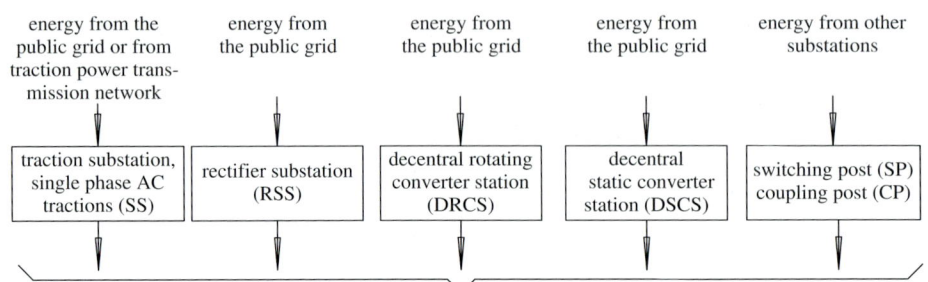

supply of the contact line network through busbars and circuit breakers

Figure 1.4: Types of substations used in traction power supply.

For design of power supply systems simulation programs like Sitras Sidytrac are used [1.12]. The interaction between power supply systems and vehicles can be verified in the Siemens testing and validation centre Wegberg-Wildenrath (PCW) [1.13]. There, all convebtional traction power supply types are available.

1.3 Basic structure of traction power supply

1.3.1 Traction power provision and transmission

Figure 1.3 depicts the common types of *traction energy generation* and provision and their connections to the public grid. With DC and AC 50 Hz single-phase traction systems, the traction energy is drawn from the public grid. The AC 15 kV 16,7 Hz contact lines are fed by a substation supplied by an AC 16,7 Hz single-phase transmission network or by:
- railway owned power plants and transmission in special grids
- central rotating or static converter stations fed by the public grid and transmission in special grids
- decentral rotating or static converter stations fed by the public network and feeding directly the contact lines

As illustrated in Figure 1.3, the DC traction systems are supplied from the three-phase network with nominal voltages between 10 kV and 66 kV. Single-phase AC 50 Hz traction systems are usually connected to grids with 110 kV or higher voltages.

1.3.2 Traction power distribution

The *traction power distribution* converts electrical energy supplied to substations into voltages and frequencies conforming with the nominal values used for traction power and the supply of this power to consumers.

Substations (SS) of various types are used to supply traction power directly into the contact line installations. As indicated in Figure 1.4, there are:
- *power transformer stations*, commonly referred to as *traction substations* (SS), which convert the voltage from the transmission network, at nominal frequency, into the nominal voltage of the contact line network as single-phase AC and supply the network with traction power

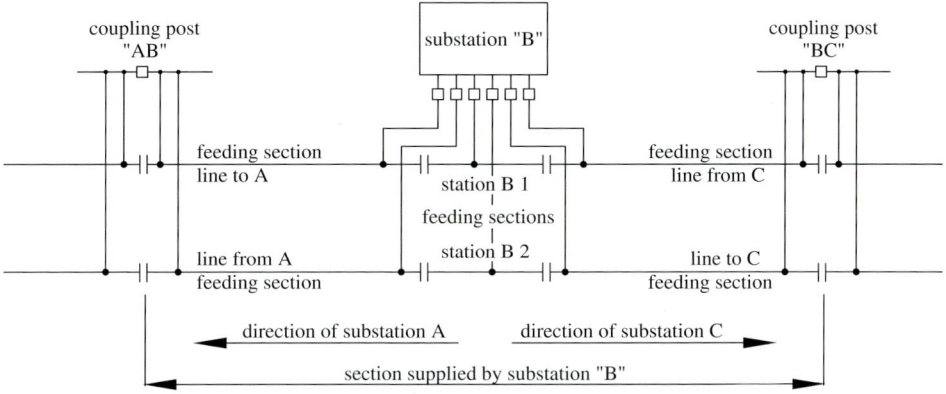

Figure 1.5: Mainline section supplied by a substation.

- *traction power rectifier substations* (RSS), which convert the applied AC three-phase electricity from the public grid into the required nominal voltage of the contact line network for direct current railways and supply that to the contact line installation
- *decentral rotating converter stations* (DRCS), in which the three-phase energy of the AC 50 Hz public grid is converted with the aid of rotating machines into AC 16,7 Hz single-phase energy for the traction network and supplied to the contact line network after conversion to the corresponding nominal values
- *decentral static converter stations* (DSCS), which have the same function as the DRCS but by means of electronic power components instead of rotating machines
- *switching posts* (SP), also switching centres and *coupling posts* (CP), which have the task of receiving the electrical energy from other substations with characteristics according to the supply system and feeding the contact line network or interconnecting different sections of contact lines and switching these sections on or off. They reduce the voltage drops and energy losses within the contact lines. By the contact line they connect substation ranges or supply lines from one side. Therefore, selective protection of contact lines and the return of breaking energy into other line sections with consumers or substations is achieved

An example of power distribution by a substation for mainline traction is shown in Figure 1.5. The function of the substation is to secure the supply of electrical energy to all trains passing through the substation supply section. The *substation supply section*, also known as the *feeding section*, designates the total of all contact line sections supplied by a substation in normal operation mode.

A *neutral section* is a contact line section which isolates adjacent feeding sections in such a way that they cannot be bridged by the pantographs of electric traction units.

Switching sections and *switching circuit groups* within the substation supply sections can be electrically separated by air insulated overlaps or section insulators, which are bridged by disconnectors during normal operation and may be bridged by the pantographs of the traction vehicles.

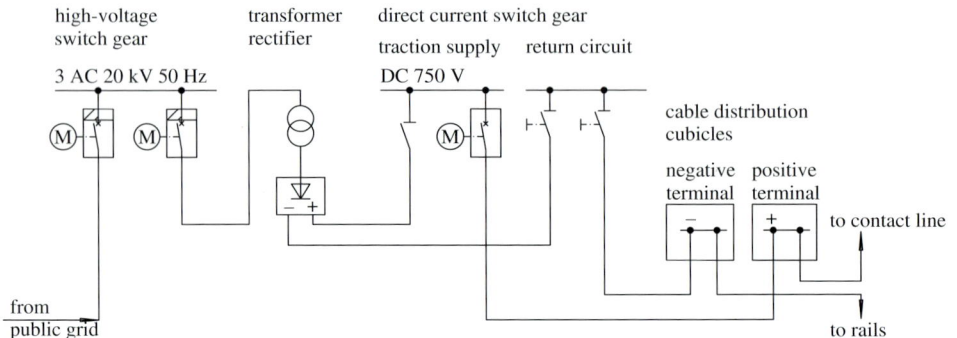

Figure 1.6: Example for direct current traction power supply system of a tramway.

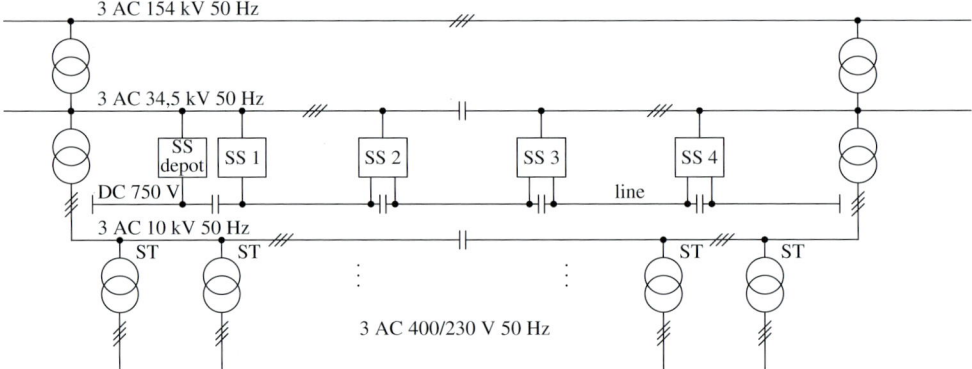

Figure 1.7: Traction power supply and railway station supply for the light rail system LRT Ankaray.

1.4 Direct current traction networks

1.4.1 General

Globally, approximately one third of all electric railways use *direct current traction*. In mass transit systems, maximum nominal voltages up to 1 500 V are used. The most common voltages are 750 V and 600 V. The distance between substations varies from 1,5 to 6 km. On some long-distance DC 1 500 V and DC 3 000 V railways the substation spacing can be up to 20 km. The power rating of direct current substations varies up to 3 MW for tramways and up to 20 MW in mass transport and main line systems.

The three-phase voltage supplied from the three-phase public grid is converted at the *rectifier substation* into direct current at the nominal voltage of the contact line network. Previously, six-pulse current converters were used, but now, mainly twelve- or twenty-four-pulse converters are used for rectifiers. The individual switching sections of DC railways are isolated vice versa by section insulators in the contact line and connected via busbars in the substation.

The switching components in rectifier substations are factory-built units usually designed for load class VI of EN 60 146-1-3. Figure 1.6 shows the basic design of a direct current traction supply for a tramway.

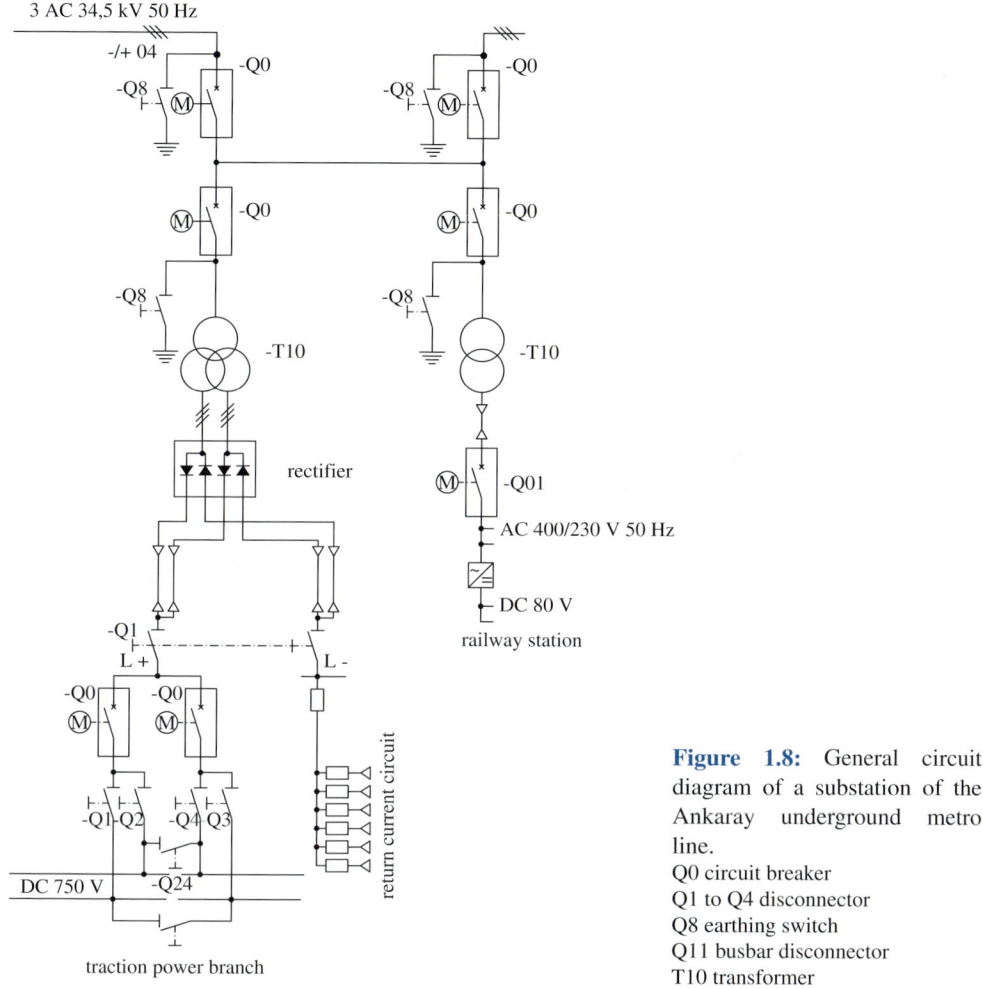

Figure 1.8: General circuit diagram of a substation of the Ankaray underground metro line.
Q0 circuit breaker
Q1 to Q4 disconnector
Q8 earthing switch
Q11 busbar disconnector
T10 transformer

In the design and operation of direct current railways, special attention has to be paid to the traction return current to minimise the hazard of *stray current corrosion*. These item is addressed in detail in Clause 6.5.

1.4.2   Metro Ankaray in Ankara

1.4.2.1   Line supply and switching

The Ankaray underground metro in Ankara/Turkey forms a typical example for a supply by DC 750 V. The line is 9 km long and includes eleven passenger stations. The vehicles travel at 120 s headway and are supplied with DC 750 V by a *third rail*. They draw currents up to 3 000 A. The installed traction power totals around 1,2 MW per km of line.

The urban energy supply provides electrical energy for the line at two feeding stations. One substation at each end of the line is fed from the 154 kV network (see Figure 1.7). The transformers 154/34,5 kV in the feeding stations feed the 34,5 kV medium-voltage line to supply the rectifier substations. With two transformers 34,5/10 kV a 10 kV medium-voltage ring is connected to supply the railway stations.

Each of the four 2,5 MW *rectifier substations* provides DC 750 V for the main line. The maximum substation spacing is around 2,8 km. For servicing or repairs, the substations can be isolated from both the DC 750 V third rail and from the AC 34,5 kV medium-voltage system.

The *open-air depot* at the end of the line is fed from a separate rectifier substation and is isolated from the main line at the tunnel entrance by insulating rail joints and gaps in the conductor rails. Consequently, it is possible to connect the rails to the protection earth of the depot earthing system in the depot and workshop.

The rails of the main line are insulated from the *tunnel earthing system* to avoid *stray currents*. The traction return current causes longitudinal voltages in the rails, which create a rail potential and thereby a potential difference to the platforms. To prevent impermissible touch voltages, which could arise when several vehicles start simultaneously, *Voltage Limiting Devices (VLD)* were installed in every station between running rails and earthing system of the station. They close if impermissible high voltages occur and thereby protect passengers from hazards. They open after approximately 10 s and reactivate their monitoring function.

1.4.2.2   Substations and components

Figure 1.8 shows an overview of the circuit diagram of a AC 34,5 kV/DC 750 V substation. They are connected through a medium-voltage substation to the 34,5 kV medium-voltage ring. The substations include two circuit breakers to connect the cable ring, the circuit breakers for the transformer for the DC traction supply and the transformer supplying the buildings of the neighbouring station and all necessary equipment for measurement recording. The transformers supplying the *rectifier* are designed as resin encapsulated types and have two secondary windings which supply voltages phase-shifted by 30 degrees. A *diode rectifier* is connected in a three-phase bridge circuit to each secondary winding, so that a *twelve-pulse direct current* results at the DC side. The protection devices reliably and selectively detect short-circuits, by measurement of the absolute current value, slew rate (di/dt) and voltage. High-speed *DC circuit breakers* with *quenching chambers* cut off the short-circuit currents on the line after being triggered by the line protection.

1.4.3   DC 3,0 kV traction power supply in Spain

1.4.3.1   Introduction

The railway network operated by the infrastructure manager ADIF in Spain is approximately 13 000 km long, whereby approximately 7 700 km are electrified. These lines with the Spanish gauge 1 668 mm are equipped with DC 3,0 kV. In total, this network is approximately 6 500 km long; 3 600 km are single-track lines and 2 900 km double-track lines. These lines are supplied by 330 substations. Fourteen remote control centres take care of the electrical operation. The substations are connected to the 50 Hz public network with voltages between 15 kV and 66 kV.

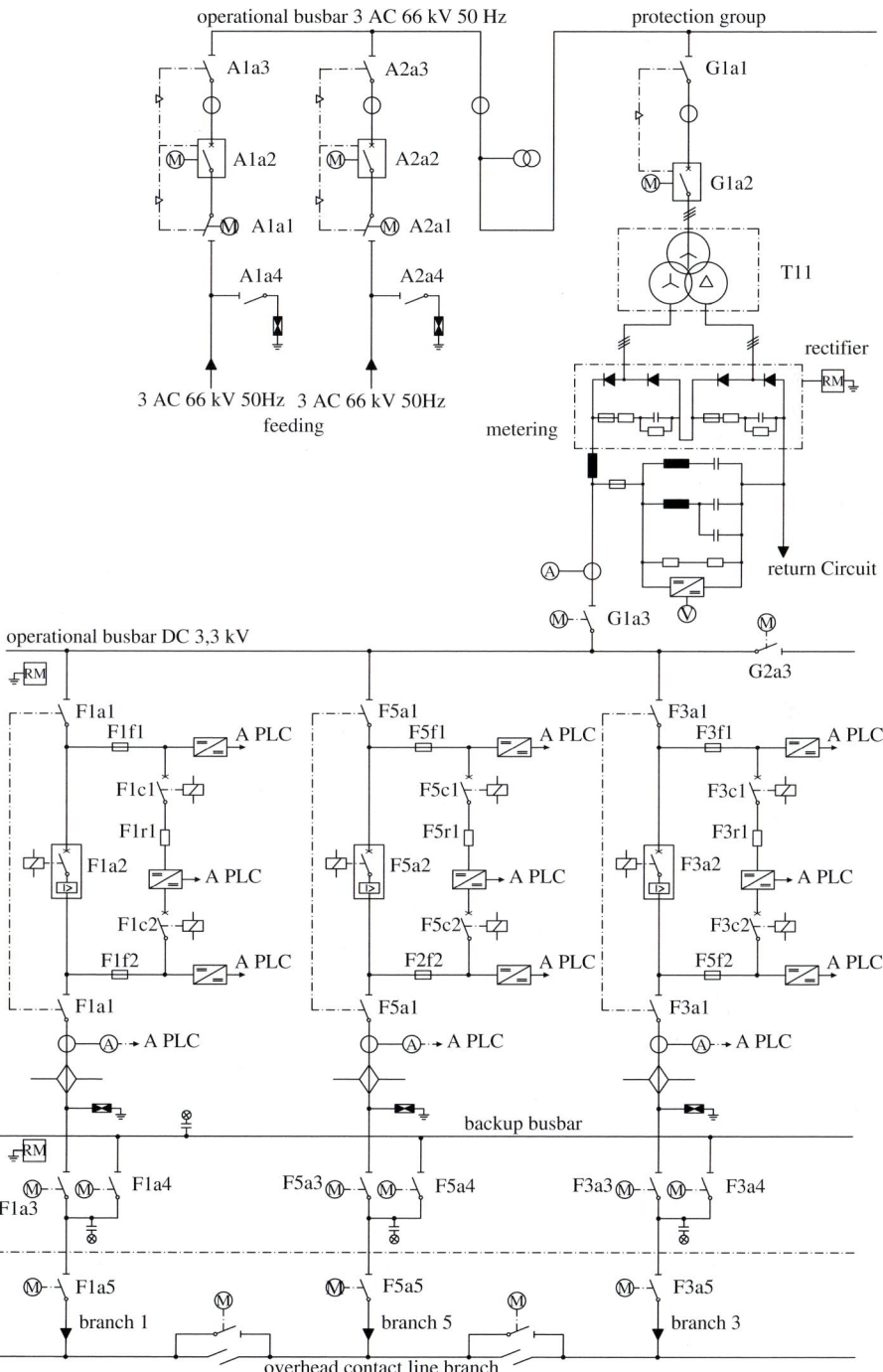

Figure 1.9: Overview circuit diagram on one half of a 3,0 kV ADIF substation.

1.4.3.2   Substations

In Figure 1.9 an overview of a rectifier substation circuit is shown supplying a single-track line by DC 3,0 kV.
The substation is equipped with:
- two feeding lines
- two transformer rectifier groups with 6,6 MVA power each
- three or six overhead contact line branches depending whether a single-track or a double-track line is supplied
- a transformer to supply the signaling by AC 50 Hz 2,2 kV and 25 kVA power as well as an auxiliary transformer which is not shown in the Figure 1.9

Each feeding line is equipped with a circuit breaker and feeds to the 66 kV busbar. A measuring system for power and energy follows. The energy supplier can monitor the energy consumption of the substations via a common transfer unit. The measuring devices, installed after the liberalization of the electric power supply, enable ADIF to negotiate more favourable contracts.

The power transformers T11 with 6,6 MVA power each comprise one primary circuit and two secondary circuits. One secondary circuit is star-connected, while the other is delta-connected. Therefore, a difference between the phase conductors of 60 degrees is achieved. Each secondary circuit supplies an uncontrolled three-phase two-way bridge rectifier with a secondary voltage DC 1,5 kV. A series connection of them results in the DC 3,0 kV secondary voltage and in a twelve-pulse rectification. To smooth the secondary current a reactance coil having 0,3 mH inductivity is arranged at the positive pole of the rectifier. In between the positive and negative pole of the rectifiers two harmonics filtres for the frequencies 600 and 1 200 Hz, respectively, are arranged, which consist of a reactance coil and a capacitor.

Every rectifier output is connected to a busbar section by a motor-driven disconnector. The individual busbar sections can be connected via disconnectors. In normal operation, one of the busbar sections supplies the line sections on the left side and the other busbar section those on the right side of the substation by overhead contact line branches by disconnectors and circuit breakers. This connection limits the short-circuit power. Each feeding branch is equipped with the necessary testing devices. Overhead contact line disconnectors to bridge the insulated overlaps and the auxiliary busbar enable auxiliary feeding in case of maintenance and disturbances.

1.4.3.3   Control and protection

Up to 1992 the substations were controlled by decentral control panels with relays and push buttons. From 1992 onwards the introduction of programmable controls at first in a master/slave operation has led to further centralizing of the substation control. In 1997 a new era of substation control started: From the central control the transition to a distributed control started, where each functional unit of a substation was equipped with programmable control units. These control units are connected with the remote control centre by a communication network with point-to-point transmission and a transmission velocity of 1 MByte/s to exchange information as well as for transmitting data with the remote control centre.

At the substations devices exist for protection against
  – overcurrent and short circuit in each feeding,
  – overvoltage, short circuit and earth faults, melting of rectifier fuses, too high temper-
    atures as well as Buchholz and oil level protectors in each of the transformer rectifier
    banks,
  – overvoltage, overcurrent and short circuit in each of the overhead contact line branches.
Within the substations there are the following functional groups:
  – input group
  – auxiliaries
  – traction power group with the transformer rectifiers
  – overhead contact line branches
  – protecting devices
Additionally, there are still other control and monitoring elements installed:
  – remote control module for the information exchange with the remote control centre
  – station control and protection
  – parameterizations and local control equipment
  – administration of the protection devices
The distributed control which supervises each functional group enhances the operational reliability of each substation.

1.5 AC 16,7 Hz single-phase traction networks

1.5.1 Traction power generation

As one option, single-phase AC with the special frequency of 16,7 Hz can be generated by
single-phase generators. The physical relationship $f = p \cdot n$ exists between the frequency f,
the number of pole pairs p and the revolutions n of a generator. The lowest possible number
of pole pairs is 1. Therefore, the highest speed at which a 16,7 Hz generator can be operated
will be $n = 16,7 \cdot \text{s}^{-1}$ or $1\,000 \cdot \text{min}^{-1}$.

A 16,7 Hz generator, therefore, runs at one third the speed of a 50 Hz generator under equal
conditions. However, the power P and the revolutions n are linked to the moment M by:
$P = M \cdot n$. Comparing 50 Hz and 16,7 Hz generators, a moment three times higher would be
required to achieve the same power at 16,7 Hz. Three times the moment means three times
the size. The generators of the public grid are three-phase generators. The generators used
in traction power supply with a frequency of 16,7 Hz are single-phase. Because of the lack
of two windings, the laminated stator core is used less efficiently at 16,7 Hz by a factor of
$\sqrt{3}$. A 16,7 Hz single-phase generator is, therefore, principally $3 \cdot \sqrt{3} \approx 5,2$ times larger in
volume than a 50 Hz three-phase generator of the same power. Practical values lie around 4,5.
The largest 16,7 Hz single-phase generator with a nominal power of 187,5 MVA, therefore,
corresponds in size to an 850 MVA generator of the 50 Hz three-phase public grid.

If 16,7 Hz single-phase generators are driven by motors supplied from the 50 Hz three-phase
network, this type of machine combination is designated, in the traction power supply, as
a *rotating converter*. Regarding the frequency ratio of 50 Hz to 16,7 Hz, elastic and *rigid
converter* are discerned.

Elastic converters are also designated as *asynchronous-synchronous converters*. By using
a variable-frequency and, thereby, revolution-variable drive of the single-phase generator

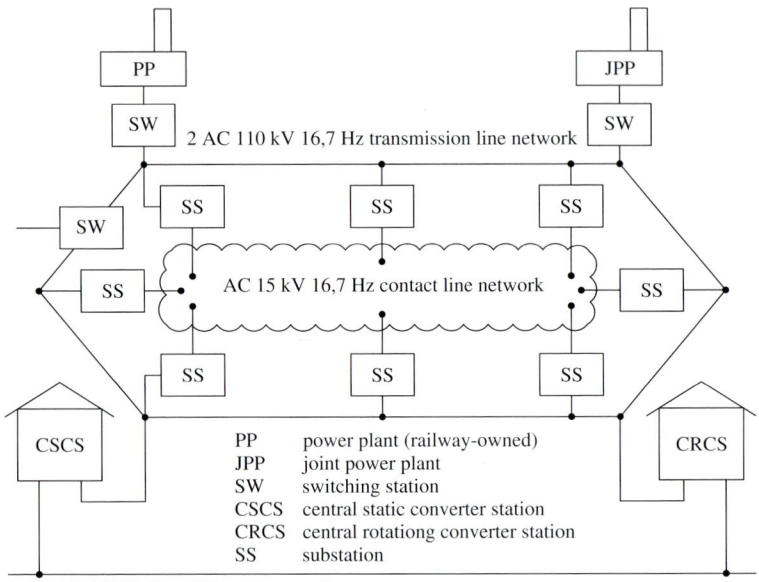

PP   power plant (railway-owned)
JPP   joint power plant
SW   switching station
CSCS   central static converter station
CRCS   central rotationg converter station
SS   substation

3 AC 110 kV 50 Hz public power supply

Figure 1.10: Basic structure of central traction power supply.

driven by an asynchronous motor, it is possible to use *elastic converters* in parallel operation with traction power plants. Elastic converters are used to cover load peaks in centrally supplied networks. The power of elastic converters lies between 10,7 and 50 MVA.

Rigid converters are *synchronous-synchronous converters*. They are designed as prefabricated and mobile units, very often (Figure 1.11). In order not to violate the gauge of railways, their power is limited to 10 MVA [1.14]. In DB's decentral network section, the single-phase power with a frequency of 16,7 Hz is generated in decentral rotating converter stations (DRCS) with the aid of synchronous-synchronous converters.

Decentral converter stations (DRCS) combine two functions in a spatial unit:
  – frequency conversion from 50 Hz to 16,7 Hz
  – distribution of the 16,7 Hz energy by busbars and line branches to the overhead contact
    line system (Figure 1.12)

Rotating converters will be replaced step-by-step by static converters [1.15, 1.16], although a certain amount of rotating masses are necessary for the network stability. There, distinction is made between converters with a DC intermediate circuit and direct converters which convert the energy in one step only without an intermediate circuit. The first type of converters erected until the beginning of the 1990's were line commutated converters. They possess a high efficiency, however, they resulted in a pulsating loading of the three-phase AC networks and coupling of harmonics. Static converters with a DC current intermediate circuit (Figure 1.13) were developed in the 1980's and 1990's and erected, for example, in Richmond, USA, with a load of 5 × 45 MVA, in Norway, Sweden and Germany. The line commutated or pulsating rectifiers (DCR) convert the voltages of the three-phase AC network into DC voltage. Four quadrant choppers (4QC) generate then the single-phase railway network voltage. The intermediate circuit converter decouples both networks by capacitors and harmonic circuits [1.17].

a) 110 kV substation

b) Converter station

c) Converter prepared for transport

e) Connection shaft

d) 15 kV line branches

Figure 1.11: Decentral rotating converter stations (DRCS) (Photos: DB Energie, T. Groh).

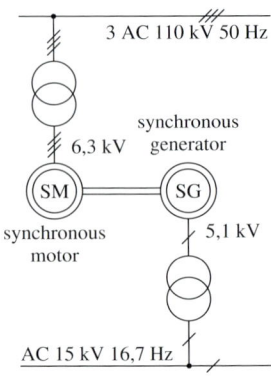

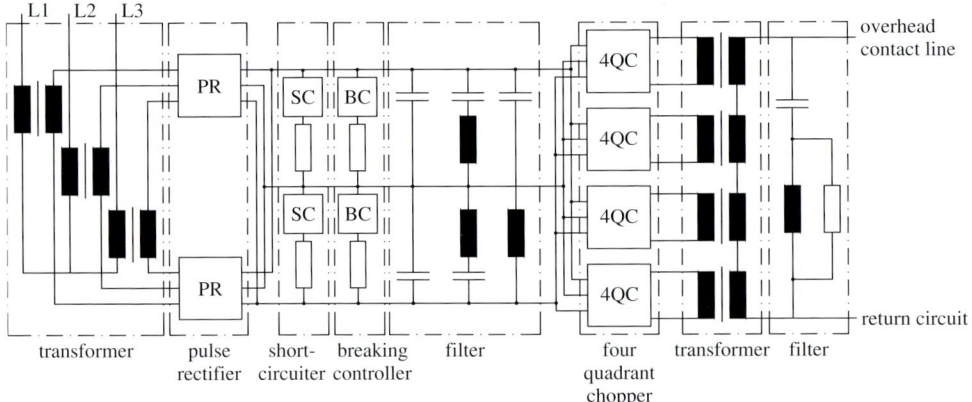

Figure 1.12: Design of a decentral rotating converter station (DRCS) with synchronous-synchronous converters, i.e. rigid converters.

Figure 1.13: Basic block diagram of a DC intermediate circuit converter in three point arrangement.

For central and decentral converter stations Siemens developed a modular, self-commutated multi-level direct converter (MMDC). In contrast to externally commutated converters known so far, this new type converts the energy practically without any system perturbations [1.18]. For this purpose several converter modules per phase are connected in series forming switchable voltage sources (Figure 1.14). They are designed as complete circuits and consist of IGBT power semi-conductors with integrated free-wheeling diodes, capacitor, control and monitoring electronics as independent units (Figure 1.16).

Due to the modular design the energy storage within the converter is distributed on many sub-modules. In case of a damage of a semi-conductor only the sub-module concerned will be short-circuited and the converter can be operated without interruption. Due to the series connection of the modules voltage-adding transformers can be waived and the overhead contact line can be supplied directly. Other advantages of an MMDC are the omission of series resonant circuits and de-coupling of the networks, advantageous control of circuit failures and limiting of short-circuit currents. They are described in [1.16]. The first MMDC with two 37,5 MWA converter units to supply into the 110 kV network of DB was erected in Nuremberg-Gebersdorf ordered by E.ON Ltd. and has been in operation since 2012. Meanwhile Siemens installed MMDC units with powers between 15 and 60 MVA in Adamsdorf, Cottbus, Frankfurt/O. and Rostock (DB), Wildenrath (Siemens), Häggvik and Eskilstuna (Trafikverket), Uttendorf (ÖBB), Winkeln (SBB) and Metuchen (AMTRAK).

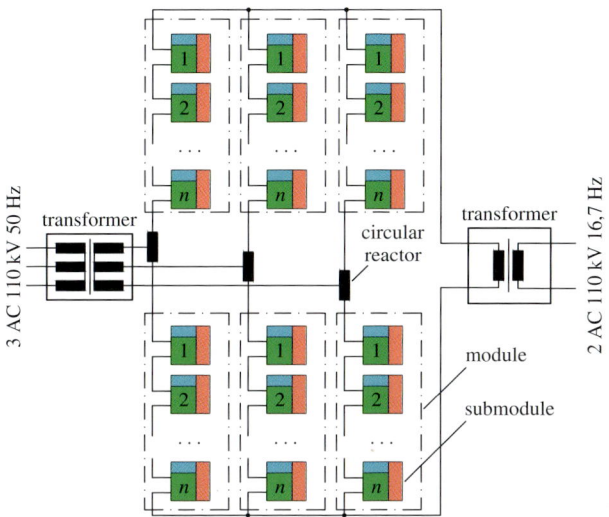

Figure 1.14: Basic block diagram of a converter in MMDC design [1.18].

a) Circuit diagram   b) Submodule unit

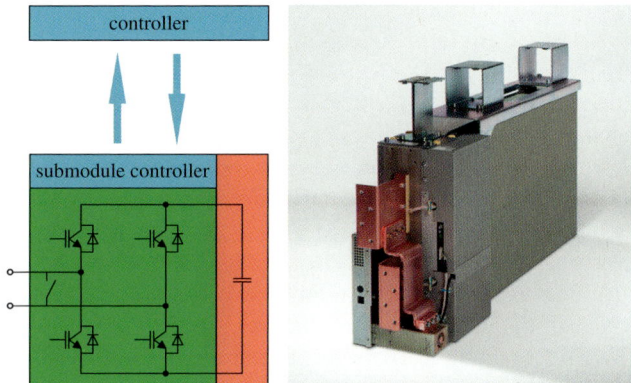

Figure 1.15: Submodule with circuit diagram [1.18].

1.5.2 Types of 16,7 Hz traction power networks in Europe

Two kinds of 16,7 Hz single-phase power supply have evolved in Europe. The *central traction power supply* (see Figure 1.15) has existed in Germany, Austria, Switzerland since 1913 and later in Norway and can be characterised by:

- power generation using 16,7 Hz single-phase generators installed in hydroelectric, thermal and nuclear power plants and driven by water or steam turbines or by rotating or static converters supplied by the public power grid. This means of generating electrical energy is known as primary energy generation
- *transmission of electrical energy* by a 110 or 132 kV overhead line network with a nominal frequency of 16,7 Hz from the power plants to the substations. This single-phase network mostly consists of two circuits and has a feed and return conductor for each circuit
- *distribution of single-phase electricity* in railway substations, where the voltage is converted from 110 or 132 kV to the nominal voltage of the contact line installation of 15 kV

Figure 1.16: Central modularized multi-level direct converter station Nuremberg.

 – feeding of the single-phase 16,7 Hz energy through circuit breakers in the substations
   to the individual feeding sections of the contact line installation

The most significant property of the *distributed traction power supply* (see Figure 1.8), which
has been operated since 1926 in Sweden and since 1968 in a network section of former
Deutsche Reichsbahn (DR) in East Germany, is the existence of *distributed rotating converter
stations* (DRCS). The DRCS has two functions:

 – generation of single-phase power with a nominal frequency of 16,7 Hz and
 – feeding of the single-phase 16,7 Hz energy through circuit breakers in the substations
   to the individual feeding sections of the contact line installation

The two types of single-phase 16,7 Hz traction power networks have different parameters. A
summary shows that both the central and the distributed traction power supplies are able to
supply trains with electricity reliably and with the required quality parameters [1.3]. Over
eighty years of electric traction transport on 20 000 km of centrally supplied traction lines
and over seventy years of electric traction on 13 000 line kilometres supplied with 16,7 Hz
electrical power from distributed equipment prove the dependability of both types of 16,7 Hz
power supply.

Due to the high short-term load peaks in the DRCS, high demand rates are currently paid
for the energy taken from the AC 50 Hz three-phase network. This economic disadvantage in
comparison with 16,7 Hz energy generation and transmission in the 110 kV network has led
to planning the extension of the centrally supplied part of the DB network.

1.5.3   16,7 Hz traction power supply of German Railway

1.5.3.1   Energy generation

In the year 2 011 [1.19], single phase 16,7 Hz *energy generation* for DB was carried out in
 – seven central thermal power plants with 988 MW,
 – five central and two decentral hydro power plants with 347 MW,

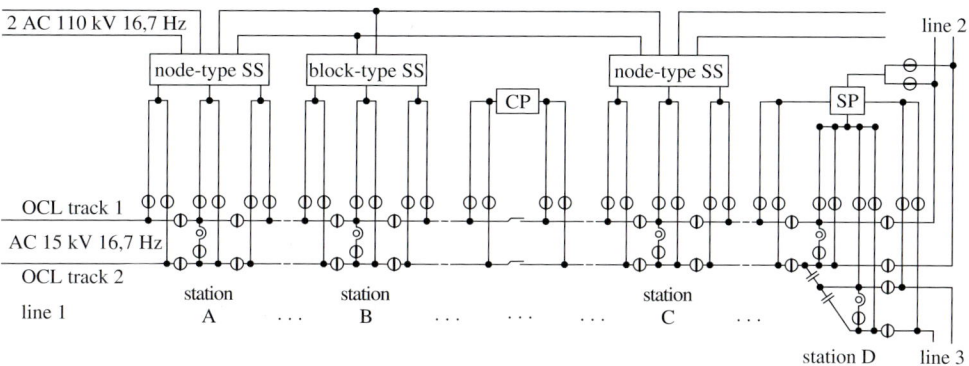

Figure 1.17: Schematic circuit diagram of a contact line supply of DB.
SS substation, SP switching post, CP coupling post, OCL overhead contact line

- ten central and twelve decentral rotating converter plants with 399 and 304 MW, respectively,
- eleven central and three decentral static converter plants with 853 and 90 MW, respectively.

1.5.3.2 Energy transmission and contact line supply

The 16,7 Hz traction power is transmitted within the central network of the DB via the 2 AC 110 kV 16,7 Hz overhead power lines. The operational voltage between phase R and T is 115 kV and the voltage between phases and midpoint 55 kV each. At the end of 2011, the 110 kV overhead line network of the DB was 7 763 km long and supplied 182 substations. At Haltingen and Singen three coupling transformers are used to connect the 110 kV system to the 132 kV traction power network of the SBB (Swiss National Railways). In Steindorf and Zirl the 110 kV DB network is connected directly to the 110 kV network of the ÖBB (Austrian National Railways). The 110 kV overhead line network of DB allows for an optimum import of energy and contributes to a high supply reliability of transport on electrified railway lines. Because it is operated as a resonant-earthed system, twelve arc suppression coils of 100 A each compensate the line capacitances.

A part of the 110 kV overhead line runs beside the main lines of the DB to supply the individual substations, which are designed as *node-type substations* with double busbars or as simple *block-type substations* (see Clause 1.5.3.3). An example of a contact line supply of this kind is shown in Figure 1.17.

1.5.3.3 Standard substations, function and types

According to standard terminology, *substations* are electrical installations with transformers, switchgear, control equipment, metering, protection and signalling facilities with the necessary instrument transformers. With these, it is possible to switch circuits on and off as required and to switch off faulty equipment quickly and selectively or to isolate it for maintenance purposes.

With the DB, substations, switching posts and coupling posts for single phase AC 15 kV 16,7 Hz are designed in accordance with DB directive 995 [1.20]. DB's *standard substations*

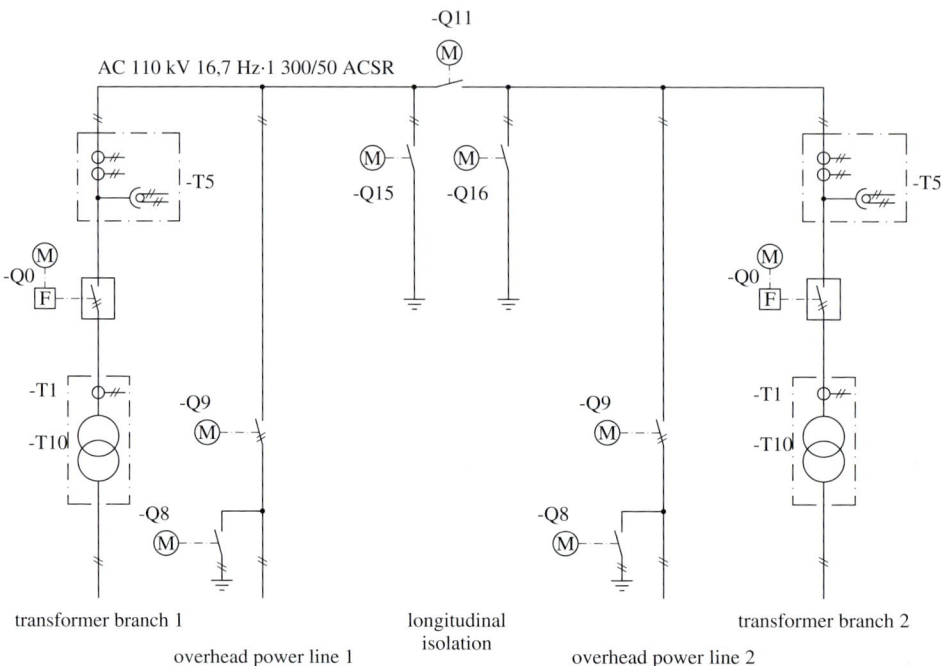

Figure 1.18: General circuit diagram of a 110 kV block substation with single busbar.
Q0 circuit breaker, T1 current transformer, T5 voltage transformer, Q9 disconnector, Q8, Q15, Q16 earthing switch (M) motor operated, [F] stored-energy spring

are unmanned in operation and consist of standardised components with standard interfaces, which can be put together and rated in a modular manner according to functional requirements [1.21]. The standard design is used for:

– substations (SS) with 110 kV and 15 kV equipment
– switching stations (SW) with 110 kV equipment only
– switching posts (SP) with several 15 kV supply branches
– coupling posts (CP) with one 15 kV circuit breaker only

Substations with transformers convert the 110 kV nominal voltage of the 16,7 Hz transmission network to the 15 kV nominal voltage of the overhead contact lines. They distribute the traction power to the individual feeding branches.

Switching stations are used to connect and branch the 110 kV electrical traction lines.

Switching posts connect the overhead contact lines and feeders of several railway lines and supply overhead contact line sections fed from one end with 15 kV power.

Coupling posts connect two feeding sections and are used especially in cases of long distances between substations or long sections fed from one end to guarantee the correct functioning of protection.

The DB substations take care of the specialties of traction power supply which are characterized by a 150 times higher frequency of short circuits and higher short-circuit currents compared with the public power supply. Therefore, special requirements need to be considered for the design of the substations, the switching and protection devices and the substation control and protection system. The fast, secure and selective switching-off of unacceptable

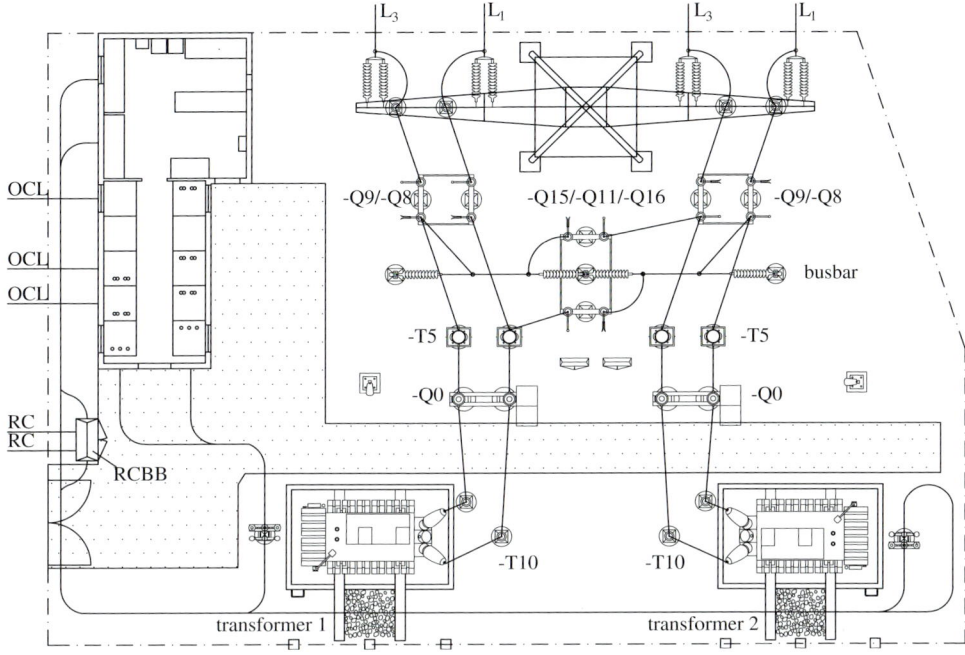

Figure 1.19: Plan view of a block-type substation.
OCL overhead contact line, RC return circuit, Q8, Q15, Q16 earthing switches, Q9, Q11 disconnector, T5 voltage transformer, T10 current transformer, RCBB return current busbar

currents and the effective processing of high numbers of information form an essential aspect. Existing technical standard solutions and devices of the public power supply are as far as possible modified and re-designed to be suitable for railway requirements. The DB standard substations of the first generation still contained pneumatically operated circuit breakers as well as control, signaling and protection technology based mostly on mechanical relays. The 15 kV vacuum circuit breakers introduced at the beginning of the 1980's, electronic information processing in SCADA systems fostered the transition to the second generation of DB's standard substations. They are characterized by a significant reduction in substation size as well as in installation and maintenance efforts [1.21]. By introducing the branch-related testing equipment AGP and division into metal-clad compartments which are insulated by air according to EN 62 271-200 the DB substations have been developed further since 2005 (see Clause 1.5.3.5).

1.5.3.4   110 kV open air equipment

DB's directive 995 [1.20] includes standard specifications for the design of the 110 kV system, based on operational requirements. The main distinguishing features are:
– 110 kV equipment with *double busbars*, two longitudinal isolations and a coupling
– 110 kV equipment with *single busbars* and two longitudinal isolations
– 110 kV equipment in block operation for *block-type substations*
Each substation consists of several branches, e. g. traction power lines, transformer and longitudinal isolation branches or block branches, which are chosen according to the local re-

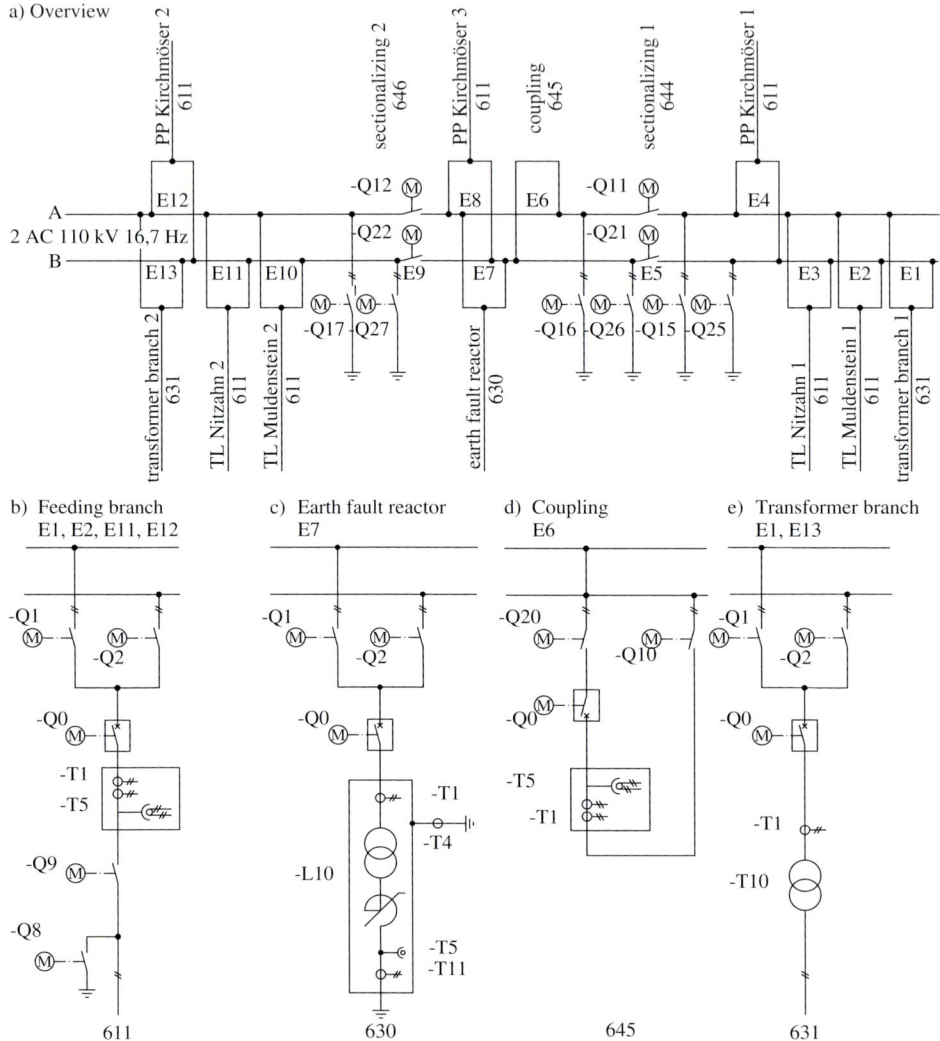

Figure 1.20: General circuit diagram of a 110 kV node-type substation with double busbar A and B. Q0 circuit breaker, Q1, Q2, Q9, Q11, Q12, Q21, Q22 disconnector, Q8, Q15-Q17, Q25-Q27 earthing switch, T1, T11 current transformer, T5 voltage transformer, T10 transformer, L10 neutral earthing reactor, Ⓜ motor operated

quirements from numerous standard branch types. A typical general *circuit diagram* of a block-type substation is shown in Figure 1.18 and the associated plan view in Figure 1.19. Whereas substations with single or double busbars in transformer and traction overhead line branches are equipped with circuit breakers, the block substation of the DB has no circuit breakers in the traction overhead line outlets.

Substations simplified in this way are used as intermediate substations between fully equipped node-type substations whose circuit breakers in the overhead power line branches switch off faulty lines including those in the vicinity of block-type substations.

Standard specifications for the electrical, mechanical and geometrical design apply to circuit breakers and disconnectors, instrument transformers and earthing coils, to achieve matching and interchangeability of equivalent types of equipment from different manufacturers.

In substations and switching stations, the individual circuits of the incoming overhead power line branches are connected with the busbars (see Figure 1.20). The conductors of the overhead power line are anchored to section supports or overhead power line end supports. They are then connected to the line disconnector (Q9) designed as a double-pole rotary earthing switches (Q8) (Figures 1.18 and 1.19). These are driven by a redundant DC 60 V supply, as are all disconnectors in standard substations of the second generation.

The twin-pole *circuit breakers* (Q0) contain SF6 as quenching gas and an electrically powered spring or pneumatic drive for actuation. Single-pole, oil-filled *combination instrument transformers* (T5) are used to measure currents and voltages.

15 MVA *single-phase oil transformers* in mobile design for outdoor installation with ONAN cooling are used as power transformers (Figure 1.19). Devices which prevent loosing of the windings caused by the high number of short-circuits [1.3] are used in power transformers for 16,7 Hz. The transformers are insulated against earth and earthed by *tank leakage protection instrument transformers*. They are also equipped with current instrument transformers (T1).

The *busbar disconnectors* (Q1, Q2) are used in substations with double busbars to switch between busbars (Figure 1.20). The longitudinal busbar disconnectors (Q11, Q21, Q12, Q22) are connected to one or two attached earth electrodes (Q15-Q17, Q25-Q27). The busbar disconnectors (Q11) with attached earthing switches (Q15, Q16) allows feeding of the transformer from other circuits if the line failed partially.

Because the 110 kV network of the DB is operated in *resonant-earthed condition, arc suppression coils* with integrated neutral point are installed in selected substations. The arc suppression coils are designed as solid core coils with step switches or, for frequency control, as plunger coils with an inductive current of 10 A to 200 A.

The mesh earth electrodes used consist of tinned copper conductors with a cross section of 95 mm^2. They are connected by loops to all steel components and with ball-type earthing studs. The *lightning protection rods* attached to the lighting masts and the earth wires above the overhead power line branches and busbars protect against lightning.

1.5.3.5 15 kV indoor equipment

At DB substations instead of previously used designs [1.3], [1.20] to [1.22] metal encapsulated, air-insulated units in accordance with EN 62 271-200 are adopted nowadays. These 15 kV indoor systems are built with the configurations:

- 15 kV installations with one *operating busbar* and two longitudinal sectionings
- 15 kV installations with one *operating busbar* (only for coupling posts)

Because of the advantageous parallel connection of the overhead contact lines of a double-track line, each feeding direction and the station in the vicinity of the substation each are usually supplied by just one overhead contact line branch. Therefore, at least three overhead contact line branches and two transformer branches result for each substation. The number of branches increases if additional lines have to be supplied (Figure 1.21). The *operating busbar* (OBB) couples the individual branches in order to distribute the current and to provide the voltage. For maintenance and clearing of disturbances the operating busbar is divided by the longitudinal isolators Q11, Q12 into three ranges. A third transformer or additionally a mobile substation can supply the middle section, if provided. By distributing the line

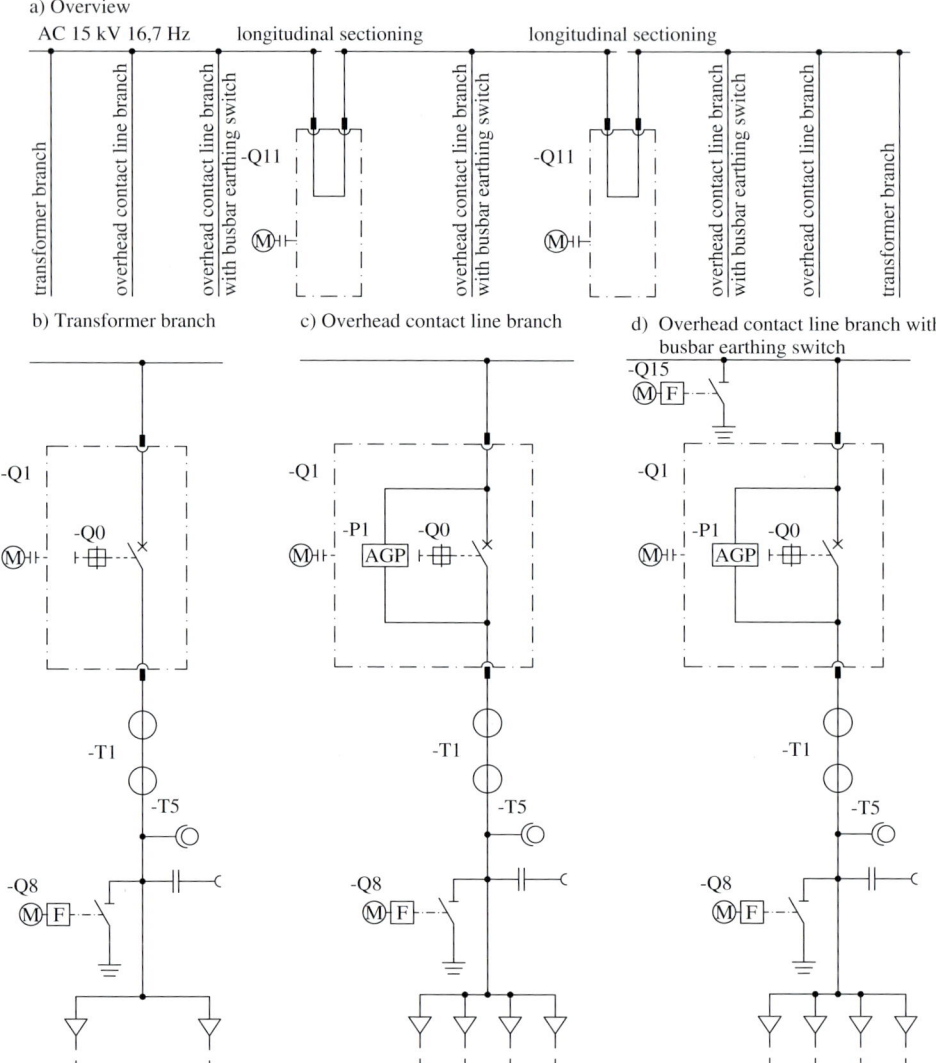

Figure 1.21: 15 kV general circuit diagram of a substation with metal-clad systems and feeder-related testing device AGP.

branches to differing busbar sections including a station branch as an auxiliary supply on the medium section a high availability of the branches especially in case of maintenance activities is achieved.

The requirements of DB on metal-clad air-insulated substations according to EN 62 271-200 are part of the DB specifications for single-pole railway traction substations [1.23]. On this basis the air-insulated medium-voltage substation Sitras ASG15 was developed (Figures 1.23 and 1.24) [1.24]. This substation is a pre-fabricated and type-tested compact substation which since 2010 has been installed in railway networks characterized by up to 40 kV short-circuit

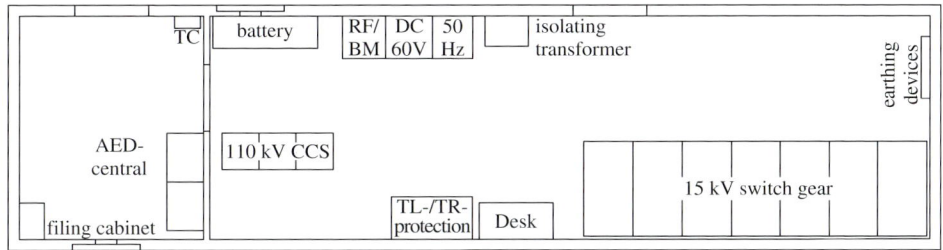

Figure 1.22: Arrangement of the high-voltage and secondary technology in a DB substation. RF/BM rectifier with battery management, CCS control and communication system, TC telecommunication, AED automatic earthing device (only for substation close to tunnel), DC 60 V power supply DC, TL/TR transmission line and transformer protection, 50 Hz power supply AC 0,4 kV 50 Hz

Table 1.4: Protection types in DB's substations.

type of protection	coupling post CP	switching post SP	block type substation BSS	substation	switching station
contact line protection	×	×	×	×	
transformer protection			×	×	
transmission line protection			×	×	×
back-up protection		×	×	×	

currents. The compact 40 kV single-walve vacuum circuit breaker 3AH4766-6 with 125 kV rated lightning impulse voltage was developed for short-cicuit currents up to 40 kA [1.25].

Each branch of this substation consists of three arc-resistant separated medium-voltage compartments: a compartment for circuit breakers, a compartment for the busbar and a compartment for the cable connections (Figure 1.23). This design avoids the spreading of a disturbing arc into the neighbouring functional compartments or to the neighbouring field. In the upper part of each compartment a pressure release channel is arranged, which is connected to each of the functional compartments. The secondary equipment such as protection and control equipment is arranged within a low-voltage compartment which can be accessed from the front. The individual compartments are equipped with a motor-driven switch gear truck which carries the circuit breakers, the branch related testing (AGP) device, an instrument transformer for currents and voltages, a fuse or a bridge. The switch gear truck takes care of function of a disconnector of the busbars by moving between operational and separation position. The switching station Sitras ASG15 can be used for

- overhead contact line branches,
- transformer and converter branches and
- longitudinal separations with and without auxiliary supply.

The overhead contact line branches contain instrument transformers T5, which measure the voltage for the overhead protection, the ACLP and the ACLRT. Voltage and current transformers in the transformer branches measure the energy. The AGP cylinder [1.25, 1.26] is arranged in parallel to the circuit breaker and the current instrument transformer (Figures 1.21 and 1.22). To be able to register the short circuits to the metal casing the individual compartments are insulated against the building and earthed by an instrument current transformer having a transformation ratio 1 000 : 1.

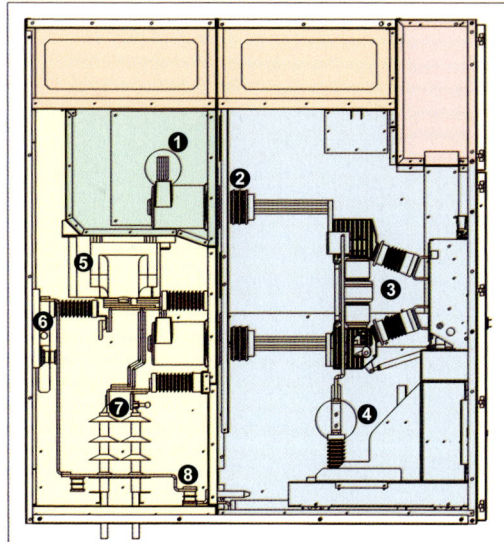

Figure 1.23: Traction substation Sitras ASG15.
1 busbar, 2 self-coupling contact, 3 vacuum circuit breaker on truck, 4 AGP, 5 instrument transformer, 6 earthing switch, 7 cable connection, 8 earthing bar
Functional compartments: *blue* circuit breaker compartment, *yellow* cable connection compartment, *green* busbar compartment, *orange* integrated pressure relief channel, *pink* low-voltage compartment

Figure 1.24: Traction substation Sitras ASG15 with open high-voltage compartment and switch truck with Siemens AG 3AH4766-6 vacuum circuit breaker.

1.5.3.6 Protection

In Table 1.4 the type of protection is shown for DB's individual substation types. Coupling posts receive *overhead contact line protection* as the only type of protection. In switching posts, the *general protection* is supplemented. Block substations have additional *transformer protection*. All other substations are equipped as shown in Figure 1.25 with general protection, overhead contact line, transformer and traction power line protection. In switching stations without transformers, only overhead power line protection is used.

The *general protection device* commands on three protective functions and, therefore, provides important backup protection:

– *busbar protection* in switching posts and substations is triggered immediately in case of more than approximately 0,5 kA short-circuit current through the frame work of the 15 kV installation and the frame work current transformer. It switches off all 15 kV circuit breakers by the main or reserve actuator and, in substations, also the 110 kV circuit breakers
– *circuit breaker monitoring*, which supervises the switching-off of circuit breakers and switches off that circuit breaker which had not switched-off within a pre-set period of time
– *total current monitoring* in substations switches off all 15 kV circuit breakers if the current measured by the total current instrument transformer exceeds an set value during a specified period

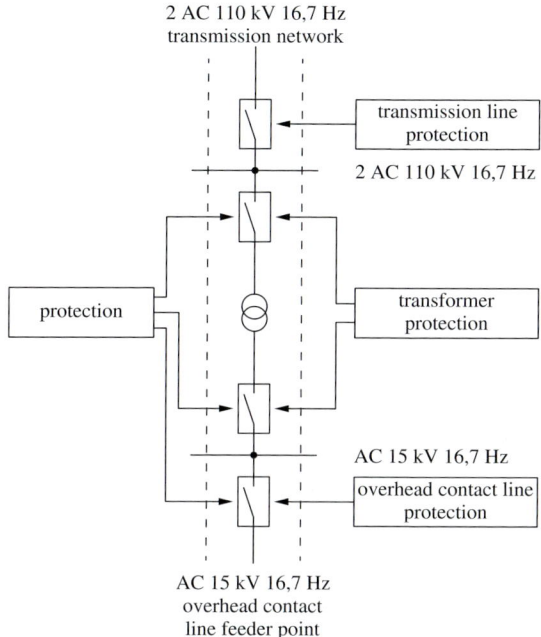

2 AC 110 kV 16,7 Hz
transmission network

transmission line
protection

2 AC 110 kV 16,7 Hz

protection

transformer
protection

AC 15 kV 16,7 Hz

overhead contact line
protection

AC 15 kV 16,7 Hz
overhead contact
line feeder point

Figure 1.25: Schematic diagram of the protection design of a DB substation.
– – coupling through SCADA or cable

For *overhead power line protection*, a static protection unit was installed in the second generation of 16,7 Hz substations. In installations constructed since 1993 a digital protection unit has been used. This protection unit is equipped with several time and direction distance steps, polygonal triggering zones, directional detection with high sensitivity, rapid activation for switching short-circuited lines, fault localisation, earth contact relays and automatic reclosing. The exchange of information with station control is possible through a serial optical glass fibre interface. To detect network faults, the impedance of the circuit is measured. If a network fault impedance is recorded a phase angle measurement would be made to determine the direction of energy flow during the short-circuit. Depending on the fault impedance and the measured angle and if a low impedance exists in both conductor-earth loops, the activation command is issued to the circuit breakers through a series of timer elements. An analogue *transient earth fault relay* used to measure earth short-circuits issues a transient earth-fault signal with the direction and the permanent earth short-circuit at the binary outputs of the protection relays. In case of a single-phase earth fault, the overhead power line and, therefore, the supplied substations can continue to operate over a limited period of time (approximately 2 hours).

For *transformer protection*, a static protection unit has been installed in standard second generation substations and, since 1995, digital protection units have also been installed. The static protection unit is equipped with high-current time protection for the high-voltage and low-voltage sides, a tank protection which measures the fault current through the tank protection, a transformer and an activation signal multiplexer for the Buchholz protection and the stepping switch of the main transformer. Because block-type substations are not equipped with overhead power line protection, additional impedance protection is used in their *transformer protection relays*. The digital transformer protection unit also incorporates differential protection, thermal overload protection and the above mentioned facilities for storing activation data. A detailed description of the *overhead contact line protection* is contained in Chapter 9. In node-type substations, central protection data units are used to store and transfer the data of all digital protection relays.

1.5.3.7 Station control and data acquisition system

The *station control* comprises control and monitoring of switchgear, automating of switching operations, signal and measured values processing and data transfer in substations, which conforms with the traction-specific requirements of the standard AC 15 kV 16,7 Hz switching equipment.

In case of metal-clad 15 kV traction current switching cubicles the protection relay and field modules of station control system have been arranged within switching cubicles since 2005.

In Figure 1.26 an overview of the SCADA system and its communication to substations, protection devices and power system control is shown.

The station control consists of the following functional parts:
- local control
- automation and booking components
- signal and measured value processing
- digital meter monitoring and processing (DMM)
- energy management
- substation communication and communication to control centers and other station control units

For local control, the data display technology employs a TFT monitor with full graphics in window technology. Each service operation is carried out in two steps with checking the plausibility.

The *automation components* secure the automatic operation of the unmanned substations and reduce the work of the operating personnel:
- automatic overhead contact line testing (ACLT)
- automatic overhead contact line residual voltage testing (ACLRT)
- automatic overhead contact line re-closing (ACLR)
- 15 kV and 110 kV automatic synchronising device (ASD)

The *automatic overhead contact line testing* (ACLT) verifies that the overhead contact line branches are free from short-circuits before the circuit breaker is switched on and after every activation of an overhead contact line protection unit. In conventional substations with a test busbar, the tested branch is connected temporarily via the test busbar and the voltage is measured at a voltage transformer of the test branch with testing resistors.

An unsatisfactory test result is reported to to DB's Central Network Control Centre, called Zes. The total testing procedure needs approximately ten seconds from the triggering of the protection to the reclosing of the circuit breaker in an overhead contact line branch of conventional systems with testing resistors in the tested branch. If several overhead contact line branches were triggered simultaneously the testing will be carried out in a defined series in order to put into operation the most important supply branches as soon as possible. In case of an AGP several branches can be tested simultaneously. The total testing period for the serial testing cycle needs only 2,5 s. The testing requirement, the testing results as well as the state of the testing device are exchanged via an optical fiber cable between the AGP control unit and the branch circuit module [1.25, 1.26].

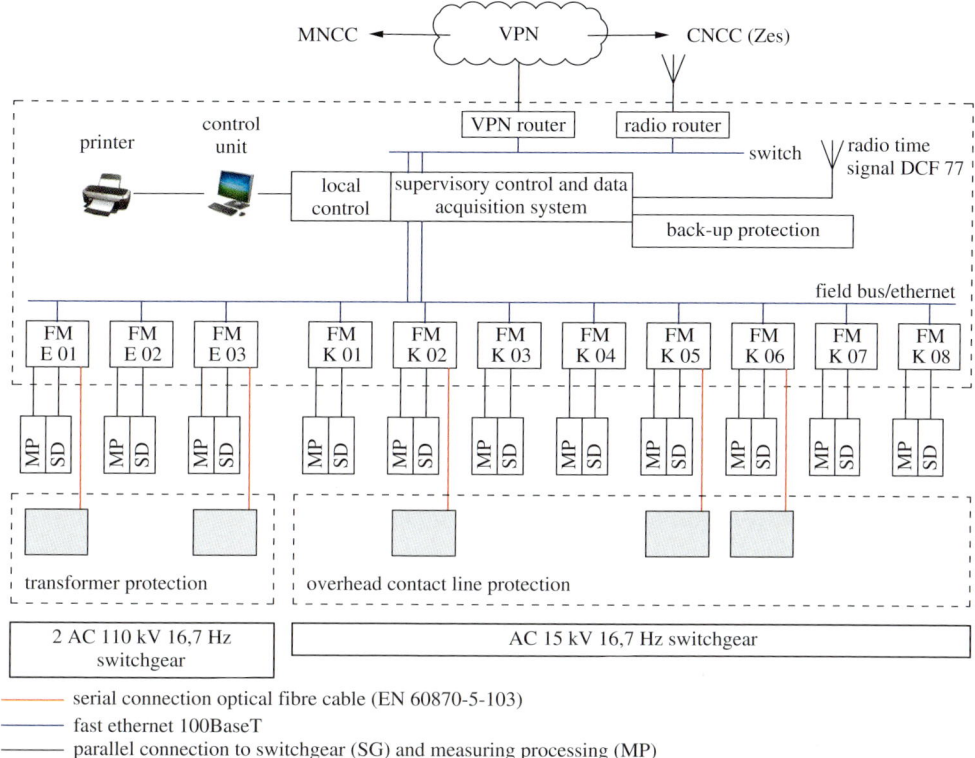

Figure 1.26: Schematic presentation local control units of DB.
FM field module, MEW measured data recording, SD switching device, VPN virtual private network, CNCC central network control centre (Zes), MNCC main network control centre

Automatic overhead contact line residual voltage testing (ACLRT) checks the residual voltage of an overhead contact line branch when a command is issued to close the earthing disconnector Q8.

The *automatic overhead contact line re-closing* (ACLR) used in standard substations without overhead contact line testing device automatically re-closes the circuit breaker Q0 after tripping of a protection unit and the operating voltage returned in a pre-set time after a successful testing of the overhead contact line by an adjacent substation.

The *automatic synchronising device* (ASD) verifies the synchronising conditions before enabling the on-command for the circuit breakers. These include the phase synchronization and equal amplitude, taking account of permissible voltage differences caused by different line loads and possible by-pass conditions if voltage is lost on one side. The interlockings are provided by the software programs. The position of the switch is monitored by two-state indication. To avoid interlocking errors only one remote-controlled switching device per branch can be operated at the same time. The switch trucks in metal-encapsulated 15 kV cubicals and the 110 kV disconnectors with attached earthing switches are additionally interlocked by mechanical systems.

The *signal and measured value processing* includes the acquisition and preparation of all *standardised operating signals* (OS), such as circuit breaker position and disturbance signals,

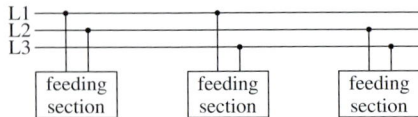

Figure 1.27: Decentral feeding of overhead contact lines by AC 50 Hz systems.

the branch currents, busbar and test voltages, reactive and effective power, which are necessary for the operation and fault analysis of an unmanned substation. The measuring data are recorded via direct inputs of 1 A, 100 V at the branching field modules. The processing of commands and output to the operation equipment comprises the testing of the interlocking conditions and can be carried out within the automation components additionally. The counting processing function records and processes the impulses coming from the counting of active and reactive power following a timely- and value-oriented algorithm which can be parametrized. The values of meter readings are transmitted to the communication interfaces using transfer cycles within minutes.

1.5.3.8   Local control equipment and remote control

In order to achieve a high availability of the energy supply even in case of disturbances or maintenance the overhead contact line is separated into numerous main and secondary switching groups. The individual overhead contact line sections are supplied via electrically driven overhead contact line disconnectors (OCLD). Short-circuit instrument transformers arranged at the type 5 disconnectors (see Clause 5.4.3.2), which connect the contact lines of parallel main tracks, serve for short-circuit localizing. To control and monitor these disconnectors, local control units (LCU) are used, which are arranged in the interlockings of each station. The overhead contact line disconnectors are connected to the drives via linkage systems. The drives are connected to the local control unit via three strands and switching components designed for this purpose. The switching components convert the 60 V control commands of the remote control modules (RCM) into the 230 V level of the motor drives and produce potential free return signals depending on the position of the drive. They also command on monitored automatic fuses, in order to protect the current circuits and the electronic components against the voltages induced in the connection cables. Special monitoring assemblies are available for measuring, displaying and acknowledging of transient impulses of short circuits.

1.6   50 Hz single-phase AC traction networks

1.6.1   Power supply with single phase AC 50 Hz

The electrical energy required for the operation of *AC 50 Hz single-phase traction networks* is obtained from one phase of the 50 Hz three-phase network of the public energy supply.

The phase conductors of the traction substations are alternatingly connected to the individual AC-three-phase conductors. Therefore, a phase difference is created between the overhead contact line sections supplied by the substations which prohibits the connection of the supply sections on the contact line level. A distributed network structure is caused being not advantageous for the operation (see Figure 1.27).

This single-phase loading of the three-phase network causes imbalance in the voltage and current of the three-phase network. The current imbalance has only a minor effect on the

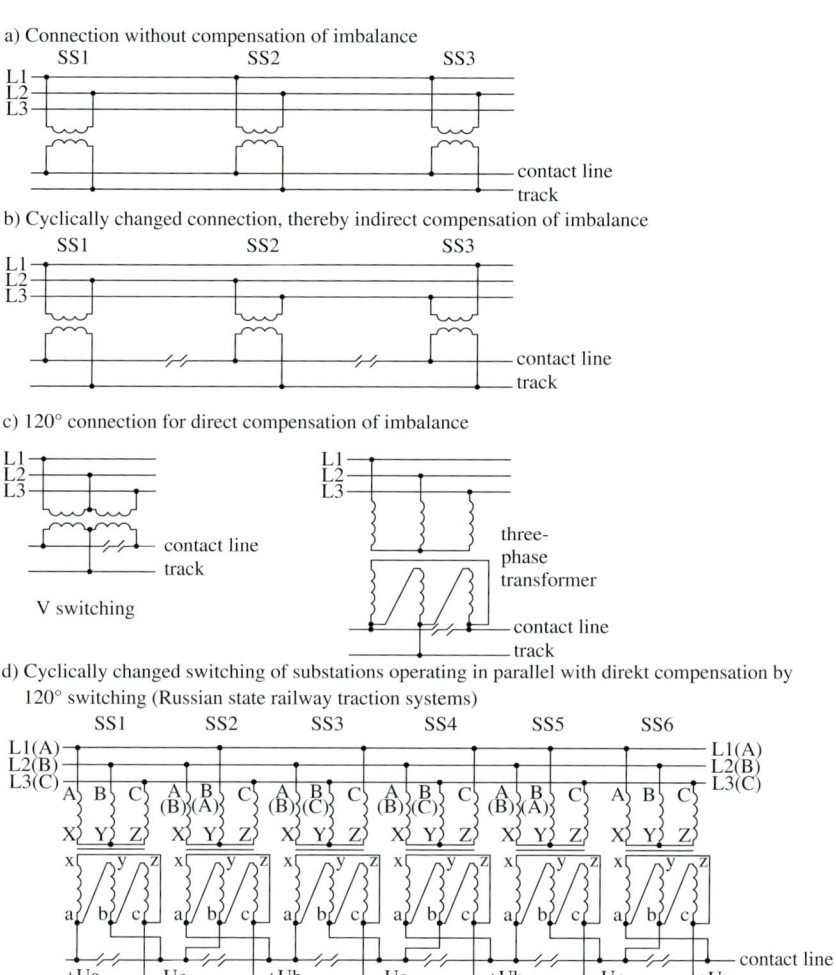

a) Connection without compensation of imbalance

b) Cyclically changed connection, thereby indirect compensation of imbalance

c) 120° connection for direct compensation of imbalance

d) Cyclically changed switching of substations operating in parallel with direkt compensation by 120° switching (Russian state railway traction systems)

Figure 1.28: Alternative methods of connecting 50 Hz single-phase traction power substations to the three-phase network.

generators, whereas the voltage imbalance has serious effects on the consumers. The *voltage imbalance* u_U is the ratio between inverse voltage U_i and direct voltage U_d. The voltage imbalance is also inversely proportional to the short-circuit power S_k'' of the three-phase network. If the traction power S_e to be drawn from one phase of the three-phase network is known, the voltage imbalance in the three-phase network at the point of supply is given by:

$$u_U = U_i/U_d \approx S_e/S_k'' \quad . \tag{1.1}$$

With short-circuit power varying between 700 MVA and 3 000 MVA in the 110 kV three-phase network and powers of the traction power substations up to 40 MVA, high values of voltage imbalance are to be expected. Voltage imbalance leads to a reduction in the life of three-phase asynchronous motors running on three-phase current. To minimise the unfavourable effects of

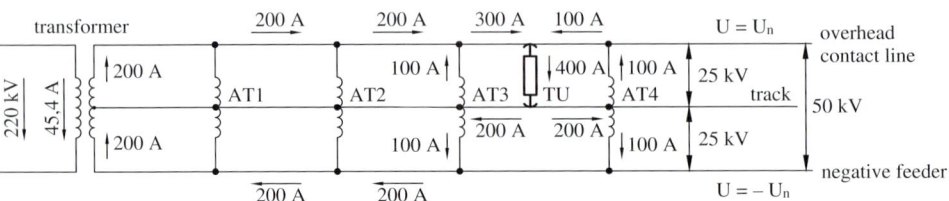

Figure 1.29: Basic design of the $2 \times U_n$ feeding system.
$U_n = 25\,\text{kV}$, feeding from the 220 kV grid, AT auto-transformer, TU traction unit, currents at $S_n = 10\,\text{MVA}$

voltage imbalance, permissible limits of u_U are specified. According to EN 60 034-1, three-phase motors may only be operated in a power supply system where the voltage imbalance does not exceed 1 % continuously or 1,5 % for only a few minutes. To comply with these stringent requirements, it is necessary to limit or compensate the imbalances [1.27].

In practice, the single-phase power is usually connected in a cyclically changed manner with the three-phase network, as can be seen in Figure 1.28 under b). However, this type of feeding leads to a compromise in the single-phase network with regard to optimum operation, which would be the case for the connection shown in Figure 1.28 a). Phase separations are necessary to avoid short circuits between adjacent line sections. At phase separations, the applied voltages have a phase shift of 120°. The voltage difference at the phase separation is $\sqrt{3} \cdot 25\,\text{kV} \approx 43,3\,\text{kV}$. Higher voltage drops result in the overhead line network. These cause unfavourable conditions for electrically regenerating traction units. Feeding as shown in Figure 1.28 b) is preferred by SNCF. This type of feeding is also used on the Madrid–Seville high-speed line [1.28]. On this line, the transformers of the individual substations are arranged with a 60° connection in such a way that the voltage differences at the phase separations correspond with the nominal voltage of 25 kV [1.29].

In Russia, where over 21 500 km (1999) of track is electrified with single phase, AC 25 kV 50 Hz, transformer connections are used, which partially correct for asymmetry (see Figure 1.28 c) and d)). However phase separations are also necessary and they are installed in the vicinity of the substations. The parallel operation of adjacent substations enables a two-side supply of vehicles which can be implemented as shown in Figure 1.28 d), however, may yield high compensating currents under certain conditions. A connection of differing high-voltage networks on the secondary networks is not permissible. Phase separations with neutral sections in the overhead contact line which require switching of the circuit breakers on the vehicles could be waived if the total line would be supplied by static converter stations connected in parallel [1.30].

1.6.2   2 AC 50/25 Hz power supply

To improve transmission properties, the 2 AC 50/25 kV system is used for high-performance traffic in France, Japan and Russia on single phase AC 25 kV 50 Hz railways. This type of feeding is characterised by additional *auto-transformers* and a return line at a potential of 25 kV. This return line is often designated as a *negative feeder*. For this reason, twin-pole switch gear is required in the overhead line network. The basic design of this type of feeding can be seen in Figure 1.29.

The line is supplied by a transformer with a centre tap. The centre tap is connected to the rails. The voltages between the negative feeder and the rails and between the overhead contact line and the rails are both 25 kV. The potential difference between the overhead contact line and the negative feeder is up to 50 kV.

The power is transmitted between the substation and the auto-transformer preceding the section on which the traction unit is collecting electric power from the contact line as in a twin-pole 50 kV line. The lower currents involved with this transmission of power result in lower voltage drops in the overhead contact line network. In the section between the substation and auto-transformer, the current flowing in the rails is low due to the almost 180° phase shift in the equally large currents in the overhead contact line and the negative feeder. The interference with adjacent lines is low, therefore.

In the section between two auto-transformers, the traction units are fed from both ends, the rails serving as return conductors in the customary manner. The interference with adjacent lines is, therefore, also lower than in single-ended feeding without auto-transformers.

In principle, this type of feeding can be used with an n-fold nominal voltage, e. g. 2 AC 75/25 kV 50 Hz. In this case, the transmission of power to the auto-transformers, between which the power consuming traction unit is located, would be performed with a voltage of 75 kV. It should be noted that this feeding principle can be used for all single-phase AC systems regardless of their nominal frequencies. Electrification of DB's Prenzlau to Stralsund line equipped with 2 AC 30/15 kV 16,7 Hz forms an example [1.31]. The requirements on the design of the insulation increase in systems with $n \cdot U_n$. For example, the larger air gaps necessary between parts with multiple nominal voltage differences need to be taken into account in the overhead contact line installations. The necessary twin-pole design of switch-gear in the overhead contact line network is a further general disadvantage of feeding systems with multiple nominal voltages.

1.6.3 Comparison of power supply with single and double phase

It is straightforward to install and to operate single-phase AC systems. The installation along the railway line consists of the contact line and the return circuit. The overhead contact line protection to be installed in the substations is specified in the design of the system. The high portion of return current flowing through earth creates unwanted interferences. They can cause electromagnetic interference in cables in the vicinity of the line if no corrective measures were taken. As a consequence, the cables need to be protected by cable sheaths. The maximum line length to be supplied single-ended from a substation is limited to approximately 25 km to comply with the tolerable voltage drops.

Since low currents flow through earth for twin-phase AC supply, in those line sections between the auto-transformers which are momentarily not traversed by trains, the electromagnetic interference will be much lower than in the case of single-phase supply. Therefore, higher currents and power can be transmitted which increases the capacity and performance of the lines. The voltage drop is lower for the same power enabling up to 50 km long supply sections. However, the enhanced performance costs more. Auto-transformers are spaced 10 to 12 km apart. An additional 25 kV feeder line is required between the auto-transformer stations and the substations. The substations need to be designed for two phases instead of one. Also the protection of the contact line is more cost-effective because of the double-phase design. The selection of a single-phase or twin-phase system depends on technical and economic aspects. Where the load is relatively low and interference does not impose constraints 1 AC 15 or 25 kV is preferred. Otherwise 2 AC 30/15 or 50/25 kV is the adequate option. Both options should be compared economically as well.

Figure 1.30: Overhead contact line and substation on the Madrid–Seville high-speed line in Spain (Picture: ADIF, T. V. Vega).

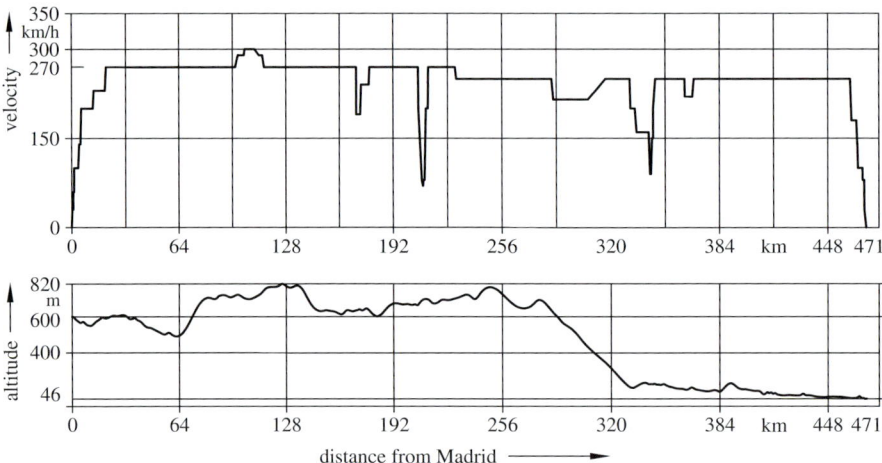

Figure 1.31: Velocity and altitude profile of the Madrid–Seville line.

1.6.4   AC 25 kV 50 Hz traction power supply of Madrid–Seville line

1.6.4.1   Description of the system

The high-speed line Madrid–Seville is in total 471 km long and commenced operation in 1992 at the beginning of the World Exhibition in Seville. The line starts in Madrid at the Puerta de Atocha station and ends in Seville at the Santa Justa station. The maximum commercial speed is 300 km/h. The line was erected with the standard European gauge of 1 435 mm instead of the Spanish standard gauche of 1 686 mm. It is traversed by the high-speed trains AVE100 having 8,8 MV power, the locomotive AVE 252 having 5,6 MV power and multiple train units AVE104 with a 4,4 MV power per unit.

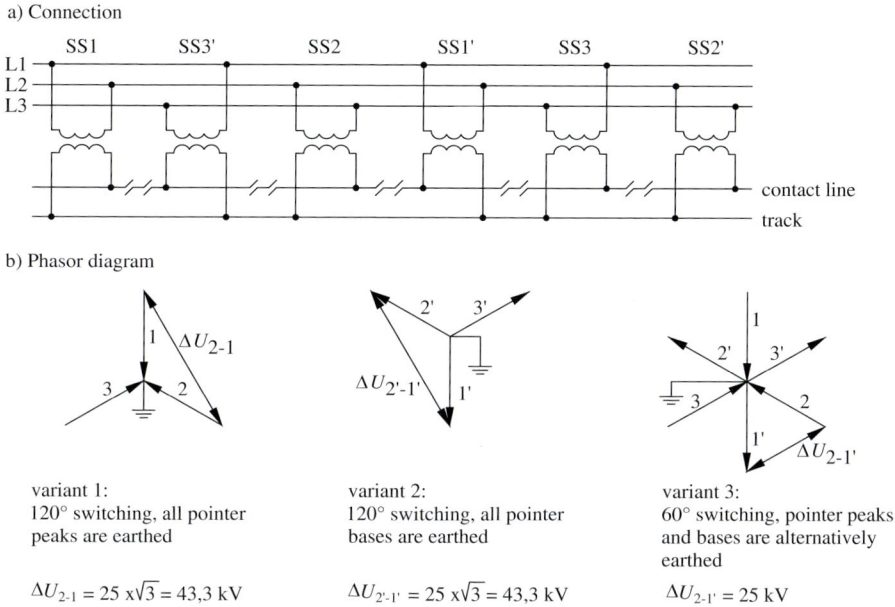

a) Connection

b) Phasor diagram

variant 1:
120° switching, all pointer
peaks are earthed

variant 2:
120° switching, all pointer
bases are earthed

variant 3:
60° switching, pointer peaks
and bases are alternatively
earthed

$\Delta U_{2\text{-}1} = 25 \text{ x}\sqrt{3} = 43{,}3 \text{ kV}$

$\Delta U_{2'\text{-}1'} = 25 \text{ x}\sqrt{3} = 43{,}3 \text{ kV}$

$\Delta U_{2\text{-}1'} = 25 \text{ kV}$

Figure 1.32: Cyclic changed switchings of the substations of the Madrid–Seville line.

The main part of the line was electrified by ADIF with 1 AC 25 kV 50 Hz and the overhead contact line type Re250 between the borders of Madrid and Seville (Figure 1.30). The last 8,5 km until the Madrid-Atocha station and 12,5 km until Seville-Santa Justa station were at first supplied by DC 3 kV. Concerns of electro-magnetic interference of signaling and communication systems at the stations in Madrid and Seville had been the reason for this decision. In 2001 also the two final sections of the line were converted to AC 25 kV [1.32]. Since then, for the trains only an equipment for AC 25 kV has been required and the system separation sections within the overhead contact line were dismantled [1.33].

Since the commission of the line, 100 million passengers had been counted until 2010. The punctuality of the trains achieved peak values of 99,8 % per year, whereby up to 20 high-speed trains run per direction and day. This proves the quality of line installation and maintenance. Figure 1.31 shows the profile of the line and the running speed. The line starts at Madrid at 640 m altitude and ends in Seville at 46 m altitude. The line is a double-track line with 4,3 m distance between the track axes. Along the line there are three stations for passengers in Ciudad Real, Puertollano and Cordoba. Rails of the type UIC60 were delivered in lengths of 288 m and thermally welded to a jointless track. Different designs of points were adopted whereby those having 17 000 m radius in the branch can be traversed at 220 km/h. The moveable frogs enable an operation with 300 km/h on the going-through track.

1.6.4.2 Electric design

The electric design assumed five minutes headway and train units with the mentioned characteristics. The power supply system AC 25 kV 50 Hz was chosen resulting in twelve substations under consideration of the possibilities of supplying them. It had to be considered that the voltage imbalance in the high-voltage network may not exceed 1,2 % continuously

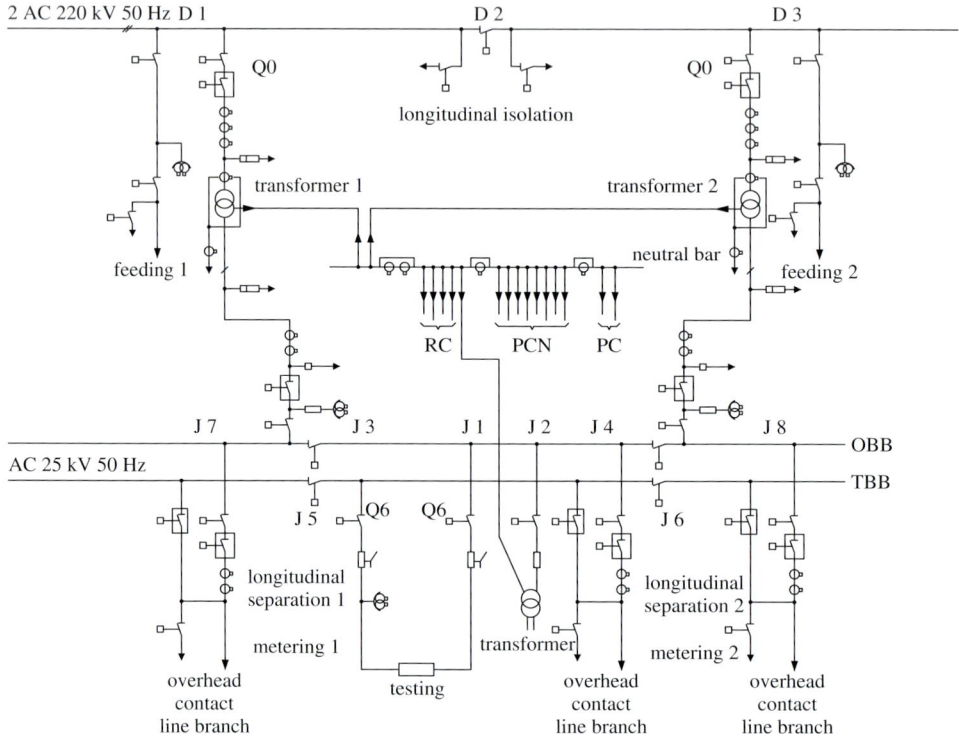

Figure 1.33: General circuit diagram for the substations of the Madrid–Seville line.
RC return current circuit, PCN potential connections of non-traction power supply equipment, PC potential connection of 25 kV traction power supply, J branch at 25 kV level, D line at 220 kV level, OBB operating busbar, TBB testing busbar

and 2 % at peak values. The distance between the substations is 40 to 50 km, whereby one substation supplies approximately 25 km of two adjacent line sections. The required phase separation sections are situated between the two adjacent substations. Nine substations of the line were connected to the 220 kV and three to the 132 kV high-voltage network arranged according to Figure 1.32, variant 3. As a result of this connection a 60° angle between the phase conductors of the neighboring substations was achieved. Therefore, the difference of potential at the phase separation section is only 25 kV and the insulation class 36 kV was always sufficient. This enabled the use of cheaper standardized components within the substations [1.28, 1.29].

1.6.4.3 Substations and their components

The traction supply substations are connected to the high-voltage network of the public power supply via AC substations erected close to each traction power substation. To these AC substations a three-phase high-voltage transmission line was constructed. The high-voltage substations are operated by the entity in charge of the public power supply. All substations have a uniform design. Their general circuit diagram as shown in Figure 1.33 has great similarity with DB block-type substations (Figure 1.18). As additional components, busbar disconnec-

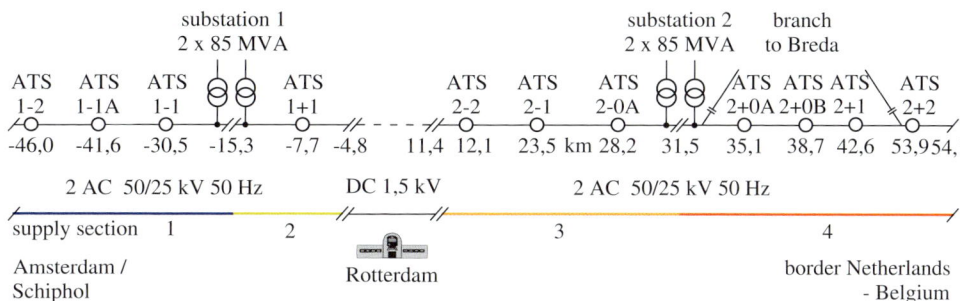

Figure 1.34: Traction power supply of the HSL Zuid line. ATS auto-transformer station

tors are installed on the high-voltage side and auxiliary transformers for the auxiliary supply of the substations at the medium-voltage end. The open-air switching equipment for the high-voltage section has different pole clearances due to the different voltages of 220 kV and 132 kV. The main transformers have a nominal power rating of 20 MVA each and are designed for a load of 150 % for 15 minutes and 200 % for 6 minutes following operation at nominal rating. The high-voltage circuit breakers with proven SF6 technology are used on the high-voltage side.

The indoor switch gear for the medium-voltage installation includes an operating busbar (OBB) and a test busbar (TBB). The vacuum circuit breakers are intended for nominal currents of 1 600 A and a cut-off capacity of 25 kA. The automatic testing of the contact line starts after triggering of a circuit breaker by the overhead contact line protection. In this case the overhead contact line is supplied at 25 kV from the test bar, the circuit breaker Q6 and a testing resistor. The testing of an overhead contact line is considered as good if the instrument voltage transformer T5 records a voltage in between 7 and 8 kV. Then, the circuit breaker Q0 recloses again. In case of a negative test, the circuit breaker stays open since it has to be assumed, that a continuous short circuit exists. The testing takes approximately 10 s.

The power transformers in each substation are protected by
- an overcurrent protection,
- a differential protection,
- a monitoring of circuit breaker failures and
- a Buchholz protector as well as temperature and a tank protection.

For each of the overhead contact line branches there is
- an undelayed overcurrent protection,
- an overcurrent time protection,
- a back-up overcurrent time protection,
- a thermal protection and
- a monitoring of circuit breaker failures.

1.6.4.4 Remote Control

The master control centre situated in the station Madrid-Atocha supervises and controls the technical processes by a SCADA system Siemens Vicos P500. A keyboard and a mouse are sufficient for the control of operations. To the remote control 41 units of the Sinaut type are connected, 12 of them concern the control of the substations. The other 29 units control

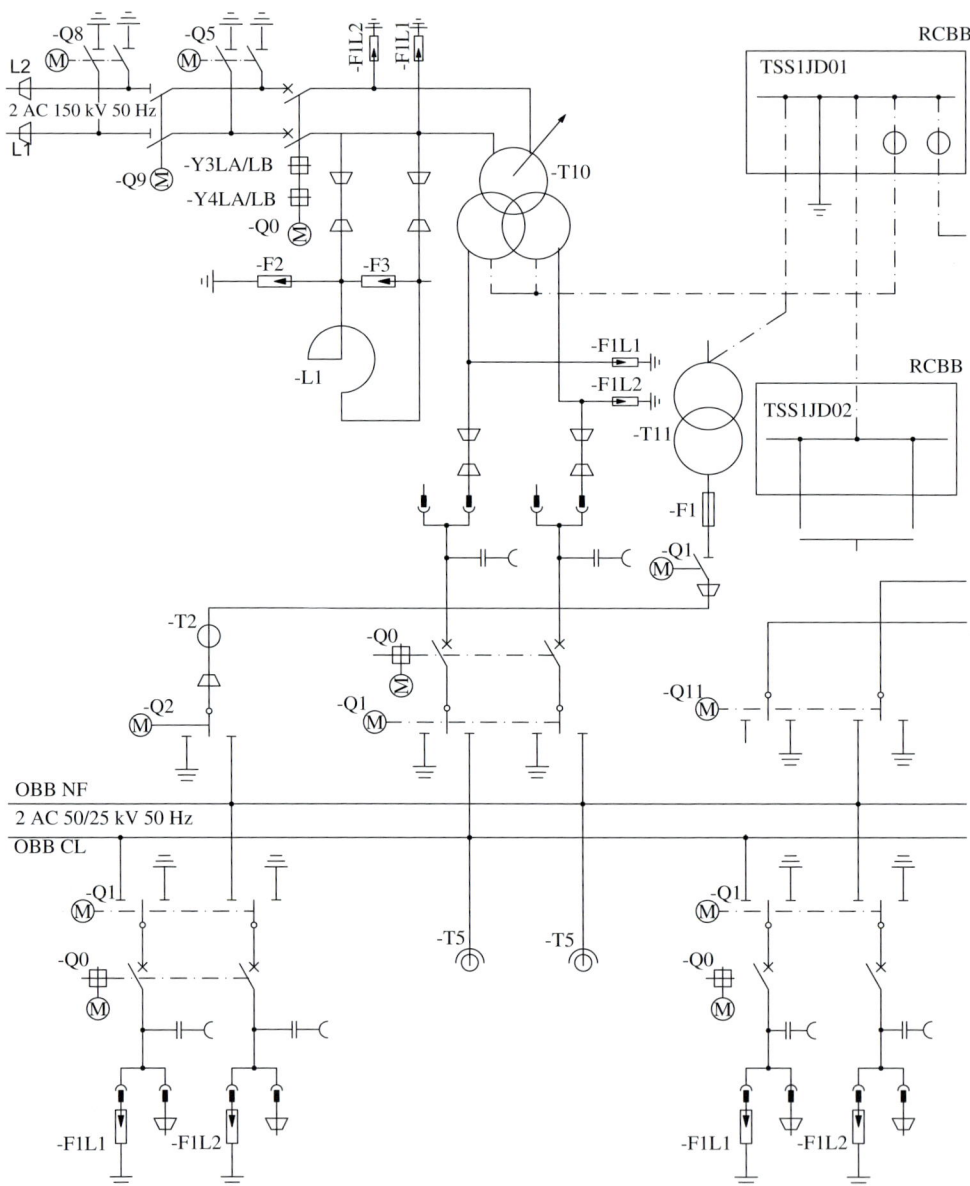

Figure 1.35: Overview switching diagram of the left part of the symmetrically designed substations AC 150 kV and 2 AC 50/25 kV of the HSL Zuid line.
Q0 Circuit breaker; Q1, Q9 disconnector; Q5, Q8 earthing switch; T1 current transformer; T5 voltage transformer; T10 power transformer; T11 auxiliary supply transformer; F1, F2, F3 surge arresters within the transmission line branch and the overhead contact line branches as well as the high-voltage fuse for the auxiliary supply; OBB operation busbar, NF negative feeder, OCL overhead contact line, RCBB return circuit busbar

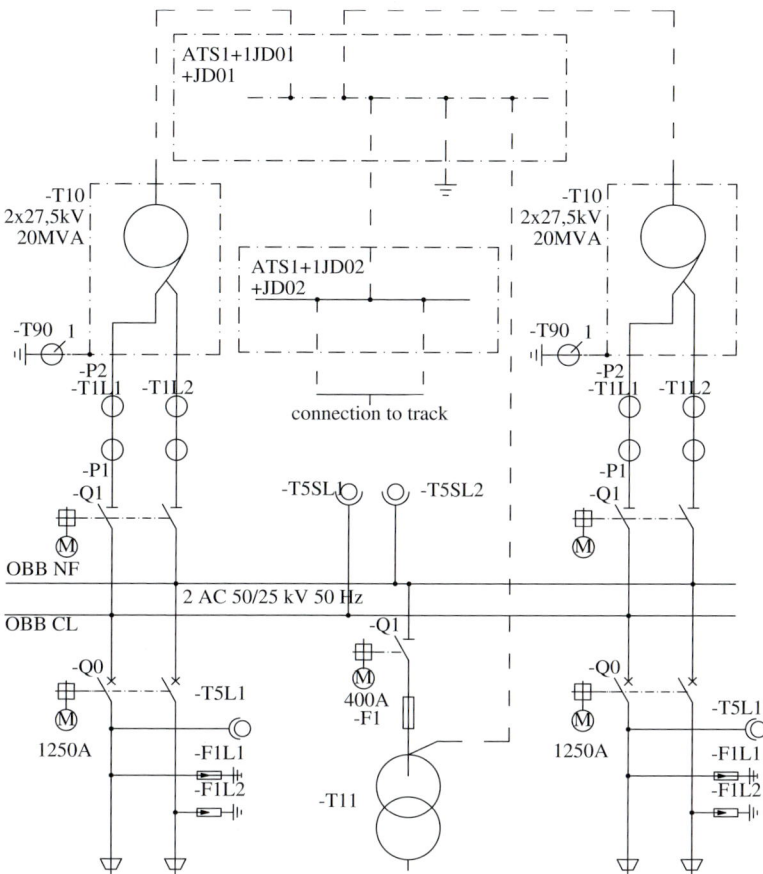

Figure 1.36: Overview of circuit diagram of an auto-transformer station of the HSL Zuid line. T10 auto-transformer; P1, P2 current instrument transformer; T90 voltage instrument transformer; other explanations see Figure 1.35

the technical systems within the buildings, the overhead contact line disconnectors and the protection installations along the line as well as the heating of the points. The remote control units within the technical buildings are connected by optical fibers. In between the substations and the technical buildings the data are exchanged via modem and cable. The system Vicos P500 is project-orientedly programmed and, therefore, flexible as far as modification of the system and the maintenance are concerned.

1.6.5 2 AC 50/25 kV 50 Hz traction power supply of HSL Zuid line

1.6.5.1 Description of the system

The high-speed line HSL Zuid in the Netherlands is a part of the trans-European railway network (TEN) and connects Amsterdam with Rotterdam and Brussels. It is a double-track, 88 km long line and starts close to the airport Amsterdam Schipol and runs via Rotterdam to the border to Belgium (Figure 1.34). The line is designed for commercial speeds up to

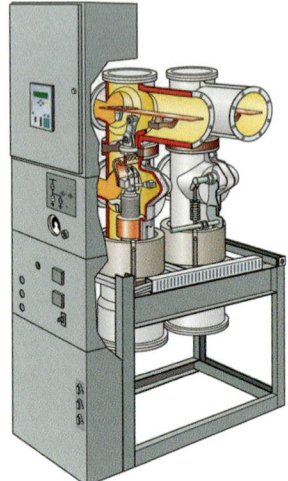

Figure 1.37: Branch of a twin-pole **Figure 1.38:** Auto-transformer station.
GIS switching station Sitras 8DA12
for 2 AC 50/25 kV systems and 1 AC
25 kV systems with testing busbar.

300 km/h and three minutes minimum head way, operated by two coupled, each 200 m long train units. The high-traction power required the electrification of the main line sections by 2 AC 50/25 kV 50 Hz. The power is supplied by two substations and seven auto-transformer stations. The connecting line sections designed for 160 km/h are supplied by DC 1,5 kV, as usual in the Netherlands. Therefore, multiple system vehicles run on the line. There are five system separation sections in between AC 25 kV and DC 1,5 kV sections and three-phase separation sections within the AC 25 kV sections.

1.6.5.2   Substations and auto-transformer stations

Both substations are supplied by the public 150 kV high-voltage network. By switching to different phase conductors of the high-voltage network the imbalance remains below the 1 % limit valid in the Netherlands for ten minutes mean values of the voltage. Half of circuit diagram of the symmetrically designed substations is shown in Figure 1.35. The feeding overhead line branch consisting of disconnectors, circuit breaker, earthing switch, instrument transformers and surge arresters supplies directly, that is without a busbar, the 150/2x27 kV 85 MVA substation transformer. The transformer is connected via the transformer branch to the 2 AC 25 kV operational busbar of the gas insulated substation Sitras 8DA12 (Figure 1.37) and with the centre tap of the return conductor busbar. The return cables lead from the transformer central tap to the track. Two overhead contact line branches supply to the overhead contact lines and the negative feeder of both running directions of the double-track line. Voltage instrument transformers record the operational busbar voltage. The left half of the busbar shown in Figure 1.35 is connected with the right half, which is not presented in the figure, via longitudinal disconnectors. An additional branch and a 50 kVA transformer 27,5/0,23 kV provides the power for the auxiliaries of the substation.

The auto-transformer stations are equipped with two 20 MVA transformers and an air-insulated 2 AC 25 kV substation (Figure 1.36 and 1.38). In Figure 1.36 the principle circuit diagram is shown for the auto-transformer station with the transformer branches, the line supply branches, the auxiliary supply and the return current bar. In the substations and auto-transformer stations digital protection devices of the SIPROTEC series are adopted. The overhead contact line protection SIPROTEC 7ST61 protects the overhead contact line as well as the negative feeder. A SCADA system of the type Telegyr 8000 in the control centre (SMC) Rotterdam controls and monitors in connection with the station control based on SIMATIC S7 the substations and the transformer stations. The SCADA system controls also the overhead contact line disconnectors.

1.7   Bibliography

1.1 *Sachs, K.*: Die ortsfesten Anlagen elektrischer Bahnen (The fixed installations of electric railways). Orell Füssli Publishing, Zürich Leipzig, 1938.

1.2 *Kummer, W.*: Die Maschinenlehre der elektrischen Zugförderung (A theory of machines for electric railway traction). Julius Springer Publishing, Berlin, 1920.

1.3 *Biesenack, H.; George, G.; Hofmann, G.; Schmieder, A. et al.*: Energieversorgung elektrischer Bahnen (Power supply of electric railways). B. G. Teubner Publishing, Stuttgart Leipzig Wiesbaden, 2006.

1.4 *Röhlig, S.; Behmann, U.*: Elektrifizierung in Europa (Electrification in Europe). In: Elektrische Bahnen 113(2015)2-3, pp. 72 to 75.

1.5 *Groh, T.; Harprecht, W.; Puschmann, R.*: Interoperabilitäat elektrischer Bahnen – 100 Jahre Vereinbarung für 16 2/3 Hz (Interoperability of electric railways – 15 kV 16 2/3 Hz agreement 100 years old). In.: Elektrische Bahnen 110(2012)12, pp. 686 to 699.

1.6 *Rossberg, R. R.*: Murnau–Oberammergau: 100 Jahre Wechselstrom mit Bahnfrequenz (100 years single phase a. c. with railroad frequency). In: Elektrische Bahnen 103(2005)1-2, pp. 45 to 50.

1.7 *Linder, C.; Heinze, R.*: Umstellung der Sollfrequenz im zentralen Bahnstromnetz von 16 2/3 Hz auf 16,7 Hz (Adjustment of frequency set value from 16 2/3 Hz to 16,7 Hz in central traction power grid). In: Elektrische Bahnen, 100(2002)12, pp. 447 to 454.

1.8 *Behmann, U.*: Nennfrequenz 16 2/3 Hz ade? (Farewell to nominal frequency 16 2/3 Hz?). In: Elektrische Bahnen, 100(2002)12, pp. 455 to 457.

1.9 *Behmann, U.*: Bahn-Normfrequenz bei IEC, CENELEC, DIN und VDE unverändert 16 2/3 Hz (Railway standard frequency with IEC, CENELEC, DIN and VDE consistently 16 2/3 Hz). In: Elektrische Bahnen 110(2012)1-2, pp. 10 to 11.

1.10 *Courtois, C.*: Fifty years of 50 Hz energy supply in France – Development and solutions. In: Elektrische Bahnen, 105(2007)4-5, pp. 232 to 240.

1.11 *Harries, K.*: Jane's World Railways 2009–2010. Jane's Information Group, London, 2011.

1.12 *Edel, R.; Schneider, E.; Schweller, M.*: Systemauslegung der Bahnstromversorgung von Gleichstrom- und Wechselstrombahnen (Design of traction power supply for DC and AC railways). In: Elektrische Bahnen 96(1998)7, pp. 213 to 221.

1.13 *Achtziger, K.-H. et al.*: Energieversorgung, Signalanlagen und Fahrleitungen für das Schienenfahrzeug-Prüfcenter Wegberg-Wildenrath (Power supply, railway signalling and contact lines for rail vehicle test centre Wegberg-Wildenrath). In: Elektrische Bahnen 96(1998)1-2, pp. 29 to 39.

1.14 *Schmidt, P.*: Energieversorgung elektrischer Bahnen (Energy supply of electric railways). transpress Publishing, Berlin, 1988.

1.15 *Lönhard, D.; Northe, J.; Wensky, D.*: Statische Bahnumrichter – Systemübersicht ausgeführter Anlagen (Static frequency converters for traction supply – System overview of realised converter stations). In: Elektrische Bahnen 93(1995)6, pp. 179 to 190.

1.16 *Pfander, J.-P.; Simons, K.*: Technik und Betrieb der Netzkupplungsanlagen 50/16,7 Hz bei der SBB (Technique and operation of network interconnection installations 50/16,7 Hz of SBB). In: Elektrische Bahnen 109(2011)1-2, pp. 55 to 62.

1.17 *Schneider, E.; Schuster, R.; Weschta, A.*: Statische Umrichter für die 15-kV-Bahnstromversorgung – Anlagenkonzept und Betriebserfahrungen (Static converters for the 15 kV railway power supply – Design of systems and operational experience). In: ETG Report No. 54 (1994), VDE Publishing, Offenbach, pp. 87 to 100.

1.18 *Halfmann, U.; Recker, W.*: Modularer Multilevel-Bahnumrichter (Modular Multilevel Converter for railway applications). In: Elektrische Bahnen 109(2011)4-5, pp. 174 to 179.

1.19 *Perschbacher, M.*: Bahnenergieversorgung der DB (Traction power supply of DB). In: Elektrische Bahnen 109(2011)1-2, pp. 50 to 54.

1.20 *DB Directive 955*: Schaltanlagen für Bahnstrom (Substations for traction energy). Deutsche Bahn AG, DB Energie GmbH, Frankfurt am Main, 2011.

1.21 *Rattmann, R.; Walter, S.*: Zweite Generation 16 2/3-Hz-Normschaltanlagen der Deutschen Bahn (Second generation of standardized 16 2/3 Hz switchgear stations at Deutsche Bahn). In: Elektrische Bahnen 96(1998)9, pp. 277 to 281.

1.22 *Schmidt, R.*: Der Standardumrichter bei der Deutschen Bahn (Standard converters used by Deutsche Bahn). In: Elektrische Bahnen 101(2003)4-5, pp. 177 to 181.

1.23 *Ebhart, S. et al.*: Lastenheft – Einpolige 15-kV-16,7-Hz-Innenraumbahnstromschaltanlagen (Specifications – Singe-phase 15 kV 16,7 Hz in-door traction power substations). DB Energie GmbH, I. EBZ5, Frankfurt am Main, 2009.

1.24 *Kinscher, J.; Jentzsch, P.*: Kompakte 40-kA-Bahnschaltanlage für 15 kV 16,7 Hz (Compact 40 kA 16,7 Hz switchgear for traction power supply). In: Elektrische Bahnen 109(2011)10, pp. 520 to 525.

1.25 *Siemens AG*: Abzweiggebundene Prüfeinrichtung AGP (Feeder-related testing device AGP). Product Information, Siemens AG, Erlangen, 2009.

1.26 *Weiland, K.; Ebhart, S.; Walter, S.*: Neue technische Entwicklungen für Bahnstromanlagen (New technological developments for traction power switchgear systems). In: Elektrische Bahnen 105(2007)4-5, pp. 206 to 212.

1.27 *Schmidt, P. et al.*: Energieversorgung elektrischer Bahnen (Energy supply of electric railways). Technik Publishing, Berlin, 1975.

1.28 *Braun, E.*: Stromversorgung der Hochgeschwindigkeitsstrecke Madrid–Sevilla (Power supply of the high-speed line Madrid–Seville). In: Elektrische Bahnen 88(1990)12, pp. 415 to 427.

1.29 *Braun, E.*: Connection of railway substations to the national three-phase power supply for the Madrid–Seville high-speed line. In: Elektrische Bahnen 88(1990)5, pp. 215 to 216.

1.30 *Behmann, U.; Rieckhoff, K.*: Umrichterwerke bei 50-Hz-Bahnen – Vorteile am Beispiel der Chinese Railways (Converter stations in 50 Hz Traction – Advantages in Case of Chinese Railways). In: Elektrische Bahnen 109(2011)1-2, pp. 63 to 74.

1.31 *Groh, T. et al.*: Elektrischer Betrieb bei der Deutschen Bahn im Jahre 2007 (Electric operation of Deutsche Bahn in 2007). In: Elektrische Bahnen, 106(2008)1-2, pp. 4 to 50.

1.32 *Vega, T.*: Schnellfahrstrecke Madrid–Sevilla durchgehend mit AC 25 kV 50 Hz betrieben (High-speed line Madrid–Seville completely operated with AC 25 kV 50 Hz). In: Elektrische Bahnen, 101(2003)3, pp. 134 to 135.

1.33 *Braun, E.; Kistner, H.*: Systemtrennstellen auf der Schnellfahrstrecke Madrid–Sevilla (Catenary sectioning devices on high-speed line Madrid–Seville). In: Elektrische Bahnen 92(1994)8, pp. 229 to 233.

2 Requirements and specifications

2.0   Symbols and abbreviations

Symbol	Definition	Unit
A	vehicle gauge A	–
B	reference gauge B	–
Bane Nor	Norwegian railway infrastructure manager	–
C	kinematic infrastructure gauge C	–
C_C	aerodynamic drag coefficient for conductors	–
CTI	Comparative Tracking Index	–
E	static effects on reference gauge B	m
F_K	dynamic contact force	N
G_{infra}	additional displacement of the vehicle in relation to gauge B	m
G_C	conductor reaction coefficient	–
G'_{ice}	length-related ice load on conductors	N/m
G1	multilateral static and kinematic vehicle gauge for international traffic	–
G2	multilateral static and kinematic vehicle gauge for national traffic	–
GA	interoperable infrastructure gauge A based on EN 15 273:2013	–
GB	interoperable infrastructure gauge B based on EN 15 273:2013	–
GB1	international infrastructure gauge GB1 based on GB	–
GB2	international infrastructure gauge GB2 based on GB	–
GC	interoperable infrastructure gauge C based on EN 15 273:2013	–
GUC	unified standard gauge	–
H	altitude of a system above sea level	m
K_1	coefficient of shape	–
L_N	frequency of indirect lightning strokes	
N	frequency of occurrence	
OV3	overvoltage category of electrical equipment in the open with overvoltage protection	–
OV4	overvoltage category of electrical equipment in the open without overvoltage protection	–
ΔP	loss of power	kW
Q'_W	length-related wind load on conductors	N/m
R	radius	m
R_{St}	aerodynamic resistance at the pantograph	kN
S	uplift of contact wire	m, mm
S'	projection of pantograph at the vehicle	m, mm
T	absolute temperature	K
ToR	Top of Rail	–
$U_{B\,max}$	peak value of lightning impulse voltage	kV
Z	surge impedance	Ω
a	height of infrastructure gauge	m

Symbol	Definition	Unit
$b_{h,ele}$	half width of electrical-kinematic pantograph gauge	m
$b_{h,mec}$	half width of mechanical-kinematic pantograph gauge	m
c	height of bevel of pantograph gauge	m
d	conductor diameter	m
e	width of bevel of pantograph gauge	m
f_u	voltage-depending distance	m
h_{CW}	contact wire height above top of rail	m
h_{SH}	system height	m
i_B	peak current of lightning stroke	kV
n	exponent	–
p	probability of exceedance, generally	–
q_b	basic wind pressure	N/m^2
$q_{b,0,02}$	wind pressure, probability of exceedance 0,02	N/m^2
$q_{b,0,10}$	wind pressure, probability of exceedance 0,10	N/m^2
$q_{b,0,20}$	wind pressure, probability of exceedance 0,20	N/m^2
$q_{b,0,33}$	wind pressure, probability of exceedance 0,33	N/m^2
$q_{b,H}$	wind pressure at the altitude H above sea level	N/m^2
q_z	reference wind pressure in height z above terrain	N/m^2
$q_{z=6}$	wind pressure in 6 m height above terrain	N/m^2
$q_{z=10}$	wind pressure in 10 m height above terrain	N/m^2
$qs'_{i,a}$	quasi-static lateral displacement of the vehicle inside the curve (i) or outside the curve (a)	m
s_0	flexibility coefficient	-
v	wind velocity	m/s
v_b	basic wind velocity in 10 m height above terrain, 10 min mean value, return period 50 a	m/s
$v_b(p)$	basic wind velocity, probability of exceedance p, deviating from 50 a	m/s
$v_{b,0,02,z=10}$	basic wind velocity, return period 50 a, 10 m above terrain	m/s
$v_{b,0,02}$	basic wind velocity, probability of exceedence 0,02	m/s
$v_{b,0,10}$	basic wind velocity, probability of exceedance 0,10	m/s
$v_{b,0,20}$	basic wind velocity, probability of exceedance 0,20	m/s
$v_{b,0,33}$	basic wind velocity, probability of exceedance 0,33	m/s
$v_{b,p}$	basic wind velocity, probability of exceedance p	m/s
v_{per}	permissible running speed	km/h
z	height above terrain	m
$\sum j$	random lateral deviation of railway vehicles	m
γ_F	partial factor for actions	–
γ_M	partial factor for materials	–
ρ	air density, at 15 °C and 0 m height above sea level: 1,225	kg/m^3

2.1 General requirements

2.1.1 Introduction

The successful operation of electric railways depends heavily on the availability and *reliability* of the traction power supply installation. The *requirements of the contact line*, whether an overhead line system or a third rail system, need to consider that the contact line is the only component in the traction power supply installation which cannot be installed redundantly. The demands on contact line systems result from their twin functions:
- *distribution line* for electrical energy from the substation to the electric traction unit and
- as a *sliding contact* for the locally varying current collection with the current collectors.

The required high availability of the contact line system necessitates thorough planning as early as practicable in the electrification planning cycle. It should make use of proven, carefully tested equipment with long service life, correct installation and effective maintenance during operation. The basic design demands made on a contact line installation are, therefore:
- persons and equipment must not be endangered by the operation of contact lines
- at all speeds up to the permissible maximum speed of the contact line type under consideration, the *dynamic interaction* of the current collector and the contact line or third rail must ensure that interruptions to power transmission do not occur under normal conditions
- it is vital that all components of the system should have a long *service life* including:
  - adequate mechanical and electrical strength.
  - resistance to loads imposed by *wind* and *ice* and *aggressive substances* in the air
  - *corrosion resistance* of all components
  - uniform, low wear of the contact wire
- during the design of overhead contact line installations in built-up areas, aesthetic and city-planning aspects should be observed
- *nature* and *environmental protection* are to be taken into account
- the *investments* for the installation and the *costs for operation* and *maintenance* should be as low as possible during the life cycle of the installation

The individual characteristics derived from these basic requirements of a contact line system can be classified into mechanical, electrical, environmental, operational and maintenance-related aspects. However, a strict distinction between the individual requirements is not always possible or necessary. In Europe, the *requirements of interoperability* across the borders of individual railway operators need to be considered additionally.

2.1.2 Mechanical Requirements

A basic *mechanical requirement* for a contact line installation is the strength of the wires, stranded conductors and other elements employed. To ensure fault free interaction of the contact line or third rail with the current collector, defined clearances between the contact line or third rail and the rails need to be maintained. With regard to overhead contact lines, the *contact wire position* is referred to the projected centreline of the super-elevated track. The *height of* the *contact wire* above rail is specified according to the type of railway and field of application. The *minimum contact wire height*, the *maximum contact wire height* and the *permissible contact wire gradient* are vital.

The forces in stranded conductors, wires and other components have to remain within permissible limits under all operational conditions. The *sag of conductors* may not exceed permissible values to ensure safety of people and the reliability of operations. It can be dangerous if the required *safety clearance* or the minimum clearance is infringed. *Minimum air gaps* to energized parts have to be maintained under all operating conditions, such as varying positions due to passing pantographs and differing sags. The wind and ice loads imposed on the conductors and components should affect railway operations as little as possible.

To ensure uniform and low wear of the pantograph collector strips and the contact wire, the *contact wire* needs to be installed with a *lateral offset* to the projected track centre line, called *stagger*. All mechanical loads acting on the overhead contact line must be carried by the poles and foundations and transmitted to the subsoil. Deformation of parts such as bending of poles or any incurred resonant vibrations should not affect the transmission of power.

Overhead contact lines, especially those for high speeds, need to comply with sophisticated quality criteria for successful power transmission. There are *static quality criteria* such as *elasticity* and its *uniformity* along the span and contact wire uplift. The *dynamic quality criteria* include *the wave propagation velocity*, the *Doppler factor* and the *reflection factor*. The contact force as a function of the running speed and its standard deviation are also significant quality features. *Overhead contact lines*, must be capable of allowing operation of trains with two or more pantographs in contact (see Chapter 10).

2.1.3 Electrical requirements

2.1.3.1 Type of current and nominal voltage

The *type of current* and the *nominal voltage*, including permissible variations (see Table 1.3) are significant characteristics of *electrical requirements*. A significant criterion for the performance of an electrified line is the *current-carrying capacity* of the contact line system. In comparison with industrial electricity distribution systems, short-circuits occur more frequently in contact line networks, therefore, the *short-circuit current capacity* of a contact line system is a determining feature.

In electric transport systems, the *voltage of the contact line network* has to be kept within given limits under all operating circumstances. The *mean useful voltage* can be considered as a scale for the suitability of a power supply system (Clause 5.2.4) The losses during power transmission must be kept within acceptable limits.

To minimise the consequences of faults on railway operations, contact line installations have to be divided into separately *fed sections*. Each installation must be designed to allow faults to be quickly and precisely localised. If conductors or other components of overhead contact line installations fail, defined fault conditions should occur to allow correct determination of the *short-circuit condition*.

2.1.3.2 Insulation coordination

Insulation co-ordination means the choice of the electric strength of the electro-technical installations depending on the voltages occurring in the contact line system. It is accomplished by the choice of associated insulating materials and their design and by respecting defined minimum air gaps, length of creepage paths that also consider the environmental conditions. *Protective measures and provisions* shall guarantee that people are protected against the possibility of *electric shock*, as stipulated in EN 50 124-1. The design voltage, which depends

Table 2.1: Allocation of the rated voltage and impulse voltage to the nominal voltage and overvoltage category according to EN 50 124-1:2005.

Nominal voltage U_n kV		Highest permanent voltage[1] $U_{max\,1}$ kV	Rated voltage U_{Nm} kV	Impulse voltage voltage U_{Ni} kV	
				Withstand level for over-voltage category	
DC	AC			OV3	OV4[2]
0,60		0,72	0,72	6	8
0,75		0,90	0,90	6	8
1,50		1,80	1,80	10	15
3,00		3,60	3,60	25	30
–	6,25[3]	8,00	8,30	45	75[1]
–	15,00	17,25	17,25	95	125
–	25,00	27,50	27,50	170	200

[1] according to EN 50 163, [2] applies to contact lines, [3] used in practice

Table 2.2: Minimum clearance for overhead contact lines according to EN 50 119, Tables 2 and 3.

Type of power supply	Clearance		Recommended clearance between differing phases		
	static voltage mm	dynamic mm	relative kV	static mm	dynamic mm
DC 0,60 kV	100	50	not applicable		
DC 0,75 kV	100	50	not applicable		
DC 1,5 kV	100	50			
DC 3,0 kV	150	50	not applicable		
AC 15 kV	–	–	26,0	260	175
	150	100	30,0	300	200
AC 25 kV	–	–	43,3	400	230
	270	150	50,0	540	300

on the nominal voltage and the utilisation of the equipment forms the criterion for electric strength. The utilisation is characterized by the overvoltage category. By selection of the design insulation level it is guaranteed that the equipment possesses the required withstand voltages. Withstand voltages are voltages with a representative wave shape that insulation withstands by a defined probability.

In case of contact line systems the insulation coordination consists of three steps:

– determination of the *impulse withstand voltage* U_{Nm} depending on the nominal voltage and the overvoltage category forming the basis for electrical testing of components. Contact lines and feeding lines of electric railways are assigned to the overvoltage category OV4 according to EN 50 124-1, Clause 2.2.2.1. However, circuits directly connected to the contact line system, protected by direct or indirect overvoltage protection devices are assigned to the overvoltage category OV3. The impulse voltage withstand level is given in Table 2.1.

– determination of minimum air clearances depending on the nominal voltage according to Tables 2 and 3 of EN 50 119:2009 (see Table 2.2). There is a distinction between permanent and temporary strains. For section insulators the permanent clearance may be 50 mm for voltages up to 3 kV, 100 mm for 15 kV and 150 mm for 25 kV

Table 2.3: Pollution severity levels and specific minimum creepage distances for insulation design according to IEC 60 815:1986 and EN 50 124-1 relating to the system voltage phase to earth.

Pollution level	Specific minimum creepage distance in mm/kV			Examples of typical environments[1]
	AC[2]	AC[3]	DC[4]	
light PD3A	28	25	28	– areas without industries and with a low density of houses equipped with heating plants – areas with a low density of industries or houses subjected to frequent wind and/or rainfall – agricultural areas – mountainous areas all these areas shall be situated at least 10 km to 20 km from the sea and not exposed to winds directly from the sea
medium PD4	35	30	36	– areas with industries not producing highly polluting smoke and/or with average density of houses with heating plants
heavy PD4A	43	40	46	– areas with heavy industrial density and the suburbs of large cities where the high density of heating systems causes contamination – areas with high density housing and /or industries subject to frequent wind and/or rainfall – areas close to the sea or exposed to relatively strong winds – winds from the sea – contact lines in tunnels
very heavy PD4B[5]	54	50	58	– areas exposed to wind from the sea but not too close to the coast (at least 10 km to 20 km distance) – areas generally of moderate extent subjected to conductive dusts and industrial smoke producing very thick conductive deposits – areas generally of moderate extent, very close to the coast and exposed to sea spray or to very strong and polluting offshore winds – desert areas, characterised by no rain for long periods, exposed to polluting winds carrying sand and salt, and subjected to regular condensation – contact lines in tunnels with heavy pollution

[1] see also EN 50 124-1, Table A.4
[2] according to IEC 60 815 multiplied by $\sqrt{3}$ since the creepage paths according to IEC 60 815 refer to the phase-to-phase voltage
[3] according to EN 50 124-1, Table A.7, these values apply to insulating material group II according to IEC 60 112
[4] recommended values due to experience
[5] in EN 50 124-1 wrongly called PD7

– determination of *required creepage path* depending on the nominal voltage according to Table 2.1 and the pollution level according to Table 2.3. In EN 50 124-1 seven pollution levels are defined. For contact lines, the levels PD3A, PD4, PD4A and PD4B need to be considered. For the creepage distances for insulation of contact lines reference is also made to IEC 60 815

Example 2.1: Insulation coordination will be explained using a 15 kV overhead contact line as an example. For the overvoltage category OV4 and 15 kV the impulse withstand voltage of 125 kV is obtained from Table 2.1. Therefore, 150 mm clearance follows from Table 2.2 for static, permanent conditions and 100 mm for temporary situations, e. g. a conductor swung out due to wind. According to Table 2.1 the rated voltage 17,25 kV corresponds to 15 kV nominal voltage. The Table 2.3 recommends a creepage path of 43 mm/kV for heavy pollution (PD4A) and the total minimum creepage path of $17,25 \cdot 43\,mm = 742\,mm$.

Electric stressing and simultaneous electrolytic pollution can create conductive paths and, therefore, creepage paths on the surface of insulating materials. Insulating materials are classified into four categories by the *comparative tracking index (CTI)* contained in EN 60 664-1. For contact lines only materials according to the classes I and II are accepted. Insulating materials are tested according to EN 61 302, EN 60 112 and EN 50 122-3. Creepage paths and air clearances may not be summed.

2.1.3.3   Interference

Unwanted interference to the supplying public network, such as harmonics and imbalances should be as low as possible. The transmission of power through the contact line network and adjacent lines of all kinds can be affected through inductive, capacitive and galvanic coupling. With direct current railways, extensive measures are necessary to limit *stray current corrosion*. When operating AC and DC vehicles on the same tracks, the requirements according to EN 50 122-3 need to be met because *rail-to-earth voltages* occurring during operations or under fault conditions may not exceed permissible limits.

2.1.4   Requirements due to interoperability

The interoperability of the European rail system is a political goal of the European Union (EU) to improve trans-European rail traffic in the Union. In view of this goal the EU ratified the following directives:
- 96/48/EC for the trans European high-speed system on the interoperability of the trans-European high-speed rail system [2.1] in 1996
- 2001/16/EC for the trans-European conventional rail system [2.2] in 2001
- 2004/50/EC for amending the directive 96/48/EC [2.3] in 2004 and
- 2008/57/EC for the interoperability directive on the modification of the mentioned directives [2.4] in 2004

The directives classify the trans-European rail system into subsystems:
- *infrastructure*
- *energy*
- *train control, train management, signalling*
- traffic management and control
- *rolling stock*
- *telematic applications for passenger and freight transport*

Technical Specifications for Interoperability (TSI) were introduced as links between the directives and the European standards (Figure 2.1). Overhead contact lines are part of the Energy subsystem. For each subsystem a TSI was established and published. The first edition of the TSI for the Energy subsystem [2.5] was published in 2002. Contents of this TSI are described in [2.6] and includes detailed goals of interoperability to tear off:
- the technical and operational borders for traffic on rails and
- the hindrances to procurement of rail equipment

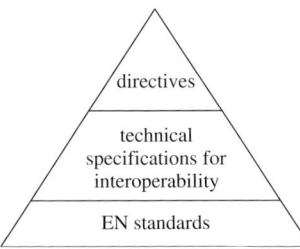

Figure 2.1: Hierarchy of European specifications for interoperability.

The high-speed advanced-technology trains must be designed to permit and take into account the technical development and uninterrupted travel:
- at a speed of at least 250 km/h on lines specially built for high speeds and enabling speeds over 300 km/h to be reached on suitable line sections
- at speeds up to 200 km/h on specially upgraded existing lines

High-speed trains are those designed for 250 km/h and above.

There are three categories of high-speed lines:
- specially built high-speed lines equipped for speeds equal to or greater than 250 km/h (high-speed lines, line category I)
- specially upgraded high-speed lines for speeds around 200 km/h (upgraded lines, line category II)
- specially upgraded high-speed lines with special features as a result of topographical, or town-planning constraints, where the speed is adapted to each case (connecting lines, line category III)

As a consequence of the TSI Energy, modification of existing standards and drafting of new standards was necessary. The existing standards EN 50 163 and EN 50 119 were involved. New standards were established:
- EN 50 388 on technical criteria for the coordination between power supply and rolling stock to achieve interoperability
- EN 50 367 on technical criteria for the interaction between pantograph and overhead contact line
- EN 50 317 on the requirements for and validation of measurements of the dynamic interaction between pantograph and contact line
- EN 50 318 on the validation of simulation of the dynamic interaction between pantograph and overhead contact line

The standards EN 50 367 and EN 50 388 refer to subjects described in the annexes to the first version of the TSI Energy [2.5]. A revised version of the TSI Energy for high-speed lines was drafted and issued in 2008 [2.7] to cover modified and new standards, revised information on the TSI Infrastructure and revised versions of directive [2.3]. The modifications were discussed in [2.8]. The TSI Energy for the conventional rail system was published in 2011 [2.9]. In 2014 the regulation [2.10] was published summarizing specifications for high-speed and conventional lines [2.7, 2.9]. The TSI Energy deals with:
- substations, that supply overhead contact lines
- sectioning posts and auto-transformers arranged in between the substations
- overhead contact lines which distribute energy to the trains
- current return circuits comprising all components conducting the operational or fault return currents
- interaction between pantographs and overhead contact lines

The TSI establish provisions for the design of subsystems and their components. These provisions need to be complied with when designing the overhead contact lines for the interoperable European rail system. The TSIs contain essential requirements concerning

– safety
– reliability, availability and maintainability
– health
– environmental protection
– technical compatibility

Strict application of the TSIs guarantee compliance with the essential requirements of the directives. The conformity of the installations needs evaluation [2.11, 2.12]. The impact of the TSI Energy on the design and installation of contact line systems is dealt with in the relevant clauses of this book. The compliance with the specifications of TSI Energy is a condition for vehicle access to DB's railway network [2.13].

2.1.5   Economic requirements

The overall investment for installation, operation and maintenance of contact line systems should be as low as possible. The components and elements should comply with the given requirements, be reliable and require little or no maintenance. Fittings, insulators and components should be easy to install and interchangeable as needed.

To minimise *wear of* the *contact wire* and pantograph *collector strips*, the contact combination of contact wire and collector strips must be made of suitable materials and designed for low wear.

If interruptions occur in railway operations, pre-planning should ensure that it is possible to ride on neighbouring tracks. At least the *electrical separation* of the contact lines of adjacent tracks and, wherever possible, the use of separate poles for each track should be considered to avoid dynamic impacts and operational dependencies between the contact lines of adjacent main line tracks. The contact line should be designed to minimise periods of line closure for planned maintenance work or contact line and track repairs.

2.2   Provisions for reliability and safety

2.2.1   Basic definitions

Within this book and in some standards, some common terms are used without being precisely defined, resulting in overlapping of terms.

Reliability is defined as the feature of a system, for example an overhead line, to carry out its purpose under given conditions, such as external loads, with a defined probability. The reliability refers to functioning of a system (see [2.14]).

Safety of people and installations exists, if there is a sufficiently improbable risk of danger. In the case of contact lines, safety means that people will not be in danger as a consequence of failure. Safety exists if the system had been installed and is being operated in compliance with the relevant standards. Therefore, safety exists or does not exist and as a consequence cannot be quantified. Generally and in technical contexts, the terms *safety* and *reliability* are not well distinguished.

The term *risk* implies the possibility of a failure. Technically speaking, risk combines the probability of a failure and the consequences of that failure and can be measured by money or inoperable periods.

Serviceability implies functionality according to given conditions that are not infringed. For example, if deflection of a pole or displacement of a contact wire under wind action occurs, however, the interaction between pantograph and contact line, up to the serviceability limit, may not be disturbed (see Figure 2.2).

The *strength* of a system is reached, if the serviceability of a system is exceeded, without damage to structures. If deflection of poles under wind action or displacement of the contact wire by more than the serviceability limit occurs and disturbs the interaction between pantographs and contact wires (see Figure 2.2), then the system is not anymore serviciable but its strength will not be inferred. Up to the failure limit, also called limit strength, no damage may occur to the structures. In the *intact state*, the system is fully functional under the design loads.

In the *damaged state* the system remains functional, however repairs might be necessary to ensure the system is capable of performing to its maximum capacity.

In the *failed state* the system is no longer functional or usable.

Availability characterizes that period of time e. g. during one year when the system can be used without limitations.

2.2.2 Standards

When installing and maintaining overhead contact line systems, many interrelated important aspects need to be considered, which have been the subject of international, regional and national standards. Annex 1 lists important standards relevant to contact lines of electrified railways which have been in force since 2015. The standardisation of contact line installation is under continuous development. In 2015, some hundred differing specifications and standards existed. Important provisions and standards for the installation and operation of contact lines are contained in the TSI Energy issued in 2014 [2.10] for the railway systems in the European Union as well as the standards EN 50 119, EN 50 121 series, EN 50 122, EN 50 163 and EN 50 367.

2.2.3 Loading and strength

During operation, contact line systems are subjected to *electrical* and *mechanical loads* resulting from electric voltage, currents and climatic environments. All elements of the overhead contact line system should withstand these effects electrically and mechanically with sufficiently high reliability. The required reliability is expressed by partial factors on the load and the strength side:

$$\text{load} \cdot \gamma_F \leq \text{strength}/\gamma_M \quad ,$$

where
γ_F is the partial factor on the load and
γ_M is the partial factor on the strength side.

If this equation is complied with, the basic design requirements are met.

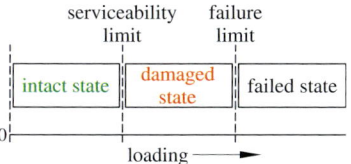

Figure 2.2: Conditions and limits of functions of systems and components.

2.2.4 Hazards due to electricity

The presence and types of electricity in the vicinity of railways can pose a hazard to life, installations and equipment. These hazards can arise from:

- *overhead contact line to rail voltage*
- *operating currents* and the *short-circuit currents*
- *electric fields*
- *magnetic fields*
- *rail to earth potentials*
- *induced* longitudinal *voltages*
- *capacitive charges*

These hazards must be limited to acceptable values by adequate design and maintenance of installations, in compliance with relevant standards.

2.2.5 Protection against electric shock

2.2.5.1 General protection against electric shock

When an electric current flows through a human body or the body of an animal, a pathophysiological effect described as an *electrical shock* or *electrical accident* can occur as a result of direct or indirect contact with live parts. In design, construction and operation of contact line installations, essential measures are required to prevent electric shocks from direct and indirect contact and are stipulated for contact line systems with nominal voltages

- up to AC 1 000 V and DC 1 500 V and
- above AC 1 000 V and DC 1 500 V each divided in
- public and restricted areas.

The protective measures are specified in EN 50 122-1, to avoid the unintended touching of live components. However, these protective measures are not always capable of preventing an intended direct contact.

2.2.5.2 Protection against electric shock by direct contact

Protection against electric shock by direct contact can be implemented by *protective clearances* or *protective barriers*, if the required clearance cannot be achieved.

Protection by clearance

Standing surfaces to which people have access shall have a minimum clearance as protection against direct contact with live parts of contact line installations or live parts of vehicles as shown in Figure 2.3. These clearances have to be met under all operating conditions.

In the case of protection by clearance, compliance with minimum height of overhead contact lines, booster and feeder lines above rails is required. As an example at road crossings with

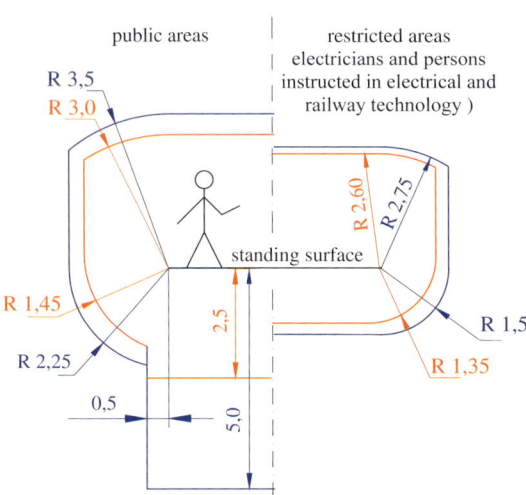

Figure 2.3: Minimal clearances to accessible live parts on the outside of vehicles as well as to live parts of overhead contact line systems from standing surfaces accessible to persons for low voltages up to AC/DC 1 kV/1,5 kV (red) and for high voltage above AC/DC 1 kV/1,5 kV (blue) (according to EN 50 122-1). Dimensions in m

Figure 2.4: Anti-climbing device at poles by Bane Nor in Norway.

an AC 15 kV overhead contact line, the minimum clearance between the road surface and the lowest point of the overhead contact line is 5,5 m (Clause 12.4.6). Furthermore, a distance of 2,5 m should be maintained between overhead contact lines and the branches of trees and bushes, under all conditions (Clause 12.7.5.2).

Protection by screening or guarding

Protection against direct contact can be achieved *by screening* energised parts to prevent contact using items such as solid walls, solid doors, gratings and grating doors made of conductive material. Gratings should have a mesh size less than 1 200 mm^2. This mesh size is required to a height of at least 1,8 m if the energised parts are higher than the standing area. Barriers should have a minimum distance of 0,6 m to live parts. Standing surfaces above live parts should be solid and have to project by at least 0,5 m from the live parts on all sides. *Anti-climbing devices* are usually unnecessary. However, some railway operators adopt anti-climbing measures (Figure 2.4).

2.2.5.3 Protection against electric shock by indirect contact

Indirect contact is contact by persons or domestic animals with conductive parts within the railway installation, also called bodies of electro-technical equipment, that are not normally energised but which may become energised under fault conditions. Therefore, in EN 50 122-1 contact line and pantograph zones are defined, within which all conductive parts need to be connected to the return circuit. Details are provided in Clause 6.2.9.

Table 2.4: Allocation of DB and Siemens overhead contact lines to maximum speeds.

Standard contact line design	v_{design} in km/h	Application
Re 100	100	through-going main lines, secondary lines and passing lines
Re 200	200	through-going main lines
Sicat S1.0	230	through-going main lines and passing lines
Re 200mod	230	through-going main lines and passing lines
Re 250	280 [1]	through-going main lines
Sicat H1.0	330 [2]	through-going main lines
Re 330	330 [2]	through-going main lines

[1]　when used by trains with two pantographs, 300 km/h with one pantograph
[2]　when used by trains with two pantographs, 350 km/h with one pantograph

2.2.5.4　Protection against electric shocks caused by the track potential

In electric railways, the tracks are used as conductors for the traction return current. The *rail potential* rises
　　－ with the power of the traction units,
　　－ due to the improved insulation features of modern permanent ways,
　　－ in the case of systems with insulated rails and
　　－ in the case of slab track.
During train operation track-earth-potentials are formed due to the transmission of power at the train and at the substation. They reach peak values at moving trains and at substations. Track-earth-potentials are also called rail potentials or rail to earth potentials in relevant standards. Track potentials depend on location and time. Protective measures are necessary to prevent electric shocks from rail potentials. The permissible rail potentials and touch voltages differ for AC and DC installations. Information on this subject including design of protective measures can be found in Clauses 6.3.4 and 6.3.5.

2.3　Requirements resulting from operations and line parameters

2.3.1　Introduction

The *purpose of the railway system*, the *operating conditions* and the type of line and track will lead to specific requirements and demands on the contact lines. The requirements resulting from the operating conditions are a function of the type of transportation required, i. e. local-area or long-distance traffic, the traffic frequency and the mass of the trains using the line. The line and track conditions particularly affecting the contact line design are the *track design*, the gauge and the geographical location of the line.

2.3.2　Operating requirements

2.3.2.1　Long-distance main-line traffic

For *long-distance traffic* the railway has to transport trains of a given mass between two stations in the network within a given time and according to a set schedule. The contact line

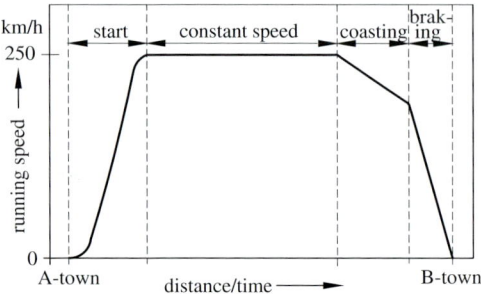

Figure 2.5: Speed-distance graph for a specific railway line.

Figure 2.6: Distance-time graph of a train on a specific railway line.

system needs to be matched to the required traffic volume capacity. The *traffic volume capacity* is a measure of the traffic that a railway line can handle. It is defined as the number of trains that actually run on the line within a given period. A train run can be subdivided into acceleration, steady-speed and braking phases (Figure 2.5). These phases will be repeated in various sections, depending on the line topography and train type. The scheduled run, the track and the geographical location of the line also determine the permissible and required speed for which the respective traction contact line needs to be designed. In long-distance traffic, the train speed is one of the essential system characteristics. As an example, DB's (German railway) and Siemens's standard contact line designs are allocated to specific running speeds as shown in Table 2.4.

Figure 2.6 shows the *distance-time graph* of a train on a specific railway line. The force required to move a train depends on the line topography and the train's traction-force characteristic, being a function of the envisaged speed (Figure 2.5) and curve radius. This function, the required speed and the related efficiency coefficients then determine the power to be transmitted via the current collector. During the acceleration phase, the force required to accelerate the train is superimposed on the force required to overcome motion resistance. Traction vehicles achieve their maximum power at speeds ranging between 80 km/h and 100 km/h due to their typical *traction-force/speed characteristics*. They can utilize the power overflowing the demand for a constant speed for further acceleration. The power supply systems and the contact line installations must be able to supply the required power for the planned *train traffic*.

The maximum train length will affect the length of platforms, of secondary and main lines in stations, of passing tracks as well as protective sections, neutral sections and location of signals. The design of the contact line installation also depends thereof.

The operating requirements and the *type of power supply* of long-distance main line traffic are the factors determining the design of overhead contact lines as traction energy supply installations for long-distance railways.

2.3.2.2 Local-area traffic

In *local-area traffic*, distinctions are made between *tramways*, *urban railways* and *metropolitan railway* systems according to their main characteristics as shown in Table 2.5. *Trolley bus lines* supplement local-area railway systems.

Whereas trams have to share the roadway with other road traffic, the tracks of urban railways run separately to some extent and metropolitan railway tracks are separate from all other

Table 2.5: Main characteristics of tramways, urban railways and metropolitan railway systems.

Characteristics	Tramway	Urban railway	Metropolitan railway
vehicle width	2,20 to 2,30 m	2,30 to 2,65 m	2,50 to 3,00 m
average speed	20 to 25 km/h	25 to 40 km/h	> 40 km/h
reserved track/roadway	none	mainly	exclusively
distance between stations	< 400 m	400 to 800 m	500 to 1 000 m

traffic. For this reason, trams and urban railways need to use overhead contact lines, while metros may use *conductor rails* or *overhead contact lines*.

The tight, close schedules of local-area traffic, particularly at peak hours, and the low voltage means that the contact lines must be able to conduct large currents. This is a characteristic of conductor rails. Overhead contact lines, formerly often installed as simple trolley-type contact wire without a catenary wire, are now mainly designed as catenary installations, providing the advantages of

- higher *speeds*,
- higher *current-carrying capacity*,
- better *collector running characteristics*,
- less *collector strip wear* and
- longer spans.

The use of *vertical catenary suspensions* is only avoided in areas where aesthetic, urban-planning and architectural aspects do not permit such systems. Then simple *trolley-type contact lines* with parallel feeder lines or twin contact wires are used instead.

To minimize *voltage drops* and the associated *power loss*, the overhead contact lines of both tracks of double-track lines can be electrically interconnected at regular intervals. Remote-controlled coupling disconnectors are only installed between the tracks on lines where single-track, two-way emergency operation is possible if a contact line fault occurred.

According to the German tramway operation regulations BOStrab (see [2.16]), the *minimum contact wire height* on lines in the open is 4,7 m. On the basis of experience gained with modern over-sized road transports, most urban communities now install overhead contact wires at heights of 5,0 to 5,5 m in the open, and approximately 4,0 m in tunnels due to the restricted space available.

The stagger of the contact wire at the supports is usually between $\pm 0,20$ and $\pm 0,40$ m. To prevent grooves from being formed in the collector strips from localised wear, the sweep of the contact wire should be at least 3 mm/m as the collector strip moves along it (see Table 4.14 and [2.17]).

In an attempt to minimise the overall line width, central supports can be used where possible, i. e. the poles are located between the tracks.

In local-area lines, the contact lines are mainly fed at both ends via *rectifier substations*, ensuring adequate distribution of the peak currents when trains are accelerating or braking.

Contact line disconnectors and section insulators should be located in the immediate vicinity of the substation to keep the feeder cables to the contact line sections as short as possible. The contact line *disconnectors* located at the *contact line section insulators* do not need to be remote-controlled. Electric separations should be arranged where the operation on one track will be continued in case of disturbances on the other.

With the ever-increasing deployment of traction vehicles able to feed braking energy back into the network, voltages exceeding the nominal traction voltage will occur frequently. This condition will need to be accommodated by the insulation coordination, if the voltages according to Table 1.1 are maintained.

The maximum permissible contact wire longitudinal inclination (gradient) and changes of gradients depend on the running speed but not on the design of the contact line. They are shown in EN 50 119, Clause 5.10.3, Table 11.

2.3.3 Requirements due to track spacing

2.3.3.1 Long-distance main-line traffic

Electric *long-distance railway networks* have their own right-of-way allowing free choice of pole location. To achieve mechanical separation of the overhead contact lines of double-track lines, the poles are placed on the outside edges of the track. The track spacing does not follow from the overhead contact line. However, separate poles may be required on lines with more than two parallel tracks to achieve *mechanical* and *electrical separation* of the contact lines of the individual tracks. In this case, enough space for the poles between the tracks is required. At DB, the spacing between main line tracks and overtaking tracks is 6,40 m.

In tunnels with rectangular cross section the overhead contact line supports can be located at the tunnel wall and in tunnels with circular cross section on the tunnel ceiling between the tracks. The *track spacing* usually used by the DB is:

– 4,00 m for train speeds up to 200 km/h
– 4,50 m for train speeds up to 350 km/h

The SNCF uses a track spacing of 4,20 m for their high-speed lines.

2.3.3.2 Urban and local-area traffic

Urban and *local-area railway systems* and trams often run on or beneath normal roadways without reserved space for the tracks. In such cases, existing buildings and structures and/or poles specially set up at suitable positions are used as supports and the contact line installations will be designed accordingly. On lines without poles between the tracks, track spacings do not effect the overhead contact line design. However, poles between the tracks are frequently used on lines running on their own reserved right-of-way. In this case, a track spacing of at least 3,60 m to 4,00 m should be used in case of tramways and railways in city centres. Transport systems that use contact rails should always have their own right-of-way.

2.3.4 Requirements due to track transverse and longitudinal gradients

2.3.4.1 Long-distance main line traffic

The running speeds determine the geometry and the layout of lines and the design of contact line installations of main lines for *long-distance traffic*, especially the curve radii and the associated cant and cant deficiency are important.

The operational limit value of cant is 180 mm. where 160 mm is acceptable for well maintained ballasted permanent way and 170 mm in the case of slab tracks. Additionally, an operational tolerance up to 150 mm is permitted for the cant deficiency [2.18, 2.19], resulting in a not-compensated centrifugal acceleration of 1 m/s^2 [2.18, 2.20, 2.21].

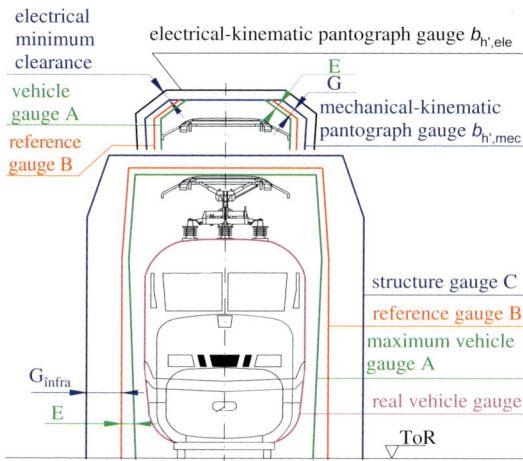

electrical minimum clearance

vehicle gauge A

reference gauge B

electrical-kinematic pantograph gauge $b_{h',ele}$

E
G

mechanical-kinematic pantograph gauge $b_{h',mec}$

structure gauge C
reference gauge B
maximum vehicle gauge A
real vehicle gauge

G_{infra}
E

ToR

Figure 2.7: Infrastructure gauge C and reference kinematic profile for the European interoperable high-speed network according to EN 15 273-2:2008, EN 15 273-3:2008 and [2.22].

Trains with tilting bodies permit a further increase in running speed of 14 % if passive tilting mechanisms are used or by as much as 30 % for active tilting mechanisms. If the train speeds in curves are increased, this will have a direct effect on the design of *contact wire stagger*. Traction vehicles with tilting bodies will experience larger lateral displacements of the pantographs due to the increased centrifugal forces at higher speeds. This means that the stagger of the contact line needs to be checked and the lateral contact wire position may need re-adjusting. It may also be necessary to replace steady arms and/or cantilevers. Alterations to the overhead contact line supporting structures are not normally needed if traction vehicles with tilting bodies and active pantograph controls are used. The maximum longitudinal gradient is generally limited to 3,5 % on future high-speed lines [2.19].

2.3.4.2 Local-area traffic

In *local-area traffic*, the operating speed is below 100 km/h. The design of the contact line installations is not only determined by the running speeds, but by the larger currents needed due to the lower traction voltage. Curve radii are smaller than on main lines, with some terminal loops of tramways in urban areas having radii of 18 m. *Line gradients* up to 11,0 % may also occur for adhesion-only vehicles but on newly-planned lines, an attempt is made to limit gradients to 5,0 %.

2.3.5 Clearance gauge related requirements

2.3.5.1 Vehicle and infrastructure gauges

To be able to calculate the usable contact wire lateral position an understanding of the terms used in EN 15 273-1 is necessary:
- static vehicle gauge A, also called loading gauge (see Figure 2.7 and Clause 4.4)
- reference gauge B
- infrastructure gauge C

The *static vehicle gauge* A (Figure 2.7) describes the envelope of the vehicle, where no vehicle component may protrude. The vehicle itself is at standstill on a straight track without cant.

Movement during running is not included in the vehicle gauge. The inclusion of movement results in the *kinematic vehicle gauge*. Starting with the static vehicle gauge and adding the following effects E results in the reference gauge B:

– horizontal displacements due to transverse play between vehicle body and wheel sets
– changes of the vehicle height due to wear, e. g. wear of the wheel rims
– vertical movements due to suspension movements caused by static and dynamic loading of the vehicles
– vertical displacements which occur due to the position of the vehicles on a convex or concave transition between gradients
– quasi-static lateral inclinations due to *cant deficiencies* or cant surplus of 50 mm
– asymmetries of the vehicle body of more than 1°, which result from production and adjustment tolerances and from the planned loading

The effects E need to be considered by the vehicle manufacturer when calculating the transverse reductions [2.19], where the gauge B is used as reference. This reference gauge forms the interface between vehicle and infrastructure.

By adding the projection S' to the reference gauge B as an exceedance of the reference gauge the limiting gauge for the statically required space for the vehicles is obtained for the case where the vehicle is situated in a curve and/or on a track with a gauge of more than 1 435 m (see Equation (4.98)), further, by adding the quasi-static lateral movement $qs'_{i,a}$ of the vehicle to the inside of the curve (i) or to the outside (a) the limiting gauge for the kinematically required space of the vehicles is obtained. If eventually the sum $\sum j$ of random lateral displacements, which result from irregular track positions is added the infrastructure gauge C is created as shown in Figure 2.7.

The vehicle may reach with the displacements E the reference gauge B and with the additional displacements G_{infra} according to

$$G_{\text{infra}} = S' + qs'_{i,a} + \sum j \tag{2.1}$$

the kinematic infrastructure gauge C. The additional displacements G_{infra} (Figure 2.7) need to be considered by the contractor and operator of the infrastructure. If the vehicle gauge A describes an outline which no vehicle component may protrude, the infrastructure gauge C forms an space into which no component of the infrastructure may project.

2.3.5.2 Pantograph gauges

The lateral movements of the pantograph determine its operational range (Figure 2.7) called in [2.10] and EN 15 273-1 the *mechanical-kinematic gauge* of the pantograph. No component of the pantograph may exceed the mechanical-kinematic gauge for both vertical and horizontal movements. On lines operated just by a single pantograph type, this type determines the mechanical-kinematic gauge. If several pantograph types with differing geometry are operated on the line then the mechanical-kinematic gauge will be determined by the type with the longest pan head.

Only the contact wire and steady arms are permitted within the pantograph mechanical-kinematic gauge. Live components may approach the mechanical-kinematic gauge but may not protrude across the mechanical-kinematic gauge into the pantograph profile.

If the electrical minimum clearance is added to the mechanical-kinematic gauge, the *electrical gauge* of the pantograph is obtained (Figure 2.7). Pantographs with insulated horns require electrical gauges that are smaller by the projection of the pan head horns.

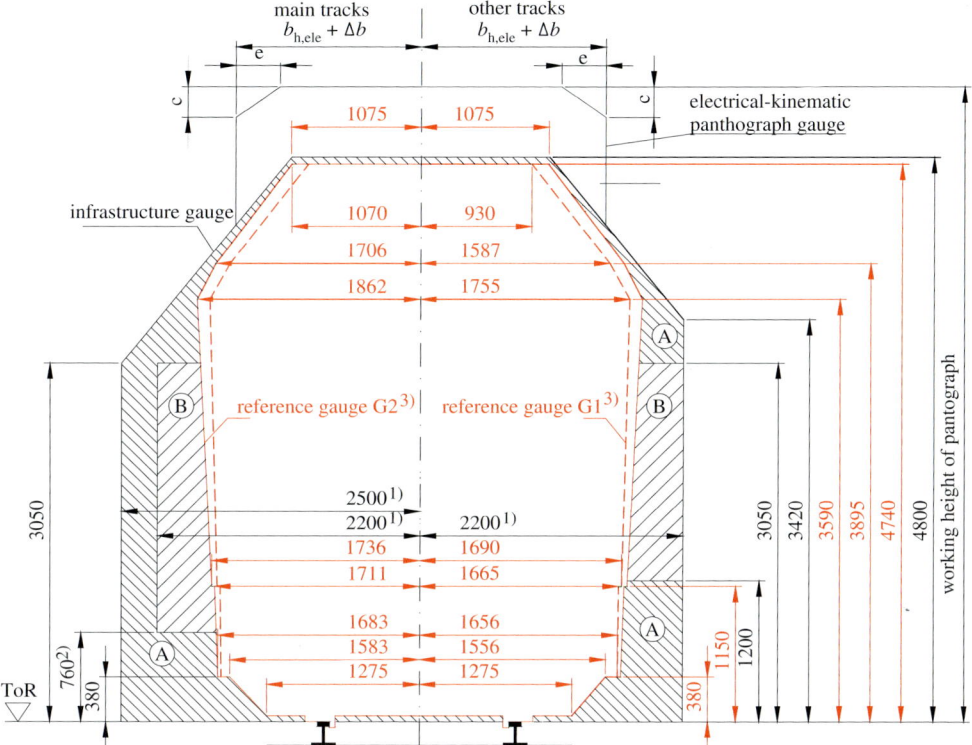

Figure 2.8: German kinematic infrastructure gauge for mainlines with 1 950 mm pantographs for $R \geq$ 250 m according to EBO [2.18]. Dimensions in mm

[1]   dimensions may be reduced by 100 mm in case of tracks only for City railway lines

[2]   dimensions may be reduced by 100 mm in case of tracks predominantly for City railway lines

[3]   determination of the electric pantograph gauge is based on the vehicle flexibility coefficient $s_0 = 0,225$ (see also EBO [2.18])

A   projections of civil engineering structures are permitted, e. g. platforms, ramps, marshalling installations, contact line foundations, signalling installations, as well as projections in case of civil works with the required security provisions

B   projections in case of civil works with the required security provisions are permitted

$b_{h,ele}$, Δb length of electrical-kinematic gauge according to TSI ENE:2014 [2.10] (see also Figure 4.27)

Pantograph type	c	e	Source
1 950	150	250	EBO[4] [2.18]
1 600	176	305	EN 15 273-3:2014, Figure C.11

[4] the dimensions 300 and 400 mm specified in EBO include 150 mm electrical clearance which need to be deduced to obtain the dimensions c and e as the electrical-kinematic pantograph gauge as shown in Figure 4.27

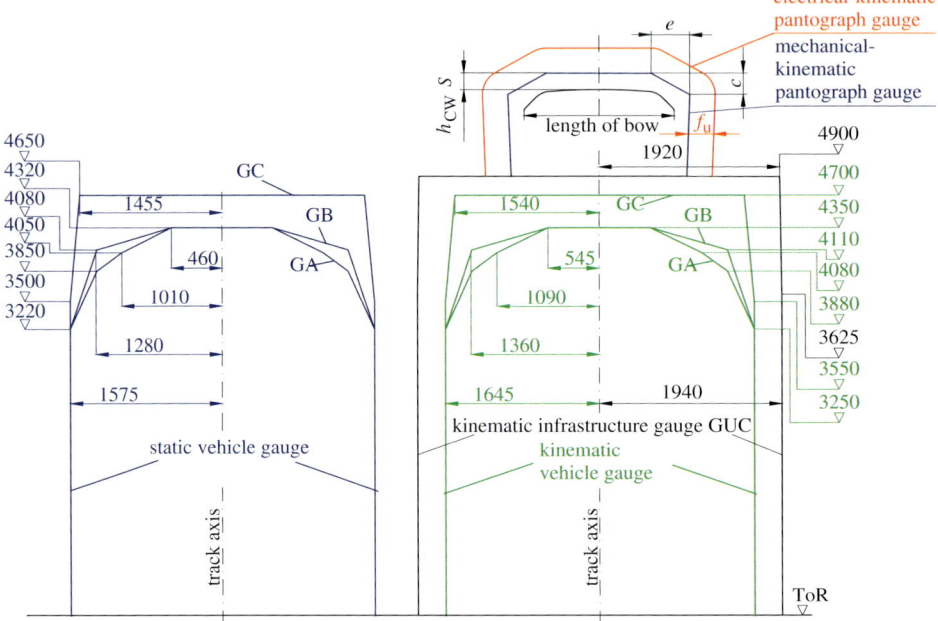

Figure 2.9: Static and kinematic vehicle gauges GA, GB, GC and kinematic infrastructure gauge GUC according to EN 15 273-1 and EN 15273-3, respectively.
Dimensions c and e for 1 600 mm and 1 950 mm long pantographs see Figure 2.8

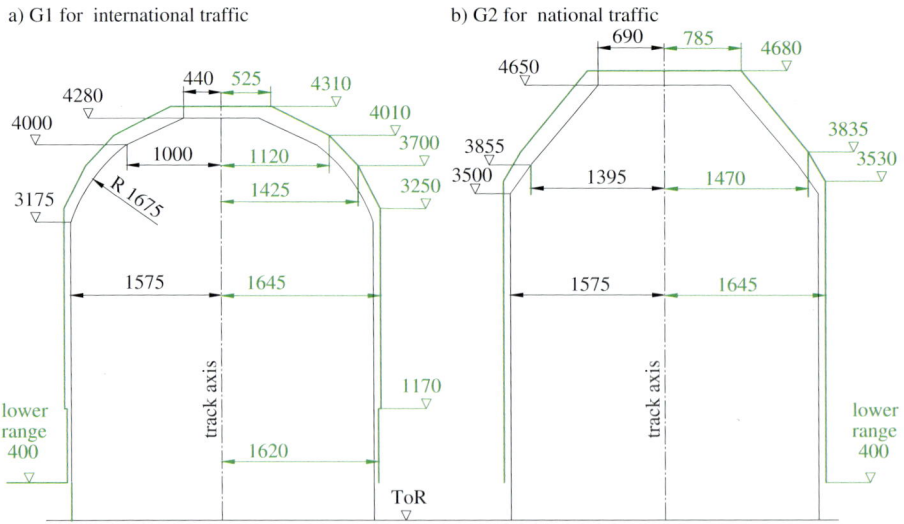

Figure 2.10: Static (black) and kinematic vehicle dimensions (green) of gauges G1 and G2 according to EN 15 273-1. Dimensions in mm

To simultaneously consider the electrical and mechanical impacts, which lead to the electrical and mechanical gauges, the standard [2.19] considers the two gauges as twin gauges, which both need to be taken into account as the infrastructure gauge by the infrastructure manager.

When transmitting energy the pantograph continuously touches the contact wire, the height of which varies. Therefore, the height of the pantograph gauge varies between the minimum and the maximum contact wire height. The uplift by the pantograph, height tolerances and other impacts are included.

2.3.5.3 Interoperable infrastructure gauge for main lines

In the context of interoperability in Europe and a common infrastructure gauge for main lines the standard EN 15 273-1, which was based on UIC 505 [2.23] and 506 [2.22], harmonized the national gauges. The allowances to half of the pantograph length include:
  - vibrations of pantograph accounting for 66 mm cant deficiency
  - displacement of the pantograph in curves
  - projection S'
  - quasi-static flexibility $qs'_{i,a}$ for the part of the cant deficiency exceeding 66 mm when a vehicle stands on a canted track or runs on a curved track
  - randomly caused lateral displacement $\sum j$
  - electrical clearance

The *pantograph gauge*, also designated as the mechanical pantograph gauge according to [2.9], needs to comply with when planning overhead contact lines.

With the exception of the contact wire and steady arms no components of the contact line may project into the pantograph gauge (Clause 2.3.5.2). The contact line foundations are permitted to project up to 0,38 m above top of rail into the lower part of the space A (Figure 2.8).

The independent and individual development of European railways led to different gauges. The existing differing gauges GA, GB and GC (Figure 2.9) were harmonized by the TSI for the subsystems infrastructure [2.19] and rolling stock [2.24] and by the International Railway Union (UIC) by the codes 505-1 [2.23], 505-4 [2.25] and 506 [2.22] targeting the goal of interoperability of the railway systems within the European Union. The small gauge GA (Gabarit A) shall be complied with on all lines. For *combined road/railway traffic*, for *heavy truck transports* etc. the larger gauges GB and GC were specified based on stipulated model loadings on special freight wagons. In Figure 2.9 the dimensions of these gauges are shown. The vehicle gauge GB was designed to accommodate the transport of standard shipping containers. In order to be able to transport 2,60 m wide containers instead of 2,50 m wide units, the gauge variant GB1 was specified. The gauge variant GB2 was designed for the *piggyback transportation* of trailer trucks on especially low wagons with a floor height of 0,27 m (Figure 2.10). Standard trucks and *articulated trucks* are transported on special wagons on certain corridor lines.

The gauge GUC defines the standard infrastructure gauge in accordance with EN 15 273-2, which was based on the interoperable gauge GC and is adopted in several European countries, especially also for the European high-speed network. The infrastructure gauge GUC was specified based on a fixed reference gauge and is limited to:
  - track radii ≥ 250 m
  - cant deficiencies ≤ 150 mm
  - change of gradients $1 : 2\,000$

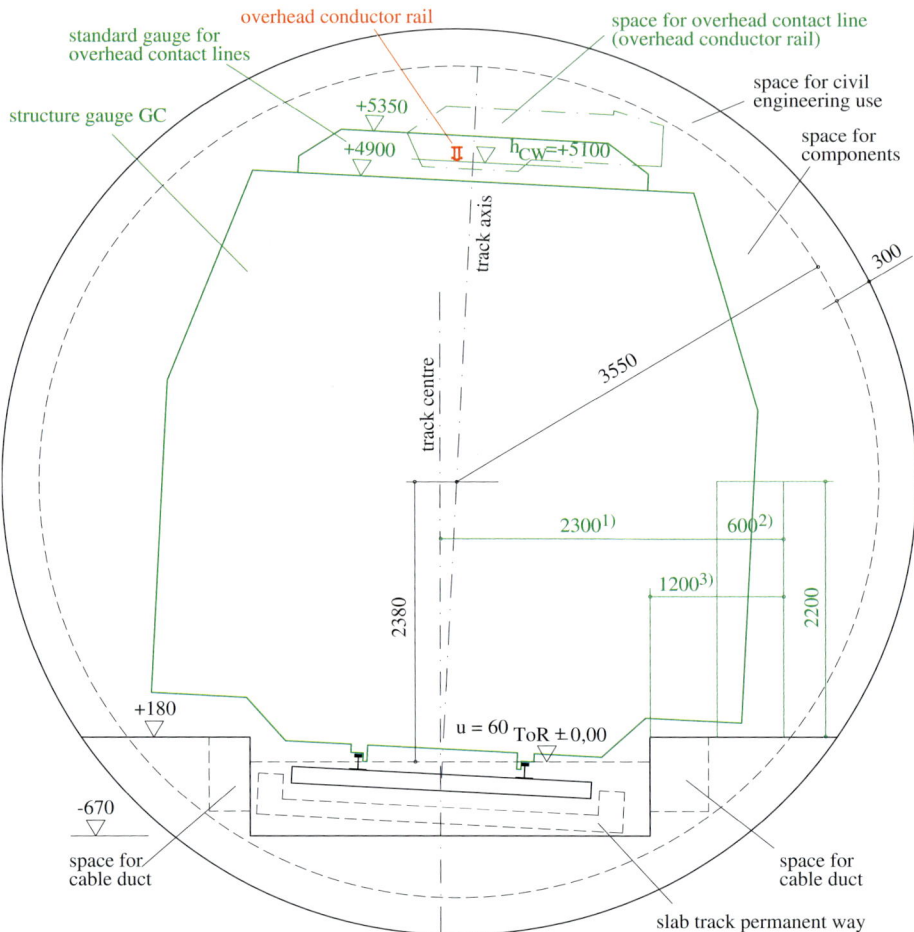

Figure 2.11: Circular tunnel cross-section with 39,6 m² space for installation for the infrastructure gauge GC with overhead conductor rail in the North-South-Tunnel Berlin [2.26].
dimensions in mm, 1) range of danger, 2) additional clearance, 3) emergency exit

The GUC infrastructure gauge includes a provision of 50 mm for track lifting. The GUC infrastructure gauge (Figure 2.9) was especially designed for the transport of heavy trucks, however, is also required to enable the use of comfortable *double decker passenger coaches* on high-speed lines. Consequently all new railway lines for the high-speed traffic in Europe will de installed utilizing the GUC infrastructure gauge. The stipulations provided in [2.19, 2.24] and EN 15 273-3, Figure B.2 and G.3 refer to:
  – new high-speed lines
  – existing high-speed lines
  – lines upgraded for high-speed traffic
  – their connecting lines
In Figure 2.8 the German infrastructure gauge as specified in EBO [2.18] is shown. This gauge corresponds to the gauges G1 and G2 where applicable. The dimensions of the pantograph gauge are shown in Figure 4.27.

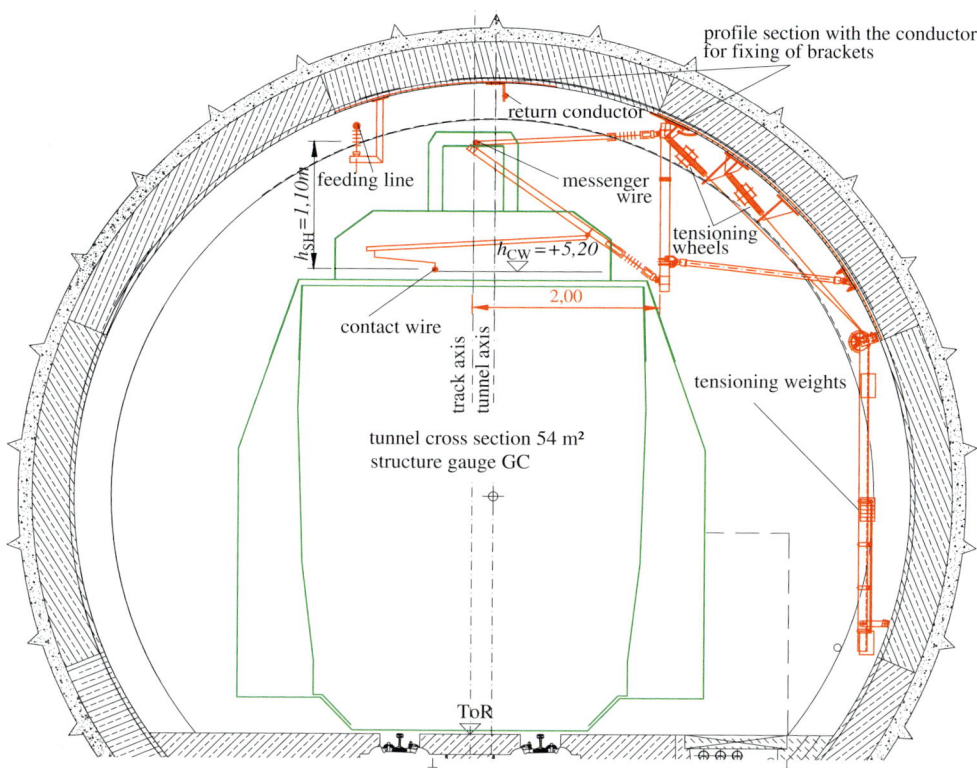

Figure 2.12: Circular tunnel cross-section with $54\,m^2$ for the structure gauge GC with a standard contact line Re 200 for 200 km/h running speed in the new Schlüchterner Tunnel close to Fulda.

2.3.5.4 Infrastructure gauges in tunnels

In general, the smallest possible cross section needs to be selected for tunnels to limit the investment. In Figure 2.11 the infrastructure gauge GC in a tunnel with $39{,}6\,m^2$ cross section is shown enabling the installation of a overhead contact line for up to 160 km/h running speed. Figure 2.12 depicts the cross-section of the Schlüchterner Tunnel having $54\,m^2$, with enough space available for the installation of cantilevers and tensioning equipment. The sections for attaching the contact line were integrated in the tubing elements made of concrete during pre-fabrication, thus reducing labour time during installation of the contact line.

An approximately constant contact wire height is required for optimum interaction between contact line and pantograph and low wear. If overhead contact lines run beneath buildings a uniform contact wire height is aimed at. However, if the clearance between contact line and building is too low, then steps need to be taken to adapt the contact line and/or to modify the permanent way (see Clause 12.4.9).

2.3.5.5 Infrastructure gauges for local railway traffic

There are much more differing infrastructure gauges in local transportation systems than in main line railways. This variety resulted from the separate development of the individual op-

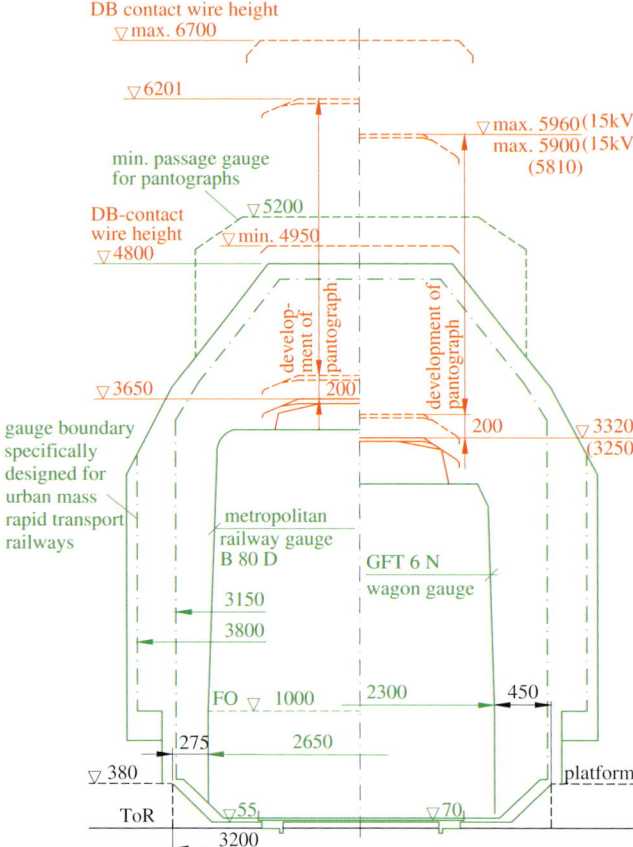

Figure 2.13: Infrastructure gauges of tramways and metropolitan railways (green) with DB's standard gauge (red) in Karlsruhe.

erating entities and their isolated networks and the increasing importance of mixed operation, which uses the same tracks for main line traffic, for regional connections and for trams, as in the cities of Karlsruhe, Chemnitz and Kassel. The infrastructure gauge adopted in Karlsruhe is shown in Figure 2.13 and there is a requirement to harmonize the infrastructure and track gauge. Figure 2.14 shows the infrastructure gauge for the Stuttgart tram.

2.4   Specifications due to pantographs

2.4.1   Design and functions

The purpose of the pantograph is to transfer electric power from the contact line to the electric traction unit. This transfer of power needs to be reliable both at standstill for the *auxiliary* and *convenience power* and for total power for the running traction vehicle. Figures 2.15 and 2.16 show the the high-performance pantographs DSA 350 S and SSS 400+ consisting of a base main frame, arms, pantograph head and drive. The design of the DSA-350 S pantograph, whose mass is approximately 100 kg, will be briefly explained. This *high-performance pantograph* is a single-arm unit [2.27, 2.28] designed for 350 km/h. The *base frame* has a mass of 53 kg including the *lifting drive* and *dampers*. The lower arm and control bar have a mass

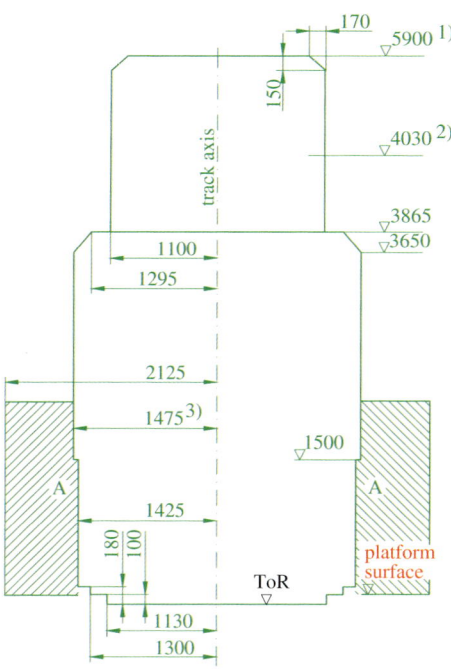

Figure 2.14: Infrastructure gauge of the tramway in Stuttgart.
[1] maximum contact wire height
[2] minimum contact wire height
[3] straight lines only
A protected space

Table 2.6: Stagger at supports and pan head length of European main line railways.

Country	Conventional lines		High-speed lines	
	stagger mm	length of pan head mm	stagger mm	length of pan head mm
Belgium	350	1 950	200	1 450 or 1 600
Denmark	275	1 950		
Germany	400	1 950	300	1 600 or 1 950
France	200	1 600 or 1 950	200	1 450 or 1 600
Great Britain	230	1 600	200	1 600
Netherlands	350	1 600 or 1 950	300	1 600
Norway	200	1 800	300	1 600 and 1 800
Italy	300	1 600	300	1 600
Austria	400	1 950	300	1 600 or 1 950
Portugal	200	1 450 or 1 600		
Schweden	200	1 800	300	1 600 and 1 800
Switzerland	150[1], 350[2]	1 450 or 1 600	150[1], 350[2]	1 450
Spain	200	1 950	300[3] or 200[4]	1 950 and 1 600

[1] straight lines; [2] track radii; [3] high-speed line Madrid–Seville
[4] high-speed line Madrid–Barcelona and succeeding high-speed lines

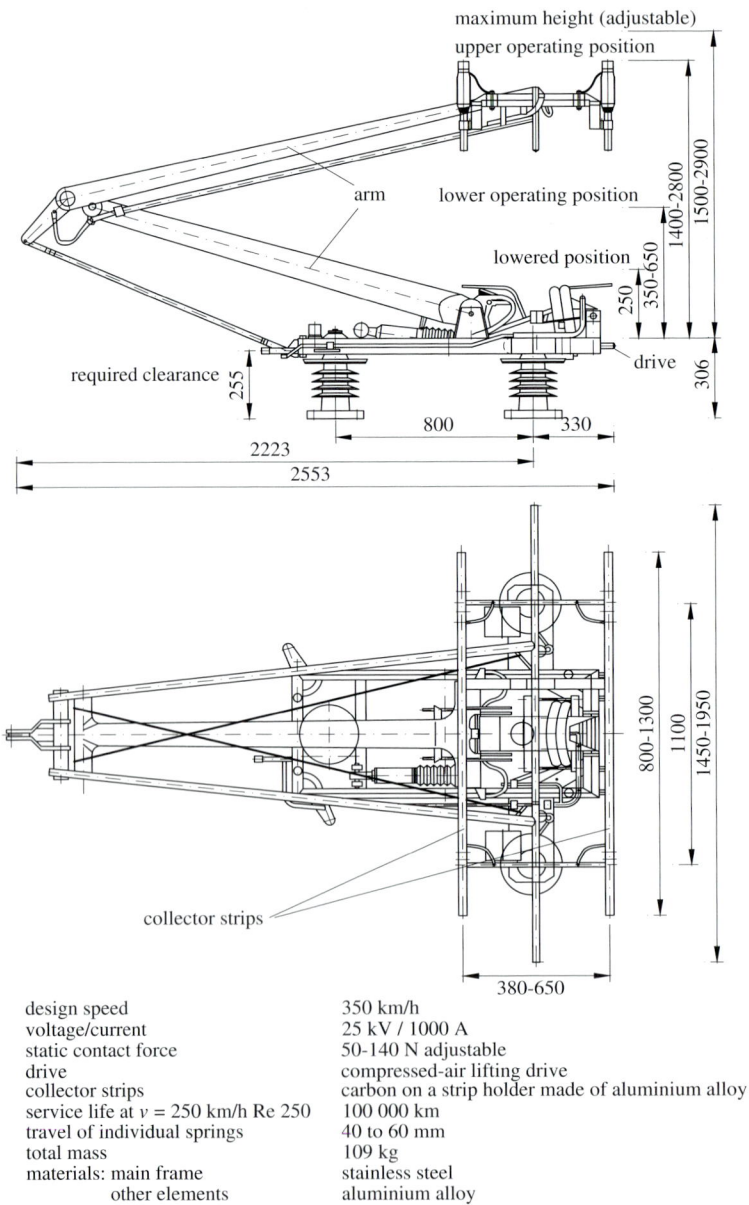

design speed	350 km/h
voltage/current	25 kV / 1000 A
static contact force	50-140 N adjustable
drive	compressed-air lifting drive
collector strips	carbon on a strip holder made of aluminium alloy
service life at v = 250 km/h Re 250	100 000 km
travel of individual springs	40 to 60 mm
total mass	109 kg
materials: main frame	stainless steel
other elements	aluminium alloy

Figure 2.15: Pantograph DSA-350 S [2.27, 2.28, 2.31, 2.38].

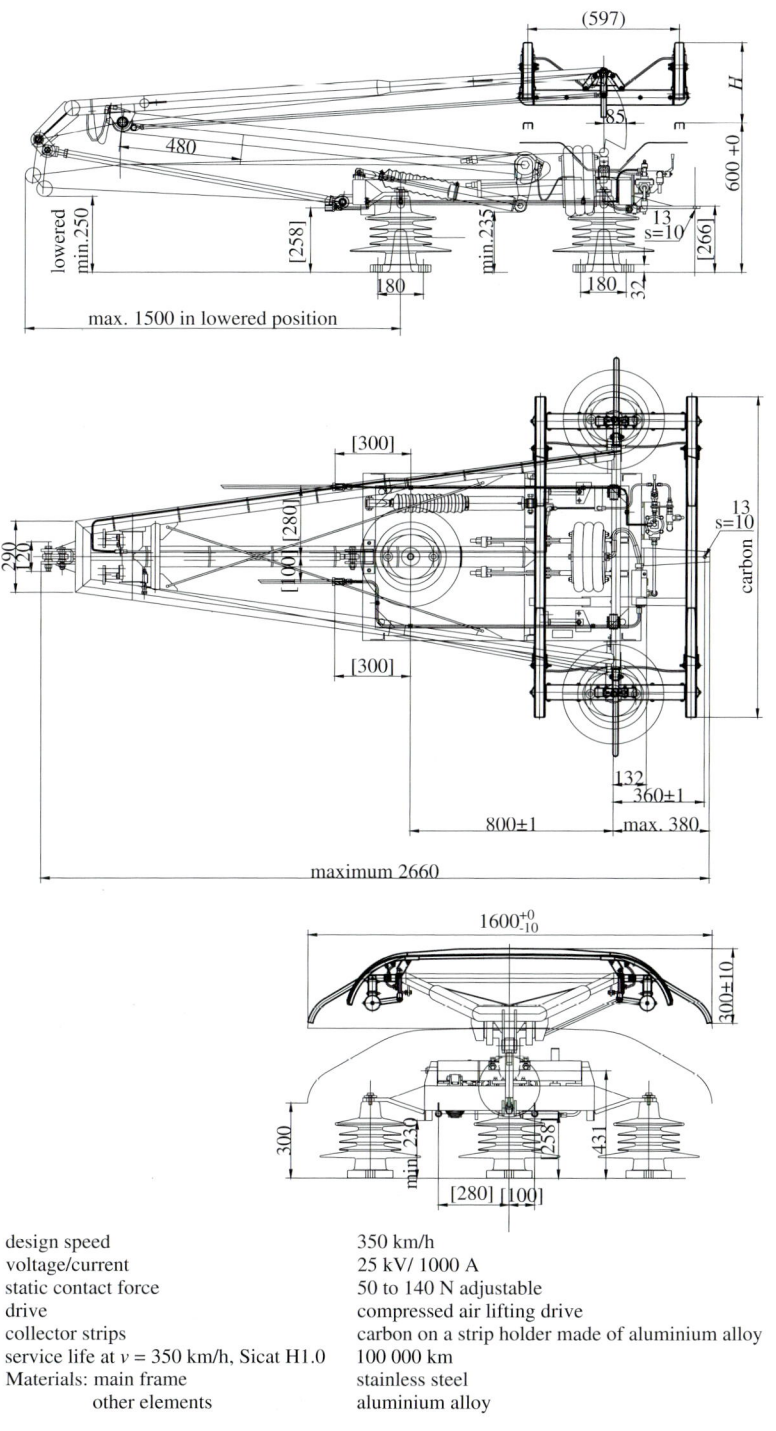

design speed	350 km/h
voltage/current	25 kV/ 1000 A
static contact force	50 to 140 N adjustable
drive	compressed air lifting drive
collector strips	carbon on a strip holder made of aluminium alloy
service life at v = 350 km/h, Sicat H1.0	100 000 km
Materials: main frame	stainless steel
other elements	aluminium alloy

Figure 2.16: Pantograph SSS 400+ [2.32, 2.33, 2.29].

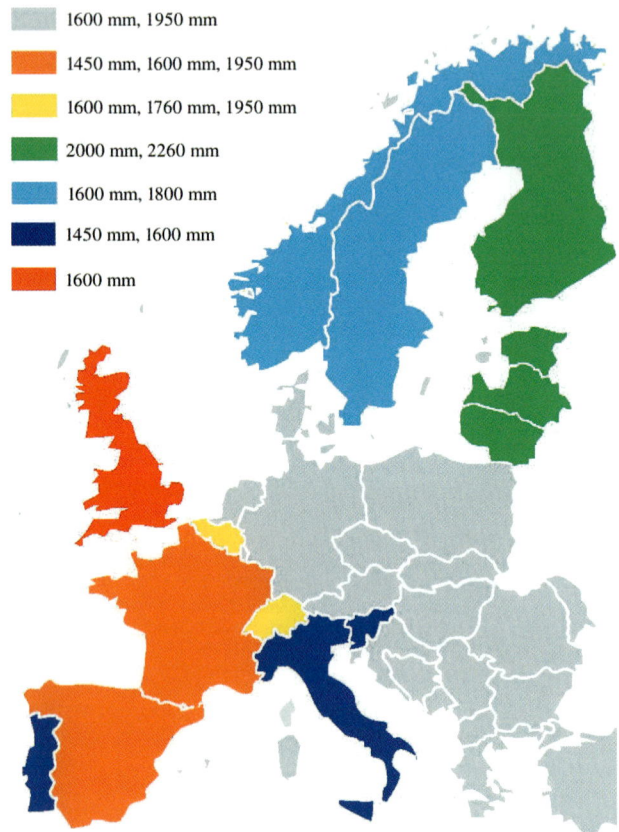

1600 mm, 1950 mm

1450 mm, 1600 mm, 1950 mm

1600 mm, 1760 mm, 1950 mm

2000 mm, 2260 mm

1600 mm, 1800 mm

1450 mm, 1600 mm

1600 mm

Figure 2.17: Preferred length of pantographs in Europe [2.37], Table 2.6. In some countries deviations are possible.

of 35 kg, the upper arm and head 9 kg. The two collector strips with holders have a mass of 3,0 kg each. The pantograph heads, consisting of collector strip holder, pantograph head guide with horns and collector strips are available for AC 25 kV 1 000 A and for DC 3 kV 2 400 A. In Figure 2.15 the *main characteristic data of* this *pantograph* are shown.

In Germany DB designs their overhead contact lines for a 1 950 mm long pan head. In 1942 this length of the pan head was introduced when merging the German and Austrian rail networks in order to be able to use all the lines in the merged networks. In [2.29, 2.30] and [2.37] and Table 2.6 information is given on pantograph profiles used in Europe. Table 2.7 contains some characteristics of pan head geometry. Figure 2.17 depicts the regions where these profiles have been adopted.

Table 2.7: Characteristics of pan head geometry.

Pan head length mm	Minimum length of collector strips mm	Working length mm	Length of horn
1 450	690	1 070	190 (SBB: 165)
1 600	800	1 200	200
1 800	–	1 394	205
1 950	1 100	1 550	200 (DB: 150)

a) 1 600 mm long pantograph according to [2.24], Clause 4.2.8.2.9.2.1

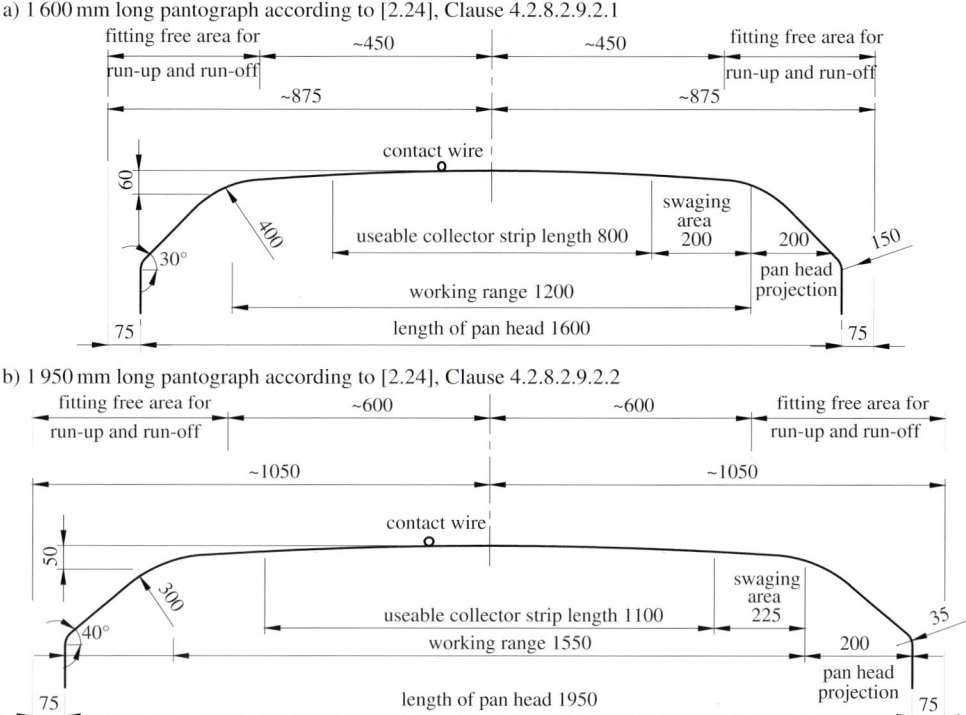

b) 1 950 mm long pantograph according to [2.24], Clause 4.2.8.2.9.2.2

Figure 2.18: Geometry of interoperable pan heads and pantographs for the static interaction between contact wire and pantographs and for the geometric interaction of contact wire and pantographs. Important dimensions in mm in a straight line.

The basic requirements on the *interaction of the pantograph with* the *contact line* are explained by Figure 2.18 a) for the 1 600 mm long EURO pantograph and Figure 2.18 b) for the 1 950 mm long pantograph. Both pantographs can be utilized in the interoperable network. The design of overhead contact lines needs to consider:

- new lines for running speeds above 250 km/h need to be designed for both pantograph types shown in Figure 2.18 a) and 2.18 b) [2.24]. If this is not possible than at least the 1 600 mm long EURO pantograph shall be considered when designing the contact line
- reconstructured or upgraded lines for speeds above 250 km/h need to be designed for at least the 1 600 mm long pantograph
- in all other cases, for lines for running speeds below 250 km/h the overhead contact line shall be designed such that at least one of the pantographs depicted in Figure 2.18 a) and 2.18 b) [2.24] can be adopted

As a basic requirement, the *pantograph head* must always protrude beyond the most unfavourable position of the contact wire due to lateral movements of the pantograph and the contact line expected during operation. Smooth operation of the system is only possible when the contact wire does not leave the *working range of the collector head* during travel. In normal operation, without high winds, it is essential that the contact wire travels on the collector strips. Therefore, the length of the collector strip needs to correspond at least to twice the useable contact wire lateral position (Clause 4.6). On straight lines that is 1 100 mm and 800 mm,

Table 2.8: Temperatures in °C exceeding which the mechanical properties of the material can be impaired according to EN 50 119.

Material	Maximum temperature		
	Up to 1 s (short circuit current)	Up to 30 min (pantograph at standstill)	Permanent
normal and high strength copper with high conductivity	170	120	80
copper silver alloy	200	150	100
copper tin alloys	170	130	100
copper magnesium alloys at least 0,2 % Mg	170	130	100
copper magnesium alloys at least 0,5 % Mg	200	150	100
aluminium alloys	130	–	80
ACSR[1] / AACSR[2]	160	–	80

[1] aluminium conductor steel reinforced, [2] aluminium alloy conductor steel reinforced

respectively, for the 1 950 mm long and the 1 600 mm long pan head. In Great Britain the Infrastructure Manager limits the working range of the 1 600 mm long pan head with insulated horns, to 1 300 mm. Each pantograph has a lower and upper working position. The range between these two positions is the *working range*. The highest and lowest working positions lie between approximately 2 800 mm and 300 mm relative to the upper edge of the main frame. On interoperable lines of the conventional European railway system only pantographs which were approved for conventional vehicles [2.24] may be adopted. These are the 1 600 mm long pantograph (Figure 2.18 a)) and the 1 950 mm long pantograph (Figure 2.18 b)). The pan heads are defined in EN 50 367, Annex A2 and B2, respectively. The Infrastructure Manager decides on the pantograph profiles to be used.

2.4.2 Properties of collector strips

The *collector strips* are part of the collector head and directly contact the contact wire to transfer power. The collector strips have to be selected and designed to cope with the requirements of current transfer when the vehicle is running and at standstill. The latter condition often governs the collector strip selection for DC applications. Stipulations for collector strip materials can be found in [2.23]. For AC lines only carbon or metal-impregnated carbon should be used [2.35, 2.36].

In modern trains, the power requirement for convenience and auxiliary can reach 1 000 kVA. This power has to be reliably transferred through the pantograph to a vehicle at standstill. The TSI Energy [2.10] requires overhead contact lines to be designed to sustain 300 A for DC 1,5 kV and 200 A for DC 3,0 kV at standstill. To avoid damage to the contact wire, temperatures may not exceed the maximum permissible values as specified in EN 50 119 and shown in Table 2.8. If higher temperatures are to be expected than those limits listed in Table 2.8, a possible reduction of the tensile strength needs to be assessed and, if required, the conducting cross section increased. The behavior of the collector strips at high temperature can be determined by testing as specified in EN 50 367, Annex A4.1. Clause 7.2.6 deals with details on design and testing of the contact wire/collector strip interface.

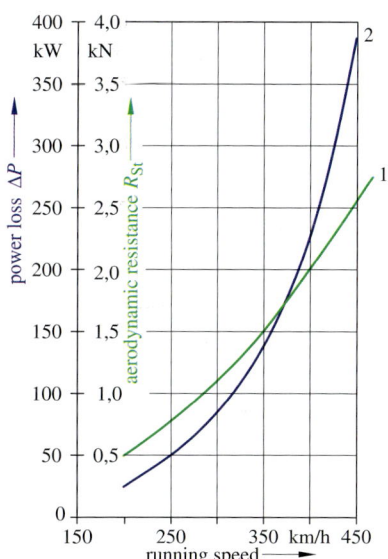

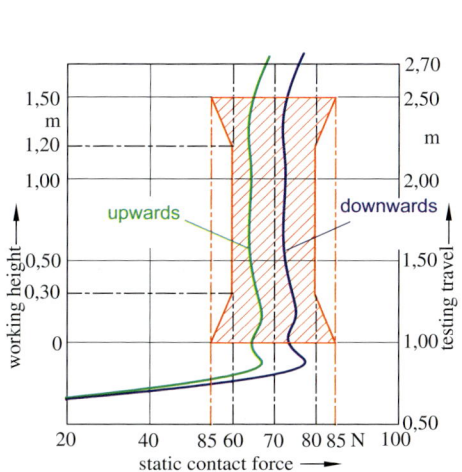

Figure 2.19: Static contact force of the pantograph DSA 250 depending on the working height [2.39] and associated temperature range (red range) according to EN 50 206-1.

Figure 2.20: Aerodynamic resistance R_{St} (1) and power losses ΔP (2) of running pantograph DSA 350 S depending on the speed according to [2.39].

2.4.3 Contact forces between the pantograph and the overhead contact line

2.4.3.1 Static contact force

The contact force governs the interaction between the pantograph and the overhead contact line. The interaction has static, aerodynamic and dynamic contact components.

The *static contact force* is the force exerted by the collector strips due to the force applied by the pantograph drive on the overhead contact line, measured at a stationary traction unit. To achieve the most consistent working conditions, this should be as constant as possible throughout the entire working range of the pantograph for both upward and downward movements. In practice, the friction in the knuckles causes differences between upward and downward motion.

According to TSI Energy [2.10] and the TSI Rolling Stock [2.23] the following static contact forces F_{K0} are recommended for the design of contact lines:

- AC: 70 N for new installations, 60 to 90 N for upgrading of existing systems
- DC 3,0 kV: 110 N for new installations, 90 to 120 N for upgrading of existing systems
- DC 1,5 kV: 90 N for new installations, 70 to 140 N for upgrading of existing systems

The static contact force may depend on the working height and whether the pantograph is rising or lowering. An example is shown in Figure 2.19 for the pantograph DSA 250.

Table 2.9: Limits for dynamic contact forces according to EN 50 119.

System	Speed in km/h	Contact forces in N minimum	maximum
AC	≤ 200	positive	300
AC	> 200	positive	350
DC	≤ 200	positive	300
DC	> 200	positive	400

2.4.3.2   Aerodynamic contact force

The sum of the static contact force and the component resulting from running speed and dependent on the aerodynamic effects is designated as the *aerodynamic contact force*. It is exerted vertically upwards and measured when the pantograph head is held still and not necessarily touching the overhead contact line. For high speeds, it is intended that the aerodynamic contact force increases only relatively slowly with the speed. The aerodynamic effect on the pantograph on the front of a train in the direction of travel is greater than on those installed on the rear of a train. Therefore, for trains or locomotives operated at high speeds it is preferred to operate pantographs mounted on the rear of the train to reduce the adverse aerodynamic effects that occur at the front of the locomotive.

The *aerodynamic resistance* of the pantograph has to be distinguished from the aerodynamic contact force. Aerodynamic resistance is exerted by the wind in a direction opposing the running direction. The main part of the aerodynamic resistance occurs at the collector head. Figure 2.20 shows the overall aerodynamic resistance depending on the wind velocity for pantograph DSA 350 S. The aerodynamic force is also called *mean contact force*. The *aerodynamic contact force* and the *resistance force* in single-arm pantographs depend on whether the knuckle is leading or trailing. Design of the pantograph can control the resistance forces and aerodynamic contact forces.

Target values for the mean contact force are stipulated by the TSI Energy [2.10] to ensure contact quality without undue arcing and to limit wear and hazards to collector strips. The requirements are defined in Clause 10.4.2. Clause 7.2.6 deals with the contact wire/collector strips interface.

2.4.3.3   Dynamic contact force

The sum of the aerodynamic contact force and the dynamic components from the interaction between overhead contact line and pantograph is designated according to EN 50 206-1 as the *dynamic contact force*. In particular, this force depends on the speed, the dynamic properties of the overhead contact line and the pantographs as well as the number of lifted pantographs on a train and their spacing. It also depends on the running behaviour of the traction unit and the quality of the tracks.

Irregularities in the overhead contact line, e. g. discrete masses such as section insulators, create peaks in the dynamic contact force. They should be avoided if possible. The quality of the contact between the overhead contact line and the pantograph can be assessed by dynamic contact forces or by the number and duration of arcs as explained in Clause 10.4.3.

To avoid arcs and also to limit the uplift of the contact line and wear of components, the dynamic contact forces should comply with the requirements stipulated in the TSI Energy [2.10] for lines of the European rail system and in EN 50 367 for lines traversed with speeds of 160 km/h and above. These requirements are dealt with in Clause 10.4.2. According to EN 50 119 the dynamic contact forces should stay within the limits given in Table 2.9.

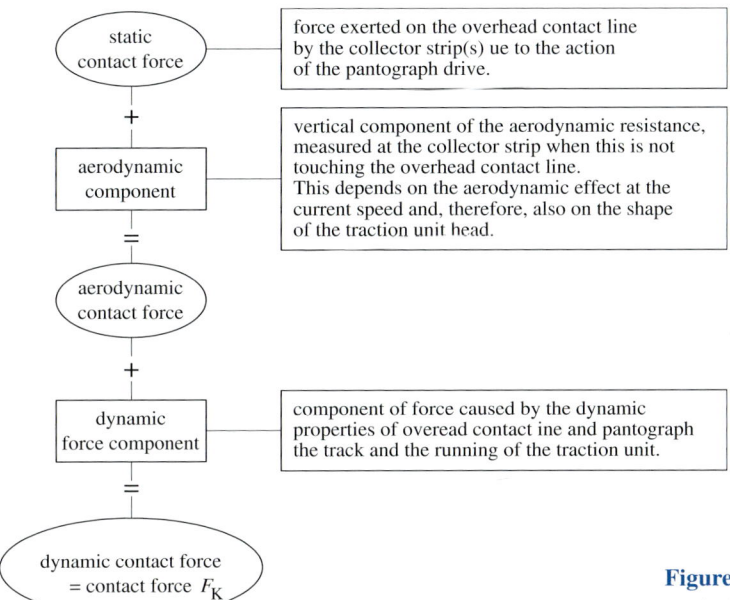

Figure 2.21: Components of the contact force.

At section insulators or other rigid components the contact force may increase up to 350 N in systems for speeds up to 200 km/h. For DC railways, a permissible value of 400 N is specified. The lower limit specified for both types of power supply is obtained by subtraction of three standard deviations from the mean contact force and should be greater than zero under all conditions. Figure 2.21 explains the individual contact force components and their relationships.

2.5 Climatic conditions

2.5.1 Temperatures

In the design of contact line systems, the *climatic conditions* applicable to the respective territory needs to be observed. In central Europe the temperature range is −30 °C to 40 °C and in Southern Europe −45 °C. The standards EN 50 341-1 and EN 50 125-2 specify guidelines to be observed for installations of overhead contact lines. The outdoor temperature range −30 °C to 40 °C shall be considered in Germany for the installation of overhead contact lines according to EN 50 119, Supplement 1:2011-04.

The valid temperature limits in Central Europe are
- *highest ambient air temperature* +40 °C
- *lowest ambient air temperature* −30 °C .

Outdoor temperatures above 35 °C occur very rarely in Central Europe. The annual averages lie between 8 °C and 10 °C. In France, the average values are approximately 15 °C. In Russia, the lowest regional outdoor temperature is around −60 °C. Equipment for outdoor systems in housings, e. g. local control facilities, should not suffer irreversible functional damage between −35 °C and +70 °C, according to EN 60 529. These requirements apply to altitudes up to 1 200 m above sea level.

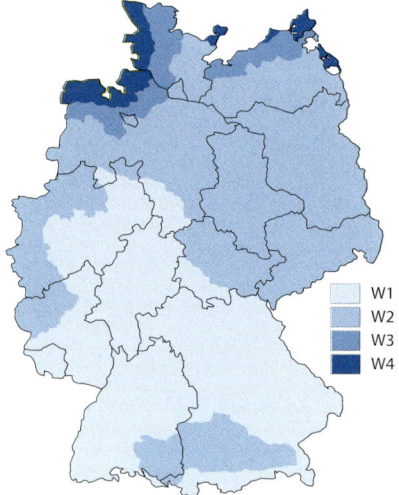

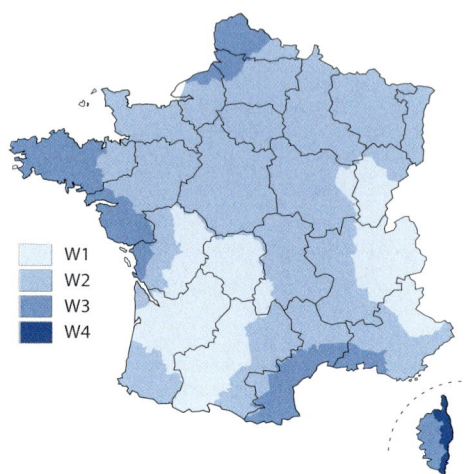

Figure 2.22: Wind zones in Germany according to DIN EN 1991-1-4.
W1: 22,5 m/s, W2: 25,0 m/s
W3: 27,5 m/s, W4: 30,0 m/s

Figure 2.23: Wind classes in France according to NF EN 1991-1-4.
W1: 22,0 m/s, W2: 24,0 m/s
W3: 26,0 m/s, W4: 27,5 m/s

2.5.2   Wind velocities and wind loads

2.5.2.1   Verification of serviceability and stability

The design of contact lines with respect to wind loads involves two main aspects
- verification of serviceability, avoidance of pantograph de-wiring caused by wind deflected contact wire and
- verification of structural strength of components to withstand the maximum wind loads which will probably occur during the life time of the installation.

The stipulations for wind loads are specified in EN 50 119, Clause 6.2.4. accordingly, the design of overhead contact lines in view of wind loads should be based on the meteorological wind velocity measured 10 m above ground over an averaging period of ten minutes in a relatively open terrain, designated as terrain category II according to EN 1991-1-4. For the structural design of supports wind velocity having a return period of 50 years shall be used, whilst, according to EN 50 119, the verification of serviceability may be carried out based on a wind velocity having a return period which may be selected by the operator of the line. EN 50 119 recommends return periods between three and ten years for this purpose.

2.5.2.2   Basic and gust wind velocities

For wind loads, design of overhead contact lines begins with the *basic wind velocity* v_b. This wind velocity is defined as:
- 10 m above ground
- averaged over 10 minutes
- in a flat, open terrain considering the altitude (terrain category II according to EN 1991-1-4, Table 4.1)

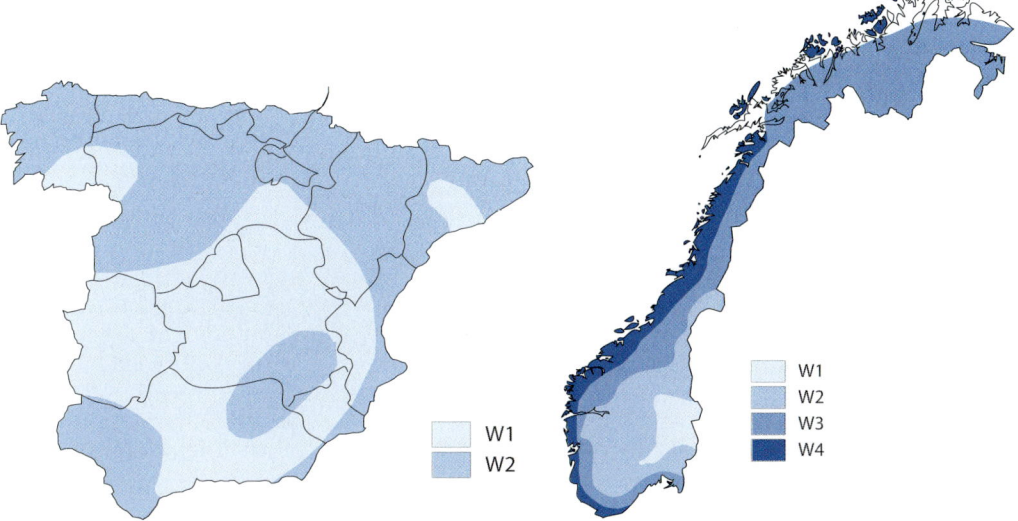

Figure 2.24: Wind classes of Spain [2.40].
W1: 24,0 m/s, W2: 28,0 m/s

Figure 2.25: Wind zones in Norway according
to NS EN 1991-1-4.
W1: 22,0 m/s, W2: 25,0 m/s
W3: 27,0 m/s, W4: 30,0 m/s

– with 50 years mean return period
– independent of the wind direction and the season

According to EN 50 341-1 a *gust* is a turbulent wind velocity. German Meteorologic Services (DWD) defines a strong wind impact as a gust, which is often combined with a sudden change of the wind direction and exceeding the 10 min average by at least 5,0 m/s.

Wind velocities have been recorded by weather services for many years. Therefore, in many countries statistical information on regional wind data is available. Data for wind loads on overhead contact lines can be taken from

– general wind load standard EN 1991-1-4,
– standard EN 50 125-2 for the environmental conditions of railways and
– project specifications of operators.

In Germany the national annex to the standard EN 1991-1-4 applies to civil engineering projects. Amendment 1 to EN 50 119:2014 stipulates that the wind pressure for the static design of contact lines can be chosen according to the standard EN 1991-1-4/NA:2010-12.

Table 2.10: Basis wind velocities v_b and basis wind pressures q_b for varification of stability according to relevant standards, 10 min mean values, return period 50 years, sea level, 10 m above ground, air dursity $\rho = 1,225\,\text{kg/m}^3$, 10 m above ground, altitude 0 m.

Standard	EN 1991-1-4/NA:2010 EN 50 341-2-4:2011		EN 50 125-2:2003	
	$v_{b,0,02,z=10}$[1]	$q_{z=10}$[1]	$v_{b,0,02,z=10}$[1]	$q_{z=10}$[2]
wind zone	m/s	N/m^2	m/s	N/m^2
W1	22,5	320	24,0	353
W2	25,0	390	27,0	463
W3	27,5	470	32,0	627
W4	30,0	560	36,0	974

[1] specified data;   [2] calculated data

Table 2.11: Basis wind velocities and wind pressures with differing return periods according to EN 1991-1-4/NA, EN 50 341-2-4, altitude 0 m, mean temperature 10 °C.

wind zone	Return period in years							
	50		10		5		3	
$v_{bp}/v_{b0,02}$	1,000		0,902		0,855		0,815	
v_{bp}	$v_{b,0,02}$	$q_{b,0,02}$	$v_{b,0,10}$	$q_{b,0,10}$	$v_{b,0,20}$	$q_{b,0,20}$	$v_{b,0,33}$	$q_{b,0,33}$
wind zone	wind velocity v in m/s s and wind pressure q in N/m^2							
W1	22,5	320	20,3	260	19,2	234	18,3	213
W2	25,0	390	22,6	317	21,4	285	20,4	259
W3	27,5	470	24,8	383	23,5	344	22,4	312
W4	30,0	560	27,1	456	25,6	410	24,5	372

Therefore, the wind pressures will depend on the region where the new line operates. Germany is divided into four wind load regions as shown in Figure 2.22. Figures 2.23, 2.24 and 2.24 show wind maps of France, Spain and Norway. The wind velocities and basic wind pressures presented in the Tables 2.10 and 2.11 can be used for contact lines. In EN 50 125-2 four wind classes are defined by the wind velocities 24,0 m/s, 27,5 m/s, 32,0 m/s and 36,0 m/s. This data represents 10 minute mean values having 2 % annual probability of being exceeded. Comparison of the wind data at a height of 10 m from EN 1991-1-4/NA:2010-12 with that from EN 50 125-2 in Table 2.10 reveals that the latter is considerably higher. In many cases railway operators prefer to specify wind velocities according to EN 50 119 for their networks.

2.5.2.3   Return period of wind velocities

The purchasing entity may specify the wind loads for verification of serviceability. It makes sense to specify wind loads having a three to ten years return period. From meteorological observations in Germany the basic wind velocities v_b, having a fifty years return period were established. The probability of exceedance during one year will be:
  – 50 years return period $1/50 = 0,02$ per year
  – 10 years return period $1/10 = 0,10$ per year
  – 5 years return period $1/5\ \ = 0,20$ per year
  – 3 years return period $1/3\ \ = 0,33$ per year
From the basic wind velocity $v_{b,0,02}$ having a 50 years return period, the basic wind velocities with the return period p can be obtained using the relationship (2.2), also see Figure 2.26, which is based on a Gumbel distribution.

$$v_{b,p} = v_{b,0,02} \cdot \left[\frac{1 - K_1 \cdot \ln\left[-\ln(1-p)\right]}{1 - K_1 \cdot \ln(-\ln 0,98)} \right]^n \quad, \tag{2.2}$$

where

$v_{b,p}$ basic wind velocity having a probability of being exceeded of p,
$v_{b,0,02}$ basic wind velocity having a probability of being exceeded of $p = 0,02$, that is once in 50 years,
K_1 coefficient of shape selected to be 0,2 according to EN 50 125-2
p probability of being exceeded during one year
n exponent, which can be selected to be 0,5 according to EN 50 125-2

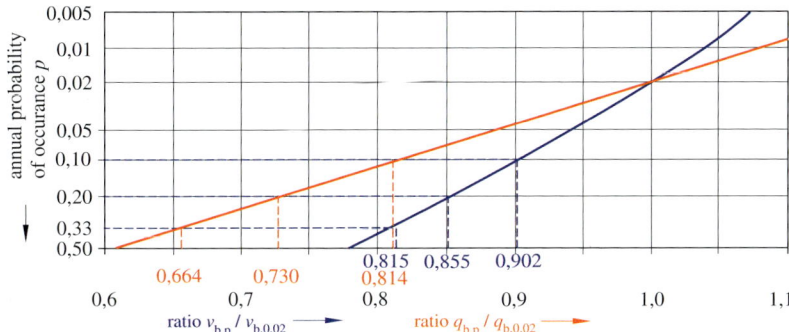

Figure 2.26: Determination of basic wind velocities $v_{b,p}$ (blue) and the basic wind pressure $q_{b,p}$ (red) with the annual probability of occurrence p from the wind velocity $v_{b,0,02}$ and the wind pressure $q_{b,0,02}$ having 2 % annual probability of occurrence (Figure 1 in EN 50 125-2).

For conversion of the basic wind pressure it applies accordingly

$$q_{b,p} = q_{b,0,02} \cdot \frac{1 - K_1 \cdot \ln\left[-\ln(1 - p)\right]}{1 - K_1 \cdot \ln(1 - \ln 0{,}98)} \quad . \tag{2.3}$$

Table 2.11 contains the basic wind velocities and pressures with return periods of 50, 10, 5 and 3 years for wind zones according to EN 1991-1-4/NA:2010-12.

The coefficient of shape K_1 depends on the standard deviation and other parameters of the Gumble distribution. In EN 50 341-1 the value 0,1 is mentioned for K_1. The value $K_1 = 0{,}2$ would lead to a conservative relation between the data for 50 and p return period in comparison with the information e, g, in EN 50 341-1.

2.5.2.4   Map of wind zones

The wind velocities depend on the individual wind region. Statistics established by the German Weather Services DWD recorded over approximately 60 years yielded the classification by the 10 min mean data and the *wind zone map* for Germany (Figure 2.22), which has been included in EN 1991-1-4/NA:2010-12 and EN 50 341-2-4:2016-04. These standards also include the detailed geographic borders.

2.5.2.5   Basic wind pressure

The relationship between *basic wind velocity* v_b at a height of $z = 10$ m and the basic wind pressure q_b is given by

$$q_b = (\rho/2) \cdot v_b^2 \quad , \tag{2.4}$$

where

ρ air density 1,225 kg/m³ at 15 °C at sea level.

v_b basic wind velocity at 10 m height z above terrain, averaged over 10 min having a return period as required for the verification (see Clauses 2.5.2.1 and 2.5.2.3).

For other values of the absolute temperature T and the altitude H above sea level the air density ρ up to 750 m altitude may be calculated from

$$\rho = 1,225 \cdot (288/T) \cdot \exp(-0,00012 \cdot H) \quad . \tag{2.5}$$

2.5.2.6 Design wind pressure depending on height

The *design wind pressure* q_z depends on the height z above ground and includes wind gusts. The design wind pressure q_z at any heights z above ground can be calculated from the basic wind pressure q_b at the height $z = 10$ m according to EN 1991-1-4/NA:2010-12, Table 4.1, for the terrain category II, including gust effects by:

$$q_z = 1,5 \cdot q_b \qquad\qquad \text{for } z \leq 7\,\text{m} \quad , \tag{2.6}$$

$$q_z = 1,7 \cdot q_b \cdot (z/10)^{0,37} \quad \text{for } 7\,\text{m} < z \leq 50\,\text{m} \quad , \tag{2.7}$$

$$q_z = 2,1 \cdot q_b \cdot (z/10)^{0,24} \quad \text{for } 50\,\text{m} < z \leq 300\,\text{m} \quad , \tag{2.8}$$

where
q_z design wind pressure at the height z above ground (including wind gusts),
q_b basic wind pressure 10 m above ground averaged over 10 min,
z height of the contact line above ground.

The relations (2.6) to (2.8) apply to altitudes up to 750 m above sea level.
For altitudes H between 750 m and 1 100 m the basic wind pressure is obtained from

$$q_{b,H} = (0,25 + H/1\,000) \cdot q_b \quad . \tag{2.9}$$

The design wind pressures q_z in (2.6) to (2.8) consider 2-sec-gusts with corresponding peak wind velocities. For sites on ridges or mountain tops and for altitudes above $H = 1\,100$ m specific advice should be obtained from local meteorological services.

2.5.2.7 Wind load on conductors

The *wind load related to length* Q'_W acting perpendicularly to the conductor or wire is obtained from

$$Q'_W = q_z \cdot G_C \cdot C_C \cdot d \quad , \tag{2.10}$$

where
q_z height dependent design wind pressure according to equations (2.6) to (2.8),
G_C conductor response factor, which considers the reaction of moveable conductors under wind load and can be assumed as $G_C = 0,75$ according to EN 50 119 the strength analysis with 50 years return period. For the verification of serviceability with wind velocities with return periods of 3 to 10 years, $G_C = 1,0$ should be assumed,
C_C drag factor for conductors with circular cross-sections
according to EN 50 119:2014 $C_C = 1,0$
according to EN 50 341-2-4:2016-04, Table 4/DE.1
– conductors with diameters up to 12,5 mm $C_C = 1,2$,
– conductors with diameters between 12,5 mm and 15,8 mm $C_C = 1,1$,
– conductors with diameters above 15,8 mm $C_C = 1,0$,
d conductor diameter.

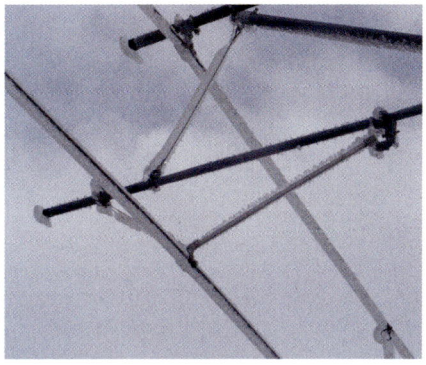

Figure 2.27: Ice and snow accretion at a contact line cantilever (Photo: SPL Power-lines GmbH, M. Goschke).

Figure 2.28: Contact line installation with heavy ice accretion in Slovenia in February 2014 [2.42].

If conductors run in parallel the wind load for the leeward conductor may be reduced to 80 % of exposed conductor, if the spacing between the conductors is less than five times their diameter. Thus, a drag factor

$$C_C = 1,2 + (0,8 \cdot 1,2) = 2,16$$

is obtained for a twin contact wire AC-100.

2.5.3 Ice accumulation

The *accumulation of ice* on conductors of overhead contact lines causes an additional load on these systems and affects their operation [2.42]. Additional details concerning this aspect are given in Clause 13.1.4.3. While in Germany, Austria and Switzerland ice loads need to be taken into account, this is unnecessary at SNCF in France. In Russian regions with *extreme ice loads* a significantly increased sag occurred on *automatically tensioned overhead contact lines*, that railway operations were temporarily suspended.

High loadings due to ice, rime or snow, summarized as *ice loads*, are relatively rare events in Central Europe. From ice observations basic assumptions for ice accretions were evaluated and used in standards for the specification of ice loads as a basis for the design of overhead power lines.

Two main types of ice accretion are distinguished

- ice accretion from precipitation: In this case hard ice having a density of $0,9\,\text{t/m}^3$ is formed from supercooled rain or drizzle at temperatures around freezing point. Ice due to *wet snow* having a density of 0,3 to $0,6\,\text{t/m}^3$ also falls into this category (Figures 2.27 and 2.28) and
- ice accretion at conductors in clouds (incloud icing) or in fog: Hard rime is formed by supercooled water droplets. The formation of hard rime is typical for lines above the lower level of clouds. Hard rime having a density of 0,4 to $0,6\,\text{t/m}^3$ and soft rime with a density of 0,2 to $0,4\,\text{t/m}^3$ are created under these conditions.

Combinations of differing types of ice also occur.

Ice loads for overhead power lines in Germany are specified in EN 50 341-2-4:2016-04. Four ice load zones E1 to E4 are distinguished as shown in Figure 2.29.

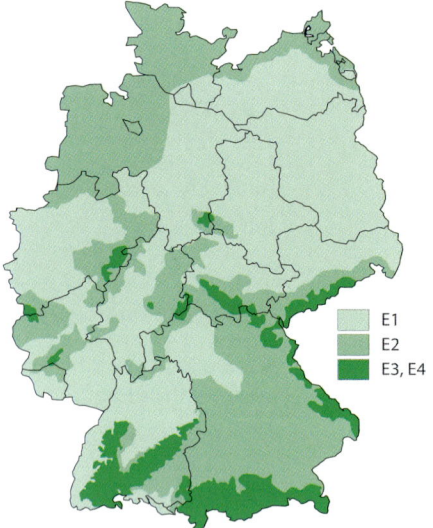

Figure 2.29: Ice load zones in Germany according to EN 50 341-2-4.

Ice load zone E1 $G'_{\text{ice}} = 5 + 0,1 \cdot d$
Ice load zone E2 $G'_{\text{ice}} = 10 + 0,2 \cdot d$
Ice load zone E3 $G'_{\text{ice}} = 15 + 0,3 \cdot d$
Ice load zone E4 $G'_{\text{ice}} = 20 + 0,4 \cdot d$

At DB and also in Germany the half of the ice load as specified for zone E1 is often used to design contact lines. Examples for the calculation of sags due to ice loads can be found in Clause 4.2.2. Procedures to avoid or remove ice loads at overhead contact lines are treated in Clause 5.5.

2.5.4   Active substances in the atmosphere

Aggressive dust, vapours, gases and extreme levels of humidity can cause rapid *contamination of insulators* and increased wear of components in contact line installations, particularly when several substances are combined. These active *airborne substances* may occur in the vicinity of production facilities and near the sea. These factors need to be accommodated in the *design of contact line systems*. These substances affect the *insulation co-ordination* described in detail in Clause 2.1.3.2.

2.5.5   Lightning surges

Lightning striking contact line installations can cause flashovers at the insulation leading to damage. From measurements made by German Railway (DB) [2.43, 2.44], it is known that one *lightning stroke* per 100 km of contact line in a year can be assumed in Central Europe. The probability of lightning is highly variable and also differs according to location. *Lightning intensity* is measured by the *keraunic level*, which is the number of days with thunderstorms per year.

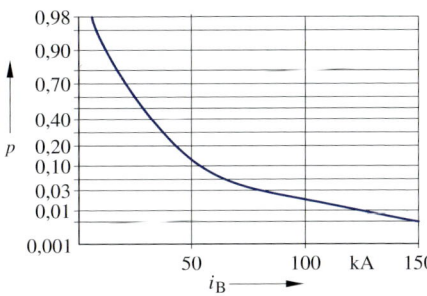

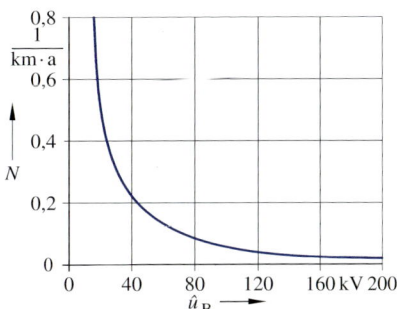

Figure 2.30: Probability p that a lightning stroke exceeds the current i_B [2.43].

Figure 2.31: Frequency N of indirect lightning surge voltages per km of electrified line and year [2.43].

A direct lightning stroke into an overhead contact line will cause *lightning voltage surges*. The voltage peak of these surges can be estimated by the empirical equation

$$U_{B\,max} = i_B \cdot Z/2 \text{ in kV} \quad , \tag{2.11}$$

where

i_B lightning current peak value in kA

Z surge impedance in Ω

The *probability of lightning currents* exceeding a given value can be seen in Figure 2.30. Indirect lightning voltage surges occur as *lightning discharges* when an overhead contact line lies in the electric field between a cloud and the earth. When a thunderstorm approaches, a field of this kind induces charges in the overhead contact line. The negative charges are drained to earth through the discharge resistance of the numerous parallel contact line insulators and the positive charges are kept by the field emitted by the cloud. If a cloud then discharges in the vicinity of an overhead contact line, the charges are released in this line and are propagated as a traveling wave along the overhead contact line. The indirect lightning impulse overvoltages are lower in magnitude than direct lightning strokes. They also rise more slowly and have less steep flanks than direct strokes. Figure 2.31 contains information on the expected indirect lightning impulse overvoltages per year and their magnitude.

In overhead contact line installations, impulse voltage limiting can be achieved by *overvoltage protection* devices. The most important overvoltage protection device is the *valve-type arrestor*. Since only limited protection is possible with overvoltage protection devices, they are not used for economical reasons unless an extreme frequency of lightning exists.

2.6 Environmental compatibility

2.6.1 General

The *climatic effects* which have to be taken into account in the design and installation of contact line systems are described in detail in Clause 2.5. Clause 2.1.3.2 addresses the problem of pollution. Other aspects of the interaction between contact line installations and the environment are explained hereafter.

Table 2.12: Environmental properties of modern transport resources [2.15]

Feature	Unit	Car	Train	Aircraft
specific energy requirement	kWh/100·P km	48,7	10,3	62,8
CO_2 – emission	kg/100·P km[1]	12,29	4,75	17,0
NO_x-emission	g/100·P km	133	3,8	88
CO-emission	g/100·P km	209	0	20
hydrocarbon emission	g/100·P km	27	0	8
soot-emission	g/100·P km	0	1,0	–
land requirement at equal performance, new installations	%	285	100	170
noise level at 25 m distance	dB(A)	73	92	–

[1] 100·P km means that the data are related to 1 Person travelling 100 km.
Notes: – car with petrol engine,
     – for the ICE, emissions in the supplying power plants are taken into account,
     – over half of the space used for electrified railway lines is an ecologically
       valuable living space because the air along the lines is not polluted

2.6.2 Environmental relevance of electric traction

The transport process consumes the lowest specific energy when performed on rails. Beyond this, transport by electric traction vehicles is the most environmentally friendly means of moving people and freight. The specific energy consumption of shipping is lower but is much slower and cannot be used in many regions, e. g. in mountainous regions. Table 2.12 compares features which characterise *environmental aspects* of transport systems. The environmental features of modern aircraft and cars are compared with those of an ICE train.

2.6.3 Land usage

Those *areas used for installation* that are concreted, asphalted, covered with gravel or otherwise surfaced can be designated as *consumed land*. Subsequently, these areas are no longer available for other purposes.

Existing railway lines have already consumed land areas. The land consumption is only increased insignificantly by the electrification of a new railway line. Additional areas can be necessary for foundations. If the poles are set on areas which have already been consumed by the construction of the railway, e. g. on railway land, no additional land area is required for the installation of the overhead contact line.

During construction of overhead contact lines, it may be necessary to use land temporarily for provisional roads, for construction and excavation work. After completion of the installation, these areas are returned to their initial state.

The construction of overhead contact lines on existing railway lines extends the clearance gauge by space required for the pantograph zone. This and the overhead contact line zone, require no additional land. The land usage for new twin-track railway lines is only 36 % of that for a four-lane motorway [2.45].

2.6.4 Nature and bird protection

When electrifying railways, the relevant regional or national directives and laws with regard to *nature and bird protection* have to be complied with. Contact line systems are often rest

Figure 2.32: Bird protection on a feeder line of the high-speed line Madrid–Valencia in Spain.

and landing places for various species of birds. This is a cause of potential danger to the birds and also to the operation of the overhead contact line installation. The danger posed to birds by contact lines and cantilevers is minor, however, tension insulators and the risk of collision with the overhead contact lines presents a particular danger. In areas where resting and landing birds are often found, the installation of *bird protection devices* has reduced the potential hazard significantly. Figure 2.32 depicts a measure to prevent birds from colliding with the feeder line of the high-speed contact line Madrid–Valencia in Spain.

2.6.5   Aesthetics

The assessment of the effects on the environment within the approval procedures is known as an *environmental impact study*. Such a study is required prior to the approval of new lines and the extension of existing railway lines. It is difficult to assess objectively the effects of a contact line on the appearance of the landscape. The layout of the railway line, the height of the overhead contact line poles, the design of cantilevers, overhead contact lines, reinforcing feeders and return lines interact in a complex manner. Assessing the effects of the electrification of a railway line on the landscape will always be subjective.

2.6.6   Electric and magnetic fields

In the accessible vicinity of contact line systems, the maximum expected *electric fields* of AC 25 kV railways is 3,0 kV/m. On lines electrified with a nominal voltage of AC 15 kV in Germany, the expected values at the edge of the railway line lie at around 2,0 kV/m. The *magnetic fields* in the vicinity of the railway are dependent upon the current and, therefore, on time and location. They can reach peak values up to 80 A/m for short duration.

Both the electric and the magnetic fields in the vicinity of electric railways are believed to be completely harmless to humans. If display monitors or other sensitive equipment are operated in the vicinity of electric railways, interference can occur. Detailed information is contained in Chapter 8.

2.7   Bibliography

2.1   *Directive 96/48/EC*: Directive on the interoperability of the trans-European high-speed rail system. In: Official Journal of European Communities, No. L235 (1996), pp. 6 to 24.

2.2   *Directive 2001/16/EC*: Directive on the interoperability of the trans-European conventional rail system. In: Official Journal of European Communities, No. L110 (2001), pp. 1 to 27.

2.3   *Directive 2004/50/EC*: Directive on amending directive 96/48/EC and directive 2001/16/EC. In: Official Journal of European Communities, No. L220 (2004), pp. 40 to 57.

2.4   *Directive 2008/57/EC*: Directive on the interoperability of the trans-European rail system. In: Official Journal of European Union, No. L191(2008) pp.1 to 45.

2.5   *Decision 2002/733/CE*: Technical specification for the interoperability relating to the energy subsystem of the trans-European high-speed rail system. In: Official Journal of European Communities, No. L245 (2002), pp. 280 to 369.

2.6   *Courtois, C.; Kiessling, F.*: Technische Spezifikation Energie und zugehörige europäische Normen (Technical Specification Energy and associated European standards). In: Elektrische Bahnen, 101(2003)4-5, pp. 144 to 153.

2.7   *Decision 2008/284/CE*: Technical specification for interoperability relating to the energy subsystem of the trans-European high-speed rail system. In: Official Journal of European Communities, No. L104 (2008), pp. 1 to 79.

2.8   *Courtois, C.; Kiessling, F.*: Überarbeitung der TSI Energie für Hochgeschwindigkeitsstrecken (Revision of the TSI Energy for high-speed lines). In: Elektrische Bahnen 103(2005)4-5, pp. 178 to 186.

2.9   *Decision 2011/274/EC*: Technical specification on the interoperability of the Energy subsystem of the European conventional rail system. In: Official Journal of European Union. No. L126 (2011), pp. 1 to 52.

2.10   *Regulation 1301/2014/EU*: Technical specification on the interoperability relating to the Energy subsystem of the rail system in the Union. In: Official Journal of European Union. No. L356 (2014), pp. 179 to 227.

2.11   *Nickel, T.; Puschmann, R.*: Technische Spezifikation Energie 2015 – Harmonisierte Auslegung der Oberleitungen (Technical specification Energy 2015 – Harmonized design of overhead contact lines). In: Elektrische Bahnen 113(2015)2-3, pp. 86 to 99.

2.12   *Behrends, D.; Brodkorb, A.; Matthes, R.*: Konformitätsbewertung und EG-Prüfverfahren für das Teilsystem Energie (Assessment of conformity and suitability of the Energy subsystem of interoperable high-speed lines). In: Elektrische Bahnen 101(2003)3-4, pp. 158 to 166.

2.13   *Resch, M.; Rusch, M.*: Zugang von Fahrzeugen zur DB-Netz-Infrastrukur (Accsess of vehicles to DB Netz infrastructure). In: Elektrische Bahnen 101(2003)3-4, pp. 167 to 171.

2.14 *Schneider, J.; Schlater, H.-P.*: Sicherheit und Zuverlässigkeit im Bauwesen. Grundwissen für Ingenieure (Safety and reliability in structural engineering. Basics for engineers). Vdf Hochschulverlag ETH Zurich, 1994.

2.15 *Bundesminister für Umwelt, Naturschutz and Reaktorsicherheit*: Beschluss der Bundesregierung zur Reduzierung der CO_2-Emission in der BRD bis zum Jahr 2005 (Federal minister for environment, nature protection and nuclear power safety: Decision of the Federal Government on the reduction of CO_2 emission within Germany until the year 2005). Druck-Service E. Böhm, Haussen, 11/1990 and 3/1991.

2.16 *BOStrab*: Verordnung über den Bau und Betrieb der Straßenbahnen (Straßenbahn-Bau- und Betriebsordnung – BOStrab) (Regulation on construction and operation of tramways – BOStrab). Bundesrepublik Deutschland, BGBl. I S. 2648, December 1987, last modification BGBl. I S. 2938, December 2016.

2.17 *VDV Leaflet 550*: Oberleitungsanlagen für Straßenbahnen und Stadtbahnen (Overhead contact line systems for tramways and urban railways). Verband Deutscher Verkehrsunternehmen, Cologne, 2003.

2.18 *EBO*: Eisenbahn-Bau und Betriebsordnung (German directive for construction and operation of railways). Bundesrepublik Deutschland, BGBl. 1967 II pp. 1 563, with last modification by chapter 1 of the regulation dated July 25, 2012, BGBl. 2012 I pp. 1 703.

2.19 *Regulation 1299/2014/EU*: Technical specification on the interoperability relating to the Infrastructure subsystem of the rail system in the Union. In: Official Journal of European Union, No. L356 (2014), pp. 1 to 109.

2.20 *Mittmann, W. et al.*: Die Dritte Verordnung zur Änderung der Eisenbahn-Bau- und Betriebsordnung (EBO) (The third amendment of the directive for construction and operation of railways). In: Die Bundesbahn (1971)7-8, pp. 759 to 770.

2.21 *Directive 800.0110*: Netzinfrastruktur; Technik entwerfen, Linienführung (Network infrastructure, design of technical system, design of railway lines). Deutsche Bahn AG, Frankfurt, 2008.

2.22 *UIC Code 506*: Rules governing the application of the enlarged GA, GB, GB1, GB2, GC and GI3 gauges. UIC, Paris, 2008.

2.23 *UIC Code 505-1*: Railway transport stock – Rolling stock construction gauge. UIC, Paris 2006.

2.24 *Regulation 1302/2014/EU*: Technical specification on the interoperability relating to the Rolling Stock – Locomotive and Passenger Rolling Stock subsystem of the rail system in the European Union. In: Official Journal of European Union, No. L356 (2014), pp. 228 to 393.

2.25 *UIC Code 505-4*: Effects of the application of the kinematic gauges defined in the 505 series of leaflets on the positioning of structures in relation to the tracks and of the tracks in relation to each other. UIC, Paris, 2007.

2.26 *Furrer, B.*: Deckenstromschiene im Berliner Nord-Süd-Fernbahntunnel (Overhead conductor rail for North-South long distance railway tunnel in Berlin). In: Elektrische Bahnen 101(2003)4-5, pp. 191 to 194.

2.27 *Bartels, S.; Herbert, W.; Seifert, R.*: Hochgeschwindigkeitsstromabnehmer für den ICE (High-speed pantograph for the ICE). In: Elektrische Bahnen 89(1991)11, pp. 436 to 441.

2.28 *Blaschko, R.; Jäger, K.*: Hochgeschwindigkeitsstromabnehmer für den ICE 3 (High-speed pan-
      tograph for the ICE 3). In: Elektrische Bahnen 98(2000)9, pp. 332 to 338.

2.29 *Zöller, H.*: Entwicklung der Pantographen der Lokomotiven der Deutschen Bundesbahn (De-
      velopment of the pantographs for locomotives of German Railway). In: Elektrische Bahnen
      49(1978)7, pp. 168 to 175.

2.30 *Brockmeyer, A.; Gerhard, Th.; Lübben, E.*: Vom ICE S zum Velaro: 10 Jahre Betriebserfahrung
      mit Hochgeschwindigkeits-Triebwagen (From ICE S to Velaro: 10 years of experience with
      high-speed motor trainsets). In: Elektrische Bahnen 105(2007)6, pp. 362 to 368.

2.31 *Horstmann, D.; Budzinski, F.; Pirwitz, J.*: Die Mehrsystemtraktionsausrüstung des Hochge-
      schwindigkeitszuges Velaro für Russland (The dual-system traction equipment of the high-
      speed train Velaro for Russia). In: ETG-Fachberichte No. 107/108, 2007, pp. 1 to 10.

2.32 *Budzinski, F.; Fischer, J.; Markowetz, H.*: Elektrische Ausruestung des Hochgeschwindigkeits-
      zuges Velaro E (The electrical equipment of the high-speed train Velaro E). In: Elektrische
      Bahnen 102(2004)3, pp. 99 to 108.

2.33 *Schunk Bahntechnik*: Full speed into the future. In: Schunk Report, Heuchelheim, Austria,
      December 2003.

2.34 *Nickel, T.*: Untersuchung zu Auswirkungen der verminderten Fahrdraht-Seitenlage auf das
      Ebs-Zeichnungswerk (Studies on the affects of a reduced contact wire lateral position on DB's
      Ebs standard drawings). TU Dresden, Diploma thesis, 2011.

2.35 *Harries, K.*: Jane's World Railways 2009–2010. Jane's Information Group, London, 2011.

2.36 *Herbert, W.*: Entwicklung und Betriebserfahrung mit den Hochgeschwindigkeitsstromab-
      nehmern DSA 350 S für den ICE (Development and operating experience with the DSA 350 S
      high-speed pantograph for the ICE). In: Eisenbahntechnische Rundschau, 41(1992)6, pp. 385
      to 390.

2.37 *Auditeau, G. et al.*: Carbon contact strip materials – Testing of wear. In: Elektrische Bahnen
      111(2013)3, pp. 186 to 195.

2.38 *UIC Code 608*: Conditions to be complied with for the pantographs of tractive units used in
      international services. UIC, Paris, 2003.

2.39 *Stemmnann Technik GmbH*: Type test for pantograph DSA350 according to EN 50 206-1.
      Fandstan Electric Group, Schüttdorf, 2005.

2.40 *Ministerio de Fomento in Spain*: Instrucción sobre las acciones a considerar en el proyecto de
      puentes de carretera (IAP) (Regulation on the effects to be considered for planning of road
      bridges). In: Orden por la que se aprueba, February, 1998.

2.41 *Behmann, U. et al.*: Elektrischer Betrieb bei der Deutschen Bahn im Jahre 2009 (Electric oper-
      ation of Deutsche Bahn in 2009). In: Elektrische Bahnen 108(2010)1-2. pp. 4 to 54.

2.42 *N. N.*: Winterschäden in Slowenien (Damage of contact lines due to ice accretion in Slovenia).
      In: Elektrische Bahnen 112(2014)3, pp. 150.

2.43 *Wilke, G.*: Neuere Untersuchungen zur Überspannungsbekämpfung in elektrischen Bahnanla-
      gen (More recent investigations to avoid overvoltages in electrical railway installations).
      In: Elektrische Bahnen 16(1940)10, pp. 161 to 170.

2.44 *Biesenack, H.; Dölling, A.; Schmieder, A.*: Schadensrisiken bei Blitzeinschlägen in Oberleitungen (Failure risk due to lightning strokes to overhead contact lines). In: Elektrische Bahnen 104(2006)4, pp. 182 to 189.

2.45 *Strebele, J.*: Zur Umweltverträglichkeit raumbedeutsamer Bahnanlagen (Environmental compatibility of railway installations important for regional planning). In: Die Bundesbahn (1986)9, pp. 701 to 705.

3 Contact line types and designs

3.0   Symbols and abbreviations

Symbol	Definition	Unit
D	length of neutral sections in phase separation sections	m
F_D	tensile force of steady arm	N
F_R	contact wire radial force at supports	N
F'_{WCW}	length related wind force on contact wire	N/m
F'_{WCA}	length related wind force on catenary wire	N/m
H_{CA}	catenary wire tensile force	kN
H_{CW}	contact wire tensile force	kN
H_Y	stitch wire tensile force	kN
$L_{neutral}$	length of neutral section	m
L_{pant}	spacing between pantographs	m
L_{tens}	length of tensioning section consisting of two halfsections	m
TGV	Train à Grande Vitesse (High-speed train in France)	–
TPL	Traction Power Line	–
ToR	Top of Rail	–
a_{SO}	distance between signal and first pole with twin cantilevers	m
a_i	length of span i	m
b_{SO}	distance between signal and last pole with twin cantilevers in the opposite track	m
b_i	stagger at support i	m
c_i	contact wire lateral position at midspan i	m
e_{max}	maximum contact wire lateral position	m
l	span length	m
$l_{D\,min}$	minimum length of dropper	m
v	running speed	km/h
ΔL_{tens}	permissible change of length in half of a tensioning section	m

3.1   History of contact line development

3.1.1   General

The use of rails insulated from earth for power transfer to electrically driven vehicles, as adopted for the first electric locomotive in 1879 (Figure 3.1) [3.1], and the use of a sliding contact, running along a contact line located in a channel below a slotted rail (Figure 3.2) [3.2], did not prove a successful solution. Electrical accidents and disturbances led to stringing of lines on wood poles beside the track.

Further development adopted a carriage running on a side line and a cable transferred power to the vehicle (Figure 3.3). The invention of the *pantograph* in 1889 by the Siemens engineer Walter Reichel [3.3, 3.4] led to the introduction of an overhead contact line above the track

Figure 3.1: First electric locomotive – worldwide (Siemens, 1879).

a) Three dimensional view   b) Cross-section

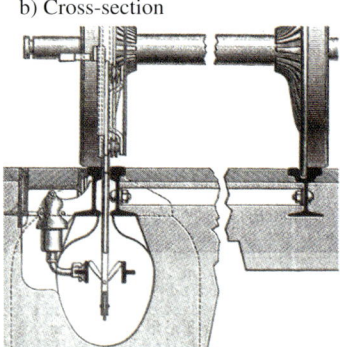

Figure 3.2: Tramway in Budapest, conductor rail guided in a duct below a slotted rail (Siemens, 1891).

(Figure 3.4). The contact wire height of 4,6 m on the tramway line from Berlin to Lichterfelde was sufficient to avoid accidents from possible contact with vehicles or persons at crossings. Since then, overhead contact lines with differing designs have been adopted worldwide. The bow-type pantograph (bow collector) invented by Reichel formed the basis for higher running speeds. For local traffic, overhead contact lines gained priority in publicly accessible areas. Conductor rails close to the running rails are often used in cases such as subways running on their own right-of-way and not open to the public.

Contact lines supplied at DC 0,6 kV and 1,5 kV were adopted for local traffic, and DC 1,5 kV and 3,0 kV as well as AC 15 kV 16,7 Hz to AC 25 kV 50 Hz for main line traffic. For voltages above AC 1,0 kV and DC 1,5 kV only overhead lines are permitted, to avoid dangers of direct contact with the contact line.

Figure 3.3: Power supply from a conductor carriage on a side line for the tramway in Berlin-Westend (Siemens, 1882).

Figure 3.4: Bow-type pantograph invented by Walter Reichel (Siemens, 1890) [3.1].

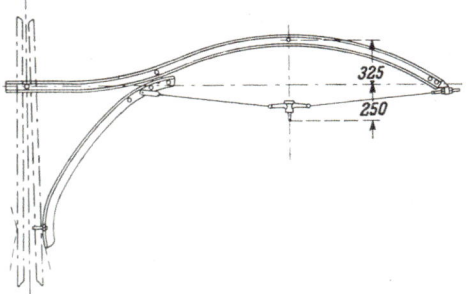

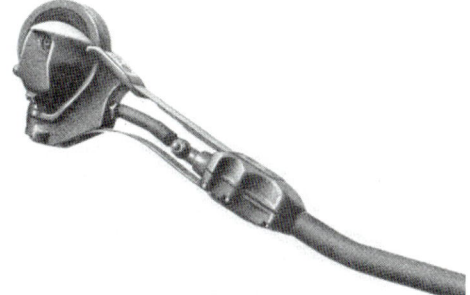

Figure 3.5: Trolley type contact line used at the tramway in Berlin-Lichterfelde (Siemens, 1890).

Figure 3.6: Siemens-wheel-type traction pole with Dickinson wheel (Siemens, 1890).

In 1890, the tramway in Berlin-Lichterfelde was equipped with a *trolley type overhead contact line* and operated by a bow-type pantograph (Clause 3.3.2). This contact line type consisted of a contact wire suspended from cantilevers (Figure 3.5) or cross-span wires spaced at 30 to 40 m in straight line sections and shorter distances in curves [3.1]. The *trolley-type contact line* used with bow-type pantographs enabled longer spans than those possible with trolley-type collectors equipped with wheels. The sag of the contact line caused vertical contact wire deviations at the suspensions where the wheel of the collectors tended to dewire. To avoid dewirements, shorter span lengths and contact wire deviations of less than 11° were required. The tensile force of the contact wire was 5 to 8 kN.

In 1894, Alfred Dickinson developed a trolley-type contact line, to be used by a trolley type collector with a wheel (Figure 3.6), installed on short cantilevers and poles located 2,0 m from the track. This design was not often employed, however, it is still in use today for trackless systems like trolley buses (see Clause 3.1.3).

Following tests using a three-phase AC supply at the Siemens premises in Berlin, a test track with a 10 kV 50 Hz three-phase AC line was established in 1901 between Marienfelde and Zossen, close to Berlin (Figure 3.7). Sponsored by both AEG and Siemens & Halske, this line was equipped with a contact line installed laterally beside the track at heights between 5 m and 7 m. An electric motor unit manufactured by AEG achieved 210,2 km/h there. At that time a world record on rails.

Figure 3.7: 3 AC 10 kV 50 Hz three-phase contact line Marienfelde–Zossen and electric motor
unit (EMU) (Siemens, 1901 to 1903).

Figure 3.8: AC 6 kV 25 Hz catenary suspended
contact line on the test line Niederschöneweide–
Spindlersfeld (AEG, 1903).

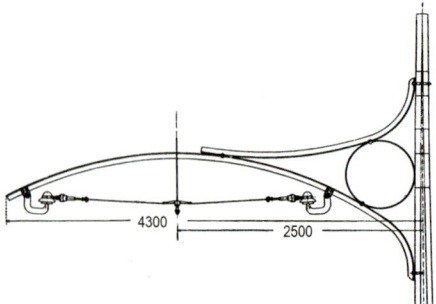

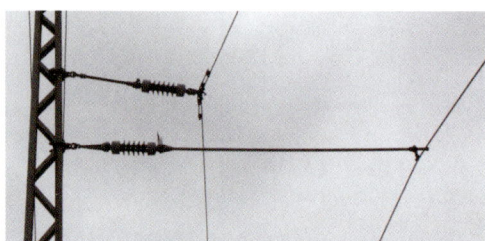

Figure 3.9: Trolley-type contact line on the
line Murnau–Oberammergau (Siemens, 1905).

Figure 3.10: Inclined overhead contact line on the
line Murnau–Oberammergau (Siemens, 1905).

AC three-phase traction supply was used on commercial lines to a limited extent only. A
1 840 km long three-phase AC 3,6 kV 16 2/3 Hz network existed in Italy between 1912 and
1976 [3.5]. The three-phase contact lines above points and crossings were not efficient and
eventually ceased to be used for main line electrification. Today, a few applications still exist
on mountain railways using three-phase AC 3,6 kV 16 2/3 Hz [3.6] supplies.

In 1903, *AEG* installed a test line between Niederschöneweide and Spindlersfeld close to
Berlin [3.7], equipped with the first contact line with catenary suspension and approximately
constant contact wire heights. It was suitable for suburban and main line railways, increased
speeds and operated at AC 6 kV 25 Hz. This design was characterized by V-type droppers
arranged at 3,0 m spacing and suspended from two steel catenary wires that supported the
contact wire (Figure 3.8). Contact and catenary wire were fixed terminated.

On 1 January 1905, the first main line electrified by AC 5 kV 16 Hz [3.8, 3.9] was commissioned for commercial services between Murnau and Oberammergau in Southern Germany
(Figure 3.9). Originally, a three-phase AC power supply had been planned, but, the three-
phase design was abandoned. The trolley-type contact line installed by *Siemens-Schuckert
AG (SSW)* comprised a 1,1 km long section with an inclined contact line (Figure 3.10) with
fixed-terminated contact and catenary wires (see Clause 3.3.3.5). At the supports, this contact
line had a low level of elasticity. At speeds of 40 km/h the pantograph touched the cantilevers
and interrupted the power supply [3.10].

a) Double-track cantilever

b) Cantilever with tube steady arm

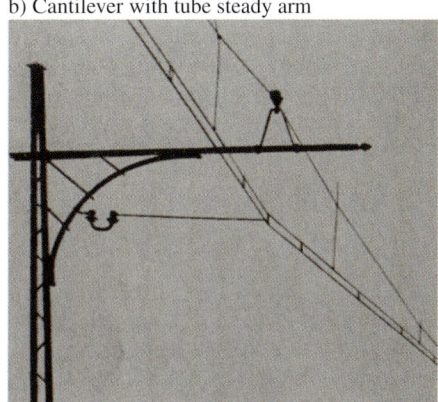

Figure 3.11: Cantilevers at Hamburg City Railway (Siemens, 1908).

In 1906, a 1,7 km long test line close to Oranienburg was electrified with a 6 kV 25 Hz contact line [3.11]. *AEG* installed a contact line with a catenary suspension where both the contact and catenary wire were auto-tensioned. In addition, an insulated lever-type tensioning device at each catenary wire support tensioned an auxiliary wire which maintained a constant contact wire height.

In 1908, *Siemens-Schuckert (SSW)* installed a 6,3 kV 25 Hz contact line, on the Hamburg city railway line Ohlsdorf–Blankenese, with catenary suspension, a fixed terminated catenary wire and a contact wire with an auxiliary catenary wire both of which were auto-tensioned by a lever-type tensioning device (Figure 3.11). Droppers at 6 m spacings carried a 100 mm^2 cross-section contact wire suspended from an auxiliary catenary wire. The auxiliary catenary wire consisting of a 35 mm^2 steel rope was supported from the catenary wire by droppers. The contact wire height was 5,2 m and the system height 1,4 m [3.12]. Cantilevers across two tracks with spacings of 40 m supported the contact line by tube-type steady arms staggered to 0,45 m. The tensioning sections were between 800 m and 1 300 m long [3.13].

3.1.2   Main line railways

AEG and *SSW* installed the contact line on the main line Dessau–Bitterfeld based on the experience gained from the test lines Ohlsdorf–Blankenese and Oranienburg [3.9, 3.14].

In April 1908, the Bavarian Traffic Ministry decided to adopt AC 15 kV 16 2/3 Hz for the operation of electric trains. This first decision to standardise the types of electrification systems eventually led to the *agreement concerning the design of electric traction of trains* [3.15] and to the development of simple contact lines with catenary suspended contact wires [3.16]. The experience gained concerning the interaction of pantographs and simple contact lines on the Murnau–Oberammergau line, resulted in multi-suspension of the contact wire from the catenary wire. This was adopted for the contact lines of the Mittenwald–Reutte and Salzburg–Berchtesgaden lines between 1908 and 1913. A horizontal tube was adopted as a steady arm, which fixed the lateral position at the supports. While the catenary wire was fixed terminated, the contact wire was auto-tensioned by a pulley- (Figure 3.12 a)) or lever-type arrangement (Figure 3.12 b)) or pulley arrangement. This design is called semi-compensated contact line. To reduce the power losses, the companies involved in electrification at that time increased the contact line voltage. In 1908, the Hamburg City railway line section Ohlsdorf–Blankenese,

a) Pulley-type tensioning device, design SSW 1939 [3.17, 3.18]

b) Lever-type tensioning device, design SSW-Hannes 1928 [3.19]

c) Wheel-type tensioning device for two times 10 kN, design AEG 1938 [3.20]

d) Wheel-type tensioning device, design SSW 1941 [3.21]

e) Mini-wheel-type tensioning device 1950 [3.22]

f) Siemens tensioning device 2006 [3.23]

Figure 3.12: Historic and current designs of tensioning devices.

similar to the test line Niederschöneweide–Spindlersfeld, commenced operations using an AC 6,3 kV 25 Hz supply. In 1910, the main line Dessau–Bitterfeld followed, with AC 10 kV 15 Hz. The differing types of power supply and contact line design hindered the installation of a common electrically operated main line network. As a consequence, in 1912 Bavaria, Baden and Prussia agreed on a common supply system of AC 15 kV 16 2/3 Hz; today 16,7 Hz [3.15]. Although, a common agreement on the frequency and voltage existed, the contact line contractors developed unique types of contact lines with new non-compatible components. The required standardisation of designs was completed in 1926, replacing the first generation of regionally designed contact lines [3.8, 3.10, 3.24, 3.25] by specified features including:

- contact line height 6,00 m
- system height 3,00 m
- optional V-type arrangement of droppers at mid-span
- spacing of droppers 12,5 m
- contact line with a catenary wire arranged vertically above the contact wire
- fixed terminated or auto-tensioned catenary wire,
- auto-tensioned groove-type contact wire
- optionally auxiliary catenary wires arranged parallel to the contact wire
- double insulation
- contact wire stagger at supports 500 mm to 600 mm
- span length 75 m
- two-span uninsulating overlaps
- single-span uninsulating overlaps
- single poles on lines outside stations
- rigid cross-spans and head-spans in stations

After the agreement on type of power supply and the first applications of catenary-suspended contact lines, further electrification was carried out in Germany [3.26]:

- Wiesental railway in Baden
- Freilassing–Berchtesgaden in Bavaria
- Garmisch–Griesen in Bavaria
- Central German lines in Prussia
- Silesian mountain railways

The lines, New York–New Haven, London–Brighton and the South Coast Railway were electrified using catenary-suspended contact lines with V-type droppers [3.25].

From 1931, and based on increased operational experience, a further unification of contact line designs was observed in Germany, leading for the first time to standard drawings for the design of contact lines, known as *second generation contact lines*:

- copper or steel catenary wire, rigidly terminated or auto-tensioned at 10 to 11 kN
- auto-tensioned copper contact wire with 100 mm^2 cross-section
- contact wire height 5,5 m
- contact wire stagger 0,5 m
- maximum span length 80 m
- cantilevers made from angle sections rigidly fixed at the poles, rope anchors and registration arms on lines between stations
- twin- and triple-span non-insulating overlaps
- triple-span insulating overlaps
- single poles on lines outside stations
- head spans in stations

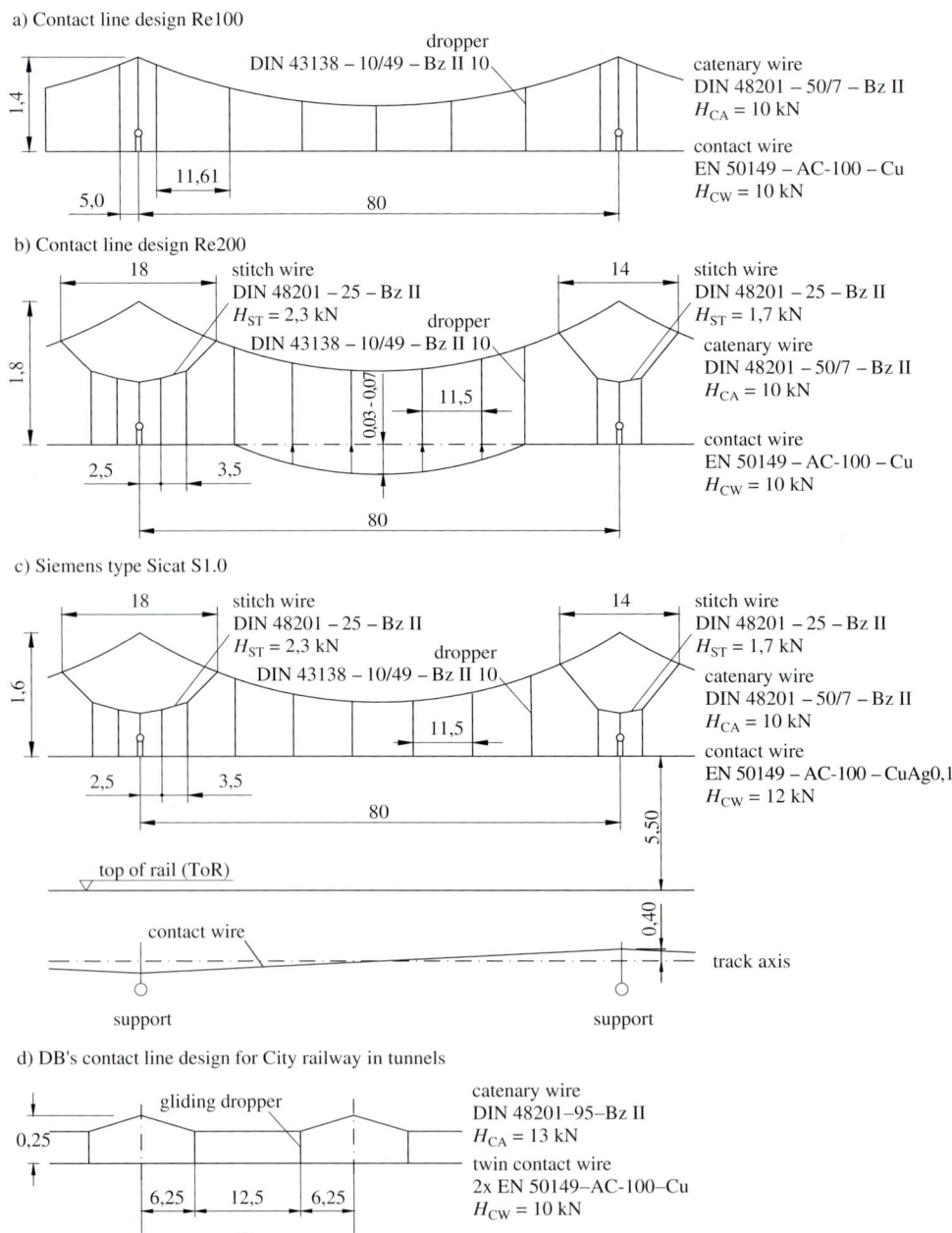

Figure 3.13: Overhead contact line types for conventional lines of DB and Siemens in 2013.

From 1926, the differing developments required standardisation of pantographs and the geometry of pan heads. Until 1926, the collector strips consisted of aluminium. The more advanced pantographs SBS 10, SBS 39 und SBS 54 were equipped with copper collector strips. In 1941, the speed could be increased under this contact line to 150 km/h by introducing auto-tensioned catenary wires, stitch wires at the supports and push-off supports where the steady arms were only tension-loaded. However, the dynamic behavior was not yet fully satisfactory. When introducing the standard contact lines at DB in 1950 [3.10], the contact line designs Re75, Re100 (Figure 3.13 a)) and Re160 were created, equipped with tensioning devices shown in Figures 3.12 a) to 3.12 d). The individual designs were distinguished by the operating speed.

The *pulley-type tensioning device*, with or without a latch-in arrangement, transferred the weight force to the contact or catenary wire with a ratio of 1 to 2 (Figure 3.12 a)). The *lever-type tensioning device* that replaced the pulley-type design in Germany, transferred the weight force with a ratio of 1 to 3 to the contact wire while the catenary wire was fixed terminated (Figure 3.12 b)). The wheel-type tensioning devices (Figure 3.12 c) and d)) were based on transferring the weight force with a ratio of 1 to 3 or 1 to 4 to the catenary or contact wire. In the next step of development small wheel-type tensioning devices (Figure 3.12 e)) tensioned the catenary wire and contact wire separately. In this case, the contact wire is fixed by two Z-type anchors close to the midpoint, approximately in the centre of the tensioning section.

The Re160 design adopted 12 m long *stitch wires*. The *light-weight steady arm* was used for the first time with this design to reduce the mass at the support. A dropper fixed the registration arm elastically to the stitch wire. Measurements of the contact forces over approximately 10 000 km of test running, confirmed good interaction of these components enabling the commercial speed to be raised to 160 km/h.

To increase the commercial speed to 200 km/h, a test line between Forchheim and Bamberg was equipped with differing contact line types consisting of 16 tensioning sections [3.27]. Designs with twin contact wires, auxiliary catenary wire, shorter span lengths and larger cross-sections were tested. As a new feature, *contact wire pre-sag* to compensate the higher uplift at mid-span was tested. The tests confirmed that expectations could be achieved with 30 mm pre-sag, an 18 m long stitch wire at pull-off supports and 14 m at push-off supports. DB applied the new contact line type, called Re200 (Figure 3.13 b)), for the first time in 1965 when upgrading the line Munich–Augsburg for higher speeds.

The City-railway in Munich and in Frankfurt/Main required a high-performance contact line for tunnels. The contact line developed for this purpose, having a 0,25 m system height, is shown in Figure 3.13 d).

On 12 March 1973, high-speed test runs started on the Gütersloh–Neubeckum line, where the recording train achieved 250 km/h. The target of these runs was to test a contact line type suitable for 250 km/h. The shorter spans, increased cross-sections and tensile forces of contact and catenary wires and additional improvements concerning the stitch wire performance created the new contact line type Re250. The most important features of this contact line, for high speeds, are shown in Figure 3.14 a). On 1 May 1988, a world record on rails was achieved under this contact line whereby the tensile force of the contact wire was increased to 21 kN [3.28].

With the target of further increasing the commercial speed, Siemens established a study that eventually led to the contact line type Re330 [3.29]. Its features are shown in Figure 3.14 b). After DB switched to functional tendering for contact line installations in 1996, the Siemens designs H1.0 for the high-speed line Köln–Rhine/Main and S1.0 for the connecting lines,

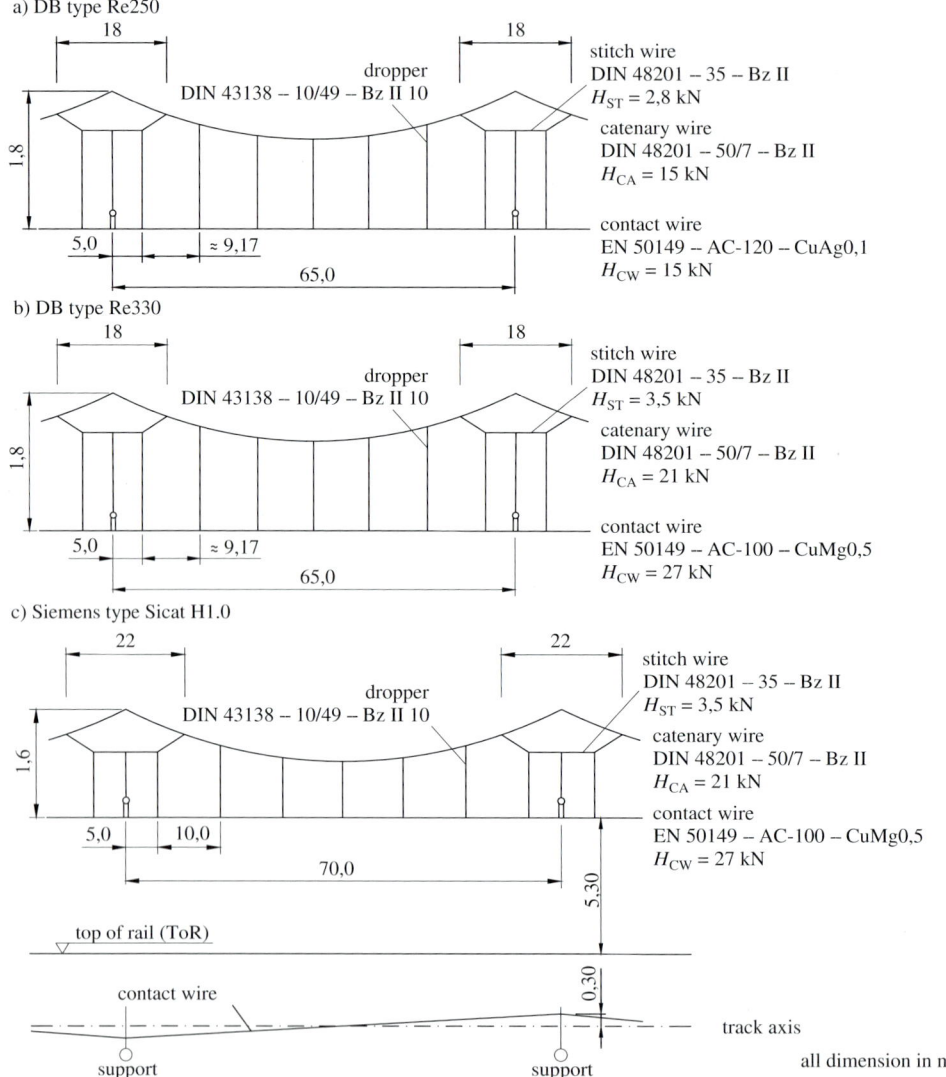

Figure 3.14: Overhead contact line types for high-speed of DB and Siemens in 2013.

were released. The design of H1.0 differs from the design of Re330 by the 70 m span length, and 1,6 m system height. The design Sicat H1.0 (Figure 3.14 c)) is suitable for running speeds up to 400 km/h with one pantograph. The characteristics of the Sicat S1.0 design which is suited for speeds up to 230 km/h, are shown in Figure 3.14 c) .

3.1.3 Overhead contact lines for tramways

At the end of the 19th century, electric tramways gradually replaced the horse-driven railways in big cities. Initially, *trolley-type contact lines* supplied the vehicles, but now modern *catenary suspended contact lines* supply them. On the first line which was commissioned in

Richmond (USA) in 1888, vehicles used *trolley-type collectors* with wheels. The single pole rigidly fixed contact wire was suspended at cross-span installations anchored on poles and buildings [3.30]. The operation of the trolley type collectors and wheels was often prone to disturbances. Dewirements often occurred at vertical contact wire bends. To reduce dewirements, shorter support spacings were required leading to increased investments.

With the introduction of *bow-type pantographs*, longer spans, simpler crossings and higher speeds could be achieved. DC power supply was used for mass transit as is still the case. The first contact lines for tramways were easy to install since they were mainly trolley-type contact lines, needing little space that did not affect the townscape. With the goal of faster commercial speeds and longer distances, catenary-suspended contact lines were increasingly used for tramways. The overhead contact line of the Berlin-Heerstraße–Spandau tramway was installed in 1927 with a catenary-suspended contact line with 120 m spacing between poles which could be traversed reliably at 50 km/h [3.31]. In view of wear and high availability, the spacing between the poles was reduced after 1927. With auto-tensioned contact lines from 1960 onwards the wear was further reduced because of the constant tensile force [3.30]. Vehicles with steadily increasing power, air conditioning and operating at shorter headways create a higher energy demand that in turn puts a higher current demand on the contact line [3.32]. Trolley- and catenary-type contact lines with twin contact or two catenary wires met these requirements. As a result of these developments, there are two main types of contact lines for tramways:

- *trolley-type contact lines* with
  - single contact wire or
  - twin contact wire having a cross-section of 80 to 150 mm^2 of E-Cu or CuAg
- *catenary-suspended contact lines* consisting of
  - one contact wire and one catenary wire or
  - twin contact wires and one catenary wire
  - twin contact wires and two catenary wires
  - each case with rigidly terminated contact wire and catenary wire or
  - auto-tensioned contact wire and auto-tensioned or fixed catenary wire
  - contact wires with 80 to 150 mm^2 E-Cu or CuAg cross-section.

3.1.4 Overhead contact lines for trackless vehicles

In 1882, Werner von Siemens developed an electrically driven trackless vehicle, called *Elektromote*. At the first application in Halensee close to Berlin an overhead contact line supplied energy to this vehicle. This first predecessor of trolley buses was supplied via a contact trolley which traveled with eight wheels on the twin contact wire line the latter being suspended on poles at fixed distances. The simple design corresponded to a simple contact line which supplied the vehicle at DC 550 V. Due to the high number of disturbances and the increasing importance of tramways, this idea was discarded [3.33].

The first trolley bus line was erected in Bielatal close to Dresden in 1901, adopting *trolley-type collectors* as is the case today. The trolley-type collector was designed by *Schiemann* and mounted on the roof of the vehicle. Both arms were spring-loaded to make good contact with the contact wires. Copper collector strips carried the power from the contact line to the vehicle. The simple design of the principle trolley-type collector and the contact line led to wide application of the *Schiemann* design. The contact line on the line Blankenese–Marienhöhe was operated with the same current collector type and consisted of a simple contact wire,

a) Third rail in Oslo (SPL Norway)

b) Conductor rail overhead contact line in Santo Domingo (Siemens AG, Germany)

Figure 3.15: Conductor rail lines.

suspended at poles spaced 50 m apart. The 440 V DC motor reached a power of 15 hp. An earthed catenary wire was arranged above the contact wire and suspended the contact wire by insulators. The *trolley contact line* used supports which differed from those for tramways. In 1935, a contact wire suspension was installed on the line Steglitz–Marienfelde, enabling the trolley pole to swing through a semi-circular sector while maintaining contact with the contact wire and allowing the vehicle to move away from directly under the contact line. The number of trolley bus lines increased in Germany up to the mid 1950's, after which cities gradually closed down trolley bus operations in favour of standard buses. In 2017, the cities of Eberswalde, Esslingen and Solingen were still operating trolley bus lines in Germany [3.34], using a DC 600 V supply [3.35]. Because of their environmental friendliness, low noise emission and economic advantages, extended trolley bus networks have been operated in large cities like Moscow and San Francisco.

3.1.5   Conductor rails for suburban railways and metros

Conductor rails are applied as *third rails* in parallel to the running rails (Figure 3.15 a)) or as *conductor rail overhead contact lines* above the tracks (Figure 3.15 b)). At the beginning of the electrification of tramways and suburban railways, cities refused the overhead installation of power supply equipment. Consequently, Siemens & Halske developed and installed a tramway from Ofen to Pest, now Budapest, with an underground power supply. A channel was constructed below one of the rails, where two T-type profiles were installed, one for the supply and one for the return current. The collector contacted both conductor rails through a slot in the rail. The collector, designed as a plate, consisted of two current shoes insulated from one another, which were pressed to the conductor rail by a spring, to supply uninterrupted power to the vehicle as far as possible. High investments, difficult design at points and expensive maintenance hindered further application of this system [3.36].

Conductor rails are well suited for railways with a high current demand. As an example, a conductor rail installed at the tunnel ceiling enabled a small tunnel cross-section for the first metro line in mainland Europe. The first time a *third rail* was adopted for a suburban line in Chicago, which was opened in 1892. The 600 V conductor rail was installed beside the running rails. An insulator supported the T-shaped steel conductor rail with a layer of copper for the contact face. A vehicle collector shoe contacted the conductor rail from the top [3.37]. Conductor rails with current collection from the top are prone to disturbances from ice and snow. Consequently, current collection from the bottom of the rail was used in the installation of the Berlin suburban railway in 1924. The H-type profile of the conductor rail enables suspension on insulated supports at a fixed horizontal and vertical position relative to the track. The original rails, made of soft steel were replaced from 1985 onwards by *compound rails made of aluminum* with the contact surface made of stainless steel [3.38]. A majority of European operators of metros and suburban railways uses this design [3.39].

The first London metro line opened in 1890, also has used the third rail close to the running rails and a fourth conductor rail as a return conductor arranged between the running rails. The traction units contact these conductor rails on the top with collector shoes. The insulation of the return current conductor rail helps contain stray current corrosion.

3.2 Definitions

As a result of the wide variety of requirements and the long period over which the contact line designs developed, different terms evolved for the same object or meaning. Some of these terms are defined in IEC 60050-811, EN 50119 and EN 50122-1. The following terms may differ from these definitions:

Contact line installation is the equipment of the electrical traction power supply between the substations and the electric traction units, the boundaries of the contact line installation, within the supply circuit, are formed by the feeding point and the contact with a *sliding current collector* to supply electrical energy to vehicles. Contact lines can be *overhead contact lines* or *conductor rails*.

A contact line installation can consist of:
- contact lines, including overhead contact lines and third rails
- feeder lines, parallel feeder and other lines
- any equipment required for the operation of the contact line installation
- return current lines and earthing installations
- any lines connected with the contact line to supply electric installations like lighting, signalling, point control and heating

Overhead contact line installations are contact lines which, according to EN 50119 use an overhead contact line for the traction power supply of traction vehicles. *Overhead contact lines* are systems of electrical conductors arranged above or laterally to the vehicle gauge used in conjunction with a *sliding current collector* to supply electrical energy to vehicles. Overhead contact line systems may consist of:
- catenary wires
- contact wires
- droppers
- clamps
- anchors
- midpoints
- terminations
- supporting structures and components, serving as lateral guidance and termination of conductors

- poles and foundations
- cross-spans and head-spans
- section insulators
- non-insulating overlaps
- insulating overlaps
- electrical connectors
- insulators
- tensioning devices
- insulating sections

Conductor rails are contact lines comprising rigid conductive rails. They are placed beside or under the vehicles as *third rails* or above the vehicle as *conductor rail overhead contact lines* also called *soffit conductor rail*.

Conductor rail overhead contact line is an overhead contact line consisting of copper profile or an aluminium profile with a U-shaped cross-section and an opening at its lower face where the contact wire is clamped in between the flanges or of solid copper profiles.

Third rails are contact lines with conductor rails installed on insulators close to the running rails.

Cross-span supports fix the lateral and vertical position of contact lines and are installed transverse to the tracks. They consist of single and multi-track cantilevers, portals and flexible cross-spans.

Poles and *soffit posts* are *structures* to carry and align the transverse supporting equipment.

Foundations anchor the poles into the subsoil and transfer the loads of the contact line into the subsoil.

Switches, in a single- or double- pole design, connect or isolate electric sections of the contact line and traction power lines to each other or to feeding lines. They include:
- disconnectors
- switch disconnectors
- earthing switch

Traction power lines are electric lines consisting of bare conductors or cables suspended by insulators above the ground or buried in the subsoil. They conduct traction power to the track installation from the substation or the switching posts to the contact line. They include:
- line feeders
- parallel feeder lines
- bypass lines
- compensation lines
- switching transverse and drop lines

The function of traction power lines is described in Clause 3.6.

Line feeders are overhead conductors installed adjacent to the contact lines on the same supporting structures and serve to supply energy to successive feed points.

Parallel feeder lines are overhead conductors installed adjacent to the contact wires and connected to these at certain intervals to increase the effective conducting cross-sectional area.

Bypass feeder lines serve to ensure unbroken energy supply while by-passing specific switching sections, e. g. stations on a single-track stretch of railway line.

Return circuits include all conducting components that form a conducting path for the traction return current in normal operation and in case of faults. The return circuits include:
- *running rails*
- *return current rails*

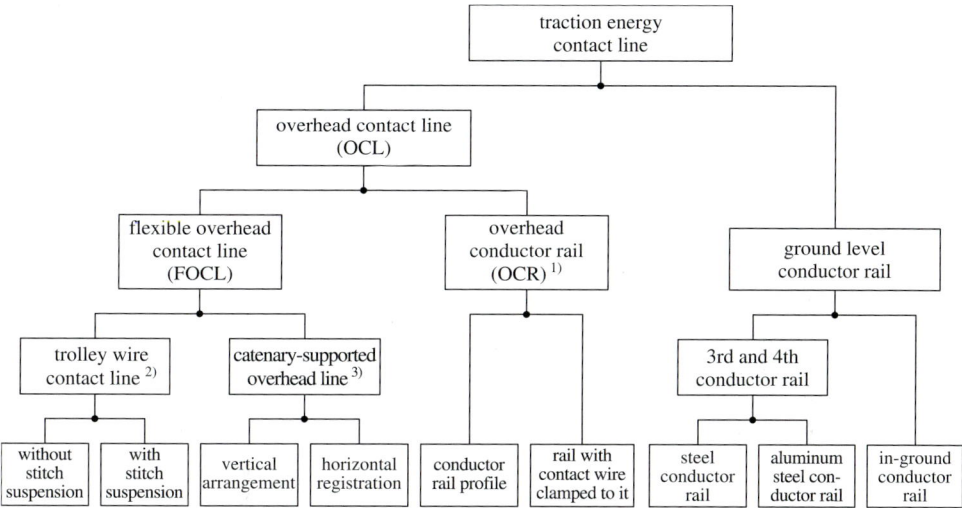

Figure 3.16: Overhead contact line designs of electric railways. 1) designated also as rigid overhead contact line (ROCL), 2) designated also as single tramway equipment, 3) designated also as catenary

- *return current conductors*
- *earthing wires*
- *return current cables* and
- all other components conducting return currents

Track return system is a system in which the running rails are used as return traction current conductors and as conductors for fault currents.

Earthing conductors are metal conductors that bond the supports to earth potential to protect people and equipment in the case of insulation faults.

Railway earthing protects against the occurrence of impermissible touch voltages and limits the *rail potential*. Rail earthing involves the connection of conductive parts with a suitable earth electrode.

Earth electrode is one or more conductive parts in intimate contact with the soil and provides a conductive connection with the soil.

Earth, as agreed, has an electric potential of zero. No direct connection to the earth is permitted in the case of DC railways because of the danger of stray current corrosion.

Overhead contact line zone and *pantograph zone* are zones within which increased requirements on the earthing of conductive parts of contact line installations are relevant.

Overhead contact line type is the description of the overhead contact line in terms of the characteristics and properties of its design, e. g. *stitched catenary supported* or horizontal arrangement. Figure 3.16 shows often used contact line types.

Overhead contact line standard design is the designation for a specific form of execution of an overhead contact line, e. g. the design Sicat H 1.0 of Siemens AG for high speeds.

Longitudinal span length, or *span* is the term used to designate the distance along track between two successive supports.

Tensioning section length is the term used to designate the distance between two consecutive terminations of an overhead contact line.

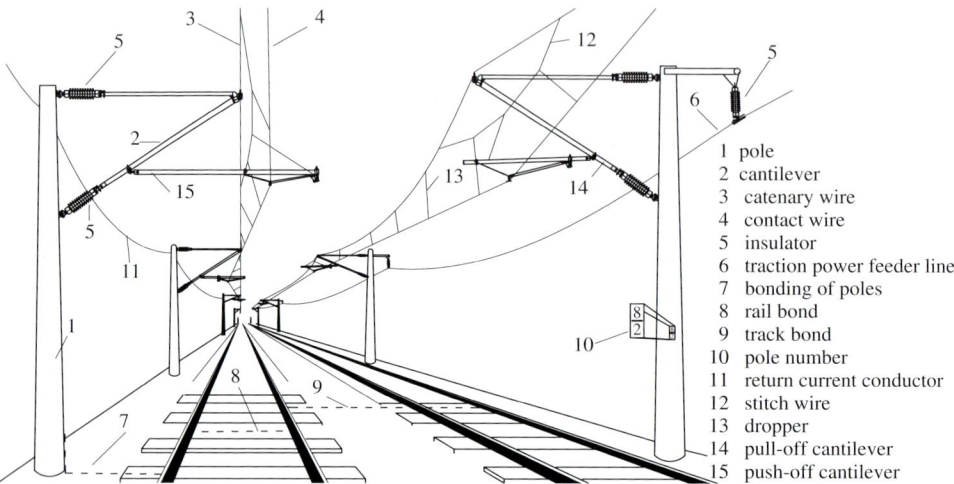

Figure 3.17: Overhead contact line on individual poles.

Automatic tensioning device is the device used to automatically maintain a constant tensile force on an overhead contact line within a specified temperature range and to compensate for contact line length variations resulting from temperature changes.

Half tensioning section length is the term used to designate the overhead contact line length between a midpoint and the tensioning equipment.

Midpoint is the term used to designate the point roughly in the middle of a tensioning section where a means of fixing the position of an overhead contact wire in the longitudinal direction relative to the running rails is installed. They are used to ensure that conductors do not migrate towards one end of the tensioning section.

Overlaps accommodate the pantograph transition from a tensioning section to the adjacent one without speed reduction and without interruption power supply (EN 50 119, Clause 5.12).

Uninsulating overlap, also called *tensioning* is bridged by electrical connectors.

Insulating overlap, also called *section insulating overlap* can only be bridged via overhead line disconnector.

3.3   Overhead contact lines

3.3.1   Structure and characteristics

The typical structure of an overhead contact line installation with individual poles is shown in Figure 3.17. This is the preferred and increasingly used structure for main line railways. The swiveling cantilevers with tubes are described in Clause 11.2.1.1.

Figure 3.18 a) shows an overhead contact line supported from a *portal*, as an alternative to the design with individual poles. Other *transverse supports* are *multi-track cantilevers* (Figure 3.18 b)) and *flexible head-spans* (Figure 3.19). Clause 11.2.1.2 describes components of head-spans.

The structure of a tensioning section is shown in Figure 3.20 a). It consists of individual spans (Figure 3.20 b)) with features corresponding to the layout of a contact line. A contact

a) Portal

b) Multi-track cantilever

Figure 3.18: Overhead contact line transverse supports in Jåttåvågen (close to Stavanger) in Norway.

line installation is divided into individual tensioning sections with fixed or constant tension terminations at their ends. A *constant tension termination* maintains the contact force of the contact and catenary wire almost constant when the temperature varies. A *midpoint* is located approximately in the middle of the tensioning section with guy anchors to prevent swiveling of the cantilever and longitudinal movements of the catenary wire. Connections, using stranded conductors between the catenary and contact wire, called Z-conductors, fix the contact wire to the catenary wire and prevent longitudinal movements of the contact wire. The overhead contact line design needs to consider all static, dynamic, thermal and electrical requirements. The following sections will describe the effect of individual parameters on the performance of the overhead contact line with reference to the terms defined in Figure 3.21, namely:

- overhead contact lines in stations
- overhead contact lines on interstation line sections
- overhead contact lines in tunnels
- overhead contact lines in the open

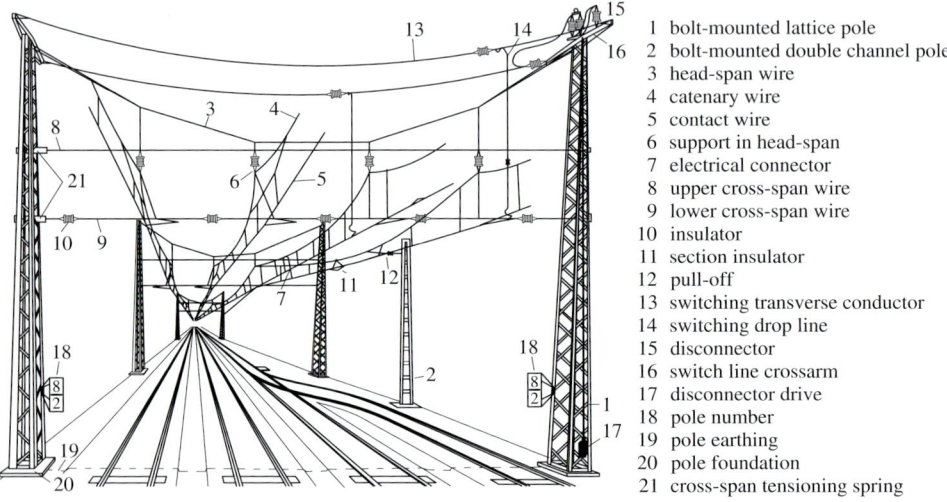

1	bolt-mounted lattice pole
2	bolt-mounted double channel pole
3	head-span wire
4	catenary wire
5	contact wire
6	support in head-span
7	electrical connector
8	upper cross-span wire
9	lower cross-span wire
10	insulator
11	section insulator
12	pull-off
13	switching transverse conductor
14	switching drop line
15	disconnector
16	switch line crossarm
17	disconnector drive
18	pole number
19	pole earthing
20	pole foundation
21	cross-span tensioning spring

Figure 3.19: Overhead contact lines at head-span structures.

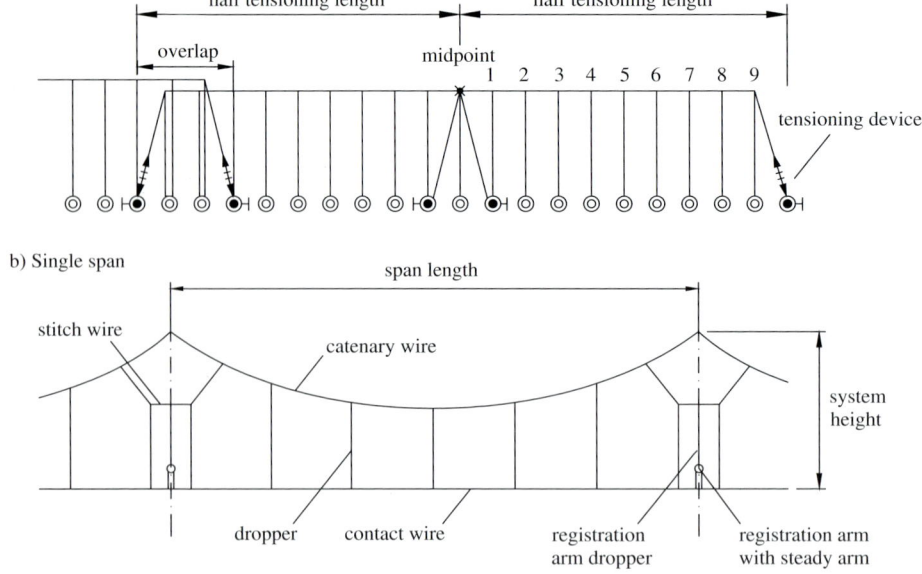

Figure 3.20: Structure of an overhead contact line.

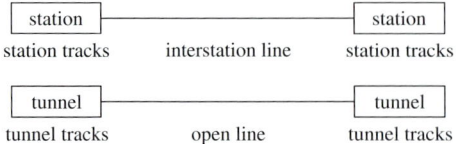

Figure 3.21: Terms for different track or line types.

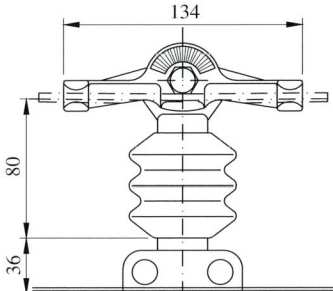

Figure 3.22: Single-point suspension.

The design of the contact lines for these line types follows from the operational requirements, implemented by contractors based on their own experience and capabilities. The resulting overhead contact lines can be classified according to the range of application or to the design characteristics, such as suspension and tensioning equipment.

3.3.2 Trolley-type contact lines

3.3.2.1 Definition and application

The term *trolley-type contact line* is applied to systems that do not have a continuous catenary wire and, therefore, have a simple structure. In comparison to catenary-type overhead contact line installations, the *contact wire sag* of systems of this kind is large and the distance between supports needs to be kept short to maintain the contact wire height as nearly constant as possible. The running speed of these systems, 80 km/h at the most, restricts their application to tramways, trolley-buses, industrial railways and turn-outs and sidings of main railway lines.

3.3.2.2 Single-point suspension with fixed anchored contact wire

With *single-point suspensions*, the contact wire is only fixed by a contact wire clip directly mounted on a cross-span wire or cantilever support (Figure 3.22). In spite of the short support spacing of approximately 30 m, a sag of up to 0,4 m is observed at mid-span due to the lack of a mechanism to compensate for temperature-dependent contact wire length variations. As it moves along the contact wire, a *pantograph-type collector* is subject to large vertical movements, while trolley collectors are subject to both horizontal and vertical movements. The sudden change of direction in vertical movement as a pantograph passes the trolley wire support can cause the pantograph to bounce, or can lead to contact separation or to excessive contact forces. The contact wire wears unevenly and is subject to premature fatigue caused by oscillations. For these reasons, the running speed of such systems is limited to 40 km/h and the design is used mainly on light-duty tramway lines.

3.3.2.3 Pendant-type suspension with and without automatic tensioning

The *pendant-type suspension* (Figure 3.23) was developed to avoid the disadvantages of the system described in Clause 3.3.2.2. In overhead contact lines of this type, the contact wire is clamped with an offset to freely swinging dropper wires fixed to the supporting points. This

a) Overview   b) Detail A

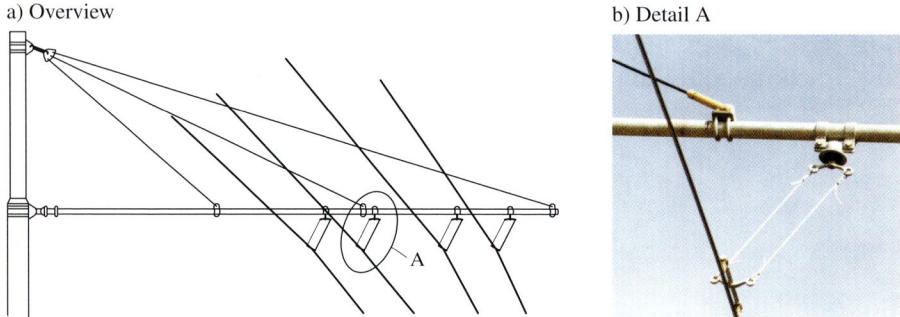

Figure 3.23: Pendant-type suspension of an overhead contact line for trolley-buses.

pull-off lever   pulley sheave   stitch rope   contact wire clip

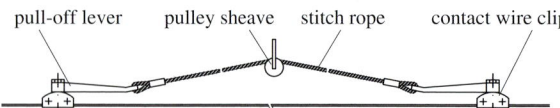

Figure 3.24: Bridle-type suspension.

improves the elasticity of the arrangement and the rate of reversal of the vertical movement of the pantograph is reduced as it moves underneath the supports.

Skew pendants are used to reduce the undesirably high sag of the contact wire. At the supports, these pull the contact wire alternately to the left and to the right. As the length of the contact wire varies with changes in temperature, its weight causes the lower end of the slanted pendants to rise or fall correspondingly, compensating for the changes in sag to some extent [3.40]. This measure allows the distance between supports to be increased to 40 m. To avoid lateral wear of the contact wire and prevent the collectors of trolley-buses from striking the clips, slanted suspensions are designed in a parallelogram or trapeze shape (Figure 3.23), enabling the contact wire to assume the desired position even if the pendant rotates. However, the resulting zig-zag path of the contact wire leads to uneven movement of the trolley-collector. The permitted running speed is up to 50 km/h.

3.3.2.4 Bridle-type suspension

With this type of overhead line design, two clips connect the contact wire with a *bridle wire* that is free to move in a longitudinal direction in a *sliding mount* or *pulley sheave* fixed to the cross-span or cantilever support (Figure 3.24). At the termination poles, the contact wire is connected to a tension adjustment mechanism, compensating the contact wire length variation. The reduction in maximum mid-span sag achieved in this way allows the support spacing to be increased to 55 m. Nevertheless, the lack of elasticity and concentration of masses at support points is a disadvantage and causes increased wear at these points, limiting the running speed to 60 km/h.

3.3.2.5 Elastic supports

Elastic supports or *elastic cantilevers* represent cantilever designs with an elastic mounting using rubber spring components (Figure 3.25) that also damp contact wire movement (see Clause 11.2.1.6). Either single contact wires or twin contact wires can be mounted in the contact wire clips. If elastic supports are the only means used to support contact wires, their

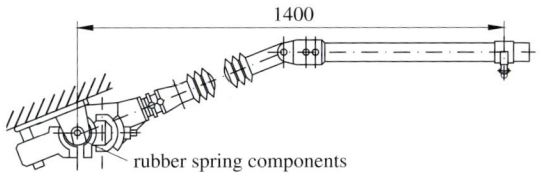

rubber spring components

Figure 3.25: Elastic support.

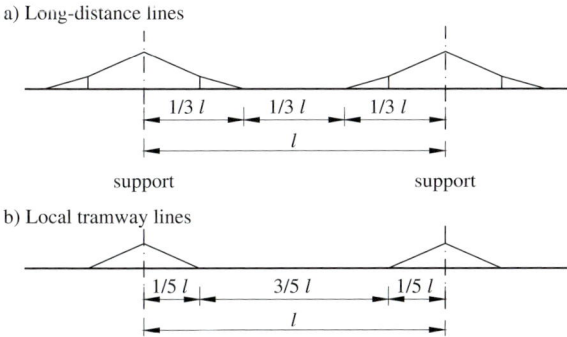

a) Long-distance lines

1/3 *l*   1/3 *l*   1/3 *l*

l

support   support

b) Local tramway lines

1/5 *l*   3/5 *l*   1/5 *l*

l

Figure 3.26: Overhead contact line with stitch suspension.

spacing should not exceed 12 m. It is possible to use this type of suspension for running speeds up to approximately 100 km/h.

An additional catenary wire clip permits the installation of a catenary wire and extending longitudinal spans to 30 m. Elastic supports are mainly used in tunnels where the space for the contact line is limited.

3.3.2.6   Trolley-type contact line with stitch suspension

A trolley-type contact line with a stitch suspension is a *simple contact line design* where the contact wire is joined to the support with a *stitch-wire* (also known as a *bridle*) and arranged in a triangular shape (Figure 3.26). The first overhead contact line installations of this type had short stitch-wires only without droppers. They were installed with a lateral pull, similar to slanted pendants that provided a certain degree of *automatic compensation* for thermal expansion and contraction of the contact wire and reduced wind deflection [3.40]. In addition, stitch wires can compensate for elasticity variations along the contact wire, which in turn improves the current collection quality. Depending on the stitch wire lengths, on the number of droppers between contact wire and stitch wire and on the tensioning method, it is possible to achieve running speeds up to 80 km/h with support spacings of 65 m.

3.3.3   Overhead contact lines with catenary suspension

3.3.3.1   Basic design

Overhead contact lines with catenary suspension are characterised by one, or in some cases two, supporting *catenary wires* located above the contact wires. The catenary wires support the contact wires by *droppers*. Because of their relatively simple design and favourable running characteristics, overhead contact line installations of the catenary design became commonly used world-wide. They permit longer support spacings than trolley-type contact lines

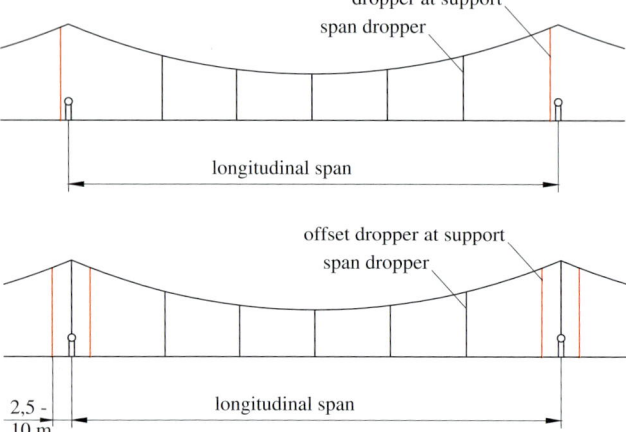

Figure 3.27: Contact line with droppers at the supports.

Figure 3.28: Contact line with offset droppers at the supports.

and reduce the wear on contact components. They are also being installed in urban mass transit transportation systems more frequently.

It is possible to classify overhead contact lines according to the design of the *tensioning system* used. A distinction is made between *completely compensated contact lines* with either combined or separate contact wire and catenary wire tensioning mechanisms and *semi-compensated contact lines* which have fixed, uncompensated catenary wires and compensated, i. e. automatically tensioned, contact wires. To suit the different applications, various catenary type contact line designs were evolved, differing mainly in the arrangement of the individual conductors and wires, in the design of supports and in the permitted running speed. Their components are described in Clause 11.2.2.

3.3.3.2   Contact line with droppers at the supports

The *simple catenary-supported overhead contact lines* used on early electrification projects were semi-compensated and characterised by a dropper connecting the contact wire to the catenary wire at or in the immediate vicinity of the support (Figure 3.27). Additional droppers were installed at spacings of 8 to 12 m along the longitudinal span. In comparison to trolley-type overhead contact lines, this system permitted the use of larger support spacings. Because of the fixed termination of the catenary wire at the ends and the rigid connection of the cantilevers to the poles, thermal expansion and contraction of the catenary wire still led to considerable variations in the height of the contact wire at this design. Whereas the catenary wire under tensile force in combination with the droppers ensures *elasticity* along the span, the elasticity at the supports is inadequate due to the droppers at the support, leading to high elasticity variations along the span. Therefore, this design cannot be recommended.

3.3.3.3   Contact lines with offset support droppers

The *contact line with offset support droppers* avoids the disadvantages described in Clause 3.3.3.2. The droppers in the immediate vicinity of the supports were eliminated and droppers at a distance of 2,5 to 10 m from the support points were introduced between catenary wire and contact wire (Figure 3.28). To reduce temperature-related changes of the contact wire

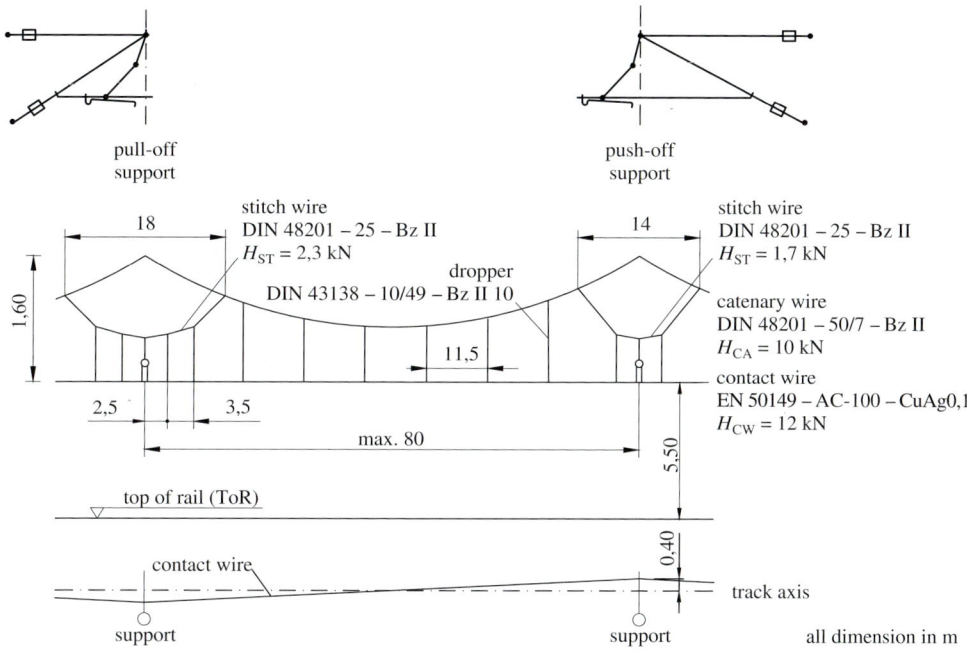

Figure 3.29: Sicat S1.0 [3.41] overhead contact line with differing stitch wire lengths.

height, completely compensated contact lines are used. In this case, the contact line is anchored approximately at the middle of the tensioning section. Instead of having cantilevers rigidly attached to the poles, as is usually the case with semi-compensated overhead contact line installations, swiveling cantilevers are used, allowing the cantilevers to follow the longitudinal contact line movement, which increases in proportion to the distance from the midpoint.

Applications of this design of contact line include main-line railways with running speeds up to 120 km/h, e. g. using DB's Re100 standard design [3.42], and tramways. Versions of this overhead contact line system with increased tensile forces and dropper spacings of approximately 6 m are also in use on high-speed lines, e. g. in France [3.43].

3.3.3.4 Contact line with Y-stitch wire suspension

The term *stitch wire* is used to designate a connecting element inserted between the catenary wire and the contact wire (Figure 3.29). At *semi-compensated overhead contact lines* stitch wires serve to compensate contact wire height differences between mid-span and the supports. When temperature changes occur, the vertical movement of the connection points of the stitch wire to the uncompensated catenary wire, in conjunction with the change in catenary wire length and tensile force, causes the contact wire at the support to be raised and lowered similarly to the height changes at the middle of the span. The spring effect of the stitch wire achieves a considerably better match of the *elasticity* at the supports to the elasticity at the mid-span. It is the latter effect which is the main reason for the current use of stitch wires in *completely compensated overhead contact line* installations. Depending on the desired

a) Semi-inclined contact line design

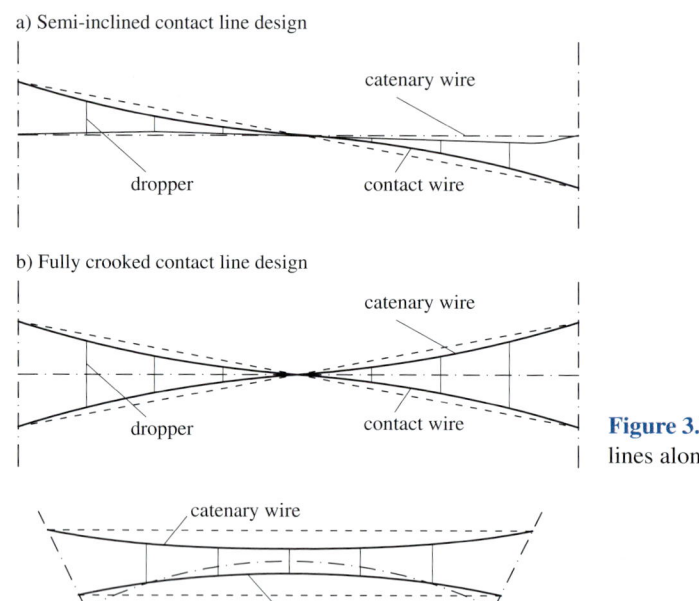

b) Fully crooked contact line design

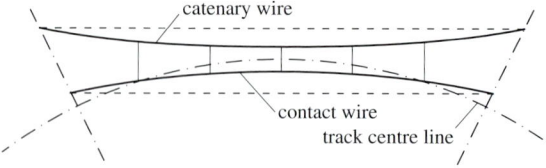

Figure 3.30: Inclined overhead contact lines along straight track.

Figure 3.31: Inclined overhead contact line in curves.

running speeds, the individual standard overhead contact line designs of DB are fitted with stitch wires with lengths of 6, 12, 14, 18 or 22 m and with one to four stitch-wire droppers [3.42]. The Siemens contact line Sicat S1.0 [3.41, 3.44], shown in Figure 3.29, has a special characteristic in that the registration arm is joined to a dropper fixed to the stitch wire. In this design, the different spring effects of short registration arms on *pull-off supports* and long registration arms on *push-off supports* is taken into account by the use of either 18 or 14 m long stitch wires with four or two droppers, respectively (Figure 3.29).

At pull-off supports, the lateral force exerted by the contact wire is directed away from the pole. At push-off supports it is directed towards the pole.

The stitch wire tension is selected with the objective of reducing variations in *elasticity* along the line. *Stitched contact lines* require careful adjustment; this can be facilitated by the use of adjusted special tools (see Clause 15.6). Properly designed stitch wires considerably improve the running characteristics of overhead contact line installations (see Chapter 10) and by allowing longer support spacings, result in lower investments.

Together with high-tensile forces on the contact and catenary wires, the use of stitch wires is one of the characteristic features of modern, low-wear *high-speed overhead contact lines*. It is possible to achieve running speeds up to 400 km/h with this type of overhead contact line installations.

3.3.3.5   Contact line with inclined suspension

In many overhead contact line designs, e. g. the standard Siemens design Sicat H1.0 [3.45], the catenary wire is located vertically above the contact wire. However, along straight lines of the track, the catenary wire is aligned with the track centre line and the contact wire can still be arranged in the usual zig-zag arrangement. The lateral position is then affected by the

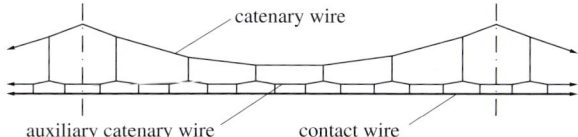

Figure 3.32: Contact line with auxiliary catenary wire – compound contact line.

alternating lateral pull. This principle is applied in the Sicat S1.0 and the design is also called a *contact line with semi-inclined suspension* (Figure 3.30 a). In *inclined catenary overhead contact lines* both the contact wire and the catenary wire are off-centre. These are pulled to opposite sides of the centre line along straight stretches (Figure 3.30 b)) but both to the same side in curves, where the catenary wire is further off-centre than the contact wire. Depending on the catenary wire arrangement, this design is also called a *semi-horizontal contact line*, also known as curvilinear. This enables adjustment of the contact wire position to match the track curvature and the use of longer support spacings. Figure 3.31 shows the arrangement of the catenary and contact wires in a curve. Overhead contact line installations of this type are relatively rare and are used mainly on mountainous stretches with tight bends. The adjustment of the contact line in such designs requires considerable effort. If higher tensile forces are exerted on the catenary wires and contact wires, the advantages of inclined overhead contact line suspensions become less pronounced. Some examples of contact lines are known in which a stitch wire is arranged on the side opposite to the contact and catenary wire offset, such as the one built on the Leipzig–Halle line (Germany) in an unsuccessful attempt to achieve longitudinal span lengths of 100 m [3.46]. As a consequence, additional poles were required.

With new materials and planing and installation tools Siemens developed the Sicat SX contact line having spans up to 100 m. Between 2010 and 2013 this design was tested successfully in Hungary and since then has been in operation at MAV (see also Clause 16.2.5, [3.49]).

3.3.3.6   Contact line with auxiliary catenary wire, compound contact line

A *compound contact line* has a second catenary wire, called the *auxiliary catenary wire* between the main catenary wire and the contact wire. It is joined to the main catenary wire and the contact wires by droppers that help to eliminate variations in elasticity (Figure 3.32). This overhead contact line design was first used by Siemens in 1912. Currently it is used for DC 1,5 kV overhead contact lines in France and for high-speed railways in Japan (see Clause 16.3.12). However, the good running characteristics of this type of installation are offset by the increased material requirements and significantly greater installation effort.

3.3.4   Horizontal catenary overhead contact lines

In *horizontal catenary contact lines*, the individual suspension wires and contact wires are in a more horizontal position relative to each other. These systems are not as strongly deflected by wind enabling lower structure heights and longer support spacings (Figure 3.33). However, they require greater planning, installation and maintenance effort than comparable vertically oriented catenary supported overhead contact lines.

There are various schemes for horizontal contact lines (Figure 3.33). In earlier designs with direct connections to the supports, larger contact wire height variations occur when the temperature changes (Figure 3.33 a)). In modern designs with a suspension similar to a *horizontal*

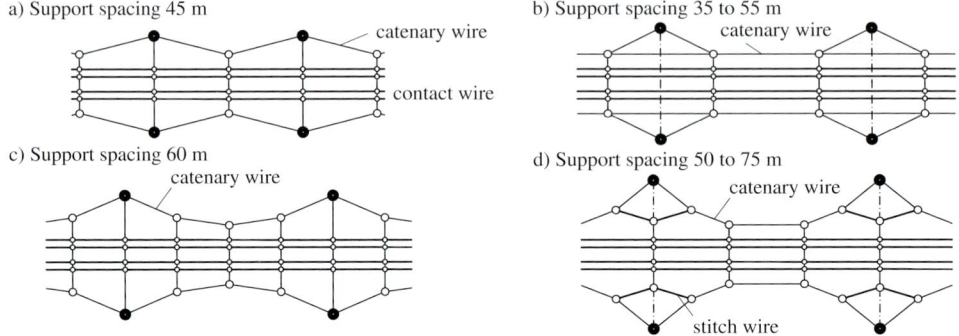

a) Support spacing 45 m

catenary wire

contact wire

b) Support spacing 35 to 55 m

catenary wire

c) Support spacing 60 m

catenary wire

d) Support spacing 50 to 75 m

catenary wire

stitch wire

Figure 3.33: Fully horizontal contact line for two trolley bus lines.

stitch wire and no cross-span wire at the support points, the contact wire clip positions are arranged so that they all rise and fall to virtually the same extent, as the temperature changes (Figure 3.33 b)). Mechanical calculations for horizontal contact lines were first carried out and discussed in a doctoral thesis in 1927 [3.46, 3.47, 3.48]. Apart from demonstrating the advantages described above, horizontal catenary contact lines lead to automatic compensation of thermal expansion and contraction and achieve almost completely uniform elasticity. As the temperature changes, the contact wires rise or fall to an equal extent along their entire length. Other known applications include long stretches of tunnels in Russia. The maximum running speed is 100 km/h. Typical criteria for the use of this design include special installation conditions or requirements with respect to clearance above ground or wind loading. Horizontal catenary overhead contact lines fixed at walls of buildings reduce the number of poles in urban environments.

3.4   Selection of the overhead contact line design

3.4.1   Basic considerations

Selection of a *type of overhead contact line* requires knowledge of the operating parameters and must take the requirements described in Chapter 2 into account.

An overhead contact line type is defined by the design and thus by the configuration of its components for a given application. Therefore, the overhead contact line should also be configured to provide minimum *operational life cycle costs*. Verification of the suitability of a design of overhead contact line for a given application can be performed by simulation of the interaction between the overhead line and the pantograph or by carrying out a track test. Table 3.2 contains applications for typical overhead contact line designs.

3.4.2   Selection of conductor cross-sections and tensile forces

The contact wire and catenary wire cross-sections should be designed and kept as small as possible, for economic reasons, but satisfy the assumed requirements. The traction power supply system, the traffic and the route profile determine the magnitude of the current flowing through the overhead contact lines. The *electrical rating of the contact line* is described in

Table 3.2: Overhead contact line designs and their applications.

No.	Design	Properties	Application
1	simple overhead contact line without continuous catenary wire, fixed termination or flexible tensioning, horizontal registration arrangements	contact wire height changes with temperature, limited span length and current-carrying capacity	light rail systems (tramways) with low electrical load, sidings on main-line railways, speed up to 100 km/h
2	vertical contact line without stitch wire, tensioned contact wire, catenary wire fixed or tensioned	contact wire height independent of temperature (tensioned catenary), span lengths up to 80 m are possible, current-carrying capacity can be adapted by selecting suitable catenary wire and contact wire cross-sections, large variation of elasticity between mid-span and support	tramways with high electrical load, main-line railways at speeds up to 120 km/h, two parallel contact wires are often employed with DC traction supplies
3	as (2), but with or without stitch wire, automatically tensioned contact and catenary wire	as (2), however lower elasticity differences between mid-span and support	main-line railways with high electrical loading and speeds up to 350 km/h
4	vertical contact line with auxiliary catenary wire automatically tensioned	as (3), however higher current-carrying capacity and more uniform elasticity	main-line railways with very high electrical loading and very high speeds

Chapter 7. Table 7.12 contains information regarding the continuous current carrying capacity of overhead contact lines for frequently adopted combinations of catenary and contact wire for a contact wire wear of 20 % operated at 16,7 or 50 Hz.

The *mechanical rating* is aimed at the maximum possible span lengths that can be implemented by adopting high tensile forces and a wide working width of the pantograph head. This reduces the number of supports and, therefore, the investment. The selection and stress analysis of the catenary wire and contact wire will be discussed in Clause 4.2.

The *elasticity* of the overhead contact line system and its variation depends mainly on the span length and the tensile forces in the contact wire and catenary wire. Clause 10.5.3.1 deals with the determination of elasticity.

Materials and tensile forces are selected for high-speed overhead contact lines so that the resulting wave propagation velocity on the contact wire is sufficiently high, as a dynamic characteristic (see Clause 10.2.1). According to EN 50 119:2013, the maximum *operating speed* should not exceed 70 % of the wave propagation velocity.

The suitability of a contact line for a given speed can also be assessed by the *Doppler factor* according to Clause 10.2.6. The non-dimensional Doppler factor should be equal to at least 0,15, or preferably 0,20.

The *contact wire lift* should be limited for high-speed traffic to 100 mm at the supports and an additional 80 mm in mid-span. The *elasticity* should, therefore, be small and evenly distributed. A increased tensile force can be achieved by increasing cross-sections and tensile stresses. However, the cross-section of the contact wires should not exceed 120 mm^2 to avoid discontinuities, microwaves and kinks during stringing.

Parallel feeder lines are installed parallel to the overhead contact lines when their cross-sections alone cannot guarantee the required current capacity and voltage stability. Clauses 7.1.2.2 and 7.1.2.3 deal with the design of contact lines with regard to current capacity.

3.4.3 Selection of span lengths

Long span lengths are desirable with a view to keeping investment down. Contact wires displaced by wind from their still air position have to guarantee an uninterrupted power transfer. Determination of *span length* must consider the wind loading per unit length for the contact wire F'_{WCW} and the catenary wire F'_{WCA} in accordance with Clause 4.10 for the anticipated *regional wind velocity*. The height of the overhead contact lines above the surrounding terrain, together with maximum regional wind velocity, determines the wind velocity to be applied.

The *maximum permissible contact wire lateral position* under wind e_{max} depends on the pantograph working range and decisively influences the span length and *contact wire stagger* at the supports. Pantographs with a narrow lateral working range require shorter span lengths. Small track radii also lead to a shortening of the maximum practical span lengths. The relationships between stagger, wind load, tensile forces, curve radius, span lengths and vehicle sway are considered in Clauses 4.9 and 4.10. For span lengths calculation see Clause 12.3.4. A radial force is created at the support because of deviation in the contact wire (see Clause 4.1.5 and 4.10), which, for example, should be within the range $80\,\mathrm{N} < F_D < 2500\,\mathrm{N}$ for light-weight steady arms. If a minimum radial force is not achieved, the connection will be loose resulting in excessive wear of the steady arm linkage hook on the drop bracket. If the radial force exceeds the permitted value because of excessive deviation of the contact line, the steady arm may be damaged. It is possible to alter the deflection of the contact wire at the support and to control the contact wire radial force by choice of span lengths and stagger. As explained in Clause 3.4.2, the span length affects the elasticity according to Equation (10.73). *Span lengths* should be adjusted to ensure that the minimum dropper length $l_{D\,min}$ is achieved on catenaries with reduced system heights. The minimum dropper length depends on the bending resistance of the dropper material.

DB, ADIF and REB can install larger span lengths on conventional lines because of their pantograph working length of 1,55 m reducing investments and maintenance costs. The 1 600 mm long European pantograph pan head requires shorter spans, than those possible with 1 950 mm long pantographs.

3.4.4 Selection of system height

The regular *system height*, being the distance between the contact wire and the catenary wire at supports should allow for the installation of minimum design length droppers at mid-span. The minimum length of flexible droppers $l_{D\,min}$ depends on running speed. Recommended minimum dropper lengths for design in Figure 11.40 are:

- $v \leq 120\,\mathrm{km/h}$ $l_{D\,min} = 300\,\mathrm{mm}$
- $120\,\mathrm{km/h} < v \leq 230\,\mathrm{km/h}$ $l_{D\,min} = 500\,\mathrm{mm}$
- $v > 230\,\mathrm{km/h}$ $l_{D\,min} = 600\,\mathrm{mm}$

If this is not possible, shorter flexible droppers, and finally *sliding droppers* have to be employed. These transfer the contact wire lift directly and less elastically to the catenary wire generating force peaks in the contact force profile. Observance of the minimum dropper length is important in view of dynamic behaviour. Shorter droppers in combination with inflexible material, increase the probability of dropper failures especially at higher speeds and greater contact wire lift.

In the case of a catenary wire DIN 48 201 – 50/7 – Bz II, tensioned at 15 kN, and a contact wire EN 50 149 – AC-120 – CuAg0,1 a sag of 0,63 m is obtained from (4.65) at the centre of a 65 m span. A system height of 1,23 m may be used in conjunction with a minimum dropper length of 0,6 m. System heights in stations are usually greater than those employed on the interstation line because of the installation of section insulators and the need to avoid electrical clearance problems between crossing contact lines of different electrical sections, especially under dynamic uplift conditions.

3.4.5 Design of contact lines in tunnels

There is a need to minimise the installation space for *tunnel overhead contact lines* in addition to the general requirements. Therefore, *contact wire height* should be kept as low as possible to minimise the tunnel cross-section and the associated construction investment. However, contact wire gradients are not permitted on high-speed lines, so the same contact wire heights need to be chosen in tunnels as found on adjacent open line sections, e. g. 5,3 m or less. To keep the *tunnel cross-section* low, a low system height should be provided within the tunnel using shorter span lengths, e. g. 55 m. Dependent upon the operating speed, alternative contact line arrangements are possible, e. g. overhead contact lines with elastic supports (see Clause 3.3.2.5) or *conductor rail overhead contact line* (see Clause 3.5) with short support intervals, e. g.10 m.

3.4.6 Adoption of contact wire pre-sag

For some overhead contact line designs, the contact wire is not strung at a constant height above the top of rail. E. g. with design Sicat S1.0, shown in Figure 3.29 and the TGV overhead line for SNCF, the contact wire is provided with a *contact wire pre-sag* of, for example, 0,1 % of the span length, so that the contact wire is lower at the centre of the span than at the supports. The provision of pre-sag is based on the premise that the overhead contact line has lower resilience at the supports than at mid-span and the pantograph lifts the contact wire at the supports to a lesser degree than at mid-span. To achieve an almost constant pantograph operating height during the passage of a train, a pre-sag is provided at mid-span. This should compensate for the difference in lift at the support compared to that at mid-span. The dynamic components of contact wire lift, however, increase with increasing speed and the pantograph is pressed downwards by the pre-sag at mid-span. Tests (see Clause 10.5.3.3) performed during the development of the overhead contact lines for high-speed line sections at DB showed that a pre-sag is not necessary for high-speed overhead contact line systems and can even be detrimental from the perspective of running characteristics.

According to SNCF's experience, a pre-sag is also useful for high-speed contact lines without stitch wires as proven by studies and tests. An initial pre-sag can also contribute to compensating for the effect of wear which could result in a negative sag.

A *pre-sag* can definitely provide better running quality for overhead contact line systems up to 200 km/h with their relatively large elasticity differences along the line, where static behaviour in the interaction between the overhead line and the pantograph is predominant.

3.4.7 Use of stitch wires

In the case of contact lines with a fixed catenary wire, stitch wires compensate for changes in contact wire height at the supports and at mid-span caused by varying temperatures. In the case of automatically tensioned contact and catenary wires, the stitch wires equalise the elasticity at the supports and in mid-span. The elasticity of truly sagged wires is considerably lower close to the supports than at mid-span. This also applies to catenary-type contact lines without stitch wires. However, the elasticity of a contact line should, as far as possible, be equal along the span so that the pantograph statically lifts the contact wire to the same extent along the line achieving a level contact wire without height changes. The stitch wires, to which droppers are fixed, are attached between two points of the catenary wire in the vicinity of the supports having a relatively low tensile force, equalising the elasticity. The stitch wires suspend the contact wire at the supports. The contact wire lift caused by the pantograph relieves the tensile load of the stitch wire and causes it to lift. The vertical movement of the contact point should be as low as possible within a span when the collector strip touches the contact wire. Low differences in the elasticity result in nearly constant contact point height and a better current transmission quality. According to [3.50], the difference between the highest and lowest dynamic contact point within a span should be:

For AC contact lines:
  – 80 mm for line design speeds of 250 km/h and above
  – 100 mm for line design speeds below 250 km/h

For DC contact lines:
  – 80 mm for line design speeds of 250 km/h and above
  – 150 mm for line design speeds below 250 km/h

The compliance of the contact wire height variation within a span should be verified on the installed system (Clause 10.4.6). This requirement is no longer defined in TSI ENE:2014 [3.51]. Figure 10.55 illustrates the effect of tensile force and length of the stitch wire on the elasticity of the overhead contact line at pull-off and push-off supports.

The suitability of an overhead contact line for a specific application can be evaluated by the *degree of non-uniformity* (see Equation (10.74)). Recommendations for degrees of non-uniformity to be envisaged are provided in Table 10.12, depending on the operating speed. SNCF used stitch wires on the Paris–Lyon high-speed line and found them difficult to install and maintain and have not used stitch wires on high-speed lines since.

3.4.8   Tensioning sections

3.4.8.1   Tensioning section lengths

The *tensioning section length* L_{tens} is determined from the temperature range of the overhead contact line, the operating range of the tensioning device, variation of the contact wire tensile force along the section and the permitted tolerances for the contact wire stagger and contact wire height.

The *operating range of the tensioning device* restricts the permitted length variation ΔL_{tens} given in Clause 4.3.2. The example in Figure 11.46 a) shows the variables that determine the operating range of a tensioning device using weights. The operating range depends on the pulley ratio, installation height, length of the weight stack and clearance above ground. Higher tensile forces require longer weight stacks, reducing the operating range. Tensioning equipment design with a pulley ratio of 3 : 1 has proved to be favourable and is widely used in Europe. Newly developed and tested tensioning equipment designed with a pulley ratio of 1,5 : 1 enables a longer contact wire section to be tensioned and unrestricted use of less expensive concrete weights instead of malleable cast iron weights [3.52]. This improvement was achieved because the working range of the weights due to the smaller pulley ratio, is reduced

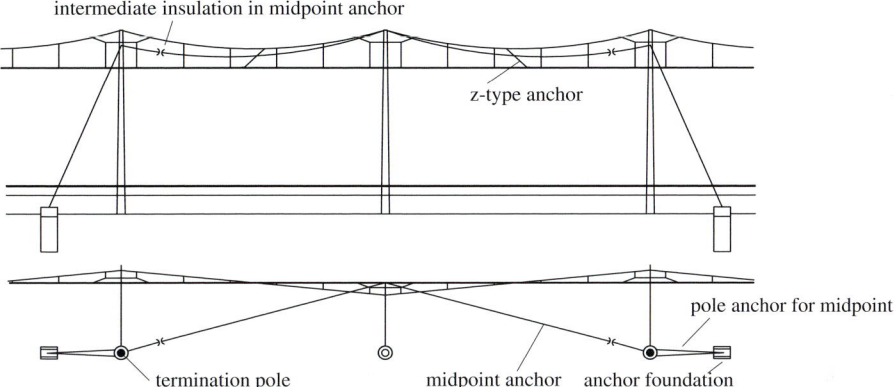

intermediate insulation in midpoint anchor

z-type anchor

pole anchor for midpoint

termination pole   midpoint anchor   anchor foundation

Figure 3.34: Mid point with hinged tubular cantilever.

more than the height of the weight stack is increased because of the use of concrete weights. Pulley ratios between 2 : 1 and 5 : 1 are employed for installations at various railways.

The greater density of cast iron, compared to concrete, reduces the dimensions of the weight stack and therefore reduces the space required for installation. A space-saving design for a tensioning devices in tunnels is essential (see Figure 11.44).

The determination of the tensioning section length has to also consider the loss of tensile force in the contact wire in curves as a result of resetting forces, to allow optimum running characteristics over the whole temperature range. The loss of tensile force should not exceed 8 % for the described contact line designs. In total, the variation in tensile forces should be less than 11 %.

Length changes caused by temperature variations in the contact line induce swiveling of the cantilevers, which in turn causes a displacement of the contact wire stagger perpendicular to the track (see Clauses 4.1.5.6 and 12.3.5).

The swiveling movements of double cantilevers at overlaps form an additional criterion for the determination of tension lengths. The cantilevers approach each other because of their travel in opposite directions in overlaps. In the case of *insulating overlaps* in accordance with Clause 3.4.8.5 the minimum electrical clearance between components on the two adjacent contact lines and their support devices also needs to be considered in extreme positions.

Tension lengths less than 750 m are tensioned at one end only and fixed at the other end; they are called half tension lengths. Such half tension lengths are formed for example:

– between a station and the interstation section for compensation
– at tunnel portals to a avoid a midpoint at the site and
– at points and overtaking section.

Because of the differing temperatures in tunnels and sections in the open, there is a difference in the change of length at the ends of the tension length that results in differing restoring forces within the two parts of the tensioning section. At the midpoint the tensile forces will vary and differing tensile forces will be found at the midpoint anchor. Such differing forces can be avoided by half tensioning sections. Half tensioning sections also have advantages for contact lines above points and at the transition between line sections in the open and in tunnels. Otherwise, longitudinal displacements and loads of the contact line could affect the midpoints. Such effects can be caused by differing temperatures in tunnels and in the open when tensioning devices are arranged on both ends of the section.

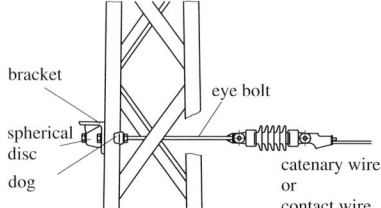

bracket

eye bolt

spherical
disc

dog

catenary wire
or
contact wire

Figure 3.35: Insulated termination.

3.4.8.2 Mechanical midpoints

Mechanical midpoints, fixed anchor points or restraints should be used in a tensioning section of an overhead contact line that is tensioned automatically at both ends, to ensure the conductors do not migrate towards either end of the tensioning section with changes in the loading conditions. This may be provided for by the installation of an anchoring arrangement at approximately the midpoint of the tensioning section as shown in Figure 3.20 a) or at a location which balances the along track forces at the midpoint.

The *midpoint* anchor as shown in Figure 3.34 fixes two-ended automatically terminated overhead contact lines. Two types of midpoint designs are considered, those with a hinged tubular cantilever and those in cross-spans. For cantilever designs, a midpoint anchor manufactured from bronze or steel conductors fixes both sides of the catenary wire support as shown in Figure 3.34. The midpoint anchors are stabilised by the adjacent poles, which are often provided with guy anchors. The contact wire is anchored to the catenary wire on both sides of the midpoint to ensure the conductors do not migrate. The cross-span wires also provide stabilisation of the midpoint at cross-spans. The rating of the midpoint equals the sum of the contact wire and catenary wire tensions and the rating of the Z-type anchor rope is equal to the contact wire tension. There are two types of midpoints: Those on swiveling cantilevers and those at flexible head-spans or on rigid portals. Figure 3.34 shows a midpoint on a swiveling cantilever made of tubes, which fixes an overhead contact line, tensioned to both sides. In the case of midpoints at head-spans, the upper cross-span wire has to withstand the longitudinal forces. As an alternative, several adjacent head-spans can act together as a midpoint.

3.4.8.3 Automatic tensioning

The *tensioning devices* serve to maintain the magnitude of the tensile forces in the contact line, and, therefore, the position of the contact wire, as constant as possible after length changes in the contact wire and catenary wire as a result of temperature variations. The device should be designed to achieve a specified efficiency over the specified design temperature range of the contact line. The efficiency, measured as the ratio of the actual to the intended tensile force, should be as high as possible, so that the horizontal tensile forces do not vary by more than 3 % [3.46]. The efficiency measured as the relationship between the actual and planned tensile force should be close to 1,0 but not below 0,97 [3.46]. The tensioning devices are dealt within the Clauses 3.1.2 and 11.2.3.

3.4.8.4 Fixed terminations

Fixed terminations in contact line systems secure catenary wires and contact wires directly to poles. The insulation is located close to the pole (Figure 3.35) taking into consideration a minimum clearance to allow authorised personnel to climb the pole.

3.4.8.5   Electrically connected and insulating overlaps

Overlaps with parallel contact lines are designed as electrically connected or insulating transitions and are installed between the tensioning sections and traversed without interruption of the energy supply or loss of contact quality.

Insulating overlaps are called *air-gap section insulations* and isolate the contact line sections electrically. There are designs with one-, two-, three-, four- and five-span overlaps. Figure 3.36 shows the designs of insulating contact line overlaps on a straight track. The designs of connected contact line overlaps on a straight track section are shown in Figure 3.37.

The pantograph contacts two contact wires at the support in twin- and four-span overlaps. The unfavourable force peaks that occur in this case favour the application of one-, three- and five-span overlaps, whereby the transition from one contact line section to the next is carried out at mid-span. Contact lines for speeds up to 300 km/h use *three-span* or *five-span overlaps* for open lines and five-span overlaps for tunnel lines. In the central span, the pantograph contacts both contact wires over approximately one third of the span. Related to the negotiated contact wire, the contact wire of the terminating contact line is raised at the supports by approximately

- 0,50 m for contact lines for speeds up to 200 km/h and
- 0,20 m for contact lines for speeds above 200 km/h.

As a result, the pantograph has contact with only one contact wire at the supports. The supports of the terminating contact lines have low elasticity due to the deviation in the contact wire and the large radial forces occurring as a consequence.

Reduced span lengths and higher contact wire tensile forces allow the contact wire to be lifted by only 0,15 m to 0,20 m on overhead contact lines for speeds above 200 km/h. The arrangement of tensioning devices directly after the transition span would increase the deflection forces at the supports, because of the high tensile forces, further reducing the elasticity with negative effects on the contact forces. The contact line is, therefore, continued to the following poles where the contact wire is raised by 0,50 m above nominal height to be able to install insulators. This design leads to five-span overlaps. Only with smaller distances between the contact lines longer spans would be possible, which would result in three-span overlaps and a more reasonable design. In Germany, the distance between parallel contact wires is 0,45 m at insulating overlaps, however, it is only 0,30 m in Switzerland. The distance selected by former Deutsche Reichsbahn was based on an electrical clearance of 0,30 m in open air, which since 1952 has been reduced to 0,15 m at 15 kV installations in accordance with DIN VDE 0115, Part 3, following operational experience. For non-insulating overlaps, the spacing between the contact lines is 0,20 m in Germany and in Switzerland. This clearance avoids clashing of the contact lines. If using five-span overlaps is not possible, three span overlaps with a negotiable plastic insulating rod instead of an insulator can be adopted (Figure 11.55).

The same conditions occur with all designs of overhead contact line systems with small track radii and short span lengths. In these cases, five-span overlaps are also employed.

Overlaps used for electrical insulation can be found, for example, at the boundaries of stations (Figure 3.38). Closed disconnectors are used to shunt the overlaps during normal operations. They may also be open with differing potentials occurring at the overlaps. Significant potential differences can lead to arcing with consequential contact line damage when a pantograph passes. To protect *insulating overlaps*, signals must be located at a minimum distance in front of the overlap with a certain distance to the first pole with twin cantilevers. This distance should be long enough that traction units accelerating behind the signal do not draw maximum currents when they enter the overlaps. Then, even in an open insulating overlap, no

a) Single-span overlap with termination portal structures

b) Two-span overlap

c) Three-span overlap

d) Four-span overlap

e) Five-span overlap

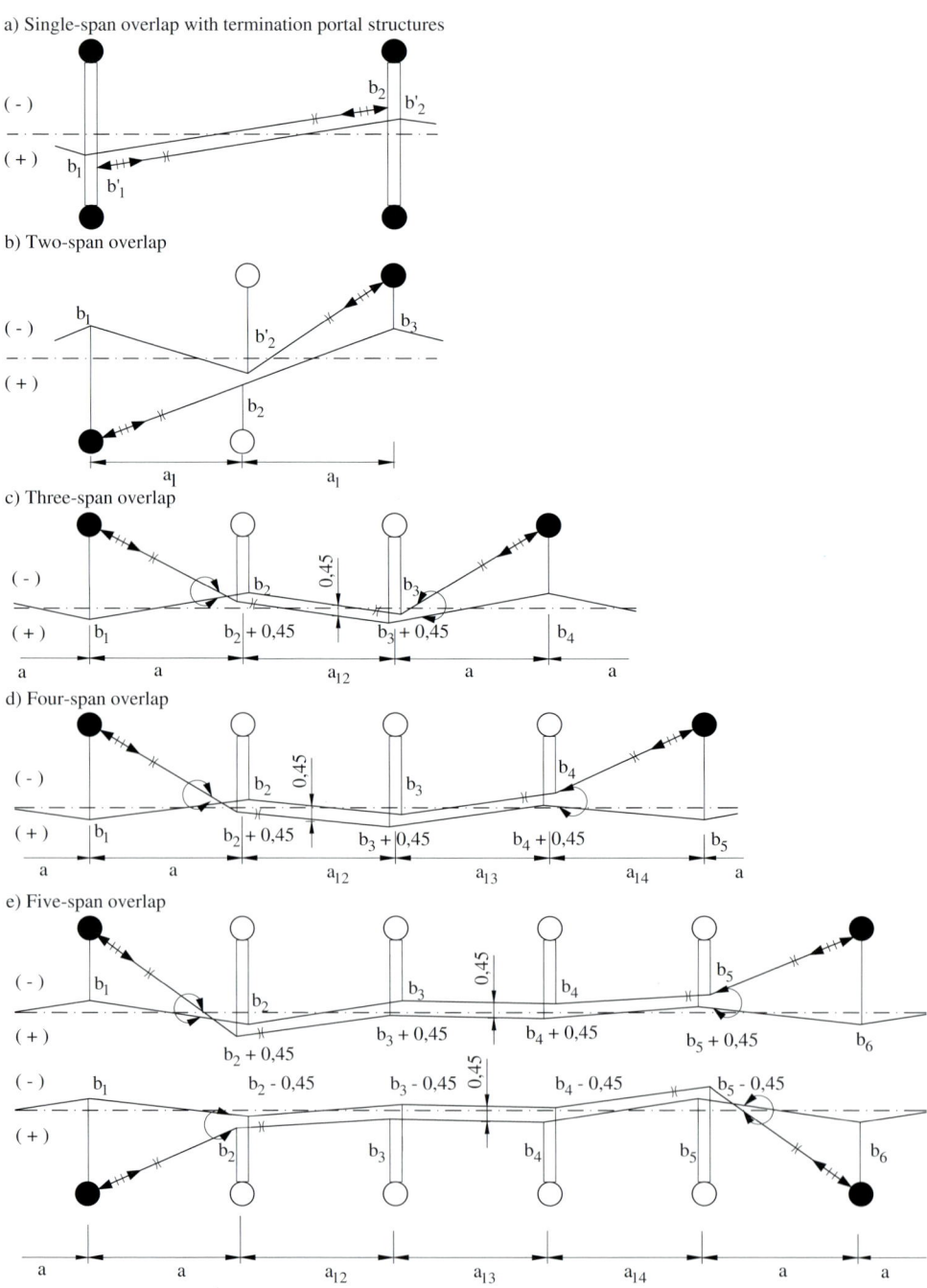

Figure 3.36: Designs of insulating overlaps on straight lines for 15 kV overhead contact line.

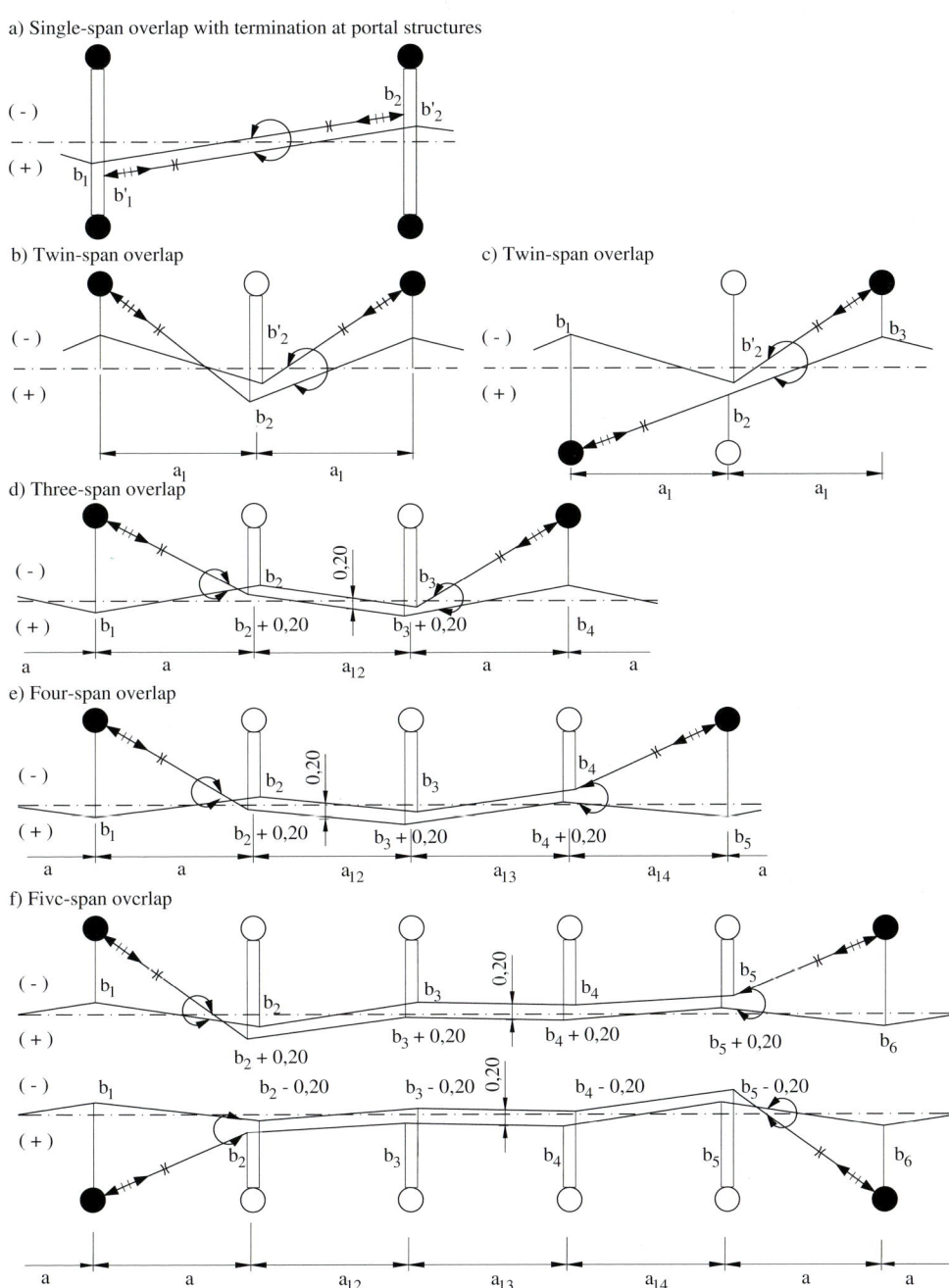

Figure 3.37: Designs of connected uninsulating overlaps on straight lines.

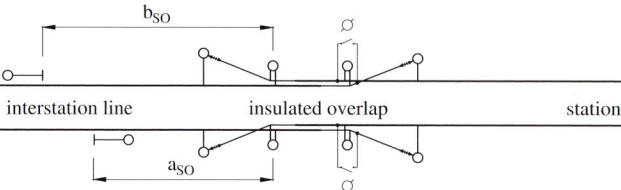

interstation line   insulated overlap   station

Figure 3.38: Insulating overlap with disconnector.

a) Neutral section shorter than the distance between pantographs

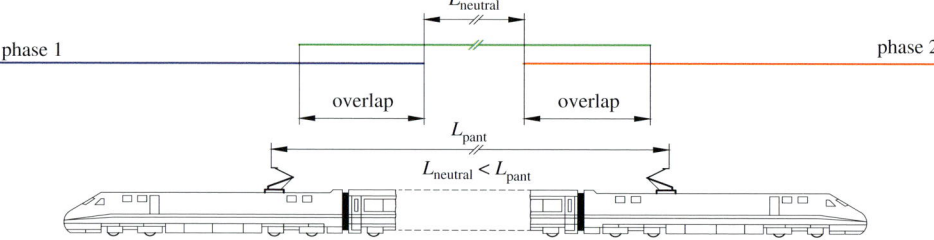

b) Neutral section longer than the distance between pantographs

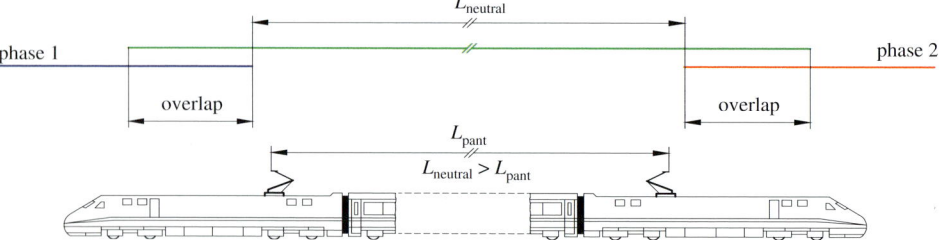

Figure 3.39: Variants of neutral section design.
$L_{neutral}$ length of neutral section, L_{pant} distance between pantographs

contact wire burn is created by voltage differences between the switching sections separated by the overlap. The minimum distance, designated as a_{SO} and b_{SO} on opposite track in in Figure 3.38, depends on the operational use of the line (see Clause 12.3.6.1).

3.4.8.6 Electrical sectioning

Electrical isolations are necessary to subdivide the contact line installation into different electrical sections or circuits. Depending on the operating speed, *sectioning devices* or *section insulators* are used for this purpose in stations and at speeds up to 160 km/h, in some cases up to 200 km/h. *Insulating overlaps* are provided in the overhead contact line, on main line tracks and at speeds above 160 km/h as discussed in Clause 3.4.8.5. Section insulators are dealt with in Clause 11.2.4.

3.4.8.7 Electrical connections

Permanent and switched *electrical connections* are used in the overhead contact line system to provide electrical current transfer. *Permanent electrical connections*, also known as current or electrical connectors, conduct the operating and short-circuit currents between the contact wire, catenary wire and contact line systems of different tensioning sections and between the

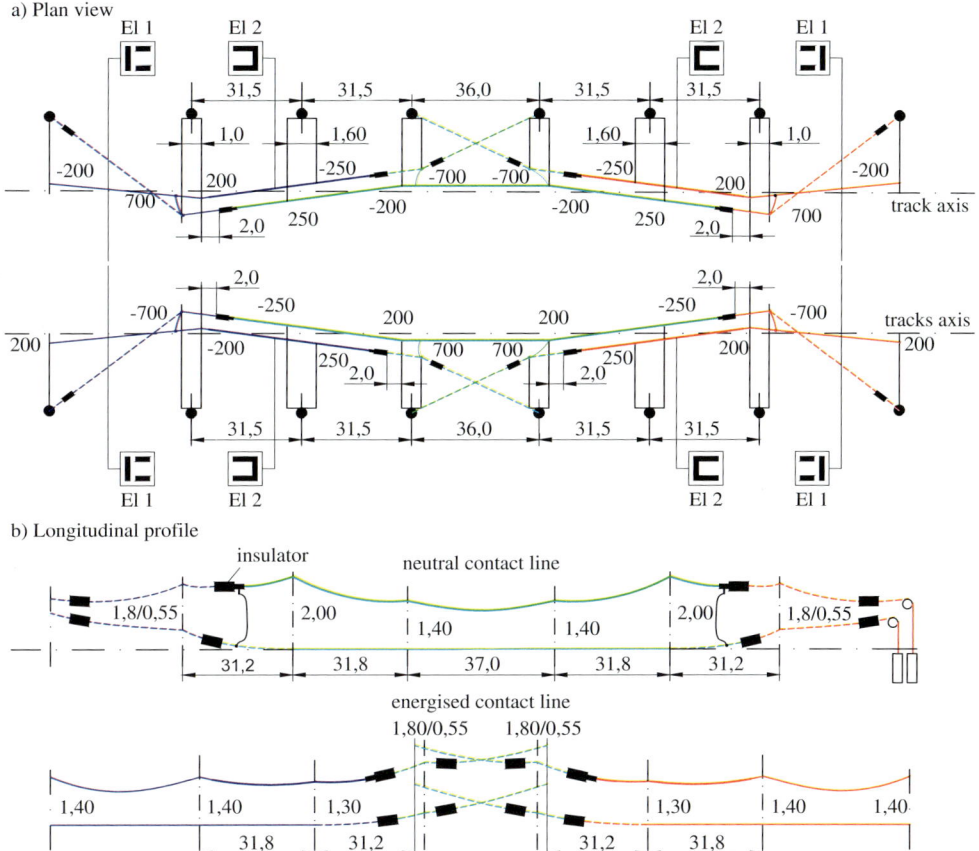

Figure 3.40: Phase separation on the French high-speed line TGV Nord.

traction feeder lines and the overhead contact line. Details regarding the design of electrical connections are given in Clause 11.2.2.8.

3.4.8.8 Neutral sections and phase separations

Neutral sections separate adjacent sections of a contact line so that the sections are not shunted by pantograph(s) during the passage of an electric traction vehicle.
Neutral sections are employed as boundaries
- of areas with different *energy supply systems*, e. g. between DC 3 kV and AC 25 kV 50 Hz or 15 kV 16,7 Hz,
- of *feeder sections with different phases*, e. g. feeder sections in AC 25 kV networks, supplied from different phases of the national public power grid. This neutral section design is also known as a *phase separation*.
- of feeder sections that can have different phases, e. g. to isolate overhead contact line sections fed from decentralised converter stations,
- of continuously earthed overhead contact line sections, e. g. under structures where there is insufficient clearance for an energised overhead contact line section.

Figure 3.41: Short neutral section with reduced length Siemens 8WL5544-4D, MAV Hungary (Photo: A. Wolf).

Neutral sections within overhead contact lines for speeds greater than 160 km/h consist of adjacent insulating overlap spans with an intermediate neutral overhead contact line section. There are two variants of this type (Figure 3.39):
- the whole neutral section is designed to be shorter than the shortest spacing between two pantographs on traction vehicles or
- the neutral section is designed to be longer than the longest distance between the extreme pantographs on a multiple unit train

An example for the layout of a neutral section based on the principle mentioned first (Figure 3.39 a)) is the French TGV Nord line (Figure 3.40) [3.53], while the AVE line Madrid–Seville is an example of the second principle (Figure 3.39 a)).

The technical TSI Energy [3.51] assumes that the pantographs are not lowered, however, the power consumption of the train should be brought to zero when entering a phase separation. In the TSI Energy three designs are specified:
- a phase separation where all the pantographs of the longest interoperable trains are inside the neutral section. In this case, there is no restriction on the arrangement and spacing of the pantographs on the trains. The length of the neutral section should be at least 402 m (Figure 3.39)
- a short phase separation with a restriction on pantograph arrangement on trains. The overall length of this design is less than 142 m. When using this design the distance between three consecutive pantographs in service must be more than 143 m
- neutral sections with reduced length consist of two neutral sections with an earthed part in between (Figure 3.41)

Neutral sections should not be installed immediately before signals, in tight curves or on steep ramps, where trains may come to a stop and need to restart again. If this should happen, then the neutral section can be connected to the contact line located in the direction of travel to enable the trapped train to pull out of the neutral section under its own power.

3.5 Overhead conductor rail systems

Overhead conductor rail systems, also called *soffit conductor rails*, consist of rigid compound or solid profiles [3.54] arranged above or beside the upper vehicle gauge. Supports carry the staggered conductor rails, as with other other overhead contact lines, to achieve uniform wear of the collector strips and supply the vehicles with electric energy by a current collector device arranged on the vehicle roof (see EN 50 119).

The quality of current collection and, therefore, the permissible running speed of rigid conductor rails depends much on the horizontal position of the contact surface. Consequently, the spacing between the supports needs to be shortened with increasing running speed to reduce variations in the contact line level. The difference in height between adjacent supports may vary by only a few tenths of a millimeter for running speeds above 140 km/h. The larger cross-sections of conductor rails improve current capacity and short-circuit resistance and reduce voltage drop and power losses.

When selecting the insulators at supports and minimum electrical clearances, the requirements for the nominal voltages of DC 750 V to 3 000 V and AC 15/25 kV need to be considered. In 25 kV systems, the required installation height at the supports is 600 mm. However, the installation height may be reduced for lower voltages and when using specially designed supports. The conductor rail requires a range to compensate for expansion caused by temperature changes. It is, therefore, designed to move at the supports and is divided into sections to compensate for the thermal expansion. Expansion joints serve the same purpose. The change of sections and section insulators divide the conductor rail system into electric switching groups and supply sections. There are midpoints in the middle of the individual sections and the length between the end of a conductor rail section and the midpoint is limited to 500 m. Rigid conductor rails reduce the required installation space because:

- the nominal height of the conductor rail contact line needs to be not more than the minimum contact wire height plus the tolerances of the track position and the installation height of the conductor rail
- its structural height is relatively low and
- tensioning devices are not necessary

Therefore, conductor rail overhead contact lines can be adopted predominantly in narrow tunnels, below bridges and in workshops where the installation of a catenary-suspended contact line may be difficult. Fixed and swiveling contact lines on turning and lifting bridges are other fields of application (see Clause 14.5).

The decision on the use of an elastic or rigid contact line depends on the engineering requirements, the installation conditions and economic aspects. Globally up to 2 000 km of conductor rail overhead contact line were installed in 2012, mainly for running speeds up to 140 km/h [3.55]. Clause 11.3 contains details. Examples are presented in Clauses 16.2.6 and 16.8. Paper [3.54] includes reports on test runs with speeds up to approximately 250 km/h.

3.6 Third rail installations

Conductor rails are the oldest form of electric traction current supply lines for railways. They are used mainly to transmit energy to the electric traction vehicles in underground and urban railway systems. Approximately 80 % of DC public mass transit railways operating with nominal voltages up to 1 000 V use conductor rails for energy transmission, while approximately

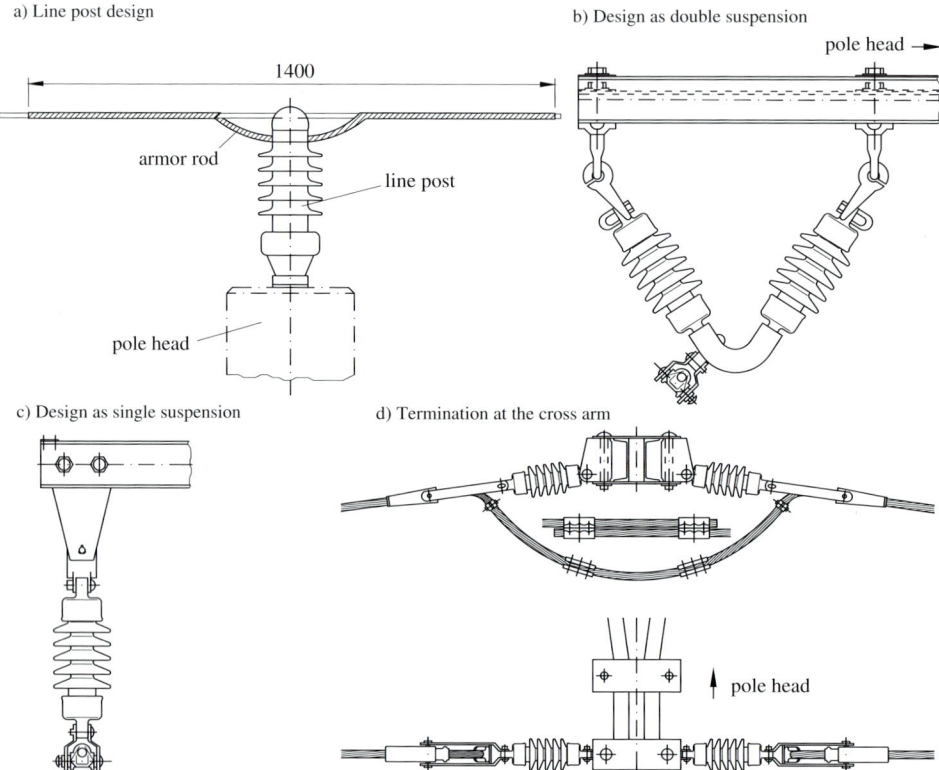

a) Line post design

1400

armor rod

line post

pole head

b) Design as double suspension

pole head →

c) Design as single suspension

d) Termination at the cross arm

pole head

Figure 3.42: Types of supports for traction power lines.

90 % of systems with nominal operating voltages above 1 000 V use overhead contact lines. For nominal voltages above 1 500 V, only experimental conductor rail systems are known.

Conductor rails are virtually *rigid conductors* installed at the side of the track on insulated mounts outside of the vehicle gauge in such a way that energy transfer in normal operation is possible while persons are protected against accidental or intentional contact.

With conductor rails, current collection can be from the top, the side or the bottom of the rail. Whereas the easier-to-construct top contact design is still used in France, United Kingdom and the USA, the design in which the bottom face of the conductor rail is used to transfer the traction energy to railway vehicles is the main type used in Germany [3.57], Russia, Austria and other European countries. The Hamburg metropolitan railway (S-Bahn) is one example of a system which uses *side-contact conductor rails*. Protection against contact with the live rail is achieved by installing insulating conductor rail cover-boards with electrical, thermal and mechanical properties suitable for the respective climatic and operating conditions [3.56]. Gaps are required in the conductor rail at points and road crossings. These are designed to match the vehicles and the installation. Ramp type end pieces guide the collector shoes into their upper end position and after passing the gap, back into the required working position for traversing the conductor rail. The conductor rail is divided into sections because of its thermal expansion, fixed at the end or in the middle by a fixed point, enabling expansion in one or both directions. At the ends of these sections expansion joints compensate thermal movement. Essential components of third rail systems are explained in Clause 11.4.

3.7 Traction power lines

The contact line supports may also carry additional *traction power lines* (TPL). Depending on the type of fixing, the line post design (Figure 3.42 a)) and the suspension design (Figure 3.42 b) and c)) can be distinguished. The line post design enables the arrangement of the TPL at the head of a concrete pole without a cross arm. However, this design is limited to one or two lines per pole. In the case of more parallel TPL, in track radii, when crossing head-spans or when clearances have to be met, a suspension design may have advantages. The suspension design can be carried out as single (Figure 3.42 c)) or double suspension (Figure 3.42 b)). The second type may be required because of standard stipulations. Suspension designs for TPLs are attached to cross-arms. Strain insulators separate individual TPL sections as shown in Figure 3.42 d). TPL tension sections start and finish at strain poles and are designed according to the standards of overhead power lines using aluminum or aluminum-steel conductors. Details for planning and installation of TPLs can be found in the Clauses 12.4.7, 12.4.8, 12.4.11 and [3.58].

3.8 Bibliography

3.1 *Höring, D. O.*: Siemens-Handbücher (Siemens Manuals), Volume 15: Electric Railways. Walter de Gruyter & Co. Publishing, Berlin and Leipzig, 1929.

3.2 *Kyser, H.*: Die elektrischen Bahnen und ihre Betriebsmittel (The electric railways and the equipment). Friedrich Vieweg und Sohn Publishing, Braunschweig, 1907.

3.3 *Reichel, W.*: Versuche über Verwendung hochgespannten Drehstromes für den Betrieb elektrischer Bahnen (Test on the use of three phase high-voltage current for the operation of electric railways). In: Elektrotechnische Zeitschrift 21(1900)23, pp. 453 to 461.

3.4 *Reichel, W.*: Über die Zuführung elektrischer Energie für größere Bahnnetze (On the transfer of electric energy at more extented railway networks). In: Elektrotechnische Zeitschrift 25(1904)23, pp. 486 to 493.

3.5 *Molino, N.*: Trifase in Italia 1902-1925 (Three phase in Italy 1902-1925). Gulliver Publishers, Torino, 1991.

3.6 *Aeschbacher, P.; Roth, G.; Schomburg, A.*: Kompakte DC-Bahnstromschaltanlage für die Jungfraubahnen (Compact DC switchgear for traction power supply at Jungfraubahnen). In: Elektrische Bahnen 110(2012)8-9, pp. 476 to 483.

3.7 *Eichenberg, E.*: Das Einphasen-Bahnsystem der Union-Elektricitätsgesellschaft, insbesondere die Versuchsbahn Niederschöneweide–Spindlersfeld (The single phase railway system of Union Electricity Company, especially the test line Niederschöneweide to Spindlersfeld). In: Z VDI (1904)9.

3.8 *Kroll, U.*: Beitrag zur Entwicklungsgeschichte der Fahrleitungen für Einphasenwechselstrom in Europa (Contribution to the development of single phase contact lines in Europe). In: Elektrische Bahnen 31(1960)6, pp. 121 to 132.

3.9 *Wiesinger, K.*: Die elektrische Zugförderung auf den Haupteisenbahnen mit besonderer Berücksichtigung der Strecke Dessau–Bitterfeld (The electric train transport on main lines with special attention to the line Dessau to Bitterfeld). In: Glasers Annalen 68(1911)4, pp. 72 to 79.

3.10 *Seifert, R.*: Zusammenwirken von Fahrleitung und Stromabnehmer (Interaction of contact line and pantographs). In: Die Eisenbahntechnik, Entwicklung und Ausblick. Hestra Publishing, Darmstadt, 1982.

3.11 *Heymann, H*: Die Versuchsbahn in Oranienburg – die maschinentechnischen Anlagen der Bahn (The test line in Oranienburg – mechanical engineering installations of railway). In: Glasers Annalen 68(1911)4, pp. 66 to 71.

3.12 *Kniffler, A.*: Die elektrische Hamburger S-Bahn, Anlagen und Betriebsführung (The electric operaton of Hamburg S-Bahn, equipment and operational control). In: Die Bundesbahn 28(1954)9/10, pp. 168 to 179.

3.13 *Kotzlott, K.*: 50 Jahre Einphasenwechselstrom mit 25 Hz und 6 300 V auf der Hamburg–Altonaer Stadt- und Vorortbahn (50 years single phase AC supply 25 Hz and 6 300 V on the Hamburg to Altona City and Suburban Railway). In: Elektrische Bahnen 26(1955)5, pp. 97 to 104.

3.14 *Usbeck, W.*: Die Fahrleitung der Allgemeinen Elektricitäts-Gesellschaft auf der Strecke Dessau–Bitterfeld (The contact line of Allgemeinen Elektricitäts-Gesellschaft on the line Dessau to Bitterleld). In: Elektrotechnische Zeitschrift 32(1911)25, pp. 609 to 612.

3.15 *Preusisch-heßische, bayerische und badische Staatseisenbahnen*: Übereinkommen betreffend die Ausführung der elektrischen Zugförderung (Agreement on the design of electric train operation). Berlin, München und Karlsruhe, 1912/1913.

3.16 *Groh, T.; Harprecht, W.; Puschmann, R.*: Interoperabilität elektrischer Bahnen – 100 Jahre Vereinbarung für 15 kV 16 2/3 Hz (Interoperability of railways – 15 kV 16 2/3 Hz agreement 100 years old). In: Elektrische Bahnen 110(2012)12, pp. 686 to 699.

3.17 *Deutsche Reichsbahn Ezs 823*: Rollen-Fahrdrahtspanner für Nebengleise (Pulley-type tensioning device for secondary tracks). München, 1939.

3.18 *Sachs, K.*: Die ortsfesten Anlagen elektrischer Bahnen (The fixed installations of electric railways). Orell Füssli Publishing, Zürich-Leipzig, 1938.

3.19 *Deutsche Reichsbahn Ezs 828*: Grundsätzliche Anordnung der Hebel-Fahrdrahtspanner (Principle arrangement of lever-type contact wire tensioning devices). München, 1931.

3.20 *Deutsche Reichsbahn EzsN 188, Bl. 1*: Radspanner für Fahrdraht und Tragseil oder für Doppelfahrdraht, ü = 1 : 3, Abspannzug maximal 2 500 kg (Tensioning device for contact and catenary wire, ü = 1 : 3, tensioning force 2 500 kg at maximum). München, 1946.

3.21 *Deutsche Reichsbahn EzsN 187, Bl. 1*: Radspanner für Fahrdraht, ü = 1 : 3, Abspannzug maximal 2 500 kg (Tensioning device for contact wire, ü = 1 : 3, tensioning force 2500 kg at maximum). München, Juli 1941.

3.22 *Wagner, R.*: Die Einheitsfahrleitung 1950 der Deutschen Bundesbahn (The standard contact line 1950 of Deutsche Bundesbahn. In: Elektrische Bahnen 25(1954)7, pp. 177 to 180.

3.23 *Abst, S.; Fiegl, B.; Fihlon, M.; Puschmann, R.*: Elektrifizierung der Hochgeschwindigkeitsstrecke HSL Zuid in den Niederlanden (Electrification of high-speed line HSL Zuid in the Netherlands). In: Der Eisenbahningenieur 58(2007)11, pp. 46 to 57.

3.24 *Pforr, P.*: Werdegang des elektrischen Zugbetriebs bei der vormals Preußisch-Hessischen Eisenbahn und bei der Reichsbahn (Development of electric train operation at the former Prussian-Hessian railways and at German state railways). In: Elektrische Bahnen 11(1935)11, pp. 310 to 314.

3.25 *Schwach, G.*: Oberleitungen für hochgespannten Einphasenwechselstrom in Deutschland, Österreich und der Schweiz (Overhead contact lines for high-voltage single phase AC in Germany, Austria and Switzerland). Wetzel-Druck KG Publishing, Villingen-Schwenningen, 1989.

3.26 *Stockklausner, H.*: 50 Jahre Elektro-Vollbahnlokomotiven in Österreich und Deutschland (50 years of electic main line locomotives in Austria and Germany). In: Sonderheft (Special edition), Eisenbahn, Wien, 1952.

3.27 *Dorenberg, O.*: Versuche der deutschen Bundesbahn zur Entwicklung einer Fahrleitung für sehr hohe Geschwindigkeiten (Tests of Deutsche Bundesbahn to develop a contact line for very high speeds). In: Elektrische Bahnen 63(1965)6, pp. 148 to 155.

3.28 *Harprecht, W.; Kießling, F.; Seifert, R.*: "406,9 km/h" Energieübertragung bei der Weltrekordfahrt des ICE ("406,9 km/h" power transmission in the record run of DB's ICE). In: Elektrische Bahnen 86(1988)9, pp. 268 to 289.

3.29 *Kießling, F. et al.*: Die neue Hochgeschwindigkeitsoberleitung Bauart Re 330 der Deutschen Bahn (New high-performance overhead contact line type Re 330 of Deutsche Bahn). In: Elektrische Bahnen 92(1994)8, pp. 234 to 240.

3.30 *Becker, W. et al.*: Entwicklung der Bahnenergieversorgung für Gleichstrom-Nahverkehrsbahnen (Development of traction power supply for DC light vehicles). In: Elektrische Bahnen 101(2003)6. pp. 276 to 282.

3.31 *N. N.*: Entwicklung der Fahrleitung der elektrischen Straßenbahn (Development of contact lines of electric tramways). In: Die Fahrt 13(1930) pp. 299 to 304.

3.32 *Thiede, J.*: Strombelastbarkeit von Oberleitungen bei Straßen- und Stadtbahnen (Current capacity of catenary systems of trams and suburban railways). In: Elektrische Bahnen 95(1997)5, pp. 123 to 129.

3.33 *Vuchic, V. R.*: Urban Transit. Systems and Technology. Wiley & Sons, Inc. Publishing, Hoboken, New Jersey (USA), 2007.

3.34 *Zabel, R.*: Trolleybusse in Schweizer und anderen Ballungsräumen, Stand und Perspektiven – Teil 1 (Trolley bus systems in Swiss and other agglomerations, situation and perspectives – Part 1). In: Elektrische Bahnen 103(2005)8, pp. 390 to 400.

3.35 *Röhlig, S.*: DC-Bahnen in Deutschland (DC railways in Germany). In: Elektrische Bahnen 104(2006)3, pp. 145 to 147.

3.36 *N. N.*: Meyers großes Konversations-Lexikon (Meyers great Encyclopaedia). Volume 5. Max Herzig Publishing, Vienna and Leipzig, pp. 605 to 609.

3.37 *Gerry, M. H.*: Electric Traction. Notes on the Application of Electric Motive Power to Railway Service, with Illustration from the Practice of the Metropolitan Elevated Road of Chicago. Paper from 14th Meeting of the AIEE, New York (USA), 1897.

3.38 *Mahlke, D.*: Die Aluminium-Stahl-Stromschiene als wichtiger Bestandteil der Stromschienen-anlage der S-Bahn Berlin (The aluminum-steel ccompound conductor rail, an important component of the traction power system of suburban railway Berlin). In: Der Eisenbahningenieur 56(2005)2, pp. 24 to 29.

3.39 *Lerner, F. et al.*: Bahnleitungsbau der AEG. Reminiszenzen, Daten und Fakten (Overhead contact and power lines at AEG. Resemblances, data and facts). Special print, 1997.

3.40 *Eichenberger, M. et al.*: Pendelaufhängungen für Einfachfahrleitungen (Pendular suspension for simple overhead contact lines). In: Elektrische Bahnen 105(2007)7, pp. 391 to 396.

3.41 *Grimrath, H.; Reuen, H.*: Elektrifizierung der Strecke Elmshorn–Itzehoe mit der Oberleitung SICAT S1.0 (Electrification of the Elmshorn – Itzehoe line section with the catenary type SICAT S1.0). In: Elektrische Bahnen 96(1998)10, pp. 320 to 326.

3.42 *N. N.*: Die Regelfahrleitung der Deutschen Bundesbahn (Standard contact lines of German railway). In: Elektrische Bahnen 77(1979)6, pp. 175 to 180 and pp. 207 to 208.

3.43 *Gourdon, C.*: Die TGV-Oberleitungsanlage der SNCF (The TGV overhead contact line system of SNCF). In: Elektrische Bahnen 88(1990)7, pp. 285 to 290.

3.44 *Matthes, R.*: Oberleitung für die Modernisierung von Gleichstromstrecken in Russland (Overhead contact line for refurbishment of DC railways in Russia). In: Elektrische Bahnen 102(2004)10, pp. 433 to 438.

3.45 *Schwab, H.-J.; Ungvari, S.*: Development and design of new overhead contact line systems. In: Elektrische Bahnen, 104(2006)5, pp. 238 to 248.

3.46 *Süberkrüb, M.*: Technik der Bahnstrom-Leitungen (Technology of traction power lines). Wilhelm Ernst & Sohn Publishing, Berlin-München-Düsseldorf, 1971.

3.47 Willenberg, N.: Die Flachkette, eine Aufhängung für Straßenbahn-Fahrleitungen (The overhead contact lines with catenary suspension, a suspension for tram overhead contact lines). In: Verkehrstechnik 11(1930)45, pp. 605 to 606.

3.48 Düskow, A.: Berechnung der Flachkettenaufhängung für Straßenbahn-Fahrleitungen (Calculation of overhead contact lines with catenary suspension for tram overhead contact lines). In: Verkehrstechnik 12(1931)10, pp. 121 to 124.

3.49 *Kökenyési, M.; Kunz, D.*: Oberleitung Sicat SX – Zulassung und Betriebserfahrungen in Ungarn (Overhead contact line Sicat SX – Apporval and operational experience in Hungary). In: Elektrische Bahnen 111(2013)6-7, pp. 440 to 444.

3.50 *Decision 2002/733/CE*: Decision concerning the technical specification for the interoperability relating to the energy subsystem of the trans-European high-speed rail system. In: Official Journal of European Communities, No. L245(2002), pp. 280 to 369.

3.51 *Regulation 1301/2014/EU*: Regulation concerning the technical specification for the interoperability relating to the energy subsystem of the rail system in the Union. In: Official Journal of European Union, No. L356(2014), pp. 179 to 207.

3.52 *Siemens AG*: Fahrleitungsmaterial für Nah- und Fernverkehr (Contact line equipment for mass transit and main-line railways). Product Catalogue, Siemens AG, Erlangen, 2017.

3.53 *SNCF*: Principes d'equipement. Dossier EF 7B 24.3, sectionnements à lame d'air, Sections de séparation, isolateurs de section (Basics of equipment. Document EF 7B 24.3, Insulating and not insulating overlappings, section insulators). Internal SNCF Standard VZC 21400/300100 Part 28.

3.54 *Kurzweil, F.; Furrer, B.*: Deckenstromschienen für hohe Fahrgeschwindigkeiten (Overhead conductor rail for high speeds). In: Elektrische Bahnen 109(2011)8, pp. 398 to 403.

3.55 *Lörtscher, M.; Urs, W.; Furrer, B.*: Stromschienenoberleitungen (Conductor rail overhead contact lines). In: Elektrische Bahnen 92(1994)9, pp. 249 to 259.

3.56 *Rosenke, D.; Uyanik, A.*: Neuentwicklung einer Stromschienenoberleitung für Tunnelstrecken (Developmemt of an overhead conductor rail contact line for tunnels). In: Verkehr und Technik (1985)5, pp. 136 to 138.

3.57 *Janetschke, K.; Freidhofer, H.; Mier, G.*: Einführung von neuen Stromschienenanlagen mit Aluminium-Verbundstromschienen bei der Berliner S-Bahn (Introduction of new conductor rail installations with aluminum-steel composite conductor rails at Berlin suburban railway). In: Elektrische Bahnen 80(1982)1, pp. 17 to 23.

3.58 *Kiessling, F.; Nefzger, P.; Nolasco, J. F.; Kaintzyk, U.*: Overhead power lines – Planning, design, construction. Springer Publishing, Berlin-Heidelberg-New York, 2003.

4 Rating of overhead contact lines

4.0   Symbols and abbreviations

Symbol	Definition	Unit
A	cross-sectional area	mm^2
A	kinematic vehicle gauge	–
A'	kinematic pantograph gauge	–
A_{CA}	cross-sectional area of catenary wire	mm^2
A_{wear}	worn contact wire cross-section	mm^2
ACSR	**A**luminum **C**onductor **S**teel **R**einforced	–
AACSR	**A**luminum **A**lloy **C**onductor **S**teel **R**einforced	–
B	reference vehicle gauge	–
B'	reference pantograph gauge	–
BA	start of track curve	–
BE	end of track curve	–
C	kinematic infrastructure gauge	–
CA	catenary wire	–
C_C	drag factor for conductors with circular cross section	–
CW	contact wire	–
C_i	integration constant	–
D	swaying distance: 0,200 m / 0,225 m for 1 600 mm / 1 950 mm pan heads	m
D_k	point of contact between pan head and contact wire	–
E	reference vehicle gauge	–
E'	reference pantograph gauge	–
E	modulus of elasticity	kN/mm^2
E_{CA}	modulus of clasticity of catenary wire	kN/mm^2
FEM	Finite Element Method	–
F_{ult}	minimum failing load	kN
F_{clamp}	tensile force acting on contact or catenary wire clamp	kN
F_{per}	permissible conductor tensile force	kN
F_{Ri1}	force acting radially at a steady arm at the support i from span 1	N
F_{Ri2}	force acting radially at a steady arm at the support i from span 2	N
F_{Ri}	force acting radially at steady arm or catenary wire clamp of support i	N
$F_{Ri\,left}$	radial force component at steady arm i on its left	N
$F_{Ri\,right}$	radial force component at steady arm i on its right	N
$F_{Ri\,anchor}$	force acting radially at catenary wire clamp of midpoint cantilever i, which is composed from components F_{Ri1} and F_{Ri2}	N
F_{RiCW}	radial force at the contact wire clamp at support i	N
F_{RiCA}	radial force at the catenary wire clamp at support i	N
$F_{RiCW\,fail}$	radial force at the catenary wire clamp of the midpoint cantilever i in case of a contact wire failure	N

Symbol	Definition	Unit
$F_{RiCA\,fail}$	radial force at the catenary wire clamp at the midpoint cantilever i in case of a catenary wire failure	N
F'_W	wind load per unit length acting perpendicularly to a conductor or wire	N/m
$F'_{W\,CA\,tot}$	load per unit length resulting in complete displacement of the catenary wire due to wind	N/m
$F'_{W\,CW\,tot}$	load per unit length resulting in complete displacement of the contact wire due to wind	N/m
$F'_{W\,CA}$	force per unit length due to wind on catenary wire	N/m
$F'_{W\,CW}$	force per unit length due to wind on contact wire	N/m
$F'_{W\,CW\,CA}$	force per unit length transferred by droppers between contact and catenary wire	N/m
$F'_{W\,Cl}$	force per unit length due to wind load on clamps	N/m
$F'_{W\,OCL}$	force per unit length due to wind on contact line	N/m
F_{per}	permissible tensile force of a conductor	kN
G	weight or weight force of a conductor	kg or N
G'	specific weight of conductor, weight per unit length	N/m
G'_{ice}	specific weight of ice and snow, weight per unit length	N/m
$G'_{ice\,CA}$	ice load on catenary wire	N/m
$G'_{ice\,CW}$	ice load on contact wire	N/m
G_C	conductor reaction coefficient considering the reaction of conductor on wind gusts according to EN 50 119 Beiblatt 1:2011 $G_C = 0,75$	–
G'_{CA}	specific weight of catenary wire, weight per unit length	N/m
G'_{CW}	specific weight of contact wire, weight per unit length	N/m
G'_{OCL}	specific weight of overhead contact line consisting of contact wire, catenary wire, droppers, stitch wire and clamps, weight per unit length	N/m
$G'_{OCL\,ice}$	specific weight of overhead contact line with ice	N/m
G'_0	specific weight of a conductor at the condition Zero	N/m
G'_x	specific weight of a conductor at the condition x	N/m
G_{infra}	additional lateral displacement whereby the vehicle reaches a limit of the infrastructure gauge beyond the reference line	m
H	conductor tensile force	kN
H_{anchor}	conductor tensile force of the anchor rope at mid point	kN
H_{CA}	conductor tensile force of catenary wire	kN
$H_{CA\,1\,or\,2}$	conductor tensile force of catenary wire in half tensioning section 1 or 2	kN
H_{CW}	conductor tensile force of contact wire	kN
$H_{CW\,1\,or\,2}$	conductor tensile force of contact wire of half tensioning section 1 or 2	kN
H_{OCL}	conductor tensile force of overhead contact line consisting of the tensile force of contact and catenary wire	kN
H_{OD}	height of the line above sea level, reference altitude of the line is ToR	m
H_0	conductor tensile force a conductor at the condition 0	kN
H_x	conductor tensile force at the condition x	kN
H_{10}	everyday force at 10 °C	kN
H_{80}	conductor force at 80 °C	kN
H_{-30}	conductor force at -30 °C	kN
L	length of a conductor	m
l_{CA}	length of catenary wire	m

Symbol	Definition	Unit
L_{CW}	length of contact wire	m
L_N	length of an overhead contact line between midpoint and tensioning device	m
L_S	distance between the centre lines of the rails of a track	m
L'_h	kinematic limit of mechanical pantograph gauge at verification height h	m
L_{imp}	distance between the cantilever i an midpoint	m
L_o	kinematic limit at the verification height $h_o = 6,5$ m	m
L_0	length of a conductor at the condition 0	m
L_x	length of a conductor at the condition x	m
L_u	kinematic limit at the verification height $h_u = 5,0$ m	m
MVK	distance between track axis and front face of pole	m
NN_i	height at a support i related to a reference height	m
P	provision for contact wire lateral displacements (tolerances)	N
P_M	provision for variation of contact wire position due to pole inclination under wind	m
P_S	provision for installation tolerances of contact wire stagger at support	m
P_T	provision for changes of contact wire lateral position due to temperature-dependent variation of contact wire length	m
P_i	point of contact wire support i	
R	track radius	m
R'	perpendicular connection of secant and centre	m
S	projection	m
S'	permissible additional projection	m
S	apex of the conductor within a span	–
S_{Bo}	margin added to the mechanical-kinematic gauge at upper verification height $h_o = 6,5$ m	m
S_{Bu}	margin added to the mechanical-kinematic gauge at lower verification height $h_u = 5,0$ m	m
S_i	support i	
S	conductor tensile force	N
S_0	flexibility of vehicle, $S_0 = 0{,}225$	–
T_{charge}	vehicle asymmetry angle η_0 due to unequal distribution of loads	m
T_D	error of transverse height of track which can result between two track maintenance activities	m
T_{osc}	error of transverse height of track to take care of the displacements from oscillation due to track position deficiencies	m
T_{susp}	vehicle asymmetry angle η_0 for inadmissible adjustments of suspension	m
T_{track}	transverse displacement of track which can occur between two track maintenance activities	m
TE	Transition curve End	–
ToF	Top of Foundation	–
ToR	Top of Rail	–
TS	Transition curve Start	–
V_b	reaction force at point B	N
V_{CA}	vertical force of catenary wire	N
V_{CW}	vertical force of catenary wire	N
V_i	vertical component of force at the support i	N

Symbol	Definition	Unit
$V_{i\,\mathrm{right}}$	vertical component of force at the support i from the span on the right l_{i+1}	N
$V_{i\,\mathrm{left}}$	vertical component of force at the support i from the span on the left l_i	N
a	partial length in the span	m
a_K	width of contact wire contact mirror	m
a_r	coefficient of circle function	1/m
b_i	stagger at support i	m
$b_{i\mathrm{CW}}$	contact wire stagger at support i	m
$b_{i\mathrm{CA}}$	catenary wire stagger at support i	m
b_r	coefficient of circle function	–
b_T	parameter of clothoide	–
$b_{\mathrm{y}\,i}$	lateral position of flexible termination at support i	m
$b_{\mathrm{h'\,ele}}$	electrical-kinematic pantograph gauge	m
$b_{\mathrm{h'\,mec}}$	mechanical-kinematic pantograph gauge	m
c	distance between contact wire and perpendicular to the middle of the line connecting top of rails	m
c_m	suspension coefficient	m/N
c_F	spring constant	m/N
c_V	lateral position at apex	–
$c_{\mathrm{T}(x)}$	track axis in a transition curve	–
c_r	coefficient of circle function	–
c_T	parameter of clothoide	–
d	conductor diameter	m
$d_{\mathrm{case\,A(B,C)}}$	distance between start of transition curve and the first support $P_{(i-1)}$	m
d_CA	catenary wire diameter	m
d_CW	contact wire diameter	m
d_T	parameter of the clothoide	–
$d_{\mathrm{pole-TS}}$	distance between pole and beginning of transition curve	m
d_i	distance between contact wire axis and secant in parallel to the perpendicular on the secant at support i (see Figure 4.6)	m
d_{i+1}	distance between contact wire axis and secant in parallel to the perpendicular on the secant at support $i+1$ (see Figure 4.6)	m
e	contact wire lateral position	m
e_WCA	displacement of catenary wire caused by wind load	m
e_WCW	displacement of contact wire caused by wind load	m
e_K	versine of track	m
$e_\mathrm{R}(x)$	mathematical presentation of track curve as a circle	m
$e_\mathrm{SR}(x)$	contact wire lateral position in curves in still air	–
e_TSk	secant in curve between support P_{i-1} and support P_i	m
e_T	parameter of the clothoide	–
$e_\mathrm{T}(x)$	contact wire position in a transition curve in still air	m
$e_\mathrm{TS}(x)$	contact wire position in still air in transition curve at position x	m
$e_\mathrm{Ttrack}(x)$	track axis in a transition curve at position x	m
$e_\mathrm{S}(x)$	contact wire position in a straight line in still air	m
$e(x)$	contact wire position in curves under wind	m
$e_\mathrm{W}(x)$	contact wire deflection under wind	m

Symbol	Definition	Unit
e_{lim}	limiting position of contact wire	m
e_{max}	maximum displacement of contact wire related to track axis	m
e_{use}	usable contact wire lateral position	m
e_{p}	swaying movement of pantograph	m
e_{po}	overthrow of pantograph due to vehicle features, 0,170 m at $h_{\text{o}} = 6,5$ m	m
e_{pu}	overthrow of pantograph due to vehicle features, 0,110 m at $h_{\text{u}} = 5,0$ m	m
$e_{1\,\text{max}\,1}$	contact wire displacement on right side in viewing direction in span 1	m
$e_{2\,\text{max}\,1}$	contact wire displacement on left side in viewing direction in span 1	m
$e_{1\,\text{max}\,2}$	contact wire displacement on right side in viewing direction in span 2	m
$e_{2\,\text{max}\,2}$	contact wire displacement on left side in viewing direction in span 2	m
f	conductor sag	m
f_{a}	conductor sag at distance a from support	m
f_{e}	conductor sag of the complementary span	m
$f_{\text{e max}}$	maximum conductor sag of the complementary span	m
f_{max}	maximum conductor sag	m
$f_{\text{CW ice max}}$	maximum contact wire sag under ice load	m
$f_{\text{CW D}-\text{D ice max}}$	maximum contact wire sag under ice load between droppers	m
$f_{\text{CW S}-\text{S ice max}}$	maximum contact wire sag under ice load between supports	m
f_i	conductor sag in span i	m
f_{id}	conductor sag in equivalent span	m
f_{CA}	catenary wire sag	m
$f_{\text{CA max}}$	maximum catenary wire sag	m
$f_{\text{CW max}}$	maximum contact wire sag	m
$f_{\text{u m}}$	uplift of contact wire due to passing pantograph at midspan	m
$f_{\text{u s}}$	uplift of contact wire due to passing pantograph at support	m
$f_{\text{W a}}$	wear of collector strip	m
f_{WS}	inclination of pantograph pan head	m
$f_{\text{W P}}$	flexibility of pan head	m
g	gravitational acceleration $g = 9,81$	m/s^2
h_{OCL}	nominal contact line height above terrain, reference is ToR	m
h	verification height above top of rail	m
h_{CW}	nominal contact wire height above top of rail	m
$h_{\text{CW S}}$	contact wire height above top of rail at support	m
h_{LH}	clearance height of building	m
h_{re}	residual contact wire thickness	m
h_{SH}	system height, defined as vertical distance between contact wire lower face and centre line of catenary wire at support	m
h_{CA}	catenary wire nominal height above top of rail	m
$h_{\text{CA Su}}$	catenary wire height above top of rail at support	m
h_{wear}	residual height of worn contact wire	m
h_{c0}	rolling centre height of reference vehicle, $h_{\text{c0}} = 0,5$ m	m
h_{o}	upper verification height above top of rail $h_{\text{o}} = 6,5$ m	m
h_{t}	height of lower pantograph knuckle above ToR	m
h_{u}	lower verification height above top of rail $h_{\text{o}} = 6,5$ m	m
k	coefficient according to EN 15 273-3, $k = 1,0$	–

Symbol	Definition	Unit
k_{wear}	factor accounting for the permissible wear of contact wire, where x_w expresses the wear in perecent	–
$k_{ice(wind)}$	factor accounting for the effect of ice and wind loads	–
k_{clamp}	factor accounting for the efficiency of dead end clamps	–
k_{load}	factor accounting for additional loads at catenary wire	–
k_{joint}	factor accounting for joints in the contact wire	–
k_{temp}	factor accounting for the relation between maximum operational temperature and tensile strength	–
k_{eff}	factor accounting for the efficiency of the tensioning device	–
l_A	working length of pantograph being $- l_A = 1,200$ m for the 1 600 mm pan head $- l_A = 1,550$ m for the 1 950 mm pan head	m
$l_{Al i}$	length of cantilever at support i	m
$l_{F(i)}$	distance between cantilever i and midpoint	m
l_S	minimum collector strip length	m
l_{ga}	local track gauge	m
l_W	length of pantograph pan head	m
l_e	length of complementary span	m
l_g	weight span	m
l_h	length of pantograph horn	m
l, l_i	span between two poles	m
l_{id}	equivalent span	m
l_{max}	maximum span between two contact line poles	m
l_o	limiting gauge of track being the distance between the running edges of rails according to the infrastructure manager: – secondary lines and tracks $l_o = 1,470$ m – tracks with $v \leq 160$ km/h $l_o = 1,465$ m – tracks with $v > 160$ km/h $l_o = 1,463$ m	m
l_0	distance between the edges of the rails of a track	m
l_{sec}	length of secant	m
m'	mass per unit length	kg/m
n	number of strands of a conductor	–
p	annual probability of occurrence	–
q_b	reference wind pressure at 10 m above ground surface	N/m^2
q_r	bearing play of the reference vehicle	m
q_z	design wind pressure at the height z above ground surface	N/m^2
$qs'_{i(a)}$	quasi-static lateral movement of the vehicle, i inside curve, a outside curve	m
s_E	supplement against dewirement	m
s'_0	flexibility of vehicles with pantographs on the roof	–
s_U	flexibility of vehicles without pantographs on the roof	–
s'_U	flexibility of vehicles with pantographs on the roof	–
sp	bogie and bearing play of the reference vehicle	m
t	flexibility of the pantograph raised to a maximum of 6,5 m and up to 300 N lateral force, $t = 0,03$ m	m
t_r	lateral displacement of the pantograph under the load of 300 m	m
t_0	transverse movement of the track	m

Symbol	Definition	Unit
t_{12}	quasi-static effect of an erroneous cant	m
$t_{3i(3a)}$	vehicle sway caused by uneven track, i inside curve, a outside curve	m
t_4	asymmetry caused by unequally loaded vehicles	m
t_5	asymmetry caused by suspension tolerances	m
u	cant of track in curves	m
u'_0	reference value for cant for vehicles with pantographs, $u'_0 = 0,066\,\text{m}$	m
u_f	cant deficiency of track in curves	m
u_fe	cant error	m
u'_f0	reference value for cant deficiency, $u'_\text{f0} = 0,066\,\text{m}$	m
$v_\text{b0,02}$	reference wind velocity with occurrence probability $p = 0,02$, i. e. once in 50 years	m/s
$v_\text{b0,10}$	reference wind velocity with occurrence probability $p = 0,10$, i. e. once in 10 years	m/s
$v_\text{b0,33}$	reference wind velocity with occurrence probability $p = 0,33$, i. e. once in 3 years	m/s
v_g	gust wind velocity	m/s
v_b	reference wind velocity at 10 m height above terrain, 10 min average and with a return period as required for the verification	m/s
$v_{\text{b}(p)}$	reference wind velocity with a probability of occurrence p, differing from 50 years	m/s
w_T	play between bogie and body of vehicle	m
w_r	bogie play of reference vehicle	m
w_ao	projection of pan head	m
x	longitudinal coordinate	m
x_max	longitudinal coordinate of maximum offset	m
x_2max	longitudinal coordinate of maximum offset in span 2	m
x_w	wear factor for the contact wire	
x_S	coordinate of apex S	m
y	coordinate transverse to track	m
y'_ia	displacement due to quasi-static inclination inside curve i / outside curve a	m
$y_{i-1\text{CA}}$	distance between track axis and catenary wire anchoring at the pole $i-1$	m
z	vertical coordinate	m
z'	displacement due to quasi-static inclination on the vehcicle side	m
z_1	sag related to support	m
z_CW	contact wire sag	m
z_CWice	contact wire sag due to ice load at midspan	m
z_CA	catenary wire sag	m
z_CAice	catenary wire sag due to ice	m
z_OCL	distance between catenary wire centre and lower surface of contact wire at midspan	m
$z_\text{u/o}$	quasi-static flexibility of the vehicle for the lower and upper point of verification	m
ΔF_R	difference in lateral forces at adjacent supports	N
$\Delta H_\text{R}((i)$	resetting force at support i	N
ΔL_E	change of conductor length caused by elastic elongation	m
ΔL_F	change of conductor length resulting from an installed spring	m

Symbol	Definition	Unit
ΔL_T	change of conductor length due to thermal expansion	m
ΔL_i	distance between cantilever i and midpoint	m
ΔT	half temperature range of the contact line	K
Δb_i	change of contact wire lateral position because of cantilever swivel i	m
Δb_m	minimum sweep (change of stagger)	mm/m
Δb_1	transverse displacement of track	m
Δb_2	transverse height error of track	m
Δb_3	vehicle oscillation due track irregularities	m
Δb_4	adjustment tolerance of suspension	degree
Δb_5	asymmetry of loading	degree
Δe	difference in lateral displacement between contact and catenary wire under wind	m
Δh_{AB}	difference in height of conductor attachments between supports A and B	m
Δh_{CWT}	installation tolerance of contact wire height	m
Δf_c	maximum change of height of contact point in a span	m
$\sum j$	sum of horizontal supplements to account for random phenomena, see Clause 4.9.2)	m
$\sum j_o$	sum of horizontal supplements to take care of random phenomena at the upper verification height $h_o = 6,5\,\mathrm{m}$	m
$\sum j_u$	sum of horizontal supplements to take care of random phenomena at the lower verification point $h_u = 5,0\,\mathrm{m}$	m
α	coefficient of thermal expansion of conductors	1/K
θ_r	angle due to vehicle asymmetry	degree
$\alpha_{1,2}$	bending angle of contact or catenary wire at support	degree
α'	angle between direction of radius and contact wire at support	degree
α''	angle between direction of radius and secant at support	degree
$\beta_{1,2}$	half of angle between contact wire directions at support	degree
δ	angle between radial force and connecting line between pole and support	degree
η_{CW}	efficiency of tensioning device for contact wires	–
η_{CA}	efficiency of tensioning device for catenary wires	–
γ_C	partial factor on conductor forces	–
γ_M	partial factor on materials	–
σ	tensile stress of a conductor	N/mm^2
$\sigma_{C\,max}$	maximum tensile stress of a conductor	N/mm^2
σ_{min}	minimum tensile stress of a conductor	N/mm^2
$\sigma_{min\,CA}$	minimum tensile strength of catenary wire	N/mm^2
$\sigma_{min\,CW}$	minimum tensile strength of contact wire	N/mm^2
$\sigma_{min\,joint}$	minimum tensile strength of conductor joints	N/mm^2
σ_{avail}	available stress of a conductor	N/mm^2
σ_{per}	permissible tensile stress of a conductor	N/mm^2
$\sigma_{per\,CA}$	permissible tensile stress of catenary wire	N/mm^2
$\sigma_{per\,CW}$	permissible tensile stress of contact wire	N/mm^2
σ_{10}	everyday stress of a conductor	N/mm^2
σ_{80}	conductor tensile stress at 80 °C	N/mm^2
σ_{-30}	conductor tensile stress at −30 °C	N/mm^2

Symbol	Definition	Unit
τ_r	pantograph production and installation tolerances	m
ϑ	displacement due to transverse swing, position tolerance of pantograph	m
$\vartheta_{o,u}$	pantograph transverse sway, installation tolerances and vehicle asymmetry at upper o and lower u verification height	m
ϑ_0	conductor temperature at condition 0	degree
ϑ_x	conductor temperature at condition x	degree
ρ	air density 15 °C and sea level 0 m equal to 1,225	kg/m^3
ξ	general symbol for a variable	–

4.1 Loads and strengths

4.1.1 Introduction

Electric traction contact lines are subjected to *mechanical, electrical, climatic* and *chemical stresses*. They need to withstand such stresses within given limits by fulfilling the requirements of relevant standards such as EN 50 119, EN 50 122 and the directives of the railway infrastructure and operating entities.

The design of *contactor rails* or *third rail installations* with large cross-sections needs to consider variations in length due to temperature changes. In the case of conductor rails installed in tunnels close to the surface beside the track or at the ceiling, climatic ambient conditions do not play an important role. However, flexible overhead contact lines are arranged approximately 6 m above the terrain and are more prone to climatic stresses and consequential longitudinal, vertical and transverse movement. External loads and movements create forces that need to be counteracted by the *supports*.

According to EN 50 119 the loads on overhead contact lines consist of:
  – *dead loads* of all conductors and components
  – *tensile forces* of conductors and their components
  – *wind loads* on conductors, poles and cantilevers
  – *additional loads* due to *construction activities* and *ice* as well as
  – *transient loads* caused by breaking conductors or by sudden changes in forces acting
    on conductors and wires

4.1.2 Coordinate system

For force and movement directions as well as geometric magnitudes, the coordinate system in Figure 4.1 applies to the following Clauses with the exception of Clause 4.8 and 4.9 for which the symbols of TSI ENE:2014 were used.

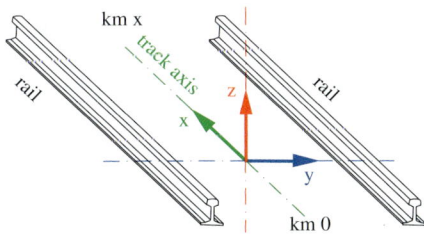

Figure 4.1: Coordinate system for force and movement directions and geometric magnitudes.

Table 4.1: Loads G' in N/m of Siemens and German Railway overhead contact line types.

Signification	G'						
Contact line type	Siemens AG – Sicat			Deutsche Bahn AG – Re			
	L1.0	S1.0	H1.0	100	200	250	330
contact wire	8,46	8,46	10,15	8,46	8,46	10,15	10,15
catenary	4,38	4,38	10,40	4,38	4,38	5,85	10,40
droppers	0,15	0,24	0,19[1]	0,15	0,27	0,21	0,21
clamps	0,24[2]	0,43[3]	0,37[4]	0,24[2]	0,39[3]	0,37[5]	0,37[4]
stitch wire	0,00	0,53	0,84	0,00	0,53	0,75	0,84
sum	13,22	14,02	21,96	13,22	14,02	17,33	21,97
practically used data for planning	$\approx 14,00$	$\approx 15,00$	$\approx 22,00$	$\approx 14,00$	$\approx 15,00$	$\approx 18,00$	$\approx 22,00$

[1] dropper wire Siemens 8WL7060-2 Bronze 10×49 mass per unit length 0,09 kg/m

[2] 1 clamps 16R Siemens 8WL4517-1K mass per piece 0,30 kg and
   12 dropper clips Siemens 50 8WL4620-0 mass per piece 0,11 kg

[3] 1 contact wire clip 16R Siemens 8WL4517-1K 16R mass per piece 0,30 kg
   13 dropper clips 50 Siemens 8WL4620-0 mass per piece 0,11 kg
   3 dropper clips 25 Siemens 8WL4620-1 mass per piece 0,16 kg and
   2 stitch wire clamps Siemens 8WL4505-7 mass per piece 0,31 kg

[4] 1 contact wire clip 16R Siemens 8WL4517-1K 16R mass per piece 0,30 kg
   5 dropper clips 120 Siemens 8WL4624-4 mass per piece 0,11 kg
   7 dropper clips 50 Siemens 8WL4620-0 mass per piece 0,11 kg
   2 dropper clips Siemens 8WL4624-2 mass per piece 0,11 kg and
   2 stitch wire clamps Siemens 8WL4505-7 mass per piece 0,31 kg

[5] 1 contact wire clip 16R Siemens 8WL4517-1K 16R mass per piece 0,30 kg
   5 dropper clips 70 Siemens 8WL4624-2 mass per piece 0,11 kg
   7 dropper clips 50 Siemens 8WL4620-0 mass per piece 0,11 kg
   2 dropper clips Siemens 8WL4624-2 mass per piece 0,11 kg and
   2 stitch wire clamps Siemens 8WL4505-7 mass per piece 0,31 kg

4.1.3 Dead loads

The *dead loads* on overhead contact lines result from the dead weights of conductors, insulators and fittings. They are described as a whole by the *mass per unit length m'* calculated
relative to the mean support spacing. Expressed in general terms, the force caused by gravity
g acting on a conductor's dead mass in relation to its length is termed *load per unit length G'*

$$G' = m' \cdot g \quad . \tag{4.1}$$

The specific mass of contact wires and conductors are listed in Tables 11.3 and 11.6. The
length related dead weights of overhead contact lines are listed in Table 11.7.

In stranded conductors, the individual strands of the conductors are up to 3 % longer than the
conductor itself. Because of the different number of droppers and fittings in different length
spans, the dead weights related to the length may be different for the individual spans. For
practical design purposes, the dead weights of the components are assumed for a representative span length, e. g. 65 m with stitch wires and 1,6 m system height in straight lines and
converted to length related loads. The rounded values for German overhead contact line types
are listed in Table 11.7.

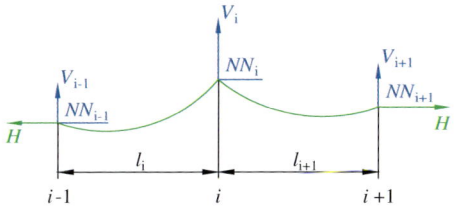

Figure 4.2: Effect of different suspension heights on vertical reaction forces - influence of unequal support heights.

4.1.4 Vertical components of loads

Corresponding to their direction of action, the tensile forces of conductors are distinguished as vertical and horizontal components. The weight force G of a conductor with the length L is calculated from the weight force G' per unit length of the conductor by $G = G'L$. For overhead contact lines, the distance between the supports l can be used instead of the conductor length L with an error of approximately $0{,}1\,\%$:

$$G = G' \cdot l \quad . \tag{4.2}$$

The portion of the *reaction forces* from one span can be obtained from the balance of moments according to Figure 4.2. The vertical component V_i results from

$$V_i = V_{i\,\mathrm{right}} + V_{i\,\mathrm{left}} \quad .$$

From the span l_{i+1} the vertical load component results by

$$V_{i\,\mathrm{right}} = G' l_{i+1}/2 + H\,(NN_i - NN_{i+1})\,/\,l_{i+1},$$

where H is the horizontal tensile force. From the span l_i it results

$$V_{i\,\mathrm{left}} = G' l_i/2 + H\,(NN_i - NN_{i-1})\,/\,l_i \quad .$$

There, NN_i is the relative height related to a common reference altitude. The total reaction force follows

$$V_i = G'\,(l_i + l_{i+1})\,/\,2 + H\,[(NN_i - NN_{i-1})\,/\,l_i + (NN_i - NN_{i+1})\,/\,l_{i+1}] \quad . \tag{4.3}$$

If the neighboring supports have a higher altitude than the one considered, the reaction force is reduced compared with spans with equal altitudes and increased in the opposite case.

4.1.5 Radial components of forces

4.1.5.1 General

Changes in the direction of conductors effect *radial forces*. Changes in conductor directions occur because of:
- *curves in the track* served by the contact or catenary wires
- *lateral offset* to achieve stagger in contact wires and cables
- *lateral deflection* towards anchoring or tensioning equipment
- *track radii*
- *wind loads*

The radial components receive a positive sign if they act towards the pole.

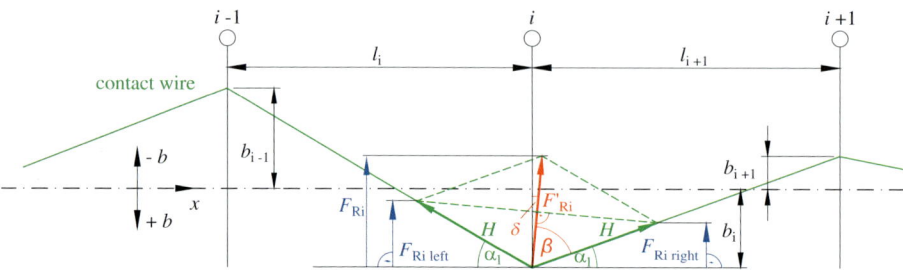

Figure 4.3: Determination of the radial components F_{Ri} of the conductor tensile force H due to stagger b_i in a straight track.

4.1.5.2 Contact line supports on straight tracks

From Figure 4.3 the geometric relations follow

$$\beta = (180 - \alpha_1 - \alpha_2)/2, \ \cos\beta = F_R/(2H)$$

and

$$\sin\alpha_1 = (|b_{i-1}| + |b_i|)/l_i, \ \sin\alpha_2 = (|b_i| + |b_{i+1}|)/l_{i+1} \quad .$$

Therefore, the radial force F_{Ri}' can be calculated from

$$F_{Ri}' = 2 \cdot H \cdot \cos\left[(180 - \sin^{-1}(|b_{i-1}| + |b_i|)/l_i - \sin^{-1}(|b_i| + |b_{i+1}|)/l_{i+1})/2\right] \quad .$$

Since the angle δ is small, F_R' can be replaced by F_R resulting in

$$F_{Ri}' = F_{Ri\,\text{left}} + F_{Ri\,\text{right}} \quad .$$

The radial components follow, therefore, from

$$F_{Ri\,\text{left}} = H \cdot \sin\alpha_1 = H \cdot \frac{|b_{i-1}| + |b_i|}{l_i}, \ F_{Ri\,\text{right}} = H \cdot \sin\alpha_2 = H \cdot \frac{|b_i| + |b_{i+1}|}{l_{i+1}} \quad .$$

The horizontal force F_{Ri} can be calculated from the absolute values of the staggers by

$$F_{Ri} = H \cdot \left(\frac{|b_{i-1}| + |b_i|}{l_i} + \frac{|b_i| + |b_{i+1}|}{l_{i+1}} \right) \quad . \tag{4.4}$$

For the calculation of the radial force F_{Ri} from the *direction related stagger* the approximations follow from Figure 4.3

$$\tan\alpha_1 = (b_i - b_{i+1})/l_{i+1} \approx \sin\alpha_1 = F_{Ri1}/H \quad \text{and}$$
$$\tan\alpha_2 = (b_i - b_{i-1})/l_i \approx \sin\alpha_2 = F_{Ri2}/H \quad . \tag{4.5}$$

The approximation $\tan\alpha \approx \sin\alpha$ applies to small angles. For example, the error is 1,5 % for an angle of $\alpha = 10°$. The force

$$F_{Ri} = F_{Ri1} + F_{Ri2} = H\left[(b_i - b_{i-1})/l_i + (b_i - b_{i+1})/l_{i+1}\right] \tag{4.6}$$

is the *horizontal component of the tensile force on the conductor* required to pull it away from the centre line. For the situation that $b_i = b$ and $b_{i-1} = b_{i+1} = -b$ and $l_{i+1} = l_i = l$ as is the case along straight line sections the equation is simplified to

$$F_R = 4H \cdot b/l \quad . \tag{4.7}$$

As agreed the radial force is positive if it acts in the direction of the pole at push-off supports. It is negative if it acts in a direction away from the pole, as at pull-off supports.

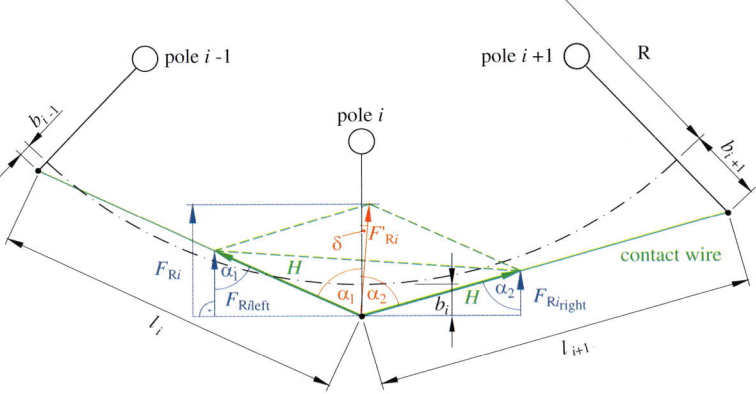

Figure 4.4: Radial force components F_{Ri} on curved tracks with differing spans.

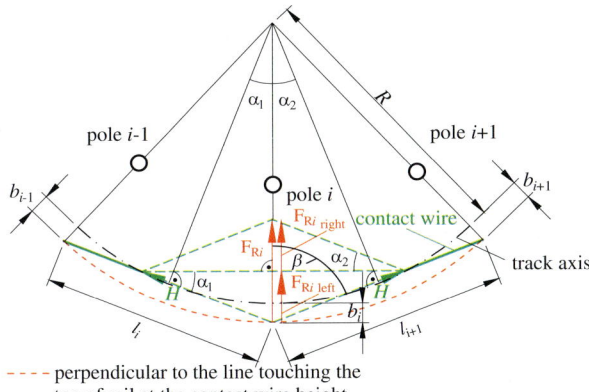

- - - - perpendicular to the line touching the
top of rail at the contact wire height

Figure 4.5: Radial force components F_{Ri} on curved tracks with equal spans $l_i = l_{i+1}$.

Example 4.1:   Assuming $H = 10\,\text{kN}$, $b_{i-1} = -0,3\,\text{m}$, $b_i = 0,1\,\text{m}$, $b_{i+1} = -0,3\,\text{m}$, $l_i = 64\,\text{m}$ and $l_{i+1} = 53\,\text{m}$ the radial force follows from Equation (4.6) to be 138 N corresponding to 1,4 % of the conductor tensile force. The radial force is positive and acts, therefore, in the direction to the pole.
Since the line will be traversed by 1 600 mm long pantographs and speeds up to 200 km/h in the future, the deviation due to wind needs to be reduced. The contact wire tensile force of the contact line is intended to be increased to 27 kN by a contact wire AC-120 – CuMg. Therefore, the radial force at the steady arm will be increased to $27/10 \cdot 138 = 373\,\text{N}$, meaning that the standard components of a cantilever for the contact line type Re200 can be used.

In Equation (4.4), the absolute data of the stagger at the supports needs to be used. However, for Equation (4.7) the existing values of the stagger need to be used with their signs. Both methods of calculation yield the same results.

4.1.5.3   Contact wire supports on curved tracks

From Figure 4.4 for curved track with the radius R follows

$$F_{Ri\,\text{left}} = H \cdot \cos\alpha_1 \text{ and } F_{Ri\,\text{right}} = H \cdot \cos\alpha_2 \quad . \tag{4.8}$$

On curved track, radial components of the conductor tensile force are also created from the deviation due to the track radii. Since the angle δ is small, with sufficient approximation $F'_{Ri} = F_{Ri}$ can be applied and the *radial force at the support i* can be taken as the sum of the radial force components of the conductor tensile force from the neighbouring spans on the left and right side to be

$$F_{Ri} = F_{Ri\,\text{left}} + F_{Ri\,\text{right}} \quad .$$

Using the cosine law, the angles α_1 and α_2 can be obtained from

$$\cos\alpha_1 = \frac{(R+b_i)^2 + l_i^2 - (R+b_{i-1})^2}{2 \cdot (R+b_i) \cdot l_i}, \quad \cos\alpha_2 = \frac{(R+b_i)^2 + l_{i+1}^2 - (R+b_{i+1})^2}{2 \cdot (R+b_{i+1}) \cdot l_{i+1}} \quad .$$

Therefore, for the radial force at the support i it follows

$$F_{Ri} = \frac{H \cdot (R+b_i)^2}{2} \cdot \left[\frac{l_i^2 - (R+b_{i-1})^2}{(R+b_i) \cdot l_i} + \frac{l_{i+1}^2 - (R+b_{i+1})^2}{(R+b_{i+1}) \cdot l_{i+1}} \right] \quad . \tag{4.9}$$

The following relationships can also be deduced from Figure 4.5

$$\sin\alpha_1 \approx l_i/(2R) = F_{Ri1}/H \text{ and } \sin\alpha_2 \approx l_{i+1}/(2R) = F_{Ri2}/H$$

$$F_{Ri} = F_{Ri1} + F_{Ri2} = \pm H \cdot (l_i + l_{i+1})/(2R) \quad . \tag{4.10}$$

In the case of equally long spans from Equation (4.10) it follows

$$F_{Ri} = H\,l/R \quad . \tag{4.11}$$

The radial force will be positive if it acts towards the pole (push-off support) and negative if it acts away from the pole (pull-off support). Equation (4.11) represents the total radial force at the support only if $b_{i-1} = b_i = b_{i+1}$, otherwise it applies

$$F_{Ri} = H \left[\frac{(b_i - b_{i-1})}{l_i} + \frac{(b_i - b_{i+1})}{l_{i+1}} \right] \pm H \cdot \frac{(l_i + l_{i+1})}{2 \cdot R} \quad . \tag{4.12}$$

Example 4.2: With $H = 10\,\text{kN}$, $b_{i-1} = 0,3\,\text{m}$, $b_i = 0,1\,\text{m}$, $b_{i+1} = -0,3\,\text{m}$, $l_i = 64\,\text{m}$, $l_{i+1} = 53\,\text{m}$ and $R = -2\,000\,\text{m}$ the radial force equates to $430,5\,\text{N}$ according to Equations 4.9) and (4.12). The direction of force is towards the pole and has a positive sign in Equation (4.10). Both equations lead to the same results.

The radial forces in straight lines and curved tracks can be obtained by means of superposition. From Figure 4.6 the trigonometric function can be derived

$$\alpha_1 = \alpha'' - \alpha' \quad .$$

The cosine of α_1 equates to

$$\cos\alpha_1 = \cos(\alpha'' - \alpha') = \cos\alpha' \cdot \cos\alpha'' + \sin\alpha' \cdot \sin\alpha'' = F_{Ri}/H \quad . \tag{4.13}$$

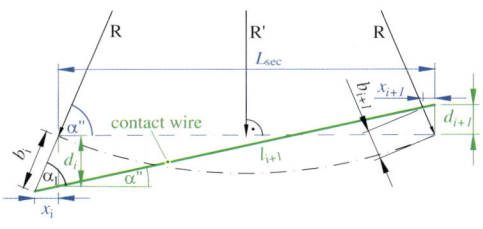

Figure 4.6: Determination of the radial force component F_{Ri} on curved tracks using superposition.

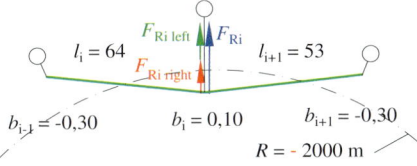

Figure 4.7: Designations in curved tracks according to Example 4.2. All dimensions in m.

Since the radius R is much bigger than the span length l_{i+1}, from Figure 4.6, $L_{sec} \approx l_{i+1}$

$$x_{i+1} \approx 0 \text{ and, therefore, } d_i \approx b_i \text{ and } R' \approx R \quad .$$

Therefore, it is obtained

$$\cos\alpha' = \frac{l_{i+1}}{L_{sec}} = 1, \ \cos\alpha'' = \frac{l_{i+1}}{2 \cdot R}, \ \sin\alpha' = \frac{b_i + b_{i+1}}{l_{i+1}} \text{ and } \sin\alpha'' = \frac{R'}{R} \approx 1 \quad .$$

The Equation (4.13) can be simplified to

$$\cos\alpha_1 = \frac{l_{i+1}}{2 \cdot R} + \frac{|b_i| + |b_{i+1}|}{l_i} = \frac{F_{Ri}}{H} \quad .$$

In span i it applies similarity

$$\frac{F_{Ri}}{H} = \frac{l_i}{2 \cdot R} + \frac{|b_i| + |b_{i-1}|}{l_i} \quad .$$

The total *radial force* F_{Ri} can be obtained by

$$F_{Ri} = H \cdot \left(\pm \frac{l_i}{2 \cdot R} + \frac{|b_i| + |b_{i-1}|}{l_i} \mp \frac{l_{i+1}}{2 \cdot R} + \frac{|b_i| + |b_{i+1}|}{l_{i+1}} \right) \tag{4.14}$$

or after rearrangement of the Equation (4.14) by

$$F_{Ri} = H \cdot \left(\frac{|b_i| + |b_{i-1}|}{l_i} + \frac{|b_i| + |b_{i+1}|}{l_{i+1}} \right) \pm H \cdot \left(\frac{l_i + l_{i+1}}{2 \cdot R} \right) \quad . \tag{4.15}$$

The *radial force* $\pm H \cdot [(l_i + l_{i+1})/(2 \cdot R)]$ is positive, if it acts towards the pole (push-off support) and negative, if it acts away from the pole (pull-off support).

4.1.5.4 Dead end anchoring

As shown in Figure 4.8 the resulting *radial component* can be obtained from Equation (4.10). By replacing b_{i+1} by y_{i+1}, where y_{i+1} assumes a negative value viewed in direction of increasing kilometrage, the radial force at the catenary wire clamp is as follows

$$F_{Ri} = H_{CA} \cdot [(b_{iCA} - b_{i-1CA})/l_i + (b_{iCA} - y_{i+1})/l_{i+1}] \quad . \tag{4.16}$$

If H_{CA} is replaced by H_{CW} in Equation (4.16) and b_{iCA} by b_{iCW} the radial force F_{RiCW} at the contact wire clamp at the registration arm is obtained.

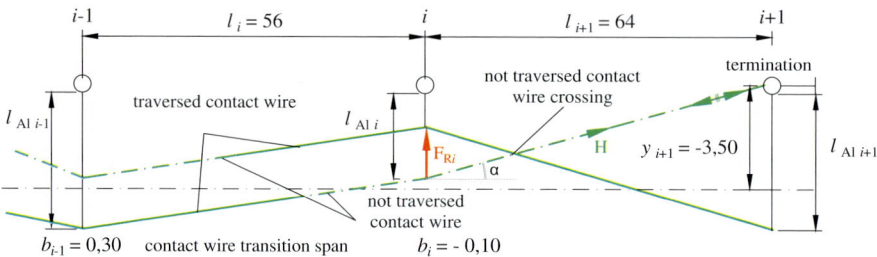

Figure 4.8: Arrangement of an overlap with a flexible termination at a pole according to Example 4.3.

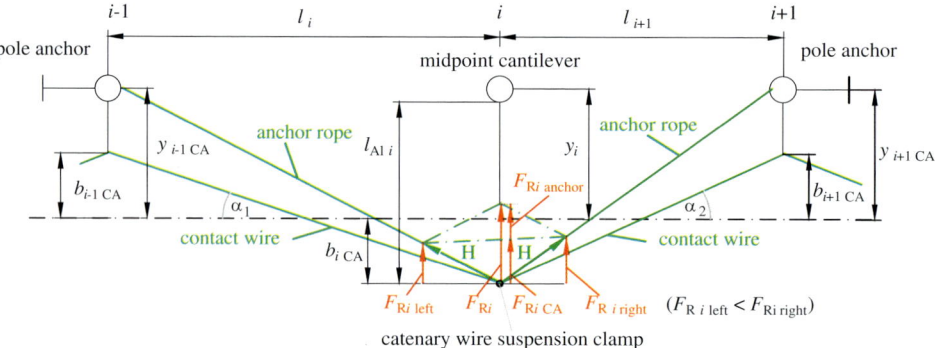

Figure 4.9: Radial component $F_{Ri\text{left/right}}$ at midpoint cantilever.

Example 4.3: For a straight line the radial force F_{Ri} at the support i according to Figure 4.8 is to be obtained.

The contact wire staggers are $b_{i-1} = 0,3$ m, $b_i = -0,10$ m and at the termination $y_{i+1} = -3,5$ m. The spans result from the interoperable overhead contact line with a 10 kN contact wire tensile force at the transition span $l_i = 56$ m and for the adjacent span $l_{i+1} = 64$ m. The radial force F_{Ri} can be obtained from (4.5) as

$$F_{Ri} = 10\,000 \cdot \left[\frac{(-0,10) - (+0,30)}{56} + \frac{(-0,10) - (-3,50)}{64} \right] = 460\,\text{N} \quad .$$

The force will be positive and acts towards the pole.

4.1.5.5 Midpoint anchoring

In the case of a sound contact line, the *radial force* F_{Ri} acts on the midpoint cantilever i at the catenary wire clamp

$$F_{Ri} = F_{Ri\text{anchor}} + F_{RiCA} \quad ,$$

where the radial force F_{Ri} at the catenary wire clamp consists of the anchor radial force $F_{Ri\text{anchor}}$ and the catenary wire radial force F_{RiCA} (Figure 4.9). For the midpoint cantilever (Figure 4.9) the anchor force F_{Ri} is obtained from Equations (4.5) and (4.8) with $H = H_{\text{anchor}}$,

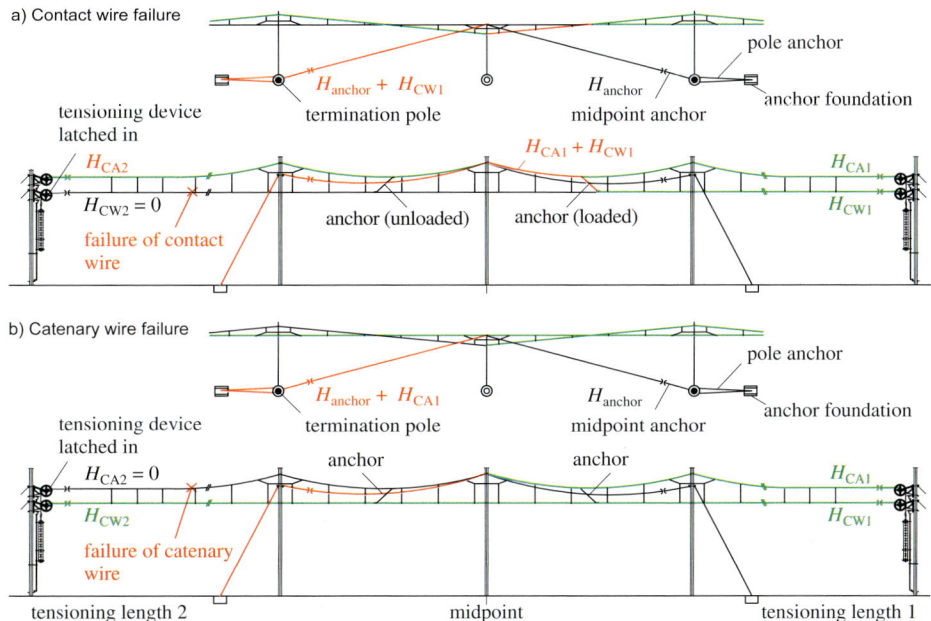

Figure 4.10: Conductor failure within the overhead contact line.

the catenary wire stagger b_{iCA} at the midpoint cantilever and the contact wire staggers at the anchor poles $b_{i-1} = y_{i-1CA}$ and $b_{i+1} = y_{i+1CA}$ to be

$$F_{Ri\,anchor} = H_{anchor} \left(\frac{y_{i-1CA} - b_{iCA}}{l_i} + \frac{y_{i+1CA} - b_{iCA}}{l_{i+1}} \pm \frac{l_i + l_{i+1}}{2R} \right) \quad . \tag{4.17}$$

If the radial force acts towards the pole (push-off support) the positive sign in (4.17) should be used. At pull-off supports, the force acts away from the pole and has a negative sign.

German Railway DB assumes 80 % of the catenary wire tensile force H_{CA} at $-30\,°C$ as anchor force H_{anchor}, e. g. for an anchor rope BzII 50 mm² [4.1].

The *catenary wire radial force* F_{RiCA} according to (4.6) acts at catenary wire clamp resulting in total radial force F_{Ri} under consideration of catenary wire staggers at the next supports

$$\begin{aligned} F_{RiCA} = \ & H_{anchor} \cdot \left(\frac{b_{iCA} - y_{i-1CA}}{l_i} + \frac{b_{iCA} - y_{i+1CA}}{l_{i+1}} \pm \frac{l_i + l_{i+1}}{2R} \right) \\ & + H_{CA} \left(\frac{b_{iCA} - b_{i-1CA}}{l_i} + \frac{b_{iCA} - b_{i+1CA}}{l_{i+1}} \pm \frac{l_i + l_{i+1}}{2R} \right) \quad . \end{aligned} \tag{4.18}$$

In the case of disturbances such as a tree falling onto the contact line, it is possible that the contact wire or the catenary wire or both will fail (Figure 4.10). Normally, the tensioning devices of the failed contact or catenary wire latch in after a short rotary motion of the tensioning wheel and lock the wheel to prevent major damage to the contact line.

In the case of a contact wire failure within the half tensioning length 2, the tensile force within tensioning section 2 will fall to zero (Figure 4.10 a)) and the midpoint anchor of the tensioning half 1 will be loaded. A radial force $F_{RiCW\,fail}$ at the catenary wire clamp of the midpoint cantilever i results after loss of the contact wire tensile force H_{CW2} according to Equation (4.18) as

$$
\begin{aligned}
F_{RiCW\,fail} &= F_{Ri\,anchor\,12} + F_{RiCA\,12} + F_{RiCW\,1\,rail} \\
&= H_{anchor\,12} \cdot \left(\frac{b_{iCA} - y_{i-1\,CA}}{l_i} + \frac{b_{iCA} - y_{i+1\,CA}}{l_{i+1}} \pm \frac{l_i + l_{i+1}}{2R} \right) \\
&\quad + H_{CA\,12} \cdot \left(\frac{b_{iCA} - b_{i-1\,CA}}{l_i} + \frac{b_{iCA} - b_{i+1\,CA}}{l_{i+1}} \pm \frac{l_i + l_{i+1}}{2R} \right) \\
&\quad + H_{CW\,1} \cdot \left(\frac{y_{i-1} - b_{iCW}}{l_i} \pm \frac{l_i + l_{i+1}}{2 \cdot R} \right) \quad .
\end{aligned}
\tag{4.19}
$$

For the component of force $F_{RiCW\,1}$ if there is a contact wire failure in the half tensioning section 2, b_{iCA} is to be replaced by the contact wire stagger b_{iCW} in (4.19). The lateral position of the contact wire b_{iCW} and the catenary wire b_{iCA} can differ.

If a catenary wire failure occurs within the half tensioning section 1, the radial force of the midpoint anchor at the catenary wire clamp results from

$$
\begin{aligned}
F_{RiCA\,fail} &= F_{Ri\,anchor\,12} + F_{RiCW\,12} + F_{RiCA\,1\,fail} \\
&= H_{anchor\,12} \cdot \left(\frac{b_{iCA} - y_{i-1}}{l_i} + \frac{b_{iCA} - y_{i+1}}{l_{i+1}} \pm \frac{l_i + l_{i+1}}{2R} \right) \\
&\quad + H_{CW\,12} \cdot \left(\frac{b_{iCW} - b_{i-1\,CW}}{l_i} + \frac{b_{iCW} - b_{i+1\,CW}}{l_{i+1}} \pm \frac{l_i + l_{i+1}}{2R} \right) \\
&\quad + H_{CA\,1} \cdot \left(\frac{y_{i-1} - b_{iCA}}{l_i} + \frac{b_{i+1\,CA} - b_i}{l_{i+1}} \pm \frac{l_i + l_{i+1}}{2 \cdot R} \right) \quad .
\end{aligned}
\tag{4.20}
$$

Figure 4.10 shows the force diagram for a conduct or catenary wire failure. The catenary wire between the anchor clamp and contact wire clamp at the midpoint cantilever within the intact tensioning section carries the highest loaded. Within this section, the tension of the catenary and the contact wire are summed resulting in a tensile force of 22 kN for a contact line type Sicat S1.0. However, the tensile strength of 28,6 kN according to Example 4.7 will not be exceeded and therefore, a failure will not occur, assuming a sufficiently rated midpoint anchor rope.

Example 4.4: Determine the radial force F_{Ri} at the catenary wire clamp of midpoint cantilever i according to Figure 4.11. Given are:

contact wire staggers $\qquad\qquad$ $b_{i-1\,CW} = -0,30\,\text{m}$, $b_{iCW} = 0,30\,\text{m}$, $b_{i+1\,CW} = -0,30\,\text{m}$
catenary wire staggers $\qquad\qquad$ $b_{i-1\,CA} = -0,30\,\text{m}$, $b_{iCA} = 0,30\,\text{m}$, $b_{i+1\,CA} = -0,30\,\text{m}$
distance between anchor points and track axis $y_{i-1} = y_{i+1} = -3,5\,\text{m}$
tensile force of the contact wire $\quad$ 10 kN
tensile force of the catenary wire $\quad$ 10 kN
span lengths $\qquad\qquad\qquad\qquad$ $l_i = l_{i+1} = 64\,\text{m}$
tensile force in the anchor ropes $\quad$ $H_{anchor} = 8\,\text{kN}$ at $-30\,^{\circ}\text{C}$

The radial force F_{Ri} at the catenary wire clamp of the midpoint at $-30\,^{\circ}\text{C}$ can be obtained from Equations (4.6) and (4.17)

$$
F_{Ri} = F_{Ri\,anchor} + F_{RiCA} \quad .
$$

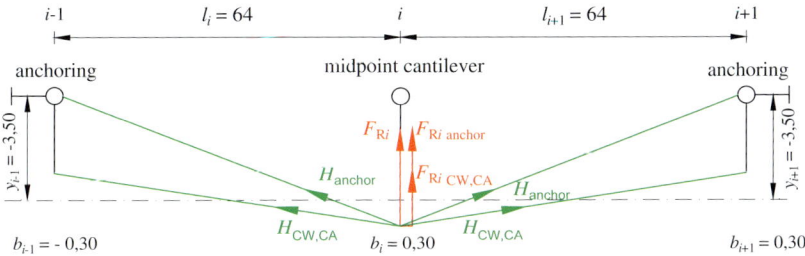

Figure 4.11: Contact and catenary stagger and spans in example 4.4. All dimensions in m.

For the intact contact line

$$
\begin{aligned}
F_{\mathrm{R}i} &= 8\,000\,\{[0,3-(-3,5)]+[0,3-(-3,5)]\}\,/64 \\
&+ 10\,000\,\{[0,3-(-0,3)]+[0,3-(-0,3)]\}\,/64 = 950 + 188 = 1\,138\,\mathrm{N} \quad .
\end{aligned}
$$

The force $F_{\mathrm{R}i}$ will be positive and acts towards the pole. In the case of a contact wire failure within the half tensioning section 2, it applies according to (4.20) and Figure 4.10 a)

$$
\begin{aligned}
F_{\mathrm{R}i\,\mathrm{CW\,fail}} &= F_{\mathrm{R}i\,\mathrm{anchor}} + F_{\mathrm{R}i,\mathrm{CA}\,12} + F_{\mathrm{R}i\,\mathrm{CA}\,1\,\mathrm{fail}} \\
&= 8\,000 \cdot \{[0,3-(-3,5)]+[0,3-(-3,5)]\}\,/64 \\
&+ 10\,000\,\{[0,3-(-0,3)]+[0,3-(-0,3)]\}\,/64 \\
&+ 10\,000\,[(-3,5)-0,3]\,/64 = 950 + 188 - 594 = 544\,\mathrm{N} \quad .
\end{aligned}
$$

The force $F_{\mathrm{R}i\,\mathrm{CW\,fail}}$ is positive and acts in direction to the pole. In case of a catenary wire failure within the half tensioning section 2 it applies according to (4.20) and Figure 4.10 b)

$$
\begin{aligned}
F_{\mathrm{R}i\,\mathrm{CA\,fail}} &= F_{\mathrm{R}i\,\mathrm{anchor}\,12} + F_{\mathrm{R}i\,\mathrm{CW}\,12} + F_{\mathrm{R}i\,\mathrm{CA}\,1\,\mathrm{fail}} \\
&= 8\,000 \cdot \{[0,3-(-3,5)]+[0,3-(-3,5)]\}\,/64 \\
&+ 10\,000\,\{[0,3-(-0,3)]+[0,3-(-0,3)]\}\,/64 \\
&+ 10\,000\,\{[(-3,5)-0,3]+[(-0,3)-0,3]\}\,/64 = 950 + 188 - 688 = 450\,\mathrm{N}.
\end{aligned}
$$

The force $F_{\mathrm{R}i\,\mathrm{CA\,fail}}$ is positive and acts towards the pole. In the case of an intact overhead contact line the radial force at the midpoint cantilever will be a maximum. In case of contact or catenary wire failure the radial force at the midpoint cantilever will be reduced.

4.1.5.6 Resetting forces

The *resetting forces*, also known as *cantilever drag*, are created by thermally induced changes in length of contact and catenary wires and consequential changes to the conductor tensile forces. A variation Δb_i of the lateral offset can be caused by a thermal change of length of the contact or catenary wire by ΔL_i, as presented in Figure 4.12. By the use of

$$
l_{\mathrm{A}li}^2 = (l_{\mathrm{A}li} - \Delta b_i)^2 + \Delta L_i^2 \quad \text{and} \quad l_{\mathrm{A}li}^2 = l_{\mathrm{A}li}^2 - 2 \cdot l_{\mathrm{A}li} \cdot \Delta b_i + \Delta b_i^2 + \Delta L_i^2
$$

and by neglecting the term Δb_i^2 an approximation for the variation of the lateral offset will be

$$
\Delta b_i = \Delta L_i^2 / 2\,l_{\mathrm{A}li} \quad \text{and} \quad \Delta L_i = \alpha \cdot l_{\mathrm{F}i} \cdot \Delta T \quad . \tag{4.21}
$$

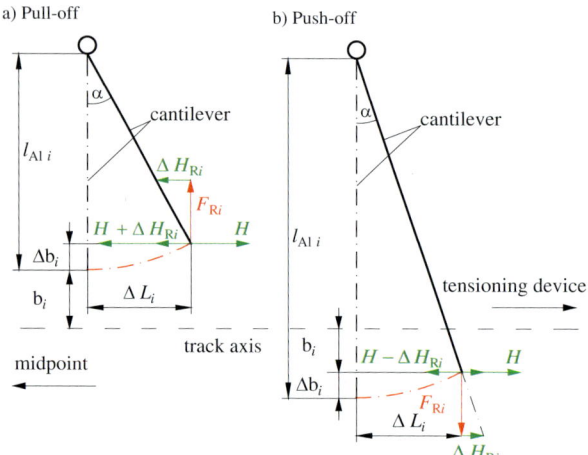

Figure 4.12: Resetting forces ΔH_{Ri} at a cantilever of the length $l_{\mathrm{Al}i}$ due to change in the lateral offset Δb by pivoting of the cantilever with resulting ΔL_i.

where

Δb_i change of the catenary or contact wire lateral position in m

$l_{\mathrm{Al}i}$ distance between pole front face and middle of the catenary wire clamp for calculation of the catenary wire resetting forces or distance between the pole front face to the middle of the contact wire clamp when calculating the contact wire resetting forces at the cantilever i in m

$l_{\mathrm{F}i}$ distance between the cantilever i and the midpoint in m

ΔL_i thermal change of length of the contact or catenary wire

α thermal expansion coefficient for copper contact wire or bronze catenary wire in K^{-1}

ΔT half temperature range of the overhead contact line in K

Example 4.5: Calculate the change of stagger Δb_i for contact wire at last cantilever in front of contact wire termination.

Given are:

$l_{\mathrm{F}i}$	half length of tensioning section	700 m
α	coefficient of thermal expansion	$17 \cdot 10^{-6}$ K^{-1} for contact wire AC-100 – Cu
ΔT	temperature range	40 K for Re200
$l_{\mathrm{Al}i}$	length of cantilever	3,6 m

It is obtained for

$$\Delta L_i = 700 \cdot 17 \cdot 10^{-6} \cdot 40 = 0,476\,\mathrm{m} \text{ and } \Delta b_i = 0,476^2/(2 \cdot 3,6) = 0,031\,\mathrm{m} \quad .$$

The variation in contact wire stagger Δb_i will be 31 mm at the last cantilever.

When considering a temperature difference of 20 K, a change in the contact wire lateral position Δb_i of 0,008 m will result at the last cantilever.

The variation ΔL_i of the position of the cantilever results in radial or longitudinal tensile forces which act as a moment around the cantilever pivot. This moment is counteracted by the moment resulting from the difference in the horizontal contact wire forces, designated as a *resetting force*. A part of the horizontal tensile force H is deviated in the direction of the pole.

With the approximation for small angles

$$\tan \alpha \approx \Delta H_{Ri}/F_{Ri} \approx \Delta L_i/l_{Ali}$$

the resetting force is obtained from Figure 4.12

$$\Delta H_{Ri} = F_{Ri} \Delta L_i/l_{Ali} \quad . \tag{4.22}$$

This calculation of the resetting forces assumes that
  – the cantilevers are installed for a certain ambient temperature and
  – the steady arms can rotate around their pivots.

Additional resetting forces result from friction within the cantilever pivots and within the tensioning devices. Measurements carried out at new tensioning devices according to Figure 11.43 for AC installations, demonstrated that the forces caused by friction amount to 3 % of the tensile force approximately. The sum of resetting forces should not exceed 11 %.

4.2 Tensile stresses and forces in conductors

4.2.1 Invariable tensile stresses and forces

The *tensile forces* within conductors are determined by the function and design of the overhead contact line. For automatically tensioned contact lines for speeds up to 230 km/h the contact and catenary wire tensile forces are selected between 10 and 12 kN, or in the case of high-speed contact lines between 15 and 31 kN [4.2, 4.3, 4.4]. For the SNCF record runs in 2007, the contact wire tensile force was 40 kN [4.5].

The strength of materials forms the basis for determining permissible tensile forces. The Tables 11.4, 11.6 and 11.7 list the essential characteristics of contact wires and conductors. The permissible tensile stress σ_{per} and the nominal cross-section A of a conductor result in the *permissible tensile force* F_{per} according to

$$F_{per} = \sigma_{per} \cdot A \quad . \tag{4.23}$$

According to EN 50 119, the permissible tensile force for contact and catenary wires under operational conditions results from

$$\sigma_{per\,CW,CA} = \sigma_{min} \cdot 0,65 \cdot k_{temp} \cdot k_{wear} \cdot k_{wind} \cdot k_{ice} \cdot k_{eff} \cdot k_{clamp} \cdot k_{joint} \cdot k_{load} \quad , \tag{4.24}$$

where the factor 0,65 represents the permissible tensile stress related to the minimum tensile strength. The other factors in Equation (4.24) represent:

$\sigma_{per\,CW,CA}$	*permissible tensile stress* of the contact wire or catenary wire
$\sigma_{min\,CW,CA}$	*minimum rated tensile strength* of the contact wire or catenary wire according to Table 4.3
k_{temp}	factor, which represents the relation ship between the maximum operating temperature and the permissible tensile stress, see Table 4.2,
k_{wear}	factor taking care of the permissible wear of the contact wire, see Table 4.2, where x_W expresses the wear of the contact wire as a percentage
k_{wind}, k_{ice}	factor taking accounting for the effect of wind or ice loads
k_{eff}	factor accounting for the efficiency of the tensioning device, where the value provided by the manufacturer, e. g. 0,97, should be used. In the case of rigid terminations $k_{eff} = 1,0$

Table 4.2: Factors for rating of contact and catenary wires according to EN 50 119.

	Contact wires (CW)		Catenary wires (CA)	
Effect of temperature k_{temp}				
material	maximum operational temperature			
	80 °C	100 °C	80 °C	100 °C
Cu	1,00	0,80	1,00	0,80
aluminum alloy	–	–	1,00	0,80
CuAg0,1	1,00	1,00	1,00	1,00
CuSn0,4	1,00	1,00	1,00	1,00
CuMg	1,00	1,00	1,00	1,00
steel	–	–	1,00	1,00
ACSR, AACSR	–	–	1,00	0,80
Wear k_{wear}, permissible wear x_{w} in %				
local mass transit	$k_{\text{wear}} = 1 - x_{\text{w}}/100,\ x_{\text{w}} \leq 30$		$k_{\text{wear}} = 1,00$	
main lines	$k_{\text{wear}} = 1 - x_{\text{w}}/100,\ x_{\text{w}} \leq 20$		$k_{\text{wear}} = 1,00$	
Effect of wind and ice loads k_{ice} and k_{wind}				
type of termination	wind and ice load k_{ice}	wind load k_{wind}	ice load k_{ice}	wind load k_{wind}
				≤ 100 km/h / > 100 km/h
CW and CA flexible	0,95	1,00	1,00	1,00 / 0,95
CW flexible, CA fixed	0,90	0,95	0,95	0,95 / 0,90
CW and CA fixed	0,70	0,80	0,95	0,95 / 0,90
CW flexible (trolley type contact line)	0,90	0,95	–	– / –
Efficiency of tensioning device k_{eff}				
CW and CA flexible	η_{CW}		η_{CA}	
CW flexible, CA fixed	η_{CW}		1,00	
CW and CA fixed	1,00		1,00	
trolley type contact line				
Effect of termination fittings k_{clamp}				
clamping force F_{clamp}				
without and with clamp: $F_{\text{clamp}} \geq 0,95\,\sigma_{\text{min CW,CA}} \cdot A$	1,00		1,00	
with clamp: $F_{\text{clamp}} < 0,95\,\sigma_{\text{min CW,CA}} \cdot A$	$F_{\text{clamp}}/(\sigma_{\text{min CW}} \cdot A)$		$F_{\text{clamp}}/(\sigma_{\text{min CA}} \cdot A)$	
Effect of welded or soldered joints k_{joint}				
without joints	1,00		1,00	
with joints	$\sigma_{\text{min joint}}/\sigma_{\text{min CW}}$		1,00	
Effect of single loads at the catenary wire k_{load}				
with loads	1,00		0,80	
without loads [1]	1,00		1,00	

A contact or catenary wire cross-section, η_{CW}, η_{CA} efficiency of tensioning device for CW and CA according to information from the manufacturer, [1] loads due to droppers are not considered for the factor k_{load}

k_{clamp} factor accounting for the efficiency of the dead end clamps. Should be assumed to be 1,0 where the clamping force is greater than 95 % of the nominal tensile strength of the contact wire or taken to be the ratio between the existing clamping force to the nominal strength of the contact wire if this relation is less than 95 %

k_{joint} factor accounting for the possible reduction in strength of a contact wire joint. Without joints this factor can be set to 1,00, otherwise the factor k_{joint} should be taken as the ratio of the tensile strength of the joint to the tensile strength of the contact wire

k_{load} factor accounting for additional loads on the catenary wire, however, dropper loads do not need to be considered

Example 4.6: Determine the permissible tensile stress of contact wire AC-100–Cu-ETP in an automatically tensioned overhead contact line without contact wire joints. What is the relation between permissible tensile force F_{per} and the minimum failing load F_{ult} of the contact wire?

From Table 4.3, $\sigma_{min} = 355$ N/mm^2 according to EN 50 149 and from Table 4.2, $k_{temp} = 1,00$ for $\vartheta_{max} = 80\,°C$, $k_{wear} = 0,80$ for a maximum wear of 20 %, $k_{wind} = 1,00$; $k_{ice} = 0,95$, $k_{eff} = 0,97$ (according to the manufacturer's data), $k_{clamp} = 1,00$, because the clamping force of the dead end clamp is more than 95 % of the nominal contact wire tensile strength of the contact wire and $k_{joint} = 1,00$, because no joints exist in the contact wire.

Therefore, Equation (4.24) yields the permissible tensile stress of the contact wire type AC-100–Cu-ETP to be 170,1 N/mm^2 and the permissible tensile force of the contact wire is

$$F_{per} = 170,1 \cdot 100 = 17011\,\text{N},$$

This value is marked in red in Table 4.3. The relation between the minimum tensile strength σ_{min} and the permissible tensile stress $\sigma_{per\,CW}$ of the contact wire worn to 80 % amounts to 2,09.

Since the wear of the contact wire is already considered in Equation (4.24) by the factor k_{wear}, the nominal cross-sectional area of the contact wire should be inserted into Equation (4.23), that is the cross-section of 100 mm^2 in example 4.6.

Example 4.7: What is the permissible tensile stress of a seven strand, 50 mm^2 bronze Bz II catenary wire in an overhead contact line installation according to EN 50 119 with a section insulator and 20,3 m/s wind velocity (Table 4.2)?

From Table 4.3, $\sigma_{min} = 572$ N/mm^2 and from Table 4.2 k_{temp} 1,0; k_{wear} 1,0; k_{wind} 1,0 with $v_{wind} = 73,1$ km/h < 100 km/h; k_{ice} 1,0; k_{eff} 0,97 according to the manufacturers data are obtained; k_{clamp} 1,0 and k_{load} 0,8 are taken to consider the section insulators.

Equation (4.24) yields the permissible tensile stress to be 288,5 mm^2. For the factor $k_{load} = 1,0$, as is the usual case, the permissible tensile stress would rise to 360,6 mm^2. Depending on whether vertical loads act at the catenary wire, the permissible tensile forces for this conductor amount to 14,4 kN or 18,0 kN, respectively (see values marked in blue in Table 4.3). For $k_{load} = 1,0$ the relation to the ultimate tensile strength is 1,59. For $k_{load} = 0,8$ this relation will be 1,98.

The Equation (4.24) can also be used to determine the permissible tensile stresses in conductors for railway traction lines because these conductors have a similar structure as conductors used for catenary wires. For rigidly terminated catenary wires, the conductor stress varies with temperature and when wind and/ice loads act, EN 50 119 requires the use of Equation (4.24) for these conductors, if the tensile force exceeds 40 % of their rated tensile force. EN 50 119 does not exclude the use of Equation (4.24) for calculating the tensile force for conductors

Table 4.3: Parameters and permissible tensile forces F_{per} for contact and catenary wires.

type of wire and conductor	n	E kN/mm^2	A mm^2	σ_{min} N/mm^2	σ_{per} N/mm^2	F_{per} kN
Contact wires according to EN 50 149:2013						
AC-80–Cu-ETP	1	120	80	355	170,1	13,6
AC-100–Cu-ETP	1	120	100	355	170,1	17,0
AC-107–Cu-ETP	1	120	107	350	167,7	17,9
AC-120–Cu-ETP	1	120	120	330	158,1	19,0
AC-150–Cu-ETP	1	120	150	310	148,5	22,3
AC-80–CuAg	1	120	80	365	174,9	14,0
AC-100–CuAg	1	120	100	360	172,5	17,3
AC-107–CuAg	1	120	107	350	167,7	17,9
AC-120–CuAg	1	120	120	350	167,7	20,1
AC-150–CuAg	1	120	150	350	167,7	25,2
AC-80–CuMg	1	120	80	520	249,2	19,9
AC-100–CuMg	1	120	100	510	244,4	24,4
AC-107–CuMg	1	120	107	500	239,6	25,6
AC-120–CuMg	1	120	120	490	234,8	28,23
AC-150–CuMg	1	120	150	470	225,2	33,8
AC-80–CuSn	1	120	80	460	220,4	17,6
AC-100–CuSn	1	120	100	450	215,6	21,6
AC-107–CuSn	1	120	107	430	206,0	22,0
AC-120–CuSn	1	120	120	420	201,3	24,2
AC-150–CuSn	1	120	150	420	201,3	30,2
Catenary wires according to DIN 48 201-1:1981 or DIN 48 201-2:1981						
DIN 48 201–50–E-Cu	7	113	50	397	250,3	12,5
DIN 48 201–70–E-Cu	19	105	70	377	237,7	16,6
DIN 48 201–95–E-Cu	19	105	95	394	248,4	23,6
DIN 48 201–120–E-Cu	19	105	120	391	246,5	29,6
DIN 48 201–150–E-Cu	37	105	150	392	247,2	37,1
DIN 48 201–50–Bz II	7	113	50	572	360,6	18,0
DIN 48 201–70–Bz II	19	105	70	552	348,0	24,4
DIN 48 201–95–Bz II	19	105	95	576	363,2	34,5
DIN 48 201–120–Bz II	19	105	120	563	355,0	42,61
DIN 48 201–150–Bz II	37	105	150	576	363,2	54,5

n number of strands E elasticity module
A nominal cross-section σ_{min} minimum tensile strength
σ_{per} permissible tensile stress F_{per} permissible tensile force
k_{wind} 1,00, $v_{wind} \leq 100$ km/h
k_{eff} 0,80 for contact wires

of less than 40 % of the rated tensile stress. However, where the permissible tensile force can also be determined according to EN 50 341-2-4.

The assumptions for wind and ice loads follow from the same climatic conditions as the static design of overhead transmission lines. The factors used in (4.24) can be taken from Table 4.2. Singular loads like section insulators within the overhead contact line are not present in this case.

Example 4.8:   What is the permissible tensile force for a railway traction supply line with the conductor 243-AL1 according to EN 50 182 at 20,3 m/s wind velocity?
From EN 50 182 $\sigma_{min} = 180$ N/mm^2 and from Table 4.2 $k_{temp} = 1,0$; $k_{wear} = 1,0$; $k_{wind} = 0,95$ at $v_{wind} = 73,1$ km/h < 100 km/h, fixed terminated line; $k_{ice} = 0,95$; $k_{eff} = 1,0$ for fixed termination; $k_{clamp} = 1,0$; $k_{joint} = 1,0$ and $k_{load} = 1,0$ without additional load result.
From Equation (4.24) the permissible tensile stress is found to be 106,8 N/mm^2. Considering the nominal cross-section of the conductor with 242,5 mm^2 (Table 4.16) the permissible tensile force for this conductor is 25,96 kN.

Table 4.3 contains the permissible tensile forces according to (4.24) for commonly used contact wires and catenary wires.

4.2.2   Change of tensile stresses and forces

4.2.2.1   Single conductors

The conductors used in overhead contact line installations vary in length due to thermal and elastic elongation. A conductor of the length L expands linearly by

$$\Delta L_T = \alpha \cdot L \cdot (\vartheta_x - \vartheta_0) \quad , \tag{4.25}$$

when its temperature rises from ϑ_0 to ϑ_x, where α is the linear *coefficient of thermal expansion*. Tables 11.2, 11.5 and 11.6 contain the thermal expansion coefficients of materials, which are used in contact line installations.

Example 4.9:   What variations of length occur if the temperature of the conductor is varied from $\vartheta_0 = -30$ °C to $\vartheta_x = +70$ °C? With the values of α according to Tables 11.2 and 11.5 it is obtained:
- $\Delta L_T = 12,0 \cdot 10^{-6}K^{-1} \cdot 15$ m $\cdot [70 - (-30)]$ K $= 0,018$ m for 15 m long conductor rails made of soft iron
- $\Delta L_T = 23,1 \cdot 10^{-6}K^{-1} \cdot 18$ m $\cdot [70 - (-30)]$ K $= 0,042$ m for 18 m long aluminum/steel compound third rails
- $\Delta L_T = 17 \cdot 10^{-6}K^{-1} \cdot 700$ m $\cdot [70 - (-30)]$ K $= 1,190$ m for 700 m long contact wire sections made of Cu-ETP

In the case of soft iron or aluminum compound conductor rails it is, therefore, necessary to arrange overlaps or expansion joints every 90 to 120 m and to use supports which do not prevent elongation with temperature.

Longitudinal forces result in elastic elongation of conductors. If the longitudinal forces do not exceed the elastic limits, the conductors resume their original length after releasing the load. The length variation of a conductor due to elastic behaviour can be described with the modulus of elasticity E according to Tables 11.2, 11.5 and 11.6, also designated as E-modulus. If the tensile force varies from H_0 to H_x, the change in length will be

$$\Delta L_E = (H_x - H_0) \cdot L / (E \cdot A) \quad . \tag{4.26}$$

Table 4.4 contains sags and tensile stresses in a not automatically tensioned conductor 243-AL1 with respect to temperature.

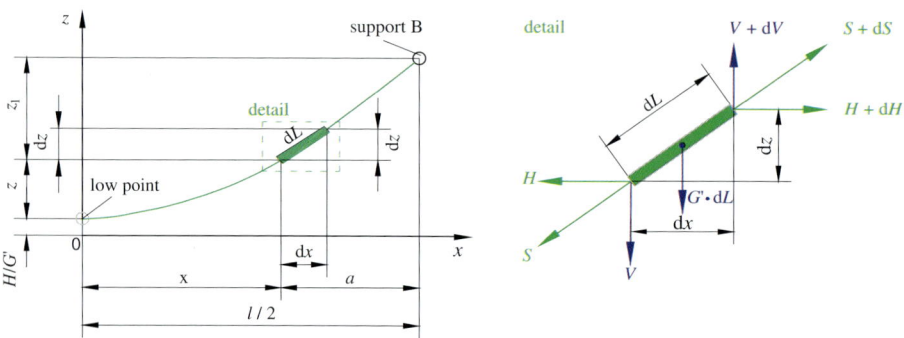

Figure 4.13: Sag of a strung conductor.

Table 4.4: Tensile stresses σ and sags f in a not automatically tensioned conductor 243-AL1 depending on the temperature for different spans in m.

Span	65		67		69		71		73		75	
ϑ	f	σ	f	σ	f	σ	f	σ	f	σ	f	σ
°C	m	N/mm²	m	N/mm²	m	N/mm²	m	N/mm²	m	N/mm²	m	N/mm²
−30	1,05	13,7	1,14	13,3	1,24	13,0	1,31	12,8	1,44	12,6	1,54	12,4
−20	1,19	12,0	1,28	11,9	1,38	11,7	1,48	11,6	1,58	11,4	1,68	11,3
−10	1,32	10,8	1,41	10,8	1,51	10,7	1,61	10,6	1,71	10,6	1,81	10,5
− 5[1]	1,48	20,0	1,57	20,0	1,67	20,0	1,77	20,0	1,87	20,0	1,97	20,0
0	1,44	9,9	1,54	9,9	1,63	9,9	1,73	9,9	1,83	9,9	1,94	9,9
10	1,56	9,2	1,65	9,2	1,75	9,2	1,85	9,3	1,95	9,3	2,05	9,3
20	1,67	8,6	1,76	8,6	1,86	8,7	1,96	8,7	2,06	8,8	2,16	8,8
30	1,77	8,1	1,86	8,2	1,96	8,2	2,06	8,3	2,16	8,3	2,27	8,4
40	1,87	7,7	1,96	7,8	2,06	7,8	2,16	7,9	2,27	8,0	2,37	8,0
50	1,96	7,3	2,06	7,4	2,16	7,5	2,26	7,6	2,36	7,6	2,47	7,7
60	2,05	7,0	2,15	7,1	2,25	7,2	2,35	7,3	2,46	7,4	2,56	7,4

[1] with ice load $G'_{ice} = 5 + 0,1 \cdot 20,3 = 7,0\,\text{N/m}$ maximum tensile stress limited to 20 N/mm²

Example 4.10: What is the change in length of a 780 m long contact wire AC-100 – Cu, if its tensile force is raised by 10 kN ?

$$\Delta L_E = 750 \cdot 10000/(120 \cdot 10^3 \cdot 100) = 0,625\,\text{m}.$$

A parallel feeder line 243-AL1 with a 242,5 mm² cross-section changes its length by $(75 \cdot 10000)/(60 \cdot 10^3 \cdot 242,5) = 0,052$ m within a 75 m span, when the load is increased by 10 kN.

Tensioning devices absorb the length variations of contact and catenary wires automatically and keep the tensile force approximately constant.

The conductors are prone to external loads, such as ice loads. These additional loads change the axial tensile force of the not automatically tensioned conductors.

Figure 4.13 shows the relations between the length dL of a conductor element and its horizontal and vertical components dx and dy. From Figure 4.13 it can be concluded that

$$dL^2 = dx^2 + dz^2 \text{ and } (dL/dx)^2 = 1 + (dz/dx)^2 \text{ and further } dL = \sqrt{1 + (dz/dx)^2} \cdot dx \quad .$$

From Figure 4.13 it can be further concluded that $dz/dx = G' \cdot x/H$. Since $(G'x/H)^2 \ll 1$

applies to conductors in overhead contact line systems, because of

$$\sqrt{1+\xi} \approx 1 + \xi/2 \text{ it can be written } dL = \left[1 + (G'x/H)^2/2\right] dx \quad .$$

By integration over the conductor span the length of the sagged conductor is obtained

$$L = l + (G'/H)^2 \cdot (l^3/24) \tag{4.27}$$

or, expressed relative to the *maximum sag* f_{max} according to (4.66),

$$L = l + (8/3) \cdot (f_{max}^2/l) \quad . \tag{4.28}$$

Example 4.11: The length of a catenary wire of the contact line Sicat S1.0 tensioned at 10 kN, weight of 15 N/m according to Table 11.7 and a 75 m span between supports is

$$L = 75 + (15/10\,000)^2 \cdot (75^3/24) = 75 + 0{,}040 = 75{,}040 \,\text{m}.$$

The catenary wire is 40 mm or 0,5 ‰ longer than the distance between the supports.

If the weight per unit length changes from state 0 to state x, e. g. because of ice loads, the tensile force in a wire without automatic tension control will also change accordingly. This variation is described by (4.27) as

$$L_x - L_0 = \left[(G'_x/H_x)^2 - (G'_0/H_0)^2\right] \cdot (l^3/24) \quad , \tag{4.29}$$

where L_x and L_0 are inserted according to the Equation (4.27). The *variation in conductor length* of a fixed terminated conductor in transition from condition 0 to condition x is equal to the thermal and elastic elongation. Therefore, it applies

$$L_x - L_0 = \Delta L_T + \Delta L_E$$

or, if the individual terms are expressed in full

$$\left[(G'_x/H_x)^2 - (G'_0/H_0)^2\right] \cdot (l^3/24) = \alpha L (\vartheta_x - \vartheta_0) + \left[(H_x - H_0)/(EA)\right] L \quad .$$

Since $L \approx l$ applies to overhead contact lines, it is obtained

$$\left[(G'_x/H_x)^2 - (G'_0/H_0)^2\right] \cdot (l^2/24) = \alpha (\vartheta_x - \vartheta_0) + (H_x - H_0)/(EA) \quad . \tag{4.30}$$

Equation (4.30) is the *equation of state change* that can be used to determine the force in wires and conductors fixed at both ends. For practical applications this equation can be solved either for ϑ_x or for H_x. If solved to H_x an equation of a third order is obtained. Its analytic solution is impractical so Equation (4.30) is solved iteratively with computer programs. As an example, the values specified in Table 4.6 apply to a traction power supply line 243-AL1, which is strung at overhead contact line poles as a parallel or bypass feeder. The values calculated according to Equation (4.30) and listed in Table 4.6 are used in Clause 4.2.2.3 to determine the sags of power traction lines as a basis for verification of the clearances.

The application of Equation (4.30) will be explained with the example of rigidly terminated conductors i. e. not automatically tensioned contact wires of a tramway contact line. By solving Equation (4.30) to the required tensile force H_x it is obtained

$$H_x^2 \left[H_x - H_0 + E A G_0'^2 \, l^2 / (24 H_0^2) + E A \, \alpha (\vartheta_x - \vartheta_0) \right] = E A G_x'^2 \, l^2 / 24 \quad . \tag{4.31}$$

For a *tensioning section* with n spans of different lengths l_i, the length l can be substituted by the *equivalent span length* l_{id} as explained in [4.6]:

$$l_{id} = \sqrt{\sum_{i=1}^{n} l_i^3 \Big/ \sum_{i=1}^{n} l_i} \quad . \tag{4.32}$$

Example 4.12: Determine the tensile force at $-30\,°C$ for the rigidly terminated contact wire AC-100 – Cu with an equivalent span of 30 m. The contact wire was installed with a tensile force of 8 kN at $+10\,°C$. According to the Tables 11.3 and 11.5 it applies: $A = 100\,\text{mm}^2$; $E = 120\,\text{kN/mm}^2$; $\alpha = 17 \cdot 10^{-6}\,\text{K}^{-1}$; $G_0' = G_x' = 0{,}862\,\text{kg/m} \cdot 9{,}81\,\text{m/s}^2 = 8{,}46\,\text{N/m}$, whereby the specific weight of the conductor is derived from the minimum mass according to Table 11.3. By inserting these data into (4.31) it is obtained

$$H_x^2 \cdot \left[H_x - 8\,000 + \frac{120 \cdot 10^3 \cdot 100 \cdot 8{,}46^2 \cdot 30^2}{24 \cdot 8\,000^2} + 120 \cdot 10^3 \cdot 100 \cdot 17 \cdot 10^{-6} (-30 - 10) \right]$$

$$= 120 \cdot 10^3 \cdot 100 \cdot 8{,}46^2 \cdot 45^2 / 24 \quad ,$$

$$H_x^2 \left(H_x - 15\,657\,\text{N} \right) = 32{,}21 \cdot 10^9\,\text{N}^3 \quad .$$

The equation for H_x is solved by iteration:

$H_x = 16\,000\,\text{N}$ results in $H_x^2 \left(H_x - 15\,657\,\text{N} \right) = 87{,}87 \cdot 10^9\,\text{N}^3$
$H_x = 15\,800\,\text{N}$ results in $H_x^2 \left(H_x - 15\,657\,\text{N} \right) = 35{,}76 \cdot 10^9\,\text{N}^3$
$H_x = 15\,786\,\text{N}$ results in $H_x^2 \left(H_x - 15\,657\,\text{N} \right) = 32{,}21 \cdot 10^9\,\text{N}^3$

The tensile force at $-30\,°C$ is, therefore, 15 786 N. The increase in temperature to $+70\,°C$ reduces the tensile force to 2 160 N. The corresponding sags are 0,060 m at $-30\,°C$ and 0,440 m at $+70\,°C$. The difference is 0,380 m.

The variation in tensile force can be reduced by inserting a spring. The length variation of the spring will be

$$\Delta L_F = (H_x - H_0) / c_F \quad , \tag{4.33}$$

where c_F is the *spring constant*. Since the change in length from the different factors are summed, it applies

$$L_x - L_0 = \Delta L_T + \Delta L_E + \Delta L_F \quad .$$

With the relations (4.16), (4.25), (4.29) and (4.33) the state equation for a tensioning section with a spring follows, which is similar to Equation (4.31)

$$H_x^2 \left[(H_x - H_0) \cdot \left(\frac{1 + (E \cdot A)}{c_F \cdot \sum_{i=1}^{n} l_i} \right) + \frac{E \cdot A \cdot G_0'^2 \cdot l_{id}^2}{24 \cdot H_0^2} + E \cdot A \cdot \alpha \cdot (\vartheta_x - \vartheta_0) \right]$$

$$= E \cdot A \cdot G_x'^2 \cdot l_{id}^2 / 24 \quad , \tag{4.34}$$

where

H_0	tensile force at condition 0 in N
H_x	tensile force at condition x in N,
E	modulus of elasticity in N/mm^2,
A	cross-sectional area in mm^2,
G_0'	length related weight force at condition 0 in N/m,
G_x'	length related weight force at condition x in N/m,
l_{id}	equivalent span in m,
l_i	span length of the span i in m,
c_F	spring constant,
α	linear thermal expansion coefficient in K^{-1},
ϑ_0	temperature of the conductor at condition $0 = 10\,°C$ in the Example 4.12 and
ϑ_x	temperature of the conductor at condition $x = x\,°C$.

Example 4.13: A spring with a spring constant $c_F = 10\,\mathrm{kN/m}$ is installed into a tensioning section comprising ten spans as described in example 4.12 having an equivalent span $l_{\mathrm{id}} = 30\,\mathrm{m}$ and the complete tensioning section length $\sum_{i=1}^{n} l_i = 300\,\mathrm{m}$. It applies $G_0' = G_x'$ since no ice load needs to be considered. The tensile force H_x at $-30\,°C$ results from Equation (4.34)

$$H_x^2 \cdot \left[(H_x - 8\,000) \cdot \frac{1 + (120 \cdot 10^3 \cdot 100)}{10 \cdot 10^3 \cdot 300} + \frac{120 \cdot 10^3 \cdot 100 \cdot 8,46^2 \cdot 30^2}{24 \cdot 8\,000^2} \right.$$

$$\left. + 120 \cdot 10^3 \cdot 100 \cdot 17 \cdot 10^{-6} \cdot (-30 - 10) \right] = 120 \cdot 10^3 \cdot 100 \cdot 8,46^2 \cdot 30^2 / 24 \quad ,$$

$$H_x^2 \cdot [(H_x - 8\,000) \cdot 4,0 + 503,2 - 8\,160,0] = 32,21 \cdot 10^9\,\mathrm{N}^3$$

to be $H_x = 9\,994,8\,\mathrm{N}$.

At $+70\,°C$ the tensile force will be $5\,121,2\,\mathrm{N}$ and, therefore, the variation in tensile force is $4\,844\,\mathrm{N}$. The installation of a spring reduces the variation of the tensile force from $13\,626\,\mathrm{N}$ of example 4.12 to $4\,844\,\mathrm{N}$ for example 4.13 and the sags at temperature of $70\,°C$ considerably from $0,380\,\mathrm{m}$ to $0,091\,\mathrm{m}$.

The examples illustrate the large tensile force and sag variations with rigidly installed contact wires. The difference will increase with decreasing equivalent span lengths. The length of the total tensile section and the number of spans has no effect. Tensioning devices are ideal for avoiding the large variations in tensile force as described in examples 4.12 and 4.13.

In the case of semi-compensated contact lines with rigidly terminated catenary wires, the stresses and sags of the catenary wire can be determined using Equation (4.31). According to [4.7] the following relation for the catenary wire sags in the case of the given condition x applies:

$$f_{\mathrm{CA\,max}} = \frac{l^2}{8} \cdot \frac{G_{\mathrm{OCL}x}' + G_{\mathrm{OCL}}' \cdot H_{\mathrm{CW}} / H_{\mathrm{CA0}}}{H_{\mathrm{CW}} + H_{\mathrm{CA0}} - \alpha_{\mathrm{CA}} \cdot E_{\mathrm{CA}} \cdot A_{\mathrm{CA}} \cdot (\vartheta_x - \vartheta_0)} \quad \tag{4.35}$$

In the case of ice load and $-5\,°C$, $G_{\mathrm{OCL}x}' = G_{\mathrm{OCL}}' + G_{\mathrm{ice}}'$ applies and in all other cases $G_{\mathrm{OCL}x}' = G_{\mathrm{OCL0}}' = G_{\mathrm{CA}}' + G_{\mathrm{CW}}'$. The catenary wire tensile force H_{CA0} needs to be determined by numerically solving the given equation, if the catenary wire sag $f_{\mathrm{CA\,max}}$ is limited.

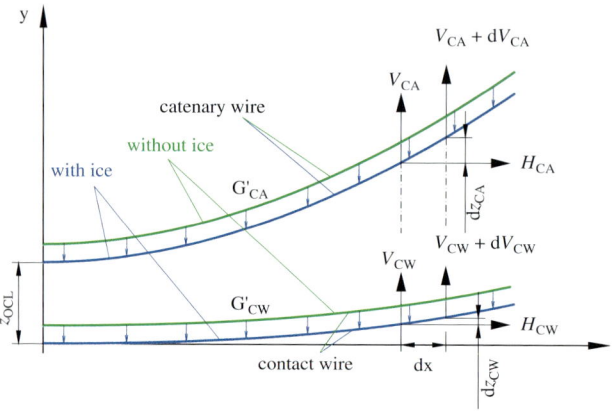

Figure 4.14: Sag in a contact line.

4.2.2.2 Catenary suspended contact lines

In this context, the term catenary type contact line is used to describe a system, where the contact wire is suspended from a catenary wire by droppers in such a way that specified distances between the catenary and the contact wire are maintained by the length of the individual droppers [4.7]. The following relationships can be deduced from Figure 4.14:

$$dV_{CA} = G'_{CA} \cdot dL_{CA} \quad , \qquad\qquad dV_{CW} = G'_{CW} \cdot dL_{CW} \quad ,$$
$$dz_{CA}/dx = V_{CA}/H_{CA} \quad \text{and} \qquad dz_{CW}/dx = V_{CW}/H_{CW} \quad .$$

The subscripts CA and CW indicate the values for the catenary wire CA and the contact wire CW, respectively. By differentiation, the following expressions are obtained

$$H_{CA}\, d^2 z_{CA}/dx^2 \ = G'_{CA} \cdot dL_{CA}/dx \quad \text{and}$$
$$H_{CW}\, d^2 z_{CW}/dx^2 \ = G'_{CW} \cdot dL_{CW}/dx \quad . \tag{4.36}$$

It follows from (4.36), by inserting the total mass per unit length of the contact line $G'_{OCL} = G'_{CA} + G'_{CW}$ and the application of the approximation $dL_{CA} \approx dL_{CW} \approx dx$

$$H_{CA}\, d^2 z_{CA}/dx^2 + H_{CW}\, d^2 z_{CW}/dx^2 = G'_{OCL} \quad . \tag{4.37}$$

Integrating this equation twice with respect to x produces the relationship

$$H_{CA}\, z_{CA} + H_{CW}\, z_{CW} = G'_{OCL}\, x^2/2 + C_1 x + C_2 \quad .$$

The integration constants are derived from the boundary conditions shown in Figure 4.14:

$$z'_{CA}(0) = 0 \quad \text{and} \quad z'_{CW}(0) = 0 \longrightarrow C_1 = 0 \, ,$$
$$z_{CA}(0) = z_{OCL} \quad \text{and} \quad z_{CW}(0) = 0 \longrightarrow C_2 = H_{CA}\, z_{OCL},$$

which yields the equation

$$H_{CA}\, z_{CA} + H_{CW}\, z_{CW} = G'_{OCL}\, x^2/2 + H_{CA}\, z_{OCL} \quad . \tag{4.38}$$

Without ice, the contact wire has no sag. Therefore, $z_{CW} = 0$ and

$$z_{CA} = (G'_{OCL}/H_{CA}) \cdot (x^2/2) + z_{OCL} \quad . \tag{4.39}$$

With ice, the sag of the catenary wire will be

$$z_{CA\,ice} = z_{CA} + z_{CW\,ice} \quad .$$

Using Equation (4.39) yields

$$z_{CA\,ice} = \left(G'_{OCL}/H_{CA} \right) \cdot \left(x^2/2 \right) + z_{CL} + z_{CW\,ice} \quad . \tag{4.40}$$

The length related load with ice is

$$G'_{OCL\,ice} = G'_{OCL} + G'_{ice} \quad .$$

From (4.38) it is obtained

$$H_{CA}\,z_{CA\,ice} + H_{CW}\,z_{CW\,ice} = \left(G'_{OCL} + G'_{ice} \right) x^2/2 + H_{CA}\,z_{OCL} \quad .$$

Inserting $z_{CW\,ice}$ according to (4.40) yields

$$\begin{aligned}
H_{CA}\,z_{CA\,ice} + H_{CW}\,y_{CA\,ice} - G'_{OCL}\,(H_{CW}/H_{CA})\,(x^2/2) - H_{CW}\,z_{OCL} \\
= \left(G'_{OCL} + G'_{ice} \right)(x^2/2) + H_{CA}\,z_{OCL}
\end{aligned} \tag{4.41}$$

and

$$\begin{aligned}
z_{CA\,ice}\,(H_{CW} + H_{CA}) \;=\;& G'_{OCL}\left[(H_{CW} + H_{CA})/H_{CA} \right](x^2/2) \\
& + \left(G'_{ice} \right)(x^2/2) + (H_{CW} + H_{CA})\,z_{OCL} \quad .
\end{aligned} \tag{4.42}$$

The sags of the catenary and contact wire under ice load $z_{CA\,ice}$ and $z_{CW\,ice}$ can then be determined individually from (4.41) and (4.42). Thus:

$$\begin{aligned}
z_{CA\,ice} \;=\;& (x^2/2)\left[(G'_{OCL}/H_{CA}) + G'_{ice}/(H_{CA} + H_{CW}) \right] + z_{OCL} \quad\text{and} \tag{4.43} \\
z_{CW\,ice} \;=\;& (x^2/2)\left[(G'_{OCL}/H_{CA}) + G'_{ice}/(H_{CA} + H_{CW}) \right] \\
& + z_{OCL} - G'_{OCL}/H_{CA}(x^2/2) - z_{OCL} \qquad\text{and} \\
z_{CW\,ice} \;=\;& (x^2/2)\left[G'_{ice}/(H_{CA} + H_{CW}) \right] \quad . \tag{4.44}
\end{aligned}$$

If the sag is expressed in relation to the supports and the variable x replaced by the variable a, equations similar to Equation (4.64) result:

$$\begin{aligned}
z_{CA\,ice} \;=\;& [a(l-a)/2] \cdot \left[G'_{OCL}/H_{CA} + G'_{ice}/(H_{CA} + H_{CW}) \right] + z_{OCL} \quad\text{and} \\
z_{CW\,ice} \;=\;& [a(l-a)/2] \cdot \left[G'_{ice}/(H_{CA} + H_{CW}) \right] \quad .
\end{aligned}$$

The *maximum contact wire sag* with ice load is obtained at the position $a = l/2$:

$$f_{CW\,ice\,max} = \left[G'_{ice}/(H_{CA} + H_{CW}) \right](l^2/8) \quad . \tag{4.45}$$

Example 4.14: Determine the contact wire sag between the supports and between the droppers of an automatically tensioned contact line, Sicat S1.0 for
- a national application with a 1 950 mm long pantograph and
- for an interoperable application with a 1 600 mm long pantograph

due to ice load in zone 1. Given are

ice load zone 1	$G'_{ice} = 0,5 \cdot (5 + 0,1 \cdot d)$ according to Clause 2.4.3,
diameter of the catenary wire Bz II 50 mm^2	$d_{CA} = 9,0$ mm according to DIN 48 201,
diameter of the contact wire AC-100–Cu	$d_{CW} = 12,0$ mm according to EN 50 149,
span length	$l_{max} = 71,4$ m (for the 1 600 mm long pantograph
	$l_{max} = 80,0$ m (for the 1 950 mm long pantograph)
length of pantograph	1 600 mm, 1 950 mm,
tensile force within the contact wire	$H_{CW} = 12$ kN,
tensile force of the catenary wire	$H_{CA} = 10$ kN.

Without ice load the contact wire of this overhead contact line type has no sag. The ice load for the overhead contact line type Sicat S1.0 at the catenary wire is

$$G'_{ice\,CA} = 0,5 \cdot (5 + 0,1 \cdot d) = 0,5 \cdot (5 + 0,1 \cdot 9,0) = 3,0\,\text{N/m}$$

and at the contact wire

$$G'_{ice\,CW} = 0,5 \cdot (5 + 0,1 \cdot d) = 0,5 \cdot (5 + 0,1 \cdot 12,0) = 3,1\,\text{N/m} \quad .$$

The contact wire height reduction between the supports is calculated according to Equation (4.45) for an 80,0 m span length operated with a 1 950 mm long pan head

$$f_{CWS-S\,ice\,max} = [6,1/(12\,000 + 10\,000)] \cdot 80,0^2/8 = 0,222\,\text{m}.$$

In the case of an 80,0 m span the maximum distance between droppers will be 10,20 m (Figure 4.15). An additional contact wire sag between the droppers (D) results

$$f_{CW\,D-D\,ice\,max} = [3,1/(12\,000)] \cdot 10,20^2/8 = 0,003\,4\,\text{m}.$$

The total sag of the contact wire with ice and snow, therefore amounts to 0,225 m.
On interoperable lines for operation with a 1 600 mm long pantograph the useable contact wire lateral position is less and, therefore, the span length shorter. The calculation of the contact wire sag with ice and snow for the maximum span length is

$$f_{CWS-S\,ice\,max} = [6,1/(12\,000 + 10\,000)] \cdot 71,4^2/8 = 0,177\,\text{m}.$$

An additional contact wire sag is formed between the droppers (D) spaced at 8,99 m distance according to Figure 4.15

$$f_{CW\,D-D\,ice\,max} = [3,1/(12\,000)] \cdot 8,99^2/8 = 0,003\,\text{m}.$$

The total sag of the contact wire with ice is, therefore, 0,180 m.

Example 4.15: The wheel tensioning devices used for overhead contact line type Re 330 are equipped with sleeve bearings with at least 97 % efficiency. For the catenary wire 120-Bz II with 19 strands (Table 11.6) the tensile force is 21 kN. The modulus of elasticity E is 105 kN/mm^2.
Determine the variation in tensile force in front of the tensioning devices of the catenary wire? What temperature difference corresponds to the variation in tensile force? What contact wire sag will occur due to the tensioning devices with 97 % efficiency?
For a catenary wire strung at 21 000 N a loss of 3 % contact wire force corresponds to 1 050 N variation in tensile force. According to (4.30) this corresponds to a temperature change

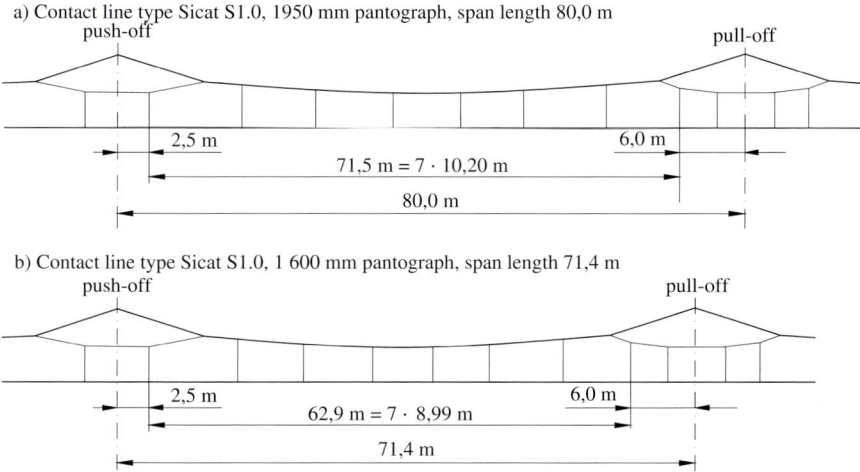

a) Contact line type Sicat S1.0, 1950 mm pantograph, span length 80,0 m

b) Contact line type Sicat S1.0, 1 600 mm pantograph, span length 71,4 m

Figure 4.15: Distances between droppers from Table 12.29 for Example 4.14.

- from $-26,2\,°C$ to $-30\,°C$ with $3{,}8\,K$ or
- from $80\,°C$ to $76{,}2\,°C$ with $3{,}8\,K$.

The temperature difference of only $3{,}8\,K$ is low. The *maximum contact wire sag* results in the middle of the span according to (4.45) at $a = l/2$. There, it is obtained

$$f_{CW\,max} = \left[G'_{OCL}/(H_{CA0}+H_{CW0})\right](l^2/8) - \left[G'_{OCL}/(H_{CAx}+H_{CWx})\right]\cdot(l^2/8)$$
$$= \left[23/(21\,000+27\,000)\right]\cdot(65^2/8) - \left[23/(21\,630+27\,810)\right]\cdot(65^2/8) = 0{,}012\,m \quad .$$

The sag of $0{,}012\,m$ is acceptable.

4.2.2.3 Traction power lines

The calculation of tensile stress and conductor sag e. g. of traction power lines at suspension insulators is based on the *nominal stress* at the annual mean temperature of $+10\,°C$ without wind load, designated as σ_{10}. For the selected conductor type, the nominal stress can be determined from Table 9/DE.2, column 5 of EN 50 341-2-4. For the conductor tensile force and the resulting sags, the following *load cases* need to be considered:

- $-20\,°C$ without ice load
- $-5\,°C$ and ice load corresponding to ice load zones 1 to 4 according to Clause 2.4.3 by
  - ice load zone E1: $G'_{ice} = 5+0{,}1\cdot d$
  - ice load zone E2: $G'_{ice} = 10+0{,}2\cdot d$
  - ice load zone E3: $G'_{ice} = 15+0{,}3\cdot d$
  - ice load zone E4: $G'_{ice} = 20+0{,}4\cdot d$
- $+5\,°C$ and maximum wind load according to the relevant wind load zone according to Clause 2.5.2
- $+40\,°C$ with swung conductors in wind with a three year return period
- maximum permissible conductor temperature $80\,°C$ for conductors made of copper, AL1 and AL1/ST1A

The verification of the clearances at simultaneous actions of extreme wind and ice load is not required according to EN 50 341-2-4:2016.

The horizontal component of the conductor tensile force is to be calculated for each loading case and may not exceed the limits according to EN 50 341-2-4, Clause 9/DE.4.

According to EN 50 341-2-4, Clause 9/DE.4, the tensile stress at

- $-20\,°C$ or $-30\,°C$ without ice load,
- $-5\,°C$ with ice load,
- $-5\,°C$ with ice load and wind load,
- $+5\,°C$ with wind load.

multiplied by the partial factor $\gamma_C = 1,35$ may not exceed the permissible stress at maximum loads. The maximum tensile stress is obtained to be 95 % of the rated tensile stress of the conductor divided by the material partial factor $\gamma_M = 1,25$. Therefore, the permissible tensile stress is

$$\sigma_{C\,per} = 0,95 \cdot \sigma_{min} / (1,35 \cdot 1,25) \quad . \tag{4.46}$$

For a conductor 243-AL1 σ_{min} equates to $180\,N/mm^2$ and $\sigma_{C\,per} = 101 \approx 100\,N/mm^2$. For approximately the same suspension heights and fixed suspensions, the tensile force H_{10} will be the same in the spans l_1, l_2 and l_3:

$$H_{10} = A \cdot \sigma_0 = H_1 = H_2 = H_3 \quad . \tag{4.47}$$

The condition of a common tensile stress applies to the *equivalent span* a_{id}

$$l_{id} = \sqrt{\sum l_i^3 / \sum l_i} \quad , \tag{4.48}$$

where a_i means the individual span lengths of the considered line section.

The sag of the equivalent span length is

$$f_{id} = \frac{G' \cdot l_{id}^2}{8 \cdot H_{id}} \quad . \tag{4.49}$$

and for the individual spans it is obtained

$$f_i = f_{id} \cdot \frac{l_i^2}{l_{id}^2} = \frac{G' \cdot l_{id}^2}{8 \cdot H_{id}} \cdot \frac{l_i^2}{l_{id}^2} = \frac{G' \cdot l_i^2}{8 \cdot H_{id}} \quad . \tag{4.50}$$

When calculating the uplift at the support B (Figure 4.17) the equivalent span a_{id} according to Equation (4.48) needs to be calculated as the first step. In the second step the sag f_{id} of the equivalent span can be determined with the Equation (4.49). The sags f_1 and f_2 for different span lengths can be calculated using Equation (4.50).

Example 4.16:   A traction power line must be designed for the conductor 243-AL1 according to EN 50 182, Table F.17. According to EN 50 341-2-4, Table 9/DE.2, the permissible nominal stress is $30\,N/mm^2$. The conductors will be strung at $20\,N/mm^2$ at $10\,°C$.

Determine the sags of the conductors in spans $a_1 = 60\,m$, $a_2 = 80\,m$ and $a_3 = 65\,m$ of the tensioning section at $-30\,°C$ and at $80\,°C$?

The equivalent span a_{id} follows from (4.48) as

$$l_{id} = \sqrt{\frac{60^3 + 80^3 + 65^3}{60 + 80 + 65}} = 69,9\,m \quad .$$

Table 4.5: Conductor sags in cm for Example 4.16.

temperature	l_{id}	Span lengths in m		
		60	65	80
$-30\,^\circ$C	31	23	27	41
$80\,^\circ$C	180	133	156	237

The state change Equation (4.30) is at $-30\,^\circ$C with $E = 55 \cdot 10^3$ N/mm^2 and $\alpha = 23 \cdot 10^{-6}$:

$$H_{-30}^2 \left[H_{-30} - 4850 + 55\,000 \cdot 242{,}5 \cdot 6{,}58^2 \cdot 69{,}9^2 / (24 \cdot 4850^2) \right.$$
$$\left. + 55\,000 \cdot 242{,}5 \cdot 23 \cdot 10^{-6} \cdot (-30 - 10) \right] = 55\,000 \cdot 242{,}5 \cdot 6{,}58^2 \cdot 69{,}9^2 / 24$$

and, therefore,

$$H_{-30}^2 \left[H_{-30} - 4850 + 4997{,}9 - 12\,270{,}5 \right] = 117{,}6 \cdot 10^9$$

$$H_{-30}^2 \left[H_{-30} - 12\,122{,}6 \right] = 117{,}6 \cdot 10^9 \quad .$$

By iteration it is obtained

$H_{-30} = 12\,610{,}0\,$N   yields to $H_{-30}^2 \, (H_{-30} - 12\,122{,}6\,\text{N}) = \;77{,}5 \cdot 10^9 \, \text{N}^3$,
$H_{-30} = 12\,850{,}0\,$N   yields to $H_{-30}^2 \, (H_{-30} - 12\,122{,}6\,\text{N}) = 120{,}1 \cdot 10^9 \, \text{N}^3$,
$H_{-30} = 12\,836{,}1\,$N   yields to $H_{-30}^2 \, (H_{-30} - 12\,122{,}6\,\text{N}) = 117{,}6 \cdot 10^9 \, \text{N}^3$.

Table 4.5 contains the conductor sags in cm for this example.

The result is $\sigma_{-30} = 12\,836{,}1\,\text{N} / 242{,}5\,\text{mm}^2 = 52{,}9\,\text{N/mm}^2$. The state change Equation (4.30) for 80 $^\circ$C is:

$$H_{80}^2 \left[H_{80} - 4850 + 55\,000 \cdot 242{,}5 \cdot 6{,}58^2 \cdot 69{,}9^2 / (24 \cdot 4850^2) \right.$$
$$\left. + 55\,000 \cdot 242{,}5 \cdot 23 \cdot 10^{-6} \cdot (80 - 10) \right] = 55\,000 \cdot 242{,}5 \cdot 6{,}58^2 \cdot 69{,}9^2 / 24$$
$$H_{80}^2 \left[H_{80} + 21\,621{,}3 \right] = 117{,}6 \cdot 10^9 \, N^3$$

$H_{80} = 2\,220{,}6\,$N   yields $H_{80}^2 \, (H_{80} + 21\,621{,}3) = 117{,}6 \cdot 10^9 \, \text{N}^3$

The stress σ_{80} will, therefore, be $\sigma_{80} = 2\,220{,}6\,\text{N} / 242{,}5\,\text{mm}^2 = 9{,}16\,\text{N/mm}^2$. From Table 4.6 it is obtained by iteration for $l_{\text{id}} = 69{,}9\,$m from 70 m to 65 m $\sigma_{-30} = 52{,}79\,\text{N/mm}^2$ and $\sigma_{80} = 9{,}16\,\text{N/mm}^2$. The sags in the individual spans are listed in Table 4.5. The stress at $-30\,^\circ$C is less than 100 N/mm^2, which would be permissible as the maximum stress according to EN 50 341-2-4:2016.

4.3 Sags and tensile stresses

4.3.1 Conductors with equal suspension heights

This section deals with the *sag* of contact wires and conductors prone to a given linear load and subjected to a constant tensile force S.

The horizontal component of S is designated as H. Since the bending stiffness of the conductors used for overhead contact lines is relatively small, consideration of the tensile forces acting in the conductors is sufficient. The conductors are supported pivoted at the suspension, therefore longitudinal movements are not possible. The balance of forces at a conductor element of the length ΔL results according to Figure 4.13 for the horizontal forces as

$$H + \mathrm{d}H - H = 0, \qquad\qquad \longrightarrow \mathrm{d}H = 0 \qquad\qquad (4.51)$$

Table 4.6: Sags and tensile stresses and of the conductor 243-AL1 according to EN 50 182:2001.

longitudinal span	temperature in °C						
m	-30	-20	0	10	20	40	80
30	5	6	10	15	24	42	69
	67,14	54,66	30,49	20,00	12,81	7,24	4,45
35	6	8	14	21	31	51	81
	65,91	53,50	29,81	20,00	13,51	8,13	5,12
40	8	10	19	27	38	61	94
	64,50	52,19	29,09	20,00	14,13	8,96	5,77
45	11	14	24	34	47	71	107
	62,92	50,73	28,35	20,00	14,68	9,72	6,40
50	14	17	31	42	56	81	121
	61,17	49,14	27,63	20,00	15,16	10,41	7,00
55	17	22	38	51	66	93	136
	59,28	47,45	26,93	20,00	15,59	11,06	7,58
60	21	27	47	61	76	105	150
	57,26	45,67	26,26	20,00	15,97	11,65	8,13
65	26	33	56	72	88	118	166
	56,11	43,83	25,65	20,0	16,32	12,19	8,66
70	31	40	66	83	100	131	181
	52,87	41,96	25,10	20,00	16,62	12,70	9,17
75	38	48	78	95	113	145	198
	50,57	40,11	24,60	20,00	16,89	13,16	9,65
80	45	57	90	109	127	160	215
	48,23,0	38,30	24,15	20,00	17,14	13,59	10,11
85	53	67	103	123	141	175	233
	45,89	36,57	23,76	20,00	17,35	13,98	10,55

data in red: sags f in cm; data in blue: tensile stress σ in N/mm^2
length-related load G' 6,58 N/m
selected every day stress G_{10} 20 N/mm^2 at 10 °C,
coefficient of linear expansion α $23{,}0 \cdot 10^{-6} \cdot 1/\mathrm{K}$
conductor cross-section A 242,5 mm^2
diameter d 20,3 mm
modulus of elasticity E 55 kN/mm^2 (see Table 11.6)
permissible tensile stress σ_per 182 N/mm^2 according to EN 50 182

and after integration $H = $ constant results. The balance of the vertical forces is

$$V + \mathrm{d}V - V - G' \cdot \mathrm{d}L = 0, \quad \longrightarrow \mathrm{d}V = G' \cdot \mathrm{d}L \quad . \tag{4.52}$$

With

$$\mathrm{d}L = \mathrm{d}x \sqrt{1 + \left(\frac{\mathrm{d}z}{\mathrm{d}x}\right)^2} \tag{4.53}$$

and the relation developed from Figure 4.13

$$\frac{\mathrm{d}z}{\mathrm{d}x} = \frac{V}{H} \tag{4.54}$$

it is obtained by inserting of (4.53) into (4.52)

$$dV = G' \cdot dx \cdot \sqrt{1 + \left(\frac{dz}{dx}\right)^2} \qquad (4.55)$$

and after transformation

$$\frac{dV}{dx} = G' \cdot \sqrt{1 + \left(\frac{dz}{dx}\right)^2} \quad . \qquad (4.56)$$

The derivation of (4.54) to x yields

$$\frac{1}{H} \cdot \frac{dV}{dx} = \frac{d^2z}{dx^2} = \frac{1}{4} \cdot \frac{dV}{dx} \text{ and } \frac{dV}{dx} = H \cdot \frac{d^2z}{dx^2} \quad . \qquad (4.57)$$

By inserting of (4.57) into (4.56), the differential equation for the conductor sagging line is obtained

$$\frac{d^2z}{dx^2} = \frac{G'}{H} \cdot \sqrt{1 + \left(\frac{dz}{dx}\right)^2} \quad . \qquad (4.58)$$

The solution of (4.58) is known generally as the *catenary curve*

$$z = \frac{H}{G'} \cdot \cosh \frac{G' \cdot x}{H} \quad , \qquad (4.59)$$

which is explained in detail in [4.6]. The solution (4.59) can be verified by inserting dz/dx and d^2z/dx^2 into Equation (4.58). The derivation of (4.59) yields

$$\frac{dz}{dx} = \sinh\left(\frac{G' \cdot x}{H}\right) \text{ and } \frac{d^2z}{dx^2} = \frac{G'}{H} \cdot \cosh\left(\frac{G' \cdot x}{H}\right) \qquad (4.60)$$

Inserting in (4.58) demonstrates the correctness of the solution

$$\frac{G'}{H} \cdot \cosh\left(\frac{G' \cdot x}{H}\right) = \frac{G'}{H} \cdot \sqrt{1 + \sinh^2\left(\frac{G' \cdot x}{H}\right)} = \frac{G'}{H} \cosh\left(\frac{G' \cdot x}{H}\right) \quad . \qquad (4.61)$$

For $x = 0$ the Equation (4.59) yields $z = H/G'$ (Figure 4.13).
For overhead contact lines the length of the conductor L is only 0,5 ‰ to 1,0 ‰ longer than the spacing of the supports l. Therefore, the approximation $dL \approx dx$ can be used, whereby $\sqrt{1 + (dz/dx)^2} = 1$ follows from (4.51) and, therefore, (4.58) yields

$$\frac{d^2z}{dx^2} = \frac{G'}{H} \quad . \qquad (4.62)$$

The solution of the differential Equation (4.62) for the *conductor sag z* is

$$z = \frac{G'}{H} \cdot \left[\left(\frac{x^2}{2}\right) + 1\right] \quad . \qquad (4.63)$$

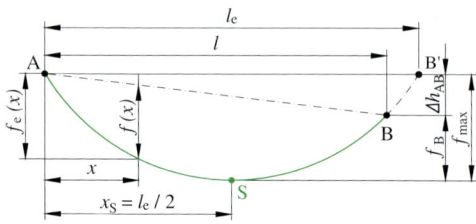

Figure 4.16: Sag in a line with supports at different heights.

The sag related to the support point B is assumed to be z_1 (Figure 4.13). At the distance a from the support it applies

$$z_1(a) = f_a = \frac{G'}{2 \cdot H} \cdot a\,(l-a) \quad . \tag{4.64}$$

The sag z_1 as a function of the distance x from the midpoint of the sag is obtained to be

$$z_1(x) = \frac{G'}{2 \cdot H} \cdot \left(\frac{l^2}{4} - x^2 \right) \quad . \tag{4.65}$$

The maximum sag is obtained for $a = l/2$ (Equation (4.64)) and for $x = 0$ (Equation (4.65)), respectively, and will be

$$z_{max} = f_{max} = \frac{G' \cdot l^2}{8 \cdot H} \cdot \quad . \tag{4.66}$$

The sag z_1 at a distance a from the support can be expressed by the maximum sag f_{max} as

$$z_1 = f_a = \frac{4 \cdot f_{max} \cdot a \cdot (l-a)}{l^2} = \frac{G' \cdot a \cdot (l-a)}{2 \cdot H} \quad . \tag{4.67}$$

Example 4.17: The maximum sag f_{max} of a contact wire type AC-100–Cu subject to a tensile force of $10\,$kN at $l = 40\,$m will be

$$f_{max} = \frac{8{,}46\,\text{N/m} \cdot 40^2\,\text{m}^2}{8 \cdot 10000\,\text{N}} = 0{,}169\,\text{m} \quad .$$

The same contact wire would have a sag of approximately $0{,}015\,$m between two droppers in a catenary type contact line with a dropper spacing of $12\,$m.

4.3.2 Conductors with different suspension heights

In a span of length l with a difference in height Δh_{AB} (Figure 4.16) the conductor curve is determined by Equations (4.66) and (4.67):

$$z = \frac{G' \cdot l^2}{2 \cdot H} \left(\frac{x}{l} - 1 \right) \frac{x}{l} - \frac{\Delta h_{AB}}{l} \cdot x \quad . \tag{4.68}$$

The position x_S of the low point S of the conductor curve can be determined from

$$\frac{dz}{dx} = \frac{G' \cdot l^2}{2 \cdot H} \left(\frac{2x_S}{l^2} - \frac{1}{l} \right) - \frac{\Delta h_{AB}}{l} = 0 \quad \text{to} \quad x_S = \frac{l}{2} + \frac{H}{G'} \cdot \frac{\Delta h_{AB}}{l} \quad . \tag{4.69}$$

The *complementary span* l_e will then be determined from

$$l_e = 2 \cdot x_S = l + \frac{2 \cdot H \cdot \Delta h_{AB}}{G' \cdot l} \tag{4.70}$$

with:

H conductor tensile force in N
Δh_{AB} difference in height of conductor fixings at A and B in m
l span length in m
G' specific weight force of the conductors in N/m

The sag in the middle of the complementary span l_e at the low point S can be expressed as

$$f_{e\,\mathrm{max}} = \frac{G' \cdot l_e^2}{8 \cdot H} = \frac{G'}{8 \cdot H} \cdot \left(l + \frac{2 \cdot H \cdot \Delta h_{AB}}{G' \cdot l} \right)^2 \quad . \tag{4.71}$$

The sag curve of the complementary span is described by

$$f_e(x) = 4 \cdot f_{e\,\mathrm{max}} \cdot \left(1 - \frac{x}{l_e} \right) \cdot \frac{x}{l_e} \quad . \tag{4.72}$$

At point B with $x = l$

$$f = 4 \cdot \frac{G'}{8 \cdot H} \cdot l_e^2 \cdot \left(1 - \frac{l}{l_e} \right) \cdot \frac{l}{l_e} = \frac{G'}{2 \cdot H} \cdot (l_e - l) \cdot l = \frac{G'}{2 \cdot H} \cdot \left(l + \frac{2 \cdot H \cdot \Delta h_{AB}}{G' \cdot l} - l \right) \cdot l = \Delta h_{AB} \quad .$$

The sag in span AB (Figure 4.16) in relation to the line connecting points A and B is

$$f(x) = 4 \cdot f_{e\,\mathrm{max}} \cdot \left(1 - \frac{x}{l_e} \right) \cdot \frac{x}{l_e} - \frac{\Delta h_{AB}}{l} \cdot x \tag{4.73}$$

and from (4.71)

$$f_{e,\mathrm{max}} = \frac{G'}{8 \cdot H} \cdot \left(l + \frac{2 \cdot H \cdot \Delta h_{AB}}{G' \cdot l} \right)^2 \tag{4.74}$$

results as

$$f(x) = \frac{G'}{8 \cdot H} \cdot (l - x) \cdot x \tag{4.75}$$

as in the case of the span with no height differences. In this case, it was assumed that the difference in height Δh_{AB} is small compared to the span length l.

Example 4.18: The longitudinal profile of the line reveals that the parallel feeder line with the conductor 243-AL1 is suspended at pole B 2 m lower than at pole A. The span length l is 75 m. According to Table 11.6 the length-related weight force is $G' = 6,58$ N/m and the conductor cross-section $A = 242,5$ mm^2. Determine the sag of the line at $-30\,°\mathrm{C}$. At the lowest temperature, the tensile stress should not exceed 20 N/mm^2.

The maximum tensile force is obtained at $-30\,°\mathrm{C}$ with $H_{-30} = 20 \cdot 242,5 = 4850$ N. From Equation (4.74) the maximum sag of the complementary span l_e is

$$f_{e\,\mathrm{max}} = \frac{6,58}{8 \cdot 4850} \cdot \left(75 + \frac{2 \cdot 4850 \cdot 2}{6,58 \cdot 75} \right)^2 = 2,22\,\mathrm{m}$$

as viewed from the support with the greater height.

a) Meaning of symbols of sag curve for unequal height suspension points

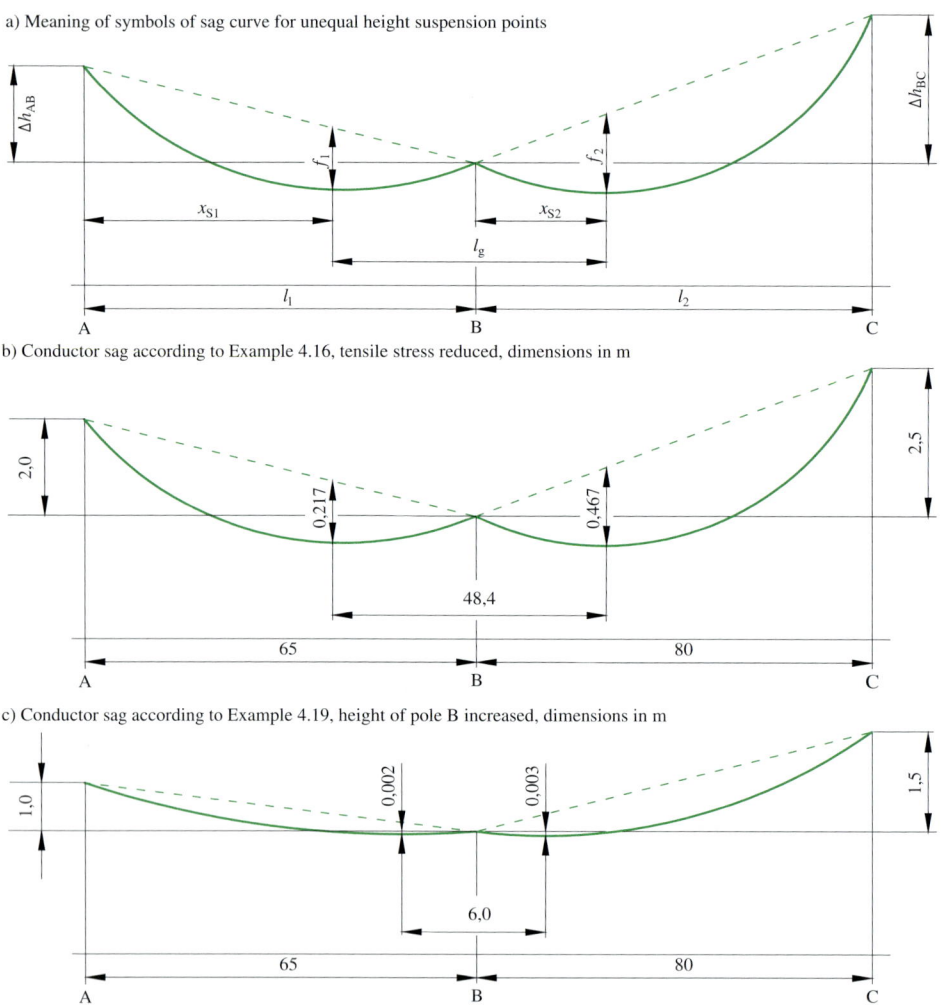

b) Conductor sag according to Example 4.16, tensile stress reduced, dimensions in m

c) Conductor sag according to Example 4.19, height of pole B increased, dimensions in m

Figure 4.17: Distorted scale diagram of sag curve with f as the conductor sag and unequal height suspension points.

4.3.3   Uplift force at the support

If the line is installed on poles with different support heights, a force directed upwards, described as an uplift force can be created at the lower support (Figure 4.17, support B). In this case, earth faults and damage to insulators and conductors can occur. Therefore, in the case of lines with varying suspension heights, a longitudinal profile needs to be established to review the consequences of height variations and to calculate the vertical forces at the supports.

In Figure 4.17 a), the force at support B must be greater than zero to avoid an uplift force. As a simplification, only the conductor load needs be considered. The weight of insulators and other loads can be disregarded since they result in downward directed loads. The vertical load at support B corresponds to the weight force of the conductors between the low points of the adjacent spans. The position of the low point in the left span of Figure 4.17 a) follows from

Equation (4.69) using the established symbols according to Figure 4.17 a)

$$x_{S1} = \frac{l_1}{2} + \frac{H \cdot \Delta h_{AB}}{G' \cdot l_1} \tag{4.76}$$

and for the span on the right side

$$x_{S2} = \frac{l_2}{2} + \frac{H \cdot \Delta h_{BC}}{G' \cdot l_2}, \tag{4.77}$$

where:

x_{S1}, x_{S2}	distances between poles A and B and the low points in spans 1 and 2, respectively, in m
l_1, l_2	span lengths 1 and 2 in m
$\Delta h_{AB}, \Delta h_{BC}$	difference in height within the spans AB and BC in m
H	conductor tension at the minimum temperature in N, $H = \sigma_{max} \cdot A$
σ_{max}	maximum conductor tensile force at the lowest temperature in N/mm^2
A	cross-section of the conductor in mm^2
G'	weight force related to the length of the conductor in N/m

The distance between the adjacent low points is called the weight span l_g (Figure 4.17 a))

$$l_g = (l_1 - x_{S1} + x_{S2}) \quad . \tag{4.78}$$

Inserting (4.77) in (4.78) and the relation $H = A \cdot \sigma_{avail}$ yields the weight span l_g

$$l_g = \frac{l_1 + l_2}{2} - \frac{A \cdot \sigma_{avail}}{G'} \left(\frac{\Delta h_{AB}}{l_1} + \frac{\Delta h_{BC}}{l_2} \right) \quad . \tag{4.79}$$

The vertical load V_B at the support B is then

$$V_B = l_g \cdot G' = G' \cdot \frac{l_1 + l_2}{2} - A \cdot \sigma_{avail} \cdot \left(\frac{\Delta h_{AB}}{l_1} + \frac{\Delta h_{BC}}{l_2} \right) \quad . \tag{4.80}$$

Since no uplift force may occur at support B, the terms V_B, according to Equation (4.80) and l_g according to Equation (4.78) need to be positive.

Example 4.19: From an overhead line longitudinal profile (Figure 4.17 a)) pole B is the lowest suspension of a parallel feeder line with the conductor 243-AL1. Given are the weight per unit length $G' = 6,58$ N/m, the nominal (everyday) stress 20 N/mm^2 at $10\,°C$, the available tensile stress $\sigma_{avail} = 52,9$ N/mm^2 at the lowest temperature of $-30\,°C$, the differences in heights $\Delta h_{AB} = 2,0$ m and $\Delta h_{BC} = 2,5$ m according to Figure 4.17. Check for uplift force at pole B.
Equation (4.79) yields the weight span at the pole B

$$l_g = \frac{65 + 80}{2} - \frac{242,5 \cdot 52,9}{6,58} \cdot \left(\frac{2,0}{65} + \frac{2,5}{80} \right) = -48,4\,\text{m} \quad .$$

From (4.80) it is calculated

$$V_B = 6,58 \cdot \frac{65 + 80}{2} - 242,5 \cdot 52,9 \cdot \left(\frac{2,0}{65} + \frac{2,5}{80} \right) = -318,5\,\text{N} \quad .$$

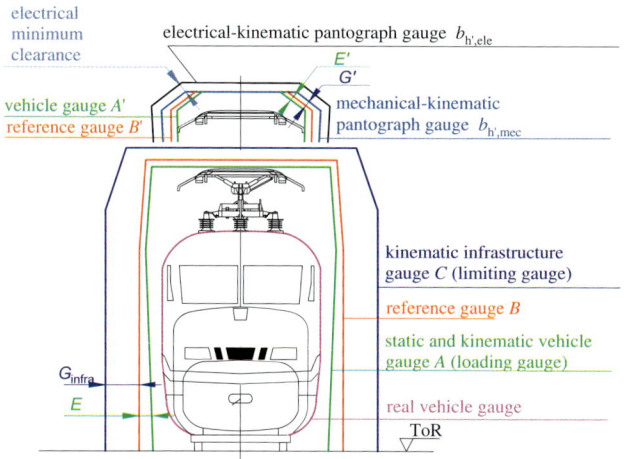

Figure 4.18: Rolling stock gauge, reference gauge and structure gauge.

At pole B, a force of $-318{,}5\,\text{N}$ acting upwards would occur, so the design needs to be revised. To avoid this the nominal stress is reduced to $10\,\text{N/mm}^2$. Then at $-30\,^\circ\text{C}$ the available tensile stress is $\sigma_{\text{avail}} = 15\,\text{N/mm}^2$. The vertical force at support B results from (4.80) to be

$$V_{\text{B}} = 6{,}58 \cdot \frac{65+80}{2} - 242{,}5 \cdot 15 \cdot \left(\frac{2{,}0}{65} + \frac{2{,}5}{80}\right) = 251.5\,\text{N} \quad .$$

By reducing the conductor stress to $\sigma_{\text{avail}} = 15\,\text{N/mm}^2$ a load of $+251{,}5\,\text{N}$ results at support B, eliminating uplift. Figure 4.17 b) shows the sag curve without uplift at support B.

In Example 4.19, the tensile force was reduced to avoid uplift at support B. Alternatively, the heights Δh_{AB} and Δh_{BC} can be reduced as shown in example 4.20.

Example 4.20:   To avoid uplift at pole B, the suspension height at pole B will be increased by $1{,}5\,\text{m}$. From (4.79) the weight span at $-30\,^\circ\text{C}$ results as

$$l_{\text{g}} = \frac{65+80}{2} - \frac{242{,}5 \cdot 52{,}9}{6{,}58} \cdot \left(\frac{1{,}0}{65} + \frac{1{,}5}{80}\right) = 6{,}0\,\text{m}$$

and from (4.80)

$$V_{\text{B}} = 6{,}58 \cdot \frac{65+80}{2} - 52{,}9 \cdot 242{,}5 \cdot \left(\frac{1{,}0}{65} + \frac{1{,}5}{80}\right) = 39\,\text{N}.$$

Therefore, at support B, uplift is eliminated as shown by the sag curve in Figure 4.17 c).

4.4   Rolling stock gauges and structure gauges

Rolling stock and structure gauges determine where pantographs may operate and contact wire may be positioned. A description of these gauges follows:
- *maximum vehicle gauge A*, also called *loading gauge*
- *reference gauge B* and
- *infrastructure gauge C*

The *static vehicle gauge A* (Figure 4.18) describes the vehicle envelope from which no component of the vehicle may protrude. The vehicle is considered to be at standstill on a straight track without cant. Vehicle movements during running are not considered in the vehicle gauge. Consideration of vehicle movements yields the *kinematic vehicle gauge*. The following vehicle movements and effects are captured under a factor called E:

- horizontal displacements due to transverse plays between the vehicle body and the wheel sets, e. g. axle bearings play as well as the position of the wheel sets within gauge channel
- variations in the vehicle height due to wear, e. g. wear of the wheel rims
- vertical peaks resulting from spring travels, caused by static and dynamic loading of the vehicle
- vertical displacements resulting from the position of the vehicle within convex as well as concave transitions between track gradients
- quasi-static lateral inclination due to *cant deficiency* or cant excesses of 0,050 m for vehicles without a pantograph on the roof and of 0,066 m for vehicles with pantographs on the roof
- imbalances of the vehicle body exceeding 1°, resulting from manufacturing and adjustment tolerances of the vehicle and planned loadings

The combination of the vehicle gauge A with the effects summarised under E result in the reference gauge B as shown in Figure 4.18: The effects E need to be considered by the manufacturer of the vehicle for limitation calculations [4.8], for which gauge B serves as reference. This reference forms the interface between vehicle and infrastructure.

By adding the projection S, the reference gauge B is exceeded when the vehicle runs on a curved track or on a track with a gauge of more than 1,435 m, arriving at the reference gauge for the static space requirement of the vehicles. By further adding the *quasi-static lateral movement* qs'_{ia} of the vehicle inside the curve (i) or outside the curve (a), the gauge for the kinematic space demand of the vehicle is obtained. By also adding the *random lateral displacements* $\sum j$, resulting from the irregular position of track, the *infrastructure gauge C* is obtained as shown in Figure 4.18.

With the displacements E the vehicle can reach the reference gauge B and with the additional displacements G_{infra} according to

$$G_{\mathrm{infra}} = S + qs'_{ia} + \sum j \tag{4.81}$$

the *kinematic infrastructure gauge C*, which is also designated as *the limiting gauge C* for fixed installations. The additional displacements G_{infra} (Figure 4.18) need to be considered by the infrastructure manager and operator.

While the maximum vehicle gauge A describes the contour from which no part of the vehicle may protrude, the kinematic infrastructure gauge C forms a contour line into which no component of the infrastructure may intrude.

4.5   Mechanical and electrical gauge

The lateral movements of the pantograph determine its operational range (Figure 4.18), which is called the *mechanical-kinematic gauge* of the pantograph in [4.9]. This gauge can be calculated according to the stipulations of EN 15 273 and TSI ENE. On lines operated by one type of pantograph only that type determines the mechanical gauge, which is also called the structure gauge of the pantograph. If on a line, vehicles with pantographs of differing geometry operate, the mechanical-kinematic structure line results from the longest pan head.

Only the pantograph, the contact wire and the steady arm are permitted within the mechanical-kinematic gauge. Only live components of the overhead contact line system may approach the mechanical-kinematic gauge. They must not cross the mechanical-kinematic gauge into the profile of the pantograph.

If the minimum electrical clearance is added to the mechanical gauge, then the *electric-kinematic gauge* of the pantograph is obtained (Figure 4.18). Pantographs with insulated horns enable the application of smaller electrical gauges equivalent to the length of the horns. The infrastructure manager needs to consider the mechanical and electrical effects that lead to the mechanical and electrical gauge.

During energy transfer, the pantograph continuously touches the contact wire and its height is variable. Therefore, the height of the pantograph gauges are also variable between the lowest and highest contact wire heights being considered, including the uplift caused by the pantograph, the height tolerances and other effects as discussed in the following clauses. The lateral movements of the pantograph, also called sway, depend on the height. Consequently, the mechanical-kinematic pantograph gauge depends on the contact wire height but also on the track radius.

4.6   Working length of the pantograph

The geometry of the pantograph of the Deutschen Reichsbahn with the useable range for contact with the contact wire, also designated as working length l_A, was the result of agreements between German provincial railways [4.10, 4.11]. Because of the hazard of dewirement, the contact wire must not leave the working range of the pantograph (Figure 4.19). The working length l_A can be separated into two parts, being the range of the collector strip l_S, where the lateral movements of the contact wire occur and two times the range of sway D, which the contact wire can reach when the pantograph sways (Figure 4.19).

To determine the working lengths of interoperable pantographs, the stipulations of TSI LOC & PAS [4.12] are decisive. These stipulations refer to EN 50 367, where the working lengths are specified in Figure A.6 for the 1 600 mm long pan head with 1 200 mm [4.13] and in Figure A.7 for the 1 950 mm long pan head with 1 550 mm. It is also specified that the working length must consist of conducting material.

Deviating from these specifications, the DB Directive 997.0101 [4.14] specifies the working range of the 1 950 mm long pan head as 1 450 mm. The contact range defining the range of the pan head in which the contact wire may move under consideration of all effects, is defined by the Directive 997.0101 [4.14] as half of the total length of the pan head minus a 150 mm length for the horn. The contact range, therefore, equates to 1 650 mm.

If the working range of the pan head consists of a conductive material, the working range and the contact range are identical. For the 1 950 mm long pan head, half of the working range is 775 mm according to EN 50 367, corresponding to a 200 mm long horn. This value should be used for calculations since this length of the horn also corresponds to the 1 650 mm long pan head.

If vehicle sway and strong wind occur simultaneously, it is permissible, that on a short line section, the contact wire reaches the limit of the working range, also called *contact wire*

a) Designations at the pantograph pan head

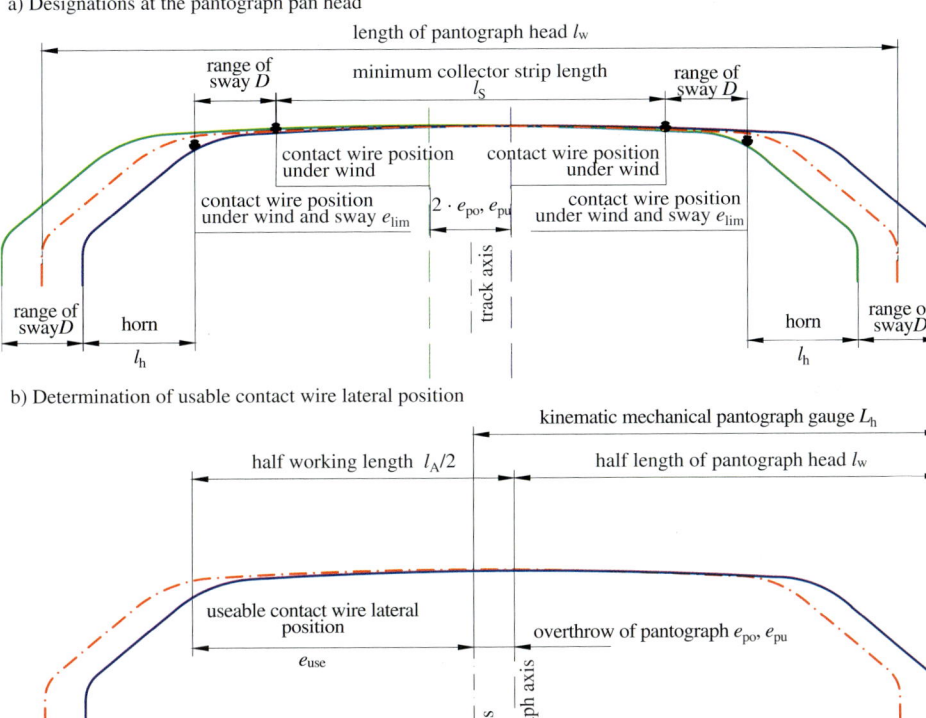

b) Determination of usable contact wire lateral position

Figure 4.19: Designations at the pantograph pan head.

lateral limit position e_{lim}. In curves, the sway of the pantograph may be more than the 225 mm or 200 mm provided, respectively. Exceeding the sway range D leads to a reduction of the useable contact wire lateral position. The *useable contact wire lateral position* e_{use} results from

$$e_{use} = l_A/2 - D \quad . \tag{4.82}$$

While e_{use} is determined by the geometry of the pantograph and its swaying movement, the *maximum contact wire lateral position* e_{max} depends on the stagger of the contact wire at the supports and its displacement relative to the canted track centreline under the action of wind. The maximum contact wire lateral position needs to be less than the useable contact wire lateral position $e_{max} \leq e_{use}$.

The maximum contact wire lateral position e_{max} follows from the lateral movement of the contact wire and provision for tolerances

$$e_{max} = e_{1,2} + P \quad , \tag{4.83}$$

where
$e_{1,2}$ contact wire lateral position due to contact wire stringing and wind action in m,
P provisional value for tolerances in m.

4.7 Provisions for the contact wire lateral position

Consideration of *provisional values* for determining the *maximum contact wire lateral position* is necessary because of

P_T temperature-caused variations to stagger due to rotation of the cantilever,
P_S tolerance of the contact wire lateral position at the supports and
P_M bending of poles under wind action.

The geometric mean of these components results in the provisional value P

$$P = \sqrt{P_T{}^2 + P_S{}^2 + P_M{}^2} \quad . \tag{4.84}$$

Example 4.21: Determine the contact wire lateral displacement at the last cantilever before the tensioning device where the cantilevers are fixed to poles made of HEB profiles.

Assuming that the cantilevers reach their mean position at 20 °C the temperature range to be considered will be 20 K between the extreme values 0 °C and 20 °C. To calculate the maximum contact wire lateral displacements at the cantilevers at a distance of 700 m from midpoint, the components of the provisional values are:

P_T = 0,008 m for 20 K (see example 4.5)
P_S = 0,030 m [4.15] and
P_M = 0,025 m for poles made of HEB profiles [4.16].

From (4.84), $P = 0,040$ m. When using concrete poles with $P_M = 0,000$ m, $P = 0,031$ m would result from (4.84) .

The provision used at German Railway since 1931 has been 0,030 m [4.17, 4.18]. Long-term experience confirms this value. The provisional value with its components considers the combination of tolerances and will be recommended in this book.

4.8 Contact wire lateral position of conventional lines

In 1939, Deutsche Reichsbahn harmonized the working length l_A for the 1 950 mm long pantograph pan head with the issue of drawing Ezs 837 [4.18]. The contact wire must not exceed half of the working length $l_A/2 = 750$ mm viewed from the middle of the pantograph. The *useable contact wire lateral position*, already designated as e at that time, was stipulated as 550 mm on straight lines. An addition s_E, as a provision against dewirement, was stipulated and defined as $s_E = l_A/2 - e$ by 200 mm in straight lines. This value corresponds to the swaying range D. The considered length of the horn l_h is obtained by

$$l_h = l_S/2 - l_A/2 = 975 - 750 = 225 \text{ mm.} \tag{4.85}$$

The useable contact wire lateral position e (Figure 4.20) specified in Ezs 837 related to a contact wire height of 6 250 mm [4.18] was applied, with small changes from 1967, until 2013 at German Railway DB (Figure 4.20). Figures 4.21 a) to c) show the changes to pan head geometry. The length of the collector strips was specified between 1 000 mm and 1 300 mm from 1931 to 1971. UIC 608-1 edition 1971 [4.19] specified the *minimum length of the collector strips* as 1 030 mm, adopted by German Railway DB in their Directive 997.0101 in 1995 [4.14]. The minimum length of collector strip should be two times the *useable lateral position* as shown in Figure 4.21 d) . While the envelope of the pantograph pan head

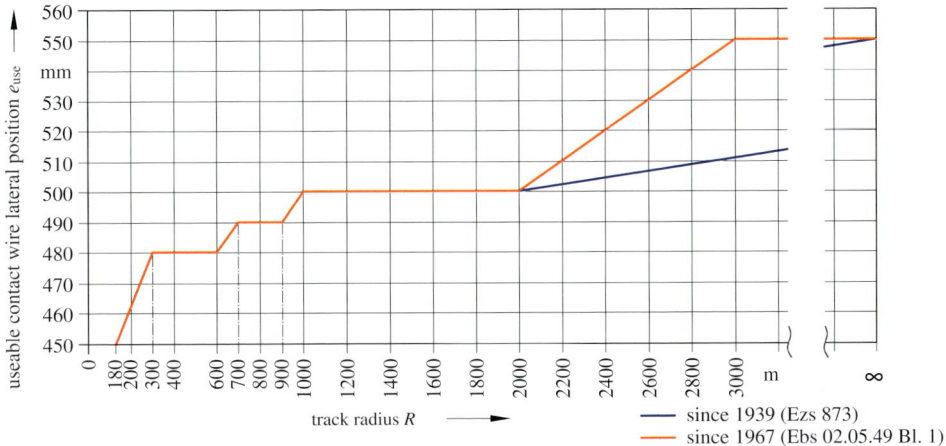

Figure 4.20: Useable contact wire lateral position e_{use} depending on the radius R at DB .

established in 1931 was not changed, the length of the horn varied from 225 mm to 150 mm [4.20]. EN 50 367:2013 and TSI LOC & PAS [4.12] specify the realistic length of the horn as 200 mm.

4.9 Contact wire lateral position for interoperable lines

4.9.1 Types of pantographs on interoperable lines

For interoperable lines listed on the Trans-European Transport Network (TEN-T) and the national infrastructure register [4.21], the contact line needs to be designed according to the requirements of TSI ENE [4.9] (see also Clause 2.4.1) for:
- new lines with speeds above 250 km/h for 1 600 mm long pan heads and for 1 950 mm long pan heads where required
- lines to be renovated or upgraded to speeds above 250 km/h for 1 600 mm and for 1 950 mm pan heads minimum, however for 1 600 mm long pan heads,
- all other lines for at least one of those pan heads, as far as practical

The overhead contact line should be designed for both pan heads but at least, the 1 600 mm long pan head. According to TSI LOC & PAS: 2014 [4.12] the 1 600 mm long and 1 950 mm long pan head are interoperable pan heads. The relevant pan head geometry determines the useable contact wire lateral position. For dual operation with both pan heads the shorter pan head determines the useable contact wire lateral position and the longer pan head, the space to be kept clear for the passage of pantographs.

4.9.2 Mechanical-kinematic gauge

The calculation of the useable contact wire lateral position e_{use} is described in EN 15 273-3 and in [4.9] and is based on the mechanical-kinematic pantograph gauge L_o at the upper verification height $h_o = 6{,}5$ m:

$$L'_o = l_W/2 + e_{po} + S' + (qs'_{i/a})_{max} + \sum j_o \tag{4.86}$$

a) Pantograph of German Railway since 1931 [4.18]

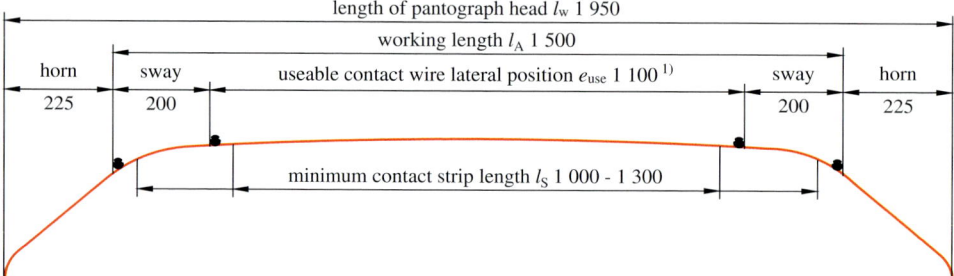

b) Pantograph of German Railway since 1956 [4.20]

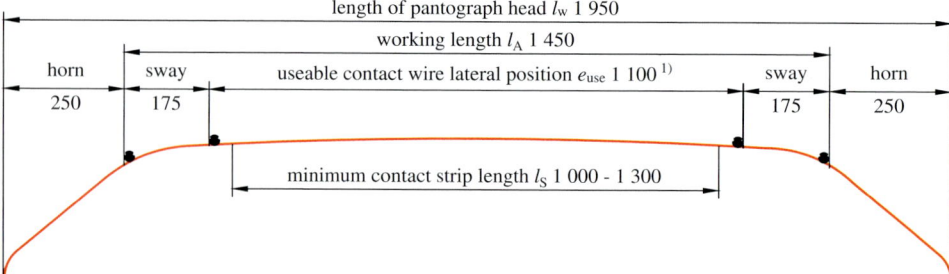

c) Pantograph of German Railway since 2001 [4.22] - [4.23]

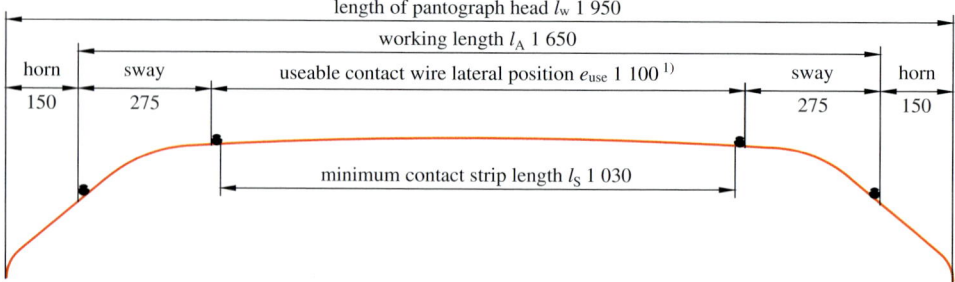

d) Interoperable pantograph of German Railway since 2014 [4.12]

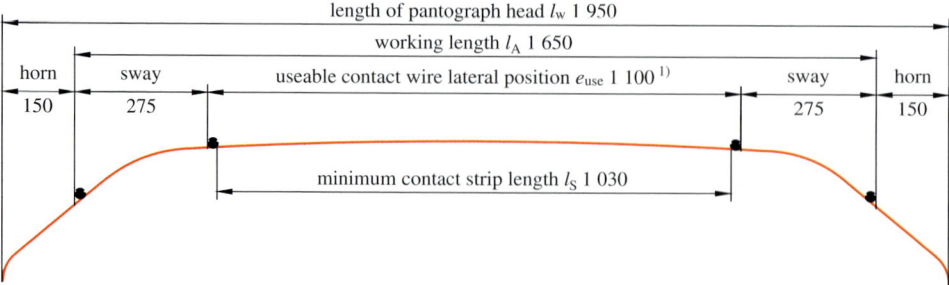

Figure 4.21: Dimensions of 1 950 mm long pantograph pan head used from 1931 to 2017.

and on the mechanical-kinematic gauge L_u at the lower verification height $h_u = 5,0\,\text{m}$

$$L'_u = l_W/2 + e_{pu} + S' + (qs'_{i/a})_{max} + \sum j_u. \tag{4.87}$$

For an intermediate height h' it is obtained

$$L'_h = L_u + \frac{(h' - h'_u)}{(h'_o - h'_u)} \cdot (L'_o - L'_u) \tag{4.88}$$

For the meaning of the variables in these equations and in the chapter are:

e_{use} useable contact wire lateral position

$l_A/2$ half working length of the pan head

being 0,600 m for the 1 600 mm long pan head and 0,775 m for the 1 950 mm pan head

$l_W/2$ half pan head length

being 0,800 m for the 1 600 mm long pan head and 0,975 m for the 1 950 mm pan head

L'_h kinematic limit at the verification height h'

L'_o kinematic limit at the upper verification height $h'_o = 6,5\,\text{m}$

L'_u kinematic limit at the lower verification height $h'_u = 5,0\,\text{m}$

e_{po} swaying movement of the pantograph with 0,170 m at $h'_o = 6,5\,\text{m}$

e_{pu} swaying movement of the pantograph with 0,110 m at $h'_u = 5,0\,\text{m}$

S' permissible additional projection

$qs'_{i,a}$ quasi-static movement into the inside (i) or outside (a) curve

$\sum j$ sum of the horizontal supplements to consider random phenomena:

     – asymmetry of the loading in m

     – transverse displacement of the track in m

     – tolerance of cant and oscillations due to track unevenness, the value of which are stipulated by the infrastructure operator

h'_u lower verification height 5,0 m

h'_o upper verification height 6,5 m

h' intermediate height for verification

4.9.3 Verification height

The permissible contact wire lateral position depends on the contact wire height. Therefore, the interaction of the pantograph with the contact line should be demonstrated for the maximum height which is planned for the contact line. This height is called the upper verification height. Verification may also be established at other heights. The verification height varies between 5,0 m and 6,5 m. The actual verification height h' comprises

$$h' = h_{CW} + f_u + f_{ws} + f_{wa} \quad , \tag{4.89}$$

where:

h_{CW} nominal contact wire height

f_u uplift of the contact wire due to passing pantographs at

f_{us} – support

f_{um} – midspan

f_{ws} inclination of the pantograph pan head

f_{wa} wear of the collector strips

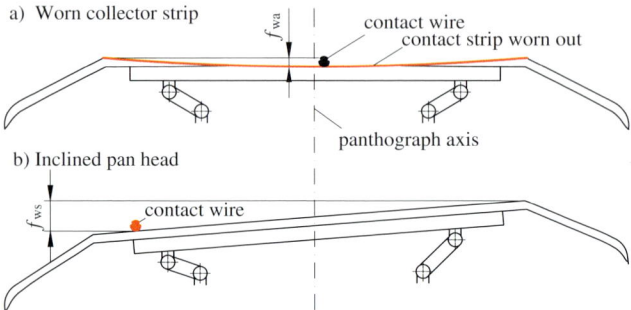

Figure 4.22: Wear of the collector strip and inclination of the pantograph pan head.

Table 4.7: Verification heights for DB's contact line types

Components of the verification height	Symbol	Verification height h' in m					
Type		Re200		Re250		Re330	
		support	midspan	support	midspan	support	midspan
nominal contact wire height	h_{CW}	5,500	5,500	5,300	5,300	5,300	5,300
installation tolerance	Δh_{CWT}	0,100	0,100	0,030	0,030	0,030	0,030
contact wire uplift	f_{us} / f_{um}	0,099	0,199	0,101	0,150	0,074	0,138
inclination of pan head	$f_{ws} + f_{wa}$	0,060	0,060	0,060	0,060	0,060	0,060
sum		5,759[1]	5,859[1]	5,491[2]	5,540[2]	5,464[3]	5,528[3]
rounded sum		**5,76**	**5,86**	**5,50**	**5,54**	**5,47**	**5,53**

[1] [4.24], [2] [4.25] with one pantograph in contact, [3] [4.26]

The locally measured contact wire height should be used as the nominal contact wire height h_{CW}. If no data are available, the planned nominal contact wire height, including the vertical installation tolerance, may be used. As an example, this tolerance is at maximum 100 mm for the contact line type Re200 [4.15]. The lift of the contact wire by passing pantographs has to be determined by measurements or simulations. The lift at supports depends on the overhead line design. For standard contact line type Re200 the simulated maximum lift at the support is approximately 100 mm [4.24]. The inclination of the pantograph head at a staggered contact wire position and the wear of the collector strip together may not exceed 60 mm according to EN 50 367 (Figure 4.22 and Table 4.7). The manufacturer of the pantograph and the rolling stock manager are responsible for meeting these requirements. Therefore, for the contact line Re200 with 5,50 m nominal contact wire height [4.14], the verification height is 5,76 m. After considering 100 mm vertical movement of the contact point at mid span [4.24] a verification height of 5,86 m results for the calculation of the useable contact wire lateral position of the contact line type Re200 [4.27].

In Table 4.7 a breakdown of the components of the verification heights is given for the contact lines Re200, Re250 and Re330 at the support and at midspan used to calculate the pantograph gauge and the useable contact wire lateral position e_{use} [4.27].

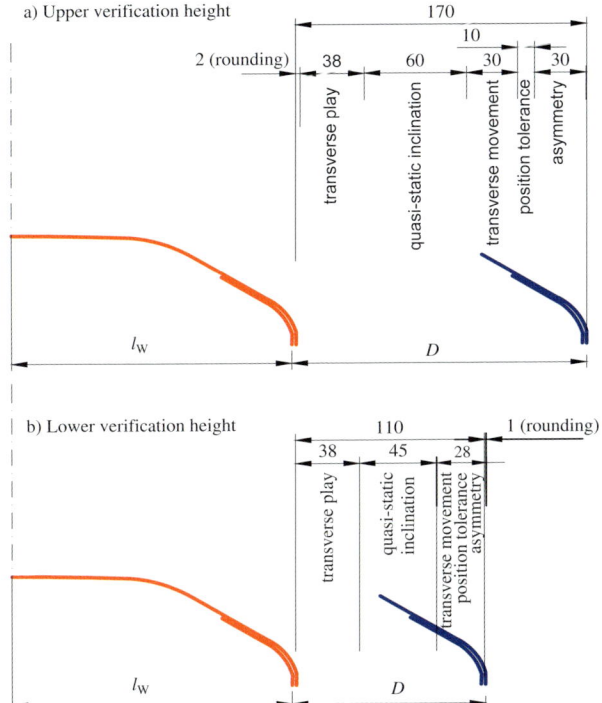

a) Upper verification height

b) Lower verification height

Figure 4.23: Parameters affecting swaying of pantographs.

4.9.4 Pantograph sway

The sway of pantographs (Figure 4.23) produces the limit position of the pantograph on the vehicle. Verification for vehicles is to be by calculation that the movement of the pantograph will not be more than the permitted overall values and the mechanical parts of the pantograph according to EN 15 273-1 are always situated within the reference line. The pantograph sway results from

$$e_{\mathrm{p}} = sp + z' + \vartheta \quad , \tag{4.90}$$

where:
e_{p} swaying movement of the pantograph
sp transverse clearance between wheel set and body of the vehicle
z' displacement due to quasi-static inclination to be considered on the vehicle side
ϑ displacement due to transverse swing, position tolerance and asymmetry of pantograph

The transverse play by bearing and bogie for the reference vehicle is defined in EN 15 273-1:2014 as an overall value

$$sp = q_{\mathrm{r}} + w_{\mathrm{r}} = 0{,}0375 \text{ m,} \tag{4.91}$$

where:
q_{r} transverse clearance between wheel set and bogy frame of the reference vehicle
w_{r} play between bogy and body of the vehicle

According to EN 15273-1:2014 the displacement from the quasi-static inclination of the vehicle results for an outside curve (a) by

$$y'_a = s'_0/L_S \cdot (u_f - u'_{f0})_{>0} \cdot (h - h_{c0}) \tag{4.92}$$

and for an inside curve (i) by

$$y'_i = s'_0/L \cdot (u - u'_0)_{>0} \cdot (h - h_{c0}) \quad , \tag{4.93}$$

where:
$y'_{a/i}$ displacement on the vehicle side to be considered due to quasi-static inclination
        at outside curve (a) or inside curve (i) of vehicles with pantographs on the roof
s'_0 flexibility for the gauge of the pantograph
L_S distance between the centre lines of the rails of a track
u cant
u'_0 reference cant for vehicles with pantographs on the roof
u_f cant deficiency
u'_{f0} reference cant deficiency for vehicles with pantographs on the roof
h_{c0} rolling centre height above top of rail

Reference cant and reference cant deficiency are 0,066 m each, whereby the same values for the quasi-static inclination of the vehicle follow from EN 15273-1 and EN 15273-2 for inside curve and outside curve. The displacement to be considered on the vehicle because of quasi-static inclinations is 0,045 m at the lower verification height and 0,060 m at the upper verification height. At the lower verification height, EN 15273-1 and UIC 505-5: 2010 [4.29] use the geometric sum of transverse swing, tolerance of position and asymmetry of the vehicle for calculating

$$\vartheta_u = \sqrt{\left(t_r \cdot \frac{h'_u - h_t}{h'_0 - h_t}\right)^2 + \tau_r^2 + [\Theta_r \cdot (h'_u - h_{c0})]^2} = 0,027 \text{ m} \quad , \tag{4.94}$$

whereas at the upper verifications height, the displacement is taken from UIC 505-5: 2010 [4.29] to be

$$\vartheta_0 = t_r + \tau_r + \Theta_r \cdot (h'_0 - h_{c0}) = 0,070 \text{ m} \quad , \tag{4.95}$$

where:
$\vartheta_{u/o}$ displacement due to transverse swing, tolerance of position and asymmetry of the
        pantograph at the lower and upper verification height
t_r lateral displacement of the pantograph under a load of 300 N
h_t installation height of the lower pantograph joint above top of rail
τ_r manufacture and installation tolerance of the pantograph
Θ_r asymmetry due to the suspension adjustment of the vehicle and
h_{c0} height of the rolling centre of the vehicle above top of rail

The data for the parameters of the reference vehicle used for the calculation are summarised in Table 4.8. According to (4.91) the sway of the reference pantograph (Figure 4.23) at the lower verification height considering the geometrical sum of the effects amounts to

$$e_{pu} = 0,038 + 0,045 + 0,027 = 0,110 \text{ m} \quad . \tag{4.96}$$

Table 4.8: Reference parameters according to [4.27].

Symbol	Meaning	Value	Source
L_S	distance between centre lines of the rails of a track	1,500 m	EN 15 273-3, Table G.1
e_{po}	swaying of pantograph at upper verification height	0,170 m	TSI ENE:2014, 4.2.10
e_{pu}	swaying of pantograph at lower verification height	0,110 m	TSI ENE:2014, 4.2.10
h'_{c0}	height of rolling centre above top of rail	0,500 m	TSI ENE:2014, 4.2.10
h'_o	height of upper verification point	6,500 m	TSI ENE:2014, 4.2.10
h_t	installation height of pantograph articulation above top of rail	4,005 m	UIC 505-1 [4.28], A.1.2
h'_u	height of lower verification point	5,000 m	TSI ENE:2014, 4.2.10
l_0	gauge of main tracks with agreed tolerances and wear effects	1,465 m	EBO, §5 [4.8]
$q_r + w_r$	transverse play between wheel set and bogie and play between bogie and body of reference vehicle	0,0375 m	EN 15 273-3, Table G.1
s_0	vehicle flexibility coefficient without pantographs	0,400	UIC 505-5 [4.29]
s'_0	vehicle flexibility coefficient with pantographs on roofs	0,225	TSI ENE:2014, 4.2.10
t_r	lateral displacement of pantograph and action of a force 300 N	0,030 m	EN 15 273-3, Table G.1
u_0	reference cant of the track for vehicles without pantographs on the roof	0,050 m	UIC 505-5 [4.29]
u'_0	reference cant of track for vehicles with pantographs on the roof	0,066 m	TSI ENE:2014, 4.2.10
u_{f0}	reference cant deficiency of track for vehicles without pantographs on the roof	0,050 m	UIC 505-5 [4.29]
u'_{f0}	reference cant deficiency of track for vehicles with pantographs on the roof	0,066 m	TSI ENE: 2014, 4.2.10
Θ_r	asymmetry due suspension adjustment the vehicle	0,005 rad	EN 15 273-3, Table G.1
τ_r	manufacturing and installation tolerance of pantographs	0,010 m	EN 15 273-3, Table G.1

Adding of effects at the upper verification height according to UIC 505-5:2010 in (4.95) it is obtained

$$e_{po} = 0,038 + 0,060 + 0,070 = 0,168 \text{ m} \quad . \tag{4.97}$$

After rounding, these values coincide with the reference parameter given in TSI ENE:2014 and EBO for e_{pu} and e_{po}.

However, the calculation of the sway of the pantograph according to UIC 505-5 [4.29] does not fully comply with the calculation methods according to EN 15 273-1: 2013, since in EN 15 273-1: 2013 a probabilistic mathematical approach is used for the effects of transverse swing, position tolerance and vehicle asymmetry. Consequently, it results $\vartheta_0 = 0,044$ m with a smaller value $e_{po} = 0,142$ m and longer span lengths.

4.9.5 Additional projection

An additional projection results from the position of the vehicle on a curved track (Figure 4.24) and on the track because of locally existing extensions of the track gauge because of tolerances and occurrence of wear (Figure 4.25).

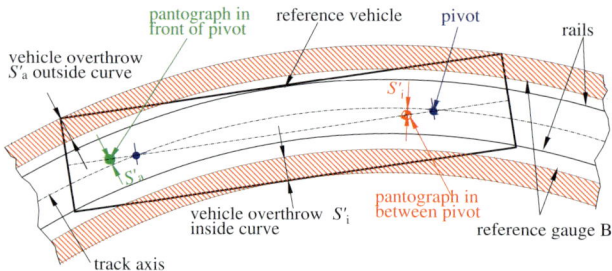

Figure 4.24: Position of the vehicle on a curved track.

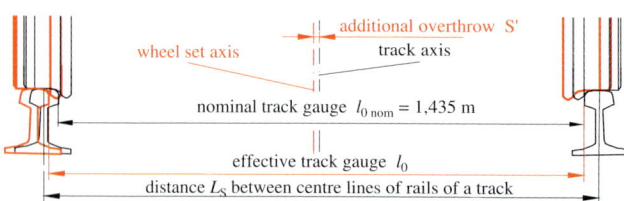

Figure 4.25: Position of the wheel axis on a track with expanded track gauge.

The permissible additional projection S' within the Equations (4.86) and (4.87) is

$$S' = 2,5/R + (l_0 - 1,435)/2 \quad . \tag{4.98}$$

where

R radius in m and

l_0 distance between the edges of the rails of a track in m (Table 4.9).

The permissible maximum values for a track gauge depend on the maintenance rules of the infrastructure manager and are not standardized within Europe. In Germany, according to EBO [4.8] the gauge may not exceed 1 465 mm at main line tracks and 1 470 mm at secondary tracks (Table 4.9).

The *additional projection* S' according to (4.98) is dependent on the deflection factor value of the radius $2,5/R$ and from the locally existing gauge extension $(l_0 - 1,435)/2$. According to EN 15 273-3, the additional projection is related to the beginning, the middle and the end of the vehicle. If the pantographs are arranged above the bogie axis, the value $2,5/R$ can be disregarded. Because of the operation of differing vehicles, the pantographs of which are not arranged exactly over the axis of the bogies, the most unfavourable case needs to be considered when calculating the additional projection. The effects of the radius and the gauge extension also need to be considered.

Table 4.9: Gauge according to [4.8].

Radius in m	Gauge l_0 in mm		
	minimum	standard	maximum
main tracks			
$\infty > R \geq 150$	1 430	1 435	1 465
secondary tracks			
$\infty > R \geq 150$	1 430	1 435	1 470
$150 > R \geq 125$		1 440	
$125 > R \geq 100$		1 445	

a) Vehicle running in outside
curve with cant deficiency

b) Vehicle at standstill in inside
curve with cant deficiency

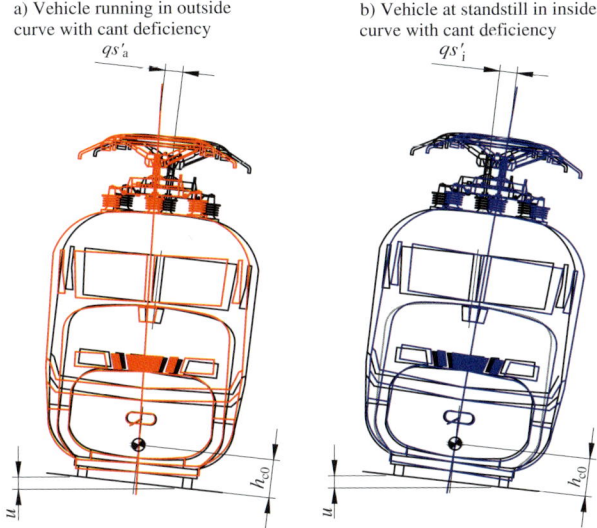

Figure 4.26: Displacement due to quasi-static inclination.

4.9.6 Quasi-static effect

The quasi-static effect results from the sway of the vehicle at standstill in a curve with a cant or in case of maximum speed in curves with a cant deficiency (Figure 4.26). This effect is considered a function of cant or cant deficiency and of values more than the reference data. The *displacements due to quasi-static effects* can be calculated for inside curve (i) by

$$qs'_i = s'_0 \cdot (u - u_0)_{>0} \cdot (h - h'_{c0})/L \tag{4.99}$$

and outside curve (a) by

$$qs'_a = s'_0 \cdot (u_f - u_{f0})_{>0} \cdot (h - h'_{c0})/L \quad, \tag{4.100}$$

where:

s'_0 flexibility taken into account for the pantograph gauge
L_S distance between the centre lines of the rails of a track
u cant
u'_0 reference value of cant of a track for vehicles with pantographs on the roof
u_f cant deficiency of the track
u'_{f0} reference value of the cant deficiency for vehicles with pantographs on the roof
h verification height and
h'_{c0} rolling centre height above top of rail

According to TSI ENE the larger of the two values from (4.99) or (4.100) is adopted. According to [4.30] the cant u is predominantly more than the *cant deficiency* u_f and therefore, the calculation of qs'_a can be omitted and the term qs'_i will be qs'.

4.9.7 Random displacements

4.9.7.1 Calculation of the random displacements

According to EN 15 273-3 the sum of the random displacements accounts for the displacements resulting from random phenomena such as:

- track transverse displacement between two consecutive maintenance actions
- transverse track height defects between two consecutive maintenance actions
- vehicle vibrations due to irregularities in the railway track
- asymmetry due to installation tolerances of the vehicle suspension and non-uniform distribution of loading

Due to the low probability of coincidence of the maximum tolerances, the sum of the random lateral displacements $\sum j$ is calculated as the geometric average of the individual coefficients

$$\sum j = k \cdot \sqrt{\Delta b_1^2 + \Delta b_2^2 + \Delta b_3^2 + \Delta b_4^2 + \Delta b_5^2} \quad , \tag{4.101}$$

where:

$\sum j$ sum of the random lateral displacements
k coefficient
Δb_1 lateral displacement of track
Δb_2 displacement due to a transverse height deficiency of the track
Δb_3 displacement due to vehicle vibrations because of irregularities of the track
Δb_4 vehicle displacement due to asymmetry caused by tolerances of the suspension
Δb_5 vehicle displacement due to asymmetric distribution of loading

4.9.7.2 Coefficient k

The coefficient k accounts for the low probability of a simultaneous occurrence of limit values of vibrations and asymmetries. Because of EN 15 273-1, it is not probable that the sum of the limit values considered within the calculation will be exceeded by 20 %. The violation of the pantograph gauge is considered as less important than the violation of the vehicle gauge, therefore, the coefficient k is taken as $k = 1,2$ for the calculation of the sum of random displacements of the vehicle and as $k = 1,0$ for the displacements of the pantograph.

4.9.7.3 Transverse displacement of track Δb_1

The track position may change between two inspections because of the effects of train running. The maximum transverse displacement depends on the maintenance conditions

$$\Delta b_1 = T_{\text{track}} \tag{4.102}$$

with T_{track} transverse displacement of track in m.
This effect may be neglected if the permanent way does not allow displacements towards structural systems as is the case with slab tracks.

4.9.7.4 Transverse track height deficiency Δb_2

The transverse height of the track may deviate from the planned value because of maintenance tolerances and the operational load of trains. The transverse height deficiency results in a geometric effect, caused by rotating the reference line with a certain angle relative to the track axis and a quasi-static effect due to suspension elasticity. Taken together, the displacement is

$$\Delta b_2 = \frac{T_{\text{D}}'}{L_{\text{S}}} \cdot h + \frac{s_0'}{L_{\text{S}}} \cdot T_{\text{D}} \cdot (h - h_{\text{co}}') \tag{4.103}$$

with T_{D} being the transverse height deficiency of the track in m occuring between two consecutive track maintenances.

Table 4.10: Sum of random displacements according to EN 15 273-3.

Parameter	Unit	Ballasted track				Slab track
		excellent track quality		other tracks		
		speed in km/h				
		$v \leq 80$	$v > 80$	$v \leq 80$	$v > 80$	
T_{track}	m	0,025	0,025	0,025	0,025	0,005
T_D	m	0,020	0,015	0,020	0,015	0,005
T_{osc}	m	0,039	0,039	0,065	0,065	0,039
T_{charge}	°	0,77	0,77	0,77	0,77	0,77
T_{susp}	°	0,23	0,23	0,23	0,23	0,23
k	–	1	1	1	1	1
$\sum j_u$	m	0,108	0,094	0,114	0,101	0,071
$\sum j_o$	m	0,141	0,123	0,149	0,132	0,095

4.9.7.5 Vehicle vibrations generated by track unevenness Δb_3

Vehicle vibrations are generated by unevenness of the track position, mainly with ballasted tracks. Their amplitude depends on the track quality, the suspension characteristics and the speed of the vehicle. As far as these effects need to be considered structurally, the vibrations are expressed by corresponding cant deficiencies

$$\Delta b_3 = \frac{s_0'}{L} \cdot T_{osc} \cdot (h - h_{c0}') \tag{4.104}$$

with T_{osc} theoretical transverse track height deficiency as a basis to calculate the displacements due to vibrations generated by track position irregularities in m.

4.9.7.6 Asymmetry Δb_4 and Δb_5

The vehicle is not positioned symmetrically above the track axis but is displaced because of suspension tolerances and non-uniform loading distribution. The resulting displacements are

$$\Delta b_4 = \tan(T_{susp}) \cdot (h - h_{c0}') \tag{4.105}$$

and

$$\Delta b_5 = \tan(T_{charge}) \cdot (h - h_{c0}') \quad , \tag{4.106}$$

where:
T_{susp} asymmetric angle due to irregularities of the suspension in degrees
T_{charge} asymmetric angle due to unfavourable loading distribution in degrees

4.9.7.7 Data

Since the sum of the random displacements depends on the maintenance regime of the infrastructure manager, standard values do not exist in Europe. EN 15 273-3 contains recommendations for the coefficients to determine the allowances for calculation of the mechanical-kinematic reference gauge. In this case, it is distinguished between ballasted and slab tracks as well as between two ranges of speeds and track qualities. The resulting values and random displacements at the lower and upper verification height are summarized in Table 4.10.

Table 4.11: Random displacements according to EBO.

Parameter	Unit	not fixed track	fixed track	slab track with cant or error in transverse height ≤ 5 mm
$\sum j_{\mathrm{u}}$	m	0,079	0,073	0,025
$\sum j_{\mathrm{o}}$	m	0,099	0,095	0,032

In Germany, EBO specifies values for the sum of the random displacements (Table 4.11). A comparison of the data based on the recommended coefficients according to EN 15 273-3 and the values according to EBO reveals, that the values according to EN 15 273-3 are higher by at least a factor of 1,2. EBO does not give reasons for the values, however, they are confirmed by experience.

For operational tolerances of other permanent way, values according to DB Directive 821 [4.31] and those recommended by EN 15 273-3 differ. In [4.31] it is specified, that differences within ± 15 mm between the axial cant and the design value are acceptable for all types of tracks and speeds. According to EBO, the consideration of vehicle vibrations caused by track unevenness is based on an excellent track quality. According to this regulation a cant deficiency of 39 mm needs to be assumed so the effect of track unevenness can be modeled. The sum of the asymmetry angles of the vehicle due to manufacturing tolerances, suspension adjustment and non-uniform vehicle loading is not considered by the reference value of 1°, since vehicles with pantographs are equipped with a stiffer suspension. The flexibility will be 0,225 for vehicles with pantographs on the roof. Consequently, the angles to determine the asymmetry are

$$T'_{\mathrm{susp}} = T_{\mathrm{susp}} \cdot \frac{s'_0}{s_0} = 0,23° \cdot \frac{0,225}{0,400} \approx 0,13° \tag{4.107}$$

and

$$T'_{\mathrm{charge}} = T_{\mathrm{charge}} \cdot \frac{s'_0}{s_0} = 0,77° \cdot \frac{0,225}{0,400} \approx 0,43° \quad . \tag{4.108}$$

This procedure coincides with the requirements of EN 15 273-1: *When considering a more rigid vehicle, the maximum values of T_{osc}, T_{charge} or T_{susp} for the reference vehicle may not be considered with s_0, but rather the proportional intermediate values for s/s_0, as a flexible vehicle will oscillate and roll more than a rigid vehicle.*

After consideration of the before mentioned coefficients, the random displacements for a ballasted track at the lower verification point are:

$$\Delta b_1 = 0,025 \text{ m},$$
$$\Delta b_2 = \frac{T'_{\mathrm{D}}}{L_{\mathrm{S}}} \cdot h + \frac{s'_0}{L_{\mathrm{S}}} \cdot T_{\mathrm{D}} \cdot (h - h'_{\mathrm{c}0}) = \frac{0,015}{1,500} \cdot 5,0 + \frac{0,225}{1,500} \cdot 0,015 \cdot (4,5) = 0,060 \text{ m},$$
$$\Delta b_3 = \frac{s'_0}{L_{\mathrm{S}}} \cdot T_{\mathrm{osc}} \cdot (h - h'_{\mathrm{c}0}) = \frac{0,225}{1,500} \cdot 0,039 \cdot (4,5) = 0,010 \text{ m},$$
$$\Delta b_4 = \tan(T_{\mathrm{susp}}) \cdot (h - h'_{\mathrm{c}0}) = \tan(0,13°) \cdot (4,5) = 0,010 \text{ m},$$
$$\Delta b_5 = \tan(T_{\mathrm{charge}}) \cdot (h - h'_{\mathrm{c}0}) = \tan(0,43°) \cdot (4,5) = 0,034 \text{ m},$$
$$\sum j_{\mathrm{u}} = 1,0 \cdot \sqrt{0,025^2 + 0,060^2 + 0,026^2 + 0,010^2 + 0,034^2} = 0,079 \text{ m}$$

Table 4.12: Dimensions of gauge L_{hEBO} according to EBO, all dimensions in m.

Working height of pantograph in m	$h \leq 5,3$	$5,3 < h \leq 5,5$	$5,5 < h \leq 5,9$	$5,9 < h \leq 6,5$
L_{hEBO} in m	1,430	1,440	1,470	1,510

and summarised for the upper verification height point

$$\sum j_o = 1,0 \cdot \sqrt{0,025^2 + 0,079^2 + 0,035^2 + 0,014^2 + 0,045^2} = 0,101 \text{ m}.$$

These data for randomly caused lateral displacements according to EBO apply to speeds up to 160 km/h. According to EN 15 273-3 the values can be used without speed limitations.

4.9.8 Pantograph gauge

Using the impacts described in Clause 4.9.2 to 4.9.7 for the determination of the mechanical-kinematic pantograph gauge and assuming, that the cant is always greater than the cant deficiency, the quasi-static displacements will be

$$qs' = s'_0 \cdot (u - u'_0)_{>0} \cdot (h - h'_{c0})/L_S = 0,15 \cdot (u - 0,066)_{>0} \cdot (h - 0,5) \tag{4.109}$$

and half width of the mechanical-kinematic pantograph gauge L_h results from (4.88):

$$L_h = \frac{l_W}{2} + \frac{2,5}{R} + \frac{l - 1,435}{2} + 0,15 \cdot (u - 0,066)_{<0} \cdot (h - 0,5) +$$
$$\left\{ (0,11 + \sum j_u) + \frac{h - 5,0}{6,5 - 5,0} \cdot \left[(0,17 + \sum j_o) - (0,11 + \sum j_u) \right] \right\}. \tag{4.110}$$

4.9.9 Standard gauge of the pantograph

The *pantograph standard gauge*, also called *unit gauge*, is the gauge for infrastructure with a constant cross-section and designed for the most unfavourable case. The limiting lines of the standard gauge profile GC according to EN 15 273-3 and also to EBO [4.8] were calculated for track radii of more than 250 m and less than 160 mm cant.

The widths of the standard gauge for electrified lines according to EBO are given in Table 4.12 depending on the working height of the pantograph. In this case, it needs to be observed that the electrical minimum clearances for 15 kV had already been considered.

The standard gauge must be kept free of obstructions for unhindered passage of the pantograph. At constraints, the pantograph gauge can be determined for the individual case if some components such as the hook end fitting of the steady arm, the drop bracket or a displaced insulator approach the standard gauge. A violation of a pantograph gauge limits by infrastructure components is not permitted.

4.9.10 Bevelling the pantograph gauge corners

The geometry of the pantograph pan head with contact horns is accommodated by the pantograph gauge by bevelling the corners. The bevelling starts in each case at the outer limit line at the verification height in the direction of the pantograph axis and the downwards from this reference point.

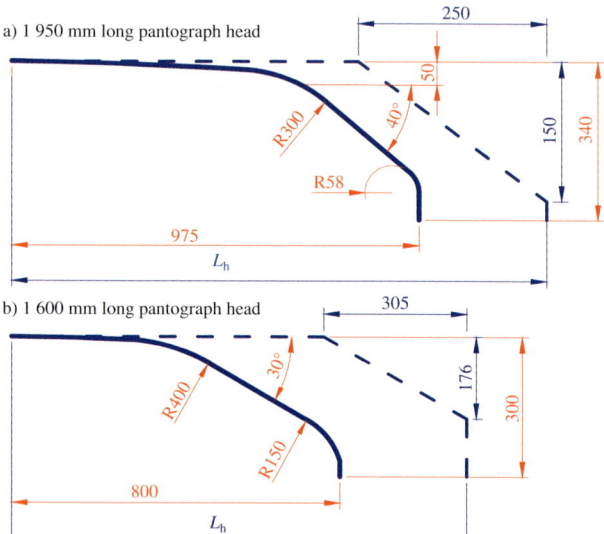

a) 1 950 mm long pantograph head

b) 1 600 mm long pantograph head

Figure 4.27: Bevelling of the pantograph gauges.

Figure 4.27 shows the bevelling for pantograph lengths of 1 600 mm and 1 950 mm. The dimensions in blue are valid for the 1 600 mm long pantograph and defined in EN 15 273-3 and for the 1 950 mm long pantograph in EBO. The dimensions shown in red result from the real geometry of the pantograph head as defined in EN 50 367.

4.9.11 Useable contact wire lateral position

4.9.11.1 Calculation procedures

To minimize the wear of contact strips, the contact wire is installed with a changing lateral position relative to the track axis. Under operational conditions, the contact wire moves horizontally on the pantograph collector strips. In Germany, there is no stipulation for this minimum lateral movement of the pantograph related to the movement of the vehicle. Under the given conditions and mechanical tolerances, the horizontal contact wire displacement and the pantograph must never lead to a pantograph head dewirement. Also, the contact wire must not leave the working length of the pantograph under the action of wind. If the working length of the pantograph is reduced by the total horizontal movement of the pantograph as sum of E' and G'_{infra} the useable contact wire lateral position results (Figures 4.18 and 4.28). This is also called maximum contact wire horizontal displacement in TSI ENE.

In Europe, there are differing rules to determine the useable contact wire lateral position. In TSI ENE a standardized, harmonized calculating procedure is now stipulated.

The useable contact wire lateral position depends on the horizontal movement of the pantograph head relative to the track axis and, therefore, is directly related to the pantograph gauge which already considers displacements of the pantograph.

The useable contact wire lateral position e_{use} is, consequently calculated by considering the total horizontal displacement of the pantograph and its working length according to TSI ENE:

$$e_{\text{use}} = \frac{l_{\text{A}}}{2} + \frac{l_{\text{W}}}{2} - L'_{\text{h}}$$

(4.111)

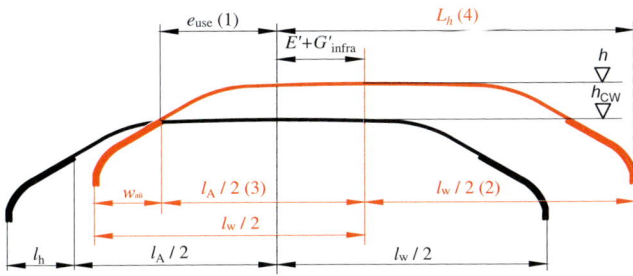

Figure 4.28: Relation between the pantograph gauge and the useable contact wire lateral position according to (4.111).

where:

e_{use} useable contact wire lateral position in m
l_A working length of the pantograph head in m
l_W length of the pantograph head in m
L'_h mechanical-kinematic pantograph gauge in m

The useable contact wire lateral position depends, therefore, on the radius, the cant and the cant deficiency and the working height of the pantograph as shown by the calculation of the mechanical-kinematic pantograph gauge. When determining the working height, the uplift at midspan needs to be considered. In addition, the useable contact wire lateral position is limited to 400 mm for the 1 600 mm long pantograph and to 550 mm for the 1 950 mm long pantograph. In Germany, the maximum useable contact wire lateral positions 400 mm and 550 mm can be used for 5 300 mm and 5 500 mm, respectively, nominal contact wire height. If the maximum useable contact wire lateral positions 400 mm and 550 mm are adopted at a lower contact wire height, the useable contact wire lateral positions at the nominal contact wire height are reduced and as a consequence, also the span length.

4.9.11.2   Useable contact wire lateral position with and without reference cant

For $u_0 = u_{f0} = 0{,}066$ m according to TSI ENE the useable contact wire lateral position follows from

$$e_{use} = \frac{l_A}{2} - \frac{2{,}5}{R} - \frac{l_0 - 1{,}435}{2} + 0{,}15 \cdot (u - 0{,}066)_{>0} \cdot (h - 0{,}5) - \left\{ \left(0{,}11 + \sum j_u\right) + \frac{h - 5{,}0}{6{,}5 - 5{,}0} \cdot \left[\left(0{,}17 + \sum j_o\right) - \left(0{,}11 + \sum j_u\right)\right] \right\}. \tag{4.112}$$

For the case $u'_0 = u'_{f0} = 0{,}0$ m the useable contact wire lateral position is obtained from

$$e_{use} = \frac{l_A}{2} - \frac{2{,}5}{R} - \frac{l_0 - 1{,}435}{2} + 0{,}15 \cdot u \cdot h + 0{,}075 \cdot u - \left\{ \left(0{,}065 + \sum j_u\right) + \frac{h - 5{,}0}{6{,}5 - 5{,}0} \cdot \left[\left(0{,}11 + \sum j_o\right) - \left(0{,}065 + \sum j_u\right)\right] \right\}. \tag{4.113}$$

The infrastructure manager decides whether to use the reference cant $u'_0 = u'_{f0} = 0{,}066$ m according to TSI ENE or $u'_0 = u'_{f0} = 0{,}0$ m corresponding as recommended in Clause 4.9.11.4.

Table 4.13: Useable contact wire lateral position with and without reference cant, parameter according to Table 4.8, random-related displacements due to EN 15 273 for speeds > 80 km/h and EBO (Table 4.11) for not fixed tracks, all dimensions in m.

Veri-fication height h	Pantograph length 1 600 mm					Pantograph length 1 950 mm				
	reference value				Δ	reference value				Δ
	0,066		0,000			0,066		0,000		
according to EN 15 273										
		l		l			l		l	
5,00	0,374	67,5	0,419	74,0	0,045	0,549	84,0	0,594	89,0	0,045
5,52	0,342	62,5	0,392	70,5	0,050	0,517	80,5	0,567	86,5	0,050
5,86	0,322	57,5	0,375	67,5	0,053	0,497	77,5	0,550	84,0	0,053
6,50	0,283	–	0,342	–	0,057	0,458	–	0,517	–	0,059
according to EBO										
5,00	0,396	71,0	0,441	76,5	0,045	0,571	86,5	0,616	91,0	0,045
5,52	0,368	67,0	0,418	74,0	0,050	0,543	84,0	0,593	89,0	0,050
5,86	0,350	63,5	0,404	71,5	0,054	0,525	81,0	0,579	87,0	0,054
6,50	0,316	–	0,376	–	0,060	0,491	–	0,551	–	0,060

Δ difference between e_{use} with and without reference cant, l longitudinal span in wind load zone II for DB contact line type Re200, verification heights according to TSI ENE and Clause 4.9.3

4.9.11.3 Impact of the reference cant

When calculating the reference gauge, standard EN 15 273-1 distinguishes between the responsibilities on the vehicle side and on the infrastructure side. The portion E' of the displacement is considered when establishing the vehicle gauge and the portion G_{infra} for the infrastructure gauge (Figure 4.18). The interface between the two responsibilities is represented by reference gauge B.

Displacements, which occur because of the inclination of the vehicle resulting from the quasi-static effects, are part of the rolling stock manager's responsibility and in part should also be considered by the infrastructure manager. The reference cant and the reference cant deficiency form the interface between responsibilities. The rolling stock manager considers all displacements due to quasi-static effects up to the reference cant and reference cant deficiency, while the infrastructure manager takes responsibility only for cants and cant deficiencies above the reference cant and the reference cant deficiency as additional quasi-static displacements.

According to agreed commitments, the rolling stock manager is responsible for verifying analytically that no part of the vehicle exceeds the reference gauge. The mechanical-kinematic reference gauge also considers the displacements from quasi-static inclination, the transverse clearance and installation tolerances (see Clause 4.10.5). As a basis for calculating the kinematic reference gauge, the data for a reference vehicle according to EN 15 273-3, Table F.2 applies.

When designing a new vehicle, the manufacturer of the vehicle may choose which parameters, e. g. for transverse clearance or flexibility may be adopted, as long as the sum of the vehicle body width and displacements do not exceed the reference gauge. For example, the vehicle manufacturer may develop a relatively weakly suspended vehicle with high flexibility and accept a reduced vehicle body width. Alternatively the manufacturer of the vehicle may intend to design an relatively wide vehicle body where the lateral displacements are reduced by an especially stiff suspension in accordance with UIC 505-5 [4.29].

This freedom concerning the development of rolling stock is only possible by applying the kinematic calculation methodology. In the case of the static calculation method used so far, the infrastructure manager had to take into account all dynamic effects when calculating the infrastructure gauge. The feature of the suspension had no effect on the rolling stock gauge and each vehicle, therefore, had approximately the same vehicle body width. An example of the optimum use of the kinematic reference gauge is the development of the gauges for the first ICE generation [4.32].

For the reference cant and the reference cant deficiency overall values of 0,050 m for the calculation of the infrastructure gauge and the value 0,066 m for the calculation of the pantograph gauge were specified.

The value of 0,050 m for the infrastructure gauge results from its historical use in the Gotthard-Base-Tunnel, as a higher value would have led to a wider reference gauge and consequently, a wider infrastructure gauge which would have required a greater distance between the adjacent tracks [4.33]. To not limit the use of the Gotthard-Base-Tunnel as an important international transit route and to avoid expensive re-construction a fixed value of 0,050 m was accepted. To calculate the additional quasi-static displacement and the pantograph gauge, the TSI ENE uses a fixed cant value and a fixed cant deficiency value of 0,066 m. The historical development and a detailed derivation of the reference values can be found in [4.34].

This specification resulted in the mechanical-kinematic pantograph gauge always considering a quasi-static displacement of the vehicle bodies resulting from cant and cant deficiency equal to the agreed overall values of 0,066 m including if the vehicle runs on a straight track without cant where a track deficiency cannot occur.

In the case of a mechanical-kinematic pantograph gauge, the real overall movement of the pantograph is not considered but a vehicle inclined at least by the reference cant is always considered. The consequence is, that the calculation of the useable contact wire lateral position is not based on the real lateral movement of the pantograph but on the theoretical displacement after considering the agreed overall values of the reference cant and the reference cant deficiency. These data were chosen for the definition of the structure gauge and the vehicle gauge.

With line sections having cants and/or cant deficiencies less than the reference cant or the reference cant deficiency, the useable contact wire lateral position is unnecessarily limited. To consider the real displacements of the pantograph for calculating the useable contact wire lateral position, the reference cant and the reference cant deficiency should be ignored and both values should be defined as 0,0 m in this case. This stipulation leads to the fact that the pantograph sway, according to Equations (4.96) and (4.97), can be reduced by the displacement considered from the vehicle side due to quasi-static inclinations. According to (4.96) and (4.97) the displacement due to quasi-static inclination can be reduced by 0,045 m at the lower verification height and 0,060 m at the upper verification height, where sway of the pantograph at the upper verification height would be reduced to $e_{po} = 0,110$ m and at the lower verification height to $e_{pu} = 0,065$ m.

4.9.11.4 Impacts of the TSI ENE on overhead contact line systems

From January 1, 2015 the TSI ENE [4.9] became the basis for planning all overhead contact lines on the trans-European railway system. By harmonizing the calculation rules for conventional and high-speed railway systems the TSI ENE took an important step towards further harmonizing the European railway systems. From this date, the calculation of the mechanical-kinematic pantograph gauge and the useable contact wire lateral position should be carried out according to TSI ENE and EN 15 273-1.

The pantograph gauge and the useable contact wire lateral position depend on parameters not yet harmonized within Europe. This applies especially to the permitted tolerances and limits for track position established by individual infrastructure managers.

For future system design, a distinction should be made between the calculation of the mechanical-kinematic pantograph gauge using the reference value 0,066 m, and the useable contact wire lateral position which does not need to consider this value. The simplified formula (4.88) for the mechanical-kinematic pantograph gauge and Equation (4.113) for the determination of the useable contact wire lateral position should be used.

The calculation of the useable contact wire lateral position relates to the individual contact wire height. The lower the verification height, the higher the permissible useable contact wire lateral position. In TSI ENE, the useable contact wire lateral position is limited to 0,400 m for operation with the 1 600 mm long pantograph pan head and to 0,550 m for operation with the 1 950 mm long pantograph pan head, independently of the contact wire height which does not make sense, since at lower contact wire heights higher values of the lateral position would be possible as shown in Table 4.13. Without this limitation, longer span lengths would be possible resulting in reduced installation and operational costs.

Example 4.22:   Design an overhead contact line of type Re200 for a conventional railway line, operated with 1 600 mm and 1 950 mm long pantographs which, according to the infrastructure register, belong to the conventional trans-European network (TEN). What useable contact wire lateral position is possible in straight line sections at the verification heights $h'_o = 6,5$ m, $h_{5,86} = 5,86$ m and $h'_u = 5,0$ m according to TSI ENE, EN 15 273-3 and EBO [4.8]? The useable contact wire lateral position should be calculated for the reference values $u'_0 = 0,066$ m and $u'_0 = 0,0$ m.

Given are:

 - straight line
 - maximum gauge $l = 1,465$ m for main tracks (Table 4.9)
 - operation with 1 600 mm and 1 950 mm long pantographs
 - ballasted permanent way as for 'other tracks' for speeds > 80 km/h with the random lateral displacements for the lower verification *height* $\sum j_u = 0,101$ m and for the upper verification height $\sum j_o = 0,132$ m (Table 4.10) according EN 15 273-3
 - ballasted permanent way for *not fixed ballasted track* according to EBO [4.8] and the random lateral displacements for the lower verification height $\sum j_u = 0,079$ m and for the upper verification height $\sum j_o = 0,099$ m.

When operating with several pantographs, the pantograph with the shorter pan head determines the *useable contact wire lateral position* e_{use}. According to Equation (4.112) for the case $u'_0 = u'_{f0} = 0,066$ m and the random lateral displacements according to EN 15 273-3 for *other tracks* at the verification height $h_{5,0}$ the useable contact wire lateral position to be

$$e_{use} = \frac{l_A}{2} - \frac{2,5}{R} - \frac{l_0 - 1,435}{2} + 0,15 \cdot (u - 0,066)_{>0} \cdot (h - 0,5) - \left\{ (0,11 + \sum j_u) + \frac{h - 5,0}{6,5 - 5,0} \cdot \left[(0,17 + \sum j_o) - (0,11 + \sum j_u) \right] \right\}$$

$$= 0,600 - 0 - \frac{1,465 - 1,435}{2} + 0 -$$
$$\left\{ (0,11 + 0,101) + \frac{5,0 - 5,0}{6,5 - 5,0} \cdot [(0,17 + 0,132) - (0,11 + 0,101)] \right\}$$
$$= 0,374\,\text{m (blue value marked in Table 4.13)}.$$

For the verification height $h_{5,86}$ it results $e_{\text{use}} = 0,322\,\text{m}$ and for $h'_{6,5} = 0,283\,\text{m}$.
For the case $u'_0 = u'_{f0} = 0,0\,\text{m}$ the useable contact wire lateral position is obtained from

$$e_{\text{use}} = \frac{l_A}{2} - \frac{2,5}{R} - \frac{l_0 - 1,435}{2} + 0,15 \cdot u \cdot h + 0,075 \cdot u -$$
$$\left\{ \left(0,065 + \sum j_u\right) + \frac{h - 5,0}{6,5 - 5,0} \cdot \left[\left(0,11 + \sum j_o\right) - \left(0,065 + \sum j_u\right)\right] \right\}$$
$$= 0,600 - 0 - \frac{1,465 - 1,435}{2} + 0 + 0 -$$
$$\left\{ (0,065 + 0,101) + \frac{5,0 - 5,0}{6,5 - 5,0} \cdot [(0,11 + 0,132) - (0,065 + 0,101)] \right\}$$
$$= 0,441\,\text{m (green value marked in Table 4.13)}.$$

For the verification height $h_{5,86}$ it results e_{use} 0,375 m and for $h'_{6,5}$ 0,342 m.
According to Equation (4.112) it results for the case $u'_0 = u'_{f0} = 0,066\,\text{m}$ for the useable contact wire lateral position e_{use} with the random lateral displacements according to EBO for a *not fixed ballasted track* for the verification height $h'_{5,0}$

$$e_{\text{use}} = \frac{l_A}{2} - \frac{2,5}{R} - \frac{l_0 - 1,435}{2} + 0,15 \cdot (u - 0,066)_{>0} \cdot (h - 0,5) -$$
$$\left\{ \left(0,11 + \sum j_u\right) + \frac{h - 5,0}{6,5 - 5,0} \cdot \left[\left(0,17 + \sum j_o\right) - \left(0,11 + \sum j_u\right)\right] \right\}$$
$$= 0,600 - 0 - \frac{1,465 - 1,435}{2} + 0 -$$
$$\left\{ (0,11 + 0,079) + \frac{5,0 - 5,0}{6,5 - 5,0} \cdot [(0,17 + 0,099) - (0,11 + 0,079)] \right\}$$
$$= 0,396\,\text{m (red value marked in Table 4.13)}.$$

For verification height $h_{5,86}$ yields 0,350 m and $h'_{6,5}$ 0,316 m.
For the case $u'_0 = u'_{f0} = 0,0\,\text{m}$ the useable contact wire lateral position is found to be

$$e_{\text{use}} = \frac{l_A}{2} - \frac{2,5}{R} - \frac{l_0 - 1,435}{2} + 0,15 \cdot u \cdot h + 0,075 \cdot u -$$
$$\left\{ \left(0,065 + \sum j_u\right) + \frac{h - 5,0}{6,5 - 5,0} \cdot \left[\left(0,11 + \sum j_o\right) - \left(0,065 + \sum j_u\right)\right] \right\}$$
$$= 0,600 - 0 - \frac{1,465 - 1,435}{2} + 0 + 0 -$$
$$\left\{ (0,065 + 0,079) + \frac{5,0 - 5,0}{6,5 - 5,0} \cdot [(0,11 + 0,099) - (0,065 + 0,079)] \right\}$$
$$= 0,441\,\text{m (magenta value marked in Table 4.13)}.$$

For verification height $h_{5,86}$ yields 0,404 m and $h_{6,5}$ 0,376 m.
As can be seen from Table 4.13, the span lengths vary in the case of the 1 600 mm pan head up to 8 m at the verification height 5,86 m, corresponding to a contact wire nominal height of 5,50 m, to 6 m for the 1 950 mm long pan head.

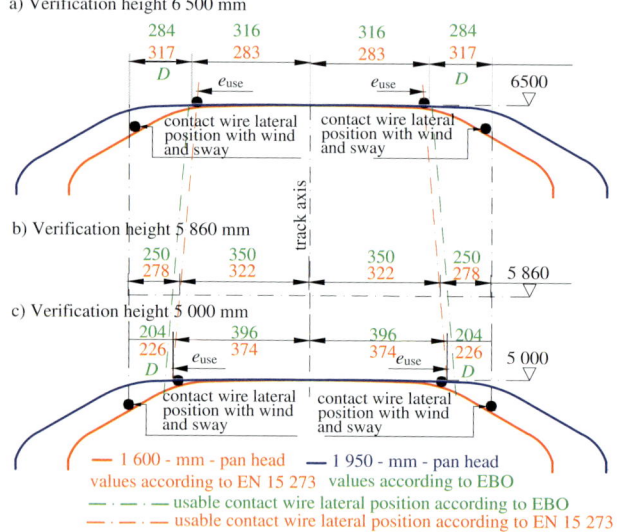

a) Verification height 6 500 mm

b) Verification height 5 860 mm

c) Verification height 5 000 mm

Figure 4.29: Useable contact wire lateral position according to example 4.22 in a straight line section with random supplements according to EN 15 273-3:2014 (red) and EBO (green) in each case with a reference value of 0,066 m (Table 4.13) for conventional lines.

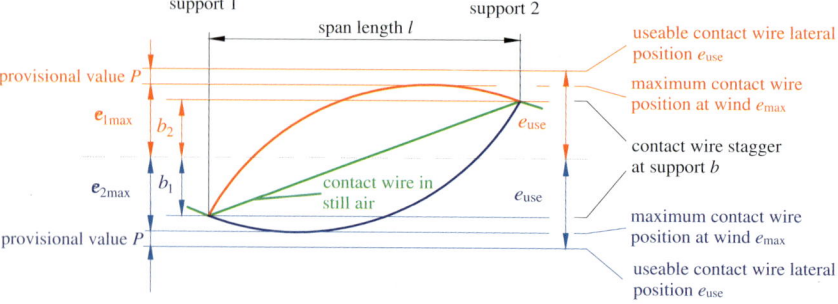

Figure 4.30: Designations within a span.

The useable contact wire lateral positions with random lateral displacements according to EBO are bigger than that with random lateral displacements according to EN 15 273-3 for the same quality of permanent way for conventional lines. Figure 4.29 shows the useable contact wire lateral positions.

4.10 Contact wire lateral position

4.10.1 Introduction

Figure 4.30 illustrates the variables in *spans with contact wire lateral position* under wind action, where in the following subscript 1 is used for a contact wire displaced to the right as seen when viewing along the line, from left to right and subscript 2 for the contact wire displaced to the left. Thus the contact wire can wear the collector strips equally. However, the *wear of the collector strips* is less important than the wear of the contact wire. Replacing collector strips is cheaper than replacing a contact wire and if the rolling stock manager does not replace worn collector strips, contact wires can be damaged.

Table 4.14: Minimum variation of lateral displacement Δb_m of contact wires.

Transport entity	Voltage kV	Displacement mm/m	Source
Mass transit Germany (VDV)	DC 0,75	$\geq 10,0$	VDV 550 [4.35]
Mass Transit Railway (MTR) Hong Kong / China	DC 1,50	$\geq 1,5$	Consultant Kennedy & Donkin
Córas Iompair Éireann (CIÉ) Dublin / Ireland	DC 1,50	$\geq 3,0$	Consultant Mott Hay & Anderson
Perth Electric Perth / Australien	AC 25	$\geq 3,0$	Consultant ELRail
Main lines internationall	AC	$> 5,0$	[4.36] for $v > 200\,\mathrm{km/h}$
West coast route Great Britain	AC 25	$\geq 2,5$	WCRM Ole Alliance Design Group
British Railway Board Great Britain	AC 25	$\geq 3,0$	Network Rail

Because of bending (change of direction) of the contact line at the supports, radial forces are created at the contact and catenary wire. The wear of overhead contact wire and collector strips depend on the radial forces as well. Therefore, according to the stipulations of German Railway the *contact wire radial force* F_R should lie within $80\,\mathrm{N} < F_R < 2\,000\,\mathrm{N}$ for light-weight steady arms. Falling below this value results in high wear of the joint between the drop bracket and steady arm (see also Clause 17.2.3). Low radial forces are caused by too small changes of direction of the contact line.

To ensure even wear of collector strips the contact wire must achieve a minimum sweep across the collector strips. Table 4.14 lists the specifications for sweep values. The selection of *contact stagger for tramway systems in cities* and on *high-speed lines* should as far as possible aim at achieving a minimum sweep of the contact wire lateral position across the collector strips to avoid local overheating and produce wear that is as even as possible. In mass transit systems, 1,5 mm/m to 10 mm/m *minimum sweep of the contact wire* relative to the line length is well proven practice (Table 4.14) [4.37]. For high-speed lines, the sweep of contact wire should not fall below 1,5 mm/m. Over one third of the span length in curves, a contact wire sweep less than 1,5 mm/m cannot always be avoided so the section concerned should be situated in the middle of the span. Table 4.15 presents an overview on contact wire height, contact wire stagger, system height, catenary wire height and stagger of catenary wire as used by some main line railway entities in Europe.

4.10.2 Wind loads on conductors

This section verifies the *serviceability of contact lines* under the action of wind. The assumptions for wind loads on conductors are dealt with in Clause 2.5.2. The wind load on conductors follows from

$$F'_W = q_z \cdot G_C \cdot C_C \cdot d \quad . \tag{4.114}$$

To verify serviceability, it is not necessary to assume the same wind conditions as for static stability of the system since the simultaneous action of extreme wind load at a specific site and a traction unit at the same place on a line is improbable. For serviceability, therefore, the

Table 4.15: Contact wire height h_{CW}, contact wire stagger b_{CW}, system height h_{SH}, catenary wire heights (h_{CA}) and catenary wire lateral position b_{CA} of European AC railways. dimensions in m.

Railway entity	h_{CW}	b_{CW}		h_{SH}	h_{CA}	b_{CA}	
		straight line	curves			straight line	curves
German Railway (DB)							
– operation with 1 950 mm pan head							
$v \le 200$ km/h[1]	5,50	$\le 0,4$	$\le 0,4$	1,80	7,30	0	$\le 0,4$
$v \le 230$ km/h[2]	5,50	$\le 0,4$	$\le 0,4$	1,60	7,10	0	$\le 0,4$
– operation with 1 600 mm pan head							
$v \le 230$ km/h[1],[2]	5,50	$\le 0,25$	$\le 0,3$	1,80	7,30	$\le 0,25$	$\le 0,3$
$v > 230$ km/h[3]	5,30	$\le 0,3$	$\le 0,3$	1,80	7,30	$\le 0,3$	$\le 0,3$
$v > 230$ km/h[4]	5,30	$\le 0,3$	$\le 0,3$	1,60	6,90	$\le 0,3$	$\le 0,3$
French Railway (SNCF)							
$v \le 230$ km/h[5]	5,50	$\le 0,2$	$\le 0,2$	1,25/1,40	6,75/7,15	$\le 0,2$	$\le 0,2$
$v > 230$ km/h[6]	5,08	$\le 0,2$	$\le 0,2$	1,40	6,48	$\le 0,2$	$\le 0,2$
Spanish Railway (ADIF)							
$v \le 230$ km/h[7]	5,30	$\le 0,3$	$\le 0,3$	1,80	7,10	$\le 0,3$	$\le 0,3$
$v > 230$ km/h[8]	5,30	$\le 0,2$	$\le 0,2$	1,40	6,70	$\le 0,2$	$\le 0,2$
Dutch Railway[9]							
$v \le 230$ km/h[10]	5,30	$\le 0,3$	$\le 0,3$	1,60	6,90	$\le 0,3$	$\le 0,3$
$v > 230$ km/h[11]	5,30	$\le 0,3$	$\le 0,3$	1,60	6,90	$\le 0,3$	$\le 0,3$
Norwegian Railway (JBV)							
$v \le 230$ km/h[12]	5,60	$\le 0,3$	$\le 0,4$	1,60	7,20	$\le 0,3$	$\le 0,4$
$v > 230$ km/h[13]	5,30	$\le 0,3$	$\le 0,3$	1,80	7,10	$\le 0,3$	$\le 0,3$

[1] DB's contact lines Re100, Re200   [2] Siemens contact line Sicat S1.0
[3] DB contact line Re330   [4] Siemens contact line Sicat H1.0
[5] contact wire tensile force 12 kN/15 kN   [6] contact wire tensile force 14 kN
     catenary wire tensile force 10 kN/14 kN   catenary wire tensile force 20 kN
[7] contact wire tensile force 15 kN   [8] ADIF contact line EAC 350
     catenary wire tensile force 15 kN   [9] HSL Zuid
[10] Siemens contact line Sicat S1.0   [11] Siemens contact line Sicat H1.0
[12] JBV contact line S20 und S35   [13] JBV contact line S25

assumption of a ten year return period of the wind load is sufficient. The meanings of the variables in Equation (4.114) are:

q_z wind pressure for the relevant wind load zone with 10 year return period at the height z. This value is obtained from the corresponding wind velocity with fifty year return period $v_{0,02}$ according to Figure 2.26 by $q_{z10} = q_{z50} \cdot (0,902)^2 = q_{z50} \cdot 0,813$

G_C conductor reaction coefficient, depending on the span length and the dynamic performance of the conductor. This factor should be selected according to according to national conditions. In accordance with German standard EN 50 341-2-4 G_C can be selected for spans up to 200 m within the wind load zones W1 and W2 as $G_C = 0,75$ and for the wind load zones W3 and W4 as 0,67 and 0,60, respectively

C_C drag factor for conductors with circular cross-section; according to EN 50 119 $C_C = 1,0$; EN 50 341-2-4:2016, Table 4.3.2/DE.1, following values are recommended
         – conductors less than 12,5 mm diameter $C_C = 1,2$
         – conductors above 12,5 mm to 15,8 mm diameter $C_C = 1,1$
         – conductors above 15,8 mm diameter $C_C = 1,0$

d conductor diameter.

If conductors are run in parallel and if the distance between the conductor axes is less than five times the conductor diameter, a reduction in wind load, to 80 % can be assumed for the leeward conductor, when compared with the windward conductor. Thus, for a twin contact wire AC-100 the drag factor equates to

$$C_C = 1,2 + (0,8 \cdot d) = 1,2 \cdot (1 + 0,8) = 2,16 \quad .$$

Example 4.23:   It is planned, to reconstruct the overhead contact line of the Munich–Augsburg line to an interoperable contact line design. The line is situated in the wind region W2 characterised by $q_b = 390\,\text{N/m}^2$ basic wind pressure and $v_b = 25,0\,\text{m/s}$ basic wind velocity. Calculate the wind loads for return periods of three, five and ten years for verification of serviceability for the line.
According to Figure 2.26 for the three, five and ten year return periods, $v_{b,0,33}/v_{b,0,02} = 0,815$, for $v_{b,0,20}/v_{b,0,02} = 0,855$ and $v_{b,0,10}/v_{b,0,02} = 0,902$ and, therefore, the basic wind velocities are $v_{b,0,33} = 20,4\,\text{m/s}$, $v_{b,0,20} = 21,4\,\text{m/s}$ and $v_{b,0,10} = 22,6\,\text{m/s}$, respectively.

Example 4.24:   Given is the basic wind velocity $v_b = 22,6\,\text{m/s}$ for wind load zone W2 for a ten years return period for verification of serviceability of an overhead contact line according to example 4.23. For the Munich–Augsburg line $T = 10\,°\text{C}$ (283 K) and $H = 520\,\text{m}$ can be assumed. Determine the basic wind pressure q_b. For values T and H the air density ρ can be taken from Equation (2.5) as

$$
\begin{aligned}
\rho & = & 1,225 \cdot (288/T) \cdot \exp(-0,00012 \cdot H) \\
& = & 1,225 \cdot (288/283) \cdot \exp(-0,00012 \cdot 520) = 1,171\,\text{kg/m}^3 \quad .
\end{aligned}
$$

The basic wind pressure for verification of serviceability with a ten year return period follows from Equation (2.4) to be

$$q_b = (\rho/2) \cdot v_b^2 = (1,171/2) \cdot 22,6^2 = 299\,\text{N/m}^2 \quad .$$

The result $q_b = 299\,\text{N/m}^2$ can be rounded to $300\,\text{N/m}^2$.

Example 4.25:   The overhead contact line on the Munich–Augsburg line is planned to be reconstructed to an interoperable design. On this line there is a bridge over the river Lech 126 m above terrain. For the contact line, the height above terrain will be $h_z = 132\,\text{m}$. Given is the basic wind pressure for the wind load zone W2 with $q_b = 300\,\text{N/m}^2$ for a return period of ten years for verification of serviceability (see Example 4.24). Find the wind pressure at a height of 132 m above terrain, for verification of serviceability.
According to Equation (2.8) the wind pressure at the height $h_z = 132\,\text{m}$ above terrain will be

$$q_{z=132} = 2,1 \cdot q_z \cdot \left(\frac{h_G}{10}\right)^{0,24} = 2,1 \cdot 300 \cdot \left(\frac{132}{10}\right)^{0,24} = 1\,170\,\text{N/m}^2 \quad .$$

Example 4.26:   Given is the wind pressure $q_z = 1\,170\,\text{N/m}$ from example 4.25 for an overhead contact line, that is planned to be installed at a height $h_z = 132\,\text{m}$ above the terrain. The contact line type Re200 will be used and consists of a contact wire AC-100 – Cu and a catenary wire BzII 50. The diameter of the contact wire is $d = 12\,\text{mm}$. $C_C = 1,2$ applies to the contact wire and the $d = 9\,\text{mm}$ diameter catenary wire. Find the wind load F'_{WOCL} for the contact line. The wind load on the contact wire results from Equation (4.114):

$$F'_{\text{WCW}} = q_z \cdot G_C \cdot C_C \cdot d = 1\,170 \cdot 0,75 \cdot 1,2 \cdot 0,012 = 12,6\,\text{N/m}$$

Table 4.16: Length related wind loads F'_W in N/m for the verification of serviceability for contact line components for a ten year return period according to EN 1991-1-4/NA:2010-12 in open terrain, height 7 m (see also Example 4.27).

Element	Application	Cross-section	Dia-meter	Drag factor	Wind zone			
					W1	W2	W3	W4
					wind velocity in m/s			
					20,3	22,6	24,8	27,1
		A	d	C_C [6]	basic wind pressure q_b N/m²			
					260	317	383	456
					design wind pressure q_h			
					390	476	575	684
		mm²	mm	–	wind load F'_W N/m			
AC-80	contact wire	80	10,6	1,2	3,72	4,54	5,48	6,53
AC-100	contact wire	100	12,0	1,2	4,21	5,14	6,20	7,39
AC-107	contact wire	107	12,3	1,2	4,32	5,26	6,36	7,57
AC-120	contact wire	120	13,2	1,1	4,25	5,18	6,26	7,45
AC-150	contact wire	150	14,8	1,1	4,76	5,81	7,01	8,35
2 ×AC-80	contact wire	80	10,6	2,16	6,70	8,17	9,87	11,75
2 × AC-100	contact wire	100	12,0	2,16	7,58	9,24	11,17	13,30
2 × AC-107	contact wire	107	12,3	2,16	7,77	9,47	11,45	13,63
2 × AC-120	contact wire	120	13,2	1,98	7,64	9,32	11,26	13,41
2 × AC-150	contact wire	150	14,8	1,98	8,57	10,45	12,63	15,03
rope 10 × 49[1]	dropper	10	4,50	1,2	1,58	1,93	2,33	2,77
rope 10 × 49[2]	dropper	10	4,65	1,2	1,63	1,99	2,40	2,86
rope 16 × 84[1]	dropper	16	6,20	1,2	2,18	2,65	3,21	3,82
rope 25 × 133[1]	dropper	25	7,50	1,2	2,63	3,21	3,88	4,62
rope 25 × 7[3]	stitch wire	25	6,30	1,2	2,21	2,70	3,26	3,88
rope 35 × 7[3]	stitch wire	35	7,50	1,2	2,63	3,21	3,88	4,62
rope 50 × 7[3]	catenary wire	50	9,00	1,2	3,16	3,85	4,65	5,54
rope 70 × 19[3]	catenary wire	70	10,5	1,2	3,69	4,49	5,43	6,46
rope 95 × 19[3]	catenary wire	95	12,5	1,2	4,39	5,35	6,46	7,70
rope 120× 19[3]	catenary wire	120	14,0	1,1	4,50	5,49	6,64	7,90
182-AL1[4]	feeder line	181,6	17,5	1,0	5,12	6,24	7,54	8,98
243-AL1[4]	feeder line	242,5	20,3	1,0	5,94	7,24	8,75	10,41
299-AL1[4]	feeder line	299,4	22,5	1,0	6,58	8,02	9,69	11,54
122-AL1/71-ST1A[5]	S feeder line	193,4	18,0	1,0	5,27	6,42	7,76	9,23
243-AL1/39-ST1A[5]	feeder line	234,1	21,8	1,0	6,38	7,77	9,39	11,18
CW AC- 80+CA 50	contact line	80/50	–	–	7,91	9,65	11,65	13,88
CW AC-100+CA 50	contact line	100/50	–	–	8,48	10,33	12,49	14,87
CW AC-107+TS 50	contact line	107/50	–	–	8,60	10,48	12,67	15,08
CW AC-120+CA 70	contact line	120/70	–	–	9,12	11,12	13,44	16,00
CW AC-120+CA 120	contact line	120/120	–	–	10,06	12,27	14,83	17,65
CW AC-150+CA 95	contact line	150/95	–	–	10,52	12,83	15,50	18,45
CW AC-150+CA 120	contact line	150/120	–	–	10,66	12,99	15,70	18,69

[1] acc. to DIN 43138 made of bronze or copper; [2] acc. to 8WL7060-2 of Siemens AG, [3] acc. to DIN 48201-2 made of bronze or copper [4] acc. to EN 50182 aluminum conductor [5] acc. to EN 50182 ACSR [6] acc. to EN 50341-2-4:2016 and EN 50119, Clause 6.2.4.3, for twin contact wire

and for the catenary wire BzII 50:

$$F'_{WCA} = q_z \cdot G_C \cdot C_C \cdot d = 1\,170 \cdot 0,75 \cdot 1,2 \cdot 0,009 = 9,5\,N/m \quad ,$$

The wind load for the total contact line is obtained by applying the factor 1,15, to account for clamps and droppers etc. in the contact line:

$$F'_{WOCL} = (F'_{WCW} + F'_{WCA}) \cdot 1,15 = (12,6 + 9,5) \cdot 1,15 = 25,4\,N/m \quad .$$

Example 4.27: An interoperable railway line is to be electrified. It is situated in the wind load zone W1, where $q_b = 320\,N/m^2$ basic wind pressure and $v_b = 22,5\,m/s$ basic wind velocity according to EN 1991-1-4/NA:2010 applicable to a 50 year return period. To verify the serviceability, a return period of 10 years is specified by the infrastructure manager. An interoperable contact line design with a 12 mm diameter contact wire AC-100–Cu and a 9 mm diameter catenary wire BzII 50 is planned to be used. The drag factor is $C_C = 1,2$ for contact and catenary wire. The mean temperature is 10 °C (283 K) and the mean altitude above sea level $H = 0\,m$, where according to (2.5) $\rho = 1,25\,kg/m^3$ results for the air density. Determine the length related wind load on the overhead contact line at a height above terrain $h_z = 7\,m$.

According to Figure 2.26, $v_{b,0,10}/v_{b,0,02} = 0,902$ for a ten year return period and therefore, the basic wind velocity $v_{b,0,10} = 20,3\,m/s$, which is shown in magenta in Table 4.16. The basic wind pressure for verification of serviceability follows from Equation (2.4)

$$q_b = (\rho/2) \cdot v_b^2 = (1,25/2) \cdot 20,3^2 = 258\,N/m^2 \quad .$$

The result $q_b = 258\,N/m^2$ is rounded to $260\,N/m^2$ and shown in blue in Table 4.16. The design wind pressure follows from Equation (2.8) at a height $h_z = 7\,m$ above surface as

$$q_z = 1,5 \cdot q_b = 1,5 \cdot 260 = 390\,N/m^2 \quad .$$

The result $q_z = 390\,N/m^2$ can be found as a value shown in brown in Table 4.16. The wind load F'_W on the contact wire within the wind load zone W1 is obtained from Equation (4.114) as

$$F'_W = q_z \cdot G_C \cdot C_C \cdot d = 390 \cdot 0,75 \cdot 1,2 \cdot 0,012 = 4,21\,N/m \quad .$$

The result $F'_W = 4,21\,N/m$ is shown in red in Table 4.16. For the catenary wire in the wind load zone W1 it is obtained

$$F'_W = q_h \cdot G_C \cdot C_C \cdot d = 390 \cdot 0,75 \cdot 1,2 \cdot 0,009 = 3,16\,N/m \quad .$$

To consider the effect of clamps, droppers, stitch wires etc. the values F'_W in Table 4.16 will be multiplied by 1,15, whereby the wind load F'_{WOCL} of the contact line is found to be

$$F'_{WOCL} = (F'_{WCW} + F'_{WCA}) \cdot 1,15 = (4,21 + 3,16) \cdot 1,15 = 8,48\,N/m \quad .$$

The result $F'_{WOCL} = 8,48\,N/m$ is shown in cyan in Table 4.16.

In the following examples, the basic data from example 4.27 will be used.
Table 4.16 contains the length related wind loads according to example 4.27, for the wind loads according to EN 1991-1-4/NA:2010 which apply to:
- up to 7 m height above sea level
- 10 minutes mean wind velocity

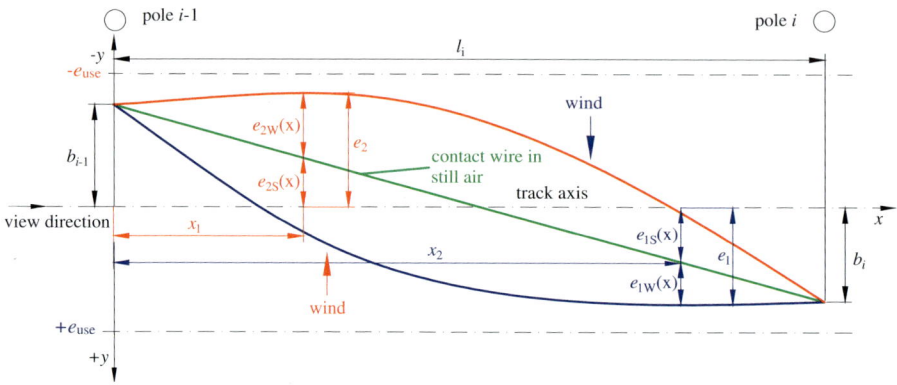

Figure 4.31: Displacement of the contact wire under wind action with stagger on a straight track.

- 10 m high above terrain level
- independent of the wind direction
- flat open terrain of category II
- reference altitude is at sea level
- 50 year return period for the wind velocity

Because of difficulties when applying the wind data according to the Supplement 1, addition to EN 50 119, Clause 6.2.4.1, the use of the specifications according to EN 1991-1-4/NA:2010-12 is recommended for verification of serviceability and strength analysis under wind action. Consequently civil engineering and overhead power line design wind standards may be applied to contact line design in Germany. The wind loads specified in EN 1991-1-4/NA and EN 50 341-2-4 are based on data gained from meteorological observations by German weather services (DWD).

4.10.3   Displacement in straight line sections

Using the coordinate system in Figure 4.31 the contact wire lateral position on the right side of the track axis, viewing in the direction with increasing pole designations (chainage), is defined as positive. Without wind action, the position of the contact wire is described by

$$e_S(x) = (b_i - b_{i-1})x/l_i + b_{i-1} \quad . \tag{4.115}$$

The wind force on the wires and conductors of a contact line displaces these horizontally in a transverse direction. The displacement is directly proportional to the wind load and inversely proportional to the horizontal tensile force of the conductor. The maximum allowable *displacement* of the contact wire is limited by the working range of the pantographs and the resulting useable contact wire lateral position. The design of the contact line should guarantee that during operation, the maximum displacement will be lower than the useable contact wire lateral position (see Clause 4.6). The wind load on a conductor is to be determined according to Clause 4.10.2. As with (4.64) the displacement of a single wire by wind, e. g. of a contact wire of a trolley contact line at point x measured from the reference support, as shown in Figure 4.31, is described by

$$e_W(x) = \pm F'_W \cdot x(l_i - x)/(2H) \quad . \tag{4.116}$$

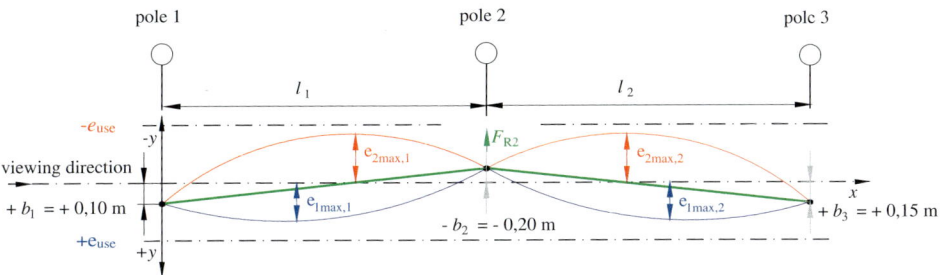

Figure 4.32: Wind deflection in straight line section of Example 4.28.

The *lateral position under wind action* is determined by superimposing both factors of influence $e_W(x)$ and $e_S(x)$ because of conductor stringing. The equation describing the lateral position of contact wire under wind action is

$$
\begin{aligned}
e = e_S(x) + e_W(x) &= \left[\pm F'_W x/(2H) + (b_{i-1} - b_i)/l_i\right] \cdot (l_i - x) + b_i \\
&= \left[\pm F'_W (l_i - x)/(2H) + (b_i - b_{i-1})/l_i\right] \cdot x + b_{i-1} \quad .
\end{aligned}
\tag{4.117}
$$

By differentiating and equating the differential to zero, the position with the *greatest lateral offset* can be found

$$
x_{max} = l_i/2 + (b_i - b_{i-1}) \cdot H/(\pm F'_W \cdot l_i) \quad .
\tag{4.118}
$$

The *maximum offset under wind action* will, therefore, be

$$
e_{max} = \pm F'_W l_i^2/(8H) + (b_{i-1} - b_i)^2 \cdot H/(\pm 2 \cdot F'_W l_i^2) + (b_{i-1} + b_i)/2 \quad .
\tag{4.119}
$$

Equation (4.119) applies if wind load F'_W exceeds $2|b_{i-1} - b_i|H/l_i^2$. If this is not the case, the mathematical maximum of the displacement under wind would be outside the span being considered. In practical applications $b_{i-1} = -b$ and $b_i = +b$ applies to straight line sections:

$$
e_{max} = \pm F'_W l^2/(8H) + 2Hb^2/(\pm F'_W l^2) \quad .
\tag{4.120}
$$

Example 4.28: On straight line sections, the maximum contact wire lateral position e_{max} due to wind is to be found according to Figure 4.32. The effect of droppers, clamps and stitch wires on the wind load should be considered. The given data are:

– span length	$l_1 = l_2 = 70\,\text{m}$
– system height	$h_{SH} = 1,6\,\text{m}$
– contact wire type and diameter	AC-120, $d = 0,013\,2\,\text{m}$
– wind load zone	W1
– height of contact line above terrain	$z \leq 7\,\text{m}$
– return period	ten years
– contact wire tensile force H_{CW}	27 kN
– contact wire stagger	$b_1 = +0,10\,\text{m}, b_2 = -0,20\,\text{m}, b_3 = +0,15\,\text{m}$

Determine the following: Radial force F_{R2} at the support 2, wind load F'_{WCW} and maximum contact wire offsets $e_{1\,max\,1}$, $e_{2\,max\,1}$, $e_{1\,max\,2}$ and $e_{2\,max\,2}$ because of wind.

Calculation of the wind load F'_{WCW}:

The wind load F'_{WCW} is calculated using Equation (4.114)

$$F'_{WCW} = q_z \cdot G_C \cdot C_C \cdot d \quad ,$$

where $G_C = 0,75$ and $C_C = 1,1$, because $12,5 < d < 15,8$.
From Equation (2.8) it is determined with $q_{b,0,10} = 320 \cdot (0,902)^2 = 260\,\mathrm{N/m^2}$

$$q_z = 1,5 \cdot q_{b0,10} = 1,5 \cdot 260 = 390\,\mathrm{N/m^2} \quad .$$

The result is

$$F'_{WCW} = 390 \cdot 0,75 \cdot 1,1 \cdot 0,0132 = 4,25\,\mathrm{N/m} \quad .$$

To include the impact of clamps, droppers and stitch wires etc. this result is multiplied by 1,15. Therefore, the wind load on the contact wire F'_{WCW} is found to be

$$F'_{WCW} = 4,25 \cdot 1,15 = 4,88\,\mathrm{N/m} \quad .$$

Calculation of the radial force F_{R2} at support 2:

The radial force F_{R2} at support 2 with differing staggers $b_1 \neq b_2 \neq b_3$ follows from Equation (4.6)

$$F_{R2} = H_{CW} \cdot [(b_1 - b_2)/l_1 + (b_3 - b_2)/l_2] \quad ,$$

$$
\begin{aligned}
F_{R2} = {} & 27\,000 \cdot \{[(+0,10) - (-0,20)]/70 + [(+0,15) \\
& - (-0,20)]/70\} + 27\,000 \cdot (70 + 70) \cdot 0 = 251\,\mathrm{N} \quad .
\end{aligned}
$$

Calculation of the contact wire offsets $e_{1\max 1}$, $e_{2\max 1}$, $e_{1\max 2}$ and $e_{2\max 2}$:

Equation 4.119 yields for the span between pole 1 and pole 2 with positive wind load F'_W

$$
\begin{aligned}
e_{1\max 1} = {} & +4,88 \cdot 70^2/(8 \cdot 27\,000) + [(+0,1) + (-0,2)]/2 \\
& + [(-0,2) - (+0,1)]^2 \cdot 27\,000/[2 \cdot 70^2 \cdot (+4,88)] \\
= {} & +0,111 - 0,050 + 0,051 = +0,112\,\mathrm{m}
\end{aligned}
$$

and with of negative wind load F'_W

$$
\begin{aligned}
e_{2\max 1} = {} & -4,88 \cdot 70^2/(8 \cdot 27\,000) + [(-0,2) + (+0,1)]/2 \\
& + [(-0,2) - (+0,1)]^2 \cdot 27\,000/[2 \cdot 70^2 \cdot (-4,88)] \\
= {} & -0,111 - 0,050 - 0,051 = -0,212\,\mathrm{m} \quad .
\end{aligned}
$$

Equation (4.119) is also used to calculate $e_{1\max 2}$ and $e_{2\max 2}$ in the span l_2 between pole 2 and pole 3

$$
\begin{aligned}
e_{1\max 2} = {} & +4,88 \cdot 70^2/(8 \cdot 27\,000) + [(-0,2) + (+0,15)]/2 \\
& + [(+0,15) - (-0,2)]^2 \cdot 27\,000/[2 \cdot 70^2 \cdot (+4,88)] \\
= {} & +0,111 - 0,025 + 0,069 = +0,155\,\mathrm{m} \quad ,
\end{aligned}
$$

$$
\begin{aligned}
e_{2\max 2} = {} & -4,88 \cdot 70^2/(8 \cdot 27\,000) + [(+0,15) + (-0,2)]/2 \\
& + [(+0,15) - (-0,2)]^2 \cdot 27\,000/[2 \cdot 70^2 \cdot (-4,88)] \\
= {} & -0,111 - 0,025 - 0,069 = -0,205\,\mathrm{m} \quad .
\end{aligned}
$$

The calculated contact wire offsets $e_{\max}$ due to wind in spans 1 and 2 are shown in Figure 4.32.

The contact wire offsets $e_{1\max}$ and $e_{2\max}$, because of wind load on straight lines can also be calculated using Equations (4.133) and (4.137) for curved track. Thereby, the curvature $1/R \sim 0$ as assumed by $R = 10^6$ m is sufficiently low.

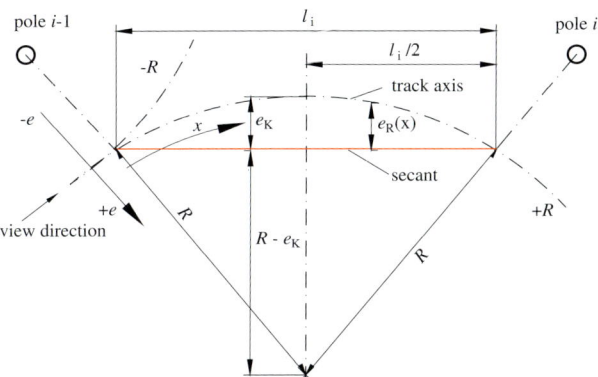

Figure 4.33: Designations of variables in a curved track.

4.10.4 Deflection of contact wire due to wind in curves

4.10.4.1 Contact wire offset in still air

Along curves, the effect of the *contact wire lateral position* is calculated relative to the position of the canted track centre line, since this line also represents the running path of the pantograph centreline. Within the (x, e)-coordinate system of Figure 4.33, the lateral position of the contact wire $e_{R(x)}$ is assigned a positive sign to the right of the track centreline and a negative sign on the left side as viewed in the direction of increasing chainage of stations and poles. The track axis is a circular line and the radius is assumed to be a constant. The versine e_K results from

$$R^2 = (R - e_K)^2 + (l_i/2)^2$$

and solved to e_K

$$e_K = R - \sqrt{R^2 - (l_i/2)^2} = R \cdot \left(1 - \sqrt{1 - [l_i/(2 \cdot R)]^2}\right) \quad . \tag{4.121}$$

It applies $\sqrt{1 - \xi_1} \approx 1 - \xi_1/2$ for $\xi_1 \ll 1$ and for contact lines $[(l_i/(2 \cdot R)]^2 \ll 1$. Then, with $\xi_1 = [l_i/(2 \cdot R)]^2$ using (4.121)

$$e_K = R \cdot \left[1 - \left(1 - \frac{1}{2} \cdot \frac{l_i^2}{(2 \cdot R)^2}\right)\right] = \frac{l_i^2}{8 \cdot R} \quad . \tag{4.122}$$

The circular line is approximated by a parabola to which $e_R(0) = e_R(l_i) = 0$ applies. The parabola assumes the general format

$$e_R(x) = a_r \cdot x^2 + b_r \cdot x + c_r \quad . \tag{4.123}$$

Because $e_R(0) = 0$ then $c = 0$. For $e_R(l_i) = 0$ it is obtained $a_r \cdot l_i^2 + b_r \cdot l_i = 0$, and also $b_r = -a_r \cdot l_i$. The coefficient a_r follows from

$$e_R(l_i/2) = e_K = \frac{l_i^2}{8 \cdot R} = a_r \cdot \left(\frac{l_i}{2}\right)^2 + b_r \cdot \frac{l_i}{2} = a_r \cdot \left(\frac{l_i}{2}\right)^2 - a_r \cdot \frac{l_i^2}{2} = a_r \cdot \frac{l_i^2}{4} - a_r \cdot \frac{l_i^2}{2}$$

to be

$$a_r = -\frac{1}{2 \cdot R} \quad .$$

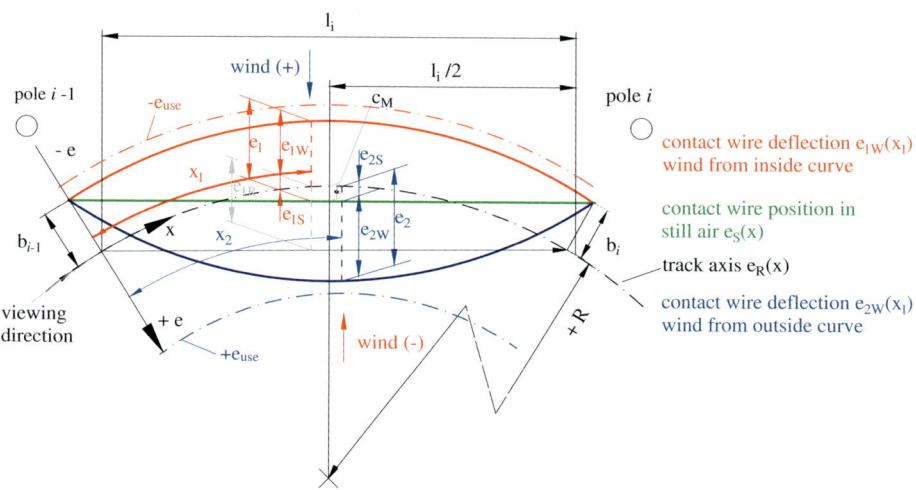

Figure 4.34: Deflection of contact or catenary wire along a curved track.

$e_R(x)$ is determined from (4.123) by inserting a_r, b_r and c_r

$$e_R(x) = -\frac{1}{2 \cdot R} \cdot x^2 + \frac{l_i}{2 \cdot R} \cdot x = \frac{x \cdot (l_i - x)}{2 \cdot R} \quad . \tag{4.124}$$

By assigning a curve radius to the right as $(+)$ and a curve to the left as $(-)$ in Figure 4.33 Equation (4.125) is obtained from (4.124) for the track axis in the chosen coordinate system

$$e_R(x) = \frac{x \cdot (l_i - x)}{2 \cdot (\pm R)} \quad . \tag{4.125}$$

The error, when applying this approximation for the deflection e_K is 0,2 % for a radius of 180 m and a span between poles of 33,4 m.

Using the staggers b_{i-1} and b_i at the supports $i-1$ and i the position in still air of the contact wire within the coordinate system of the Figure 4.34 follows

$$e_{SR}(x) = e_S(x) + e_R(x) = (b_i - b_{i-1}) \cdot x/l_i + b_{i-1} + x \cdot (l_i - x)/(2 \cdot (\pm R)) \quad . \tag{4.126}$$

At midspan at $x = l_i/2$ the contact wire lateral position in still air c_M is determined:

$$e_{SR}(l_i/2) = c_M = (b_i + b_{i-1})/2 + l^2/(8 \cdot (\pm R)) \quad . \tag{4.127}$$

For a curve to the right using $R = +R$, $b_{i-1} = b_i = -b$ and $l_i = l$

$$c_{M+} = -b + l^2/(8 \cdot R) \tag{4.128}$$

and for a curve to the left using $R = -R$, $b_{i-1} = b_i = +b$

$$c_{M-} = +b - l^2/(8 \cdot R) \quad . \tag{4.129}$$

The lateral displacement c_M in midspan is positive in a curve to the right if $l^2/(8R) > b$.

4.10.4.2 Contact wire lateral position under wind action

When determining the *contact wire position under wind action* it is necessary to distinguish between offset because of wind on the contact wire to the right $(+)$ or to the left $(-)$ (Figure 4.34). As with Equations (4.64) and (4.116) the contact wire displacement e_W of the contact wire under wind action is determined by

$$e_W(x) = \pm F'_W \cdot x \cdot (l_i - x) / (2 \cdot H) \quad . \tag{4.130}$$

The positive sign applies to wind according to Figure 4.34 from the left side, which displaces the contact wire to the right $(+)$ in the viewing direction and the negative sign for wind acting from the right displacing the contact wire in the viewing direction to the left side $(-)$. The lateral position of the contact wire e is obtained relative to the track axis

$$e(x) = e_{SR} + e_W = e_S(x) + e_R(x) + e_W(x) \quad .$$

The maximum displacement of the contact wire position relative to the track axis and the wind acting from outside the curve $(+)$ in a curve to the right $(+)$ (Figure 4.35) is obtained by

$$\begin{aligned}
e(x) &= e_S(x_1) + e_R(x_1) + e_{1W}(x_1) \\
e(x) &= \frac{(b_i - b_{i-1}) \cdot x}{l_i} + b_{i-1} + \frac{x \cdot (l_i - x)}{2 \cdot R} + \frac{F'_W \cdot x \cdot (l_i - x)}{2 \cdot H} \\
e(x) &= \frac{(b_i - b_{i-1}) \cdot x}{l_i} + b_{i-1} + \frac{x \cdot (l_i - x)}{2} \cdot \left(\frac{1}{R} + \frac{F'_W}{H} \right) \quad .
\end{aligned} \tag{4.131}$$

In order to determine the *maximum lateral position*, Equation (4.131) is differentiated and $de(x)/dx$ is set to be zero. Then

$$x_{1\max} = \frac{l_i}{2} + \frac{b_i - b_{i-1}}{l_i \cdot (1/R + F'_W/H)} \quad , \tag{4.132}$$

and the maximum lateral displacement of the contact wire under wind action from outside the curve is found to be

$$e_{1\max} = \frac{l_i^2}{8} \cdot \left(\frac{1}{R} + \frac{F'_W}{H} \right) + \frac{(b_i + b_{i-1})}{2} + \frac{(b_i - b_{i-1})^2}{2 \cdot l_i^2 \cdot (1/R + + F'_W/H)} \quad . \tag{4.133}$$

For case $b_{i-1} = b_i = -b$ it is determined by considering the signs

$$e_{1\max} = \frac{l_i^2}{8} \cdot \left(\frac{1}{R} + \frac{F'_W}{H} \right) \quad . \tag{4.134}$$

The wind acting from inside the curve for a curve to the right $(+)$ displaces the contact wire to the outside of the curve $(-)$ as shown in Figures 4.35 and 4.36. Giving

$$\begin{aligned}
e_2(x) &= e_S(x_2) + e_R(x_2) + e_{1W}(x_2) \\
e_2(x) &= \frac{(b_i - b_{i-1}) \cdot x}{l_i} + b_{i-1} + \frac{x \cdot (l_i - x)}{2 \cdot R} - \frac{F'_W \cdot x \cdot (l_i - x)/H)}{2 \cdot H} \quad .
\end{aligned} \tag{4.135}$$

The *maximum lateral position* will be

$$x_{2\max} = \frac{l_i}{2} + \frac{b_i - b_{i-1}}{l_i \cdot (1/R - F'_W/H)} \quad , \tag{4.136}$$

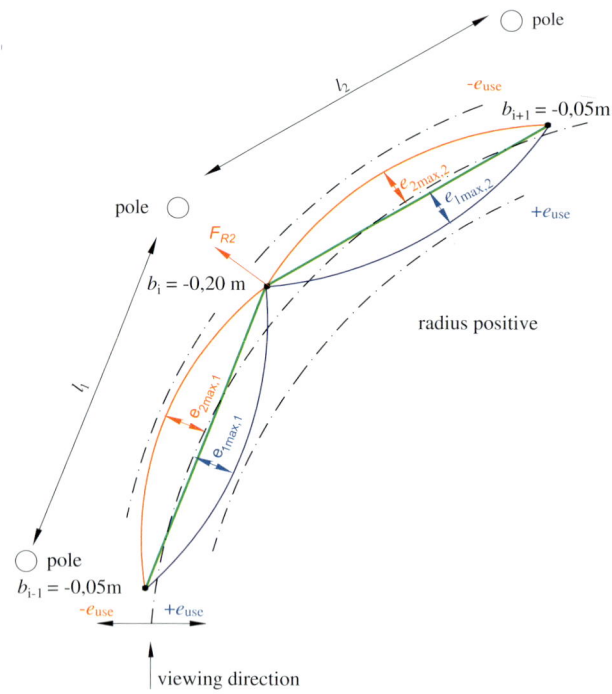

Figure 4.35: Wind deflection in right hand bends with positive radius.

and the maximum contact wire lateral position under wind action obtained is

$$e_{2\max} = \frac{l_i^2}{8}\left(\frac{1}{R} - \frac{F_W'}{H}\right) + \frac{b_i + b_{i-1}}{2} + \frac{(b_i - b_{i-1})^2}{2\cdot l_i^2 \cdot (1/R - F_W'/H)} \quad . \tag{4.137}$$

For the frequent case $b_{i-1} = b_i = -b$ in curves to the right, from Equation (4.137) yields

$$e_{2\max} = \frac{l^2}{8}\cdot\left(\frac{1}{R} - \frac{F_W'}{H}\right) - b \quad . \tag{4.138}$$

If the radius R tends to ∞, Equation (4.137) is transformed to Equation (4.107).
The following examples demonstrate the calculation of contact wire radial forces at supports in curves and its offsets due to wind.

Example 4.29:   On right hand bend line sections, the maximum contact wire lateral position $e_{\max}$ due to wind is to be found according to Figure 4.35. The effect of droppers, clamps and stitch wires on the wind load should be considered.

Given data

$R = +\,3\,400\,\text{m}$, contact wire AC-120, $H_{CW} = 27\,\text{kN}$, $l_1 = l_2 = 70\,\text{m}$, $b_1 = -\,0{,}05\,\text{m}$, $b_2 = -\,0{,}20\,\text{m}$, $b_3 = -\,0{,}05\,\text{m}$ and F_W' according to Table 4.16 multiplied by 1,15 to consider clamps etc. $F_W' = +\,4{,}88\,\text{N/m}$.
Data to be determined:
F_{R2}, $e_{1\max 1}$, $e_{2\max 1}$, $e_{1\max 2}$ and $e_{2\max 2}$.

Calculation of the radial force F_{R2} at the support i with different staggers $b_0 \neq b_1 \neq b_2$:

$$F_{R2} = H_{CW} \cdot \left[\frac{b_1 - b_2}{l_1} + \frac{b_3 - b_2}{l_2} \right] + H_{CW} \cdot \left[\frac{l_1 + l_2}{2 \cdot R} \right]$$

$$F_{H2} = 27\,000 \cdot \left[\frac{(-0,05) - (-0,2)}{70} + \frac{(-0,05) - (-0,2)}{70} \right] + 27\,000 \cdot \left[\frac{70 + 70}{2 \cdot 3400} \right]$$

$$= 27\,000 \cdot 0,0043 + 27\,000 \cdot 0,0206 = +672\,N \quad .$$

Calculation of $e_{1\max 1}$, $e_{2\max 1}$, $e_{1\max 2}$ and $e_{2\max 2}$ in the spans l_1 and l_2:

Within the span l_1 it results

$$e_{1\max 1} = \left(\frac{1}{3400} + \frac{4,88}{27\,000} \right) \cdot \frac{70^2}{8} + \frac{(-0,05) + (-0,2)}{2} + \frac{[(-0,2) - (-0,05)]^2}{2 \cdot 70^2 \cdot (1/3400 + 4,88/27\,000)}$$

$$= 0,291 - 0,125 + 0,006 = +0,172\,m \quad .$$

The position of the maximum lateral displacement is

$$x_{1\max 1} = \frac{70}{2} + \frac{(-0,20) - (-0,05)}{70 \cdot (1/3400 + 1/27\,000)} = 31,00\,m \quad .$$

In the span l_1 with wind acting from inside the curve and $R = +3400\,m$ and $F'_W = -4,88\,N/m$ it results

$$e_{2\max 1} = \left(\frac{1}{3400} - \frac{4,88}{27\,000} \right) \cdot \frac{70^2}{8} + \frac{(-0,2) + (-0,05)}{2} + \frac{[(-0,2) - (-0,05)]^2}{2 \cdot 70^2 \cdot (1/3400 - 4,88/27\,000)}$$

$$= 0,070 - 0,125 + 0,020 = -0,035\,m \quad .$$

The negative sign of $e_{2\max 1}$ means, that $e_{2\max 1}$ is situated on the negative side in the viewing direction relative to the track axis (see Figure 4.35). For the position of the maximum lateral displacement the result is $x_{2\max 1} = -5,4\,m$. The mathematical maximum, therefore, is not situated within the span l_1, but in the preceding span l_0.

In span l_2 with wind from outside the curve $R = +3400\,m$ and $F'_W = +4,88\,N/m$ the result is

$$e_{1\max 2} = \left(\frac{1}{3400} + \frac{4,88}{27\,000} \right) \cdot \frac{70^2}{8} + \frac{(-0,2) + (-0,05)}{2} + \frac{[(-0,05) - (-0,2)]^2}{2 \cdot 70^2 \cdot (1/3400 + 4,88/27\,000)}$$

$$= +0,291 - 0,125 + 0,005\,m = +0,171 \quad .$$

The position of the maximum lateral displacement is at $x_{1\max 2} = 39,00\,m$. The maximum is therefore situated within span l_2. Within span l_2 with wind from inside the curve, $R = +3400\,m$ and $F'_W = -4,88\,N/m$ results in

$$e_{2\max 2} = \left(\frac{1}{3400} - \frac{4,88}{27\,000} \right) \cdot \frac{70^2}{8} + \frac{(-0,05) + (-0,2)}{2} + \frac{[(-0,05) - (-0,2)]^2}{2 \cdot 70^2 \cdot (1/3400 - 4,88/27\,000)}$$

$$= 0,070 - 0,125 + 0,020 = -0,035\,m \quad .$$

The result means, that $e_{2\max 2}$ is situated on the negative side of the track axis (see Figure 4.35). The position of the maximum offset is $x_{2\max 2} = 74,5\,m$. The mathematical maximum is not situated within span l_2, but within the adjacent span l_3.

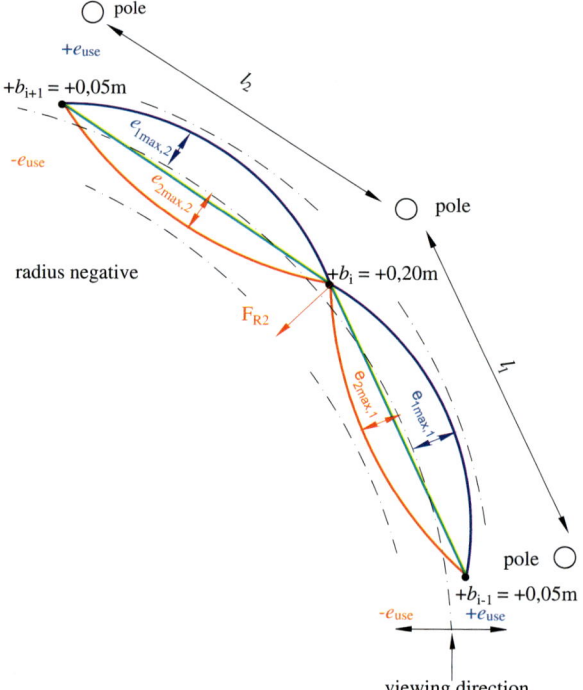

Example 4.30:   On left hand bend line sections, the maximum contact wire lateral position e_{max} due to wind is to be found according to Figure 4.36. The effect of droppers, clamps and stitch wires on the wind load should be considered.

Given data

Contact wire AC-120 , $H_{CW} = 27\,\text{kN}$, $R = -3\,400\,\text{m}$, $l_1 = 70\,\text{m}$, $l_2 = 70\,\text{m}$, $b_1 = +0{,}05\,\text{m}$, $b_2 = +0{,}20\,\text{m}$, $b_3 = +0{,}05\,\text{m}$, $e_{use} = 0{,}40\,\text{m}$ and F'_W according to Table 4.16 multiplied by 1,15 to consider clamps etc. $F'_W = +\,4{,}88\,\text{N/m}$.

Data to be determined: F_{R2} $e_{1\,max\,1}$, $e_{2\,max\,1}$, $e_{1\,max\,2}$ and $e_{2\,max\,2}$.

Calculation of radial force F_{R2} at support i with differing lateral displacements $b_1 \neq b_2 \neq b_3$:

$$F_{R2} = H_{CW} \cdot \left(\frac{b_1 - b_2}{l_1} + \frac{b_3 - b_2}{l_2}\right) - H_{CW} \cdot \left(\frac{l_1 + l_2}{2 \cdot R}\right)$$

$$F_{R2} = 27\,000 \cdot \left(\frac{0{,}05 - 0{,}2}{70} + \frac{0{,}05 - 0{,}2}{70}\right) - 27\,000 \cdot \left[\frac{70 + 70}{2 \cdot (-3\,400)}\right]$$

$$= 27\,000 \cdot (-0{,}0043) + 27\,000 \cdot (-0{,}0206) = -440\,\text{N} \quad .$$

Calculation of $e_{1\,max\,1}$, $e_{2\,max\,1}$, $e_{1\,max\,2}$ and $e_{2\,max\,2}$ within spans l_1 and l_2:

In span l_1 with wind from the inside curve, $R = -3\,400\,\text{m}$ and $F'_W = +\,4{,}88\,\text{N/m}$ it is obtained

$$e_{1\,max\,1} = \left(\frac{1}{-3\,400} + \frac{4{,}88}{27\,000}\right) \cdot \frac{70^2}{8} + \frac{0{,}2 + 0{,}05}{2} + \frac{[0{,}2 - 0{,}05]^2}{2 \cdot 70^2 \cdot [1/(-3\,400) + 4{,}88/27\,000]}$$

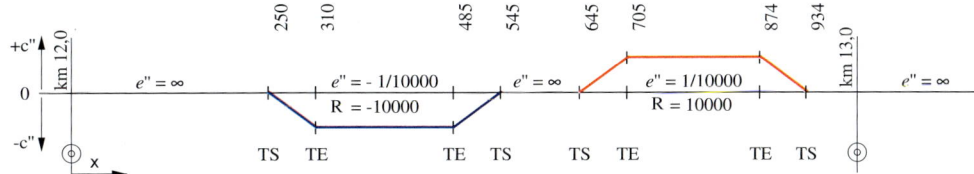

Figure 4.37: Curvature flow of a typical line section of the Cologne–Aachen high-speed line in Germany. e'' curvature, R corresponding radius, TS start of transition curve, TE end of transition curve

$$= -0,069 + 0,125 - 0,020 = +0,035\,\mathrm{m} \quad .$$

Within span l_1, with wind from outside the curve, $R = -3400\,\mathrm{m}$ and $F'_W = -4,88\,\mathrm{m}$ it results

$$e_{2\,\mathrm{max}\,1} = \left[\frac{1}{-3400} - \frac{4,88}{27000}\right]\cdot\frac{70^2}{8} + \frac{0,05+0,2}{2} + \frac{[0,2-0,05]^2}{2\cdot 70^2\cdot[1/(-3400)-4,88/27000]}$$

$$= -0,291 + 0,125 - 0,005 = -0,171\,\mathrm{m} \quad .$$

In span l_2, with wind from inside the curve, $R = -3400\,\mathrm{m}$ and $F'_W = +4,88\,\mathrm{N/m}$ it results

$$e_{1\,\mathrm{max}\,2} = \left[\frac{1}{-3400} + \frac{4,88}{27000}\right]\cdot\frac{70^2}{8} + \frac{0,05+0,2}{2} + \frac{[0,05-0,2]^2}{2\cdot 70^2\cdot[1/(-3400)+4,88/27000]}$$

$$= -0,069 + 0,125 - 0,020 = +0,035\,\mathrm{m} \quad .$$

In span l_2 with wind from outside the curve, $R = -3400\,\mathrm{m}$ and $F'_W = -4,88\,\mathrm{N/m}$ it results

$$e_{2\,\mathrm{max}\,2} = \left[\frac{1}{-3400} - \frac{4,88}{27000}\right]\cdot\frac{70^2}{8} + \frac{0,2+0,05}{2} + \frac{[0,05-0,2]^2}{2\cdot 70^2\cdot[1/(-3400)-4,88/27000]}$$

$$= -0,291 + 0,125 - 0,005 = -0,171\,\mathrm{m} \quad .$$

4.10.5 Contact wire lateral position in transition curves

4.10.5.1 Contact wire lateral position without wind

Transition curves connect straight line sections with curved line sections and curved line sections with different radii. These transition curves are intended to increase the curvature linearly to allow a smooth transition between straight and curved sections. A typical diagram of a line section is shown in Figure 4.37.

Mathematically, a curve with linearly increasing curvature forms a *clothoide*, which can be approximated by a cubic parabola. The second derivation of that is given as

$$e''_T(x) = c_T \cdot x + b_T \quad . \tag{4.139}$$

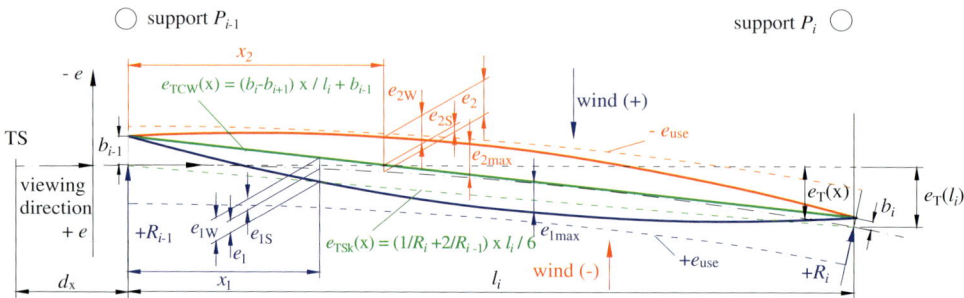

Figure 4.38: Wind deflection in a transition curve for the case B in Figure 4.39.
R_{i-1} and R_i radii at the supports P_{i-1} and P_i with the lateral displacement b_{i-1} and b_i, respectively,
d distance between the start of the transition TS and the support P_{i-1} (Figure 4.39, case B, l_i length of span i)

With definition of $1/R_{i-1}$ as the curvature at support P_{i-1} at $x = 0$, and $b_T = 1/R_i$, as the curvature at support P_i at $x = l_i$, it is determined that

$$c_T = (1/l_i) \cdot (1/R_i - 1/R_{i-1}) \quad . \tag{4.140}$$

The integration of the second derivation e_T'' yields the first derivation

$$e_T'(x) = c_T \cdot x^2/2 + b_T \cdot x + d_T \quad . \tag{4.141}$$

Eventually, the integration of e_T' results in the cubic parabola

$$e_T(x) = (1/6) \cdot c_T \cdot x^3 + (1/2) \cdot b_T \cdot x^2 + d_T \cdot x + e_T \quad , \tag{4.142}$$

as it is used at railways for mathematical presentation of the track axis in transition curves. Because the coordinate and the first derivation are at the curvature start $x = 0$ the integration constant d_T and e_T will be zero. Therefore

$$e_T = (1/6) \cdot c_T \cdot x^3 + (1/2) \cdot b_T \cdot x^2 \text{ and} \tag{4.143}$$

$$e_T'(x) = c_T \cdot x^2/2 + b_T \cdot x \quad . \tag{4.144}$$

With the curvature $b_T = 1/R_{i-1}$ at the support P_{i-1} and $c_T = (1/l_i) \cdot (1/R_i - 1/R_{i-1})$ according to Equation (4.140) it results

$$e_T(x) = \frac{x^3}{6 \cdot l_i} \cdot \left(\frac{1}{R_i} - \frac{1}{R_{i-1}} \right) + \frac{x^2}{2 \cdot R_{i-1}} \quad . \tag{4.145}$$

In Equation (4.145) and in Figure 4.38 the following symbols apply:
$e_T(x)$ track axis in the transition curve as a function of the variable x
l_i span length in the considered span in m
R_{i-1} radius at the support P_{i-1} in m
R_i radius at the support P_i in m and
x variable with $x = 0$ at the support P_{i-1}

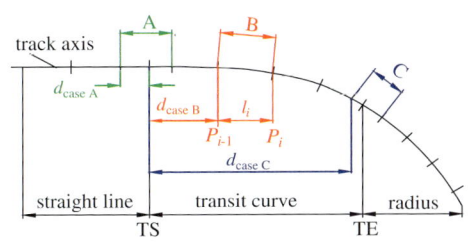

A   span in partly straight line and partly in transition curve

B   span completely in transition curve

C   span in partly transition curve and partly in curve

TS   start of the transition curve

TE   end of the transition curve

P_{i-1} first support of analyzed span l_i

P_i second support of analyzed span l_i

d distance between start of the transition curve and the first support P_{i-1} of the analyzed span l_i

Figure 4.39: Arrangement of spans in transition curves.

At the position $x = l_i$ the coordinate of the track axis assumes the value

$$e_{\mathrm{T}}(l_i) = \pm \frac{l_i^2}{6} \cdot \left(\frac{1}{R_i} - \frac{1}{R_{i-1}} \right) + \frac{l_i^2}{2 \cdot R_{i-1}} = \frac{l_i^2}{6} \cdot \left(\frac{1}{R_i} + \frac{2}{R_{i-1}} \right) \quad , \tag{4.146}$$

from which the secant equation results as

$$e_{\mathrm{TSk}}(x) = \pm \frac{l_i \cdot x}{6} \cdot \left(\frac{1}{R_i} + \frac{2}{R_{i-1}} \right) \quad . \tag{4.147}$$

Figure 4.38 shows the straight line e_{TSk} as the secant between the radius R_{i-1} at the support P_{i-1} and the radius R_i at the support P_i.

The coordinate of the track axis at the position $x = l_i$ is positive in curves to the right $(+)$ and negative in curves to the left $(-)$, consequently the secant is situated to the right of the x axis in curves to the right and left of the x axis in curves to the left, whereby the radii may assume positive and negative values, respectively (Figure 4.38).

Figure 4.39 shows three cases for the position of spans in transition curves. In case B the span is situated completely within the transition curve. The following calculations deal with case B. For the cases A and C, the track axis needs to be described section by section, with the connecting line between the track central points at the supports used as a reference line. Imaginary supports are assumed at TS and TE.

To determine the transition curve within a span, it is necessary to know the distance d between the start of the transition curve TS to the support P_{i-1} (see Figure 4.38 and Figure 4.39, case B). The Equation (4.148) represents the relation with the radii R_{i-1} and R_i:

$$\frac{1}{R_{i-1}} = e_{\mathrm{T}}''(d) \text{ and } \frac{1}{R_i} = e_{\mathrm{T}}''(d + l_i) \quad . \tag{4.148}$$

Within the transition curve, the contact wire position in still air, e_{TCW} without wind, is identical to the connecting line between the contact wire support $P_{i-1}(0; b_{i-1})$ and $P_i(l_i; e_{\mathrm{T}}(l_i) + b_i)$ with the contact wire staggers b_{i-1} and b_i, respectively

$$e_{\mathrm{TCW}}(x) = \frac{(b_i - b_{i-1}) \cdot x}{l_i} + b_{i-1} \quad . \tag{4.149}$$

Equation (4.145) represents the track axis using the coordinate system shown in Figure 4.38. In this coordinate system, the contact wire position in still air $e_{\mathrm{TSCW}}(x)$ is represented by the linear connection of the contact wire supports P_{i-1} and P_i and results from Equations (4.149) and (4.147) as:

$$e_{\mathrm{TSCW}}(x) = e_{\mathrm{TCW}}(x) + e_{\mathrm{TSk}}(x) = \frac{(b_i - b_{i-1}) \cdot x}{l_i} + b_{i-1} \pm \frac{l_i \cdot x}{6} \cdot \left(\frac{1}{R_i} + \frac{2}{R_{i-1}} \right) \quad . \tag{4.150}$$

a) Contact wire position related to track axis as clothoide

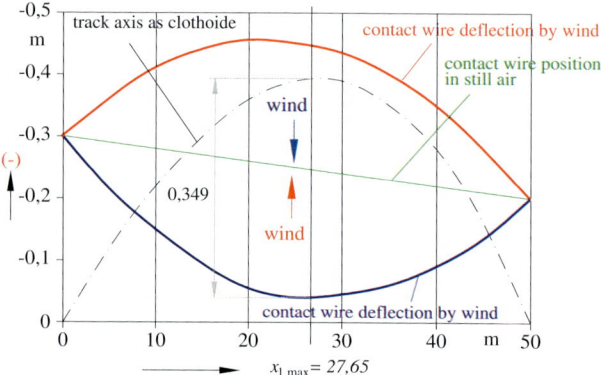

b) Contact wire position related to linearised track axis

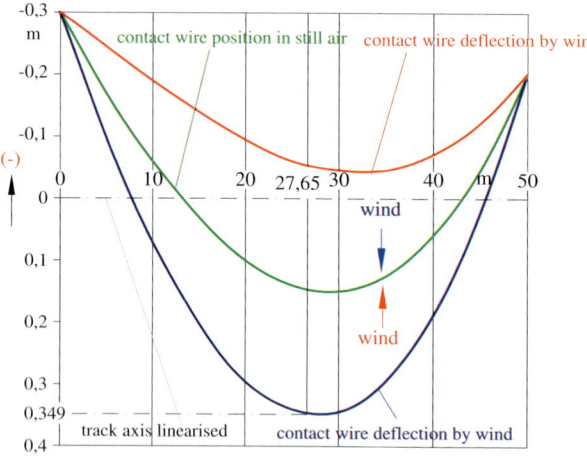

Figure 4.40: Wind deflection in transition curve with and without wind action according to Example 4.31.

From the Equations (4.150) and (4.146) the contact wire position in still air within the transition curve $e_T(x)$ in relation to the tack axis is determined by

$$
\begin{aligned}
e_T(x) &= e_{TSCW} - e_{Ttrack} \\
&= x \cdot \left[\frac{(b_i - b_{i-1})}{l_i} + \left(\frac{1}{R_i} + \frac{2}{R_{i-1}} \right) \cdot \frac{l_i}{6} \right] + b_{i-1} \\
&\quad - \left(\frac{1}{R_i} - \frac{1}{R_{i-1}} \right) \cdot \frac{x^3}{6 \cdot l_i} - \frac{x^2}{2 \cdot R_{i-1}} \quad .
\end{aligned}
\tag{4.151}
$$

In curves to the right, the radii are positive $(+)$ and in curves to the left they are negative $(-)$.

4.10.5.2 Contact wire lateral position under wind action

Similarly to (4.116) and with the agreed assumptions in Clauses 4.10.2 and 4.10.3, the lateral displacement e_W of the contact wire under the action of wind follows from

$$
e_W(x) = \pm F'_W \cdot x \cdot (l_i - x)/(2 \cdot H) \quad .
\tag{4.152}
$$

The contact wire position under wind actions in transition curves results from Equations (4.151) and (4.152) as

$$e_{TW}(x) = e_T(x) + e_W(x)$$

$$
\begin{aligned}
e_{TW}(x) = {} & x \cdot \left[\frac{(b_i - b_{i-1})}{l_i} + \left(\frac{1}{R_i} + \frac{2}{R_{i-1}} \right) \cdot \frac{l_i}{6} \right] + b_{i-1} \\
& - \left(\frac{1}{R_i} - \frac{1}{R_{i-1}} \right) \cdot \frac{x^3}{6 \cdot l_i} - \frac{x^2}{2 \cdot R_{i-1}} \pm \frac{F'_W \cdot x \cdot (l_i - x)}{2 \cdot H}
\end{aligned}
\qquad (4.153)
$$

To determine the *maximum lateral contact wire position* $x_{1\,max}$, the derivation $de_1(x)/dx$ is set to zero. The position of the maximum lateral contact wire position is determined by

$$
\begin{aligned}
x_{1\,max} = {} & \left[-\frac{1}{R_{i-1}} - \frac{F'_W}{H} + \sqrt{ \left(\frac{1}{R_{i-1}} + \frac{F_W}{H} \right)^2 + 2 \cdot \left(\frac{1}{R_i} - \frac{1}{R_{i-1}} \right) } \cdot \right. \\
& \left. \cdot \left\{ \left(\frac{1}{R_i} + \frac{2}{R_{i-1}} \right) \cdot \frac{l}{6} + \frac{b_i - b_{i-1}}{l^2} + \frac{F'_W}{2 \cdot H} \right\} \right] \Big/ \left[\frac{1/R_i - 1/R_{i-1}}{l} \right].
\end{aligned}
\qquad (4.154)
$$

With $x_{1\,max}$ the maximum offset by wind can be obtained from (4.153).

Equation (4.153) can be transformed into the Equation (4.131) for calculating the wind deflection in curves with constant radius by using $R_{i-1} = R_i = R$, whereby the radius R and the wind action F'_W get specific signs. In the straight line sections, R trends to infinity and from (4.153) the Equation (4.117) for the calculation of the wind deflection in straight line sections is determined.

Example 4.31: Find the contact wire displacement in a transition curve (Figure 4.40 a) and b)). Given data:
Stagger $b_{i-1} = -0,3$ m, $b_i = -0,2$ m; radius $R_{i-1} = 2\,000$ m, $R_i = 500$ m; span length $l = 50$ m; $e_{use} = 0,4$ m; contact wire AC-100; contact wire tensile force $H_{CW} = 10$ kN and wind load $F'_W = 4,21 \cdot 1,15 = 4,84$ N/m (Table 4.16, wind load zone W1 multiplied by 1,15 to account for droppers, clamps etc.).
Data to be determined: $x_{1\,max}$ and $e_{1\,max}$
Find the position at support 2 for the location $x = l$ according to Equation (4.146):

$$e_T(50) = \frac{(1/500 + 2/2\,000) \cdot 50^2}{6} = 1,25 \text{ m} \quad .$$

Then, the position of the maximum deflection x_{1max} according to Equation (4.154) is determined:

$$
\begin{aligned}
x_{1\,max} = {} & \left[-\frac{1}{2\,000} - \frac{4,84}{10\,000} + \sqrt{ \left(\frac{1}{2\,000} + \frac{4,84}{10\,000} \right)^2 + 2 \cdot \left(\frac{1}{500} - \frac{1}{2\,000} \right) } \cdot \right. \\
& \left. \cdot \left\{ \left(\frac{1}{500} + \frac{2}{2\,000} \right) \cdot \frac{1}{6} + \frac{-0,2 - (-0,3)}{50^2} + \frac{4,84}{2 \cdot 10\,000} \right\} \right] \Big/ \left[\frac{1/500 - 1/2\,000}{50} \right]
\end{aligned}
$$

$$x_{1\,max} = [-0,000984 + 0,001821] / [0,00003] = 27,88 \text{ m} \quad .$$

In Figure 4.40, the track axis and the contact wire positions with and without wind are shown. In the case of wind action from the left side in the direction of increasing kilometrage, no mathematical maximum of the contact wire lateral position is obtained within the discussed span.

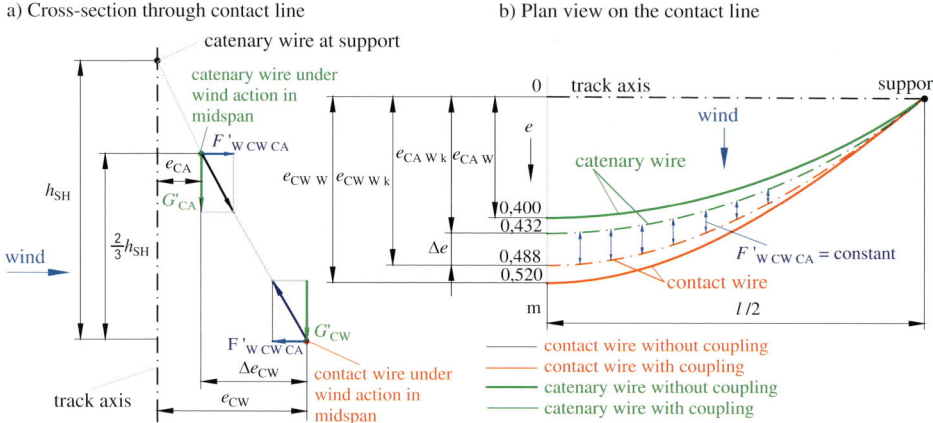

a) Cross-section through contact line   b) Plan view on the contact line

Figure 4.41: Deflection of the overhead contact line, where the contact wire is deflected further than the catenary wire and the droppers transmit coupling forces F'_{WCWCA} from the catenary wire on to the contact wire. The contact and catenary wire lateral position are as zero for approximation (see also values according to (4.159) in Table 4.18). dimensions in m

Inserting $x_{1\max}$ into Equation (4.153) gives

$$
\begin{aligned}
e_1(27,88) &= 27,88 \cdot \left[\frac{(-0,20-(-0,30))}{50} + \left(\frac{1}{500} + \frac{2}{2\,000} \right) \cdot \frac{50}{6} \right] + (-0,30) \\
&\quad - \left(\frac{1}{500} - \frac{1}{2\,000} \right) \cdot \frac{27,88^3}{6 \cdot 50} - \frac{27,88^2}{2 \cdot 2\,000} + \frac{4,84 \cdot 27,88 \cdot (50-27,88)}{2 \cdot 10\,000} \\
&= 0,299\,\text{m} \quad .
\end{aligned}
$$

The contact wire lateral position $e_1(27,88) = 0,299\,\text{m}$ is marked in Figures 4.40 a) and 4.40 b).

4.10.6   Wind deflection of overhead contact lines

If Equations (4.107), (4.108), (4.126) and (4.130) are applied individually to the catenary or contact wire, different lateral offsets are obtained under wind action. When the contact and catenary wires are deflected differently, they exert force components upon each other because of their connection via the droppers.

Some earlier methods applied in practical calculations ignored this fact and calculated *wind deflection of an overhead contact line* based on the assumption that the entire overhead contact line is deflected by the same offset when exposed to wind loads. The force exerted by the wind was calculated for the complete overhead contact line strung with a force corresponding to the sum of the tensile forces of the contact and catenary wire (see Equations (4.126) and (4.130)). Depending on the individual conditions, displacements obtained were either too small or to high when compared with the results obtained for detailed assumptions.

An alternative to this approximation is the individual determination of displacement under wind for the contact and catenary wire. In Table 4.18, these values are given in the columns *Approximation*. If the contact and catenary wire assume different lateral positions under wind action, the droppers assume an inclined position and transfer part of the wind force from

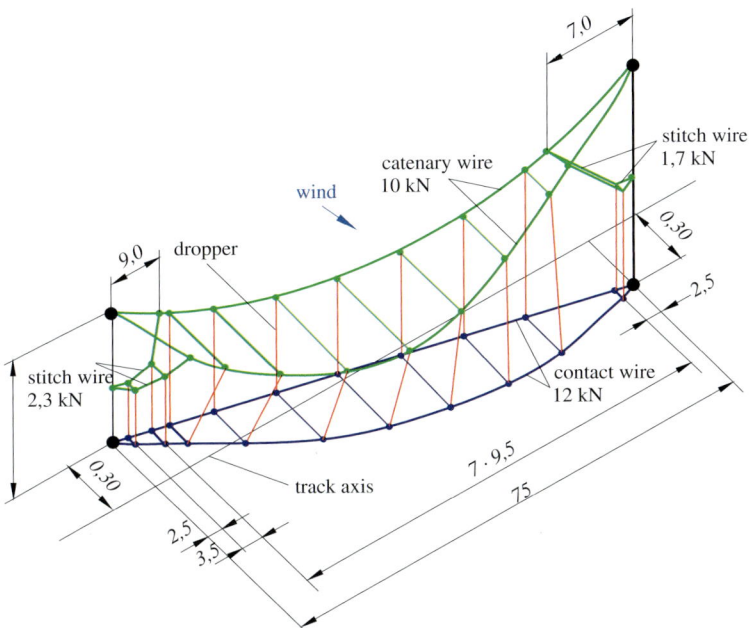

Figure 4.42: Design of overhead contact line type Re200 with wind action according to FEA analysis. dimensions in m

the component with the lower wind displacement to the other component. Based on [4.38] an approach was developed that includes the interaction between contact and catenary wire via the droppers. The forces transferred between contact and catenary wire because of wind action are length-related forces, designated by F'_{WCWCA}. As an approximation, it is assumed that the dropper lengths along the span are all equal to an average. This means, as shown in Figure 4.41, that all droppers in the span have the same deflection angle to a line transverse to the track. This assumption is permissible for system heights greater than 1,4 m. Table 4.18 demonstrates that these assumptions for the wind deflection at midspan lead to values, that practically coincide with the results of a more exact Finite Element Analysis. The values calculated using this method are shown in Table 4.18 within the columns designated by (4.159).

As stated, this assumption is permissible for system heights above 1,4 m. However below that value this procedure can be considered as a useable approximation to mutual impacts as comparisons with FEM have demonstrated.

A parabolic function is assumed to describe the horizontal deflection of the catenary wire due to the force per unit length F'_{WCWCA} exerted by the contact wire. Expressed in relation to the half span shown in Figure 4.41 b), this function can be described as

$$y = F'_{\text{WCWCA}} x^2 / (2H) \quad . \tag{4.155}$$

To the deflection of catenary and contact wire due to wind loads, the following wind loads per unit lengths must be taken into account

$$-\text{on catenary wire}: F'_{\text{WCA tot}} = F'_{\text{WCA}} + F'_{\text{WCWCA}}$$
$$-\text{on contact wire}: \ F'_{\text{WCW tot}} = F'_{\text{WCW}} - F'_{\text{WCWCA}} \quad . \tag{4.156}$$

Table 4.17: Data for overhead contact line types Sicat S1.0 and Sicat H1.0.

Parameter	Symbol	Unit	Sicat S	Sicat H
length related weight force of the contact wire (Table 11.7)	G'_{CW}	N/m	8,46	10,15
tensile force of the contact wire	H_{CW}	kN	12	27
tensile force of the catenary wire	H_{CA}	kN	10	21
contact wire lateral position (stagger) at the support	$-b_{i-1} = +b_i$	m	0,30	0,30
basic wind pressure for wind load zone W1 (Table 4.16)	q_b	N/m	260	260
wind load on the contact line with clamps etc. (Table 4.16)	$F'_{W OCL}$	N/m	8,48	10,06
wind load on the contact wire type AC-100/120 (Table 4.16)	$F'_{W CW}$	N/m	4,21	4,25
wind load on the catenary wire Bz50/120 (Table 4.16)	$F'_{W CA}$	N/m	3,16	4,50
span length between the supports	l	m	75	70

There, it has been assumed, that if the tensile forces of catenary and contact wire are equal, the contact wire will be deflected further than the catenary wire because the former has a larger diameter. In such cases, the catenary wire reduces the wind deflection of the contact wire. The difference in lateral offset because of the different deflections of the contact and catenary wire at the middle of the span is assigned the term Δe. It is calculated by

$$\Delta e = e_{CW} - e_{CA} = \frac{\left(F'_{W CW} - F'_{W CW CA}\right) l^2}{8 H_{CW}} - \frac{\left(F'_{W CA} + F'_{W CW CA}\right) l^2}{8 H_{CA}} \quad . \tag{4.157}$$

Further, from Figure 4.41 the following relation can be obtained

$$\Delta e / (2 h_{SH}/3) = F'_{W,CW CA}/G'_{CW} \quad ,$$

where h_{SH} is the system height. It can be resolved to give

$$\Delta e = 2 F'_{W CW CA} h_{SH}/(3 G'_{CW}) \quad . \tag{4.158}$$

By elimination of Δe in Equations (4.157) and (4.158) an equation describing the *length related coupling force* between contact and catenary wire under wind loads is obtained

$$F'_{W CW CA} = \frac{F'_{W CW} \cdot H_{CA} - F'_{W CA} \cdot H_{CW}}{H_{CW} + H_{CA} + (16 \cdot H_{CW} \cdot H_{CA} \cdot h_{SH})/(3 \cdot l^2 \cdot G'_{CW})} \quad . \tag{4.159}$$

A realistic calculation of *overhead contact line equipment lateral position* is possible with computer programs using *Finite Element Analysis* (FEA). The application to the discussed problem was demonstrated by several examples. Figure 4.42 shows the relative positions of the contact and catenary wire of a standard overhead contact line with a span of 75 m, as determined by FEA.

Example 4.32: Calculate the wind deflections of the contact and catenary wire of standard overhead contact lines Sicat S1.0 and Sicat H1.0. The data are given in Table 4.17.
The calculation is carried out using the coupling force $F'_{W CW CA}$ in accordance with Equation (4.159) with and without taking the stagger b into consideration. The results of the calculation are shown in Table 4.18. The results obtained with Equation (4.159) coincide well with the results of the FEA analysis.

The example using contact line Sicat S1.0 illustrates that if the deflection of the contact wire alone by wind is taken into account, the calculated deflection values are larger than those

Table 4.18: Displacement of contact lines Sicat S1.0 and Sicat H1.0 obtained by conventional approximations (columns 3 and 6), by finite elment analysis (column 4 and 7) and according to equation (4.159) with coupling factor (columns 5 and 8), dimensions in m.

Type	Component	without consideration of value b			with consideration of value b		
		Approximation	FEA	(4.159)	Approximation	FEA	(4.159)
1	2	3	4	5	6	7	8
Sicat S1.0	contact wire e_{CW}	0,520	0,481	0,488	0,597	0,570	0,570
	catenary wire e_{CA}	0,400	0,437	0,432	0,400	0,450	0,434
	contact line e_{OCL}	0,460	–	–	0,547	–	–
Sicat H1.0	contact wire e_{CW}	0,268	0,282	0,277	0,352	0,363	0,358
	catenary wire e_{CA}	0,345	0,342	0,335	0,410	0,406	0,402
	contact line e_{OCL}	0,302	–	–	0,369	–	–

obtained if the complete overhead contact line equipment is considered. The reason is, that the wind has less impact on the catenary wire than on the contact wire subjected to the same tensile force. The results of FEA and the results obtained using Equation (4.159) coincide quite well.

The Sicat H1.0 example also provides good correlation of the values obtained by FEA to those obtained according to Equation (4.159). In this case, calculating the wind-dependent deflection of the contact wire alone leads to lower deflection values. This is because the tensile force on the catenary wire is lower than that on the contact wire, when exposed to the same wind load. In Table 4.18, the determined values for these two examples are given to an accuracy of three decimal places to show the differences between the individual methods discussed here. In practical applications, two decimal places are sufficient because of the assumptions made.

4.11 Span lengths

4.11.1 General

The longitudinal *span length* and tensioning section length have a considerable effect on the investment required for overhead contact line installations. They also affect the the quality parameters such as uniformity of elasticity and contact force performance. Long spans effectively reduce costs. However, for overhead contact lines with running speeds above 230 km/h, it is necessary to limit the span length to comply with the dynamic interaction requirements between pantographs and overhead contact lines (Clause 10.5.3.2).

4.11.2 Impacts on the span length

The *maximum possible span length* is the maximum distance between two adjacent overhead contact line supports, where, after consideration of vehicle movements, tolerances and specified wind loads, the contact wire will not leave the working range of the pantograph pan head. The interaction between track, vehicle, pantograph and overhead contact line determines the *useable contact wire lateral position* e_{use}. The following impacts need to be considered when determining e_{use}

— the geometry of the track consisting of

  - radius R
  - cant u
  - reference cant u_0
  - limit of track gauge l_o
  - distance between the centre lines of the tracks L_S
  - sum of the horizontal supplements $\sum j$ at the lower verification height h_u and at the upper verification height h_o
- the vehicle characteristics with
  - flexibility coefficient s_0
  - reference height of the rolling centre h_{c0}
- the pantograph with
  - length of pan head l_W
  - working length l_A
  - sway D
- the overhead contact line with
  - nominal contact wire height h_{CW}
  - contact wire uplift at the support f_{us}
  - maximum change of height of contact point Δf_c within the span.

When determining e_{max} the following impacts need to be considered
- the wind load with
  - wind velocities, in Germany wind load zones W1 to W4
  - height of the line above sea level, reference is ToR
  - basic and gust wind pressure q_b and q_z
  - the return period of the wind velocity $v_{b,p}$
  - nominal contact line height h_{OCL} above terrain, reference height is ToR
- the overhead contact line with
  - areas prone to wind of conductors and clamps
  - reaction coefficient G_C
  - drag coefficient C_C

The maximum contact wire lateral position e_{max}, which is not determined by the pantograph but by the type of contact line, is analyzed in Clause 4.3.3.

In comparison with the 1 950 mm pan head, the shorter working length of the 1 600 mm pan head leads to a smaller useable contact wire lateral position. At a given wind velocity it leads to shorter span lengths and consequently higher investments when installing new or modifying existing contact lines to interoperable operation with a 1 600 mm long pantograph. To reduce the additional investment, the following possibilities exist:
- reduction of the contact wire height to 5,00 m according to TSI ENE for conventional lines and to 5,09 m for high-speed lines
- waiving the limitation of e_{use} to 400 mm according to TSI ENE
- choosing wind velocities (Clause 2.5.2) according to the local conditions
- selection of random lateral displacements (Clause 4.9.7) corresponding to the quality and type of track and
- increasing the contact wire tensile force

The combination of these measures leads to an increase of e_{use}, e_{max} and consequently to longer span lengths. Further considerations are necessary when designing the insulating and non insulating overlaps regarding the distance between the parallel contact lines. SBB takes advantage of these possibilities in design; the example 4.34 targets these possibilities.

4.11.3 Permissible span lengths

The wind deflection is a decisive factor in determining the permissible *span length* of over-head contact lines. If the maximum contact wire lateral position e_{max} is known, the Equations derived in Clause 4.3.3 can be used as a basis for the calculation of the span length. The maximum possible span length can be found using the symbols in Figure 4.34 and (4.133) as

$$l_{max} = \sqrt{\frac{2 \cdot H_{CW}}{F'_{WCW} + \frac{H_{CW}}{R}} \left(2 e_{max} + b_{i-1} + b_i + \sqrt{(2 e_{max} + b_{i-1} + b_i)^2 - (b_{i-1} - b_i)^2} \right)} \quad (4.160)$$

and can be also used for whole contact lines. Instead of the contact wire tensile force H_{CW}, in this case it is used

$$H_{OCL} = H_{CW} + H_{CA}$$

and instead of F'_{WCW}

$$F'_{WOCL} = F'_{WCW} + F'_{WCA} \quad .$$

Whether this approximation can be accepted depends on the relative wind displacements of the contact and catenary wire. For a theoretically more exact calculation of the permissible span length, the mechanical coupling of the catenary wire with the contact wire as described in Clause 4.10.6 needs to be considered. In a straight line section, the radius R tends to infinity and often it occurs that $b_{i-1} = -b_i = b$. Then from (4.160) it is obtained

$$l_{max} = 2 \sqrt{\frac{H_{OCL}}{F'_{WOCL}} \left(e_{max} + \sqrt{e_{max}^2 - b^2} \right)} \quad . \quad (4.161)$$

Example 4.33: What is the maximum possible span length of the interoperable overhead contact line Re200, traversed by a 1 600 m long pantograph?

Given data of the overhead contact line type Re200 are:

- – verification height for 5,5 m nominal contact wire height $h = 5,86\,m$
- – lower verification height $h'_u = 5,0\,m$
- – provisional value $V = 0,03\,m$
- – tensile force of contact wire $H_{CW} = 10\,kN$
- – tensile force of catenary wire $H_{CA} = 10\,kN$
- – contact wire lateral position at the support $-b_{i-1} = +b_i = 0,30\,m$
- – basic wind pressure for wind load zone W2 (Table 4.16) $q_b = 317\,N/m^2$
- – reference cant and cant deficiency according to TSI ENE $u'_0 = u'_{f0} = 0,066\,m$
- – design wind pressure up to 7 m height (Table 4.16) $q_z = 476\,N/m^2$
- – wind load on the contact line (Table 4.16) $F'_{WOCL} = 10,33\,N/m$

The *useable contact wire lateral position* is according to Table 4.13 for 5,5 m contact wire nominal height and 5,0 m lower verification height $e_{use\,5,86\,m} = 0,350\,m$ and $e_{use\,5,0\,m} = 0,369\,m$, respectively. After deducing the provisional value $P = 0,03\,m$ then $e_{max\,5,86\,m} = 0,320\,m$ and $e_{max\,5,0m} = 0,339\,m$. The tensile forces of the catenary and contact wire are 10 kN each. At the verification height of 5,86 m it is obtained for $b = 0,30\,m$

$$l_{max} = 2 \sqrt{\frac{20000\,N \cdot m}{10,33\,N} \left(0,320\,m + \sqrt{0,320^2\,m^2 - 0,3^2\,m^2} \right)} = 57,8\,m$$

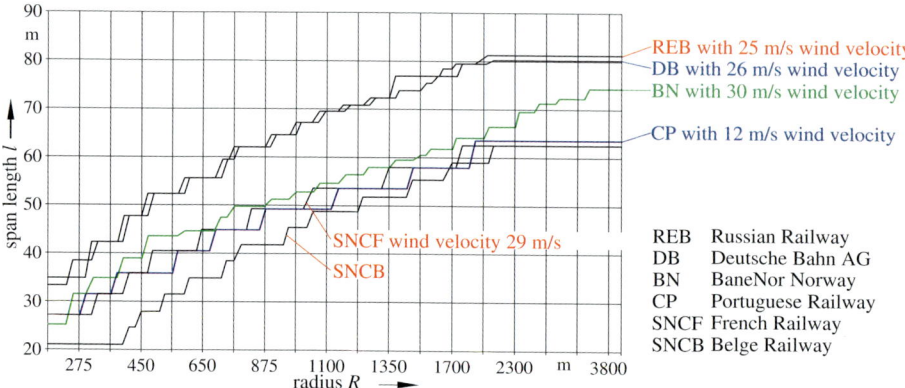

Figure 4.43: Span length depending on the track radii for European railway entities.

and for the contact wire height 5,0 m $l_{max} = 66,8$ m. Because of the 0,86 m lower verification height, a 9 m longer span length can be achieved.

For an interoperable overhead contact line Re200, planned for operation with the 1 950 mm m long pantograph, the maximum possible span length should be determined.

Given are $-b_{i-1} = b_i = 0,40$ m instead of 0,30 m.

The useable contact wire lateral position, therefore, is obtained from Table 4.13 for the nominal contact wire height of 5,5 m $e_{use\,5,86\,m} = 0,525$ m and for the lower verification height 5,0 m $e_{use\,5,0\,m} = 0,571$ m. After subtracting the provision value $P = 0,03$ m it is obtained $e_{max\,5,86\,m} = 0,495$ m and $e_{max\,5,0\,m} = 0,541$ m.

The tensile forces for catenary and contact wire are 10 kN each. For $b = 0,40$ m and verification height 5,86 m the following results

$$l_{max} = 2 \sqrt{\frac{20\,000\,\text{N} \cdot \text{m}}{10,33\,\text{N}} \left(0,495\,\text{m} + \sqrt{0,495^2\,\text{m}^2 - 0,4^2\,\text{m}^2}\right)} = 78,0\,\text{m}$$

and for the contact wire height 5,0 m $l_{max} = 83,5$ m. Because of the 0,86 m lower verification height, 5,0 m longer span lengths can be achieved. In the wind load zone W2, it is possible to have a maximum 78,0 m span length according to the discussed wind deflection calculations and TSI ENE. If the recommendation in accordance with Clause 4.9.11.3 is accepted and $u'_0 = u'_{f0} = 0,0$ m is used, then a span length of 84,5 m can be achieved.

For economic reasons, the design for *pole locations* aims at maximum possible span lengths, considering the minimum sweep $\Delta b_m \geq 1,5$ mm/m of the contact wire (Table 4.14). Figure 4.43 shows the relation between the track radius and the span lengths for several European railway entities.

Example 4.34: The railway line on the bridge across the Elster Valley in Saxony, which is the second largest bridge made of bricks worldwide (Figure 4.44), should be electrified with an interoperable contact line. According to the infrastructure register, this line belongs to the conventional Trans-European Network (TEN) and design needs to consider operation with 1 600 mm and 1 950 mm long pantographs. Consequently, the 1 600 mm long pantograph determines the useable contact wire lateral position and the 1 950 mm pantograph the mechanical-kinematic gauge for the unhindered passage of the pantograph. Given are:

Figure 4.44: Bridge across the Elster Valley without overhead contact wire.

- straight line with laterally fixed ballasted track according to EBO [4.8] with random lateral displacements for the lower verification height $\sum j_u = 0,073$ m and for the upper verification height $\sum j_o = 0,095$ m (Table 4.11)
- wind load zone W2, basic wind pressure $317 \, N/m^2$, 10 year return period (Table 4.16)
- altitude H of the line 412 m above sea level
- annual average temperature 9,25 °C or 282,25 K
- height h_{OCL} of the contact line above the valley floor 68 m
- overhead contact line type Re200, consisting of a contact wire AC-100–CuAg with 10 kN, 13 kN or 15 kN tensile force and a catenary wire Bz II 50 mm^2 with 10 kN tensile force
- cant of track $u = 0,0$ m and reference values $u'_0 = u'_{f0} = 0,066$ m
- provision for an 100 K overhead contact line, with 3,5 m long cantilevers mounted on H-type poles and $P_T = 0,008$ m, $P_S = 0,030$ m and $P_M = 0,025$ m

Find the following data:

- useable contact wire lateral position e_{use} for verification heights of 5,0 m and 5,85 m without limitation on the useable contact wire lateral position to $\leq 0,4$ m
- three spans on the bridge
- radial force at the support in the middle of the three spans

Useable contact wire lateral position e_{usc}

For the laterally fixed track according to [4.8], as installed on the Elster Valley Bridge, with random lateral displacements, results are, for the lower verification height $\sum j_u = 0,073$ m and for the upper verification height $\sum j_o = 0,095$ m and for the useable contact wire lateral positions e_{use} according to Equation (4.86) for the lower verification height $h_{5,0}$ with $u_0 = u_{f0} = 0,066$ m according to TSI ENE

$$
\begin{aligned}
e_{use} &= \frac{l_A}{2} - \frac{2,5}{R} - \frac{l - 1,435}{2} + 0,15 \cdot (u - 0,066)_{>0} \cdot (h' - 0,5) - \\
&\quad \left\{ (0,11 + \sum j_u) + \frac{h' - 5,0}{6,5 - 5,0} \cdot [(0,17 + \sum j_o) - (0,11 + \sum j_u)] \right\} \cdot \\
&= \frac{1,200}{2} - \frac{2,5}{\infty} - \frac{1,465 - 1,435}{2} + 0,15 \cdot (0 - 0,066)_{>0} \cdot (5,0 - 0,5) - \\
&\quad \left\{ (0,11 + 0,073) + \frac{5,0 - 5,0}{6,5 - 5,0} \cdot [(0,17 + 0,095) - (0,11 + 0,073)] \right\} \cdot \\
&= 0,600 - 0 - 0,015 + 0 - 0,183 = 0,402 \text{ m}
\end{aligned}
$$

For the upper verification height of $h_{5,86}$ and using the same approach $e_{use} = 0,355$ m results.

Table 4.19: Useable contact wire lateral position e_{use}, maximum contact wire lateral position e_{max} and span length l according to 4.34. dimensions in m

Specification		EBO		
verification height		$\sum j$	e_{use}	e_{max}
6,50		0,095	0,320	0,290
5,86		–	0,355	0,325
5,00		0,073	0,402	0,372
tensile force $H_{CW} + H_{CA}$ kN	verification height m	span length m		
10 +10	5,86	39,5		
10 +10	5,00	45,0		
13 +10	5,86	42,5		
13 +10	5,00	48,5		
15 +10	5,86	44,0		
15 +10	5,00	50,5		

Provisional value P

The *provisional value P* varies within a half tensioning section length. Starting with the fixed cantilever at the midpoint with $P_T = 0,000\,$m to the last traversed support before the tensioning device, which is approximately 700 m from the midpoint, the temperature-related value $P_T =$ rises to 0,008 m. For the calculation, the maximum value $P_T = 0,008\,$m is chosen resulting in

$$P = \sqrt{P_T{}^2 + P_S{}^2 + P_M{}^2} = \sqrt{0,008^2 + 0,030^2 + 0,025^2} = 0,040\,\text{m} \quad .$$

The provisional value P is assumed to apply at the heights 5,0 m to 6,5 m.

Maximum contact wire lateral position e_{max}

As $e_{max} \leq e_{use} - V$ the maximum contact wire lateral position e_{max} as given in Table 4.19 results.

Basic wind velocity v_b for ten years return period

The bridge across the Elster valley is situated in the wind load zone W2 with $390\,$N/m^2 basic wind pressure for a 50 year return period according to EN 1991-1 NA:2010. Considering the diagram in Figure 2.26 the basic wind pressure $q_{b;0,02} = 390\,$N/m^2 can be transformed to the basic design wind pressure for the a 10 year return period for verification of serviceability to

$$q_{b0.10} = 390 \cdot 0,902^2 = 317\,\text{N/m}^2 \quad .$$

Design wind pressure q_z depending on height

The *design wind pressure, depending on height* q_z can be obtained by considering the height of the contact line above the valley floor $z = 68\,$m according to Equation (2.8) as

$$q_{z=68} = 2,1 \cdot q_b \cdot \left(\frac{z}{10}\right)^{0,24} = 2,1 \cdot 317 \cdot \left(\frac{68}{10}\right)^{0,24} = 1055\,\text{N/m}^2 \quad .$$

Wind load F'_{WCW}

The wind load F'_{WCW} can be calculated from q_b and q_h according to Equation (4.114) for the contact wire AC-100 with $G_C = 0,75$, $C_C = 1,2$ and $d = 0,012$ to be

$$F'_{WCW} = q_z \cdot G_C \cdot C_C \cdot d = 1055 \cdot 0,75 \cdot 1,2 \cdot 0,012 = 11,4\,\text{N/m}$$

and for the catenary wire with a diameter $d = 0,009$ m as

$$F'_{WCA} = q_z \cdot G_C \cdot C_C \cdot d = 1055 \cdot 0,75 \cdot 1,2 \cdot 0,009 = 8,6 \, \text{N/m} \quad .$$

For the contact line, the wind load is determined by adding the wind loads on contact and catenary wire and by multiplying with the factor 1,15 to consider clamps etc.

$$F'_{WOCL} = (11,4 + 8,6) \cdot 1,15 = 22,9 \, \text{N/m} \quad .$$

Maximum span length l_{max}

To calculate the maximum possible span length on straight track, the maximum contact wire lateral position e_{max} after consideration of the contact wire tensile force (10 kN, 13 kN and 15 kN) and the catenary wire tensile force (10 kN) and the contact wire lateral position at the supports are decisive. The staggers are assumed as $b = \pm 0,30$ m. The maximum possible span length l can be obtained using the symbols from Figure 4.31 and Equation (4.107) as

$$l_{max} = 2 \sqrt{\frac{H_{OCL}}{F'_{WOCL}} \left(e_{max} + \sqrt{e_{max}^2 - b^2} \right)} \quad .$$

With the given data, the verification height equates to 5,86 m and $e_{max} = 0,325$ m and the maximum span length is

$$l_{max} = 2 \sqrt{\frac{20\,000}{22,9} \left(0,325 + \sqrt{0,325^2 - 0,3^2} \right)} = 39,5 \, \text{m} \quad ,$$

the length is shown in red in Table 4.19. All other calculated span lengths l are listed in Table 4.19.

Sweep or rate of lateral displacement of the contact wire

The sweep of the contact wire results from the staggers at the supports and the span length l. For the maximum span length l, according to Table 4.19 it follows

$$\Delta b_m = (|b_1| + |b_2|) \cdot 1000/l = (|0,3| + |0,3|) \cdot 1000/50,5 = 12 \, \text{mm/m}$$

The *minimum sweep* of the contact wire $\Delta b_m \geq 1,5$ mm/m is complied with in each span mentioned in Table 4.19 for the overhead contact line on the bridge across the Elster Valley.

Radial forces at the supports

The radial force at both supports in the middle of the bridge is calculated according to Equation (4.7) with the lateral contact wire positions $b_{i-1} = -b_i = 0,3$ m as

$$F_R = 4 \cdot H_{CW} \cdot b/l$$

and using the numerical data for the longest span length in Table 4.19, a contact wire tensile force of 15 kN is obtained

$$F_R = 4 \cdot 15\,000 \cdot 0,3/47,2 = 381 \, \text{N} \quad .$$

The calculated radial force is used for the calculation of the poles and lies within the specified range $80 \, \text{N} \leq F_R \leq 2\,000 \, \text{N}$.

Assessment

Consideration of the bridge height above the terrain, as with the bridge across the Elster Valley, demonstrates its considerable impact. According to the design assumptions of Deutsche Bahn, which specify for conventional contact lines in wind zone W2, 26 m/s wind velocity and, therefore, 11,44 N/m wind load for the interoperable contact line Re200, span lengths up to 63,5 m would be possible on the bridge. For type Re200 with a 10 kN contact wire tension and catenary wire with 10 kN tension, only a 39,5 m maximum span length can be achieved according to Example 4.34, on the bridge. The approach of Deutsche Bahn to useable span lengths exceeds the limiting span length according to Example 4.34 by 24 m.

Span lengths by up to 11 m longer would be possible if a verification height of 5,0 m was used instead of 5,85 m and the contact wire tension was increased from 10 kN to 15 kN.

4.12 Process for verification of serviceability

The following steps are necessary for verification of overhead contact line serviceability:
1. calculation of e_{use}
2. determination of the provisional value v
3. calculation of e_{max}
4. determination of the wind load zones W1 to W4 for the geographic region
5. determination of the basic wind velocity with a 50 year return period $v_{b,0,02}$
6. determination of the basic wind velocity with a 10 year return period $v_{b,0.10}$
7. calculation of the basic pressure q_b
8. calculation of the design wind pressure q_z depending on the height
9. calculation of the wind load on conductors F'_w
10. calculation of resetting forces and their affect on contact and catenary wire tensile force
11. calculation of the longitudinal span l

4.13 Bibliography

4.1 *Ebs 02.05.61*: Spann- und Durchhangskurven zum Einregulieren der Festpunktverankerungsseile (Höchstspannung 8 000 N bei −30 °C) (Tension and sagging diagrams for adjusting the midpoint anchor ropes (maximum tensile force 8 000 N at −30 °C)). Deutsche Bahn AG, Frankfurt, 1997.

4.2 *Ungvari, S.; Paul, G.*: Oberleitung Sicat H1.0 für die Neubaustrecke Köln–Rhein/Main (Overhead contact line Sicat H1.0 for the new high-speed line Cologne–Rhine/Main). In: Elektrische Bahnen 96(1998)7, pp. 23 to 242.

4.3 *Bausch, J.; Kießling, F.; Semrau, M.*: Hochfester Fahrdraht aus Kupfer-Magnesiumlegierungen (High-strength contact wire made of copper magnesium alloy). In: Elektrische Bahnen 92(1994)11, pp. 295 to 300.

4.4 *Payan Cuevas, F.; Puschmann, R.; Vega, T.*: Overhead contact line maintenance for the Madrid–Lérida high-speed line. In: Elektrische Bahnen 106(2008)5, pp. 211 to 221.

4.5 *Bobillot, A.; Mentel, J.-P.*: World record − 574,8 km/h on rails. In: Elektrische Bahnen 107(2009)9, pp. 396 to 375.

4.6 *Kiessling, F.; Nefzger, P.; Nolasco, J. F.; Kaintzyk, U.*: Overhead power lines – Planning, design, construction. Springer Publishing, Berlin-Heidelberg-New York, 2003.

4.7 *Schmidt, P. et al.*: VEM Handbuch: Energieversorgung elektrischer Bahnen (VEM Manual: Power supply of electric railways). VEB Technik Publishing, Berlin, 1975.

4.8 *EBO*: Eisenbahn-Bau- und Betriebsordnung (railway construction and operation ordinance). Bundesrepublik Deutschland, BGBl. 1967 II p. 1563, last version in: BGBl. 2012 I p. 173.

4.9 *Regulation 1301/2014/EU*: Technical specification for the interoperability of the energy subsystem of the rail system in the Union. In: Official Journal of the European Union, No. L 356 (2014), pp. 179 to 227.

4.10 *Prussian-Hessian, Bavarian and Badinian state railways*: Übereinkommen betreffend die Ausführung elektrischer Zugförderung (Agreement concerning the design of electric train operation). Berlin-Munich-Karlsruhe, 1912/1913.

4.11 *Groh, T.; Harprecht, W.; Puschmann, R.*: Interoperabilität elektrischer Bahnen – 100 Jahre Vereinbarung für 15 kV 16 2/3 Hz (Interoperability of electric railways 15 kV 16 2/3 Hz agreement 100 years old). In: Elektrische Bahnen 10(2012)12, pp. 686 to 699.

4.12 *Regulation 1302/2014/EU*: Technical specification for interoperability relating to the rolling stock – locomotives and passenger rolling stock subsystem of the rail system in the European Union. In: Official Journal of the European Union, No. L 356 (2014), pp. 228 to 393.

4.13 *Wili, U.*: Vereinheitlichte Stromabnehmerwippe – die Europawippe (Standardized current collector pan – the European pan). In: Elektrische Bahnen 113(1994)11, pp. 301 to 304.

4.14 *Directive 997.0101*: Oberleitungsanlagen; Allgemeine Grundsätze (Overhead contact line installations; general requirements). Deutsche Bahn AG, Frankfurt, 1995.

4.15 *Ebs 02.05.29*: Toleranzen für Oberleitungen (Tolerances of overhead contact lines). Deutsche Bahn AG, Munich, 1982.

4.16 *Ebs 02.03.20*: Verwendung der IPB-Maste auf der freien Strecke (Use of IPB poles on interstation lines). Deutsche Bahn AG, Munich 1974.

4.17 *Deutsche Reichsbahn*: Dienstvorschrift für die Ausführung und die Festigkeitsberechnung der Fahrleitungen für Wechselstrombahnen mit 15 kV (Specification for execution of strength analysis for 15 kV alternating current railways). Deutsche Reichsbahn, Munich, 1931.

4.18 *Ezs 837*: Seitliche Festlegung des Fahrdrahts, abhängig vom Bogenhalbmesser R und der Windgeschwindigkeit w für Reichsstromabnehmer 1950 (Lateral fixing of the contact wire depending on the radius R and the running speed w for the pantograph type 1950). Deutsche Reichsbahn, Munich, 1931.

4.19 *UIC Code 608*: Conditions to be complied with for the pantographs of tractive units used in international services. UIC, Paris, 3rd edition 2003.

4.20 *Directive 997*: Oberleitungsanlagen; Allgemeine Grundsätze (Overhead contact lines; general principles). Deutsche Bahn AG, Frankfurt, 2001.

4.21 *Deutsche Bahn*: Infrastrukturregister (Register of infrastructures).
Internet: http://fahrweg.dbnetze.com/fahrweg-de, 2013.

4.22 *Deutsche Bahn*: Richtlinien für die Errichtung von Fahrleitungen für 15 kV und 25 kV Nenn-
      spannung und Regelstromabnehmer (Fahrleitungsrichtlinien) (Directives for installation of
      contact lines for 15 kV and 25 kV nominal voltage and standard pantographs). Munich, 1953.

4.23 *Olv 1*: Vorschrift für Oberleitungsanlagen, Teilheft 1: Errichtung und Instandhaltung von Ober-
      leitungsanlagen, Oberleitungsvorschrift (Entwurf)(Specification for overhead contact lines,
      Part 1: Installation and maintenance of overhead contact lines. Overhead contact line speci-
      fication (draft)). Deutsche Bahn, Frankfurt, 1986.

4.24 *Railway CERT*: Technisches Dossier für Regeloberleitung Re200, Re200i und seitenhalterlose
      Re200 (Technical dossier on the standard overhead contact lines Re200, Re200i and Re200
      without steady arms). Railway CERT, Bonn, 2011.

4.25 *Railway CERT*: Technisches Dossier für die Interoperabilitätskomponente Regeloberleitung
      Re 250 (Technical dossier on the interoperability component standard overhead contact line
      Re250). Railway CERT, Bonn, 2011.

4.26 *Deutsche Bahn*: Anhubmesswerte SFS Nürnberg–Ingolstadt, km 38,4 (Measurement data on
      contact line uplift on the high-speed line Nuremberg–Ingolstadt). Deutsche Bahn Systemtech-
      nik, Test report: 06-P-002803-TZF74.1-PR-0012, Munich, 2006.

4.27 *Nickel, T.; Puschmann, R.*: Technical Specification Energy 2015 – Harmonized design of over-
      head contact lines. In: Elektrische Bahnen 113(2015)INT2, pp. 33 to 45.

4.28 *UIC Code 505-1*: Railway transport stock – Rolling stock construction gauge. UIC, Paris, 2006.

4.29 *UIC Code 505-5*: History, justification and commentaries on the elaboration and development
      of UIC leaflets of the series 505 and 506 on gauges. UIC, Paris, 2010.

4.30 *Directive 800.0110*: Netzinfrastruktur; Technik entwerfen, Linienführung (Network infrastruc-
      ture; technological design, line layout). Deutsche Bahn, Frankfurt, 2008.

4.31 *Directive 821*: Oberbau inspizieren (Inspection of permanent way). Deutsche Bahn, Frankfurt,
      2010.

4.32 *Jänsch, E.*: Die Abmessungen der Mittelwagen des Intercity Express (Dimensions of interme-
      diate cars of Intercity Express). In: ETR 37(1988)4, pp. 197 to 204.

4.33 *Jacobs, K.; Mittmann, W.*: Neue Lichtraumbestimmungen auf kinematischer Grundlage –
      Regeln, Auswirkungen, Perspektiven (New specifications on gauges on kinematic basis –
      Rules, effects, perspectives). In: Heinisch, R.; Koch, P.; Kracke, R.; Rahn, T. (Editors), Edi-
      tion ETR: Erstellen und Instandhalten von Bahnanlagen (Construction and Maintenance of
      railway installations), pp. 241 to 249, Hestra Publishing, Darmstadt, 1993.

4.34 *Nickel, T.; Puschmann, R.*: TSI Energy 2015 – Reference parameters for overhead contact
      lines. In: Elektrische Bahnen 113(2015)INT2, pp. 46 to 50.

4.35 *VDV Publication 550*: Oberleitungsanlagen für Straßen und Stadtbahnen. (Overhead contact
      line installations for trams and mass transit railways), Association of German Public Transport
      Entities (VDV), Köln, 2003.

4.36 *UIC Code 799*: Characteristics of alternative current overhead contact systems for lines oper-
      ated at speeds over 200 km/h. UIC, Paris, 2002.

4.37 *Puschmann, R.*: Zulässige Fahrdrahtseitenlage für interoperable Strecken (Permissible lateral deviation of contact wires on interoperable lines). In: Elektrische Bahnen 110(2012)6, pp. 270 to 279.

4.38 *Wlassow, I. I.*: Fahrleitungsnetz (Overhead contact line network). Technical books Publishing, Leipzig, 1955.

5 Currents and voltages in the contact line system

5.0   Symbols and abbreviations

Any symbols and abbreviations followed by an apostrophe ' indicate a magnitude per unit length, given in 1/km or 1/m

Symbol	Definition	Unit
A	cross-section	mm^2
C	capacitance, capacitor	nF
C_E	capacitance versus earth	nF
CA	CAtenary wire, alternatively messenger wire	–
CW	Contact Wire	–
D_m	mean distance between two conductors	m
$F(\lambda)$	Gaussian standard distribution	–
FL	Feeder Line	–
G	admittance	S
I	current	A
I_{1s}	maximum 1-sec mean current	A
I_{10s}	maximum 10-sec mean current	A
I_{12}	current between point 1 and 2	A
I_{3min}	maximum 3-min mean current	A
I_{CW}	contact wire current	A
I_{OCL}	current in the overhead contact line	A
I'_{OCL}	distributed line load in a feeding section	A/m
I_P	short-circuit impulse current	kA
$|\underline{I}_{Pi}|$	absolute value of the average current of the traction unit i	A
I_T	test current	A
I_a	mean annual load current	A
I_c	balancing current between substations	A
$\underline{I}_b$	known current	A
$\underline{I}_c$	balancing current between substations	A
$\underline{I}_e$	current to be determined	A
I_k	short-circuit current	kA
$I_{k\,min}$	minimum short-circuit current	kA
I''_k	initial symmetrical short-circuit current	kA
I_{max}	maximum current	A
I_{min}	minimum current	A
I_p	peak short-circuit current	kA
I''_{rG}	generator current rating	kA
I_{th}	thermal equivalent short-circuit current	kA
I_{tot}	total current	A

Symbol	Definition	Unit
I_{trc}	traction current of a train	A
$\underline{I}_{trc}^*$	conjugated complex traction current	A
L	inductance, inductivity	mH
L	length of a feeding section	km
L_C	characteristic length	km
L'_{ex}	external inductance of a conductor	mH/m
L'_{in}	internal inductance of a conductor	mH/m
L_{trans}	transition length	km
M	number of calculation steps	–
N	number of simulation steps	–
OCL	Overhead Contact Line	–
P	power	kW
P_L	power limit	kW
P_a	annually mean power	kW
P_d	daily mean power	kW
P_h	hourly mean power	kW
$P_{h\,max}$	maximum mean power per hour	kW
R	resistance, real part of the impedance	Ω
R_C	resistance of a conductor	Ω
R'_E	longitudinal resistance of the earth	Ω/km
R_{Req}	equivalent rail resistance	Ω
R_{loop}	short-circuit loop resistance	Ω
R_{trc}	resistance of a traction unit	Ω
RC	Return Conductor	–
S_k''	initial short-circuit AC power	MVA
S_{trc}	apparent traction power	kW
SP	Switching Post	–
SS	Substation	–
T	Track	–
T	time period	s
T_i	integration period	s, min
U	voltage	V, kV
U_{12}	source voltage	kV
$\underline{U}_{Pi}$	momentary rms voltage at the pantograph of train i	V, kV
U_S	longitudinal structure voltage	V
U_S	source voltage	V, kV
U_{SS}	voltage at the substation busbar	V, kV
U_T	test voltage	V
$U_{j,k}(t)$	instantaneous voltage for evaluation $U_{mean\,useful}$	V
$U_{mean\,useful}$	mean useful voltage	V
U_{trc}	voltage of a traction vehicle	kV
U_x	voltage at position x	V
W_P	real energy per train cycle	kWh
W_Q	reactive energy per train cycle	kWh
W_a	annual energy demand	kWh

Symbol	Definition	Unit
X	reactive part of the impedance	Ω
X_{ex}	external reactance of a conductor	mΩ
X_{in}	inner reactance of a conductor	mΩ
X_{inR}	inner reactance of a rail	mΩ
X_{ik}	mutual reactance between the conductors i and k	mΩ
X_{trc}	reactance of a traction unit	Ω
$\underline{Z}'$	line impedance per unit length	Ω/km
$\underline{Z}_k$	short-circuit impedance	Ω
$\underline{Z}_{line}$	line impedance	Ω
$\underline{Z}_{12}$	impedance between point 1 and 2	Ω
$\underline{Z}_T$	measured impedance	Ω
$\underline{Z}_{ii}$	self impedance of the loop i	Ω
$\underline{Z}_{ik}$	mutual impedance between the conductors i and k	Ω
$\underline{Z}_{inR}$	inner rail impedance per unit length	Ω
a_{ik}	mutual distance between the conductors i and k	m
c_a	annually load coefficient	–
c_d	daily load coefficient	–
c_h	hourly load coefficient	–
f	frequency	Hz
h_{CW}	contact wire height	m
h_m	mean height of the catenary	m
k_{CW}	contact wire current distribution coefficient	–
k_{FL}	feeder line current distribution coefficient	–
l	length	km, m
l_T	length of test circuit	km, m
l_{max}	maximum distance between substations	km
l_{min}	minimum distance between substations	km
n	number of trains per section	–
n_1	number of parallel conductors	–
r	radius of a conductor	mm
r_{eq}	equivalent radius	mm
r_{eqA}	equivalent radius of cross-section for a rail	mm
t	time	s
t^*	time window for a time-weighted equivalent continuous load curve	s
t_D	time step for a time-weighted equivalent continuous load curve	s
t_{RE}	response or operating time of a protective relay	s
t_{SA}	break-time of a power circuit breaker	s
t_k	duration of a short circuit	s
t_m	time period for effective current evaluation	s
t_{zz}	time quotient for evaluation of mean useful voltage	–
v_W	wind velocity	m/s
v_p	variation coefficient of the power per hour	–
ΔP	line losses	W
$\underline{\Delta U}$	voltage drop	V
$\underline{\Delta U}_l$	voltage drop along the length l	V

Symbol	Definition	Unit
$\Delta \underline{U}_{\mathrm{max}}$	maximum voltage drop	V
$\Delta \underline{U}_{\mathrm{q}}$	transversal voltage drop	V
$\Delta \underline{U}_{\mathrm{t}}$	time period	s
$\Delta \underline{U}_{\mathrm{xe}}$	voltage drop occurring at single-end feed	V
α_{R}	temperature coefficient of a resistor	1/K
δ_{E}	current penetration into earth	m
ε_0	electric field constant $8,85 \cdot 10^{-9}$	F/km
ε_{r}	relative permittivity	–
γ	propagation constant	1/km
κ_{P}	short-circuit current impulse factor	–
λ_{L}	variable of the Gaussian standard distribution	–
λ_{P}	factor according to EN 60 865-1	–
φ	phase angle	°
φ_{Z}	phase angle of the line impedance	°
μ	soil permeability	V s/(A m)
μ_0	permeability constant	V s/(A m)
μ_{P}	decay factor for short circuits	–
μ_{r}	relative permeability	–
ω	circular frequency	s^{-1}
ρ_{20}	specific resistance at 20 °C	$\Omega \cdot \mathrm{m}$
ρ_E	specific soil resistivity	$\Omega \cdot \mathrm{m}$
ϑ	conductor temperature	°C
ρ	specific resistivity	$\Omega \cdot \mathrm{m}$,
κ	specific conductivity	$1/(\Omega \cdot \mathrm{m})$
σ_{p}	standard deviation of the power per hour	kW
θ_{air}	air temperature at de-icing procedure	°C

5.1   Electrical characteristics of contact lines

5.1.1   Basic relationships

Electrical characteristics such as the impedance, current distribution and current capacity determine the *energy transmission behavior* of a contact line system. The electrical characteristics of a contact line and the corresponding protection required for the electric installations and operating equipment are designed in view of the current to be transmitted. Once the transmission characteristics and currents are known, it is possible to evaluate the *electromagnetic interferences* being emitted by an electric railway line. The contact line system can be assumed to act as a long conductor installed above ground. Figure 5.1 shows the schematic circuit diagram of the traction power supply:

- the substation supplies the energy with a source voltage $\underline{U}_{12}$ and the current $\underline{I}_{\mathrm{trc}}$
- the energy is transmitted from the substation to the traction vehicles via the contact line system. The line impedance $\underline{Z}$ in case of AC supplies or the resistance R in case of DC supplies causes a voltage drop $\Delta \underline{U}$ and a power loss ΔP along the contact line
- the electric power depends on the status of the train at the respective time, either traction or regenerative breaking mode

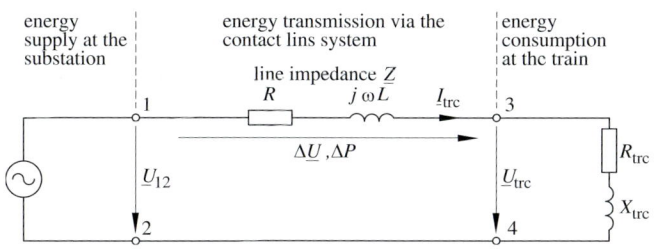

Figure 5.1: Schematic circuit diagram of traction power supply.

– the traction current $\underline{I}_{\text{trc}}$ returns to the substation through the return circuit consisting of the rails and return conductors. In AC supplies, the earth is part of the return circuit

The following equations apply to single-phase AC railways. Since there is no reactive component effective in direct-current systems, the simpler relationships applying to DC railways can be deduced from the AC-related equations by replacing the impedance by the resistance. The power $\underline{S}_{\text{trc}}$ of the train has to be transferred to the train via the collectors under the respective conditions and amounts to the complex value

$$\underline{S}_{\text{trc}} = \underline{U}_{\text{trc}} \cdot \underline{I}_{\text{trc}}^* \quad , \tag{5.1}$$

where $\underline{U}_{\text{trc}}$ is the voltage at the pantograph, $\underline{I}_{\text{trc}}^*$ is the conjugated complex value of the current. This power is supplied by the substation and transferred to the train via the contact line system. The contact line and the return circuit form an impedance for power transmission. This impedance can be measured by applying a voltage between points 1 and 2 according to Figure 5.1 and short-circuiting points 3 and 4. The impedance $\underline{Z}_{12}$ is then determined from the voltage $\underline{U}_{12}$, applied between points 1 and 2 and the resulting current $\underline{I}_{\text{trc}}$

$$\underline{Z}_{12} = \underline{U}_{12}/\underline{I}_{\text{trc}} \quad . \tag{5.2}$$

The impedance has a real component R and a *reactive component X*

$$X = \omega L \quad . \tag{5.3}$$

There, ω is the *angular frequency* $\omega = 2\pi f$ with f as basic frequency of the traction power network.

In (5.3), L is the *inductance* of the system between points 1 and 2 and can be measured when points 3 and 4 are short-circuited. The real component R of the impedance is the effective resistance of the contact line and the return circuit.

The total impedance is thus

$$\underline{Z}_{12} = R + j\omega L \quad . \tag{5.4}$$

If expressed in the commonly used formats, $\underline{Z} = \underline{Z}_{12}$ can be written as

$$\underline{Z} = R + j\omega L = R + jX = |\underline{Z}| \angle \arctan(X/R) = |\underline{Z}| \angle \varphi_Z \quad . \tag{5.5}$$

In Equation (5.5), $|\underline{Z}|$ is the absolute value of the impedance and $\arctan(X/R)$ is the *phase angle* φ_Z. The term $\angle \varphi_Z$ can be written as $e^{j\varphi_Z} = \exp(j\varphi_Z) = \cos\varphi_Z + j\sin\varphi_Z$.

Table 5.1: Resistance per unit length R'_C of conductors in mΩ/km for different cross-sections A and temperatures.

Conductor	A mm^2	R'_C at 20°C new	R'_C at 20°C 20 % worn	R'_C at 40°C new	R'_C at 40°C 20 % worn
Contact wires					
AC-80–Cu	80	223	278	240	300
AC-100–Cu	100	179	223	193	240
AC-120–Cu	120	149	186	160	200
AC-150–Cu	150	119	149	128	160
Catenary wires					
Cu	50[1]	360	–	390	–
Cu	70	271	–	292	–
Cu	95	191	–	206	–
Cu	120	153	–	165	–
Cu	150	121	–	131	–
Bz II	50[1]	561	–	605	–
Bz II	70	422	–	455	–
Bz II	95	298	–	321	–
Bz II	120	237	–	255	–
Bz II	150	189	–	204	–
Steel	50[1]	3 880	–	4 230	–
Conductor rails					
soft iron	5 100	22,5	–	25,2	–
soft iron	7 625	15,0	–	16,8	–
composite[2]	5 100	6,8	–	7,3	–
composite[2]	2 100	16,4	–	17,6	–
Reinforcing wires and feeding wires					
243-AL1	240	118	–	126	–
625-AL1	625	45	–	48	–

[1] seven strand, [2] aluminum-steel composite rail

5.1.2   Impedances

5.1.2.1   Components

The impedance of the loop comprising the contact line and the return circuit is commonly called the *line impedance*. In DC railway installations, the line impedance is obtained from the resistances of all parallel contact lines, reinforcing feeder conductors or cables and the return circuit comprising the track resistance including all parallel return conductors. The impedances of contact lines are usually expressed in relation to the length.

5.1.2.2   Resistance per unit length

The *resistance per unit length* of conductors, wires, cables and rails is determined by the electrical properties of the materials from which these components are made. A summary of material properties specified in standards, publications on contact line materials and in research reports on new contact line materials, has been listed in Tables 5.1 to 5.4. The resistance per unit length of wires, conductors, rails and earth is determined in the following.

Table 5.2: Resistance per unit length R' of overhead contact lines in mΩ/km at 20 °C and 40 °C.

Contact line configuration		Catenary wire cross-sectional area in mm^2									
		50		70		95		120		150	
Contact wire	Catenary wire	20 °C	40 °C	20 °C	40 °C	20 °C	40 °C	20 °C	40 °C	20 °C	40 °C
AC-100–Cu[1]	Bz II	136	147	126	135	112	121	102	110	92	99
AC-100–Cu[1]	Cu	119	129	108	116	92	100	82	89	72	78
AC-100–Cu[2]	Bz II	160	172	146	157	128	137	115	124	102	110
AC-100–Cu[2]	Cu	138	148	122	132	103	111	91	98	78	85
AC-120–Cu[1]	Bz II	118	127	110	119	99	107	91	99	83	90
AC-120–Cu[1]	Cu	105	114	96	104	84	90	75	81	67	72
AC-120–Cu[2]	Bz II	140	151	129	139	115	123	104	112	94	101
AC-120–Cu[2]	Cu	123	132	110	119	94	102	84	90	73	79
AC-150–Cu[1]	Bz II	98	106	93	100	85	92	79	85	73	79
AC-150–Cu[1]	Cu	89	96	83	89	73	79	67	72	60	65
AC-150–Cu[2]	Bz II	118	127	110	118	99	107	92	98	83	90
AC-150–Cu[2]	Cu	105	114	96	103	84	90	76	81	67	72
2 AC-120–Cu[1]	Bz II	66	71	63	68	60	64	57	61	53	57
2 AC-120–Cu[1]	Cu	62	67	58	63	54	58	50	54	46	50
2 AC-120–Cu[2]	Bz II	80	86	76	82	71	76	67	72	62	67
2 AC-120–Cu[2]	Cu	74	80	69	75	63	67	58	62	53	57
2 AC-120–Cu[1]	2Bz II	59	63	55	59	50	54	46	49	42	45
2 AC-120–Cu[1]	2Cu	53	57	48	52	42	45	38	41	33	36
2 AC-120–Cu[2]	2Bz II	70	75	65	70	57	62	52	56	47	51
2 AC-120–Cu[2]	2Cu	61	66	55	59	47	51	42	45	37	40

[1] new contact wire, [2] contact wire 20 % worn

Wires and conductors

The resistance per unit length of wires and conductors is calculated by

$$R'_C = R/l = \rho \cdot l/(A \cdot l) = \rho/A = 1/(\kappa \cdot A) \quad , \tag{5.6}$$

with

ρ specific resistivity in Ω·m,
κ specific conductivity $1/\rho = 1/\Omega$·m,
l length in m or km,
A cross section in mm^2

The *resistivity* ρ of the conductor material is a function of the temperature. Up to 200 °C, the following applies

$$\rho = \rho(\vartheta) = \rho_{20} \cdot [1 + \alpha_R \cdot (\vartheta - 20)] \quad , \tag{5.7}$$

where ϑ is the conductor temperature in °C and α_R the temperature coefficient of resistivity. In the Tables 11.2, 11.5 and 11.6 the specific properties at 20 °C and the *temperature coefficients* are presented for contact line conductor materials.

The total resistance per unit length of n_l parallel conductors is obtained from

$$R'_C = 1 \Big/ \sum_{i=1}^{n_l} (1/R'_{Ci}) \quad . \tag{5.8}$$

Table 5.3: Characteristic properties of commonly used running rail types.

Rail type	m'	H	F_w	A	U	r_{eqA}	R' mΩ/km	
	kg/m	mm	mm	mm²	mm	mm	wear 0 %	wear 15 %
S 49	49,43	149	125	6297	600	44,77	35,7	42,0
R 50	50,50	152	132	6450	620	45,31	34,5	40,6
S 54	54,54	154	125	6948	630	47,03	32,0	37,6
UIC 54	54,40	159	140	6934	630	46,98	32,0	37,6
S 60	60,30	172	150	7650	680	49,35	28,9	34,0
UIC 60	60,34	172	150	7686	680	49,46	28,9	34,0
R 65	65,10	180	150	8288	700	51,36	25,2	29,9

m' mass per unit length
H height of rail
F_w foot width
A cross section
r_{eqA} cross section-area-equivalent radius $r_{eqA} = \sqrt{A/\pi}$
R' longitudinal resistance at 20 °C

Table 5.4: Resistance per unit length R' of conductor rail types in mΩ/km.

ϑ_{rail}	Type of conductor rail					
°C	soft iron 5 100 mm² degree of wear		soft iron 7 625 mm² degree of wear		Al-composite 5 100 mm²	Al-composite 2 100 mm²
	0 %	20 %	0 %	20 %	degree of wear 0 %	degree of wear 0 %
−30	15,8	19,7	10,5	13,2	5,6	13,4
20	22,5	28,1	15,0	18,8	6,8	16,4
40	25,2	31,5	16,8	21,0	7,3	17,6

Table 5.1 presents the DC resistance of conductors often used in contact line installations. Equation (5.8) can be used to calculate the resistance of overhead contact line designs given in Table 5.2.

Running rails, conductor rails

The resistance of steel running rails can be obtained from Equations (5.6) and (5.7) with $\rho_{20} = 0,222\ \Omega\text{mm}^2/\text{m}$ and $\alpha_R = 0,0047\ \text{K}^{-1}$. Table 5.3 gives the characteristic properties of commonly used running rails. The resistance of single-track and double-track lines is one half or one quarter respectively. Where rail joints are used, each rail joint has to be calculated by an additional resistance corresponding to 2,5 m rail length. The resistance of steel running rails can be obtained from (5.6) and (5.7) with $\rho_{20} = 0,115\ \Omega\text{mm}^2/\text{m}$ and $\alpha_R = 0,006\ \text{K}^{-1}$. The conductivity of an aluminum composite conductor rail is governed by the electrical properties of aluminum with $\rho_{20} = 0,0345\ \Omega\text{mm}^2/\text{m}$ and $\alpha_R = 0,0036\ \text{K}^{-1}$ including the steel content. Wear of the steel plate does not need to be considered when determining the resistance. Table 5.4 shows the resistance per unit length of conductor rails.
For example, a conductor rail made of soft iron with 5 100 mm² without wear and 20 joints per km has a resistance of $0,115 \cdot (1\,000 + 20 \cdot 2,5)/5\,100 = 23,7\ \text{mΩ/km}$ at 20 °C.

Earth return path

Although the differing types of soil show a great variety of resistivities, the resistance of earth to DC currents is zero due to the huge cross-section involved. However, in case of AC currents, the earth possesses a resistance. The longitonal resistance of the earth return path

R'_E is a function of the frequency of the power supply. According to [5.1],

$$R'_E = (\pi/4)\mu_0 \cdot \mu_r \cdot f = (\pi/4) \cdot \mu_0 \cdot f \quad , \tag{5.9}$$

with

μ_0 magnetic space constant with

$$4\pi \cdot 10^{-7}\,\mathrm{H/m} = 4\pi \cdot 10^{-4}\,\mathrm{Vs/(A\,km)} \quad , \tag{5.10}$$

μ_r relative permeability, for soil $\mu_r = 1{,}0$ [5.2],
f frequency.
Inserting μ_0 from Equation (5.10) into (5.9) leads

$$R'_E = 10^{-4} \cdot \pi^2 \cdot f \ \ \Omega/\mathrm{km} \quad . \tag{5.11}$$

Therefore, the longitudinal resistance of earth is calculated to be $16{,}4\,\mathrm{m\Omega/km}$ for $16{,}7\,\mathrm{Hz}$ and $49{,}3\,\mathrm{m\Omega/km}$ for $50\,\mathrm{Hz}$.

5.1.2.3   Inductance, reactance and impedance

The impedance of a *conductor-earth loop* consists of its resistance and reactance. The reactance depends on the inductance L and the frequency f. The self-impedance of the conductor-earth loop is composed of the resistance, the inner self-inductance and the external inductance. The *self-impedance* of the loop i can be expressed by

$$\underline{Z}'_{ii} = R' + R'_E + \mathrm{j}\,(X'_{ex} + X'_{in}) \tag{5.12}$$

with

R' resistance in Ω/km, see Clause 5.1.2.2,
R'_E longitudinal resistance of the earth return path as per Equation (5.11),
X'_{ex} external reactance (5.13),
X'_{in} inner reactance of the conductor (5.14).
The *external reactance* can be obtained from

$$X'_{ex} = 2\pi \cdot f \cdot L'_{ex} = 4\pi \cdot 10^{-4} \cdot f \cdot \ln(\delta_E/r) \ \ \Omega/\mathrm{km} \tag{5.13}$$

with

L'_{ex} external inductance,
f power supply frequency,
δ_E penetration depth of the current in the earth and
r conductor radius.
The *inner reactance* X'_{in} is obtained from

$$X'_{in} = 2\pi \cdot f \cdot L'_{in} = 4 \cdot \pi \cdot 10^{-4} \cdot f \cdot \ln(r/r_{eq\,A}) \ \ \Omega/\mathrm{km} \tag{5.14}$$

with L'_{in} inner inductance, $r_{eq\,A}$ equivalent radius.
The *penetration depth* δ_E of a current flowing through earth can be taken from

$$\delta_E = 0{,}738/\sqrt{f \cdot \mu_0/\rho_E} \quad , \tag{5.15}$$

with ρ_E resistivity of the soil in $\Omega\mathrm{m}$.

Table 5.5: Equivalent radii, inner inductances L'_{in} and inner reactances X'_{in} for conductors at 16,7 and 50 Hz.

Conductor	r_{eq}/r	L'_{in} mH/km	X'_{in} mΩ/km 16,7 Hz	X'_{in} mΩ/km 50 Hz
contact wires	0,7788	0,0500	5,24	15,71
conductors (droppers, catenary wires, reinforcing feeder wires, earth wires)				
7 strands 10… 50 mm²	0,726	0,0640	6,72	20,12
19 strands 70…120 mm²	0,758	0,0554	5,81	17,41
37 strands 150…185 mm²	0,768	0,0528	5,54	18,59
61 strands 240…500 mm²	0,772	0,0518	5,43	16,26
91 strands 630 mm²	0,774	0,0512	5,38	16,10

Equation (5.15) yields

$$\delta_E \approx 160\sqrt{\rho_E} \quad \text{for} \quad f = 16,7\,\text{Hz} \quad \text{and} \tag{5.16}$$
$$\delta_E \approx 90\sqrt{\rho_E} \quad \text{for} \quad f = 50\,\text{Hz} \quad .$$

For calculation of the penetration depth, the simplifying assumption is made that the earth is a homogeneous body having a semi-circular cross section located under the electrified railway line. This is an approximation since the soil resistivity changes with the depth.

The inner inductance L'_{in} was found to be $\mu/(8\pi)$ for solid conductors with circular cross sections independent of the conductor radius. Therefore,

$$L'_{in} = 2 \cdot 10^{-4} \cdot \ln(r/r_{eq}) = 4\pi \cdot 10^{-4}/(8\pi)\ \text{H/km} \tag{5.17}$$

results. From this equation it is obtained

$$\ln(r/r_{eq}) = 1/4 = 0,25 \quad \text{and} \quad r_{eq} = r \cdot e^{-0,25} = 0,7788 \cdot r \quad .$$

In Table 5.5 the inner reactances X'_{in} are given for conductor types used for contact lines. The *coupling impedance* $\underline{Z}'_{ik}$ of two conductor-earth loops i and k can be expressed by

$$\underline{Z}'_{ik} = R'_E + \mathrm{j}X'_{ik} \quad . \tag{5.18}$$

In Equation (5.18) the *mutual reactance* is

$$X'_{ik} = 4\pi \cdot 10^{-4} \cdot f \cdot \ln(\delta_E/a_{ik})\ \Omega/\text{km} \quad , \tag{5.19}$$

where a_{ik} is the mutual distance of the conductors in the loops.

5.1.2.4 Impedance of running rails

The equations mentioned in Clause 5.1.2.3 are not appropriate to calculate the inner self-impedance of running rails because of the skin effect and the current and frequency-dependent relative permeability μ_r of the steel. Skin effect cannot be neglected because of the large cross section of the rail. This effect increases with the frequency and is very distinct at operational frequencies. Therefore, it is not advisable to calculate the inner self-inductance of the rail based on the rail dimensions and the permeability. It is more reliable to measure the resistance and the self-reactance of rails at operational frequencies depending on the current. In the literature, some results of measurements are reported, e. g. in [5.3]. The results of the measurements vary depending on the different materials and differences in the measurement arrangement. For calculations, the following data can be used for 50 Hz:

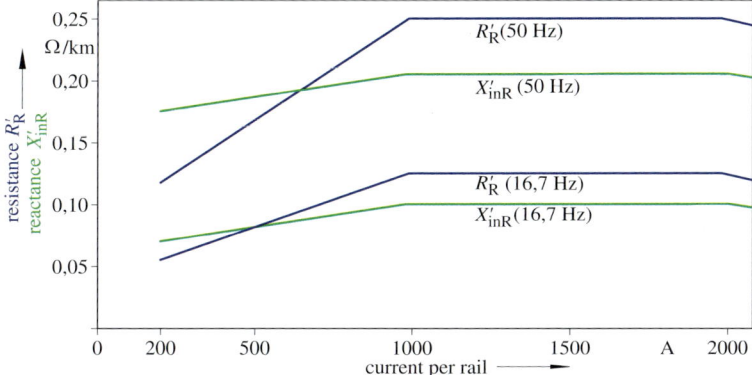

Figure 5.2: Resistance R'_R and inner self-reactance X'_{inR} per unit length for running rails.

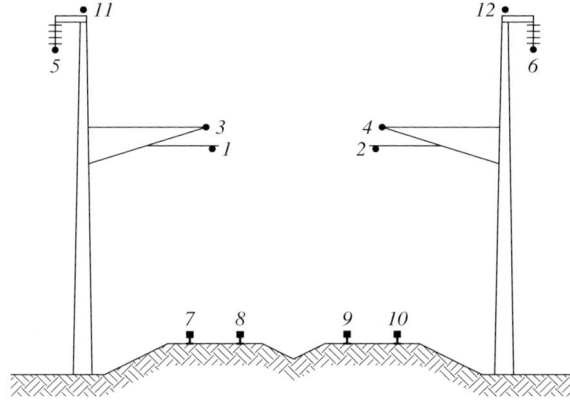

Figure 5.3: Presentation of a double-track railway line by a n-conductor system.
1, 2 contact wires
3, 4 catenary wires
5, 6 parallel feeder lines
11, 12 return conductors
7, 8, 9, 10 rails

– resistance $\qquad R'_R$ 0,12 to 0,25 Ω/km
– inner self-reactance X'_{inR} 0,15 to 0,22 Ω/km

The values apply to currents between 100 and 1 000 A per rail, where the lower data is associated with the lower current and the higher data with the higher ones.

For 16,7 Hz systems the data can be assumed to be half that given for 50 Hz system. Figure 5.2 depicts the rail resistances and reactances for both frequencies.

The resistance and reactance rise linearly with increasing current up to approximately 1 kA in the given range. Between 1 kA and 2 kA, they remain constant and drop again with the rising current. The measured resistances and reactances can be used for calculating the *line impedances* as the values for the inner self-impedance of a running rail $\underline{Z}'_{inR} = R'_R + j X'_{inR}$.

5.1.2.5 Impedance of AC overhead contact lines

According to [5.4] the line impedance of any overhead contact line configuration can be calculated by establishing and solving equations for the voltages and currents of the individual conductors forming the contact line and the return circuit. By utilizing adequate computer software, all conductors involved can be considered with their specific characteristics.

Any overhead contact line system can be represented by a n-conductor system as shown in Figure 5.3. Within such a n-conductor system, some conductors form a loop that allows

Table 5.6: Example to calculate line impedance, self-impedance for 50 Hz.

No	Conductor type	Resistance R'_i Ω/km	Equivalent radius r/r_{eq} –	Reactance X'_{iE} Ω/km	X'_{ii} Ω/km	Self impedance $\underline{Z}'_{ii}$ Ω/km
1	contact wire AC-120–Cu	0,15	0,7788	0,765	0,016	$0,20+\mathrm{j}\cdot 0,78$
2	contact wire AC-120–Cu	0,15	0,7788	0,765	0,016	$0,20+\mathrm{j}\cdot 0,78$
3	catenary wire BzII70	0,42	0,758	0,778	0,017	$0,32+\mathrm{j}\cdot 0,80$
4	catenary wire BzII70	0,42	0,758	0,778	0,017	$0,32+\mathrm{j}\cdot 0,80$
7	rail UIC60	0,03	–	–	–	$0,20+\mathrm{j}\cdot 0,20$
8	rail UIC60	0,03	–	–	–	$0,20+\mathrm{j}\cdot 0,20$
9	rail UIC60	0,03	–	–	–	$0,20+\mathrm{j}\cdot 0,20$
10	rail UIC60	0,03	–	–	–	$0,20+\mathrm{j}\cdot 0,20$
11	return conductor 240-AL1	0,12	0,772	0,738	0,016	$0,17+\mathrm{j}\cdot 0,75$
12	return conductor 240-AL1	0,12	0,772	0,738	0,016	$0,17+\mathrm{j}\cdot 0,75$

current to flow through the earth. The currents in the supplying conductors are designated as I_1 to I_h and those following in the return circuit I_{h+1} to I_n.

Based on the voltage/current relation of a differentially small section of the n-conductor system, the Equation (5.20) can be obtained for the conductor-earth loops. The impedances in Equation (5.20) are related to the common current return through earth. The conductor-earth loops are represented by the coupling impedances $\underline{Z}_{ik}$ in the impedance matrix. The self-impedances $\underline{Z}_{ii}$ of the conductor loops form the diagonal elements of the impedance matrix.

$$
\begin{pmatrix} \mathrm{d}\underline{U}_{1E} \\ \mathrm{d}\underline{U}_{2E} \\ \vdots \\ \mathrm{d}\underline{U}_{hE} \\ \vdots \\ \mathrm{d}\underline{U}_{nE} \end{pmatrix} = \mathrm{d}x
\begin{pmatrix}
\underline{Z}_{11} & \cdots & \underline{Z}_{1h} & \underline{Z}_{1h+1} & \cdots & \underline{Z}_{1n} \\
\underline{Z}_{21} & \cdots & \underline{Z}_{2h} & \underline{Z}_{2h+1} & \cdots & \underline{Z}_{2n} \\
\vdots & \ddots & \vdots & \vdots & \ddots & \vdots \\
\underline{Z}_{h1} & \cdots & \underline{Z}_{hh} & \underline{Z}_{hh+1} & \cdots & \underline{Z}_{hn} \\
\underline{Z}_{(h+1)1} & \cdots & \underline{Z}_{(h+1)h} & \underline{Z}_{(h+1)(h+1)} & \cdots & \underline{Z}_{(h+1)n} \\
\vdots & \ddots & \vdots & \vdots & \ddots & \vdots \\
\underline{Z}_{n1} & \cdots & \underline{Z}_{nh} & \underline{Z}_{n(h+1)} & \cdots & \underline{Z}_{nn}
\end{pmatrix}
\cdot
\begin{pmatrix} I_1 \\ I_2 \\ \vdots \\ I_h \\ I_{h+1} \\ \vdots \\ I_n \end{pmatrix}
\quad (5.20)
$$

Further calculations can be found in [5.4]. Example 5.1 shows the calculations and results.

Example 5.1: Calculate the impedance of the double-track line Madrid–Seville with two transversely coupled contact lines and two return conductors. Figure 5.3 shows the arrangement of the conductors. The soil resistivity is 200 Ωm resulting in a penetration depth of

$$\delta_E = 90\sqrt{200} = 1\,270\,\mathrm{m}.$$

The longitudinal resistance of earth is $R'_E = 49,3\,\mathrm{m}\Omega/\mathrm{km} \approx 0,05\,\Omega/\mathrm{km}$, see Clause 5.1.2.2, Equation (5.11).

Table 5.6 gives the self-impedance as calculated from Equations (5.12), (5.14) and (5.17). The self-impedance of the rails was taken as $0,20+\mathrm{j}0,20\,\Omega/\mathrm{km}$, see Clause 5.1.2.4. Table 5.7 contains the mutual reactances calculated with Equation (5.19).

The result of the calculation is $\underline{Z}' = 0,08+\mathrm{j}0,20\,\Omega/\mathrm{km}$ [5.5], whereas the measured value is $\underline{Z}' = 0,07+\mathrm{j}0,20\,\Omega/\mathrm{km}$. Calculated and measured data agree to an acceptable extent.

Table 5.7: Example to calculate line impedance and mutual reactance X'_{ik} in Ω/km for 50 Hz according to Example 5.1 and mutual distance a_{ik} between the conductors in m.

	Conductor	1	2	3	4	7	8	9	10	11	12
a_{ik}	1	–									
X_{ik}		–									
a_{ik}	2	4,60	–								
X_{ik}		0,35	–								
a_{ik}	3	0,70	4,65	–							
X_{ik}		0,47	0,35	–							
a_{ik}	4	4,65	0,70	4,60	–						
X_{ik}		0,35	0,47	0,35	–						
a_{ik}	7	5,30	7,50	6,00	8,00	–					
X_{ik}		0,34	0,32	0,34	0,32	–					
a_{ik}	8	5,30	6,60	6,00	7,20	1,50	–				
X_{ik}		0,34	0,33	0,34	0,33	0,43	–				
a_{ik}	9	6,60	5,30	7,20	6,00	4,60	3,20	–			
X_{ik}		0,33	0,34	0,33	0,34	0,35	0,38	–			
a_{ik}	10	7,50	5,30	8,00	6,00	6,10	4,60	1,50	–		
X_{ik}		0,32	0,34	0,32	0,34	0,34	0,35	0,43	–		
a_{ik}	11	4,1	8,60	4,00	8,60	6,80	7,65	9,70	11,10	–	
X_{ik}		0,36	0,31	0,36	0,31	0,33	0,32	0,30	0,30	–	
a_{ik}	12	8,60	4,10	8,60	4,00	11,10	9,70	7,65	6,80	12,00	–
X_{ik}		0,31	0,36	0,31	0,36	0,30	0,30	0,32	0,33	0,29	–

5.1.2.6 Measuring of line impedances

The impedance of a line section can be determined by measuring currents and voltages. The principle of an *impedance measurement* is shown in Figure 5.4, see [5.5]. For the measurement it is recommended that:

- the impedances of the line be measured several times in order to obtain statistical certainty of the results
- all measurements be carried out under the same conditions, i. e. the method used to establish the short-circuit loop and the magnitude of the current through it
- effects of the feeder line sections be taken into consideration
- if the contact line in question is an AC traction contact line, the reactive power and the effective power be measured in addition to the current and voltage in order to obtain realistic validation and comparison values

Measurements are quite simple on DC traction contact lines. All that is needed is to short-circuit the line under test and apply a measuring test voltage U_T between the overhead contact line and the running rails at a distance l_T from the short circuit and then measure the resulting current I_T. The *line resistance per unit length* is then calculated as the quotient of the applied voltage and the measured current

$$Z'_T = R'_T = U_T/(I_T \cdot l_T) \quad . \tag{5.21}$$

Reliable impedance values of single-phase AC railway lines can be obtained if the length l_T of the measured section is considerably longer than the *transition length* L_{trans}. The transition range or length describes the region within which currents are observed to pass into and out of the earth due to electromagnetic inductive coupling processes. The term *transition length*

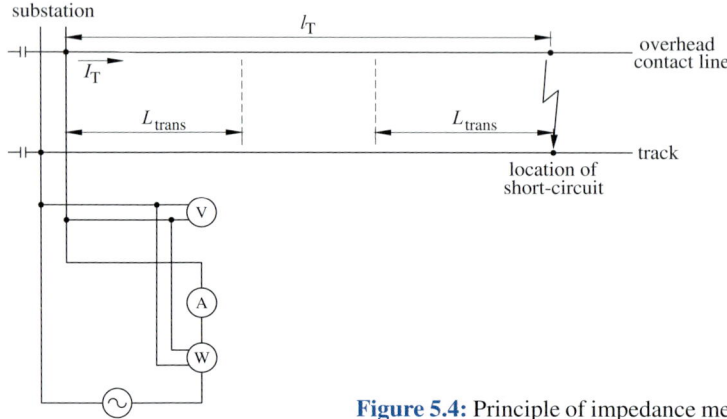

Figure 5.4: Principle of impedance measurements of an overhead contact line. l_T length of the measuring section, L_{trans} transition length

L_{trans} is explained in Clause 6.4.3.3 normally being between 5 km and 8 km. It is advisable to use measuring currents as high as feasible and close to the operating currents. Once the values of the voltage U_T, the apparent current I_T and the effective power P_T have been measured and the length l_T of the section is known, the impedance per unit length is calculated using the following equations:

$$\varphi = \arccos\left[P_T / (\underline{U}_T \cdot \underline{I}_T)\right]$$

$$\underline{Z}'_T = |\underline{U}_T/\underline{I}_T| \cdot (\cos\varphi + \mathrm{j}\sin\varphi) \text{ and } \underline{Z}' = |\underline{U}/(\underline{I} \cdot l_T)| \cdot (\cos\varphi + \mathrm{j}\sin\varphi) \tag{5.22}$$

Where $R'_T = |\underline{U}_T/(\underline{I}_T \cdot l_T)| \cdot \cos\varphi$ and $X'_T = |\underline{U}_T/(\underline{I}_T \cdot l_T)| \cdot \sin\varphi$.
If the single-phase AC line measurements are carried out under the condition that the measuring length l_T is less than $2 \cdot L_{trans}$, there will be a tendency to obtain a higher line resistance and a lower line reactance. If the length of the section being tested were very short, the self-impedance of the contact line to track circuit alone would be measured.

Example 5.2:   On a 50 Hz single-phase AC railway line, measurements were taken on a 3,58 km section and the following values obtained: $U_T = 23,8$ V; $I_T = 19$ A and $P_T = 190$ W
Therefore, the measured line impedance is given by:
- $\varphi = \arccos\left[190/(23,8 \cdot 19)\right] = \arccos 0,4202 = 65,2°$
- $|\underline{Z}'| = 23,8/(19 \cdot 3,58) = 0,35 \ \Omega/\text{km}$
- $\underline{Z}' = 0,35 \ \angle 65,2° = 0,35 \cdot (0,419 + \mathrm{j}0,908) = (0,147 + \mathrm{j}0,318) \ \Omega/\text{km}$
- $R' = 0,147 \ \Omega/\text{km}; \ X' = 0,318 \ \Omega/\text{km}$

The impedances per unit length can also be determined using a traction vehicle an measuring the aforementioned quantities I, U and P at the substation while simultaneously measuring the voltage U_{trc} at the traction vehicle's pantograph and the effective traction power P_{trc} consumed by the traction vehicle. This means that the *voltage drop* $\Delta \underline{U}$ and the *power loss* $\Delta P = P - P_{trc}$ along the overhead contact line can be determined by measurement. Likewise it applies to Equation (5.30) and with reference to Figure 5.5:

$$\Delta \underline{U} = \underline{U}_{12} - \underline{U}_{trc} \approx |\Delta \underline{U}| \qquad\qquad \underline{Z} = \Delta \underline{U}/\underline{I}_{trc}$$

a) Equivalent circuit

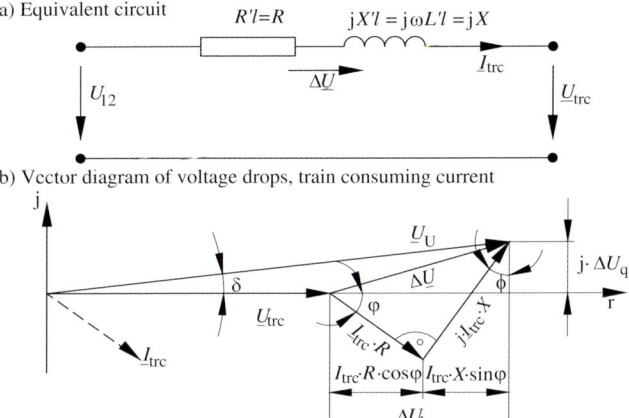

b) Vector diagram of voltage drops, train consuming current

c) Vector diagram of voltage drops, train braking

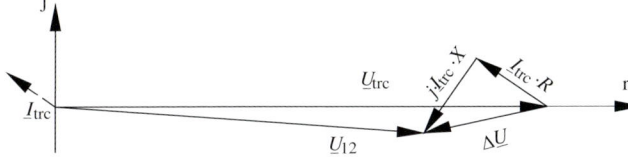

Figure 5.5: Voltage relationships in a traction power supply contact line network.

The resistance R is calculated now using

$$R = \Delta P / I_{trc}^2 \quad .$$

The impedance per unit length is determined by

$$\underline{Z}' = \Delta \underline{U} / (\underline{I}_{trc} \cdot l_T) \quad . \tag{5.23}$$

The *impedance angle* can be calculated using R' and Z':

$$\varphi = \arccos(R'/Z') \quad .$$

This method was used to determine the impedances per unit length of the overhead contact line installation of the Magdeburg–Marienborn line in 1993 [5.6] and Madrid–Seville line in 1994 [5.5].

5.1.2.7 Calculated and measured impedances per unit length

The values shown in Tables 5.8 to 5.10 were calculated or obtained by measurement. There is a wide variety of factors affecting the impedances per unit length of single-phase AC railways that can lead to differences between calculated and measured values. Table 5.8 shows the calculated values of resistance, reactance and *impedance per unit length* for five standardized overhead contact line designs used for 16,7 Hz systems, including the current distributions in various configurations. In Table 5.9, impedances per unit length are shown for 50 Hz single-phase AC railway lines. In Table 5.10, calculated and measured impedances of AC railway lines are compared. With a few exceptions the differences between measured and calculated data are less than 10 %. Therefore, the calculated data can be considered as confirmed by measurements. For setting of protection relays of line circuit breakers, the effective impedance should be measured on site so that the real conditions such as number of tracks and return conductor configuration are considered.

Table 5.8: Calculated line impedances in Ω/km of double-track 16,7 Hz single-phase AC railway lines according to [5.4] and current distribution among the individual conductors.

Overhead contact line	No. of OCLs	FL	RC	Impedances per unit length			Current distribution (in %)				
				R'	X'	$\underline{Z}'$	CW	CA	FL	T	RC
Re 200	1	n	n	0,148	0,140	0,206 $\angle 45°$	0,74	0,27		0,70	
	1	n	y	0,152	0,127	0,198 $\angle 40°$	0,74	0,27		0,50	0,32
	1	y	n	0,073	0,105	0,127 $\angle 55°$	0,39	0,14	0,49	0,68	
	1	y	y	0,078	0,085	0,115 $\angle 47°$	0,38	0,14	0,49	0,47	0,38
Re 200	2	n	n	0,077	0,091	0,119 $\angle 50°$	0,74	0,27		0,70	
	2	n	y	0,080	0,079	0,112 $\angle 45°$	0,74	0,27		0,50	0,32
	2	y	n	0,038	0,070	0,080 $\angle 61°$	0,38	0,14	0,50	0,68	
	2	y	y	0,043	0,052	0,068 $\angle 50°$	0,37	0,13	0,51	0,45	0,38
Re 250	1	n	n	0,122	0,135	0,182 $\angle 48°$	0,71	0,30		0,71	
	1	n	y	0,125	0,123	0,176 $\angle 44°$	0,71	0,30		0,52	0,32
	1	y	n	0,065	0,101	0,121 $\angle 57°$	0,40	0,17	0,45	0,68	
	1	y	y	0,071	0,082	0,109 $\angle 49°$	0,39	0,16	0,45	0,46	0,38
Re 250	2	n	n	0,064	0,087	0,108 $\angle 54°$	0,71	0,30		0,71	
	2	n	y	0,067	0,075	0,100 $\angle 49°$	0,71	0,30		0,51	0,32
	2	y	n	0,035	0,068	0,076 $\angle 63°$	0,39	0,16	0,45	0,68	
	2	y	y	0,040	0,051	0,064 $\angle 52°$	0,38	0,16	0,47	0,46	0,38
Re 330	1	n	n	0,126	0,126	0,178 $\angle 45°$	0,51	0,49		0,70	
	1	n	y	0,127	0,113	0,171 $\angle 41°$	0,51	0,49		0,51	0,32
	1	y	n	0,064	0,099	0,120 $\angle 56°$	0,29	0,27	0,44	0,68	
	1	y	y	0,073	0,080	0,108 $\angle 48°$	0,28	0,27	0,45	0,45	0,38
Re 330	2	n	n	0,066	0,083	0,105 $\angle 52°$	0,51	0,49		0,70	
	2	n	y	0,068	0,071	0,098 $\angle 46°$	0,51	0,49		0,51	0,32
	2	y	n	0,036	0,067	0,076 $\angle 62°$	0,29	0,27	0,45	0,68	
	2	y	y	0,040	0,049	0,064 $\angle 51°$	0,28	0,26	0,46	0,45	0,39

n = no, y = yes
Re 200: contact wire AC-100 – Cu, new; catenary wire BzII 50 mm^2; rails UIC 60
Re 250: contact wire AC-120 – CuAg, new; catenary wire BzII 70 mm^2; rails UIC 60
Re 330: contact wire AC-120 – CuMg, new; catenary wire BzII 120 mm^2; rails UIC 60
OCL overhead contact line, CW contact wire, CA catenary wire, FL feeder line 243-AL1, RC return conductor 243-AL1, T track
Note: The data for two overhead contact lines apply to lines connected in parallel. The data apply to one or two overhead contact line installations, each in conjunction with two tracks and two return conductors. Where the sum of the partial current components differs from 1,00 this is due to the phase differences between the individual components.

5.1.3   Capacitances

Harmonic oscillations in a contact line system are affected by the *capacitance per unit length*. Every conductor in a contact line installation constitutes a capacitance relative to earth. This characteristic depends on the shape and the dimension of the conductors and on the dielectric media within the electrical field under consideration. As a result, overhead contact lines, conductor rails and even tracks will have a specific capacitance related to the earth.

Table 5.9: Calculated line impedances per unit length in Ω/km of double-track 50 Hz single-phase AC railway lines according to [5.5] and current distribution among the individual conductors.

Overhead contact line	No. of OCLs	FL	RC	Impedances per unit length			Current distribution (in %)				
				R'	X'	$\underline{Z}'$	CW	CA	FL	T	RC
Cu AC-100	1 [1]			0,148	0,422	0,447 $\angle71°$					
+	1 [2]			0,139	0,414	0,437 $\angle74°$					
Cu 95	2 [1]			0,110	0,297	0,317 $\angle70°$					
	2 [2]			0,092	0,289	0,303 $\angle72°$					
Cu AC-100	1 [1]			0,139	0,422	0,444 $\angle72°$					
+	1 [2]			0,130	0,414	0,434 $\angle73°$					
Cu 120	2 [1]			0,097	0,297	0,312 $\angle72°$					
	2 [2]			0,088	0,289	0,302 $\angle73°$					
Re 200	1	n	n	0,170	0,396	0,431 $\angle67°$	0,66	0,37		0,70	
	1	n	y	0,172	0,355	0,394 $\angle64°$	0,66	0,38		0,47	0,35
	1	y	n	0,087	0,297	0,309 $\angle74°$	0,39	0,22	0,42	0,68	
	1	y	y	0,088	0,233	0,249 $\angle65°$	0,36	0,20	0,46	0,40	0,42
Re 200	2	n	n	0,090	0,269	0,274 $\angle71°$	0,66	0,37		0,70	
	2	n	y	0,091	0,220	0,237 $\angle68°$	0,66	0,37		0,47	0,35
	2	y	n	0,047	0,199	0,204 $\angle77°$	0,38	0,20	0,44	0,68	
	2	y	y	0,048	0,142	0,150 $\angle71°$	0,34	0,18	0,49	0,40	0,43
Re 250	1	n	n	0,141	0,382	0,407 $\angle70°$	0,62	0,40		0,71	
	1	n	y	0,142	0,342	0,371 $\angle68°$	0,62	0,41		0,48	0,35
	1	y	n	0,077	0,289	0,299 $\angle75°$	0,38	0,24	0,40	0,69	
	1	y	y	0,079	0,227	0,247 $\angle71°$	0,35	0,23	0,44	0,41	0,42
Re 250	2	n	n	0,075	0,246	0,257 $\angle73°$	0,62	0,40		0,71	
	2	n	y	0,076	0,209	0,222 $\angle70°$	0,62	0,40		0,48	0,35
	2	y	n	0,043	0,192	0,197 $\angle77°$	0,37	0,22	0,42	0,69	
	2	y	y	0,044	0,138	0,145 $\angle72°$	0,34	0,21	0,47	0,41	0,42
Re 330	1	n	n	0,139	0,366	0,391 $\angle70°$	0,52	0,48		0,71	
	1	n	y	0,132	0,329	0,354 $\angle68°$	0,51	0,49		0,48	0,35
	1	y	n	0,075	0,284	0,294 $\angle75°$	0,33	0,30	0,38	0,69	
	1	y	y	0,077	0,223	0,236 $\angle71°$	0,30	0,28	0,42	0,41	0,42
Re 330	2	n	n	0,071	0,240	0,250 $\angle74°$	0,53	0,48		0,71	
	2	n	y	0,071	0,202	0,214 $\angle71°$	0,52	0,48		0,48	0,35
	2	y	n	0,042	0,190	0,195 $\angle77°$	0,33	0,28	0,40	0,68	
	2	y	y	0,043	0,136	0,143 $\angle72°$	0,29	0,25	0,46	0,41	0,42

[1] rails R 50, [2] rails R 65;   Note: designations and assumptions as for Table 5.8

Capacitance of overhead contact lines to earth

The capacitance per unit length of a single track overhead contact line to earth can be described as follows [5.7]:

$$C'_{\mathrm{E}} = 2\pi\left(\varepsilon_0 \cdot \varepsilon_{\mathrm{r}}\right) / \ln(2\,h_{\mathrm{m}}/r_{\mathrm{eq}}) \qquad (5.24)$$

with
ε_{r} relative permittivity ≈ 1 for air,
ε_0 electric field constant $8,85 \cdot 10^{-9}$ F/km,
h_{m} mean height of the contact line above ground $\approx 6,1$ m $= h_{\mathrm{CW}} + D_{\mathrm{m}}/2$,
D_{m} mean distance between contact wire and catenary wire and
r_{eq} equivalent radius of the contact line consisting of contact wire and catenary wire.

Table 5.10: Measured and calculated line impedances per unit length.

No. of OCLs	CW mm²	CA mm²	FL mm²	RC mm²	calculated Ω/km	measured Ω/km	Remarks, reference
\multicolumn{5}{OCL configuration}					Impedance per unit length		

No. of OCLs	CW mm²	CA mm²	FL mm²	RC mm²	calculated Ω/km	measured Ω/km	Remarks, reference
AC 15 kV 16,7 Hz							
1	100	50	–	–	0,206 ∠45°	0,215 ∠58°	DR [5.7, 5.8]
						0,230 ∠45°	DB [5.8]
2	100	50	–	–	0,119 ∠50°	0,117 ∠54°	DR [5.7, 5.8]
						0,130 ∠48°	DB [5.8]
2	100	50	240	–	0,080 ∠61°	0,118 ∠60°	DB [5.8]
						0,088 ∠48°	Swiss Railway
2	100	50	240 [1]	240	0,068 ∠50°	0,077 ∠40°	DB [5.9]
1	2 × 100	2 × 95	–	–		0,150 ∠53°	DB [5.8], Urban mass transit in tunnel
1	120	70	–	–	0,182 ∠48°	0,172 ∠47°	DB [5.8], open line
1	120	70	–	–	0,182 ∠48°	0,165 ∠46°	DB [5.8], tunnel
1	120	70	240	–	0,121 ∠57°	0,110 ∠59°	DB [5.8]
2	120	70	–	–	0,108 ∠54°	0,106 ∠52°	DB [5.8]
						0,096 ∠48°	DB [5.8], tunnel
2	120	70	240	–	0,076 ∠63°	0,070 ∠63°	DB [5.8]
AC 25 kV 50 Hz							
1	120	70	–	–	0,407 ∠70°	0,420 ∠69°	Hambachbahn, Germany
1	120	70	–	240	0,371 ∠62°	0,330 ∠69°	ADIF [5.9]
2	120	70	–	–	0,257 ∠73°	0,280 ∠71°	Hambachbahn, Germany
2	120	70	–	240	0,222 ∠66°	0,210 ∠71°	ADIF [5.9]

[1] one track only equipped with feeder line

The *equivalent radius* r_{eq} is calculated using the equation

$$r_{eq} = \sqrt{D_m \cdot r} \quad , \tag{5.25}$$

where r is the radius of the contact wire.

For $r = 0,0056$ m for a contact wire AC-100 – Cu, $D_m = 1,2$ m, $h_{CW} = 5,5$ m and $h_m = 6,1$ m mean height above rails the capacitance per unit length of the two parallel overhead contact lines of a double-track line, with respect to earth, is calculated to be $2 \cdot 11$ nF/km $= 22$ nF/km.

Capacitance of conductor rails to earth (third rails)

The capacitance per unit length of a conductor rail with respect to earth can be approximated using Equation (5.24). The equivalent radius r_{eqA} of a conductor rail in such a calculation is derived from cross section A by

$$r_{eqA} = \sqrt{A/\pi} \quad . \tag{5.26}$$

If the permittivity $\varepsilon_r = 2,5$ to take the ballast bed below the rail into account, the capacitance per unit length of a conductor rail of cross-sectional area $A = 5\,100$ mm² and $h_m = 500$ mm height above surface of the earth is found from (5.24) to be 43 nF/km. Measurements carried out by the former Deutsche Reichsbahn (DR) with another conductor arrangement, have shown a capacitance per unit length to be between 70 and 100 nF/km.

Capacitance of track to earth

According to (5.24) and (5.26) the *capacitance* between the two rails of a track to earth is calculated as

$$C'_{TE} = 2 \cdot C'_{RE} \quad .$$ (5.27)

where $h_m = 0,3$ m, $A = 7680\,\text{mm}^2$ for the rail UIC60, $\varepsilon_r = 2,5$ for dry ballast.

The capacitance per rail is calculated as 56 nF/km, so that a value of 223 nF/km results for two tracks. The difference to the measurements in [5.8] with values of 120 nF/km and double track may be caused by other height h_m or other ε_r than in the calculation.

5.2 Voltages in contact line networks

5.2.1 Basic requirements and principles

When electric power is transmitted from substations to traction vehicles moving along the contact lines, *voltage drops* will occur along the contact lines. Conversely, if a vehicle feeds braking energy into the network, the voltage at the traction vehicle position will rise and the transfer of the braking energy to the contact line network will be possible.

Thus the voltage at the pantograph of a traction vehicle depends on the electrical characteristics of the contact line installation, the present power consumption of all electric traction vehicles in the supply section and their respective distance from the feed point. Under normal operating conditions, the voltages should never exceed or drop below the nominal voltage limits given in Table 1.3. In railway lines for high-speed traffic and heavy traffic, the recommendations are stricter [5.9] and state that the voltage of the electric traction contact line network should never drop below the nominal voltage at any point of the network in normal operation.

In the European TSI Energy [5.10, 5.11] and EN 50 388 a quality index for the power supply is defined and expressed as the *mean useful voltage* at the current collector or at the substation busbar. Details are discussed in Clause 5.2.4.

Figure 5.5 shows a simplified equivalent circuit diagram of a contact line section together with the corresponding voltage and current vector diagrams. In this illustration, the *longitudinal voltage drop* ΔU_l due to the traction current I_{trc} flowing through the resistance and reactance can be deduced as

$$\Delta U_l = l \left(R' I_{trc} \cos \varphi + X' I_{trc} \sin \varphi \right) = I_{trc} \cdot l \left(R' \cos \varphi + X' \sin \varphi \right) \quad .$$ (5.28)

The *transversal voltage drop per unit length* is

$$\Delta U_q = l \left(X' I_{trc} \cos \varphi - R' I_{trc} \sin \varphi \right) = I_{trc} \cdot l \left(X' \cos \varphi - R' \sin \varphi \right) \quad ,$$ (5.29)

yielding the *overall voltage drop* $\Delta \underline{U}$

$$\Delta \underline{U} = (\underline{U}_{12} - \underline{U}_{trc}) = l \cdot \underline{I}_{trc} \left(R' + \mathrm{j} X' \right) \quad .$$ (5.30)

For all practical applications, the lateral voltage drop caused by the traction current is negligible, so that the longitudinal voltage drop ΔU_l may be used to consider the overall voltage drop $\Delta \underline{U}$.

In an AC traction energy supply network, the voltage drop between the substation and a traction vehicle located l km away and drawing a current $\underline{I}_{trc}$ can be described with sufficient accuracy by the equations

$$\Delta U = \mathrm{Re}\{\underline{\Delta U}\} = I_{trc}\,l\,|\underline{Z}'|\,,\ \text{where} \tag{5.31}$$

$$\Delta U \approx \Delta U_1 = I_{trc}\,l\,\left(R'\cos\varphi + X'\sin\varphi\right)\quad.$$

In DC traction energy supply networks, the corresponding equation is

$$\Delta U = I_{trc}\,l\,R'\quad. \tag{5.32}$$

A comparison of the diagrams in Figures 5.5 b) and 5.5 c) illustrates that when braking electrically with *energy regeneration*, the voltage U_{trc} is increased to feed the energy back into the contact line network. The braking energy is used to supply other traction vehicles in the same feed section or fed back into the electric power supply network that feeds the railway traction network. The voltage at the traction vehicle returning energy to the network is determined by the respective energy recovery conditions.

Equations (5.31) and (5.32) differ with respect to the resistances. As already explained in Clause 5.1.2.2, the sum of the resistances per unit length of the contact line and the return current conductors is relevant in DC traction systems, whereas in single-phase AC traction systems, the complex value of the line impedance $\underline{Z}'$ determines the voltage drop, see Clause 5.1.2.7.

The voltage drops and the currents flowing in the contact line installations are associated with corresponding energy dissipation, i. e. power losses. The *power losses*

$$\Delta P = I^2 \cdot R \tag{5.33}$$

are caused by the resistances of the contact line network and increase the temperature of the conductors. Corresponding to this, the power losses $\Delta P'$ per unit length along an AC or DC traction contact line are

$$\Delta P' = I^2 \cdot R/l\quad. \tag{5.34}$$

5.2.2   Voltage drop calculations

5.2.2.1   Introduction

This clause deals with the *voltage drop* occurring between the feeding substation and the current position of one or several trains within the same feeding section. Apart from the traction current, distance and impedance per unit length, the feeding arrangement will determine the voltage drops to be expected, see Figure 5.24.

5.2.2.2   Single-end feed

One train in the feeding section

In Figure 5.6 , the voltage drop from the substation up to a position x can be read off as

$$\underline{\Delta U}_x = \underline{I}_{trc}\,\underline{Z}'\,x\quad.$$

The *maximum voltage drop* $\underline{\Delta U}_{max}$ in the section will occur when the train reaches the far end of the feeding section. In this case

$$\underline{\Delta U}_{max} = \underline{I}_{trc}\,\underline{Z}'\,l\quad.$$

Several trains in the feeding section

If the designations from Figure 5.6 b) are used, the voltage drop between the substation and the third train can be calculated as

$$\Delta \underline{U}_3 = \underline{Z}' \left(\underline{I}_1 l_1 + \underline{I}_2 l_2 + \underline{I}_3 l_3 \right) = \underline{Z}' \left(\underline{I}_{\text{trc }1} x_1 + \underline{I}_{\text{trc }2} x_2 + \underline{I}_{\text{trc }3} x_3 \right) \quad .$$

or, if this is generalized to describe n trains in a feed section

$$\Delta \underline{U}_n = \underline{Z}' \sum_{i=1}^{n} \underline{I}_i l_i = \underline{Z}' \sum_{i=1}^{n} \underline{I}_{\text{trc }i} x_i \quad . \tag{5.35}$$

If the number of trains within a feeding section is very large, the boundary case of a *uniformly distributed line load* as shown in Figure 5.6 c) will be reached. The line load $\underline{I}'_{\text{OCL}}$ in a contact line feeding section can be defined in relation to the length as follows

$$\underline{I}'_{\text{OCL}} = \frac{1}{l} \sum_{i=1}^{n} \underline{I}_{\text{trc }i} \quad . \tag{5.36}$$

The current $\underline{I}_{\text{OCL}}(x)$ flowing in the contact line section at a distance x from the feed point is then

$$\underline{I}_{\text{OCL}}(x) = \underline{I}'_{\text{OCL}} (l - x) \quad ,$$

as can be seen in Figure 5.6 c).

This expression enables the equation for the voltage drop $\Delta \underline{U}_x$ between the substation and the point x to be defined for a uniform line load distribution:

$$\Delta \underline{U}_x = \int_0^x \underline{I}_x \underline{Z}' \, dx = \frac{\underline{Z}'}{l} \left(lx - x^2/2 \right) \sum_{i=1}^{n} \underline{I}_{\text{trc }i} \quad . \tag{5.37}$$

For the special case of n trains drawing equal currents $\underline{I}_{\text{trc}}$,

$$\Delta \underline{U}_x = n \underline{I}_{\text{trc}} \underline{Z}' \left(lx - x^2/2 \right) / l \quad . \tag{5.38}$$

If it is assumed that all trains travel along the section at a constant speed, the maximum voltage drop is

$$\Delta \underline{U}_{\text{max}} = (1/2) n \underline{I}_{\text{trc}} \underline{Z}' l \quad . \tag{5.39}$$

Reference [5.12] specifies an equation that produces adequate results for the mean voltage drop for n trains. This equation is

$$\Delta \underline{U} = (1/3) \underline{I}_{\text{trc}} \underline{Z}' l \left(n + 1{,}5 t_{zz} - 1 \right) \quad , \tag{5.40}$$

where t_{zz} is the quotient of the period between two subsequent accelerations of a train and the period of time in between where power is drawn from the contact line network. Values of t_{zz} obtained by empirical methods range from two for conventional railways to approximately six for metropolitan mass-transit train traffic, see Clause 4.4.1 of [5.13].

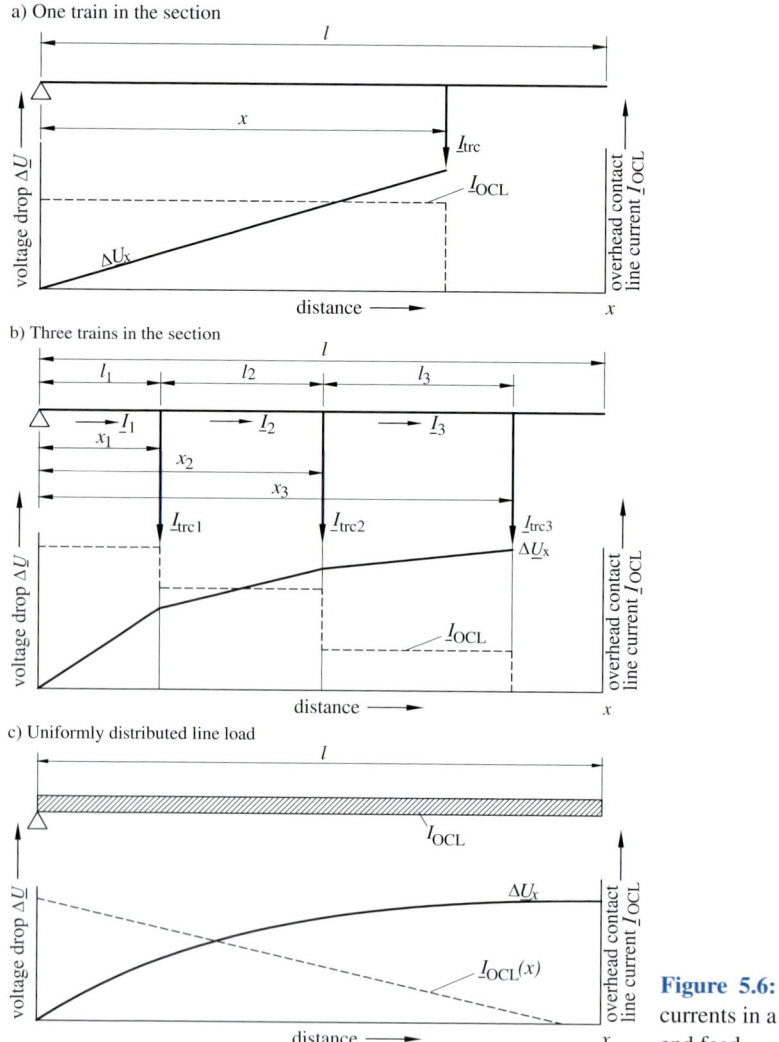

a) One train in the section

b) Three trains in the section

c) Uniformly distributed line load

Figure 5.6: Voltage drops and currents in a section with single-end feed.

5.2.2.3 Double-end feeding

One train in the feeding section

The feeding conditions can be seen in Figure 5.7. The length l_2 is the distance between the two feeding substations A and B. In *double-end feedings*, the length of a feeding section is defined as $l_2 = 2 \cdot l$. Assuming that $U_A = U_B = U$ and that $\underline{Z}'$ is constant between the two substations, the voltage divider rule leads to the equation

$$(\underline{I}_A / \underline{I}_{trc}) = \underline{Z}' \left(l_2 - x\right) / \left(\underline{Z}' l_2\right)   .$$

Taking substation A as a reference point, the above considerations lead to an expression for the *voltage drop* between the substation and a point x

$$\Delta \underline{U}_x = \underline{I}_{trc} \underline{Z}' \left(x - x^2 / l_2\right)$$

(5.41)

a) One train in the section

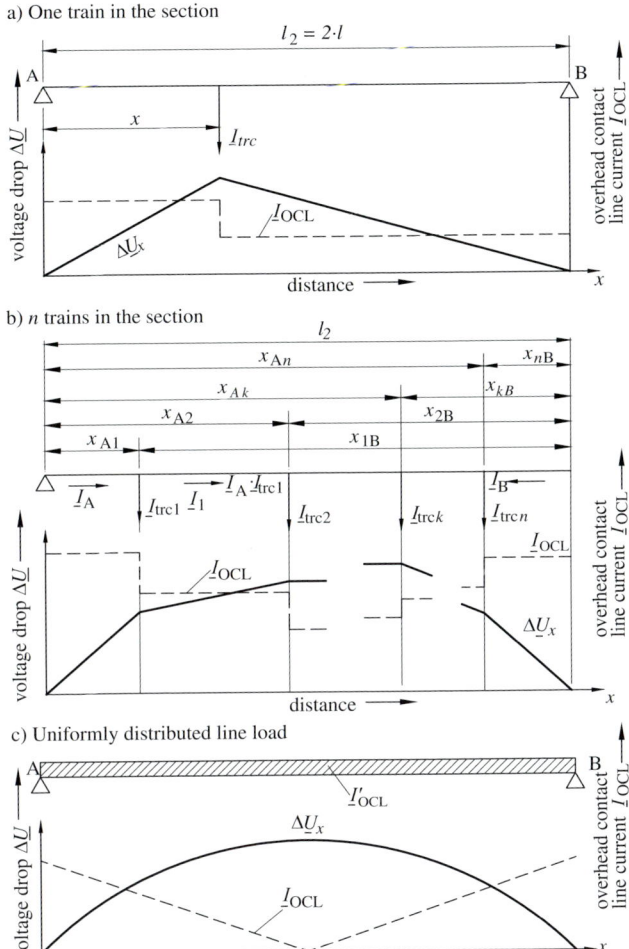

b) n trains in the section

c) Uniformly distributed line load

Figure 5.7: Voltage drops and overhead contact line currents in sections with double-end feed.

and for the maximum that will occur at the point $x = l_2/2$,

$$\Delta \underline{U}_{\text{max}} = (1/4)\underline{I}_{\text{trc}}\underline{Z}'\, l_2 \quad .$$

Using the above assumption, i. e. $l_2 = 2l$, it is obtained

$$\Delta \underline{U}_{\text{max}} = (1/2)\underline{I}_{\text{trc}}\underline{Z}'\, l \quad .$$

If the contact line installations of a double-track line are cross-coupled at the mid-point of the feeding section, the voltage drop caused by a train travelling on one of the parallel sections is described by the following equations [5.13]

$$\Delta \underline{U}_x = \underline{I}_{\text{trc}}\underline{Z}'\,\left[x - 3x^2/(2\,l_2)\right] \quad ,$$

$$\Delta \underline{U}_{\text{max}} = (1/6)\underline{I}_{\text{trc}}\underline{Z}'\, l_2 = (1/3)\underline{I}_{\text{trc}}\underline{Z}'\, l \quad ,$$

$$\Delta \underline{U} = (1/8)\underline{I}_{\text{trc}}\underline{Z}'\, l_2 = (1/4)\underline{I}_{\text{trc}}\underline{Z}'\, l \quad . \tag{5.42}$$

Table 5.11: Voltage drops in contact line network feeding sections [5.13, 5.14].

Type of feed	Number of trains n in the section	Instantaneous value ΔU_x	Mean value ΔU	Maximum value ΔU_{max}
	1	x	$l/2$	l
single-side	uniform load		$l \cdot n/3$	$l \cdot n/2$
	n	$\left(\sum\limits_{i=1}^{n} I_{trci} x_i \right) / I_{trc}$	$l \cdot (n+1,5t_{zz}-1)/3$	$l \cdot (n+1,5t_{zz}-1)/2$
	1	$x[1-x/(2l)]$	$l/3$	$l/2$
double-sided	uniform load		$l \cdot n/6$	$l \cdot n/4$
	n	see (5.37)	$l(n+2t_{zz}-1)/6$	$l \cdot (n+2t_{zz}-1)/4$
double-sided with	1	$x-3x^2/(4l)$	$l/4$	$l/3$
cross-coupling	uniform load		$l \cdot n/12$	$l \cdot n/8$
connection	n		$l \cdot (n+3t_{zz}-1)/12$	$l \cdot (n+3t_{zz}-1)/8$
Note:	all formulae and expressions need to be multiplied by $I_{trc}R'$ for DC installations and by $\underline{I}_{trc}\underline{Z}'$ for AC installations			

Several trains in the feeding section

Using the assumptions and conditions described above, the instantaneous value of the voltage drop between substation A and the train number k is given by

$$\Delta \underline{U}_{A,k} = \frac{\underline{Z}'}{l_2} \left[(l_2 - x_{A,k}) \sum_{i=1}^{k} \underline{I}_{trci} x_{A,i} + x_{A,k} \sum_{i=k+1}^{n} \underline{I}_{trci}(l_2 - x_{A,i}) \right] \quad . \tag{5.43}$$

If the voltages $\underline{U}_A$ and $\underline{U}_B$ of the two substations are not equal, the right-hand side of Equation (5.43) must be supplemented by the sum $x_{A,k}(\underline{U}_A - \underline{U}_B)/l_2$. In this case, a compensating current $\underline{I}_c$ will flow through the contact line section from one substation to the other if there is no load along the section. The value of this no-load *compensation current* will be

$$\underline{I}_c = (\underline{U}_A - \underline{U}_B)/(\underline{Z}' l_2). \tag{5.44}$$

Assuming all traction currents to be equal and applying (5.37) it is obtained:

$$\Delta \underline{U} = (\underline{Z}' l_2/12) \sum_{i=1}^{n} \underline{I}_{trc\,i} = (\underline{Z}' l/6) \sum_{i=1}^{n} \underline{I}_{trc\,i}$$

and for the maximum voltage drop

$$\Delta \underline{U}_{max} = (1/8)\, n\, \underline{I}_{trc}\, \underline{Z}'\, l_2 = (1/4)\, n\, \underline{I}_{trc}\, \underline{Z}'\, l \quad .$$

The parameter t_{zz} is defined in Clause 5.2.2.2 in context of Equation (5.40). The expressions required for calculating the *voltage drops* for different types of contact line feeding sections are summarized in Table 5.11. To allow easier comparison of the results, all lengths are expressed relative to a contact line section of length l_2, whereby the length of the section under consideration in the case of double-ended feed is defined as half the distance l_2 between the substations, i. e. $l_2 = 2 \cdot l$.

A feeding section of length l is thus considered to extend from the feed point up to the coupling point between substations or, for terminating sections, up to the end of the section.

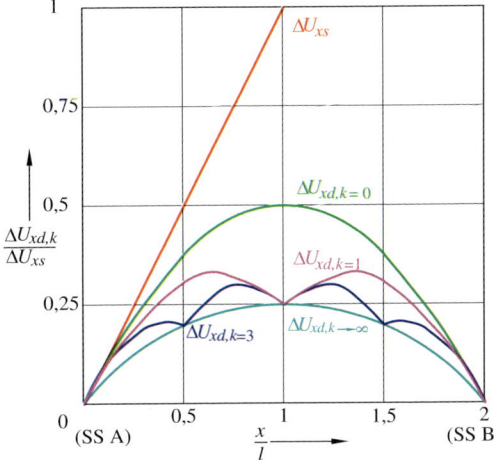

Figure 5.8: Voltage drop graphs of cases with equal total loads, different types of feed, without cross-coupling, with one cross-coupling and with three cross-couplings, relative to the voltage drop $\Delta \underline{U}_{xs}$ occurring with single-end feed and with one train in the feeding section. meaning of the indices:
s single-end feed, d double-end feed
k number of cross-couplings between the substations
SS A, SS B substations

The theoretical limit $n \rightarrow \infty$ and $\underline{I}_{\mathrm{trc}} \rightarrow 0$ gives a value for the uniform line load which can be calculated using (5.36).

Figure 5.8 is a graphic comparison of the different voltage drop situations for equal total loads in the section between two substations. This graph also demonstrates that increasing the number of *cross-couplings* within a feeding section improves the voltage conditions in a traction power contact line network.

5.2.3   Other calculation algorithms

The train schedule of high-speed lines cannot be described as a stochastic process. Therefore, it is advisable to calculate the instantaneous OCL voltages for the actual train schedules for such lines. In [5.15] an algorithm is described for the *calculation of the voltage drops*. The *train simulation* and the simultaneous network calculations are detailed in Clause 7.6 of [5.13]. The electrical data and the description of the voltage situation in the traction network derived with this technique are represented in [5.16].

Figure 5.9 shows an example of the voltage at a high-speed train calculated on the basis of the method described in [5.15]. The supply arrangement for AC 15 kV 16,7 Hz is shown in Figure 5.9 a) and the voltage at the pantographs for 4, 10 and 30 minutes headway in Figure 5.9 c). Figure 5.10 contains the same information for AC 25 kV 50 Hz. Single-end feeds are used in the AC 25 kV 50 Hz system as opposed to double-end feeds in the AC 15 kV 16,7 Hz system. In parts c) of Figures 5.9 and 5.10 the voltages at the traction vehicles are shown as they travel along a 200 km long line.

Other methods have also been developed for assessing the voltage conditions in contact line networks of conventional, normal-load railways. These include

– calculating voltage drops for mixed traffic conditions
– calculating voltage drops using stochastic methods and
– estimated calculation of the maximum voltage drops.

These methods are also described in detail in [5.13].

a) Structure of traction power supply for 1 AC 15 kV 16,7 Hz

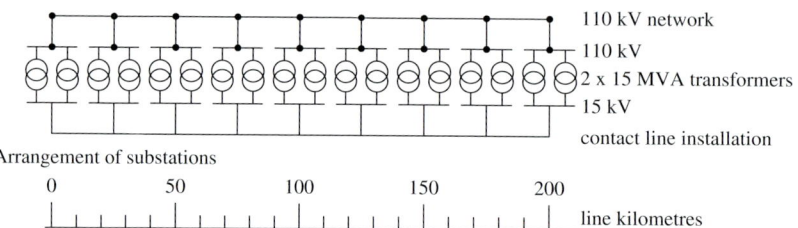

b) Arrangement of substations

c) Voltage at pantographs for headways 4, 10 and 30 minutes

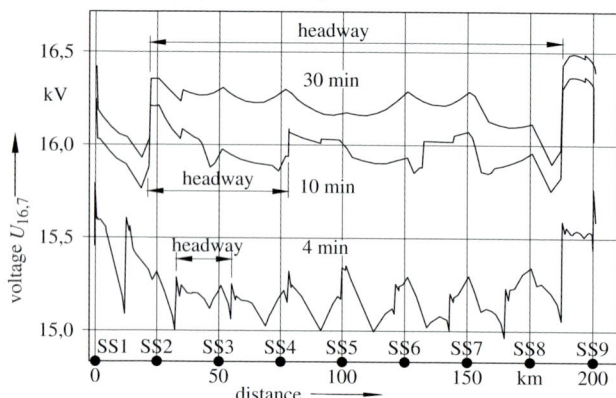

Figure 5.9: Calculated pantograph voltages for high-speed trains with AC 15 kV 16,7 Hz supply.

a) Atructure of traction power supply for 1 AC 25 kV 50 Hz

b) Arrangement of substations

c) Voltage at pantograph for a headway 4 minutes

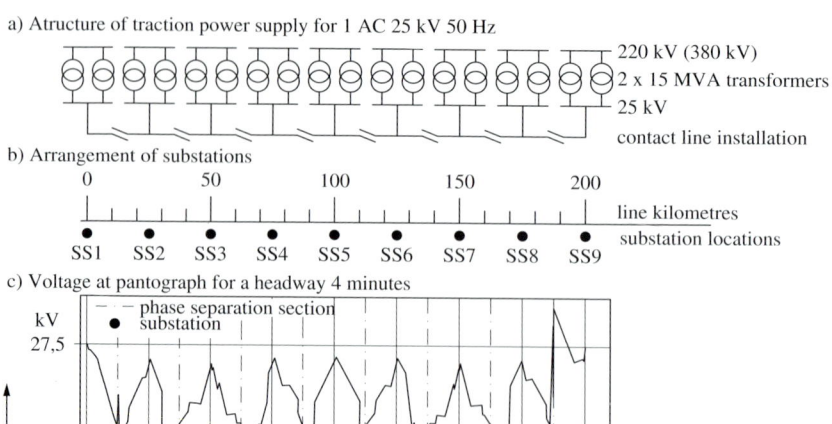

Figure 5.10: Calculated pantograph voltages for high-speed trains with AC 25 kV 50 Hz supply.

Table 5.12: Minimum mean useful voltages $U_{\mathrm{mean\,useful}}$ at pantograph.

Power supply system	High-speed lines zone and train	Conventional high-speed lines and classic lines zone and trains
1 500 V	1 300 V	1 300 V
3 000 V	2 800 V	2 700 V
15 000 V 16,7 Hz	14 200 V	13 500 V
25 000 V 50 Hz	22 500 V	22 000 V

5.2.4 Mean useful voltage

5.2.4.1 Requirements and definitions

Substations and overhead contact lines should be able to accommodate the most severe conditions:

– at the densest operating period in the timetable, corresponding to peak traffic
– by the characteristics of the different types of train involved, taking into account selected traction units.

The *quality index mean useful voltage* $U_{\mathrm{mean\,useful}}$ was defined in EN 50 388 to assess the suitability of the installations with respect to the traffic requirements.

The mean useful voltage $U_{\mathrm{mean\,useful}}$ of a geographic zone is calculated by computer simulation, taking into account all trains scheduled to pass through the zone in an appropriate period of time corresponding to the peak traffic period in the timetable. This period of time should be sufficient to include the highest load on each electrical section in the geographic zone.

All trains in this geographic zone, considered over the peak traffic period, are included in this analysis whether they are in traction mode or not (stationary, traction, regeneration, coasting) at each simulation time step.

The voltage $U_{\mathrm{mean\,useful\,train}}$ is the mean value of all voltages in the same simulation run as for the geographic zone study but only analysing the voltages for one particular train at each time step. Precondition is, that the train is taking traction load ignoring steps when the train is stationary, regenerating or coasting mode.

The mean value of these voltages indicates the performance of each train in the simulation and as a result, identifies the governing train, the train most constrained by low voltage, under acceleration.

The minimum values for mean useful voltages are given in Table 5.12 in accordance with EN 50 388, Table 4. The power supply should be designed in such a way that the simulations of $U_{\mathrm{mean\,useful}}$ under normal operating conditions never generate momentary voltage values at the pantograph of any train lower than the limit $U_{\mathrm{min\,1}}$ specified in Table 4 of EN 50 163, see also Table 1.1 for traffic corresponding to the type of line concerned. The electrification system design should ensure the ability of the power supply to achieve the specified performance. The correct mean useful voltage

– allows the tractive units to function close to their nominal voltage, hence optimising efficiency and performance,
– ensures that the values of minimum voltage specified by the standards are respected,
– reflects the fact that the fixed installations for electric traction have the correct power capability and that, as a result, increased traffic volumes can be considered and
– allows to fulfill certain impaired traffic situations.

5.2.4.2 Calculation

The mean useful voltage at the pantograph can be obtained from

$$U_{\mathrm{mean\,useful}} = \sum_{i=1}^{n} \frac{1}{T_i} \int\limits_{0}^{T_i} U_{\mathrm{P}_i} \cdot |I_{\mathrm{p}i}| \cdot dt \left/ \left(\sum_{i=1}^{n} \frac{1}{T_i} \int\limits_{0}^{T_i} |I_{\mathrm{p}i}| \cdot dt \right) \right. ,$$
(5.45)

where
T_i integration or study period on train i,
n number of trains considered in the simulation.

For AC electrification
$U_{\mathrm{P}i}$ momentary rms voltage at fundamental frequency at the pantograph of train i,
$|I_{\mathrm{P}i}|$ absolute value of momentary rms current at fundamental frequency flowing through the pantograph of train i.

For DC electrification
$U_{\mathrm{P}i}$ momentary average DC voltage at the pantograph of train i,
$|I_{\mathrm{P}i}|$ absolute value of momentary average DC current passing through the pantograph of train i.

The mean useful voltage represents the relationship between mean power calculated for the trains during their traction sequences and the corresponding mean current.
An equivalent result is obtained with

$$U_{\mathrm{mean\,useful}} = \frac{1}{n} \sum_{i=1}^{n} \left[\left(\frac{1}{M \cdot N \cdot \Delta t} \right) \sum_{j=1}^{N} \sum_{k=1}^{M} \left(U_{j,k}(t) \cdot \Delta t \right) \right]_i ,$$
(5.46)

where
n number of trains considered in the simulation
$U_{j,k}(t)$ voltage (AC: rms value of fundamental frequency, DC: average value)
M number of calculation steps in the integration period
N number of integration over the simulation
Δt time during which each calculation step M is simulated.

The period Δt must be short enough to include all events in the timetable including the short-term values of the maximum currents and minimum voltages.
The formulae (5.45) and (5.46) can be used to study
   – a geographic zone i. e. the part of the network under consideration during a given period, accounting taken for all trains passing through the zone, whether they are in a traction mode or not. The value of $U_{\mathrm{mean\,useful}}$, therefore, can be taken as an indicator of the quality of the power supply for the entire zone
   – the mean useful voltage at the pantograph of train; only the traction periods of the train are taken into account. In this case, n is equal to 1 in the Equations (5.45) and (5.46).

Example 5.3: In Figure 5.11 the mean useful voltage $U_{\mathrm{mean\,useful}}$ is shown at a train running along the high-speed line HSL-Zuid in the Netherlands supplied by 25 kV 50 Hz. The calculation is based on operation of high-speed trains with 3 min headway. The voltage at the train varies between 21 300 V and 27 250 V and the calculated mean useful voltage is 25 300 V, which is above the minimum required value of 22 500 V.

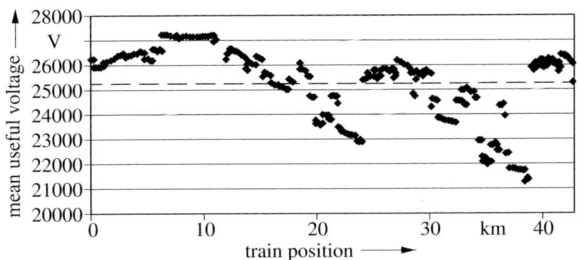

Figure 5.11: Mean useful voltage at a train, high-speed line HSL Zuid, Netherlands.
◆ train voltage
– – $U_{\mathrm{mean\,useful}}$

5.3 Electric traction loads

5.3.1 Introduction

The *power requirement* of a railway and the related currents result from the physical power necessary to fulfill the transportation purpose. The transportation process itself differs greatly over time and in geographical locations. The physical power required to achieve a specific line transport demand and the currents resulting currents depend on many parameters. The most essential are:

- the *speed*: the power required being proportional to the cube of the speed
- the *mass of the trains*
- the aerodynamic features
- the *frequency of service*
- the line topography
- the frequency of restarts
- the availability of regenerative braking
- the driving style of train drivers and
- the electrification system.

To achieve an adequate power supply and contact line design, the characteristics of traction loads should be determined and described as accurately as possible.

In power supply systems, the rated power of the traction vehicle will be the quantity on which the operating current calculations is based. The time function of traction currents drawn by moving trains can be analysed by simulating train runs and is determined by the parameters of the respective run. In a contact line installation, the traction currents of all trains travelling in the same supply sections at a given time will be superimposed.

The train loads and, therefore, the *load currents* of contact line supply sections of general-purpose railway lines can be described as stochastic functions, as demonstrated in Clause 5.3.3. However, in high-speed traffic, there is often only one train in each feeding section. Consequently the current load is intermittent. This case is detailed in Clause 5.3.4.

5.3.2 Time-weighted equivalent continuous load

Accurate calculation of the thermal stress on overhead contact line installations subjected to currents represented by $I_{\mathrm{trc}}(t)$ requires considerable effort. However, *time-weighted equivalent continuous load curves* [5.17] are realistic models of the effective currents determining the thermal stresses. Thereby, the time-dependent characteristics of the currents will not be lost [5.18].

In the following, modelling of the time-weighted equivalent continuous load curves is discussed for the arithmetical mean values:

- the real *time graph of the load current* $I_{trc}(t)$ within a reference period T forms the basis. This is usually available in the form of a time-discrete sequence of values within a defined time interval t_D as shown in Figure 5.12
- the next step is to define a time window t^*. This variable time window is moved across the entire load current graph in steps of t_D, over the complete period of the load current $T - t^*$, see Figure 5.12. The time step t_D should be chosen between 10 and 20 s
- the mean load current is then calculated for every possible position of the time window of width t^*. The maximum mean current value I_{max} subsequently determined, is stored relative to the current window width
- this is repeated with varying window widths ranging from the smallest possible value, i. e. t_D, right up to the largest possible value i. e. $t^* = T$
- consequently, a function of the *maximum mean currents* in relation to the load duration is obtained as represented by the window width and can be called the *time-weighted equivalent continuous load curve* of the mean values. In reference [5.17] this is also given the designation "peak value graph"

For discrete-interval load current value sequences $I(t)$ related to a time interval of t_D, the rule for calculating the time-weighted equivalent continuous load curve of the arithmetic mean value is

$$I_{max}(t^*) = \max \left(\frac{1}{t^*} \sum_{i=t}^{t+t^*} |I_i| \cdot t_D \right) \quad .$$

For the heating characteristics and thermal load calculations, however, the effective values of the load current are decisive. The effective current value, which is the root-mean square value of the current, corresponds to the equivalent direct current which would generate the same heat in an electric resistance over a period t_m as the time-variable current under consideration. The general equation for the *effective current value* I_{eff} is:

$$I_{eff}(t_m) = \sqrt{\frac{1}{t_m} \int_0^{t_m} I(t)^2 \cdot dt} \quad .$$

The modelling rule for calculating the time-weighted equivalent continuous load curve of the effective values is analogous to the arithmetic mean values. For a given time-discrete sequence of load current values I_i, the equation is:

$$I_{eff\,max}(t^*) = \max \left(\sqrt{\frac{1}{t^*} \sum_{i=t}^{t+t^*} I_i^2 \cdot t_D} \right) \quad , \tag{5.47}$$

where $0 \leq t \leq (T - t^*)$ and $t_D \leq t^* \leq T$.

In Figure 5.12 c) the time-weighted equivalent continuous load curves are shown for 6/7 min headway (on left side) and for 14/15 min headway (right side). These were calculated using the above algorithms.

The recognised *standardized equivalent continuous load curve* for general traffic railway lines and the time-weighted equivalent continuous load curve of high-speed or heavy-duty railway lines, form the basis for rating the *thermal load capacity* of contact line installations, see Chapter 7.

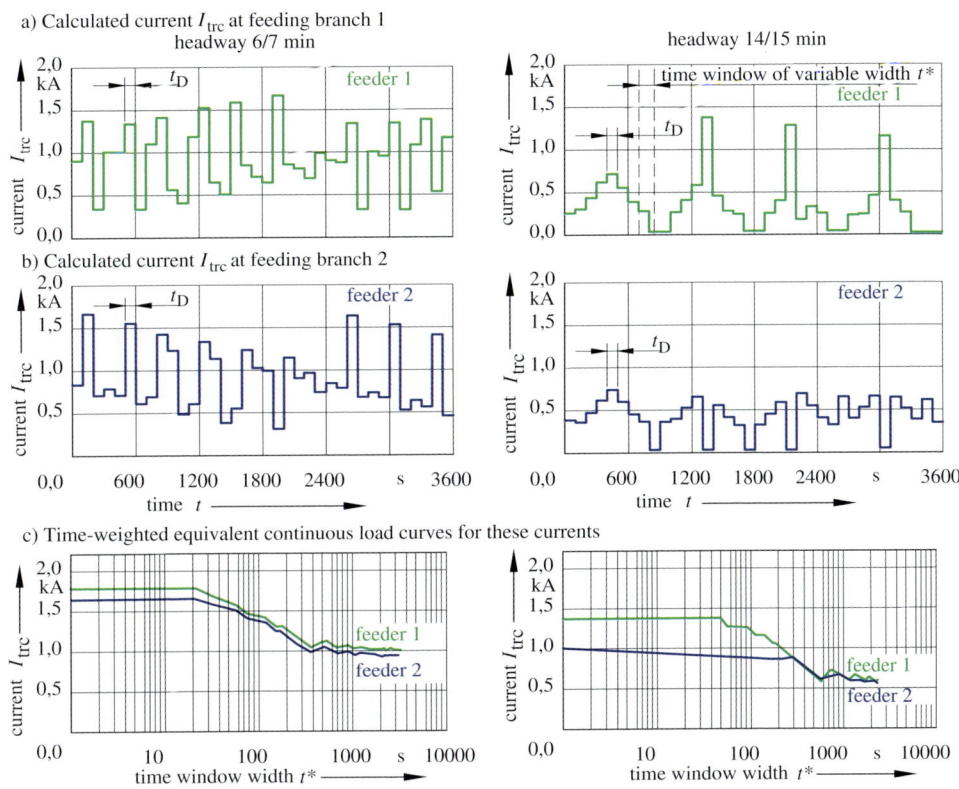

Figure 5.12: Load currents of two line feeders of a substation with traction currents of 1 130 A drawn by high-speed trains travelling at constant speeds of 330 km/h [5.18].

5.3.3 Railways for general traffic

Railways for general traffic are used by trains of various categories at speeds up to 200 km/h; resulting in power demands up to approximately 300 kW/km. Their load distribution characteristic can be described with stochastic functions [5.19]. The components describing the loads on *main line railways* for general traffic are shown in Figure 5.13.

Graph a) in Figure 5.13 shows the variation of *monthly mean load* of a substation during a year. The annual load variation is determined, for example, by the need to heat passenger trains in the winter, by holiday traffic in summer months and by other seasonal mass transportation demands. For design calculations, the statistically determined *day's load coefficient* c_d is used:

$$c_d = P_{d\,max}/P_a. \tag{5.48}$$

$P_{d\,max}$ is the maximum daily average load occurring in the entire year and P_a is the *annual mean load*. Experience has shown that the daily load coefficient is virtually only dependent on the annual mean value. In Figure 5.14 the relationship is shown between c_d and P_a.

Figure 5.13 b) shows the typical variation in the load values of a main line substation in the course of a day. The variation in the load is characterized by peaks of rush-hour traffic in

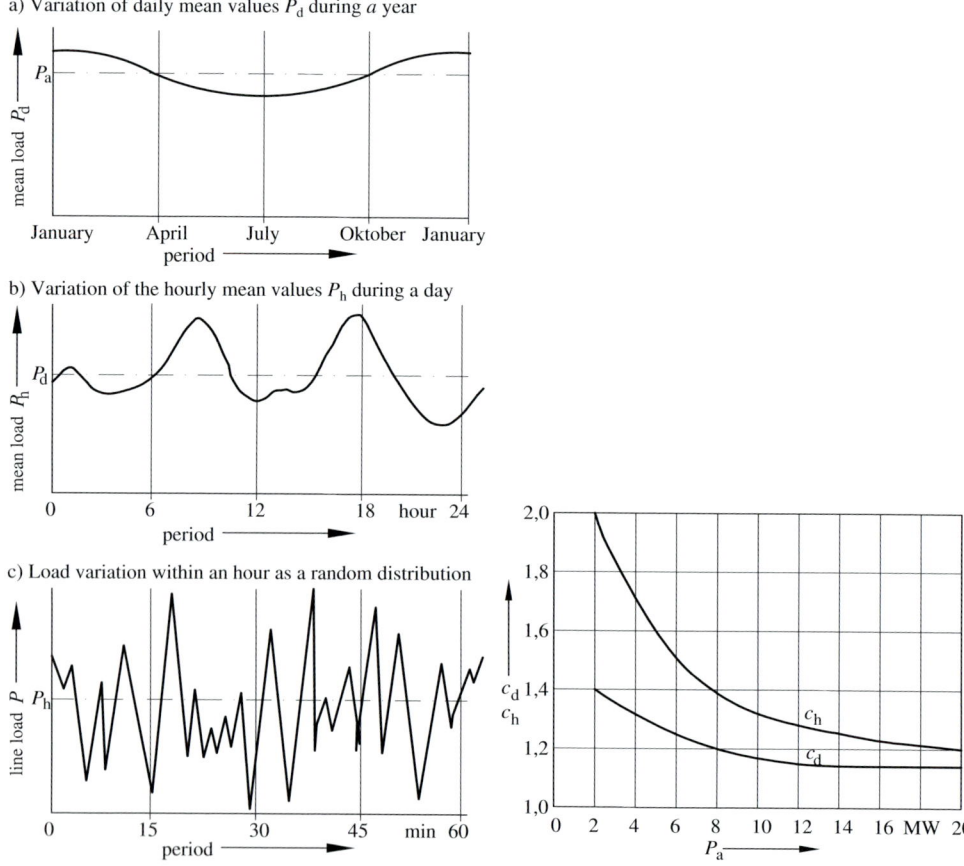

a) Variation of daily mean values P_d during *a* year

b) Variation of the hourly mean values P_h during a day

c) Load variation within an hour as a random distribution

Figure 5.13: Components of the idealised random functions describing railway line loads.
P_a annual mean load, P_d daily mean load, P_h hourly mean load

Figure 5.14: Daily load coefficient c_d and hourly load coefficient c_h as functions of the annual mean load P_a, as given in [5.19].

the mornings and evenings and by low-load periods at midday and at night. By statistical evaluation of a large number of implemented installations, it is also possible to determine an *hour's load coefficient* c_h defined as

$$c_h = P_h/P_d = P_{h\,max}/P_{d\,max} \quad . \tag{5.49}$$

where P_h is the maximum hourly mean power consumption during a day and P_d is the corresponding daily mean power. The hour's load coefficient c_h is also related to the annual mean power. This relationship is also depicted in Figure 5.14.

The variation of the power load within an hour represents the sum of the respective loads occurring on the system due to the individual trains travelling on the section of a line under consideration. This power load can be described as a random function. Using Equations (5.48) and (5.49), the mean value of the power $P_{h\,max}$ of a feeding section during the hour with the highest load within the entire year is

$$P_{h\,max} = c_d \cdot c_h \cdot P_a \quad . \tag{5.50}$$

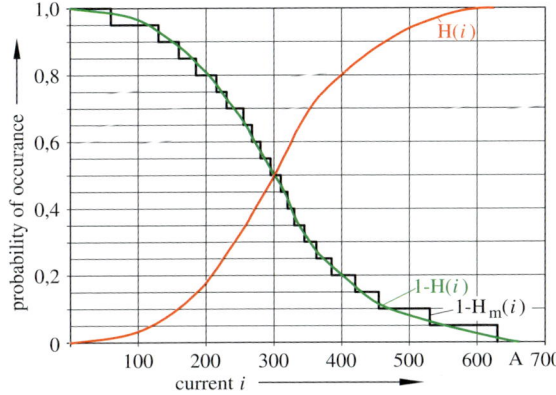

Figure 5.15: Measured $1 - H_m(i)$ histogram and $1 - H(i)$ continuous load curve, as well as the cumulative distribution curve $H(i)$ of a normally distributed load current drawn from a traction substation.

The mean annual load P_a can be calculated from the total annual energy demand W_a of the section. These idealized components simplify the description of the load by a random function. The same is true for *tramway line loads* [5.20]. In many cases, it is also possible to describe the traction power load as a variable not dependent on time. For supply power to the contact lines, the currents at high loads can also be described as non-time-dependent variables.

It is acceptable to assume that the traction power in the hour of the highest load follows a *normal distribution* function, where $P_{h\,max}$ is the mean value of the distribution and σ_p the standard deviation. With this hypothesis, a forecast of the probability $F(P)$ with which the load P will remain below a specified value P_L is possible:

$$F(P \leq P_L) = \frac{1}{\sigma_p \sqrt{2\pi}} \int\limits_{-\infty}^{P_L} \exp\left[-(P - P_{h\,max})^2 / 2\sigma_p^2\right] dP \quad . \tag{5.51}$$

The probability of $F(P \leq P_L)$ can be expressed with (5.51) because $P_{h\,max} \gg \sigma_p > 0$. Equation (5.51) is also called the distribution function of the random variable P. The *standard deviation* σ_p can be expressed in terms of the *coefficient of variation* v_p and the mean value $P_{h\,max}$

$$\sigma_p = v_p \cdot P_{h\,max} \tag{5.52}$$

and the power P as the sum of the mean value of the one-hour power $P_{h\,max}$ and a multiple λ_L of the standard deviation σ_p

$$P_L = P_{h\,max} + \lambda_L \sigma_p = P_{h\,max} (1 + \lambda_L v_p) \quad . \tag{5.53}$$

Equation (5.51) can be transformed into the standardized form of the distribution function $F(\lambda_L)$, being the *Gaussian standard distribution*

$$F(\lambda_L) = \frac{1}{\sqrt{2\pi}} \int\limits_{-\infty}^{\lambda_L} \exp(-\lambda^2/2) \, d\lambda \quad . \tag{5.54}$$

This standardized form of distribution is described in tabular form in relevant engineering handbooks. If $F(P \leq P_L)$ or $F(\lambda_L)$ is the probability of a random load P remaining below the given limit value P_L, as discussed above, the converse

$$G(\lambda_L) = 1 - F(\lambda_L) \tag{5.55}$$

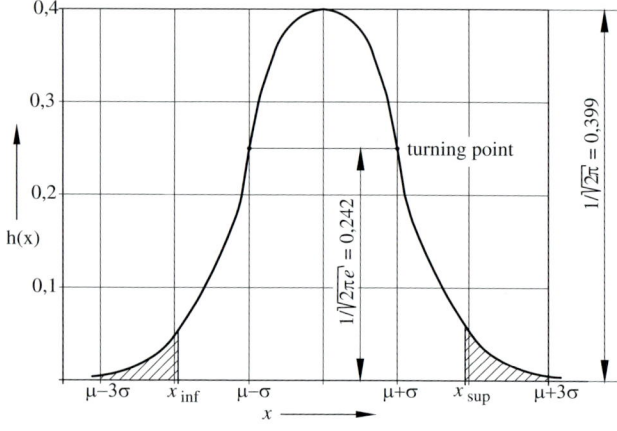

Figure 5.16: Standardised density $h(x)$ for $\sigma = 1$ and mean value $\mu = 1$ of a normally distributed quantity, e. g. power or current of a substation.

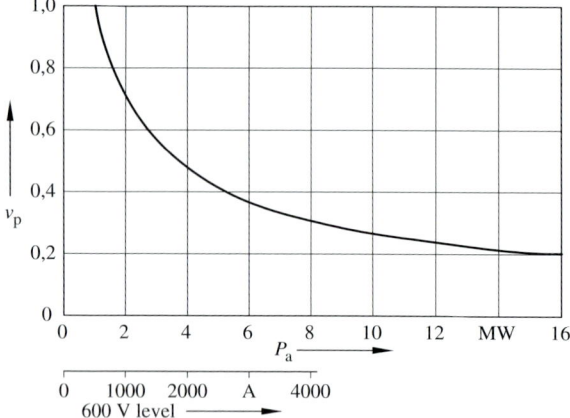

Figure 5.17: Coefficient of variation v_{p} as a function of mean annual power consumption P_{a}.

can also be determined as the *probability of occurrence of a load* that exceeds the limit P_{L}. In electrical energy engineering, this function is called the *continuous loading diagram* or *standardised loading diagram*. Within a given period T, the load occurrences will remain below the limit P_{L} with a probability of $(T - t)/T = \mathrm{F}(\lambda_{\mathrm{L}})$ and conversely, will exceed the limit P_{L} with a probability of t/T.

In Figure 5.15, the measured *loading histogram* $1 - \mathrm{H_m}(i)$ and the continuous loading graph $1 - \mathrm{H}(i)$ derived from this are shown, as well as the distribution function $\mathrm{H}(i)$ of the load current of a railroad traction substation. To illustrate the explanations given above, the *density function of the standardized normal distribution* is shown in Figure 5.16.

In Figure 5.17, the *coefficient of variation* v_{p} is shown in relation to the *mean annual power consumption* P_{a} of a DC 0,6 kV mass transit system. The relationship shown has been determined empirically from a large number of measurements taken in several European railway networks [5.13].

The annual mean load current can be obtained from

$$I_{\mathrm{a}} = P_{\mathrm{a}} / (U \cdot \cos \varphi) \quad . \tag{5.56}$$

Table 5.13: Load peaks of duration t within a load observation period $T = 1\,\mathrm{h}$ in a normally distributed batch.

t	$(T-t)/T = \mathrm{F}(\lambda_{\mathrm{L}})$	λ_{L}
3 min	0,9500	1,645
1 min	0,9833	2,13
10 s	0,9972	2,77
1 s	0,9997	3,44

If the mean annual value of the power P_{a} is known, the current occurring in the hour of maximum load in the course of an entire year is

$$I_{\mathrm{h\,max}} = c_{\mathrm{d}} \cdot c_{\mathrm{h}} \cdot P_{\mathrm{a}} \,/\, (U \cdot \cos\varphi) \quad , \tag{5.57}$$

where U is the rated voltage of the traction energy supply network and $\cos\varphi$ the mean value of the *power factor* in the traction power network.

To estimate the load peaks of a defined duration within the maximum load hour, the peak values corresponding to the λ_{L} values are taken from Table 5.13, which is based on a normal distribution function for $\mathrm{F}(\lambda_{\mathrm{L}}) = (T-t)/T$, in relation to $T = 1$ hour. This will be illustrated by Example 5.4.

Example 5.4: The mean annual line load of a heavily loaded feeding section is $P_{\mathrm{a}} = 3\,\mathrm{MW}$. Determine the peak load currents in this feeding section, with the voltage $U = 25\,\mathrm{kV}$ and $\cos\varphi = 0,83$?
From Figures 5.14 and 5.17, $c_{\mathrm{d}} = 1,36$, $c_{\mathrm{h}} = 1,86$ and $v_{\mathrm{p}} = 0,58$ are obtained. These data can be used to calculate
- the mean annual value of the load current, using Equation (5.56)

$$I_{\mathrm{a}} = 3\,000\,\mathrm{kW}/(25\,\mathrm{kV} \cdot 0,83) = 145\,\mathrm{A}$$

- and the maximum hour's mean value on the day with the maximum load occurring in the entire year, using Equation (5.57)

$$I_{\mathrm{h\,max}} = 1,36 \cdot 1,86 \cdot 145\,\mathrm{A} = 367\,\mathrm{A} \quad .$$

Using Equation (5.53) and Table 5.13, it is deduced:
- the maximum 3 minute peak value in the entire year: Since $(60-3)/60 = 0,95$, see Table 5.13, and, therefore, $\lambda_{\mathrm{L}} = 1,645$ and $v_{\mathrm{p}} = 0,58$:

$$I_{\mathrm{3min}} = 367\,\mathrm{A} \cdot (1 + 1,645 \cdot 0,58) = 717\,\mathrm{A},$$

- the maximum 10 second peak value: Since $(3\,600 - 10)/3\,600 = 0,9972$, see Table 5.13, and, therefore, $\lambda_{\mathrm{L}} = 2,77$:

$$I_{\mathrm{10s}} = 367\,\mathrm{A} \cdot (1 + 2,77 \cdot 0,58) = 957\,\mathrm{A} \text{ and}$$

- the maximum 1 second peak value: Since $(3\,600 - 1)/3\,600 = 0,9997$, see Table 5.13, and, therefore, $\lambda_{\mathrm{L}} = 3,44$:

$$I_{\mathrm{1s}} = 367\,\mathrm{A} \cdot (1 + 3,44 \cdot 0,58) = 1\,099\,\mathrm{A}.$$

To be able to determine the electrical parameters, data is required on the expected load currents. In Table 5.14, guideline data is presented on expected operating currents in various power supply systems. Since the instantaneous load cases, as shown in Figure 5.7, only apply to a particular moment in time, the use of the *line current loads* I'_{OCL} per unit length in

Table 5.14: Guideline values of expected maximum operating currents in various power supply systems.

Vehicle/ train type	Power supply system	Rated power	Auxiliaries	Probable maximum currents		
				individual vehicle/train	double traction	contact line section
		kW	kW	A	A	A
T4D Dresden	DC 600 V	172	70	600	1 200	3 000
GT6N Mannheim	DC 600 V	1 480	80	780	1 700	4 000
AEL Hong Kong	DC 1 500 V	5 300	800	4 500		4 500
Munich subway	DC 750 V	2 340		1 050	3 000 [1)]	4 500
Berlin heavy rail	DC 750 V	2 400		800	3 200 [2)]	4 500
DB, BR 420	AC 15 kV	2 400	110	250	500	1 200
DB, BR 120	AC 15 kV	6 400	800	460	800	1 800
DB, BR 112/143	AC 15 kV	3 720	600	290	550	1 000
DB, ICE	AC 15 kV	4 800	500	420 [3)]	840	1 500
DB, ICE 3	AC 15 kV	8 000	500	725	1 450	2 000
SNCF, Thalys	AC 25 kV	4 440	500	200	400	800
	DC 1 500 V	1 840	500	1 500	3 000	

[1)] triple train, [2)] $4 \times$ Br 481+482, [3)] per traction unit

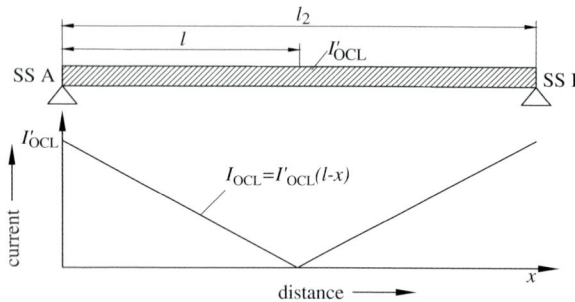

Figure 5.18: Contact line currents I between two substations assuming a uniformly distributed line load I'_{OCL} according to (5.36).

calculations represents an often adopted approach. The line current can be deduced from the power per unit length P' on which the design of the electric railway system was originally based:

$$I'_{OCL} = P'/(U_n \cdot \cos \varphi) \quad . \tag{5.58}$$

As shown in Figure 5.18, the current flowing at any point a distance x away from the left-hand feed point of a contact line installation is

$$I_{OCL} = P' \cdot (l - x)/(U_n \cdot \cos \varphi) \quad . \tag{5.59}$$

At the feed point x, the current flowing into the line section is

$$I_{OCL}(x = 0) = P' \cdot l/(U_n \cdot \cos \varphi) \quad . \tag{5.60}$$

The values shown in Table 5.14 can serve as a realistic guide to typical values of power loads per unit length. In the UIC leaflet No. 795-0 [5.21] a value of 3 MVA/km is mentioned as the power per unit length to be installed for double-track high-speed traffic lines (Table 5.15). This value is extremely high. The value 5,5 MVA/km mentioned in the same publication as being the power requirement of a line with a train headway of 2 minutes and speeds of up to 200 km/h applies only to exceptional cases.

Table 5.15: Guideline values for installed power per unit length P' on double-track electric railway lines, values given in kW/km.

Type of railway and traffic	P' in kW/km
– Lines with little traffic, trains at up to 120 km/h	up to 300
– Lines with heavier traffic loads, trains at up to 160 km/h	up to 500
– Lines with very heavy traffic loads, trains at up to 200 km/h	up to 1 000
– Local-area railways, 10 000 passengers per hour and direction, trains at up to 80 km/h, starting acceleration 1,1 m/s²	up to 750
– Local-area railways, 40 000 passengers per hour and direction, trains at up to 80 km/h, starting acceleration 1,1 m/s²	up to 3 000
– Madrid–Seville, 5 minute headway, speed 300 km/h	up to 1 065
– Madrid–Lerida, 5 minute headway, speed 350 km/h	up to 1 065

Table 5.16: Total inductive power factor of a train.

Instantaneous train power P in MW at the pantograph	$\cos\varphi$
$P > 2$	$\geq 0,95$
$0 \leq P \leq 2$	$\geq 0,85$[1)]

[1)] To control the total power of the auxiliary load of a train during the coasting phases, the overall average $\cos\varphi$ (traction and auxiliaries) defined by simulation and/or measurement should be higher than 0,85 over a complete timetable journey

Example 5.5: Determine the line load and current in a contact line feeding section of railway line with very heavy traffic, 25 kV nominal voltage where trains travel up to 200 km/h?

In Table 5.15, the power per unit length of double track line required for this type of traffic is given as 1 000 kW/km. With an assumed mean power factor of 0,76, the current per unit length of the double track is

$$I'_{\text{OCL}} = 1\,000\,\text{kW/km} / (25\,\text{kV} \cdot 0,76) = 52,6\,\text{A/km} \quad .$$

If the distance between substations on this line with double-end feeding is 50 km, a contact line feeding section length of 25 km must be used in the calculations.

The feeding current for the double track section is then calculated as

$$I_{\text{OCL}} = I'_{\text{OCL}} \cdot l = 52,6\,\text{A/km} \cdot 25\,\text{km} = 1316\,\text{A} \quad .$$

The currents calculated using Equation (5.60) can be considered to be the currents $I_{h\,\text{max}}$ drawn in the peak-load hour according to (5.57). They can then be used to calculate the currents for defined periods using Equation (5.53) by replacing the power P by the current I.

5.3.4 Power factor

The power factor of trains significantly affects the voltage along the line and the current load of substation and line equipment. EN 50 388 specifies the total inductive power factor for high-speed trains.

For yards or depots, when a train is stationary, with traction power-off, and the active power taken from the overhead contact line is greater than 10 kW per vehicle, the total power factor resulting from the train load should not be less than 0,8, and should have a target value of 0,9. The overall average $\cos\varphi$ for a train journey, including the stops, is calculated from the active energy W_P and reactive energy W_Q given by a computer simulation of a train journey or metered on an actual train.

$$\cos\varphi = \sqrt{1/(1 + W_Q/W_P)^2} \quad . \tag{5.61}$$

Table 5.17: Maximum allowable train current in A (excerpt from EN 50 388, Table 2).

Power supply system	TSI lines			Conventional TSI lines and classical lines			
	High-speed lines	Upgraded lines	Connecting lines	Target	Austria Germany Switzerland	Spain	Great Britain
DC 0,75 kV	–	–	6 800	–	–	–	–
DC 1,5 kV[1]	–	5 000	5 000	5 000	–	–	–
DC 3 kV[1]	4 000	4 000	4 000	4 000	–	2 500 [2]	–
AC 15 kV[1] 16,7 Hz	1 500	900	900	900	900	–	–
AC 25 kV[1] 50 Hz	1 500	600	500	800	–	–	300

[1] on special lines, e. g. freight lines in mountainous areas and suburban networks these values may be exceeded
[2] 3 200 A for upgraded TSI lines

The data given in Table 5.16 must be met for trains in accordance with the TSI for high-speed lines. They are recommended for all other lines. Modern trains achieve values of 0,98 or even 1,00. During regeneration, inductive power factor is allowed to decrease freely to keep voltage within limits.

5.3.5   High-speed and heavy-duty railway lines

High-speed and heavy-duty *railway lines*, e. g. underground and metropolitan railway lines with short headways between trains, have totally different traction power load characteristics to those of general railway traffic. These types of railway are characterised by an impulse-like load on contact line installations, feedings and substations. Studies have shown that the specific energy demand of high-speed railway lines can be as high as 1,0 to 1,3 MW/km and that of heavy-duty railway lines even as high as 1,7 to 2,5 MW/km, both cases are valid for double-track lines [5.18].

Figure 5.12 shows the load currents of supply sub-sections of a substation on a high-speed railway line. The high-speed trains draw traction currents of 1 130 A at collector strips, travelling at constant speeds of 330 km/h. On the left-hand side, the graphs of the contact line currents are shown for a case where all trains travelling in one direction pass at 6 minute intervals and trains travelling in the other direction pass at 7 minute intervals. The right-hand side shows the corresponding load current curves for 14/15 min intervals.

Although the load currents in double-end feeding sections of heavy-traffic lines with large loads are intermittent, they can be described in a simplified form with the aid of Equations (5.58) to (5.60). Because of high train frequency on such heavily travelled lines, the load currents exhibit a low statistic variation. Expressed quantitatively, a Gaussian distribution, i. e. a coefficient of variation, of less than 0,1 can be expected in these cases. Therefore, the load current per unit length can be used as an acceptable basis for estimating the required substation capacity.

5.3.6   Maximum train currents

To achieve compatibility between traction power supply and rolling stock for interoperable lines in the European railway networks the standard EN 50 388 specifies maximum allowable train currents including auxiliaries as set out in Table 5.17. The levels apply both in tractive

Table 5.18: Characteristic parameters of short-circuit currents according to standards EN 60 865-1.

Definition	Formula
Initial symmetrical short-circuit current I''_K: effective (rms) value of the symmetrical alternating component of a short-circuit current at the moment the short circuit occurs if the short-circuit impedance remains constant and equal to that existing at time $t = 0$.	$I''_k = c \cdot U_n / Z_k$ [1)
Peak short-circuit current i_p: maximum absolute value of the expected short-circuit current.	$i_p = \kappa_P \cdot \sqrt{2} \cdot I''_k$ [2)
Symmetrical short-circuit breaking current I_a: effective (rms) value of a short-circuit alternating current at the moment the circuit is opened by the circuit-breaker.	$I_a = \mu \cdot I''_k$ [3)
Sustained short-circuit current I_k: effective (rms) value of a short-circuit alternating current that would remain at a constant value after all transient processes have decayed.	$I_k = \lambda \cdot I''_{rG}$ [4)
Thermally equivalent short-circuit current I_{th}: the effective (rms) value of a current which would have the same thermal effect in the same time as the actual short-circuit current that might have direct-current component and decay with time.	$I_{th} = I''_k \cdot \sqrt{m+n}$ [5)
Initial symmetrical short-circuit AC power S''_k: the product of the initial symmetrical short-circuit current and the nominal voltage. These quantities are not quantities of power in the physical sense, only factors used in calculations.	$S''^{6)}_k = U_n \cdot I''_k$ or $S''^{7)}_k = \sqrt{3} \cdot U_n \cdot I''_k$

[1) c voltage factor $= 1{,}03$ to $1{,}1$ in railway networks
    Z_k network short-circuit impedance
[2) κ_P impulse factor according to Figure 5.19 [5.23]
[3) μ decay factor according to Figure 5.20 [5.23] for AC 16,7 Hz, $\mu = 1$ for AC 50 Hz
[4) λ factor for calculating steady-state short-circuit currents according to EN 60 865-1
    I''_{rG} generator current rating
[5) m, n factors describing the heating effect of direct and alternating current components in accordance with EN 60 865-1 ($n \approx 0{,}95$ in the centrally fed railway network)
[6)   in railway traction power networks
[7)   in three-phase AC power distribution networks

and regenerative modes. The trains need automatic devices that adapt the level of power consumption depending on the overhead line voltage in steady-state operation.

5.3.7   Short-circuit currents

Short circuits in traction contact line installations are caused by damage to or faults in the insulating components installed between conductive components having different electric potentials. Short circuits in contact line systems occur more often than in three-phase installations. Under adverse conditions such short circuits can lead to damage at the contact wires and/or catenary wires.

The occurrence rate of short circuits in German railways's network is 0,8 to 1,2 per kilometre and year. For the Madrid–Seville line it is 0,25 to 0,30 per kilometre and year. By also considering the frequency of trains, it can be concluded that short circuits occur less frequently in sections with lower traffic than in those with very frequent traffic. By comparison, only 0,02 faults per km line per annum occur in a comparable 3 AC 30 kV 50 Hz power supply network [5.22].

Table 5.19: Maximum contact line – short-circuit currents according to EN 50 388.

Power supply system	Substations connected in parallel Y / N	Short-circuit current kA
AC 25 kV 50 Hz	N	15
AC 15 V 16,7 Hz	Y	40
DC 3,0 kV	Y	50*
DC 1,5 kV	Y	100*
DC 0,750 kV	Y	100*

* for definition see EN 50 123-1

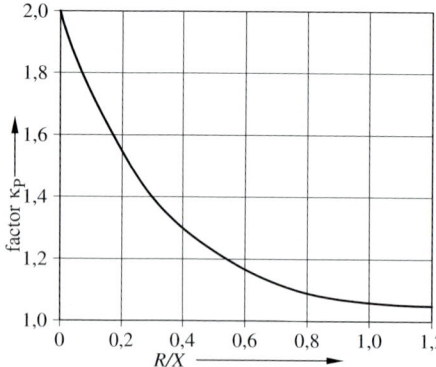

Figure 5.19: Impulse factor κ_P as a function of R/X in accordance with IEC 60 865-1.

In a traction power contact line network, every short circuit interrupts operations. However, less than five percent of all circuit-breaker trippings occurring in DB's electric railway traction network are caused by *sustained short circuits*. The most important factors leading to the high annual frequency of short circuits in traction contact line networks are:

- third-party interference e. g.
    - parts of loads, such as wagon awnings
    - birds or other animals bridging insulators
- faults due to electric railway operation *per se*
    - defects on traction vehicles
    - collector strip and pantograph damage
    - switching errors in traction power network operation
- meteorological impacts
    - lightning
    - storms with strong wind gusts
- bad maintenance condition of contact wire installations
    - wear and tear
    - material defects

Short circuits lead to high mechanical and thermal stresses in the affected electric installations. Energy supply cuts and dangerous situations can result where the rating of the installations is inadequate. To select the components, especially circuit-breakers, and to set up protective equipment correctly, requires knowledge of the magnitude of the expected *short-circuit currents*. Short-circuit currents in railway installations can also induce dangerous voltages in cables and metal structures parallel to the railway line.

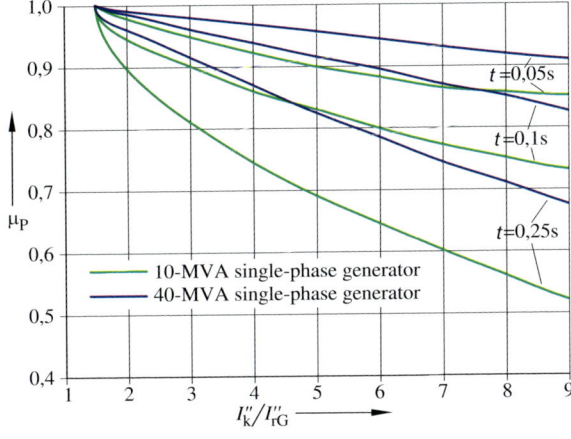

Figure 5.20: Decay factor μ_P applicable to a 16,7 Hz railway traction power network [5.23]. Decay factor μ_P can be taken as 1,0 for AC 50 Hz.
t duration of short circuit

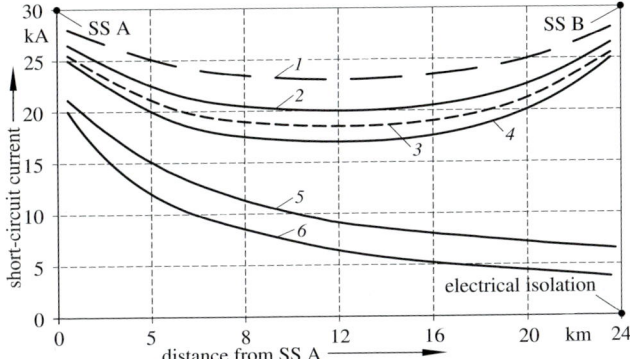

1 with feeder line and return conductors
2 with feeder line without return conductors
3 without feeder line with return conductors
4 without feeder line and return conductors
5 with feeder line and return conductors
6 without feeder line and return conductors

Figure 5.21: Maximum short-circuit currents in various power supply systems and traction contact line configurations *1, 2, 3, 4* for AC 15 kV 16,7 Hz, double-end feeding; *5, 6* for AC 25 kV 50 Hz

In an AC traction contact line system, any earth connection will constitute a short circuit. The short-circuit current in single-phase AC railway traction systems can be calculated using the formulae in Table 5.18. Table 5.19 lists typical values for short-circuit currents taken from EN 50 388. The values depend on the type of power supply and are considerably higher in DC and AC 15 kV 16,7 Hz systems than in AC 25 kV 50 Hz.

In Figure 5.21, the maximum short-circuit currents of a 24 km long high-speed railway section are shown with various overhead contact line configurations and different types of feeding, namely AC 15 kV 16,7 Hz and AC 25 kV 50 Hz [5.24]. The double-end feeding used in 16,7 Hz overhead contact lines leads to noticeably higher short-circuit currents.

If the short-circuit currents are calculated using the formulae given in Table 5.18, then the values obtained are higher than those obtained by measurements in actual practice. In reference [5.25], it has been established, by probability-based methods of all factors that can be expected in real-life applications, that the real maximum short-circuit currents are approximately 0,8 times the values obtained by using the formulae given in Table 5.18.

Figure 5.22 depicts an example for the *cumulative frequency distribution of short-circuit currents*. Reference [5.26] also explains an alternative method of calculating the expected short-circuit currents in electric railway traction power networks.

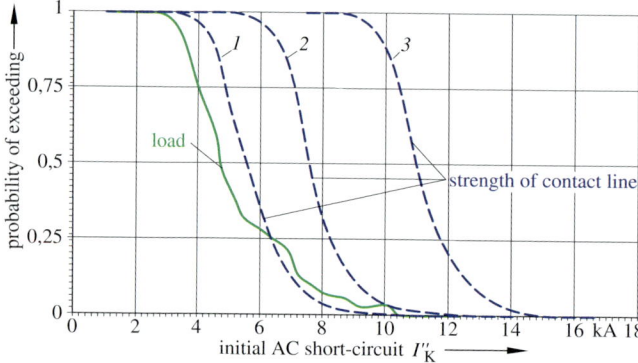

Figure 5.22: Stochastic rating for short circuits.
1 design too weak
2 optimum design
3 design too strong

In direct-current (DC) traction power systems, short circuits occurring in the contact line installation are also relevant for the design of the *rectifier equipment*. Such short circuits have a characteristic current behaviour, as shown in Figure 5.23. The short-circuit impulse current I_P is the main parameter affecting the dynamic short-circuit load. Thermal short-circuit loads are caused by the continuous short-circuit current I_{Kd}. In DC railway supplies without *current impulse suppression chokes* in the DC circuits, an I_P/I_{Kd} ratio of around 1,2 can be assumed as a reasonable approximation. The steepest *short-circuit current rise*, $(dI_k/dt)_{max}$ is the parameter used as a basis for determining the required circuit-breaker operating times. Furthermore, in DC railway networks, the *minimum short-circuit current* is an important factor in the setting of protective equipment. In practice, this current value is calculated frequently, using the following approximation:

$$I_{k\,min} = (U_{SS} - 0,15 \cdot U_n)/R_{loop} \tag{5.62}$$

In this equation, U_{SS} is the busbar voltage in the substation, which is usually assumed to be 1,1 times the rated voltage U_n. R_{loop} is the *loop resistance of the contact line* and track, and reaches its maximum value, when the short circuit occurs at the maximum possible distance from the substation.

The duration t_k of the short-circuit currents in traction power networks is determined by the response or operating time of the protective relays and the break-time of the power circuit breakers used. The following values may be used as guidelines for short-circuit durations t_K:

$t_K \approx 10$ to $25\,ms$ in DC systems

$t_K \approx 20$ to $45\,ms$ in single-phase AC systems where vacuum circuit breakers are used

$t_K \approx 45$ to $75\,ms$ in single-phase AC systems where compressed-air circuit breakers and minimum-oil-content circuit-breakers are used

5.4 Line feeding circuits

5.4.1 Basic requirements

To ensure reliable operation of electric railway lines, the contact line installation is subdivided into *electrical sections* that can be switched on or off or isolated in such a way that it can still be operated in the case of faults or planned disconnection of particular sections. Specific requirements are set out in the TSI Energy [5.27].

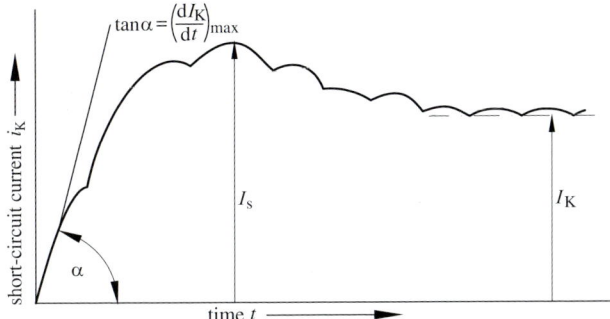

Figure 5.23: Characteristic graph of short-circuit currents in DC railways.

When drawing up configurations, designing and constructing contact line installations, the following aspects should be taken into consideration with regard to the *contact line circuits*:

- the circuit should enable *optimum efficiency* of the contact line installation with the least possible voltage and power losses in regular operation
- the contact line installation circuit should enable distinct, localized sections of the contact line installation to be isolated in case of necessary maintenance or repair work or in case of short circuits. Electric rail traffic should be able to continue in the unaffected sections. This circuit design principle, in conjunction with the respective protection design, is also called *selectivity configuration*
- the contact line circuit arrangement should be clearly understood and easy to monitor to prevent erroneous connections and work accidents. Consequently, the circuits used in public railway networks should be designed in accordance with standardized operational practices and aspects
- the number of contact line network components necessitated for the selectivity configuration, such as circuit-breakers, *section disconnectors*, *section insulators* and insulated overlaps should be restricted to the minimum

Therefore, the design of the contact line circuit configuration requires a sensible balance between the requirements of electrical engineering, protection, railway operations, maintenance and economics.

Town-planning aspects also have to be taken into consideration when drafting contact line circuit configurations for urban mass-transit networks. The criteria listed above, the calculated power requirements, the location of the power supply lines from the main energy sources, the railway line profile and the location of the fixed railway installations forms the basis of the line feeding configurations, also called *line feed plans*. These plans are then used as a basis for the design of contact line circuit diagrams.

5.4.2   Basic types of feeding circuits

Substations supply electric power to the electric traction vehicles within a specific section of the contact line network, which is also termed a *substation supply section*. A *contact line feeding section* receives its power from a section feed branch or section terminal of a substation via a particular feed line. In the case of main line railways, the overhead contact line feeding sections between two substations are subdivided into *switching sections* and these, in turn, are further divided into *switching groups*. This principle is detailed in Clause 5.5.3.

a) Single-end feeding

b) Double-end feeding with longitudinal coupling

c) Double-end feeding with longitudinal and cross-coupling

d) Double-end feeding with a large number of cross-couplings

e) Cross-connection circuit

f) Feeding scheme Madrid-Seville

g) Double-end feeding with reinforcing conductors and one cross-coupling

h) Double-end feeding with reinforcing conductors on one track only
   and with cross-couplings

i) Double-end feeding with reinforcing conductors along part-sections of
   of both tracks

j) Distributed feedings

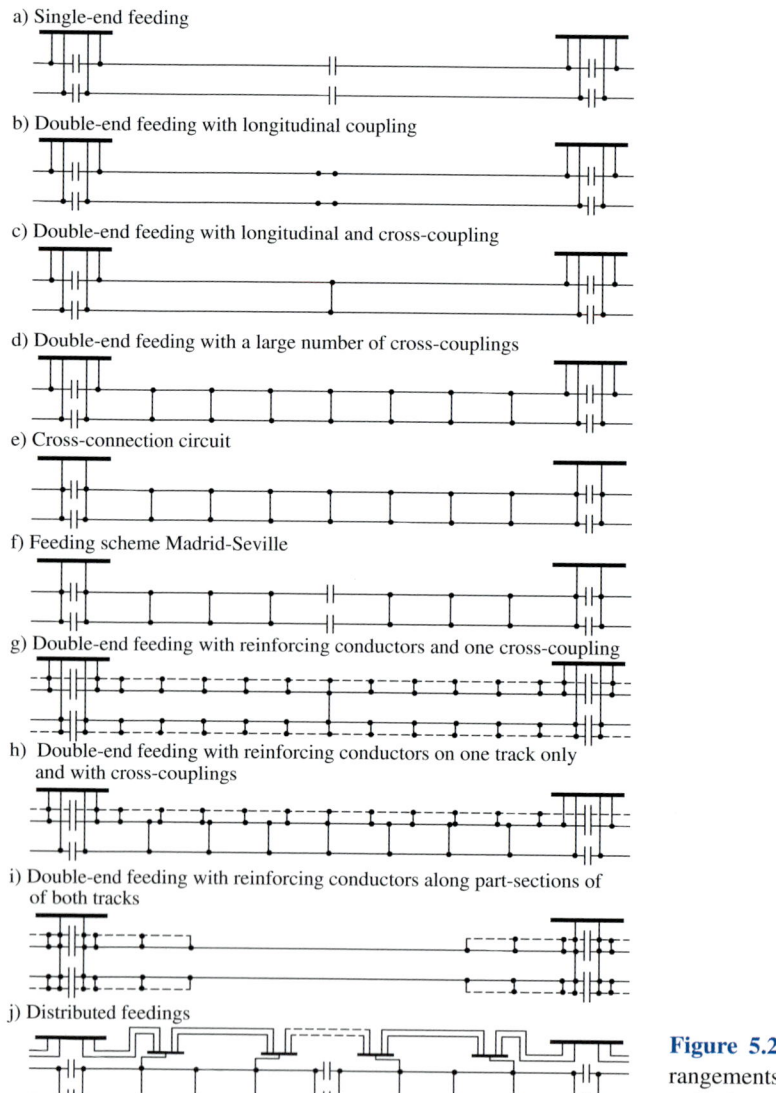

Figure 5.24: Basic feeding arrangements of contact line installations.

In Figure 5.24 the *basic feeding arrangements* of contact line installations are shown. To ensure clarity, no switchgear has been shown. The important basic circuits are described below.

- *Single-end feeding* (**unidirectional feeding**), Figure 5.24 a)
  Power is supplied to each feeding section via a separate circuit-breaker. From the protection aspect, this type of circuit is easy to control. It is sometimes used for local-area railways and frequently in AC 25 kV 50 Hz systems with and without cross-couplings.
- *Double-end feeding* **with longitudinal coupling,** Figure 5.24 b)
  At the ends of the feeding sections of the respective substations, the contact lines are coupled together via circuit breakers or disconnectors located in a switchgear housing.

As has been shown in Clause 5.2, double-end feedings considerably reduce voltage drops and power losses. Depending on the distance between the substations and the protection equipment, the coupling post with circuit breakers or disconnectors can be dispensed with.

– *Double-end feeding with cross-coupling(s)*, Figures 5.24 c) and 5.24 d)

Cross-coupling the contact line installations of the two tracks further reduces the voltage drops and power losses. In the case of a fault, all four circuit-breakers are tripped, meaning that initially both tracks of the sections between the substations will be isolated. After this, the disconnectors in the cross-couplings are opened. If only one of the switching groups has a sustained short circuit or other fault, power is returned to the unaffected sections approximately one to two minutes after the required testing and switching processes have been carried out, permitting trains to resume running on that section.

– *Cross-connection of double-end feeding installations*, Figure 5.24 e)

In this case, both tracks are supplied with power from the substation via a common circuit-breaker. This circuit design has to be introduced in contact line installations of railway lines with extremely high power demands to prevent excessive *potential differences* building up at section insulators. Where potential differences of more than 800 V occurred at section insulators, arcing and overhead contact line disturbances frequently occurred when traction vehicles travelled across the section insulators. Cross-bonding the overhead contact lines in the vicinity of railway stations reduces the potential differences at the section insulators. In case of disturbances, selectivity is achieved by disconnecting the faulty basic sections. The time required to do this hardly affects train traffic on the sections that are not faulty.

– *Feeding schematic Madrid–Seville*, Figure 5.24 f)

The Madrid–Seville line [5.28] is fed by one circuit breaker for the two OCLs. These are connected by disconnectors. Phase separation sections are installed midway between the substations.

– *Circuits with reinforcing feeder lines*, Figures 5.24 g), h), i)

The reinforcing feeder lines are permanently electrically connected to the contact line at specific intervals and may be installed parallel to both or just one of the contact lines of a double-track line. In practice, the variant shown in Figure 5.24 i), which only has reinforcing feeders in the vicinity of the substations, is also used.

– **Distributed feedings,** Figure 5.24 j)

This circuit is found in some tram traction power supplies. The circuit is characterized by feeding cables installed parallel with and connected to the overhead contact line via cable distribution junctions as shown in the diagram.

5.4.3   Overhead contact line circuits of 16,7 Hz railways

5.4.3.1   History

The basic circuits of 16,7 Hz railways have been well known since the beginning of 16,7 Hz electrification at the beginning of the 20th century and summarised in [5.29]. At first, single-sided supplies were used because the performance of protection technology was very limited. In the middle of two supply sections, electric separations caused disturbances because of voltage differences. Therefore, both sections were connected by a circuit breaker in the lon-

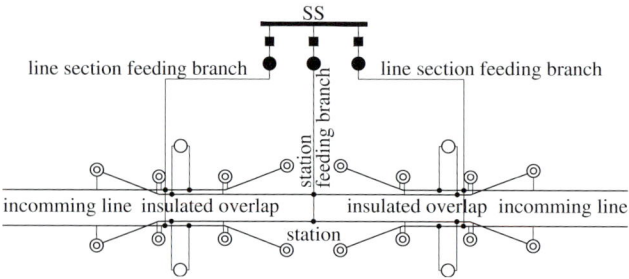

Figure 5.25: Line section feeding branches with a station feeding branch.

● disconnectors closed in normal position

○ disconnectors open in normal position

■ circuit-breakers on

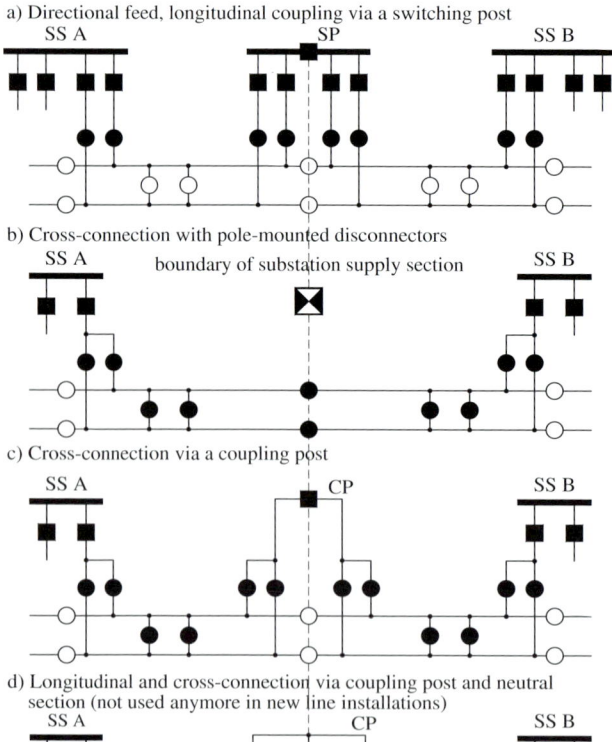

a) Directional feed, longitudinal coupling via a switching post

b) Cross-connection with pole-mounted disconnectors

c) Cross-connection via a coupling post

d) Longitudinal and cross-connection via coupling post and neutral section (not used anymore in new line installations)

─○─ pole-mounted disconnector open   ─□─ circuit breaker OFF

─●─ pole-mounted disconnector closed   ─■─ circuit breaker ON

Figure 5.26: Overhead contact line circuits used by the DB.
SS substation
SP switching post
CP coupling post

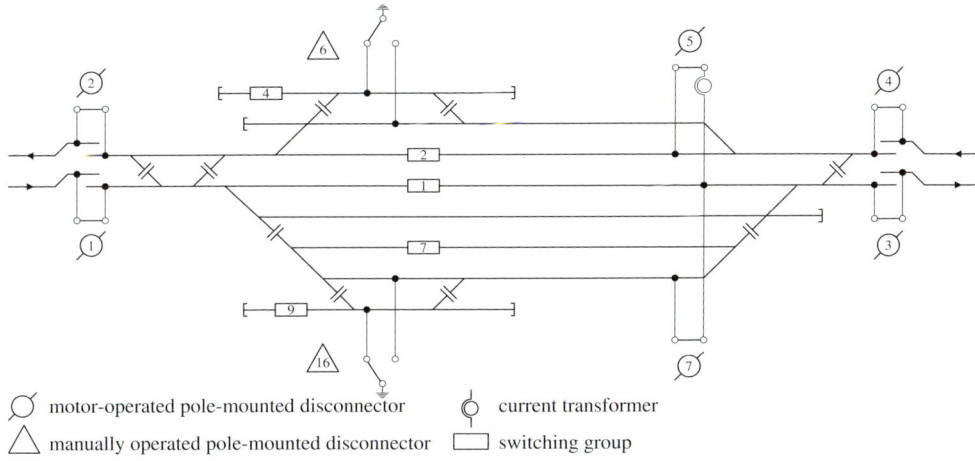

Symbol	Description	Symbol	Description
⌀	motor-operated pole-mounted disconnector	⌀	current transformer
△	manually operated pole-mounted disconnector	☐	switching group

Figure 5.27: Simplified group section circuit diagram of a station.

gitudinal direction. In addition, the two parallel contact lines were connected transversely. By monitoring currents and voltages at the couplings, directional supply could be restored after short circuits. A further increase of power was counteracted by additional transverse connections equipped with short-circuit detector relais. Figures 5.24 d) to g) show circuits commonly used nowadays.

5.4.3.2 Contact line circuits used by the German Railways, DB

The contact line circuits used by the DB are described in [5.30]. A typical *supply section* includes all overhead contact lines, feeder lines and reinforcing feeders, all being connected by not more than two *circuit breakers*. Electric power is fed to a supply section either by a *substation* or a switching post. An overhead contact line branch can supply power to *line branches*, *station branches* or *substitute feeder branches*, as can be seen in Figure 5.25. The boundaries of the supply sections are at the sectioning overlaps or at the coupling points, as shown in Figure 5.26.

In a longitudinal direction, the supply sections are subdivided into *sub-sections* that can be isolated electrically. A distinction is made between the *sub-sections of inter-station section* of railway lines and *sub-sections of stations*. The boundary of a sub-section normally coincides with the *insulated overlaps*. Signals should be located so that an electric traction vehicle with a raised pantograph does not stop directly under the insulated overlap.

In Figure 5.27, a simplified group section circuit diagram is shown for a station. At the beginning and the end, insulated overlaps are arranged separating the station from the inter-station sections. In normal operation, these insulated overlaps are closed by disconnectors ①, ②, ③ and ④. There is a connection in the station between the two section groups by the disconnector ⑤. The siding group ⑦ is connected to the section groups ② by the disconnector ⑦. The loading sidings are supplied by the disconnectors ⑥ and ⑯, earthed in normal operation. Figure 5.28 a) shows the switching of an overtaking station in a high-speed line. The switching circuit diagram of cross-over sections is depicted in Figures 5.28 b) and c). The arrangement permits switching off any of the four adjacent sections and operating the others. *Station sub-sections* are further divided into individual *switching groups* that are contact line

a) Circuit diagram of an overtaking station

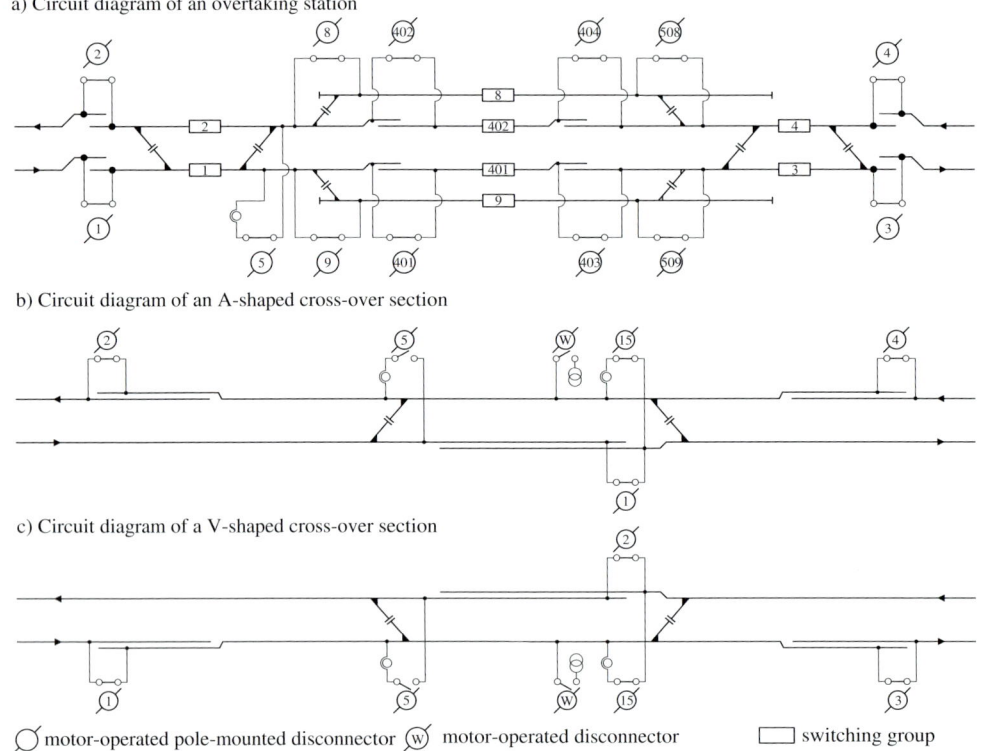

b) Circuit diagram of an A-shaped cross-over section

c) Circuit diagram of a V-shaped cross-over section

○ motor-operated pole-mounted disconnector Ⓦ motor-operated disconnector   ▭ switching group

Figure 5.28: Simplified group section circuit diagrams of overtaking stations and cross-over sections on new line projects.

sections that can be switched off, i. e. individually isolated. It is usual to set up separate switching groups for main lines and secondary lines. In long stations, the main switching groups are sub-divided longitudinally. Electrically, the switching groups are linked by *section insulators* or *insulated overlaps*. Section insulators are used only for operational speeds until 160 km/h, exceptionally also until 200 km/h. Disconnectors are used in overhead contact line installations for the following purposes:

– *section disconnectors* link sub-sections
– *connector disconnectors* connect auxiliary equipment to the overhead contact line
– group disconnectors connect one switching group to another
– *loading siding disconnectors* disconnect the overhead contact lines of loading sidings from the installation and connect them to the return circuit
– *longitudinal disconnectors* are all disconnectors used to link longitudinal sections
– *cross-connecting disconnectors* link the overhead contact lines of the main line tracks of a substation supply section
– *cross-connecting disconnectors* link the overhead contact lines of the main line tracks of a substation supply section
– *feeder disconnectors* connect the overhead contact line to the feeders or connecting lines
– *bypass feeder disconnectors* connect overhead contact lines to bypass feeders

– substation section link connectors connect the overhead contact lines of different sub-station feeding sections

5.4.3.3 Codes used in circuit diagrams

The German Railway DB uses a numerical code to identify the disconnector application as specified in the operating directive Gbr 997.0302.

Figures 5.27 and 5.28 show simplified group section circuit diagrams and switch codes of a station and operating facilities of new line installations.

Units:

1 section disconnector, South or West side, on double-track lines the arrival track,
2 section disconnector – only on double-track lines – South or West side, departure track,
3 section disconnector, North or East side, on double-track lines the departure track,
4 section disconnector – only on double-track lines – North or East side, arrival track,
5 cross-connecting disconnector, usually with short-circuit signalling trans-former,
6 loading siding disconnector, workshop shed isolator disconnector with earthing contact,
7 group section disconnector of tracks on the station side with odd section disconnector numbers,
8 group section disconnector of tracks on the station side with even section disconnector numbers,
9 group section disconnector, where required, preferably for special applications
0 substation section link switch, only in conjunction with corresponding tens digits.

Tens (in combination with the unit-digits described above):

1 to 9 supplementary, serial numbers if required.

For *section disconnectors* of arriving or departing lines odd-numbered tens-digits are used in conjunction with odd-numbered units and even-numbered tens-digits are used in conjunction with even-numbered units for the through-going main line.

Hundreds:

1 *group section disconnector*, in cases where the tens-digits are not sufficient,
2 disconnectors of operating facilities of open lines,
3 special cases, e. g. private sidings, repair or vehicle maintenance workshops, auxiliary longitudinal sectioning, secondary connections to railway power systems, system conversion switchgear, special designations to prevent confusion etc.,
4 longitudinal disconnectors in stations,
5 group section disconnector, for secondary connections in stations,

6 to 9 as for 3.

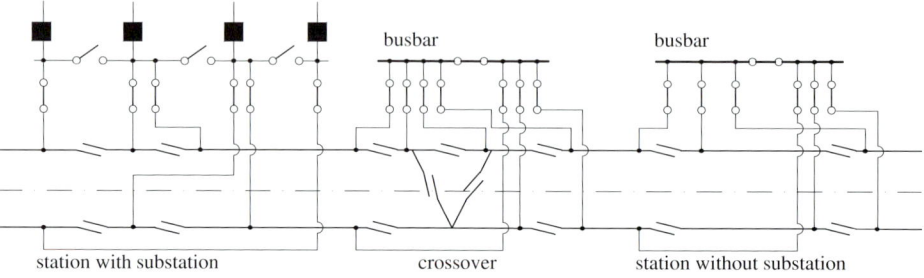

Figure 5.29: Switching scheme of Austrian railway.

Code letters used to identify switches and circuit-breakers:

A   protective neutral section disconnector, South or West side; on double-track lines the track with odd-numbered section switches,

B   protective neutral section disconnector – only on double-track lines – South or West side, the track with even-numbered section disconnectors,

C   protective neutral section disconnector, North or East side, on double-track lines the track with odd-numbered section disconnectors,

D   protective neutral section disconnector – only open lines – North or East side, the track with section disconnectors,

E   earthing switch, only for special cases, e. g. flood gate doors,

F   disconnector for connecting basic sections to bypass feeder lines,

G   disconnector for connecting main-line overhead contact lines to bypass feeder lines,

L   disconnector for loading facilities

Q   disconnector for third-party facilities,

R   feeder disconnector of substitute feed branches with supplementary feed busbar on the outside,

S   feeder disconnector of station feed branches and protective feed branches,

T   disconnector for longitudinal subdivision of feeders, bypass feeders and connecting lines,

U   feeder disconnector of line section feed branches,

V   feeder disconnector connecting overhead contact lines to connecting lines,

W   connector disconnector for switch-point heaters

Z   Connector disconnector for train heating facilities.

5.4.3.4   Contact line circuits of European 16,7 Hz railways

Austrian railway substations are supplied through a railway-owned 110/55 kV transmission network fed by power stations and other sources. The substations are connected in parallel through the contact lines enhancing the performance and stabilising the voltages. The circuitry is described in [5.31]. Substations are arranged at spacings of 25 to 60 km depending on the requirements of the line. Coupling posts are installed where required depending on the line protection to cope with all types of short circuits.

In Figure 5.29, the switching circuitry of an Austrian railway is depicted. Busbars arranged on special gantries characterise this design. The feeders to the overhead contact line sections

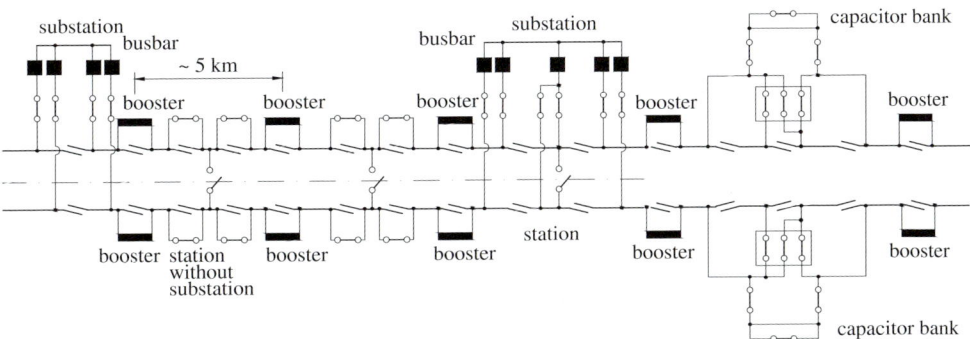

Figure 5.30: Basic circuitry of Norwegian railway.

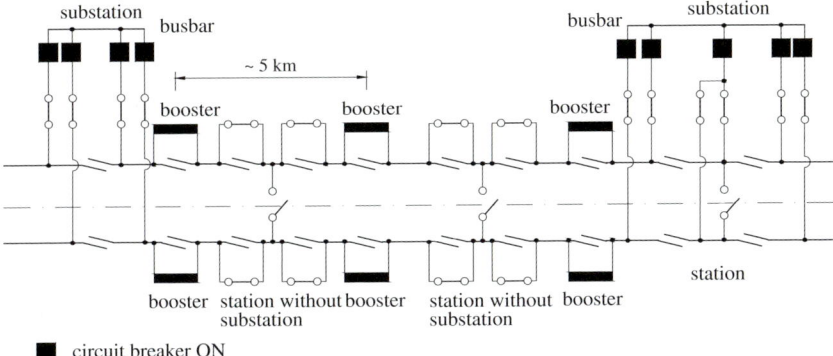

Figure 5.31: Basic circuitry of Swedish railway.

are guided via disconnectors to a busbar. A cross connection of parallel tracks is achieved by connecting the busbars.

The circuitry of overhead contact lines for Norwegian railways is described in [5.32]. The network is supplied from the 50 Hz public power network by means of rotating and static converters. The majority of the network consists of single-track lines. Booster transformers are installed on most lines to reduce rail potentials and electromagnetic interference to railway-owned and other installations close to the line, because of the high resistivity of the earth. Figure 5.30 shows typical circuitry of a single-track line including a station. The overhead lines at stations are electrically separated from the surrounding lines by insulated sections. The overhead contact lines of the adjacent inter-station line sections are connected via a station bypass line to continue electrical supply on these sections when the station has been de-energized. Coupling posts are located midway between two substations. These posts automatically separate failed sections and are also effective in cases where short circuits are not detected by the substation protection. In Figure 5.30, the booster transformers and capacitor banks are clearly visible.

The Swedish railway is mainly supplied from the national high-voltage 50 Hz grid by means of rotating and static converters [5.33]. To cope with increasing power demand, the distances between substations were decreased and auto-transformer systems installed. Figure 5.31 shows the basic circuitry of overhead contact lines. The substations are equipped with

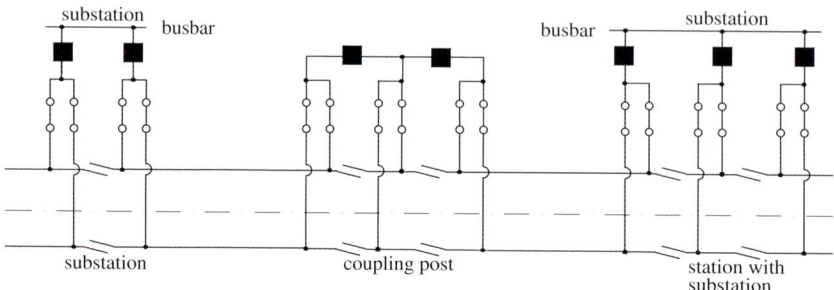

Figure 5.32: Basic circuitry of Swiss railway.

separate circuit breakers for each track in both directions. Couplings between the two tracks exist only at stations or crossovers. These disconnectors are open in normal operation. Therefore, the two contact lines are not coupled permanently. Booster transformers are installed with approximately 5 km spacing. The stations are supplied by a separate circuit breaker where a substation is arranged at a station. As described in [5.34], the existing booster transformer feeding is being more and more replaced by auto-transformer feeding.

Swiss railways supply the substations from a 132/66 kV interconnected overhead transmission network fed by power plants. The main features of the overhead contact line circuitry are described in [5.35]. The contact line conductors on a line are supplied from both adjacent substations, see Figure 5.32. The contact lines on parallel tracks are coupled only at stations or coupling posts. The arrangement of a coupling post depends on local conditions.

5.5 Icing on overhead contact lines

5.5.1 Introduction

In common with public power supply overhead transmission line conductors, *ice layers* up to ten millimetres thick can form on the conductors of overhead traction lines [5.36, 5.37]. Ice accumulation only affects operations when the headway between electric trains is more than 15 min. Lines without overnight traffic can accumulate sufficient ice to cause disturbing effects. At temperatures below freezing and with sufficient atmospheric humidity, hoar frost can form on conductors when wind blows across them. Super-cooled rain and wet snow can also cause considerable ice accumulation, see Figures 2.27 and 13.2. This increases the vertical loads and horizontal loads due to wind which need to be considered in the relevant planning Standards see Clauses 2.4.3 and 13.1.4.4.

Heavy ice loads typically arise in the Nordic countries, in mountainous areas and in the vicinity of lakes and rivers. They can disturb train operations by:

- large sags on automatically tensioned overhead contact lines
- reduced clearances between conductors compared to the design calculations
- short circuits and conductor oscillations with large amplitudes, so-called *galloping*, occurring with the simultaneous impact of wind and ice
- extreme stress at and possible failure of the supports, especially with a combination of ice accumulation and wind
- damage to the collector strips caused by severe arcing and mechanical impacts

Extraordinarily thick ice layers up to 35 mm causing interruption to train operations for three days on the line Paris–Le Mans occured in France in the winter of 1941/42 [5.38]. On New Years eve 1978/1979, ice layers 10 mm thick were formed on the overhead contact lines of Rheinische Braunkohlenwerke AG, in the lignite coal mining area of West Germany. Traffic and production had to be stopped for 30 hours to remove the ice layers. During Christmas 2002 ice deposits on overhead contact lines caused delays and cancellation of many trains especially in the northern part of Germany. For the same reason there was considerable interruption to train traffic in the last days of the year 2009 in the Göttingen–Osnabrück area in northern Germany [5.39]. Ice accumulation on contact wires forms an insulating layer which cannot be penetrated by pantographs. However, the current is not be completely interrupted but flows across arcs between the contact wire and the collector strips and such arcs, cause erosion and current burn marks at the edge of the contact elements on the contact wire, on the collector strips and their fixings.

To limit these adverse effects, procedures, depending on the intensity and the effects of the ice accumulation, are required on susceptible railway lines to avoid or minimize impacts on train operations. It is especially important on high-speed lines where a high availability is expected even after over night operational breaks. To avoid, or to remove the ice accumulation mechanical, chemical and electric techniques are used.

5.5.2 Mechanical techniques

Mechanical removal of the ice covering is possible with lines men standing on the tracks beating against the contact wire with insulated bars causing wave-like oscillations. This results in ice shedding over the whole span or in case of hard ice only in the vicinity of the strokes [5.40, 5.41]. Currently, vibrating pantographs are used as a mechanical technique to remove ice accumulation by a few railway operators. The collector strips of the pantographs cause the contact wire to vibrate. Locomotives prepared for this purpose run with two raised pantographs with a speed of 40 to 80 km/h.

A further mechanical device to clear ice from contact wires is installed on a working platform and uses an electrically driven rotating drum, that strikes the lower part of the ice cover with metal rods. Such a device can be used for an ice thickness of up to four millimetres. For thicker ice layers, electrical de-icing methods are preferred, see Clause 5.5.4 and [5.42]. Mechanical de-icing methods can damage the contact wires and should only be used in exceptional circumstances, for example, to avoid even more severe damage from the icing.

5.5.3 Chemical techniques

Chemical removal of ice accumulation can avoid or reduce ice covering at current collectors and contact line disconnectors. This can be achieved by using grease or other hydrophobic liquids that prevent the adhesion and the accumulation of ice. To achieve this, the treatment must be applied at 3-day intervals or less. Overhead contact line disconnectors using coated contacts should be treated at least once a year. Coating the contact wires with chemically active substances is considered as uneconomical because of the high labour effort and the short period of effectiveness.

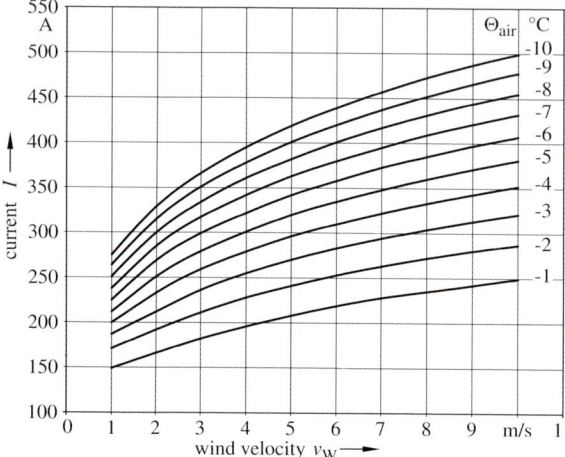

Figure 5.33: Minimum required currents for melting ice on contact wire AC-120 – CuMg0,5.

5.5.4 Electrical techniques

Ice covering can be avoided or melted by two electrical removal techniques:

– For *contact wire preheating* a current is injected guaranteeing a temperature of at least 2 °C thus avoiding ice adhesion. Depending on the weather conditions, for this purpose a current density of 1 to 3,5 A/mm^2 is sufficient [5.43, 5.44]. Preheating starts when weather conditions promote the formation of ice accumulation. The supply circuits used may or may not require ceasing of train services. Procedures for the calculation of the required currents are given in Clause 7.1.2.4. Figure 5.33 shows the pre-heating currents I_{CW} in the catenary for a contact wire AC-120 – CuMg0,5 to 5 °C depending on ambient temperature and wind speeds. Heating currents required for the complete catenary I_{OCL} can be calculated using Equation (5.63), considering the current distribution coefficient k_{CW} for contact wires for AC systems, see Tables 5.8 and 5.9.

$$I_{OCL} = I_{CW}/k_{CW} \quad .\tag{5.63}$$

– *Melting ice accregation* is achieved by current densities from 3,5 to 8 A/mm^2. This allows the return of train operations within one hour. For determination of the current value and the corresponding melting time, line-related tables and diagrams can be used, adopted to the applicable installations. They consider the size and density of the ice layers, ambient temperature and wind velocity. Melting starts with an ice thickness of three to five millimetres defined by ice parameters. After removal of the accumulated ice, an additional 10 to 15 minutes period is required to dry out the contact wire surface. The current required to melt the ice must not exceed the current-carrying capacity of the overhead contact line under the given ambient temperature, wind force and duration of the current flow and the corresponding material-dependent limits for these conditions, see Clause 2.3.2. To calculate the permissible melting currents according to Clause 7.1.2.4 a lower temperature and 1 m/s for the lowest wind velocity is considered to avoid overheating due to locally different and temporary changing wind velocities and variation over the time of temporary application and conditions in tunnel sections. Higher conductor temperatures than those normally permitted can be used for a short

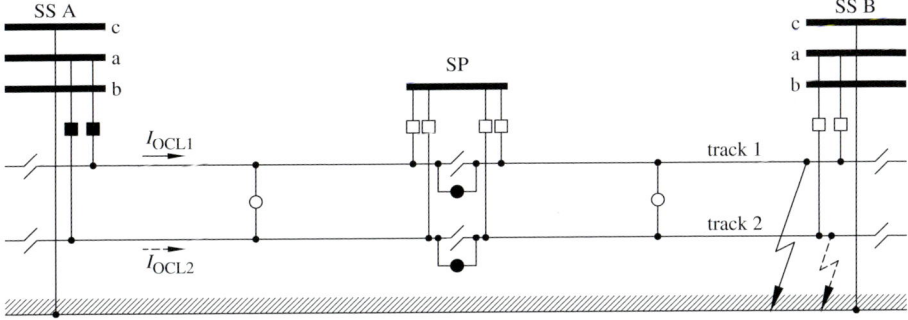

Figure 5.34: Circuit diagram for melting ice accregation on overhead contact lines between the feeding substation SS A and the temporarily applied short-circuit connection at the adjacent substation SS B or, alternatively at the switching post SP. Applicable for one or two tracks.
Note: The circuit breakers and disconnectors showing filled symbols are closed
SS A substation A, SS B substation B, SP switching post

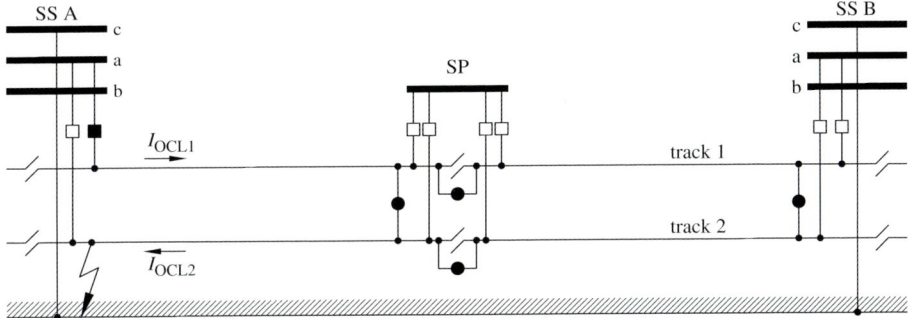

Figure 5.35: Circuit diagram for melting ice accregation on overhead contact lines in a closed loop starting at the feeding substation SS A, track 1, to the switching post SP or to the SS B with a deliberate short-circuit connection at the feeding substation SS A, track 2, alternatively at the SP. Note: abbreviations see Figure 5.34

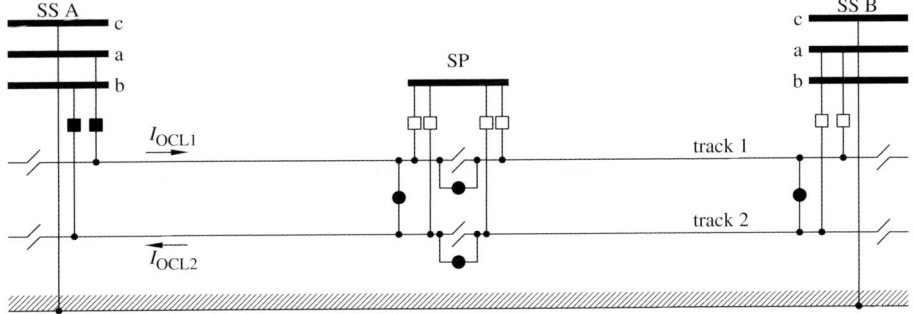

Figure 5.36: Circuit diagram for melting ice accregation on overhead contact lines in a closed loop fed by different phases a and b of the tree-phase transformer in substation SS A. Loop length given by switching post SP or adjacent substation SS B.
Note: abbreviations see Figure 5.34

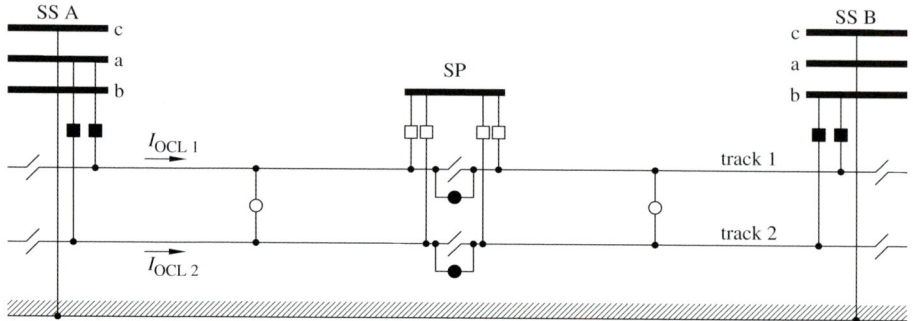

Figure 5.37: Circuit diagram for melting ice accregation on overhead contact lines fed by phase a in substation SS A via switching post SP and phase b in substation SS B. Applicable for one or two tracks. Note: abbreviations see Figure 5.34

duration, e. g. up to 30 min, see Table 2.9, if the tensioning devices, cantilevers and contact lines can handle the additional conductor lengthening.

It is also necessary to consider the current carrying capacity of all power supply devices, e. g. disconnectors, current transformers, cabling and electrical connectors and to adopt them if applicable.

Two basically different *ice melting circuits* can be used [5.45]:

– application of a temporary deliberate short circuit between the contact line and the rails shown in Figures 5.34 and 5.35. In this case, train operation is not possible. This method is used for de-icing of overhead contact lines during a short period. The short-circuit location is selected depending on the line section to be de-iced and on the required current. After de-icing of the first line section, it is switched over to the following one and so on.

– supplying the overhead contact line from two phases of one or two three-phase transformers shown in Figures 5.36 and 5.37 creating a phase to phase short circuit. A simultaneous train operation is possible. As a result the overhead contact line protection must ensure that the sum of traction and heating currents does not exceed the permissible line loading under the given ambient conditions. This connection can be used depending on the length of the supply section and the current required for melting as well as for pre-heating of the contact line. Since the de-icing current only flows via the overhead contact line, no additional losses arise in the traction return circuit.

The contact wires in overlaps should be connected not in parallel but in series by a purpose made electrical connection to utilize the whole current for heating or de-icing. Electrical techniques, circuits and equipment are designed to accommodate the regional climatic conditions and required heating currents. In areas of frequent and intensive ice formation, for example in the northern regions of Russia, the spacing between substations is also designed to accommodate the additional heating power of the *de-icing circuits*. Thus, after calculating the maximum permissible currents, the optimum de-icing method is selected from Figures 5.34 to 5.37. A controlling of the current density during heating is complicated and not cost-effective and, therefore, usually not implemented. Instead, design work is based on available transformer power, the permissible operational voltage, the line length and impedance and the method of line connection.

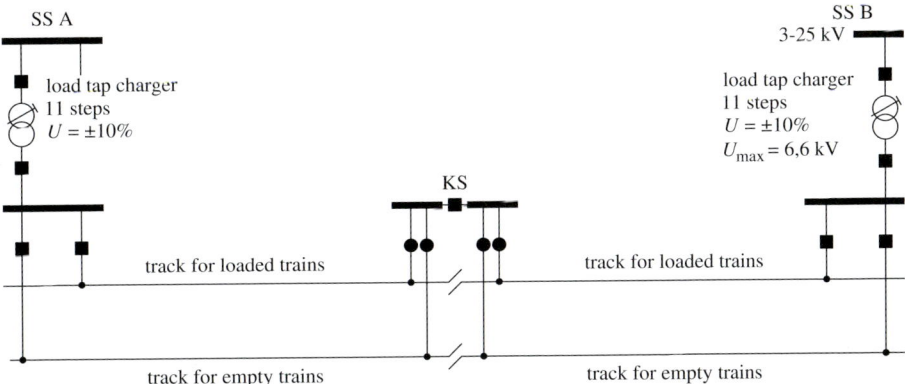

Figure 5.38: Contact line heating connection used by Rheinische Braunkohlenwerke AG, Cologne, Germany. Note: abbreviations see Figure 5.34

In cases where the calculated heating current is too high or too low, the length of the feeding section must be increased or reduced respectively. This can be achieved by an alternative circuit arrangement, (Figures 5.34 to 5.37) or by modifying the distance between the substations or feeding points where no other possibilities are available. For longer tunnels either bypass lines or increased cross-sections achieved by cross-connections can be used, where a requirement to limit the current in the tunnel would reduce the heating current on the lines in the open. For this purpose, disconnectors must be installed in advance.

By connecting the overhead contact line sections to the installed de-icing circuits protection requirements change. The protection relays have to be set according to the new parameters by remote control. The modelling of the contact wire temperature within the protection methods should also consider the wind velocity in addition to the ambient temperatures.

The approaches and electric connections described are predominantly used in the northern parts of Russia, Scandinavia, France and the Netherlands. German Railway DB pre-heats the contact line of the high-speed line Cologne–Frankfort when weather conditions are prone to ice accumulation during the overnight operational by supplying several supply sections connected in series. The Rheinische Braunkohlenwerke AG use the voltage difference controlled by the transformer load tap changer (Figure 5.38) for this purpose. The maximally possible voltage differences between $U_{\max 1}$ and $U_{\min 1}$ are shown in Table 1.1.

Example 5.6: Calculate the minimum required de-icing current I_{OCL} and the maximum ice melting currents of the overhead contact line system Sicat H1.0 for a two-track open-air line with a new contact wire AC-120–CuMg, messenger wire BzII 120, feeder line and return conductors 243-AL1. The power supply frequency is AC 50 Hz, air temperature $\vartheta = -5\,°\mathrm{C}$ and wind velocity $v_{\mathrm{W}} = 3$ m/s. Current distribution coefficients for AC overhead contact lines are given in Table 5.9 with $k_{\mathrm{CW}} = 0,30$ and $k_{\mathrm{FL}} = 0,42$. The minimum required de-icing current for the contact wire is taken from Figure 5.33:

$$I_{\mathrm{CW}} = 280\,\mathrm{A} \text{ or } 2,3\,\mathrm{A/mm}^2 \quad .$$

The minimum required de-icing current for the overhead contact line system Sicat H1.0 is calculated by Equation (5.63):

$$I_{\mathrm{OCL}} = 280/0,30 = 930\,\mathrm{A} \quad .$$

The permissible de-icing current of the overhead contact line can be derived from Table 7.3 by inter-polation of the contact wire AC-120 – CuMg values for $0\,°C$ and $-10\,°C$

$$I_{CW} = (631\,A + 593\,A)/2 = 612\,A \quad .$$

The maximum ice melting current of the overhead contact line system Sicat H1.0 can also be calculated by applying Equation (5.63):

$$I_{OCL} = 612\,A/0,30 = 2\,040\,A \quad .$$

However, it is also necessary to check, in the same way, the maximum current capacity and distribution coefficients of the other conductors using the Tables 5.8, 5.9 and 7.3 and to determine the lowest value for the maximum contact line current. In this example the lowest value is $2\,040\,A$, governed by the contact wire current. The calculated temperature of the overhead contact line according Chapter 7 is $57\,°C$. Assuming that the heated contact wire must melt 50 % of the attached ice before it sheds, the calculated temperature should be sufficient to remove a 3 mm thick ice layer within 20 minutes. In the following example, the minimum and maximum supply section length is determined for the overhead contact line system Sicat H1.0, at which the required currents are found. The voltage at the substation busbar is 27,5 kV. The approximate minimum and maximum impedance is calculated by considering the minimum and maximum contact line currents by means of the absolute values:

$$Z_{max} = U/I_{min} = 27\,500\,V \,/\, 930\,A = 18\,\Omega \quad \text{and}$$

$$Z_{min} = U/I_{max} = 27\,500\,V \,/\, 2\,040\,A = 8\,\Omega \quad .$$

The minimum and maximum supply section lengths of the overhead contact lines can be estimated with a related impedance of $0,236\,\Omega/km$ according to Table 5.9:

$$l_{max} = Z_{max}/Z^{`} = 18\,\Omega \,/\, 0,236\,\Omega/km = 75\,km \quad \text{and}$$

$$l_{min} = Z_{min}/Z^{`} = 8\,\Omega \,/\, 0,236\,\Omega/km = 34\,km \quad .$$

With a distance of approximately 35 km between adjacent sub-stations the contact line can be fed for melting the ice with the connections shown in Figure 5.36 or 5.37 from the various branches of one or both sub-stations. For de-icing heating, the contact lines of both main tracks can be connected in series at sub-station B and supplied by different phases of sub-station A.

5.5.5 Combination of different de-icing techniques

Each of the de-icing methods presented above can have advantages and disadvantages. The vibration drums only remove locally accumulated ice and with low effectiveness. They can also damage the contact wire. Chemical procedures are short term and in-efficient. When melting existing accumulated ice there is a danger of annealing conductors, however, the energy consumption is low. With preventive pre-heating there is no danger of annealing the conductors, however, the energy consumption is high.

The type of *de-icing technique* used depends on economic considerations including the fre-quency of the ice accumulation and the operational conditions, e. g. operational breaks during the night and the envisaged availability of the railway line. Where there is a low probability of ice accumulation it is likely to be more economic, during icing periods, to run trains more frequently than to install specific de-icing counter-measures. Where requirements to cope with ice accumulation are necessary, they should be taken into consideration when planning and installing the electric traction equipment to avoid high investments for later retrofitting.

5.6   Bibliography

5.1   *Eichhorn, K. F.*: Stromverdrängung und Stromleitung über Erde (Current displacement and conduction through earth). In: Elektrische Bahnen 95(1997)3, pp. 74 to 81.

5.2   *Carson, J. R.*: Wave propagation in overhead wires with ground-return. In: Bell System Technical Journal 5(1926), pp. 539 to 555.

5.3   *Mariscoti, A.; Pozzobon, P.*: Measurement of the internal impedances of traction rails at 50 Hz. In: IEEE Transactions on Instrumentation and measurement Vol.49(2000)2.

5.4   *Behrends, D.; Brodkorb, A.; Hofmann, G.*: Berechnungsverfahren für Fahrleitungsimpedanzen (Calculation of overhead contact line impedances). In: Elektrische Bahnen 92(1994)4, pp. 114 to 122.

5.5   *Kießling, F.; Schneider, E.*: Verwendung von Bahnstromrückleitern an der Schnellfahrstrecke Madrid–Sevilla (Use of return conductors at the high-speed line Madrid–Seville). In: Elektrische Bahnen 92(1994)4, pp. 112 to 116.

5.6   *Zimmert, G. et. al.*: Rückleiter in der Oberleitungsanlage der Strecke Magdeburg–Marienborn (Return conductors in catenary system on the Magdeburg–Marienborn line). In: Elektrische Bahnen 92(1994)4, pp. 105 to 115.

5.7   *Brüderlink, R.*: Induktivität und Kapazität der Starkstrom-Freileitung (Inductivity and capacity of power lines). Braun Publishing, Karlsruhe, 1954.

5.8   *Zimmert, G.*: Bericht über Ableitungsbeläge moderner Oberbauarten (Report on the leakance of modern superstructures). Frankfurt, 1993.

5.9   *Milz, K.*: Elektrifizierungssysteme für den Hochgeschwindigkeitsverkehr (Electrification systems for the high-speed railway traffic). In: Elektrische Bahnen 89(1991)11, pp. 323 to 325.

5.10 *Decision 2002/733/EC*: Decision concerning the technical specification for the interoperability relating to the energy subsystem of the trans-European high-speed rail system. In: Official Journal of European Communities, (2002), pp. L245/280 to L245/369.

5.11 *Decision 2008/284/CE*: Decision concerning a technical specification for interoperability relating to the energy sub-system of the trans-European high-speed rail system. In: Official Journal of European Communities, (2008), pp. L104/1 to L104/79.

5.12 *Schmidt, P.*: Berechnung der Spannungsabfälle zweigleisiger elektrischer Bahnen bei beliebiger Anzahl von Querkuppelstellen (Calculation of the voltage drops on double-track electric railways with any number of cross-coupling). In: Scientific Journal of HfV Dresden 22(1975)2, pp. 401 to 427.

5.13 *Schmidt, P.*: Energieversorgung elektrischer Bahnen (Power supply of electric railways). Edition Transpress, Berlin, 1988.

5.14 *Irsigler, M.*: Aktuelle Fragen der Bahnenergieversorgung für Hochleistungsstrecken (Current problems of traction power supply of high-capacity lines). In: Elektrische Bahnen 90(1992)6, pp. 189 to 196.

5.15 *Brodkorb, A.*: Ein Modell der elektrischen Bahnbelastung auf der Grundlage der digitalen Simulation der Zugfahrten (Modelling of elctrical railway loads based on digital simulation of train runnings). HfV Dresden, doctoral thesis, 1986.

5.16 *Biesenack, H.; Hauptmann, A.; Müller, K.; Schmidt, P.*: Bahnbelastung und Spannungshaltung im Hochgeschwindigkeitsverkehr (Electrical railway load and voltage stability in case of high-speed traffic). In: Elektrische Bahnen 50(1996)9-11, pp. 324 to 333.

5.17 *Röhlig, S.*: Beschreibung und Berechnung der Bahnbelastung von Gleichstrom-Nahverkehrsbahnen (Description and calculation of the electrical load of DC load railways). HfV Dresden, doctoral thesis, 1992.

5.18 *Lingen, J.; Schmidt, P.*: Strombelastbarkeit von Oberleitungen des Hochgeschwindigkeitsverkehrs (Current capacity of overhead contact lines for high-speed traffic). In: Elektrische Bahnen 94(1996)1-2, pp. 38 to 44.

5.19 *Schmidt, P.*: Elektrische Belastung als Zufallsgröße und thermische Belastbarkeit von Leitungen bei mitteleuropäischen Bahnen (Electric load as a random magnitude and thermal strength of contact lines of Central European railways). In: Elektrische Bahnen 90(1992)6, pp. 204 to 212.

5.20 *Hellige, B.*: Beitrag zur Untersuchung der Belastung von Energieversorgungsanlagen bei Straßenbahnen (Investigation of the electrical loading of tramway power supply installations). HfV Dresden, doctoral thesis, 1971.

5.21 *UIC Code 795*: Minimal installierte Leistung – Streckenkategorien (Minimum installed power – Line categories). UIC, Paris, 1996.

5.22 *Pundt, H.*: Elektroenergiesysteme (Electrical power supply systems, manuscript). TU Dresden, 1980.

5.23 *Heide, S.*: Ein Beitrag zur Berechnung von Kurzschlußströmen im 15-kV-Fahrleitungsnetz der DR unter besonderer Beachtung ausgewählter Probleme des Fahrleitungsschutzes (Calculation of short-circuit currents in the 15 kV contact line network of the East German Railways DR with respect to specific problems of overhead contact line protection). HfV Dresden, doctoral thesis, 1980.

5.24 *Lingen, J. v.; Schmidt, P.*: Methodik einer zuverlässigen und ressourcensparenden Bemessung elektrotechnischer Betriebsmittel des Hochgeschwindigkeitsverkehrs (Procedures for a reliable and economic design of electrotechnical operational equipment for high-speed traffic). In: Scientific Journal of TU Dresden 45(1996)5, pp. 30 to 39.

5.25 *Lingen, J. v.*: Kurzschlussberechnung im Fahrleitungsnetz (Short-circuit calculation for contact line networks). TU Dresden, doctoral thesis, 1995.

5.26 *Kontcha, A.*: Analyse elektromagnetischer Verhältnisse in Mehrleiterfahrleitungssystemen bei Einphasenwechselstrombahnen (Analysis of electromagnetic conditions in multi-conductor overhead contact line systems at single-phase AC railways). TU Dresden, doctoral thesis, 1996.

5.27 *Regulation 1301/2014/EU*: Technical specification on the interoperability relating to the Energy subsystem of the rail system in the Union. In: Official Journal of European Union. No. L356 (2014), pp. 179 to 227.

5.28 *Behmann, U.*: Operating and power supply concept of Madrid–Seville high-speed line. In: Elektrische Bahnen 88(1990)5, pp. 207 to 214.

5.29 *Lörtscher, M.*: Vergleich der Fahrleitungsschaltungen bei 16,7-Hz-Bahnen (Comparison between the overhead contact line circuits of railways powered with 16,7Hz) In: Elektrische Bahnen 103(2005)4-5, pp. 164 to 170.

5.30 *Ebhart, S.; Ruch, M.; Hunger, W.*: Schaltungsaufbau im 16,7-Hz-Oberleitungsnetz bei DB Netz (Circuitry of 16,7 Hz overhead contact line network at DB). In: Elektrische Bahnen 102(2004)4, pp. 152 to 163.

5.31 *Punz, G.*: Schaltungsaufbau im Oberleitungsnetz der ÖBB (Circuitry of 16,7Hz overhead contact line network at ÖBB). In: Elektrische Bahnen 102(2004)4, pp. 174 to 183.

5.32 *Johnsen, F.; Nyebak, M.*: Schaltungsaufbau im Oberleitungsnetz der Norwegischen Eisenbahn Jernbaneverket (Circuitry of overhead contact line network at Norwegian Railway Jernbaneverket). In: Elektrische Bahnen 102(2004)4, pp. 195 to 200.

5.33 *Bülund, A.; Deutschmann, P.; Lindahl, B.*: Schaltungsaufbau im Oberleitungsnetz der Schwedischen Eisenbahn Banverket (Circuitry of overhead contact line network at Swedish Railway Banverket). In: Elektrische Bahnen 102(2004)4, pp. 184 to 194.

5.34 *Deutschmann, P.; Marquass, J.-P.*: Electrification by AC50/15 kV 16,7 Hz in Sweden. In: Elektrische bahnen 112(2014)INT, pp. 55 to 65.

5.35 *Basler, E.*: Schaltungsaufbau im 16,7-Hz-Oberleitungsnetz der SBB (Circuitry of 16,7Hz overhead contact line network at SBB). In: Elektrische Bahnen 102(2004)4, pp. 164 to 172.

5.36 *Kiessling, F.; Nefzger, P.; Nolasco, J. F.; Kaintzyk, U.*: Overhead power lines – Planning, Design, Construction. Springer Publishing, Berlin-Heidelberg-New York, 1st edition 2003.

5.37 *Porzelan, A. A.; Pavlov, I. V.; Neganov, A. A.*: Controlling of ice accretion at electrified railways (in Russian). Moscow, Transport, 1970.

5.38 *Zorn, W.*: Das Abtauen von raureifbelegten und vereisten Fahrleitungen (De-icing of hoar frost and clear ice on overhead contact lines). In: Elektrische Bahnen 41(1943)4-5, pp. 74 to 84.

5.39 *N. N.*: Elektrischer Betrieb bei der Deutschen Bahn im Jahre 2009 (Electric operation of German Railway DB in 2009). In: Elektrische Bahnen 108(2010)1-2, pp. 4 to 54.

5.40 *Serdinov, S. M.*: Improvement of the reliability of the power supply installations of electrified railways (Russian language). Transport Publishing, Moskau, 2nd edition 1985.

5.41 *German patent 2324287, Class B60L 5/02.*: Vorrichtung zur Entfernung eines Eisbelags von langgestreckten Teilen wie Drahtseilen, Fahrdrähten, Stromschienen oder dergleichen (Device to remove ice accretion from long-stretched components such as stranded wires, contact wires, contact rails or similar subjects).

5.42 *Russian Railways*: Planned controlling of ice accretion on and galloping of overhead contact lines (in Russian language). Moscow, Department of electrification and power supply of Russian Railways, 2004.

5.43 *Heide, E.*: Der Fahrleitungsbau: Handbuch für Bau und Unterhaltung (Overhead contact line installation – Manual for installation and maintenance). E. Schmidt Publishing, Berlin, 1956.

5.44 *Wlassow, I. I.*: Fahrleitungsnetz (Overhead contact line network). Technical Publishing, Leipzig, 1955.

5.45 *Russian Railways: Russian Scientific Research Institut of Railways (RSRIR)*: Calculation methods for power supply systems with electrical ice fighting techniques. Moscow, 2005.

6 Current return circuit and earthing

6.0   Symbols and abbreviations

Any symbols and abbreviations followed by an apostrophe ' indicate a magnitude per unit length, given in 1/km or 1/m

Symbol	Definition	Unit
AC	Alternating Current	A, kA
AT	Auto transformer	–
A, B	integration constants	A
BT	Booster Transformer	–
C	electrochemical equivalent	$kg/(A \cdot a)$
DC	Direct Current	A, kA
G	admittance, 1/resistance	S
G_M	conductance to earth of a mast foundation	S
G_{RE}	conductance between rail and earth	S
G_{REeff}	effective conductance between rail and earth	S
G_{RS}	conductance between return circuit and structure	S
G_{SE}	conductance between structure and earth	S
G_{TE}	conductance between track and earth	S
I	current	A
I'	current density	A/mm^2
I_{C1}	body current (IEC)	A
I_E	earth current	A, kA
I'_{OCL}	distributed line load	A/m
I_R	rail current	kA
I_{RC}	return conductor current	kA
I_S	stray current	A
I_T	track current	kA
I_{an}	anodic current	A
I_k	short-circuit current	kA
I_{ka}	cathodic current	A
I_p	peak value of the lightning current	kA
I_{tot}	total current	A, kA
I_{trc}	traction current	A
K	rail fixing, metallic plates	
L	length of a section in consideration	km
L_C	characteristic length	km
L_{trans}	transition length	km
ME	Main Equipotential busbar	–
OCZ	Overhead Contact Line Zone	–
R_C	characteristic resistance	Ω, Ω/km
R_E	resistance to earth	Ω

Symbol	Definition	Unit
R_{ESS}	resistance to earth of a substation	Ω
R_R	longitudinal resistance of the return circuit	Ω
RR	Rail Return	–
RRR	Rail Return with Return conductors	–
R_{Req}	equivalent track resistance	Ω
R_S	longitudinal resistance of structure	Ω
R_a	additional resistance	Ω
R_{a1}	additional resistance for shoes	Ω
R_{a2}	additional resistance of standing surface	Ω
R_{dis}	impulse earthing resistance	Ω
R_{rail}	longitudinal resistance of a rail	Ω
$T1, T2, T3$	current transformers	–
U_{C1}	body voltage (IEC)	V
U_{PE}	voltage between point P and earth	V
U_R	rail potential	V
U_{RE}	rail potential	V
U_{RP}	voltage between rails and a point P	V
U_{RS}	voltage between return circuit and structure	V
U_S	source voltage	V
U_S	longitudinal structure voltage	V
U_{SE}	voltage between structure and earth	V
U_b	body voltage	V
U_{ins}	impulse withstand voltage of the insulation	V
U_{sub}	source voltage substation	kV
U_{te}	effective touch voltage	V
$U_{te,max}$	permissible effective touch voltage	V
U_{tp}	prospective touch voltage	V
V_G	volume of a foundation	m^3
VLD	Voltage Limiting Device	–
W	rail fixing, plastic lining	
WK	rail fixing, thick plastic lining	
Z_A	terminating impedance	Ω
Z_E	impedance versus earth	Ω
Z_{KE}	coupling impedance of the track-to-earth circuit	Ω
Z_{RE}	self-impedance of the track-to-earth circuit	Ω
Z_b	body impedance	Ω
Z_o	surge impedance	Ω
a	distance between measuring electrodes	m
h_m	the mean height of contact line above track	m
k	coupling factor	–
k_a	bridging factor rail potential	–
n_R	number of parallel rails	–
pH	pH-value	–
r_{eq}	equivalent radius	m
m'	metal wear per year	kg/a

Symbol	Definition	Unit
t	time duration	sec
w	width of a flat steel	mm
x_B	boundary distance between current leaving and entering	km
α	attenuation constant	1/km
β	phase constant	1/km
δ_E	penetration depth	m
γ	propagation constant	1/km
ρ_E	soil resistivity	$\Omega \cdot m$
μ_r	relative permeability	–
μ_o	permeability constant	$V \cdot s/(A \cdot m)$

6.1 Introduction

The contact line system and the return circuit form the traction current circuit of electrified railways. The traction current flows via the contact line system to the trains and via the return circuit back to the substation. Wherever the return path of traction currents is discussed in the following clauses, the currents are considered to also include braking currents. The running rails serve as the primary conductors for the return current. As the resistance between the rails and earth is finite and the rails have a longitudinal resistance, a portion of the return current will flow to earth and back to the substation through earth. Near the substation, this current flows back into the running rails and in case of AC systems, into the substation earthing system. The sum total of the currents flowing through the rails, earth and any metal objects running parallel to the track in the railway track area, such as cable sheaths and pipelines, is identical to the current flowing in the contact line system.

Up to several thousand amperes can flow in the return circuit and cause voltages at the running rails and conductive parts of vehicles during operation. To avoid voltages which could be potentially dangerous when bridged by passengers and staff, the return circuits need to be adequately designed.

Compared with conventional three-phase power transmission and distribution systems, where hazardous voltages at accessible parts can only arise during fault conditions, electrified railways also require provisions to ensure the safety of persons and protection of installations during operation. In the case of short circuits, the conditions are the same as for short-time voltage impacts in other electric transmission systems.

Some aspects of the return conductor arrangements are common to both direct current and alternating current traction systems. However, there are also fundamental differences between the two. In DC railways, the coupling between the rails and earth is found to be completely *galvanic* in nature, whereas in single-phase AC railways, the *inductive coupling* between all conductors, i. e. between the rails, earth, contact line, reinforcing feeder lines and return conductors, affects the way the return current is distributed among the individual conductive paths.

In DC railway systems, the current flowing through earth can cause *stray current corrosion*, so this portion of the return current must be minimised. The standards IEC 62 128-2 and corresponding EN 50 122-2 deal with stray currents in DC traction installations and specifie that adequate insulation must be arranged between track and earth. Consequently, in principle the return circuit of DC railways must not be connected to earth or earthing installations.

When operating or short-circuit currents flow through the track of electric traction railways, voltages can reach their maximum values at the feed or load points. In both the operating state and during short circuits, the voltage between the rails and earth must not exceed permissible values as specified in the relevant standards. In AC railways, the rail-to-earth voltages are reduced by bonding other metallic, conducting elements to the rails, eliminating any possibility of affecting people and ensuring that the entire system can be switched off safely in case of faults.

To reduce the rail-to-earth voltage in or near direct current railway installations, other measures are required, e. g. installation of parallel return conductors and/or *voltage limiting devices* (VDL).

Track release systems for railway operations often use the tracks as part of their electrical circuits. The return circuit has to be designed in such a way that its electrical characteristics are suitable for safe return current conduction and simultaneously serve as part of the electric circuits of the track release systems.

In both AC and DC railway traction systems, inductive, capacitive and galvanic coupling with the traction current circuit can cause adverse effects on technical devices and equipment in the vicinity of energy being transmitted from the substation to the trains. The design of the return circuit has to take these effects into consideration and reduce them to a tolerable level.

6.2   Terms and Definitions

6.2.1   Introduction

Standards and publications related to earthing and bonding sometimes use terms with different meanings so that definitions and comments are necessary for a common understanding [6.1]. In the main, these are derived from the International Standard series IEC 62 128 and the corresponding European Standard series EN 50 122, which were elaborated for protective provisions related to electrical safety and earthing and the effects of stray currents caused by railway traction systems.

6.2.2   Return circuit

With conventional AC and DC traction power supply, the operating current flows through the contact line to the vehicle. The return current, for traction and regenerative braking, flows from the vehicle via the *return circuit* to the substation.

As defined in IEC 62 128-1 and EN 50 122-1, the return circuit includes all conductors that form the path provided for the return current during operation and in the case of faults. They include:

– running rails, that conduct the return current
– *return conductors*, laid parallel to the running rails and connected to the running rails at regular intervals.
– on DC lines cables laid parallel to the running rails and insulated against earth reduce the *longitudinal rail voltages* and the rail potentials
– on AC lines, this function can be fulfilled by *return conductors* suspended on the contact line poles or earth strips alongside the track

- return conductors of twin-wire supply systems like trolley buses and *return conductor rails* of metro systems using a separate fourth rail insulated from the running rails. These conductors are treated as live conductors and no voltage arises at the running rails and car bodies during normal operation
- on AC lines, the soil is a part of the return circuit, as a portion of the return current flows because of the earthing of the running rails and inductive coupling

6.2.3 Earth

The *earth* from an electrical point of view is defined as the *conductive soil*, whose electric potential at any point is taken conventionally as equal to zero, see IEC 62 128-1 and relating EN 50 122-1. Often the terms *reference earth, neutral earth, separate earth* or *remote earth* are used. Earth in the context of this definition is found outside the area of interference of electrical installations, where no potential difference can be detected between different points as a result of earth currents.

The distance between earthing installations of energy supply facilities and *earth* can be several tens of metres up to one kilometre and depends upon the dimensions of the installations, the soil composition and the magnitude of the earth current. The earth is taken as reference for determining the *rail potential*, i. e. rail-to-earth voltage.

6.2.4 Earth electrode

Earth electrodes are one or more conductive parts in intimate contact with soil, providing an electrical connection with the earth. It is advantageous to use metallic or steel reinforced structures as earth electrodes that primarily serve other purposes including foundations for buildings and poles. Early project planning is required to ensure provisions of adequate electrical cross-bonding and terminals.

6.2.5 Soil resistivity and resistance to earth

The electrical characteristics of earth electrodes depend on their design and the conductivity of the surrounding soil. The *soil resistivity* indicates the electrical conductivity of the soil. Normally, it is measured in $\Omega \cdot m$. Its numeric value represents the resistance of a cube of soil with edge lengths of 1 m between two opposing cube surfaces. The *resistance to earth* of an earth electrode or an earthing system can be calculated with sufficient accuracy for design purposes from the geometric dimensions of the electrode and the local soil resistivity [6.2].

6.2.6 Structure earth, tunnel earth

An *earthing system* consists of several earth electrodes that are connected to each other by conductors. The conductive, interconnected reinforcement of concrete structures and the metallic components or other structures are defined as *structure earth* [6.1]. This includes passenger stations and technical buildings, bridges, viaducts, concrete slab permanent way and tunnels. The structure earth of tunnels is also known as *tunnel earth*.

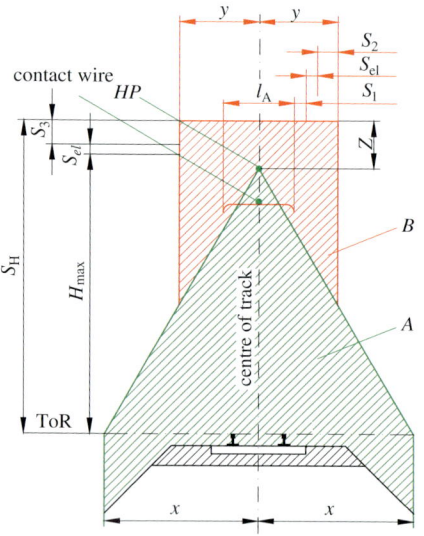

Figure 6.1: Overhead contact line zone and current collector zone according to EN 50 122-1.

ToR top of rail
HP   highest point of the overhead contact line
A   overhead contact line zone
B   current collector zone or pantograph zone
X   maximum unidirectional (half) horizontal A, top of rail level
Y   maximum unidirectional (half) horizontal B
Z   distance between HP and S_H
S_1 width of lateral movement of the pantograph
S_2 lateral safety distance for the broken or dewired pantograph
S_3 vertical safety distance for the broken or dewired pantograph
S_{el} electrical clearance in accordance with EN 50 119
S_H maximum height of pantograph zone
l_A pantograph width
H_{max} maximum height of the fully extended pantograph

6.2.7 Rail potential and track-to-earth voltage

The resistance to earth of an earth electrode and the current through the electrode result in a voltage rise versus far earth. In the same way the traction return current in the running rails causes a voltage rise designated as the *rail potential* or *track-to-earth voltage*. The rail potential arises at the running rails and the conductive parts connected to them during both operational and fault situations.

6.2.8 Touch voltage

The voltage between two conductive parts, which can be bridged by a human being, is known as the *touch voltage*. Consideration is given to direct and indirect touch voltage.
Direct touch voltage (IEC 60 364-1) refers to possible touch contact to live parts. Protection measures against direct touch contact are obstacles such as insulating enclosures, covers or barriers or sufficiently large clearances to accessible surfaces.
Indirect touch voltage refers to conductive parts energized only under fault conditions. The metallic enclosures of switchgear, earthing connections and steel-reinforced concrete structures that can carry a voltage are categorised as touchable parts. The voltage to earth of an earth electrode, that can be bridged by a person, also falls within the term of a touch voltage. Protective devices normally switch-off within a short period so that the permissible touch voltage can affect a human being for only a short period. Standard EN 50 122-1 indicates the permissible touch voltage versus time for railway applications, see Figure 6.5.

6.2.9 Overhead contact line zone and pantograph zone

For traction systems with overhead contact lines, EN 50 122-1, Clause 4.1, defines *overhead contact line* and *current collector zones* and requires special protective provisions there. In this book the term *pantograph zone* is used instead of current collector zone. In Figure 6.1 the dimensions of these zones are defined. In EN 50 122-1 it is assumed that structures and equipment in these zones could accidentally become live and, therefore, need to be connected to the return circuit. The authors of this book would like to emphasize that broken contact lines or dewired pantographs are extreme rare events which are caused by external impacts. The connection of conductive parts situated in these zones to the return circuit is in most cases anyway necessary because of equipotential bonding and protective earthing purposes. The parameters X, Y, Z in Figure 6.1 must be defined by national rules. Guiding values are $X = 4{,}0$ m; $Y = 2{,}0$ m; $Z = 2{,}0$ m. In Germany, X is specified as $4{,}0$ m . Y and Z must be fixed by the operator of the infrastructure. The value Y of the current collector area is calculated by

$$Y = l_A/2 + S_1 + S_{el} + S_2 \quad . \tag{6.1}$$

Figure 6.1 depicts the abbreviations and definitions.
The collector head dimensions of the European pantograph results in $Y = 800 + 345 + 270 + 530 = 1\,945$ mm for $S_1 = 345$ mm in a height of $5{,}50$ m above rail, $S_{el} = 270$ mm and a safety margin $S_2 = 530$ mm at a voltage AC 25 kV. For a pantograph head with a length of $1\,950$ mm $Y = 975 + 345 + 150 + 530 = 2\,000$ mm at AC 15 kV.
The height of the current collecting zone Z is given by:

$$Z = H_{max} + S_{el} + S_3 - HP \quad . \tag{6.2}$$

For a maximum height of $6\,500$ mm of the extended pantograph, a safety margin of $1\,350$ mm and $HP = 5\,500 + 1\,800 = 7\,300$ mm (6.2) at AC 15 kV results in

$$Z = 6\,500 + 150 + 1\,300 - 7\,300 = 700 \text{ mm} \quad .$$

For trolley systems EN 50 122-1, Clause 4.3. indicates the specific dimensions. Conductor rail systems do not require a contact line zone.
The protective provisions for exposed conductive parts or conductive parts in this zone are described in 6.3.4, while the specific precautions for DC-traction systems are given in 6.5.2. AC-traction systems are noted in 6.6.2.

6.2.10   Stray Current

Because perfect insulation from earth of the return circuit of DC lines can never be achieved in practice, part of the return current leaks from the running rails into the structure or earth. This current component, that does not flow in the intended return circuit, is defined as *stray current*.

6.3   Design principles and requirements

6.3.1   Principles of AC and DC railways

The traction current flows back to the feeding substation via the return circuit. From the electrical engineering aspect, contact line and return circuit constitute an inseparable unit.
Use of the running rails as a part of the return circuit is a common feature of AC and DC railway systems. However measures for earthing and bonding applied to the return circuit differ fundamentally. In DC railway electrification systems, the running rails are laid with a high resistance to earth and to earthing systems, to avoid return current leaving the running

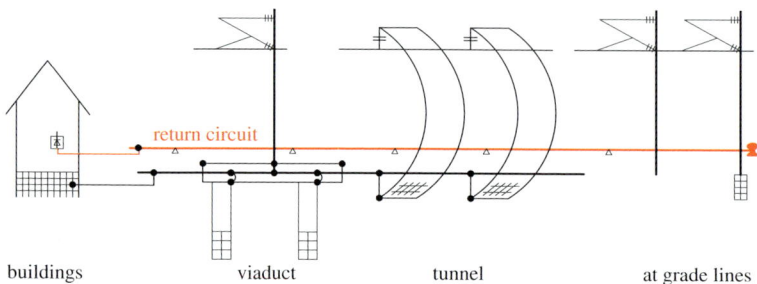

Figure 6.2: Simplified circuit diagram of return circuit and earthing of DC traction systems.

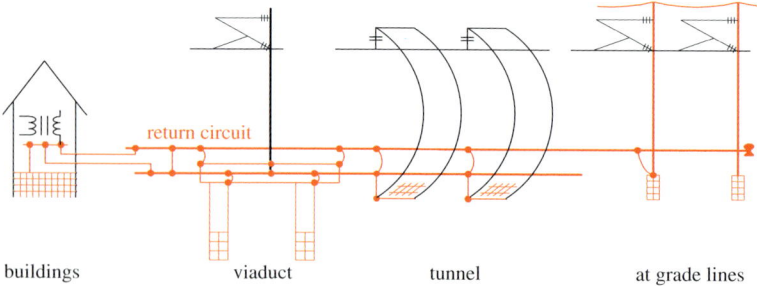

Figure 6.3: Simplified circuit diagram of return circuit and earthing of AC traction systems.

rails as stray currents, causing *stray current corrosion* of metallic components in close contact with the earth. Items such as pipelines, cable screens, steel reinforced foundations of buildings or poles, reinforced tunnel structures, bridges and viaducts are at risk. The strict separation of the track return circuit from the earthing system is demonstrated in Figure 6.2. Voltage drops occur in the running rails along the line, causing track-to-earth voltages during normal operation and in short circuits. Since no earthing connections are present, there is a risk that the permissible *touch voltage* will be exceeded with high currents and long feeding sections. The danger arises on surface lines in the open against earth and in tunnels, on viaducts and in stations and substations against the structure earth. Suitable measures for the arrangement of the *track return circuit* for DC railways are specified in IEC 60 128-2 and EN 50 122-2 and detailed in Clause 6.5.

Beside to *ohmic voltage drops* arising in DC railways, alternating current of AC railways causes additional inductive voltage drops, which are almost the same magnitude as the ohmic component at the operating frequency of 16,7 Hz and more than double that value at 50/60 Hz. This, together with the longer feeding sections, leads to significantly higher *rail potentials* than with DC railways, in spite of the smaller operating currents. To restrict the rail potentials to permissible values, it is necessary to connect the return circuit to earth, i. e. to connect the running rails and additional return conductors along the track and in the substation. Figure 6.3 illustrates the necessary connections between the return circuit and the earthing systems for AC railways, which are described in more detail in IEC 62 128-1 and EN 50 122-1 and Clause 6.6.

The earthing of the running rails is independent of the type of AC power supply. By contrast to the DC return circuit, current flows through the earthing installations of the buildings and

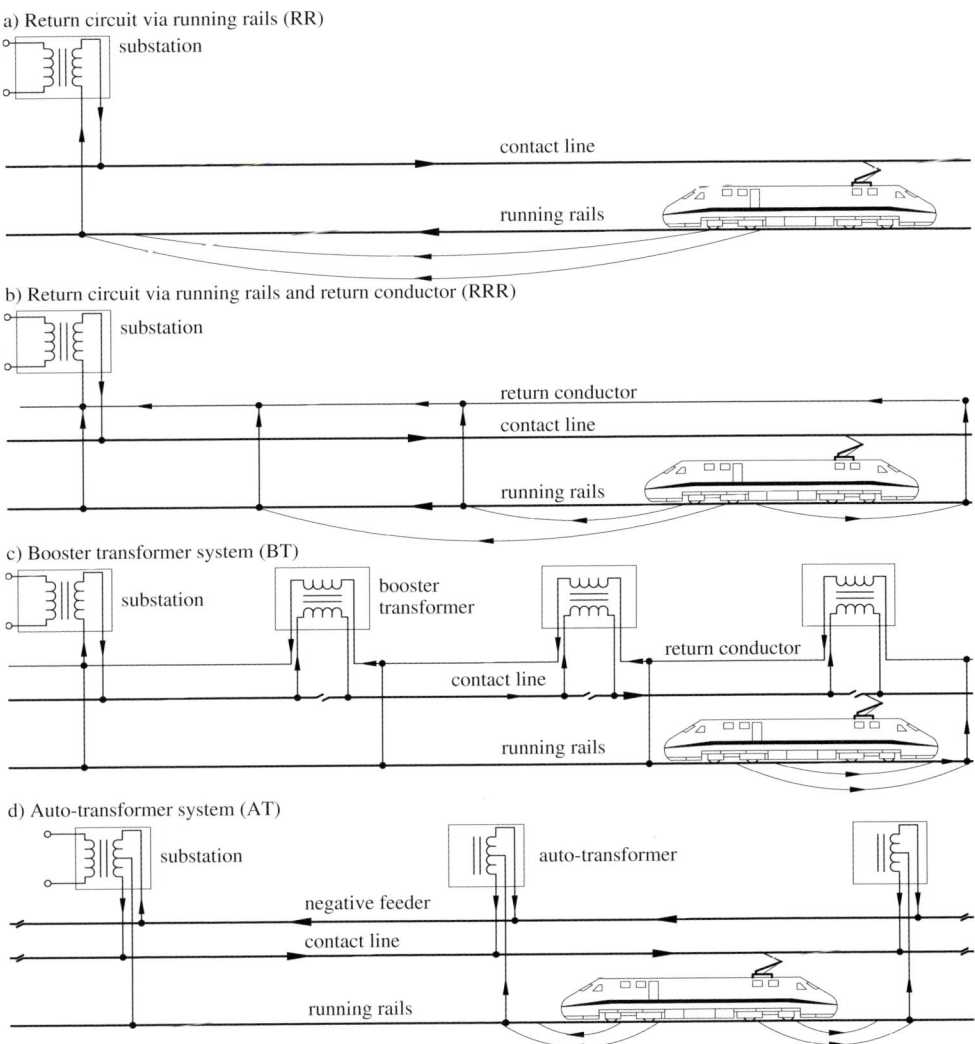

Figure 6.4: Track return current on AC railways.

through the earth, due to the earthing connections of the return circuit. This current produces undesirable inductive and magnetic field interference on equipment alongside the railway line and can cause disturbances in electronic equipment.

6.3.2   Return circuit of DC railways

Strict separation of the return circuit and earthing systems is a principle of DC railways. Practical applications can differ because of the installation of additional return conductors parallel to the tracks, of stray current collecting nets or of differing design of the earthing system. However, the design principle as shown in Figure 6.2 can be used for all applications. The design of the return circuit, earthing installations and the distances between substations are

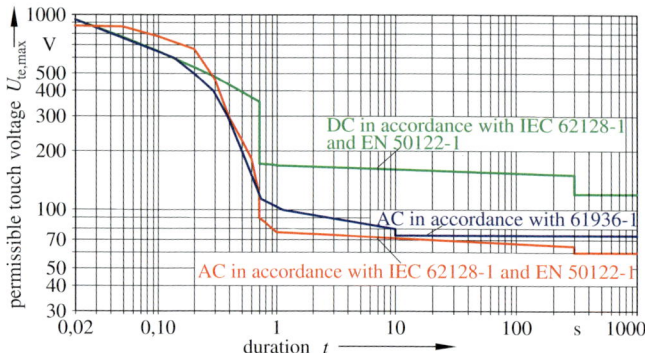

Figure 6.5: Permissible touch voltage $U_{te,max}$ for AC and DC railway systems as a function of duration t according to IEC 63 128-1, EN 50 122-1 and IEC 61 936-1.

determined by protective provisions against electric shock and overvoltages and by protective measures against stray current corrosion.

6.3.3   Return circuit of AC railways

The main configurations of AC traction power supply systems are shown in Figure 6.4 a–d). The rail-return system (RR) is commonly used for conventional lines. Booster transformer (BT) and auto-transformer (AT) systems are used to reduce the return currents through rails and earth where high electric currents and difficult earthing conditions could lead to unacceptable interference and rail potentials. In BT systems, see Figure 6.4 c), the return conductors are connected to the running rails midway between BT locations. They are almost at rail potential and carry the largest part of the return current. AC lines with AT adopt a double or multiple system voltage using an energized return circuit known as a negative feeder, see Figure 6.4 d) and Clauses 1.6.2 and 6.6.3.3. In traction sections with railway traffic, the running rails and the soil conduct the traction return current.

Theoretically, outside the *auto-transformer section* currently under load, no current should flow through the running rails. In practice however, studies have established that up to 10 % of the load current will flow there.

As explained in Clause 6.3.4 the running rails form a reliable return current path and so provide protection against *indirect electrical contact*. For this purpose conductive parts are connected to the running rails (Figure 6.3). Depending on the configuration of the return circuit and the cross sections, the portion of the current returning via earth will be 5 % to 40 % of the traction current.

6.3.4   Protection against electric shock and human safety

The protection of people against *electric shock* has highest priority. This involves protection against remaining dangerous voltages in the case of faults in the overhead contact zone and pantograph zone by tripping and by limiting rail potential during operations to permissible levels. To guarantee the *safety of persons*, the *touch voltages* during operations and under fault conditions must not exceed permissible voltages in accordance with IEC 62 128-1 and EN 50 122-1. To fulfill the criteria for electrical protection, a satisfactory rating of the return circuit and the earthing system is necessary. The specific measures for AC traction systems are given in Clause 6.5.2 and for DC-traction systems in Clause 6.6.2.

Table 6.1: Permissible effective touch voltage $U_{te,max}$ in V for AC and DC railways depending on duration of exposure in s.

Long-term			Short-term		
Duration t	AC railways	DC railways	Duration t	AC railways	DC railways
permanent	60	120	$< 0,7$	155	350
300	65	150	0,6	180	360
1	75	160	0,5	220	385
0,9	80	165	0,4	295	420
0,8	85	170	0,3	480	460
0,7	90	175	0,2	645	520
			0,1	785	625
			0,05	835	735
			0,02	865	870

The *return circuit* must conduct the traction and regenerative braking currents, as well as the short-circuit currents during faults, to the substation at low impedance. This limits *longitudinal rail voltages* and, therefore, the *track-to-earth voltages* and the requirements concerning permissible touch voltages can be met.

The running rails used to conduct the return current should be through-connected at low impedance. Rail bonds, track bonds and track release circuit bonds that conduct the return current serve this purpose. For upgrading existing systems, cables could be laid parallel to the running rails to supplement the return circuit. *Interruptions* to the *return circuit* are not permitted, because parts of it could become live.

In the area of the overhead contact line zone and pantograph zone, wholly or partially conductive structures require connection to the return circuit. The dimensions of this zone are defined in Clause 6.2.9. This requirement, for example, applies to metal fences, pipes, metal and steel reinforced bridges. Similarly, the metal covers of electrical equipment in the overhead contact line zone and pantograph zone must be connected to the return circuit. Therefore, no remaining high voltage can arise at metal structures within the overhead contact line and pantograph zones after failures. Exceptions from direct connection to the return circuit for small conductive parts are allowed for items such as, pit covers, signal posts, level crossing posts, individual masts, warning labels, trash cans, fences, mesh constructions and metallic structures with a maximum length as given in Table 1 of EN 50 122-1. Clause 6 of the Standard also indicates exceptions for tunnel sections and for low voltage traction systems.

The *permissible voltage values* specified in IEC 62 128-1/EN 50 122-1 and IEC 61 936-1 against electric shock are based on comprehensive examinations of body resistance and the effects of body currents (IEC 60 479-1). The mentioned standards specify different values for permissible touch voltages because of the consideration of varying footwear, insulation of the location and probability of ventricular fibrillation. Other values have been derived for low voltage applications in IEC 60 364-4-41. Figure 6.5 depicts the permissible touch voltages for railway applications and for high-voltage three-phase applications, depending upon the duration of exposure.

6.3.5 Permissible touch voltages

6.3.5.1 Requirements

EN 50 122-1 gives the limits for the permissible touch voltages for electric traction systems, which includes contact line systems. For three-phase AC installations however, the standard

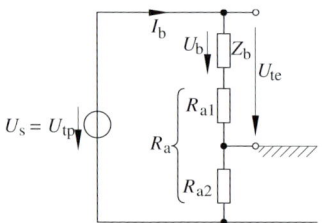

Figure 6.6: Equivalent circuit diagram of a touch circuit.
U_s source voltage
U_{tp} prospective touch voltage
U_b body voltage
U_{te} effective touch voltage
Z_b body impedance
R_a additional resistances
R_{a1} additional resistance for shoes
R_{a2} additional resistance of standing surface

IEC 61 936-1 has to be considered. The permissible *effective touch voltage* $U_{te,max}$ depends on the duration of the current, giving rise to the potential at the running rails, at contact lines poles and other components. Table 6.1 indicates the values $U_{te,max}$ for railway applications according to IEC 62 128-1 and EN 50 122-1, Tables 4 and 6.

The permissible touch voltages are categorized into long-term values for operations and short-term values for fault situations. In the case of floating voltages, the most severe case has to be considered. The values assume a current path through the human body from one hand to both feet, also covering the path from hand to hand. The voltages are indicated as root-mean-square values in the considered time interval. In workshops and similar locations, the touch voltages must not exceed the let-go limit of 25 V for AC installations and 60 V for DC-installations.

6.3.5.2   Body current, body voltage and touch voltage

The permissible touch voltages specified in EN 50 122-1 are derived from IEC/TC 60 479 with the following assumptions:
– current path from one hand to both feed
– body impedance for large touch areas and dry conditions
– 50 % probability for a body impedance to be higher than the assumed value

Table 6.2: Permissible body currents in mA and body voltages in V according to IEC 62 128-1 and EN 50 122-1, Table D.3 and D.4, depending on duration of impact.

Duration t s	Body current I_{C1}		Body voltage U_{C1}		Maximum body voltage $U_{b,max}$	
	AC	DC	AC	DC	AC	DC
> 300	37	140	62	153	60	120
300	38	140	64	153	65	150
1,0	50	150	75	160	75	160
0,9	52	160	77	167	80	165
0,8	58	165	83	170	85	170
0,7	66	175	91	177	90	175
< 0,7	66	175	91	177	90	175
0,6	78	180	101	180	100	180
0,5	100	195	119	191	120	190
0,4	145	215	152	204	155	205
0,3	252	240	230	222	230	220
0,2	350	275	293	246	295	245
0,1	440	380	343	287	345	285
0,05	475	410	361	327	360	325
0,02	495	500	370	372	370	370

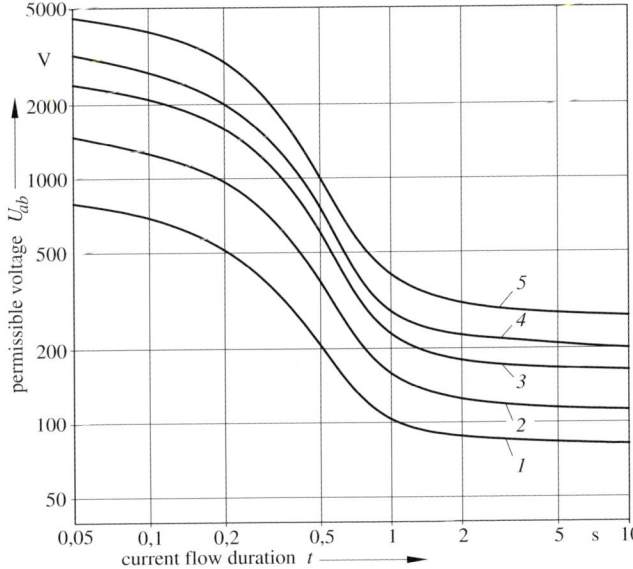

Figure 6.7: Maximum permissible voltage U_{ab} according to IEC 61936-1 assuming additional resistances in the touch circuit, plotted as a function of the current flow duration t.
1 $R_a = 0\,\Omega$, d. h. $U_{te} = U_t$
2 $R_a = 750\,\Omega$ with $R_{a1} = 710\,\Omega$ and $\rho_E = 27\,\Omega m$
3 $R_a = 1\,750\,\Omega$ with $R_{a1} = 1\,315\,\Omega$ and $\rho_E = 290\,\Omega m$
4 $R_a = 2\,500\,\Omega$ with $R_{a1} = 1\,000\,\Omega$ and $\rho_E = 1\,500\,\Omega m$
5 $R_a = 4\,000\,\Omega$ with $R_{a1} = 3\,960\,\Omega$ and $\rho_E = 27\,\Omega m$

 – zero % probability for heart fibrillation
 – for short-term conditions $1\,000\,\Omega$ resistance R_{a1} for old wet shoes additional to U_{C1}
The resistance of the standing surface may be considered for all time periods. Figure 6.6 shows additional resistances in the touch circuit. The current path from one hand to both feet according to IEC 62128-1 is used to determine the body impedance, which is 0,75 times the impedance of the hand-to-hand path. Table 6.2 gives an example of the permissible touch voltages for $R_{a1} = 1\,000\,\Omega$ and $R_{a2} = 150\,\Omega$. It also indicates the body current I_{C1} according to the curve C1 in IEC/TC 60479-1 as well as U_{C1} according to I_{C1} as a basis for the evaluation of the permissible touch voltages. Based on experience, the maximum permissible body voltages $U_{b,max}$ for AC and DC railway systems differ insignificantly from U_{C1}. They are identical with the long-term values of the permissible touch voltage. The maximum permissible short-term touch voltage is calculated as

$$U_{te,max} = U_{C1} + R_{a1} \cdot I_{C1} \cdot 10^{-3} \tag{6.3}$$

with $R_{a1} = 1\,000\,\Omega$ for old wet shoes.
Unlike IEC 61936-1, which specifies permissible touch voltages of 80 V for a 10 second current flow and 75 V for longer durations, IEC 62128-1 and EN 50122-1 define the permissible touch voltages for current flow durations exceeding 300 s as 60 V for AC and 120 V for DC railway traction systems.
In practice, every current circuit, where parts with higher potential can be touched, contains additional resistances as shown in Figure 6.6. For example, the additional resistance R_a for persons working in the railway environment comprises of R_{a1}, e. g. resistance of shoes, and the local resistance to earth R_{a2}. Figure 6.7 shows the permissible touch voltages for three-phase high-voltage installations as a function of the duration of current flow. In this case, additional resistances are taken into account, e. g. the resistance of shoes and the resistance to earth. Those voltages are considerably higher than the touch voltage $U_{te\,max}$ where there is no additional resistance.

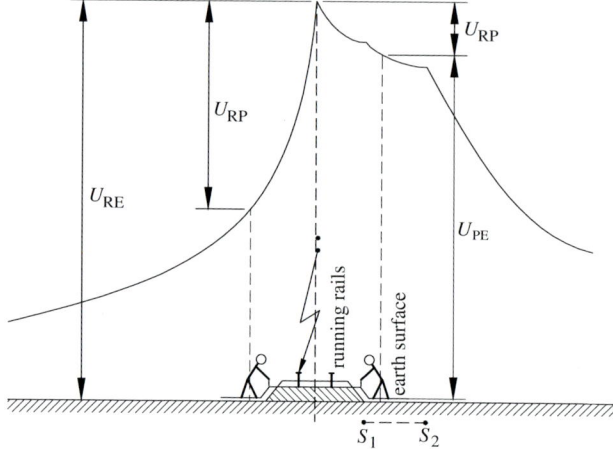

Figure 6.8: Voltage difference versus earth U_{PE} and touch voltage in a perpendicular direction to the track.
U_{RE} rail potential, U_{RP} voltage shunted by persons, S_1, S_2 potential grading, voltage-limiting provisions, e. g. earthing electrode strip bonded to the rails

A hazard to persons due to high rail potential could only occur in railway installations if the shunted portion exceeded the permissible touch voltage. This portion of the rail potential is shown as U_{RP} in Figure 6.8 and as U_{RP}/U_{RE} in Figure 6.17, where U_{RE} is the rail potential or rail-to-earth voltage. The diagram shows that the relation U_{RP}/U_{RE} will be well below one in almost all cases. In worst-case situations, U_{RP} can become equal and identical to U_{RE}.

For AC railways, a factor k_a is used in [6.3] to define that portion of the rail potential or other potentials that could affect persons. This factor considers most cases where *additional resistances* are present in the *touch circuit*. In practice, only a part of the rail potential can be shunted by persons. This factor is expressed by the ratio:

$$k_a = U_{RP}/U_{RE} \tag{6.4}$$

In the quoted references, values of 0,3 to 0,8 are given for k_a with 0,5 being a recommended value for design purposes. In individual cases, it can be necessary to introduce *equipotential bonding measures* to reduce the expected voltage U_{RP}. Such equipotential bonding is achieved by electrical connections which ensure that external conducting parts are kept at the same or nearly the same potential as the accessible conducting parts of electrical equipment, where potentials arise during operation or in case of a fault. Some electric railway operators require mandatory equipotential bonding provisions, see Figure 6.8.

6.3.5.3 Touch voltage measurement

In cases where it is necessary to measure the touch voltage, the procedure is given in the normative Annex E of IEC 62 128-1 and EN 50 122-1. A resistor is used that corresponds to the human body resistance Z_b and to the additional resistance R_{a1}, see Figure 6.6. For practical applications, a value of $2\,200\,\Omega$ covers short-term and long-term conditions.

The measuring electrode for the simulation of the foot resistance
  – shall have a total area of $400\,cm^2$ and shall be pressed to the earth with a minimum force of $500\,N$,
  – alternatively, a measuring electrode with $2\,cm$ diameter and $30\,cm$ length can be used. This corresponds to an earth electrode with $2,2\,\Omega/(\Omega m)$.
  – shall be placed at a distance not less than $1\,m$ from the measured part,

– shall be a tip electrode. In this case, paint coatings, but not insulation, shall be pierced reliably.

One clamp of the voltmeter is connected to the hand-electrode, the other clamp to the feet electrode. It is sufficient to carry out such measurements by random checks of an installation. At measuring points where the resistance to earth of the measuring electrode for the simulation of feet does not exceed several hundred ohms, a measurement with and without parallel resistance is recommended. The resistance represents the human body resistance Z_b and the additional resistance R_{a1}. If the voltage breaks down when using the parallel resistance, it can be concluded that effective touch voltage is considerably lower than the unaffected prospective touch voltage, e. g. the rail potential.

6.3.6 Interference

Malfunction can arise from interference by railway circuits on railway-owned and third-party installations. With respect to the interference caused by traction supply currents, the following coupling mechanisms should be considered:

– ohmic or *galvanic interference*
– *inductive* and *capacitive interference*
– *electric* and *magnetic fields*

The galvanic interference is caused by conductive connections to the return circuit. Voltage and currents influenced by the conductor capacitance are called capacitive interference. They only arise in insulated conductors close to the contact line systems of AC railways. Inductive interference is important in the case of AC power supply systems. The magnitude of interference depends on the self and mutual impedances of the overhead contact line arrangements. The magnitude of interference depends on the distribution of the currents. This means that, the *current* flowing *through earth* indicates the degree of interference. The design of contact line configuration aims at limiting the return current through earth and reducing interference in the vicinity of the railway. The interference concerns railway owned as well as third-party installations. Depending on the sensitivity of the devices, operational impairment can occur. Details on analysis and acceptable levels are discussed in Clause 6.6 and Chapter 8.

6.3.7 Stray current corrosion

Metals in contact with an electrolyte such as humid soil show chemical reactions if currents leave the metal parts. Therefore, the DC currents flowing from the tracks to the earth and returning to the substations – called stray currents – can cause *stray current corrosion* at the running rails and at metal structures in the vicinity of the DC railway. One aim of DC traction power installations design is to avoid stray current corrosion at railway-owned and third-party installations. This can be achieved by limiting stray currents through adequate design of the return circuit, in particular by insulating the tracks from the earth or structures, e. g. tunnels and viaducts and by planned maintenance to identify rail to earth connections and repairing such defects, see EN 50 122-2, VDV 501 Part 1 to 3 [6.4], VDV 507 [6.5] and [6.6]–[6.11].

A low longitudinal voltage drop in the return circuit and good insulation of the rails from earth also reduces stray currents substantially. Since the longitudinal voltage drop depends on the resistance of the return circuit and the distance between the substations, stray current protection can also determine the number of substations.

During the operation of DC railways it is not possible to avoid stray currents completely. If, however, the compatibility criteria according to Clause 5 of EN 50 122-2 are fulfilled, adverse effects caused by stray currents can be neglected. One assessment criterion concerns the return circuit, meaning the running rails and is given in EN 50 122-2, Clause 5.2. For stray current protection purposes, the average stray current leaving the running rails should not exceed 2,5 mA/m for a single track line. This value is based on 25 years experience and is assumed correct, if the conductance per unit length G'_{RE} of the running rails and the average rail potential fulfills the following conditions:

- $G'_{RE} = 0,5$ S/km per track and $U_{RE} \leq +5$ V for open formation
- $G'_{RE} = 2,5$ S/km per track and $U_{RE} \leq +1$ V for closed formation

In case this simplified verification does not meet the postulated stray current values, an alternative conductance per unit length can be used according to IEC 62 128-2 or EN 50 122-2, Clause 5.2

For design of the rail insulation, EN 50 122-2, Annex C.1. gives a worst-case estimation for U_{RE} in the case of dead-end lines or line extensions. The rail potential is calculated as

$$U_{RE} = 0,5 \cdot I \cdot R_C \cdot [1 - \exp(-L/L_C)] \tag{6.5}$$

with the characteristic resistance

$$R_C = \sqrt{R'_R / G'_{RE}} \tag{6.6}$$

and the characteristic length

$$L_C = 1/\sqrt{R'_R \cdot G'_{RE}} \quad , \tag{6.7}$$

with the following symbols:

U_{RE} rail potential
I average value of the traction return current in the considered section in the hour of the highest load
R_C characteristic resistance of the circuit between rails and structure
L_C characteristic length of the circuit between rails and structure
L length of the considered section
R'_R longitudinal resistance of the running rails, including parallel return conductors per length
G'_{RE} conductance per length of the running rails versus earth

Using the rail potential and the conductance of the running rails versus earth, the stray current per length is calculated according to Equation (6.8)

$$I'_S = 0,5 \cdot (L/L_C) \cdot [1 - \exp(-L/L_C)] \quad . \tag{6.8}$$

If the stray current I'_S, divided by the number of parallel tracks is less than 2,5 mA/m, the conditions for the permissible conductance according to Clause 5.2 of IEC 62 128-2 and EN 50 122-2 are fulfilled.

The other requirement concerns the steel reinforced installations of railways and third parties, e. g. foundations of buildings and metal installations like pipelines. This is given in Clause 5.3 of IEC 62 128-2 and EN 50 122-2, and 6.5.4 of this book.

Protection against stray current corrosion of steel and steel reinforced concrete installations uses the voltage rise with respect to earth as a compatibility criterion. The average of the potential rise of steel reinforced concrete structures during the highest traffic should not exceed $+200$ mV. Further details and values for metal buried in soil are given in the specific engineering standard for stray current corrosion EN 50 162. In general, good insulation of the running rails from earth and strict electrical separation from earthing installations is the precondition to fulfill the above criteria. For further details see Clause 6.5.3.

6.3.8 Measurements for earthing installations and return circuit

In many cases, reliable information for design of electrical installations is only available from *measurements*. The *design of earthing installations* requires information relating to the soil resistivity to enable calculation of the resistance to earth of foundations or earth electrodes. If existing earthing systems are used, it is recommended that direct measurement of resistance to earth is undertaken. Together with the design values for the operational and short-circuit currents, the *touch voltages* can be calculated. This is required as a design value for the assessment of the safety of persons.

During the construction phase, the planned earth connections must be inspected before they are covered with concrete. The measurement of the earth resistance of subsystems is recommended if the design provides critical values, so that corrective measures are possible in due course.

During the commissioning phase, *verification of the safety of persons* and operational reliability of the installations are necessary. Measurements provide meaningful information and will provide a significant contribution to the rapid technical approval of the installations.

During this phase, measurements are also necessary to testify the effectiveness of stray current protection measures, see IEC 62 128-2 and EN 50 122-2, Appendix A.

During operation of DC railway systems, reasonable monitoring is also necessary to confirm that the proposed measures against stray current corrosion perform as required. See IEC 62 128-2 and EN 50 122-2 Annex B. Such monitoring or equivalent measurements can also be used to support the *permanent supervision of safety of persons* [6.10].

6.4 Return currents and rail potentials

6.4.1 Soil resistivity and conductivity

Earth is considered to include all types of soil and rock that make up the Earth's external crust that contribute towards conducting currents. The soil presents a conductivity and resistance to the circulation of currents, depending on its physical and chemical properties. When a voltage is applied to a conductor with uniform cross section and homogeneous material, the determination of its resistivity and resistance is a simple task. However, when dealing with current conduction through the earth, the analysis becomes complex, because of the huge dimensions of the earth as compared to the metallic conductors and the great variation of its characteristics.

For example, experimental tests made with red clay soil indicated that with only 10 % moisture content, the resistivity was over 30 times that of the same soil having a moisture content of about 20 %. For values above 20 %, the resistivity is not affected too much but below 20 %,

Table 6.3: Typical soil resistivities.

Type of soil	Soil resistivity $\Omega \cdot m$
sea water	1
marshy soil	5–40
loam, clay, humus	50–350
sand	200–2 500
gravel	2 000–3 000
lime stone	350
sand stone	2 000–3 000
weathered rock	bis 1 000
granite	$\sim 3\,000$–50 000
moraine	bis 30 000

it increases rapidly with the decrease in moisture content. As defined in Clause 6.2.5 *soil resistivity* is expressed in $\Omega \cdot m$, the *soil conductivity* in S/m. Resistivities of typical soils are around the values indicated in Table 6.3.

Figure 6.9 shows a histogram of soil resistivity values measured along 6 000 km of railway lines in Germany [6.12]. The majority of measured data is below $50\,\Omega \cdot m$, the mean value being $25\,\Omega \cdot m$. From Equation (6.9), for soil resistivity of $25\,\Omega \cdot m$, a current penetration depth of 800 m results for AC 16,7 Hz and a depth of 450 m for AC 50 Hz.

The most frequently used method to determine the *soil resistivity*, depending on the depth, is the *four-point method*, also called *Wenner method* [6.13] where an *earth megger* [6.14] is used (see Figure 6.10). The four rods are arranged with the same spacing a; five measurements with the spacing $a = 2$, 4, 8, 16 and 32 m are carried out. For each measurement a current I is injected between the probes C_1 and C_2 and the voltage between the points P_1 and P_2 is measured. With increasing spacing a the measured soil resistivity applies to greater depths since the current flows through soil strata in greater depth.

The soil resistivity ρ_E results from

$$\rho_E = 2\,\pi \cdot a \cdot R_E \quad , \tag{6.9}$$

where a is the distance between the probes and R_E is the recorded resistance.

6.4.2   Earth electrodes in the vicinity of railways

6.4.2.1   Earth resistance of electrodes and pole earthing

An *earth electrode* is a bare conductor or other conductive component which is in electrical conductive contact with the earth, or a bare conductor or other conductive component embedded in a concrete structure which, in turn, has a large contact area with the earth. Earth electrodes in a railway installation may include

- contact line support foundations,
- *earthing strips* installed parallel to the track and
- natural earth contacts, such as metal pipes, cable sheaths, parts of steel structures, foundations of buildings and *substation earthing systems*.

Earth electrodes installed in the vicinity of railways and connected to the track increase the conductance per unit length between the track and earth. The connection to the track is commonly used for AC Systems, but not for DC systems. Earth electrodes are characterized by

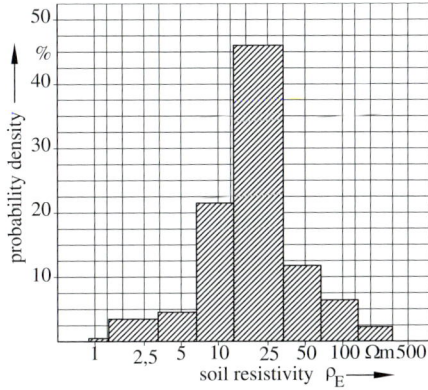

Figure 6.9: Histogram of soil resistivities in the vicinity of railway lines according to [6.12].

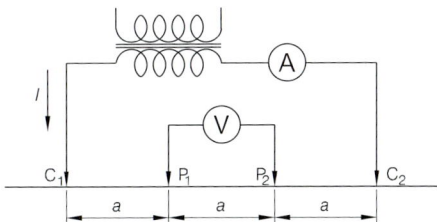

Figure 6.10: Arrangement for soil resistivity measurements according to Wenner [6.13].

an earth resistance defined as the effective resistance between the earth electrode and the reference or remote earth. The *earth resistance* depends on the soil resistivity ρ_E, on the geometrical dimensions of the electrode and on its arrangement.

The *earthing strips* installed along sections of electric railway lines are *surface earth electrodes* normally buried at a depth of 1 m. The resistance to earth of an earthing strip of a width w and length L is given by

$$R_E = \rho_E/(\pi L) \cdot \ln(4L/w) \quad . \tag{6.10}$$

Example 6.1: What is the earth resistance of an earthing strip of galvanized steel, 1 km long and 30 mm diameter with earth resistivities of $27\,\Omega\text{m}$ and $290\,\Omega\text{m}$ respectively?

For $\rho_E = 27\,\Omega\text{m}$, the earth resistance of this 1 km long earthing strip is approximately $R_E = 27/(\pi \cdot 1\,000) \cdot \ln(4 \cdot 1\,000/0,03) = 0,1\,\Omega$. For $\rho_E = 290\,\Omega\text{m}$, it is $1,06\,\Omega$.

Earth rods are earth electrodes buried or driven deeper than surface earth electrodes. Overhead contact line pole foundations can be considered as earth rods. As explained in Clause 13.7.4, poles are frequently set up on steel piles or pipes that have been driven into the ground to a depth of several metres.

The *earth resistances of pole foundations* form an important part of traction earth systems. To calculate the expected earth resistance of a pole foundation, it is treated as an earth rod. This permits the use of the following equation for calculating the earth resistance R_E for a circular-shaped pole foundation of depth L and diameter d:

$$R_E = \rho_E/(2\,\pi L) \cdot \ln(4L/d) \tag{6.11}$$

For foundations with a rectangular cross section, a good approximation is obtained by substituting the diameter by the shorter edge of the rectangle.

As indicated, in addition to the foundation geometry, the soil resistivity above all, has a decisive effect on R_E. Poles set in in-situ cast concrete may have values of several hundred ohms in dry locations because of the high resistivity of concrete, see Table 11.2. In comparison,

Table 6.4: Guideline values of earth resistance and conductance of earth electrodes in railway applications for $\rho_E \approx 100\,\Omega$m.

Type of pole, type of natural earthing	R_E Ω	G_E S
concrete pole with concrete foundation	50	0,02
steel pole on in-situ concrete foundation	40	0,025
pole with conductive connection to steel pile	14,3	0,07
earthing strip electrodes, double-track line, per km	0,167	6,00
lighting pole	50 to 100	0,01 to 0,02
bridge railings	30 to 60	0,03 to 0,07
roof drain with drainpipe	125	0,008
water supply pipeline network, buried 2 m deep, pipes having diameter between 40 mm and 150 mm[1]	0,2 to 0,4	2,5 to 5
water pipelines, 3 km long, diameter 150[1] mm	2,3	0,43

1) according to [6.15]

pole foundations on steel piles driven into the ground have earth resistances between 8 and 15 Ω. Similar low values are found for *driven steel pipes*. Earth resistances of 2 to 13 Ω have been measured on driven steel pipes with an external diameter of 508 mm. The length of such pipes commonly varies between 3,5 and 6,0 m. Table 6.4 shows guideline values of pole to earth resistances commonly occurring in the DB lines, based on the DB Directive 997.204.

Example 6.2: What is the earth resistance of a pipe of diameter 0,508 m driven into the earth to a depth of 5 m, assuming the earth resistivity to be 27 Ωm and 290 Ωm? The earth resistance is $R_E = 27/(2 \cdot \pi \cdot 5,0) \cdot \ln(4 \cdot 5,0/0,508) = 3,2\,\Omega$ for $\rho_E = 27\,\Omega\cdot$m and 33,9 Ω for $\rho_E = 290\,\Omega\cdot$m.

According to EN 50 341-1 and DB Directive 997.204, the following equation can be used to calculate the earth resistance of concrete foundations with steel reinforcement

$$R_M = \rho_E/(\pi \cdot d) \quad . \tag{6.12}$$

In this equation, d is the diameter of a hemisphere with a volume equal to the volume V of the foundation: $d = 1,57\,V^{1/3}$.

Table 6.5 shows the resistance of foundations for sizes of 1, 2 and 3 m^3. The pole earth resistance is determined by the soil resistivity, while the volume of the foundation has a minor effect.

It should be noted that Equation (6.12) applies to concrete foundations with steel reinforcement. However in sandy soils without ground water contact, the earth resistance of in-situ concrete foundations without reinforcement can reach values up to 300 Ω.

6.4.2.2 Effective conductance between tracks and earth

The electrical resistance between the rails of a track and earth is called the *rail-to-earth resistance*. This resistance which describes the *galvanic* or conductive *coupling* of track and earth depends on the properties and condition of the superstructure between the running rails and the earth. The essential characteristics of the superstructure include:
  – the type of *superstructure*, i. e. type of sleepers and track fasteners used e. g. sole plates, including insulating pads between rails and sleepers

Table 6.5: Pole earth resistances R_E of steel-reinforced concrete foundations in soils with different soil resistivities. Values given in Ω.

Volume	Soil resistivity ρ_E in Ωm		
m^3	27	100	290
1	5,6	20,3	58,9
2	4,3	16,1	46,7
3	3,8	14,1	40,9

- the bedding of the sleepers, e. g. in gravel or sand ballast, in a road, on concrete or, as is now used for tramways, in turf
- concrete track slab

The condition of the track embedding is mainly determined, from the electrical engineering aspect, by:

- the degree of contamination
- weather conditions such as damp, rain and frost

Measurements have proven that the characteristic variations of the rail-to-earth resistance of *tracks with concrete sleepers* were in the range of 0,4 to 2,5 Ω·km corresponding to a conductance per unit length of 2,5 to 0,4 S/km in summer weather conditions and 1,5 to 17,5 Ω·km corresponding to a conductance per unit length of 0,67 to 0,06 S/km in winter conditions. Measurements and analytical studies of concrete-sleeper track superstructures have shown that the rail-to-earth resistance is determined to an extent of 90 % by the type of sleepers and ballast. The remaining 10 % are a function of the substructure and the subsoil in the vicinity of the line.

Rail-to-earth resistance measurements carried out under varying conditions with normal operating currents and with short-circuit currents have also led to the conclusion that the rail-to-earth resistance is virtually independent of the currents flowing to and from the track to earth within the entire range of currents possible in an electric traction network. This means that the *galvanic coupling* of any given superstructure is also independent of whether the railway is powered by direct current or by alternating current. The *rail-to-earth impedance* of single-phase AC railway systems is a complex vectorial quantity with a phase angle of between 1° and 3°. Because of this, the very small reactive component is ignored in practice and the resistance is also assumed to apply, as a purely ohmic quantity, in calculations for single-phase AC railways.

The resistance between the two running rails of a track is the *rail-to-rail resistance*. High rail-to-rail resistances are required to ensure reliable operation of *track release systems*. The rail-to-rail resistance is affected by the type of *insulating pads* placed between the rail and the sole plates. High rail-to-rail resistances can be achieved by installing high quality insulating pads. If the insulating pads of both running rails have the same electrical characteristics, the *superstructure* is considered to be *symmetrical* from the electrical engineering aspect. Superstructure with different insulation characteristics of each running rail of a track is termed an *asymmetric superstructure*. To ensure reliable functioning of the track release systems, the rail-to-rail resistance should not be permitted to drop below permissible values given by the signalling system.

As an example, the DB Directive 997.204 specifies a rail-to-rail resistance of at least 1,5 Ω·km for symmetrical superstructure and at least 2,5 Ω·km for asymmetric superstructure if it is to be used for audio frequency track release circuits.

The reciprocal value of the resistance between the rails or the track and earth is the *conductance* G_{RE} and is expressed in S. The length related value G'_{RE} is expressed in S/km. It has

Table 6.6: Conductance per unit length G'_{RE} of tracks versus earth (guideline values) according to data from [6.12, 6.16], values given in S/km.

Construction and condition of the track ballast	Single-track line	Double-track line
impregnated wood or concrete sleepers, clean gravel ballast, heavy frost	0,02 to 0,04	0,04 to 0,08
ditto, but no frost	0,5 to 1,0	1,0 to 2,0
ditto, but contaminated gravel ballast	1,0 to 2,2	2,0 to 4,4
ditto, but clean sand ballast	1,5 to 3,3	3,0 to 6,7
long-distance line track on gravel ballast	1,5 to 4,0	3,0 to 8,0
concrete slab track on an insulating layer of bitumenized stone chippings	0,25 to 5,0	0,5 to 10,0
impregnated wood or concrete sleepers on sand ballast with clay content	3,2 to 5,0	6,0 to 10,0
wood sleepers in lignite open-cast mines	2,5 to 8,0	6,0 to 16,0
concrete sleepers on gravel ballast with stone paving	2,0 to 5,0	4,0 to 10,0
concrete sleepers on sand ballast with stone paving	3,5 to 10,0	7,0 to 20,0
concrete slab track on sand bed	10,0 to 25,0	20,0 to 50,0
track in tunnel, well-insulated, dry bed	0,3 to 1,3	0,6 to 2,5
track in tunnel, old insulation, wet bed	2,0 to 8,0	4,0 to 17,0
tracks in roads	9,5 to 23,0	19,0 to 45,0
WK type superstructure, new, dry	0,005	0,01
WK type superstructure, older, dry	0,02	0,04
WK type superstructure, older, damp	0,23	0,5
W type superstructure, new, dry	0,05	0,1
W type superstructure, older, dry	0,1	0,2
W type superstructure, older, damp	0,4	0,8
K type superstructure, older, dry	0,5 to 1,0	1,0 to 2,0
K type superstructure, older, damp	1,5 to 3,0	3,0 to 6,0
slab track	$\approx 0,01$	$\approx 0,02$

Where no specific information is shown in the table, the values apply to normal humid beds. In the case of very dirty ballast and extreme *damp*, the G'_{RE} values should be multiplied by a factor of 1,5 to 2,2 or a factor of 0,1 to 0,3 in the case of frost.

K: Rail fixing on wood, steel or concrete sleepers by means of metallic plates
W: Rail fixing on concrete sleepers by means of pre-stressed plates with plastic lining
WK: Rail fixing on concrete sleepers by means of pre-stressed plates with thick plastic lining

a significant effect on the return current conduction and on the *track-to-earth voltage*, as explained in Clause 6.4.3. Therefore, rails and tracks are thus characterized by a longitudinal resistance and a conductance to the reference earth potential with both properties depending on the length of the track system.

The *track-to-earth conductance* depends on the following factors:
- the permanent way structure
- the sub-grade structure
- the degree of pollution of the superstructure
- the weather conditions and
- the specific soil resistivity

Table 6.6 contains a list of *conductance per unit length* values measured on single-track and double-track railway lines. The effect of some important factors can be observed in this table and some of these factors will be discussed briefly. Important parameters are water content, frost and temperature changes. Refrence [6.6] reports on rail-earth conductances per unit length of 0,1 S/km being measured at temperatures under 0 °C and 0,5 S/km being measured at the same location when the temperature had risen above 0 °C. The poles, the foundations of which each have an earth resistance R_E, are earth electrodes connected parallel

Table 6.7: Effective conductance $G'_{RE\,eff}$ per unit length per single track for different superstructures and for different pole earth resistances, assuming 16 poles per kilometre, all values given in S/km.

Design of single track permanent way	Track conductance S/km	Effective conductance pole earth resistance					
		$10\,\Omega$	$20\,\Omega$	$50\,\Omega$	$100\,\Omega$	$200\,\Omega$	$500\,\Omega$
concrete slab track [1]	0,01	1,61	0,81	0,33	0,17	0,09	0,042
WK type superstructure	0,05	1,65	0,85	0,37	0,21	0,13	0,082
W type superstructure	0,10	1,70	0,90	0,42	0,27	0,18	0,132
K type superstructure	1,00	2,60	1,80	1,32	1,16	1,08	1,032

[1] In practice, the concrete slab track is cross-connected to the pole foundations. The conductance versus earth is 10 to 20 S/km or even higher and, Therefore, the overall value of the effective conductance predominates.

to the track. These parallel earth electrodes represent a significant contribution to the *effective conductance* $G'_{RE\,eff}$. The effect of the pole to earth resistances on the effective conductance per unit length of a track was calculated for typical track conductances and various types of *track superstructure*. A R_E value range of $10\,\Omega$ to $500\,\Omega$, which is realistic in practice, was chosen. The calculation was carried out for a line with 16 poles per kilometre being connected electrically to the track. The results are shown in Table 6.7. The track conductance and the effective conductance per unit length are to be considered as switched in parallel. In double-track lines, the potential sinks of the rail-to-earth voltage overlap when two trains meet. As a consequence, the rail potential doubles. In the vicinity of stations and buildings or non-railway metal structures, the standards IEC 62 128-1 and EN 50 122-1 require that, for AC traction systems, all conductive parts e. g. handrails of bridges, signal masts etc. be directly connected to the return circuit, i. e. with the running rails. This additional traction earthing leads to the *effective conductance per unit length* in such areas being higher than the ones calculated above. In stations, the effective conductance per unit length is further increased by other tracks running parallel to the main track.

For example, in a railway station with four tracks with W type superstructure according to Table 6.7 a track conductance of 0,10 S/km per one row of poles is obtained from Table 6.7. Assuming a pole resistance to earth of $50\,\Omega$, the total effective conductance per unit length is $G'_{RE\,eff} = 2 \cdot 0,42 + 4 \cdot 0,10 = 1,24$ S/km. Because of superimposing effects, effective conductances per unit length of 10 S/km and higher can be observed in large stations with many tracks in parallel.

6.4.3 Track-to-earth circuit

6.4.3.1 General

In Figure 6.11, a circuit diagram is shown of the galvanic coupling between track and earth. In this model, the distributed or continuous quantities *longitudinal resistance per unit length* R'_R and *conductance per unit length* G'_{RE} between the rails and earth are represented as discrete resistors. By accepted definition, the resistance of the subsoil between the individual connecting points of the resistors with the soil has been assumed to be zero.

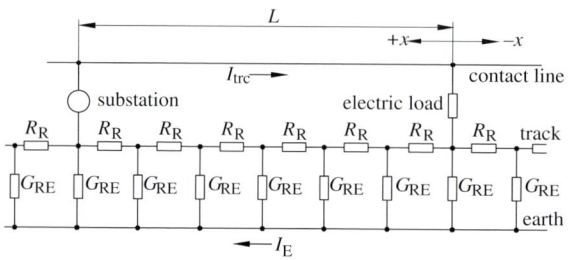

Figure 6.11: Model of the galvanic coupling between a railway track and earth.

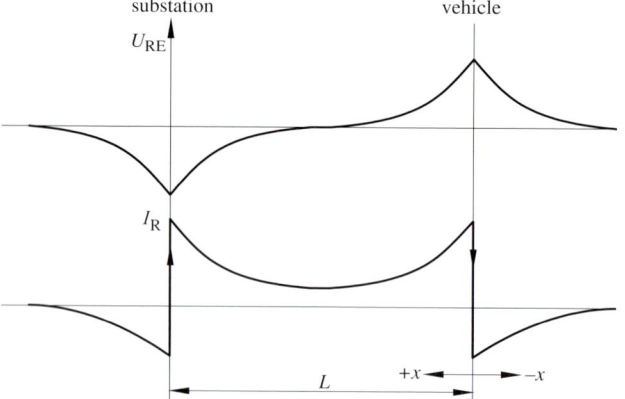

Figure 6.12: Track-to-earth voltages U_{RE} and track currents I_R in a DC railway line with a single feed substation and a single load.

6.4.3.2 Track-to-earth circuit of DC systems

In DC systems, the rails are intentionally insulated from earth to avoid stray currents as far as possible. However, depending on the actual condition of the rail insulation and the resistance of the superstructure, a part of the traction current I_{trc} flows through the earth back to the substation.

In Figure 6.11, a substation supplies energy to an electric traction vehicle. In reality, an electric railway system with a larger number of trains running simultaneously, will receive its energy supply from a multitude of substations. For this reason, either the individual loads or the railway line load per unit length are taken into consideration when discussing the currents and voltages between track and earth. The railway line load per unit length is defined by Equation (5.58). Calculating earth currents and track-to-earth voltages is a complex procedure – the results depend on train and load combinations existing at any particular time. For this reason, only some fundamental conclusions will be drawn based on the model shown in Figure 6.11.

With the assumption that the track leading up to the substation and away from the traction vehicle is of infinite length and using the coordinate designations given in Figure 6.12, the rail potential U_{RE} is calculated from

$$U_{RE}(x) = (Z_0 I_{trc}/2) \cdot \left(e^{-\alpha x} - e^{-\alpha(L-x)}\right) = Z_0 I_{trc}\, e^{-\alpha(L/2)} \sinh\left[\alpha\,(L/2 - x)\right] \qquad (6.13)$$

and the rail current from

$$I_R(x) = (I_{trc}/2)\left(e^{-\alpha x} + e^{-\alpha(L-x)}\right) = I_{trc}\, e^{-\alpha(L/2)} \cdot \cosh\left[\alpha\,(L/2 - x)\right] \qquad . \qquad (6.14)$$

Table 6.8: Earth current and track-to-earth voltage for a DC line and an example with UIC 60 rails for a traction current of 1 000 A.

Conductance G'_{RE} S/km	Surge impedance Z_0 Ω	Propagation constant α 1/km	Earth current $L = 10\,\text{km}$ A	$L = 5\,\text{km}$ A	Track-to-earth voltage $L = 10\,\text{km}$ V	$L = 5\,\text{km}$ V
0,1	0,3873	0,0387	176	92	62	34
1	0,1225	0,1225	459	264	43	28
2	0,0866	0,1732	579	351	36	25

In (6.13) and (6.14), α is the *propagation constant* of the dimension $(\text{length})^{-1}$. It is determined by

$$\alpha = \sqrt{R'_R \cdot G'_{RE}} \quad , \tag{6.15}$$

where R'_R is the track resistance per unit length and G'_{RE} the conductance to earth per unit length.
The value Z_0 is the surge impedance in Ω, which is calculated as

$$Z_0 = \sqrt{R'_R / G'_{RE}} \quad . \tag{6.16}$$

At the position $x = 0$ at the electric load, the voltage U_{RE} between track and earth is

$$U_{RE,x=0} = Z_0 \cdot I_{trc}/2 \cdot \left(1 - e^{-\alpha L}\right) \quad . \tag{6.17}$$

The earth current I_E midway between the load and substation is

$$I_{E,x=L/2} = I_{trc} - I_R = I_{trc}\left(1 - e^{-\alpha L/2}\right) \quad . \tag{6.18}$$

Example 6.3: For a single-track DC railway line, assuming the conductance per unit length to be 2; 1 and 0,1 S/km, determine the current through earth midway between the traction vehicle and the substation for $L = 5\,\text{km}$ and $L = 10\,\text{km}$. The rails are type UIC 60. In addition, the track-to-earth voltage should be calculated at the traction vehicle location if the traction current drawn is 1 000 A. From Table 5.6 $R' = R'_R = 0,015\,\Omega/\text{km}$ is obtained for a single track with UIC 60 rails. This leads to the results presented in Table 6.8. The example demonstrates the importance of a low conductance to limit the current through earth. The potentials will not cause any risks since they are well below the touch voltage permissible for 300 s, which is 150 V.

The graphs shown in Figure 6.12 are obtained by calculation of the entire range of values of the track current and the *track-to-earth voltage* between the substation and the point where the energy is consumed and these values plotted as a function of the distance. For practical applications, the effect of the conductance on the effective resistance of the track is of significance. This resistance value, which is also termed the *equivalent track resistance* R_{Req}, can be defined as

$$R_{Req} = Z_0\left(1 - e^{-\alpha L}\right) \quad . \tag{6.19}$$

For high values of αL, the equivalent track resistance approaches the value of the surge impedance Z_0. In practice, this applies to substation-load distances between 13 and 15 km if the conductance per unit length is 2 S/km. However, if the conductance is as low as 0,1 S/km, the corresponding distance reaches 65 to 70 km.

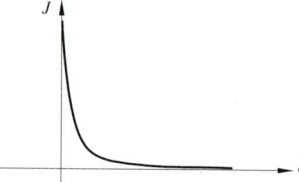

Figure 6.13: Graph of the current density J in the earth plotted as a function of the depth δ.

6.4.3.3 Track-to-earth circuit of AC systems

If the soil is assumed to be homogeneous below a *single-phase AC railway line*, the current density in the soil will decrease exponentially as a function of the depth. The penetration depth δ_E as defined by equation (5.15) is the equivalent depth of the current through earth used for calculation of inductance. The electromagnetic coupling of the current within an area close to the line causes the longitidinal resistance of earth to achieve a value not equal to zero and proportional to the frequency, as described by Equation (5.11). The penetration depth enables the effective inductance and *resistance of the earth* in the track-to-earth circuit to be determined. Figure 6.13 shows the current density in homogeneous subsoil below a track. This model is based on reference [6.17] and has been discussed in greater depth in [6.18]. However, as the earth is composed of many layers of differing properties and thickness, the conclusions drawn from this model merely provide a basis for estimating the order of magnitude of the penetration depth. Measurements described in [6.19] demonstrated that voltages were induced in a conductor loop located in a mine 400 m below an AC railway line. The measured voltages depended on the current flowing in the contact line.

The *longitudinal profile of rail potentials and currents* in a single-phase AC railway track is shown in Figure 6.14. Here, too, single end feeding and a load at a distance L from the feeding point is assumed to obtain a simplified model. If the track on both sides of the load exceeds 5 km, which is a normal situation, the curves shown are applicable for the currents flowing in the track and in earth and for the transition of currents between track and earth. The following basic statements and conclusions can be drawn from Figure 6.14:

– the traction current I_{trc} flows to the track at the location of the traction vehicle
– the major part of this current flows towards the substation via the track. The remainder

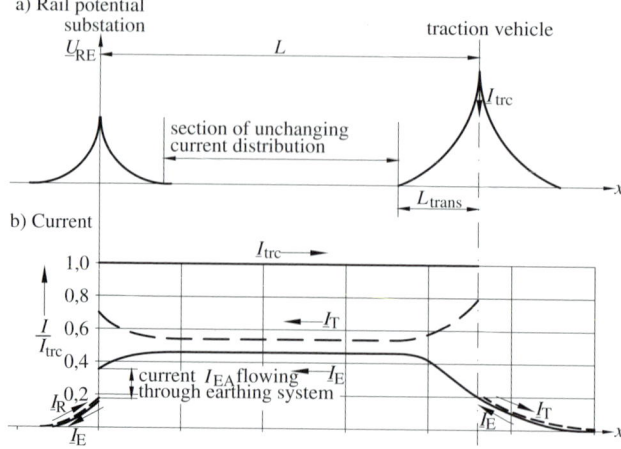

Figure 6.14: Rail potential (a) and currents (b) of a single-phase AC railway with single-ended feed to the traction vehicle by one substation. I_R current through track I_E current through earth I_{EA} current through substation earthing I_{trc} traction current

Table 6.9: Earth resistances R_{ESS} of substation earthing systems and portions I_{EA} of the return currents flowing to the substation via the earthing system, in relation to the total traction current.

Substation	Substation locations	R_{ESS} Ω	I_{EA}/I_{trc} %
Dresden–Stetzsch	along double-track line	0,12	21
Riesa	large railway station area	0,23	9
Chemnitz	along double-track line	0,07	51
Gößnitz	large railway station area	0,10	15

- flows through the track in the opposite direction, i. e. to the right-hand side of the load in Figure 6.14
- currents flow from the rails to earth on both sides of the load location. This section in which the rail-to-earth currents are observed is called the *transition section* with the *transition length* L_{trans}
- in the section close to the substation, a portion of the return current in the earth flows back into the track, whereby a certain fraction of the earth current flows back to the substation through the substation earthing system. The magnitude of this fraction depends mainly on the earth resistance of the substation foundations. Table 6.9 contains examples for earth resistance of substation earth electrodes and the associated earth currents.
- a *rail potential* occurs within the transition ranges near the substation and near the load location

Inductive coupling of two conductive loops is effective in the case of *earth return current* of AC traction systems. The current flowing through earth is determined mainly by the inductive coupling between the conductive loops and only to a minor extent by the *galvanic coupling*, which is a function of the *conductance per unit length*. As a result of the *inductive coupling*, there will be a section of *unchanging current distribution* where the transition processes have already decayed. In this section, no return currents will flow from rails to earth or vice verse and no rail potential occurs. The *impedances per unit length* given in Clause 5.1.2.3 apply to this clause.

Applying the model used in [6.20] and assuming an infinitely long, i. e. longer than 5 km, electric railway line, according to Figure 6.14, the current flowing through earth to the left of the feed point and to the right-hand side of the traction vehicle location is

$$\underline{I}_E = -\underline{I}_{trc}\left(1 - \underline{k}\right)\left[1 - 1/(2\,\underline{\gamma}L)\left(e^{-\underline{\gamma}(L-x)} + e^{-\underline{\gamma}x}\right)\right] \quad . \tag{6.20}$$

Equation (6.20) describes the earth current, comprising two components: the first is a constant component observed in the section of balanced current distribution. The second, variable component describes the transition current depending on the distance of the relative points. Correspondingly, the *rail potential* or *track-to-earth voltage* is (Figure 6.14 a))

$$\underline{U}_{RE} = \underline{I}_{trc}\left(1 - \underline{k}\right)\left(e^{-\underline{\gamma}(L-x)} - e^{-\underline{\gamma}x}\right)\underline{Z}_0/2 \quad , \tag{6.21}$$

with
$\underline{k}$ coupling factor,
$\underline{\gamma}$ *propagation constant* of the track-earth circuit,

$\underline{Z}_0$ *surge impedance* of the track-earth circuit and
L distance between substation and load.

These quantities are determined by the following relationships:

$$\underline{k} = \underline{Z}'_{KE}/\underline{Z}'_{RE}   \quad, \tag{6.22}$$

where $\underline{Z}'_{KE} = R' + \mathrm{j}f\mu \cdot \ln(\delta_E/h_c)$ is the *coupling impedance per unit length* acting between the overhead contact line-to-earth circuit and the track-to-earth circuit according to (5.18) and (5.19) and $\underline{Z}'_{RE} = R'_R + R'_E + \mathrm{j} \cdot f \cdot \mu \ln(\delta_E/r_{eq})$ is the *self-impedance per unit length* of the track-to-earth circuit in analogy to equations (5.12) to (5.14). There
 - $R'_E = \pi^2 \cdot 10^{-4} \cdot f$ Ω/km (5.11) is the earth resistance per unit length along the line,
 - $\mu = \mu_0 \cdot \mu_r = 4 \cdot \pi \cdot 10^{-4} \cdot 1{,}0$ Vs/(A·km) (5.10) is the soil permeability,
 - $\delta_E = 160 \cdot \sqrt{\rho_E}$ (m) for 16,7 Hz and $90\sqrt{\rho_E}$ (m) für 50 Hz is the penetration depth of currents into earth (5.15), with ρ_E specific soil resistivity in Ω·m,
 - h_m is the mean height of contact line above track in m,
 - r_{eq} is the equivalent rail radius in m according to Table 5.3,
 - $R'_R = R'_{rail}/n_R$ is the track resistance in Ω/m. Values for R'_{rail} are given in Table 5.3. $R'_R = 0{,}030/2 = 0{,}015$ Ω/m for UIC 60 and
 - n_R is the number of parallel rails.
The *coupling factor* $\underline{k}$ is calculated by the equation

$$\underline{k} = \frac{R'_E + \mathrm{j}\mu f \ln(\delta_E/h_c)}{R'_E + R'_R + \mathrm{j}\mu f \ln(\delta_E/r_{eq})}   \quad. \tag{6.23}$$

The propagation constant γ is

$$\underline{\gamma} = \alpha + \mathrm{j}\beta = \sqrt{\underline{Z}'_{RE} \cdot G'_{RE}}   \quad. \tag{6.24}$$

The conductance per unit length G'_{RE} of the track can be assumed to be a purely ohmic property, α is the *attenuation constant* and β the *phase constant*.
Lastly, the *surge impedance* of the track-earth circuit is

$$\underline{Z}_0 = \sqrt{\underline{Z}'_{RE}/G'_{RE}}   \quad. \tag{6.25}$$

The *transition length* L_{trans} is defined as the distance over which the transition processes and values have decayed to approximately 5 % of their maximum value. This is the case for $e^{-\alpha L_{trans}} \leq 0{,}05$ or $\alpha L_{trans} = -\ln(0{,}05) \approx 3{,}0$. Therefore,

$$L_{trans} = 3/\alpha   \quad. \tag{6.26}$$

The maximum rail potentials occur at the substation ($x = 0$) or at the traction vehicle $x = L$, where according to Equation (6.21)

$$\underline{U}_{RE} = \underline{I}_{trc}(1 - \underline{k})\left(1 - e^{-\gamma L}\right)\underline{Z}_0/2   \quad. \tag{6.27}$$

For sufficiently long sections $e^{-\gamma L}$ can be neglected relative to 1 and it is obtained

$$\underline{U}_{RE} = \underline{I}_{trc} \cdot (1 - \underline{k})\underline{Z}_0/2   \quad. \tag{6.28}$$

Table 6.10: Transition length and rail potential depending on the conductance per unit length, frequency 50 Hz and soil resistivity 290 Ωm.

Conductance G'_{RE} S/km	Phase constant γ 1/km	Surge impedance Z_0 Ω	Transition length L_{trans} km	Rail potential U_{RE} V/kA	Current through earth for $I_{trc} = 1$ kA A
0,5	$0,452 + j\,0,361$	$0,904 + j\,0,721$	6,6	279	480
1,0	$0,639 + j\,0,510$	$0,639 + j\,0,510$	4,7	197	480
2,0	$0,904 + j\,0,721$	$0,452 + j\,0,360$	3,3	140	480
4,0	$1,278 + j\,1,020$	$0,320 + j\,0,255$	2,3	100	480
8,0	$1,807 + j\,1,443$	$0,226 + j\,0,180$	1,7	87	480

Equation (6.28) assumes that the line is terminated with an impedance equal to the surge impedance Z_0. However, if the line is terminated by $\underline{Z}_A$, then Equation (6.28) is transformed to

$$\underline{U}_{RE} = \underline{I}_{trc} \frac{\underline{Z}_A\,\underline{Z}_0}{\underline{Z}_A + \underline{Z}_0}\,(1 - \underline{k}) \quad . \tag{6.29}$$

At a substation with an earthing system of earth resistance $\underline{Z}_E$, the voltage is calculated by

$$\underline{U}_{RE} = \underline{I}_{trc} \frac{\underline{Z}_A \cdot \underline{Z}_0 \cdot \underline{Z}_E}{\underline{Z}_A\,\underline{Z}_0 + \underline{Z}_E\,(\underline{Z}_A + \underline{Z}_0)}\,(1 - \underline{k}) \quad , \tag{6.30}$$

assuming the line is terminated by the impedance $\underline{Z}_A$.

Example 6.4: How do different conductance data affect the earth currents and the rail potential of a single-track railway line?
The line is equipped with a contact line type Re 200, rails UIC 60 and operated with 50 Hz. The considered conductances per unit length should be 0.5, 1, 2, 4 and 8 S/km, the traction current is $I_{trc} = 1$ kA. The tracks are terminated by the surge impedance Z_0. The line data are:
- average contact line height $h_m = 6,5$ m
- two rails UIC 60 (Table 5.6) $R'_R \sim 0,030/2 = 15$ mΩ/km
- equivalent rail radius $49,46 \sim 50$ mm according to Table 5.3
- longitudinal earth resistance, $R'_E = \pi^2 \cdot 50 \cdot 10^{-4} = 0,0493$ Ω/km
- mean soil resistivity 290 Ωm and 27 Ωm

The penetration depth is, therefore, $\delta_E = 90\sqrt{290} \approx 1\,530$ m for $\rho_E = 290$ Ωm and ≈ 470 m for $\rho_E = 27$ Ωm, respectively. The coupling factor k is obtained for $\rho_E = 290$ Ωm from Equation (6.22):

$$\underline{k} = \frac{0,0493 + j \cdot 10^{-3} \cdot 0,4 \cdot \pi \cdot 50 \ln(1\,530/6,5)}{0,015 + 0,0493 + j \cdot 10^{-3} \cdot 0,4 \cdot \pi \cdot 50 \ln(1\,530/0,05)} = \frac{0,0493 + j\,0,343}{0,0643 + j\,0,649} =$$

$$= 0,526 - j\,0,130$$

The absolute value of k is 0,52. For $\rho_E = 27$ Ωm, $|k| = 0,46$ is obtained. The current flowing through earth in the section with unchanging current distribution, see Figure 6.14, can be calculated using Equation (6.20): $I_E = I_{trc}\,(1 - \underline{k}) = I_{trc}\,(1 - 0,52) \sim 0,48\,I_{trc}$ for $\rho_E = 290$ Ωm and $\sim 0,54\,I_{trc}$ for 27 Ωm. The current through earth increases with dropping soil resistivity. These ratios do not depend on the conductance.
The self-impedance per unit length of the track-to-earth circuit is given by the denominator of Equation (6.23). It resolves to be for $\rho = 290$ Ωm

$$\underline{Z}'_{RE} = (0,064 + j\,0,649)\,\Omega/\text{km} \quad .$$

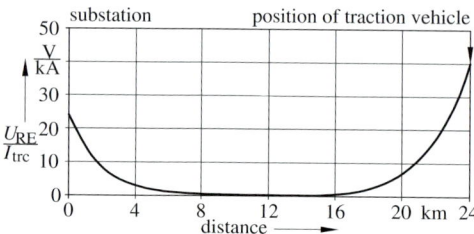

Figure 6.15: Rail potential U_{RE} for 1 kA I_{trc} along an AC 25 kV 50 Hz line: no return conductors, effective conductance per unit length of 2 S/km and a substation earth resistance of 0,2 Ω.

Since G'_{RE} is purely ohmic, equations (6.24) and (6.25) can be transposed to

$$\underline{\gamma} = \sqrt{G'_{RE}} \cdot \sqrt{\underline{Z}'_{RE}} \quad \text{and} \quad \underline{Z}_0 = \sqrt{\underline{Z}'_{RE}} / \sqrt{G'_{RE}} \quad .$$

$$\sqrt{\underline{Z}'_{RE}} = \sqrt{0,064 + j0,649} = \sqrt{0,652 \cdot e^{j85}} = 0,808 \cdot e^{j\,42,5°} = 0,595 + j0,546 \quad .$$

For $G'_{RE} = 1$ S/km Equation (6.25) yields $\underline{Z}_0 = 0,595 + j0,546\,\Omega$ and (6.24) $\underline{\gamma} = 0,595 + j0,546\,\text{km}^{-1}$. γ and Z_0 are presented in Table 6.10 for various values of conductance G'_{RE}.
Assuming $G'_{RE} = 1$ S/km and applying the equations (6.24) and (6.25) there is

$$\underline{\gamma} = \underline{Z}_0 = \sqrt{\underline{Z}'_{RE}} = \sqrt{0,064 + j0,649} = \sqrt{0,652 \cdot e^{j84,4°}} = 0,807 \cdot e^{j42,2°} = 0,593 + j0,543 \quad .$$

The values $\underline{\gamma}$ and $\underline{Z}_0$ are presented in Table 6.10 depending on the conductance G'_{RE}. The transition length L_{trans} follows from Equation (6.26) with α real part of γ and varies between 7,1 and 1,8 km. The rail potential $\underline{U}_{RE}$ can be calculated using Equation (6.21) at $x = 0$ or $x = L$. Since $e^0 = 1$ and $e^{-\alpha L}$ is very low for $L \geq 10$ km, Equation (6.21) results in

$$\underline{U}_{RE} = \underline{I}_{trc}\,(1 - \underline{k})\,\underline{Z}_0/2.$$

Assuming $G'_{RE} = 1$ S/km there is

$$\underline{U}_{RE}/\underline{I}_{trc} = (1 - 0,526 + j0,130)\,(0,593 + j0,543)\,/2 = 0,106 + j0,167\,\text{V/A}$$

and the absolute value $|\underline{U}_{RE}/\underline{I}_{trc}| = 0,198$ V/A. The absolute value of $\underline{U}_{RE}$ is given in Table 6.10 for the current $I_{trc} = 1$ kA, depending on the conductance G'_{RE}.

Example 6.5: How will the rail potential change at the substation, if the substation earthing system has an earth resistance $\underline{Z}_E = R_{ESS} = 0,1\,\Omega$?
For $Z_A = Z_0$ and $G'_{RE} = 1,0$ S/km, Equation (6.30) yields for

$$\underline{U}_{RE} = \frac{(0,593 + j \cdot 0,543) \cdot 0,1}{(0,593 + j \cdot 0,543) + 2 \cdot 0,1}\,(1 - 0,526 + j0,130) = (-0,006 + j0,070)\,\text{V/A}$$

$$|\underline{U}_{RE}| = 0,070\,\text{V/kA} = 70\,\text{V/A}$$

With a current of $I_{trc} = 1\,000$ A the rail potential will drop from 197 V to 40 V if the substation earth resistance R_{ESS} of 0,1 Ω is considered.

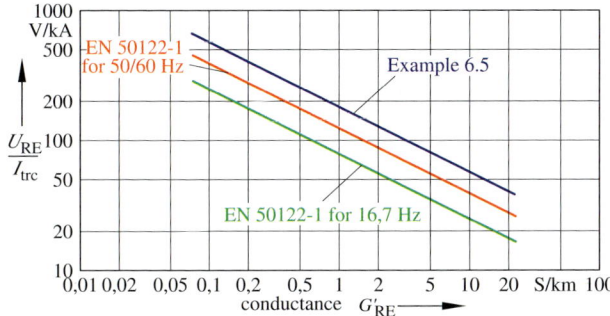

Figure 6.16: Guide values for the specific rail U_{RE}/I_{trc} for AC double track sections according to EN 50 122-1:1997.

6.4.4   Rail potentials

6.4.4.1   AC traction systems

The *rail potential* or the *rail-to-earth voltage* U_{RE} is defined as the voltage between the track and earth during operations and during fault conditions such as short circuits. As can be concluded from the equations and the example in Clause 6.4.3.3, the value of U_{RE} is determined by the traction current, the effective conductance per unit length, the rail resistance, the soil resistivity, the frequency and the geometry of the conductor arrangement.

Equation (6.21) shows that the peak values of the rail potential are located at the substation and at the load. Along the tracks, the rail potential drops according to the transition length to practically zero in the section midway between the substation and the load.

In Figure 6.15 the rail potential for AC lines is shown along the line between the substation and a train at a distance of 24 km. The voltage U_{RE} is related to a traction current I_{trc} of 1 kA. In Figure 6.16 the rail potential for AC lines related to a 1 kA traction current is shown for varying conductance and frequency according to EN 50 122-1:1997, Annex C. These values apply to soil resistivity between 40 Ωm and 200 Ωm and double tracks. The conductance is indicated as

- 0,4 to 1,7 S/km for at-grade sections with concrete sleepers,
- 1,7 to 7,0 S/km for at-grade sections with wood or steel sleepers and
- 7,0 to 15,0 S/km for tunnel sections with electrical connections between tracks and tunnel reinforcement.

The results obtained from the example in Clause 6.4.3.3 are higher. The voltage U_{PE} measurable between a point P on the Earth's surface and reference earth E decreases with increasing distance a from the track as shown in Figure 6.17. The voltage U_{RP} is measured between the track and point P. The ratio $U_{RP}/U_{RE} = 1 - U_{PE}/U_{RE}$ is also plotted in Figure 6.17. This

Table 6.11: Guiding values for the potential U_{PE}/U_{RE} for AC double track sections at right angle according to EN 50 122-1, Annex C.

U_{PE} voltage between measuring point and earth
U_{RP} voltage between rails and measuring point
U_{RE} rail potential
a distance from outer rail

Distance a m	Ratio U_{PE}/U_{RE} %	Ratio U_{RP}/U_{RE} %
1	70	30
2	50	50
5	30	70
10	20	80
20	10	90
50	5	95
100	0	100

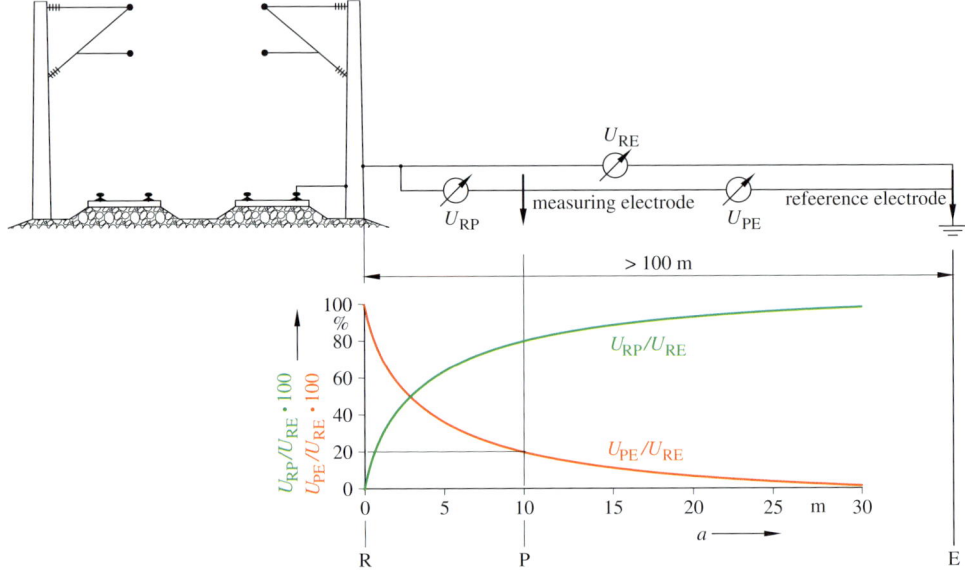

Figure 6.17: Guideline values for the characteristic curves of the voltage U_{PE} between a point P and earth reference potential and of the voltage U_{TP} between the track and a point P on the Earth's surface, at right angles to the rails. The values are related to U_{RE} and the specific soil resistivity is $\rho_E \approx 100\,\Omega m$. Examples of practical relevance for accessible voltage:
– U_{RP}/U_{RE} at a_{1m}: touch voltage between the rail and a point on the Earth's surface at a distance of 1 m.
– U_{RP}/U_{RE} at $a_{4,5m}$: accessible voltage between rails and a point on the Earth's surface outside the overhead contact line zone, at a distance of 4,5 m.

value increases with distance from the track and reaches its maximum value, which is equal to U_{RE}, at the reference earth. In Table 6.11, guideline values are given for the ratio U_{PE}/U_{RE} and U_{RP}/U_{RE} of a double track AC line according to EN 50 122-1: Annex C.

6.4.4.2   DC traction systems

The rail potential can be calculated using equation (6.11). The parameters affecting this potential are the conductance G'_{RE} and the number of parallel rails and their cross section and the distance between the substation and the traction vehicle. The potential decreases along the track and at right angles as for AC systems. Thus, basically Figure 6.17 is valid for DC systems as well. More precise calculations of rail potentials can be carried out using appropriate computer programs [6.1, 6.3, 6.21].

6.4.4.3   Rail potential under operational conditions

Rail potentials under operational conditions are determined by the traction vehicle currents I_{trc} with values up to 1 500 A in AC installations and 5 000 A in DC installations. The highest rail potential will occur where two trains meet on a double-track line, each train drawing its maximum traction current. Where two trains accelerating in two directions meet, the resulting rail potential can last one minute or even longer. In such cases, the permissible voltage will be $U_{RE} = 65$ V for AC systems and $U_{RE} = 150$ V for DC systems, as stated in Table 6.1 for

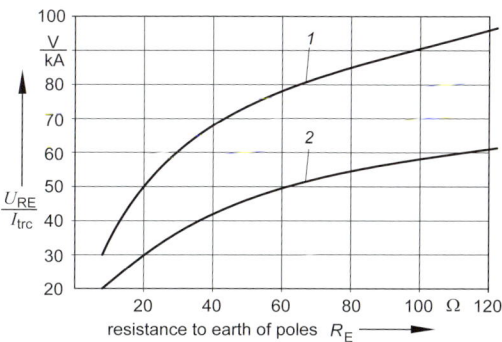

Figure 6.18: Specific rail potential U_{RE}/I_{trc} relative to the traction current at the load location on a double-track AC 16,7 Hz line with a track conductance per unit length of 0,1 S/km per track, plotted as a function of the resistance to earth of the pole R_E, assuming 16 poles per km of railway line.
1 without return conductor
2 with return conductor type 243-AL1

a duration of up to 300 s. As a consequence, the rail potential must not exceed $U_{RE} = 65/k_a$ in V where k_a is defined in Clause 6.3.5, Equation (6.4). For $k_a = 0,5$, the permissible rail potential would be 130 V for AC systems. Because of the insulation of running rails against earth for DC systems, the whole rail potential can act as touch voltage, see EN 50 122-1, Annex C.2. From Figure 6.16, specific rail potentials of 125 V/kA for 50 Hz and 79 V/kA for 16,7 Hz are obtained with $G'_{RE} = 1$ S/km. Then the permissible currents would be 1 040 A for 50 Hz and 1 710 A for 16,7 Hz systems.

If the conductance between tracks and earth decreases due to improved insulation between the tracks and earth, then the currents necessary to reach the permissible accessible voltage will also drop. A high conductance is favourable with regard to touch voltages. However, the hazard of stray current corrosion in the case of DC lines would increase.

The effective conductance is established by the conductance of the tracks and the conductivity of the poles and return conductors. Return conductors considerably reduce the rail potential as shown in Figure 6.18. In each case, the rail potentials need to be checked, taking all relevant data into account guaranteeing the safety of persons. Examples are given in [6.3].

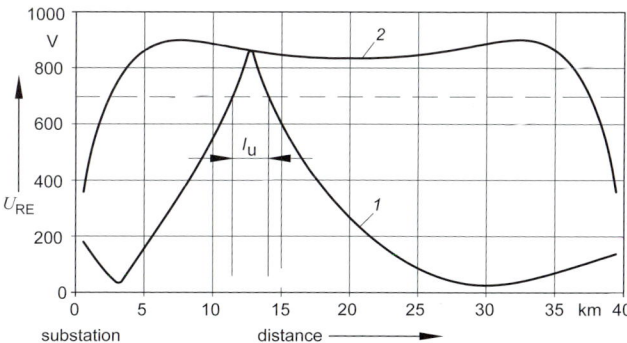

Figure 6.19: Rail potential U_{RE} along a 40 km double track AC 15 kV line section between two substations supplied from both sides. Short-circuit current 35 kA; leakance of track per unit length 0,01 S/km; Resistance to earth of poles 100 Ω; line section l_u, where U_{RE} exceeds 700 V.
1 voltage along the line section for a short circuit at 12,5 km
2 U_{RE} at location of the short circuit, when the location moves from one substation to the next.

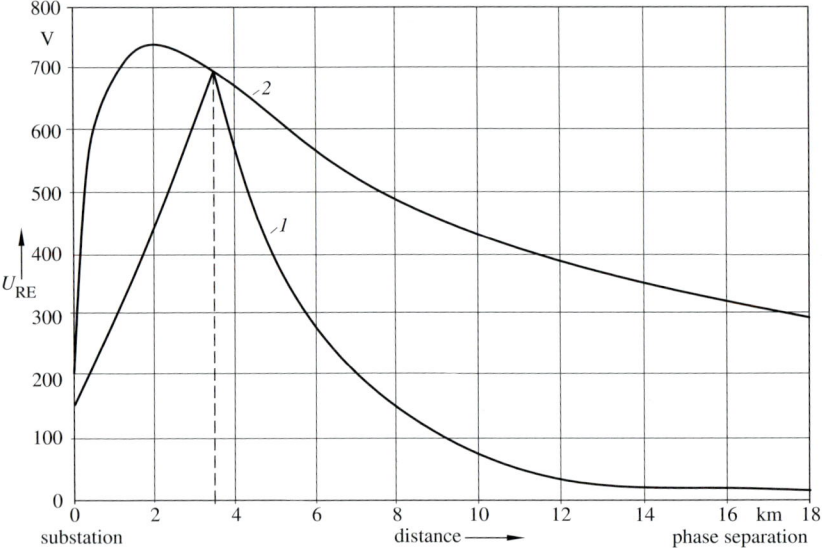

Figure 6.20: Rail potential U_{RE} along an 18 km double-track AC 25 kV line section fed from one end. Track leakance 0,15 S/km; rail leakance 0,01 S/km; pole earthing resistance 100 Ω; U_{RE} does not exceed 700 V.

1 voltage U_{RE} along the line section for a short-circuit current of 5,5 kA at km 3,5
2 voltage U_{RE} at the short-circuit location moving from the substation to the phase separation section

6.4.4.4 Track-to-earth voltage in case of short circuits

The *short-circuit current* is determined mainly by the impedances of the contact line and the substation transformers. The related rail potential depends on the short-circuit current, the track conductance per unit length and the pole earthing resistances.

In Figure 6.19 the rail potential U_{RE} is shown for a 35 kA short-circuit current on a 40 km long line section between two substations of an AC 15 kV 16,7 Hz line:

- (*1*) along the line between the two substations when the short-circuit occurs 12,5 km from the substation
- (*2*) voltage at the short-circuit location, moving from one substation to the next

In Figure 6.20, the rail potential U_{RE} is shown for a 5,5 kA short-circuit current acting on a 18 km long AC 25 kV 50 Hz line section between the substation and the phase separation section.

The relatively low rail potentials at the substation, despite the high short-circuit currents occurring in the vicinity of the substation, are due to the low resistance of the substation earthing system, being assumed as 0,2 Ω in the example.

In reference [6.3] the hazards to be expected in the track area in the case of a short circuit are discussed and illustrated in Figure 6.19. If the short-circuit duration is 0,07 s, a value not normally exceeded in AC operations, ensures the probability of an accident due to electric shock is zero, irrespective of whether return conductors are installed or not. For a short-circuit duration of 0,1 s and assuming $G'_{RE} = 0,1$ S/km and $R_E = 200$ Ω, the probability of electric shock is $1,3 \cdot 10^{-5}$ for people working in the contact line zone for four hours a day, twenty days per year.

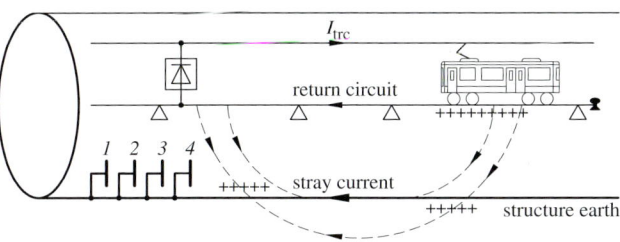

Figure 6.21: Return circuit and earthing of DC railways.
1 high- and medium-voltage protective earthing
2 low-voltage protective earthing
3 earthing of telecommunications and signalling systems
4 lightning protection earthing
+++ possible stray current corrosion areas
△ insulated arrangement of rails

6.5 Design of DC traction systems

6.5.1 Basic rules for the design of the return circuit

Protective provisions against *stray currents* significantly determine the *design of the traction return circuit* and *earthing installations* of DC supplied railways. Using the information given in Clauses 6.3 and 6.4, this clause deals with the configurations of the return circuit, its planning and implementation. The strict separation between the return circuit and structure earth complies with the stipulations of standards IEC 62 128-2 and EN 50 122-2. These design principles described in [6.8] have proven essential for mass transit systems.

Power supply for DC railways includes the three-phase AC feeding network on the medium- or high-voltage side, the traction power supply system and the auxiliary low-voltage supply for technical equipment and buildings. Various configurations for the traction return circuit and earthing and bonding exist and are also suggested for new installations. They cover requirements for both safety of persons, see EN 50 122-1 and IEC 61 936-1 and for *protection against* the effects of *stray currents*, see [6.8], EN 50 122-2, VDV 501 [6.4] Part 1 to 3 and EN 50 162. In addition, they must also ensure protection of electrical equipment and lightning protection, see VDV 525 [6.9]. Where provisions for safety of persons conflict with stray current protection, then safety must be given highest priority. Practical applications require coherent solutions that can be implemented into the overall configuration in a simple and economic manner.

The standards EN 50 122-1 and EN 50 122-2 deal with the addressed set of issues and contain stipulations for earthing of

– structures of buildings and tunnels,
– three-phase high-voltage power supply,
– DC traction power supply,
– signalling and telecommunications installations and
– low-voltage power supplies.

The standards are the basis for the *system configuration* described below. Figure 6.21 illustrates the main elements of the *return circuit* and earthing using the example of a DC system in a tunnel. The return currents flow through running rails and return cables to the feeding rectifiers. Running rails form part of the return circuit, however, due to varying track voltage and insulation along the line, currents from the running rails stray into the soil and flow through soil and metallic conductors in contact with the soil. Stray current corrosion occurs at the position of current transition from metallic conductors to an electrolyte. Figure 6.21 shows the possible *stray current corrosion areas* for the case where a vehicle is fed only from

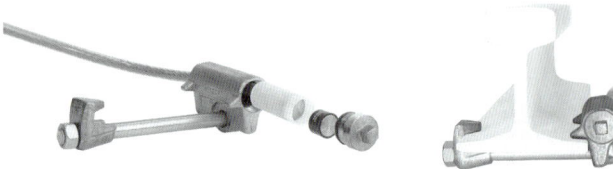

Figure 6.22: Example of a voltage limiting device VLD-F, called also voltage fuse (Photo: Siemens AG).

one substation. The degree of metal erosion depends upon the current, the type of metal and the duration of exposure, see Clause 6.5.3.

The *structure earth*, also known as tunnel earth, see Figure 6.21, is not connected to the return circuit and serves as protective earth for all equipment components such as the three-phase high-voltage and medium-voltage installations, signalling and telecommunication installations.

6.5.2　Safety of persons

Both the *touch voltage* following faults in the three-phase feeding system and the *potentials on running rails* must not exceed the permitted values in accordance with EN 50 122-1 to guarantee human safety. During faults with earth contact in the three-phase AC system, the fault current flows to earth via the earthing system and causes a voltage between the earthing installations and earth. The resistance to earth then determines the voltage to earth and the touch voltage. The operating and short-circuit currents cause longitudinal voltage drops in the rails. In certain cases the rail potential, measured relative to earth, can reach the value of the voltage drop. It must not exceed the permissible touch voltage according to Table 6 of EN 50 122-1, see also Figure 6.5.

The rail potentials for the operating and short-circuit cases should be calculated during the design of DC railway systems. In many cases, full operation must also be guaranteed after the outage of a substation by supplying traction power from neighbouring substations through extended feeding sections. In this situation, the rail potentials limit the maximum distance between substations.

To guarantee safety, protective provisions against electric shock must be provided in the overhead contact line area of electric railway lines. Protection against indirect contact is relevant to issues concerning return current conduction through the running rails of the tracks. *Indirect contact* is the contact of persons with exposed conductive parts that are not normally live but could become live under fault conditions. To avoid hazardous voltage at conductive parts, in the case of electric faults, the preferred and generally recommended method for AC systems is to directly connect all conductive parts in the overhead line zone to the running rails. However, to prevent or at least counteract, in DC systems, stray current corrosion, conductive metal parts or installations that are not insulated from earth must not be directly connected to the return circuit, in particular the running rails.

Standards EN 50 122-1 and EN 50 122-2 are based on the fundamental principle that protective provisions against electric shock must be given higher priority than provisions against stray current corrosion. The standards specify that for DC systems the resistance between the return conductors and conductive installations not insulated against earth should be as high as possible. Therefore, *voltage limiting devices* (VLD) are to be installed between the equipment to be earthed and the return circuit, see EN 50 122-2, Clause 6.2 and Annex F. They form a short-circuit path to the return circuit after a threshold voltage is exceeded and limit potential

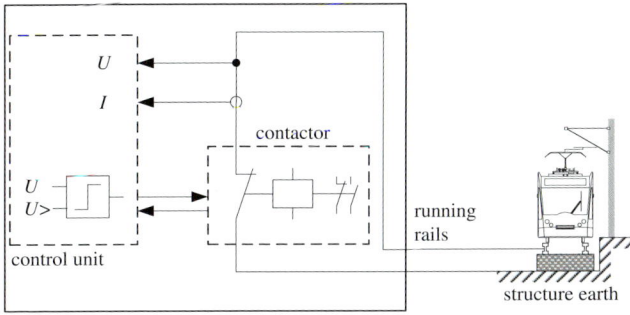

Figure 6.23: Principle circuit diagram of the voltage limiting device Sitras SCD-C.

differences that may arise in fault situations. These devices, also known as *voltage fuses*, are normally open connections between the conductive parts and the running rails. Alternatively, *electronic voltage limiters* can be used, as described in [6.22]. They take care of the safety of persons and the protection of equipment against overvoltages, using for example anti-parallel thyristors.

Depending on the requirements of the individual application, alternative designs of the voltage limiting devices are used:

– *voltage limiting devices* VLD-O type connect the rails and structural earth in railway stations if the rail potential exceeds the permissible value. After a short duration, typically 60 s, they open automatically under the condition to be able to switch off the currents flowing through it. If the device does not return to its idle state, methods need to be implemented for documenting the cause of the fault, which must be remedied immediately

– voltage-triggered permanent connections – so-called voltage fuses VLD-F types – establish a permanent conducting connection between the conductive parts and the return circuit in case of voltages between the return circuit and the conductive part being above a certain threshold value. A typical device is shown in Figure 6.22

– electronic voltage limiting devices using for example two anti-parallel thyristors. They are triggered if the specified voltage level is exceeded. The thyristors isolate the circuit when the current has decreased to zero

Voltage limiting devices VLD-O type are offered by Siemens Rail Electrification, for example, with the product series Sitras SCD for DC traction systems. They form an electrical connection between the return circuit, i. e. the running rails and the structure earth in case of undue high rail potentials. Automatic reopening avoids long-duration stray currents flowing through this connection. The devices protect a remaining impermissible touch voltage and fulfill the requirements of EN 50 122-1 and IEC 62 128-1.

Figure 6.23 shows the principle circuit diagram with connection to the rails. In normal state, the main contactor is open. The voltage between the return circuit and structure earth is monitored. When the preselected voltage is exceeded, the contactor closes. Sitras SCD reopens automatically after a minimum time period to ensure the current does not exceed a safety limit for the maximum current.

6.5.3 Stray current protection

6.5.3.1 Stray current corrosion process

The rails are installed on sleepers, which in turn are placed on ballast, the sub-ballast and sometimes an insulating layer and finally the earth. A concrete slab permanent way is an alternative to sleepers and ballast. A high rail-to-earth resistance results when a new track is laid with high quality insulating ballast, on exceedingly dry sandy soils or where the rails are specifically insulated from the sleepers. In many cases, however, the track-to-earth resistance is low, so that a part of the return current can flow through earth, whereby the soil acts like an electrolyte. Currents leaving metal such as the running rails or metal pipes and other underground metallic installations in the vicinity of DC traction railways can cause electrolytic corrosion known as *stray current corrosion.*

Every metal object in an electrolyte is subject to an osmotic pressure and a solution pressure, which are normally in equilibrium. If this equilibrium is disturbed by an electric current, e. g. due to currents passing from the rails into earth, *electro-chemical corrosion* takes place. In such cases, two parallel processes occur. Using steel as an example, these two concurrent processes are the:

- anodic reaction

$$Fe \rightarrow Fe^{++} + 2\,e \qquad\qquad \text{and the}$$

- cathodic reaction

$$1/2\,O_2 + H_2O + 2\,e \rightarrow 2\,OH^- \qquad \text{at pH} > 7\,, \text{and}$$

$$2\,H^+ + 2\,e \rightarrow H_2 \qquad\qquad \text{at pH} < 7\,,$$

where pH is the pH-value, with a range of 0 to 14. In the anodic reaction, an *anodic current* component I_{an} flows from the metal into the electrolyte. In the cathodic reaction, a *cathodic current* component I_{ka} flows from the electrolyte to the metal. When no external current is imposed, there is an equilibrium between I_{an} and I_{ka}.

If this equilibrium is disturbed by an externally imposed current, two cases can occur:

- $I_{tot} > 0$ i. e. increase of the anodic reaction, in which case there will be *stray current corrosion*
- $I_{tot} < 0$ i. e. increase of the cathodic reaction. This is the principle of *cathodic protection*

At the point where the current leaves the conductor, stray current corrosion occurs and metal dissipates. The mass m of *metal erosion* can be calculated according to *Faraday's first law* of electrolysis:

$$m' = C \int_{t_1}^{t_2} I(t)\,dt \quad . \tag{6.31}$$

C is the *electrochemical equivalent* of the metal and $I(t)$ is the current flowing in the time interval between t_1 and t_2. The metal masses eroded by a current of 1 A within one year is 9,1 kg iron, 33,4 kg lead or 10,4 kg copper. To be able to calculate the equipment dimensions to prevent this, it is necessary to know how high the proportion of the traction current flowing into earth is, as well as the resulting rail potential. Examples are given in Clause 6.4.3.2.

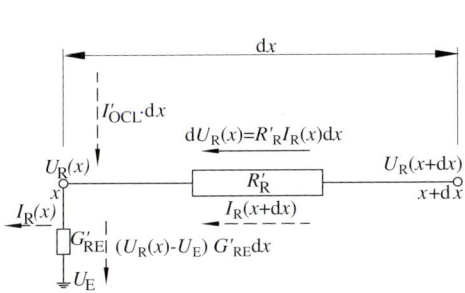

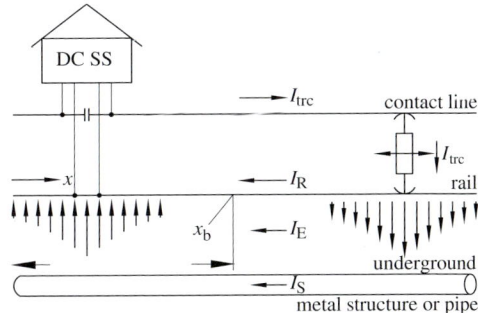

Figure 6.24: Potential drop along a track element of length dx.

Figure 6.25: Stray current corrosion areas. Contact line having positive polarity

6.5.3.2 Voltage and current relations in the return circuit

On the basis of the equivalent circuit in Figure 6.24, the current and voltage calculations for the return circuit can be deduced from the potential gradient along a track element, assuming a *distributed line load* I'_{OCL} according to (5.36)

$$dU_R(x)/dx = I_R(x) \cdot R'_R \tag{6.32}$$

and from *Kirchhoff's law* of currents

$$dI_R(x)/dx + I'_{OCL} = (U_R(x) - U_E) \cdot G'_{RE} \quad . \tag{6.33}$$

By inserting the *propagation constant* α according to (6.15) the following equation results for the current $I_R(x)$ flowing in the track:

$$I_R(x) = A \cdot \exp[-\alpha (L - x)] + B \cdot \exp[\alpha (L - x)] \quad , \tag{6.34}$$

where L is the length of the considered line section, x starts at the feeding substation, see Figure 6.25. With this equation and the related boundary conditions that are the voltages and currents at the feeding point and at the load point, it is obtained

$$I_R(x) = \frac{I'_{OCL} \cdot L}{\sinh(\alpha L)} \cdot \sinh[\alpha (L - x)] \quad . \tag{6.35}$$

The current $I_E(x)$ flowing through earth at a distance x from the substation can be obtained from

$$I_E(x) = I'_{OCL} \cdot (L - x) - I_R(x)$$

or

$$I_E(x) = I'_{OCL} \cdot L \cdot \left(\frac{L - x}{L} - \frac{\sinh[\alpha (L - x)]}{\sinh(\alpha L)} \right) \quad . \tag{6.36}$$

For comparison, the current resulting from a single load I_{trc} at the point L in a feed section is

$$I_E(x) = I_{trc} \left(1 - \frac{\cosh[\alpha (L/2 - x)]}{\cosh(\alpha L/2)} \right) \quad . \tag{6.37}$$

Furthermore, the *rail potential* or *track-to-earth voltage*, assuming a uniformly distributed load I'_{OCL} is given by

$$U_{RE}(x) = U_R(x) - U_E = \frac{I'_{OCL}}{G'_{RE}} \left(1 - \alpha L \frac{\cosh[\alpha(L-x)]}{\sinh(\alpha L)}\right) \quad . \tag{6.38}$$

For a single load I_{trc} at point L in a feed section, the rail potential is

$$U_{RE}(x) = \frac{I_{trc} \cdot \alpha}{G'_{RE}} \cdot \frac{\sinh[\alpha(L/2 - x)]}{\cosh(\alpha L/2)} \quad . \tag{6.39}$$

The latter equation corresponds to Equation (6.13).

For practical applications, it is important to distinguish between areas liable to stray current corrosion and areas where there is no such danger. As shown in Figure 6.25, the boundary between the area where current flows out of the track and where current flows from earth back into the track is at the point x_B. This point is also the boundary between sections with positive and with negative rail potentials. This is also the point where the largest stray current will occur within the section under consideration. By inserting $x = 0$ in Equation (6.38) it is obtained

$$U_{RE}(0) = I'_{OCL} \cdot L/G'_{RE} \cdot [1 - \alpha L \coth(\alpha L)] \quad .$$

The term $\alpha L \coth(\alpha L)$ is always greater than 1, i.e. U_{RE} is negative if the contact wire polarity is positive. The boundary between the *anodic area* and the *cathodic area* is termed the *boundary distance* x_B. At this point $U_{RE}(x_B) = 0$ and Equation (6.38) is then transformed to

$$\sinh(\alpha L) = \alpha L \cdot \cosh[\alpha(L - x_B)] \quad .$$

For small values of α, both sides of the equation can be developed to series and taking two terms only, yields

$$x_B \approx L\left(1 - \sqrt{3}/3\right) = 0,42 \cdot L \quad . \tag{6.40}$$

In real applications, however, the total load on a traction system comprises discrete, moving individual loads because of trains moving along the lines. Therefore, the boundary between the anodic and the cathodic areas varies during operation and x_B only gives an indication for the extension of both areas.

6.5.3.3 Protective provisions against stray current corrosion

The main objective of *protective provisions against* the effects of *stray currents* is to avoid the danger of corrosion on third-party and railway-owned installations. A distinction should be made between passive and active protective measures. *Passive protection* involves coating the relevant metal installations with an insulating material or a corrosion-resistant metal. Stray current protection can also be achieved by reducing the stray currents to an acceptable level, to avoid a reduction of the installations' service life (see EN 50 122-2, VDV 501, [6.6] and [6.7]). For this purpose, it can be necessary to
- reduce the distance between substations,
- reduce the length of the track return system by moving the track return connection away from the substation,

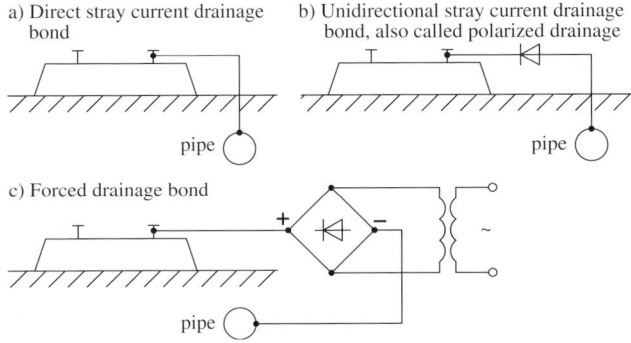

a) Direct stray current drainage bond

b) Unidirectional stray current drainage bond, also called polarized drainage

pipe

pipe

c) Forced drainage bond

pipe

Figure 6.26: Active protective measures against stray current corrosion.

- reduce the conductance per unit length between running rails and earth,
- reduce the resistance per unit length of the current return system, and
- install *return conductors*, i. e. conductors running in parallel to the track and connected to the rails at short intervals, being insulated versus earth.

As stray current corrosion directly corresponds with the duration of action, quick detection and repair of deficiencies is necessary. See EN 50 122-2, Chapter 10 and Annex B and Clause 6.5.5 for possible solutions.

Protective measures against stray currents are necessary to protect the running rails and rail fastenings. Additionally, steel reinforced track bed, railway-owned steel-reinforced tunnel and viaduct structures and third party installations are endangered and need specific provisions.

The traction return current must flow in the intended return circuit and not as stray current through earth and earthing installations. To achieve this, the longitudinal voltage drop in the running rails should be as low as possible. This can be achieved by using rails with a large cross section and cross-bonding of parallel rails and tracks. Since the longitudinal voltage drop depends on the distance between substations and the resistance of the return circuit, stray current protection also influences the required number of substations and, consequently project investment.

Another important factor is a high insulation level between the running rails and earth, when the running rails are used as part of the return circuit. This can be achieved by effectively draining water from the track substructure, proper ballast, wooden sleepers and insulating rail fixings. The return cables from the rails to the substation must be insulated.

The tracks of DC railways must be insulated from the tracks of other railways, especially AC railways. EN 50 122-3 and Clause 6.7. provide requirements for the operation of adjacent AC and DC railways.

In some cases, conventional active *cathodic protection* was adopted, using drainage bonds. According to EN 50 122-2, Chapter 8, the connection of any metallic structure to the return bus bar in a substation, even via polarized electric drainage bond will increase the overall stray current. Therefore, any metallic structure should be connected to the return busbar only with due consideration to the effect on the running rails and other structures. Generally, polarized electric drainage is applicable only when the structure to be protected is remote from other structures.

The cathodic protection principle is based on preventing *anodic reactions* on the metal to be protected. Figure 6.26 shows several cathodic protection methods:

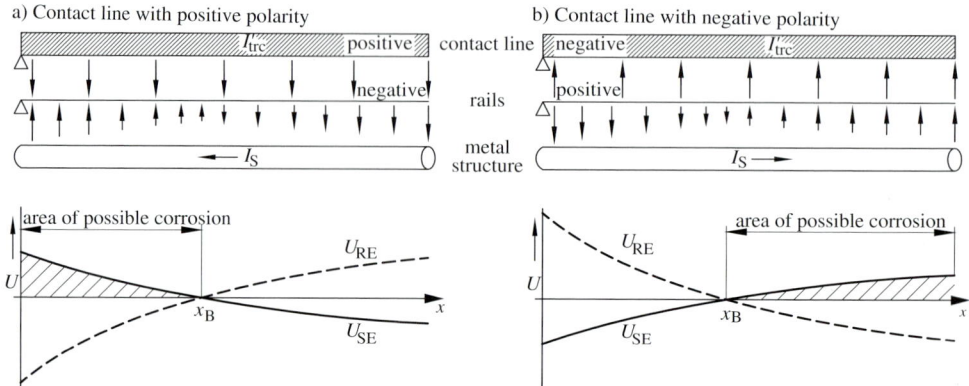

Figure 6.27: Effect of polarity on the location of the area with possible corrosion.
SS substation, U_{RE} rail potential, U_{SE} potential of pipeline or metal structures in contact with earth

- *direct drainage of stray currents* from earth-buried metallic installations to the tracks can be used when currents would also flow to the tracks without bonding. The direct stray current drainage as shown in Figure 6.26 a) may only be adopted where a sufficiently high negative rail potential is guaranteed and a reversal of the current flow in the bonding connection can be excluded
- *unidirectional stray current drainage*, also called *polarized drainage*, is required where reversal of current direction is expected. A rectifier or a diode hinders the stray current flow to the earth-buried metal installation to be protected (Figure 6.26 b))
- if the negative rail potential is not high enough, the mean diverted current could be so low that sufficient cathodic protection would not be achieved. As shown in Figure 6.26 c) a more effective *stray current drainage*, also called *soutirage*, can be achieved by inserting a direct current source

Associated issues are found in Standards EN 50 162 and EN 50 122-2.

6.5.3.4   Effect of the polarity

The *polarity* of the *contact line* will affect the position and size of the area where stray current corrosion can occur. The *polarity* of the *contact line* will affect the position and size of the area in which stray current corrosion can occur.

In Figure 6.27 the *track-to-earth voltages* U_{RE} and the voltages U_{SE} between metal structures and earth are shown for both positive and negative contact line voltages, assuming a continuously distributed load along the respective line section. In case of negative polarity of the contact line, according to (6.38), the cathodic area of the underground metal installation is at the far end of the line and, according to (6.40) it is $0,58 \cdot l$ long. This situation is described as *diffuse stray current corrosion*. If the contact line voltage is positive, the area where corrosion is likely to occur is at the substation end and it is smaller. In this case, the effect is termed *concentrated stray current corrosion*. The positive polarity of the contact line is the preferred and most common version.

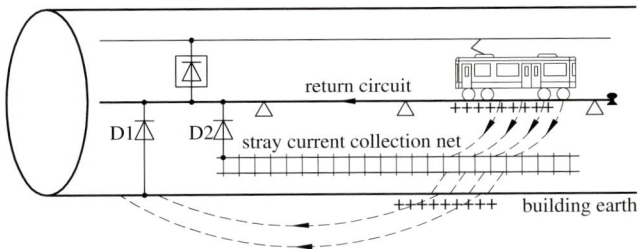

Figure 6.28: Increasing stray currents with collecting nets and drainage diodes in DC railway systems.

6.5.3.5 Stray current collecting nets

The system design according to EN 50 122-2 is based on insulated running rails and continuity of the earthing installations for protective earthing and protection against stray currents. However, designs also exist for polarized drainage or *stray current collecting nets* for tunnel and viaduct structures and for systems with steel-reinforced rigid superstructure to protect against stray current corrosion.

The *stray current drainage* via diode D1 in Figure 6.28 forms a metallic unilateral connection between the tunnel and running rails and avoids corrosion at metallic installations in the vicinity of the connection. However, this connection reduces the resistance of the stray current path between the running rails and structure increasing the overall amount of stray currents. The high DC currents flowing through such drainage installations evidently show this unintentional effect.

In a further alternative, additional reinforcement rods in the concrete layer under the running rails form a stray current collecting net that is connected to the running rails with the *stray current drainage diode* D2 (Figure 6.28). In reality, the stray current collecting net cannot be insulated from the structure satisfactorily, or only with immense effort. In case of faulty conductive connections during construction work, it is difficult to localize and then to correct them. The current densities and, in consequence, stray current corrosion will be increased by a stray current drainage diode D2.

A *stray current collecting net* without stray current drainage reduces the stray currents outside the railway facilities. However, no defined potential can be associated to this collecting net, so that induced voltages could also interfere with the track release systems. Furthermore, protective tripping must be ensured during a short circuit between the contact wire and the stray current collecting system to guarantee safety.

Extensive comparative calculations on the effects of stray current collecting nets for the European Standard EN 50 122-2 demonstrated that stray current drainage increases the rail potentials by a factor of up to two and the stray currents by a factor of four to ten. Measurements in a tunnel system confirmed the theoretical investigation [6.23].

The BTS Mass Transit System in Bangkok, built as a viaduct, was designed and constructed under Siemens leadership [6.24]. The stray currents for a design with through-connected structure reinforcement was compared with a design adopting a stray current collecting net with drainage diodes. The stray current drainage would have increased stray currents through the structure reinforcement by a factor of ten. Consequently the BTS System was installed with continuously connected structure reinforcement and without a separate stray current collecting net and without diodes.

The results of the investigations illustrate the technical problems involved and lead to the conclusion that drainage diodes cannot be recommended as a stray current protection measure.

6.5.4   Earthing and bonding with respect to stray currents

6.5.4.1   Basic recommendations

The design of the return circuit and earthing installations must satisfy both *safety* and *stray current protection* requirements [6.8]. In Figure 6.29 provisions are shown in an overall circuit diagram that satisfy these requirements. Particular requirements arise for the design of the superstructure [6.25]. The running rails determine the longitudinal resistance and the rail potential of the return circuit. For a low rail potential and consequently low stray currents, running rails with a large cross-section should be used. Welded rail connections are preferred and points should be longitudinally connected with rail connectors to ensure that the total resistance is not increased by more than 5 %.

As far as possible, all running rails should share the return currents. To achieve this there should be frequent cross bonding of rails and tracks. If necessary additional return conductors insulated from earth can be laid parallel to the running rails. Track release systems using one rail insulated are not appropriate, because only one rail per track conducts the return current. The low conductance per unit length of the running rails to earth is an important factor in limiting stray currents. It should not exceed 0,2 S/km in tunnel sections. Such values can only be achieved by appropriate high-insulating rail fixings. Furthermore, all installations and equipment connected to the rails, e. g. track release systems, point machines or point heating need to have a correspondingly high insulation level from earth. If connections of the return circuit to earth are necessary for safety of persons in case of undue high touch voltages, voltage limiting devices are required, see [6.26] and Clause 6.5.2. They provide a temporary or permanent electrical connection if a pre-set voltage limit is exceeded.

6.5.4.2   Railway-owned earthing systems

Building foundations, tunnel structures and the foundations of viaducts form the *railway-owned earthing system* generally known as *structure earth*, see Figure 6.29. The overall resistance to earth must be so low, that the permissible touch voltage is not exceeded during earth faults in the three-phase supply system.

Tunnels or viaducts form a through-connected earthing system where specific rules exist to reduce stray current corrosion effects, see Clause 6.5.4.7.

Usually, there is no through-connected earthing system for at-grade lines. Stations, substations, industrial buildings, buildings housing railway equipment such as interlockings and all contact line supports act as independent earthing systems.

6.5.4.3   Three-phase power supply

In the case of single-phase short circuits, the earth current in combination with the resistance to earth causes a voltage rise of the earthing system, which can cause impermissible touch voltages. A favourable way to achieve low values of voltage to earth in case of single-phase faults is to limit the short-circuit current by using a star-point resistor at the feeding transformer of the *three-phase power supply*. The star-point resistor needs to be rated according to the resistance to earth of the installations.

The *resistance to earth of tunnel* and viaduct systems is usually below 100 mΩ, so that the voltage to earth in case of faults is likely to be low and no further measures are required.

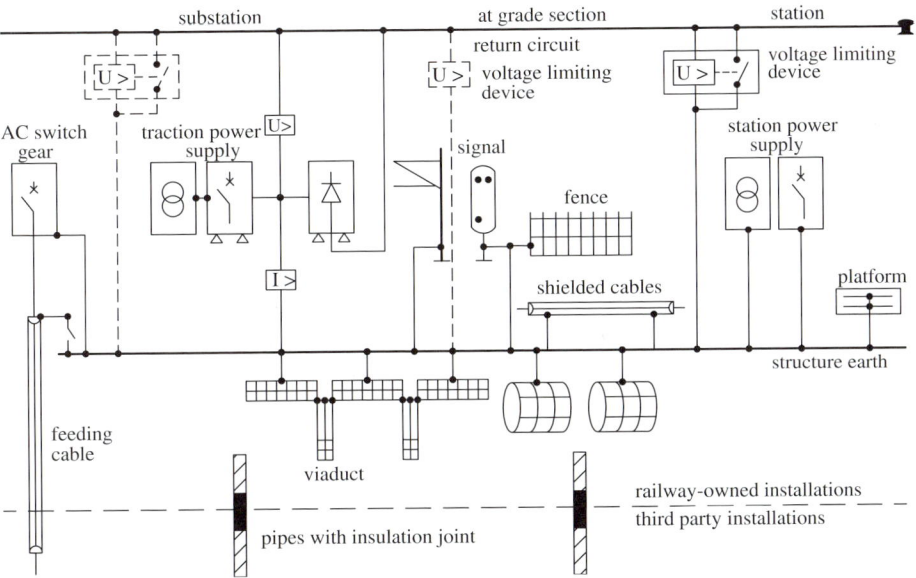

Figure 6.29: Schematic diagram of the return circuit and related measures for earthing and bonding for DC railway systems.

For at-grade structures with foundations of minor size, however, it can be necessary to add additional earth rods to comply with the *permissible touch voltage* requirements.

For *stray current protection* purposes the earthing installations of the three-phase supply network from outside should be separated from the DC railway earthing system. This separation can only be achieved if the cable sheaths or earth wires are not connected to the DC railway earthing installations. Since dangerous voltages can occur at the open end of sheaths, the endings of them should be protected against touch contact and labeled accordingly. However, during work on the medium-voltage installations, the open cable sheaths must be connected to the railway-structure earth to prevent impermissible touch voltages.

In many cases, the substations and passenger stations are fed by the railway-owned medium-voltage network. In this case, the medium-voltage cable sheaths are to be connected directly to the structure earth.

If the station supply is provided from the public low-voltage network, neither the neutral conductor nor the protective earth conductor are allowed to transfer the potential to third-party installations. In this case, the low-voltage protection must be ensured by other methods, e. g. *residual-current circuit breakers*, see EN 50 122-1, Clause 7.4.

6.5.4.4   Traction substations

Usually, the frames of the DC-switchgear in *DC traction substations*, e. g. rectifier and DC switchgear, are erected on insulating layers. Earth fault detection is provided by a low resistive connection versus structure earth at one point, as shown in Figure 6.29. A current detection in the earth connection and an optional voltage supervision between the return circuit and structure earth trip the medium-voltage transformer in case of insulation faults or unacceptable touch voltages. In combination with an optional voltage relay between the DC switchgear structure and the return circuit, the medium-voltage transformer is switched off in case of a frame fault. The frames of the medium-voltage AC switchgear and the rectifier transformers are directly connected to the structure earth.

The return cables from the running rails to the rectifier substation must be insulated from structure earth to avoid stray currents. To keep the *running rail potentials* as low as possible, the running rails should be sufficiently cross-bonded at the connection point of the return cables. In case of high rail potentials voltage limiting devices can be necessary, see Clause 6.5.2.

6.5.4.5 Passenger stations

During simultaneous starts of serveral trains, and also in case of the through connection of feeding sections due to a substation outage, the *rail potential* can reach or exceed the maximum permissible touch voltage. To ensure the *safety of persons*, *voltage limiting devices* [6.26] according to Clause 6.5.2 and EN 50 122-1, Annex F, are employed, particularly in stations with heavy duty suburban, regional and metro traffic.

These devices register the voltage between the return circuit and the structure earth at stations and connect both for a short period of time if the rail potential reaches excessively high values. The connection automatically reopens. A suggested value for this time is 60 s. The tripping of the *voltage limiting devices* should be registered or signalled to a control centre for monitoring and indication of unusually frequent switching operations prompting investigation of the cause.

If a rectifier substation is located in the passenger station building, then setting of the frame fault detection devices and the response time of the voltage limiting devices must be co-ordinated.

6.5.4.6 At-grade line sections

According to EN 50 122-1, Chapter 6, provisions for protection against indirect contact shall be provided for exposed conductive parts and components of overhead contact line systems. In addition, protective provisions shall be taken against any remaining dangerous touch voltages for wholly or partially conductive structures in the overhead contact line or pantograph zone. These include steel structures, reinforced concrete structures, metallic poles, metallic fences, drainpipes and running rails of non-electric traction systems, that will become live when there are faults.

Direct earthing of the running rails as is the practice with AC traction systems is not allowed for stray current reasons. Therefore, it is also not allowed to connect exposed conductive parts with the running rails, when they are in contact with earth. In this case, voltage limiting devices shall be used to make an open connection from the exposed conductive parts to the return circuit as shown in Figure 6.29 to cause immediate tripping of the circuit breakers and interrupt the current to limit touch voltages to tolerable values.

To avoid connecting every pole or every exposed conductive component in the overhead contact line zone via a voltage limiting device type F according EN 50 122-1, Annex F, it is common practice, especially for DC 3 kV contact line systems, to install earth wires on the poles that are connected via voltage limiting devices to the running rails in distances of several hundred metres. The earth wire should be subdivided into sections not longer than some kilometres to reduce the hazard of stray current corrosion at the pole foundations. An example for the interconnection of individual devices to common earthing points along a tramway line is given in [6.27].

To protect the poles against the effects of lightning strokes the metallic parts of the poles should be connected to the reinforcement of the pole foundations. For practical application the following guidelines should be considered:

- the cubicles and frames of equipment with traction voltage inside should be insulated from supports or foundations and be directly earthed at the *structure earthing system* or connected to the return circuit by means of voltage limiting VLD-F
- metallic supports of conductor rails (third rails) need not to be earthed if they are installed on insulated base plates from earth

According to EN 50 122-1, supporting structures for overhead contact lines with voltages up to DC 1,5 kV do not need to be connected to the return circuit via voltage limiting devices if the insulation of the contact line has been doubled or reinforced. This applies to many tramway and urban mass transit systems.

6.5.4.7 Tunnel and viaducts

Stray currents can flow from the running rails into reinforced concrete structures of tunnels and viaducts and cause voltage drop and voltage shift versus earth. To avoid the stray currents flowing via the outer reinforcement shell, the *electrical bonding of conductive metal reinforcement* and all other metal parts is required

- to provide protection against electric shock,
- to provide protection against the hazards of the rail potential and
- to reduce hazards associated with stray currents.

Installing *earth conductors* in parallel to the structure proved to be advantageous, so that the structure segments can be connected to it, as shown in Figure 6.29. The installation of this type of through-connection is simple and can be checked easily with respect to the criterion of 200 mV. This criterion replaces the former 100 mV criterion from EN 50 122-2:1996 and is also used for assessment of cathodic protection systems. EN 50 122-2 states, that for stray current protection purposes only, it is possible to achieve adequate electrical conductivity of reinforcing bars within a structure section by means of conventional steel wire wrapping.

In consideration of stray currents, EN 50 122-2, Clause 5.3, specifies the calculated maximum longitudinal voltage between any two points of the entire tunnel structure. Experience confirm,s that there is no cause for concern, if the average value of the potential shift between the structure and earth in the hour of highest traffic does not exceed +200 mV for steel in concrete structures. For details refer also to EN 50 162, Table 1.

Using the track resistance per unit length R'_R, which can be taken from Table 5.3, the *longitudinal voltage drop* in the tunnel is assessed by a worst-case study according to EN 50 122-2, Annex C.2. The formula assumes an infinitely long tunnel on each side of the considered section. Furthermore, it does not take into account the reducing effects of train movements in adjacent sections and the conductance per unit length of the tunnel structure against earth. Therefore, the calculated values for the longitudinal voltage drop according to Equation (6.41) are likely to be higher than in reality. In case the calculated value exceeds 200 mV by less than double, a more detailed calculation model can be applied, which usually results in lower values. the voltage drop in the metallic structure of the tunnel or viaduct is

$$U_S = 0,5 \cdot I \cdot L \cdot \frac{R'_R \cdot R'_S}{R'_R + R'_S} \cdot \left\{ 1 - \frac{L_C}{L} \cdot [1 - \exp(-L/L_C)] \right\} \quad , \tag{6.41}$$

where

$$L_C = 1/\sqrt{(R'_R + R'_S) \cdot G'_{RE}} \quad , \tag{6.42}$$

with:

U_S longitudinal voltage in the reinforced structure, in V
G'_{RE} conductance of the rails to earth, in S/km
I average value of the traction return current in the hour of the highest traffic, in A
L length of the considered line section, in km
L_C characteristic length of the system running rails/structure, in km
R'_R longitudinal resistance of the running rails per unit length, in Ω/km
R'_S longitudinal resistance of the structure per unit length, in Ω/km.

Figure 6.30 shows the *electrical circuit diagram* for the exact calculation. The maximum longitudinal voltage U_S occurring between any two points in the structure depends upon the following parameters (see EN 50 122-2, VDV 501 Part 1-3 [6.4]):

- length of a supply section
- resistance of the tracks
- resistance of the tunnel structure
- conductance per unit length G'_{RS} between the return circuit and the structure, in case of tunnels, the tunnel reinforcement
- conductance per unit length G'_{SE} between the structure and earth
- maximum one-hour average value of the traction current

Example 6.6:   The longitudinal voltage U_S in a tunnel structure for a 1 km long double-track tunnel section is calculated according to Equation (6.41). The hourly mean value of the traction current is 1 000 A. The other required parameters are:

- $R'_R = 0,01\ \Omega$/km according to Table 5.3 for four rails UIC 60 and reducing margin for wear and rail joints
- $G'_{RE} = 0,05$ S/km superstructure with one rail per track insulated, after a long period of use
- $R'_S = 0,05\ \Omega$/km for eight reinforcing rods with a cross section of 400 mm^2 each.

The characteristic length is determined as L_C from (6.42) with $L_C = 1/\sqrt{(0,01+0,05)\cdot 0,05} = 18,3$ km.
The longitudinal voltage of the structure U_S results using (6.41) as
$U_S = 0,5 \cdot 1\,000 \cdot 1,0 \cdot 0,01 \cdot 0,05 \cdot [1 - (18,3/1,0) \cdot (1 - \exp(-1/18,3))](0,01 + 0,05) = 0,112$ V
which is below 200 mV as the basis for the stray current assessment. This example also shows that U_S can exceed 200 mV with higher conductance G'_{RE}. In this case, exact calculations would be necessary.

Other requirements for the design of electrical installations in tunnels are:

- metallic, conductive connections between the running rails and the tunnel reinforcement or other steel components are not allowed
- metal pipes which lead into the tunnels must be insulated electrically from the sections of pipe outside of the tunnel
- cable sheaths and armouring also need to be insulated by insulating joints where they lead into tunnels

The installation of *return conductors parallel to the running rails* is an effective way of limiting track-to-earth voltages and reducing the hazards due to stray currents, simultaneously achieving favourable conditions for implementing *protection* against *electric shock*.

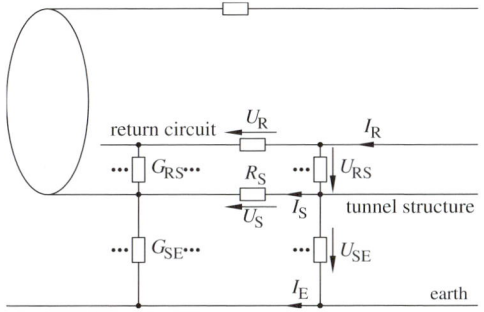

Figure 6.30: Equivalent electrical circuit diagram for a DC railway system in a steel-reinforced concrete tunnel.
R Rails, *S* Structure, *E* Earth

As an example, computer simulations were used to determine the track-to-earth voltages and stray currents at the location of a traction vehicle. The vehicle was on a double-track DC 750 V metro line with a heavy traffic load. Headway between trains was 5 minutes, traction and braking power consumption was up to 4 000 kW per train and track-to-earth conductance values were 0,02 to 2 S/km. The highest *track-to-earth voltage*, $U_{RE} = 210$ V, was found for $G'_{RE} = 0,02$ S/km. An increase of the conductance per unit length to 2 S/km led to a reduction of this voltage to 140 V. At the same time, the stray currents increased by a factor of 50. The solution found for this problem, meeting both the demand for a reduction of the track-to-earth voltage and of the stray currents, was to install a 1 000 mm² cross section copper return cable parallel to the running rails. On a new permanent way, with one rail insulated and a conductance per unit length of 0,02 S/km, this supplementary return conductor reduces the maximum track-to-earth voltage from 210 V to 120 V and lowers the stray currents by more than 60 %. A model for computer simulation is presented in [6.28].

6.5.4.8 Third party earthing installations

Cable sheaths, pipelines and metallic and steel reinforced structures in *third party systems* can transfer potentials and also cause stray current corrosion on third party equipment. External pipelines entering tunnels or viaducts in an electric traction system, must be laid with insulation against the structure earth, or be separated electrically with insulating sections at the entry points into buildings, see Figure 6.29. This also reduces corrosion due to different *open-circuit potentials* in the earthing system.

The screens of communication cables that lead to the railway system from outside must also be insulated from the structure earth. Despite this, to take advantage of the reduction effect, the cable sheaths can be connected to the structure earth via low-inductive capacitors.

Between a DC railway installation and underground pipelines or cables a minimum distance of 1 m should be maintained according to EN 50 122-2, where in Annex A.4. a measuring procedure is also given for further tests.

The *earthing installations* of *third party* systems should be insulated from DC railway systems. If such insulation is not possible, e. g. where the DC railway system and third party systems are integrated into commonly used buildings, then the earthing systems of the building need to be grouped together with the structure earth of the DC railway.

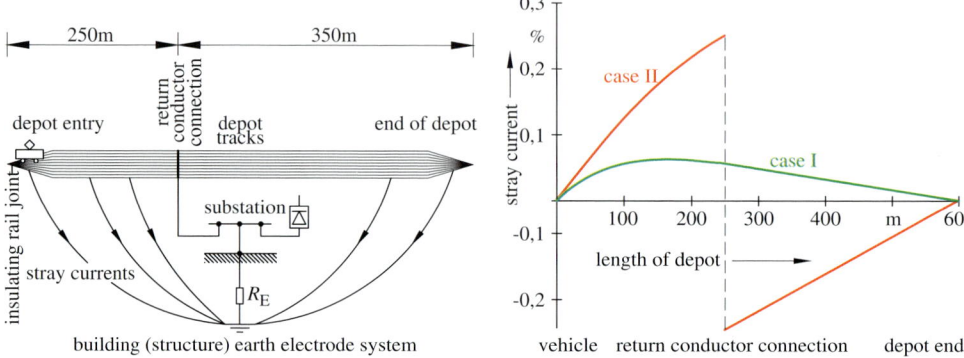

Figure 6.31: Return conductor arrangement in a depot area.

Figure 6.32: Stray currents in a depot when the rails and the structure earth system are separated (case I) and when they are connected (case II).

6.5.4.9 Depot and workshop areas

Voltage differences between structure earth and return circuit can be a hazard for staff and equipment while working. In such areas the permissible values for the short-term touch voltages apply as for other track areas. For long-term touch voltages with a duration of 0,7 s and longer, only 60 V is permissible in workshops and similar locations, see also Clause 6.3.5.1. This value has to be considered, when high currents for air conditioning and pre-heating of trains occur. To avoid dangerous rail potentials, it is allowable to interconnect the return circuit and structure earth in *depots* and *works hop areas*, see also EN 50 122-1, Clause 6.4.2 and EN 50 122-2, Chapter 9. These provisions also help to avoid damage to electrical tools. If the running rails are connected to the main line, direct earthing is not allowed and the rail insulation from earth must be the same as on the main line.

The conditions under which direct connections between the return circuit and earth are permissible are:

– separate traction substations for workshop area and main line sections
– isolation from the main line track by *insulated rail joints* and
– provision of insulating joints at all cable sheaths and pipelines entering the depot or workshop area.

In Figure 6.31, the running rails of the depot are connected to the earthing system and separated from the main line by insulating rail joints. The depot and workshop area is supplied by a separate traction power supply substation. Because of short feeding sections and low operating currents in these areas, stray currents are low.

In [6.29], an investigation was carried out to determine how stray currents are affected by connecting the running rails in a depot with the structure earth. The depot under investigation had a separate feed and the rails were isolated from the main line track by insulated rail joints. The depot parameters are:

– 10 tracks
– track resistance per unit length of 22 mΩ/km
– track conductance per unit length of 0,5 S/km
– structure earth electrode resistance $R_E = 0,33\ \Omega$ and
– return conductor resistance of 1,5 mΩ

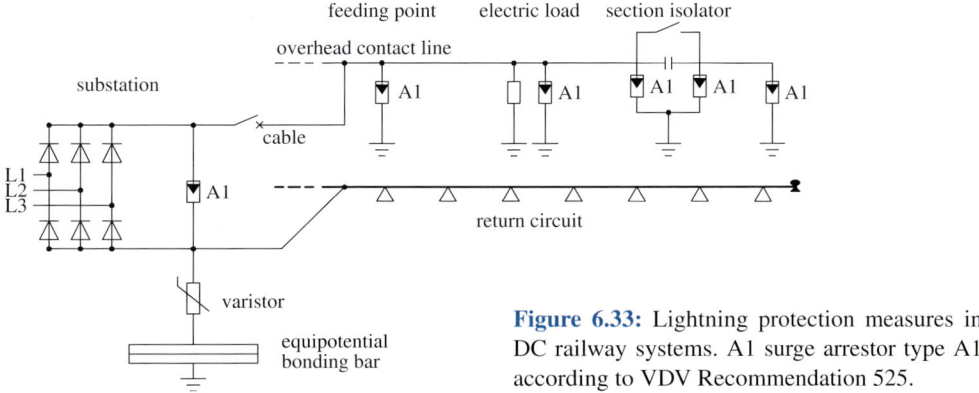

Figure 6.33: Lightning protection measures in DC railway systems. A1 surge arrestor type A1 according to VDV Recommendation 525.

Assuming that a traction vehicle is drawing current at the depot entrance, the stray currents were calculated as shown in Figure 6.32. The average current, observed over a longer period, must be taken as a basis for evaluating the corrosive effect of stray currents. In the case in question, the advantages of having tracks and structure earth at the same potential during repair and maintenance work outweigh the disadvantages resulting from the higher stray currents.

The return circuit and the structure earth should be connected in the centre of the depot tracks only at one point, to keep the *longitudinal rail voltages* in the depot as low as possible. Further connections to wheel lathes, vehicle lifting devices and crane systems often cannot be avoided during working. It is advantageous to install them close to the central connection of depot tracks and structure earth. A survey carried out on 22 public transport operators in Germany showed that 15 had chosen this form of separate feeding and interconnection of the structure and traction earth for existing, projected or planned workshop or depot installations [6.16].

6.5.4.10 Signalling and telecommunications installations

Since electrically conductive connections between running rails and structure earth or earth are not permitted, low-voltage equipment like *signalling installations, track release installations*, point machines and other installations connected to the running rails must also be insulated from the structure earth and earth to avoid stray currents. The sheaths of railway-owned telecommunication and signalling cables not directly connected to the rails can be connected on both sides to the structure earth in stations and along the line to support the reduction effect.

6.5.4.11 Lightning protection

Lightning strokes cannot be avoided on at-grade or elevated lines. They can create hazards to DC installations with considerable consequences. [6.30]. Direct strokes to the overhead contact lines and strokes to neighbouring installations result in overvoltages on the contact lines. Strokes to poles can lead to back-flashovers and to overvoltages. Such overvoltages can also result from strokes to running rails, especially in the case of third rail installations. The overvoltages can reach values up to 2 000 kV. The equipment used for DC lines cannot be designed economically to withstand such overvoltages. The overvoltages need to be limited to values not hazardous to DC installations. The most recent and most effective protection is achieved with metal-oxide arrestors [6.31, 6.32].

According to the recommendations given in VDV 525 [6.9], outdoor-suited arrestors should be installed at each feeding point, at the ends of feeding sections and lines, at line couplings and at the connections of electric loads according to Figure 6.33. In VDV 525 the recommended arrestors are called type A1 arrestors. Additional A1 arrestors should be installed on long at-grade sections and on bridges.

The A1 arrestors should be installed close to the contact line and connected to the contact line by an insulated conductor. In the case of ballast superstructure with insulated rails, the A1 arrestors should be connected to an earthing rod having an earthing resistance not more than $10\,\Omega$ according to IEC 62 305-3. Further details can be found in [6.31]. The terminals of the arrestors should be as close as possible to the equipment to be protected, e. g. the arrestor should be directly connected to the cable termination at a feeding point.

To protect the substation, VDV 525 recommends the installation of A1 arrestors in the substation at each feeding cable and additionally a varistor (type A2 arrestor) between the return conductor and the structure earth, connected to the equipotential bonding bar, see Figure 6.33.

6.5.4.12 Implementation of electrification projects

Earthing and *stray current protection* for DC railways are important for the installation of each electrification project. Electrical connections in the reinforcement of buildings, bridges, tunnels and pole foundations need to be defined early, before the first construction activities. If the earthing connections and through-connections have not been provided in the structures, then alternative solutions need to be provided later, resulting in considerable additional expenditures. It is especially important to agree early on the materials, cross-sections and connection techniques to be employed to fulfil the requirements for the earthing of DC electric railway systems. The *minimum cross-sections of earthing conductors* are specified IEC EN 61 936-1 with respect to corrosion and mechanical strength, for example $50\,\text{mm}^2$ for steel and $16\,\text{mm}^2$ for copper. According to IEC 61 000-5, welded connections are preferred to clamp connections for earth conductors, since the electrical resistance at the connection could be increased by corrosion of the clamps.

6.5.5 Measurements for design verification and maintenance

6.5.5.1 Measurements for design, construction and approval

An overall plan of how and when to measure the essential parameters for electrical safety and stray current provisions is a precondition for the installation of electric railway systems.

Measurements during the design phase for evaluation of design parameters such as the soil resistivity and resistance to earth of existing earthing structures provide the data for the system design calculations.

During the construction phase of tunnels and viaducts, the electrical interconnection of the steel reinforcement has to be checked visually, to be measured and documented to prove the proper execution of civil works and electrical installations. Short-comings in this procedure can only be corrected before the reinforcement is cast in concrete. The correct design and implementation of earthing connections and earthing terminals for the electrical installations is also important.

The continuity of the return circuit and the insulation of the trackwork from earth must be confirmed by measuring relevant values. It is expedient to do this for short sections before connecting them together.

During the commissioning phase, measurements must be performed for verification of electrical safety and stray current provisions. Measurements include rail potentials, insulation of the rails from earth or voltage shift of structure earth versus remote earth due to train operations. The results are essential for the system approval.

With the beginning of passenger services, reference measurements are recommended as a basis for comparative measurements for maintenance, so changes and deficiencies during operation e. g. in the rail insulation can be detected and countermeasures initiated.

The rail potential as a basis for electrical safety is measured during the commissioning phase and periodically during operation. To test worst-case conditions, the measurements should be performed for the highest train load as well as for substation outages, taking into account a reduced traffic load.

6.5.5.2 Measurements for maintenance purposes

For DC traction systems, in addition to electrical safety, acceptable stray current conditions are important for maintenance work. The permissible touch voltage is governed by the rail potential. In case of high values, voltage limiting devices register the rail potentials and limit them if impermissible high values arise. If no voltage limiting devices are installed, rail potentials during operation should be recorded.

To avoid damage to the return circuit and nearby metal installations, the stray current conditions must be surveyed during operation. A direct measurement of stray currents through earth is not possible. The conductance per unit length of the running rails, however, is representative of the stray currents leaving them. A local change of the conductance per unit length of the running rails or an overall deterioration is reflected in the rail potential. Therefore, the rail potential is used for assessment of the stray current situation. The danger of *stray current corrosion* would increase if the average value of the rail potential changed, under the same operational conditions, compared to the initial measurements during the commissioning phase. The reason could be low-resistance connections between the return circuit and the structure earth, which should be located by further measurements. A low rail potential indicates near-by low-resistance connections between the running rails and structure earth. At large distances from the faulty connection, the rail potential increases to double the value compared to undisturbed operation. The faulty connections should be repaired without delay. EN 50 122-2, Chapter 10, proposes a time interval of five years for *repetitive stray current measurements*. Annex B.2. of the standard describes the measuring procedure. Such measurements could also be performed at the terminals of voltage limiting devices, if installed, during train operation.

In the case of *continuous stray current monitoring* according to EN 50 122-2, Annex B.1, repetitive measurements are not necessary. For this, the installation of measuring points for the rail potential along the line is required. The values are recorded at a centralized data acquisition system. Due to the varying traffic loads, an averaging is required. If there is a major change in the average rail potential along the line, a change in the rail to earth conductance is likely and indicates increased stray currents.

If the repetitive stray current measurements or the continuous stray current monitoring show extraordinary large deviations from the reference measurements, a test of the rail insulation

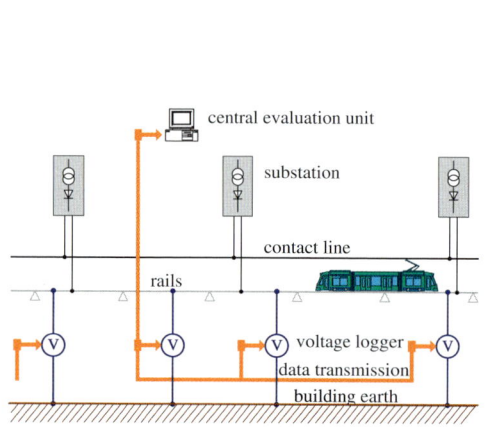

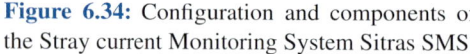

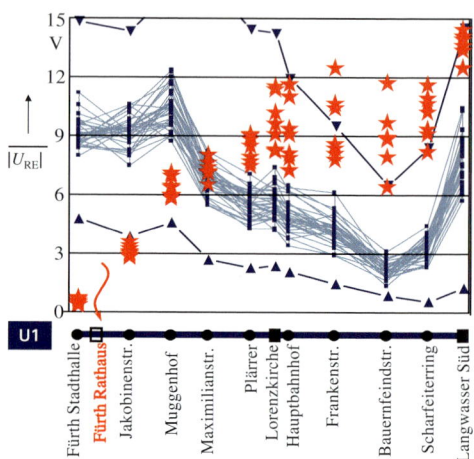

Figure 6.34: Configuration and components of the Stray current Monitoring System Sitras SMS

Figure 6.35: Fingerprint of the rail potential along the line U1. 15-minute mean absolute values of the track potentials in blue and in red, a rail to earth connection in Fürth Rathaus station.

and the structure/earth potential in accordance with EN 50 122-2, Annex A.2, A.3 or A.4 is recommended to localise the cause, see also [6.33, 6.34]. The effort required for such measurements is high and requires stopping of train operation.

Continuous stray current monitoring was developed by Siemens Transportation Systems, called Stray Current Monitoring System Sitras SMS [6.11]. The experience in practical application at the Metro Nuremberg in 2002 has been published in [6.10]. Since then, Sitras SMS has been installed in numerous Metro Projects worldwide. The operator VAG had recorded rail potentials manually over a long period and evaluated the average values for stray current assessment. Installing Sitras SMS, the continuous monitoring reduced this effort almost to zero. Figure 6.34 shows the configuration and the components involved. The central evaluation unit of Sitras SMS is located in the DC substation in the main station, where the data from eleven measuring points on the line U1 is recorded. A distance between measuring points of 2 km is sufficient. The main features are:

- continuous data acquisition of the rail potentials during operation
- continuous averaging and graphical processing of the rail potential along the line
- evaluation and localisation of rail to earth deficiencies and faults
- remote control of the operation and archiving of the measured values

Figure 6.35 shows an example of the situation along the line U1. The upper limit and the lower limit form a reference band where the rail potential is likely during operation. The 15 minute average values are given by the blue lines for operation during the day. Such a fingerprint is typical for the line in operation. The red stars demonstrate the situation of a rail to earth connection in the station Fürth Rathaus. As expected, the rail potential is close to zero at this location and increases at the other end of the line. It exceeds both the upper and the lower limit band. Sitras SMS reports the fault and its location to the control centre for further action.

6.5.6 Earthing and bonding for Ankaray LRT system

6.5.6.1 Description of the project

The provisions described above were implemented for earthing and bonding of the Ankaray LRT in Ankara, Turkey, as illustrated in Figure 6.29. The project was presented in [6.8]. During construction and commissioning phases, measurements were carried out to verify the design parameters and the results provided representative results for similar electrification projects. The measurements at the Ankaray metro system demonstrate that an *overall strategy for earthing and bonding* and the arrangement of the return circuit does not only accelerate project progress but also simplifies the maintenance of the return circuit and earthing system with respect to safety of persons and the effectiveness of stray current protection.

6.5.6.2 Resistance-to-earth of the tunnel structure

The *resistance to earth* of all stations and substations was measured during the construction phase using the *three-point method*. The maximum value measured was $0,35\,\Omega$. This was significantly lower than the value of $0,9\,\Omega$ required for compliance with the permissible touch voltage in the case of earth faults in the three-phase supply.

6.5.6.3 Rail potentials

During the system trial run, the rail potentials were measured at the stations, the trains operating at the shortest permissible train headway and the maximum train load. During normal operation, *maximum rail potentials* of $\pm 60\,\mathrm{V}$ occurred. Feeding sections were through connected to investigate the effect of substation outages. As expected, these extraordinary conditions resulted in higher rail potentials during multiple starting, causing tripping of some voltage limiting devices in the stations in some cases due to increased rail potentials.

6.5.6.4 Rail insulation

To test the insulation of rails, the conductivity between the running rails and structure earth was measured using the method described in EN 50 122-2, Annex A.2. The measured values were close to $0,02\,\mathrm{S/km}$ per track, significantly lower than the $0,1\,\mathrm{S/km}$, which is recommended in EN 50 122-2 for the design of tunnel sections.

During the measurements, the *longitudinal resistance* of the running rails was also measured. The values were found to be between 36 and $40\,\mathrm{m\Omega/km}$ for one running rail and they correspond quite well with the values specified in documents for the rail type S 49, see Table 5.3.

6.5.6.5 Voltage between structure earth and reference earth

For assessment of the danger of stray current corrosion the *potential of the tunnel structure* was measured against a $Cu/CuSO_4$ reference electrode, without vehicle operation and during maximum operational load. In Figure 6.36, a printout is shown of a typical result of potential measurement. The average potential between the tunnel structure and remote earth during train operation is insignificantly higher than that occurring without operation. Only short duration voltage peaks of up to $50\,\mathrm{mV}$ occurred because of train operation. Since the average value of the measured voltage shift was far below $200\,\mathrm{mV}$, in accordance with EN 50 122-2 and EN 50 162 respectively, there is no danger of stray current corrosion.

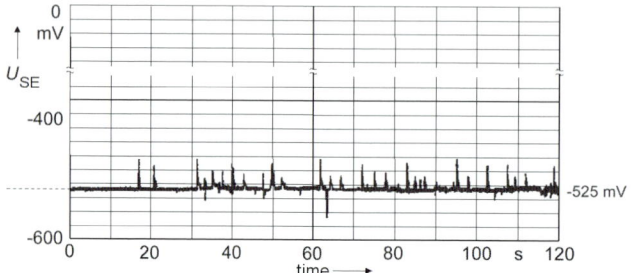

Figure 6.36: Ankaray LRT system: Potential between tunnel structure and remote earth during train operation.

6.5.6.6 Voltage limiting devices for stray current assessment

A quantitative assessment of stray currents is possible basing on a current measurement with voltage limiting devices closed for testing purpose. They connect the running rails with the structure earth and have an effect similar to *stray current drainage*. Even with this undesirable case of an earth connection, stray currents smaller than 10 A were measured at the voltage limiting device. This low value confirmed the high quality of rail insulation used for the Ankaray metro system, especially when compared to other DC railway systems.

Without the intentional connection, the values were assumed to be lower by a factor of 10. If several voltage limiting devices were closed simultaneously, then considerably higher currents – up to 500 A – would flow through the connecting conductors. This situation is equivalent to a through-connected tunnel reinforcement switched in parallel to the running rails.

6.5.7 Conclusions for return circuit and earthing in DC traction systems

The system design of the traction return circuit, including earthing and bonding described in Clause 6.5, is based on insulated running rails and a continuous and integrated earthing system. It complies with the relevant international standards. The example of the Ankaray LRT system demonstrates that this design proved itself in practical applications. The configurations with *stray current collecting nets* also discussed, would cause technical disadvantages and require additional expenditure for installation and maintenance. Stray current collecting nets and stray current drainage can not be recommended as methods for stray current protection.

The measures for earthing and bonding also affect civil engineering works and need to be defined at an early stage of railway projects to decide on the necessary provisions, prior to construction, to avoid more expensive alternative measures.

In summary, in Figure 6.29 the recommended system is shown for earthing and bonding of DC lines to provide protection against electric shock, overvoltages and stray current corrosion on at-grade sections, in tunnels and on viaducts.

6.6 Design of AC traction systems

6.6.1 Basic rules for the design of the return circuit

The systems used for the power supplies of AC railways are directly correlated to the *return circuit*. In simple *track return systems* (Figure 6.4 a)) using the running rails only as the

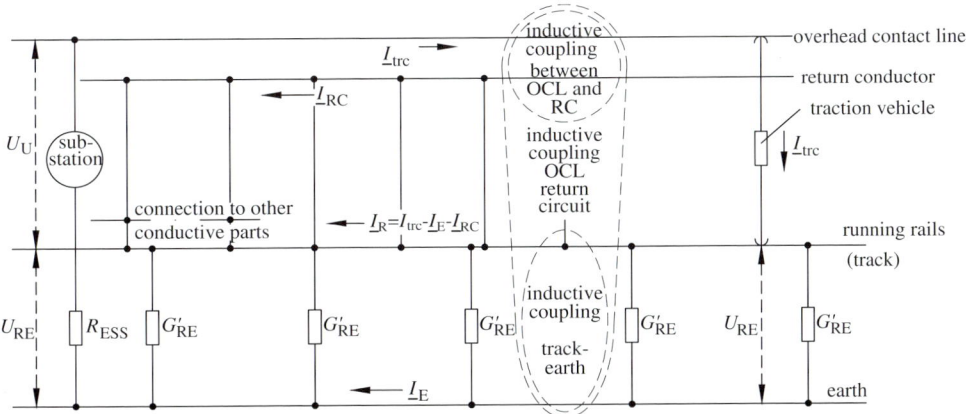

Figure 6.37: Current return path and traction earthing of single-phase AC railways.
R_{ESS} earthing resistance of the protective earthing system of the substation

return circuit, 30 to 40 % of the return current flows through the earth. This proportion can be reduced to 15 to 20 % by installing *return conductors* at the poles as shown in Figure 6.4 b). The *auto-transformer system*, Figure 6.4 d), feeds the railway line at a higher transmission voltage between the overhead line and the *negative feeder*. Auto-transformers, arranged at intervals of 10 to 20 km, transform the transmission voltage to the contact line voltage. Two neighbouring auto-transformers function like substations on track sections supplied at both ends.

The *booster transformer system*, Figure 6.4 c), employs current transformers with a transformation ratio of 1 : 1 connected into the overhead line at intervals of 3 to 5 km. The secondary winding sucks the return current from the running rails via connections to a return conductor, suspended from insulators. The return current flows back to the substation in close proximity to the contact line. The current flowing through the running rails and the earth is low over large sections of the line.

The basics for the traction return circuit and its earthing and bonding are explained in [6.35] and [6.36]. A schematic circuit diagram of the current return and traction earthing system, in accordance with the basic principles of Figure 6.29 is shown in Figure 6.37. The traction current I_{trc} flows from the overhead contact line through the traction vehicle to the running rails that are directly bonded to the return conductor. There is a galvanic connection to the earth characterized by the conductance G'_{RE}. In addition, the contact line is inductively coupled with the return conductor, the track and the earth. Tracks and earth are also coupled inductively.

An electric potential exists between the rail and remote earth, which has maxima at the vehicle position and at the substation, see also Figure 6.14. The return current flows through the return conductor, the rails and the earth. For sufficiently long distances between substation and vehicle, there is a balanced current distribution in the middle section, where the rail potential is zero. The return circuit on AC railways, unlike that on DC lines, is connected to the earthing system, see Figures 6.3 and 6.37. The earthing system includes large area earth electrodes such as building foundations, bridge and viaduct foundations, tunnel reinforcements and piling foundations for overhead contact line poles along the track. Their interconnection via the return circuit line forms the 'railway earth' to which the following are connected:

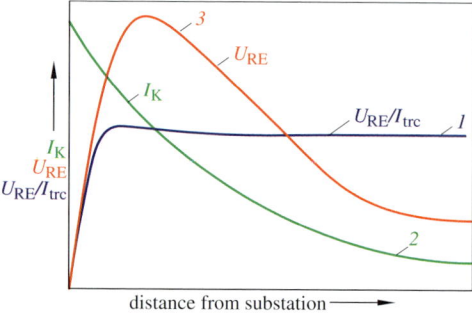

Figure 6.38: Short-circuit current and rail potential in the operating case and in the short-circuit
case depending on the distance to the substation.
1 related rail potential U_{RE}/I_{trc} at the train
2 short-circuit current I_K depending on the short-
circuit distance from substation
3 rail potential U_{RE} at the location of the short circuit

- *medium-voltage protection earth*
- *low-voltage protection earth*
- earthing of *telecommunications and signalling* and
- earthing of lightning protection devices

With AC traction systems, the earth is part of the traction current return path due to the inductive and ohmic coupling with the tracks. Parts of the track return current flow through the connected earthing system and through earth. This results in an extended area within which non-railway installations can be affected by the railway system. The stronger the current flowing through the earth, the higher the risk of other installations, pipes, cables and devices in the vicinity of the railway being affected by inductive and galvanic coupling.

The following types of *interference* are examined for AC lines:

- *galvanic interference*
- *inductive* and capacitive *interference*
- *electric* and *magnetic fields*

The galvanic interference arises from conductive connections of any systems with the re-turn circuit. Capacitive interference caused by influenced voltage is insignificant in railway applications.

Inductive interference and magnetic fields are important in AC railway systems. Their magnitudes depend upon the self-impedance and coupling impedance of the overhead line arrangement in the same manner as the return current distribution. For this, the return current through the earth represents a measure of the interference to be expected. Additional return conductors, booster transformer or auto-transformers reduce the return current flowing through the earth and, therefore, the interference in the vicinity of the electric railway.

The interference affects railway-owned and third party electrical devices in the direct neighbourhood. Impairments and disturbances can occur, depending on the sensitivity of the equipment. Chapter 8 deals with interference issues.

6.6.2   Safety of persons

Primarily, the earthing and bonding of AC railways must prevent hazards of *electric shock* and guarantee the *safety of persons* [6.1]. To achieve this, electrical equipment and components of the overhead contact line system, that could become energized at contact line voltage under fault conditions, are bonded directly to the *return circuit*. This especially applies to components that lie within the overhead contact line and pantograph zones, see Clause 6.2.9. The connection of these components with the return circuit results in reliable *protection tripping*,

Table 6.12: Permissible touch voltage $U_{te,max}$ according to EN 50 122-1 and rail potential U_{RE} for AC railways.

	permissible voltage $U_{te,max}$ in V	permissible rail potential U_{RE} in V
operational case $t > 300\,s$	60	120
operational case $t = 300\,s$	65	130
fault case $t_F = 0, 1\,s$	842	1 648

e. g. during insulator flash-over or short circuits between the overhead contact line and poles. If a direct connection to the return circuit is not possible, for example, because the parts to be earthed are part of a return circuit belonging to a DC railway, then they are to be connected to the return circuit of the AC line using voltage limiting devices. Small conducting components, with horizontal dimensions not exceeding the values given EN 50 122-1, Clause 6.3.1.2, Table 1, and which do not support electrical equipment, are excepted from the bonding.

The *rail potentials* must satisfy the requirements for permissible *touch voltage*. Feeding traction currents into the return circuit at the location of the vehicle or short circuits causes a local potential increase of the return circuit against earth, see Figure 6.38. This potential difference, the *rail potential* depends on the operating or short-circuit currents, the conductance of the track to earth and the distance of the vehicles or the earth fault from the substation, see Clause 6.4.4.1. For convenience, the rail potential is referred to 100 A as a specific value.

The maximum rail potentials need to be determined for the operational and short-circuit cases to assess the hazards caused by the rail potential, see Clause 6.4.4. For a constant operational traction current, the *rail potential* rises with increasing distance of the vehicle from the substation up to a distance of 0,5 to 5 km and then remains almost constant, depending on the earthing conditions. Figure 6.38 shows the corresponding trend of the related rail potential at the traction vehicle.

The short-circuit current is largest for a short circuit at the substation but in this case the rail potential is zero. The rail potential reaches its maximum with a short circuit at a distance of several kilometres from the substation in the transitional area shown in Figure 6.38.

To assess the hazard caused by the rail potential, the shape of the potential against earth must be considered perpendicular to the track. In Figure 6.8 and 6.17 the voltage U_{RP} between the track and point P and that of voltage U_{PE} between P and the reference earth are shown relative to the rail potential U_{RE}. The voltage U_{RE} cannot be shunted by a person at a distance of 1 m. However, the part of the rail potential between the outer rail of the track and a point 1 m away must not exceed the permissible voltage according to EN 50 122-1, where only approximately 20 % of it can be shunted. For the earthing of high voltage systems, IEC 61 936-1 specifies that the permissible accessible or touch voltage is considered to be compliant if the potential does not exceed twice the *permissible touch voltage*. This relation applied to the rail potential is taken into account in Table 6.12.

The permissible touch voltages specified in EN 50 122-1 also apply to power supply installations, where faults with earth contact within the three-phase medium- and low-voltage systems must be taken into account. For this, the potential increase of the earthing system is to be treated in the same manner as the rail potential.

The values for permissible touch voltages and rail potentials in the case of short circuits depend on the time of action and are shown in Table 6.12. They take into account the fault

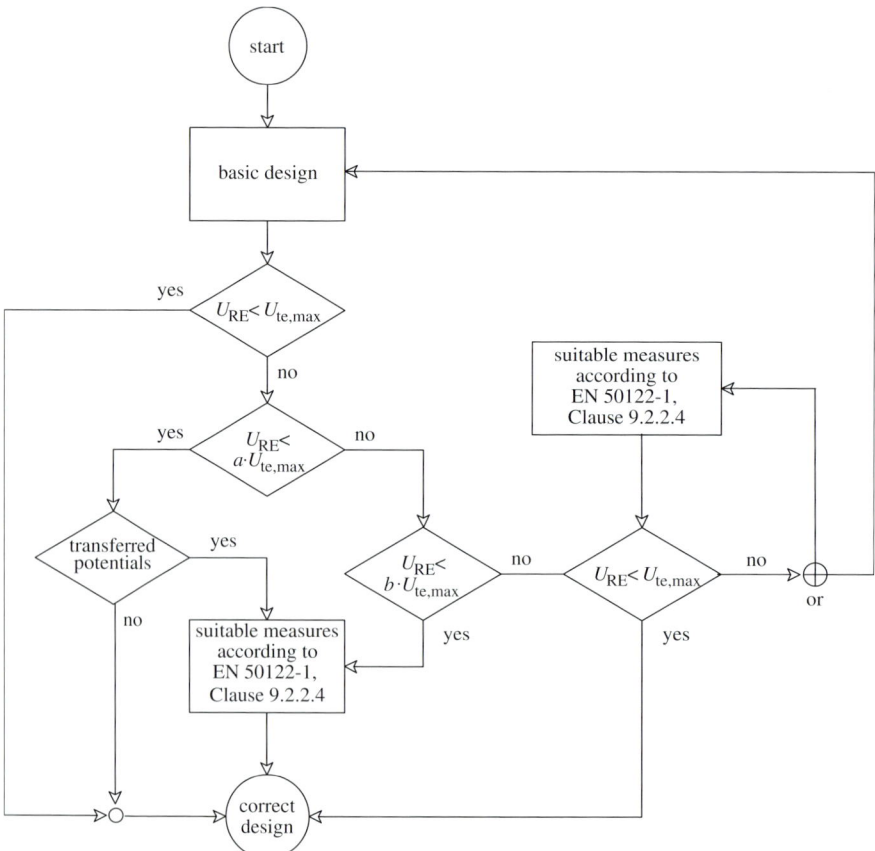

Figure 6.39: Decision diagram to achieve safety relating to permissible touch voltages according to EN 50 122-1, Figure 26.

disconnection time of modern protection devices, which are less than 100 ms. For further values see Table 6.1.

In case the permissible touch voltages shown in Table 6.1 are exceeded, direct countermeasures, rail potential reduction measures or management derived according to Figure 6.39 from EN 50 122-1, Clause 9.2.2.4, should be employed. These include:

- reduction of the rail to earth resistance, e. g. by means of improved or additional earth electrodes
- equipotential bonding
- improvement of the return circuit taking electromagnetic coupling into account
- insulation of the standing surface
- potential grading by means of appropriate surface earth electrodes
- obstacles or insulated accessible parts
- access restrictions (fences) including instructions for maintenance staff
- reduction of fault and/or operation currents
- installation of voltage limiting devices (VLD)
- reduction of the tripping time needed to interrupt the short-circuit current

Table 6.13: Electrical characteristics of AC lines.

	AC 25 kV 50 Hz			AC 15 kV 16,7 Hz		
	A	B	C	A	B	C
Distribution of return current (%)						
– tracks	70	65	40	70	65	50
– earth	30	30	20	30	30	15
– buried strips	–	5	–	–	5	–
– return conductors	–	–	40	–	–	35
Rail potential at load (V/kA)						
– track conductance 1 S/km	130	70	60	85	45	40
– track conductance 10 S/km	40	25	20	25	15	12
Induced voltage in 10 m distance (V/(kA·km))						
– earth resistivity 100 Ω·m	140	130	85	50	47	30
Magnetic flux density (μT/kA)[1]	75	75	50	75	75	50

A rails connected to earth, B rails and buried strips, C rails and return conductors
[1] measured under the contact line 1 m above rail head

Paper [6.37] gives an example for improvement of the return circuit to reduce the rail potential. Paper [6.38] reports on the improvement of the Lötschberg mountain section by additional return conductors, reducing the rail potential to one third.

6.6.3   Return circuit

6.6.3.1   Return current through rails and earth-buried return conductors

Lines with high electrical loads and, as a consequence, high currents, using the tracks together with the earth as a return path for the currents may not be sufficient to limit the rail potential to tolerable values and at the same time, keep interferences within acceptable limits. Additional measures need to be taken to achieve the design goals.

The conductance to earth can be improved by earth buried conductors. Clause 6.4.2 describes the effect of *strip-type earth electrodes*. On some DB lines one earthing strip made of galvanized steel with a cross section of $30 \times 4 \, mm^2$ is buried approximately 1 m underground for each track. Because of their underground installation and the cross sections used, the earthing strips only slightly improve the return current distribution. For DB overhead contact lines, it was calculated that in comparison to lines using the tracks only

- the line impedance is reduced by approximately 2 to 3 %,
- the longitudinal voltages induced in conductive parts located at a distance of 3,5 m from the track centre line and 0,1 m above the rail head are reduced by roughly 7 %, and
- the rail potential is reduced by approximately 53 %.

The shape and the field strength of the *electric and magnetic fields* in the area surrounding the railway line barely changes when earthing strips are installed. The main advantage is the reduction of *rail potential*.

In Table 6.13 the effect of earth-buried return strips on the distribution of return current, rail potential and induced voltages is shown in comparision with the data from a system with rails only.

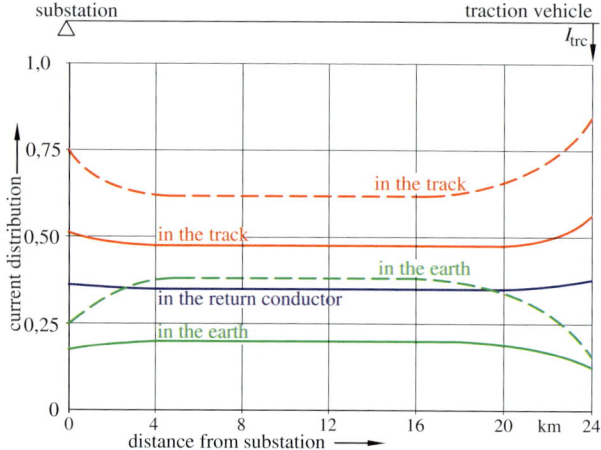

Figure 6.40: Distribution of the return currents among the individual return paths of a double-track AC line with an effective conductance between track and earth of 2 S/km.
--- without return conductors
—— with 243-AL1 return conductors

6.6.3.2 Return conductors

Installing parallel *return conductors* in the vicinity of the overhead contact line is a simple and effective way of reducing the proportion of return current flowing through track and earth, rail potential and induced voltages [6.39]. Parallel return conductors, with a close inductive coupling with the overhead contact line, offer the following measurable effects:

– they reduce the proportions of *return current* flowing through track and earth considerably as shown in Figure 6.40 and Table 6.13
– *track-to-earth voltages* are also lowered considerably. Calculations have shown that a reduction of rail potentials by 50 to 55 %, can be expected, when compared to systems without return conductors. Reference [6.40] reports that a 53 % reduction was confirmed by measurement after return conductors had been installed
– *longitudinal voltages* induced in conductors installed parallel to the railway line are reduced by 40 to 50 %. For a DB standard contact line of type Re 250 with return conductors, it was found that the induced longitudinal voltage in a conductor located 3,5 m from the track centre line was almost 45 % lower than without parallel return conductors. Measurement of induced voltages on the AC 25 kV 50 Hz Madrid–Seville railway line gave evidence that a reduction of the induced voltage of approximately 40 % had been achieved by return conductors, see Figure 6.41. In [6.40] the interference voltage is shown to be almost 45 % lower when parallel return conductors are installed
– the *magnetic field* in the vicinity of the railway line is reduced considerably. In [6.40] it was shown that with a current of $2 \times 1\,000$ A the magnetic field strength 1 m above rail height directly under a contact line of type Re 200 is 54 A/m if parallel return conductors are installed and 75 A/m when no return conductors are installed. The reduced magnetic field in case of parallel return conductors can also be seen in Figure 8.11
– the *impedance* per unit length is reduced. Paper [6.40] reports that at the DB line Magdeburg–Marienborn, AC 15 kV 16,7 Hz, a reduction of the impedance per unit length by 9 % relative to the variant without return conductors was measured. The resistance per unit length was increased by approximately 8 % relative to the design without return conductors and the reactance per unit length was decreased by approximately 18 %. As a result, the phase angle changed from 56° to 48°

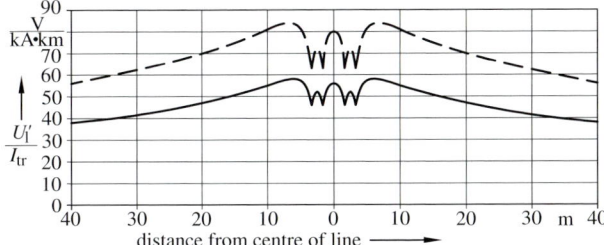

Figure 6.41: Induced longitudinal voltage per km and per kA traction current in a conductor installed in parallel to the line for $\rho_E = 100\,\mathrm{m\Omega}$.
--- without return conductor
—— with 243-AL1 return conductors

Figure 6.40 shows the *distribution of the return current* in the rails and earth. If a return conductor, at earth potential, is installed along the poles at the same height as the overhead contact line, then almost half the current that normally flows back through earth is uncoupled from earth and flows back via the return conductor. The values shown in Tables 5.8 and 5.9 show the same results. In Table 6.13, the effect of return conductors is shown in comparison with current return through rails only.

The *installation* of parallel *return conductors* requires only 5 % additional expenditure. The noticeable reduction in the magnetic field strength in the vicinity of the railway, the interference voltage reduction and the reduction of track-to-earth voltages and impedance justify this additional investment. In reference [6.41], it was concluded that for the Austrian railway company ÖBB, a return current configuration involving parallel return conductors along high-traffic-load lines was an economically and technically sound way to cope with the return traction currents and the associated issues of interference and disturbance. Contact line designs with return conductors are used e. g. on the AVE Madrid–Seville high-speed line [6.42] and on the high-speed line Berlin–Hannover in Germany [6.43].

6.6.3.3 Autotransformers

The *auto-transformer* system or 2 AC system is used for AC lines worldwide [6.44] to [6.45] mainly for AC 25 kV 50 Hz supplies, where the load and currents are high or the distances between locations to connect the substations to the public power grid are long. The principle of this system is shown in Figure 6.4 b). In the 2 AC system, an additional conductor called *negative feeder* is included. This conductor has the same voltage to earth as the contact line and is in phase opposition. Therefore, twice the contact line voltage is present between the contact line and the negative feeder. Contact line and negative feeder line are connected through the auto-transformers having a transformation ratio of 1 : 1. The centre tap is connected to the rails. The design of the windings ensures the same currents flow in both winding but in opposite directions. Outside the section between two auto-transformers, the current flows mainly through the contact and feeder lines while in between them it flows through the contact line, the traction unit and tracks and the return circuit to the auto-transformers.

In the sections where trains are running, the distribution of the return current on tracks, earth and other components is similar to that without auto-transformers, but with traction return through rails and return conductors. Provided the train power consumption is the same, the rail potential, the induced voltage in parallel conductors and the magnetic flux density are lower than with conventional designs. The amount of reduction depends to a large extent on the specific situation and generalisations cannot be made. Measurements of the current distribution are reported in [6.46] and [6.47].

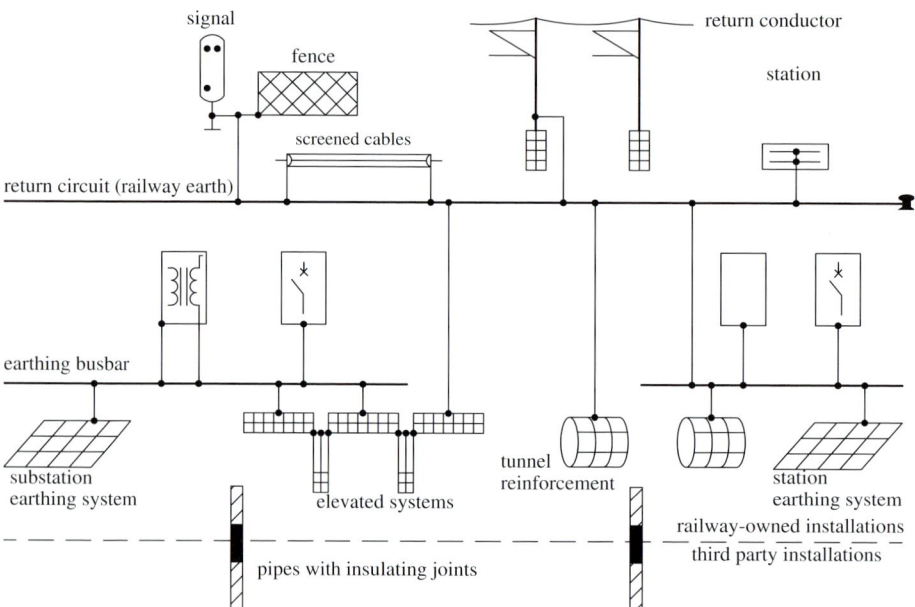

Figure 6.42: Schematic circuit diagram of the return circuit and earthing and bonding for AC railway systems.

Further detailed explanations on auto-transformer systems are to be found in references [6.21]. Auto-transformer systems effectively reduce interference and magnetic fields.

6.6.3.4   Booster transformers

Figure 6.4 c) illustrates the principle of operation of *booster transformers* (BT system). From the feeding substation transformer onwards, the contact line is interrupted at 3 to 8 km intervals and the traction current is passed through the primary winding of a booster transformer with a transformation ratio of 1 : 1. This principle is also described in [6.48]. The secondary winding of the transformer is connected to the *return feeder* and passes the traction current back to the feeding substation. Booster transformers act as inductive coupling between overhead contact line and running rails and force the traction current from the rails and earth to pass into the return feeder, i. e. they drain the current. The booster transformer system, in most train positions along the line, reduces interference effects.

The disadvantages of this design are
- the high cost of installing and operating a large number of booster transformers, return feeders and switchgear, especially taking into account that each track of a multiple-track line has to be equipped with booster transformers,
- increased effective line impedance due to the booster transformers, this being associated with increased voltage drops and power losses,
- arcing across electric isolation gaps in the contact line equipment, leading to faster contact wire and collector strip wear, as well as radio frequency interference and
- the *train-in-section effect*, i. e. when a traction vehicle is travelling and drawing a current between two booster transformers, the reduction of interferences is not fully effective.

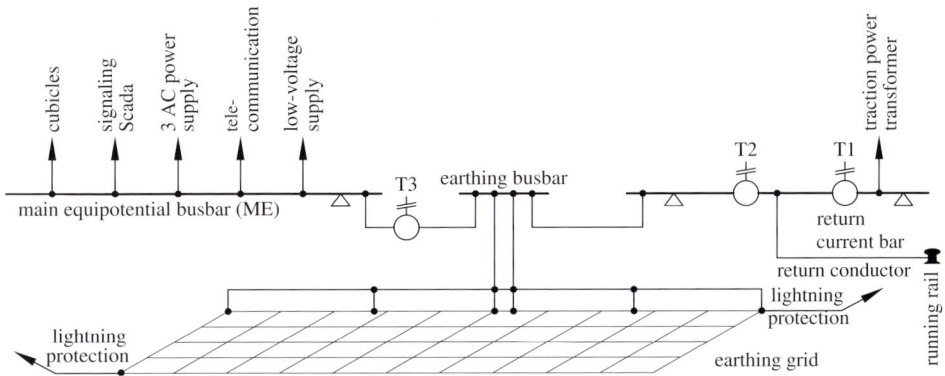

Figure 6.43: Circuit diagram for return circuit in AC substations.
T1, T2, T3 are current transformers

Because of these disadvantages, for new lines booster transformers are only occasionally used. However, BT systems are used in Sweden and Norway [6.49, 6.50] because of high soil resistivity. Paper [6.51] describes the testing of a combination of auto- and booster transformers in Sweden. As in the case of auto-transformer systems, the current return path and the effects differ essentially between the section with the traction unit location and others.
The current distribution is affected by the distance between adjacent booster transformers. In the case of shorter sections, less current flows through earth and other components and also the rail potential and the induced voltage are lower for the same train current. The substantial reduction of induced voltage is the main advantage of a booster transformer system. However, the magnetic flux density is higher than for other alternative designs.

6.6.4 AC specific safety rules for earthing and bonding

6.6.4.1 Basic requirements

The *return circuit*, the electrical equipment enclosures and the conducting components in the overhead contact line zone, see EN 50 122-1, are connected to the return circuit to avoid hazardous touch voltages during operation and during short-circuit faults. Figure 6.42 shows provisions that satisfy the safety requirements in a simplified circuit diagram. The individual earthing systems for bridges, tunnel segments, substations and pole foundations are connected to the return circuit and form the *overall earthing system* for an AC railway system.
The running rails, the return circuit conductors and the connections to the substation form the return circuit. To achieve as low a voltage drop as possible, welded rail connections are preferred and the track points are bonded longitudinally with low-resistance joints. To distribute the return currents evenly among all parallel tracks and return conductors, they are bonded to each other. The intervals between the cross bonds are defined depending on the earthing characteristics and the permitted touch voltage. The intervals normally vary between 300 and 1 200 m. Longer transverse cross-bonding intervals can be selected for sections supplied by substations equipped with *power electronic converters*, since the converters limit the short-circuit currents. The requirements of the *track-release circuits* should also to be taken into account for the spacing of the cross bonds.

The return currents in the substation flow through the return circuit and earth connections to the return current bar, see Figure 6.43 according to DB Directive 954.0107. A minimum of two return conductors must be provided between the track and the substation and rated so that the intact ones can carry the whole current after a failure of one of the conductors. Measurement of the return currents, as provided in the design of the return current system shown in Figure 6.43, permits testing of the return circuit. The current transformer T1 measures the whole railway return current if the cables from the return current bar to the transformer are insulated against earth. The portion of the return current through the substation earthing system is measured using the current transformer T2. T3 measures the residual current through the cubicles of the medium-voltage switch gear.

6.6.4.2   Substations and passenger stations

The traction power to *substations* is supplied decentralized from the public energy supply network or from railway-owned high-voltage networks in Austria, Germany and Switzerland for example. The high-voltage supply and the railway substations use a common earthing system. All operating assets in the high-, medium- and low-voltage supplies are connected to it for *equipotential bonding* and reliable protection tripping, as shown in Figure 6.42. The sheaths of cables used to connect the substation with the contact line can be earthed at both ends only if they can carry the whole traction current. In the case of earthing at one end, high voltages can occur at the free end of the cable sheaths. In this case, the ends of the cable sheaths need to be insulated against accidental touching. When the low-voltage supply is provided from the public network, the protective earth and the neutral conductor of the low-voltage system should not be connected to the AC railway earth, as they could be damaged by the railway return currents.

A portion of the return current flows through the earthing system of the substation and leads to a potential increase at the earthing system. When earth short circuits occur in the high-voltage supply, the fault currents flow through the earthing system. Therefore, it is necessary to have an especially low earthing resistance in the substation earthing system, to achieve a low earthing voltage.

Figure 6.43 illustrates the earthing provisions for the return circuit in the substation. The *main equipotential busbar* (ME) does not carry return currents to avoid interference in the connected SCADA system, telecommunications system, three-phase supply system, signalling system or the operating equipment cubicles. The ME busbar is connected to the *earthing busbar* of the earthing system at one end only. The foundation earth electrodes are connected to each other and attached twice to the earthing busbar. These connections are also designed to carry the maximum operating and short-circuit currents in the event of a failure of one of the earthing connections.

Since station platforms are located in the overhead contact line zone, the reinforcement of their concrete structures are connected to the return circuit. The platform foundations should be designed as earth electrodes to reduce the resistance to earth and to provide equipotential bonding, see Figure 6.42. Paper [6.52] describes the return circuit and earthing and bonding in an AC substation.

6.6.4.3 At-grade line sections

Contact line poles and conducting structures within the contact line zone are connected to the return circuit to allow rapid clearing of short circuits. Alternatively, the rails are earthed via the foundations of the poles to increase the conductance to earth and, therefore, reduce the rail potential. Adopting aerial return conductors connected to the steel poles or the reinforcement of concrete poles is preferred, because it is not necessary for it to be directly connected to the rails. Not having to connect each pole individually at the track offers significant savings during maintenance of the superstructure and earthing system. The connection to the aerial return conductor is also more reliable than a connection to the rails, because of the risk of damage by track maintenance works [6.39, 6.42]. The return conductors are connected to the tracks at spacings of 300 m to 600 m.

Concrete slab track, often used for high-speed-lines, requires additional provisions as described in [6.53]. These comprise:

- longitudinal interconnection of the concrete reinforcement of the surface layers
- cross connection at the position of the poles
- earthing sockets at the concrete surface
- earth connections at the location of the track bonds
- 50 mm minimum concrete cover

6.6.4.4 Tunnel sections

The reinforcement of tunnels is used as an earth electrode along the line. Both the tunnel reinforcement and the overhead contact line components are connected to the return circuit to reduce rail potentials and guarantee a reliable connection to the return circuit in the case of earth faults.

Components, such as supports of the contact line, are fixed to the tunnel ceiling. In certain cases, they are connected to unistrut brackets. If there is a reliable connection between the unistruts and the tunnel earth, no additional electrical connections are required. Measurements of the short-circuit resistance of fixing bolts and profiles manufactured from stainless steel have shown that the permitted heating was not exceeded for thermally effective short-circuit currents of 33 kA over a fault period of 350 ms. It is sufficient to connect just the unistruts to the return circuit.

Tunnels with sealing systems against ground water penetration, consisting e. g. of welded PVC sheaths between the inner and outer tunnel shells, lose contact with the soil and the *earthing effect of the tunnel*. Because of the danger of potential transfers at the emergency exits, which, in accordance with national tunnel safety codes, may not be more than 1 000 m apart, additional earthing measures are necessary if the rail potentials can cause unacceptable touch voltages.

Traction currents of up to 1,5 kA per train in the tunnels on the Cologne–Rhine/Main high-speed line and the sealing of the tunnels against penetration by ground water was in unfavourable conditions for the earthing of the contact line. With increasing tunnel length, the *rail potentials* also increased and exceed the permissible values by a factor of two if no counteractions had been taken. As long as no potentials are transferred into the tunnels from outside, humans cannot bridge the voltages to earth in the tunnel. The design of the tunnels, with emergency exits, leads to the possibility that the potential differences at the exits can be bridged by members of staff, for example. The measures taken to lower the rail potentials to permitted levels consisted of the following:

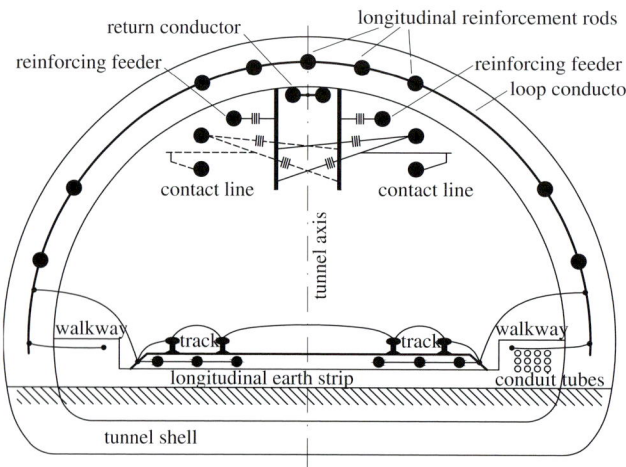

Figure 6.44: Conductor arrangement and earthing rods in a DB tunnel.

- an earth strip electrode laid in the outer tunnel shell, which is led through the sealing into the tunnel every 500 m and connected to the return circuit
- loop earth electrodes, arranged around the emergency exits to control the potential and reduce the earth electrode resistance

Figure 6.44 shows the arrangement of conductors and earthing system in a tunnel of DB's high-speed lines. The tunnel floor forms an earth grid to ensure good earth contact. Longitudinal earth strips are installed at a horizontal spacing of not more than 1,5 m across the tunnel floor to ensure that a broken contact wire touches one of the strips and trips the circuit breaker in a case of a short circuit. However, the running rails also form longitudinal conductors and are considered as sufficient for conducting short-circuit current if there is a broken contact wire. It should be kept in mind that a contact wire breakage cannot occur if there are no trains running.

DB also installs so called bouncing contact strips along the tunnel wall approximately 1,5 m above top of rail to ensure a short circuit if a broken contact wire touches the tunnel wall. As discussed in the context of longitudinal strips along the tunnel floor, these strips are considered as superfluous by other railway operators and engineers.

In the majority of cases, return conductors are installed above each track and serve as an earth connection for all components in the tunnel. Longitudinal bonding of the tunnel reinforcement is not necessary in this case. Figure 6.44 shows the conductor arrangement, including return conductors and the earthing rods. Paper [6.54] describes the earthing arrangement and earthing installations in Berlin Centre.

6.6.4.5 Viaducts

The foundations of *viaducts* form earth electrodes along the railway line. To utilise their earth electrode effect, the reinforcement of the individual viaduct segments is connected electrically via the supports down to the base of the foundations. Contact line poles on viaducts should be earthed like those on at-grade sections by connecting the poles to the viaduct reinforcement. The electrically interconnected reinforcement also forms an earth terminal for lightning protection for the viaduct. These connections should be kept as short as possible, to keep the surge inductance in the path of the lightning current as low as possible.

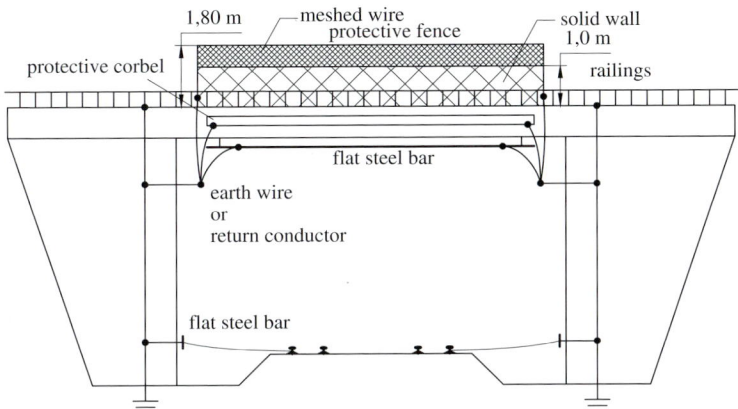

Figure 6.45: Earthing measures at road bridges (DB AG, Directive 954.0107).

6.6.4.6 Depot and workshop areas

No railway specific earthing measures are necessary in the *depot and workshop areas* of AC railways. In such areas, permissible values for the short-term touch voltages apply as permitted for the other track areas. For long-term touch voltages with a duration of 0,7 s and longer, only 25 V are permissible in workshops and similar locations, see also Clause 6.3.5.1. This value has to be considered when there are high currents for air conditioning and pre-heating of trains.

6.6.4.7 Signalling and telecommunications installations

If components of the signalling and track release system are placed within the overhead contact line zone, they will be connected to the running rails. The connections are designed to withstand short-circuit currents. Telecommunication and signalling cables can be affected by the traction power supply system. The cable sheaths of *signalling* and *telecommunication installations* can propagate voltages over long distances. Voltages caused by *inductive interference* can also arise. The cable sheaths are connected to the earthing systems at both ends in the substations and along the track to reduce interference. Since operational currents from the traction power supply flow through the cable sheaths, attention needs to be paid to providing the cable sheaths with sufficient current capacity.

6.6.4.8 Third-party earthing installations

Third-party earthing installations in the vicinity of the track should not be connected to the railway earthing system because of the danger of voltage and current propagation. For this reason, pipework from outside should be manufactured from non-conducting materials or interrupted at the site boundary by an insulating segment as illustrated in Figure 6.42.

If a separation between the railway and public supply network earthing systems is not possible due to lack of space, the return circuit should be interconnected with the neighbouring earthing system of the public networks. A satisfactory cross section for conducting railway return currents must be provided for this. As an example, Germany Railway DB permits the

Table 6.14: Required earthing resistance for AC railways for voltage impulses.

nominal voltage kV	15		25	
overvoltage category	III	IV	III	IV
withstand surge voltage kV	75	95	145	170
impulse earthing resistance Ω	1,9	2,4	3,6	4,3

operation of a three-phase star point conductor without special protection measures only over a distance of less than 1,5 km along the AC railway system [6.55].

If components of crossing *road bridges* lie within the overhead contact line and pantograph zone, then EN 50 122-1 and the DB Directive 954.0107 specify additional earthing provisions to ensure *personal safety*. Figure 6.45 illustrates recommended treatments:

- galvanized steel strip on both bridge walls, if these are located in the overhead contact line zone
- galvanized steel strip or angle section-profile above the overhead line at the start and end of the bridge if the bridge ceiling is within the pantograph zone
- protective fence or projecting device to avoid touching components possibly live on the bridge sides.

The metallic parts are connected at two points to the return circuit. It is recommended that the reinforcement of new bridges is also interconnected electrically and then connected to the return circuit for lightning protection. The bridge foundations can also be employed as earth electrodes in this case.

6.6.4.9 Lightning protection

Railway systems require high reliability and availability. Consequently, it is necessary to determine the measures to be applied to protect them against damage from lightning strokes, according to the risk management regime in IEC 62 305-2. Lightning strokes into the overhead contact line system cause a flash-over of the neighbouring insulators and the current is conducted via the poles and the earthing installations into the earth. It is possible for insulators to be damaged, however, a fracture is unlikely. The earth connections should be kept as short as possible, to keep the impulse resistance and inductance in the arrester path to a minimum. Suitable *overvoltage protection circuits*, in addition to the external *lightning protection*, are adopted to protect sensitive equipment assets.

Evaluations of the frequency of lightning current in accordance with IEC 62 305-2 revealed that 95 % of all lightning currents are lower than 40 kA and 99 % lower than 60 kA. Back flashovers should not be expected if the following condition is fulfilled:

$$R_{dis} \leq U_{ins}/I_p \tag{6.43}$$

R_{dis} *impulse earthing resistance*
U_{ins} impulse withstand voltage of the insulation
I_p peak value of the lightning current at pole or steel structure

For earth electrodes of small dimensions, such as pole foundations, the surge earthing resistance corresponds approximately to the earth electrode resistance at nominal frequency.

The necessary *impulse earthing resistance* for a lightning current I_p of 40 kA with respect to the overvoltage category for AC railways is shown in Table 6.14.

The permissible values for the impulse earthing resistance increase due to the withstand impulse voltage. Consequently, the requirements for the impulse earthing resistance for AC 25 kV railways can be satisfied more easily than for 15 kV.

In Europe, the keraunic level that indicates the lightning activity is quite low. Therefore, specific protection measures are not generally applied. In case of higher exposure, e. g. on bridges, surge arresters are installed. Their sphere of influence is limited to a short distance.

6.6.4.10   Implementation procedure for earthing provisions

The return circuit and earthing provisions have an impact on steel-reinforced concrete structures for railway purposes and should be defined at an early stage prior to construction. Preparation of *electrical connections* of the reinforcement, the provision of additional reinforcement rods and earthing connections in the foundation and the lead-out of the earthing connections need to be carried out during the first *implementation* phase and initiated much earlier than the detailed planning of the electrical system, especially for railways on viaducts and in tunnels with long lead times for the construction work.

This includes the timely agreement of materials to be employed, cross sections and connection technology for the structure earth. If earthing connections and electrical through-connections were not provided on the structures and were missing during the installation of the electrical systems, then alternative solutions would be required later. This can mean considerable additional investigations for implementation of alternatives.

The electrical connections between the reinforcement rods should preferably be welded, because the resistance of clamp connections can increase with corrosion at the connection point, see DB Directive 997.0101. The defined earthing provisions should be monitored by *visual inspection* during construction, because errors during construction are difficult and expensive to correct.

6.6.4.11   Measurements for design verification

The operational reliability of the earth return current path and the safety provisions for persons must be verified during the *commissioning of the contact line installation*. Based on calculations made during the design phase, the *verification of earthing provisions* can be performed by measurements made after completion of the installation. It is expedient to measure the earth *electrode resistance*, *rail potentials* and *induced voltages* to check the parameters upon which the calculations were based. These measurements also serve as a reference for the subsequent operation of the system.

The earth electrode resistance of the earthing system determines the *touch voltages* and rail potentials. The rail potential is measured by feeding a constant current into the running rails, between two rail connectors. The potential of the running rails is measured against remote earth. The distance of the measuring point to the nearest cross bond of the return circuit should be as long as possible, to ensure examination of worst case conditions. Feeding current to one track represents the operational case, while the supply to one rail represents the short-circuit case. The measured rail potentials can be converted by calculating an indicative potential due to operational and short-circuit currents. The measurements can be used to verify that the permitted touch voltages will not be exceeded during operational and short-circuit conditions.

6.6.5 Earthing and bonding for DB lines

6.6.5.1 Design of traction current return circuit

German Railway DB is used as an example to explain some aspects to be taken into consideration regarding earthing and bonding when planning, constructing and operating overhead contact line installations for AC 15 kV 16,7 Hz installations. DB Directives DS 997.0201 to 997.0224 deal with the respective details. Other railway entities operating single-phase alternating current railways own similar internal specifications and rules. On DB's network, return circuits via running rails and earth, see Figure 6.4 a) are adopted for conventional lines and with additional return conductors for lines where higher currents are expected, see Figure 6.4 b).

The running rails are used to conduct the majority of the traction current back to the feeding substation, however, they are also used as part of the signalling and train control system as well as for the track release system using *track circuits*. Depending on the type of track circuits, additional measures or devices are required to conduct the return current without disturbing the track circuit. Track circuits are operated at frequencies of 42 Hz or 100 Hz and *audio-frequency track release circuits* at frequencies of 4 to 6 kHz and 9 to 17 kHz.

The terms of mutual utilization of the running rails must be agreed upon and co-ordinated by the responsible technical departments with strict adherence to the agreed rules. From the electrical engineering aspect, track design for return circuit and traction earthing distinguishes between:

- *uninsulated track* without track circuits. Both rails per track conduct the return current
- *track with one rail insulated* and with track circuits. One rail per track conducts the return current
- *track with both rails insulated* and with track circuits. Both rails are insulated from the next section and connected via impedance bonds
- track with audio-frequency track release circuits

The type of track release system determines the location and number of insulating rail joints. Figure 6.46 shows various track circuits installations and the permitted earthing connections. Basically, both rails and all tracks are used as return current conductors. The rails of electric railway lines should be connected electrically both longitudinally and across all tracks wherever it is permissible to distribute the return currents more evenly and to ensure an equal potential. Welded rail connections provide the best electrical connection. Generally, *fishplate joints* between rails are adequate for the longitudinal connection. Fishplate joints in sections with track circuits need to be bridged by an additional *longitudinal bond*. Transverse rail bonds or *crossbonds* are used to connect both rails of a track but the design and arrangement depends on the requirements of the track circuits.

6.6.5.2 Tracks with uninsulated rails

In tracks with no track circuits, including tracks with *axle counting equipment*, no specific requirements exist from the signalling viewpoint for the insulation of the running rails from earth. Both rails are used continuously to conduct the return traction current and may also be used for protective earthing. The running rails of a track without track circuits are bonded at intervals of approximately 150 m on conventional lines or approximately 75 m on lines carrying heavy traffic loads. The tracks of multi-track railway lines are cross-bonded by *track cross-bonds*. These are installed at intervals of approximately 300 m on conventional lines or

a) Track release circuit with single-rail insulation

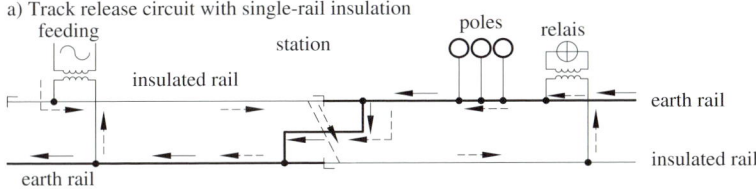

b) Track release circuit with double-rail insulation

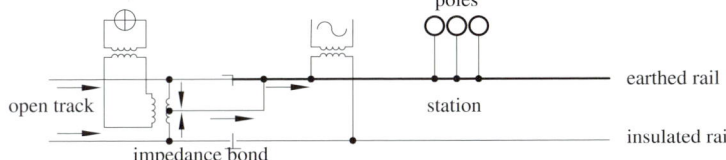

c) Track release circuit with transition from single-rail to double-rail insulation

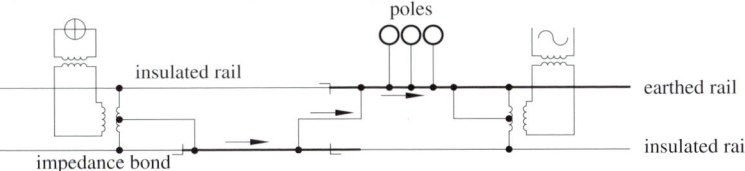

d) Track release circuit with special centre-tapped reactor

Figure 6.46: Track release circuits.

approximately 150 m on lines carrying heavy traffic loads. The individual distances depend on the expected rail potentials during operation and short circuits. The rail next to the contact line poles is used for the earthing connection.

6.6.5.3 Tracks with single-rail insulation

On tracks with one rail insulated from earth for track release purposes, the other rail is used for the return circuit and earthing purposes and is called earth rail. (Figure 6.46 a)). Earth connections are only permitted to the earth rail. The earth rails of adjacent tracks are cross bonded at intervals of approximately 300 m on conventional lines or approximately 150 m on lines carrying heavy traffic loads. The insulated rail must have a voltage limiting device, also called a voltage fuse, see Clause 6.5.2, Figure 6.22, that is connected to the earth rail and forms a short-circuit path in the case of contact with the overhead contact line system.

Single-rail insulation can be used with 42 Hz or 100 Hz track release circuits. The version according to Figure 6.46 a) is mostly applied in railway stations, where the earth rail is provided with at least two connections to the return circuit. Figure 6.46 c) shows the transition from double-rail insulation to single-rail insulation and Figure 6.46 d) the change of the insulated rail via a rail joint.

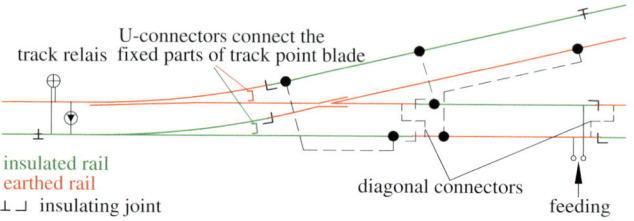

U-connectors connect the
track relais   fixed parts of track point blade

insulated rail
earthed rail   diagonal connectors
⊥⌐⌐ insulating joint   feeding

Figure 6.47: Track release circuit at a track point

a) Connection of components with $R_E < 4\,\Omega$

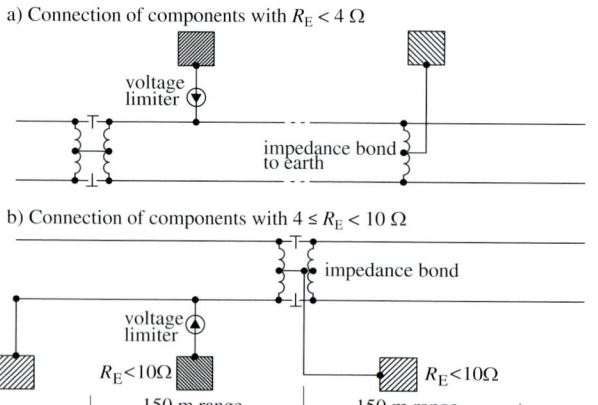

voltage
limiter

impedance bond
to earth

b) Connection of components with $4 \leq R_E < 10\,\Omega$

impedance bond

voltage
limiter

$R_E < 10\Omega$ $R_E < 10\Omega$

150 m range   150 m range

Figure 6.48: Earthing of tracks to poles and other components with a low resistance to earth, both rails insulated.

Insulating rail joints must be taken into account for the design of the return circuit, especially at track points, see Figure 6.47. At passenger stations, where single-rail insulation is common practice, the earth rail must be connected to the return circuit via two cables. The earth rails of parallel tracks are cross-bonded with a spacing of about 300 m for conventional lines and 150 m for high-speed lines.

6.6.5.4   Tracks with both rails insulated

Tracks with both rails insulated from earth are used with 42 Hz or 100 Hz track circuits. Then, as shown in Figure 6.46 b), both rails are used as return conductors for the traction current. To enable reliable operation of the track circuits, each track release section is isolated from the adjoining one by *insulating rail joints* in conjunction with *impedance bonds*. Figure 6.46 b) shows where track joints are installed along a line. The traction current flowing back to the substation passes through the impedance bonds, comprising two reactors with connected centre taps. The current flows through the reactors in such a way that the inductive effects cancel one another. The centre tap connections of the reactors can also be used as connection points for earthing purposes.

In tracks with both rails insulated, one of the rails is defined as being the *earth rail*. Following DB's rules, any components with earthing resistances $\geq 10\,\Omega$ may be connected to this rail without any restrictions. Any components having earthing resistances between $4\,\Omega$ and $10\,\Omega$ may only be connected with the earth rail at distances up to 150 m in front of and more than 150 m behind the *impedance bonds*. Within the range of 150 m to either side of the impedance bond, all such components must be earthed to the centre-tap connection, either via impedance bonds or voltage fuses. Figure 6.48 illustrates aspects to be considered when connecting components to the running rails.

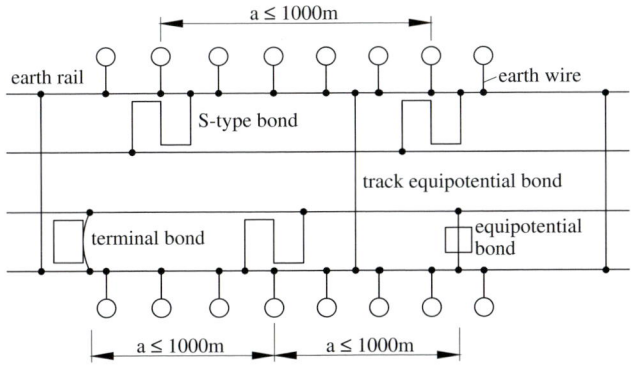

Figure 6.49: Track bonds and S-type rail bonds for audio-frequency track release circuits.

It is not permissible to install track or rail equipotential bonds in tracks with both rails insulated. In this case, the tracks are bonded by connecting the centre-taps of the impedance bonds.

6.6.5.5 Tracks with audio-frequency circuits

Like all other types of track circuits, audio-frequency track release circuits operate on the principle of axle shunt sensing. Remotely fed audio-frequency track release circuits are circuits using a frequency of 4 to 6 kHz for track release systems on main lines and a frequency of 9 to 17 kHz for track release systems in railway station areas.

Both rails are used to conduct the return current. Usually, the rail next to the contact line poles is defined as the *earth rail*. On at-grade main line sections, the two rails are interconnected by S-type bonds, terminal bonds, short-circuit or equipotential bond connections at intervals of less than 1 000 m. As indicated in Figure 6.49, S-type bonds, short-circuit bonds and terminal bonds form the control and command circuit termination of the respective audio-frequency track circuits. In control and command engineering, they are called *electrically insulating rail joints*. The bonds use copper wires of cross sections between 50 and 600 mm². Insulating rail joints are not necessary. The earth rails of parallel tracks are bonded by track bonds at distances depending on the audio-frequency circuits.

6.6.5.6 Cabling requirements for bonds

Usually, equipotential bonds are made of insulated copper cable type NYY-O, normally with a cross section of 50 mm². If short-circuit currents of $I_k'' > 25$ kA are to be expected at the respective location, then cross sections of 70 mm² are installed. If the bonds are embedded in concrete, a minimum cross section of 70 mm² is required and for $I_k'' > 25$ kA a 95 mm² cable is used. The cross sections of bonds for audio-frequency lines need to be selected individually. The bonds are permanently connected to the rails by welding, soldering, brazing or other adequate methods. Care must be taken not to damage the rails.

6.6.5.7 Crossbonding between the return circuit and steel reinforcement

The DB Directive 997.0223 requires that all *reinforcement* of concrete structures on or within which tracks are laid are bonded to the return circuit. This ensures effective equipotential

Figure 6.50: Overhead contact line with return conductors at the high-speed line Madrid–Seville.

bonding and a definite short circuit that will cause the corresponding circuit breaker to trip if an unintentional contact between the contact line and the concrete structure occurred.

The reinforcing rods and all corresponding longitudinal conductive parts are interconnected electrically and connected with the earth rail or the *return conductors* at intervals of not more than 100 m. The connections between the steel reinforcement embedded in the concrete are welded. However, it is not permissible to interconnect and bond steel rods used for pre-stressed concrete components. Here, additional steel rods need to be installed and bonded to the return circuit. Poles, railings and noise-reduction barriers installed on railway bridges are connected with the reinforcement of the respective structures. Insulating rail joints for track release circuits must not be shunted by parallel return conductors or earthing installations, e. g. noise protection walls. In such cases a separation is required.

In structures longer than 100 m, additional continuous steel bars with a cross section of at least 120 mm^2 or additional continuous reinforcement rods of at least 16 mm in diameter are placed in the top concrete layer under each track. In the case of track release systems with both rails insulated, it is not permissible to connect the rails with the reinforcement of bridges because of their low resistance to earth. Earth busbars are installed in such cases and all components to be bonded to the return circuit are connected to these busbars. The busbars are connected to the centre-taps of impedance bonds, see Figure 6.48. The DB Directive 997.02 contains further details on the *design of earthing and bonding* for DB lines.

6.6.6 Earthing and bonding of the Madrid–Seville AC 25 kV 50 Hz high-speed line

Examples of the application of the earthing design described in Clause 6.6.3 for AC systems are represented by:
- Madrid–Seville and Madrid–Toledo high-speed lines in Spain
- high-speed line HSL Zuid, (Netherlands)
- ERL Express Rail Link in Kuala Lumpur (Malaysia)

As an example, details for earthing and bonding will be follow for the Madrid–Seville line. The line is supplied by AC 25 kV 50 Hz [6.42]. The electrification scheme was discussed in

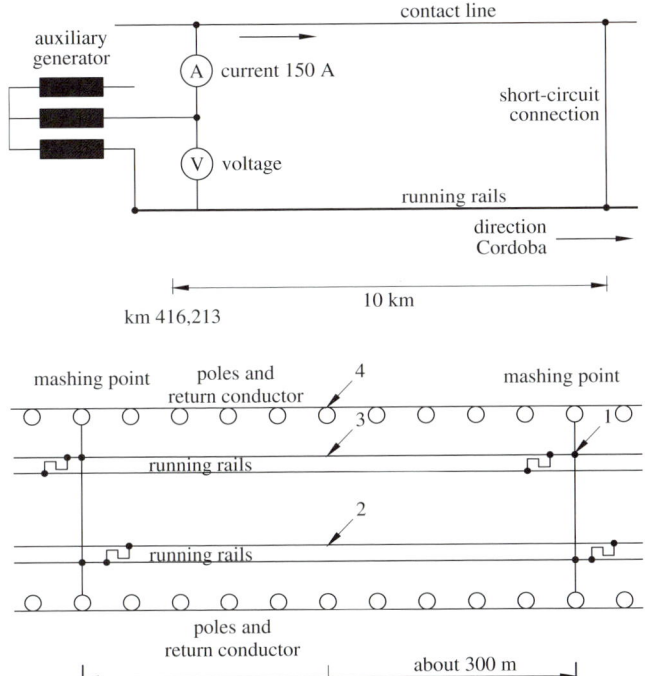

Figure 6.51: Circuit diagram for measurements on the high-speed line Madrid–Seville.

Figure 6.52: Return circuit arrangement and feeding points to measure the rail potential.

Clause 1.6. Comparisons of several alternatives resulted in utilisation of *return conductors* Al 240 installed on the contact line poles. Figure 6.50 shows the line with the return conductors. Calculations carried out during the design confirmed that this solution is favourable considering line impedance and keeping rail potentials and *interference* at acceptable levels. The investment is relatively low compared to other alternatives such as the use of autotransformers. The design calculations were verified by measurements during the commissioning phase.

The verification measurements required a section which represented typical conditions of the line as found in the southern part of the line between Córdoba and Seville close to the Lora del Rio substation. The length of the measuring section was set at 10 km. Boundary effects in the vicinity of the substation and at the load site are, therefore, negligible. The test circuit is shown in Figure 6.51. A diesel generator set in the Lora del Rio substation provided currents of 150 A. Measurements taken along the section before tests started showed that the soil resistivity was 30 Ωm.

The rail potential was measured for four situations as shown in Figure 6.52:

- current fed into one rail at the bonding point of tracks and return conductors (1)
- current fed into two rails between the bonding point of tracks and return conductors (2 and 3)
- current fed into one rail between the bonding point of tracks and return conductors (3)
- short circuit at an insulator at a pole midway between two adjacent bonding points of tracks and return conductors (4).

Table 6.15 lists the measured *rail potentials* observed in all four cases. In case 4, a touch voltage of 5,8 V/100 A resulted from the measurements between the pole and a location 1 m

Table 6.15: Rail potential.

Feeding arrangemant	Rail potential V/100 A
case 1	2,6
case 2	3,5
case 3	5,8
case 4	5,8

Table 6.16: Induced voltages in unshielded conductors along the line.

Distance perpendicularly to the centre of track in m	Measurement V/(kA/km)	Calculation V/(kA/km)
6	34	42
11	40	43
20	41	39
120	13	20

distant from the side of the pole away from the track. The voltage difference between pole and rail was 7 V/100 A.

In case 1, calculations and measurements yielded the same results based on an earthing resistance of 5 Ω for each pole. When the earthing resistance was assumed to be 15 Ω per pole, the rail potential was computed to be 50 % higher revealing the close dependency between the earthing characteristics of the poles and the rail potential.

Without return conductors, the calculations for case 1 yielded rail potentials 50 % higher than for the installed system.

In addition to the magnitude of the return current, the current distribution within the return circuit is responsible for the induced longitudinal voltages in cables laid parallel to the track. The specific *induced voltage* related to the cable length is highest at the midpoint between substation and vehicle or substation and short-circuit location, because, there, the proportion of current returning through earth is at its maximum. Consequently, the measurements were taken in the middle of the test section. At the measuring position, unsheathed cables were laid out at various distances away from the track centre line, and the induced longitudinal voltages were measured. The measured and calculated results are listed in Table 6.16. The relatively close correlation between calculations and measurements validated the calculations as a reliable planning tool.

For a section without return feeders, calculations with the same basic parameters yielded interference voltages of 70 V/(kA·km), i. e. values about 70 % higher, for an unsheathed cable 6 m away from the track centreline.

In the contact line system and return circuit of electrified railway lines many individual conductors are connected in parallel. Unlike DC railways, in AC railways the currents are distributed according to the self and coupling impedances and not just the resistances. Near the substations and load locations, conductance between the return conductors and earth also has to be considered. In the middle part of sufficiently long sections, a constant current distribution establishes itself in the return feeder system because no current is exchanged between the return feeder system and the earth. That is why the current distribution was measured at the midpoint of the measuring section, as shown in Figure 6.52. At the measuring position current transformers were fitted in the feeding side to the contact lines and to rails, to return conductors and to traction-earthed cable sheaths. Table 6.17 lists the calculated and measured results for the conductors concerned. The return current component flowing through earth cannot be measured, so only the calculated value is given. The measured values yield an earth current of about 20 %.

Compared with systems without them, the return conductors reduce the return current component flowing within the soil by approximately 40 % and through rails by approximately

Table 6.17: Current distribution related to the total traction current.

| | With return conductor | | Without return conductor |
	calculation	measurement	calculation
	%	%	%
Contact line equipment			
contact wire A	30,6	29,7	31,0
catenary wire A	20,4	20,7	20,0
contact wire B	30,6	29,5	31,0
catenary wire B	20,4	20,2	20,0
Return conductor system			
rails A	23,2	20,4	34,2
return conductor A	18,5	17,7	–
cable shield A	–	2,8	–
rails B	23,2	19,5	34,2
return conductor B	18,5	16,1	–
cable shield B	–	3,0	–
earth	20,6	not possible	34,4

35 %, as shown in Table 6.17. This is also the reason for the favourable effect on interference voltages.

6.6.7 Conclusion for design of AC return circuit and earthing

The use of the running rails for the traction return circuit is sufficient for conventional lines with moderate traffic. The inductive interference is tolerable for AC 16,7 Hz as well as for AC 50/60 Hz systems

Additional return conductors on the poles are recommended for increased traffic and power demand. They reduce the rail potential and the induced voltage along the line by the half. In Clause 6.6.6 the advantages of return conductors are shown for the Madrid–Seville high-speed line.

Auto-transformer systems generally allow higher traffic loads and longer substation spacing with respect to interference and rail potentials. The complexity of the system, the high power demand at the feeding connection to the high-voltage network and operating effort must also be considered. Auto-transformer systems e. g. are often used for high-speed lines in France, Japan and Russia.

Booster-transformer systems reduce the return currents in the rails and earth by far more than return conductors, especially important for regions with high soil resistivity. The voltage drop along the line, however, and the insulating sections in the contact line allow only low-power traffic. Therefore, BT Systems are used only under specific conditions, e. g. in Norway and Sweden.

6.7 Interaction between AC and DC traction systems

6.7.1 Mutual electrical interaction situations and interference processes

Where electric traction systems with AC and DC power supply approach, both systems interact in principle. Interference situations arise:

parallel crossing system common
                    change buildings

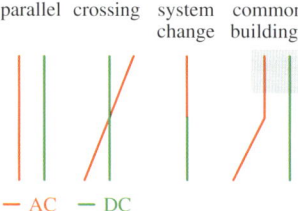

— AC   — DC

Figure 6.53: Typical situations for a mutual approach between AC and DC lines.

a) AC 25 kV High-speed line AVE Madrid–Seville and DC 3 kV mainline

b) 2 AC 50/25 kV High-speed line HSL Zuid and DC 1,5 kV mainline

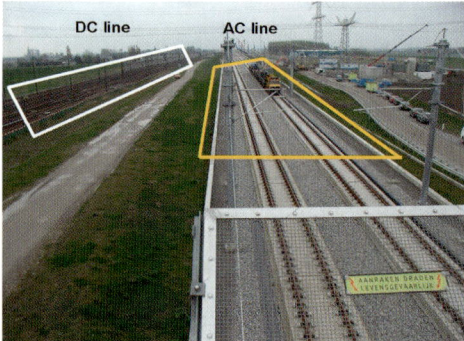

Figure 6.54: Examples for AC and DC railway lines close together.

– where AC and DC lines run in parallel, caused by inductive or capacitive coupling from the AC side and galvanic interaction due to the rail potentials of both systems
– where AC and DC lines cross, often combined with common structures
– at system separation sections between AC and DC lines and at commonly used tracks for AC and DC operation
– where railway stations contain AC mainlines and DC mass transit systems, which is the case in cities with mass transit systems and incoming railway lines.

Figure 6.53 shows the main interference situations, Figure 6.54 gives two examples for parallel AC and DC lines in Spain and the Netherlands.

The technical interaction processes between AC and DC traction systems can be deduced from the three basic relations:

– galvanic coupling
– inductive coupling
– capacitive coupling

Standard EN 50 122-3 describes these effects and [6.56] gives further information.

The AC interaction can cause AC currents flowing in the DC traction system and cause voltages which can endanger people and technical equipment. The DC interaction can cause DC currents flowing in the AC traction system with the risk of stray current corrosion and increased touch voltage. Examples of the consequences of such interactions are:

– electric shock from AC and DC and mixed AC-DC touch voltages
– transformer saturation in substations and trains caused by DC current components
– stray current corrosion in the AC return circuit and earthing installations
– malfunction of telecommunication and signalling circuits.

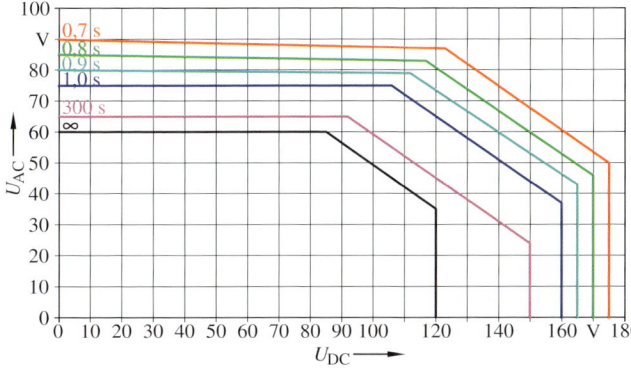

Figure 6.55: Limiting curves for permissible mixed AC and DC touch voltages for long-term situation.

6.7.2 Zone of mutual interaction

Interference effects only need to be considered in a zone where adverse effects are likely. EN 50 122-3 specifies this zone in Clause 6 and gives calculated examples in Clause 6.2 and Annex A.

Where there is a distance between AC and DC lines of less than 50 m, a mutual interaction is likely and has to be considered. An AC line e. g. can affect a DC line with high induced voltage as indicated in the following calculation. The running rails in the DC line are affected up to 1 000 m in case of:

– a double track line with running rails as return circuit
– inducing current of 500 A per overhead contact line (1 000 A in total)
– a length of parallelism between AC and DC railway of 4 km

The DC-running rails are assumed to be insulated from earth. Standard EN 50 122-3, Annex A gives calculated support for different line parameters.

The effects of DC railway systems on AC railway systems can be neglected, if the insulation of the running rails from earth accords with the requirements of EN 50 122-2. Only the voltage transfer needs to be considered for the assessment of the touch voltage.

6.7.3 Permissible values for combined AC-DC touch voltages

The permissible touch voltages for AC or DC traction systems are given in EN 50 122-1, see also Clause 6.3.5. The permissible touch voltages for mixed AC and DC voltages, which is relevant for AC-DC interaction, are specified in EN 50 122-3. The limits depend on the time duration of both types of voltage:

– AC long-term and DC long-term conditions
– AC short-term DC long-term conditions
– AC long-term and DC short-term conditions
– AC short-term and DC short-term conditions
– specific requirements for workshops and similar locations

Figure 6.55, taken from EN 50 122-3, Figure 1, shows the permissible combined effective touch voltages, depending on the duration of impact. The diagram shows that, e. g. for a duration of 1 s and an AC voltage of 75 V, an additional DC component of up to 105 V is allowed. Permissible values for mixed voltages with other conditions of duration are given in EN 50 122-3, Chapter 7 and Annex B.

6.7.4   Technical requirements and implementation measures

If investigations state a zone of mutual interaction, countermeasures are necessary for the safety of persons and the protection of installations.

Where there are overlapping overhead contact line zones of AC and DC lines, both return circuits need to be connected via voltage limiting devices VLD, see EN 50 122-3, Figure 5. They provide a permanent direct connection of both return circuits and ensure short-circuit tripping if there is a connection of an overhead contact line with the other return circuit. Both return circuits are not connected during operation.

A separation of the structure earth associated with the AC- and the DC-line respectively needs to be considered depending on the feasibility. If there is no possibility of a proper separation, both structures have to be connected together with sufficient cross section of conductors so that fault currents can also be conducted. Separation from outside earthing systems is also required to block stray currents.

Where there is separated structure earth of AC and DC lines, insulating gaps are necessary in the whole structures and no other installations must bridge these gaps. Clause 8.2 of EN 50 122-3 gives details.

The common use of running rails for AC and DC is a specific operational situation, where the running rails must not be connected with the structure earth because of the risk of stray current corrosion. Specific precautions are given Clause 8.3 of EN 50 122-3, mainly to avoid the risk of stray current corrosion.

System separation sections between AC and DC require specific operational procedures to avoid interaction between AC and DC system. The main requirements are, to avoid the current collector of the trains coming into contact with other power supply systems and vice versa. From the traction power supply;

- no connection between AC- and DC-overhead contact lines are allowed
- reliable current return path is necessary to the connected AC or DC traction power substations for operation and in case of short circuits.
- limitation of current exchange between the return circuit of AC and DC, e. g. by insulation rail joints, where the voltage between them must be limited to the permissible touch voltage.

Examples for system separation sections are given in Clause 14.4 and in [6.57, 6.58, 6.59].

6.8   Bibliography

6.1 *Schneider, E.*: Bahnrückstromführung und Erdung – Teil 1: Grundsätze (Earthing and bonding and return circuit in railway installations – Part 1: Principles). In: Elektrische Bahnen 96(1998)4, pp. 85 to 90.

6.2 *Kießling, F.; Nefzger, P.; Nolasco, J. F.; Kaintzyk, U.*: Overhead power lines – Planning, Design, Construction. Springer-Publishing, Berlin – Heidelberg – New York, 2003.

6.3 *Kontcha, A.; Schmidt, P.*: Elektrosicherheit im Bereich von Oberleitungen elektrischer Bahnen (Electrical safety within the overhead contact line zone of electric railways). In: Elektrische Bahnen 94(1996)10, pp. 297 to 303.

6.4 VDV Recommendation 501, Part 1 to 3: Reduction of the corrosion danger due to stray currents in tunnels of DC traction systems with return current via running rails. Verband Deutscher Verkehrsunternehmen (VDV), Köln 06/2012.

6.5 VDV Recommendation 507: Design and protective provisions for electrical power installations along DC mass transit lines.Verband Deutscher Verkehrsunternehmen (VDV), Köln 06/2005.

6.6 *Bette, U.*: Verringerung der Streustromkorrosionsgefahr an Bauwerken von Gleichstrombahnen (Reduction of the stray current corrosion hazard in buildings of DC railway installations). In: Nahverkehrspraxis (1994)9, pp. 312 to 316.

6.7 *Bette, U.*: Maßnahmen zur Verringerung der Korrosionsgefahr durch Streuströme und Erdungsmaßnahmen bei Gleichstrombahnen (Measures to reduce the corrosion hazard by stray currents and earthing for DC railways). In: ETG Report, Part 30, vde Publishing, Berlin – Offenbach.

6.8 *Schneider, E.; Zachmeier, M.*: Rückstromführung und Erdung bei Bahnanlagen – Teil 3: Gleichstrombahnen (Earthing and bonding and return circuit in railway installations – Part 3: DC traction systems). In: Elektrische Bahnen, 96(1998)4, pp. 99 to 106.

6.9 VDV Recommendation 525: Overvoltage protection for traction supply systems of DC urban rail systems. Verband Deutscher Verkehrsunternehmen (VDV), Köln 06/2012.

6.10 *Altmann, M. et al.*: Streustromüberwachung bei der U-Bahn Nürnberg (Stray Current Monitoring at Nuremberg subway). In: Elektrische Bahnen 102(2004)5, pp. 223 to 230.

6.11 *Siemens AG*: Sitras SMS – Streustrom-Monitoring-System für die DC-Bahnstromversorgung (Sitras SMS – Stray Current Monitoring for DC traction power supplies). Product information, 2013.

6.12 *Feydt, M.*: Vorschläge zur Verwendung der Kabelmäntel metallener Rohrleitungen, der Gleise und der Erdseil-Maste-Kettenleiter als natürliche Erder (Proposals to use cable sheets, metallic pipelines, tracks and earthwire pole recurrent network as natural earth electrodes). Report of the Institute for Energy Suppl, Dresden, 1982.

6.13 *Wenner, F.*: A method of measuring earth resistivity. Scientific papers of the Bureau of Standards 258(1917) pp. 469 to478.

6.14 Digital earth tester MEGGER DET/3R & DET5/3D: User Guide, AVO-International, Kent CT179EN, England.

6.15 *Nitsch, K.*: Ergebnisse der Untersuchung des Isolationswiderstandes von Stahlbetonschwellen (Results of investigations of the insulation resistance of steel-reinforced concrete sleepers). In: Signal und Schiene, 10(1966)9, pp. 376 to 383.

6.16 *Hellige, B.; Hampel, H.*: Untersuchung des Überganswiderstandes von Straßenbahngleisen (Investigation of the transition resistance of tramway tracks). In: VESK Information, Dresden, 4(1971)4, pp. 28 to 34.

6.17 *Ollendorf, F.*: Erdströme (Currents through the earth). Birkhäuser-Publishing, Basel – Stuttgart, 1969.

6.18 *Eichhorn, K. F.*: Stromverdrängung und Stromleitung über Erde (Current displacement and current conduction through earth). In: Elektrische Bahnen, 95(1997)3, pp. 74 to 81.

6.19 *Schaller, K.-P.*: Untersuchung über das Verhalten der Rückströme im Erdreich bei Einphasenwechselstrombahnen (Investigation on the behaviour of return currents through earth for single-phase AC railways). HfV Dresden, thesis for diploma, 1965.

6.20 *Schmidt, P. et al.*: VEM Handbuch Energieversorgung elektrischer Bahnen (VEM Manual Power supply of electrical railways). VEB-Publishing Technik, Berlin, 1975.

6.21 *Kontcha, A.*: Analyse elektromagnetischer Verhältnisse in Mehrleiterfahrleitungssystemen bei Einphasenwechselstrombahnen (Analysis of electromagnetic processes in multi-conductor overhead contact line systems of single phase AC railways). TU Dresden, doctoral thesis, 1996.

6.22 *Thiede, J.; Zeller, P.*: Niederspannungsbegrenzer für Gleichstrombahnen (Low voltage limiters at DC railways). In: Elektrische Bahnen, 100(2002)10, pp. 399 to 403.

6.23 *Bette, U.*: Messungen in Betriebshöfen und an Verkehrsbauwerken (Measurements in depots and general traffic facilities). In: Reports and information of HTW Dresden, 4(1996)1, pp. 89 to 101.

6.24 *Weitlaner, E.; Schneider, E.*: Bahnstromversorgung für die Stadtbahn BTS in Bangkok (Railway electrification system of MRT system BTS in Bangkok). In: Glasers Annalen, 123(1999)6, pp. 253 to 260.

6.25 *Röhlig, S.*: Streuströme bei DC-Bahnen und elektrotechnische Anforderungen an den Gleisbau (Stray currents at DC traction systems and requirements for the trackwork). In: Elektrische Bahnen 99(2001)1-2, pp. 84 to 89.

6.26 *Altmann, M.; Schneider, E.*: Spannungen und Überspannungen in der Rückleitung von Gleichstrombahnen (Voltages and over-voltages in the return circuit of DC traction systems). In: Elektrische Bahnen 104(2006)3, pp. 129 to 136.

6.27 *Schneider, S.*: Erdung und Potenzialausgleich an oberirdischen Bestandsstrecken (Earthing and bonding at existing at-grade lines). In: Elektrische Bahnen 109(2012)4, pp. 152 to 157.

6.28 *Röhlig, S.; Rothe, M.*: Dynamische Berechnung von Streuströmen und Gleis-Erde-Spannungen (Dynamical calculation of stray currents and track to earth potentials). In: Reports and information of HTW Dresden, 2(1994)1, pp. 59 to 64.

6.29 *Schneider, E.*: Streustromberechnung bei geerdetem und nicht geerdetem Rückleiteranschluss von Gleichstrombahn-Unterwerken (Stray current analysis for earthed and earth-free return conductor connection at DC substations). In: Reports and information of HTW Dresden, 2(1994)1, pp. 65 to 71.

6.30 *Biesenack, H.; Dölling, A.; Schmieder, A.*: Schadensrisiken bei Blitzeinschlägen in Oberleitungen (Risk of damage in case of lightning strokes at overhead contact lines). In: Elektrische Bahnen 104(2006)4, pp. 182 to 189.

6.31 *Lingohr, H.; Stahlberg, U.; Richter, B.; Hinrichsen, V.*: Überspannungsschutzkonzept für DC-Bahnanlagen (Overvoltage protection concept for DC railways). In: Elektrische Bahnen 101(2003)7, pp. 315 to 320.

6.32 *Bette, U.; Galow, M.*: Ableiter und Spannungsbegrenzungseinrichtungen für Netze DC 750 V (Surge arresters and voltage limiting devices for DC 750 V lines). In: Elektrische Bahnen 104(2006)3, pp. 137 to 144.

6.33 *Bette, U.; Sons, W.*: Streustrombewertungen gemäß DIN EN 50 122-2 (Stray current assessment according to DIN EN 50 122-2). In: Elektrische Bahnen 106(2008)1-2, pp. 66 to 67.

6.34 *Fischer, Ch.; Thiede, J.*: Erfahrungen mit der Streustrombewertung gemäß DIN EN 50 122-2 (Experience with the stray current assessment according to DIN EN 50 122-2). In: Elektrische Bahnen 106(2008)11,pp. 501 to 507.

6.35 *Deutschmann, P.; Schneider, E.; Zachmeier, M.*: Rückstromführung und Erdung bei Bahnanlagen – Teil 2: Wechselstrombahnen (Earthing and bonding and return circuit in railway installations – Part 2: AC traction systems) . In: Elektrische Bahnen 96(1998)4, pp. 91 to 98.

6.36 *Braun, W.; Schneider, E.*: Konzepte für Rückstromführung und Erdung bei AC-Bahnen (Conceptions for the traction return circuit and earthing for AC traction systems). In: Elektrische Bahnen 103(2005)4-5, pp. 219 to 224.

6.37 *Behrends, D.; Fischer, Ch.*: Berechnungen nach DIN EN 50 122-1 – Erdung im Katzenbergtunnel (Calculations according to DIN EN 50 122-1 for the earthing installations in the Katzenbergtunnel). In: Elektrische Bahnen 109(2011)11, pp. 592 to 600 and 12, pp. 680 to 684.

6.38 *Aeberhard, M.; Kocher, M.; Koch, M.*: Ausbau der Bahnstromrückleitung auf der Lötschberg-Bergstrecke (Construction of the installation for the traction current return at the Lötschberg mountain railway). In: Elektrische Bahnen 101(2003)8, pp. 377 to 386.

6.39 *Tischer, G.*: 20 Jahre Einsatz von Bahnrückstromleitern (20 years application of return conductors). In: Elektrische Bahnen 92(1994)4, pp. 97 to 104.

6.40 *Zimmert, G. et al.*: Rückleiterseile in Oberleitungsanlagen auf der Strecke Magdeburg–Marienborn (Return conductors used for overhead contact line installations on the Magdeburg–Marienborn line). In: Elektrische Bahnen, 92(1994)4, pp. 105 to 111.

6.41 *Gruber, A.*: Rückstromführung auf ÖBB-Hochleitungsstrecken (Traction current return on Austrian Railway's high-performance lines). In: Elektrische Bahnen, 89(1991)11, pp. 404 to 408.

6.42 *Kießling, F.; Schneider, E.*: Verwendung von Bahnstromrückleitern an der Schnellfahrstrecke Madrid–Sevilla (Utilization of traction return conductors on the high-speed line Madrid–Seville). In: Elektrische Bahnen 92(1994)4, pp. 112 to 116.

6.43 *Knüpfer, S.; Christoph, L.*: Hochgeschwindigkeitsstrecke Hannover–Berlin 1998 in Betrieb (Hannover – Berlin high-speed line in 1998 in operation). In: ETR, 46(1997)9, pp. 531 to 532, pp. 535 to 540.

6.44 *Courtois, C.*: Bahnenergieversorgung in Frankreich (Traction power supply in France). In: Elektrische Bahnen 92(1994)6, pp. 167 to 170 and 7, pp. 202 to 205.

6.45 *Klinge, R. et al.*: Hochgeschwindigkeitsverkehr in Italien am Beispiel Rom–Neapel (Rome–Naples – part of Italy´s future high-speed network). In: Elektrische Bahnen, 103(2005)4-5, pp. 253 to 256.

6.46 *Alphen, G.-J.; Smulders, E.*: Messungen im Autotransformatornetz der CFL und Überprüfung mit SIMSPOG (Measurements on the autotransformer system of CFL and verification with SIMSPOG). In: Elektrische Bahnen, 98(2000)7, pp. 242 to 248.

6.47 *Levermann-Vollmer, D.; Thiede, J.*: Messungen am Mehrspannungssystem Prenzlau–Stralsund (Measurements at the multi-voltage traction power supply Prenzlau–Stralsund). In: Elektrische Bahnen 100(2002)10,pp. 385 to 389.

6.48 *Hofmann, G.; Kontcha, A.*: Boostertranformatoren auf AC-Bahnen (Booster transformers used for AC traction systems). In: Elektrische Bahnen 98(2000)7, pp. 233 to 237.

6.49 *Bühlund, A.; Deutschmann, P.; Lindahl, B.*: Schaltungsaufbau im Oberleitungsnetz der schwedischen Eisenbahn Banverket (Circuitry of overhead contact line network at Swedish railway contact line Banverket). In: Elektrische Bahnen, 102(2004)4, pp. 184 to 194.

6.50 *Johnson, F.; Nyebak, M.*: Schaltungsaufbau im Oberleitungsnetz der norwegischen Eisenbahn Jernbaneverket (Circuitry of overhead contact line network at Norwegian railway Jembaneverket). In: Elektrische Bahnen, 102(2004)4, pp. 195 to 200.

6.51 *Schütte, T.; Tiede, J.*: Kombinierte Streckenspeisung mit Auto- und Saugtransformatoren (Line supply by combined auto- and booster transformers). In: Elektrische Bahnen, 98(2000)7, pp. 249 to 253.

6.52 *Lörtscher, M.; Voegeli, H.*: Bahnstromrückführung und Erdung beim Unterwerk Zürich (Traction return circuit and earthing at the substation Zurich). In: Elektrische Bahnen 99(2001)1-2, pp. 51 to 63.

6.53 *Braun, W.*: Feste Fahrbahn und AC-Bahnenergieversorgung (Concrete track slab and AC traction power supply): In: Elektrische Bahnen 101(2003)4-5, pp. 213 to 216.

6.54 *Tschiedel, H.; König, F.; Kuypers, K.-H.*: Erdungskonzept für Verkehrsanlagen im zentralen Bereich Berlins (conception for earthing and bonding for traffic installations in centre of Berlin). In: Elektrische Bahnen 104(2006)6, pp. 290 to 296.

6.55 *Zimmert, G.*: Erdung von Oberleitungsanlagen (Earthing of overhead contact line installations). In: Eisenbahningenieur, 43(1992)2, pp. 86 to 90.

6.56 *Deutschmann, P.; Röhlig, S.; Smulders, E.*: Parallelbetrieb von AC-und DC-Bahnen : Ziele der neuen DIN EN 50 122-3 (Operation of AC and DC railways in parallel: Goals of the new DIN EN 50 122-3). In: Elektrische Bahnen 103(2005)4-5, pp. 191 to 197.

6.57 *Braun, E.; Kistner, H.*: Systemtrennstellen auf der Schnellfahrstrecke Madrid–Sevilla (System separation sections at the High-Speed-Line Madrid–Seville). In: Elektrische Bahnen 92(1994)8, pp. 229 to 233.

6.58 *Cinieri, E. et al.*: Interference assessment at the interface between 2 AC 25 kV 50 Hz and DC 3 kV systems. In: Elektrische Bahnen 102(2004)12, pp. 551 to 557.

6.59 *Cinieri, E. et al.*: Compatibility problems of AC and DC electric traction lines. In: UIC Spoornet Meeting on Alternative Traction Technologies. Johannesburg, April 2000.

7 Thermal rating of conductors

7.0 Symbols and abbreviations

Symbol	Definition	Unit
A	conductor cross section	mm^2
A_S	cross section of conductor rail	mm^2
B	thermal constant	m
D	conductor diameter	m
I	current	A
I_{CA}	current in catenary wire	A
I_{CW}	current in contact wire	A
I_{FL}	current in parallel feeder line	A
I''_K	initial short circuit current	A
$I_{K\,max}$	limiting current of an element of the OCL	A
I_{OCL}	current in overhead contact line	A
I_{dOCL}	current capacity overhead contact line	A
I_d	ampacity of a conductor	A
$I_{eff\,max}$	time-weighted loading of a contact line	A
$I_{h\,max}$	maximum hourly mean value of current	A
I_{dlim}	maximum hourly mean current	A
I_{max}	maximum operational current	A
I_{conper}	permissible current at a contact strip at standstill	A
I_{th}	thermally equivalent short-circuit current	A
N_C	energy losses by convection	W/m
N_J	energy impact by Joule's heat	W/m
N_M	energy impact by magnetic losses	W/m
N_R	energy losses by radiation	W/m
N_S	energy input by solar radiation	W/m
N_{in}	energy input along the conductor	W
N_{out}	energy losses along the conductor	W
N_{Sh}	energy input by the sun	W/m^2
Nu	Nusselt number	–
OCL	Overhead Contact Line	–
R'_{20}	resistance related to length at 20 °C	Ω/m
R'_{CA}	resistance of catenary related to length	Ω/m
R'_{CW}	resistance of contact wire related to length	Ω/m
Re	Reynolds number	–
R'_{FL}	resistance related to length of parallel feeder line	Ω/m
R'_T	resistance related to length at temperature T	Ω/m
R_{tot}	total resistance of OCL	Ω
R_{tv}	transition resistance between contact wire and contact strip	Ω
T	temperature of conductor	°C

Symbol	Definition	Unit
T_1	initial temperature	°C
T_2	short-circuit final temperature	°C
T_∞	contact wire temperature far away from point with elevated temperature	°C
T_{CW}	temperature at contact wire	°C
T_{Fa}	temperature at contact strip holder	°C
T_{am}	ambient temperature	°C
T_{lim}	short circuit final temperature	°C
T_m	asymptotic limit temperature	°C
U_{CW}	circumference of contact wire	m
U_S	circumference of conductor rail	m
c	specific heat	Ws/(K·kg)
c_d	daily load coefficient	–
c_h	hourly load coefficient	–
d	contact wire diameter	m
f	frequency	1/s
h_{al}	altitude above sea level	m
k_a	absorption coefficient	–
k_e	emission coefficient	–
k_s	Stefan-Boltzmann constant	W/(m² K⁴)
l	length of damaged contact wire section	m
m_C	conductor mass related to length	kg/m
m_P	DC portion of short-circuit current	–
$n_{2,3}$	portion on limit loading of a contact line element	–
n_{CW}	portion on limit loading of contact wire	–
n_{CA}	portion on limit loading of catenary wire	–
n_{FL}	portion on limit loading of parallel feeder line	–
n_P	AC portion of short-circuit current	–
n_{lim}	portion on limit loading of a contact line element	–
t	time	s
t^*	averaging period	s
t_k	duration of short circuit	s
t_{sc}	duration of fusing current	s
v_w	wind velocity	m/s
ΔP	loss of power	W
α_R	temperature coefficient of resistance	1/K
α_S	thermal heat transfer coefficient	W/(K·m²)
α_{Scon}	thermal heat transfer coefficient in case of free convection	W/(K m²)
α_{rd}	thermal heat transfer coefficient due to radiation	W/(K·m²)
γ	specific mass of air	kg/m³
γ_0	specific mass of air at sea level	kg/m³
γ_C	specific mass of conductor	kg/dm³
η	dynamic viscosity of air	W/(K·m)
θ	absolute temperature	K
θ_{am}	absolute ambient temperature	K
ϑ_{CW}	temperature at contact wire	°C

Symbol	Definition	Unit
ϑ_{Fa}	temperature at collector strip holder	°C
λ_{C}	thermal conductivity of contact wire	W/(K·m)
λ_{L}	variable of frequency distribution	–
ρ_{20}	specific resistance at 20 °C	$\Omega\,\mathrm{mm}^2/\mathrm{m}$
σ	contact wire stress	$\mathrm{N/mm}^2$
σ_u	ultimate strength	$\mathrm{N/mm}^2$
$\sigma_{0,2}$	0,2 % yield strength	$\mathrm{N/mm}^2$
τ	thermal time parameter	min

7.1 Current-carrying capacity

7.1.1 Introduction

The current loads on contact lines have been analysed and discussed in Chapter 5. To be able to withstand different types of loads, the contact line needs an adequate *current-carrying capacity* determined by the maximum permissible temperature of the conductors and the operating range of tensioning equipments. The current-carrying capacity is also termed *thermal resistance*, *thermal loading capability* or *ampacity*.

The current-carrying capacity characterizes the thermal design of contact lines and is used to compare the capability of various contact lines. As described in Chapter 5 the electrical loading is not constant but represented by time-dependent values. Therefore, the current-carrying capacity should also be presented by corresponding parameters.

7.1.2 Single conductors

7.1.2.1 Basic relationships

Although a contact line is composed of several conductors in parallel, the basic equations for the ampacity refer to a single conductor. The ampacity of contact lines can be obtained from the ampacities of individual conductors. The evaluation of the *conductor temperature* and ampacity is based on the heat balance at the conductor which, according to [7.1] to [7.4], is affected by

- the energy input by *Joule's heat* N_{J} due to conductor resistance,
- the energy input by *solar radiation* N_{S},
- the energy input by magnetic losses N_{M},
- energy loss by radiation N_{R} and
- energy loss by *convection* N_{C}.

In the case of reinforced cables and steel cored conductors as used for overhead power lines, the magnetic losses contribute to increases in conductor temperature. For the elements of contact lines they may be neglected. Therefore, the heat balance related to the unit length of a single conductor can be established by

$$m_{\mathrm{C}} \cdot c \cdot \mathrm{d}T/\mathrm{d}t = N_{\mathrm{J}} + N_{\mathrm{S}} - N_{\mathrm{R}} - N_{\mathrm{C}} \quad , \tag{7.1}$$

where

- m_{C} conductor mass per unit length,
- c specific heat,

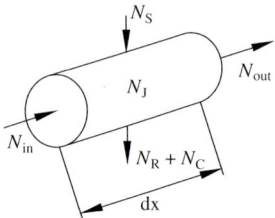

Figure 7.1: Energy balance of a bare wire element.

  – T conductor temperature,
  – dt derivative in terms of time.
The terms N_J, N_R and N_C also depend on the conductor temperature.
Figure 7.1 represents the energy balance of a bare wire. In Equation (7.1) it is assumed that there is no energy flow along the conductor:

$$N_{in} = N_{out} = 0 \quad .$$

Three cases of substantial practical interest include:
  – varying operational loads because of starting and braking vehicles (see Clause 7.1.2.4) where temperature also varies. This general case represents an unsteady state
  – loads due to short circuits with durations up to a maximum of approximately 100 ms. Because of the large magnitude of the term N_J, the terms N_S, N_R and N_C may be disregarded in Equation (7.1). Since there is no external heat exchange, the process is adiabatic (see Clause 7.1.2.3)
  – permanent, long-term operational loads acting for half an hour or more (see Clause 7.1.2.2). This leads to a steady-state condition with a constant conductor temperature

7.1.2.2   Long-term operational loads

The literature describes several approaches to determine the individual terms of Equation (7.1). In [7.1] a summary of approaches is given that also forms the basis for the equations used in IEC 61 597. A detailed study of the steady-state condition can be found in [7.2].

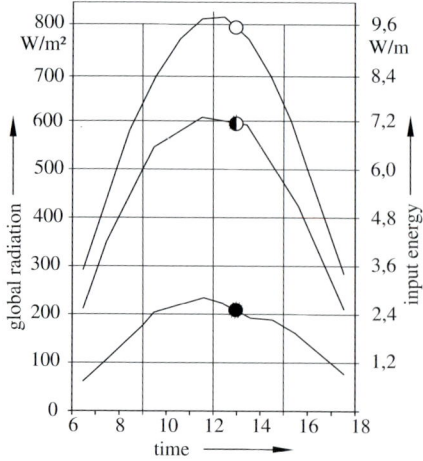

Figure 7.2: Typical daily variation of global radiation in mid-summer as a function of the cloud coverage, showing the values of the resulting radiation heat per unit length acting on a type AC-100 – Cu contact wire.
○   hazy to clear sky
◑   50 % cloud cover
●   100 % cloud cover

Table 7.1: Solar absorption and emission coefficients k_a and k_e of metallic surfaces according to [7.2] and [7.4].

[1] cast iron

Surface	Copper	Aluminum	Iron
semi-polished	0,15	0,08	
matted, smooth	0,24	0,23	0,45[1]
oxidized, slightly dirty	0,60	0,50	
heavily oxidized	0,75	0,70	0,96[1]
heavily oxidized, dirty	0,85 to 0,95	0,88 to 0,93	
rolling-mill skin			0,65
sand-blasted			0,67
rusty			0,61 to 0,85

The losses due to Joule's heat are obtained from

$$N_J = R'_T \cdot I^2, \qquad (7.2)$$

where I is the effective current in A and R'_T is the resistance at temperature T in Ω/m. In general, the resistance R'_T is different for AC and DC currents due to skin, spiral and magnetic effects [7.2]. However, for the dimensions and composition of conductors used for contact lines, the DC resistance may also be adopted for AC applications. The resistance is therefore:

$$R'_T = \frac{\rho_{20}}{A} \left[1 + \alpha_R (T - 20)\right] = R'_{20} \left[1 + \alpha_R (T - 20)\right] \qquad , \qquad (7.3)$$

where R'_{20} is the DC resistance at $20\,°C$ in Ω/m and α_R the temperature coefficient of resistivity in K^{-1}. Values for α_R can be found in Tables 11.2, 11.5 and 11.6.

Overhead contact lines and conductor rails are heated externally by *solar radiation* and *diffuse sky radiation*. The sum of the effect of both types of radiation is called *global radiation*. Representative surveys carried out in Germany have shown that the following values can be assumed for global radiation on a contact wire of type AC-100 – Cu:

    1,2 W/m   annual average
    2,3 W/m   in mid-summer at 100 % cloud coverage
    6,1 W/m   in mid-summer at 50 % cloud coverage
    8,2 W/m   in mid-summer with clear skies

Measurements have shown that the temperatures of contact wires without current loads can be 6 K to 8 K higher than the surrounding air in cases of exposure to sun shine.

In Figure 7.2, the typical daily variation of heat radiation is shown acting on a type AC-100 – Cu contact wire in the summer in Central Europe. The corresponding values for AC-120 – Cu are 10 % higher.

In accordance with IEC 61 597 the *solar radiation* is taken from

$$N_S = k_a \cdot D \cdot N_{Sh} \qquad , \qquad (7.4)$$

where N_{Sh} is the standard solar radiation, which is between 850 and 1 350 W/m^2 at maximum depending on the latitude of the site, sun position, air pollution and time of year or day. A typical maximum value for Central Europe is 900 W/m^2. D is the conductor diameter in m and k_a is the absorption coefficient being 0,5 for conductors used for overhead contact lines (Table 7.1).

According to IEC 61 597 the *energy loss by radiation N_R* is given by

$$N_R = k_s \cdot k_e \cdot D \cdot \pi \cdot \left(\theta^4 - \theta_{am}^4\right) \qquad , \qquad (7.5)$$

where

Table 7.2: Material constants of air.

Temperature T °C	Specific mass γ kg/m^3	Thermal conductivity λ W/(K·m)	Dynamic viscosity η Ns/m^2
0	1,290	0,0243	$0,175 \cdot 10^{-4}$
10	1,250	0,0250	$0,180 \cdot 10^{-4}$
20	1,200	0,0257	$0,184 \cdot 10^{-4}$
30	1,170	0,0265	$0,189 \cdot 10^{-4}$
40	1,13	0,0272	$0,194 \cdot 10^{-4}$
50	1,09	0,0280	$0,199 \cdot 10^{-4}$
60	1,06	0,0287	$0,203 \cdot 10^{-4}$
70	1,04	0,0294	$0,208 \cdot 10^{-4}$
80	1,01	0,0301	$0,213 \cdot 10^{-4}$
90	0,97	0,0309	$0,217 \cdot 10^{-4}$
100	0,95	0,0316	$0,222 \cdot 10^{-4}$

- θ absolute temperature of the conductor,
- θ_{am} absolute ambient temperature,
- k_s *Stefan-Boltzmann constant* equal to $5,67 \, \mathrm{W}/(\mathrm{m}^2\mathrm{K}^4) \cdot 10^{-8}$ and
- k_e emission coefficient which can be taken from Table 7.1.

The absolute temperature θ in K is obtained by $\theta = T + 273$.

The *energy loss by convection* N_C can be calculated from

$$N_C = \pi \cdot \lambda \cdot Nu \cdot (T - T_{am}) \quad , \tag{7.6}$$

where

- λ is the thermal conductivity of air in W/(K·m) (see Table 7.2) and
- Nu is the *Nusselt number*, which, in case of forced convection, depends on the *Reynolds number* according to IEC 61 597

$$Nu = 0,65 \cdot Re^{0,2} + 0,23 \cdot Re^{0,61} \quad . \tag{7.7}$$

The Reynolds number Re is given by

$$Re = v_w \cdot D \cdot \gamma / \eta \quad , \tag{7.8}$$

where

- v_w wind velocity in m/s,
- γ specific mass of air in kg/m^3 and
- η dynamic viscosity in N·s/m^2.

These data depend on the temperature and air pressure. At sea level the data given in Table 7.2 apply.

In the case of free convection, the wind velocity is zero and the approach presented above cannot be applied directly. Then, the Nusselt number can be calculated by using procedures presented in [7.2]. The condition of free convection can arise in tunnels. However, because of the lower ambient temperature and the lack of solar radiation, the ampacities calculated for lines in the open based on wind velocities of 1,0 m/s or less may also be used for lines in tunnels.

The specific mass of air depends on the absolute ambient temperature θ_{am} and the altitude above sea level h_{al} by [7.1]

$$\gamma = \gamma_0 (288 / \theta_{am}) \exp(-0,0001 \cdot h_{al}) \quad , \tag{7.9}$$

where γ_0 is 1,225 kg/m^3 at sea level.

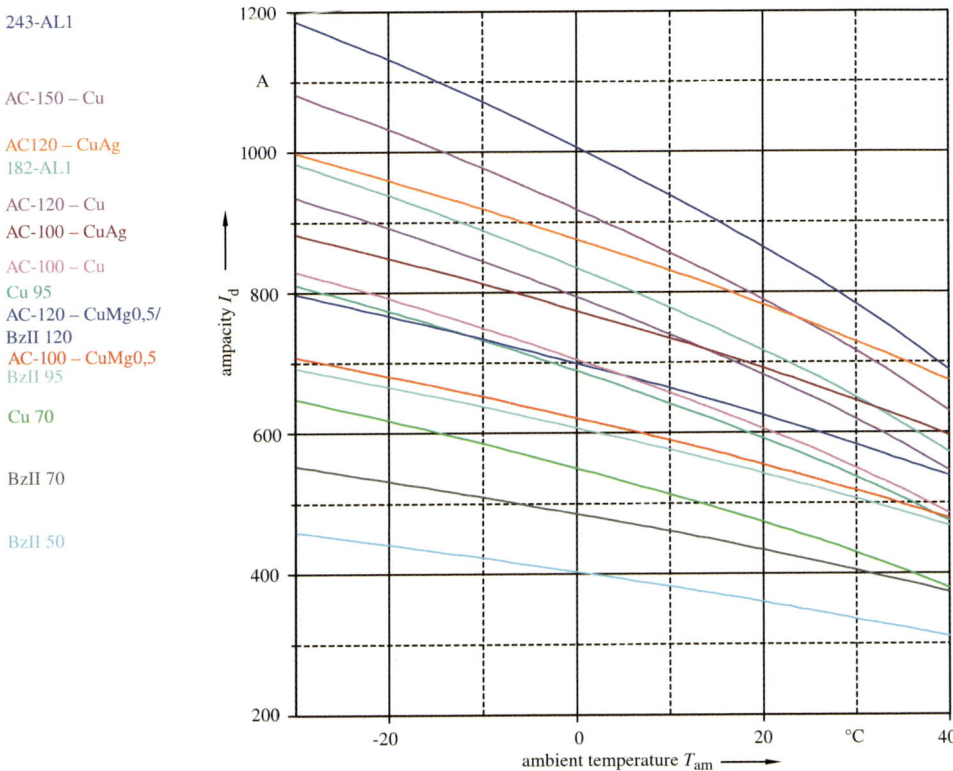

243-AL1

AC-150 – Cu

AC120 – CuAg
182-AL1

AC-120 – Cu
AC-100 – CuAg

AC-100 – Cu
Cu 95
AC-120 – CuMg0,5/
BzII 120
AC-100 – CuMg0,5
BzII 95

Cu 70

BzII 70

BzII 50

Figure 7.3: Ampacities I_d of conductors and contact wires.

For practical calculations, the characteristics of air should be evaluated for the mean value $(T + T_{am})/2$. The ampacity can be obtained from Equation (7.1) with $dT/dt = 0$:

$$I_d = \sqrt{(N_C + N_R - N_S)/R'_T} \quad .$$ (7.10)

To determine the ampacity of conductors used in overhead contact lines, an ambient temperature of 40 °C and a wind velocity of 1,0 m/s are often assumed.

Example 7.1: To calculate the ampacity of a contact wire AC-100 – Cu at 1,0 m/s wind velocity and 80 °C temperature. The ambient temperature is assumed to be 40 °C. The solar radiation is assumed as 900 W/m².

 – The resistance at 80 °C is obtained from (7.3) with $\rho_{20} = 0,01777\,\Omega mm^2/m$; $A = 100\,mm^2$; $\alpha_R = 0,00380$ (Table 11.5) to be $R'_{80} = 0,01777/100 \cdot [1 + 0,00380 \cdot (80 - 20)] = 0,219 \cdot 10^{-3}\,\Omega/m$.
 – The solar radiation follows from (7.4) with $k_a = 0,75$ from Table 7.1 (heavily oxidized): $N_S = 0,75 \cdot 0,012 \cdot 900 = 8,1\,W/m$.
 – The energy loss by radiation is obtained from (7.5): $N_R = 0,75 \cdot 5,67 \cdot 10^{-8} \cdot 0,012 \cdot \pi (353^4 - 313^4) = 9,5\,W/m$
 – The Reynolds number for the wind speed of 1,0 m/s is obtained from (7.8) with γ and η taken from Table 7.2 for 60 °C: $Re = 1,0 \cdot 0,012 \cdot 1,06/(0,203 \cdot 10^{-4}) = 627$.
 – The Nusselt number follows from (7.7): $Nu = 0,65 \cdot (627)^{0,2} + 0,23 \cdot (627)^{0,61} = 14,0$.

Table 7.3: Permanent current-carrying capacity (ampacity) I_d of conductors used in contact lines with a wind velocity of 1,0 m/s, values given in A.

Type of conductor	Conductor temperature	Ambient temperature °C							
		−30	−20	−10	0	10	20	30	40
AC-100−Cu[1]	80	830	791	749	704	656	605	550	485
AC-120−Cu	80	935	891	844	794	739	682	619	546
AC-150−Cu	80	1 081	1 031	976	918	855	788	716	630
AC-100−CuAg[1]	100	883	848	812	774	735	691	645	596
AC-120−CuAg	100	999	959	919	875	831	782	729	674
AC-100−CuMg0,5	100	708	681	652	621	589	555	517	478
AC-120−CuMg0,5	100	798	766	734	699	664	625	583	538
Cu70[2]	80	648	618	585	550	513	473	430	379
Cu95	80	810	773	732	688	641	591	543	473
Cu120	80	939	895	847	797	742	684	621	548
Cu150	80	1 090	1 040	984	925	862	795	721	635
BzII50[3]	100	459	441	422	402	382	360	335	310
BzII70	100	553	531	508	484	460	433	404	373
BzII95	100	693	665	637	607	576	542	506	468
BzII120	100	803	771	738	704	668	629	586	541
182-AL1[4]	80	984	938	888	835	777	716	650	572
243-AL1	80	1 186	1 131	1 071	1 007	937	863	783	689
626-AL1	80	2 195	2 093	1 982	1 862	1 733	1 594	1 443	1 265

[1] Contact wires AC-Cu and AC-CuAg according to EN 50 149, [2] Cu conductors according to DIN 48 201-1, [3] BzII conductors according to DIN 48 201-2, [4] AL1 conductors according to EN 50 182

- The energy loss by convection amounts to (Equation (7.6)): $N_C = \pi \cdot 0,0287 \cdot 14,0(80 - 40) = 50,5$ W/m.

Therefore, the ampacity will be obtained from Equation (7.10)

$$I_d = \sqrt{(9,5 + 50,5 - 8,1)/(0,219 \cdot 10^{-3})} = 487\,A \approx 485\,A.$$

Table 2.8 corresponding to EN 50 119, Table 1, contains conductor temperatures above which, the mechanical properties of the material could be impaired. The permanently acceptable temperature should not be exceeded when rating the conductors.

In Table 7.3 and Figure 7.3 the continuous current-carrying capacity of conductors frequently used for overhead contact lines is given without wear at 1,0 m/s wind velocity and with 900 W/m² solar radiation at various ambient temperatures.

In Table 7.4, the dependence of ampacities on wind velocity is demonstrated for two frequently used conductors: Contact wire AC-120−Cu and catenary wire BzII 70. At a wind velocity of 2,0 m/s the ampacity is 20 % greater than at 1,0 m/s and 55 % higher at 5,0 m/s. A thermal rating based on 1,0 m/s wind velocity is conservative, as this wind velocity is frequently exceeded.

Long-term weather statistics for Germany confirm that 35 °C was exceeded in only 0,01 % of the records. When 35 °C was exceeded, the wind velocity was at least 1,8 m/s. From [7.5] it can be concluded that the probability of simultaneous occurrence of a temperature higher than 30 °C and a wind velocity less than 1,0 m/s is practically zero. However, the maximum ambient temperature and simultaneous wind velocity must be specified for each project based on local conditions. In Table 7.5 the ampacity of contact wire AC-120−CuAg depending on the conductor temperature is listed.

Table 7.4: Permanent current-carrying capacity (ampacity) I_d of conductors used in contact lines at varying wind velocities, values given in A.

Conductor	Wind velocity m/s	Ambient temperature °C							
		−30	−20	−10	0	10	20	30	40
AC-120−Cu	0,6	824	786	744	700	652	600	544	478
80 °C	1,0	935	892	844	794	739	682	619	546
	2,0	1 119	1 067	1 010	950	886	817	743	657
	3,0	1 248	1 190	1 126	1 059	988	912	830	736
	5,0	1 437	1 369	1 296	1 218	1 137	1 049	957	849
BzII70	0,6	490	471	451	430	407	384	358	330
100 °C	1,0	553	531	508	484	460	433	404	373
	2,0	660	635	608	578	548	516	483	445
	3,0	736	707	676	643	609	574	537	496
	5,0	845	812	776	739	699	659	616	569

Table 7.5: Permanent current-carrying capacity (ampacity) I_d depending on the conductor temperature, contact wire AC-120−CuAg, wind velocity 1,0 m/s, values given in A.

Conductor temperature °C	Ambient temperature °C							
	−30	−20	−10	0	10	20	30	40
100	999	959	919	875	831	782	729	674
80	938	895	981	794	739	682	619	546
60	863	811	752	693	627	553	467	359
40	778	715	637	561	473	364	198	0

Table 7.6: Final steady-state conductor temperature in °C, ambient temperature 40 °C, wind velocity 1,0 m/s.

Type of conductor	Current in A										
	200	300	400	500	600	700	800	900	1 000	1 100	1 200
AC-100−Cu[1]	−	58	69	82	101	125	155	−	−	−	−
AC-120−Cu	−	55	62	72	87	104	126	153	−	-	−
AC-150−Cu	−	52	58	67	77	90	106	125	148	−	−
AC-100−CuMg0,5	−	66	82	107	139	−	−	−	−	−	−
AC-120−CuMg0,5	−	61	75	93	117	149	−	−	−	−	−
Cu70[2]	56	68	85	110	144	189	−	−	−	−	−
Cu95	51	59	71	85	104	130	162	−	−	−	−
Cu120	−	55	63	75	88	106	128	155	−	−	−
Cu150	−	52	58	66	76	89	104	124	147	−	−
BzII50[3]	71	100	151	−	−	−	−	−	−	−	−
BzII70	61	80	110	154	−	−	−	−	−	−	−
BzII95	55	67	84	110	144	−	−	−	−	−	−
BzII120	51	60	74	92	116	148	−	−	−	−	−
182-AL1[4]	−	54	61	72	84	100	120	145	174	−	−
243-AL1	−	50	56	62	72	82	95	111	130	151	−
626-AL1	−	−	−	−	51	54	58	62	66	70	76

Notes:
[1] contact wires AC-Cu and AC-CuAg according to EN 50 149, [2] Cu conductors according to DIN 48 201-1,
[3] BzII conductors according to DIN 48 201-2, [4] AL1 conductors according to EN 50 182

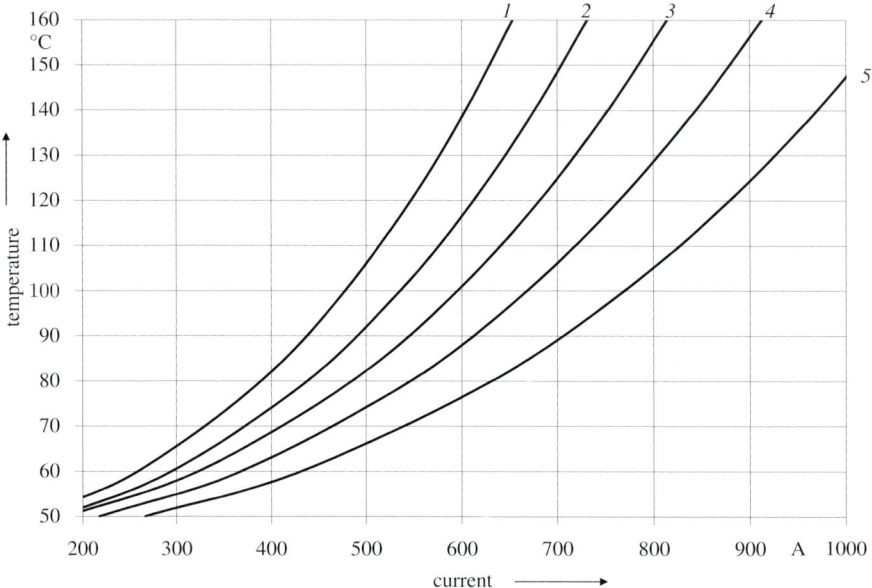

Figure 7.4: Steady-state temperatures of contact wires, ambient temperature 40 °C, wind velocity 1,0 m/s. Limit temperatures according to Table 2.8. *1* CuMg0,5 AC-100, *2* CuMg0,5 AC-120, *3* AC-100 – Cu and AC-100 – CuAg, *4* AC-120 – Cu and AC-120 – CuAg, *5* Cu AC-150

For a constant current I, the asymptotic temperature T_m can be obtained by solving Equation (7.1) with $\mathrm{d}T/\mathrm{d}t = 0$. With (7.3) to (7.6) it is obtained

$$I^2 \cdot R'_{20}\left[1 + \alpha_R(T_m - 20)\right] + k_a \cdot D \cdot N_{Sh} - \pi \cdot \lambda \cdot Nu \cdot (T_m - T_{am})$$
$$- k_s \cdot k_e \cdot D \cdot \pi \left[(T_m + 273)^4 - (T_{am} + 273)^4\right] = 0 \quad . \tag{7.11}$$

In Table 7.6, the final steady-state conductor temperature in °C is shown for an ambient temperature of 40 °C and a wind velocity 1,0 m/s depending on the current for contact wires and other conductors, for materials often used in overhead contact lines. Figures 7.4 and 7.5 show the final temperatures for these conductors.

7.1.2.3 Short-circuit current-carrying capacity

Methods of determining the short-circuit current-carrying capacity of overhead contact lines and their main components are explained in this clause. The *short-circuit current-carrying capacity*, also termed *short-circuit capability* or *short-circuit rating*, is important for the thermal design considerations of overhead contact line installations. If, in Equation (7.1) the heat applied by external sources is ignored and it is assumed that no heat is dissipated from the wire because of the rapid rise of a short-circuit current, then all energy will heat the conductor and if the protective measures fail, the conductor may eventually melt.

In accordance with [7.6], it is obtained from (7.1) with $m_C = A \cdot \gamma_C$

$$A \cdot \gamma_C \cdot c \cdot \mathrm{d}T/\mathrm{d}t = N_J \quad . \tag{7.12}$$

The Joule's heat follows from (7.2) and (7.3) with $R'_{20} = \rho/A$:

$$N_J = I^2[1 + \alpha_R(T - 20)] \cdot \rho_{20}/A \quad . \tag{7.13}$$

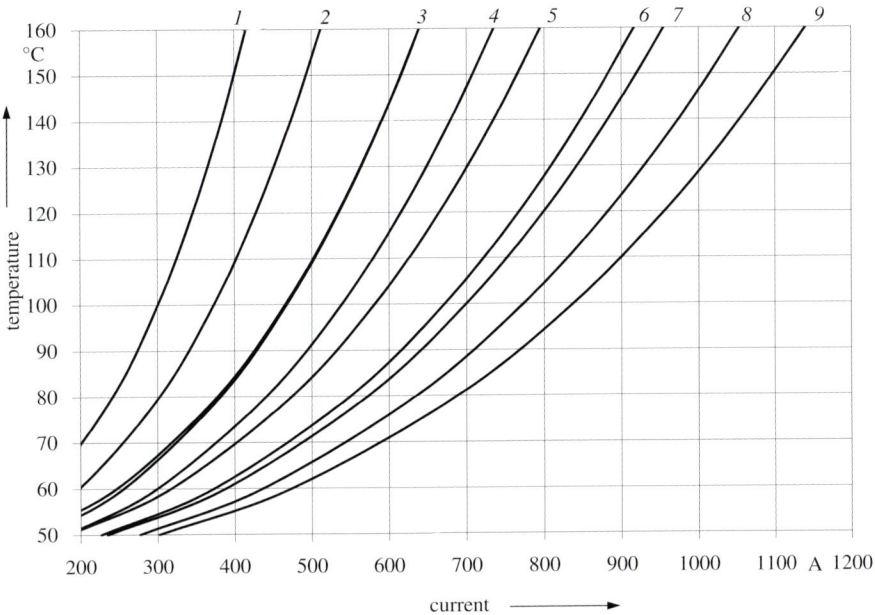

Figure 7.5: Steady-state temperatures of conductors, ambient temperature 40 °C, wind velocity 1,0 m/s. Limit temperatures in accordance with Table 2.8. *1* BzII50, *2* BzII70, *3* BzII95 and Cu70, *4* BzII120, *5* Cu95, *6* Cu120, *7* 182-AL1, *8* Cu150, *9* 243-AL1

Therefore,

$$dT/dt = I^2 \left[1 + \alpha_R (T - 20) \right] \cdot \rho_{20} / (A^2 \cdot \gamma_C \cdot c) \quad , \tag{7.14}$$

where I represents the current, ρ_{20} the resistivity of the conductor at 20 °C, α_R the thermal coefficient of resistance, A the cross-sectional area, γ_C the specific mass, c the specific heat. Equation (7.14) can be integrated after being transformed to

$$\int_{T_1}^{T_2} \frac{dT}{1 + \alpha_R (T - 20)} = \int_0^{t_k} \frac{I^2 \cdot \rho_{20} \, dt}{A^2 \cdot \gamma_C \cdot c} \quad . \tag{7.15}$$

Integration yields

$$\frac{1}{\alpha_R} \ln \frac{1 + \alpha_R (T_2 - 20)}{1 + \alpha_R (T_1 - 20)} = \frac{I_{th}^2 \cdot \rho_{20} \cdot t_k}{A^2 \cdot \gamma_C \cdot c} \quad , \tag{7.16}$$

where T_1 and T_2 are the initial and final temperatures, t_k the short-circuit duration and I_{th} is the thermal effective short-circuit current. The final temperature for a thermally equivalent current I_{th} is

$$T_2 = 20 + \frac{1}{\alpha_R} \left\{ \left[1 + \alpha_R (T_1 - 20) \right] \exp \left(\frac{I_{th}^2 \cdot \rho_{20} \cdot \alpha_R \cdot t_k}{A^2 \cdot \gamma_C \cdot c} \right) - 1 \right\} \quad . \tag{7.17}$$

The *short-circuit capacity* depends on the permissible limit temperature T_{lim} and can be calculated from

$$I_{th} = A \cdot \sqrt{\frac{c \cdot \gamma_C}{\rho_{20} \cdot \alpha_R \cdot t_k} \cdot \ln \left(\frac{1 + \alpha_R \cdot (T_{lim} - 20)}{1 + \alpha_R \cdot (T_1 - 20)} \right)} \quad . \tag{7.18}$$

Table 7.7: Permissible initial short circuit currents I_K'' in kA.

Type of conductor	Permissible temperature °C	Short-circuit current duration					
		double side feeding (16,7 Hz) and DC			single side feeding (50 Hz)		
		0,1 s	0,5 s	1,0 s	0,1 s	0,5 s	1,0 s
AC-100–Cu	170	43,3	19,4	13,7	34,8	15,5	11,0
AC-120–Cu	170	51,9	23,3	16,4	41,4	18,6	13,1
AC-150–Cu	170	65,1	29,0	20,6	52,2	23,2	16,5
AC-100–CuAg	200	47,1	21,1	14,9	37,6	16,9	11,9
AC-120–CuAg	200	56,6	25,3	17,9	45,2	20,2	14,3
AC-100–CuMg0,5	200	37,0	16,6	11,7	29,7	13,3	9,4
AC-120–CuMg0,5	200	45,6	19,9	14,1	35,4	15,9	11,2
Cu 70	170	28,5	12,8	9,0	22,8	10,2	7,2
Cu 95	170	40,5	18,1	12,8	32,3	14,5	10,2
Cu 120	170	50,6	22,7	16,0	40,5	18,2	12,8
Cu 150	170	63,9	28,5	20,2	51,2	22,8	16,2
BzII 50	200	18,3	8,2	5,8	14,5	6,6	4,6
BzII 70	200	24,3	10,9	7,7	19,6	8,7	6,2
BzII 95	200	34,5	15,5	10,9	25,9	12,4	8,2
BzII 120	200	43,3	19,4	13,7	34,8	15,5	11,0
182-AL1	130	45,2	20,2	14,3	36,1	16,2	11,4
243-AL1	130	60,4	27,0	19,1	48,4	21,6	15,3
626-AL1	130	155,3	69,6	49,1	124,3	55,7	39,3
BzII 10	300	4,4	2,0	1,4	3,5	1,6	1,1
BzII 16	300	7,0	3,2	2,2	5,6	2,6	1,8

With the often used units c in $\mathrm{W \cdot s / kg \cdot K}$, γ_C in $\mathrm{kg / dm^3}$, ρ_{20} in $\mathrm{\Omega \cdot mm^2 / m}$, α_R in $\mathrm{1 / K}$, t_k in s, T in K, A in $\mathrm{mm^2}$ the Equation (7.18) will be

$$I_{th} = A \cdot \sqrt{\frac{c \cdot \gamma_C \cdot 10^{-3}}{\rho_{20} \cdot \alpha_R \cdot t_k} \cdot \ln\left(\frac{1 + \alpha_R \cdot (T_{lim} - 20)}{1 + \alpha_R \cdot (T_1 - 20)}\right)} \quad . \tag{7.19}$$

In Equation (7.18), t_k is the *duration of the short-circuit current*, T_1 the initial temperature of the conductor when the short circuit occurs and T_{lim} is the permissible maximum temperature of the conductor in case of a short circuit. For grooved contact wires of electrolytic copper, a value of 170 °C is specified in EN 50 119, Table 1 (see also Table 2.8), as permissible maximum temperature. A temperature of 200 °C is permitted for contact wires made of CuAg0,1 and CuMg0,5 alloys. Some railway operators permit a final temperature of 300 °C for bronze catenary wires and 600 °C for dropper wires.

The short-circuit capability values determined using (7.18) or (7.19) refer to the thermally equivalent short-circuit currents. However, the permissible *initial short-circuit alternating currents* I_K'', differ from the thermally equivalent short-circuit currents I_{th}. According to standard EN 60 865-1 it applies:

$$I_K'' = I_{th} / \sqrt{m_P + n_P} \quad , \tag{7.20}$$

where m_P describes the heat generated by the DC component and n_P the heat generated by the AC component.

EN 60 865-1 gives factors m_P and n_P as functions of the duration t_k of the short circuit situation and of the product of the short-circuit duration and frequency, $t_k \cdot f$.

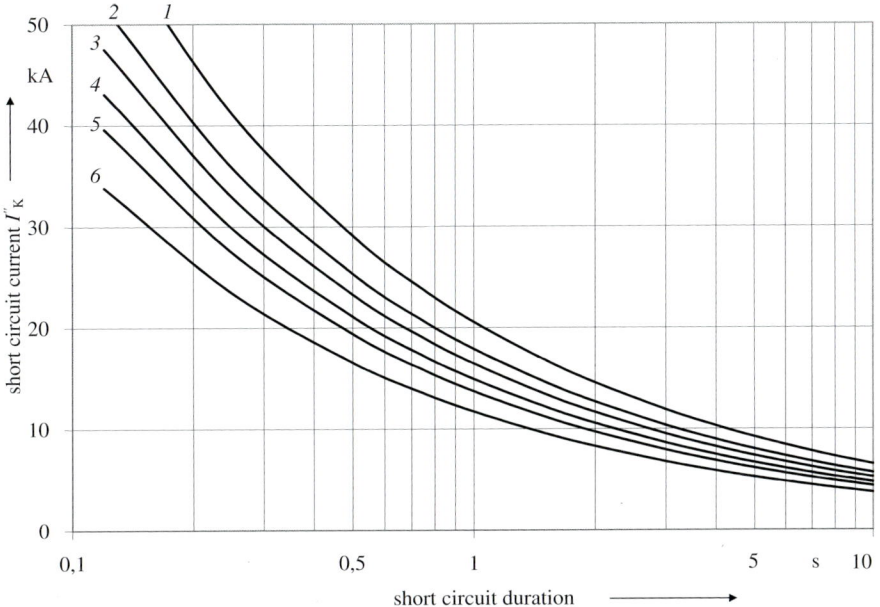

Figure 7.6: Permissible short-circuit currents I_K'' in kA of contact wires for double-sided fed AC systems and DC systems. *1* AC-150 – Cu, *2* AC-120 – CuAg, *3* AC-120 – Cu, *4* CuAg AC-100, *5* AC-100 – Cu and AC-120 – CuMg0,5, *6* AC-100 – CuMg0,5

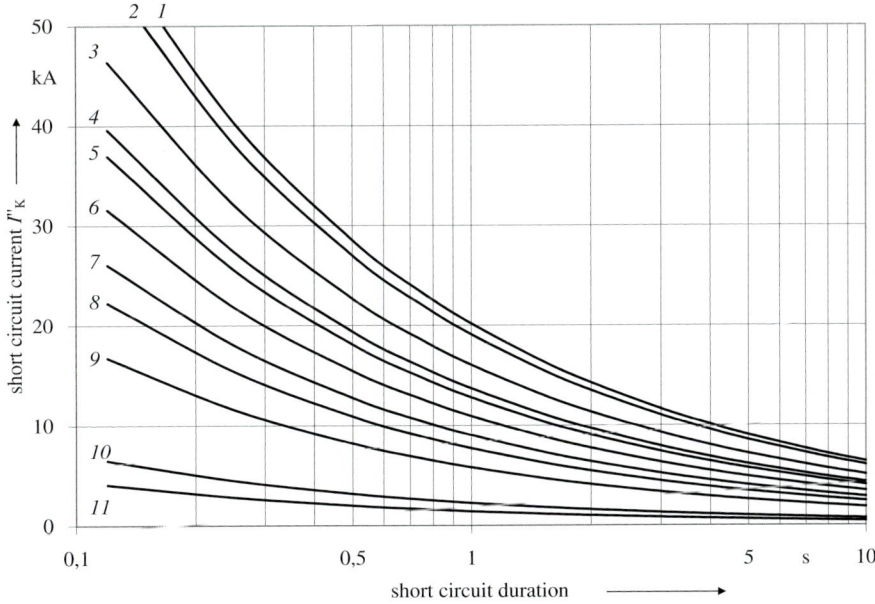

Figure 7.7: Permissible short-circuit currents I_K'' in kA of conductors for double-sided fed AC systems and DC systems. *1* Cu 150, *2* 243-AL1, *3* Cu 120, *4* 182-AL1 and BzII120, *5* Cu 95, *6* BzII95, *7* Cu 70, *8* BzII70, *9* BzII50, *10* BzII16, *11* BzII10

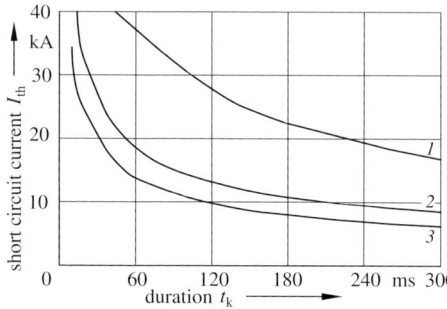

Figure 7.8: Short-circuit current carrying capability of a contact line, 10 % worn contact wire type AC-100 – Cu, 50 mm² catenary wire type BzII, initial temperature $T_1 = 70\,°C$; $T_{lim} = 200\,°C$, $300\,°C$ and $600\,°C$ for contact wire, catenary wire and dropper wire, respectively.

1 contact wire, short circuit at a dropper at mid span
2 catenary wire, short circuit 0,5 m away from a dropper at mid span
3 catenary wire, short circuit at the dropper at mid span

As a default value $\sqrt{m_P + n_P} = 1$ can be used for networks with high short-circuit currents as is the case for the 16,7 Hz network of German Railway. The 50 Hz traction systems are supplied by an overlying integrated grid. Consequently, the traction supply system is not interconnected and the short-circuit currents are relatively low. In this case, $\sqrt{m_P + n_P} = 1$ can be assumed. The permissible initial short-circuit alternating current in 50 Hz systems is 80 % of the limits in 16,7 Hz systems. Therefore, it applies

$I_K'' \approx 1 \cdot I_{th}$ in the centrally supplied networks (16,7 Hz) with high short-circuit currents,
$I_K'' \approx 0,8 \cdot I_{th}$ in the decentrally supplied networks (50 Hz) with low short-circuit currents.

In Table 7.7 and in Figures 7.6 and 7.7 the permissible short-circuit currents are presented for contact wires and other conductors.

Example 7.2: Determine the short-circuit current capability I_K''. Given are the contact wire AC-120 – CuMg0,5, the permissible temperature of 200 °C, the short-circuit duration of 1 s, the initial temperature of 20 °C, and double-sided supply. The other data are

specific heat (Table 11.5)	c = 380 W·s / kg·K
specific mass (Table 11.5)	γ_C = 8,9 kg / dm³
specific resistivity (Table 11.5)	ρ_{20} = 0,02778 Ω·mm² / m
coefficient of resistivity (EN 50 149:2013)	α_R = 0,00270 K⁻¹
cross-section	A = 120 mm²

$$I_{th} = 120 \cdot \sqrt{\frac{380 \cdot 8,9 \cdot 10^{-3}}{0,02778 \cdot 0,00270 \cdot 1} \cdot \ln\left(\frac{1 + 0,00270 \cdot (200 - 20)}{1 + 0,00270 \cdot (40 - 20)}\right)} = 15\,kA \quad .$$

For the centrally supplied network of DB $\sqrt{m_P + n_P}$ can be assumed as 1,0, therefore $I_K'' = I_{th}$. For a one-sided supply (also referred to as single ended) $\sqrt{m_P + n_P} = 1,25$ can be assumed, therefore $I_K'' = I_{th} / 1,25$. Table 7.7 shows the calculated short-circuit currents. In Figure 7.8, the short-circuit capability is given depending on the duration, for a contact line made of a contact wire AC-100 – Cu and a messenger wire BzII 50, as is often used for conventional contact lines.

In Figure 7.8, the short-circuit current carrying capacity is shown as a function of the circuit breaking duration for a contact line composed of a contact wire AC-100 – Cu and a catenary wire BzII 50, which is used frequently for conventional contact lines.

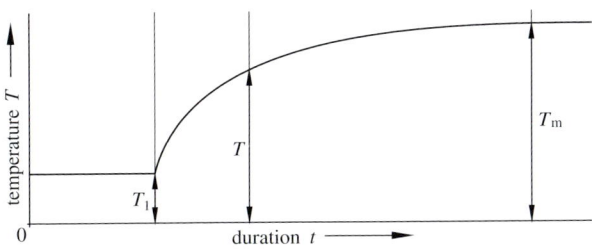

Figure 7.9: Temperature rise of a conductor under action of current I.
T_1 temperature at start of current I
T_m temperature in steady-state condition

7.1.2.4 Varying operational loads

In the differential Equation (7.1) the
- Joule's heat (Equations (7.2) and (7.3)),
- energy loss by radiation (Equation (7.4)) and
- energy loss by convection (Equation (7.5))

depend on the conductor temperature, while the solar radiation can be assumed as independent of the conductor temperature.

The specific heat c also varies with the temperature, but may be assumed as constant for the temperature ranges discussed here. The Joule's heat and the energy loss by convection depend linearly on the temperature but the energy loss by radiation is non-linear with temperature. If the radiation loss is small, relative to the convective loss, it can be linearised and a general solution of Equation (7.1) in the unsteady condition according to [7.7] is

$$ t = \frac{-m_C \cdot c \cdot T_m}{I^2 \cdot R'_{20} + N_S} \ln \left(\frac{T_m - T}{T_m - T_1} \right) \quad , \tag{7.21} $$

where
- R'_{20} resistance per unit length at 20 °C,
- T_m conductor temperature in the steady-state condition for the current I and
- t time elapsed after commencement of current I.

Figure 7.9 explains the parameters T_1, T and T_m. Equation (7.21) can be transformed to

$$ T = T_m - (T_m - T_1) \exp(-t/\tau) \quad , \tag{7.22} $$

where τ is the *thermal time parameter*

$$ \tau = (m_C \cdot c \cdot T_m) / (I^2 R'_{20} + N_S) \quad , \tag{7.23} $$

which depends on the data of the conductor, on the current I and on the steady-state temperature T_m. In Table 7.8, the thermal time patameters are given for contact wires and other conductors.

From Equation (7.22) it can be concluded that after the time of one time parameter has elapsed, the temperature reaches 63 % of the final temperature and after four times this parameter, the final temperature will be reached. In Figure 7.10, these relations can be seen for the contact wire AC-100 – Cu. The ampacity of the contact wire AC-100 – Cu is 485 A. The time after which the temperature limit 80 °C will be reached can be obtained from (7.20).

Table 7.8: Thermal time parameter τ in minutes of contact wires and conductors, ambient temperature 40 °C, wind velocity 1,0 m/s.

Type of conductor	Current in A									
	200	300	400	500	600	700	800	900	1000	1100
AC-100 – CuAg	–	–	11,0	9,1	8,1	7,5	7,4	–	–	–
AC-120 – CuAg	–	–	13,3	11,0	9,8	8,9	8,5	8,3	–	–
AC-150 – CuAg	–	–	17,6	14,8	12,8	11,6	10,8	10,3	10,0	–
AC-100 – CuMg0,5	–	11,1	8,7	7,7	7,2	–	–	–	–	–
AC-120 – CuMg0,5	–	12,1	9,5	9,0	7,2	6,9	–	–	–	–
Cu70	12,3	8,5	6,6	5,8	5,4	5,3	–	–	–	–
Cu95	17,6	12,8	10,1	8,4	7,5	7,1	6,9	–	–	–
Cu120	–	16,5	12,9	10,8	9,5	8,7	8,3	8,1	–	–
Cu150	–	23,3	18,6	15,6	13,5	12,3	11,4	11,0	10,8	–
BzII50	6,9	4,9	4,4	–	–	–	–	–	–	–
BzII70	9,6	6,7	5,6	5,2	–	–	–	–	–	–
BzII95	14,5	10,2	8,0	7,1	6,6	–	–	–	–	–
BzII120	17,9	13,0	10,4	8,9	8,1	7,8	–	–	–	–
182-AL1	–	15,3	12,1	10,4	9,1	8,3	7,9	7,7	7,6	–
243-AL1	–	20,4	17,1	14,3	12,8	11,5	10,7	10,2	9,9	9,7

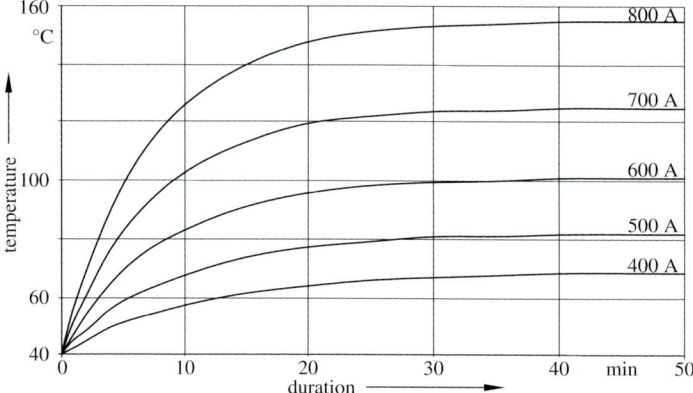

Figure 7.10: Temperature rise of a contact wire AC-100 – Cu, ambient temperature 40 °C, wind velocity 1 m/s, time parameter according to Table 7.8.

Example 7.3:　Determine the time it takes a contact wire AC-100–Cu to reach 80 °C when currents of 700 A and 800 A act. The thermal time parameters obtained from Table 7.8 are 7,5 min and 7,4 min. The steady-state temperatures would be 125 °C for 700 A and 155 °C for 800 A, see Table 7.6. Equation (7.20) yields

$$\tau = -7,5 \ln\left[(125 - 80)/(125 - 40)\right] = 4,8 \text{ min for } 700\,\text{A and}$$

$$\tau = -7,4 \ln\left[(155 - 80)/(155 - 40)\right] = 3,2 \text{ min for } 800\,\text{A}.$$

Figure 7.10 confirms this data.

Table 7.9: Current-carrying capacity of conductor rails at various ambient temperatures, conductor rail temperature 85 °C.

Type of rail	Wind velocity	Ambient temperature °C				
	m/s	−20	0	20	35	40
Fe5100	0,0	4 120	3 710	3 240	2 830	2 700
	0,6	5 620	5 160	4 510	3 660	3 750
AL5100	0,0	7 460	6 710	5 870	5 150	4 880
	0,6	10 550	9 490	8 300	7 240	6 900
AL2100	0,0	4 790	4 300	3 760	3 300	3 130
	0,6	6 770	6 090	5 320	4 670	4 430

7.1.2.5 Conductor rails

The current-carrying capacity of conductor rails (Clause 11.4) can be obtained in principle from Equation (7.10). However, the case of very low wind velocities, being practically zero, has to be considered, since wind speeds are low close to the ground level and in tunnels. In this case, the Nusselt number in Equation (7.6) cannot be calculated from Equation (7.7), which is based on the Reynolds number. The energy loss by radiation is low and solar radiation is not present in tunnels and Equation (7.10) can be transformed to

$$I = \sqrt{A \cdot \alpha_S \cdot U_S (T - T_{am})/\rho_T} \quad , \tag{7.24}$$

where A is the cross section, U_S the circumference, α_S the heat transfer coefficient and ρ_T the resistivity at the rail temperature T. The resistivity ρ_T can be taken from $\rho_T = \rho_{20} [1 + \alpha_R (T - 20)]$ with ρ_{20} and α_R from Table 11.2
  – steel $\rho_{20} = 0,12060 \, \Omega mm^2/m$ and $\alpha_R = 5 \cdot 10^{-3} \, K^{-1}$ and
  – aluminum $\rho_{20} = 0,03268 \, \Omega mm^2/m$ and $\alpha_R = 3,82 \cdot 10^{-3} \, K^{-1}$.
The heat transfer coefficient α_S consists of the components α_{Scon} because of convection and α_{rd} because of radiation: $\alpha_S = \alpha_{Scon} + \alpha_{rd}$.
Measurements carried out on *conductor rails* made of AlMgSi0,5 and with a cross-sectional area of 3 578 mm², have shown that the heat transmission coefficient for free convection α_{Scon} is approximately 5,3 W/(K·m²). The radiation heat transmission coefficient α_{rd} is calculated according to Stefan-Boltzmann's law for a conductor rail temperature of 85 °C. With an ambient air temperature of 40 °C it was found $\alpha_{rd} = k_e \cdot 8,62$ W/(K·m²). An emissivity k_e of 0,75 to 0,85 can be assumed for steel conductor rails and 0,6 for aluminum after being in place for two years and as high as 0,8 after four years [7.8].
In still air, the heat transmission coefficient of aluminum composite conductor rails is assumed to be 9 W/(K·m²) and that of iron conductor rails to be 12,2 W/(K·m²). Considering an analogy to the heating of contact wires heat transfer coefficients, values of 18 and 24 W/(K·m²) were assumed for a wind velocity of $v_W = 0,6$ m/s. These values were then used to calculate the continuous current-carrying capacities shown in Table 7.9. A cross-sectional area reduction due to wear of the iron conductor rail by 10 % of the nominal value was taken into account.
For comparison purposes, the continuous current-carrying capacity of an aluminum conductor rail with a cross-sectional area of 5 100 mm² was calculated according to

$$I_d = 5,75 \cdot \sqrt{A} \cdot U_S^{0,39} \quad , \tag{7.25}$$

as presented in [7.9]. If the circumference $U_S = 450\,\text{mm}$ is inserted, a continuous *current-carrying capacity* of $4\,430\,\text{A}$ is obtained. This value is lower than the value of $5\,150\,\text{A}$ at $v_W = 0\,\text{m/s}$ shown in Table 7.9 because the emissivity of a clean, brilliant rail was assumed in Equation (7.25).

Berlin metropolitan railway has specified the permissible continuous current-carrying capacity to be $2\,800\,\text{A}$ for iron conductor rails and $4\,700\,\text{A}$ for aluminum composite conductor rails with a cross-sectional area of $5\,100\,\text{mm}^2$, based on a maximum temperature of $85\,°\text{C}$ and an ambient temperature of $40\,°\text{C}$. The ampacity of overhead conductor rails is included in Clause 11.3.

7.1.3 Overhead contact lines

The current-carrying capacity I_{dOCL} of an overhead contact line is the sum of the currents flowing through the contact wire CW, the catenary wire CA and the parallel feeders FL in the limit state.

$$I_{dOCL} = I_{CW} + I_{CA} + I_{FL} \quad . \tag{7.26}$$

In the case of DC supply, the portions of the total current flowing through the individual components depend on the conductance of the components.

Contact wire

$$I_{CW} = I_{dOCL} \cdot R_{tot}/R_{CW} = n_{CW} I_{dOCL} \quad .$$

Catenary wire

$$I_{CA} = I_{dOCL} \cdot R_{tot}/R_{CA} = n_{CA} I_{dOCL} \quad . \tag{7.27}$$

Feeder line

$$I_{FL} = I_{dOCL} \cdot R_{tot}/R_{FL} = n_{FL} I_{dOCL} \quad .$$

The total resistance R_{tot} follows from

$$1/R_{tot} = 1/R_{CW} + 1/R_{CA} + 1/R_{FL} \quad ,$$

therefore,

$$R_{tot} = \frac{R_{CW} \cdot R_{CA} \cdot R_{FL}}{R_{CW} \cdot R_{FL} + R_{CW} \cdot R_{CA} + R_{CA} \cdot R_{FL}} \quad . \tag{7.28}$$

The current-carrying capacity of a contact line I_{dOCL} is determined by the component reaching first its thermal limit, which is called I_{dlim} and its portion of the total current is n_{lim}. The current capacity of the contact line can be obtained from

$$I_{dOCL} = I_{dlim} (1 + n_2/n_{lim} + n_3/n_{lim}) \quad . \tag{7.29}$$

Provided, that the contact wire reaches its thermal capacity first, than

$$n_{lim} = n_{CW} \quad ; \quad n_2 = n_{CA} \text{ and } n_3 = n_{FL} \quad .$$

Table 7.10: Current-carrying capacities of DC overhead contact lines at various ambient temperatures, values given in A, wind velocity 1,0 m/s.

Composition	T_{am} °C			
	−20	0	20	40
AC-100 – Cu + Bz II 50	1 060	945	810	650
AC-120 – Cu + Cu 70	1 370	1 220	1 050	840
2 AC-120 – Cu + Cu 70	2 285	2 030	1 750	1 400
2 AC-120 – Cu + Cu 150	2 740	2 430	2 090	1 670
2 AC-120 – Cu + 2 · Cu 150	3 780	3 360	2 890	2 310

Example 7.4: Determine the current-carrying capacity of a Sicat S1.0 line consisting of a contact wire AC-100 – Cu and a catenary wire BzII 50, at 40 °C ambient temperature and 1,0 m/s wind velocity.

Resistance of the contact wire at 80 °C

$$R_{CW} = \rho_{20}/A \cdot [1 + \alpha_R(t-20)] = 0,0179/100\,[1+0,00294 \cdot 60] = 0,222 \cdot 10^{-3}\,\Omega/m$$

Resistance of the catenary wire at 80 °C

$$R_{CA} = 0,0278/49,5\,[1+0,00377 \cdot 60] = 0,689 \cdot 10^{-3}\,\Omega/m$$

Total resistance

$$R_{tot} = (0,222 \cdot 0,689 \cdot 10^{-3})/(0,222+0,689) = 0,168 \cdot 10^{-3}\,\Omega/m$$

$$n_{CW} = 0,168/0,222 = 0,76$$

$$n_{CA} = 0,168/0,689 = 0,24$$

The contact wire limits the current capacity: $I_{d\,lim} = 485$ A (see Table 7.3), $n_{lim} = 0,76$

$$I_{dOCL} = 485\,(1+0,24/0,76) = 640\,A.$$

Current through contact wire

$$I_{CW} = 640 \cdot 0,76 = 485\,A$$

Current through catenary wire

$$I_{CA} = 640 \cdot 0,24 = 155\,A$$

The ampacity of the catenary wire BzII 50 was calculated to be 310 A at 100 °C conductor temperature. The current of 155 A will only heat the catenary wire to approximately 60 °C. Therefore, its resistance needs to be adjusted.

$$R_{CA} = 0,0278/49,5\,[1+0,00377 \cdot 40] = 0,646 \cdot 10^{-3}\,\Omega/m$$

$$R_{tot} = (0,222 \cdot 0,646 \cdot 10^{-3})/(0,222+0,646) = 0,165 \cdot 10^{-3}\,\Omega/m$$

$$n_{CW} = 0,165/0,222 = 0,744$$

$$n_{CA} = 0,165/0,689 = 0,256$$

The adjusted current capacity is

$$I_{dOCL} = 485(1+0,256/0,744) = 652 \approx 650\,A.$$

Table 7.11: Calculated current distribution on the individual components of 50 Hz and 16,7 Hz overhead contact lines.

Overhead line type	Feeder line	Return conductor	Power supply					
			50 Hz			16,7 Hz		
			n_{CW}	n_{CA}	n_{FL}	n_{CW}	n_{CA}	n_{FL}
Sicat S1.0 [1]	no	no	0,66	0,37	–	0,74	0,27	–
	yes	no	0,38	0,20	0,44	0,39	0,14	0,49
	yes	yes	0,36	0,18	0,49	0,38	0,14	0,49
Madrid–Seville [2]	no	no	0,62	0,40	–	0,71	0,30	–
	no	yes	0,62	0,40	–	0,71	0,30	–
	yes	no	0,37	0,22	0,42	0,39	0,16	0,45
	yes	yes	0,34	0,21	0,47	0,38	0,16	0,47
Sicat H1.0 [3]	no	no	0,53	0,48	–	0,51	0,49	–
	yes	no	0,33	0,28	0,40	0,29	0,27	0,45
	yes	yes	0,29	0,25	0,46	0,28	0,26	0,46

[1] Contact wire AC-100–Cu, catenary wire BzII 50, feeder line 243-AL1, return conductor 243-AL1, [2] Contact wire AC-120–CuAg, catenary wire BzII 70, feeder line 243-AL1, return conductor 243-AL1, [3] Contact wire AC-120–CuMg0,5, catenary wire BzII 120, feeder line 243-AL1, return conductor 243-AL1

Current through contact wire

$$I_{CW} = 652 \cdot 0,744 = 485 \, A.$$

Current through catenary wire

$$I_{CA} = 652 \cdot 0,256 = 167 \, A.$$

Table 7.10 shows *ampacities* of DC overhead contact lines which were calculated using Equations (7.26) and (7.28).

In the case of AC supply systems, the individual portions of current also depend on the *inductive coupling* and, therefore, on the frequency of the power supply and the arrangement of conductors. In Chapter 5 information is given on the calculation of current distribution in an AC overhead contact line.

In Table 7.11, the current distribution is summarized for some frequently used contact line types. The presence of a return conductor has only a minor effect on the current distribution in the feeding system. The sum of the current components can be more than one because of the phase difference between the currents. For the Madrid–Seville line, equipped with a return conductor, without a feeder line, the following is obtained:

- $n_{CW} = 0,62$ for the contact wire and
- $n_{CA} = 0,40$ for the catenary wire.

In Table 7.12, the current-carrying capacity (ampacity) of contact lines frequently used for AC installations is presented for 50 Hz and 16,7 Hz power supply. There, the effect of the frequency on the permissible current can clearly been observed.

7.1.4 Thermal design calculations

7.1.4.1 Design alternatives

Contact lines of electric railways need to be designed to exclude detrimental overloading. However, it is also desirable to achieve a coordinated utilization of the contact line. These

Table 7.12: Current-carrying capacity in A of AC contact lines at various ambient temperatures, wind velocity 1,0 m/s, new contact wires, feeder line and return conductor 243-AL1.

Contact line arrangement	Feeder line	Return conductor	Frequency Hz	Contact wire temperature °C	Ambient temperature °C			
					−20	0	20	40
Sicat S1.0	no	no	50		1 230	1 100	944	755
contact wire:	no	no	16,7		1 080	960	820	660
AC-100–CuAg	yes	no	50	80	2 120	1 890	1 620	1 300
catenary wire:	yes	no	16,7		1 950	1 610	1 490	1 190
BzII 50	yes	yes	50		1 970	1 750	1 500	1 200
	yes	yes	16,7		1 950	1 610	1 490	1 190
Re250	no	no	50		1 350	1 230	1 100	950
contact wire:	no	no	16,7		1 270	1 130	970	770
AC-120–CuAg	no	yes	50		1 350	1 230	1 100	950
catenary wire:	no	yes	16,7	100	1 270	1 130	970	770
BzII 70	yes	no	50		2 250	2 000	1 720	1 370
	yes	no	16,7		2 080	1 850	1 590	1 270
	yes	yes	50		2 030	1 810	1 550	1 240
	yes	yes	16,7		2 010	1 790	1 540	1 230
Sicat H1.0	no	no	50		1 620	1 480	1 300	1 040
contact wire:	no	no	16,7		1 570	1 430	1 280	1 030
AC-120–CuMg0,5	yes	no	50	100	2 360	2 100	1 800	1 440
catenary wire:	yes	no	16,7		2 110	1 870	1 600	1 280
BzII 120	yes	yes	50		2 040	1 810	1 550	1 240
	yes	yes	16,7		2 040	1 810	1 550	1 240

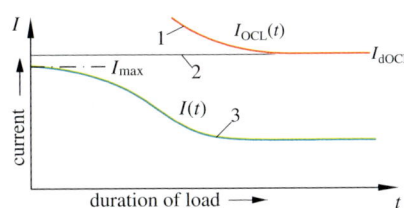

Figure 7.11: Rating based on continuous current-carrying capacity.

— load current in contact line section $I(t)$
— continuous current-carrying capacity of contact line $I_{OCL}(t)$

contradicting requirements can be met by applying methods, which consider the real characteristics of loadings and of the features of the installation.

7.1.4.2 Maximum principle

In the *maximum principle*, which is sufficient for many cases, the continuous current-carrying capacity I_{dOCL} is taken as the basis of all calculations. In this case, the criterion assumed is that this value must be equal to or higher than the *maximum expected load current* I_{max} (Figure 7.11) at all times.

$$I_{dOCL} \geq I_{max} \qquad (7.30)$$

In Figure 7.11, the currents are related to the period of their action are called *time-weighted load currents*. Using this procedure, the contact line never reaches the permissible maximum temperature during normal operation because the peak loads only occur over short periods however, they are assumed to be acting permanently. For this reason, this method is uneconomic and not recommended if the time-dependent loads are known.

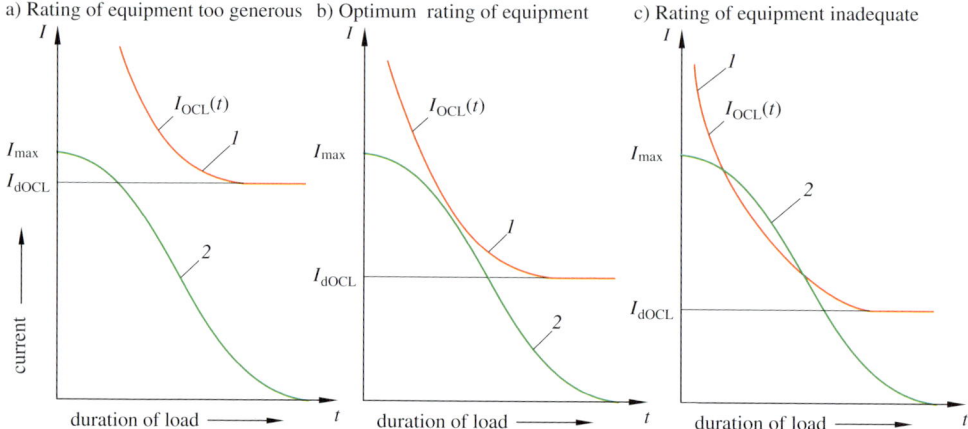

a) Rating of equipment too generous　b) Optimum rating of equipment　c) Rating of equipment inadequate

Figure 7.12: Design calculation principle: matching load characteristics (—) and current-carrying capacity characteristics (—).

7.1.4.3　Matching load and ampacity characteristics

Figure 7.12 illustrates the principle of the *thermal rating design method* [7.10, 7.11]. The graph of the time-weighted load current $I(t)$, representing the load, is matched with the graph of the thermal characteristics $I_{OCL}(t)$ of the equipment, i. e. the traction contact line. The objective is to achieve the best possible match of the graphs of $I_{OCL}(t)$ and $I(t)$ as shown in Figure 7.12 b). The thermally determined ampacity of overhead contact lines can be calculated using Equations (7.10), (7.19) and (7.22), where $I_{OCL}(t) = I$. As demonstrated in Figure 5.22, this principle is also applicable to short-circuit design considerations.

Railway lines for general traffic

The loading current of the feeding section of a railway line which is classified as general traffic line can be considered as a stochastic variable. The Gaussian distribution can be used to describe its characteristics. The distribution is determined by the maximum annual hourly mean value $I_{h\,max}$ of currents and its standard deviation which can be obtained from the annual power consumption. The time-dependant current $I(t)$ is obtained from

$$I(t) = I_{h\,max}\left(1 + \lambda_L \cdot v_p\right) \quad , \tag{7.31}$$

where λ_L is the variable of the Gaussian distribution $F(\lambda_L)$ and v_p the coefficient of variation of the distribution which depends on the annual mean power consumption P_a according to Figure 5.17. The relation between the Gaussian distribution $F(\lambda_L)$ and the duration t is $F(\lambda_L) = T - t/T$, where T is one hour or 3 600 s. In Table 5.13 the relation of λ_L and t is presented. The principle to be applied in order to achieve optimum contact line dimensions can be expressed as:

$$I_{OCL}(t) - I(t) \longrightarrow \text{Minimum} \quad . \tag{7.32}$$

Table 7.13: Time-depending ampacity of the contact line Sicat S1.0 with and without parallel feeder line, temperature 80 °C.

	Without feeder line			With feeder line 243-AL1	
Current in contact wire	Current in catenary	Duration		Current in catenary	Duration
A	A	s		A	s
485	660	∞		1 270	∞
500	680	1 660		1 310	1 660
600	820	720		1 570	720
700	960	290		1 830	290
800	1 090	190		2 090	190

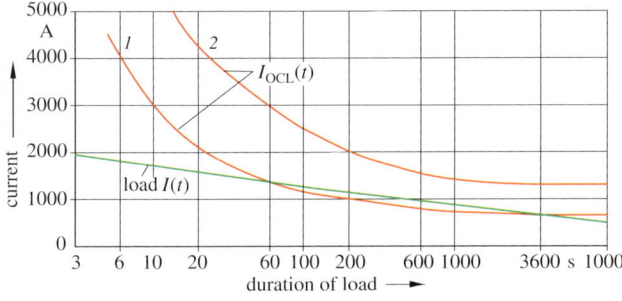

Figure 7.13: Comparison of a normally distributed load with a maximum annual hourly mean of 610 A with the current-carrying capacity of a contact line system of type Sicat S1.0; wind speed 1,0 m/s, ambient temperature 40 °C. *1* contact line without parallel feeder line, *2* contact line with a 243-AL1 parallel feeder line

Example 7.5: The suitability of the contact line type Sicat S1.0 is to be checked for a railway of the general traffic with an annual mean power consumption of 3 500 MW. The mean current for the hour with the maximum load during one year is obtained from Equation (5.57) for $U = 15$ kV, $\cos = 0,9$, $c_d = 1,35$, $c_h = 1,75$ (see Figure 5.14):

$$I_{h\,max} = c_d \cdot c_h / (15 \cdot \cos \varphi) = 1,35 \cdot 1,75 \cdot 3\,500 / (15 \cdot 0,9) = 613\,A \approx 610\,A$$

The time-dependant current carrying capacity for the contact line type Sicat S1.0 with and without parallel feeder line is presented in Table 7.13 and Figure 7.13. The example reveals that the contact line without a parallel feeder line does not comply with the requirements in the range of current durations between 50 s and one hour.

High-speed railway lines

For *high-speed railways*, the rating should be based on the *time-weighted parameters*, see Clause 5.3.2, the principle to be applied being:

$$I_{OCL}(t^*) - I_{eff\,max}(t^*) \longrightarrow \text{Minimum} \quad . \tag{7.33}$$

In Figure 7.14 the time-weighted ampacity of an overhead contact line type Sicat H1.0 is shown in comparison with the load of high-speed trains as calculated in Clause 5.3.2. The example is taken from [7.12].

By comparing the time-weighted load with the time weighted ampacity of the contact line, conclusions can be drawn about the real thermal behaviour of overhead contact line installations. Assuming the load situation discussed in detail in Clause 5.3.2, the condition shown in Figure 7.15 is obtained for a contact line type Sicat H1.0. In this case, the ambient temperature in the tunnel was assumed to be $T_{am} = 30$ °C and the wind speed was assumed to be

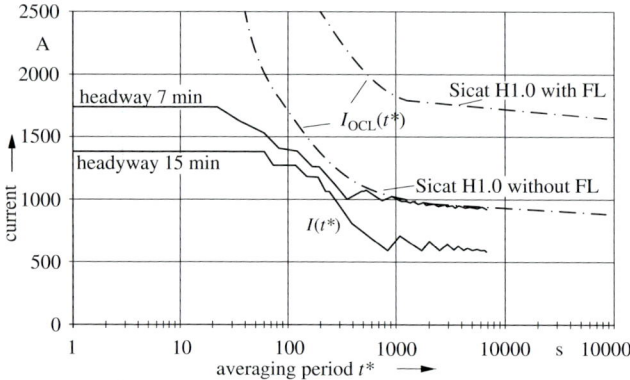

Figure 7.14: Comparison of the time-weighted ampacity $I_{OCL}(t^*)$ of a contact line Sicat H1.0, with and without parallel feeder line (FL), with the load $I(t^*)$ on line sections used by high-speed trains, as determined in Clauses 5.3.2 and 5.3.7.

zero. The contact lines of the two tracks under consideration are connected at a distance of 10 km from the feed point and at 5 km intervals thereafter. For 15 min headway, a contact line Sicat H1.0 without a parallel feeder would comply with the requirements, however, for 7 min headway, a parallel feeder line should be installed.

7.2 Effects of temperature on contact wire characteristics

7.2.1 Introduction

Clause 7.1 describes the basis for determining current-carrying capacity and *thermal rating of contact lines*. This clause identifies the basis for permissible temperature limits and presents the consequences of contact wire operation at elevated temperatures, which may occur from increased power consumption, after short circuits and in the case of failures of protective devices or circuit breakers. Local temperature rises may be caused, for example, by damaged connecting fittings or defects in the contact wire.

Localized and short-term temperature rises can occur where the collector strips touch the contact wire at standstill over a longer period. At these locations, the *melting temperature* of the contact wire material may even be reached. The associated reduction of the tensile strength of the contact wire and high collector strip wear can limit the capacity of DC railway traction power supply systems. Currently, the maximum economically and technically manageable current to flow through a contact wire-collector strip transition is deemed to be 500 to 700 A for a running pantograph.

Increasing contact wire temperatures tends to increase *permanent elongation* and reduce tensile strength. Additionally, the mechanical properties of the wire change, depending on the tensile stress in the contact wire and on the time it has been in operation. The wire production process and the cross section (see Table 2.8) also affect the behaviour under temperature changes. The effects of these parameters on the contact wire characteristics will be discussed in this clause. They are also essential when assessing the residual life of a contact wire. The behaviour of contact wires under elevated temperatures determines the operational temperature limits.

a) Variation of temperature of a type Sicat H1.0 overhead contact line for various tunnel air temperatures T_{am}

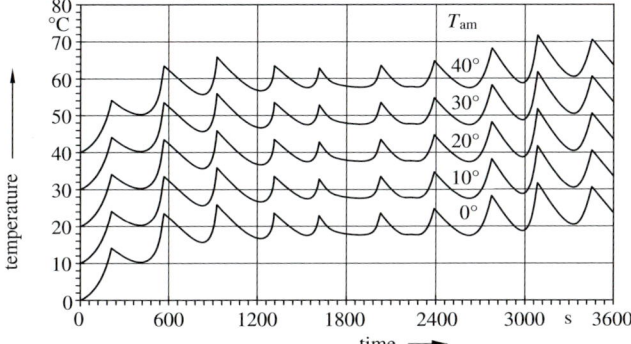

b) Comparison of a time-weighted current-carrying capacity $I_{dOCL}(t*)$ for T_{am} = 30°C with the load that trains require traveling at 6/7 min headway through the tunnel and consuming 1130 A per train

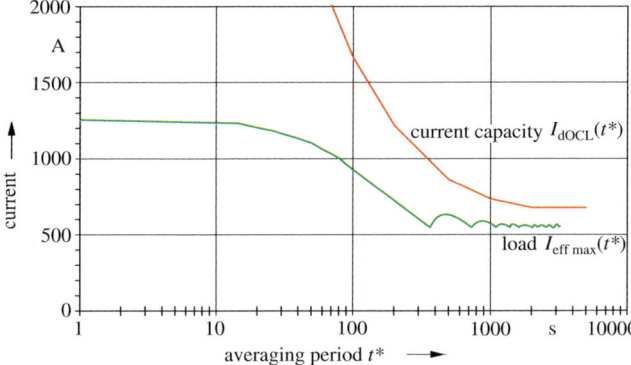

Figure 7.15: Rating of an overhead contact line for use by high-speed trains in a tunnel, without parallel feeder line.

7.2.2 Characteristics of contact wire material

Traditionally, contact wires were mainly made of hard-drawn electrolytic copper, because they achieve adequate mechanical strength and provide, in a pure form, low resistivity and a good conducting surface for current transfer to the collector strips. However, improved contact wire properties are required to permit the running speeds of modern high-speed trains. For this reason, contact wires made of *copper alloys* containing silver, cadmium, magnesium, tin, nickel or zinc were introduced. The standard EN 50149 contains relevant data. While alloys with silver, magnesium and tin are of practical relevance for high-speed lines, alloys containing cadmium are prohibited by law in many European countries because of its adverse environmental toxicity.

Electrolytic copper is also called Cu-ETP (electrolytic tough-pitched) in engineering publications and the alloy with 0,1 % silver Cu-LSTP (low silver tough-pitched).

Alloying additives to copper leads to isomorphous crystalline structures that achieve higher tensile strengths and temperature resistances than pure copper. The electrical conductivity is not affected by the addition of silver since both have similar electrical properties. However, while the electrical conductivity of magnesium and zinc alloys is lower, their tensile strength is higher. The physical properties of contact wire materials can be obtained from Tables 11.2, 11.5 and 11.6.

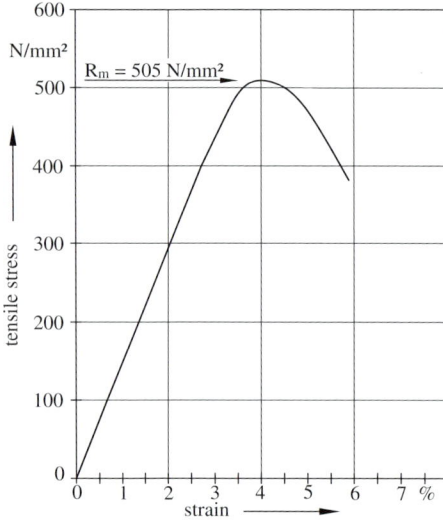

Figure 7.16: Contact wire material made of CuMg0,5 tested in accordance with EN 10 002-1.

Contact wires are shaped from 18 to 24 mm diameter drawing stock produced by a rolling or extruding process. The stock is drawn through several circular dies in a sequence followed by a grooving die and then a final finishing die. The wire is cold-drawn into its final dimensions and profile. This changes the material micro-structure from almost round into an elongated shape and finally into a fibrous-like structure aligned in the direction that the wire was drawn. Discontinuities in the crystalline structure increase the shear resistance and harden the material. As copper is subjected to several cold-drawing processes, the resistance to deformation increases while the electrical conductivity and plasticity decrease.

The reduction of the cross-sectional area caused by drawing in relation to the original stock cross section is called the conversion ratio. It is expressed as a percentage and it is equal to the ratio of the reduced cross-sectional area to the original cross-sectional area. For example, the optimum conversion ratio for producing a grooved contact wire CuMg0,5, that contains approximately 0,5 % magnesium, was determined to be about 75 %.

Under normal operating conditions, automatically tensioned contact wires are subjected to a nearly constant tensile force. The mechanical load can be assumed as being static. Therefore, permissible stress may be defined in relation to the minimum strength. Figure 7.16 shows the stress-strain diagram obtained by tensile testing of a contact wire made of CuMg0,5. Over a wide range, the strain is directly proportional to the stress.

7.2.3   Effect of heating on the tensile strength

Depending on the temperature and the composition of contact wire material, operation of contact wires at elevated temperatures can reduce their strength. Long-term heating of cold-drawn copper wire causes the crystalline micro-structure to regain the condition it had before the cold-drawing process. This transition to the stable crystalline micro-structure is called *recrystallization* and is accompanied by the resetting of physical characteristics typical of the cold-drawn contact wire, to those of annealed copper. Figure 7.17 shows how the tensile strength of contact wires made of Cu, CuAg0,1, CuMg0,2 and CuMg0,5 decreases because of recrystallization. As the recrystallization temperature is exceeded, the micro-structure starts

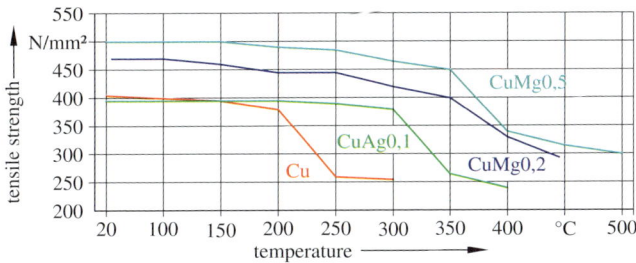

Figure 7.17: Tensile strength of various contact wire alloys due to recrystallization at rising temperatures.

to change and is accompanied by a loss of tensile strength. In this process, the crystalline grains re-assume the stable round shape and the micro-structure created by cold drawing is almost totally converted.

The reduction in tensile strength can be evaluated on the basis of the *annealing point*. This is the temperature at which the material can be kept for one hour, when its tensile strength drops by half the difference between the original high-tensile strength and the final, low-tensile strength of the material, resulting from exposure to high temperatures for long periods. The process is a function of both the temperature and the period of time the material is kept at that temperature. For example, for a material conversion ratio of 60 % and an exposure of one hour, the annealing point is 215 °C for copper and 340 °C for CuAg0,1 contact wires. For a conversion ratio of 85 %, the corresponding values drop to 180 °C and 300 °C, respectively. Figure 7.18 shows a graph for determining the annealing point of CuMg0,5 with a conversion ratio of 85 %.

The *loss of tensile strength* of copper wires because of heating increases with the duration that the material is kept at high temperatures, with the conversion factor and with purity of the copper. Alloying copper with silver dramatically delays the tensile strength reduction [7.13], enabling contact wires made of CuAg0,1 to be operated at higher temperatures and tensile loads than those made of E-Cu, although, their strength is the same (see Table 7.13).

In [7.14] the effect of heating on the loss of tensile strength of copper was studied when subjected to periodic temperature changes and when kept at constant temperatures in the 100 °C to 150 °C range. It was discovered that a series of short-term exposures to higher temperatures did not affect the tensile strength. The same conclusions were made from studies reported in [7.15].

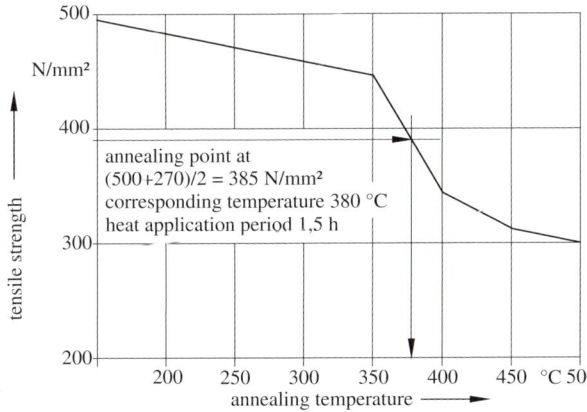

Figure 7.18: Annealing point of CuMg0,5.

Table 7.14: Reduction in percent of contact wire strength because of exposure to raised temperatures.

Contact wire	Stress N/mm²	Temperature °C	Exposure period in h				
			100	200	300	400	500
AC-100 – Cu and AC-120 – Cu	100	120	0,8	1,0	1,2	1,4	1,6
		140	1,5	1,8	2,2	2,8	3,6
		160	2,2	2,7	3,3	4,0	4,8
AC-100 – CuAg	150	120	1,6	2,0	2,4	2,8	3,2
AC-120 – CuAg	100	170	3,0	3,4	3,9	4,4	5,0

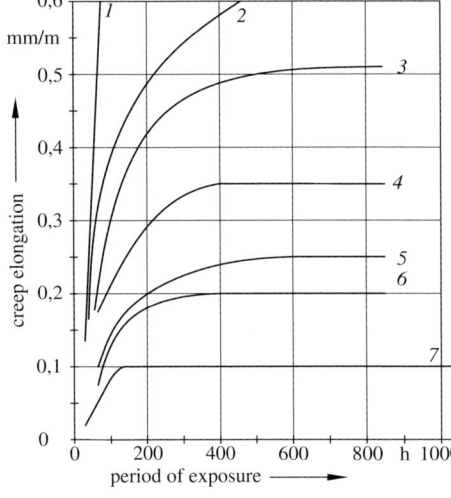

Figure 7.19: Time/elongation graphs of contact wires measured in mm/m.
1 AC-120 – Cu, $T = 120°C$, $\sigma = 150\,\text{N/mm}^2$
2 AC-120 – CuAg, $T = 170°C$, $\sigma = 150\,\text{N/mm}^2$
3 AC-120 – Cu, $T = 120°C$, $\sigma = 100\,\text{N/mm}^2$
4 AC-120 – CuAg, $T = 170°C$, $\sigma = 100\,\text{N/mm}^2$
5 AC-120 – Cu, $T = 120°C$, $\sigma = 50\,\text{N/mm}^2$
6 AC-120 – CuAg, $T = 170°C$, $\sigma = 50\,\text{N/mm}^2$
7 AC-120 – CuMg, $T = 150°C$, $\sigma = 225\,\text{N/mm}^2$

7.2.4   Effect of period at increased temperatures on tensile strength

This clause summarizes tests and studies to determine the effect of the exposure period to increased temperatures on the tensile strength of contact wires. Papers [7.16, 7.17, 7.18] deal with this subject. A measurable effect of the exposure period on the strength was recorded only at temperatures above 120 °C. The findings are presented in Table 7.14. At 160 °C (AC-100 – Cu) and 170 °C (AC-120 – CuAg), the reduction is approximately 5 % after 500 h of exposure. Since temperatures above 80 °C or 100 °C rarely occur, this effect can be ignored. In [7.18] the effect of the exposure period on *permanent elongation* was studied. The results are summarized in Figure 7.19; e. g., the permanent elongation of a contact wire AC-120 – Cu subject to 100 N/mm² at 120 °C for 600 h was measured as 0,5 mm/m. This result may also be applied to other contact wire types, when the same stress and temperature are assumed.

7.2.5   Heating of contact wire at locations subject to increased wear

Figure 7.20 b) shows a contact wire that was worn unevenly and locally. The temperature distribution along this section of contact wire was calculated using methods that accurately model the variable heating characteristics. Figure 7.21 shows the calculated temperatures for a current of 1 000 A. Cross-section reductions of 25 % and 35 %, an ambient temperature of 35 °C and a wind speed of 1,0 m/s were assumed in the calculations. If the worn section is

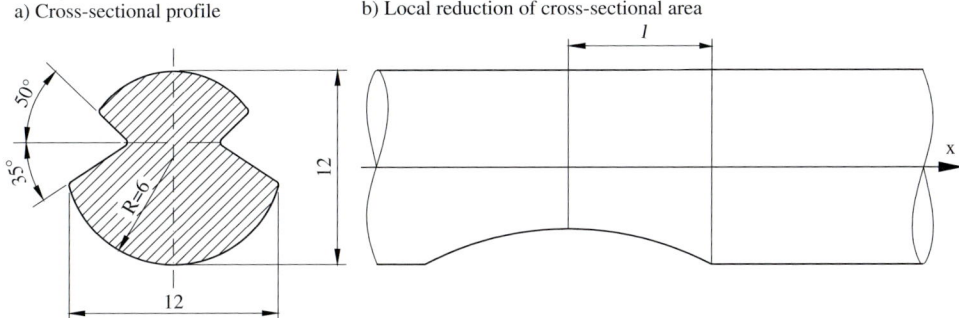

Figure 7.20: Contact wire AC-100–Cu, dimensions in mm.

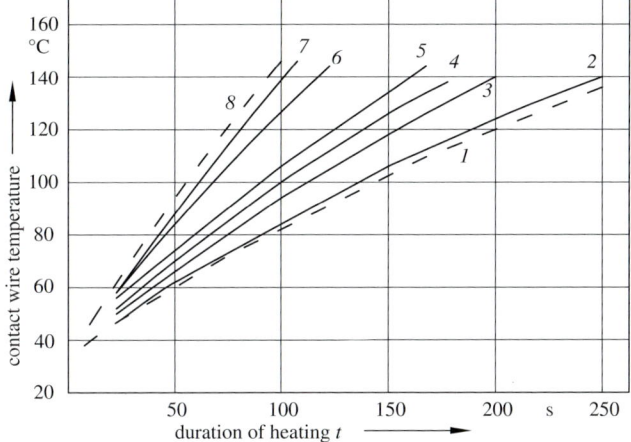

Figure 7.21: Temperature of the section of contact wire experiencing the greatest temperature increase depending on the heating duration [7.19], current 1 000 A.
1 contact wire without wear AC-100–Cu
2 at large distances from position of local wear
3 local wear 25 %, $l = 0,1$ m
4 local wear 25 %, $l = 0,2$ m
5 local wear 35 %, $l = 0,1$ m
6 local wear 35 %, $l = 0,2$ m
7 local wear 35 %, $l = 0,4$ m
8 contact wire AC-100–Cu, evenly worn by 35 %

longer than 0,4 m, the contact wire temperature increases with the cross section reduction and the amount of wear. Then, the condition of the contact wire should only be assessed on the basis of the remaining minimum cross-section.

To calculate the effect of varying cross-sections, the power transmitted along the contact wire because of thermal conduction is added to the energy balance Equation (7.1) for an element with the length dx

$$m_C \cdot c \cdot dT/dt \cdot dx = (N_j - N_S - N_R - N_C) \cdot dx + N_{in} - N_{out} \quad . \tag{7.34}$$

The energies N_{in} and N_{out} can be described by

$$N_{in} = \lambda_c A(x) \frac{\partial}{\partial x} T(x) \quad \text{for the energy conducted into the element and} \tag{7.35}$$

$$N_{out} = \lambda_c A(x + dx) \frac{\partial}{\partial x} T(x + dx) \quad \text{for the energy lost by conduction,} \tag{7.36}$$

where λ_c is the thermal conductivity of a contact wire and $A(x)$ the variable cross-section. If $dT/dt = 0$ for stationary conditions and (7.35) and (7.36) are considered in (7.34), the resulting differential equation describes the temperatures along the conductor axis [7.20].

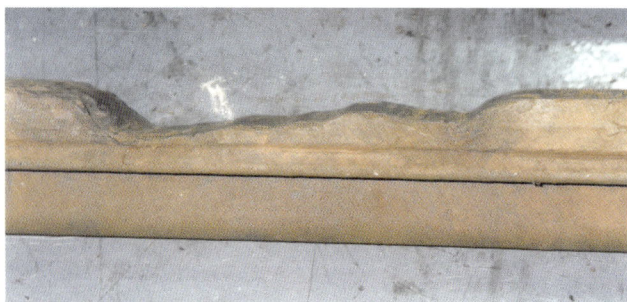

Figure 7.22: Collector strip of a DC 0,75 kV urban mass transit vehicle showing severe wear due to contact at the same position over an extended line section.

The solution of the equation can be written as

$$T(x) = (T_K - T_\infty)\, e^{-|x|/B} + T_\infty \quad . \tag{7.37}$$

In (7.37), $T_K = T(0)$ is the increased temperature occurring at a position where the cross-sectional area is reduced locally or where a faulty fitting is placed and $T_\infty = T(x \to \pm\infty)$ is the temperature at points far away from the position of the reduced cross-section. The thermal constant B in (7.37) is

$$B = \sqrt{\lambda_C A/(\alpha_S U_S)} \quad , \tag{7.38}$$

where λ_C is the thermal conductivity of the contact wire (see Table 7.13), A the sound contact wire cross-section, U_S its circumference and α_S the heat transmission coefficient, which depends mainly on the wind velocity. Table 7.15 contains data for α_S relevant for conventional contact wires.

The parameter B having the unit of length determines the length affected by changes in cross-sectional areas or by local heat sources. It can be assumed that conductor temperature is increased on a section of $\pm 3B$, in total on a length of $6B$.

Example 7.6:   From Tables 7.13 and 11.5 the following values are obtained for a contact wire AC-100 – Cu: $\lambda_C = 377\,\text{W/(K·m)}$; $U_S = 0{,}0412\,\text{m}$, $A = 10^{-4}\,\text{m}^2$; $\alpha_S = 36\,\text{W/(K·m}^2)$ at $v_W = 1\,\text{m/s}$. Inserting these data into (7.38) yields $B = 0{,}16\,\text{m}$. This means that the transitional state virtually decayed at a distance of $3B = 0{,}5\,\text{m}$.

Under the same wind conditions and $U_S = 0{,}0454\,\text{m}$ and $\lambda_C = 245\,\text{W/(K·m)}$, $B = 0{,}13\,\text{m}$ is obtained for a grooved contact wire type AC-120 – CuMg0,5.

These examples confirm that, in cases of local reductions of a cross section, only the remaining, reduced cross-sectional area may be used to assess the ampacity of the AC-100 – Cu contact wire if the worn section is longer than 1,0 m; for CuMg AC-120 contact wires this length is 0,8 m. A comprehensive series of measurements described in [7.21] and [7.22] confirms that the temperature of contact wires of type AC-100 – Cu in the vicinity of a connector fitting differs slightly from the fitting temperature.

Table 7.15: Heat transmission coefficients α_S in W/(K·m²), ambient temperature 40°C.

Wind velocity m/s	α_S W/(K·m²)
1	36
2	52
4	75
8	112

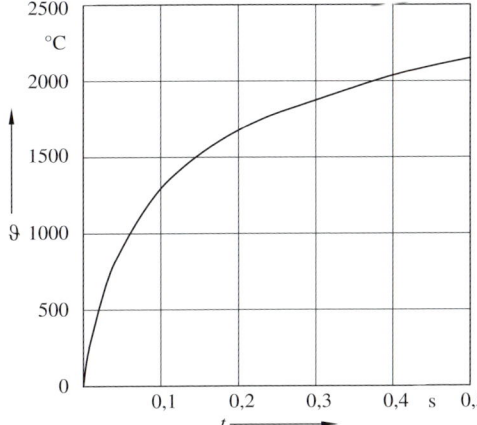

Figure 7.23: Temperature ϑ of the contact surface of a stationary carbon collector strip at the contact point as a function of current action period t, as determined by different methods [7.22], current 1 000 A.
Carbon strip properties:

γ = 1 810   kg/m^3
c = 140   Ws/(kg·K)
λ_C = 30   W/(K·m)
ρ_0 = 30 · 10^{-6} Ωm

7.2.6 Interface between contact wire and collector strips

Design of overhead contact lines and pantograph needs especially to consider the interface between contact wire and collector strips, especially when a train is stationary and the auxiliaries are in operation. In the case of DC power supply the current to be transmitted at standstill can determine the selection of contact wires and collector strips. In Clause 2.4.2 some basic requirements are presented.

Maximum current is drawn by an accelerating train in motion when the full power is utilized. The current can reach 1 500 A per train in the case of AC 15 kV 16,7 Hz supply and 5 000 A in the case of DC 1,5 kV (see EN 50 388).

At standstill of train, the power for air conditioning and auxiliaries flows through a fixed contact point for a longer period of time. The TSI Energy [7.23] specifies a current of 300 A per train for DC 1,5 kV and 200 A per train for DC 3,0 kV to be transmitted without local overheating of both components.

To cope with the requirements of the current transmission of a running train, an adequately selected number of collector strips is necessary. The permissible current for one carbon collector strip in motion is 500 to 700 A, the latter being adopted for DC systems. Staggering of the contact wire is assumed when applying the mentioned data. If the contact wire contacted the collector strip at the same point over a longer period of time, severe wear would occur at the collector strip because of local heating, reducing the strength. Figure 7.22 shows a severely worn carbon collector strip that was used on a vehicle in a DC 0,75 kV installation where the contact wire was installed with almost no stagger, resulting in local overheating of the strip.

In Figure 7.23, the temperature increase of the contact surface of a carbon collector strip is shown at the contact point as a function of the duration of the action of a 1 000 A current. With this load, the temperature rises very quickly to 500 °C and above. To cope with the above mentioned requirements, the collector strips should have the following properties:

- low *electric contact resistance*
- high melting point
- good *thermal conductivity*
- low dead weight
- high compressive strength

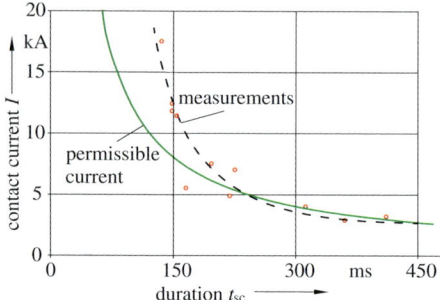

Figure 7.24: Measured fusing currents, stationary vehicle, contact wire type AC-100 – Cu, simple carbon contact strip.
  ∘ measurement made by DR
  — recommended limit values

   – high elasticity
   – low coefficient of friction at the interface with the contact wire
Carbon collector strips, i. e. made of electro-carbon or graphite with a binder, proved particularly favourable in combination with copper contact wires. Many European railways operating AC systems replaced the *metal collector strips* used in the first half of the 20th century with carbon collector strips. Paper [7.24] concludes that metalized carbon is suited for all AC railway networks in Europe. The upper limit of the *permissible operating current for a pantograph* with two collector strips is at around 1 400 A. In DC railways with lubricated copper collector strips, the value of the upper permissible operating current is 1 250 A per collector strip. If higher currents are required for the transfer of power to the traction unit, then the number of collector strips per pantograph, or the number of pantographs per vehicle, must be increased.

On high-speed trains the power demand for convenience and auxiliaries reaches 1 000 VA. This power has to be safely transferred to a pantograph on a stationary vehicle.

The *fusing current* that would cause the contact wire to melt depends, to a high degree, on the current duration t_{sc}. Figure 7.24 shows the required currents, as measured in Germany, to melt a contact wire AC-100 – Cu using one carbon collector strip depending on the duration of the current. From the measurements, recommended values for the permissible current can be expressed as a function of the duration period by

$$I_{\text{con per}} = 1\,200/t_{sc} + 100 \text{ in A} \quad , \tag{7.39}$$

where t_{sc} is the duration of current action measured in seconds. This function is also depicted in Figure 7.24. For AC-120 – Cu contact wires, the permissible values are 20 % higher. As a conclusion from Equation (7.39) a current of 100 A is permanently permissible for a AC-100 – Cu contact wire as well as 120 A for AC-120 – Cu and 150 A for AC-150 – Cu.

Verification of the correct selection and design of collector strips can be carried out in accordance with EN 50 367, Annex A.4. The test should be carried out with one pantograph equipped with the pantograph head to be checked and installed on a traction unit or on a validated test facility. The test should be performed in a protected environment and with one or two contact wires equipped with temperature sensors. A static force, as specified for the application and a current, representative of the maximum consumption of the rolling stock or according to the TSI Energy should be applied. Each test should last 30 minutes unless the temperature recorded by a sensor reaches the maximum permissible value for the contact wire as specified by EN 50 119 and presented in Table 2.8. The test can be deemed as satisfactory if the maximum temperature after 30 minutes is not higher than the value specified in Table 2.8 or a project specification.

7.3 Bibliography

7.1 *Cigre SC22-WG22-12*: The thermal behaviour of overhead conductors. Section 1 and 2: Mathematical model for evaluation of conductor temperature in the steady state and the application thereof. In: Electra, 144(1992), pp. 107 to 125.

7.2 *Webs, A.*: Dauerstrombelastbarkeit von nach DIN 48201 gefertigten Freileitungsseilen aus Kupfer, Aluminium und Aldrey (Current carrying capacity of overhead line conductors made from copper, aluminium and aluminium alloy). In: Elektrizitätswirtschaft, 62(1963)23, pp. 861 to 872.

7.3 *Gorub, J. C.; Wolf, N. F.*: Load capability of ASCR and aluminum conductors based on longtime outdoor temperature rise tests. American Institute of Electrical engineers, 1963, pp. 63 to 81.

7.4 *Kiessling, F.; Nefzger, P.; Nolasco, J. F.; Kaintzyk, U.*: Overhead power lines – Planning, design, construction. Springer Publishing, Berlin – Heidelberg – New York, 2003.

7.5 *Bencard, R.*: Querschnittsauswahl von Freileitungsseilen bei zufällig variablen Betriebsströmen und Umgebungsbedingungen nach thermischen und ökonomischen Kriterien (Selection of cross sections of overhead power line conductors at randomly variable operating currents and ambient conditions using thermal and economic criteria). Ingenieurhochschule Wismar, doctoral thesis, 1985.

7.6 *Cigre SC22-WG22-12*: The thermal behaviour of overhead conductors. Section 4: Mathematical model for evaluation of conductor temperature in the adiabatic state. In: Electra, 185(1999), pp. 75 to 87.

7.7 *Cigre SC22-WG22-12*: The thermal behaviour of overhead conductors. Section 3: Mathematical model for evaluating of conductor temperature in the unsteady state. In: Electra, 174(1997), pp. 59 to 69.

7.8 *Rigdon, W. S. et al.*: Emissivity of weathered conductors after service in rural and industrial environments. In: American Institute of Electrical Engineers. Transactions, Part III, Power Apparatus and Systems, Vol 81, 1962.

7.9 *Mier, G.*: Herstellung und Anwendung von Aluminium-Stromschienen (Production and use of aluminium conductor rails). In: Schweizer Aluminium Rundschau, (1984)3.

7.10 *Lingen J. v.; Schmidt, P.*: Wärmeübergang und Strombelastbarkeit von Hochgeschwindigkeitsoberleitungen im Tunnel (Heat transfer and current capacity of overhead contact lines in tunnels). In: Elektrische Bahnen, 94(1996)4, pp. 110 to 114.

7.11 *Röhlig, S.; Rothe, M.; Schmidt, P.; Weschta, A.*: Höhere Leistungsfähigkeit der Bahnenergieversorgung bei modernen Stadt- und U-Bahnen (Higher capacity of power energy supply for modern city and underground railways). In: Elektrische Bahnen, 91(1993)11, pp. 359 to 365.

7.12 *Lingen, J. v.; Schmidt, P.*: Strombelastbarkeit von Oberleitungen des Hochgeschwindigkeitsverkehrs (Current carrying capacity of contact lines for high-speed transport). In: Elektrische Bahnen 94(1996)1-2, pp. 38 to 44.

7.13 *Freudiger, E. et al.*: Erweichung verschiedener Kupferarten während dreizehn 1/2 Jahren bei 100 °C (Softening of various copper types at 100 °C over a period of 13,5 years). In: Schweizer Archiv für angewandte Wissenschaft und Technik, 36(1970)9, pp. 357 to 359.

7.14 *Roggen, F.:* Erweichung von Kupfer bei zyklischer Erwärmung (Annealing of copper during cyclic heating). In: Schweizer Archiv für angewandte Wissenschaft und Technik, 36(1970)9, pp. 360 to 362.

7.15 *Flink, J. V.:* Effect of the conductor heating in the overhead contact line network on its stability (in Russian language). In: Activities of MIIT, part 104, Moscow 1959.

7.16 *Busche, N. A.; Berent, W. J.; Porcelan, A. A.; Alechin, W. J.:* Annealing of various copper alloys during heating (in Russian language). In: Increase of life-cycle period of non-ferrons metals, Edition Transport, Moscow 1972.

7.17 *Szepek, B.:* Beitrag zur Ermittlung der Belastbarkeit und Zuverlässigkeit elektrotechnischer Betriebsmittel von Industriegleichstrombahnen (Contribution to the determination of the capacity and reliability of electrical equipment for industrial DC railways).
HfV Dresden, doctoral thesis, 1974.

7.18 *Merz, H.; Roggen, F.; Zürrer, Th.:* Erwärmung und Belastbarkeit von Fahrleitungen (Heating and load capacity of overhead contact lines). In: Schweizer Archiv für angewandte Wissenschaftund Technik, 33(1967)7, pp. 189 to 215.

7.19 *Tschutschew, A. P.:* Result of studies on mechanical characteristics of wires and conductors in overhead contact lines (in Russian language). In: Improvement of design and analysis of electric traction installations, Edition Transport, Moscow, 1985.

7.20 *Löbl, H.:* Zur Dauerstrombelastbarkeit und Lebensdauer der Geräte der Elektroenergieübertragung (Current carrying capacity and life cycle period of equipment for electric power transmission). TU Dresden, lecturer thesis, 1985.

7.21 *Petrausch, D.:* Beitrag zur Anwendung der thermischen Modellierung für die Instandhaltung und Diagnose der Fahrleitungsanlage unter Berücksichtigung der Temperaturmessung mittels Infrarottechnik (Contribution to the use of thermal modelling for maintenance and diagnostics of overhead contact line installations considering temperature measurements by means of infrared technology). HfV Dresden, doctoral thesis, 1988.

7.22 *Porcelan, A. A.:* Investigation on the heating and mechanical characteristics of contact wires (in Russian language). In: Progress of the railway research institute, Edition Transport, Moscow (1968)337, pp. 44 to 63.

7.23 *Council directive 2004/50/EC:* Directive on amending directive 96/48/EC and directive 2001/16/EC. In: Official Journal of European Communities, (2004), pp. L220/40 to L220/57.

7.24 *Auditeau, G.; Avronsart, S.; Courtois, C.; Krötz, W.:* Carbon contact strip materials – Testing of wear. In: Elektrische Bahnen 111(2013)3, pp. 186 to 195.

8 Interference

8.0 Symbols and abbreviations

Any symbols and abbreviations followed by an apostrophe ' indicate a magnitude per unit length, given in 1/km or 1/m

Symbol	Definition	Unit
A_1	constant	V/Ω
B_1	constant	V/Ω
B	magnetic flux density	T
C_1, C_2	capacity versus earth of conductor 1 and 2	nF
C_{12}	capacity between conductor 1 and 2	nF
E	electric field strength	V/m
G	admittance	S
H	magnetic field strength	A/m
I, i	current	A
I_C	capacitive charging current	A
I_{trc}	traction current	A, kA
L	inductance	mH
M	coupling impedance	mH
OCL	Overhead Contact Line	–
R	resistance	Ω
U, u	voltage	V, kV
U_1, U_2	influencing voltage	V, kV
U_1'	induced longitudinal voltage	V/km
U_{int}	interference voltage	V
Z	impedance	Ω
Z_1, Z_2	terminating impedance	Ω
a_{12}	distance between the conductors 1 and 2	m
a_1, a_2	height of conductor 1 and 2 above surface	m
b_w	band width	–
f	frequency	Hz
$i(x)$	current at the location x	V
l	length	m
r	reduction factor	–
r_c	reduction factor due to the cable sheaths	–
r_r	reduction factor due to the rail current	–
r_{rc}	reduction factor due to the return conductor current	–
r_{env}	environmental reduction factor due to other earth conductors	–
$u(x)$	voltage at the location x	V
w	probability factor of the short-circuit current	–
δ_E	current penetration depth	m
γ	propagation constant	1/m

information on the variation of the operating currents along the section from the specific energy demand of the line, if the exact operating current distribution for the line is not known. The operating currents of railway lines for general traffic and lines for high-speed traffic are discussed in detail in Clauses 5.3.3 and 5.3.5 .

Short-term interference is caused by short circuit currents between the contact lines and the return circuit. Table 5.18 can be used to calculate the short-circuit currents. For further explanations reference is made to Clause 5.3.7.

8.3.2 Galvanic interference

Installations spread over a wide area such as electrified railway lines can unintentionally transfer voltages via conductive connections to remote locations. Technical equipment and lines in the vicinity of electric railways can be connected to a part of the return current path by *galvanic coupling* via the earth and/or direct metallic contact, see Figure 8.2 b).

In addition to the induced currents flowing in underground metal cable sheaths and pipes running in parallel to electric railway lines, a current will flow through these installations because of galvanic coupling with the return current. Voltage rise of substation earthing equipment is especially relevant for installations of telecommunication and signalling equipment. Vice versa the remote earth can be transferred to the tracks or substation causing voltage differences resulting in hazardous effects in the vicinity of the substation. The voltages arising must comply with the permissible touch voltages, see Clauses 6.3.4 and 6.5.2.

A special kind of galvanic interference is produced by stray currents in case of DC installations, refer to Clause 6.5.3. The design of return circuits aims to avoid or minimize stray currents by adequate insulation of the running rails and return cables.

8.3.3 Inductive interference

8.3.3.1 Inductive interference related to the power frequency

Inductive interference is caused by the loop formed by the contact line and the return circuit. The magnetic field generated by this current acts on metal installations and cables in the vicinity of the railway line and on people. The *alternating magnetic field* generated by the operating current of AC railway systems as well as by the higher harmonics occurring in both AC and DC systems, can induce voltages in the affected installations, cables and people, potentially causing damage or interferences.

Inductive interference to conductors in the vicinity of traction contact lines can be described by the *inductive coupling* between two conductor-earth circuits located in parallel to one another. As shown in Figure 8.3, it is assumed that the conductor subjected to interference is situated in a railway line section where no currents flow from the track to earth.

In Figure 8.3 a) it is shown that a current $\underline{I}$ flowing in the contact line will induce a length-dependent longitudinal voltage $\underline{U}'_1$ in the affected installation. Assuming that the length l of the affected installation is less than that of the contact line emitting the interference, the induced *longitudinal voltage per unit length* is described by the equation:

$$\underline{U}'_1 = 2\pi f \underline{M}' \cdot \underline{I} \cdot r \qquad (8.1)$$

In this equation, $\underline{M}'$ is the mutual inductance per unit length of the conductor-earth loops of the electric traction system and the system subjected to interference. The reduction factor,

a) Coupling mechanism

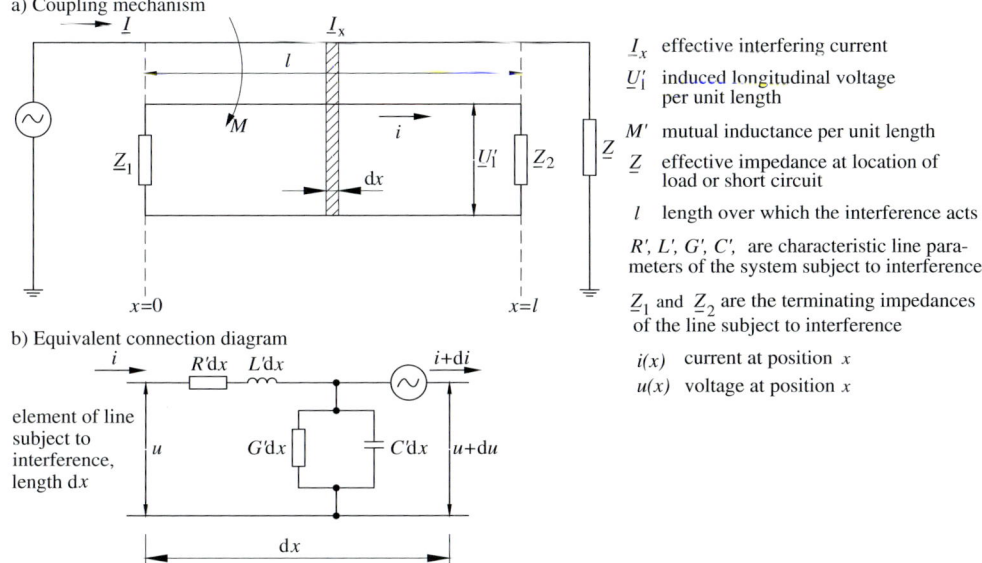

I_x effective interfering current

U_1' induced longitudinal voltage per unit length

M' mutual inductance per unit length

Z effective impedance at location of load or short circuit

l length over which the interference acts

$R', L', G', C',$ are characteristic line parameters of the system subject to interference

Z_1 and Z_2 are the terminating impedances of the line subject to interference

$i(x)$ current at position x

$u(x)$ voltage at position x

b) Equivalent connection diagram

element of line subject to interference, length dx

Figure 8.3: Inductive interference coupling mechanism.

$r < 1$, describes the effect of the current flowing in rails, cable armorings, earth wires and other conducts reducing the induced voltage. The induced longitudinal voltage per unit length is directly proportional to the frequency f of the traction power supply.

To determine the magnitudes of induced voltages and currents along the line, the circuit diagram shown in Figure 8.3 b) can be used to create differential equations. They are also known as *telegraph equations* in related references [8.2] . Their general solutions are:

$$\underline{u}(x) = -\underline{Z}_0 \left[\underline{A}_1 \exp(\underline{\gamma}x) + \underline{B}_1 \exp(-\underline{\gamma}x) \right]$$
$$\underline{i}(x) = \underline{U}_1'/\underline{Z}' + \underline{A}_1 \exp(\underline{\gamma}x) + \underline{B}_1 \exp(-\underline{\gamma}x) \tag{8.2}$$

The surge impedance $\underline{Z}_0$ and the propagation constant $\underline{\gamma}$ of the affected line can be derived from the equations (6.25) and (6.24), respectively. The parameters $\underline{A}_1$ and $\underline{B}_1$ are functions of the reflections in the line subjected to interference depending on its type of connection. The type of *connection* describes how the ends of the installation, e. g. the cable sheaths, are terminated. Figure 8.4 shows typical connections for *cable sheaths* as commonly used in the vicinity of railways.

For electric conductors with an adequate termination, including metal pipes and rails extending beyond the range of influence, and assuming that the induced longitudinal voltage per unit length $\underline{U}_1'$ is constant, the following equations apply at the end of the conductor subjected to interference:

$$\underline{u} = -\underline{U}_1' \left[1 - \exp(-\underline{\gamma}l) \right] / (2\underline{\gamma})$$
$$\underline{i} = (\underline{U}_1'/\underline{Z}') \left[1 - \exp(-0{,}5\,\underline{\gamma}l) \right] \tag{8.3}$$

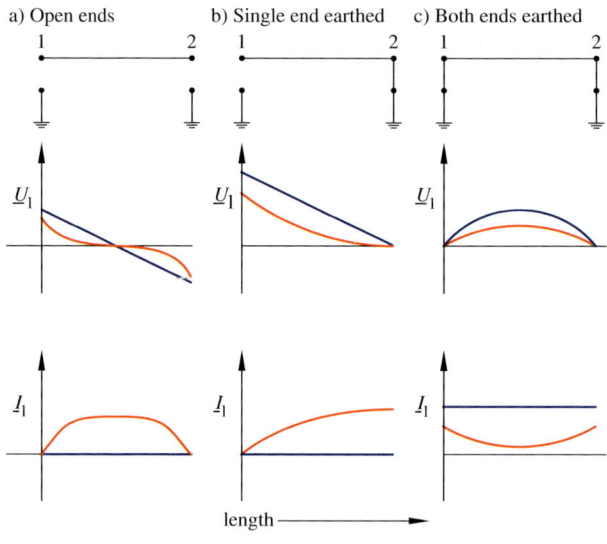

a) Open ends   b) Single end earthed   c) Both ends earthed

Figure 8.4: Effect of the earthing condition on induced longitudinal voltages and currents flowing in the interfered conductor due to these voltages.
——— insulated versus earth (cable strand, suspended conductor)
——— contact with earth (cable sheath, metal pipe)

If both ends of insulated conductors of a cable subject to interference are open, see Figure 8.4 a), then the solution of Equation (8.3) for the voltage is:

$$\underline{U} = \underline{U}'_1/2 \cdot l = \pi f \underline{M}' \underline{I}_{\text{trc}} \cdot r \cdot w \cdot l \tag{8.4}$$

In the case of affected conductors connected to earth at one end, the solution of Equation (8.3) is the *longitudinal voltage*:

$$\underline{U} = \underline{U}'_1 \cdot l \tag{8.5}$$

This earthing condition, which is the most important one in railway engineering practice, is depicted in Figure 8.4 b) . The longitudinal voltage is proportional to the longitudinal voltage per unit length and the length l, is also known as the *effective length* of the affected section. The absolute value of this length-related quantity can be used for calculations in practice. The voltage in an affected cable or installation earthed at one end is:

$$\underline{U} = 2 \pi f \cdot \underline{M}' \cdot \underline{I}_{\text{trc}} \cdot r \cdot w \cdot l \tag{8.6}$$

In Equations (8.4) and (8.6) w is the *probability factor* of a short-circuit current. It allows for worst-case conditions to be used as a basis of the calculations and renders the simultaneous occurrence of all unfavourable circumstances and events as extremely unlikely. On the basis of the studies described in [8.3] and [8.4] values 0,55 and 0,70 can be assumed for the probability factor w. The value w is 1,0 for operational currents.

The *reduction coefficient r* considers reducing effects of currents flowing in the rails, in return conductors, cable sheaths and environmental conditions and is given by

$$r = r_r \cdot r_{rc} \cdot r_c \cdot r_{env} \tag{8.7}$$

with the specific reduction coefficients of
r_r current flowing through rails,
r_{rc} return conductors,
r_c cable sheath of the affected cable and
r_{env} caused by other earthed conductors and components within the interference range.

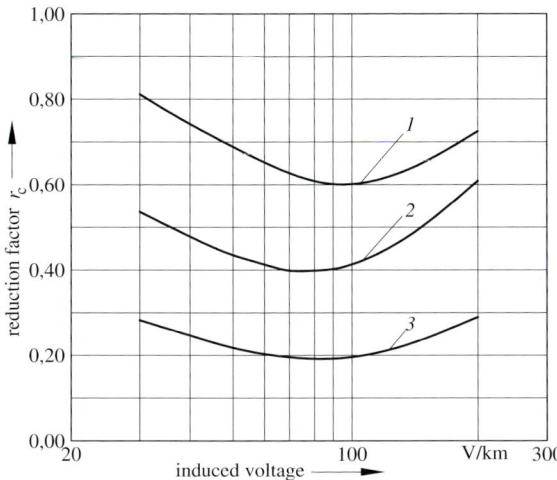

Figure 8.5: Examples for reduction factors of three typical cable classes depending on the induced voltage according to [8.6].
1 cable with low reduction effects
2 cable with medium reduction effects
3 cable with high reduction effects

If reduction coefficient measurements are not available, the application of the following values is recommended:

r_r = 0,2 near substations for double-track lines,

= 0,45 further than 2 km from substation for double-track lines,

= 0,55 further than 2 km from substation for single-track lines,

r_{rc} = 0,55 to 0,7 if return conductors are installed, depending on the position relative to the contact line system,

r_c = 0,1 to 0,5 telecommunications cables, depending on the cable design as specified by the manufacturers,

r_{env} = 0,7 to 0,8 in densely built-up areas (according to [8.5]),

= 0,9 to 1,0 in rural areas (according to [8.5]).

The reduction factors depend on the cable design and on the magnitude of the induced voltage. Figure 8.5 shows as example this dependency of the reduction factors for three typical cable classes. Further examples for reduction factors of specific cables are given in the Annexes of DB Directive 819.08.05 [8.6].

The *mutual inductance* per unit length M' can be obtained from (5.19) . Inserting $\mu_0 = 4\pi \cdot 10^{-4}\,\text{Vs}/(\text{A} \cdot \text{km})$, $\mu_r = 1,0$ and $\delta_E = 0,738/\sqrt{f \cdot \mu_0/\rho_E}$ yields the numerical equation

$$M' = 0{,}2 \cdot \ln\left[660 / \left(a\sqrt{f/\rho_E}\right)\right] \quad , \tag{8.8}$$

where the unit of M' is mH/km, a is given in m, f is Hz and ρ_E is Ωm.

The same result can be derived from

$$M' = \left\{1 + 2 \cdot \ln\left[400 / \left(a\sqrt{f/\rho_E}\right)\right] - \mathrm{j}\pi/2\right\} \cdot 10^{-4}\,\text{H/km} \tag{8.9}$$

as given by [8.7] only taking into account the real part of (8.9).

In Figure 8.6 values are shown for the mutual inductance per unit length for frequencies of 16,7 Hz and 50 Hz for typical soil resistivities. In practice, the longitudinal voltages per unit length in conductors in the immediate vicinity, i. e. at a distance of roughly four to eight metres from and parallel to the railway centre line, are of particular relevance, because signalling and telecommunication cables are located in this area. The induced longitudinal voltages per

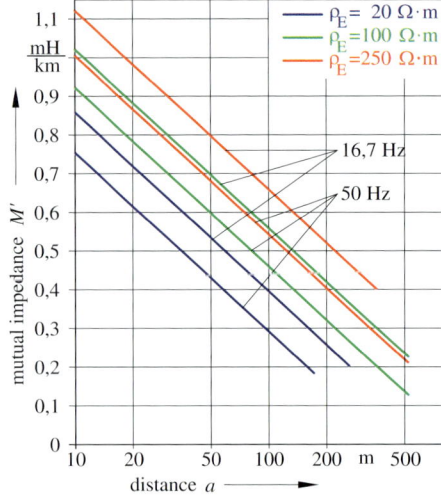

Figure 8.6: Approximate values for mutual inductance per unit length as a function of the distance between affected conductor and contact line.

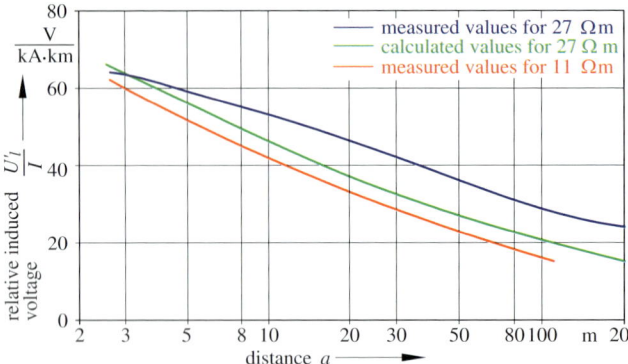

Figure 8.7: Longitudinal voltages per unit length in a cable running in parallel to the track induced by traction currents at 16,7 Hz frequency.

unit length were measured and calculated as functions of the distance from the track centre line. The results shown in Figure 8.7 apply to specific soil resistivities of $27\,\Omega\cdot\text{m}$ and $11\,\Omega\cdot\text{m}$, measured in the immediate vicinity of the track. The longitudinal voltages per unit length shown in these graphs are referenced to an inducing current of 1 kA. An example for 50 Hz is given in Clause 6.6.6.

Example 8.1: Calculation of the voltage induced by a 16,7 Hz railway line, equipped with or without return conductors on the overhead contact line poles into a parallel cable, with or without sheath. The cable is 10 km long and earthed at one end, the operating current is 1 000 A and the short-circuit current is 25 kA. The distance between the overhead contact line system and the cable is 20 m, the specific soil resistivity is assumed to be $20\,\Omega\text{m}$.

Using (8.8), the mutual inductance results in

$$M' = 0{,}2 \cdot \ln \left[660 \Big/ \left(20{,}0\sqrt{16{,}7/20}\,\right) \right] = 0{,}71\,\text{mH/km} \quad .$$

Table 8.1: Induced voltage according to Example 8.1.

Case	Reduction factor	Operation			Short-circuit		
		w	current A	voltage V	w	current kA	voltage V
1	0,41	1,0	1 000	300	0,7	25	5 280
2	0,24	1,0	1 000	180	0,7	25	3 170
3	0,20	1,0	1 000	150	0,7	25	2 650
4	0,12	1,0	1 000	90	0,7	25	1 590

The reduction factors used are:

r_r = 0,45 (double track line at a distance of more than 2 km from the substation)

r_{rc} = 1,0 without return conductors

r_{rc} = 0,60 with return conductors

r_c = 1,0 cable without sheath

r_c = 0,5 cable with sheath

r_{env} = 0,9 other conductors with contact to earth, close to the affected cable

This yields to the total reduction factor r according to Table 8.1

$r = 0,45 \cdot 1,0 \cdot 1,0 \cdot 0,9 = 0,41$ without return conductors and a cable without sheath (case 1)

$r = 0,45 \cdot 1,0 \cdot 0,6 \cdot 0,9 = 0,24$ with return conductors and a cable without sheath (case 2)

$r = 0,45 \cdot 0,5 \cdot 1,0 \cdot 0,9 = 0,20$ without return conductors and a cable with sheath (case 3)

$r = 0,45 \cdot 0,5 \cdot 0,6 \cdot 0,9 = 0,12$ with return conductors and a cable with sheath (case 4)

Table 8.1 indicates the induced voltages derived from (8.6).

Example 8.2: It should be checked whether a cable without sheath ($r_c = 1,0$) can be used for a 2,9 km connection between the signal box and the electronics control cabinet of an audio frequency track release circuit if the traction current is 800 A and a short-circuit current of 15 kA is expected. The line in question is a double-track line operated with 50 Hz single-phase AC, equipped with return conductors. Furthermore, $r_r = 0,45$, $r_{rc} = 0,6$, $r_c = 1,0$, $r_{env} = 0,8$, $w = 0,7$, $a = 10$ m and $\rho_E = 100\,\Omega\cdot$m. From Figure 8.6, the coupling inductance per unit length between contact line and the affected cable can be found to 0,90 mH/km. Equation (8.8) yields 0,91 mH/km. Using equation (8.4) the voltage at the cable ends is

$$U = \pi \cdot 50 \cdot 0,9 \cdot 10^{-3} \cdot 800 \cdot (0,45 \cdot 0,60 \cdot 0,80) \cdot 2,9 = 70,8\,\text{V} \quad .$$

This value is higher than the permanently permissible voltage of 65 V according to EN 50 122-1. Therefore, it is not possible to use a cable without a sheath. In case of a short-circuit current of 15 kA the interference voltage would be

$$U = \pi \cdot 50 \cdot 0,9 \cdot 10^{-3} \cdot 15 \cdot 10^3 (0,45 \cdot 0,60 \cdot 0,80) \cdot 0,7 \cdot 2,9 = 930\,\text{V} \quad .$$

The ITU-T Directive [8.8] specifies a permissible voltage of 430 V up to 0,5 s duration, EN 50 122-1 permits 500 V for short circuits up to 120 ms. Therefore, a cable with a sheath is required for this application.

8.3.3.2 Inductive interference related to harmonics

Harmonics of currents and voltages can occur in AC and DC railway systems and can cause interference. They are mainly caused by frequency converters on vehicles and in substations

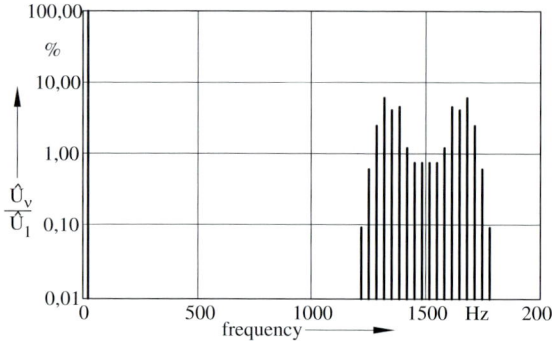

Figure 8.8: Voltage spectrum, relative to the fundamental voltage $\hat{U}_1$, of the input voltage of a four-quadrant power control circuit.

and by transformers. Interference caused by harmonics depends above all on the power control scheme used in the power converters. Overhead contact lines do not produce harmonics but transfer them. Resonance effects can be created by harmonics on the contact line system. In transformers, the saturation effects in the magnetic material lead to a magnetic flux that deviates from an exact sine wave. The frequencies of the harmonics are integer multiples of the basic frequency, the amplitudes of which decrease almost exponentially with rising frequencies characterising transformers as low-pass components.

In power electronics circuits, non-sine-wave currents and voltages are the result of the switching actions of the power electronic components that produce harmonics with various frequencies. The voltage spectrum, relative to the fundamental wave shape, shown in Figure 8.8 is an example for a four-quadrant drive control.

The voltage harmonics produce current harmonics that can affect compatibility with signalling circuits, such as track release circuits. Therefore, propagation of current harmonics in the overhead contact line and resonance phenomena need to be investigated by network calculation models for medium frequencies up to 20 kHz.

For this purpose, different calculation models are used and verified by measurements, see [8.9] to [8.13]. The results indicate that the overhead contact line system does not change the harmonic behavior of the line by more than 5 % eliminating the need to take into account such interference aspects for design, construction and implementation of overhead contact line systems.

8.3.4   Capacitive interference

The electric field generated by the live parts of contact line installations of AC railways can electrically charge conductors and system components located in the interference range of the contact lines by *influence* effects. However, this charge would only lead to measurable voltages if the respective conductors were insulated relative to earth. Buried cables or systems are not subject to *capacitive interference*.

By applying the potential divider rule to the circuit shown in Figure 8.9, the voltage U_2 on a conductor running parallel to the overhead contact line can be expressed as:

$$U_2 = U_1 C'_{12}/(C'_2 + C'_{12}) \quad . \tag{8.10}$$

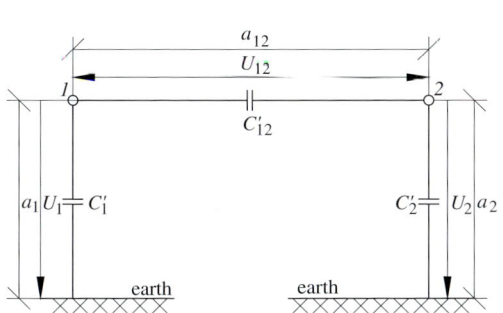

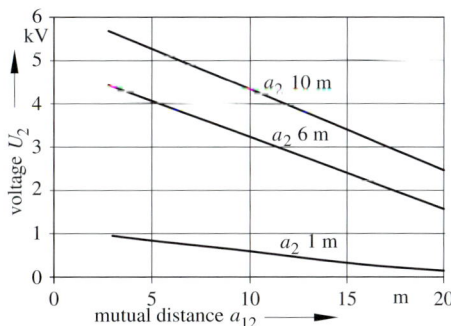

Figure 8.9: Capacitive interference on conductor *2* caused by overhead contact line *1*.
1 overhead contact line
2 influenced conductor

Figure 8.10: Voltage generated in a suspended conductor because of the influence of a contact line voltage of 25 kV, as a function of the mutual distance a_{12} between the parallel sections and for various heights a_2 of the affected conductor above ground.

C'_2 is calculated according to Equation (5.24) . Reference [8.14] explains that the capacitance per unit length C'_{12} can be approximated using the following empirical formula

$$C'_{12} = \frac{54\,a_2}{144 + a_{12}^2 + a_2^2} \quad , \tag{8.11}$$

where the unit of C' is nF/km in case of a_2 and a_{12} are given in m. a_2 und a_{12} are indicated in Figure 8.9. For double-track lines, the expected capacitance per unit length is approximately 1,5 times the value obtained by Equation (8.11).

The *influenced voltage*, which is independent of the length of the sections of a conductor running parallel to a 25 kV overhead contact line, is shown in Figure 8.10. A conclusion that can be drawn from this graph is that unearthed lines and metal objects as low as 1 m above ground level and also near the track, can achieve voltages of up to 1 kV. The possible hazards due to capacitive influence include the danger of *electric shock* to people who touch the high-voltage metal surfaces. The charging current is calculated as

$$I_C = 2\pi f C'_{12} U_1 l \quad , \tag{8.12}$$

where *l* is the length of parallelism. Hazards to people are unlikely for a section of insulated cable 2 km long running in parallel to and at a distance of 10 m from an overhead contact line. According to IEC 60 479-1, the physological effects depend on the duration of current flow. The limits for harmful effects are below 10 mA to 200 mA. This also applies to telecommunication cables in a cable duct along the railway line. Since C'_{12} is extremely low for these cables, I_c cannot reach hazardous values.

Example 8.3:   What is the amount of the capacitive interference of a non-earthed aerial conductor in a distance of 10 m from a 25 kV overhead contact line, situated 6 m above ground and a diameter of 20 mm?
The first step is the calculation of C'_2 versus earth of the conductor under influence and the mutual capacity C'_{12} of both conductors. C'_2 results from (5.24):

$$C'_2 = 2 \cdot \pi \cdot 8,85 \cdot 10^{-9} / \ln(2 \cdot 6/0,01) = 7,85\,\text{nF/km}$$

The mutual capacity C'_{12} results from (8.11):

$$C'_{12} = \frac{54 \cdot 6}{144 + 10^2 + 6^2} = 1,16 \text{nF/km}$$

The influenced voltage results from (8.10):

$$U_2 = 25 \cdot 1,16/(7,85 + 1,16) = 3,22 \text{ kV}$$

This value also can be derived from Figure 8.10. The capacitive current of a conductor 2,5 km long is obtained from (8.12)

$$I_C = 2 \cdot \pi \cdot 50 \cdot 1,16 \cdot 10^{-9} \cdot 25 \cdot 10^3 \cdot 2,5 = 23 \text{mA} \quad ,$$

which is not a hazardous value.

8.4 Electric and magnetic fields close to contact lines

8.4.1 Basics

Electric energy to be transmitted to traction vehicles requires the operating voltage between the contact line and the return circuit and a current to flow through the contact line. The consequences are
- an *electric field E* which exists as long as the contact line is energized, and
- a *magnetic field H* which varies with the flowing current, time and location.

With reference to media reports on so-called *electric smog* (electromagnetic pollution), the question on adverse effects of electric and magnetic fields on persons in the vicinity of railway installations has been raised repeatedly.

Electric fields are so-called source fields. They arise where there is a separation of charges between two poles. The field lines begin at the positive electrode and end at the negative electrode.

The *electric field strength E* is measured in kV/m and defines the force exerted on an electrical charge carrier. It is measured as a constant value or as an rms value in the case of alternating fields. Natural electric fields exist in the atmosphere and vary widely in time and space as a result of varying atmospheric conditions and in thunderstorms reach values up to 20 kV/m.

Electric fields generated by technical equipment depend on the voltage, the arrangement of conductors and distance from the live conductors. The fields under a single-phase overhead contact line can reach values up to 1,7 kV/m for AC 15 kV and up to 2,7 kV/m for AC 25 kV. Electric fields exist as long as contact lines are energized and can be easily screened by metal parts, that are in contact with earth.

Magnetic fields are defined as a circuital vector field caused by magnetic poles or surround electrical conductors when currents are flowing through them. The *magnetic field strength H* is measured in A/m and is not dependent on material characteristics. It can be measured as a constant value or as an rms value in the case of alternating currents.

The *magnetic flux density B*, also called *magnetic induction*, defines the effect of magnetic fields in materials and media. The unit of magnetic flux density is 1 Tesla = 1 Vs/m^2. The relation between the flux density B and the field strength H is

$$B = \mu_0 \cdot \mu_\mathrm{r} \cdot H \quad , \tag{8.13}$$

Table 8.2: Effect of low-frequency electric and magnetic fields on the human organism according to results given in [8.20].

Current density threshold values mA/m²	Consequences if threshold is exceeded	Current in body mA	Thresholds occur at			
			$f = 50\,\text{Hz}$		$f = 16{,}7\,\text{Hz}$	
			E kV/m	B µT	E kV/m	B µT
1	measurable effects	0,07	4 to 5	100	12 to 15	300
10	stimulation (flicker felt in eyes)	0,7	40 to 50	1 000	120 to 150	3 000
100	muscle and nerve stimulation (potentially dangerous)	7	400 to 500	10 000	1 200 to 1 500	30 000
1 000	injury (possibly lethal, ventricular fibrillation)	70	4 000 to 5 000	100 000	[1]	[1]

[1] no values available

where the permeability constant $\mu_0 = 1{,}25664 \cdot 10^{-6}$ Vs/Am. With the relative permeability in air $\mu_r = 1$ is

$$1\,\text{A/m} = 1{,}26\,\text{µT} \quad \text{and} \quad 1\,\text{µT} = 0{,}80\,\text{A/m} \quad .$$

Since the flux density B is easier to measure, it is often taken as a reference value instead of the *magnetic field strength H*.

The natural DC magnetic field of the Earth varies between 30 and 60 µT with geographic position and slightly as a result of variations in the atmosphere and inside the Earth.

Technical magnetic fields are a function of the current flowing through a conductor, the conductor configuration and the distance from the conductors. Strong DC fields of 0,5 T are generated in magnetic resonance imaging for medical diagnostics. Field strengths of up to 100 µT can be expected under overhead contact lines of single-phase AC railways. In DC railway systems with a conductor rail, DC fields up to 500 µT can be caused by the high operating currents in DC 750 V systems.

Magnetic fields vary with traction currents in time and location. The screening of emitted magnetic fields is complicated in the case of wide spread installations such as traction power supply systems. In principle local protection measures such as screening of monitors with high-permeability materials are possible. In contrast to high-voltage transmission systems, electric traction systems use low voltages and high currents. Therefore, the effects of magnetic fields are more important than those of electric fields.

The duration of magnetic fields caused by short-circuit currents is short. Therefore affects on persons and on technical equipment are unlikely.

8.4.2   Effects of fields on people

The *permissible values of electric field and magnetic field strengths* in high-voltage installations accessible to the general public are legislated in several countries. The German Federal Minister of the Environment (BMU) established reference data in [8.15] and [8.16] taking into account the values of ICNIRP recommendations [8.17] and [8.18].

The AC *effects of electric and magnetic fields on human beings* are described in [8.19] and [8.20]. On the surface of the body, the AC electric field creates a charge which in turn can lead

Table 8.3: Precaution values for electric fields and magnetic flux density for technical frequencies according to ICNIRP and 26th BImSchV.

	ICNIRP				26. BImSchV			
	16,7 Hz		50 Hz		16,7 Hz		50 Hz	
	E kV/m	B µT	E kV/m	B µT	E kV/m	B µT	E kV/m	B µT
occupation	20 (20)	1 500 (1 500)	10 (10)	1 000 (500)				
public	5 (10)	300 (300)	5 (5)	200 (100)	5 (10)	300 (300)	5 (5)	200 (100)

ICNIRP   values from 2010, in brackets from 1998
BImSchV   values from 2013, in brackets from 1996, Germany
ICNIRP   International Commission on Non-Ionizing Radiation Protection
BImSchV   Bundes-Immissionsschutzverordnung, German directive for electromagnetic fields

Table 8.4: Electric and magnetic field strengths measured in the vicinity of electric railway systems.

Traction power supply	measured at	E kV/m	$B^{1)}$ µT
DC 600 V	edge of station platform, 1 m above rails, 7 m apart from track centerline	0,07 0,05	100 25
DC 3 000 V	edge of station platform, 1 m above rails, 7 m apart from track centerline	0,3 0,2	100 25
AC 16,7 Hz 15 kV	edge of station platform, 1 m above rails, 7 m apart from track centerline	1,6 1,1	100 25
AC 50 Hz 25 kV	edge of station platform, 1 m above rails, 7 m apart from track centerline	2,7 1,8	100 25

1)   the magnetic flux densities B are measured for a current of 1 000 A flowing in each overhead contact line

to currents passing through the body. Large numbers of experimental studies have shown that an electric field strength of 1 kV/m will lead to a current of approximately 0,015 mA in the human body. The corresponding current densities are between $0,2\,\text{mA/m}^2$ and $0,3\,\text{mA/m}^2$. The currents resulting from the electric field are neither a function of the conductivity of the body nor of the person's size.

In contrast, the magnetic field induces body currents that are functions of both the person's size and the body's conductivity. The average cross-section area of the human body is between 0,06 and 0,07 m². An induction of 1 µT at a frequency of 50 Hz will lead to a current density of roughly 0,01 mA/m².

It was also shown that current densities up to $0,1\,\text{mA/m}^2$ cause no discernible effects on the human body. Current densities of $10\,\text{mA/m}^2$ and above can lead to a flickering sensation in the eyes, and current densities of $100\,\text{mA/m}^2$ lead to nerve and muscle stimulation. The danger threshold is $100\,\text{mA/m}^2$.

The impact on the human body for the typical railway traction energy frequencies of 16,7 Hz and 50 Hz are summarized in Table 8.2. Table 8.3 gives reference values stated in applicable German and international standards and recommendations. Table 8.4 shows electric field strength and induction values measured in railway environments.

Figure 8.11 shows the characteristic graph of measured magnetic field strength of an electric railway line as a function of the distance from the centreline of a double-track line. The

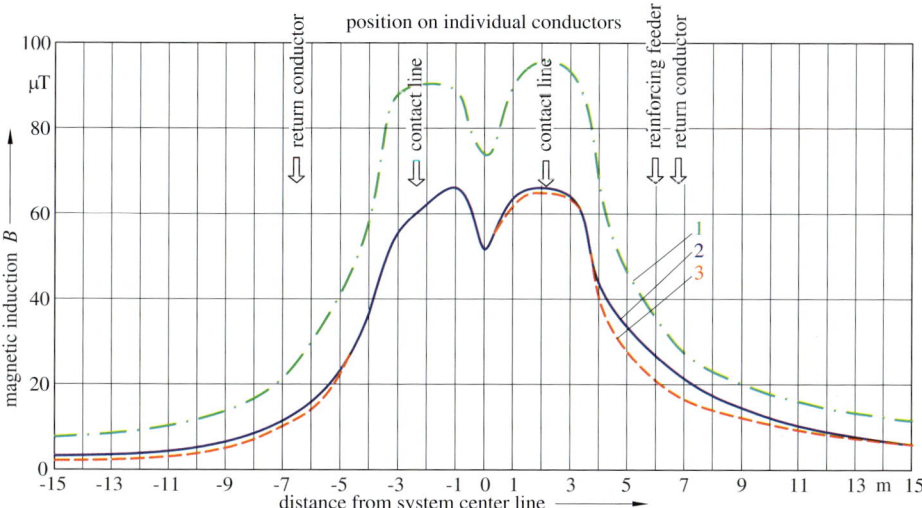

Figure 8.11: Magnetic induction B, 1 m above rail head, comparison of measured and calculated values, traction current $I = 2 \cdot 1\,000$ A [8.21].

1 without return conductor, calculated; 2 with return conductor, measured; 3 with return conductor, calculated

Table 8.5: Precaution values for pace makers with reduced immunity against electromagnetic impact.

Frequency Hz	Electric field strength kV/m	Magnetic induction µT
16,7	4,1	65
50	10	300

magnetic field strength magnitudes refer to an overhead contact line current of 1 kA per track. If return conductors are installed, the flux density is reduced by one third in close vicinity to the tracks and by 40 % at a distance of 7 m. Tables 8.2 to 8.4 demonstrate that the stringent precautionary limits set by ICNIRP and German legislation for both electric and magnetic fields are not exceeded in railway applications. Therefore, electric or electromagnetic fields caused by railway operating equipment pose no danger to people.

8.4.3   Effects of fields on equipment

8.4.3.1   Effects in general

Electric and electromagnetic fields can affect apparatus and installations in the vicinity of railways. Persons with implanted cardiac pacemakers or other similar implants can be affected adversely, but there are no occurrences reported. The operation of *information-technology equipment*, especially visual display units (VDU), can be impaired. In addition, electric traction systems emit *radio frequency interference* with intensities that can disturb equipment in the vicinity of the railway.

Table 8.6: Summary of EN 50 121 series.

EN 50 121	Railway Applications – Electromagnetic compatibility
EN 50 121-1	General
	– General overview of all parts of the standard. – Description of the electromagnetic behaviour of a railway. – Specification of the performance criteria for the whole set. – Reference to a management process to achieve EMC at the interface between the railway infrastructure and trains.
EN 50 121-2	Emission of the whole railway system to the outside world
	– Methods of measuring radio frequency interference due to passing trains (peak detection). – Limits for radio frequency interference in the range of 9 kHz to 1 GHz. – Typical field strength for power frequency and high frequencies (cartography).
EN 50121-3-1	Rolling stock – Train and complete vehicle
	– Emission and immunity requirements for all types of rolling stock. It covers traction stock and train-sets, as well as independent hauled stock. – Immunity to interference including the equipment to be installed on rolling stock energy inputs and outputs.
EN 50121-3-2	Rolling stock – Apparatus
	– Emission and immunity aspects of EMC for electrical and electronic apparatus intended for use on railway rolling stock. – Means of dealing with the impracticality of immunity testing a complete vehicle.
EN 50121-4	Emissions and immunity of signalling and telecommunication apparatus
	– Specification of limits for electromagnetic emission and immunity for signalling and telecommunications apparatus.
EN 50121-5	Emissions and immunity of installations and apparatus of electric traction systems
	– Radio frequency interference emissions by substations, contact lines and feeder lines. – Emission and immunity aspects for electrical and electronic apparatus and components of electric traction systems.

8.4.3.2 Persons with implanted cardiac pacemakers

Implanted cardiac pacemakers with reduced immunity require the highest restrictions for magnetic fields. German Standard VDE 0848-3-1 stipulates precausion values for fields with the frequency range 9 Hz to 300 Hz which are listed in Table 8.5.

In practice, interferences below 300 µT on *cardiac pacemakers* are unlikely even for a frequency of 50 Hz because of the non-uniform magnetic field distribution and the lower susceptibility of the absorbing signal circuits. Refrence [8.22] reports, that it was not possible to detect any impacts on implanted cardiac pacemakers caused by magnetic flux densities up 500 µT. This is supported by the fact that there have been no reported problems by people with pacemakers.

8.4.3.3 Information technology and electronic data processing equipment

The magnetic fields in the vicinity of railway installations can cause *interference to cathode-ray tube monitors*. Other susceptible equipment can also experience influences. The paper [8.23], for example, reports interference to an electron microscope due to the power cable of a DC railway system located at a distance of 70 m. A current of 1 400 A flowing through this

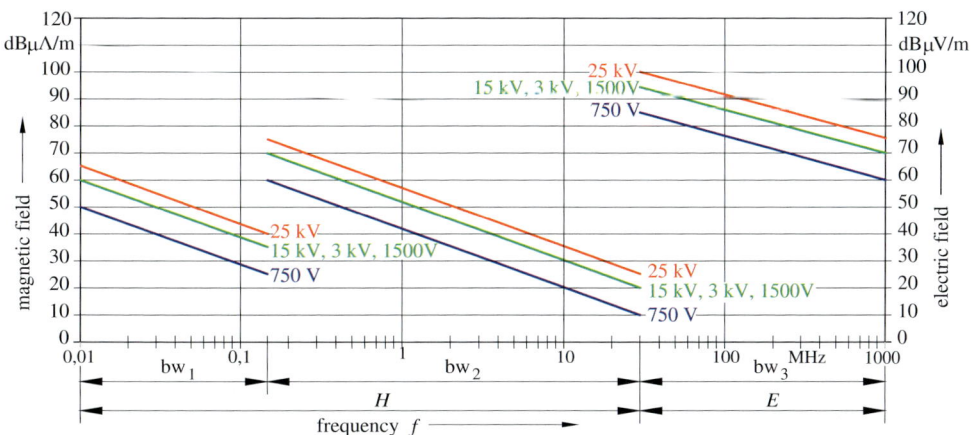

Figure 8.12: Permissible values of radio frequency influence levels according to EN 50 121-2:2015. bw_1: band width at measuring receiver 0,2 kHz, bw_2: band width at measuring receiver 9,0 kHz, bw_3: band width at measuring receiver 120,0 kHz

cable, the electron microscope was exposed to a 4 µT magnetic field, making it impossible to use the device.

8.4.3.4   Electric traction systems as sources of radio-frequency interference

Electric railways can emit *radio-frequency interference* (RFI). The standard series EN 50 121 was drawn up to deal with this issue. Table 8.6 shows a summary of the contents of the six parts of this standard. The main sources of RFI caused by electric railway systems include:

- *spark discharges* in the traction contact line network, e. g. across droppers that are inadequate for the currents
- *loss of contact* between the contact wire and the pantograph collector strip, with subsequent arcing
- *commutation processes* in the power electronic of electric traction vehicles
- *switching and control transients* in electric railway switchgear and vehicles

The graphs in Figure 8.12, showing the permissible maximum values of radio frequency influence levels for frequencies between 9 kHz and 1 GHz were taken from standard EN 50 121-2. The stepped characteristics result from the different methods of measurement used. For instance, between 150 kHz and 30 MHz, the level is measured as a magnetic field with the aid of a coil antenna. At frequencies above 30 MHz the electric field strength is measured using a dipole antenna. The measurements are carried out using the 10 m peak detection method. The values from 9 kHz to 150 kHz were indicated as informative in EN 50 121-2:2015-03 because of poor reproducibility of measured values for frequencies below 150 kHz.

8.5   Conclusions

A main characteristic of electric railways is that electric traction currents return to the substation via the running rails. In the case of single-phase AC railway systems, the running rails are connected to earth intentionally. For this reason, a portion of the return current will also

flow through earth to the respective feeding substation. In single-phase AC railway systems, inductive coupling creates a line-to-earth current loop in addition to the *galvanic coupling* of the rails to earth. This characteristic of electric railway traction systems, also termed *unbalanced or asymmetrical* with respect to earth, coincides with the widespread and large area in which technical and biological systems could be affected.

From the deductions made and the discussions set forth in this chapter, it can be concluded that the *electric* and *electromagnetic fields* in the vicinity of railways

- do not lead to any organic stimulation or pose any danger to people,
- do not endanger persons with implanted cardiac pacemakers,
- can disturb the performance of information equipment and other highly susceptible devices. The cause of such influences is mainly magnetic field strengths in the region of 1 to 30 μT.

Possible *corrosion* of underground metal parts caused *by stray currents* is an adverse affect of DC railway systems on other installations located underground.

Standard EN 50 122-2 describes protective measures to eliminate the effects of stray currents from DC traction power supply systems, see Clause 6.5. Well managed co-operation of the operators of installations with underground components, cables and pipelines with the operator of the DC railway is essential to avoid rapid degradation of the infrastructure.

In single-phase AC railway traction systems, the capacitive interference must be counteracted by *earthing all metal parts* that might otherwise become electrically charged.

Galvanic interference in the vicinity of single-phase AC railway traction systems, that means transfer of remote voltages, can be prevented by installing insulating joints in potentially susceptible conductive systems within the range of influence, e. g. in cable sheaths and metal pipes that lead into the substations.

The *inductive interference* must be accounted for when designing and operating technical systems and devices in the vicinity of railway lines. The inductive interference of the fundamental frequency can affect equipment and installations in the vicinity of railway traction power installations. Harmonics occurring in railway traction networks are sources of interference, especially in telecommunications systems. Because the technical equipment has a wide range of immunity level, no binding international standards could be fixed. In Germany the Technical Rules No. 1 to 3 of the arbitration body in Germany, comprising of DB, Deutsche Telekom and the umbrella organization of German electric power utilities are the basis of mutual agreements. However, with the increasing adoption of glass-fibre cables for telecommunication purposes this issue loses importance.

8.6 Bibliography

8.1 *Habiger, E.*: Elektromagnetische Verträglichkeit. Grundzüge ihrer Sicherstellung in der Geräte- und Anlagentechnik (Electromagnetic compatibility, principles of guaranteeing in equipment and installation technology). Hüthig Publishing, Heidelberg, 1996.

8.2 *Koettnitz, H.; Pundt, H.*: Berechnung elektrischer Energieversorgungsnetze, Mathematische Grundlagen und Netzparameter (Calculation of electrical energy supply networks, mathematical basics and network parameters). Grundstoffindustrie Publishing, Leipzig, 1968.

8.3 *Koch, H.*: Ein Beitrag zur Gewährleistung der elektromagnetischen Verträglichkeit von Anlagen der Sicherungs- und Fernmeldetechnik mit eisenbahntypischen elektrischen Systemen

hoher Leistung (Contribution to secure the electromagnetic compatibility of installations for signalling and telecommunication technology with railway-typical electric systems of high power). HfV Dresden, doctoral thesis, 1986.

8.4 *Lingen, J. v.*: Kurzschlussberechnung im Fahrleitungsnetz (Short-circuit calculation for contact line networks). TU Dresden, doctoral thesis, 1995.

8.5 *Feydt, M.*: Vorschläge zur Verwendung der Kabelmäntel, metallener Rohrleitungen, der Gleise und der Erdseil-Maste-Kettenleiter als natürliche Erder (Proposals to use cable sheaths, metallic pipelines, tracks and earthwire-pole recurrent networks as natural earth electrodes). Report of the Institute for Energy Supply, Dresden, 1982.

8.6 *Deutsche Bahn AG Directive 819.0805*: LST-Anlagen planen – Beeinflussung und Schutzmaßnahmen – Induktive Beeinflussung (Planning of control and signalling installations – interference and relating protective provision – inductive interference). Deutsche Bahn AG, Frankfurt, 1997.

8.7 *Pollaczek, F.*: Über das Feld einer unendlich langen, wechselstromdurchflossenen Einfachleitung (On the field of an infinitely long single conductor line used by AC current). In: Elektrische Nachrichten-Technik 3(1926), pp. 339 to 359.

8.8 *ITU-T*: Directives concerning the protection of telecommunication lines against harmful effects from electric power and electrified railways. In: ITU (1999), Vol. 1 to 9.

8.9 *Putz, R.*: Über Streckenwiderstände und Gleisströme bei Einphasenbahnen (About line resistances and track currents at single-phase railway lines). In: Elektrische Bahnen 20(1944), pp. 74 to 92.

8.10 *Behrends, D.; Brodkorb, A.; Hofmann, G.*: Berechnungsverfahren für Fahrleitungsimpedanzen (Calculation of catenary impedance values). In: Elektrische Bahnen 92(1994)4, pp. 117 to 122.

8.11 *Kontcha, A.*: Mehrpolverfahren für Berechnngen in Mehrleitersystemen bei Einphasenwechselstrombahnen (Multipolar method for calculations of multi-conductor contact line systems for railways with sinlge-phase AC). In: Elektrische Bahnen 94(1996)4, pp. 97 to 102.

8.12 *Xie, J. et al.*: Berechnung hochfrequenter Oberschwingungen in Oberleitungsnetzen (Calculation of high-frequency harmonics in catenary networks). In: Elektrische Bahnen 103(2005)6, pp. 286 to 290.

8.13 *Zynovchenko, A. et al.*: Oberleitungsimpedanzen und Ausbreitung von Oberschwingungen (Impedances of contact lines and propagation of harmonics). In: Elektrische Bahnen 104(2006)5, pp. 222 to 227.

8.14 *Schmidt, P.*: VEW-Handbuch Energieversorgung elektrischer Bahnen (Power supply of electrical railways). Technik publishing, Berlin, 1975.

8.15 *26. Bundesimmissionsschutzverordnung (BImSchV)*: Verordnung über elektromagnetische Felder (26th directive on the German Federal immission protection law: Directive on electromagnetic fields). In: Bundesgesetzblatt 1996, Part I, December 16, 1996, p. 1966.

8.16 *26. Verordnung zur Durchführung des Bundes-Immissionsschutzgesetzes*: Verordnung über elektromagnetische Felder - 26. BImSchV (26th directive on the German Federal immission protection law: Directive on electromagnetic fields – 26. BImSchV). In: Bundesgesetzblatt 2013, New edition by Bek. Part I, August 14, 2013, p. 3266.

8.17 *International Commission on Non-Ionizing Radiation Protection (ICNIRP)* Guidelines for limiting exposure to time-varying electric, magnetic and electromagnetic fields (up to 300 GHz). In: Health Physics 74(1998)4, pp. 494 to 522.

8.18 *International Commission on Non-Ionizing Radiation Protection (ICNIRP)* Guidelines for limiting exposure to electric fields induced by movement of the human body in a static magnetic field and by time-varying magnetic fields below 1 Hz. In: Health Physics 106(2014)3, pp. 418 to 425.

8.19 *David, E.*: Elektrische und elektromagnetische Felder im Nahbereich von Freileitungen (Electric and electromagnetic fields in the vicinity of overhead power lines). In: Deutsches Ärzteblatt (1986)12.

8.20 *David, E.*: Wirkungen der Elektrizität auf den menschlichen Organismus (Effects of electricity on the human organism). Speech at TU Dresden, November 1993.

8.21 *Zimmert, G. et al.*: Rückleiter in Oberleitungsanlagen auf der Strecke Magdeburg–Marienborn (Return conductors in catenary system on the Magdeburg–Marienborn line). In: Elektrische Bahnen 92(1994)4, pp. 105 to 111.

8.22 *Wahl, H.-P.*: Messungen von elektrischen und elektromagnetischen Feldern bei Nahverkehrsbahnen (Measurement of electric and magnetic fields of mass transit systems). In: Berichte und Informationen HTW Dresden 4(1996)1, pp. 39 to 41.

8.23 *Fischer, C.*: Diskussionsbeitrag auf dem 2. Symposium des Fachbereiches Elektrotechnik der HTW Dresden (Contribution to the 2nd symposium of the electrotechnical department of HTW Dresden). November 1995, pp. 89 to 101.

9 Line protection and fault location

9.0 Symbols and abbreviations

Symbol	Definition	Unit
I	current	A
I_E	threshold current	A
I_K	short-circuit current	A
I_e	setting value of maximum operational current	kA
$I_{e\,max}$	maximum operational current	kA
$I_{e\,min}$	minimum of maximum operational current	kA
U	voltage	V
$U_{B\,min}$	minimum supply voltage	kV
Z_C	impedance of section C	Ω
$Z_{CA,CB}$	impedance of section CA and CB	Ω
Z_q	source impedance	Ω
Z_1, Z_2, Z_{e1}	impedance	Ω/km
t	time	s
ϑ	temperature	°C
ϑ_{am}	ambient temperature	°C

9.1 Contact line protection

9.1.1 Purpose, requirements and functioning

Protective provisions for contact line systems serve the basic purpose of sensing and evaluating the occurrence of any electrical faults and overloads to:
- prevent or keep to a minimum any damage to the contact line and its components
- minimise hazards to persons directly or indirectly exposed to impermissible voltages
- maintain the best possible availability of the traction power supply and
- provide and process information that assists fault analysis

To achieve these aims, the protective installations must be able to switch off all impermissible loads safely, quickly and selectively. Examples of such loads are:
- all types of short circuits occurring in the network
- operating currents that cause the permissible temperature to be exceeded

Standard EN 50 633 contains specific protection principles for railways and guided mass transport systems. Protection systems must be able to distinguish fault currents from the maximum operating currents and the compensating currents caused by commutating processes when changing between supply sections. The operating currents may momentarily reach 8 kA in the case of DC and 2 kA in the case of AC on many line sections. These currents would lead to excessive heating of the contact line if they acted over a long period. Since the overhead contact line is susceptible to excessive temperatures because of relatively small cross sections and its mechanical configuration, a correctly designed contact line protection

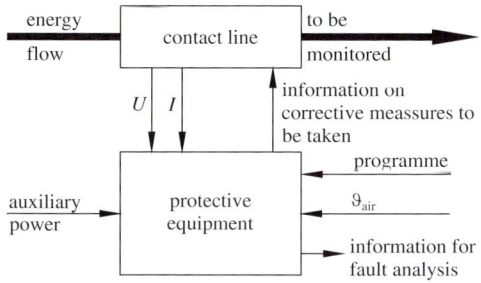

Figure 9.1: Purpose and objectives of protective provisions for contact lines.
U voltage, I current, ϑ_{air} temperature of air

scheme is essential for optimum utilisation of the contact line thermal characteristics. The permanent temperature limits of $80\,°C$ to $100\,°C$ according to EN 50 119 (see Table 2.7) are determined by the higher tensile stress in comparison with overhead power lines, the small permissible sags and the exact lateral position, by the operating range of the tensioning devices and by the temperature compatibility of the conductors and wires used. For higher thermal loadings special measures are required to comply with the requirements on the position of the contact line and the strength of conductors and wires (see EN 50 119). Excessive high and long-duration currents would reduce the contact line strength and cause deterioration of the overhead contact line within a short period.

A suitably designed protective system for contact line installations enables the thermal characteristics of contact lines to be fully utilised. This, in turn, is a prerequisite for optimum operation of adequately designed contact line systems. Contact line installations have the lowest thermal load bearing capacity in railway traction energy supply systems. All other operating components connected in series with the contact lines are less sensitive to short-term overloads.

In traction power supply installations, faults occur more frequently than in public power supply grids. The purpose, objectives and principle of operation of contact line protection provisions are provided in Figure 9.1.

Clauses 1.5 and 1.6 describe, how power is supplied independently to each feed section via a circuit breaker. Each circuit breaker has associated protection circuits and equipment.

The protective equipment that directly handles the protection is called the *primary protection*. Primary protection equipment must be able to recognise whether a short circuit has occurred in the associated feed section and distinguish it from faults in other line sections. If the protection relay responsible for a feed section or the associated power circuit breaker fails, the fault current would not be switched off immediately. In such cases, the *backup protection* provisions must ensure that the current is cut off. It is usual in such arrangements to distinguish:

- back-up protection level 1, which is meant to come into effect if protection relays or power circuit-breakers fail
- back-up protection level 2, which is used as master protection as described in Clause 1.5.3.6 for a DB substation.

When combined with the other protection components of the complete system, overhead contact line protection contributes to ensuring maximum possible availability of the traction power supply. In this context, selectivity (see Clause 9.2) is essential in ensuring that only directly involved sections of the power supply system are switched off in case of a short circuit or excessive loading. This enables continued operation of the railway on those sections not affected by the disturbance.

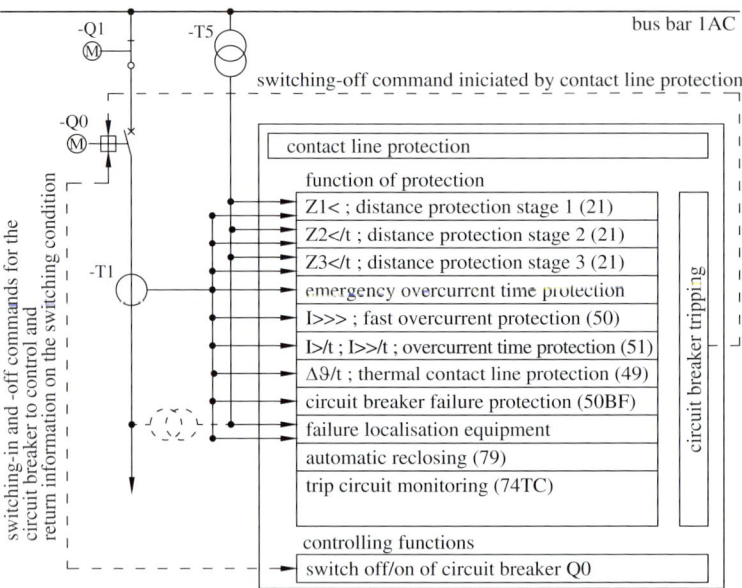

Figure 9.2: Connection of the overhead contact line protection to the voltage and current transformers in case of an AC single-voltage system. Note: the ANSI code numbers to designate the protection function are given in parenthesis

9.1.2   Design and components

9.1.2.1   Overview

Railway operations are characterised by distinct peaks in operational current and short-circuit currents up to 120 kA in the case of DC systems and 45 kA in the case of AC systems. To adequately service the special requirements of railways, contact line protection systems consist of a combination of protection levels as described below each supply branch, depending on the specific requirements [9.1, 9.2]:

- *overcurrent protection* ($I\ggg$), also called high-speed high-current protection
- *multi-step overcurrent protection* ($I\gg/t, I>/t$), partly equipped with an in-rush current recognition system
- *multi-step distance protection* system, also called *impedance protection* ($Z<, Z</t$)
- protection step to supervise the rate of current rise ($\Delta I/\Delta t$), of the voltage variation ($\Delta U/\Delta t$) or impedance variation ($\Delta Z/\Delta t$) related to the time unit to distinguish between the operational and short-circuit currents - also called *starting protection*
- sudden current variation (ΔI)
- thermal overcurrent protection ($\delta>$)
- *overcurrent time protection* ($I>/t$), used as back-up protection if the distance protection stages fail
- back-up protection

Figure 9.2 shows the connection of the contact line protection to the voltage and current transformers for an AC single-voltage system. In Figure 9.3 the limits of current protection ranges are shown depending on the duration of currents and applied protection stages. Additional

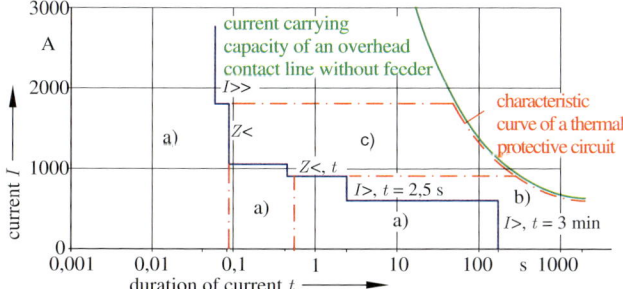

Figure 9.3: Overload protective provisions for overhead contact lines.
a) protection range with a power circuit-breaker break time of 60 ms
b) extended protection by thermal protective provisions $I^2 \cdot t$
c) range which can be covered by $I^2 \cdot t$ and starting-current limitation circuitry

functions of modern contact line protection systems include response to circuit breaker failures, automatic re-closing and fault localisation as well as recording protection data and triggering criteria. The short-circuit protection being the main function must detect short circuits without any transition resistances and trigger a switch-off command to the circuit breakers in control. In the case of short circuits with transition resistances, the total loop impedance must not be higher than the resistance in the case of a short circuit at the end of the protected line section.

9.1.2.2 High-current and overcurrent time protection stages

Because of the impedance between the feeding point and the short-circuit position, maximum short-circuit currents occur close to a substation. To switch-off high short-circuit currents without delay, fast acting overcurrent protection is used as *high-current protection*. The trigger time of the *high-current protection* is 1 ms to 8 ms depending on the type of device.

The high-current time protection, which supplements the high-current protection, is designed as an independent maximum current/time protection and initiates the switching-off only after a settable time has elapsed. When combined with low current setting values, it serves to switch-off excessively high operating currents, remote short circuits and similarly with transition resistances.

9.1.2.3 Distance protection

The *distance protection*, also called *impedance protection*, recognises remote short circuits with low currents and triggers their selective disconnection.

The distance protection usually consists of several stages:
- the first impedance stage ($Z1<$), also called the high-speed stage, is operated directionally and un-delayed, this protects the contact line up to the adjacent supply section, for example between the substation and a switching post. To avoid recognition of short circuits outside the feeding section, the first impedance stage is set to 90 to 95 % of the line impedance with consideration of measuring tolerances in favour of selectivity. The switch-off command to the corresponding circuit breaker is initiated after a delay of approximately 30 ms (high-speed time) which is necessary to locate the short circuit.
- the second impedance stage ($Z2</t$) operates as main protection for the final 5 to 10 % of the feeding section and for back-up protection for the adjacent feeding sections when their protection relays or circuit breakers fail. This stage is also used to switch-off a circuit breaker that feeds a short circuit in an already switched-off feeding section where

an electric traction unit runs into this section. For this reason, the second impedance stage is set to a higher impedance value that also considers the adjacent feeding section. The second impedance stage can be operated directionally or non-directionally and operates with a delay of 150 to 500 ms. The delay can be extended to 60 s. As an example, the German Railway DB operates the second impedance stage directionally and in all substations with a 200 to 300 ms short-delay period. In this case the starting stage governs whether a switching operation is initiated within a millisecond range or a minutes range depending on an operational or short-circuit current. If a starting current is recognised, the distance protection switches the second distance stage to the longer period as mentioned above.

- modern protection relays control a third impedance stage which enables resolution of several protection functions more effectively. The non-reset measuring voltage is a prerequisite for appropriate functioning of impedance protection. In the case of a longer interruption of the measuring voltage or automatic triggering of the voltage transformer, a standardised independent backup *maximum current-time protection* will be activated. In [9.3] the application of distance protection on an AC 25 kV 50 Hz system is described.

9.1.2.4 Starting stage

On lines with high starting currents for electrical traction units, this protection stage is used to distinguish between operational starting and short-circuit currents. Within recent AC protection systems the starting stage does not represent a stand-alone protection relay - it is usually combined with the second and third impedance stage.

This stage can be designed:

- as a $\mathrm{d}I/\mathrm{d}t$ stage in case of DC supply and $\Delta I/\Delta t$ stage in case of AC supply when the contact line is supplied directly or indirectly by rotating units or rectifiers with potentially high short-circuit currents
- as a $\mathrm{d}U/\mathrm{d}t$ stage in case of DC supply and $\Delta U/\Delta t$ stage in case of AC supply to recognise a voltage dip, if the contact line is supplied by converters or rectifiers, which, due to their characteristics, cannot supply a correspondingly high short-circuit current that could be detected by a $\mathrm{d}I/\mathrm{d}t$ or $\Delta I/\Delta t$ stage or
- as a $\Delta Z/\Delta t$ stage that avoids a supply-dependent operating range because of individually locally and timely varying short-circuit powers

The starting stage switches the second and third impedance stage to a longer operating time, if a starting current is detected as a consequence of a slower current rise, a slower voltage drop or impedance reduction.

9.1.2.5 Overload protection

Thermal overload protection ensures optimum utilisation of overhead contact lines up to the pre-defined limit capacities according to Clause 7.1.3. Since continuous measurement of the conductor temperature is difficult, the thermal overload protection measures the operating current flowing through the circuit and models the current-related heating of the overhead contact line.

In addition to the operating current, the ambient temperature is measured at the north outside wall of the substation building. It is incorporated in the calculation algorithm of the protec-

tion relay as a voltage signal between zero and 10 V or as a current signal between zero and 20 mA. Using these signals, the temperature characteristics of the conductor are adjusted to the current ambient temperature and consequently the current capacity of the overhead contact line will be increased in many cases. The acting wind velocity affects the contact wire temperature and wind velocity can be recorded using a heated temperature sensor. Unlike the outside ambient temperature, a wind velocity measured close to the substation building would not be representative of the feeding section as a whole. However, starting at approximately 60 °C the inherently produced thermal load may be considered in conjunction with an assumed wind velocity of 1,0 m/s when calculating the permissible currents or setting the time constant τ within the protection system. Thermal protection systems with a direct measurement of the contact wire temperature are used on some industrial railways operated at DC 2,4 kV [9.4].

Thermal overload protection is not required in coupling posts equipped with only one circuit breaker if summing of currents is not possible there and the overhead contact line is protected by corresponding systems installed in the feeding substations.

Thermal overload protection is associated with only one individual branch of the substation. To guarantee appropriate operation of the thermal overload protection, feeding of a particular contact line section via several substation branches must be avoided because the total current flowing and the resulting temperatures would not be detected by the operationally necessary arrangements and settings of the protection system.

On certain railway lines, additional parallel feeders are only installed on a part of the total length, e. g. on two thirds, since the current capacity of the contact line alone is sufficient to supply the traction units in the remaining sections. The thermal overload protection is set in this case to a higher current capacity together with the parallel feeder line to fully utilise them.

Under these conditions overloads occurring on the final section without a parallel feeder line cannot be completely detected with rare events of high-ohmic failures such as long arcs and current flow through ballast and concrete poles with loose connections to earth. Further, thermal overloads of individual components such as droppers, current connectors and clamps cannot be completely excluded in unfavourable cases such as short circuits across adjacent pantographs. Possible damage can be minimised by a corresponding rating of cross sections, lengths and spacings of the components. Additional arrangements may be required for vehicles with electrical regenerative braking.

9.1.2.6 Other components in digital protection equipment

Some railway operators use *digital protection equipment* with a two-stage circuit breaker failure protection that checks whether the current flow was interrupted after a switch-off command in a certain branch [9.2]. If that is not the case, reserve (also known as backup) tripping of the circuit breaker will be initiated after a settable time period, e. g. 100 ms. If a current still flows after approximately 150 ms, the relay sends a command to the master protection to initiate an immediate switch-off of all medium-voltage circuit breakers feeding the operating busbar section and the circuit breakers on the high-voltage-side transformers.

The digital protection equipment provides special signals in the case of a disturbance, thus allowing the dispatch engineer to initiate reserve feeding. To avoid sections without power, automatic switch-off should be avoided. Such switch-off sections could lead to damage from arcing caused by an electric traction unit running across the separation between the section

under voltage and the section without voltage. If contact line protection failed in the sub-station, a short circuit could be detected and switched-off by the reserve protection of the substation.

The contact line protection comprises remote-control setting possibilities for the impedance stages and the thermal protection. Adjustment to higher impedance or lower current values for thermal loading is used in the case of double-track sections temporarily used for single-track operation during repairs or disturbances.

Several remotely adjustable parameter sets for all protection stages in digital protection equip-ment accommodate the substitute switching variations for sections with deviating impedances and permissible load currents as well as the number of feeding transformers. This digital protection equipment, increasingly used in electrical railway systems nowadays, offers the possibility of extending functionalities and precision with faster processors, more exact ana-log/digital converters and high-performance storage chips [9.5, 9.2]. The functions already described also control:

- extended possibilities to set parameters and operation via an interactive PC and an integrated operating panel on site
- the display of the operational measurement values
- calculation of the failure position using the reactance values and display the failure position on a computer screen
- *automatic reclosing* for the switch-on of the circuit breakers after protection releases
- the logging and output of data and measurement values with time recognition for sev-eral disturbance cases
- the recording of operational and disturbance data and operational diagnostics for the analysis of disturbances
- high reliability by self-monitoring of hardware, software and external measurement transformer circuits and
- the data transfer to the substation control equipment via serial interfaces.

Screening measures and protection for transient overvoltages are necessary to protect the micro processors against electro-magnetic effects in the substations [9.6].

The functions of *contact line protection* and *station control technology* can be integrated into one device or designed with separate hardware for redundancy. A distinctive functional and equipment separation on site between control technology and protection technology exists at many railway operators for reliability reasons. The contact line protection equipment is often arranged in separate cubicles and connected to the auxiliary services and the substation con-trol technology with screened copper cables. For transmission of information to the system control equipment (see Chapter 1), such as measured values for disturbances which are not needed immediately, the coupling of digital protection uses serial interfaces or data buses us-ing optical cables to the station control equipment. This arrangement, which has been in use with some railway operators for 10 years, enables a reduction in cabling expenditure [9.7].

9.2 Protection settings

9.2.1 Introduction

The function of overhead contact line protection and the procedures for selection of *protection settings* is explained in the following. As an example taken from [9.1] a double-track main

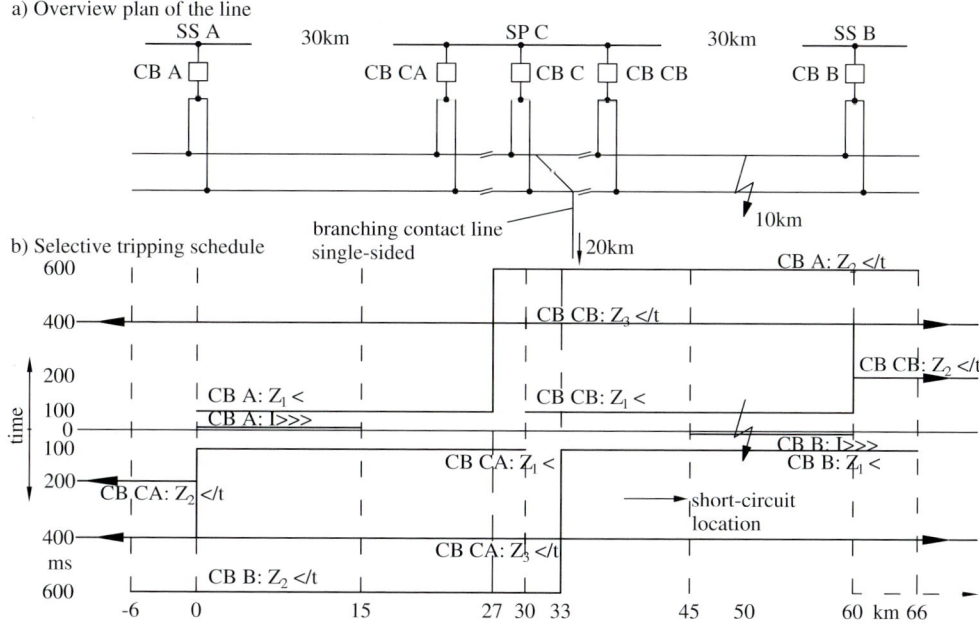

Figure 9.4: Time settings of the overhead contact line protection, where the protection settings starting from substation A to substation B are shown upwards and those starting from substation B downwards from time 0. SS substation, SP switching post, CB circuit breaker

line of an AC operated railway 60 km long with transversely coupled contact lines between two substations A and B was selected where a switching post C is arranged exactly in between the two substations (Figure 9.4 a)). Switching post C provides a single-ended supply for a 20 km long single-track secondary line and the station groups.

Using the selective tripping schedule presented in Figure 9.4 b) the protection function for a short circuit, without transition resistance, at a distance of 10 km to substation B will be explained. The selective tripping schedule is only shown for circuit breakers A, B, CA and CB.

9.2.2   Distance and high-current stages

An overhead contact line protection relay is provided for each of the circuit breakers within the line sections considered. The protection relays of the circuit breakers CA and CB at the switching post for the branching line contain three-impedance stages.

The primary settings of the impedances of all relays with the exception of the circuit breaker C are:

- for the first impedance stage: in the substation $Z_1 = 0,124\,\Omega/\text{km} \cdot 0,9 \cdot 30\,\text{km} = 3,35\,\Omega$, directional, undelayed; at the switching post $Z_1 = 0,124\,\Omega/\text{km} \cdot 1,0 \cdot 30\,\text{km} = 3,72\,\Omega$, directional, undelayed
- for the second impedance stage, which is operated directionally in both the substation and the switching post: $Z_2 = 0,124\,\Omega/\text{km} \cdot 1,1 \cdot 60\,\text{km} = 8,18\,\Omega$

- in the substation if there is a positive short-circuit detection by the starting stage to e. g. 600 ms delayed, in the switching post to 200 ms. If the starting stage does not detect a short-circuit, tripping is initiated after approximately 60 s
- the third impedance stage within the switching post is operated non-directionally and also set to $8,18\,\Omega$ with 400 ms delay and equipped with a starting stage

These settings are based on the impedance of $0,124\,\Omega/\text{km}$ for the selected symmetrical arrangement and the contact line Sicat S,1.0 without a parallel feeder line and without return conductors.

The first impedance stage within the substation is set for example to 90 % (factor 0,9). The second impedance stage is factored by 1,1 to ensure detection of all failures in the whole feeding section.

The selected Z_2 setting would permit a long-duration maximum operational current of

$$I_{B\,\text{max}} = 17,25\,\text{kV}/(8,18\,\Omega + 1,5\,\Omega) = 1,78\,\text{kA}$$

in the case of the minimum operational voltage

$$U_{B\,\text{min}} = 17,25\,\text{kV} - (1,5\,\Omega \cdot 1,78\,\text{kA}) = 14,77\,\text{kV}$$

where a source impedance of $Z_q = 1,5\,\Omega$ considering the supplying devices are taken into account. The threshold current of high-current protection should be

$$I_E = 17,25\,\text{kV}/(1,5\,\Omega + 0,5 \cdot 30\,\text{km} \cdot 0,124\,\Omega/\text{km}) = 5,13\,\text{kA}.$$

This protection stage uses a factor of 0,5 to detect short circuits without a transition resistance which occurs in the first half of the line to be protected. Factors of 0,9 to 0,95 extend the protection range to 90 % or 95 % of the line, however, they lead to corresponding lower threshold currents being 3,56 kA or 3,34 kA respectively. In the case of excessively low threshold currents, the magnetising inrush current of the traction unit transformers could lead to tripping with a returning voltage or a short circuit with DC components outside the section to be protected.

If the actual short-circuit current is lower than the set value, the high-current protection of the relay of circuit breaker B does not react. If $I_K > 5,13\,\text{kA}$ this being higher than the set value in the discussed example, then a switch-off command will be initiated by the high-current stage to circuit breaker B.

The switching-off of short circuits on the remaining feeding sections is initiated by the first impedance stage having a reaction period of approximately 30 ms at a frequency of 16,7 Hz or with the corresponding selective tripping period. In cases of short circuits without a transition resistance, the first impedance stage takes care of:
- localising all failures which are closer than 90 % of the line length that is 27 km from the substation A or B
- detection of all failures on the lines to be protected in switching post C since the recorded impedance will always be below the selected set value.

In the example with $I_K > 5,13\,\text{kA}$ in substation B, the first impedance stage initiates the switching-off of circuit breaker CB, arranged in the switching post C within a command period of approximately 30 ms and in addition to the high-current stage, the switching-off of circuit breaker B installed in substation B. The contact line between substation B and switching post C has been selectively switched-off, therefore, the station and line C remain live.

If short circuits occur with a transition resistance, high-speed tripping must not be carried out, especially in the case of failures at the end of the overhead contact line being protected. Switching-off the circuit breakers in this case is done with a selective tripping period using the second impedance stage, provided that the sum of the line impedance and the transition resistance does not exceed the set value. Impedance measuring principles, including an arcing provision, have not been implemented for overhead contact line protection because of the complex calculations required and the resulting extension of the relay response time.

The second impedance stages in branches A and CA record an impedance lower than the set impedance of $8,18\,\Omega$. Since the first impedance stage had switched-off the circuit breaker CB without delay, a tripping of the circuit breakers A and CA does not occur because of the preceding switching-off by CB and the wrong direction of energy flow.

If a short circuit occurs between 27 and 30 km from substation B the second impedance stage would initiate the switching-off of the circuit breaker B after 600 ms. If, on the other hand, the short circuit occurs close to substation B with a transition resistance, the switching-off of circuit breaker C would be initiated by the selective tripping stage of CB. However, the relay CA experiences the same current and is switched to the same voltage. To avoid un- wanted, non-selective switching-off of CA the second impedance stage of CA and, therefore, that of CB, must be operated directionally with the tripping direction: energy flow from the busbar. Then only the circuit breaker CB will be tripped and CA remains switched-on with the consequence that station C and the contact line beginning there remain live. If the relay or the circuit breaker CB failed in the case of the original failure or the failure now described, only the circuit breaker CA should trip to keep the line section A to CA, fed from circuit breaker A, live. Tripping of circuit breaker CA within the first selective tripping stage or even with high-speed tripping is not possible because of the wrong direction of energy flow. How- ever, to cope with this failure case as well, the third impedance stage will be effective for circuit breaker CA which switches off the failure undirectionally after 400 ms. Consequently selective tripping period of Z_2 in A and B must be set to at least 600 ms, to give priority to switching-off within the switching post.

If the second impedance stage of the substations and the switching post were operated di- rectionally and with the same selective tripping periods, a non-selective switching-off of the total line section between substations A and B would result since circuit breaker A would also be tripped. In the line supply scheme described here, the switching post C acts mainly as a coupling post since the line section C does not feed-in a current. The undirected *high-speed overcurrent stages* $I\ggg$ will not be activated for the circuit breakers CA and CB because they conduct the same current in case of short circuits on the sections CA to A and CB to B and, therefore, a circuit breaker would always switch-off un-selectively. Accordingly, the same issue occurs with the emergency maximum current-time protection and reserve protec- tion. For the circuit breaker C the high-speed overcurrent stage $I\ggg$ is adequate since, in the case of failures within section C the currents from substations A and B are added there.

The setting impedances of the relay for circuit breaker C of the single-track, single-end fed section has to consider a length-related impedance of approximately $0,214\,\Omega/\mathrm{km}$ for the con- tact line Re 200 without a parallel feeder or a return conductor and the factor 1,1 for both, the first and the second impedance stage: $Z_{\mathrm{e1}} = 1,1 \cdot 0,214\,\Omega/\mathrm{km} \cdot 20\,\mathrm{km} = 4,71\,\Omega$.

If there is a failure of protection or circuit breaker C, the third impedance stages of circuit breakers CA and CB trip after 400 ms for example. The impedance stages of the substations can detect failures on the single-end fed section under the selected setting values only close to switching post C. However, this requires an increased selective tripping period of 600 ms.

Longer impedance setting values are often not acceptable because of the high operational currents. In the case of coupling posts, the third impedance stage can make superfluous the second protection relay, which is necessary for directional operation with selective short-circuit detection. The first and second impedance stage is then directed, without delay to one or other line sections and the third impedance stage is operated non-directionally and with a third selective tripping period.

The advantage of the separate station supply at switching post C between the substations is demonstrated by the case of switching-off circuit breaker CB, because of a permanent short circuit and the running of an electric traction unit into the short-circuit section. The short-circuit current flowing anew would be switched off by circuit breaker CA and not by circuit breaker A. The section between substation A and switching post C would remain live.

In the example above, if the relay of circuit breaker B failed, the branch-related back up protection described below would be effective as a *maximum current time protection* or otherwise the total current protection as described in Clause 1.5.3.6. These higher level protection schemes switch off all the circuit breakers connected to the busbar of substation B and consequently the circuit breakers of the transformers as well. This rare switching-off operation has the disadvantage of making the whole substation dead.

9.2.3 Emergency maximum current time protection and back up protection

If there is a failure of the supply voltage of the transformer instrument required for distance protection, an *emergency maximum current time protection* stage will be activated which does not depend on the voltage. This protection stage is designed as an independent overcurrent time protection. An adequate protection setting can only be found as a compromise between what is an absolutely necessary and desirable failure detection similar to the previously described $Z<$ selective tripping stages. It is necessary that all failures within the complete supply section of the protection relay are detected. This determines the maximum value of the $I>$ setting. However, it would be advantageous if the emergency maximum current time protection stage also recognised failures on the adjacent line sections if they were not switched-off. The limitation is given by the maximum operational current to be expected.

As an example, a primary maximum setting value of

$$I_{e\,max} = \frac{U_{B\,min}}{1{,}25 \cdot Z_1} = \frac{14{,}77\,\text{kV}}{1{,}25 \cdot 30\,\text{km} \cdot 0{,}124\,\Omega/\text{km}} = 3{,}18\,\text{kA}$$

results based on the values described in the context of settings for distance and high-current stages for the substations and with a safety margin of 25 %. As a minimum value the maximum operational current calculated for the same line section will be used:

$$I_{e\,min} = I_{B\,max} = 1{,}78\,\text{kA} \quad .$$

A setting value of $I_e = 2{,}5\,\text{kA}$ is selected as a mean value of $I_{e\,max}$ and $I_{e\,min}$. The selective tripping period will be set to 400 ms or 600 ms, therefore, above the selective tripping stages within the switching post. When setting the emergency current time protection stage in the switching post it must be distinguished between the protection relays of the line CA and CB set equally and the protection relay for the single-end fed line section C.

The following applies to protection relays CA and CB.

$$I_{e\,max} - \frac{U_{B\,min}}{1,25\cdot(Z_{CA}+Z_{CB})} = \frac{14,77\,kV}{1,25\cdot(30\,km+30\,km)\cdot0,124\,\Omega/km} = 1,59\,kA,$$

because they must be able to detect failures even near the substations. This value will be selected for the setting, it is below the current $I_{B\,max} = 1,78\,kA$ calculated for the substations, however, this can be accepted since the load peaks are more likely taken from the adjacent substation than across the switching post from the far distant feeding point.

The $I\gg$ setting of the emergency maximum current time protection for the branch C is obtained from the failure current for a short circuit at the line end:

$$I_{e\,max} = \frac{U_0}{0,5\cdot(Z_q+Z_{CA})+Z_C} = \frac{17,25}{0,5\cdot(1,5+30\cdot0,124)+20\cdot0,214} = 2,5\,kA.$$

For consistency of settings, a threshold value of 1,59 kA is also selected for this relay. The time delay of the relays CA and CB is set to 400 ms. These settings assign priority to the second protection relay for selective failure tripping by the high-speed protection stage or the first selective tripping stage. However, if because of failure of the common current transformer and if the distance stages of all protection relays were ineffective, selectivity between CA and CB cannot be achieved. In this case, both circuit breakers trip simultaneously. However, for branch C, the emergency maximum current time protection can trip undelayed because selectivity problems do not exist.

For branch-oriented backup protection, maximum current time protection relays are used when required. The current values are calculated and set, as in the case of the emergency current time protection. The following considerations apply to the selective tripping periods:

 – if backup protection is activated only when the main protection has failed, the same command periods are selected as in the case of the emergency maximum current time protection stage
 – if the back-up protection is operated permanently – that is independently of the availability of the main protection – the command periods have to be increased by at least 200 ms compared to the emergency maximum current time protection, maintaining priority for the main protection stages.

9.2.4 Overload protection

According to Table 7.12 the permissible permanent current of a contact line type Sicat S1.0 is 660 A for 16,7 Hz operation with a new contact wire and is 1 320 A for a double-track line. Overloading the overhead contact lines cannot be completely excluded with the short-circuit current values as calculated above and the relay settings for the short-circuit protection derived thereof. There will be an unprotected current range within the substations between the permissible load of 1 320 A and the threshold value of the second impedance stage of 1 780 A. A remedy for this situation is overload protection as described in Clause 9.1.2.5. This protection is set to the permissible final temperature of the contact wire, which is 70 °C in the example, which resulted from the permissible permanent currents described above. In addition, the time constant valid for heating and cooling must be considered. This constant can be taken roughly as 9 min for a 100 mm^2 copper contact wire at 40 °C with 1,0 m/s wind velocity (see Table 7.8).

The tripping command activated by the overload protection will last long enough for the contact wire to cool down to the settable reclosing value. During this period lasting approximately 10 to 60 s, which together with a time constant depends on the ambient temperature, the circuit breaker cannot be re-closed. It can be recognised from the example that within switching post C, the overload protection is essential for single-end supplied line sections. However, for the other two branches the overload protection installed in the substations will be sufficient since no additional intermediate in-feed exists. In the case of regeneration by traction units, this needs to be reviewed.

9.2.5   Fault localisation

The location of a sustained short circuit must be found as soon as possible and isolated with circuit breakers and overhead contact line disconnectors. This enables the continuation of electric railway operation on the remaining sound line sections, the elimination of the cause of the fault and allows repairs to be carried out. Accurate and reliable *fault localisation* is important and is accomplished with a short-circuit tracing system based on current transformers. On double-track lines, these current transformers, which have a transformation ratio of 600 : 1 or 1 200 : 1, are installed mainly at the cross-coupling disconnectors, i. e. in the circuit connecting two overhead contact line main groups. If a short circuit occurs, a current considerably higher than in normal operating conditions will flow through the current transformers. They are located to one side of the fault location in single-end feed sections, or both sides of the fault location in double-end feed sections. The short-circuit sensing relays connected to the secondary windings of the current transformers will register this high current and send a signal to the control centre in charge of that section, via the local control unit and the associated telemetry module. This, in conjunction with the information on the tripped circuit breakers, enables a rough determination of the location of the fault between the positions of the activating current transformers.

On single-track lines with single-end feeds, the fault location will be known to lie beyond the last short-circuit sensing current transformers that detected an overcurrent. On single-track lines with double-end feeds, the fault will be located between the two transformers that detected a change in the energy flow direction. However, other procedures will have to be used to determine whether the short-circuit is in one of the secondary groups of a railway station, in one of the main station groups or somewhere along the line. This task is carried out using pole-mounted disconnectors and either by overhead contact line testing with automatic short-circuit localisation equipment at the master control centre or manually, step-by-step.

The evaluation of the measurement data collected by the digital protection circuitry (see Clause 9.1.2.5), which records the impedances before and during the short circuit, until the current is cut off, is an effective tool. Such systems can log the data of more than one fault event. To determine the fault location, the reactance is used and output is a resistance or a distance value. The system is relatively accurate, achieving a tolerance of 500 m for a section length of 30 km. However, if the high-speed high-current protection is tripped, it is technically impossible to locate the fault position. Automatic and highly accurate fault localisation, together with appropriate reporting, considerably reduces downtime as the repair crew will be able to proceed directly to the fault location. Down-times because of repeated transient short circuits will be reduced by enabling preventive and corrective measures to be carried out immediately.

9.3   Bibliography

9.1   *Biesenack, H. et al.*: Energieversorgung elektrischer Bahnen (Power supply of electric railways). B. G. Teubner Publishing, Wiesbaden, 2006.

9.2   *Braun, W.; Kinscher, J.*: Innovatives Schutz- und Steuergerät für AC-Bahnenergieversorgung. (Innovative control and protective device for use in AC traction power supplies). In: Elektrische Bahnen, 105(2007)6, pp. 440 to 447.

9.3   *Cinieri, E. et al.*: Protection of high-speed railway lines in Italy against faults. In: Elektrische Bahnen, 105(2007)1, pp. 81 to 90.

9.4   *Orzeszko, S.*: Schutztechnik bei Einphasen-Wechselstrombahnen am Beispiel der Deutschen Bundesbahn (Protection technique at single-phase alternating current railway lines shown on the example of the German Federal Railway). In: Elektrische Bahnen, 80(1982)9, pp. 264 to 270.

9.5   *Braun, H.-J.; Liebach, T.; Schneerson, E.*: Digitalschutz für Bahnenergienetze (Digital protection of railway power supply grids). In: Elektrische Bahnen, 97(1999)1-2, pp. 32 to 39.

9.6   *Girbert, K.-H.; Orzeszko, S.; Schegner, P.*: Digitaler Schutz für 16 2/3-Hz-Oberleitungsnetze (Digital protection of 16 2/3 Hz overhead contact line systems). In: Elektrische Bahnen, 92(1994)6, pp. 183 to 190.

9.7   *Rattmann, R.; Walter, S.*: Zweite Generation 16 2/3-Hz-Normschaltanlagen der Deutschen Bahn (Second generation of 16 2/3 Hz standard substations of German railway (DB)). In: Elektrische Bahnen, 96(1998)9, pp. 277 to 284.

10 Interaction of pantograph and overhead contact line

10.0   Symbols and abbreviations

Symbol	Definition	Unit
A	cross-section	mm^2
A_n	coefficient of a series	–
C_A	basic parameter	N
C_D	damping	Ns/m
C_V	parameter of speed	N/(km/h)^2
C_{1n}	coefficient of a Fourier series	mm
C_{2n}	coefficient of a Fourier series	mm
C_n	coefficient of a Fourier series	–
D_v	damping element	%
F	contact force	N
F_0	contact force between pantograph and contact wire	N
$F_{1,2,3,4}$	reaction forces of the collector strips	N
F_C	total contact force	N
F_{CA}	reaction force at catenary wire	N
F_{CW}	reaction force at contact wire	N
$F_{C\,dyn}$	dynamic contact force	N
F_{Ci}	contact force at leading (1) and at trailing (2) pantograph	N
$F_{C\,stat}$	contact force exerted by the pantograph drive at standstill	N
F_{Ri}	friction force within the pantograph model	N
F_S	reaction force at a pantograph	N
F_{SI}	reaction force at leading collector strip	N
F_{SII}	reaction force at trailing collector strip	N
F_a	acceleration force of a contact line element	N
F_{aero}	aerodynamic component on force on collector strip	N
$F_{contact}$	corrected contact force	N
F_i	friction force at pantograph	N
F_i	measured contact force i	N
F_m	mean contact force	N
$F_{m\,max}$	maximum contact force	N
$F_{recorded}$	measured contact force	N
$F_{m\,min}$	minimum contact force	N
F_{msz}	acceleration force of collector strip	N
F_r	reactive contact force	N
F_p	reaction force at a mass	N
F_v	reaction force	N
F_z	contact force of collector strip of a pantograph	N

Symbol	Definition	Unit
F_{zr}	restoring force of a contact wire element	N
H_0	tensile force in a wire	N, kN
H_{CA}	messenger wire tensile force	N, kN
H_{CW}	contact wire tensile force	N, kN
K_i	spring modulus within the pantograph model	mm/N
L_y	length of stitch wire	m
M	concentrated mass; dropper mass	kg
M_S	mass of pantograph head	kg
MP	midpoint anchor	
$M(y)$	bending moment of contact strip	Nm
NQ	arcing period at maximum speed	%
O	overlapping section	
$Q(z)$	transverse force at collector strip	N
S_0	steady arm force at support	mm
W	point	
Y_S	coordinate along the collector strip	mm
c	modulus of spring elasticity	mm/N
$\bar{c}$	mean wave propagation speed	m/s
c_{CA}	wave propagation speed of messenger wire	m/s
c_{CW}	wave propagation speed of contact wire	m/s
c_p	wave propagation speed	m/s
d	damping within the pantograph	kg/s
e	elasticity of contact line	mm/N
$f_{0,1,1,2}$	function of wave movement	–
f_n	frequency of order n	Hz
$f_{z\,stat}$	static uplift of contact wire	m, mm
g	gravitational acceleration	m/s^2
h_{CW}	contact wire height	mm
k	numeric factor for contact line elasticity	–
k_S	calibration quantity for contact wire position	mm
k_a	number of acceleration sensors	–
k_e	numeric constant for elasticity	–
k_f	number of contact force sensors	–
l	length of the considered line section or span	m
l_1	distance between droppers in a span	m
l_{SS}	distance of collector strip bearings	m
l_d	distance between droppers	m
l_{fz}	length of the contact wire section lifted by the pantograph	m
l_{sp}	distance between supports	m
m'	length related contact line mass	kg/m
m'_{CA}	length related messenger wire mass	kg/m
m'_{CW}	length related contact wire mass	kg/m
m_S	mass of collector strip	kg
$m_{SI,SII}$	mass of collector strips I and II	kg
m_i	pantograph partial mass	kg

Symbol	Definition	Unit
m_{min}	minimum modul mass	kg/m
m_n	modal mass of contact line model	kg/m
n	index	–
r	reflection coefficient	–
q'	load per unit length	N/m
s_m	standard deviation of contact force	N
s_{max}	standard deviation of contact force at maximum speed	N
t	time	s
t_{arc}	duration of arcs	s
u	non-uniformity degree of elasticity	–
$u(t)$	time-dependent step function	–
v	running speed	m/s
v_α	limiting speed	m/s
x	contact wire longitudinal coordinate	m
$\bar{x}$	mean value of measured contact forces	N
x_r	position of a contact wire discontinuity	m
x_t	position of pantograph at the time t	m
z	contact wire vertical coordinate and deflection	mm
$\ddot{z}_S$	measured mean acceleration of the collector strip mass	m/s^2
$\ddot{z}_{1,2,3,4}$	mean acceleration measured at the sensors 1, 2, 3 and 4	m/s^2
$\hat{z}$	amplitude of a sinusoidal wave	mm
z_0	initial motion of a point	mm
$\ddot{z}_i$	acceleration measured at the sensor i	m/s^2
z_r	additional motion of a point	mm
z_{stat}	static contact wire uplift	mm
$z(x,t)$	function of contact wire uplift	–
ΔF_0	contact force difference	N
ΔF_1	contact force increase	N
Δt	time step	s
α	angle of contact wire uplift	degree
α_D	Doppler factor	–
γ	specific mass	kg/m^3
γ_A	amplification coefficient	–
γ_{CW}	specific contact wire mass	kg/m^3
$\delta(x)$	Dirac delta function	–
θ	constant	s
ϑ	auxiliary variable	m^2/s
ε	interval limit	–
$\nu_{1,2}$	first, second fundamental frequency	Hz
σ	standard deviation	–
σ_{CW}	contact wire tensile stress	N/mm^2
τ	variable of time	s
ω	angular frequency	1/s

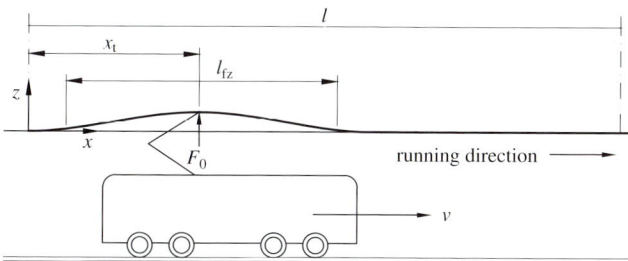

Figure 10.1: Traction vehicle with pantograph moving along a contact wire.

10.1 Introduction

The interaction of the overhead contact line with the pantograph determines the reliability and quality of the energy transmission to traction units. This interaction depends on the design of the pantograph and the overhead contact line and thus depends on a large number of parameters. High-speed trials have shown that the *pantograph/contact line interaction* is of extreme importance because energy transmission is a factor that could limit maximum achievable speeds [10.1]. Objective criteria, that can be calculated and empirically confirmed by on-track tests, are required for the evaluation and prediction of contact characteristics. Considerable progress has been made in understanding the theory of contact behaviour and the respective findings have been supplemented by simulation processes and advanced measurement methods. Simulation processes are particularly useful for developing new systems with enhanced performance requirements because there is a limit to the extent of field tests and trials that can be carried out.

The system overhead contact line-pantograph is supposed to supply energy to the traction vehicle via continuous electrical and mechanical contact, i. e. without interruptions, whilst simultaneously keeping the wear on the contact wire and the collector strips as low as possible. The energy transmission system, in this case the overhead contact line because it involves high expenditures, is expected to achieve long service life with minimum maintenance. Checking the *contact behaviour* of existing overhead contact line installations is a method of reviewing and assessing such lines. These checks also detect *localized irregularities* which can then be eliminated [10.2].

10.2 Physical basis

10.2.1 Pantograph running along an overhead contact line

A pantograph running along the contact wire of an overhead contact line (Figure 10.1) exerts a force F_0 on the elastic contact line. Therefore, the contact wire will be lifted. If the traction vehicle bearing the pantograph is at standstill or travels slowly, the dynamic effects can be neglected and the uplift $f_{z\,\text{stat}}$ at the point of contact is equal to the product of the force F_0 acting on the contact wire and the elasticity e of the contact line

$$f_{z\,\text{stat}} = F_0 \cdot e \quad . \tag{10.1}$$

Typical values for F_0 and e are 70 N and 0,5 mm/N for high-speed lines, resulting in a static uplift of 35 mm. As shown in Figure 10.1 the uplift curve of the contact wire is symmetrical

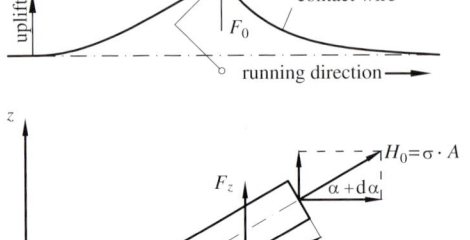

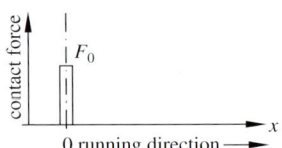

Figure 10.2: Schematic of uplift of a contact wire above a running pantograph.

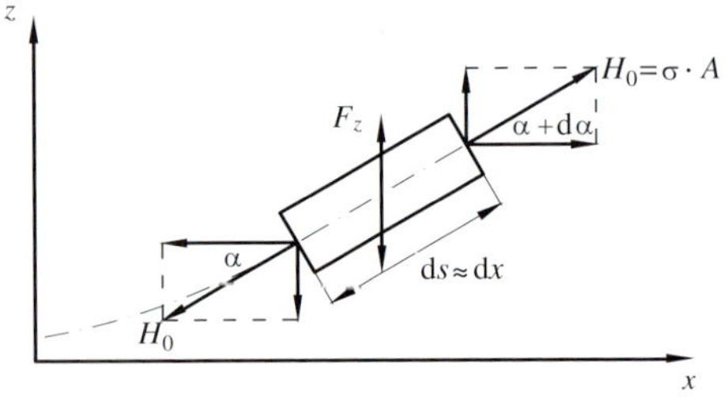

Figure 10.3: Equilibrium of forces acting on a contact wire element with the mass $m'_{CW} \cdot dx$.

Figure 10.4: Contact force exerted by a running pantograph.

about the position of the pantograph and the length of the uplifted contact wire is

$$l_{fz} = F_0/(m'_{CW} \cdot g) \tag{10.2}$$

A static force of 70 N would, therefore, affect a contact wire AC-120 over a length of 6,5 m. With increasing running speed stronger dynamic forces are exerted, that run along the contact wire in both directions, as seen from the instantaneous position of the pantograph. The overhead contact line section affected by the running pantograph will extend far beyond the static range determined above and the pattern of the uplift curve shown in Figure 10.2 is no longer symmetrical.

The objective is to describe the behaviour of the contact wire when a pantograph pressing against it with a force F_0 travels along it at a speed of v (Figure 10.1). A useful model of the processes involved is achieved by treating the contact wire as a tensioned string without bending stiffness. Alternatively, in [10.3] the contact wire was treated as a flexible beam.

To study the *propagation of a transverse impulse*, i. e. the local vertical uplift caused by the pantograph moving along the contact wire, the wire is assumed to have a negligible stiffness with a longitudinal stress σ_{CW} and a specific mass γ_{CW} (Figure 10.3). When a wire, which is subjected to a longitudinal force $H_0 = \sigma_{CW}A$, is deflected transversely, each wire element of length dx experiences a restoring force F_{zr}:

$$F_{zr} = H_0 \sin(\alpha + d\alpha) - H_0 \sin\alpha \approx H_0 d\alpha \quad . \tag{10.3}$$

With $\alpha \sim \tan\alpha = \partial z/\partial x$, $d\alpha \approx dx \cdot (\partial^2 z/\partial x^2)$ is obtained, which results in the restoring force

$$F_{zr} = H_0 \cdot dx \cdot (\partial^2 z / \partial x^2) = \sigma_{CW} \cdot A \cdot dx (\partial^2 z / \partial x^2) \quad . \tag{10.4}$$

The mass of a wire element of length dx is $m'_{CW}dx = \gamma_{CW}A\,dx$ and the acceleration force is

$$F_a = m'_{CW}dx (\partial^2 z / \partial t^2) = \gamma_{CW}A\,dx (\partial^2 z / \partial t^2) \quad . \tag{10.5}$$

The equilibrium of forces of (10.4) and (10.5) yields

$$\sigma_{CW}A \cdot dx (\partial^2 z / \partial x^2) - \gamma_{CW}A\,dx (\partial^2 z / \partial t^2) = 0 \tag{10.6}$$

and eventually the differential equation describing the motion of the tensioned contact wire is obtained:

$$\frac{\partial^2 z}{\partial x^2} - \frac{\gamma_{CW}}{\sigma_{CW}} \frac{\partial^2 z}{\partial t^2} = 0 \quad . \tag{10.7}$$

This equation is known in engineering mechanics as the *wave equation* of a taut wire or string. The general solution of this equation is given by all functions having the format

$$z = f(x \pm c_p \cdot t) \quad , \tag{10.8}$$

whereby

$$c_p = \sqrt{\sigma_{CW}/\gamma_{CW}} = \sqrt{H_0/m'_{CW}} \tag{10.9}$$

is the *wave propagation speed*. For an AC-100 copper contact wire subject to a force of 10 kN, the wave propagation speed is found to be $\sqrt{10\,000/0,89} = 106$ m/s, which is roughly equal to 380 km/h.

10.2.2   Contact wire subjected to a moving pantograph exerting a constant contact force

As described in Clause 10.2.1 the contact force between a running pantograph and a contact line is not constant but affected by reflections of the oscillating contact wire. Nevertheless some basic characteristics of the pantograph/contact line interaction can be demonstrated by assuming a constant contact force F_0 which acts at the time-dependent position x_t (Figure 10.2). Mathematically, this contact force can be described using the *Dirac delta function*

$$F_z = F_0 \cdot \delta(x - x_t) \quad . \tag{10.10}$$

The *Dirac delta function* is defined with the characteristics $\delta(0) = 1$ and $\delta(x \neq 0) = 0$. In Figure 10.4 the contact force is shown. The contact force is F_0 at $x = x_t$ and 0 in the other contact line ranges. By adding the term $F_z \cdot dx = F_0 \cdot dx \, \delta(x - x_t)$ to the equation of equilibrium of forces (10.7) it is obtained

$$\frac{\partial^2 z}{\partial t^2} = c_p^2 \frac{\partial^2 z}{\partial x^2} + \frac{F_0}{m'_{CW}} \delta(x - x_t) \quad . \tag{10.11}$$

If the collector strip is located at point $x = 0$ at the time $t = 0$, its location at time t is given by

$$x_t = v \cdot t \quad . \tag{10.12}$$

The delta function can be replaced by a *Fourier series*, where relation (10.12) is applied as a boundary condition

$$\delta(x - x_t) = \sum_{n=1}^{\infty} C_n \cdot \sin(n\pi x/l) \quad , \tag{10.13}$$

where l is the length of the section considered and

$$C_n = 2/l \cdot \sin(n\pi x_t/l) = 2/l \cdot \sin(n\pi v t/l) \quad . \tag{10.14}$$

By inserting (10.13) and (10.14), Equation (10.11) is transformed to

$$\frac{\partial^2 z}{\partial t^2} = c_p^2 \frac{\partial^2 z}{\partial x^2} + \frac{2F_0}{m'_{CW}l} \cdot \sum_{n=1}^{\infty} \sin(n\pi x/l) \cdot \sin(n\pi vt/l) \quad . \tag{10.15}$$

As a solution for Equation (10.15), the function

$$z(x,t) = \sum_{n=1}^{\infty} z_n(t) \cdot \sin(n\pi x/l) \tag{10.16}$$

can be used. If (10.16) is inserted into (10.15), a set of second-order linear differential equations is obtained for the functions $z_n(t)$

$$\ddot{z}_n(t) + c_p^2 (n\pi/l)^2 z_n(t) = (2F_0/m'_{CW}l) \sin(n\pi vt/l) \quad . \tag{10.17}$$

The general solutions for equations of this format are

$$z_n(t) = C_{1n} \cos(n\pi c_p t/l) + C_{2n} \sin(n\pi c_p t/l) + A_n \sin(n\pi vt/l) \quad . \tag{10.18}$$

From (10.17) and (10.7) it is obtained

$$A_n = 2F_0 l \ / \ \left[m'_{CW} (n\pi)^2 \left(c_p^2 - v^2 \right) \right] \quad . \tag{10.19}$$

The coefficients C_{1n} and C_{2n} are deduced from the boundary conditions $z_n(0) = 0$ and $\dot{z}_n(0) = 0$. The first condition results in $C_{1n} = 0$. Then Equation (10.18) yields

$$\dot{z}_n(0) = C_{2n} \left(n\pi c_p/l \right) + 2F_0'l \ / \ \left[m'_{CW} (n\pi)^2 \left(c_p^2 - v^2 \right) \right] (\pi n v/l) = 0 \tag{10.20}$$

and eventually

$$C_{2n} = -\frac{2F_0 l}{m'_{CW} (n\pi)^2 \left(c_p^2 - v^2 \right)} \cdot \frac{v}{c_p} \quad . \tag{10.21}$$

With this result, the solution for the differential Equation (10.15) is found to be:

$$z(x,t) = \frac{2F_0 l}{m'_{CW} \pi^2 \left(c_p^2 - v^2 \right)} \cdot \sum_{n=1}^{\infty} \frac{1}{n^2} \sin\frac{n\pi x}{l} \left(\sin\frac{n\pi vt}{l} - \frac{v}{c_p} \sin\frac{n\pi c_p t}{l} \right) \quad . \tag{10.22}$$

This solution presents the fundamental *resonance characteristic* that will be effective when the train's speed v approaches the *wave propagation speed* c_p. In this case, the uplift of the contact wire would tend towards infinity, making it impossible to draw current from the contact wire. The wave propagation speed is a physical limit to energy transmission between an overhead contact wire and a pantograph. This theoretical deduction was confirmed in practice during high-speed trials. As the trains approached the wave propagation speed, the contact wire uplift increased to unacceptable values and prevented further speed increases [10.4]. The contact wire design and the tensile stress applied to the contact wire need to be selected to ensure that the difference between the *maximum operating speed* and this limit is sufficient to ensure continuous contact. Further details regarding this subject are explained in Clauses 10.4.2 and 10.9.2. Practical experience has shown that the wave propagation speed should be at least 1,4 and 1,5 times the envisaged maximum train speed.

10.2.3 Contact wire uplift at high speeds

As an initial condition for determining *contact wire uplift* at high train speeds, reference [10.5] assumes that at the time $t = 0$, a constant force F_0 is acting on point $x = 0$ of a stationary contact wire (Figure 10.4). By multiplying this with the contact wire cross section and adding the term $q'(x,t)$, Equation (10.7) is transformed to

$$m'_{CW}\frac{\partial^2 z}{\partial t^2} - H_0\frac{\partial^2 z}{\partial x^2} = q'(x,t) \quad , \tag{10.23}$$

where $q'(x,t)$ is a time-variable load per unit length. The force F_0 can be formally expressed as a load per unit length by

$$q'(x,t) = F_0 \cdot \delta(x)u(t) \quad . \tag{10.24}$$

Here again, $\delta(x)$ is the *Dirac delta function* and $u(t)$ is a time-dependent step function of the type

$$u(t < 0) = 0 \;;\; u(0) = 0,5 \;;\; u(t > 0) = 1 \quad . \tag{10.25}$$

Since $q'(x,t) = 0$ for $x \neq 0$, because $\delta(x) = 0$ for $x \neq 0$, Equation (10.23) yields

$$\partial^2 z/\partial t^2 = c_p^2 \partial^2 z/\partial x^2 \quad . \tag{10.26}$$

Integration of Equation (10.23) over any small interval of $-\varepsilon \leq x \leq \varepsilon$, and consideration of Equation (10.24) as well as the fact that $\partial^2 z/\partial t^2$ must be a continuous function, yield

$$-H_0 \cdot [z'(\varepsilon,t) - z'(-\varepsilon,t)] + 2m'_{CW} \cdot \varepsilon \frac{\partial^2 z(0,t)}{\partial t^2} = F_0 \cdot u(t) \quad . \tag{10.27}$$

Due to the symmetry at the point, where the force F_0 acts, $z'(\varepsilon,t)$ must be equal to $-z'(-\varepsilon,t)$. For the boundary condition $\varepsilon \to 0$, Equation (10.27) is transformed to

$$z'(0,t) = -F_0/(2H_0) \cdot u(t) \quad . \tag{10.28}$$

In the range $x > 0$, only one wave $z(x,t) = f_1(x - c_p t)$ can occur, and, for this reason, there is

$$\dot{z}(0,t) = -c_p z'(0,t) = c_p F_0/(2H_0) \cdot u(t) = F_0/(2m'_{CW}c_p)u(t) \quad . \tag{10.29}$$

Integration yields

$$z(0,t) = F_0 c_p t/(2H_0) = f_0(-c_p \cdot t) \quad . \tag{10.30}$$

The solutions of this equation for $x \neq 0$ are

$$\begin{aligned}
z(x,t) &= 0 & &\text{for } |x| > c_p t, \\
z(x,t) &= F_0(c_p t - x)/(2H_0) & &\text{for } 0 \leq x \leq c_p t, \\
z(x,t) &= F_0(c_p t + x)/(2H_0) & &\text{for } -c_p t \leq x \leq 0.
\end{aligned} \tag{10.31}$$

Therefore, at any time $t > 0$, there are mirror-symmetrical, straight-line wave fronts with gradients $-F_0/(2H_0)$ and $+F_0/(2H_0)$ on the right-hand side and on the left-hand side of the point at which the force acts. Starting at time $t = 0$, the point $x = 0$ at which the force acts will be lifted with a speed of

$$\dot{z} = F_0 \cdot c_p/(2 \cdot H_0) = F_0/(2 m'_{CW} c_p) = F_0 \left/ \left(2\sqrt{m'_{CW} H_0} \right) \right. . \tag{10.32}$$

This lifting speed can be considered as a signal that is generated by the concentrated force F_0 and moves along the contact wire with the propagation speed c_p. At a point $|x|$ distant from the force, the lifting motion will start at a time $|x|/c_p$. At any time $t > 0$, a contact wire section of length $2c_p t$ will move in the direction of the y axis (vertically) at a speed of $F_0/(m'_{CW} c_p)$. This element will have a total inertia of $2F_0 m'_{CW} c_p t/(2 m'_{CW} c_p) = F_0 t$, i.e. it is exactly equal to the impulse of the force applied.

The considerations above apply to a constant force F_0. These considerations can be generalised for a force $F_0(t)$ that varies with time. This force is represented by a load related to the unit length $q'(x,t)$ according to:

$$q'(x,t) = F_0(t)\delta(x) \cdot u(t) \quad . \tag{10.33}$$

Instead of (10.30), the corresponding solution is

$$z(0,t) = c_p/(2H_0) \cdot \int_0^t F_0(\tau)d\tau \tag{10.34}$$

and instead of (10.31), it is obtained

$$z(x,t) = c_p/(2H_0) \cdot \int_0^{t-|x/c|} F_0(\tau)d\tau \qquad \text{for } |x| \leq c_p t \quad ,$$

$$z(x,t) = 0 \qquad \text{for } |x| \geq c_p t \tag{10.35}$$

and thus

$$\dot{z}(x,t) = c_p/(2H_0) \cdot F_0 \cdot (t - |x|/c_p) \cdot u(t - |x|/c_p) \quad . \tag{10.36}$$

The uplift at the point of force application, $x = 0$, of a contact wire that is stationary at the beginning, is proportional to the total impulse transmitted up to the time t. The *uplift speed* $\dot{z}(0,t)$ is proportional to the force currently acting on the contact wire. Some studies are based on the erroneous assumption that the uplift is proportional to the contact force. Especially where high speeds are concerned, this assumption leads to wrong conclusions (For more details see [10.5]).

It can also be stated that the bending angle of the contact wire is exactly the same as would be caused by a force F_0 acting on the centre of the contact wire anchored at both ends and subject to a constant tensile force under equilibrium conditions. The reactive force is the sum of the vertical components of the tensile forces acting at this point in both directions.

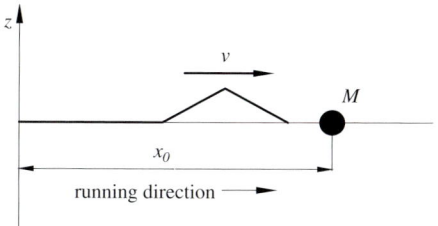

10.2.4 Reflection of transverse impulses travelling along a contact wire, at a concentrated mass

An impulse moving along a contact wire may be blocked, i. e. stopped, at a point x_0 where the motion is prevented or compensated by a force acting at this point. The motion to be compensated at the point x_0 can be termed $z_0(t)$ and described by

$$z_0(t) = z(x_0, t) = f_1(x_0 - c_p t) + f_2(x_0 + c_p t) \quad , \tag{10.37}$$

where the wave arriving from the left of this point is $f_1(x_0 - c_p t)$ and that arriving from the right is $f_2(x_0 + c_p t)$ [10.5]. According to Equation (10.29), a *concentrated reactive force*

$$F_r(t) = -2m'_{CW} c_p \dot{z}_0 = -(2H_0/c_p) \cdot \dot{z}_0 \tag{10.38}$$

would be needed at this point. A corresponding reactive force would be exerted because of the elastic reaction at any point where the contact wire is fixed. As a result of this reactive force, reflected waves moving in the opposite direction to the original waves are generated. In mathematical terms, the reflection of waves in the contact wire is treated by applying the boundary condition $z(x_0, t) \equiv 0$ to Equation (10.23), which leads to *d'Alembert's principle* of the reflection of waves by a fixed point. The method described above leads to the identical solution with the additional advantage that it can be applied relatively easily and more generally to any type of reflection, e. g. at points of concentrated masses, springs or droppers. The example given below demonstrates this by using a concentrated mass at which the wave is reflected.

Assume that a *concentrated mass M* is fixed rigidly to the contact wire at a point $x = x_0$ (Figure 10.5). As a result of the wave coming from the left, $z_0(t) = f(x_0 - c_p t)$, a reactive force $F_r(t)$, the magnitude of which is not yet known, will act on the contact wire and, in the opposite direction, on the mass. According to Equation (10.29) the contact wire at point $x = x_0$ will assume the speed

$$\dot{z}_r(t) = F_r(t)/(2m'_{CW} c_p) \tag{10.39}$$

because of the action of this force. This motion will be superimposed on the motion of this point because of the incoming wave. The total speed of this point will, therefore, be equal to

$$\dot{z}(t) = \dot{z}_0(t) + \dot{z}_r(t) = \dot{z}_0(t) + F_r(t)/(2m'_{CW} c_p) \quad . \tag{10.40}$$

The contact wire movement $z(t)$ at the point $x = x_0$ is of course identical to that of the point mass M. The movement of the mass M is described by the differential equation

$$M\ddot{z} = -F_r(t) \quad . \tag{10.41}$$

By eliminating the reaction force $F_r(t)$ from the differential Equations (10.40) and (10.41), an equation is obtained which describes the motion of the point x_0:

$$M\ddot{z} + 2m'_{CW}c_p\dot{z} = 2m'_{CW}c_p\dot{z}_0 \quad . \tag{10.42}$$

Equation (10.42) can be integrated immediately since the incoming wave $z_0(t)$ is known. From the overall motion $z(t)$ of the concentrated mass M, Equation (10.40) can be used to calculate the reactive force

$$F_r(t) = 2m'_{CW}c_p\left(\dot{z}(t) - \dot{z}_0(t)\right) \quad .$$

Therefore, the additional speed component is

$$\dot{z}_r(t) = F_r(t)/(2m'_{CW}c_p) = \dot{z} - \dot{z}_0$$

and the additional motion component

$$z_r(t) = z(t) - z_0(t) \quad .$$

This additional motion $z_r(t)$ is imparted to the left-hand section (i. e. for $x \leq x_0$) of the contact wire in the form of a *reflected wave* $z_r[t + (x - x_0)/c_p]$.
The wave transmitted to the contact wire section to the right of the mass, i. e. $x \geq x_0$, is:

$$z_t\left[t - (x - x_0)/c_p\right] = z_0\left[t - (x - x_0)/c_p\right] + z_r\left[t - (x - x_0)/c_p\right] \quad . \tag{10.43}$$

For an incoming sine wave $z_0(t) = \hat{z}_0 e^{j\omega t}$, the solution of (10.42) is

$$z(t) = \hat{z}_0(1 - \vartheta j\omega)/(1 + \vartheta^2\omega^2)\cdot e^{j\omega t} \quad , \tag{10.44}$$

where

$$\vartheta = M/(2m'_{CW}c_p) \quad . \tag{10.45}$$

The additional motion $z_r(t)$ is thus calculated as

$$z_r(t) = z(t) - z_0(t) = -\hat{z}_0\vartheta\omega(j + \vartheta\omega)/(1 + \vartheta^2\omega^2)\cdot e^{j\omega t} = \hat{z}_r e^{j\omega t} \quad , \tag{10.46}$$

and from this equation, the reaction force

$$\begin{aligned}
F_r(t) &= 2m'_{CW}c_p\dot{z}_r(t) = -\hat{z}_02j m'_{CW}c_p\vartheta\omega^2(j + \vartheta\omega)/\left(1 + \vartheta^2\omega^2\right)\cdot e^{j\omega t} \\
&= \hat{z}_0M\omega^2(1 - j\vartheta\omega)/\left(1 + \vartheta^2\omega^2\right)\cdot e^{j\omega t} \tag{10.47}
\end{aligned}$$

can be calculated. The *reflection coefficient* is defined as

$$r = \hat{z}_r/\hat{z}_0 = -\vartheta\omega(j + \vartheta\omega)/(1 + \vartheta^2\omega^2) \quad . \tag{10.48}$$

From (10.47) it can be deduced that the amplitude of the reaction force $F_r(t)$ will be $\hat{z}_0M\omega^2$ at low frequencies. At higher frequencies, this amplitude will be $\hat{z}_0M\omega/\vartheta = 2\hat{z}_0m'_{CW}c_p\omega$, and, therefore, proportional to ω. In this situation, the reflection coefficient will be -1, i. e. for short wave lengths the concentrated mass has the same effect as fixing the contact wire at that point.

10.2.5   Reflection of transverse impulses at a dropper

In an overhead contact line, the contact wire and the catenary wire are connected to each other by droppers. The contact wire is subject to a tensile force H_{CW} and the catenary wire is subject to a tensile force H_{CA}. The masses per unit length are m'_{CW} and m'_{CA} respectively. The mass of the dropper is M.

The wave on the contact wire, $z_0(t - x/c_{CW})$, reaches this dropper, which is at position $x = 0$, from the left-hand side (Figure 10.6 a) and tends to impart a motion $z_0(t)$ to the dropper. The dropper reacts to this wave by carrying out a motion $z(t)$, the catenary wire which is considered to be stationary, exerts a reaction force $F_{CA} = -2m'_{CA}c_{CA}\dot{z}$ and the contact wire exerts the reaction force $F_{CW} = -2m'_{CW}c_{CW}(\dot{z} - \dot{z}_0)$. In addition, an inertia reaction force $-M\ddot{z}$ will occur. The equation of motion of the dropper is, therefore, [10.5]

$$2\left(m'_{CA}c_{CA} + m'_{CW}c_{CW}\right)\dot{z} + M\ddot{z} = 2m'_{CW}c_{CW}\dot{z}_0 \quad , \tag{10.49}$$

where $c_{CA} = \sqrt{H_{CA}/m'_{CA}}$ and $c_{CW} = \sqrt{H_{CW}/m'_{CW}}$ are the *wave propagation speeds* along the catenary wire and the contact wire, respectively. Equation (10.49) is of the same type as (10.42). For an incoming sine wave $z_0(t) = \hat{z}_0 \cdot e^{j\omega t}$, the solution

$$z(t) = 2m'_{CW}c_{CW} \cdot \hat{z}_0 \cdot e^{j\omega t} / \left[2\left(m'_{CW}c_{CW} + m'_{CA}c_{CA}\right) + Mj\omega\right] \quad . \tag{10.50}$$

is obtained. With this, the *reflected wave* along the contact wire can be described as follows

$$\begin{aligned}
z_r(t) &= z(t) - z_0(t) \\
&= -\left(2m'_{CA}c_{CA} + Mj\omega\right) \cdot \hat{y}_0 \cdot e^{j\omega t} / \left[2\left(m'_{CW}c_{CW} + m'_{CA}c_{CA}\right) + Mj\omega\right]. \tag{10.51}
\end{aligned}$$

Since the mass of the dropper and that of the clips at both ends is low, it is possible to ignore $Mj\omega$ in (10.50) for frequencies which are not too high. With this assumption, the *reflection coefficient r* for the reflection of contact wire waves by a mass-free dropper is:

$$\begin{aligned}
-(z_r/z_0) &= r = m'_{CA}c_{CA}/(m'_{CA}c_{CA} + m'_{CW}c_{CW}) \\
&= \sqrt{H_{CA}m'_{CA}} \Big/ \left(\sqrt{H_{CA}m'_{CA}} + \sqrt{H_{CW}m'_{CW}}\right) \quad . \tag{10.52}
\end{aligned}$$

In (10.52) the sign which expresses the phase reversal was eliminated.

Usually, a dropper is made of a thin, highly flexible stranded wire that is subjected to a load equal to half the weight of the adjacent contact wire segments. If the dropper is lifted by a wave moving along a contact wire, this dropper tensile force is reduced by $m'_{CA}c_{CA}\dot{z} - m'_{CW}c_{CW}(\dot{z} - \dot{z}_0)$. The dropper will become slack when the resulting tensile force is negative. If the distance between adjacent droppers is l_d, the initial tensile force on the dropper is $m'_{CW}g \cdot l_d$ and the dropper will become slack if

$$\left(m'_{CA}c_{CA} - m'_{CW}c_{CW}\right)\dot{z} + m'_{CW}c_{CW}\dot{z}_0 \geq m'_{CW}g \cdot l_d \quad . \tag{10.53}$$

In conjunction with (10.49), the assumption $M = 0$ and (10.52), the condition for the *slackening of the dropper* is found to be

$$2m'_{CW}c_{CW}\dot{z}_0 \geq m'_{CW}g \cdot l_d/r \quad . \tag{10.54}$$

The lower the reflection factor, the lower will be the tendency of the droppers to go slack due to a contact wire wave.

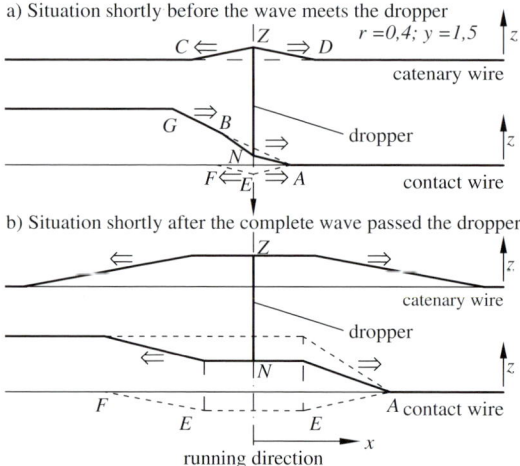

a) Situation shortly before the wave meets the dropper

b) Situation shortly after the complete wave passed the dropper

running direction

Figure 10.6: Reflection of a wave front at a dropper.
GBA: primary wave, NA: transmitted wave, EF and EA: secondary waves, NZ: dropper

The *reflection* factor or *coefficient* is a characteristic quality parameter of an overhead contact line. For a standard overhead contact installation of type Sicat S1.0 [10.6] with an AC-100–CuAg 0,1 contact wire and type Bz 50 catenary wire, subject to tensile forces of 12 kN and 10 kN, respectively, the reflection factor is found to be $r = 0,39$; for the standard design Sicat H1.0 with an AC-120–CuMg0,5 contact wire at a tensile force of 27 kN and a catenary wire Bz 120 tensioned to 21 kN, the value is $r = 0,47$. The reflection factor is lower when the catenary wire's mass and tensile force are lower in relation to the contact wire's mass and tensile force.

In Figure 10.6 a schematic representation is shown of how a wave is reflected by a dropper NZ, in this case by a straight-line wave front GA generated in the contact wire by a rectangular impulse $F_0 \cdot \Delta t$. This figure also demonstrates the transmission of this wave in the contact wire and the catenary wire. The reflection factor is 0,4. The uplift $z(t)$ of the dropper NZ generates the wave front NA in the contact wire section to the right of the dropper. In the catenary wire, it generates the wave front ZD, travelling to the right, and the wave front ZC, which is symmetrical to ZD, but moving to the left. The reflected wave EF in the contact wire will be superimposed on the incoming wave front GA, leading to steeper contact wire slope section BN. The wave front NA can be considered as a superposition of the primary wave front BA and the wave front EA, which is symmetrical to the wave front EF.

10.2.6 Doppler factor

Reflections of transverse waves by stationary passive masses or by other non-homogeneous components of contact lines do not lead to an increase in amplitudes. However, increased amplitudes can occur at a pantograph moving towards the transverse waves [10.7, 10.8]. The pantograph moves along the contact wire at a speed v. Provided that the contact force acting between the collector and the contact wire is increased by ΔF_0 due to some disturbance, e. g. due to an impulse on the wire, the effect of this contact force increase is superimposed linearly on the other contact wire and pantograph motion components. According to Equation (10.32), the speed of the resulting vertical motion is

$$\dot{z}_0 = \Delta F_0 / 2\, m'_{CW} \cdot c_{CW} \quad . \tag{10.55}$$

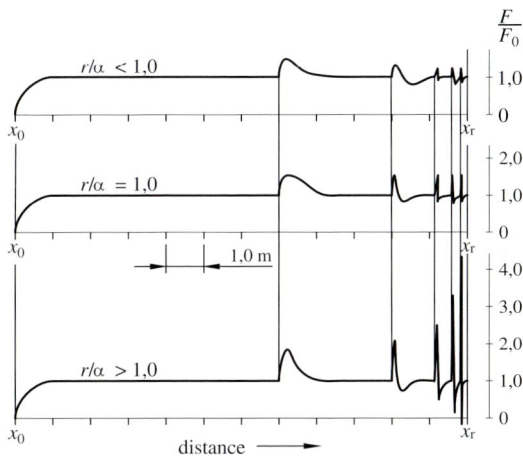

Figure 10.7: Contact force F of a mass of 1 kg that is pressed against the contact wire with a force of F_0 and moves towards a dropper at a speed $v = 44,4$ m/s. The wave propagation speed is 106 m/s.

This generates a wave front moving in the direction of the pantograph travel. Due to the *Doppler effect* of a moving source, the gradient of this wave front is

$$z_0' = \dot{z}_0 / (c_{CW} - v) \quad . \tag{10.56}$$

This wave front is reflected with a factor $r < 1$ by the next dropper and then moves towards the pantograph with a gradient of

$$z_r' = r \cdot z_0' = r \cdot \dot{z}_0 / (c_{CW} - v) \tag{10.57}$$

and forces the collector head to move vertically at a speed of

$$\dot{z}_1 = z_r' \cdot (c_{CW} + v) = \dot{z}_0 \cdot r \cdot (c_{CW} + v) / (c_{CW} - v) \quad , \tag{10.58}$$

since the wave meets a moving receiver object. The factor $(c_{CW} + v)$ describes the motion of the *receiver object*. According to Equation (10.32), the inertia of the collector head causes a sudden steep increase of the contact force by

$$\Delta F_1 = 2 m_{CW}' \cdot c_{CW} \cdot \dot{z}_1 = \Delta F_0 \cdot r / \alpha_D \quad , \tag{10.59}$$

where

$$\alpha_D = (c_{CW} - v) / (c_{CW} + v) \tag{10.60}$$

is the *Doppler factor* for the interaction of the overhead contact line and pantograph. The collector head with a mass M_S is subsequently subjected to an impulse of

$$M_S \dot{z}_1 = M_S \Delta F_1 / (2 m_{CW} c_{CW}) = M_S \cdot \Delta F_0 \cdot (r / \alpha_D) / (2 m_{CW} c_{CW}) \quad , \tag{10.61}$$

from which it can be deduced

$$\Delta F_1 = (r / \alpha_D) \cdot \Delta F_0 \quad . \tag{10.62}$$

If $r / \alpha_D > 1$, the contact force step ΔF_1 is greater than the initial ΔF_0. Figure 10.7 shows the result of such an effect for the simple case of a tensioned contact wire stationary at time 0

and having a discontinuity at the point x_r. For mechanical waves, the *reflection factor* of this discontinuity is r. A collector head of mass M_S is assumed to be moving towards the discontinuity at a speed v, keeping in contact with the contact wire, but not exerting any force upon it. A force F_0 suddenly occurs after point x_0 was passed. From x_0 onwards, it will lift the contact wire. The upward motion of the contact wire precedes the pantograph at the wave propagation speed c_{CW} and is reflected by that discontinuity. The reflected wave front travels back towards the pantograph and is reflected by this, whereby additional energy is imparted to the wire due to the pantograph's motion. This procedure is repeated continually until the pantograph has reached point x_r. In Figure 10.7, it can be seen that each consecutive force increase will be greater than the initial one if $r/\alpha_D > 1$. The amplitudes of the system will increase until point x_r is reached. If, $r/\alpha_D < 1$, the contact force variations will decay. The ratio r/α_D is called the *amplification coefficient*

$$\gamma_A = r/\alpha_D \quad . \tag{10.63}$$

As the Doppler factor α_D is a function of the train speed v, the condition $\gamma_A = 1$ defines the *limiting speed* v_α, below which the consecutive force amplitudes are not amplified:

$$v_\alpha = c_{CW}(1-r)/(1+r) \quad . \tag{10.64}$$

The limiting speed is always lower than the wave propagation speed c_{CW} along the contact wire. The *reflection coefficient* r is deduced from Equation (10.52) to be:

$$r = 1 \left/ \left(1 + \sqrt{(H_{CW}m'_{CW})/(H_{CA}m'_{CA})}\right)\right. \quad . \tag{10.65}$$

Example 10.1:   If the characteristic values of the contact line system Re 250 [10.9] are inserted in Equation (10.64), the result obtained with $H_{CW} = H_{CA} = 15\,\text{kN}$, $m'_{CW} = 1,08\,\text{kg/m}$, $m'_{CA} = 0,59\,\text{kg/m}$, $c_{CW} = 422\,\text{km/h}$ and $r = 0,425$ is $v_\alpha = 170\,\text{km/h}$. This value is far lower than the design speed $250\,\text{km/h}$. For $v = 250\,\text{km/h}$, it is obtained $\alpha_D = 0,26$ and an amplification coefficient $\gamma_A = 1,63$.

10.2.7   Natural frequencies of an overhead contact line

An overhead contact line is a mechanical system which can oscillate with a large number of degrees of freedom and has numerous *natural frequencies*. Figures 10.19 and 10.20 show frequency *spectra of overhead contact line designs*. According to [10.5], an overhead contact line suspended between equally spaced poles will exhibit *symmetrical* and *non-symmetrical oscillation* modes. In the first case, two points spaced symmetrically relative to a reference point will oscillate in phase. In the latter case, they will oscillate at opposite phases. In the symmetrical mode, there will be an oscillation peak or antinode at the axis of symmetry, i. e. the reference point, and in the non-symmetrical mode, there will be an oscillation node. In an overhead contact line comprising an even number of pole intervals, the axis of symmetry is at the support.

With *non-symmetrical oscillations*, the wave length of the basic natural frequency is equal to twice the support spacing. If the oscillation is assumed to be a stationary wave, the frequency can be calculated as

$$\nu_1 = \bar{c}(2l) = \sqrt{(H_{CW} + H_{CA})/(m'_{CW} + m'_{CA})} \left/ (2l) \right. , \tag{10.66}$$

Table 10.1: Dynamic characteristics of standard overhead contact line installations [10.8] to [10.11].

Contact line design	Units	Sicat S1.0	Re 250	Sicat H1.0
contact wire		AC-100 – Cu	AC-120 – CuAg	AC-120 – CuMg
– tensile force	kN	10	15	27
catenary wire		Bz 50	Bz 70	Bz 120
– tensile force	kN	10	15	21
wave propagation speed	km/h	382	427	572
non-uniformity	%	20	10	8
reflection coefficient	–	0,413	0,425	0,465
doppler factor	–	0,41	0,26	0,27
at	km/h	200	250	330
amplification factor	–	1,01	1,63	1,72
natural frequencies	Hz	0,74/0,76	0,96/1,02	1,06/1,15

where $\bar{c}$ is the *mean wave propagation speed* along the overhead contact line. With symmetrical oscillations, the section up to the first dropper is also taken into account. This means that the frequency is given by the equation:

$$v_2 = \bar{c}/(2l + l_1) = \sqrt{(H_{CW} + H_{CA}) / (m'_{CW} + m'_{CA})} / (2l + l_1) \quad , \tag{10.67}$$

in which l_1 is the distance between the two droppers nearest to the support. In this simplified model, the frequency of the first harmonic is double the natural frequency. For all other frequencies, it will be necessary to take into account the respective oscillation modes (see [10.10]).

For an overhead contact line Re 250 with $l = 65$ m and $l_1 = 10$ m, the natural frequencies are found to be $v_1 = 1,02$ Hz and $v_2 = 0,96$ Hz, as can also be seen in Figure 10.19.

10.2.8 Dynamic characteristics of typical overhead contact lines

Table 10.1 contains the dynamic characteristics of the standard overhead contact line designs Sicat S1.0, Re 250 and Sicat H1.0. The *wave propagation speeds* are between 382 km/h and 572 km/h. They form the main component determining the Doppler factor, which is 0,41 for Sicat S1,0 at 200 km/h and 0,26 for Re 250 at 250 km/h. The reflection coefficients of all three designs are almost equal. It is not possible to choose the catenary wire specifications solely with the objective of minimising the reflection coefficient, since the *current-carrying capacity* and *elasticity* are of equal importance.

The *Doppler factor* and the *amplification coefficient* are functions of the train speed as shown in Figure 10.8. The amplification coefficient tends asymptotically towards infinity as the running speed approaches the *wave propagation speed*. Because of this, it is not possible to operate trains at speeds near the wave propagation speed of the catenary wire. Experience in practical applications has shown that it is possible to operate overhead contact lines at amplification coefficients of up to 2,5. In the course of test runs for experimental purposes, it was also observed that energy transmission is still possible at amplification coefficients up to 5,0 (see Clause 10.5.3).

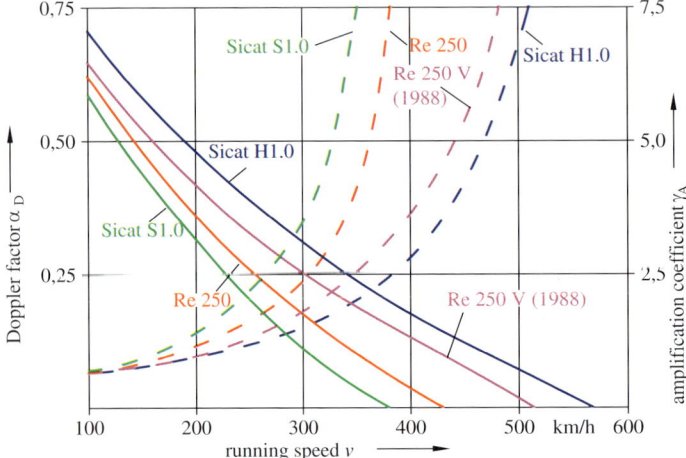

Figure 10.8: Doppler factor α_D and amplification coefficient γ_A.
— — amplification coefficient γ_A, ——— Doppler factor α_D

10.3 Simulation of pantograph interaction with overhead contact lines

10.3.1 Objectives

Use of empirical methods in the development of new overhead contact line designs, which were common in the early period of electric traction systems, is no longer feasible for systems intended to supply high-speed vehicles. The practical experience required for further empirical developments is not available for the effects occurring at such speeds. At the same time, the interaction of contact line systems and pantographs becomes more pronounced with increasing speeds, so that a useful design strategy can only be achieved using various models of the overall system comprising the two components, overhead contact line and pantograph assembly. *Computer simulation* is a powerful modelling tool for the interaction of these components. Mathematical simulation models can be used to illustrate the effects of parameter variations and to evaluate the interaction of different contact line and pantograph designs. Simulation of the pantograph/overhead contact line interaction is also important to assess the interaction characteristics gained from test runs on interoperable lines. This assessment is also required by the Energy TSI [10.12] for the high-speed system and by [10.13] for conventional lines. In 2014 the new TSI Energy [10.14] was published and replaces former TSI HS [10.12] for high-speed and the TSI CR [10.13] for conventional lines. The new edition of the TSI RST [10.15] for rolling stock was also published and contains stipulations for pantographs.

The objective of dynamic simulations is to determine the time-related behaviour of the moving contact *force exerted* by the collector strips *on the contact wire* and of the associated *lifting of the contact wire*. In this process, it should be possible to analyse the interaction of multiple contact points simultaneously, e. g. when studying the use of individually sprung collector strips or train consists with several pantographs active simultaneously. To enable validation of the models used, it should also be possible to calculate other characteristics that are easier to measure than the contact force, e. g. the motion of the contact line system [10.16]. Requirements for validation are given in EN 50 318.

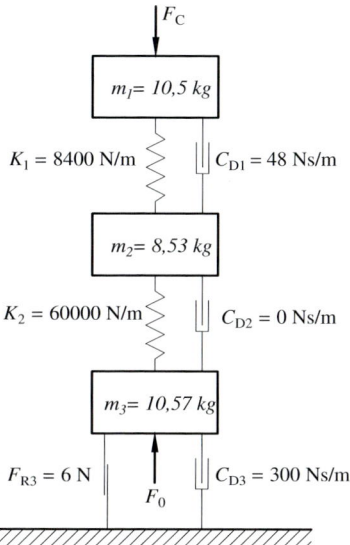

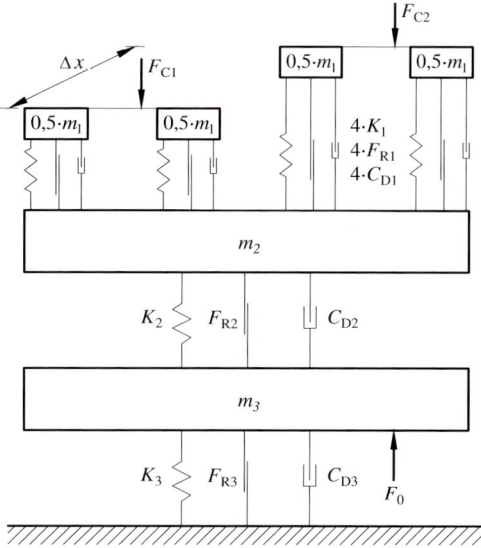

Figure 10.9: Three-mass-model of a pantograph type SBS 81 [10.16] (Symbols as in Figure 10.10).

Figure 10.10: Six-mass-model of pantographs with individually sprung collector strips.
m_1 to m_3 partial masses; K_1 to K_3 modulus of elasticity of the springs; C_{D1} to C_{D3} damping units, F_{R1} to F_{R3} friction force; F_0 static force of pantograph drive; F_C contact force

To adequately take consider all relevant characteristics, a model should be able to simulate the following characteristics of the overhead contact line:

- all types of contact wires, catenary wires, stitch wires and droppers, including their material characteristics and installation conditions
- different overhead contact line designs, e. g. *stitched contact lines*, contact lines with *auxiliary catenary wires* or with varying dropper spacing
- the dynamic characteristics of all supports, i. e. of the steady arms, cantilever supports
- discontinuities such as section insulators, overlap sections, variation of contact line height and contact line installation above points and
- complete tensioning sections

The pantograph and collector strip model should also consider the following essential parameters:

- different types of *pantograph mechanisms* and their respective characteristics, e. g. single-arm pantographs, twin-arm pantographs and
- different types of contact elements, e. g. pan heads, individual collector strips

The parameters of subsystems should be easy to change permitting their optimization. The accuracy of the pantograph and overhead contact line models should prevent misleading calculation results.

The pantograph and the contact line are two independent systems capable of oscillating and are coupled to each other at the point of contact. Pantographs having multiple collector strips will have multiple points of contact at short distances apart. The simulation is used to establish the coupling between the partial models via the contact force and the position of the contact point.

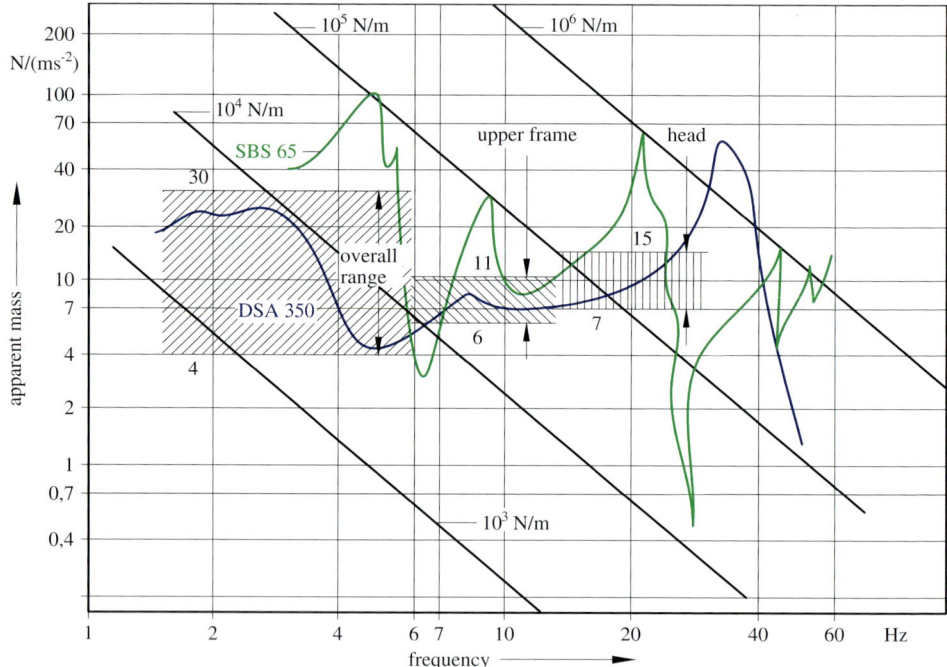

Figure 10.11: Dynamic modal mass of pantograph designs SBS 65 and DSA 350 used by German Railway [10.1].

10.3.2   Model of the pantograph system

The point on the collector strip contacting the overhead contact line couples the two systems and suitable simulation of the behaviour and interaction is required at this point. The contact force and the vertical motion of the contact point also need to be calculated. If a plane model of the system is used, the force is considered to always act on the same point of the collector strip. The spatial effect of lateral contact wire shift can be considered by assuming a linear lateral shift of the point at which the force is applied to the collector strip.

A simple model represents the pantograph by *substitute masses* coupled to one another through springs and dampers. The oscillation behaviour of such systems is described by a system of second-order differential equations.

The number of equations is determined by the number of substitute masses, i. e. by the number of *degrees of freedom* of the system - models with three substitute masses are often used. The masses represent the lower frame, the upper frame and the collector pan head. Figure 10.9 shows the data of a pantograph type SBS 81 as represented in a three-mass model [10.16].

The relatively small number of substitute masses means that only selected *pantograph os-cillation modes* will be taken into consideration. For instance, the flexural oscillations of the upper frame members are not covered by this type of model, nor does a model as the one shown in Figure 10.9 take into account individually sprung collector strips.

A six-mass-model of the type shown in Figure 10.10 is used to study pantographs with *in-dividually sprung collector strips*. Here, the masses of the collector strips are modelled as separate units arranged on the respective supports. The excitation force along the collector

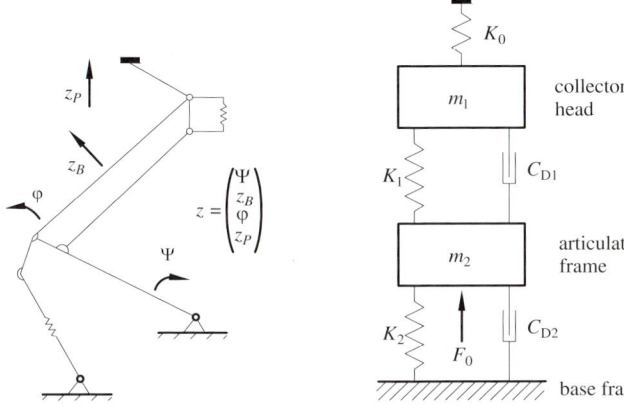

Figure 10.12: Analytical pantograph model.

Figure 10.13: Pantograph model
according to EN 50 318.

strips is subdivided linearly among the two respective part-masses corresponding to the position of the contact point.

It is also possible to model pantographs using measured, frequency-dependent, dynamic modal masses and dynamic elasticities. Figure 10.11 from [10.1] represents an example. The modal mass expresses the resistance with which an elastic system reacts to an acceleration. In this case, the excitation and the response of the pantograph are taken into consideration in the calculation as superimposed individual responses at the observed frequencies. In this model, the use of frequency-dependent calculation algorithms is an advantage but with other models, excitation patterns can be determined by carrying out harmonic analysis. In addition to the above measurements, the phase responses of the *dynamic apparent masses* are also determined to take into account the inertia of transmission at the individual frequencies.

In reference [10.17], an analytical model was developed for single-arm pantographs with pan-mounted collector strips. This pantograph model, which has four degrees of freedom (Figure 10.12), considers the vertical motion of the pan springs and the angular motion of the middle and lower pivots and the bending of the upper frame section. The parameters inserted in the mathematical model are derived from the geometry and material data of the pantograph components. Unfortunately, none of the analytical models are applicable universally, as every minor change in the design, e. g. introduction of individually sprung collector strips, will require modified calculation algorithms.

Models of any desired accuracy can be obtained by applying *finite element modelling*. In reference [10.18], calculations have been presented where the pantograph of the ICE was modelled using finite element methods with 480 degrees of freedom. The calculation effort required for such solutions is high and there is little improvement in the precision of the model. For this reason, the authors of the paper [10.18] only used a simple three-mass model to optimize the contact line design.

In [10.19] a pantograph model is adopted consisting of three masses m_0 to m_2 arranged vertically (Figure 10.13) and connected by springs having only one degree of freedom. The spring elements are characterized by the spring rigidity K, the damping C and the friction F. The static force acts on the mass m_2, the aerodynamic force on the mass m_1 and the aerodynamic correction force on the mass m_0. The model complies with the requirements of EN 50 318.

Table 10.2: Data of pantograph model according to EN 50318.

	Effective dynamic mass m in kg	Stiffness K N/m	Damping C_D Ns/m
contact spring	–	$K_0 = 50\,000$[1]	–
collector head	$m_1 = 7,2$	$K_1 = 4\,200$	$C_{D1} = 10$
articulation frame	$m_2 = 15$	$K_2 = 50$	$C_{D2} = 90$

[1] The contact spring is not a part of the pantograph but necessary for a correct comparison result

In EN 50318, a pantograph reference model is specified for the validation of simulation calculations. The pantograph is defined as a discrete mass-spring-damper-model as shown in Figure 10.13. The data are given in Table 10.2. For this validation step, this simple one-dimensional pantograph model can be used with a multi-dimensional overhead contact line model. A constant force F_0 is applied to the mass m_2 so that the static force is equal to 120 N.

10.3.3 Contact line models

10.3.3.1 Basic considerations

Frequently, simple *contact line models* are used when analysing pantograph behaviour. A model used for optimising the high-speed pantograph for the ICE is described in reference [10.20]. In this model, the contact wire is treated as a taut string of zero mass stretched between the droppers. The contact wire masses are assumed to be concentrated at the dropper positions. The droppers are modelled as dampers at the contact wire suspension points and the steady arms as springs and dampers, however, the catenary wire is not taken into account in this model. The model does not permit the overhead contact line installation to be analysed per se because it does not consider the dynamic behaviour of the catenary and stitch wires. Consequently, the contact force functions deduced using such simplified models, do not describe any responses because of the catenary and stitch wires and are not useful for predicting behaviour at high speeds. In the following, some contact line system models are presented that cover all essential parameters.

10.3.3.2 Modelling with the aid of the finite-element method

When using the *finite element method* [10.18, 10.19] the overhead contact line is subdivided into individual elements linked by coupling mechanisms described in mathematical terms. The result is a system of differential equations that permit modelling of the overhead contact line installation with any desired level of accuracy.

However, since the elements selected need to be small enough to permit the study of the dynamic processes, systems of approximately 2 000 to 3 000 differential equations need to be solved for a complete tensioning section. The moving excitation point means that the system description matrices are time-dependent and need to be re-established for each point of contact.

Up to the year 2000, models of this kind were used successfully for calculating stationary processes, e. g. for *calculations of elasticity* [10.21]. They were also used for dynamic calculations e. g. as reported in references [10.22] and [10.23]. However, to reduce the calculation time in these cases, the contact wire between the droppers was not further broken down into

individual elements but substituted as a whole by rod or string elements. The application of a finite-element-model is described in [10.19].

10.3.3.3   Analytical solution in the frequency area

Analytical *solution in the frequency area* [10.17] is based on a contact line of infinite length, supported by jointed supports at finite spacing. The individual sections are allowed to oscillate independently. The associated *Lagrange's equations* are solved by using the Ritz approximation method. The accuracy of the resulting solution is determined by the order of the approximation approaches. An order of 90 was used in calculation examples for three spans in [10.17]. Computer calculations for a complete tensioning section take a long time. Fourier series of higher orders are used to model the coupling between the contact wire and the pantograph to calculate the contact forces. This further increases calculation complexity and the computer requirements. The model was used for optimizing pantograph parameters. Any modification of contact line parameters is difficult.

10.3.3.4   Method using frequency-dependent finite elements

The *frequency-dependent finite element* [10.16] method was introduced with the objective of reducing the order of the matrices needed in the analysis of the higher frequency processes, while retaining the universal applicability of the method. The taut string equation is solved analytically at the element level so that it is not necessary to subdivide the sections of the contact wire between dropper locations. For the entire overhead contact line installation, the *frequency-dependent matrices* derived account for additional elements, e. g. steady arms, clips, cantilevers etc. as individual masses or oscillating elements. This enables modelling of any type of contact line design.

To begin with, the *natural frequencies* and the corresponding *natural vectors* of the overhead contact line are calculated. The reaction of the contact line to excitation by the pantograph can be determined by superimposing the independent responses at individual natural frequencies. Using this method, the majority of the effort involves calculating the natural frequencies and vectors but this has to be done only once for a given overhead contact line configuration. An iterative approach is used to calculate the reactions to a force acting on the analytically modelled contact wire sections.

10.3.3.5   Modelling on the basis of d'Alembert's wave equations

The contact force deflects the contact wire and this deflection is propagated along the contact wire at the wave propagation velocity. It is reflected by discontinuities such as droppers and is also transmitted to other wires by these components (also refer to Clause 10.2.5). At the point of contact, the pantograph responds to impulses in a specific way. This model is based on *d'Alembert's wave equation* [10.5] and available as a computer model and can be used for simulation [10.3]. The motions of the elements of the contact line are obtained by superposition of the individual waves. The realistic representation of the dropper wires by string elements, which can only exert tensile forces, is a major advantage here. However, the significance of dropper wires going slack and the consequent effect on the contact force is assessed differently by various authors in their respective publications. The calculation effort for this model is high for complex contact lines and increases even further if varying dropper spacings must be considered. An application is described in [10.24].

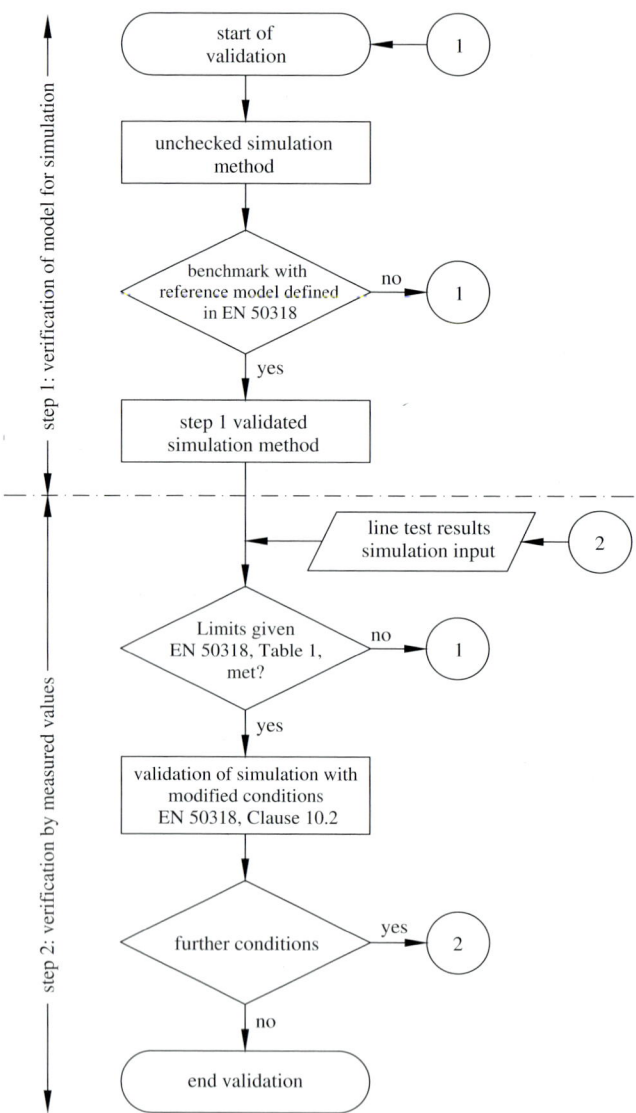

Figure 10.14: Flow chart of simulation validation according to EN 50 318.

10.3.4 Validation of simulation methods

10.3.4.1 Introduction

To ensure confidence, each simulation method should be validated. The standard EN 50 318 specifies functional requirements for the validation of simulation methods to ensure the acceptance of:

- input and output parameters
- simulation methods
- comparisons with measurements
- comparisons between simulation methods

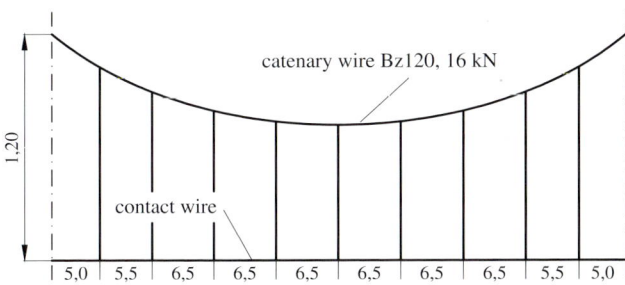

Figure 10.15: Overhead contact line model according to EN 50 318.

As shown in Figure 10.14, the validation is carried out in two steps:

- the first step consists of checking the simulation method by using a reference model. The data of the reference model are specified in the standard together with limits of the simulation results to be complied with
- the second step refers to the comparison of results obtained by simulation with measured data from one or several line tests

10.3.4.2   Requirements on simulation methods

The simulation models used should describe the pantograph and overhead contact line characteristics in the frequency range of interest. The pantograph can be described by a discrete mass-spring-damper-model, a multi-body-system, a finite-element-model or the transfer function based on the dynamic characteristics of the active pantographs. The models require information on kinematics, mass distribution, degree of freedom of joints, damping and spring characteristics, component stiffness, bump stops, application of static force, action of aerodynamic forces that may depend on orientation of the pantograph towards the running direction, operational height and position on and their type of trains.

The overhead contact line may be modelled with two or three dimensional geometry and should include the tensioning equipment as well as any discrete components such as section insulators and connectors. The model should at least consider a minimum length being at least three times the spacing between the first and last pantograph but not less than ten spans, the actual length of each span, the position of droppers, the contact wire position at rest, pre-sag, gradients, the system height at the supports, the geometry and mass distribution of steady arms, the stagger, the arrangement and characteristics of contact wire, catenary wire, auxiliary wires, stitch wires, droppers as well as the mass of clamps.

The simulation should consider the train speed, the number and spacing between active pantographs, the static force and aerodynamic force of each pantograph, the operational height and the frequency range of interest.

The simulation should calculate the contact forces and the wire and pantograph movements of each active pantograph when passing along the contact line model. The output parameters may be filtered to exclude frequencies outside the range of interest. The required outputs relating to contact forces, contact wire and pantograph displacements and/or loss of contact are:

- mean value F_{m} of contact force
- standard deviation s_{m} of contact force
- actual and statistical maximum and minimum of contact forces
- time history of the contact force
- variation of the contact force along the line for each pantograph

Table 10.3: Range of results acceptable from reference model.

	Unit	Range of results	
speed	km/h	250	300
F_m	N	110 – 120	110 – 120
s_m	N	26 – 31	32 – 40
statistical maximum of contact force	N	190 – 210	210 – 230
statistical minimum of contact force	N	20 – 40	−5 – 20
actual maximum of contact force	N	175 – 210	190 – 225
actual minimum of contact force	N	50 – 75	30 – 55
maximum uplift at support	mm	48 – 55	55 – 65
percentage of loss of contact	%	0	0

Note: The values in the table are based on results from five independent simulation models. These methods had been checked with results from line tests.

- histogram of contact forces (statistical distribution)
- variation of the contact wire uplift along the line for each pantograph
- maximum uplift of the contact wire at a support
- time history of the uplift at any specific point
- trajectory of the contact point along the line
- maximum and minimum uplift of contact points
- positions and lengths of loss of contact
- percentage of time of loss of contact

10.3.4.3 Validation by comparison with the reference model

In EN 50 318, reference models are defined for a pantograph and an overhead contact line which should be used to check and validate the simulation method. The pantograph model is a discrete mass-spring-model with two masses and three springs as shown in Figure 10.13. Table 10.2 contains the data for the pantograph model.

The overhead contact line model is defined as a contact line system with a single contact wire and identical spans without stitch wires as shown in Figure 10.15. The stagger is ±0,2 m and the steady arms are modelled by an 1,0 m long rod having a mass of 1,0 kg. The support of the catenary wire and the end of the steady arm are fixed points. The dropper stiffness is defined as 100 000 N/m under tension and zero under compression. The mass of the droppers and fittings are taken as zero.

The simulation should be carried out for speeds of 250 km/h and 300 km/h using one pantograph. The frequency range of interest is 0 to 20 Hz. If the outputs of simulating the reference model are within the ranges given in Table 10.3 for the individual parameters, the simulation method can be used for the second step of validation. Otherwise, the method should be rejected. An example of the application of the reference model is presented in Clause 10.3.6.

10.3.4.4 Validation with measured values

The second step of validation is to compare simulated data with values obtained from measurements on lines. It is assumed that the line tests are carried out using the equipment and procedures according to EN 50 317. The test results should be available as time histories. The data to be compared are the standard deviation of the contact force, the uplift at the support and the maximum and minimum uplift at the point of contact. The acceptable tol-

Table 10.4: Acceptable differences between simulated and measured data.

Parameter	Required accuracy (%)
standard deviation s_m of contact force	± 20
maximum uplift at support	± 20
range of vertical position of contact point	± 20

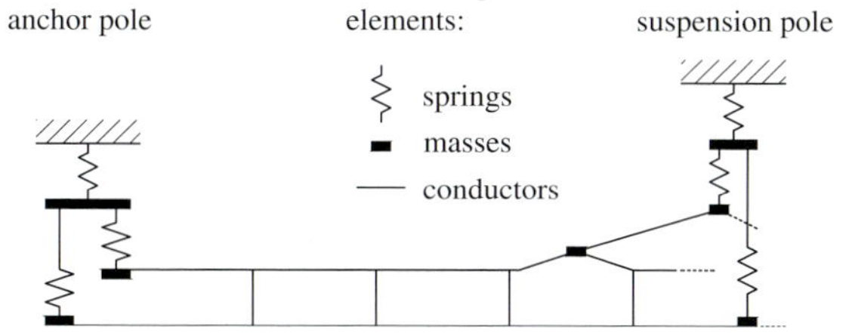

anchor pole   elements: suspension pole

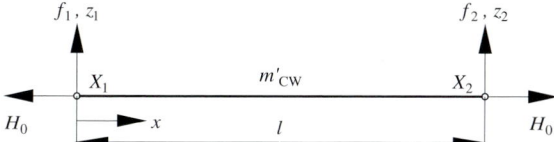

Figure 10.16: Equivalent model schematic describing the oscillation of a catenary contact line installation.

erances between measured and simulated values are defined in Table 10.4. The comparison should be made on a sufficiently long line section of the contact line. The comparison can only be considered as acceptable if the configuration of the pantographs and their number and arrangement on the train are the same. No differences to the main overhead contact line characteristics such as contact wires, catenary wires, auxiliary wires and stitch wires can be tolerated.

Changes in the simulation speed can be accepted up to the validation speed increased by 5 % of the wave propagation speed of the contact wire. Also, minor differences in the pantograph spacing can be accepted, as well as in the static and aerodynamic forces of the pantograph, in damping characteristics, in the height of contact wire. Differences in the frequency range and the number of active pantographs cannot be accepted.

The simulation model is validated if the calculated data do not deviate from the measured values by more than the tolerances specified in Table 10.4. An example is described in Clause 10.4.3.

10.3.5   Interaction simulation with frequency-dependent finite elements

The simulation of the interaction of pantographs and overhead contact lines, using *frequency-dependent finite elements* is detailed in [10.16] and [10.25]. With regards to oscillation characteristics, a vertical contact line is modelled by a plane system comprising individual masses and springs as shown in Figure 10.16.

The conductor elements connect the masses at nodes and are described by their mass per unit length m' and their tensile force H_0 (Figure 10.17). Their stiffness is considered to be negligible [10.16] and all other elements can be modelled by spring elements and masses as oscillating finite elements. The excitation can be applied at any point along the contact wire,

Figure 10.17: Differential element of conductors with degrees of freedom z_1 and z_2 at the boundaries.

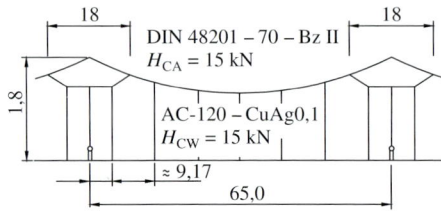

Figure 10.18: Overhead contact line of standard type Re 250, dimensions in metres.

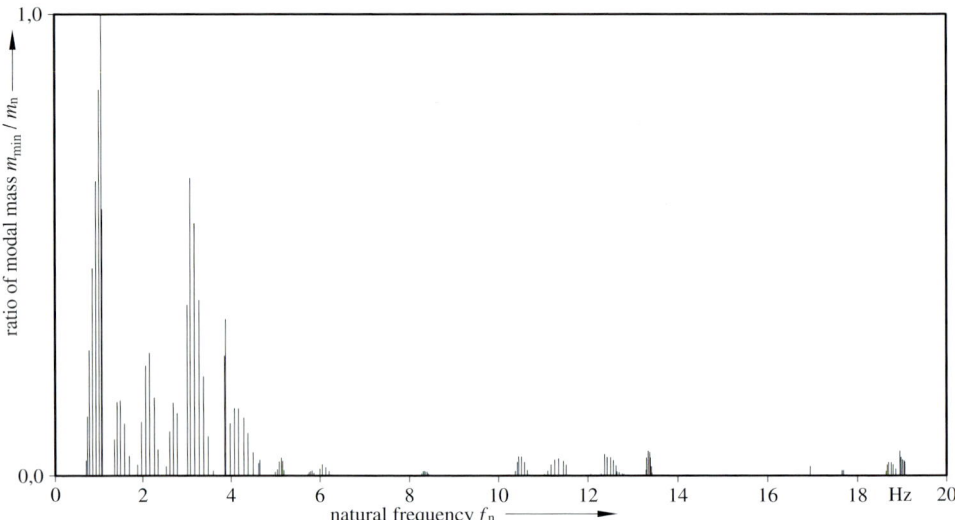

Figure 10.19: Natural frequencies f_n and ratio of modal masses m_{min}/m_n for the Re 250 overhead contact line type, m_n modal mass at frequency f_n.

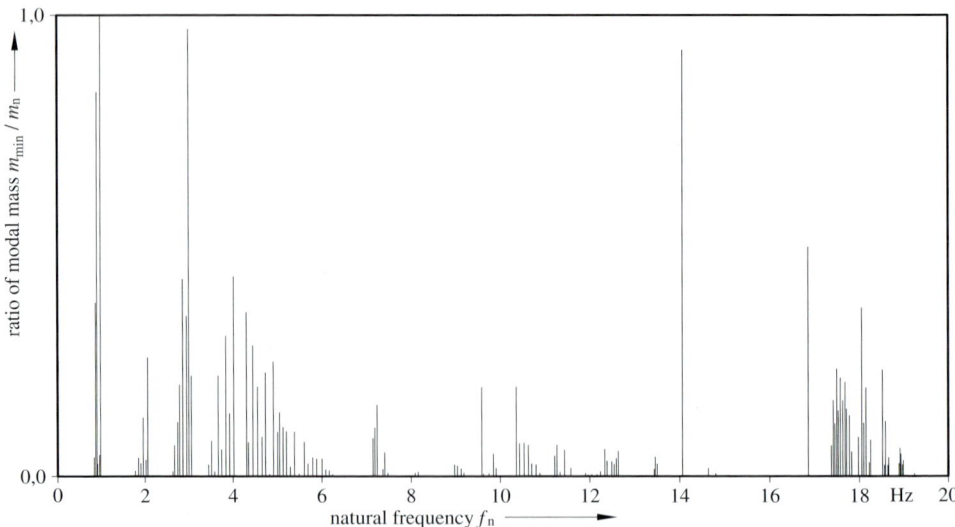

Figure 10.20: Natural frequencies f_n and ratio of modal masses m_{min}/m_n for the Re 250 overhead contact line type, but without stitch wires, m_n modal mass at frequency f_n.

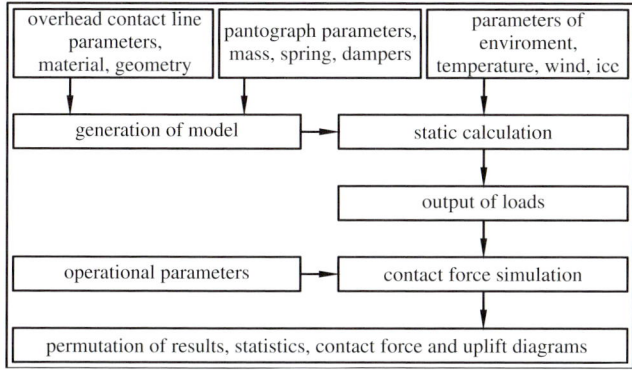

Figure 10.21: Flow chart of simulation with the finite element method.

either along a wire element or at a dropper position. The motion behaviour of the wires is described in terms of the frequency [10.25].

The method of frequency-dependent finite elements can be used to calculate the *natural frequencies* and associated modal masses of contact lines. This approach was applied to the contact line Re 250 shown in Figure 10.18. In Figure 10.19 the modal masses m_n calculated for the natural frequencies shown relate to the minimum value m_{min} as experienced at approximately 1 Hz. The first 200 frequencies that occur below 20 Hz were considered. The modal mass expresses the resistance with which an oscillating system reacts to an acceleration from outside at a given frequency. A low ratio of m_{min} and m_n expresses a low impact of the frequency f_n on the motion of the contact line. Figure 10.19 demonstrates that this overhead contact line type will only be slightly affected by natural frequencies above 6 Hz.

In Figure 10.20 the same information is given for a contact line design as shown in Figure 10.18 but without stitch wires. Comparison with Figure 10.19 reveals that frequencies up to 20 Hz affect the dynamic behaviour. The apparently slight modification, results in a considerable impact at higher frequencies. The difference in contact forces is presented in Figures 10.56 and 10.57. Simulation with frequency-dependent finite elements is irrelevant in 2017.

10.3.6   Simulation with commercially available finite element programs

The paper [10.19] reports on the use of standard finite element programs for the simulation of the overhead contact line/pantograph interaction at Siemens. The program needs to include elements specifically required for the simulation model:

- – a non-linear contact element to model the contact between contact wire and pantograph
- – a wire element permitting tensile forces only and buckling under compression.

The model can be adjusted easily to special contact line designs and project requirements and does not need to be very sophisticated when used to study the interaction and the overhead contact line. The pantograph model described in [10.19] consists of three masses m_1 to m_3 (Figure 10.9) representing the collector strips, the pantograph head and the base frame. The masses are connected by spring, damping and friction elements. Since this model is often used for simulation purposes, parameter sets are available for many pantograph designs. The model is compliant with EN 50 318.

Each finite element model consists of nodes connected by differing elements that also enable the simulation of overhead contact lines. The model used in [10.19] consists of elastic elements having a mass as well. It describes all elements and sets of elements of an overhead

Table 10.5: Statistical results of simulation of EN 50 318 reference model.

		250 km/h		300 km/h	
		Permissible range	Simulation	Permissible range	Simulation
mean value of contact force	N	110 – 120	115,4	110 – 120	114,9
standard deviation of contact force	N	26 – 31	26,4	32 – 40	32,6
statistical maximum of contact force	N	190 – 210	194,4	210 – 230	212,7
statistical minimum of contact force	N	20 – 40	36	−5 – 20	17,1
recorded maximum of contact force	N	175 – 210	197,2	190 – 225	196,5
recorded minimum of contact force	N	50 – 75	62,8	30 – 55	33,8
maximum uplift at support	mm	48 – 55	49,0	55 – 65	55,3

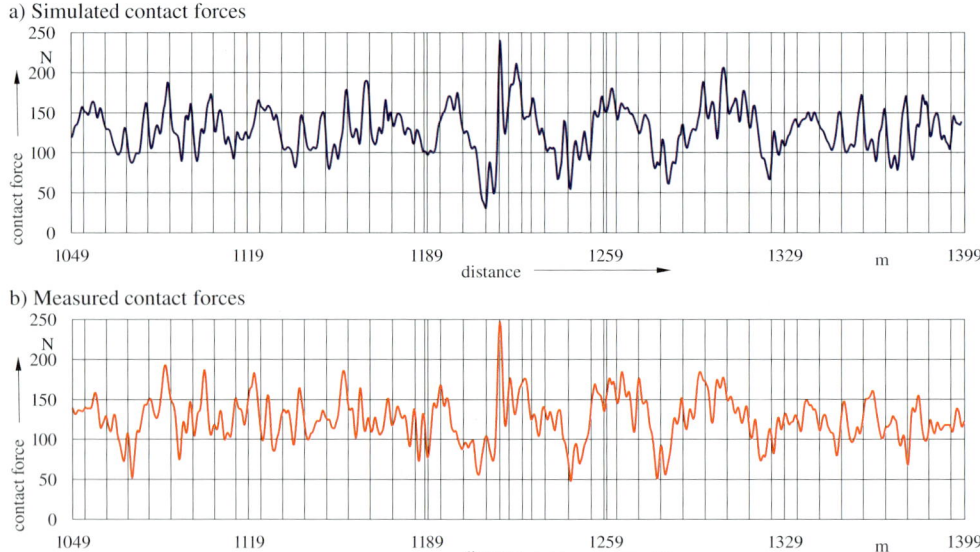

a) Simulated contact forces

b) Measured contact forces

Figure 10.22: Comparison of simulated and measured contact forces.

contact line exactly, e. g. the dropper supporting the registration arm suspended at the stitch wire.

Figure 10.21 shows a flow chart of the contact force simulation. The contact line geometry and the material properties are established using program macros. The contact line fixed point coordinates are checked by the program and adjusted if needed. The passage of the pantographs is then simulated for given conditions, e. g. a specified running speed. In addition to contact forces and contact wire uplift, the dynamic loading of the droppers and supports can be calculated for example. To achieve confidence in the simulation, a validation was carried out as stipulated in EN 50 318. Validation is a two step approval process as described in Clause 10.3.4, with the first step being an analysis of the contact line model given in EN 50 318 with the *simulation method* to be validated. In Table 10.5, the results for the reference model, as defined in EN 50 318, obtained by the finite-element-simulation using the method described in [10.19] are compared with the reference data. All results must be within the permitted ranges. In the second step, a comparison is made between data measured on a

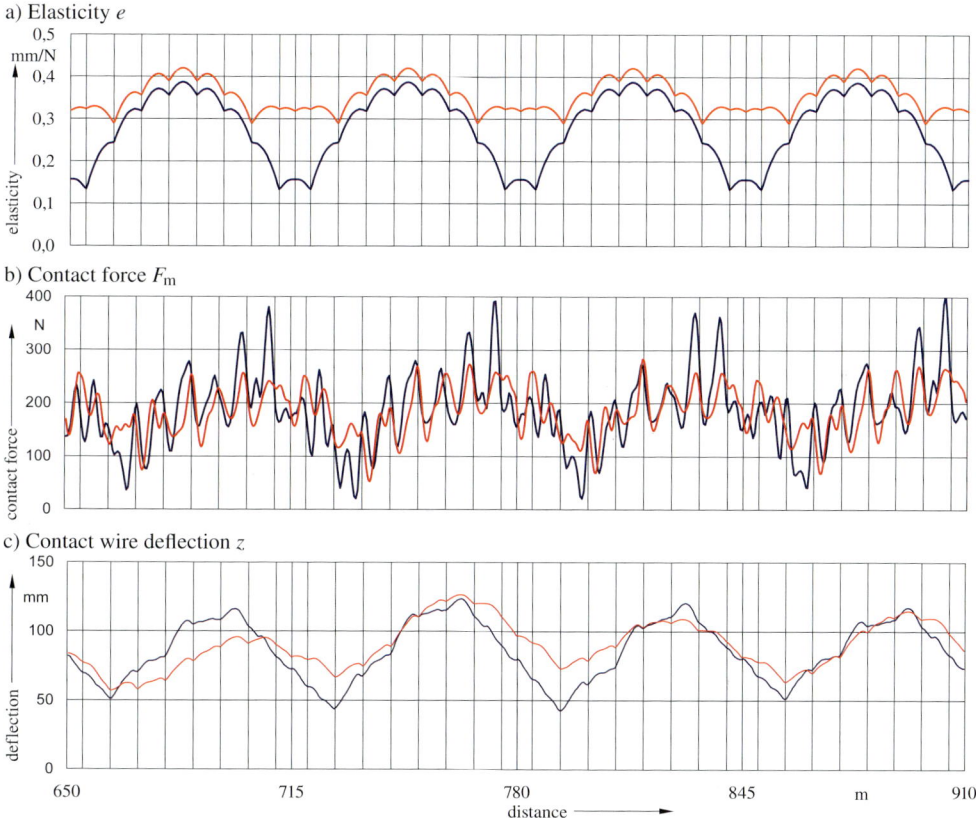

a) Elasticity *e*

b) Contact force F_m

c) Contact wire deflection *z*

Figure 10.23: Simulation of a contact line with and without stitch wires, span length 65 m.

real line and the results obtained by simulation of the line. Figure 10.22 shows the comparison. The measured and calculated standard deviations of contact forces differ by 10,2 % and the uplift at supports by 9,5 % are well within the specified limits. The two tests complied well with the specified requirements and approved the simulation method. The application of the simulation can be demonstrated at an overhead contact line with and without stitch wires. The results of the comparison are shown in Figure 10.23. The diagram above shows the elasticity calculated with 100 N contact force without stitch wires. For both designs, the elasticity at midspan is approximately 0,4 mm/N, however, the non-uniformity is 48 % without and 18 % with stitch wires.

The diagram in the middle, for the overhead contact line with stitch wire shows that the more uniform elasticity reduces the band width of the contact forces considerably. The maximum is 400 N without stitch wires and 280 N with stitch wires, the minimum being between close to zero and 100 N. Because of the reduction of the maximum contact forces, the contact wire wear will be reduced. The diagram on the bottom of Figure 10.23 demonstrates that the uplift at the supports is higher without stitch wires, the maxima, however, reach nearly the same values. An application of the simulation method for the acceptance of traction units is described in [10.26].

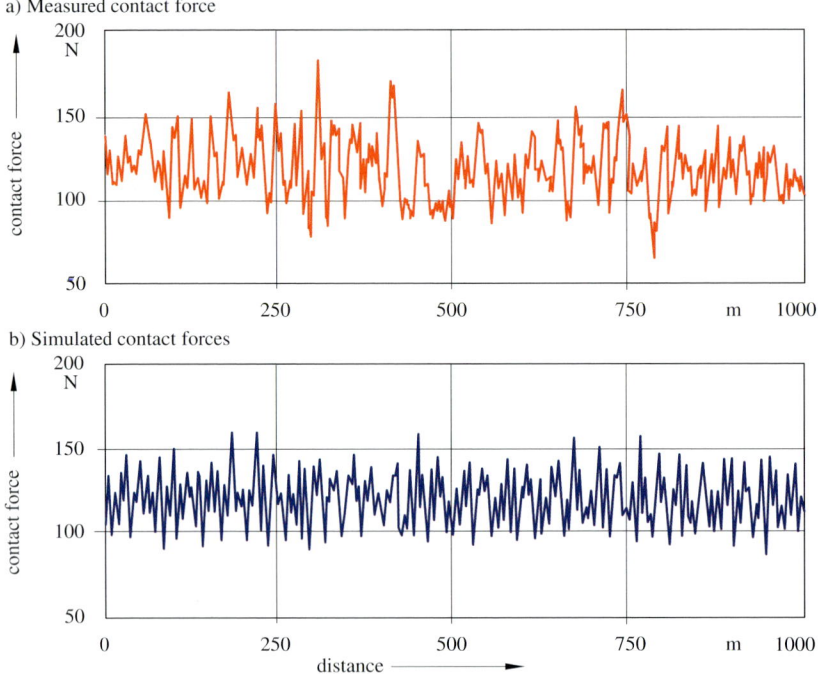

Figure 10.24: Contact force for contact line type Re 250 at 200 km/h.

10.3.7 Simulation with a program tailored to overhead contact line/pantograph interaction

The paper [10.27] describes a computer program system to analyse the interaction of overhead contact lines and pantographs. Besides simple multiple-mass models for pantographs, this program system can handle more sophisticated pantograph models and especially, closed-loop controlled pantographs as described in [10.28]. The contact lines are composed of standardized elements like catenary wire, contact wire, stitch wire, droppers, steady arms, insulators and overlaps. Mechanical models with mathematical descriptions are associated to the individual elements. Material and design characteristics are defined by the elements. E. g., the droppers can be modelled with and without resistance against compression forces. A total line section can be composed of models of individual spans. Overlaps are formed by contact lines running in parallel but not necessarily designed equally. The program can be used to calculate the static contact wire position, the elasticity of the contact line and the contact wire displacement due to wind action. The simulation of the contact line/pantograph interaction comprises running one or several pantographs along a contact line section, the consideration of installation tolerances and the effects of lateral displacements.

The program system includes detailed modelling of pantographs. The parameters can be obtained directly from the pantograph design. The non-linear characteristics of springs and dampers can be considered. The finite element method is integrated to model elements capable of oscillations, like collector strips. Control circuits can implement actively controlled pantographs. The results of measurements are shown in Figures 10.24 and the associated sta-

Table 10.6: Statistical evaluation contact forces of contact line type Re 250, with the simulation model described by [10.27], speed 200 km/h.

		Measurements	Simulation
mean value	N	120,3	120,7
minimum	N	60,6	69,4
maximum	N	169,0	168,0
standard deviation	N	26,5	17,4

tistical data are presented in Table 10.6. The differences between the measured and simulated standard deviations are higher than the values specified in EN 50 318 (see Table 10.4).

10.4   Measurements and tests

10.4.1   Introduction

In parallel to theoretical studies of interaction between pantographs and overhead contact lines, *measurement techniques* were developed in parallel for *assessing the quality of current transmission*. Three aspects can be defined:

– *assessment of the contact line* alone
– *assessment of the pantograph* alone and
– *assessment of the interaction* of these two components

Assuming compliance with safety related limits, high power transmission quality is achieved when energy is transmitted:

– continuously without *voltage* or *current drops* or losses. This means that mechanical contact exists at all times. Initially if the mechanical contact is lost, arcing occurs. An *electrical arc* is environmentally disturbing, causes interference and increased wear but it does ensure that the current flow is upheld. Consequently it is of fundamental importance for energy transmission between moving contacts. If the air gap becomes too long and the current is interrupted, the vehicle drive is switched off and traction power is lost. The number and duration of arcs is a criterion for assessing the *quality of energy transmission*

– without leading to unacceptable *environmental disturbances*. Arcing is associated with the emission of high-frequency electromagnetic waves that can interfere with amplitude-modulated radio transmissions at frequencies up to 30 MHz. At the same time, *audible noise* is generated, but usually this is blanketed by the general train noise

– without causing *wear* of the *components* involved, i. e. contact wire and contact strips, to an extent that would be economically unacceptable. Such excessive wear can be caused by arcing and/or excessive contact forces

Arcing occurs when the contact force approaches zero or is lost completely. In contrast, the contact force may not be too high as this would also lead to the contact wire being lifted too much and cause unacceptable wear. Consequently, the *contact force* is the crucial physical quantity by which interaction of the pantograph and overhead contact line can be assessed. In principle, the physical quantity used to judge the quality should meet some other general criteria:

– as far as possible, the respective characteristic quantity should provide for assessment on a continuous, graduated scale that allows for a "Yes/No" decision and a description of *quality variations* also

  – it should be possible to measure the respective quantity and to carry out *forecast calculations* on simulation models
  – measurements of the respective quantity should be reproducible and not affected by any random factors. Measurements repeated under comparable conditions should lead to comparable results
  – it should be possible to measure the assessment quantity on an active live pantograph

In earlier references [10.29, 10.30], the *quantity* and *duration of arcs* measured as *voltage losses* were the physical quantities used to assess the contact behaviour. However, these quantities do not meet the criteria listed above. If there are no arcs or only relatively few occur, this characteristic is not helpful for comparing different contact line systems. It is not possible to simulate arcing. Measurements of arcs have shown that it is not possible to reproduce the results in repeated test runs. Even under identical conditions on the same lines, repeated tests yielded different results.

The contact force couples the two mechanical systems – overhead contact line and pantograph systems, both of which are capable of oscillating and which have various masses, coefficients of elasticity, *damping coefficients* and *natural frequencies*. The pantograph lifts the overhead contact line by an amount which is a function of the contact line elasticity. The elasticity variation along the contact line leads to periodic upward and downward movements of the pantograph head and the amplitude of this motion depends on the lifting force itself. Mass inertia forces, which are a function of the change rate of the vertical motion, are superimposed on this mean lifting force (see Clause 10.3).

As speeds increase, the contact force is affected increasingly by the dynamic components. To keep the moving collector strips in continuous contact with the contact wire, the contact force values should remain within the dynamic contact force range.

The variation of the contact force with time is the most suitable characteristic quantity to evaluate the dynamic behaviour of the system components and their interaction. The TSI ENE [10.13, 10.14] stipulates requirements for contact forces (see Clause 10.4.2).

Parallel theoretical studies [10.31, 10.32] on the dynamic motion, the German Railway Research Institute, located in Munich, (Versuchsanstalt der Deutsche Bahn AG) developed a *force measurement method* [10.2, 10.33]. After successful testing, this measurement system has been used since 1980 by the DB and other railway operators and continuously developed.

In addition to the contact force, other characteristic quantities were introduced as criteria for evaluating the pantograph and overhead contact line interaction:

  – the *overhead contact line uplift*
  – the *pantograph's vertical motions*
  – the *contact behaviour* of the pantograph head or collector strips expressed in terms of the frequency and *duration of power losses* (see [10.29, 10.30]) at higher train speeds by monitoring arcing (see [10.34]).

All these are secondary quantities resulting from respective reactions to the continuous variations in the contact force that couples the two oscillating systems: overhead contact line and pantograph.

The uplift of the contact wire by the pantograph is recorded either by a stationary measuring unit installed at a support (see Clause 10.4.6.1) or on running trains by an optical measuring system installed close to the pantograph on a traction unit (see Clause 10.4.6.2). Assessment of the contact line uplift only can be carried out by recording the initial and uplift contact wire position and calculating the *contact line elasticity* related to the contact force.

Table 10.7: Contact forces in N according to TSI ENE [10.14] combined with TSI RS [10.15] and EN 50 367:2012, Table 3.

	Nominal Value	Range
AC	70	60 to 90
DC 3,0 kV	110	90 to 120
DC 1,5 kV	90	70 to 140

Table 10.8: Parameters C_V and C_A for the mean contact force F_m.

Type of supply		AC		DC 1,5 kV		DC 3,0 kV	
speed	km/h	≤ 200	> 200	≤ 200	> 200	≤ 200	> 200
EN 50 367: Maximum mean contact force $F_{m\,max}$ for the contact line design							
parameter C_V	N/(km/h)^2	0,00047	0,00097	0,00097	0,00228	0,00097	0,00097
parameter C_A	N	90	70	140	90	110	110
EN 50 367: Minimum mean contact force $F_{m\,min}$ for the acceptance of pantographs							
parameter C_V	N/(km/h)^2	0,00047	0,00047	0,00112	0,00112	0,00072	0,00072
parameter C_A	N	60	60	70	70	90	90
TSI Energy [10.13]: Maximum mean contact force $F_{m\,max}$ for the contact line design							
parameter C_V	N/(km/h)^2	0,00097	0,00097	0,00228	0,00228	0,00097	0,00097
parameter C_A	N	70	70	90	90	110	110

The components of the contact line are also checked at regular intervals, especially the position of the steady and registration arms and the clearances of live components to structures and tunnel walls. Unacceptable extremes of contact force result in increased contact wire wear determined by measuring the *residual contact wire vertical dimension*. Assessment of pantograph alone as a separate component is carried out on a *pantograph test stand* (see Clause 10.4.5).

10.4.2   Requirements on the interaction between overhead contact lines and pantographs

10.4.2.1   Introduction

TSI Energy [10.12, 10.13, 10.14] and EN 50 367 set out the basic requirements for the interoperability of the European rail system. Because this system is operated by electrically propelled trains, a reliable and safe power supply is an important condition for the interoperability of trains. Therefore, the interaction of *interoperable pantographs* with interoperable overhead contact lines requires a set of stipulated requirements concerning the static contact force, the mean contact force and the quality of current collection. The *measurement of contact forces* is one of the specified approaches for the assessment of components and subsystem.

The quality of dynamic interaction performance can be measured by:
- the contact forces characterised by mean value, standard deviation and maxima and minima or
- the percentage of arcing

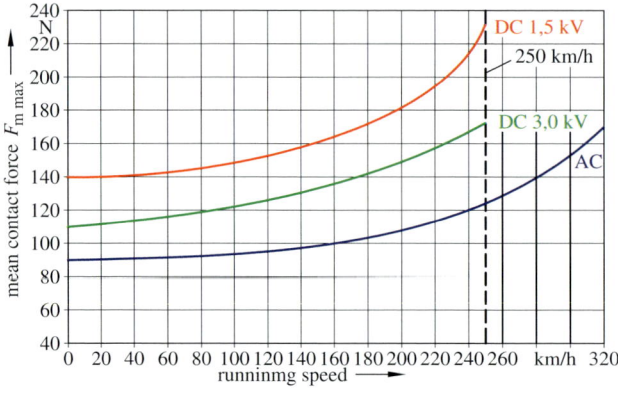

Figure 10.25: Mean contact forces $F_{m\,max}$ for contact line design of AC and DC systems as a function of speed according to TSI Energy [10.14] and EN 50 367:2016, Table 6.

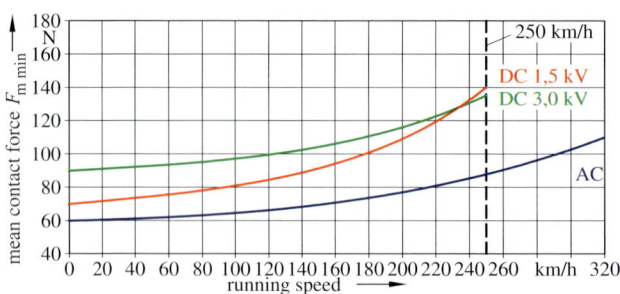

Figure 10.26: Mean contact forces $F_{m\,min}$ for acceptance test for pantographs for AC and DC systems according to TSI Energy [10.14] and EN 50 367: 2016, Table 6.

10.4.2.2 Static contact force

Following the definition in EN 50 206-1, the *static contact force* is the mean vertical force exerted upward by the collector head on the contact wire, and caused by the pantograph raising device whilst the vehicle is at a standstill. The static contact forces are specified in Table 10.7.

For DC 1,5 kV systems, the overhead contact line should be designed to withstand a static contact force of 140 N per pantograph to avoid overheating the contact wire while a train is at standstill with its auxiliaries working. The target values apply to the design of new installations while the alternative data can be used for upgrading. The pantograph design needs to permit the adjustment of contact force to the requirements.

10.4.2.3 Mean contact force

The *mean contact force* F_m is formed by the static and aerodynamic components of the pantograph contact force with dynamic correction. F_m represents a target value to be achieved that ensures a current collection quality without undue arcing and to limit wear and hazards to contact strips. For the mean contact force F_m the TSI Energy [10.14] refers to EN 50 367, Table 6. The TSI Energy [10.13] specified slightly differing data for DC 1,5 kV below 200 km/h. The data from Table 10.8 apply up to 320 km/h.

The mean contact force depends on the square of the running speed and a constant parameter according

$$F_m = C_V \cdot v^2 + C_A \tag{10.68}$$

The parameter C_V and C_A are given in Table 10.8.

Table 10.9: Minimum distances in m between pantographs for the design of contact lines according to TSI Energy [10.14], Table 4.2.13.

Type of line	AC			DC 3,0 kV			DC 1,5 kV		
	A	B	C	A	B	C	A	B	C
speed in km/h									
$v > 250$	200	200	200	200	200	200	200	200	35
$160 < v \leq 250$	200	85	35	200	115	35	200	85	35
$120 < v \leq 160$	85	85	35	20	20	20	85	35	20
$80 < v \leq 120$	20	15	15	20	15	15	35	20	15
$v \leq 80$	8	8	8	8	8	8	20	8	8

Table 10.10: Requirements for dynamic behaviour and current collection quality for contact lines of railway lines complying with TSI Energy [10.14], Table 4.2.12.

Requirement of line	250 km/h and above	160 to 250 km/h	up to 160 km/h
Space for steady arm uplift	without uplift stop $2 S_0$; with uplift stop $1,5 S_0$		
Mean contact force F_m	Equation (10.68) and Figures 10.25 and 10.26		
Standard deviation σ_{max} at maximum line speed	$0,3 F_m$		
Percentage of arcing at maximum line speed NQ (%)[1].	$\leq 0,2$	$\leq 0,1$ for AC systems $\leq 0,2$ for DC systems	$\leq 0,1$

[1] minimum duration of an arc to be taken into account 5 ms

For speeds above 320 km/h, values for the mean contact force are not detailed neither in the TSI Energy [10.14] nor in EN 50 367. Contact lines for speeds above 320 km/h are considered as innovative solutions to which Article 10 of [10.14] applies. However, experience with some installations shows that the values resulting from (10.68) can be used also above 320 km/h. The contact lines need to be designed such that, at least the mean contact forces $F_{m\,max}$ according to EN 50 367, Table 6 can be used as upper limits as shown in Figure 10.25. Table 6 in EN 50 367 also contains the lower limits of mean contact forces as depicted in Figure 10.26 for acceptance of pantographs. Contact line and pantograph test results should aim to conform with this data. The mean contact force needs to be between the limiting values of $F_{m\,max}$ (Figure 10.25) and $F_{m\,min}$ (Figure 10.26).

For AC systems, the specifications were modified slightly compared with latest TSI issue [10.12]. The previous data, according to [10.12] are as well given in Table 10.8. The data mentioned as the AC target values are mandatory. According to [10.12] new installations also need to accommodate pantographs with curves for F_m called AC/C1 and AC/C2. Existing contact line systems may also require pantographs with AC/C1 and AC/C2 curves. The curve to be applied is listed in the infrastructure register.

10.4.2.4   Minimum and maximum spacing between two pantographs

The contact line needs to be designed for the minimum spacing between two adjacent raised pantographs according to type of line A, B or C as specified in Table 10.9. The type of line is specified by the infrastructure manager.

The limits for maximum spacing are important for trouble free operation of neutral sections, section separations and section insulators. In AC systems, pantographs must not be connected

Table 10.11: Permissible range of vertical movement of the contact point within a span.

	High-speed lines 250 km/h and above	Upgraded lines 200 km/h	Other upgraded lines
AC	80 mm	100 mm	National rules apply
DC	80 mm	150 mm	National rules apply

electrically through the train. If, in the case of DC systems, an electrical connection exists between the pantographs, equipment for the interruption of this connection must be provided.

10.4.2.5 Dynamic behaviour and quality of current collection

The quality of current collection has a fundamental impact on the life of contact wire and should, therefore, comply with generally agreed and measurable parameters. Compliance with the requirements for *dynamic behaviour* may be verified in accordance with EN 50 367, Clause 7.2, by assessment of contact wire uplift and either mean contact force F_m and its standard deviation σ_{max} or the percentage of arcing.

The verification method may be selected by the entity in charge of the Energy subsystem. The requirements for compliant high-speed lines according the TSI Energy [10.14] are summarized in Table 10.10. The test methods are described in EN 50 317 and EN 50 318 (see also Clause 10.4.3.2). A similar set of requirements for the interaction performance is stipulated in EN 50 367 for all types of main line contact lines.

The magnitude S_0 is the calculated, simulated or measured *uplift of the contact wire* at a steady arm, generated in normal operating conditions with one or more pantographs with a mean contact force F_m (Clause 10.4.3.2) at the maximum line speed. When the uplift of the steady arm is physically limited because of the overhead contact line design, it is permissible for the necessary space to be reduced to 1,5 S_0 according to the [10.14] and EN 50 119 (Clause 10.4.3.2).

The standard deviation is stipulated in relation to the mean contact force F_m. Where the measured mean contact force is less than the limit specified in Figure 10.26 the permissible standard deviation s_{max} should also be reduced proportionally to the mean contact force to avoid loss of contact.

10.4.2.6 Vertical height of the contact point

The contact point is the point of the mechanical contact between a contact strip and a contact wire. The vertical height of the contact point above the track should be as uniform as possible along the span length; this is essential for a high-quality current collection. The maximum difference between the highest and the lowest *dynamic contact point height*, as specified by [10.12] and EN 50 119, Clause 5.2.3 within one span, should be less than the values shown in Table 10.11.

This should be verified by measurements according to EN 50 317 or simulations validated according to EN 50 318:
  - for the maximum line speed of the overhead contact line
  - by using the mean contact force F_m (see Clause 10.4.2.3)
  - for the longest span length

This does not need to be verified for overlap spans or for spans over points.

10.4.2.7 Conformity assessment of the interoperability constituent *Overhead Contact Line*

A new design of an overhead contact line should be assessed as an interoperability constituent by simulation according to EN 50318 and by measuring a test section of the new design according to EN 50317.

The simulations should be made using at least two TSI compliant pantographs for the appropriate system, up to the design speed of the reference pantograph and the overhead contact line proposed as an *interoperability constituent*. For simulation purposes, the control sections should be defined so that they are fully representative, including special features, for example tunnels, crossovers, neutral sections etc. To be acceptable, the simulated current collection quality should be within the limits of Table 10.10 for each of the relevant reference pantographs.

Where the simulated values are acceptable, a site test on a representative section of the new overhead contact line should be undertaken with one of the reference pantographs used in the simulation, installed on a train or locomotive producing a mean contact force for the envisaged target speeds as required by Clause 10.4.2.3. The measured current collection quality should be within the limits of Table 10.10, to be acceptable.

If all the above criteria are met, the tested overhead contact line design should be considered as compliant and may be used on lines where the characteristics of the design match the requirements of the line.

If an overhead contact line has not been assessed as an interoperability constituent and is intended to be installed in a subsystem, compliance with the requirements on dynamic behaviour can be verified at the level of the subsystem by measurements according to EN 50317. A pantograph, that is a certified interoperability constituent, should be installed on a vehicle that meets the requirements of Clause 10.4.2.3. Statistical values should be associated to the speed of the line and calculated separately for open sections and in tunnels.

10.4.2.8 Assessment of the interoperability constituent *Pantograph*

Interoperability requirements for *pantographs* are specified in [10.15]. The requirements on dynamic performance of a new design of pantograph should be assessed by tests in accordance with EN 50206-1 and by simulation in according to EN 50318. The interaction characteristics should be measured in accordance with EN 50317 on an interoperable overhead contact line.

The simulations should be made using at least two TSI compliant reference overhead contact lines for the appropriate system, up to the design speed of the pantograph. The simulated current collection quality should be within the limits of Table 10.10 for each of the relevant reference overhead contact line systems.

Where simulated values are acceptable, a site test measuring the interaction characteristics in accordance with EN 50317 should be carried out using a representative section of one of the relevant reference overhead contact lines used in the simulation.

If all the assessments are passed successfully, the tested pantograph design is considered compliant and can be used on various designs of trains provided that the mean contact force on the train for the contact wire height in question complies with the requirements of Clause 10.4.2.5.

10.4.2.9 Assessment of an overhead contact line in a newly installed line

If the overhead contact line to be installed on a new high-speed line has been certified as an *interoperability constituent*, the measurements of the interaction parameters intended to be used to check the correct installation is carried out with an interoperability constituent pantograph installed on a train or locomotive exhibiting the mean contact force characteristics as required by Clause 10.4.2.3 for the envisaged target speed. The main goal of this test is to identify installation errors, but not to assess the contact line design in principle. The installed overhead contact line can be accepted if the measurements comply with the requirements of Table 10.10.

10.4.2.10 Assessment of a pantograph installed on a new train or locomotive

When a pantograph approved as an interoperability constituent is to be installed on a new train or locomotive, testing at the required speed can be limited to the mean contact force requirements. The tests can be carried in accordance with EN 50 206-1, Clause 6.10 or EN 50 317. In the case of EN 50 206-1, the pantograph does not need to be in contact with the contact wire. The rolling stock manufacturer can decide on the type of test used. The tests should be carried out in both directions of travel at the range of nominal contact wire height, as applied for. The measured results should follow the mean contact force curve, plotted using at least five speed intervals for Class 1 trains (for speeds of 250 km/h and above) and at least three intervals for Class 2 trains (speeds up to 200 km/h). The results should comply with the curves throughout the speed range for the train within a range of $\pm 10\,\%$ for the curves specified in Figures 10.25 and 10.26. If the tests are successful, the pantograph mounted on that particular train or locomotive can be used on TSI compliant high-speed lines at the tested contact wire height.

10.4.2.11 Statistical calculations and simulations

The calculation of statistical values should be appropriate to the line speed and should be carried out separately for open sections and in tunnels. For simulation purposes, the control sections should be defined so that they are fully representative, including special features for example tunnels, crossovers and neutral sections.

10.4.3 Measurements of the interaction between overhead contact lines and pantographs

10.4.3.1 Basic principles

According to EN 50 317 the measurements of interaction between contact lines and pantographs are intended to check the reliability and the quality of the current collection system. Results of the measurements of different current collection systems should be comparable to approved components for the interoperability of railways. Measured values are also required for validation of simulation programs and other measuring systems.

To check the performance capability of the *current collection system* the following data at least should be measured:
- the contact wire uplift at the support when the pantograph passes or
- the contact force, the average contact force and standard deviation
- percentage of arcing

The contact force occurring between the contact wire and collector strip cannot be recorded directly because of the moving contact point. Due to the simpler possibilities of monitoring and because of developments carried out in the past, often the sum of the collector strip reaction forces, the so-called internal forces, is taken as an approximate quantity instead of the contact forces themselves. For *contact force measurements*, sensors are installed where the collector strip connects to its socket. The *mass inertial forces* acting at the collector strips and the running-speed-dependent *aerodynamic forces* of the collector strips are not recorded by the force sensors.

To determine the contact force, a *dynamic correction* quantity has to be added to the internal force which is evaluated from the collector strip acceleration and includes the inertial forces of the collector strips. *Aerodynamic correction* quantities, depending on the running speed must also be considered and evaluated according to the procedure described in Clause 10.4.3.8. To enable comparisons between results obtained with different pantographs or by measuring systems with differing arrangements of the force sensors and to judge them based on the same criteria, dynamic and aerodynamic corrections cannot be waived.

10.4.3.2   Requirements on contact force measurements

The standard EN 50 317 sets out the requirements for and the validation of measurements of the dynamic interaction between pantographs and overhead contact lines. The standard applies to the measurement of contact forces as well as displacements and arcing.

The contact force measurements should be carried out on the pantograph with force sensors located as near as possible to the contact points. The measurement system should record forces in the vertical direction without interference with forces from other directions. For pantographs with independently suspended contact strips, each contact strip should be measured separately. The maximum error of the measurements is required to be less than 10 %.

The inertia forces due to the effect of the mass between the sensors and the contact point need to be considered. This can be done by *measuring the acceleration* of these components.

A correction needs to be applied to allow for the influence of aerodynamic forces on the components between sensors and the contact points. Aerodynamic tests should be carried out to establish the aerodynamic corrections. The aerodynamic influence can be checked by a *tethered test* on a line as described in EN 50 206-1, ensuring that the pantograph does not touch the contact wire.

Aerodynamic tests should be carried out with nominally the same configuration of contact wire height, train configuration, measurement equipment, environmental conditions etc. as during the measuring of contact forces. The aerodynamic test may also be carried out during a line test.

The measuring system should be laboratory-calibrated to check the accuracy of the measured forces. The test should be carried out for the complete pantograph fitted with the complete force measuring devices and any accelerometers, the data transfer system and amplifiers. The ratio between the applied and the measured forces – the *transfer function* of the pantograph and instrumentation – should be determined by dynamic excitation of the pantograph at the pantograph head for a range of frequencies. If a sinusoidal force is used, a peak to peak amplitude of 30 % of the static force gives representative results. The tests should be carried out for the two cases:

- the force being applied centrally to the pantograph head
- the force being applied 205 mm from the centre line of the pantograph head, if possible.

Otherwise the point of force application should be as close as possible to this value. This test should be carried out with the mean force equal to the static force

If the pantograph contact force increases with speed, the test should also be carried out at the maximum quasi-statical force. Measurements of the applied force and the measured force should be taken at frequencies up to 20 Hz in 0,5 Hz steps with reduced intervals at resonant frequencies. These frequency steps near the resonant frequencies should be specified.

The sampling rate for the measurements on a line should be greater than 200 Hz for time sampling or smaller than 0,40 m for distance sampling. The contact force should be low-pass filtered with a cut-off frequency of 20 Hz.

The measuring range should be at least:
- for AC pantographs from 0 N to 550 N
- for DC pantographs from 0 N to 700 N

As a minimum the
- mean value F_m
- maximum value
- minimum value
- *standard deviation* (σ_m)
- histogram or probability curve

of the contact force should be calculated for a test section, which should not be shorter than a tensioning section length.

10.4.3.3 Measurement of contact wire uplift

The uplift of the contact wire should be measured with an equipment that must not have any effects on the measured displacement, which could change the result by more than 3 %. The error in *uplift* at the support should be less than 5 mm. The vertical displacement of the contact point should be measured relative to the base frame of the pantograph. The accuracy of the measurement system should be better than 10 mm.

10.4.3.4 Measurement of arcing

The detector for *arcing measurements* should be sensitive to the wave lengths of light emitted by copper materials. For copper and copper-alloy contact wires, the range should include 220 nm to 225 nm or 323 nm to 329 nm, since these two wave length ranges have substantial copper emissivity.

The detector should
- be close enough to the pantograph and to the vehicle's longitudinal axis to achieve a sufficiently high sensitivity
- be located behind the pantograph with respect to the travel direction
- be aimed at the trailing contact strip with respect to the travel direction
- be sensitive over the whole working area of the pantograph head
- have a response time to the beginning and end of an arc of less than 100 µs
- have a detection threshold, depending on the minimum arc energy intended for measurement

The detectors should be calibrated in the spectral range of interest. An adjustment of the detector is needed if the distance between the sensor and the light source differs in operation from the calibration condition. As a minimum, the system should measure and record:

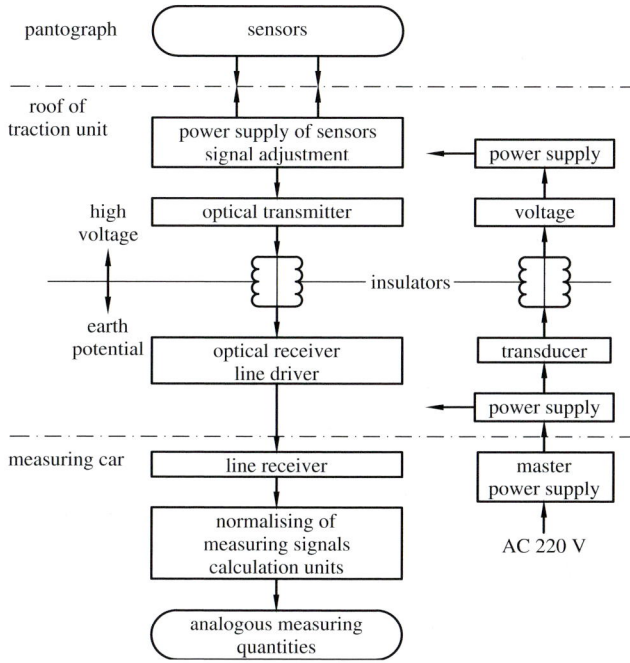

Figure 10.27: Schematic diagram of the system used to measure contact forces between contact wire and pantograph.

– the duration of each arc
– the train speed during the test
– the pantograph current
– the location of the arcs (kilometric position)

Evaluation should be carried out for a testing section not shorter than 10 km and travelled at a constant speed. For the output, only arcs lasting longer than 5 ms should be analysed. Sections with a pantograph current below 30 % of the nominal current should be disregarded. As a minimum, the following should be recorded and evaluated:

– train speed
– the number of arcs
– the sum of the duration of all arcs
– the longest arc duration
– the total period with pantograph currents greater than 30 % of the nominal current
– the total running period for the test section
– the percentage of arcing

10.4.3.5 Description of DB's contact force measuring technologies

The recording system described in [10.33] is based on proposals made in [10.35]. Figure 10.27 shows the flow chart of the *contact force measurement system*, which has been used successfully by Deutsche Bahn AG for several years.

The sensors arranged at the collector strips are the most important components of the system (see Figure 10.28). Pantograph heads equipped with two collector strips need four sensors for monitoring the internal forces. Special cables connect the sensors to the amplifiers housed in a case mounted on the pantograph base. There, the measured signals are converted into a

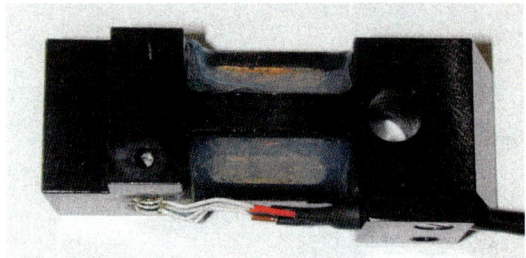

Figure 10.28: Force sensors of the contact force measuring system.

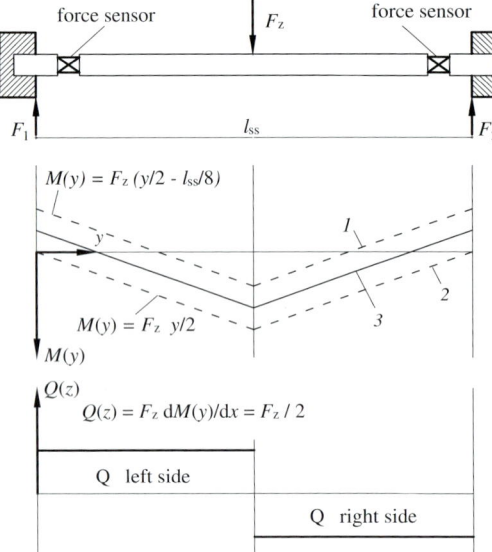

Figure 10.29: Shear forces $Q(z)$ and moments $M(y)$ acting along a collector strip.
F, F_1, F_2 forces
1 bending moment corresponding to a rigid fixing
2 bending moment corresponding to a completely flexible support
3 bending moments corresponding to the real collector strip fixing conditions

format that can be transmitted from the high-voltage potential equipment to earth potential equipment by an optical link and galvanic decoupling. Within the traction unit, the optical signals received are converted back into electrical signals and passed to the *measuring car* [10.2] for further processing.

The recording sensors comply with the following requirements:

- they have minimum effect on the dimensions or mass of the pantograph head and do not change the behaviour of the pantograph head to any unacceptable extent
- the sensors are able to measure *static and dynamic forces*
- additional sensors record the acceleration of the collector strips
- wide variations of ambient temperatures or strong electrical and electromagnetic fields at traction currents up to 1 000 A do not affect the measurements
- the vertical force components and other components are measured without being affected by each other forces acting horizontally on the collector strips

Figure 10.29 shows the forces and moments acting on a collector strip. The strains in a bending bar-type sensor are determined by the moments acting on the sensors. The bending moment curve passes through zero at two points near the sensors. The locations of the zero positions depend on the stiffness of the support fixing, and application point of the contact force. This results in complicated relationships between the applied force and the values measured by bending bar-type sensors.

Figure 10.30: Arrangement of the signal processing unit with bushing isolator (right) and voltage transmitter.

Figure 10.29 illustrates that the relationships between the *shear forces* and the applied forces are less complex. The sum of the shear forces is equal to the applied forces and is independent of the contact force position and the boundary condition at the end fixing. To ensure reliable measurements, force sensors able to measure the shear forces independently of the moments are used.

The shear forces cause shear stresses in the lateral sides of beam elements with maxima under an angle of 45° to the vertical axis. Using special strain gauges, the deformation – elongation or compression caused by the shear forces – can be recorded from which a measuring value proportional to the acting shear force can be deduced. *Strain gauge sensors* are passive sensors and require a separate supply voltage and amplifiers. For example, a force variation of 10 N results in a variation of the diagonal bridge voltage of the strain gauge of only 60 μV.

The sensors are arranged at high-voltage potential (3 kV to 25 kV). Strong electrical and electromagnetic alternating fields occur in the vicinity of the contact wire because of the high alternating currents and electrical arcing. To minimize induced interference before amplification of the signal, adequate shielding of the electrical connections to and from the sensors is provided. It is also useful to amplify the diagonal bridge signal of the strain gauge sensors as close as possible to their origin. Therefore, the *bridge amplifiers* are integrated into a casing installed on the traction unit roof above the insulators (Figure 10.30). The casing is also required for downstream electronic devices. A multi-core, 4 m long cable is used for the connection between the sensors, amplifier and power supply.

10.4.3.6   Measured quantities

Internal forces

Measured quantities are primarily the *reaction forces* at the supports of the collector strips. If the reaction forces at collector strip I are designated F_1 and F_2 (Figure 10.31) and those at collector strip II are designated F_3 and F_4 the analogous addition of the forces acting on the collector strips leads to the internal forces exerted by the contact wire

$$F_{SI} = F_1 + F_2 \quad \text{and} \quad F_{SII} = F_3 + F_4$$

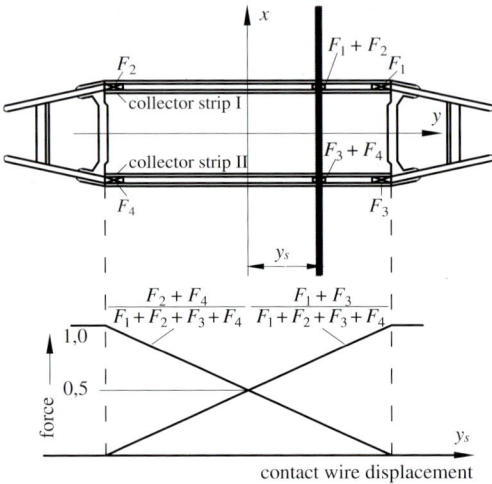

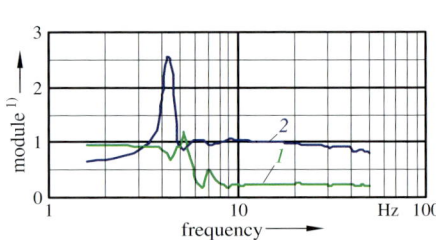

Figure 10.31: Determination of the contact wire position on the basis of the forces measured.

Figure 10.32: Amplitude of transfer function.
1 without dynamic contact force correction
2 with dynamic contact force correction
[1] Module of $F_{\text{recorded}}/F_{\text{contact}}$

as well as to a total force on both collector strips

$$F_S = F_{SI} + F_{SII} \quad .$$

The position where the *contact force* is applied can be determined from the ratio of the difference of the summed individual forces on the left and right side to the total force:

$$Y_S = k_S \frac{(F_1 + F_3) - (F_2 + F_4)}{F_1 + F_2 + F_3 + F_4} \quad , \tag{10.69}$$

where Y_S is the distance of the contact wire from an imaginary central axis on the collector strips (Figure 10.31). The factor k_S, which has the unit of length, serves as a calibration quantity.

Contact forces

The relationship between acting *contact force* and measured internal force is valid only for low frequencies because of dynamic processes (see Figure 10.32, curve *1*). The relationships between input amplitudes and recorded force and phase shift are functions of the frequencies. The relationships may vary greatly for individual pantograph systems and can be determined, for example, on a *pantograph test stand* as described in Clause 10.4.5.

When measuring the collector strip acceleration $\ddot{z}$ in the vertical (z) direction simultaneously with the internal forces and correcting them according to (10.70) with the mass inertia forces F_{msz} resulting from the collector strip masses, the amplitude curve of the transfer function and the measuring precision of the contact force will be improved significantly (see Figure 10.32, curve b).

a) Force between pantograph head and collector strip

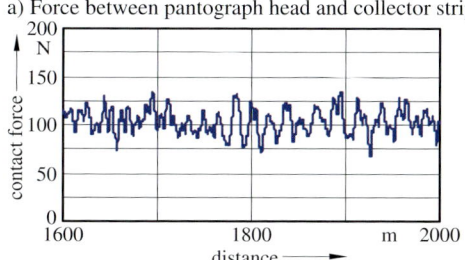

b) Contact forces, dynamically corrected

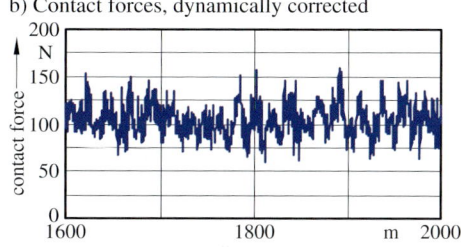

Figure 10.33: Record of forces during a test run.

The contact force F_C is obtained from

$$F_C = F_S + F_{msz} = F_S + \ddot{z}_S \cdot m_S \quad ,$$ (10.70)

where

$\ddot{z}_S$	is the recorded mean acceleration $[(\ddot{z}_1 + \ddot{z}_2 + \ddot{z}_3 + \ddot{z}_4)/4]$
m_S	the related mass of the collector strips $(m_{SI} + m_{SII})$
$\ddot{z}_{1,2,3,4}$	the acceleration at the position of the force sensors

To correct the recorded *internal force*, *acceleration sensors* are installed at the collector strips or at the force sensors already located there. The output signals of the force and acceleration sensors are available separately. The correction of signals is done by processing them according to Equation (10.70) by the signal circuitry that enables linear filtering to the phase.

In Figure 10.33, the difference is demonstrated in measured results of contact force before and after applying dynamic correction to internal force measurements. The pattern of the two curves coincides in principle. However, the contact forces show considerably more high-frequency signal components than the internal forces.

The measured parameters and the values derived along the tested track sections are logged using graphic recorders [10.2]. Figure 10.34 shows an example of such a recording. In addition to some general information on the tested section of the line, the record contains the following information:

- *running speed*
- *distance travelled* (line kilometres)
- symbols for particular characteristic points of the overhead contact line such as overlap sections (O), midpoint anchors (MP), points (W) etc.
- *vertical pantograph motion*
- arcing
- *contact forces*:
    - force of panthograph F_S
    - sum of forces on leading collector strip F_{SI}
    - sum of forces on trailing collector strip F_{SII}
    - the four individual forces F_1, F_2, F_3, F_4
- the dynamic *lateral contact wire position* relative to the collector strips, as calculated from the measured individual forces
- contact line supports identified by vertical lines in the oscillogram

The evaluation of the results by DB focuses on the graphs of the dynamic forces and particularly on the total internal force F_S and the total contact force F_K, respectively. A precondition for superior pantograph running performance is a uniform distribution of the contact forces

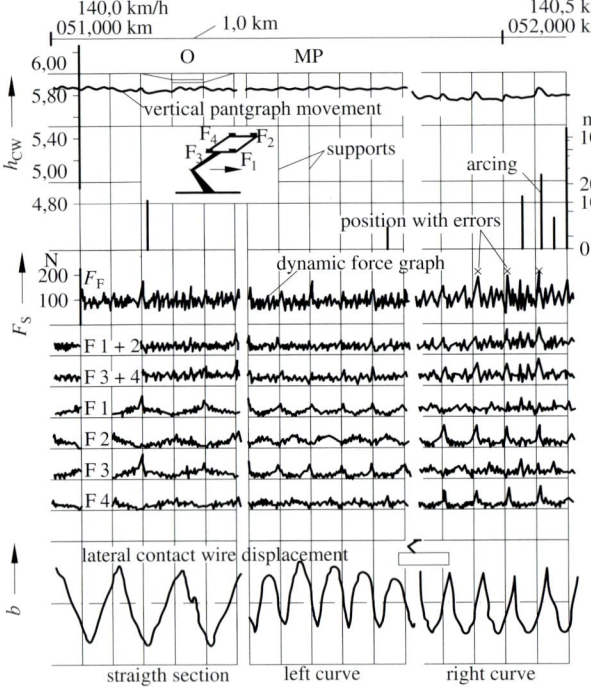

Figure 10.34: Record of a contact force test run.

F_C	contact force
MP	midpoint
O	overlapping section
b	lateral position
h_{CW}	contact wire height
t_{arc}	duration of arcs

on both collector strips. This can be seen from the graph of the forces F_{SI} and F_{SII}. Evaluation of the results is done by analysing the graphs of the dynamic forces and particularly on the total internal force F_S and the total contact force F_K. A precondition for superior pantograph running performance is a uniform distribution of the contact forces on both collector strips. This can be seen on the graph of the forces F_{SI} and F_{SII}.

The system for testing contact force performance is also able to measure the following features:

- approximate *vertical position of the pantograph top* tube as gained from the recorded support tube angle
- *horizontal forces* acting on each collector strip in track axis direction (forces due to wind and friction)
- vertical acceleration of the pantograph base frame as a quantity for assessing the effects caused by irregularities in the track superstructure

The vertical movements of the pantograph along the contact line have a close correlation to the forces. A uniform uplift pattern characterises smooth pantograph running with small dynamic force variations. With the quantities described above, a comprehensive representation of the *contact performance* and the reasons for irregularities and disturbances can be given.

10.4.3.7 Aerodynamic uplift force acting on the collector strips

Since the sensors recording the forces are mounted under the collector strips, the contact force components created by the aerodynamic effects on the collector strips cannot be detected by the force sensors. To account for these running-speed-dependent aerodynamic force components, a speed-dependent correction is applied to the recorded contact forces.

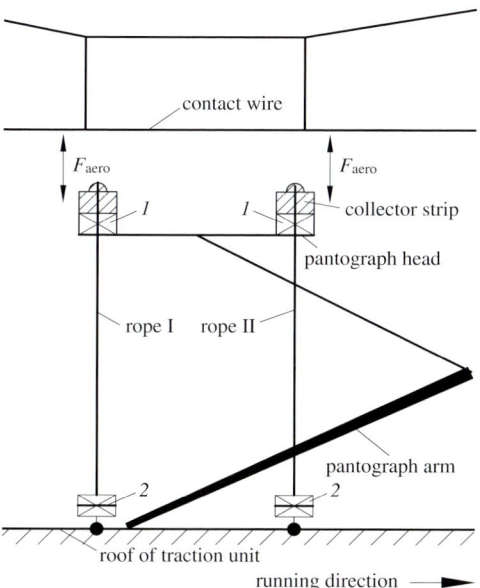

Figure 10.35: Determination of the aerodynamic force components acting at the collector strips of a pantograph by measuring the forces in ropes which fix the collector strips.
1 equipment to measure the forces between collector strips and pantograph head
2 equipment to measure the tensile forces in the strings

The *aerodynamic uplift force* component is determined by measuring the uplift force in accordance with the UIC code 608 [10.36] or EN 50 206-1. To carry out this measurement, a pantograph equipped with a contact force measuring device is fixed by two ropes attached to a collector strip. The collector strips are fixed vertically (see Figure 10.35). The collector strips do not touch the contact wire during the test run. The gap between the collector strips and the contact wire is approximately 100 mm. The test is also known as a *tethered test*.

At the lower ends of both ropes, load cells are fixed to record the forces transferred by the rope to the collector strips. Simultaneously, the internal forces underneath the collector strips are recorded by the contact force measuring system. The *aerodynamic component on the collector strip* F_{aero} is equal to the difference of the force recorded via the two ropes and the internal force F_S which is recorded by the contact force measuring system. Using this measuring procedure, the aerodynamic components acting on the collector strips are determined, depending on, the running speed, the running directions: knuckle in running direction or knuckle opposite to running direction as well as on the arrangement of the pantographs on the train. The aerodynamic force component has to be considered when evaluating test runs with pantographs raised to the contact wire.

10.4.3.8   Evaluation and assessment of the measurement results

The described *contact force measurements* are used to assess the quality of overhead contact line and pantograph designs. The following statistical criteria of the forces can be used for assessment:

- the *contact force arithmetic mean* and the *route mean square* value
- the *standard deviation*
- the deviation from the mean value
- the *extreme values*: maximum and minimum contact force values

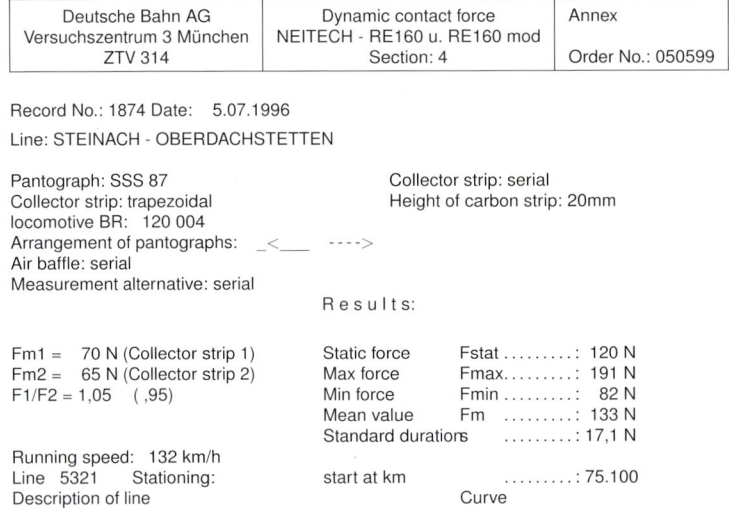

Deutsche Bahn AG Versuchszentrum 3 München ZTV 314	Dynamic contact force NEITECH - RE160 u. RE160 mod Section: 4	Annex Order No.: 050599

Record No.: 1874 Date: 5.07.1996

Line: STEINACH - OBERDACHSTETTEN

Pantograph: SSS 87
Collector strip: trapezoidal
locomotive BR: 120 004
Arrangement of pantographs: _<___ ----->
Air baffle: serial
Measurement alternative: serial

Collector strip: serial
Height of carbon strip: 20mm

R e s u l t s:

Fm1 = 70 N (Collector strip 1)
Fm2 = 65 N (Collector strip 2)
F1/F2 = 1,05 (,95)

Static force	Fstat ....:	120 N
Max force	Fmax.........:	191 N
Min force	Fmin ....:	82 N
Mean value	Fm ....:	133 N
Standard durations	.........:	17,1 N

Running speed: 132 km/h
Line 5321 Stationing:
Description of line

start at km ....: 75.100

Curve

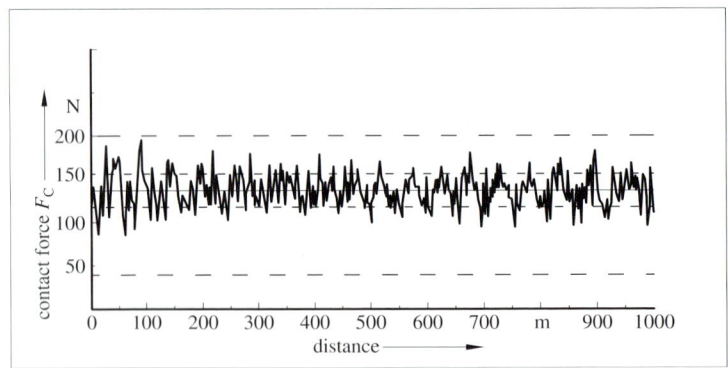

Figure 10.36: Record of a contact force F_C measuring run.

The analogous graphical force recordings obtained along the contact line do not provide any of these values except for the extreme values (Figure 10.36). However, the required quantities can be obtained by statistical methods. The evaluation of the cumulative frequency values of the measurement series has shown that forces can be assumed to follow a *Gaussian distribution*. Using this assumption, the relationships between the most important characteristics of a randomly distributed sample: mean value $\bar{x}$, standard deviation σ and the distribution of the recorded forces are determined. The *standard deviation* can be introduced as a direct criterion for judging the contact performance. As the ideal goal is that the contact force should be constant, it follows that the lower the standard deviation the better the contact performance. The standard deviation σ and the *mean recorded force* can be used to establish limits for dynamic ranges whereby the following characteristics of the Gaussian frequency distribution apply:

68,3 % of all contact force values are between $\bar{x} - \sigma$ and $\bar{x} + \sigma$,
95,5 % of all contact force values are between $\bar{x} - 2\sigma$ and $\bar{x} + 2\sigma$,
99,7 % of all contact force values are between $\bar{x} - 3\sigma$ and $\bar{x} + 3\sigma$.

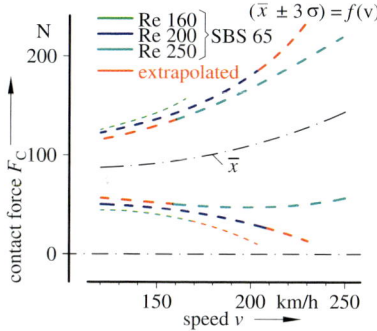

Figure 10.37: Dynamic range $\bar{x} \pm 3\sigma$ of contact forces F_C of DB overhead contact line designs depending on the running speed. Pantograph type SBS 65.

The values $\bar{x} + 3\sigma$ and $\bar{x} - 3\sigma$ form the virtual limits of the *range of dynamic effects*. Therefore, the mean values plus the standard deviations determine the total loading of the system components and their wear. The acceptable minimum is determined by the rise of the electrical contact resistance and the beginning of arcing. Where there are low standard deviations, the force mean value can be reduced by design or adjustments at the pantograph, resulting in a further reduction of contact line wear without any *contact interruptions*.

The standard deviations obtained under the same conditions can be used to compare the contact performance of various overhead contact line and pantograph designs and then optimise the *contact performance* by adjusting design characteristics accordingly.

Figure 10.36 shows the statistical evaluation of a test run with the focus on internal forces. In Figure 10.37 the *dynamic range of the internal forces*, which is $\bar{x} \pm 3\sigma$, is depicted as a function of the running speed for some DB overhead contact line types. From this figure the contribution of the overhead contact line design to the contact performance can be seen. With contact line Re 250, at 250 km/h, internal forces and standard deviations have been achieved which equal those of contact line type Re 200 at 200 km/h. This is the limit speed at which the latter contact line type is used.

Further improvements of pantograph design can contribute to reducing the standard deviation of the forces as well as the aerodynamic component of the uplift force and, therefore, the mean force and the total range of dynamic forces as well. This improves the contact performance of the pantograph accordingly, however, it is essential that the value $\bar{x} - 3\sigma$ should not tend towards zero when the running speed rises.

Apart from the mean and minimum value criteria for the internal forces, the maximum force F_{max} forms the parameter by which local wear can be assessed. The extreme values of the dynamic forces occur mainly at spots with irregularities such as:

 – extreme peaks of uplift values
 – errors during contact wire installations
 – defects in the contact wire and
 – singular masses created by concentrated loads

Consequently, they mainly indicate concentrated deviations from the specified contact wire position rather than the contact performance of the overhead contact line design. Singular extreme values of the recorded forces may fall considerably outside the statistically defined range of dynamic forces. They can be identified as local faults in the overhead contact line. Evaluation of the recordings from regularly scheduled *maintenance test runs* on the overhead contact line network focuses on locating such faults.

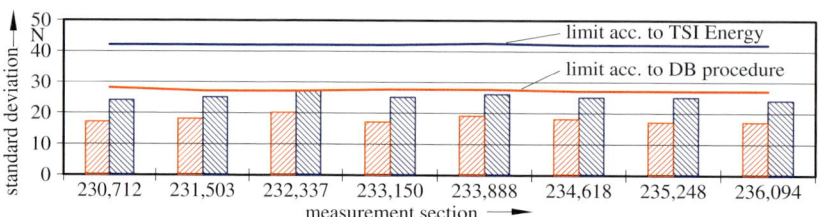

Figure 10.38: Test results obtained at contact line Re 330 with one pantograph, speed 290 km/h according to [10.37]. DB procedure, TSI Energy procedure

The faults in the overhead contact line are assessed and located by checking the force records that are available in the form of a graphical printout or a list (see Figure 10.36). The evaluated documents, containing information on identified fault locations, can be forwarded to maintenance departments immediately after the test runs, so corrective measures can be initiated.

The *evaluation of* recorded *measurements*, including faults, led to the following conclusions:
- any clearly pronounced discontinuity in the dynamic force record, as indicated by a force peak with an amplitude greater than 1,8 times the mean value, is always related to a particular cause
- increased contact wire wear is observed at every fault location, even at relatively low running speeds
- in many cases, the cause of the fault is poor adjustment of the overhead contact line during installation. Corrective measures can be specified by checking the contact line adjustment.
- other reasons may be local mass accumulations, faulty overlap sections and faulty contact line installation above points.

Before 2002, DB assessed the interaction between pantographs and contact lines using the aerodynamically corrected internal forces, however, without dynamic corrections. The TSI Energy [10.12, 10.13, 10.14], on the other hand, specifies requirements for the contact forces including a dynamic correction obtained from acceleration measurements. In EN 50 317 the contact forces are evaluated according to the TSI Energy. The contact forces follow from

$$F_{\text{C}} = \sum_{i=1}^{k_{\text{f}}} F_{\text{C}i} + \frac{m_{\text{s}}}{k_{\text{a}}} \sum_{i=1}^{k_{\text{a}}} \ddot{z}_i + F_{\text{aero}} \quad . \tag{10.71}$$

The symbols are explained in Clause 10.0.

In paper [10.37], measurements of the same line sections were evaluated with both procedures:
- the DB procedure requires a mean value of the aerodynamically corrected internal forces of 120 N with 24 N standard deviation
- the procedures of TSI Energy [10.12, 10.13, 10.14] and EN 50 317 refer to the contact forces obtained by a dynamic correction of the internal forces based on acceleration measurements. For a 250 km/h running speed, the mean contact force should be 130 N and the standard deviation not more than 39 N.

In Figure 10.38, the measured and permissible standard deviations are shown for several test sections of the contact line type Re 330 with one active pantograph at 290 km/h. The comparison demonstrates that the TSI Energy requirements are less stringent than those of the DB procedure. Unsatisfactory interaction performance leads to exceeding the stipulated limits in

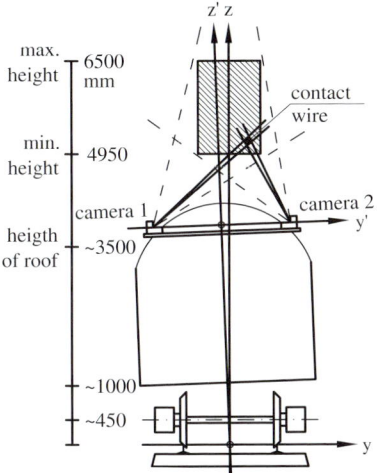

Figure 10.39: Measuring principle for recording the contact wire position.

both cases. However, using the TSI Energy criteria [10.14] should not lead to unacceptable arcing or increased wear of the contact wire and collector strip.

10.4.4   Measuring overhead contact line position and contact wire thickness

The correct *position of the overhead contact line* relative to the track is very important to contact performance and operational security. An *optical contact wire position measuring system* is used for checking the wire position before acceptance of newly installed overhead contact lines (Figure 10.39).

The method records the contact wire position with a resolution of 1 mm relative to the track position with four high-resolution diode line cameras (6 000 or 8 192 pixels) and a specifically designed evaluation computer. The measuring intervals are approximately 3,2 ms. The system uses active lighting of the contact wire and can be used under virtually all lighting conditions. The cameras are arranged on a base frame with high torsional stiffness and may be installed on any kind of vehicle. Sensors for measuring and correction of *vehicle sway* are also arranged between the car body and the wheel set bearing. The recorded data is then evaluated online, logged digitally and output graphically on a computer screen or a printer. Figure 10.40 shows a typical printout of the results.

In addition to the contact wire position, in vertical and transverse directions, various secondary information can also be displayed, for example the line kilometres, the location of poles, the location of droppers, the contact wire gradients, contact forces, the uplift and the elasticity. The pole locations are identified and recorded automatically.

The specified adjustment position of an overhead contact line is recorded during acceptance procedures of a newly installed contact line system and after re-adjusting existing installations. It would be useful to measure the deviation from the specified position in every case. But the initial evaluation system has to be provided with the entire contact wire position data requiring considerable effort. A simple and practical solution consists of a graphical presentation of the vertical and lateral position relative to the track. In this kind of recording,

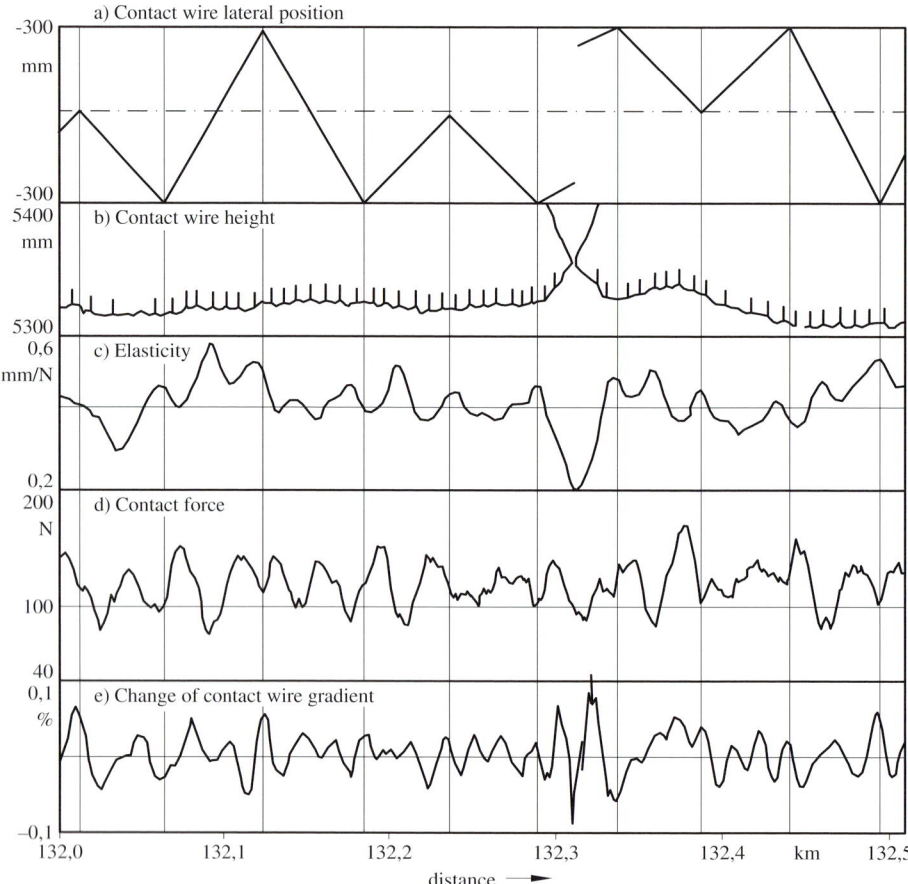

Figure 10.40: Typical printout of contact wire position measurements with lateral displacement, contact wire height, contact force and contact line gradient.

deviations from the specified position can be identified quickly and easily. The compliance of stagger with the specified limits can be checked at the same time.

As contact force measurements have shown, sudden and large force variations are often caused by discontinuities in the overhead contact line position and it is sufficient to visually inspect the contact force graphical records of the tested line section. Computer systems can also be used to detect and automatically display any abnormal situation recorded during the test runs.

The equipment to monitor the contact wire position can also be used to measure the *contact wire thickness* using high-resolution optical cameras that permit detection of the contact wire with a precision of 0,1 mm. By evaluating the width of the contact wire mirrored running surface and considering the contact wire diameter, the residual thickness of the contact wire can be determined automatically. This is done by using the data from the four cameras and calculating online the contact wire position relative to the cameras. This enables continuous monitoring of the *contact wire wear* while measuring the contact wire position. This approach will also allow detection of premature wear of the contact wire at critical spots and may

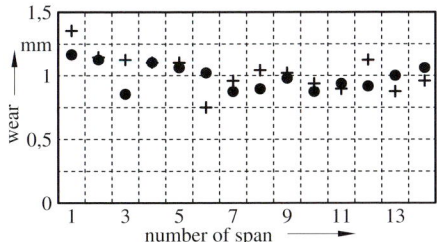

Figure 10.41: Comparison of optically measured contact wire thickness with measurements carried out by hand.

\+ measurements by hand

• measurements with optical system

initiate corresponding maintenance and adjustment procedures to extend the contact wire life cycle. Figure 10.41 shows a comparison of contact line thickness measurements by hand and by the optical detection system, with good agreement.

Reference [10.38] presents the OHV Wizard system for touch-less measurement of the contact wire position. The equipment uses ultrasonic impulses for measurements during a measurement run. Compared with optical principles, it offers the advantage of being able to operate in direct sunshine, light rain and fog. The ultrasonic impulses are reflected by the contact wire and received by a sensor. The position of the contact wire is determined with respect to location along the track and is recorded during the running period. Measurements are also possible at the live contact line. The accuracy is quoted as $\pm 2\,\%$ and software is available for the evaluation. The contact wire thickness can also be measured using this equipment.

10.4.5 Assessment of dynamic characteristics of pantographs

10.4.5.1 Introduction

An oscillation test stand can be used to analyse and assess the *dynamic characteristics of pantographs*. The DB test stand for pantographs has been described in [10.39]. For testing purposes, the pantographs are connected to a mechanical shaker via the collector strips and subjected to vibrations. Relevant parameters such as forces, accelerations and displacements are monitored by the measuring technology and then evaluated.

10.4.5.2 Measurement of frequency response

To determine the pantograph's vibration performance, in all degrees of freedom and without any undue reactions, a coupling as loose as possible between the vibration exciter and the collector strips of the pantograph is necessary during a *frequency response analysis*. From the exciter periodical or stochastic excitations of sufficient amplitude within the frequency range of approximately 0,1 to 70 Hz is transmitted to the pantograph. Depending on the pantograph equipment and force excitation system with measuring devices, a frequency analyser is used to determine:

 – the *dynamic apparent mass* graphs (an example is shown in Figure 10.11)
 – the *mechanical impedance*
 – the *transfer function of disturbances*
 – the *transfer function of contact force* recording systems

These quantities can be obtained relative to the frequency and presented by amplitude and phase response functions. Information on the dynamic performance and the operating quality of a pantograph can be deduced from the pattern of the functions.

To assess the dynamic characteristics of pantographs, the presentation of the dynamic apparent mass representing the relationship between the input force (contact force) and the sum of the resulting collector strip accelerations has proven informative. A pantograph, whose curve of apparent mass shows only a few slightly outstanding natural vibration modes as well as a low level of apparent mass in total, will also demonstrate a favourable running performance.

Similar conclusions can be drawn from the *disturbance transfer function* of a pantograph. The disturbance transfer function is given by the ratio of the contact force to the amplitude of excitation of a contact line model, presented by a mass-spring-damper system, coupled to the pantograph.

Using frequency response analysis, dynamic running characteristics of pantographs can be studied on the test stand without expensive running tests on track. Additionally, measures can be decided that will improve the pantograph dynamic.

Measurements and analysis of frequency responses yield additional important data for the establishment and validation of simulation models that describe the dynamic behaviour of pantographs in a mathematical format. These allow numerical studies of the interaction between the pantographs and contact lines in connection with simulation models for the overhead contact line.

10.4.5.3   Structural analysis

By using stroboscopic lights, simple optical *structural analysis* can be carried out on vibration test stands. Short-period, intermittent lighting of individual pantograph parts enables the vibration modes of components to be monitored. From the formation of vibration nodes and antinodes, at which fatigue failures may occur under extreme conditions of usage, information on the material stressing during operations can be obtained.

10.4.5.4   Modelling of line running

During frequency response analysis, periodical or purely stochastic excitation signals are transmitted to the pantograph. However, an assessment of the motion curve and the mechanical stresses occurring during real operation is limited.

By simulating line running in a test stand [10.39], it is possible to produce realistic motions of a pantograph interacting with a contact line with particular characteristics. Effects to be considered include changes in the contact wire height, its lateral position (*contact wire stagger*) and the highly dynamic motion effects imposed on the pantograph by interference excitation using simplified contact line models. The evaluation of relevant parameters such as internal force and contact force permits a precise assessment of the *running performances of pantographs* with various overhead contact line designs.

Figure 10.42 shows schematically, the structure of a test stand established at Deutsche Bahn AG for carrying out *simulated test runs on lines* [10.39]. A gantry (axis A3) is used to simulate the gradual changes of the contact wire position, for example, at contact wire lowerings or rises. A moveable slide, arranged horizontally on the gantry (axis A2), is used to simulate the contact wire stagger. An actuator arranged at the horizontal slide, acting in the vertical direction (axis A1), applies high-frequency excitation signals to the pantograph through a mass-spring-damper system arranged in between and used for the simplified modelling of the contact line. The signals for excitation along the different axes are deduced from data

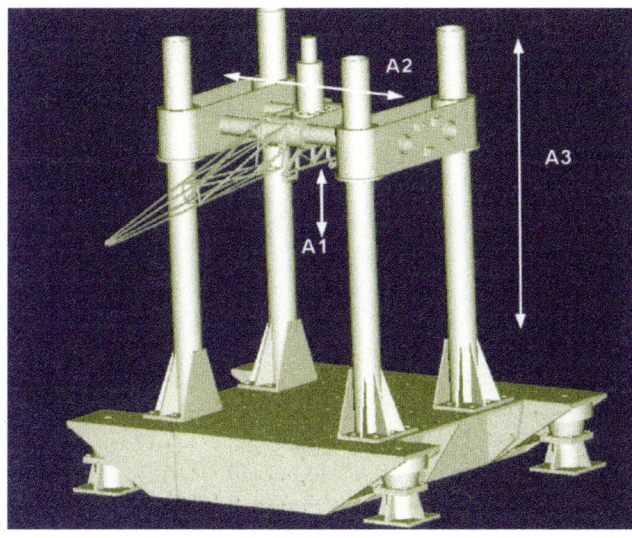

Figure 10.42: Schematic presentation of a test stand to carry out simulated pantograph test runs on lines.

recorded on test runs conducted on real lines or from computer simulations taking into account the contact line model.

10.4.6 Measurement of contact wire uplift and dynamic contact line elasticity

10.4.6.1 Stationary measurement of contact wire uplift

According to EN 50 317, the *contact wire uplift* can be determined either by stationary measuring equipment at a contact line support or by mobile measuring equipment installed on a vehicle. To monitor the development of dynamic uplift with time, at a support, stationary *uplift measuring devices* are used. They are required for:

- acceptance or determination of the maximum permitted speed of new vehicles or pantographs, in conjunction with the overhead contact line design so as not to exceed the maximum permissible uplift
- the stationary monitoring of pantographs in commercial operations

The uplift movement is recorded using a potentiometer connected to the steady arm with a pre-tensioned rope. Isolation of potentials is achieved with an insulating section within the rope (Figure 10.43). The signal is transferred by an optical coupler to the measurement amplifier, connected directly to a PC. Because the uplift is a function of the running speed, this is registered automatically by two contacts at the rails and the recorded data is transferred by GMS radio. This device has the advantage of simply monitoring the uplift of the contact wire at a support over time.

Figure 10.44 shows the *vertical movement of a contact wire* at a support for a passing train with two pantographs spaced 270 m apart. The leading pantograph lifts the contact wire by approximately 80 mm. The trailing pantograph then runs along a contact wire oscillating at one of the natural frequencies of the overhead contact line. However, the resulting uplift is almost the same as that caused by the leading pantograph. After the passage of the two pantographs, the contact line oscillates at an amplitude of ±20 mm with relatively low damping.

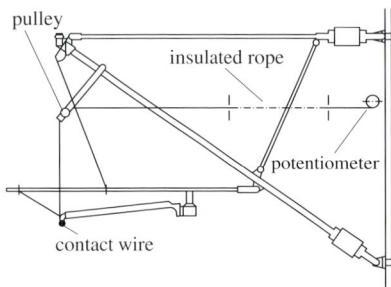

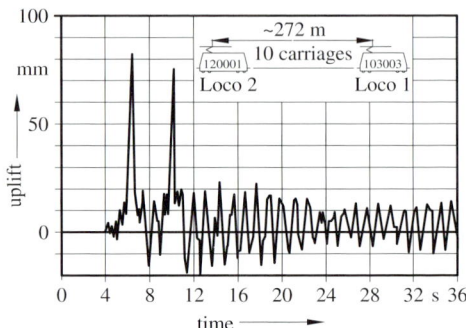

Figure 10.43: Stationary recording device for contact wire uplift.

Figure 10.44: Contact wire uplift as a function of time, train with two pantographs travelling at $v = 270$ km/h.

In view of the *pantograph diagnostics* during commercial operation, the uplift during a pantograph passage enables conclusions about the condition of the pantograph. This is possible since maximum uplift is proportional to the contact force of the pantograph at the location of the support. The contact force is the sum of three components:

$$F_C = F_{C\,stat} + F_{aero} + F_{C\,dyn} \quad , \tag{10.72}$$

where $F_{C\,stat}$ is the contact force exerted by the drive of the pantograph at standstill, F_{aero} denotes the increase in contact force due to aerodynamic effects on the pantograph and $F_{C\,dyn}$ represents the component of the contact force due to the dynamic interaction between the pantograph and overhead contact line.

For a particular train running speed, any marked increase or decrease registered in uplift indicates disturbances or *defects at the pantograph*. These can be caused by:

- a too high or too low static contact force $F_{C\,stat}$ caused by:
  - incorrectly adjusted static contact force
  - a large change in the contact strip mass, caused by a collector strip, for example, worn beyond acceptable limits
- a too high or too low *aerodynamic force* F_{aero} caused by:
  - incorrectly adjusted or damaged wind baffles
  - obliquely worn contact strips
- a too large a dynamic contact force component $F_{C\,dyn}$ caused by defective mechanical parts on the pantograph, for example dampers.

Observation of uplift is an important tool for automatic pantograph diagnostics that can effectively monitor many defects. However, the actual reasons for the defects will not be identified.

10.4.6.2 Mobile measurement of the contact wire uplift

Contact wire uplift can be measured from a moving vehicle using the measurement device described in Clause 10.4.3. The initial position of the contact wire is recorded on a run with a train hauled by a diesel engine without a raised pantograph on the train. Then, during an additional test run with an electric traction unit and a pantograph raised, the contact wire position is measured again. The optical measuring system is installed directly at the pantograph.

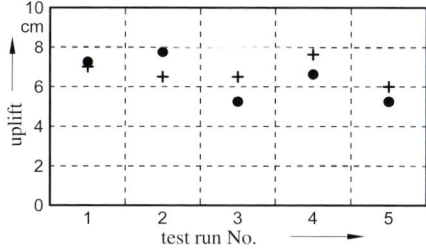

Figure 10.45: Comparison of the results of mobile and stationary contact wire uplift measurements (five test runs, one stationary measuring location).
+ stationary measurements
• mobile measurements

The contact wire uplift can be determined by subtracting the registered contact wire heights during the two measuring runs. This method requires equipment to precisely measure the kilometre position and the running distance. Only with a precise correlation of the two runs can a subtraction of the registered contact wire heights give accurate values. Comparison of the mobile contact wire uplift measurements with the data obtained from a stationary measuring installation demonstrates good compatibility as shown in Figure 10.45.

10.4.6.3   Measurement of the dynamic elasticity

Measurement of the *overhead contact line elasticity* can be achieved by supplementing the measuring equipment described in Clause 10.4.6.2. with equipment for measuring the contact force as per Clause 10.4.3. The measuring process is similar to that described in Clause 10.4.6.2. In addition to the contact wire uplift, the contact force is recorded and synchronised with the running distance. By dividing the uplift and the contact force, the dynamic contact line elasticity is obtained (Figure 10.40).

10.5   Effects of contact line design

10.5.1   Introduction

A large number of *design parameters* affect the dynamic behaviour of an overhead contact line, especially at high speeds. Theoretical studies of the overhead contact line and pantograph interaction (Clause 10.2) have led to the definition of a series of criteria, that can also be used to assess the effects of the individual parameters. Measurements of the interaction of the two components also enable empirical studies of the effects of the individual parameters on the *contact quality* [10.7, 10.40] of a moving train (see Clause 10.4). The conclusions drawn from these studies are used as a basis for structural and mechanical design of energy supply systems using overhead contact lines.

10.5.2   Criteria for overhead contact line designs

10.5.2.1   Elasticity and uplift

The uplift of an overhead contact line should be kept to a minimum in order to achieve good contact quality. However, a certain amount of elasticity is required. The mechanical design of the supports limits the possible vertical contact wire uplift at the supports. At low and medium speeds, i. e. at speeds up to approximately 50 % of the wave propagation speed, the uplift is proportional to the elasticity of the overhead contact line and the contact force exerted by

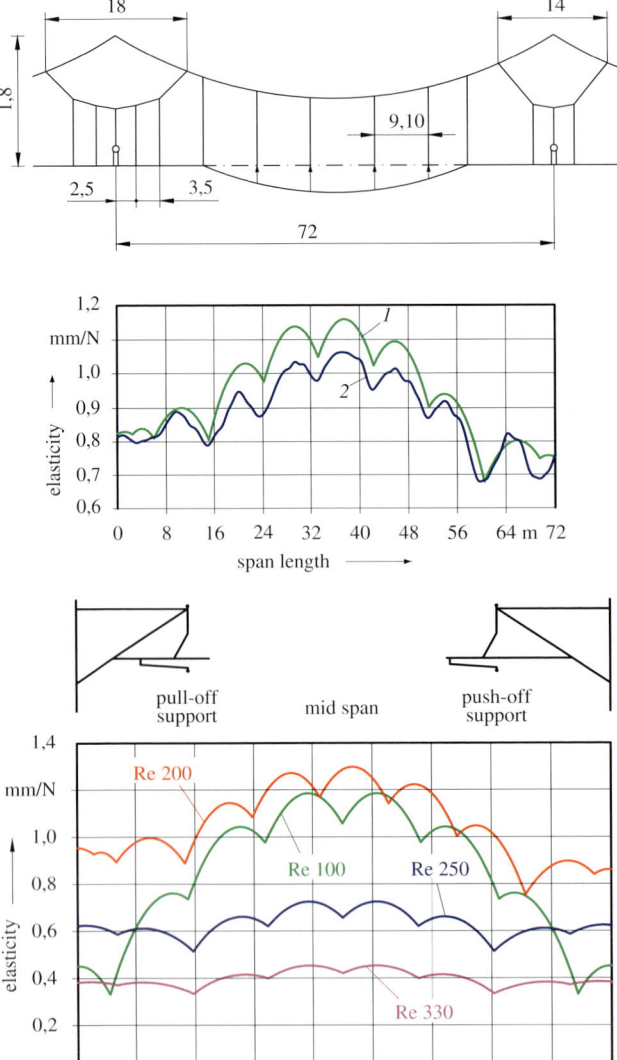

Figure 10.46: Elasticity of contact line type Re 200, span length 72 m, comparison between calculation and measurement.
1 calculation
2 measurement

Figure 10.47: Calculated elasticity of the line types Re 100, Re 200, Re 250 and Re 330. Re 100 and Re 200 span length 80 m, Re 250 and Re 330 span length 65 m.

the pantograph. To maintain favourable contact quality at increasing speeds, the contact force also needs to be increased. Therefore, the elasticity needs to be kept as low as possible to limit the resulting uplift.

The *elasticity* of an overhead contact line can be calculated with sufficient accuracy using a mathematical model based on the finite-element method (FEM) [10.19, 10.21, 10.27]. Figure 10.46 shows a calculated elasticity graph of DB's overhead contact line design Re 200. Reference [10.21] contains a description of a suitable calculation method. As a comparison, the values shown for various standard DB overhead contact lines in Figure 10.47 and Table 10.12 were obtained by an extended series of measurements.

Table 10.12: Elasticity and non-uniformity degree of contact lines.

Contact line	Running speed in km/h	Elasticity in mm/N			Non-uniformity in %	
		target	approximation (10.73)	calculation Figure 10.47	approximation (10.74)	calculation Figure 10.47
Re 100	up to 100	1,00	1,00	1,20	50	46
Re 200	up to 200	1,20	1,15	1,30	20	21
Re 250	up to 280	0,60	0,62	0,70	10	20
Re 330	above 280	0,40	0,45	0,45	10	12

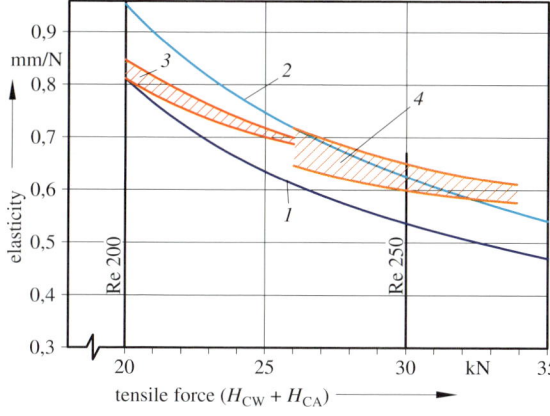

Figure 10.48: Elasticity in the middle of a span, overhead contact lines with and without stitch wires, plotted as a function of the tensile forces. Comparison of measured values with values estimated by calculation. Span length 65 m.
$1 \; l/[4,0 \cdot (H_{CW} + H_{CA})]$
$2 \; l/[3,5 \cdot (H_{CW} + H_{CA})]$
3 measured values Ri 100, BZ II 50
4 measured values RiS 100, BZ II 70

The *elasticity at the middle of a span* can be obtained using the equation

$$e = l_{sp} / [k_e \cdot (H_{CW} + H_{CA})] \qquad \text{in mm/N,} \tag{10.73}$$

where:

l_{sp} longitudinal span in m,
H_{CW} contact wire tensile force in kN,
H_{CA} catenary wire tensile force in kN,
k_e numeric factor (constant) ranging between 3,5 and 4,0

as described in [10.42]. For overhead contact lines without a stitch wire, $k_e = 4,0$, for those with a stitch wire $k_e = 3,5$.

Figure 10.48 and Table 10.12 demonstrate that Equation (10.73) gives a reliable approximation of the midspan elasticity of an overhead contact line. The elasticity at the supports depends on the structure of the overhead contact line. At the supports, contact lines without stitch wires achieve an elasticity of only 30 % to 50 % of the midspan values. However, by adding suitable stitch wire arrangements, the elasticity at the supports can be increased to approximately 90 % of the midspan values (Figure 10.47 and Table 10.12).

As train speeds increase, the *uniformity of the elasticity* becomes increasingly important. The *non-uniformity degree of elasticity u* as defined by

$$u = 100 \cdot (e_{max} - e_{min}) / (e_{max} + e_{min}) \qquad \text{in \%,} \tag{10.74}$$

where e_{max} and e_{min} are the maxima and minima of the elasticity within a span. The non-uniformity characterises the elasticity variation along a span. Values of less than 15 % as

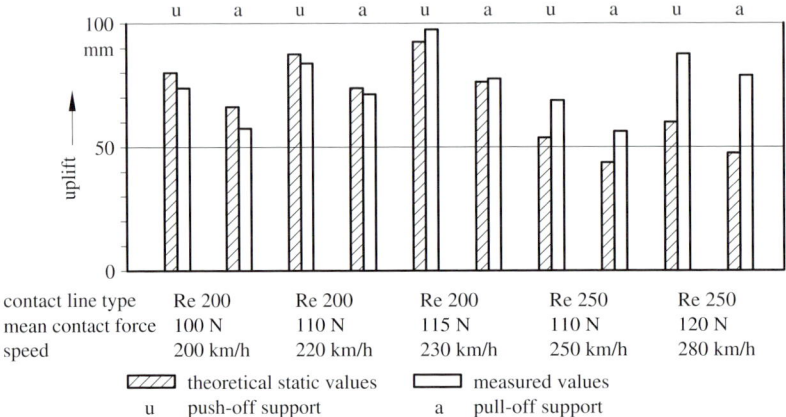

Figure 10.49: Comparison of theoretical static uplift values and measured dynamic uplift values.

defined in the 2008 edition of the TSI Energy [10.12] are desirable for overhead contact lines for high-speed traffic and can be achieved (see Table 10.12). The data given in Table 10.12 represent findings from successful applications of contact line designs in Germany and some European countries. The uniformity degree of contact line designs without stitch wires could be higher than the values recommended in Table 10.12.

The mean value of the contact force exerted by the pantograph and the elasticity of the contact line determine the contact wire uplift. At high train speeds, a *dynamic uplift component* is superimposed on the static one. The dynamic component, which increases sharply with speed, is a function of the dynamic characteristics of the overhead contact line. Figure 10.49, taken from reference [10.42], shows the development of the dynamic uplift compared with the calculated static values as a function of the train speed. The calculated static values are the product of the mean contact force exerted by the pantograph and the elasticity of the overhead contact line. For an overhead contact line installation designed for 200 km/h (Re 200), the measured dynamic uplift values exceed the calculated values only at speeds of 230 km/h and above. At higher speeds, a sharp increase in the dynamic component is noticeable.

10.5.2.2 Dynamic criteria

A series of *dynamic criteria* was derived in Clause 10.2 that can be used to formulate the specifications for design of overhead contact lines with certain desired characteristics. The *wave propagation speed* of transverse waves along the contact wire, as described by Equation (10.9), is one of the fundamental dynamic design parameters. This parameter, in relation to the train speed, enables the determination of the *Doppler factor* using (10.60). The Doppler factor approaches zero as the train speed approaches the propagation speed of mechanical waves running along the contact wire. The *reflection coefficient*, according to (10.65) is another parameter that determines the dynamic behaviour of overhead contact lines. However, it is only a function of the overhead contact line design data, i. e. it does not depend on the train speed. As explained in reference [10.7], the ratio of the reflection coefficient to the Doppler factor is called the *amplification coefficient* (see (10.63)) and depends on the train speed.

The impact of the dynamic criteria on the behaviour of an overhead contact line can be verified by measurements. During preparations for high-speed trial runs in 1988 [10.1], one of

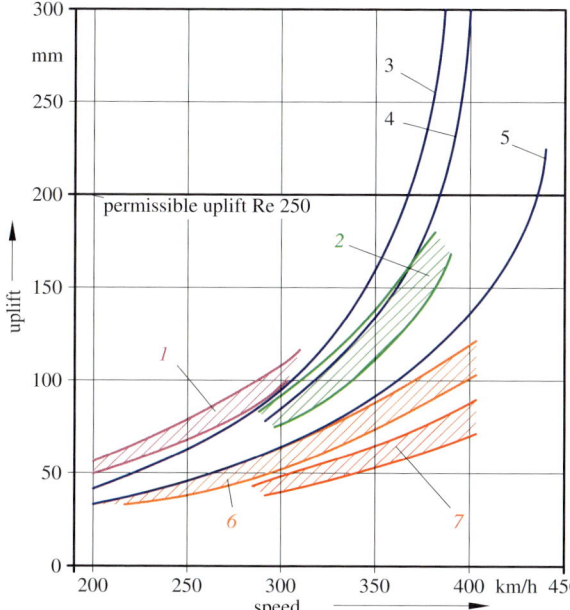

Figure 10.50: Vertical uplift of overhead contact lines as a function of the train speed [10.1].
1 measured values at Re 250
2 measured values SNCF
3 uplift expected at Re 250, contact wire AC-120–CuAg0,1, H_{CW} 15 kN, $\sigma_{CW} = 125$ N/mm^2
4 uplift expected at SNCF, contact wire AC-120–Cu, H_{CW} 20 kN, $\sigma_{CW} = 133$ N/mm^2
5 uplift expected at Re 250V, contact wire AC-120–CuAg0,1, H_{CW} 21 kN, $\sigma_{CW} = 175$ N/mm^2
6 measured values Re 250V, line in the open
7 measured values Re 250V, line in tunnels

the decisive issues was whether an overhead contact line design of type Re 250 would permit speeds in the region of 400 km/h. During trials carried out by the SNCF in 1981, the dynamic uplift of the overhead contact line limited the maximum speed to 380 km/h [10.43]. The uplift reached values of around 200 mm.

In November 1986, test runs using the ICE/V test train on a section of the Hanover–Würzburg line, the maximum uplift measured at a support was 105 mm with the maximum speed of 310 km/h. The measured uplift increased more than proportionally with the train speed (Figure 10.50). Since the mean pantograph contact force was constant at 120 N, the dynamic effects had increased the uplift considerably.

In 1981, the French railways SNCF used an experimental overhead contact line with a contact wire cross section of 150 mm^2 and a tensile force of 20 kN, i. e. a stress of 133 N/mm^2 [10.43], for which a wave propagation speed of 440 km/h can be deduced. The uplift values measured and calculated for this overhead contact line are plotted as a function of the train speed in Figure 10.50. According to reference [10.43] the measured values obtained by the SNCF at 300 km/h are lower than those monitored for the Re 250 design. At 400 km/h, uplift values of approximately 300 mm were to be expected with the SNCF overhead contact line. Since the wave propagation speed of the Re 250 design is 426 km/h and, therefore, lower than that of the SNCF's experimental overhead contact line system, the dynamic effects would be even greater. The uplift in a Re 250 system with no modifications, would be considerably higher than 300 mm at a train speed of 400 km/h. Such high values could not be accepted because the uplift range of a standard Re 250 design is limited to 200 mm by structural parameters. Consequently, it would not have been possible to achieve a train speed of 400 km/h using a standard Re 250 overhead contact line.

To reduce the *dynamic uplift*, the Doppler factor needs to be increased. So, a higher wave propagation speed of the contact wire is required. According to Equation (10.9), this can be achieved by increasing the contact wire stress but not by raising the tensile force just by in-

Table 10.13: Dynamic characteristics of overhead contact lines used for high-speed test runs.

	Units	SNCF 1981	Re 250 DB 1988	Sicat H1.0	SNCF 1991	SNCF 2007
contact wire		AC-150– Cu	AC-120– CuAg	AC-120– CuMg	AC-150– CuCd	AC-150– CuSn
tensile force	kN	20	21	27	33	40
tensile stress	N/mm^2	133	175	225	229	267
catenary wire		Bz II 65	Bz II 70	Bz II 120	Bz II 70	Bz II 116
tensile force	kN	14	15	21	15	20
wave propagation speed	km/h	440	504	572	560	623
reflection coefficient		0,363	0,392	0,469	0,314	0,38
elasticity at middle of span	mm/N	0,53	0,44	0,39	0,33	0,26
maximum speed	km/h	380	407	–	515	575
Doppler factor						
at 250 km/h	–	0,275	0,337	0,392	0,383	0,427
at 450 km/h	–	–	0,057	0,120	0,109	0,218
at maximum speed	–	0,073	0,106	–	0,042	0,040
amplification coefficient						
at 250 km/h	–	1,3	1,2	1,2	0,8	0,9
at 450 km/h	–	–	6,9	3,9	2,9	1,7
at maximum speed	–	5,0	3,7	–	7,5	9,6

creasing the cross-sectional area and retaining the same stress. According to Equation (10.60), to keep the Doppler factor above 0,1 at a train speed of 400 km/h, the wave propagation speed should be approximately 490 km/h. As per (10.9), this would correspond to a tensile force of approximately 20 kN acting on the 120 mm^2 contact wire, or a stress of 167 N/mm^2. To provide the best possible conditions for the trial runs, a force of 21 kN was applied to the contact wire [10.1]. This increased the wave propagation speed to 504 km/h, the reflection coefficient was 0,392 and, at 400 km/h, the amplification coefficient was 3,4 (see Table 10.13).

From Figure 10.8, a considerable improvement of the dynamic criteria because of the increased contact wire stress is obvious. Particularly at speeds above 350 km/h, considerably reduced overhead contact line uplift could be expected (Figure 10.50). These expectations were fully confirmed by the high-speed trial runs. The 400 km/h barrier was broken for the first time and a top speed of 407 km/h achieved. The measured values of overhead contact line uplift are plotted in Figure 10.50. The maximum value was roughly 140 mm. The calculated predictions concerning the uplift were confirmed. This example illustrates the effect of the *dynamic criteria* on the contact performance.

During the record run of SNCF in 2007, which reached a speed of 575 km/h, the Doppler factor was only 0,040 and the amplification coefficient had risen to 9,5. The gap to the wave propagation speed was very small.

10.5.3 Overhead contact line design parameters

10.5.3.1 Cross-sectional areas and tensile stress

The *cross-sectional areas of contact* and catenary *wires* can have a crucial effect on the behaviour of an overhead contact line when being transversed by a pantograph at high speeds. According to Equation (10.73), the requirement for a low and uniform *elasticity* can be satisfied by high tensile forces on the contact wire and the catenary wire. This can be achieved by

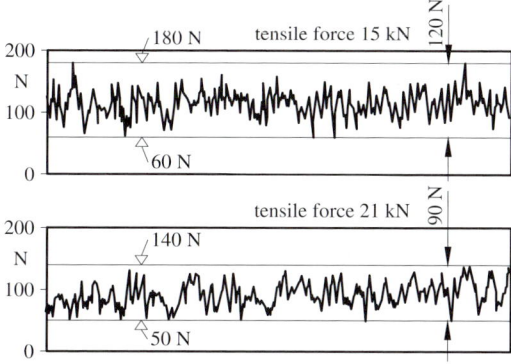

Figure 10.51: Contact forces measured at contact wires subject to tensile forces of 15 kN and 21 kN, contact wire AC-120−CuAg, overhead contact line design Re 250.

using large cross-sectional areas and corresponding stresses. The need to handle and install solid contact wires limits the maximum cross-sectional dimensions of the contact wire. As a result, contact wire cross-sections are limited to a maximum of 150 to 170 mm^2, whereby even these cross-sections incur a great risk of localized defects such as bends etc. being created during the installation work, subsequently leading to rapid *wear* at the respective locations. Special care is required when producing and installing contact wires with large cross-sections.

If the tensile stress is kept constant, increasing the contact and catenary wire cross sections leads to a linear reduction in the elasticity. For this reason, contact wires and catenary wires with as large a cross section as possible are desirable to achieve low elasticity. However, the investment increases in proportion to the cross section, and for commercial reasons, invest-ment should be kept as low as possible.

In the course of development work on the contact line type Re 250, DB also tested over-head contact lines using contact wires AC-100−CuAg and AC-120−CuAg. Both wires were subjected to a tensile stress of 125 N/mm^2 for speeds up to 280 km/h [10.44]. The bigger cross-section yielded lower standard deviations of the contact forces. As can be deduced from (10.9), equal wire stresses cause equal wave propagation speeds along the wire along with equal Doppler factors. For this reason, increasing the cross-sections while maintaining the same stress would not contribute to progress with respect to the suitability of an overhead contact line for operation close to maximum speeds.

Assuming equal cross sections, increasing the stress will reduce the elasticity of the overhead contact line as can be seen from (10.73) while at the same time increasing the wave propaga-tion speed along the wire, as demonstrated by (10.9). Equation (10.65) shows that increasing the contact wire stress will also affect the reflection coefficient.

Increasing the contact wire stress improves all significant parameters of an overhead con-tact line, as well as the *dynamic performance*. Figure 10.51 shows the contact force graphs recorded when travelling along an overhead contact line of standard design Re 250 with con-tact wire forces of 15 kN and 21 kN, respectively, at a speed of 280 km/h. The dynamic band-width is reduced considerably and the contact force peaks are lower on the wire subjected to the higher tensile force. In Figure 10.52, the observed standard deviations are shown for tensile forces of 15 kN and 21 kN as a function of the train speed. The standard deviation achieved with the tighter wire is 3 N lower on average, which is equivalent to a 15 % decrease at train speeds of 250 km/h. Consequently, increasing the stress in the contact wire is one of the most suitable measures for adapting an overhead contact line to *high-speed traffic*.

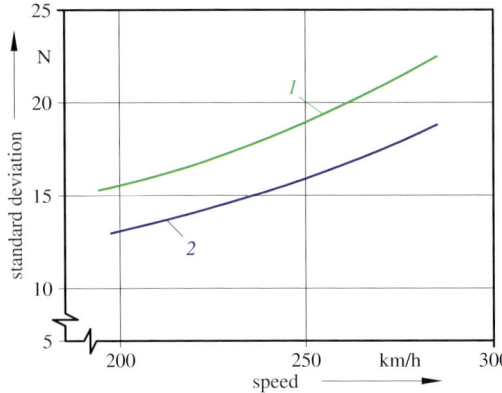

Figure 10.52: Standard deviation of contact forces plotted against running speed, contact wires subject to tensile forces of 15 kN (*1*) and 21 kN (*2*), contact wire AC-120 – CuAg, overhead contact line design Re 250.

Equation (10.65) indicates that the stress in the catenary wire also affects the reflection coefficient. A reduction of the catenary wire stress is desirable to obtain a low amplification coefficient. With the objective of achieving a low elasticity, the DB standard overhead contact line Re 250 was originally designed to operate with a catenary wire tensile force of 19 kN, which corresponds to a stress of roughly 290 N/mm². Reducing the stress to approximately 210 N/mm² lowered the reflection coefficient from approximately 0,46 to 0,42. For a train speed of 280 km/h, the *amplification coefficient* was reduced from 2,2 to 2,0, i. e. by 10 %.

For this reason, the tensile force on the catenary wire was reduced to 14 kN in some overhead contact line sections. The dynamic contact forces did not differ significantly from those observed in the sections with a tensile force of 19 kN. As a result, all other new overhead contact lines were built with contact wire and the catenary wire tensile forces of 15 kN each. Table 10.13 also includes the specifications of the overhead contact line type Sicat H1.0 [10.45]. This design was chosen to achieve favourable performance characteristics at speeds above 300 km/h.

10.5.3.2 Span lengths and system height

The *span length* affects the system elasticity as expressed in Equation (10.73) which shows the elasticity at the midpoint of the span being proportional to the span length. Reducing the span lengths will also reduce the *elasticity* and the *non-uniformity* of the overhead contact line system. Shorter spans are preferable for high speeds but the larger number of poles and foundations required imply higher expenditures for installation. The demand for longer spans to reduce investment without leading to unacceptable interaction characteristics, poses an optimization problem.

When designing the contact line Re 250 [10.44, 10.46] the target was to halve the elasticity in comparison with standard designs Re 160 and Re 200. Equation (10.73) concludes that the increase in contact and catenary wire tensile forces to 15 kN would have decreased the elasticity from 1,14 mm/N to 0,76 mm/N, i. e. to only two thirds of the Re 160 design [10.11]. To achieve the design target, the span lengths were limited to 65 m resulting in an elasticity of 0,55 N/m.

For the Sicat H1.0 design [10.42] the elasticity target was 0,40 mm/N and was complied with by increasing the contact and catenary wire tensile forces to 27 kN and 21 kN, respectively. The midspan elasticity of the Sicat H1.0 contact line is 0,39 mm/N.

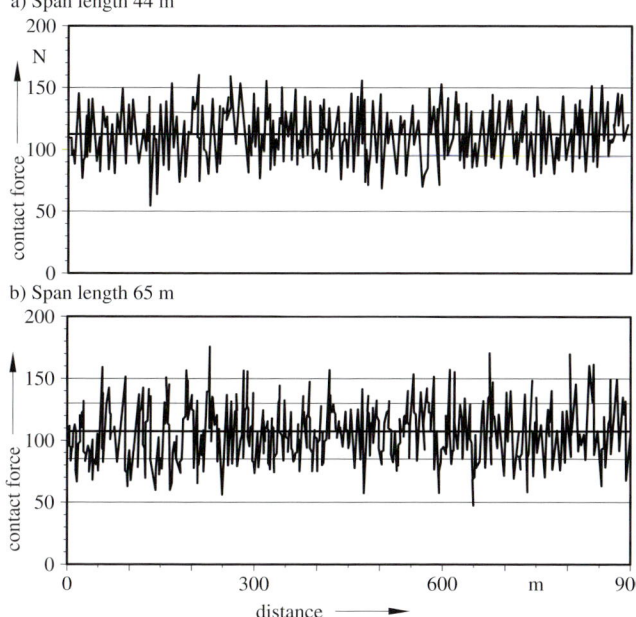

Figure 10.53: Dynamic contact forces as a function of the span length; system height 1,80 m, overhead contact line Re 250.

On real installations, the span lengths vary due to local constraints offering the possibility of carrying out contact force measurements on line sections with differing span lengths and comparable results. Figure 10.53 shows *graphs of contact forces* recorded along span lengths of 44 m and along span lengths of 65 m under otherwise equal conditions. At a train speed of 280 km/h, the standard deviation of 19 N obtained with shorter spans is clearly lower than the observed value of roughly 22 N with 65 m long spans. Therefore, reducing span lengths and consequently lowering the contact line elasticity contributes to reducing dynamic force effects.

The *system height* describes the distance between the catenary wire and the contact wire at the supports and does not occur as a parameter in any of the expressions for the characteristic properties of overhead contact lines. The system height of overhead contact lines in tunnels of German high-speed lines is 1,10 m and the span length is 44 m [10.47]. Figure 10.54 shows contact forces for lines having the same span lengths of 44 m, but with system heights of 1,80 m as installed on open line sections compared with spans in tunnels with 1,10 m system height. The higher system height appears to have more favourable dynamic characteristics. This can also be seen in the difference between the standard deviation values at a train speed of 280 km/h being reduced from 23 N to 19 N. Overhead contact lines for high-speed railways should be designed with adequate system heights that permit a minimum dropper length of 0,6 m.

10.5.3.3 Pre-sag and stitch wires

Adjusting an overhead contact line so that there is an *initial sag*, called *pre-sag* at the mid-point of a span relative to the supports, is based on the concept that if the elasticity at the midspan is higher than that at the supports, the pantograph will lift the contact wire to a greater extent at that position. The pre-sag aids in achieving contact points at a constant

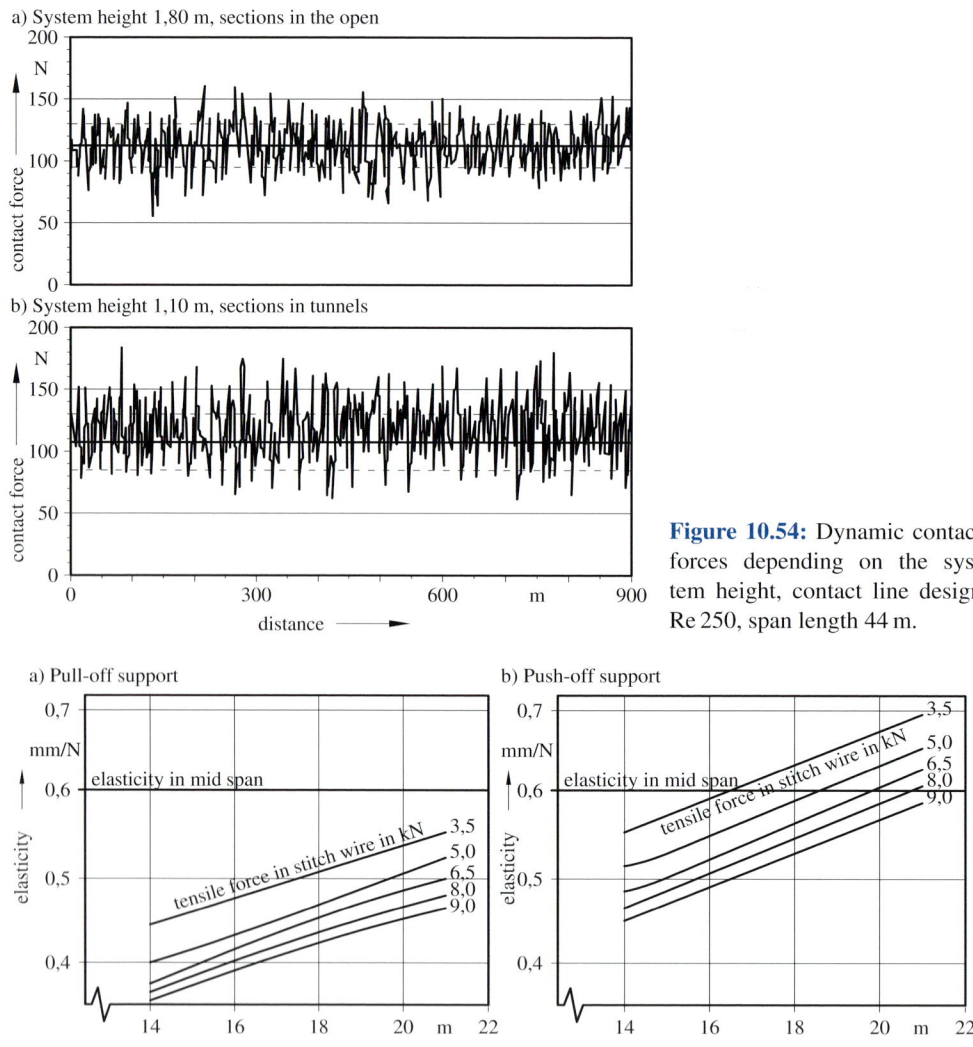

a) System height 1,80 m, sections in the open

b) System height 1,10 m, sections in tunnels

Figure 10.54: Dynamic contact forces depending on the system height, contact line design Re 250, span length 44 m.

a) Pull-off support

b) Push-off support

Figure 10.55: Effect of the length and tensile force of stitch wires on the elasticity of the overhead contact line at supports, overhead contact line design Re 250, contact wire RiS 120, tensile force 15 kN, catenary wire Bz II 70, tensile force 15 kN.

height relative to top of rail. However, this desired effect is only achieved if the contact forces exerted by the pantograph are independent of the *pantograph design* and train speed. As this is not the case, it is only possible to adjust the system for a constant contact point height for static uplift conditions and for a specific contact force.

The overhead contact line designs of Re 160 and Re 200 still show relatively great differences between the elasticity at the supports and at middle of spans. In 1962 [10.29], designs with and without pre-sag were tested and a pre-sag of approximately 50 mm improved the contact quality for 80 m long spans. *Stitch wires* installed at the supports, increase the elasticity there and can, therefore, lead to considerably more uniform elasticity along a span.

a) With stitch wires

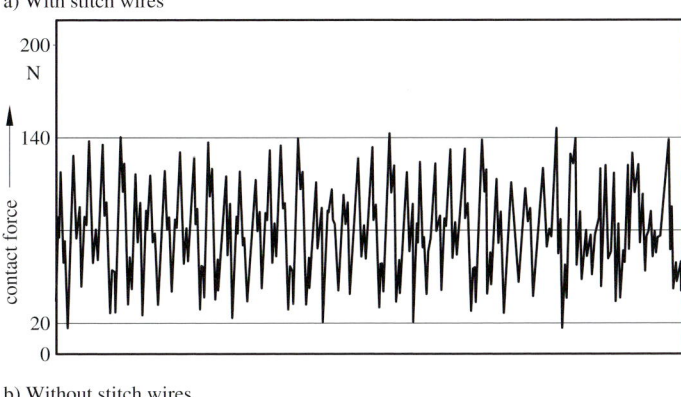

b) Without stitch wires

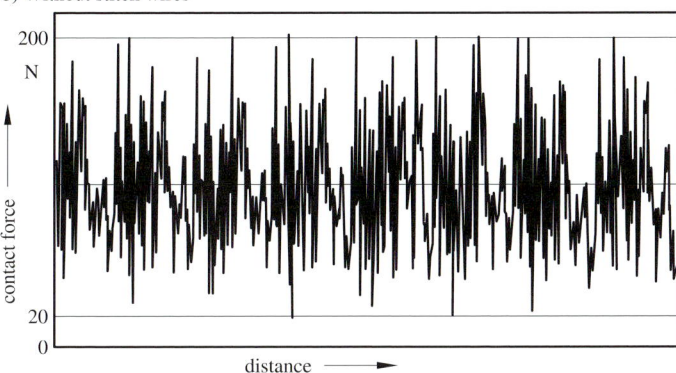

distance ⟶

Figure 10.56: Contact force simulation of an overhead contact line Re 250, span length 65 m, train speed 250 km/h.

Figure 10.55 shows the effect of stitch wire length and tensile force on the elasticity of the contact line at supports, as determined for the DB contact line Re 250. By installing 18 m long stitch wires, the elasticity at a *push-off support* can be made roughly equal to midspan points. The elasticity of *pull-off supports* with stitch wires of this length is only slightly lower. Uniform elasticity leads to a constant static uplift and causes less vertical pantograph/collector strip motion. The dynamic effects of stitch wires can be assessed both by simulation and empirically by test runs. Figure 10.56 shows the results obtained by contact force simulations. When the high-speed Hanover–Würzburg line was built, some tensioning sections were installed without stitch wires at the supports [10.48]. To compensate for the greater differences in system elasticity within a span, the contact wire was adjusted to obtain a pre-sag of approximately 50 mm, i. e. less than 0,1 % of the pole spacing. Figure 10.57 shows the results of contact force measurements at 265 km/h. With stitch wires, the dynamic range of contact forces is narrower and no pronounced contact force peaks are observed at the support positions. Without stitch wires, contact force peaks occur and the standard deviation is higher. This demonstrates the importance of *stitch wires* for superior operating characteristics of overhead contact lines even at high running speeds. It is not difficult to install stitch wires accurately if adequate tools are available and the additional effort required is negligible. The design and installation parameters of the stitch wires can be determined by *simulations of elasticity* and *contact forces*.

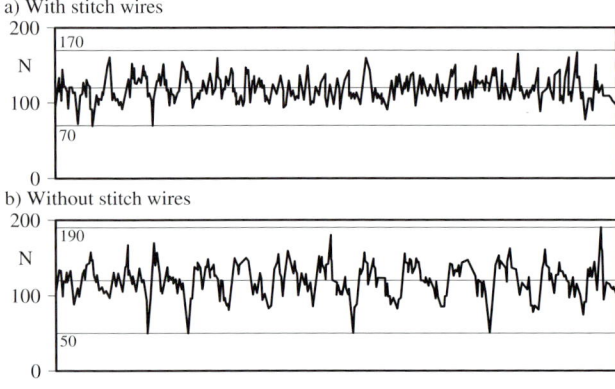

a) With stitch wires

b) Without stitch wires

Figure 10.57: Results of contact force measurements, overhead contact line design Re 250, train speed 265 km/h [10.48].

10.5.3.4 Effect of adjustment accuracy

The designed contact wire positions can only be achieved by the installation process within a more or less narrow tolerance range. Accordingly, the German Railway DB has defined *tolerance limits* for their overhead contact lines to ensure the required running qualities. These tolerance ranges are narrower for the higher-quality overhead contact line designs. Table 10.14 gives examples of these values. The following parameters are most important:
  – height differences from one dropper to the next
  – height differences from one support to the next
  – change of contact wire gradient at the supports and
  – contact wire height tolerances
The evaluation of results from test runs, has shown that the desired *contact quality* is easily achieved if the installation is within these specified tolerances. A pre-sag less than 30 mm at the middle of a span has no adverse effects. However, any substantial deviation from the stipulated tolerances, especially above points and in overlapping sections, leads to noticeable contact force effects in the form of pronounced peaks.

10.5.4 Assessment of high-speed contact lines according to TSI Energy

The Spanish railway installed a new 620 km long high-speed line between Madrid and Barcelona that was designed for 350 km/h maximum commercial speed. The line was equipped with contact line EAC 350, described in [10.49] and assessed in [10.50]. The contact line was treated as an interoperability constituent according to [10.12]. For the purpose of assessment, a 10 km long line section was equipped with the contact line for testing. The section comprised all characteristics of the line. Static and dynamic tests were carried out to assess the interaction between pantographs and the contact line by measuring the contact forces. The contact line was certified after successful testing and has been in operation at 300 km/h since 2009.

Table 10.14: Tolerances of contact wire height, stagger and gradient of DB overhead contact line designs Re 200 and Re 250 (excerpt from DB document Ebs 02.05.19, p. 3).

	Re 200	Re 250
contact wire height	±100 mm	±30 mm
support to support	1 mm/m	±20 mm
dropper to dropper	20 mm	10 mm
contact wire stagger	±30 mm	±30 mm
gradient	1 : 1 000	1 : 3 000

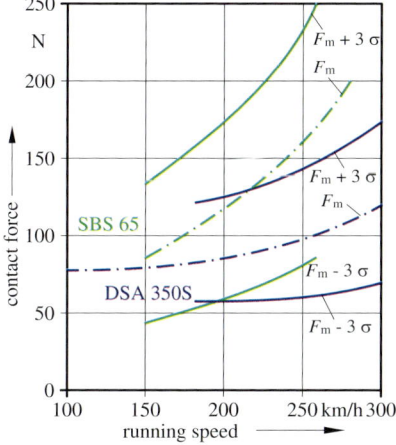

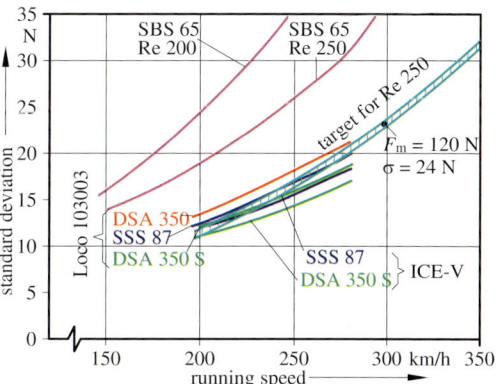

Figure 10.58: Contact forces of the pantograph types SBS 65 and DSA 350 S depending on the train speed, overhead contact line Re 250. F_m mean value, σ standard deviation of contact force.

Figure 10.59: Standard deviation of the contact force as a function of the train speed, measured for pantograph types SBS 65, WBL 85, SSS 87 and DSA 350.

The Austrian Railway ÖBB developed a contact line suited for 250 km/h [10.51] for their high-speed lines, called Type 2.1. It complies with the requirements of line category I. A line section was equipped with this contact line for testing. The contact line consists of a contact wire AC-120 – CuAg0,1 and a catenary wire CuMg 70 mm^2. The static and dynamic parameters meet the requirements and recommendations stipulated by TSI Energy [10.12]. After simulating operations a series of modifications was carried out, in particular the insulation sections were re-designed. With the improved design, contact force measurements with one raised pantograph were carried out at 300 km/h and with two pantographs spaced 200 m apart at 280 km/h. The target values were complied with in the open and in tunnels. The contact line type was certified subsequently as an interoperability constituent according to the TSI Energy [10.12].

10.6 Impact of pantograph design

10.6.1 Introduction

The *design* and the *characteristics of pantographs* have considerable effects on running quality. Running an unsuitable pantograph along a contact line that is suitable for high speeds per se, will not produce the desired result. Nor does a *pantograph* suitable *for high speeds* increase the acceptable maximum speed of a conventional overhead contact line to any great extent. Experiments carried out by the DB [10.44] have demonstrated this repeatedly for the standard overhead contact line design Re 200. Even when sophisticated pantograph designs were used, the capacity of this contact line installation was exhausted at a speed of 200 km/h. For satisfactory energy transmission to high-speed trains, suitable overhead contact lines and corresponding high-speed pantographs are essential.

10.6.2 Features of pantograph designs

The DB standard overhead contact line Re 250 was first tested using a standard pantograph of type SBS 65 [10.52]. On this single-arm pantograph, both collector strips are mounted on a frame-type pan head which is spring-mounted on the upper frame by rubber elements. Its contact force is 70 N under static conditions and increases substantially with speed due to both the aerodynamic effects and contact strip wear, as shown in Figure 10.58. Measured average contact forces were as high as 170 N on sections in the open and 200 N in tunnels at 250 km/h. Figure 10.58 also shows the wide dynamic range. At 250 km/h, the contact force peaks exceeded 300 N and the standard deviations were in the range of 26 N. Trials proved that a pantograph of type SBS 65 is not suitable for speeds above 200 km/h because of its unfavourable aerodynamic characteristics and the high dynamic forces, occurring when the pantograph moves along the overhead contact line. This deficiency was caused by the high unsprung masses of the frame-type pan head and the hard rubber torsion springs on which the pan head is mounted.

To maintain the previously achieved long service life of contact wires and pantographs at higher train speeds, it became necessary to develop new *pantograph designs*. The design specifications for these new pantographs were derived from the experience gained in the course of the overhead contact line trials. The *mean contact force* should not exceed 120 N at a running speed of 300 km/h.

Figure 10.59 shows graphs of the standard deviations for various combinations of standard overhead contact lines and pantographs and the target specifications for high-speed traffic derived from these values. The *standard deviation of* the *dynamic forces* should be less than 20 % of the mean value, i. e. 24 N. The dynamic load should also be distributed evenly on both collector strips to ensure arc-free sliding contact.

These specifications were fulfilled by several new pantograph designs [10.46, 10.48]. The test runs demonstrated that pantograph performances at high speeds are determined by the design of the pan head and the collector strips. Consequently, the head mass and head damping were reduced in comparison to earlier models. The pantograph mass as a whole also needs to be as low as possible. The DSA 350 pantograph has independently sprung *collector strips* with four spring mounts and progressive spring characteristics [10.53] reducing the unsprung mass in direct contact with the contact wire to 2,9 kg per collector strip. The total mass of the pantograph upper arm is 9 kg.

The new *single-arm pantograph* designs achieve similar contact characteristics, whether running in the usual position with the knuckle pointing away from the direction of travel or with the knuckle pointing into the direction of travel. This objective was achieved by installing suitably arranged air baffles which also have the effect of controlling the mean contact force, so that it only increases slightly with speed up to an approximate value 120 N at 300 km/h (Figure 10.58). The dynamic characteristics, evaluated by observing the *apparent mass* (Figure 10.11), also improved considerably. The apparent mass ranges from 4 to 30 Ns^2/m at frequencies of 1 to 6 Hz and from 6 to 11 Ns^2/m at frequencies of 7 to 12 Hz. By contrast, the respective values of the SBS 65 pantograph range from 0,4 to 70 Ns^2/m. A pantograph with independently sprung collector strips is shown in Figure 10.60.

With the aid of the methods described in Clause 10.4 for systematically measuring and evaluating the dynamic forces acting between the contact wire and the collector strips, it was possible to observe the effects of pantograph design parameters. For example, the spring characteristics were investigated during several test series to determine stipulations for fur-

Figure 10.60: Pantograph DSA 350 S with independently sprung collector strips.

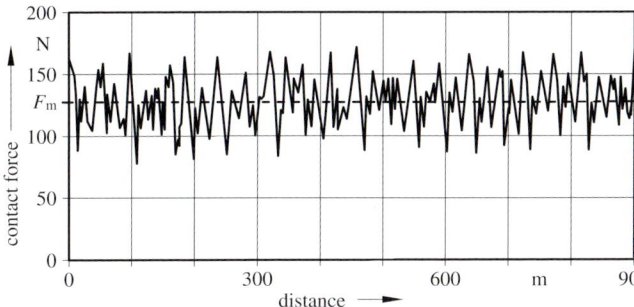

Figure 10.61: Contact force graph of a pantograph DSA 350 S, overhead contact line Re 250, train speed 310 km/h. $F_{stat} = 95$ N, $F_{max} = 176$ N, $F_{min} = 78$ N, $F_m = 128$ N, $\sigma = 18$ N
F_m mean value of F_{tot}
σ standard deviation

ther optimization work. Figure 10.61 shows a contact force recording taken at a train speed of 310 km/h. The mean contact force is 128 N and the standard deviation 18,2 N, i.e. 14 %, achieving the targets set for the contact line/pantograph system. Figure 10.59 shows the standard deviations of a pantograph SBS 65 operated under a standard overhead contact line design Re 200 and the values obtained with pantographs SBS 65, as well as the more sophisticated designs DSA 350, DSA 350 S and SSS 87, operated under an overhead contact line design Re 250. By dedicated further development work, the standard deviations observed at a running speed of 250 km/h were reduced from approximately 26 N, achieved with the SBS 65 to between 18 and 19 N, then to between 16 and 17 N and finally to a value just below 15 N achieved with more sophisticated designs. This improvement was achieved by reducing the masses, introducing independent springs under the collector strips, systematically tuning the individual structural components and optimising the aerodynamic behaviour to be as near as possible to neutral. The report [10.53] describes the pantograph for the train ICE3 running at 330 km/h, which was also designed to achieve low noise emission.

The TSI for rolling stock [10.15] of the European high-speed system specifies the pantographs as an interoperability constituent. The mean contact force should comply with the requirements shown in Figure 10.25 and 10.26 with a tolerance of ± 10 %. Pantographs must meet this requirement when installed on a train or locomotive, independently of the position on the train and the running direction. Its dynamic characteristics need to be demonstrated under a contact line which complies with the Energy TSI (see Clause 10.4.2).

Figure 10.62: Actively controlled high-speed pantograph ASP.

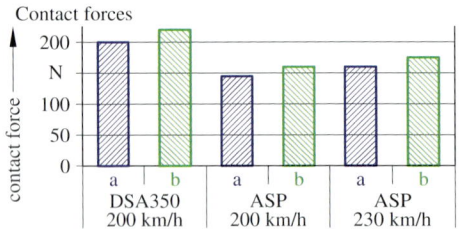
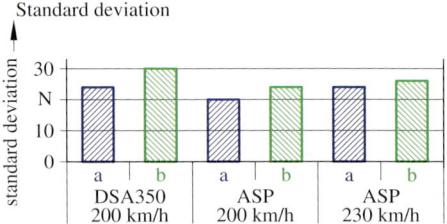

Figure 10.63: Comparison of contact forces F and standard deviation σ of a controlled pantograph ASP and a standard pantograph DSA 350 at 200 and 230 km/h for contact line DB Re 200.
a) heading pantograph, b) trailing pantograph

10.6.3 Sophisticated pantograph designs

Current standard pantographs are passive components without any equipment to actively control the contact forces. They have achieved an excellent level of performance, however, they reach their limits when increasing the operational speed under existing conventional contact lines. A new actively controlled single-arm pantograph described in [10.28] was developed by German Railway (DB).

The new ASP pantograph (Figure 10.62) is a single-arm design with an acoustically optimised head and a two-level control. Compared with conventional pantographs, the new unit produces considerably less noise, the control does not create resonance at the contact line and the unit is reliable and easily maintained. To permit operation in the existing network and mounting on existing trains, the main dimensions and installation data are the same as for standard pantographs. The pantograph head was newly designed. The number of noise-emitting small parts was reduced and the horns directly connected to the frames of collector strips sprung by torsion rods. The force and acceleration sensors are integrated into the supports of the collector strip frames.

The control aims to reduce the contact force variations by influencing the movement of the collector strips. A two-level control was implemented using the contact force as a controlled variable.

The first control level compensates slow contact force variations, e. g. because of aerodynamic forces, by adjusting the pantograph drive pressure. The first level can be used to adjust the mean contact force to the requirements of a specific line or line section. The second level controls contact force variations with frequencies up to 25 Hz via smaller pneumatic bellows arranged close to the torsion springs. The contact force is measured by four force and acceleration sensors and processed by a device mounted close to the sensors.

a) Heading pantograph

b) Trailing pantograph

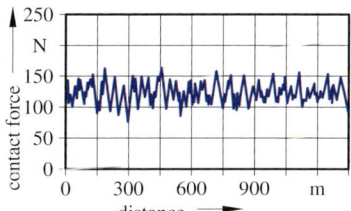

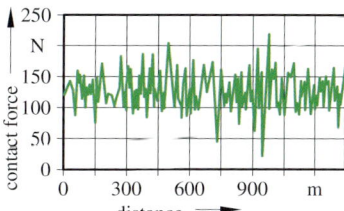

Figure 10.64: Contact force diagrams of a train with two DSA 350 pantographs, overhead contact line Re 250, train speed 275 km/h.

a)	F_{stat}	:	80 N	b)	F_{stat}	:	80 N
	F_{max}	:	162 N		F_{max}	:	215 N
	F_{min}	:	70 N		F_{min}	:	17 N
	F_m (mean value of F_{tot})	:	122 N		F_m (mean value of F_{tot})	:	124 N
	σ (standard deviation)	:	15,0 N		σ (standard deviation)	:	23,3 N

The performance was studied by simulation and comparison with the passive high-speed pantograph DSA 350. Where two active pantographs running at 200 km/h under a contact line Re 200 the maximum contact force was reduced by 50 N, the minimum raised by 30 N and the standard deviation was dropped by 30 %.

Figure 10.63 shows a comparison of contact forces and their standard deviations between the controlled pantograph and the DSA 350 type running at 200 km/h under the contact line Re 200. The dynamic range is reduced by 20 %. The standard deviation is also smaller at 230 km/h. The uplift at 230 km/h is not higher than at 200 km/h with the pantograph DSA 350. The measurements of noise emission also confirmed expectations. The new design is considerably less noisy which is essential at speeds of 250 km/h and above.

10.6.4 Trains running with multiple pantographs

High-speed trains drawn by a locomotive use only a single pantograph. The record runs on 1st of May 1988 [10.1] were carried out by a train with two traction units but with only one pantograph in contact with the contact wire. However, the traction units at both ends of the DB's high-speed ICE 1 train are supplied directly via their own pantographs. In commercial service, two active pantographs are required, spaced between 200 m and 400 m apart and according to EN 50 367, an electrical connection is not permitted with an AC power supply. Contact wire uplift measurements at a support during running of a train with *multiple pantographs* (Figure 10.44) has shown that the second pantograph always runs along an oscillating section of the contact line and is subject to less favourable conditions. This is confirmed by the contact force diagram, as can be seen in Figure 10.64. While the mean values are virtually equal, the maxima differ considerably, these being 162 N and 215 N respectively. The same applies to the minimum values, which are 70 N and 15 N. This is also visible in the standard deviations. The *contact behaviour* of the leading pantograph does not differ from that of a train with only one pantograph.

Figure 10.65 shows the standard deviations of the contact forces measured on the leading and the trailing pantograph. The values recorded on the trailing pantograph rise more sharply with speed than those of the leading pantograph, which are nearly the same as the values of single-

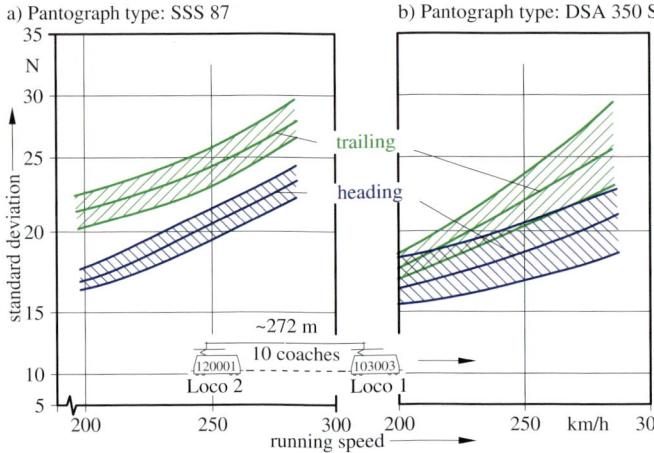

a) Pantograph type: SSS 87   b) Pantograph type: DSA 350 S

Figure 10.65: Standard deviations of the contact forces measured on a train with two pantographs, pantograph types SSS 87 and DSA 350 S, both heading and trailing.

pantograph operation. The standard deviation of the forces on the trailing pantograph reaches 24 N at a speed of 250 km/h. At 280 km/h, it rises to 28 N. Furthermore, it is not possible to limit the *mean contact force* to 120 N, which must be increased to 140 N at 280 km/h to keep arcing to an acceptable minimum. Such force increases are associated with corresponding force peaks and increased wear. In terms of arc suppression and wear, it is not possible, under these conditions, to achieve the current transmission quality normally required for single-pantograph operations. For this reason, all attempts should be made to transmit energy to *multiple traction units* only via a single pantograph. In Clause 10.4.2.4 specifications for minimum spacing of pantographs are presented which correspond to EN 50 367.

The *uplift of overhead contact lines* at the supports caused by passing trains is a parameter affecting the operational security of the system. The operating limits of standard DB overhead contact lines is close to 120 mm. In general, this value should never be exceeded because greater uplift values lead to unfavourable dynamic stresses on the overhead contact line system. When trains with two pantographs run on the lines, the value at the trailing pantograph reaches this limit at a speed far below that which would be possible with a single pantograph, even if the two pantographs are at the maximum possible distance apart. Figure 10.66 shows how the pantograph spacing affects contact wire uplift at a support.

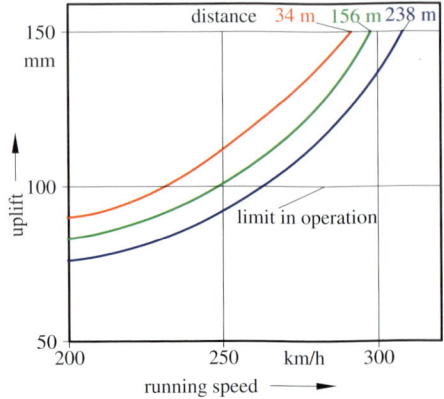

Figure 10.66: Effect of pantograph spacing on the contact wire uplift measured at a support, overhead contact line type Re 250.

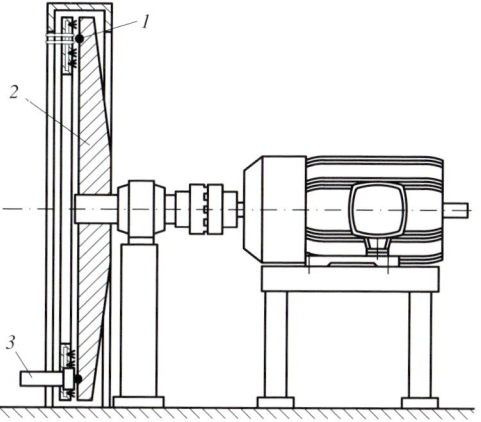

Figure 10.67: Test apparatus for measuring contact wire wear.
1 contact wire
2 mounting disc
3 collector pressure adjustment device

Whether and to what extent, multiple-pantograph operation is possible also depends on the overhead contact line design. Trial runs along an Re 330 overhead contact line have shown that it is possible to use two pantographs at speeds up to 200 km/h, even if they are spaced only 34 m apart. With a spacing of more than 240 m, it is also possible to achieve speeds of 350 km/h.

Operation with only one raised pantograph, therefore, enables high running speeds. E. g. the contact line Re 250 is traversed on the line Madrid–Sevilla at 300 km/h using one pantograph. However, because of power required for high-speed traffic and the commercial advantage, double traction unit trains are often operated with several raised pantographs.

10.7 Collector strip and contact wire materials

The *service life of contact wires* and collector strips essentially depends on:
- the contact force exerted by the pantograph on the contact wire, as discussed in Clause 10.5.3 from the overhead contact line perspective and in 10.6 from the pantograph perspective
- the *collector strip* and *contact wire* material
- the number and the dimensions of the collector strips
- the current flowing through the contact point
- the traction vehicle speed
- environmental impacts such as lines in tunnels or in the open

The last three aspects cannot be directly controlled or affected when designing energy transmission systems. They must be adequately considered when selecting the materials and calculating the dimensions of the components.

Pure copper (electrolytic copper E-Cu) and copper alloys became the primary material for contact wires (see Clause 11.2.2.2). The standard EN 50 149 specifies the following materials: E-Cu, CuAg, CuSn, CuCd and CuMg. Multi-component alloys such as CuCrZr and CuCrZrMg [10.54] have already been discussed as possible contact wire materials. Copper-clad steel wires hab been used by German Railways [10.55] and considered for use in Japan [10.56]. With respect to the contact behaviour, the latter material does not differ from pure copper. As commonly known, copper is also used as a material for *sliding contacts* in electrical motors and generators.

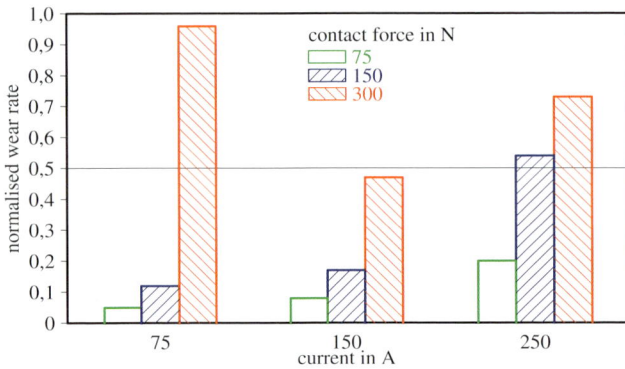

Figure 10.68: Wear rates of a CuMg0,5 contact wire at 150 km/h, measured on a test stand according to [10.59].

Depending on the environmental conditions and the contact partner material, copper will form a 5 to 20 µm thick layer of CuO and CuO_2, which may also have graphite inclusions stemming from the collector strip material. This layer is electrically conductive, hard and provides ideal conditions for sliding electric contacts.

Various attempts were made to use aluminum as a contact wire material. Since aluminum forms a hard non-conductive oxide layer that needs to be ground off every time a collector passes, energy transmission involves abrasion and continuous arcing. For this reason, aluminum is not suitable for use as a contact wire material.

Among the materials mentioned above, besides standard E-Cu, the alloys CuAg and CuMg are very suitable for contact wires, especially in high-speed and high-power applications. CuCd is no longer permitted because of the environmental contamination involved. CuSn has no decisive advantages over CuMg. The wear characteristics of these materials were the object of many studies [10.57, 10.58]. The studies were followed up systematically on a *test-stand for contact wires* designed and built in Germany [10.59]. The test wires were mounted on a 2,0 m diameter disc (Figure 10.67). With an approximate maximum 1 500 rpm, running speeds of up to 500 km/h could be simulated. The contact force of the collector could be varied from 0 to 300 N and the AC current through the contact from 0 to 1 000 A. The wear was measured by two laser sensors and the measuring circuits used enabled a direct wear measurement with a resolution in the µm-region.

The results of *wear measurements* on contact wires made of CuAg0,1 and CuMg0,5 are described in reference [10.60]. When the current is increased under otherwise unchanged parameters, the wear rate initially decreases. This can be attributed to the current's lubrication effect due to the formation of a lubricating graphite layer. This leads to a minimum wear rate at currents of 100 to 150 A at a speed of roughly 200 km/h (Figure 10.68).

As the current is increased even further, an *electric wear component* begins to take effect and the wear rate increases. The *mechanical wear component* dominates and this component definitely increases with increasing contact force (Figure 10.68) confirming the importance of achieving as uniform a contact force as possible for optimum performance of an overhead contact line.

At constant contact forces and currents, the *wear rate* initially increases with speed (Figure 10.69) to a maximum value at around 150 km/h and then decreases again. This justifies the assumption that it is still possible to achieve long contact wire service life in high-speed applications in spite of the tendency towards higher currents and contact forces. As a guideline, the *normalised wear rate* can be taken as 1 mm^2 per 10^5 pantograph passages.

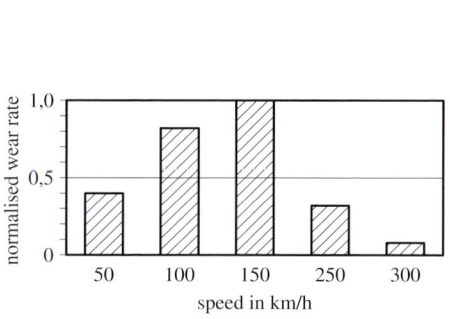
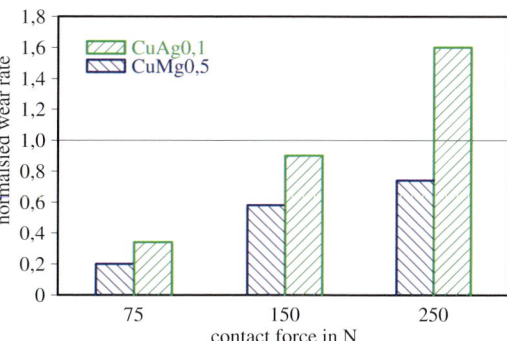

Figure 10.69: Wear rate of a CuMg0,5 contact wire depending on the running speed, contact force 250 N, current 300 A [10.60].

Figure 10.70: Comparison of wear rates of contact wires made of CuAg0,1 and CuMg0,5, running speed 150 km/h, current 300 A [10.60].

Figure 10.70 shows a comparison of wear of the materials CuAg0,1 and CuMg0,5. It can be concluded that the harder contact wire made of CuMg0,5 wears only half as quickly as contact wires made of CuAg0,1, almost irrespective of the current and contact force. CuMg0,5 is an obviously superior choice for contact wires with a greatly increased service life.

Steel, copper alloys, graphite and copper impregnated carbon have been used as *materials* for *collector strips* [10.57]. The interactions of these materials with the contact wire differ considerably. Carbon and graphite lead to a smooth, shiny surface without any visible roughness on the contact wire. However, copper and steel form a rough surface similar to that of a fine file. This roughness acts as an abrasive and leads to rapid wear, of both the contact wire and the collector strips.

Figure 10.71, taken from reference [10.61], shows the wear rates of contact wires in combination with various collector strip materials. It can be seen that the *metal collector strips* lead to wear rates almost ten times those caused by carbon collector strips. Whereas the DB uses only *carbon collector strips* as a matter of principle and is able to achieve a contact wire service life of 30 years and more, the Japanese railways and the SNCF use metal (i.e. steel) collector strips even in AC traction systems or they used them in the past. The associated wear only allows a *service life* of few years. Although this fact has been well known for a long time, these railway companies continue to use metal collector strips because they fear that the impact-sensitive, brittle carbon collector strips might shatter under mechanical impacts. The experience gained by German Railways has shown that this rarely occurs if the overhead contact lines are optimized with respect to contact force characteristics.

Metal collector strips are considerably heavier than those of carbon, leading to unfavourable dynamic characteristics. Consequently, these collector strips negatively affect the contact forces. Because of the heavy currents associated with DC traction applications, such collector strips are often used in DC railways [10.22]. For this application the copper-chromium-zirconium alloy CuCrZr has proved very suitable because of its good *thermal stability*.

Differences in the surface conditions and contact forces also affect the wear rates of collector strips. German Railway DB achieves service lives of up to 100 000 km for carbon collector strips, while the metal collector strips used in DC traction applications have to be replaced every 30 000 km.

The different *contact wire surface structures* caused by carbon collector strips and metal collector strips mean that mixed operation of carbon collector strips and metal collector strips

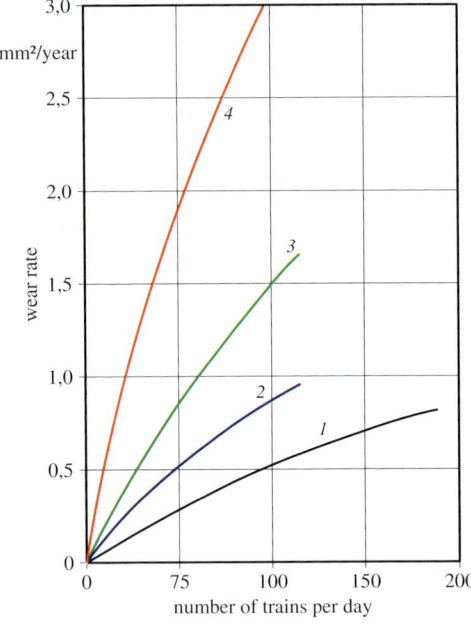

Figure 10.71: Wear rates of copper contact wires (according to [10.61]).
1 pantograph with two carbon collector strips
2 two pantographs with one carbon collector strip each
3 pantograph with two aluminum collector strips
4 pantograph with two steel collector strips

on the same contact wires is not advisable. It would lead to greatly increased wear rates, both of the contact wires and of the carbon collector strips. For this reason, the TSI Energy for the interoperability of European high-speed railway networks [10.13, 10.14] specifies carbon as the collector strip material. The paper [10.62] reports on a study to compare the effect of collector strips made of carbon and those made of copper impregnated carbon on the wear of contact wires and collector strips. According to the study, both materials are equally suitable and can be used on interoperable lines. Therefore, the TSI Energy [10.14] specifies these materials for collector strips.

The paper [10.63] reports on the development of contact lines and pantographs for high-speed lines in Japan. Copper-clad steel contact wire of $170 \, mm^2$ and $24,5 \, kN$ tensile force is adopted. The mean contact force is low in comparison with the requirements according to TSI Energy [10.13, 10.14].

High-strength contact wires are required for current transfer at high speeds. Such contact wires consist of copper-tin or copper-magnesium alloys which tend to deviations from the ideal contact wire position caused by *micro waves* [10.64, 10.65]. Such micro waviness can cause or amplify arcing. The effect will remain low if the height difference is less than 0,2 mm and will increase at greater differences. Wave lengths between 100 and 900 mm were recorded (see also Clause 15.5.2 and Figure 15.10).

As reported in [10.66], the contact wire blanks and the stringing technology are the main reasons for microwaves. Therefore, the maximum strength of the contact wire blanks should be limited, the contact wire production should be supervised to detect microwaves at manufacture, alignment devices for the contact wires should be used and the contact wire position should be monitored after stringing. Using these improved techniques, the microwaves can be reduced to amplitudes less than 0,1 mm and avoid arcing.

10.8   Examples for the assessment of interaction

10.8.1   Assessment of the interoperable overhead contact line EAC 350 in Spain

The paper [10.50] reports on the assessment of the *overhead contact line design EAC 350* in Spain, designed in compliance with the TSI Energy [10.12]. This design is provided for running speeds of 350 km/h and, therefore, has to be tested at a 10 % higher speed. The design was tested on a 10 km long section between Madrid and Lleida and the tensile forces used for contact and catenary wire are higher than with other contact line designs. This section comprised all typical line features such as track grades, curves, open sections, tunnels, cross-overs and phase separations. The static tests related to the geometry, the performance during temperature changes and to elasticity. The dynamic criteria were tested by measuring the contact forces with a pantograph DSA 380 EU. Based on the test results, this contact line type was certified as suitable for the intended application.

10.8.2   Overhead contact line for high speeds in Austria

The paper [10.51] describes *ÖBB's overhead contact line design* 2.1 for high speeds and its conformity assessment. This contact line is equipped with a contact wire AC-120 – CuAg0,1 tensioned to 15,3 kN. The 70 mm^2 bronze catenary wire has a tensile force of 10,8 kN and stitch wires at the supports. Because of official requirements, the contact line must be accepted by an authorized person. The performance of the contact line was simulated before the acceptance tests. For this simulation, the planning and design data were first used. In addition, contact force measurements using DB's test train were carried out at 300 km/h with one active pantograph DSA 380-D and at 280 km/h with two active pantographs 200 m apart. During the tests, arcing was only observed at a few positions, at insulated overlaps, tensioning sections and above points, without any effects on the energy transmission. At 250 km/h the maximum contact forces remained below the permissible 250 N and the minima above 20 N. Therefore, the contact line design 2.1 met the requirements of TSI Energy [10.12] at 250 km/h.

10.8.3   Dynamic performance of the contact line Wuhan–Guangzhou, China

The dynamic performance of the contact line from Wuhan to Guangzhou in China is reported in [10.67]. The contact line has 60 m spans with stitch wires at the supports. The contact wire consists of AC-150 – CuMg0,5 tensioned to 30 kN and the catenary wire is Bz 120 at 21 kN. The running speed of 350 km/h is higher than the range covered by the TSI Energy [10.14]. For 350 km/h the maximum contact forces were specified as 360 N and the minimum values at 20 N. According to the report, not all the specifications for contact lines and pantographs of European standards could be met. The measured contact forces, however, showed that the minima and maxima were obeyed. The assessment concluded that the contact line is suitable for envisaged targets.

Figure 10.72: TGV 150 during the world record run on the Paris–Strasbourg line, France.

10.9 Conclusions

10.9.1 Energy transmission limits for overhead contact lines

Since 1980, electric traction railways have seen great progress in terms of running speeds in commercial everyday operations and high-speed trials that test the *performance limits of the wheel-on-rail system*. In 1988, the DB's experimental train ICE/V achieved a speed of 407 km/h [10.1]. The pantograph and the overhead contact line – tensioned to 21 kN performed to expectations, indicating that these components should be able to achieve speeds up to 450 km/h in single-pantograph operations.

The paper [10.68] reports on the increased running speeds on rails in France until 2002. In May 1991, an SNCF train of an enhanced TGV-Atlantique type achieved a speed of 515 km/h on the Paris–Tours line near Tours [10.69] and set a *world speed record for railway vehicles*. Here again, the importance of the overhead contact line design, especially of contact wire stress, became apparent. During preparatory runs along contact wires tensioned to a force of 28 kN, the trials had to be aborted at a speed of approximately 480 km/h because of current interruptions caused by contact wire uplift values of more than 300 mm. The *Doppler factor* was only 0,040 and the *amplification coefficient* had already reached 8,2. Under these conditions, the current transmission had reached its limits. The final speed of 515 km/h was made possible by increasing the tensile force on the contact wire to 33 kN. Table 10.13 shows some of the parameters relating to overhead contact lines used for these trials.

On April 3, 2007 French railways SNCF, the Infrastructure Manager RFF (Réseau Ferré de France) and Alstom Transport set a new world record for railways by achieving 574,8 km/h on the Paris–Strasbourg line using the test train TGV 150 (Figure 10.72) [10.70]. A single-arm pantograph was used on an 85 km long test section with track radii of more than 12 000 m. On a 35 km long section, the speed was above 500 km/h. Figure 10.73 shows the development of the speed during the record run. The line section where the high speeds were achieved was equipped with a contact line comprising a contact wire AC-150 – CuSn and a Bz 116 catenary wire with a tensile force of 20 kN. On a 70 km long section, the tensile force of contact wire was increased progressively from 26 kN to 40 kN depending on the envisaged

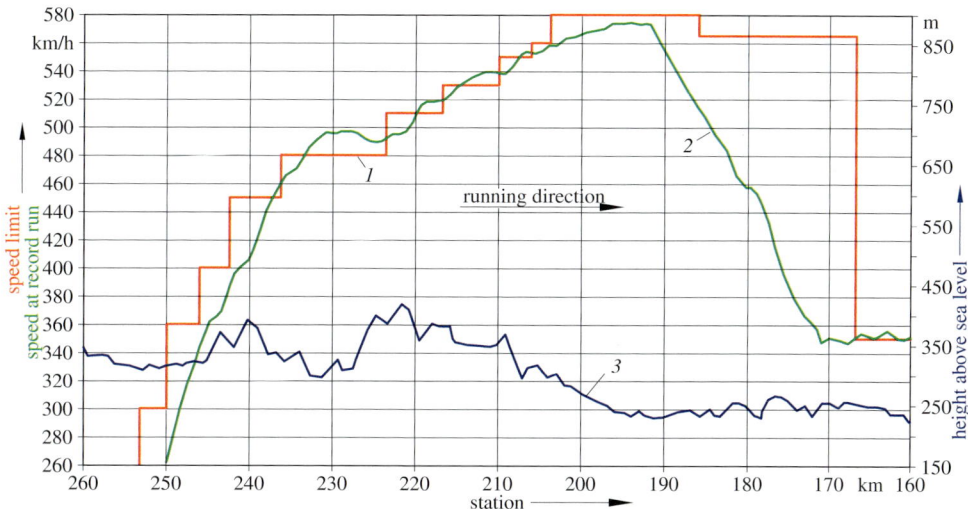

Figure 10.73: Development of speed during SNCF's world record in 2007 [10.70].
1 speed limit, *2* speed at record run, *3* height above sea level

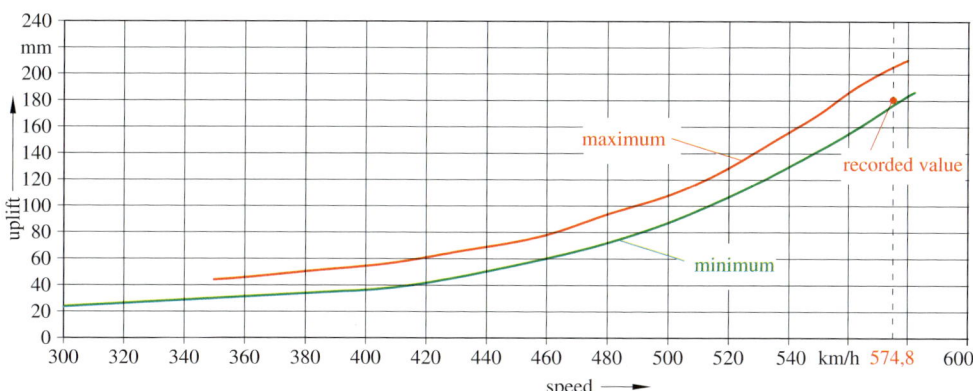

Figure 10.74: Simulated maximum and minimum uplift during the world record run.

speed. The pre-sag was 22 mm at midspan. The wave propagation speed of the contact wire reached 622 km/h resulting in a Doppler factor of 0,039 (see Table 10.13). In Figure 10.74, the simulated uplift is shown relative to the running speed. The maximum recorded uplift was 180 mm at 574,8 km/h and in total 28 runs with speeds above 500 km/h were made, achieving a cumulative distance of 720 km as well as runs above 400 km/h with a cumulative distance of 2 200 km.

10.9.2 Overhead contact line requirements

Overhead contact lines need to be capable of reliably transmitting electric current to the traction vehicles. The design methods regarding current carrying capacity are discussed in Chapter 7. They are used to determine the conductor cross-sections, especially for DC traction systems. Importantly, the mechanical design dimensions must be especially tuned to suit the

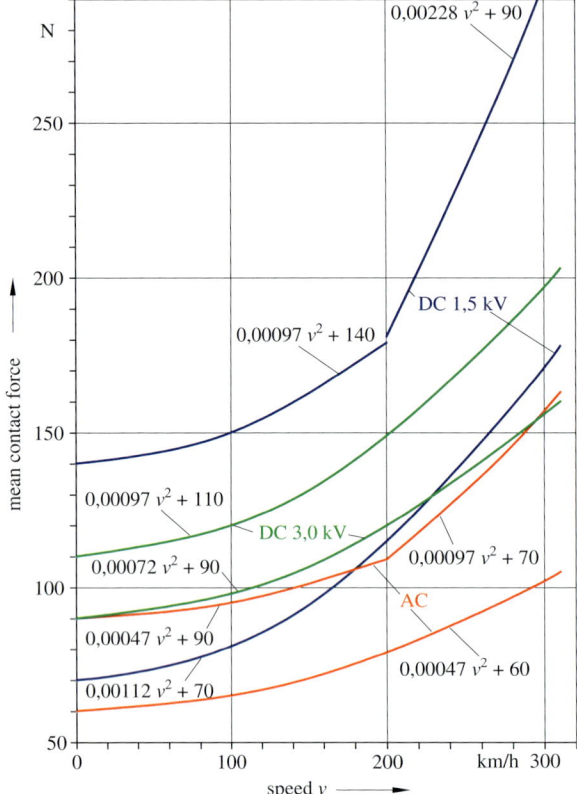

Figure 10.75: Mean contact forces and their tolerance ranges according to EN 50 367, Table 6.

running speeds. The geometric and static criteria relating to the *interaction of the contact line with pantographs* are of importance for overhead contact lines intended for running speeds up to 160 km/h. At higher speeds, the *dynamic criteria* become increasingly important and these are particularly dependent on the tensile stress in the contact wire.

EN 50 367 specifies requirements for contact wire heights and these are presented in Table 10.15. Contact lines for speeds equal to and above 250 km/h should be installed with a uniform contact wire height.

For lines in the open, span lengths, tensile forces and the design of contact lines should be selected to ensure that the parameters describing the mean contact forces presented in Table 10.8 are met.

It is possible to design an overhead contact line for a given speed range on the basis of the dynamic criteria. The wave propagation speed should be chosen in such a way that the

Table 10.15: Contact wire heights in m for AC and DC systems according to EN 50 367.

Speed km/h	$\leq$ 160	160 to 250	$\geq$ 250
nominal	5,0 to 5,75	5,0 to 5,5	5,08 to 5,30
minimum	4,95	4,95	–
maximum, design / actual	6,2 / 6,5	6,2 / 6,5	–

Doppler factor never drops below 0,2. This means that the wave propagation speed should be between 1,4 and 1,5 times the planned train running speed. EN 50119 and the TSI Energy [10.12] and [10.14] limit the operational speed to 70 % of the wave propagation velocity. The *reflection coefficient* should be designed to keep the amplification coefficient below 2,0. Reflection coefficients around 0,4 meet this requirement.

In view of dynamic interactions, overhead contact lines need to be designed to operate with pantographs exerting a speed-dependent mean contact force as per Figure 10.75. According to TSI Energy [10.14], the requirements specified in Figures 10.25 and 10.26 apply. When testing, the mean contact force F_m should be between the data presented in Figures 10.25 and 10.26.

When designing cantilevers, the space for maximum uplift of the steady arm should be a minimum of twice the calculated or simulated uplift value. If restrictions or design limitations for uplift of the steady arm are provided, then a space not less than 1,5 times the uplift will be sufficient.

10.9.3   Pantograph requirements

Experience as well as theoretical considerations have shown that it is not possible to design pantographs solely with the intention of optimizing the interaction with a specific overhead contact line design. Even standardised overhead contact line designs do not have uniform dynamic characteristics, because the span lengths, masses and tensile forces will vary under real line and operating conditions. However, pantographs need to have certain basic characteristics to make them suitable for a specific range of applications. Trial runs have demonstrated that well-designed pantographs will always achieve running performances of comparable quality under a variety of different overhead contact lines. Consequently, the following general *requirements for pantographs* should be applied:

- the *mean contact force* should be equal for both travel directions and only increase slightly with speed as described in Figure 10.75. The mean contact force should be concurrently high enough to prevent arcing but also as low as possible to keep the contact wire uplift to a minimum to avoid unnecessary dynamic excitation of the contact line (see Clause 10.4.2.3)
- to achieve a satisfactory quality of current collection the static contact force exerted by the pantograph and the mean aerodynamic contact force should obey a certain set criteria e. g. by [10.14, 10.15]
- the nominal *static contact force* should be inside the following ranges (see Table 10.7):
  - 60 N to 90 N   for AC supply systems
  - 90 N to 120 N   for DC 3 kV supply systems
  - 70 N to 140 N   for DC 1,5 kV supply systems

  To improve the contact of carbon collector strips with the contact wire in DC 1,5 kV systems, in general, 140 N, may be needed to avoid hazardous heating of the contact wire when the train is at standstill with its auxiliaries working
- the target for the *mean contact force* F_m formed by the static and aerodynamic components of the contact force with dynamic correction specified by [10.12, 10.14, 10.15] is shown in Figure 10.25 and 10.26 for AC and DC systems as a function of running speed. In this context, F_m represents a target value which should be achieved to ensure current collection without excessive undue arcing and to limit wear and hazards to current collector strips

- for trains with *multiple pantographs* in simultaneous operation, the mean contact force F_m for any pantograph should not be higher than the value given by Equation (10.68) since for each individual pantograph, the current collection criteria needs to be met
- it will not always be possible to apply the forces specified for a given running speed as shown in Figures 10.25 and 10.26. The contact force should be within the tolerances shown in Figure 10.75, depending on the chosen basic curve
- the mean contact force is the mean value of the forces due to static and aerodynamic actions. It is equal to the sum of static contact force and the *aerodynamic force* (see Clause 2.4.3) caused by the airflow on the pantograph elements at the considered speed. The mean uplift force is a characteristic of the pantograph for given rolling stock and a given development of the pantograph
- the mean contact force can be measured at the collector head, the latter not touching the contact line, according to EN 50 206-1, or by measuring the contact forces according to EN 50 317
- to comply with these specifications, the static contact force of the pantograph should be adjustable between 40 N and 120 N for AC systems and between 50 N and 150 N for DC systems
- the *mass of the collector strips* should be as low as possible to obtain optimum dynamic characteristics
- the *modal mass* should be within a relatively narrow range of values between 4 and 30 Ns^2/m, depending on the frequency and devoid of any sharply distinct peaks
- according to [10.15] pantographs should be equipped with an *automatic dropping device* that lowers the pantograph in case of a failure (see EN 50 206-1). The paper [10.71] describes such a device

10.9.4 Requirements concerning interaction of overhead contact lines and pantographs

The *interaction of a pantograph* with an *overhead contact line* can be assessed by observing the contact forces or arcing and the *contact wire uplift*. Concerning uplift, the latest issues of TSI Energy [10.12] and [10.14] specify that the vertical height of the contact point above the track should be as uniform as possible along the span length. The maximum difference between the highest and the lowest dynamic contact point height within one span should be less than the values shown in Table 10.11 at the maximum line speed for a pantograph exerting a mean contact force as per Equation (10.68). Although neither EN 50 119 or the TSI Energy [10.14] specify limits, the uplift should not exceed

- 100 mm for single and leading pantographs of a multi-pantograph train and
- 120 mm for trailing pantographs of multi-pantograph trains

According to TSI Energy [10.13, 10.14] and EN 50 367, the interaction of overhead contact lines and pantographs may be assessed by the mean contact force in connection with its standard deviation or with the percentage of *arcing*. Table 10.10 of Clause 10.4.2.5 presents the requirements.

With regard to contact forces, the German Railway DB specified that overhead contact lines should have standard deviation/speed characteristics as shown in Figure 10.59. The values shown in Table 10.7 were derived from this graph. Experience proves that these specifications result in a superior quality of current collection.

Table 10.16: DB Specifications for contact forces at the collector strip reaction and the associated deviations, in relation to the intended application of overhead contact lines.

Number of pantographs	1	2	
Speed (km/h)	300	280	
Pantograph		heading	trailing
contact force (N)	120	120	140
maximum contact force (N)	200	185	240
minimum contact force (N)	40	55	40
standard deviation (N)	22	18	28
variation coefficient (%)	18	15	20

Table 10.17: Contact force in N at the point of contact as specified in EN 50 119.

System	Speed km/h	Contact force maximum	Contact force minimum
AC	≤ 200	300	positive
AC	> 200	350	positive
DC	≤ 200	300	positive
DC	> 200	400	positive

Table 10.17 shows the contact force specifications given in EN 50 119, Table 9.4. When comparing the criteria given in Tables 10.16 and 10.17, the different definitions for contact forces have to be considered. In addition to the data given in Table 10.17, EN 50 119 requires that the mean contact force minus three standard deviations should be positive. The DB specification refers to the measured data at the collector strip reaction while EN 50 119 specifies forces between contact wire and collector strips (see also [10.37]). Compliance with these specifications can be verified by simulation calculations when designing an energy transmission system and then validated empirically by trial runs and measurements in accordance with EN 50 317.

10.10 Bibliography

10.1 *Harprecht, W.; Kießling, F.; Seifert, R.:* "406,9 km/h" Energieübertragung bei der Weltrekordfahrt des ICE ("406,9 km/h" power transmission during the world record run of ICE). In: Elektrische Bahnen 86(1988)9, pp. 268 to 289.

10.2 *Seifert, R.:* Der neue Oberleitungsmeßwagen und seine messtechnischen Möglichkeiten zur Überprüfung des Energieübertragungssystems Oberleitung-Stromnehmer (The new catenary measuring vehicle and its measuring technical facilities for the inspection of overhead contact line pantograph/power transmission system). In: Elektrische Bahnen 81(1983)11, pp. 341 to 343 and 12, pp. 370 to 374.

10.3 *Resch, U.:* Simulation des dynamischen Verhaltens von Oberleitungen und Stromnehmer bei hohen Geschwindigkeiten (Simulation of the dynamic behaviour of contact lines and pantographs at high speeds). In: Elektrische Bahnen 89(1991)11, pp. 445 to 446.

10.4 *Dupuy, J.:* 380 km/h. In: Rails of the world (1981)8, pp. 316 to 323.

10.5   *Buksch, R.*: Beitrag zum Verständnis des Schwingungsverhaltens eines Fahrdrahtkettenwerks (Contribution to understanding the vibration behaviour of an overhead contact line). In: Wissenschaftliche Berichte AEG-Telefunken 52(1979)5, pp. 250 to 262.

10.6   *Schwab, H.-J.; Ungvari, S.*: Entwicklung und Ausführung neuer Oberleitungssysteme (Development and design of new overhead contact line systems). In: Elektrische Bahnen 104(2006)5, pp. 137 to 145.

10.7   *Bauer, K.-H.; Buksch, R.; Lerner, F.; Mahrt, R.; Schneider, F.*: Dynamische Kriterien zur Auslegung von Fahrleitungen (Dynamical criteria for the design of overhead contact lines). In: ZEV-Glasers Annalen 103(1979)10, pp. 365 to 370.

10.8   *Buksch, R.*: Theorie der Wechselwirkung von Fahrdrahtwellen mit angekoppelten mechanischen Systemen (Theory of the interaction between contact line waves with coupled mechanical systems). In: Wissenschaftliche Berichte AEG-Telefunken 54(1981)3, pp. 129 to 140 and 55(1982)12, pp. 112 to 122.

10.9   *Beier, S.; Lerner, F.; Lichtenberg, A.; Spöhrer, W.*: Die Oberleitung der Deutschen Bundesbahn für ihre Neubaustrecken (German Railway's overhead contact line for their new high-speed lines). In: Elektrische Bahnen 80(1982)4, pp. 119 to 125.

10.10 *Buksch, R.*: Eigenschwingungen eines Fahrleitungs-Kettenwerks (Natural vibration modes of the overhead contact line equipment). In: Wissenschaftliche Berichte AEG-Telefunken 53(1980)4/5, pp. 186 to 199.

10.11 *N. N.*: Die Regelfahrleitung der Deutschen Bundesbahn (The standard overhead line of German Federal Railway). In: Elektrische Bahnen 77(1979)6, pp. 175 to 180 and 7, pp. 207 to 208.

10.12 *Decision 2008/284/EG*: Technical specification for interoperability of the trans-European high-speed rail system for the Energy subsystem. In: Official Journal of the European Communities (2008) No. L217, pp. 1 to 96.

10.13 *Decision 2011/274/EU*: Technical specification for interoperability of the Energy subsystem of the conventional trans-European rail system. In: Official Journal of the European Union (2011) No. L126, pp. 1 to 52.

10.14 *Regulation 1301/2014/EU*: Technical specification for interoperability of the Energy subsystem of the rail system in the Union. In: Official Journal of the European Union (2014) No. L356, pp. 179 to 207.

10.15 *Regulation 1302/2014/EU*: Technical specification for interoperability of the Rolling Stock, Locomotions and Passenger Rolling Stock subsystem of the rail system in the Union. In: Official Journal of the European Union (2014) No. L356, pp. 228 to 393.

10.16 *Brodkorb, A.; Semrau, M.*: Simulationsmodell des Systems Oberleitungskettenwerk und Stromabnehmer (Simulation model of the overhead contact line and pantograph system). In: Elektrische Bahnen 91(1993)4, pp. 105 to 113.

10.17 *Renger, A.*: Dynamische Analyse des Systems Stromabnehmer und Oberleitungskettenwerk (Dynamical analysis of the overhead contact line – pantograph system). Final report, Kombinat Engine fabrication – Electrotechnical workshop, Henningsdorf, 1987.

10.18 *Nowak, B.; Link, M.*: Zur Optimierung der dynamischen Parameter des ICE-Stromabnehmer durch Simulation der Fahrdynamik (Optimizing of dynamical parameters of the ICE pantograph by simulation of the running dynamics). VDI-Bericht Nr. 635 (1987), pp. 147 to 166.

10.19 *Reichmann, T.*: Simulation des Systems Oberleitungskettenwerk und Stromabnehmer mit der Finite-Elemente-Methode (Simulation of the interaction between overhead contact line and pantographs with the finite element method). In: Elektrische Bahnen 103(2005)1-2, pp. 69 to 75.

10.20 *Bartels, S.; Herbert, W.; Seifert, R.*: Hochgeschwindigkeitsstromabnehmer für den ICE (Highspeed pantograph for the ICE train). In: Elektrische Bahnen 89(1991)11, pp. 436 to 441.

10.21 *Buck, K. E.; von Bodisco, V.; Winkler, K.*: Berechnung der statischen Elastizität beliebiger Oberleitungskettenwerke (Calculation of the static elasticity of overhead contact lines). In: Elektrische Bahnen 89(1991)11, pp. 510 to 511.

10.22 *Bianchi, C.; Tacci, G.; Vandi, A.*: Studio dell'interazione dinamica pantografi – catenaria con programma di simulazione agli elementi finiti. Verifiche sperimentali (Study of the dynamic interaction between pantograph and catenary with simulation software of finite elements. Experimental investigation). In: Sciena e tecnica (1991)11, pp. 647 to 667.

10.23 *Hobbs, A. E. W.*: Accurate prediction of overhead line behaviour. In: Railway Gazette International (1977)9, pp. 339 to 343.

10.24 *Bader, K.; Mohrich, J.*: Oberleitung Bauart N-FL der SBB – Erweiterung der Anwendung (Overhead contact line type N-FL of Swiss rail – Extension of application). In: Elektrische Bahnen 113(2015)6-7, pp. 354 to 362.

10.25 *Link, M.*: Zur Berechnung von Fahrleitungsschwingungen mit Hilfe frequenzabhängiger finiter Elemente (Calculation of overhead contact line vibrations by means of frequency dependent finite elements). In: Ingenieur-Archiv 51(1981), pp. 45 to 60.

10.26 *Reichmann, T.; Raubold, J.*: Triebfahrzeugzulassung mit Hilfe der Simulation Fahrdraht/Stromabnehmer (Approval of multiple train units by means of the simulation of contact wire/pantograph). In: Elektrische Bahnen 109(2011)4-5, pp. 225 to 230.

10.27 *Poetsch, G.; Baldauf, W.; Schulze, T.*: Simulation der Wechselwirkung zwischen Stromabnehmer und Oberleitung (Simulating the interaction between pantographs and overhead contact lines). In: Elektrische Bahnen 99(2001)9, pp. 386 to 392.

10.28 *Baldauf, W.; Kolbe, M.; Krötz, W.*: Geregelter Stromabnehmer für Hochgeschwindigkeitsanwendungen (Closed-loop controlled pantograph for high-speed applications). In: Elektrische Bahnen 103(2005)4-5, pp. 225 to 230.

10.29 *Dorenberg, O.*: Versuche der Deutschen Bundesbahn zur Entwicklung einer Fahrleitung für sehr hohe Geschwindigkeiten (German Railway tests to develop an overhead contact line for very high speeds). In: Elektrische Bahnen 63(1965)6, pp. 148 to 155.

10.30 *Heigl, H.*: Messeinrichtungen zur Registrierung von Kontaktunterbrechungen zwischen Fahrdraht und Stromabnehmer (Measuring equipment to record the contact losses between contact wire and pantograph). In: Elektrische Bahnen 63(1965)7, pp. 171 to 174.

10.31 *Fischer, W.*: Kettenwerk und Stromabnehmer bei hohen Zuggeschwindigkeiten (Overhead contact line and pantograph at high running speeds). In: ZEV – Glasers Annalen 101(1977)5, pp. 142 to 147.

10.32 *König, A.; Resch, U.*: Numerische Simulation des Systems Stromabnehmer – Oberleitungs-kettenwerk (Numerical simulation of the pantograph / overhead contact line system). In: e&i 111(1994)4, pp. 473 to 476.

10.33 *Ostermeyer, M.; Dörfler, E.*: Die Messung der Kontaktkräfte zwischen Fahrdraht und Schleifleisten (Measuring of contact forces between contact wire and collector strips). In: Elektrische Bahnen 80(1982)2, pp. 47 to 52.

10.34 *Bethge, W.; Seifert, R.*: Messtechnische Möglichkeiten der DB zur Erprobung von Fahrleitungssystemen für 250 km/h (German Railway's equipment to carry out measurements for testing overhead contact line systems for 250 km/h). In: ETR-Eisenbahntechnische Rundschau 25(1976)3, pp. 162 to 171.

10.35 *Kluzowski, B.*: Einrichtung zur Messung der Kontaktkraft zwischen Fahrdraht und Stromabnehmer (Devices of measuring of the contact force between contact wire and pantograph). In: Elektrische Bahnen 74(1976)5, pp. 112 to 114.

10.36 UIC Codex 608: Conditions to be complied with for the pantographs of tractive units used in international services. UIC, Paris, 3rd edition April 2003.

10.37 *Koss, G.-R.; Kunz, A.; Resch, U.*: Bewertung der Kontaktkraftmessungen an Stromabnehmern (Assessment of contact force measurements at pantographs). In: Elektrische Bahnen 103(2005)7, pp. 332 to 337.

10.38 *Puschmann, R.; Wehrhahn, D.*: Fahrdrahtlagemessung mit Ultraschall (Ultrasonic measurement of contact wire position). In: Elektrische Bahnen 109(2011)7, pp. 323 to 330.

10.39 *Deml, J.; Baldauf, W.*: Prüfstand zur Untersuchung des Zusammenwirkens Stromabnehmer und Oberleitung (Test bench for examinations of the pantograph-catenary interaction). In: Elektrische Bahnen 100(2002)5, pp. 178 to 181.

10.40 *Bauer, K.-H.; Kießling, F.; Seifert, R.*: Einfluss der Konstruktionsparameter auf die Befahrung einer Oberleitung für hohe Geschwindigkeiten – Theorie und Versuch (Effect of design parameters on traversing an overhead contact line at high speeds – theory and tests). In: Elektrische Bahnen 87(1989)10, pp. 269 to 279.

10.41 *Ebeling, H.*: Stromabnahme bei hohen Geschwindigkeiten – Probleme der Fahrleitungen und Stromabnehmer (Current collection at high speeds – problems of the contact lines and pantographs). In: Elektrische Bahnen 67(1969)2, pp. 26 to 39 and 3, pp. 60 to 66.

10.42 *Kießling, F. et al.*: Die neue Hochleistungsoberleitung Bauart Re 330 der Deutschen Bahn (The new high-performance overhead contact line type Re 330 of Deutsche Bahn). In: Elektrische Bahnen 92(1994)8, pp. 234 to 240.

10.43 *Bauer, K.-H.; Kießling, F.; Seifert, R.*: Weiterentwicklung der Oberleitungen für höhere Fahrgeschwindigkeiten (Development of overhead contact lines for elevated running speeds). In: Eisenbahntechnische Rundschau 38(1989)1-2, pp. 59 to 66.

10.44 *Bauer, K.-H.; Koch, K.*: Von der Versuchsoberleitung zur Regeloberleitung Re 250 (The steps from an experimental contact line to the standard contact line Re 250). In: Die Bundesbahn 62(1986) pp. 423 to 426.

10.45 *Ungvari, S.; Paul, G.*: Oberleitungsbauart Sicat H1.0 für die Hochgeschwindigkeitsstrecke Köln–Rhein/Main (Contact line type Sicat H1.0 for high-speed line Cologne–Rhine/Main). In: Elektrische Bahnen 96(1998)7, pp. 236 to 242.

10.46 *Bauer, K.-H.; Reinold, K.*: Die Fahrleitung Re 250 für Neubaustrecken (The overhead contact line type Re 250 for new high-speed lines). In: Elsners Taschenbuch der Eisenbahntechnik (1980) pp. 199 to 216.

10.47 *Bauer, K.-H.; Kiessling, F.*: Die Regeloberleitung in den Tunneln der Neubaustrecken der DB (The standard contact line in tunnels of German Railway's high-speed lines). In: Eisenbahntechnische Rundschau, 36(1987)11, pp. 719 to 728.

10.48 *Bauer, K.-H.; Seifert, R.*: Testing of the high-speed overhead contact line Re 250 of Deutsche Bundesbahn. In: Elektrische Bahnen 89(1991)11, pp. 424 to 425.

10.49 *Ortiz, J. M. G. et al.*: Elektrifizierung der Hochgeschwindigkeitsstrecke Madrid–Lerida (Electrification of hih-speed line Madrid–Lerida). In: Elektrische Bahnen 100(2002)12, pp. 466 to 472.

10.50 *Behrends, D.; Vega, T.*: Assessment of interoperable overhead contact line system EAC 350. In: Elektrische Bahnen 103(2005)4-5, S. 237–241.

10.51 *Kurzweil, F.; Streimelweger, K.; Hofbauer, G.*: Oberleitung der ÖBB für hohe Geschwindigkeiten – Konformitätsbewertung (ÖBB overhead contact line type for high speeds – Conformity assessement). In: Elektrische Bahnen 103(2005)9, pp. 442 to 449.

10.52 *Zöller, H.*: Entwicklung der Stromabnehmer der Triebfahrzeuge der Deutschen Bundesbahn (Development of pantographs for German Railway's traction vehicles). In: Elektrische Bahnen 49(1978)7, pp. 168 to 175.

10.53 *Bartels, S.*: Versuchsstromabnehmer für ICE (Test pantograph for ICE). In: Elektrische Bahnen 86(1988)9, pp. 290 to 296.

10.54 *Ikeda, K. et al.*: Development of the new copper alloy trolley wire. In: Sunitomo Electric Technical Review. 39(1995)1, pp. 24 to 28.

10.55 *Nibler, H.*: Fahrleitung aus Heimstoffen für elektrischen Hauptbahnbetrieb (Contact line made of locally produced material for electrical main line operation). In: Elektrische Bahnen 39(1941)10, pp. 186 to 191 and 39(1941)12, pp. 258 to 259 and 40(1942)1, pp. 12 to 16.

10.56 *Nagasawa, H.*: Verwendung von Verbundwerkstoffen für Fahrleitungen (Use of composite material for overhead contact lines). In: Elektrische Bahnen 90(1992)3, pp. 92 to 96.

10.57 *Kasperowski, O.*: Kontaktwerkstoffe für Stromabnehmer elektrischer Fahrzeuge (Contact materials for pantographs of electric railway vehicles). In: Elektrische Bahnen 34(1963)8, pp. 170 to 182.

10.58 *Hinkelbein, A.*: Der Fahrdrahtverschleiß und seine Ursachen (Contact wire wear and its reasons). In: Elektrische Bahnen 40(1969)9, pp. 210 to 213.

10.59 *Becker, K.; Resch, U.; Zweig, B.-W.*: Optimierung von Hochgeschwindigkeitsoberleitungen (Optimizing high-speed overhead contact lines). In: Elektrische Bahnen 92(1994)9, pp. 243 to 248.

10.60 *Becker, K.; Resch, U.; Rukwied, A.; Zweig, B.-W.*: Lebensdauermodellierung von Ober-leitungen (Life cycle modelling of catenaries). In: Elektrische Bahnen 94(1996)11, pp. 329 to 336.

10.61 *Borgwardt, H.*: Verschleißverhalten des Fahrdrahtes der Regeloberleitung der Deutschen Bun-desbahn (Wearing behaviour of the contact wire of German Railway's standard contact lines). In: Elektrische Bahnen 87(1989)10, pp. 287 to 295.

10.62 *Auditeau, G.; Avronsart, S.; Courtois, C.; Krötz, M.*: Carbon Contact strip materials – Testing of wear. In: Elektrische Bahnen 111(2013)3, pp. 186 to 195.

10.63 *Ikeda, M.; Uzuka, T.*: Interaction of pantographs and contact lines at Shinkansen. In: Elek-trische Bahnen 109(2011)7, pp. 338 to 343.

10.64 *Nagasaka, S.; Aboshi, M.*: Measurement and estimation of contact wire uneveness. In: QR of RTRI Volume 45, No. 2, May 2004, pp. 86 to 91.

10.65 *Rux, M.; Schmieder, A.; Zweig, B.-W.*: Qualitätsgerechte Fertigung und Montage hochfester Fahrdrähte (Quality-oriented production and installation of high-strength contact wires). In: Elektrische Bahnen 105(2007)4-5, pp. 269 to 275.

10.66 *Schmidt, H.; Schmieder, A.*: Stromabnahme im Hochgeschwindigkeitsverkehr (Power trans-mission to high-speed trains). In: Elektrische Bahnen 103(2005)4-5, pp. 231 to 236.

10.67 *Zimmert, G.*: Dynamisches Verhalten der Oberleitung für 350 km/h auf der neuen Strecke Wuhan–Guangzhou (Dynamic behaviour on the contact line for 350 km/h on the new line Wuhan–Guangzhou). In: Elektrische Bahnen 108(2010)9, pp. 147 to 155.

10.68 *N. N.*: Record-smashing run completes TGV speed trials. In: Railway Gazette International (1990)7, pp. 515 to 517.

10.69 *N. N.*: Hochgeschwindigkeitsverkehr in Frankreich (High-speed traffic in France). In: Elek-trische Bahnen 100(2002)1-2, pp. 61 to 67.

10.70 *N. N.*: Rekord auf Schienen: 574,8 km/h (Record on rails: 574,8 km/h). In: Elektrische Bahnen 105(2007)3, pp. 173.

10.71 *Landwehr, B.*: Automatische Senkeinrichtung für Stromabnehmer (Automatic pantograph lowering device). In: Elektrische Bahnen 100(2002)9, pp. 172 to 177.

11 Components and functional modules

11.0 Symbols and abbreviations

Symbol	Definition	Unit
A	cross section of conductor	mm^2
A_N	nominal cross section of conductor	mm^2
A_S	reference cross section of conductor	mm^2
AES	Automatic Earthing System	–
CA	CAtenary wire	–
CW	Contact Wire	–
D	diameter of conductor	mm
E	modulus of elasticity	kN/mm^2
F_z	tensile force of dropper	N
G'	mean dead weight per unit length	N/m
$G'_{min(max)}$	minimum (maximum) dead weight per unit length	N/m
GRP	Glass-fibre Reinforced Plastic	–
H	conductor tensile force	N
M_t	tightening torque	Nm
R'	resistance per unit length	Ω/km
SPL	Powerlines Group GmbH	–
ToR	Top of Rail	–
W	weight force of weight units for tensioning devices	N
a	width of contact wire between the two grooves	mm
a	amplitude	mm
c	height between groove and contact wire head	mm
c	specific heat	Ws/(kg K)
d	height of contact wire	mm
h	distance between contact wire and catenary wire	mm
l	minimum specified length	mm
m'	mean mass per unit length	kg/m
$m'_{min(max)}$	minimum (maximum) mass per unit length	kg/m
n	number of strands	–
n	batch size	–
p	number of specimens to be tested	–
r	transmission ratio of tensioning device	–
r	radius of contact wire upper edge and inside of grooves	mm
r_1	radius of BF contact wire shoulders	mm
r_2	radius of BF contact wire bottom	mm
α	thermal expansion coefficient	K^{-1}
α_R	coefficient of resistivity	K^{-1}
γ	specific mass	kg/dm^3

Symbol	Definition	Unit
η	mechanical efficiency	–
κ_{20}	minimum conductivity	$S\,m/mm^2$
λ	coefficient of thermal conductivity	$W/(K\,m)$
λ_c	thermal conductivity	$W/(K\,m)$
ρ_{20}	resistivity at 20 °C	$\Omega\,mm^2/m$
σ	tensile strength	N/mm^2
σ_{min}	minimum tensile strength	N/mm^2

11.1 Structure of overhead contact line systems

11.1.1 Structure

According to Figure 11.1 overhead contact line systems can be classified into the following
main functional groups:
- foundations
- supporting structures
- cross-span equipment
- overhead contact line
- traction power lines
- disconnectors and accessories
- earthing, return circuit, special fittings and plates

This classification serves as an overview of the contact line system and as a basis for planning and installation. Each main functional group consists of several subgroups. Figure 11.2
shows an example of the cross-span equipment main functional group . Five functional groups
belong to the main functional group:
- single cantilevers
- cantilevers across several tracks
- cross-span equipment
- portal structures
- pull-offs

Depending on the individual application, the required functions are assigned to the individual
functional groups. Thus, a cantilever belongs to the functional group of cross-span equipment Figure 11.3, which consists of several component assemblies. In the case of a single
cantilever these are the catenary wire registration and the pole attachments of the cantilevers.
The individual component assemblies consist of the respective components, which in the case
of the catenary wire support include:
- top tube
- cantilever tube
- diagonal tube
- catenary wire support clamp

Classification into functional groups forms the basis for material selection using computer
programs during design, material procurement, on site construction management and maintenance.

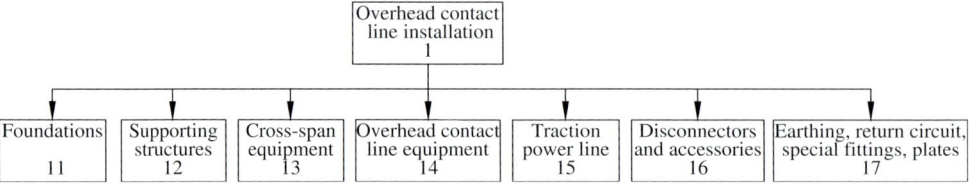

Figure 11.1: Main functional groups in an overhead contact line system.

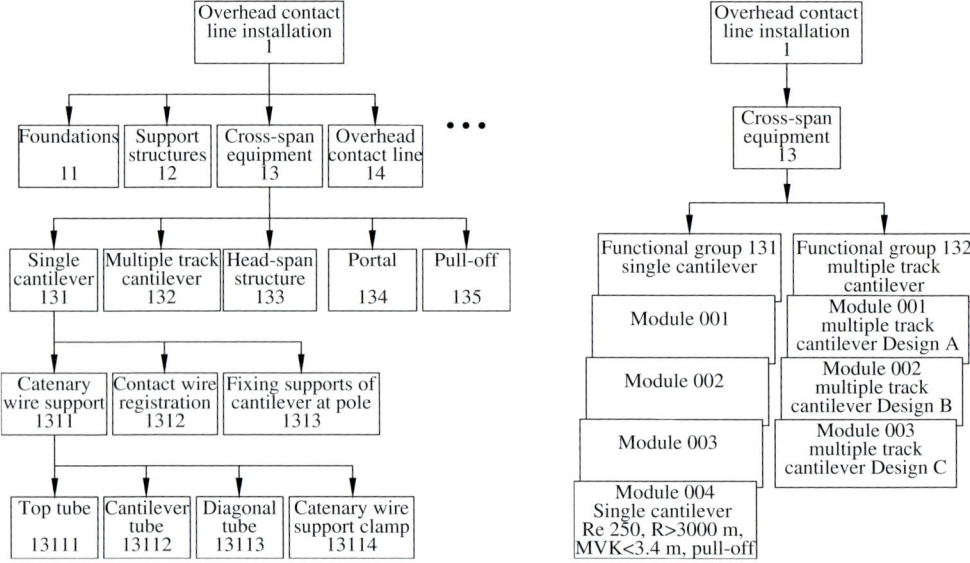

Figure 11.2: Extract from a two-dimensional structure of functional groups.

Figure 11.3: Extract from a three-dimensional structure of functional groups with component assemblies.

11.1.2 General requirements

A master material list contains all components and respective drawings for each component assembly with several thousand components available for the *contact line types* described in this book. Standard catalogues for contact line material [11.1] contain approximately 1 000 component assemblies and components. The following clauses describe frequently used components.

Requirements and testing specifications for contact line components are contained in Chapters 7 and 8 of EN 50 119. The product standards for insulators, disconnectors, contact wires etc. are mentioned there (Table 11.1).

Under most circumstances, installation, inspection and replacement of assemblies and components requires the interruption of railway operations and de-energising and earthing the contact line system. Components should have a long service life to conform to the service life of the system. Components should not require servicing and only minimal maintenance. Contact wires and the skids of section insulators wear during operation and need to be replaced after reaching their wear limits. A few components of older disconnectors, disconnec-

Table 11.1: Standards for components, see also Annex: Standards.

Component	Standards for production and testing
poles and founda-tions	EN 1992, EN 1993, EN 12843, EN 10025, EN 10204, EN 61773, EN 1997, EN ISO 14688-1, EN ISO 14689-1, EN 50119, IEC 60913
contact wires	EN 50149, IEC 62917
ropes	IEC 61089, EN 50182, EN 50183, EN 50189, EN 50326, EN 50345, EN 60889, EN 61232, DIN 48200-1, DIN 48200" 2, DIN 43138
conductor rails	DIN 17122, DIN 50142
insulators	EN 60383-1/-2, EN 60672-1/-2/-3, EN 62621, EN 61109, EN 60305, EN 61952, EN 60168, EN 60660, IEC 60273, EN 60433, EN 60437, IEC/TR2 61245, EN 61325
clamps and fittings	EN 50119, IEC 60913
section insulators	EN 50119, IEC 60913
disconnectors	EN 50119, EN 50152-2, EN 50123-4, EN 62271-102/-103, IEC 60913

tor drives and tensioning devices need to be greased periodically. According to EN 50119, Clause 7.1.2, all components shall be labeled with name of manufacturer and part number.

Components made of steel need surface protection, the composition of which depends on the local ambient conditions. Components made of corrosion resistant materials do not require surface protection. However, additional protection, in the form of greasing, needs to be provided for galvanized wires.

The design of clamps, connectors and other components should not lead to *bi-metallic corrosion* when installed on overhead line conductors. If clamps are designed to minimise ponding of water, damage from freezing will be avoided. The possibility of *stress crack corrosion* should be avoided by careful selection of materials.

Components and assemblies of a contact line system need to have sufficient mechanical and electrical strength. A knowledge of the physical properties is, therefore, imperative for the design of overhead contact lines. Tables 11.2 to 11.8 list the physical properties of materials, as the basis for design of overhead contact line systems.

11.2 Overhead contact lines

11.2.1 Cross-span equipment

11.2.1.1 Hinged cantilevers

Application, function and structural design

The functions and requirements can be found in Clause 3.3. *Hinged cantilevers*, made of tubes, follow the temperature induced longitudinal travel of catenary and contact wire and are used on fully automatically tensioned contact lines. They can consist of:
- steel tubes with components made of malleable cast iron
- aluminum tubes with components made of aluminum alloys
- stainless steel with components made of steel or aluminum alloys
- glass-fibre reinforced plastic (GRP) tubes or rods with components made of copper or aluminum alloys

Table 11.2: Physical properties of materials in overhead contact line systems.

Property	Unit	Conductor rails		Rail steel	Concrete C45/55	Source
		Al	steel			
ultimate strength σ	N/mm^2	240	290	700 ... 1 080	55	[11.2] DIN 17 122
modulus of elasticity E	kN/mm^2	70	210		30	[11.2] [11.3]
thermal coefficient of expansion α	10^{-6} K^{-1}	23,1	12	11,7	10 ... 14	[11.2] DIN 17 122, [11.3]
coefficient of resistivity α_R	10^{-3} K^{-1}	3,82	5	4,7		[11.2], DIN 17 122 [11.4]
resistivity ρ_{20}	Ωmm^2/m	0,032 68	0,120 6	0,207 / 0,228	$150 \cdot 10^6$ [1) / $2 \cdot 10^9$ [2)	[11.2], DIN 17 122 [11.5] [11.6]
conductivity κ_{20}	S m/mm^2	30,6	8,29	4,83		[11.2], DIN 17 122 [11.5]
specific mass γ	kg/dm^3	2,7	7,87	7,9	2,2 ... 2,5	
specific heat c	Ws/(kg K)	920	470	477	880	[11.2], DIN 17 122 [11.7] [11.3]
coefficient of thermal conductivity λ	W/(K m)	199	72	≈ 50	0,8 ... 1,8	[11.2] [11.3] [11.3]

[1) in moist soil   [2) in air

The last mentioned design is used mainly for mass transit applications with nominal voltages up to 1,5 kV where cast-resin material is also used for insulation.

Figure 11.4 shows a tube-type hinged cantilever for mass transit application. The cantilever can be subdivided into contact wire and catenary wire supports. The *contact wire support* comprises the registration arm, the drop bracket, the steady arm with contact wire clip and the registration arm dropper or the registration arm diagonal tube. *Pull-off supports* are characterised by the contact wire being pulled towards the pole or drop post and *push-off supports* by the contact wire being pushed away from the pole.

The *catenary wire support* consists of a cantilever tube, top anchor, catenary wire support clamp and diagonal tube, where used. In the case of high-speed overhead contact lines, tubular top anchors are often referred to as a *top tube*. A top tube anchor has a higher short-circuit resistance than a rope-type anchor. Catenary wire support clamps that are moveable along the top tube enable re-adjustment of the catenary lateral position (stagger) (Figure 11.5).

Arrangement and design of components

Figures 11.4 and 11.5 show the component parts and the assembly configuration of two cantilevers. Using hinges that swivel about a vertical axis (Figure 11.6) cantilevers can be attached to poles, drop posts and walls. Insulators (Clause 11.2.6) with eye caps or tube caps (Figure 11.8) connect the swivel hinges with the cantilever tube and the top tube. The cantilever tube is fixed to the catenary wire support clamp (Figure 11.9 a)). The *clevis end fitting* shown in Figure 11.7 connects the top tube with the *catenary wire support clamp*. In the case

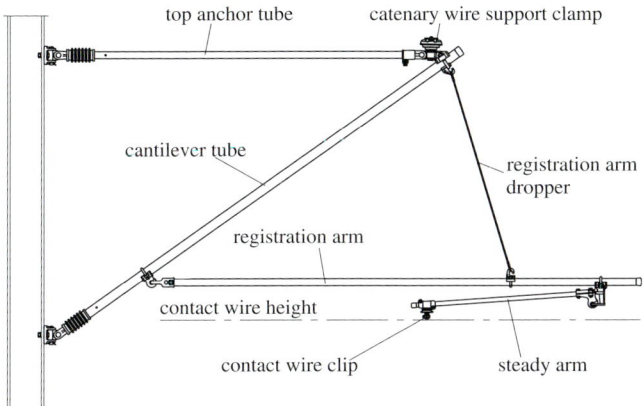

top anchor tube   catenary wire support clamp

cantilever tube

registration arm
dropper

registration arm

contact wire height

contact wire clip   steady arm

Figure 11.4: Cantilever for mass transit systems with push-off contact wire support and an insulating catenary wire support clamp moveable of the top tube.

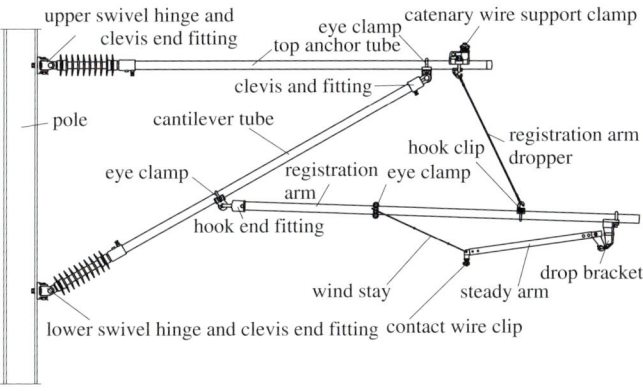

upper swivel hinge and   catenary wire support clamp
clevis end fitting   eye clamp
top anchor tube

clevis and fitting

pole   cantilever tube

registration arm
dropper

hook clip

eye clamp   registration eye clamp
arm

hook end fitting

wind stay   steady arm   drop bracket

lower swivel hinge and clevis end fitting contact wire clip

Figure 11.5: Cantilever for main lines with a catenary wire support clamp moveable along the top tube.

Figure 11.6: Swivel hinge.

Figure 11.7: Clevis end fitting.

Figure 11.8: Composite insulator.

a) At the cantilever tube   b) At the top tube   c) At the top tube

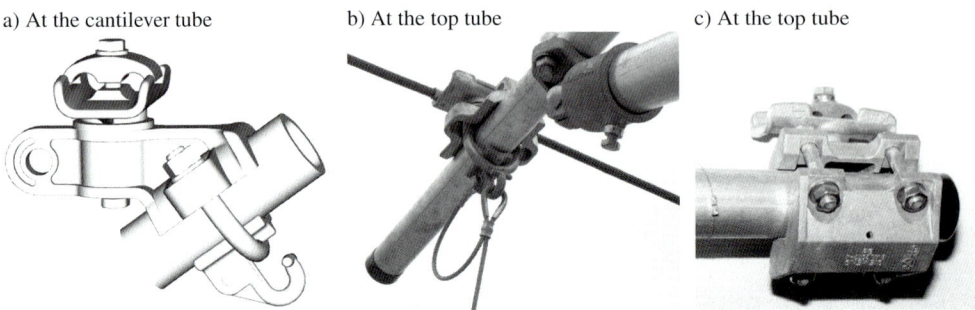

Figure 11.9: Fixed and adjustable catenary wire support clamps.

a) Hook clamp Sicat 8WL2122-6E   b) Eye clamp Sicat 8WL2112-5H   c) Hook clip Sicat 8WL2148-6

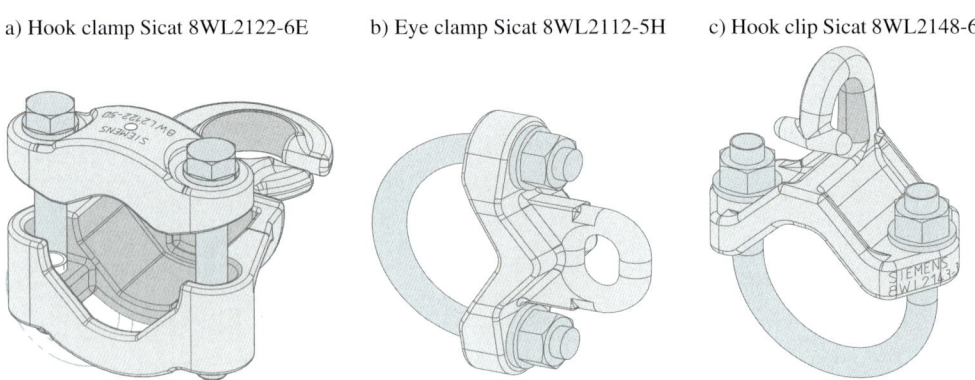

Figure 11.10: Cantilever fittings.

of a rope top anchor, *wedge-type clamps* are used for the connection between the insulator and catenary wire support clamp. When using an adjustable catenary wire support clamp (Figure 11.5, 11.9 a) and b)) the cantilever and top tube are connected to each other by an eye clamp and a clevis end fitting (Figure 11.5).

Catenary wire support clamps (Figure 11.9) support the catenary wire from the hinged cantilever tube. They allow the catenary wire to align in parallel with the track centreline, independent of the momentary cantilever position. Insulated catenary wire clamps and insulating steady arms, in conjunction with the cantilever and top tube insulators, provide double insulation for mass transit applications made of aluminum or steel. *Hook clamps* and eye clamps (Figures 11.10 a and b)) connect the top tube vertically and horizontally. They can swivel and move along the catenary tube for adjustment purposes. The requirement for adjustability is not so much a design and installation function, but a requirement by some operators to accommodate variations in track position during operation.

The registration arm is supported by a *registration arm dropper* fixed to the stitch wire by a *hook clip* (Figure 11.10 c)) and a dropper clamp (Figure 11.5) or a clevis end fitting at the catenary wire support clamp (Figure 11.9). An *eye clamp* (Figure 11.10 b)) fixes the wind stay at the registration arm.

The *steady arm* forms an essential component of the contact wire support (Figure 11.11). It is fixed to the registration arm by a drop bracket (Figure 11.12 a)) and is attached to the contact wire with a contact wire clip (Figure 11.12 b)). It restrains the contact wire horizontally against wind and radial forces and keeps the contact wire in stagger at supports. For twin

Figure 11.11: Swivelling tube cantilever with light-weight steady arm in a curve.

a) Drop bracket   b) Contact wire clip

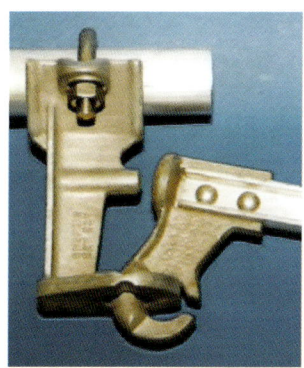 

Figure 11.12: Light-weight steady arm.

Figure 11.13: GRP cantilever on a mass transit railway.

contact wires, double drop brackets are used with twin steady arms (Figure 11.14). Drop brackets can be moved along the registration arm to adjust the lateral contact wire position. The *contact wire clip* (Figure 11.12 b)) is secured with the *groove stud* of the steady arm.

Light-weight steady arms, made from aluminum sections, which can manage contact wire lateral forces up to 2 500 N [11.1] are used to reduce concentrated masses at the contact wire. Cast resin insulated steady arms are a standard solution in mass transit installations (Figure 11.13).

The steady arm can pivot vertically at its attachment point with the drop bracket in case of contact wire lift. An adjustable uplift stop between the drop bracket and steady arm can limit this movement and avoid impacts between pantographs and support components. However, the clear space allowed for dynamic lift should be twice the expected lift under normal operation for designs without an uplift stop or 1,5 times for designs with an uplift stop. The *wind stay* prevents the registration arm and steady arm turning in different directions and secures the contact wire position in case of wind action. Wind stays are used at DB in straight line sections and curves with radii larger than 1 200 m.

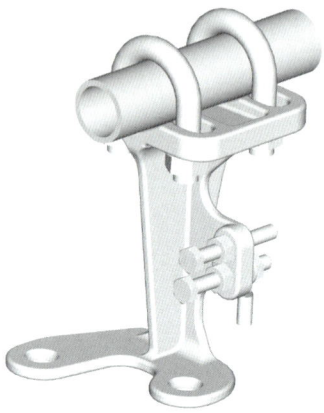

top anchor

upper
swivel bracket

insulator

catilever tube

catenary wire

registration arm strut
registration arm

steady arm
contact wire
height

drop bracket for large
contact wire uplift

contact wire and
contact wire clip

lower swivel bracket

Figure 11.14: Drop bracket for two steady arms.

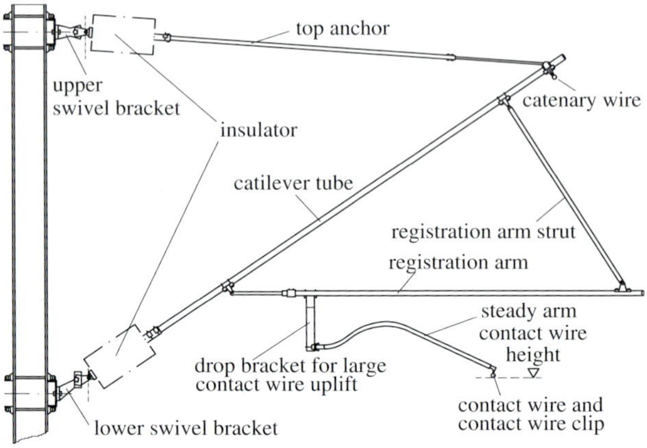

Figure 11.15: Cantilever with registration arm strut (see Clause 13.4.2.1).

Figure 11.16: Cantilever for a section with reduced system height and auxiliary catenary wire support.

Figure 11.17: Cantilever across two tracks.

At locations with long drop brackets for contact wire lifts above 150 mm, an upward turning moment can be formed. This is counteracted by a compression member between the registration arm and cantilever tube, refer to Figure 11.15.

To guide contact lines below low-height bridges, a reduction of the system height may be required. A stitch wire is used for the catenary wire suspension and an registration arm to fix it laterally in such situations (Figure 11.16).

11.2.1.2 Cantilevers across several tracks

If poles can only be installed on one side of the railway line, *cantilevers across several tracks* can be used. The example shown in Figure 11.17 uses a top-rope-stayed cantilever made of two channel sections connected together and hinged to the pole. A drop post at the track-side end of the cantilever arm, supports the catenary cantilever. For single-wire contact lines

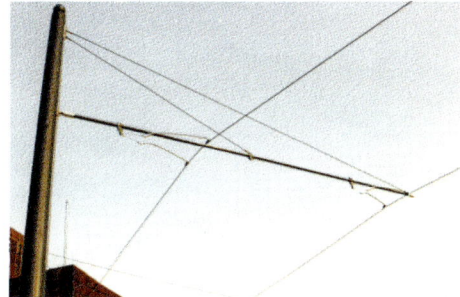

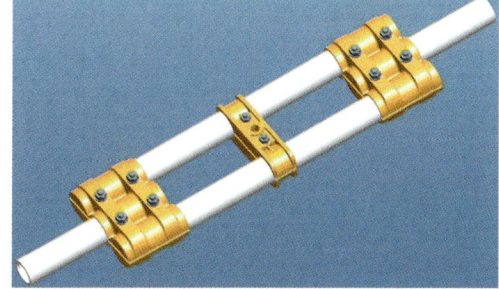

Figure 11.18: Cantilever made of GRP in a mass transit railway.

Figure 11.19: Twin (middle) and triple (end) tube clip to connect GRP sections.

on mass transit installations, cantilevers on several tracks made of GRP tubes or rods are used (Figure 11.18). The GRP profiles are connected using double or triple tube clips (Figure 11.19).

11.2.1.3   Flexible head spans

For electrification of railway systems with more than two parallel tracks, the use of flexible *head spans* is an economic solution as only two poles are needed. Some basic information is given in Clause 3.3.

It is not necessary to locate poles close to tracks for headspans because of limited cantilever lengths or to accommodate reduced clearances for poles that need to be positioned between closely spaced tracks. Head spans are usually up to 40 m long, with maximum lengths up to 80 m with poles up to 16 m high. Head spans transfer pantograph-induced movements between individual contact lines so that the contact behaviour of several pantographs and contact lines can interfere with each other during simultaneous train running on several tracks. Damage or maintenance activities at head spans may require the blocking of all tracks below the head span. Because of this, some railway operators do not use head spans for the main tracks in new installations.

The *head span wire* carries the vertical forces of the overhead contact line supports through the *head span wire droppers* (Figure 11.20). The number of head span wires and their cross section depends on the load being carried. Usually, at least two head span wires are provided in main line installations. Normally head span wire sag is specified as 10 to 15 % of the head span length (see Clause 13.4.3.2).

The *upper cross-span wire* carries the horizontal forces resulting from the catenary wire support; earthed upper cross-span wires are used where ever possible. In curves, lateral forces from both the catenary wires and the contact wires cause the support insulators in head spans to be inclined. Due to the resulting violation of the minimum clearances between the earthed upper cross-span wire and the energized insulator caps, energized upper cross-span wires are used for track radii below 800 m. This design increases the number of insulators, as the intermediate insulation for the electrical insulation of the contact lines in each switching group also needs to be provided in the upper cross-span wire.

The *lower cross-span wire* carries the horizontal forces from the contact wires and cross-span wire springs compensate for temperature-dependent length variations in the cross-span wires. A sufficiently large pre-tension is exerted by the cross-span wire spring to compensate for

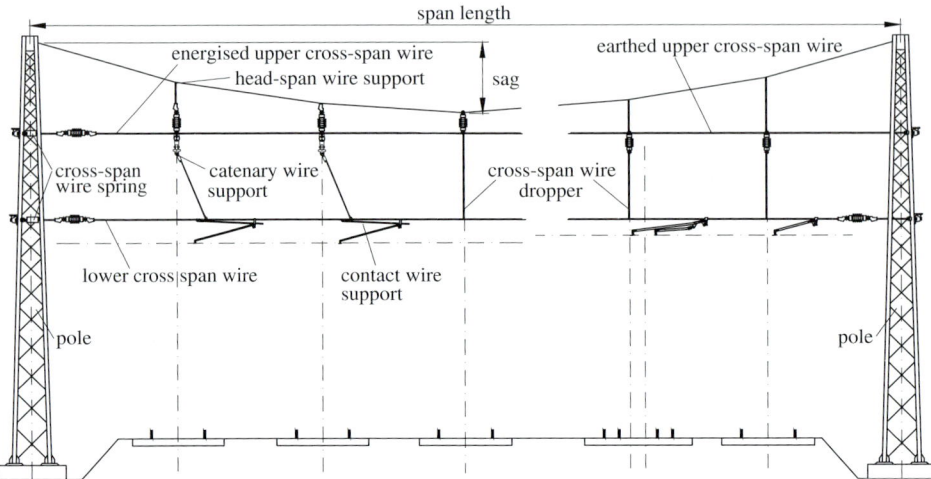

Figure 11.20: Flexible head span.

a) One contact wire   b) Two contact wires

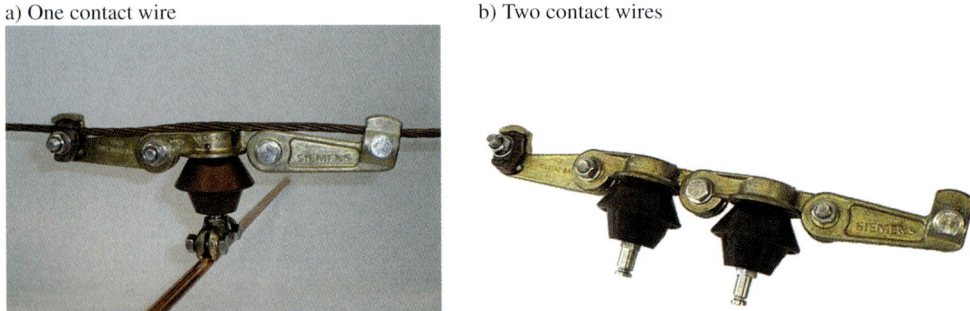

Figure 11.21: Insulated line hanger in curves.

wind forces and maintain a positive tension in all sections of the cross-span wire. Information on wires and conductors can be found in Clauses 11.2.2.2 to 11.2.2.4.

In mass transit installations simple cross-span wires may be attached to the walls of buildings [11.8] and insulated line hangers may be used as contact wire supports. They have cast resin insulators and are fixed to the cross-span wire by two adjustable arms. Due to their symmetrical design, several of these line hangers can be connected together (Figure 11.21). Steady arms can be fixed to the cross-span wire using cross-span eye clamps. The design shown in Figure 11.22 can be connected with only one bolt.

11.2.1.4   Portal structures

Instead of cross-spans, portal structures are frequently used to carry the contact line supports of more than two tracks. Multi-track portals are designed as lattice beams or solid girder structures, supported by concrete or steel poles (Figure 11.23). Due to the bending resistant design of portals, a smaller loading is imposed on the poles and foundations, than with head spans (see Chapter 13). Consequently shorter and weaker poles and smaller foundations are possible with portal structures. Lower foundation loads are especially useful for soils with

Figure 11.22: Cross-span eye clamp for the attachment of steady arms to the lower cross-span wire.

Figure 11.23: Lattice portal structure with drop posts at RENFE in Madrid-Atocha.

Figure 11.24: Pull-off arrangement.

low load-bearing capacity. Portals can be designed with *drop posts* and cantilevers or with lower cross-span wires and contact wire supports as used with head spans. The use of portals with drop posts and cantilevers effectively de-couple track to track contact line oscillations. However, this advantage is counteracted by reduced flexibility in the arrangement of the contact lines and restricted visibility of signals. Galvanized and coated steel lattice or hollow tube structures are predominantly used for portals requiring ongoing maintenance of corrosion protection, as opposed to head spans. Therefore, light-weight and less maintenance intensive aluminum structures are often employed for portal structures.

11.2.1.5   Contact line pull-offs

Pull-offs in contact lines are used near track points and other locations to secure the contact line position in between two contact line supports. Figure 11.24 shows a pull-off arrangement connecting catenary and contact wire by ropes and insulators to the pole.

11.2.1.6   Cross-span equipment in tunnels and below bridges

Cantilevers in *rectangular tunnels* are mounted on the tunnel walls, with and without recesses or on the *ceilings* using soffit posts. The supports are arranged between the tracks in tunnels with round cross sections. Cantilevers mounted individually on soffit posts as in Figure 11.25 on a double-track line enable the complete mechanical separation of the contact lines. Unistruts are provided in the tunnels on the new DB lines to attach soffit posts and supports for the feeder lines as well as for other equipment such as tensioning devices and feeder lines. Alternatively, the soffit posts are directly fixed to the tunnel walls using anchor bolts.

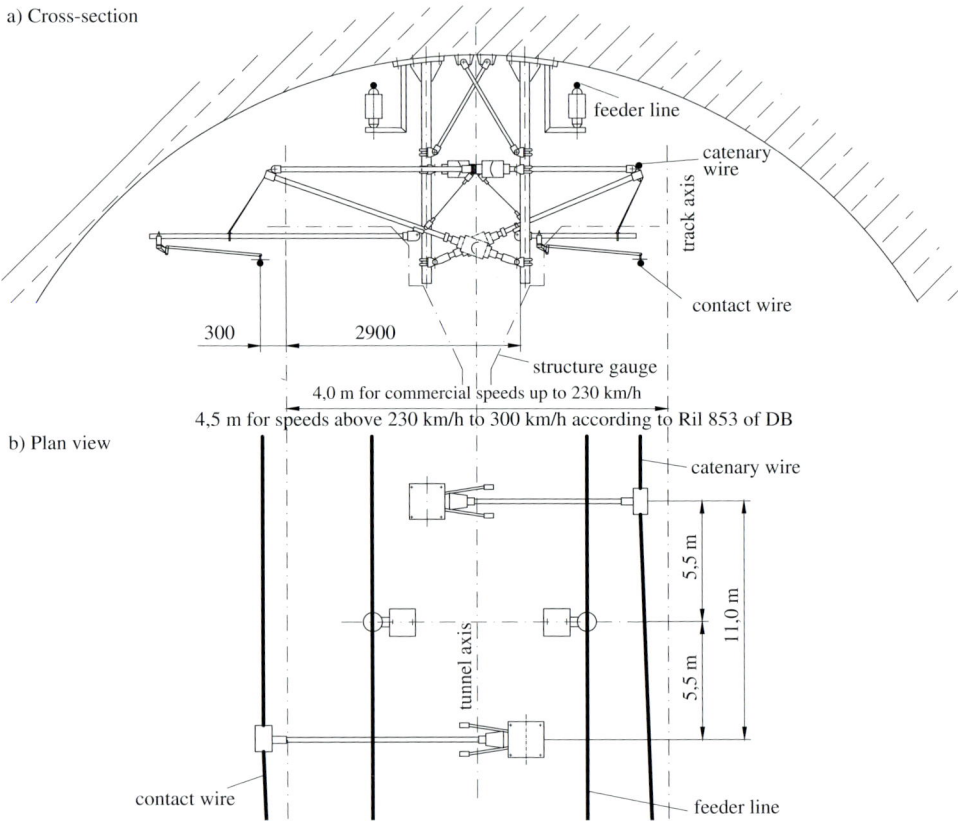

a) Cross-section

feeder line

catenary wire

track axis

contact wire

300   2900

structure gauge

4,0 m for commercial speeds up to 230 km/h

4,5 m for speeds above 230 km/h to 300 km/h according to Ril 853 of DB

b) Plan view

catenary wire

5,5 m

11,0 m

tunnel axis

5,5 m

contact wire

feeder line

Figure 11.25: Supports in a double-track circular tunnel for the contact line Sicat H1.0.

11.2.1.7 Elastic supports

For narrow tunnel cross sections and below bridges with limited clearance *elastic supports* are used (Figure 11.26). They are fixed to the tunnel wall or the building structure at spacings between 8 m and 12 m and support one or two contact wires. Because of the pivot between the insulator and cantilever arm, the arm is adjustable by $\pm 25°$, to allow adjustment to the tunnel profile and to the pantograph gauge. A metal and rubber friction element for damping of oscillations creates favourable running performance.

Other cross-span equipment for tunnels and under bridges is described in Clause 11.3 in the context of contact lines consisting of conductor rails.

11.2.2 Longitudinal contact line equipment

11.2.2.1 Functions and requirements

The main components of the *longitudinal catenary support* are the wires, conductors, clamps and insulators, serving to support the contact wire and conduct the current. Structure and design of the longitudinal catenary supports as essential distinguishing characteristics of overhead contact line designs and types are described in Clause 3.3.

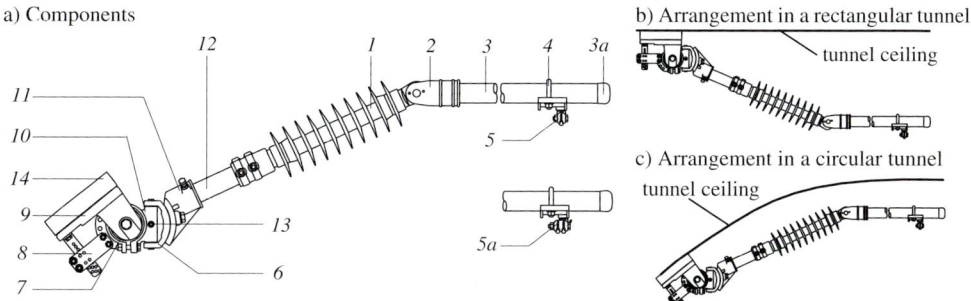

a) Components

b) Arrangement in a rectangular tunnel

c) Arrangement in a circular tunnel

Figure 11.26: Elastic support Sicat 8WL4200.
1 composite insulator, *2* tube-type cap, *3* cantilever arm (Tube 55×6), *3a* end cap, *4* clamp holder, *5* contact wire clip, *5a* double contact wire clip, *6* clevis end fitting, *7* metal rubber insert, *8* friction component, *9* basic frame, *10* central knuckle, *11* tube cap, *12* tube 55×6, *13* fixing bolt, *14* fixing base plate

Table 11.3: Dimensions of grooved contact wires according to EN 50 149.

Designation according to EN 50 149	A	Dimensions (see Figure 11.27)						m'_{min}	m'_{max}	G'_{min}	G'_{max}
		a	c	d	r	r_1	r_2				
	mm^2	mm						kg/m	kg/m	N/m	N/m
AC-80	80	5,60	3,80	10,60	0,40	–	–	0,690	0,733	6,769	7,191
AC-100	100	5,60	4,00	12,00	0,40	–	–	0,862	0,916	8,456	8,986
AC-107	107	5,60	4,00	12,30	0,40	–	–	0,923	0,980	9,055	9,614
AC-120	120	5,60	4,00	13,20	0,40	–	–	1,035	1,099	10,153	10,781
AC-150	150	5,60	4,00	14,80	0,40	–	–	1,293	1,378	12,684	13,518
BC-100	100	6,92	4,30	12,00	0,40	–	–	0,862	0,916	8,456	8,986
BC-107	107	6,92	4,30	12,24	0,40	–	–	0,923	0,980	9,055	9,614
BC-120	120	6,92	4,30	12,85	0,40	–	–	1,035	1,099	10,153	10,781
BC-150	150	6,92	4,00	14,50	0,40	–	–	1,293	1,378	12,684	13,518
BF-100	100	6,92	4,00	11,04	0,40	6,00	15,60	0,862	0,916	8,456	8,986
BF-107	107	6,92	4,24	11,35	0,40	6,43	15,85	0,923	0,980	9,055	9,614
BF-120	120	6,92	4,30	12,27	0,40	6,60	17,21	1,035	1,099	10,153	10,781
BF-150	150	6,92	3,90	13,60	0,40	7,55	20,00	1,293	1,378	12,684	13,518

Operating pantographs create oscillations in the contact line that produce mechanical wear. Operational and short-circuit currents load these components leading to aging and electrical wear in addition to that produced by temperature variations and component movement. Table 2.8 shows the limiting temperatures for conducting materials in overhead contact lines. Where temperatures exceed those mentioned in Table 2.8 the remaining tensile strength of the conductor should be established after considering the period of high temperature action and where required the conductor dimensions need to be increased or the operational load decreased.

11.2.2.2 Contact wires

The wires of overhead contact lines contacted by the pantographs are called *contact wires*. As they are equipped with two grooves on their upper part, they are also known as *grooved contact wires*. The standards EN 50 149 and IEC 62 917 set the requirements and characteristics of contact wires. There are different contact wire types depending on requirements and development, distinguished by the type of cross-section. The cross-sections and the material

a) Circular cross section type AC and BC

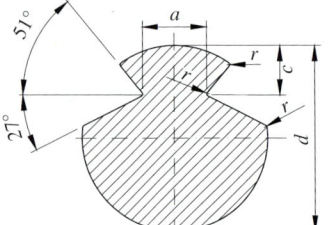

b) Cross section type BF with flat profile

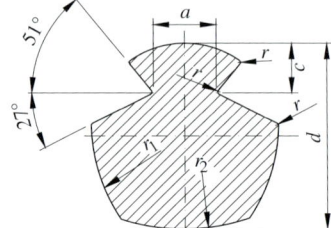

Figure 11.27: Contact wire cross-sections according to EN 50 149. Dimensions see Table 11.3

Table 11.4: Tensile strength σ_{min} in MPa of grooved contact wires according to EN 50 149.

Cross section	Contact wires made of			
	Cu-ETP	CuAg0,1	CuMg0,2	CuMg0,5
80	355	365	460	520
100	355	360	450	510
107	350	350	440	500
120	330	350	430	490
150	310	350	420	470

and are indicated by identification grooves. The dimensions of the types AC, BC and BF are shown in Figure 11.27 and Table 11.3. The circular cross-section shape is favoured for contact wires. Table 11.4 contains the minimum tensile strength and Table 11.5 details the physical properties of contact wires generally used in Europe.

The selection of the contact wire cross-section depends mainly on the required current capability, voltage regulation and the applied tensile forces.

For DC traction systems with operating voltages up to 3 kV and high traction power requirements, it is usual to install parallel contact wires, also known as *twin contact wires*.

Table 11.5: Physical properties of contact wires.

Property	Unit	Materials				Source
		Cu	CuAg0,1	CuMg0,2	CuMg0,5	
ultimate strength σ	MPa	345	350	436	495	EN 50 149: AC-100
		320	340	418	475	EN 50 149: AC-120
modulus of elasticity E	kN/mm^2	120	120	120	120	—
coefficient of thermal expansion α	$10^{-6} K^{-1}$	17	17	17	17	EN 50 149
coefficient of resistivity α_R	$10^{-3} K^{-1}$	3,8	3,8	3,1	2,7	EN 50 149
resistivity ρ_{20}	$\Omega \cdot mm^2/m$	0,01777	0,01777	0,02240	0,02778	EN 50 149
conductivity κ_{20}	$S \cdot m/mm^2$	56,3	56,3	44,6	36,0	EN 50 149
specific mass γ	kg/dm^3	8,9	8,9	8,9	8,8	DIN 43 140; [11.9]
specific heat c	$Ws/(kg \cdot K)$	394	394	378	378	EN 60 865-1
thermal conductivity λ_c	$W/(K \cdot m)$	377	375		245	[11.7]
annealing point	°C	220	320	370	370	—

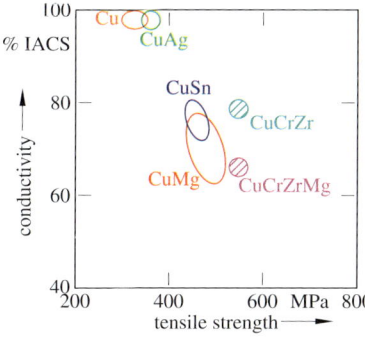

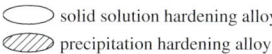

Figure 11.28: Conductivity of copper alloys plotted in relation to their tensile strength and expressed in relation to the conductivity of electrolytic copper.

⌒ solid solution hardening alloy

⧄ precipitation hardening alloy

The installation technology for contact wires requires that the cross-sectional area of contact wires has to be limited. According to EN 50 149 the maximum standardised cross-section is 150 mm^2. However, some operators use contact wires with 161 mm^2, 170 mm^2 and 193 mm^2. Because of high conductivity, tensile strength, hardness and the ability to withstand temperature changes and corrosion, *hard-drawn electrolytic copper* and *copper alloys* have become the established global contact wire material. When exposed to air, copper forms a hard but conductive oxide layer that does not prevent current flow. This is the reason why copper is a suitable material for sliding contacts, as opposed to aluminum which forms an oxide layer of poor conductivity. All attempts to use aluminum as a contact wire material have failed.

Alloy additives such as silver, tin or magnesium further improve the mechanical or thermal properties of copper wires and permit the application of higher tensile forces. These properties are important for high-speed traffic. With the exception of silver the alloying metals reduce the material's conductivity (see Figure 11.28). Lesser amounts of alloys reduce the conductivity of copper to a lesser extent, however, they raise considerably the thermal resistivity of contact wires. The use of cadmium as an additive is no longer permitted in most European countries because of the associated environmental pollution risks.

Copper-clad steel contact wires with a copper content of 45 % were installed in Germany on some lines at the beginning of the 1940s. Up to the time when the copper on the contact surfaces had worn away these wires proved to have similar mechanical characteristics as copper wires, but they wore away very quickly, impairing operating reliability. Nevertheless, copper-clad steel contact wires are currently used in Japan to enable the use of high tensile forces.

Contact wires are worn away by the collectors sliding along them (see Clause 10.7). The rates of wear of contact wire and collector strips depend, among other things, on the combination of materials for collector strips and contact wires. To date low wear rates have been achieved using a combination of copper contact wires with *carbon collector strips*. Contact strips made of steel or copper produce a considerably higher rate of wear.

Since the resulting reduction of the cross-sectional area of the contact wire reduces its current-carrying capacity and increases the tensile stress, the permitted wear is limited to between 20 % and 30 % of the original cross-sectional areas, if the tensile force is reduced in proportion to the wear. The limit of the permitted wear is reached at those parts of the contact wire experiencing locally higher wear.

The basic requirements for near uniform wear of contact wires and consequently long service life are optimum *overhead contact line and collector interaction* (Chapter 10) as well as design and installation and adequate maintenance (Chapter 17).

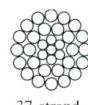

7-strand   19-strand   37-strand

Figure 11.29: Stranded wire cross sections.

11.2.2.3   Steel wires

Galvanised steel or high-grade stainless steel wires are used for pole earthing and wind stays on steady arms, respectively. Stranded steel wires are also used for head spans and catenary wires.

11.2.2.4   Stranded metal conductors

In overhead contact line installations, *stranded wires* are used both for suspension and tensioning purposes and as electrical conductors. The most common stranded wire designs are shown in Figure 11.29. The *wrought copper alloy* CuMg0,5, also called Bz II, is now widely used. In Central Europe, the majority of catenary wires, head span and cross span wires, stitch wires and droppers, which all carry heavy mechanical and electrical loads, are made of this alloy. Flexible stranded wires of *electrolytic copper* (E-Cu) are mainly used as electrical connectors between the catenary wire and the contact wire, to connect consecutive tensioning sections of contact line systems and as switch gear cable. E-Cu conductors are often used to increase the current-carrying capability of contact line systems in DC railways.

Galvanised stranded steel conductors were also used as catenary wires, head span and cross-span wires in early contact line installations but their susceptibility to corrosion was the main disadvantage of simple steel conductors. *Flexible, high-tensile strength stranded steel conductors* with bitumen protection are used for tensioning wheel ropes subjected to high mechanical loads. Stainless steel cables are not suitable for this purpose because of susceptibility to stress-crack corrosion. Electrical parallel feeder lines, bypass and other feeder cables which are only subject to loading due to their dead weight are made of *aluminum conductors*. Although aluminum has a lower conductivity and tensile strength than copper, it is cheaper and extremely corrosion resistant after the protective oxide layer has formed. Some railway operators use ASCR conductors as catenary wires as well.

In Russia, copper catenary wires comprising copper clad wires with individual strands of copper with a steel core are widely used. German Railways DB experienced negative results with *copper clad steel wires* because damage to the thin outer copper layer caused by clamps used during installation, work leads to rapid corrosion of the steel core. To date no European standards for copper and copper alloy conductors exist, so national standards should be applied. The most important specifications and technical delivery conditions for conductors and stranded wires made of copper alloys are:

- DIN 43 138: Flexible copper and copper alloy conductors
- DIN 48 201-1: Copper stranded conductors
- DIN 48 201-2: Bronze stranded conductors
- EN 50 182: Conductors for overhead lines. Round wire concentric lay stranded conductors, conductors made of aluminum, aluminum alloy and steel

Table 11.6 contains the physical properties of copper, copper alloy, aluminum and steel stranded wires which are used in overhead contact line systems.

Table 11.6: Physical properties of stranded conductors.

Property	Unit	Conductors				Source
		E-Cu	Bz II	AL1	St1A	
modulus of elasticity E	kN/mm^2	100–113[2]	100–113[2]	55–60[2]	162	DIN 48 203-1, DIN 48 203-2
coefficient of thermal expansion α	$10^{-6}\,K^{-1}$	17	17	23	11	DIN 48 203-1, DIN 48 203-2
coefficient of resistivity α_R	$10^{-3}\,K^{-1}$	3,94	3,78	4,0	4	DIN 48 203-1 DIN 48 203-2
resistivity ρ_{20}	$\Omega \cdot mm^2/m$	0,01786	0,02778	0,028264	0,192	DIN 48 203-1, -2
conductivity κ_{20}	$S \cdot m/mm^2$		36	35,4	7,25	DIN 48 203-1, -2
specific mass γ	kg/dm^3	8,9	8,9	2,7	7,8	DIN 48 200
specific heat c	$Ws/(kg \cdot K)$	394	380	897	480	[11.7]
thermal conductivity λ_c	$W/(K \cdot m)$	400	300	236	42	[11.3]

[1] dependent on the wire diameter, [2] dependent on the number of wires

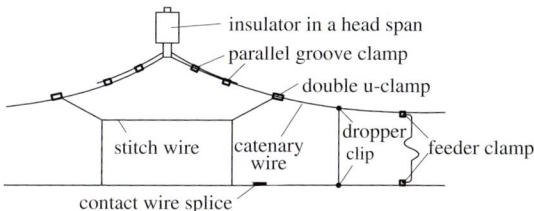

Figure 11.30: Examples of installed clamps and fittings.

11.2.2.5 Synthetic ropes

Various types of *synthetic ropes* made of polyester acrylamide fibres with a polyamide sheath, also called Minoroc rope, are used for anchors in plastic cantilevers, bridle and pulley suspensions and cross spans. These ropes fulfill mechanical as well as insulating functions. Details and testing conditions are contained in EN 50 345 and [11.1].

11.2.2.6 Clamps

Clamps are used in the contact line to connect wires and conductors and other components under tension and/or carrying load currents. The selection and rating of *clamps* follows the specifications in EN 50 119. *Anchoring clamps* or wire connections should be capable of securing conductors with a minimum of 2,5 times the working load or with 85 % of the specified tensile strength of the conductors to be supported. The lower value may be used in each case. Anchoring clamps should not induce permanent deformations that impair operation at 1,33 times the working load.

Other clamps and line fittings should have a load capacity of 2,5 times the working load. Clamps and line fittings subjected to vibration should be designed to prevent loosening over time. The mass of line fittings should also be kept to a minimum, however, they should meet the functional requirements of the component. Clamps and line fittings should provide a path for the specified normal and short-circuit current flow without failing.

The selection of materials for clamps and connection fittings depends on the required conductivity, tensile strength and long-term durability. Copper and copper alloys best fulfill the

Table 11.7: Nominal cross-section A_N, actual cross section A_S, number of strands n, diameter D, mass per unit length m', weight per unit length G' and minimum tensile strength σ_{min} of conductors according to DIN 48 201, parts 1 and 2 and EN 50 182.

Designation	A_N mm^2	A_S mm^2	n -	D mm	m' kg/m	G' N/m	σ_{min} N/mm^2
Conductors made of E-Cu according to DIN 48 201-1							
DIN 48 201–10–E-Cu	10	10,02	7	4,1	0,090	0,88	402
DIN 48 201–16–E-Cu	16	15,89	7	5,1	0,143	1,40	398
DIN 48 201–25–E-Cu	25	24,25	7	6,3	0,218	2,14	389
DIN 48 201–35–E-Cu	35	34,36	7	7,5	0,310	3,04	393
DIN 48 201–50–E-Cu	50	49,48	7	9,0	0,446	4,38	397
DIN 48 201–50–E-Cu	50	48,35	19	9,0	0,437	4,29	388
DIN 48 201–70–E-Cu	70	65,81	19	10,5	0,596	5,85	377
DIN 48 201–95–E-Cu	95	93,27	19	12,5	0,845	8,29	394
DIN 48 201–120–E-Cu	120	116,99	19	14,0	1,060	10,40	391
DIN 48 202–240–E-Cu	240	242,54	61	20,3	2,209	21,67	405
Conductors made of BzII according to DIN 48 201-2							
DIN 48 201–10–Bz II	10	10,02	7	4,1	0,090	0,88	588
DIN 48 201–16–Bz II	16	15,89	7	5,1	0,143	1,40	583
DIN 48 201–25–Bz II	25	24,25	7	6,3	0,218	2,14	570
DIN 48 201–35–Bz II	35	34,36	7	7,5	0,310	3,04	576
DIN 48 201–50–Bz II	50	49,48	7	9,0	0,446	4,38	572
DIN 48 201–50–Bz II	50	48,35	19	9,0	0,437	4,29	568
DIN 48 201–70–Bz II	70	65,81	19	10,5	0,596	5,85	552
DIN 48 201–95–Bz II	95	93,27	19	12,5	0,845	8,29	576
DIN 48 201–120–Bz II	120	116,99	19	14,0	1,060	10,40	563
DIN 48 201–240–Bz II	240	242,54	61	20,3	2,209	21,67	593
Conductors made of AL1 according to EN 50 182, Table F17							
16-AL1	16	15,9	7	5,1	0,0434	0,42	189
24-AL1	25	24,2	7	6,3	0,0663	0,65	174
34-AL1	35	34,4	7	7,5	0,0939	0,92	172
49-AL1	50	49,5	7	9,0	0,1352	1,33	168
48-AL1	50	48,3	19	9,0	0,1329	1,30	179
66-AL1	70	65,8	19	10,5	0,1809	1,77	169
93-AL1	95	93,3	19	12,5	0,2563	2,51	172
117-AL1	120	117,0	19	14,0	0,3215	3,15	166
147-AL1	150	147,1	37	15,8	0,4057	3,8	177
182-AL1	185	181,6	37	17,5	0,5009	4,91	172
243-AL1	240	242,5	61	20,3	0,6711	6,58	182
299-AL1	300	299,4	61	22,5	0,8285	8,13	175
ACSR conductors according to EN 50 182, Table 19							
44-AL1/32-ST1A	44/32	75,6	14/7	11,2	0,3693	3,62	584
97-AL1/56-ST1A	95/55	152,8	12/7	16,0	0,7068	6,93	509
106-AL1/76-ST1A	105/75	181,2	14/19	17,5	0,8853	8,68	584
184-AL1/30-ST1A	185/30	213,6	26/7	19,0	0,7410	7,27	306
243-AL1/39-ST1A	240/40	282,5	26/7	21,8	0,9801	9,21	301
304-AL1/49-ST1A	300/50	353,7	26/7	24,4	1,2273	12,04	297
653-AL1/45-ST1A	650/45	698,8	45/7	34,4	2,1595	21,19	223

Table 11.8: Material properties for clamps and fittings [11.1].

Material	Tensile strength N/mm^2	Electrical conductivity at 20 °C m/(Ω·mm^2)	Application examples
electrolyte copper	200 to 300	58	crimped connector, feeder clamp (E-clamps and C-clamps), protective sleeves
copper-nickel wrought alloy	290 to 640	15 to 18	conductor crossing clamp, contact wire clamp, contact wire splice, parallel groove clamp, sliding dropper clip, dropper clip, dead-end clamp, bolts, nuts, stud bolt, double U-clamps, bridle wire parallel clamp, dropper strap, body for contact wire end fitting, clamps for section insulators, dead-end clamps for contact and catenary wires and compression joints
copper-tin alloy	440 to 590	9	double U-clamps, bolts
copper-zinc cast alloy	440 to 490	15	contact wire clips, double contact wire clips, parallel groove clamps, stitch-wire clamps
copper-aluminum cast alloy	460 to 720	4 to 8	contact wire clips, contact wire crossing splice, sliding dropper clip, bridle wire parallel clamps, feeder clamps, conductor clamps, dropper clips, cone-type dead-end clamps
aluminum	115 to 130	37,7	pin for clevis end fitting
aluminum wrought alloy	215 to 320	30	sheet metal, crimped connector, winding tape, tubes, straps, cable dog, hollow section, pin, hook end fitting, swivel clip holder, dropper clip, wedge-type dead-end clamp, suspension clamp, cone-type dead-end clamp
aluminum cast alloy	230 to 310	25 to 30	swivel clip holders, hook clips, eye clip, spade end fittings, drop brackets, catenary wire support clamp, hook end clamp, hook end fitting, eye clamp, reducing socket, clevis end fitting, tongue end fitting, swivel bracket, swivels, dog, washer, wedge-type dead-end clamp
malleable cast iron, galvanized	~ 400	2,5 to 3,5	dog, head-span wire clamps, cross-span wire clamps, catenary wire suspension clamps, cross-span drop bracket clamps, dead-end clamps, hook end clamps, wedge-type dead-end clamps, connection clamps
stainless steel	500 to 850	1,2 to 1,7	threaded rods, bolts, nuts, washers
steel	360 to 850	5 to 6	threaded rods, bolts, nuts, washers, angle sections, flat bars, hollow sections, crossarms, swinging straps

requirements for sleeves for copper contact wires, copper and bronze catenary wires etc. as energized parts of the overhead contact line system. These materials have long-term durability. Clamps and connection fittings for fixed and flexible *dead-end devices* consist either of galvanized malleable cast iron, copper aluminum alloys or aluminum cast alloys. They completely satisfy requirements for mechanical strength and long-term durability. Table 11.8 contains mechanical and electrical properties of important materials for clamps and connection fittings.

Joining of conductors and fittings made of copper or bronze with those made of aluminum is accomplished with copper-clad bi-metallic sheets that avoid electrolytic corrosion. Aluminum and copper sheet materials are compressed together resulting in a *bi-metallic copper-*

a) Wedge-type dead-end clamps

b) Cone-type dead-end clamp

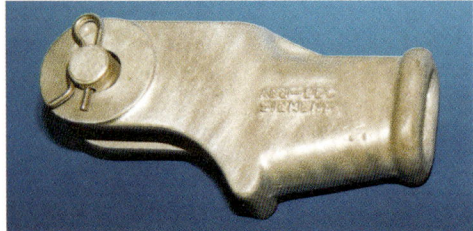

Figure 11.31: Dead-end clamps.

a) AC-80 to AC-120 made of Cu-ETP or CuAg0,1   b) AC-120 made of Cu-ETP, CuAg0,1 or CuMg0,5

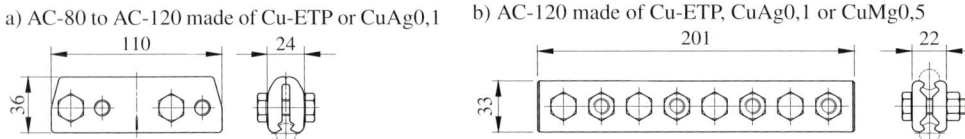

Figure 11.32: Contact wire splices for contact wires according to EN 50 149 [11.1].

clad sheet. The aluminum layer is faced towards the aluminum fitting and the copper layer towards the copper or bronze components. In this way, a copper aluminum bi-metal casing forms the transition from bronze or copper catenary wire or contact wire to a wedge-type dead-end clamp made of AlSi7Mg0,3 in accordance with EN 1706.

Wedge-type dead-end clamps and *cone-type dead-end clamps* as in Figure 11.31 connect contact wires and conductors and other components, e. g with insulators at dead-ends or intermediate terminations of all kinds, at positions clear of pantographs. These connections can be undone if required. They are considered as dead-end clamps and have a tensile strength close to the failing load of the conductors they are used to terminate. They can be re-used several times. Wedge-type dead-end clamps made of galvanized malleable iron or aluminum are suitable for copper and copper-silver alloy contact wires. For installation, the contact wire or conductor is passed through the housing of the clamp and then bent into a loop. The loop is then drawn back into the housing of the wedge-type dead-end clamp, together with a suitable wedge and tensioned. *Cone-type dead-end clamps* are used specifically for high-tensile contact wire materials like CuMg0,5 that can only be fitted with difficulty into wedge-type dead-end clamps. The end of the contact wire is as with a wedge-type dead-end clamp passed through the housing and then, with a suitable cone drawn back into the housing. After screwing together the cover, a tension and current proof connection is produced.

Contact wire splices shown in Figure 11.32 are used to connect two contact wires, e. g after failure or damage to a contact wire. They are designed predominantly as compression or bolted clamps of the copper alloy CuNi2Si. The tension and current proof connection may be traversed by pantographs.

Parallel groove clamps shown in Figure 11.33 are used to connect two contact wires, a contact wire and another conductor or two conductors. Examples are the connection of the contact wire and an auxiliary contact wire in front of or behind section insulators, at contact wire crossings and the connection of the Z-anchor conductor with the contact wire or catenary wire. Parallel groove clamps are unsuitable for tensioned conductor or contact wire connections.

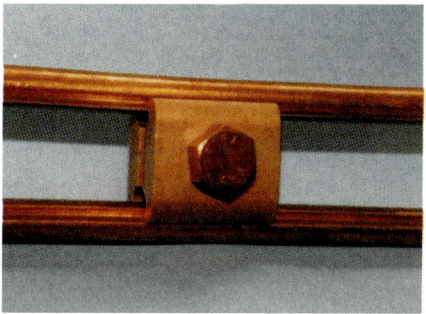

Figure 11.33: Parallel groove clamp.

Figure 11.34: Double U-clamp.

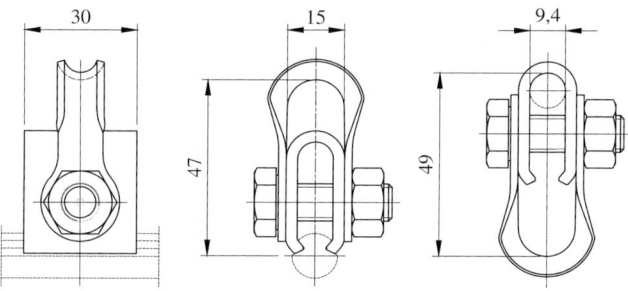

Figure 11.35: Dropper clip with clamping and suspension eye.

Feeder clamps, as shown in Figure 11.39 provide short-circuit proof connections to contact wires and between two conductors. Because of the toothed inner surface of the clamp it provides a reliable connection and can transfer tensile forces between the conductors. The clamp shown in Figure 11.34 connects the catenary wire with the stitch wire.

Dropper clips connect the droppers with the contact wire or the catenary wire. The dropper clips shown in Figure 11.35 consist of a clamping bracket, a suspension clevis, a bolt and a nut. The *thimble* with a compression connector and conductor loop is used together with this dropper clip as a current-carrying connection as shown in Figures 11.36 a) and b).

a) Fastening at contact wire

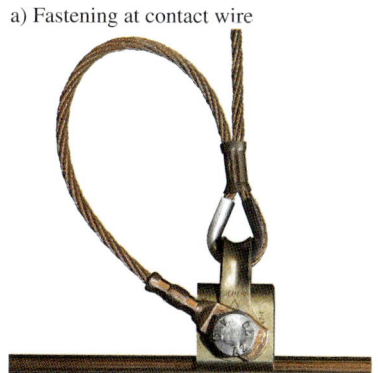

b) Fastening at catenary wire

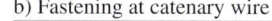

Figure 11.36: Dropper clip for a current-carrying fastening of dropper.

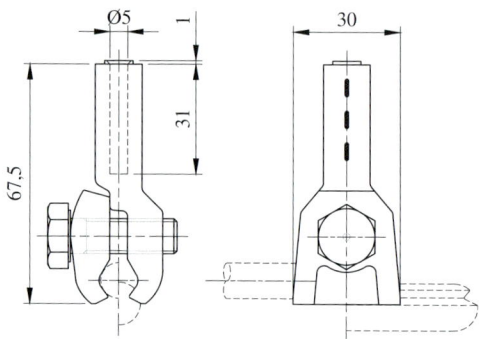

Figure 11.37: Universal dropper clip Sicat 8WL4591-6.　　**Figure 11.38:** Dropper clip (SNCF).

a) C-type clamp Sicat 8WL4550-0 for catenary wires　　b) E-type clamp Sicat 8WL4570-0 for contact wires

Figure 11.39: Feeder clamps for electrical connections.

The universal dropper clip as shown in Figure 11.37 consists of a CuAl alloy and is used predominantly on mass transit installations. The dropper wire is compressed with the clamp body. The dropper clip in Figure 11.38 consists of the same material but has two parts clamped together after assembly on the contact wire. The dropper wire is compressed with the dropper body. The complete droppers are described in Clause 11.2.2.7.

Feeder clamps connect conductors to catenary and contact wires. They are current and short circuit proof, but are not designed for tensile loads. The *E-type clamp* shown in Figure 11.39 b) is compressed with a flexible conductor and the contact wire. The *C-type clamp* in Figure 11.39 a) connects the other end of the conductor with the catenary wire. (see also Clause 11.2.2.8).

11.2.2.7 Droppers

The *droppers* support the contact wire and are attached to the contact wire and catenary wire or to the stitch wire, using thimbles and various types of dropper clips (Figure 11.40). Some dropper designs perform a mechanical suspension function and an electrical function. They conduct a portion of the operational or short-circuit current between the contact wire and the catenary or stitch wire while a current drawing operating traction unit or a short circuit occurs close to the dropper. Therefore, the droppers are mechanically and electrically loaded.

Droppers can be made rigid, flexible (Figure 11.40) or sliding. The different types should withstand the loads imposed without an adverse effect on the performance of the dropper over the life cycle of the system. The loading capacity of the dropper, including the clamps, should be 2,5 times the vertical and 1,5 times the horizontal working loads. The following

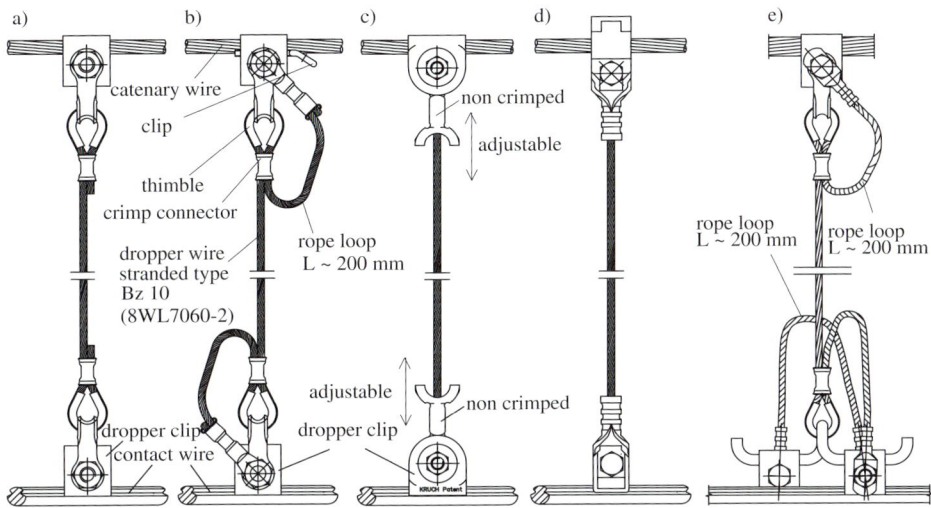

Figure 11.40: Types of droppers. b)–e) current-proof, a) – d) for one contact wire,
b) dropper Bz 10 (8WL7060-2) for running speeds up to 400 km/h, e) for twin contact wires

service loads need to be considered in the design of droppers:
- vertical loads from contact wire weight, ice loads, wind loads and loads resulting from the contact wire profile
- horizontal loads along the axis of the contact wire resulting from a dropper inclination up to 30°
- dynamic loads such as oscillation and alternate bending due to the pantograph passage, especially in the case of contact lines for speeds above 160 km/h, as the droppers are compressed when a pantograph passes

In addition to the operational loads, the following needs to be considered:
- construction loads
- loads after damage, e. g the failure of an adjacent dropper

If the additional loads are greater than 2,5 times the operating load, then these loads should be used for design of droppers instead of the operational loads. Additional loads less than 2,5 times the operating loads can be neglected. In overhead contact line sections, *reduced system height* droppers that slide along the catenary wire as shown in Figure 11.41 a) are suitable for avoiding inclinations and contact wire height variations where length changes in contact and catenary wire are not equal. The dropper type shown in Figure 11.41 a) transfers uplift rigidly to the catenary wire producing higher contact forces than elastic droppers due to passing pantographs. The East Japan Railways (JP East) adopts rigid droppers with unlimited uplift (Figure 11.41 b)) to attach the contact wire to the auxiliary catenary wire. Triangle-type droppers, as shown in Figures 11.41 c) and d) are used for low clearance as well as low running speeds.

Since 1960, several railway operators have experienced repeated dropper failures in sections where the overhead contact lines are often traversed at the limit of their maximum commercial speed [11.10]. The assumed reason for this is the high frequency compression loads in the dropper caused by high speed pantograph uplift of the contact wire and the delayed reaction of the catenary wire. For this reason, dropper design and testing methods for verification of

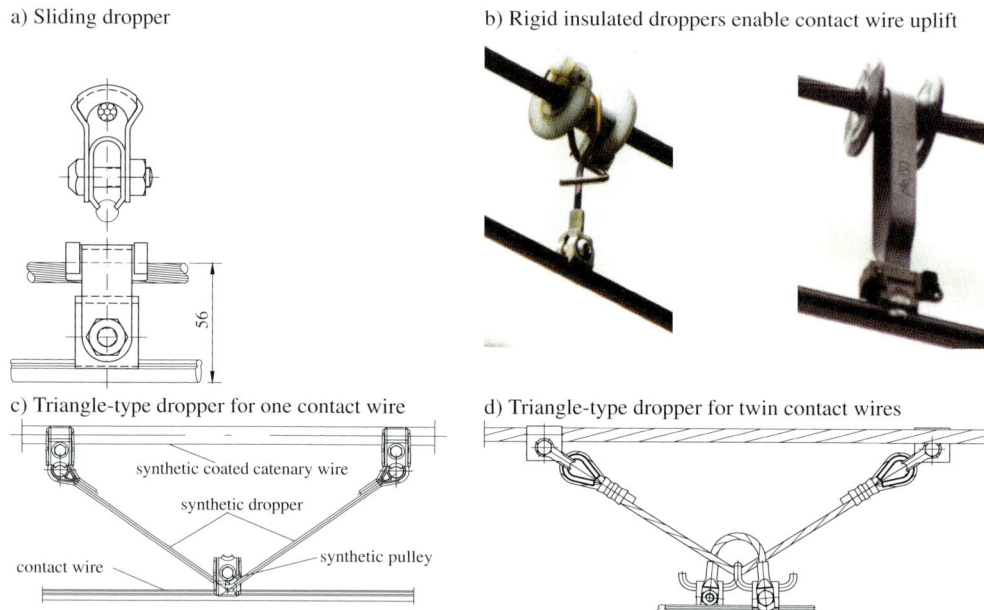

a) Sliding dropper

b) Rigid insulated droppers enable contact wire uplift

c) Triangle-type dropper for one contact wire

synthetic coated catenary wire

synthetic dropper

contact wire

synthetic pulley

d) Triangle-type dropper for twin contact wires

Figure 11.41: Droppers for reduced system height design.

alternate bending resistance were further developed. Clause 11.5.5 describes these dropper tests. For droppers of high-speed lines, vibration testing close to practice, in the range of 5 to 10 Hz, is carried out. The dropper design 8WL7060-2 shown in Figure 11.40 b) made of bending-resistant stranded wires complied with all tests at this frequency for running speeds up to 400 km/h. It is adopted in all Sicat types of contact lines for running speeds above 160 km.

Droppers can be designed as current carrying, non current carrying or insulating. Current carrying droppers should be designed to allow current to flow between the contact and catenary wire. A short circuit close to a single dropper could cause a very high short-circuit current in the dropper, which the dropper cannot sustain. Under this condition, a dropper is not expected to be short-circuit-proof. The current-carrying droppers in Figure 11.40 b) to d) are suitable for systems with high operating currents. Figure 11.40 d) shows a *dropper* used for *twin contact wires* in DC systems. The current loops, with cable lugs at their ends, enable a continuous current flow between contact wire and catenary wire or stitch wire. In the case of high currents in DC systems, additional electrical connections also known as jumpers or feeders are installed between the catenary and contact wire.

In the case of non-current-carrying droppers, currents flow across moveable connections that are subjected to vibrations by passing pantographs. This can lead to electrical erosion with frequent current flow and to radio frequency interference caused by arcs. Consequently, these non-current-carrying droppers are only used in overhead contact lines on secondary tracks or on lines with low currents. All other systems are equipped with current-carrying or insulated droppers in combination with electrical connectors.

The calculation of length and pre-fabrication of droppers avoids cost-effective adjustments during installation (see Clause 15.5).

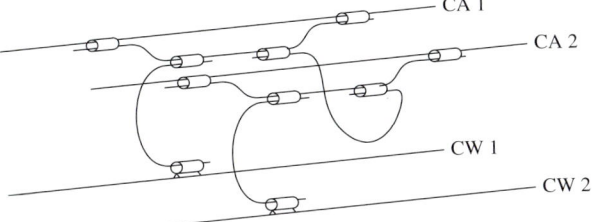

Figure 11.42: Electrical connectors between two contact lines
CA 1, CA 2 catenary wire
CW 1, CW 2 contact wires

Figure 11.43: Tensioning device Sicat 8WL5070 for tensile forces up to 40 kN with a transmission ratio 1 : 3 for contact line Sicat H1.0 with separate tensioning of contact and catenary wire.

11.2.2.8   Electrical connectors

Electrical connectors, also called *current connectors* or jumpers, provide current-carrying connections between the contact and catenary wire, between two overhead contact lines, between overhead contact lines and parallel feeder lines and between a great number of overhead contact line components. Current connectors need to be designed for the expected operational current and they need to be short-circuit-proof, unlike individual droppers. Along the contact line the electrical connectors are prone to oscillations caused by passing pantographs and, therefore, are designed in an S-type or circular shape (Figure 11.42). They are only subjected to tensile loads under the dynamic consequences of short circuits and rare conductor galloping induced by wind.

Electrical connections should have the following properties:
  – they need to resist alternating thermal loads
  – their temperature, because of operational and short-circuit currents, may not exceed the maximum permitted temperature of the connected stranded wires and conductors according to Table 2.8
  – the mechanical effectiveness of the electrical connection must be sustained even under short circuit

The current connector, made of flexible copper conductors, may be pressed with current-carrying compression clamps (Figure 11.39) or bolted after contact grease is applied to the mating surfaces.

Figure 11.44: Sicat H1.0 tensioning device in a tunnel of DB high-speed line Frankfurt–Cologne.

11.2.3   Tensioning devices

11.2.3.1   Functions and requirements

Tensioning devices serve to keep constant the tensile forces in the contact line and, therefore, to compensate the variations in length of contact and catenary wire. Tensioning devices can be categorised into devices with or without tensioning weights. Devices with tensioning weights can be found world-wide. They have a simple and reliable design and proven performance. A tensioning wheel or a pulley arrangement transmits the weight force and the migration length in a ratio of 1 : 1,5 to 1 : 5 depending on the specific design. For special cases, e. g in narrow tunnels, tensioning devices are used with either tensioning springs, hydraulic or electro-mechanical devices. Tensioning devices should be service-free and characterized by high level efficiency and should include a drop arresting device in case of conductor failure.

11.2.3.2   Tensioning devices with weights

The *wheel tensioner* consists of a tensioning wheel with two rope coils on a common axle and a drop protection device. The overhead contact line to be tensioned is attached to the small split coil by flexible steel wire ropes while a tensioning mass weight acts on the large coil. Figure 11.43 shows the wheel tensioner 8WL5070 with a transmission ratio of 1 : 3. for winding lengths up to 2,4 m and for tensioning lengths up to 2 000 m a tensioning device with a transmission ratio of 1 : 1,5 is available. Both types of tensioning devices are rated for tensioning forces up to 40 kN. At this wheel tensioner, the latch-in device consists of a conical-shaped wave-type surface on the flanges of the wheel. After conductor failure, the

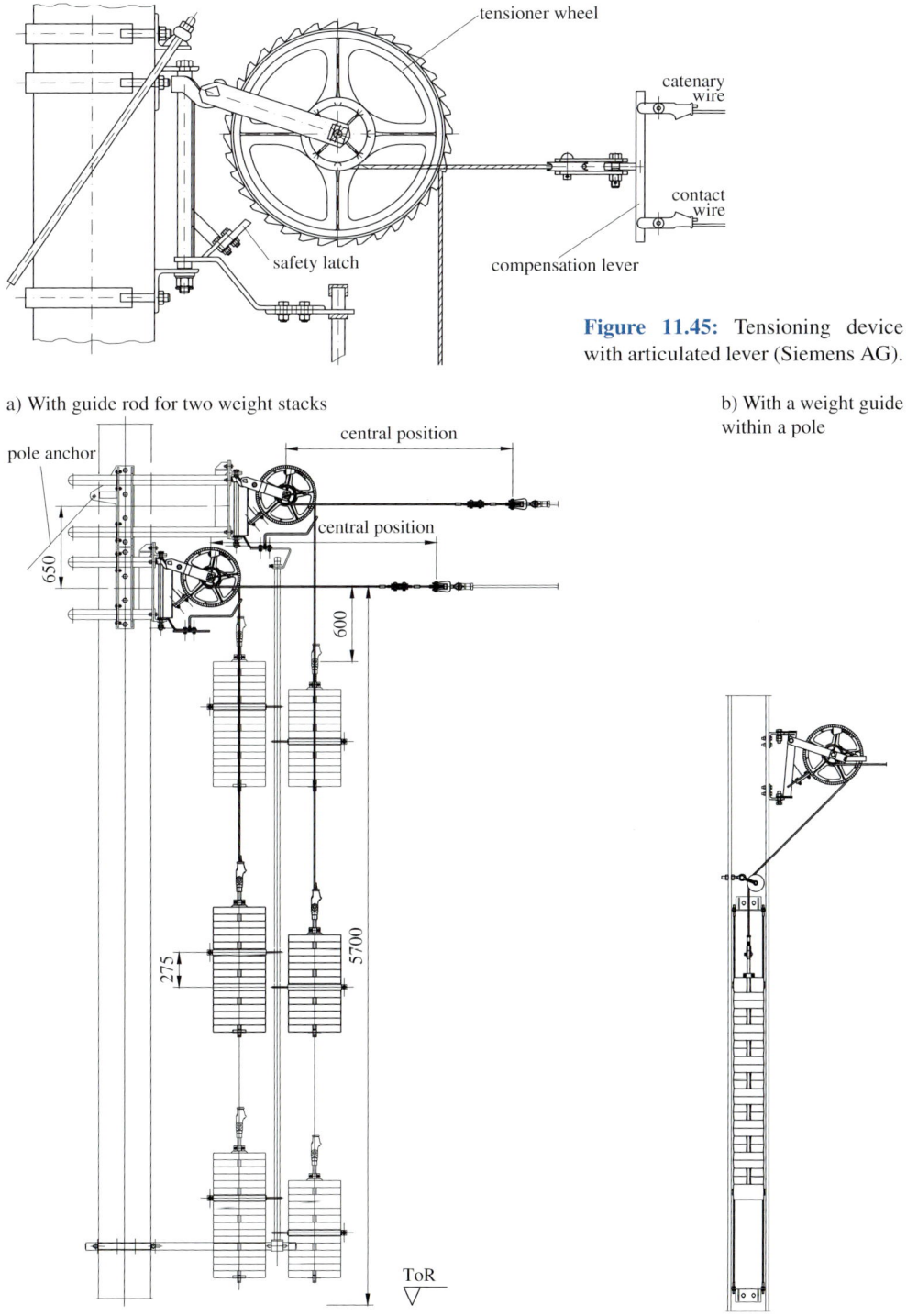

tensioner wheel

catenary wire

contact wire

safety latch

compensation lever

Figure 11.45: Tensioning device with articulated lever (Siemens AG).

a) With guide rod for two weight stacks

b) With a weight guide within a pole

pole anchor

central position

central position

650

600

275

5700

ToR

Figure 11.46: Tensioning device for the overhead line Sicat H1.0 (Siemens AG).

Figure 11.47: Pulley wheel tensioner as used on the Paris–Strasbourg line (Foto: SNCF).

wheel drops and latches into a U-shaped plate with consequent deceleration of the turning of the wheel. This system also works with large tensioning forces. This device avoids the weight stack falling to the ground and limits damage to the overhead contact line, especially failure of droppers. This is the main advantage of *wheel-type tensioners* in comparison with *pulley-type tensioners* based on the block and tackle principle. Separate tensioning of contact wire and catenary wire by individual tensioning devices as in Figures 11.43 and 11.44 guarantees constant tensile forces even when the catenary and contact wire vary in length.

For high-speed lines in tunnels, Siemens developed tensioning devices with weight guides shaped to match the tunnel cross-section (Figure 11.44) and keep emergency escapes open. An efficiency of at least 97 % guarantees a precise contact wire position and excellent running qualities of the contact line [11.11].

The combined tensioning of contact wire and catenary wire as shown in Figure 11.45 requires only one tensioning device to tension the contact line within a half tensioning section. It has been adopted by several railway operators for lines up to 200 km/h for long periods. The articulated lever compensates minor length variations in the catenary and contact wire. The drop protection device consists of a rim with teeth that drops on to a safety latch in the case of conductor failure and stops the weights dropping any further, see Figure 11.45.

The *weight stack* of the tensioning device consists of individual weights made of concrete or galvanized malleable iron weighing 12,5 kg, 25 kg or 50 kg each and guided along a rod (Figure 11.46 a)). For mass transit systems lead weights are also used. They can be inconspicuously placed between the flanges of H-type poles. At sites with public access, such as on platforms, a protective guard is installed around the weights to protect people in case of a dropping weight stack.

Pulley wheel tensioners operate on the *pulley block principle*. The weight force is transmitted to the contact line via several pulley wheels as a horizontal tensile force. Figure 11.47 shows a tensioning device used by SNCF on their Paris–Strasbourg high-speed line. In the case of conductor failure, consecutive failures within the contact line can occur with pulley-type tensioners, where a drop protection device or limiting of wire movement is not provided.

Figure 11.48: Spring-type tensioning device for cross-span wires and tensioning lengths up to 180 m (Siemens AG).

11.2.3.3   Tensioning devices without weights

Tensioning devices with springs

Simple *tensioning devices with springs* with a linear spring in a piston-type housing are suited for short tensioning sections up to 180 m in length and tensile forces up to 10 kN (Figure 11.48).

For tensioning lengths above 180 m in individual cases, e. g at very narrow spaces in tunnels or in stations, tensioning devices according to Figure 11.49 are used. Their housing contains several springs, the characteristics of which is a result of laterally arranged spiral-type springs. For this type of tensioning device, specific test conditions (Clause 11.5.4), possible reductions of the permissible conductor tensile force (Clause 4.2.2.1) caused by lower efficiencies and specified service intervals need to be considered [11.12].

Gas and hydraulic tensioning devices

The *hydraulic tensioning device* controls the tensile force of the contact line by means of a change in volume of a gas or fluid in a cylinder. This gas moves a piston axially, which adjusts the tensile force in the contact line as illustrated in Figure 11.50. The tensile force of the contact line can only be adjusted with the gas pressure within the cylinder during installation. This device only reacts to variations of the ambient temperature and not to the current-related length variations of the contact line.

Electro-mechanical tensioning devices

The elctro-mechanical tensioning device [11.13] compensates tensile force changes resulting from temperature induced contact line length variations using an electrically driven spindle whose reaction threshold can be adjusted. The *electro-mechanical tensioning device* requires an electricity supply.

11.2.4   Electrical sectioning devices

11.2.4.1   Functions and requirements

According to EN 50 119, section insulators with or without neutral sections are *sectioning devices*. A *section insulator* without a neutral section establishes a device formed by insulators, installed in the overhead contact line, and equipped with skids or similar components to enable an uninterrupted current transmission when traversed by a pantograph (IEV-IEC 60 050-811 36-15). *Section insulators* are used predominantly in main line systems at junctions and in mass transit systems in all kinds of electrical sectioning to divide the overhead contact line into individual electrical sections to enable selective switching off and earthing.

A *section insulator with a neutral section* creates a separation that avoids bridging by pantographs of two adjacent electrical sections with different voltages or different phases to act as a protective section (IEV-IEC 60 050-811 36-16). Sectioning devices with neutral sections are used in main line systems for phase and system separation sections. In mass transit systems some operators use them instead of section insulators. In a neutral section, power supply

Figure 11.49: Tensioning device with spring for a complete tensioning section in the Nevjanski station in Russia (Foto: J.N.Eberle Company, spring manufacturer GmbH).

or regeneration of vehicle current is not possible because its main switch needs to be switched off in front of the section insulator to avoid arcing.

Depending on their design, the overhead contact line type and the type of pantograph, sectioning devices may be traversed at running speeds up to 200 km/h. For higher speeds, *insulated overlaps* are adopted for main line tracks (IEV-IEC 60050-811 36-14).

Since sectioning devices need to be installed in the overhead contact line, they are designed for the tensile forces in the attached wires or conductors. The mechanical strength of the tensioning device needs to be rated such that at 1,33 times the operational load, no permanent deformation occurs. According to EN 50119, the dead-end clamps have to withstand at least 2,5 times the operational load of the fixed conductors or wires or 0,85 times the nominal strength of the conductors. The lower value have to be attained in any case. The clamps and height of the skids need to be adjusted for the contact wire used according to EN 50149 and its variation in height because of wear.

The dynamic pantograph contact force may not exceed 350 N at a section insulator, according to EN 50119 and the contact strips of the pantograph should not be damaged. This limit is necessary because of practical contact conditions to accommodate differences in elasticity and because it is simpler to replace a sectioning device than a section of contact wire. To maintain this limit requires, especially at running speeds above 130 km/h, minimal mass and smooth passage of the pantograph along the contact wire. When designing the skids and air clearances, the dimensions and spacing of the contact strips of the pantographs used and vehicle sway need to be considered.

The insulators of the sectioning devices have to comply with the relevant standards, e.g. EN 62621. In cases were the contact strips of the pantograph slide along the insulating elements of the section insulator conducting carbon or metal layers can be formed creating creepage paths. A short circuit caused by running a pantograph into an earthed line section or

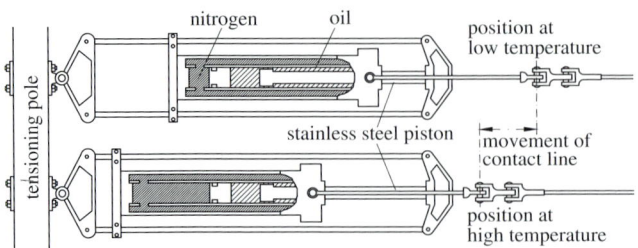

Figure 11.50: Gas tensioning device.

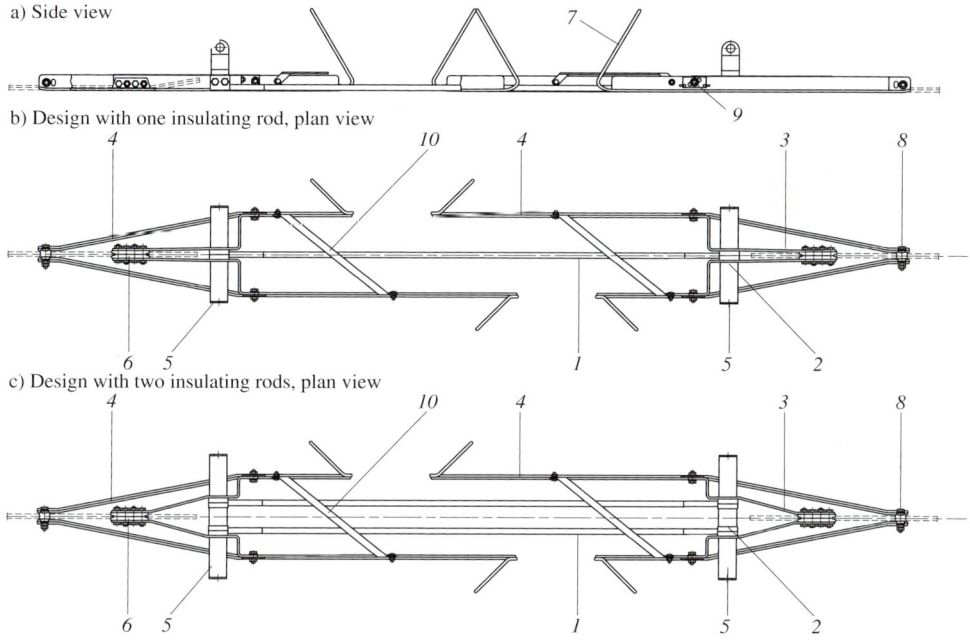

a) Side view

b) Design with one insulating rod, plan view

c) Design with two insulating rods, plan view

Figure 11.51: Light-weight section insulator Sicat 8WL5545.
1 composite insulator, *2* clamping bracket, *3* tensioning strap, *4* runner skid, *5* suspension clip, *6* contact wire clamp, *7* arcing horn, *8* runner clamp, *9* height adjustment link, *10* bar

bridging a section insulator when running into neutral sections may not mechanically impair the sectioning device. It follows that arcs should be kept away from heat sensitive components by arcing horns designed to extinguish the arc.

Sectioning devices should be capable of easy and rapid installation and adjustment to minimise line blocking periods. Damaged components prone to wear like skids and arcing horns should be easily replaced.

11.2.4.2   Section insulators

Section insulators, which separate sectioning groups and feeding sections, may be designed according to IEC 60 913 and EN 50 122-1 with lower minimum clearances in air than usually specified between live and earthed components:

    – 50 mm up to 3 kV
    – 100 mm at 15 kV
    – 150 mm at 25 kV

These smaller clearances, also permitted for short-period approaches in overhead contact lines, enable improved running performance of pantographs. In normal operation the section insulators separate adjacent sectioning groups having the same potential. When one or both separated sections are de-energised for maintenance or removal of a disturbance, earthing and short circuit devices on both sides of the work site are used to protect personnel.

For section insulators with neutral sections, longer air clearances are required, depending on the design, to avoid bridging by pantographs.

a) Sicat 8WL5545-4A (25 kV)   b) Sicat 8WL5545-8A (3 kV)

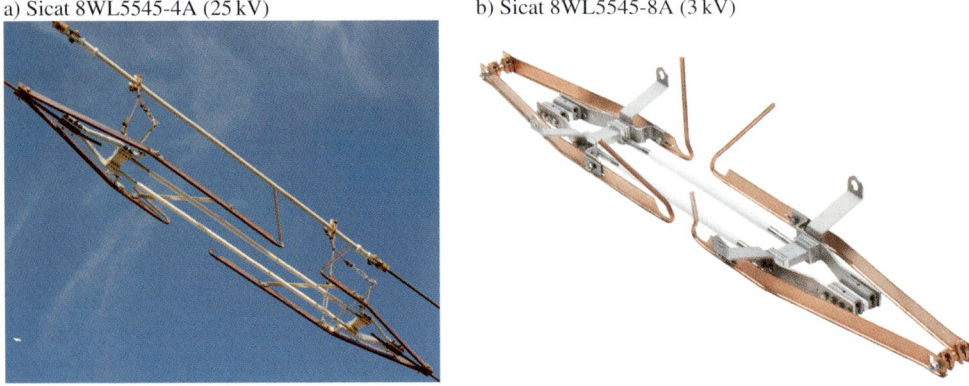

Figure 11.52: 25 kV light-weight section insulator Sicat 8WL5545-4A for one contact wire a) and Sicat 8WL5545-8A for 3 kV nominal voltage and twin contact wire b).

Table 11.9: Technical data of section insulators.

Series Sicat 8WL		5545-7A	5545-8A	5545-4A	5545-2A
nominal voltage	kV	3	3	25	25
air clearance	mm	60	60	220	220
creepage path	mm	450	450	1 200	1 200
length	mm	1 725	1 725	2 490	2 490
height	mm	208	208	238	238
width	mm	340	362	450	472
mass	kg	13	13,8	15,9	16,4
nominal strength	kN	90	90	90	90
operating force	kN	30	30	30	30
number of contact wires		1	2	1	2

Numerous designs exist for section insulators and are distinguished mainly by the shape and arrangement of the insulators and runners.

The section insulator Sicat 8WL5545, is known as a light-weight section insulator, see Figures 11.51 to 11.53 and Table 11.9, for nominal voltages up to DC 3 kV and AC 25 kV possesses the following design characteristics:

 – ensioned components are arranged along the axis of force, close to the centre of the contact wire to keep the bending moment low
 – the weight of components is low
 – the individual components form a modular system from which several types of section insulators can be assembled (see Table 11.9)
 – the clamps used are suitable for the connection of contact wires AC-/BC-80 to AC-/BC-150 according to EN 50 149
 – installation is possible on tensioned contact wires without unloading and cutting the wire first

Section insulators are provided for contact lines with catenary suspension. When adjusted for individual installation sites, they are also used in tunnels, workshops and in trolley wire contact lines. They are designed to ensure that the contact strips of the pantographs slide smoothly from the contact wire to the intake runners, which guide the contact strips at least 2 mm below the end fittings of the insulators, to avoid damage. Because of the asymmetric

a) Sicat 8WL5510-0   b) Sicat 8WL5570-1AF

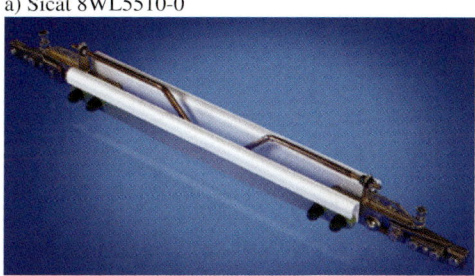
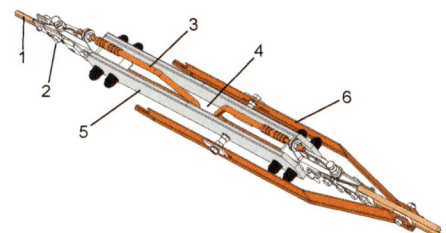

Figure 11.53: Section insulators for mass transit systems up to 1,5 kV.

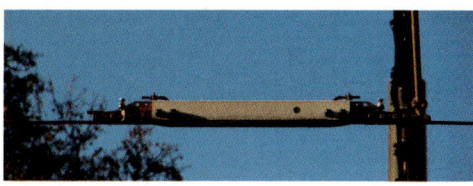

Figure 11.54: Section insulators with neutral section, Sicat 8WL5546-3 for mass transit systems up to 1,5 kV.

arrangement of the section insulator air gaps, each of the pantograph contact strips is always in contact with at least one runner of the section insulator. This ensures there is always voltage at the pantograph. The transfer of voltage to the runners is continuous. When traversing the middle part of the section insulator commutation arcs can occur because of the high power consumption of traction units. This is also the case for all other types of section insulators.

When pantographs accidentally run into neutral or earthed overhead contact line sections high power arcs result from either the operational or short-circuit currents. These arcs are directed upwards at the inner arcing horns and extinguish without damaging the section insulator. Table 11.9 contains the technical data for various section insulator designs. For superior interaction with pantographs low unit mass is important. This mass is reduced to only 6 kg/m for a 25 kV section insulator design. The light-weight section insulator Sicat 8WL5545 can be traversed at up to 200 km/h, depending on overhead contact line type it is installed in.

Section insulator designs 8WL5510-0 and 8WL5570-1AF proved their performance in mass transit installations up to DC 1,5 kV (Figure 11.53). They can be traversed at running speeds up to 80 km/h. These section insulators are installed on to the uncut contact wire (1) and fixed into the end clamp (2) at each end. After clamping, the contact wire is cut away. The contact to the pantograph clearance is set from this point with two height-adjustable brackets (3) trained upwards in the middle of the section insulator to guide any arcs upwards. The 60 mm airgap (4) is located between the two installation brackets. The special features of this design are the two insulating rods transferring the contact wire tensile force. The insulating rods are traversed by the pantograph contact strips in the middle section. The type 8WL5570-1AF (Figure 11.53 b)) has runners (6) on the out side and a main running direction, resulting in more favourable electrical characteristics at high operating currents. The wear of electrically stressed components can be reduced or eliminated by reducing the operational current when passing the section insulator.

Figure 11.55: Short 25 kV phase separation section consisting of neutral sections Sicat 8WL5545-4D.

11.2.4.3 Section insulators with neutral sections

A *section insulator with a neutral section* shown in Figure 11.54 was adopted by several mass transit operators. As distinct from section insulators shown in Figure 11.53, the air gap is 450 mm long and is not bridged by the pantograph. This type of section insulator should be traversed without drawing current to avoid arcing when running into the neutral section. The required switching-off of the circuit breaker is signalled to the tramway driver by a T-sign. The bridging of different DC substation supply sections and resulting protection problems can be avoided by this. For the driver, frequent switching operations are necessary, however, these are sometimes forgotten resulting in arcing and wear at the section insulators.

Short neutral sections as shown in Figure 11.55 are used in main line systems. They consist of two neutral sections, the design of which is described in Clause 11.2.4.2 and is similar to the section insulator 8WL5545. The length of the air gap of the neutral section, is 1 500 mm and the creepage path is 2 010 mm. This section insulator with neutral section can be traversed at 160 km/h. The section in between the two neutral sections is earthed.

The neutral sections can also be used for system separation sections. Phase separation sections and *system separation sections* with neutral sections may also be formed by several section insulators, e. g. Sicat 8WL5545 (Figure 11.51).

Usually the required switching on and off of the circuit breaker before and after the separation section when being traversed by a raised pantograph is signalled to the driver with signals EL1 and EL2, respectively. The tests to verify the arc extinguishing performance of the phase separation section in case of unintended traversing by a pantograph with the circuit breaker switched on is shown in Figure 11.110.

a) Disconnector Sicat 8WL6144-1

b) Disconnector arrangement with drive and link

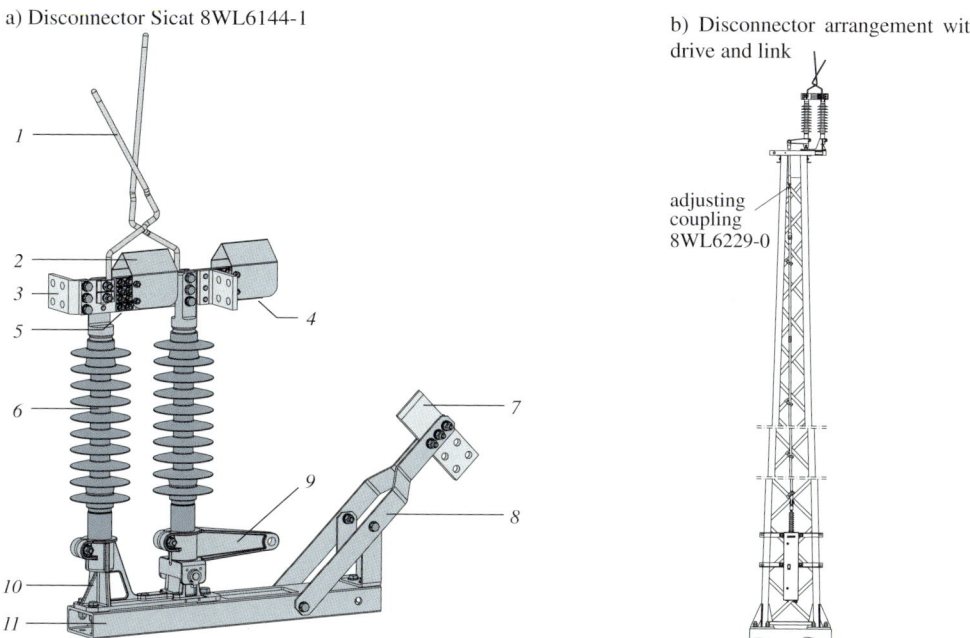

adjusting
coupling
8WL6229-0

Figure 11.56: Overhead contact line disconnector Sicat 8WL6144-1 for 15 kV and 25 kV (a) and its arrangement with drive and link at a pole (b); Explanation of figures within the text.

11.2.5   Overhead contact line disconnectors and earthing switches

11.2.5.1   Functions and requirements

Overhead contact line disconnectors connect and separate switching groups and feeding ranges. They enable disconnection and isolation of individual sections in case of faults or maintenance work, allowing operations to continue on the remaining tracks. They are usually mounted on contact line poles or structures and provide *visible confirmation of section separation*. Overhead contact line disconnectors are installed at the feeding points of the overhead contact line or in parallel to insulated overlaps and section insulators at stations, transition connections, branch-offs etc. They can be fitted with contacts for earthing purposes, for a range of applications on tracks or in service workshops. Bi-polar overhead contact line disconnectors serve the same function in multiple-voltage systems like 2 AC 15 kV or 2 AC 25 kV [11.14]. Earthing switches are used for earthing of overhead contact line sections.

Standard EN 50 119 specifies that overhead contact line disconnectors and the drives and linkages installed on site, must comply with the requirements of EN 50 123-4 for DC disconnectors and EN 50 152-2 for AC disconnectors. These standards specify requirements for single-pole disconnectors, earthing switches and load switches for use in DC and AC systems. EN 50 152-2 refers to other standards such as EN 62 271 and EN 60 060 for switch gear and high-voltage testing. Overhead contact line disconnectors need to be designed for the rated current and the nominal voltage of the system. They are unsuitable for switching full load currents but should be able to interrupt limited operational current since switching without current cannot always be guaranteed in overhead contact line systems. They are also designated as overhead contact line switches and equipped with arcing horns. EN 50 119,

a) Disconnector Sicat 8WL6144-1

b) Contact set

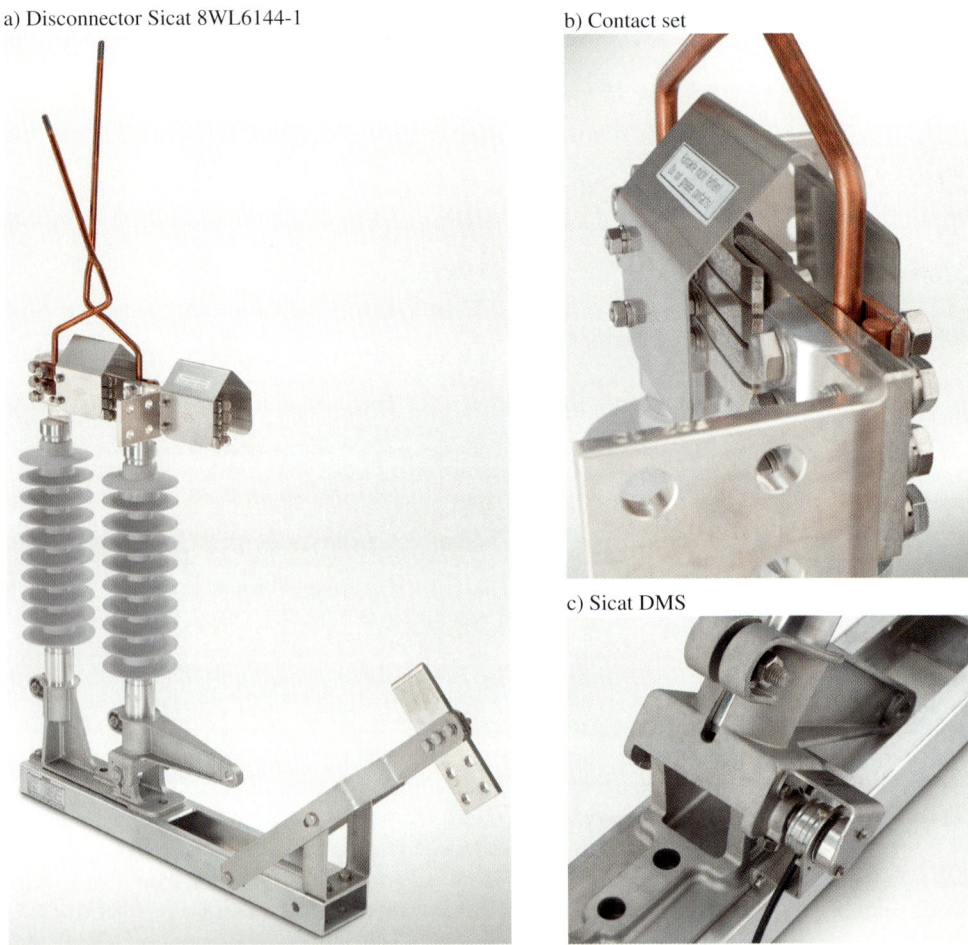

c) Sicat DMS

Figure 11.57: Overhead contact line disconnector Sicat 8WL6144-1 with earth contact.

also specifies that overhead contact line disconnectors should be able to switch-off the operational current with the number of switch cycles to be specified by the system operator. The disconnectors should be opened or closed without load or with a stipulated current. The purchaser should specify the required switch on and switch off performance as well as the mechanical and electrical service life. Since switching-off current with disconnectors causes arcs the switches should be arranged such that the arcs cannot damage other components of the system. If this is not possible, it is important to ensure that the disconnector is operated with no load current or a switch disconnector with an arc chamber should be used instead.

The drives must be able to open and close the disconnectors reliably, by a link, in all weather conditions and should indicate the switch condition to the control centre. An adjustment coupling allows incremental adjustment of the contact plate into the contact sets (Figure 11.56 b)). The testing of overhead contact line disconnectors is described in Clause 11.5.9.

Table 11.10: Technical data of the disconnector Sicat 8WL6144.

Type		8WL6144-0	8WL6144-1
earth contact		no	yes
design holding short circuit	kA	40	40/40[1]
design holding impulse current	kA	100	100/100[1]
design holding short circuit duration	s	1	1/1[1]
design impulse voltage – between mass contact – to earth	 kV kV	 290 250	 290 250
short-term power frequency withstand voltage, wet – between main contacts – to earth	 kV kV	 110 95	 110 95
dimensions – length – width – height	 mm mm mm	 760 232 1 400	 1 070 232 1 400
weight	kg	22,5	30,0
air gap – between main contacts – to earth	 mm mm	 460 420	 460 420

[1] main contact/earth contact

11.2.5.2 Disconnectors for AC railway systems

Figure 11.56 shows the *overhead contact line disconnector* Sicat 8WL6144-1 for nominal voltages 15 kV and 25 kV. The base plate (*10*) and the pivoting base (*9*) of the disconnector are mounted on a base frame (*11*) with one composite insulator (*6*), the contact piece (*5*), the connecting lug (*3*) and the arcing horn (*1*) and the fixed and the pivoting switch column. At the pivoting column, a contact plate for earthing (*4*) is attached for switches of type 8WL6144-1. This is pressed onto the earthed contact pieces (*7*) when the disconnector is opened. The contact sets are protected by covers (*2*) against weathering.

In most cases, the disconnector is mounted on a cross-arm at the top of a contact line pole and connected via links with the drive. The drive has an operating travel of 200 mm (Figure 11.56 b)) to open and close the disconnector.

The composite insulators (Figure 11.56 a)) and long air gaps of the overhead contact line disconnector results in high insulation characteristics allowing their universal use in 15 kV and 25 kV railway systems (Table 11.10). The design and the silver graphite coating of the service-free contact sets permit 40 kA design short-term current for 1 s and 2,5 kA operating current. Arcing horns with temperature-resistant end pieces can break operational currents when the disconnector is opened under load, extinguishing arcs within a few seconds. The additional earth contact is able to reliably earth the overhead contact lines of loading, inspection and tunnel tracks. The disconnector monitoring system (DMS) at the lower end of the moveable contact column (Figure 11.57 c)) produces, if required, an indication of disconnector position as an alternative and/or in addition to signalling of drive position. The DMS was certified for safety level SIL 1 and, therefore, suited for use in tunnel rescue systems.

Where disconnectors need to be installed below overhead lines or other structural systems and it is not possible to switch free of current, an enclosed overhead line switch disconnector

Figure 11.58: Switch disconnector at a cross-arm.

Table 11.11: Electric properties of 3 kV disconnectors series 8WL6134.

nominal voltage	V DC	3 000
insulation design voltage	V DC	4 800
operational current – Sicat 8WL6134-3/-3A – Sicat 8WL6134-0B/-0C/-2/-2A/-4/-4A – Sicat 8WL6134-5	 A A A	 2 000 3 000 4 000
creepage path	mm	300
air gap to earth / across separation gab – Sicat 8WL6134-0B/-0C/-2/-2A – Sicat 8WL6134-3/-3A/-4/-4A/-5	 mm A	 180 / 200 180 / 90
short-term withstand impulse voltage to earth / across separation gap	kV	40 / 80
power frequency withstand voltage, wet to earth / across separation gap	kV	18,5 / 22,2
design short-circuit current	kA	40
design short-circuit current duration Sicat 8WL6134-1 to -4	ms	250
design short-circuit current duration Sicat 8WL6134-5	s	1

according to Figure 11.58 can be used. This is equipped with a vacuum switching chamber to extinguish arcs.

11.2.5.3 Overhead contact line disconnectors for DC systems

One of the DC overhead contact line disconnector types Sicat 8WL6134 is depicted in Figure 11.59 a). Also for this series, composite insulators with high insulation resistance and silver graphite coated service-free contact sets with 3 kA current capacity are used. The electrical properties of these disconnectors are shown in Table 11.11. Disconnectors are arranged in mass transit systems below the top of towers and on cross-arms because of the use of cables. The operating link can be arranged within the pole (Figure 11.59 b)).

a) Disconnector Sicat 8WL6134-4A   b) Arrangement of linkage

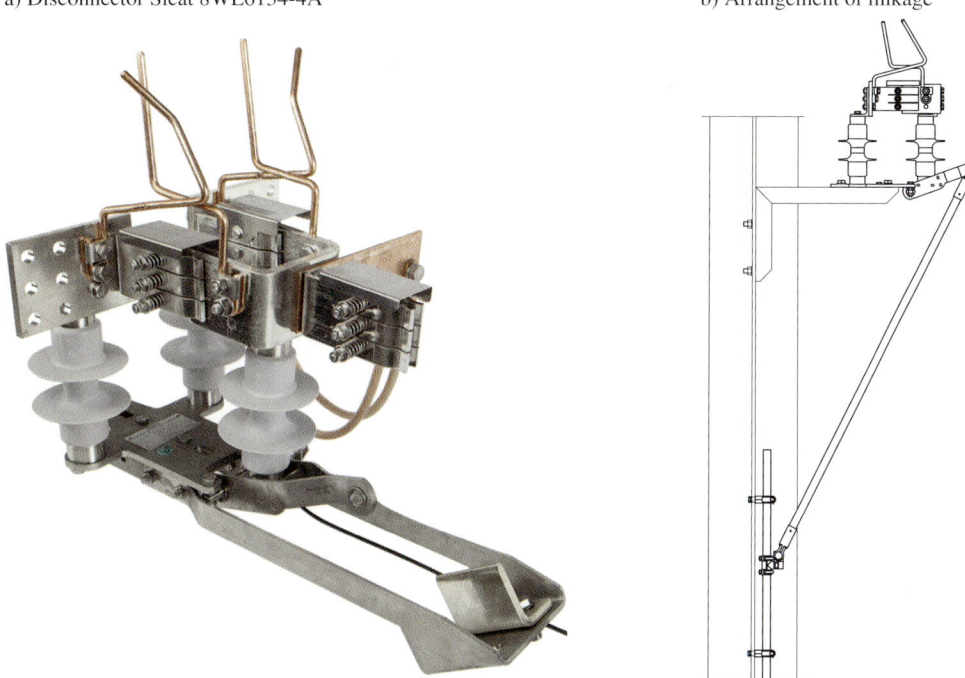

Figure 11.59: Overhead contact line disconnector Sicat 8WL6134-4A for DC systems up to 3 kV.

a) Earthing switch Sicat 8WL6144-1A   b) Arrangement on a pole

Figure 11.60: Earthing switch Sicat 8WL6144-1A a) and arrangement of an earthing switch on a pole close to a tunnel entrance b).

Figure 11.61: Electromechanical disconnector drive Sicat 8WL6253-0 in a stainless steel housing.

Figure 11.62: Disconnector drive Sicat 8WL6243 with control unit and DMS evaluating unit in an insulating GRP-housing.

11.2.5.4 Earthing switches

The *earthing switch* earths overhead contact lines in tunnels. It has only one composite insulator in the moveable contact column (Figure 11.60). The fixed contact column is installed at earth potential. The earthing switch is operated by a standard switching drive for disconnectors and is not resistant to in-rush current. The closed contact sets can manage a 40 kA design holding short-circuit current for 1 s. The position of the contacts is monitored in addition to the signal from the drive position switch. The disconnector is controlled by the automatic earthing system (AES) [11.15] (Clause 17.1.7). Other applications of earthing switches are for overhead contact lines in workshops or rolling stock maintenance.

11.2.5.5 Disconnector drives

Disconnector drives of the series Sicat 8WL6243, 8WL6244, 8WL6253 and 8WL6254 operate disconnectors and earthing switches in overhead contact lines via a linkage with 200 mm contact travel. Depending on the type, they are arranged in stainless steel (Figure 11.61) or insulating GRP housings (Figure 11.62) which can be installed into an H-section pole because of its narrow design. The GRP housing is used predominantly in mass transit systems. The switching drives can be remotely controlled by cable connections and on site electrically or manually. An optional radio control is also available. Depending on the type of design, the nominal operating voltage is DC 24 V, DC 48 V, DC 60 V, DC 220 V, AC 50/60 Hz 110–125 V or 230 V. The force travel characteristics of the reversing Geneva gearing offers the maximum

switching force at the end positions when closing or opening the contact sets. A switching cycle requires several seconds. Unintentional movement of the disconnector, e. g. under short-circuit loads, is prevented by mechanical stops at both end positions. *Signalling the position of the drive* is mostly transmitted over three or four conductor control circuits. The evaluating unit for the additional disconnector position monitoring Sicat DMS at the moving base of the disconnector or earthing switch is included as standard and does not require additional control cables [11.16].

11.2.6 Insulators

11.2.6.1 Functions and requirements

Insulators separate energized components of contact wires and traction power lines from each other and from earth. They experience mechanical and electrical loading as a result of the energised system and must satisfy both electrical and mechanical requirements.

While insulators are subjected only to tensile stress in suspended and dead-end positions they also have to withstand compression and bending loads in cantilevers and as post insulators on poles. The selection and design of insulators have to take these stresses and the local ambient environmental conditions into consideration.

Data for the design insulation voltage of overhead contact lines and the creepage path for various pollution levels is contained in EN 50 124-1.

The requirements for *composite insulators* in fixed railway systems are stipulated in EN 50 151 which also refers to general standards for this type of insulators. The standards EN 62 217, EN 61 109 and EN 61 952 also apply.

For other types of insulators, there are no specific standards for railway or contact line insulators. EN 50 119 refers to applicable standards for insulators made of ceramic or glass like EN 60 071, EN 60 305, EN 60 433, EN 60 672-1, EN 60 672-3, EN 60 273 and stipulates that:
  – the minimum tensile strength of an insulator shall at least be 95 % of the nominal tensile strength of the conductor
  – the maximum operational tensile load may not be more than 40 % of the minimum failing load of the insulator to be strained by it
  – the maximum operating load in bending or torsion may not exceed 40 % of the minimum bending and torsion strength of the insulator and
  – simultaneously occurring tensile, bending and/or torsional loading needs to be considered

11.2.6.2 Insulating material

Porcelain, glass, cast resin and glass fibre reinforced plastic with or without polymeric sheath are employed as *insulating materials* for insulators in overhead contact line systems. Porcelain insulators consist mainly of group C120 *hard porcelain* in accordance with EN 60 672-1 which is composed of china clay, feldspar and alumina. Quartz-enriched porcelain is no longer used for high-performance insulators. The quality of porcelain largely depends on the uniformity of the mineral composition within the insulator and on the manufacturing process, especially management of the firing. Porcelain is used for long-rod insulators, line posts and cap-and-pin insulators.

a) Long-rod insulator with eye caps b) Line post insulator with tie top c) Cap-and-pin insulator

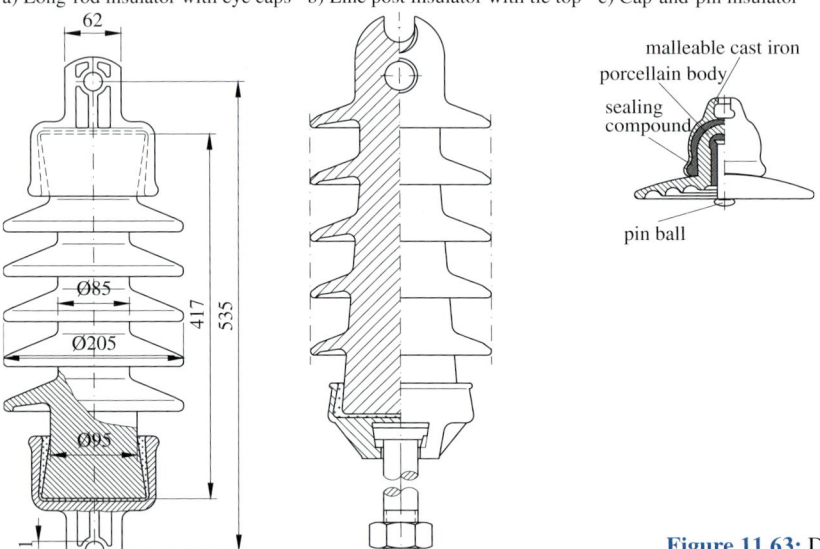

malleable cast iron
porcellain body
sealing
compound
pin ball

Figure 11.63: Designs of porcelain insulators.

Cap and pin insulators can also be made from *pre-stressed glass* in accordance with IEC 60 672-2. Alkali-lime-silicate glass is also used. It is shaped in a fluid condition and gradually cooled helping to avoid undesirable internal stresses.

Plastic insulators of various designs are manufactured from cycloaliphatic epoxy resin, polyurithane cast resin (CEP and PUR), PTFE (Teflon) and silicone rubber. High UV resistance and stability against climatic affects are required for outdoor applications, lacking in epoxy resins. Compared to ceramic and glass, there is more freedom in shaping epoxy resin insulators, which results in high dimensional precision that allows for insertion of fastening elements. However, the *leakage current resistance* is low.

Composite insulators with glass-fibre reinforced plastic cores and sheds made from various materials like silicone or PTFE and connecting fittings made of metal, are suited for high voltages and high mechanical loads.

While porcelain and glass insulators are brittle materials that are impact sensitive, plastic insulators are tough. They are vandalism resistant and because of their low weight facilitate transport and installation. Mass transit systems often use GRP materials without silicone or PTFE housings. They may contain only glass-fibres made of boron-free glass.

11.2.6.3 Designs and applications

The bodies of *long-rod insulators* (Figure 11.63 a)) are standardised in IEC 60 433. The ends of the fired and glazed porcelain bodies are provided with end fittings using lead-antimony, Portland cement or sulfur cement. Lead-antimony alloys are elastic but heat sensitive. Portland cement-sealant is hard and heat compatible, while sulfur cement-sealant is more elastic but less heat compatible. *Porcelain line post insulators* according to IEC 60 273 are used on the top of poles to carry traction power lines (Figure 11.63 b)).

Cap-and-pin insulators are manufactured either from porcelain or glass. The individual caps are provided with a pin ball and a cap (Figure 11.63 c)). The shape of the individual insu-

a) Insulators for 25 kV

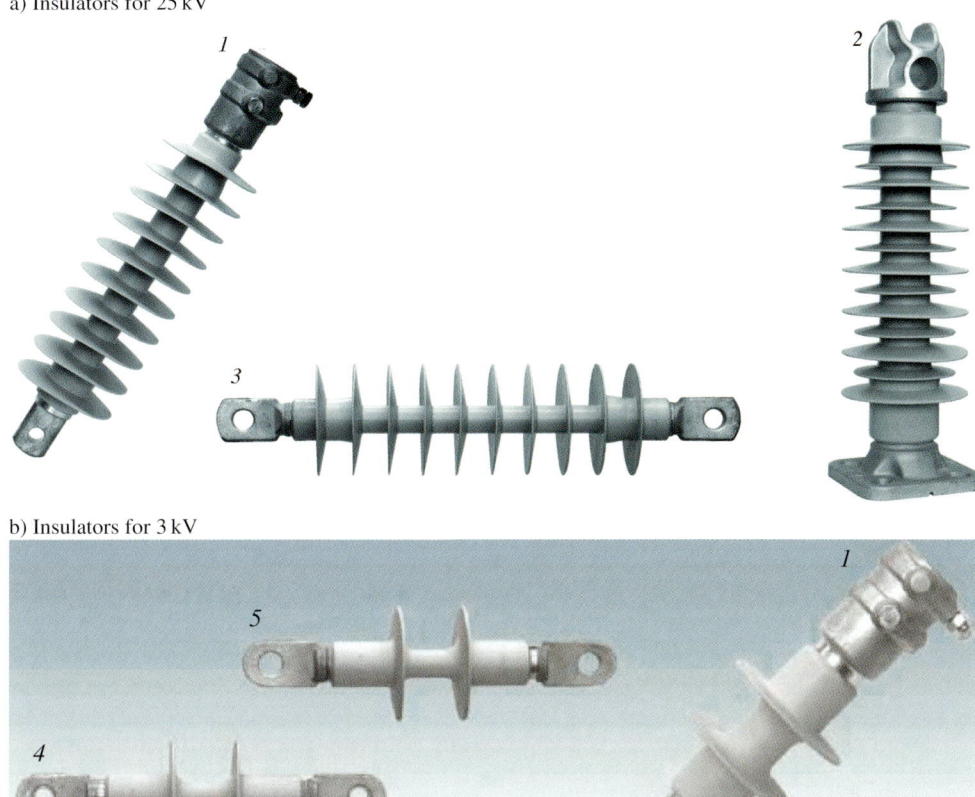

b) Insulators for 3 kV

Figure 11.64: Composite insulators.
1 insulator for cantilever tube, *2* insulator for tap tube, *3* line post insulator, *4* linepost insulator (tongue with hole diameter 17 mm), *5* linepost insulator (tongue with hole diameter 21 mm)

lators can be adopted to the specific application and the required creepage path. The annual failure rate for cap-and-pin insulators is 10^{-5} and approximately ten times higher than for long-rod insulators. A mechanical fracture of the insulator string does not occur immediately since the socket cap and pinball maintain the mechanical strength of the damaged insulators. Cap-and-pin insulators behave less favourably than long-rod insulators when contaminated. Consequently, their creepage path should be approximately 10 % higher than with long-rod insulators. Cap and pin insulators are standardised in IEC 60 305. Porcelain insulators for overhead contact lines are tested in accordance with IEC 60 383. Because of the already mentioned disadvantages of cap and pin insulators, composite insulators and porcelain long-rod insulators are increasingly being adopted world-wide.

Composite insulators (Figure 11.64) are used as tensile-loaded insulators in overhead contact lines and tension, bending (cantilever) loaded insulators and in head-span installations because of their good electrical and mechanical properties in both AC and DC systems for main lines and mass transit traffic. They are used on poles as compression and bending loaded line

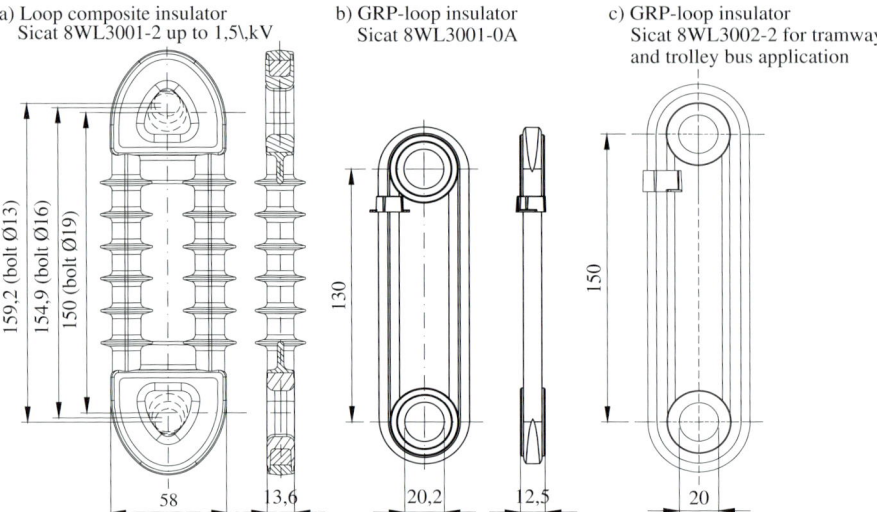

a) Loop composite insulator
   Sicat 8WL3001-2 up to 1,5\,kV

b) GRP-loop insulator
   Sicat 8WL3001-0A

c) GRP-loop insulator
   Sicat 8WL3002-2 for tramway
   and trolley bus application

Figure 11.65: Types of loop insulators.

post insulators. The fittings, partly covered with silicone, positively influence the electrical field of the insulator. [11.17].

Composite- and GRP-*loop insulators* (Figure 11.65) with GRP loop made of boron-free glass fibres wound around the fitting and filled with epoxy resin are used for overhead contact lines of tramways, city railways and trolley bus lines. The type Sicat 8WL3001-2 shown in the Figures 11.65 a) and 11.66 has a silicone sheath and complies with the requirements on composite insulators according to EN 62 621. This insulator type is especially suited for

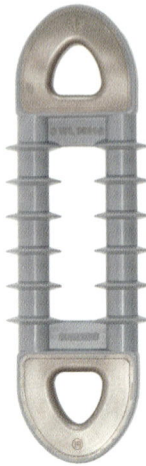

Figure 11.66: Loop composite insulator Sicat 8WL3001-2.

Figure 11.67: Cantilever with double insulation for mass transit with insulating catenary wire clamps, GRP steady arms and loop insulators.

contact wire   contact wire

Figure 11.68: Composite insulator Sicat 8WL3092-1 for overlap spans with emergency running capability for pantographs.

Table 11.12: Porcelain insulators for AC 15 kV and 25 kV main line applications.

Insulator design	Application		Electrical parameters	Mechanical parameters
eye end cap insulator	top tube / registration arm contact line/ head-span		creepage path 484 mm rated voltage 15 kV	failing load 100 kN e. g. working force up to 16 kN
eye end cap insulator	top tube contact line		creepage path 760 mm rated voltage 25 kV	Failing load 130 kN e. g. working force up to 27 kN
tube end cap insulator	cantilever tube		creepage path 420 mm rated voltage 15 kV	Working bending moment up to 1,13 kNm
tube end cap insulator	cantilever tube		creepage path 760 mm Rated voltage 25 kV	Working bending moment up to 2,8 kNm

unfavourable ambient conditions with high pollution. The shape of the thimble can be used for bolts of 13, 16 and 19 mm diameter. The other two types without a silicone sheath are manufactured with a round eye and proved their performance under normal conditions. Table 11.12 lists technical properties.

Double insulation is often used in mass transit systems in which case railway earthing may be waived. In this case, two insulators are arranged in series with a neutral section in between (as in Figure 11.67). The double insulation protects the linesmen, when working under live-

Table 11.13: Composite insulators for AC 15 kV, 25 kV and DC 3 kV main line applications.

Insulator design	Application		Electrical parameters	Mechanical parameters
eye end cap insulator	contact line/ head-span		creepage path 1 230 mm rated voltage 25 kV	SML 135 kN
eye and tube end cap insulator	top tube		creepage path 1 215 mm rated voltage 25 kV	MDCL 1,9 kN STL 60 kN
Tube end cap insulator	cantilever tube		creepage path 1 215 mm rated voltage 25 kV	MDCL 1,9 kN STL 60 kN
line post insulator with flanges	Traction power lines		creepage path 1 215 mm rated voltage 25 kV	MDCL 1,9 kN STL 20 kN
eye end cap insulator	contact line/ head span		creepage path 320 mm rated voltage DC 3 kV	SML 90 kN OML 30 kN
eye and tube end cap insulator	cantilever tube		creepage path 300 mm rated voltage DC 3 kV	MDCL 1,9 kN STL 60 kN

SML Specified Mechanical Load (IEC 61 109), MDCL Maximum Design Cantilever Load (IEC 61 952), STL Specified Tensile Load (IEC 61 952)

Table 11.14: Insulators in urban transportation systems up to DC 1,5 kV.

Insulator design	Application		Electrical parameters	Mechanical parameters
eye end and threaded tube cap insulator	top tube		creepage path 130 mm rated voltage 1,5 kV	failing load 40 kN
eye end and tube clamp cap insulator	cantilever tube		creepage path 130 mm rated voltage 1,5 kV	failing load 70 kN
loop insulator	contact line, conductors		creepage path 130 mm rated voltage 1,0 kV	SML 70 kN
loop insulator	contact line, conductors		creepage path 90 mm Rated voltage 1,0 kV	SML 30 kN
composite loop insulator	contact line, conductors		creepage path 150 mm rated voltage 1,5 kV	SML 70 kN
Insulator body	Cantilever, feeder lines		creepage path 240 mm rated voltage 3,0 kV	failing load 50 kN
GRP-tube	cantilever		creepage path ≥ 570 mm Rated voltage 1,5 kV	diameter 26 mm, 38 mm, 55 mm
GRP-rod	cantilever		creepage path ≥ 570 mm rated voltage 1,5 kV	diameter 10 mm, 26 mm, 38 mm, 55 mm

GRP Glass-fibre Reinforced Plastic; SML Specified Mechanical Load (IEC 61109)

line conditions from simultaneously touching live and earthed parts. In the case of damage, at least one of the two insulating units will function correctly.

To reduce the required contact wire lift in parallel spans (overlaps), a composite insulator with emergency running capability can be used (Figure 11.68). Like other insulators in the line, it has flat fittings and a PTFE or silicone sheath without insulator sheds.

11.2.6.4 Electrical and mechanical rating

Insulators should be rated electrically in accordance with the conditions given in Clause 2.5.4. The lightning impulse withstand voltage for insulator testing is specified in Table 2.1. The permissible operating forces and moments specified in the tables depend on the partial design factors and must not be exceeded during operation.

11.2.6.5 Selection and application

The Tables 11.12 to 11.14 provide an overview of the frequently used insulators in overhead contact line systems for AC 25 kV 50 Hz, AC 15 kV 16,7 Hz and for DC. Selection of the insulators should consider the maximum operational load and the design specifications according to Clause 11.2.6.1.

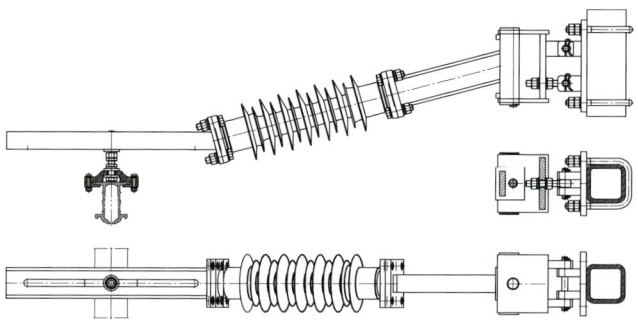

Figure 11.69: Horizontally pivoting support with a support clamp Sicat 8WL7232-3, which is directly attached to the conductor rail.

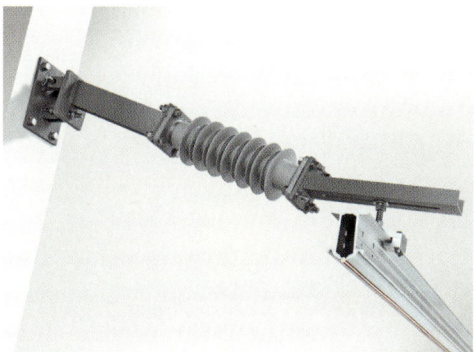

Figure 11.70: Fixed support arranged at a wall.

Figure 11.71: Adjustment with an eye end bolt.

a) Clamp 8WL7233-0 for sliding movements

b) Clamp 8WL7232-0 for midpoints

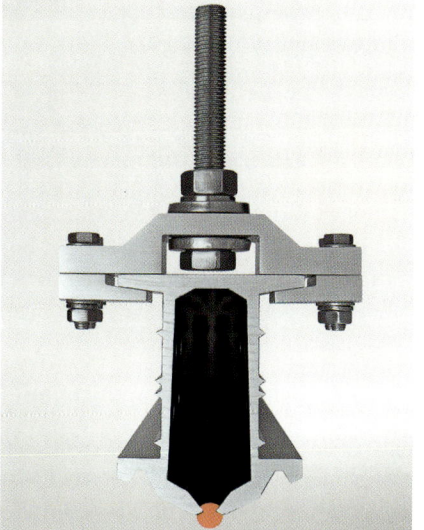

Figure 11.72: Support clamps.

a) RailTech design (Furrer+Frey)

b) Siemens design

Figure 11.73: Pivoting supports for moveable overhead conductor rails in maintenance workshops.

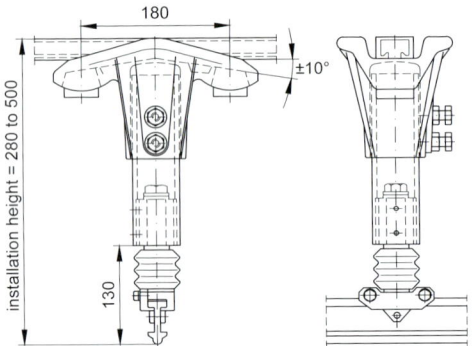

Figure 11.74: Supports Sicat 8WL3584-6 for overhead conductor rails made of solid sections.

11.3 Overhead conductor rails

11.3.1 Support assemblies

The *support assembly* carries the overhead conductor rail and aligns it in the vertical and horizontal position relative to the track. The conductor rail supports are installed at 8 to 12 m spacings depending on the permissible sag of the rail. Two types of supports permit temperature-induced longitudinal movements of the conductor rail:

- horizontally pivoting supports according to Figure 11.69. The pivoting support maintains the conductor rail at a constant height above and parallel to the track
- fixed supports as in Figure 11.70 with a suspension clamp Figure 11.72 a) in which the conductor rail slides, according to for systems with space limitations, e. g. in circular tunnels with only one track. The suspension clamp can be mounted to fixed cantilever arms or at other supports installed underneath of buildings. The contact spring serves as potential compensation in AC installations and is also recommended for DC systems with running speeds up to 130 km/h because of their damping characteristics

Figure 11.71 shows the attachment of a support arm on a wall. Fixed supports with suspension clamps according to Figure 11.72 b) serve as midpoints approximately in the middle of the section.

a) Shape and dimensions
b) Detail of contact wire clamping and
wear limits

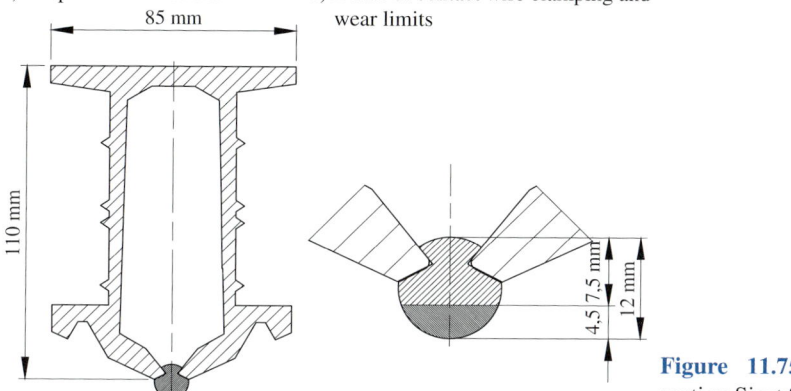

Figure 11.75: Conductor rail section Sicat 8WL7230-0.

For *pivoting overhead conductor rails* in maintenance workshops and on moveable bridges, horizontally pivoting supports are used (Figure 11.73). They enable unhindered access to the roofs of vehicles and wagon loading operations after de-energising, earthing and pivoting the installation. If the pivoting conductor rail is situated above the track in the operational position, trains can be driven electrically through the maintenance and loading sections.

Supports, as shown in Figure 11.74, carry the solid copper section conductor rails, used in city railways and tram systems as contact lines in tunnels. They are fixed at approximately four meter spacings (Clause 11.3.2.2).

11.3.2   Conductor rail profiles

11.3.2.1   Compound section

Overhead conductor rails consist of a U-shaped extruded section made of aluminum alloys with contact wire clamped into the lower part of the section. Figure 11.75 a) shows the shape and dimensions of the bending-resistant conductor rail section 8WL7230-0 with a $2\,300\,\text{mm}^2$ cross-section and a weight of 6,2 kg/m. The rail sections are up to 12 m long. The upper part has horizontally protruding flanks used to attach the suspension clamps and electrical connections. Four lateral ribs are used to locate the splice plates used to join individual sections (see Clause 11.3.3). In the lower part, all types of contact wires in accordance with EN 50 149 and non-standardised types can be clamped in. An installation carriage is used to open the split in the bottom of the section and introduce the contact wire. At this time, a special grease is applied to the contact surfaces to avoid electrical corrosion of both materials with differing electrochemical voltage potential. After the installation of the contact wire, the aluminum section maintains the contact wire within its grooves (Figure 11.75 b)). Since the contact wire is not tensioned, up to 40 % contact wire wear can be allowed. The coefficient of longitudinal expansion of the conductor rail material is $23,4 \cdot 10^{-6}\,\text{K}^{-1}$. Consequently, 100 m of conductor rail will expanded by 23,4 mm for a 10 K temperature variation. The electrical properties of the compound cross-section are discussed in Clause 11.3.4.

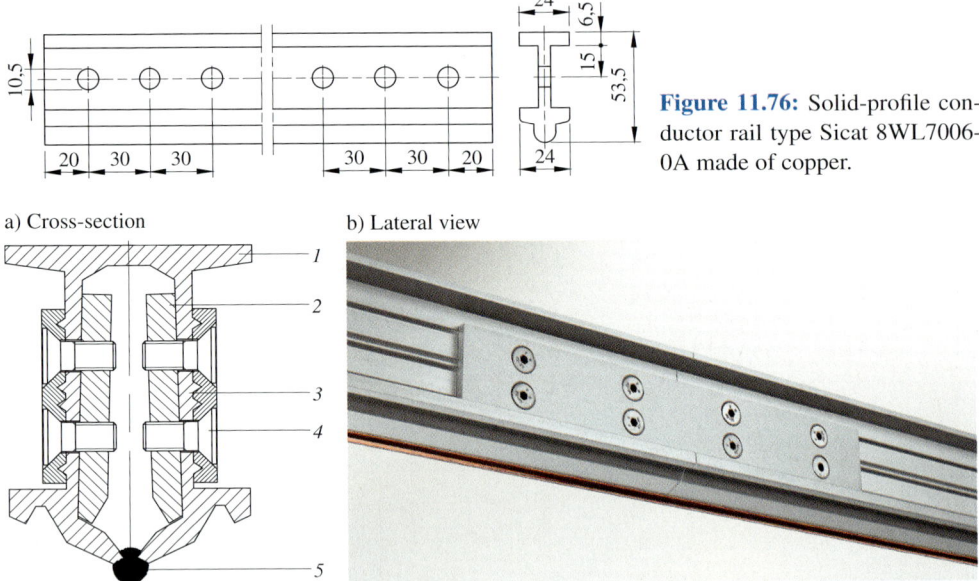

Figure 11.76: Solid-profile conductor rail type Sicat 8WL7006-0A made of copper.

a) Cross-section   b) Lateral view

Figure 11.77: Conductor rail joint Sicat 8WL7231-0. *1* U-type conductor rail cross-section with countersunk connection bolts, *2* internal splice plate, *3* external splice plate, *4* self-securing bolt, *5* contact wire

11.3.2.2 Solid-profile type

The cross-section of a copper profile for conductor rails for mass transit systems is made of a bending-resistant Cu-ETP solid section (Figure 11.76). It is installed in tunnels or under buildings on supports as shown in Figure 11.74. The rails are connected by straps and strung in an S-shape. Where there are temperature variations, the lateral deviation increases or decreases, so expansion joints are not necessary. Overhead conductor rail systems equipped with this type of section can be permanently loaded at 1 600 A and traversed at 80 km/h [11.1, 11.18].

11.3.3 Mechanical connections and components

11.3.3.1 Conductor rail joints

The *conductor rail joint* is used to connect the individual sections of conductor rails with splice plates. The Siemens conductor rail joint consists of two internal and two external plates made of extruded aluminum profiles (Figure 11.77). The external plates interlock with locating ribs on the conductor rail, fixing the relative heights of the rails being joined. The internal plates are in contact with a large surface of the aluminum hollow section (Clause 11.3.2.1) and are, therefore, able to transfer high currents. The segments of the solid Cu sections (Clause 11.3.2.2) are connected by straps made of Cu-ETP (Figure 11.78).

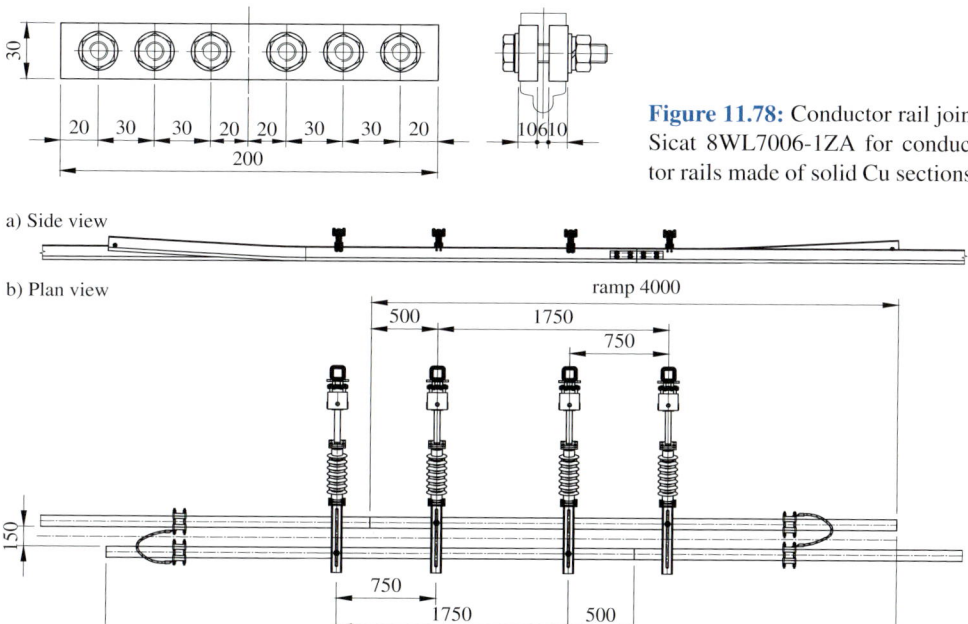

Figure 11.78: Conductor rail joint Sicat 8WL7006-1ZA for conductor rails made of solid Cu sections.

Figure 11.79: Non-insulating section transition consisting of overlapping conductor rail ramps (Sicat 8WL7230-1A).

11.3.3.2 Transition between individual sections and arrangements above points

The *transitions* between the individual *conductor rail sections* compensate for thermal expansion. At running speeds up to 140 km/h, the transitions between the individual sections of the Siemens composite conductor rail consist of overlapping conductor rail ramps. Figure 11.79 shows a non-insulating transition with the supports on one side. Two supports for each conductor rail are necessary within this section to limit the sag of the conductor rails and to adjust the heights of the ramps. By inclining the ramps at approximately 1 : 50, a smooth passage of the pantograph at the end of each section is achieved. The length of the parallel overlap of the conductor rails should be at least 6,00 m to ensure good quality running. The distance between the supports of different sections needs to be selected to ensure the temperature induced longitudinal expansion of the conductor rails is not impeded. Insulating section transitions that are also used as separations are described in Clause 11.3.5.

11.3.3.3 Expansion joints

For running speeds above 140 km/h, the ends of two adjacent sections are controlled by a mechanical connection called an *expansion joint* or dilatation joint (Figure 11.80). The expansion joint enables temperature dependent movement of the conductor rail and guarantees a uniform sliding plane for the pantograph collector strips. The joint is designed for a 1 m long migration distance on both ends of the section. Flexible copper conductors, that comply with DIN 43 138, electrically bridge the expansion joint.

a) Side view

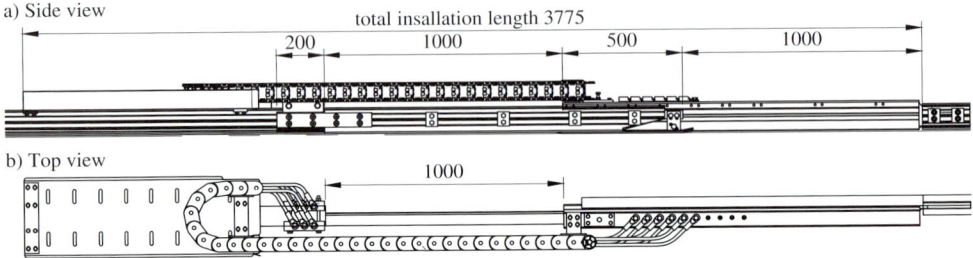

b) Top view

Figure 11.80: Expansion joint Sicat 8WL7238-0 for conductor rails with running speeds above 140 km/h.

a) Cross-section

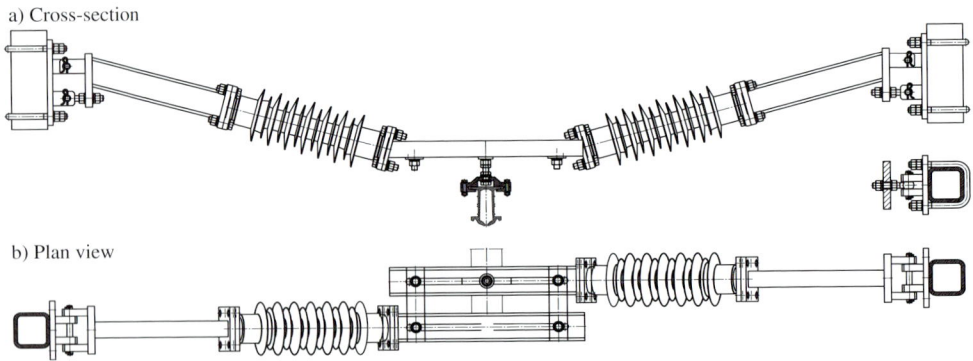

b) Plan view

Figure 11.81: Midpoint consisting of supports on both sides.

11.3.3.4 Midpoint

Located in the middle of each section a *midpoint* fixes the conductor rail ensuring that it is free to move in either direction with temperature variations, enabling the installation of longer individual sections. The midpoint may be designed with anchor ropes as with overhead contact lines. The midpoint shown in Figure 11.81 consists of supports arranged on both sides. When using supports with the same design in front of and behind the suspension clamps Sicat 8WL7233-0, additional dead-end clamps Sicat 8WL7235-0B are provided. .

11.3.4 Electrical connections

11.3.4.1 Current-carrying capacity

All *current-carrying mechanical connections* and the specific electrical connections as described in Clauses 11.3.4.2 to 11.3.4.4, have to carry the operational and short-circuit current as specified for the individual installation. They need to be rated accordingly. The components within the main current path of a conductor rail overhead line include the conductor rail Sicat 8WL7230-0, the conductor rail joint Sicat 8WL7231-0 and the feeding clamp Sicat 8WL7235-0A. Without a contact wire, these components can permanently carry at least 2 900 A with 1 m/s wind velocity and 40 °C ambient temperature reaching a permanent temperature of 90 °C. In addition there will be the current capacity of the contact wire.

Table 11.15: Resistance and ampacity of Siemens conductor rail contact lines dependent on the contact wire cross-section.

Contact wire [1] acc. to EN 50 149	Ampacity [2]		Resistance of 1 km at 20 °C			Copper-equivalent
	Contact wire at 90 °C	Conductor rail/ Contact wire	Contact wire	Conductor rail	Total	
	A	A	Ω/km	Ω/km	Ω/km	mm^2
AC-/BC-80	401	3 301	0,229	0,0143	0,0135	1 306
AC-/BC-100	465	3 365	0,183	0,0143	0,0133	1 326
AC-/BC-107	484	3 384	0,171	0,0143	0,0132	1 333
AC-/BC-120	523	3 423	0,153	0,0143	0,0131	1 346
AC-/BC-150	605	3 505	0,122	0,0143	0,0128	1 376

[1] applies to Cu-ETP and CuAg0,1; [2] at conductor initial and final temperature 40 °C and 90 °C, respectively, wind velocity 1,0 m/s, contact wire wear 20 %

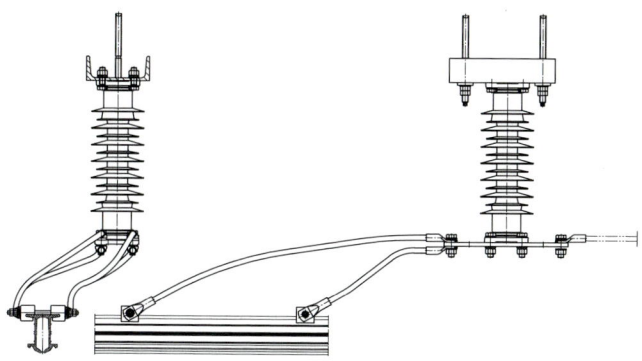

Figure 11.82: Arrangement of feeder line and feeder clamp Sicat 8WL7235-0A.

Table 11.15 contains the electrical parameters of a conductor rail overhead contact line. Several designs exist for the *expansion joint* where up to twelve flexible conductors DIN 43 138-E-Cu58-70×189 can be installed to attain the required current capacity. The feeder lines and current connectors are rated to suit.

In addition to the above components, a conductor rail overhead line includes the expansion joint Sicat 8WL7238-0. These components command on a short-circuit capacity of 45 kA for 100 ms. After the short circuit, all these components are fully functional.

The suspension clamps are part of the short-circuit current path and in case of an insulation defect can be fitted with M16 bolts to withstand up to 30 kA short-circuit current or with M 20 bolts up to 40 kA short-circuit current. If a short-circuit resistance of 45 kA should be required additional current connectors made of E-Cu35f will be installed between the suspension clamp and the cantilever suspension arm.

11.3.4.2 Feeding

The current is fed into the conductor rail by *feeders* as shown in Figure 11.82. The feeder line is fixed to a copper plate at an insulator above the conductor rail.

11.3.4.3 Current connector between individual sections

Current connectors made of flexible copper conductors 95×259-Cu-ETP and feeder clamps electrically connect the individual sections of the overhead conductor rail. Figure 11.83 shows

Symbol	Definition	Unit
μ_0	permeability constant	V s/(A km)
μ_r	relative permeability	–
ρ_E	specific soil resistivity	Ωm

8.1   Introduction

The physical laws for transmission and distribution of electrical energy are accompanied by several processes of interference within the industrial and public environment. In Figure 8.1 these interference processes are shown together with the consequences:

- *conductive interference* is caused by galvanic coupling via conductive connections and results in rail potential transferring voltages with return current flow in installations other than the return circuit as stray currents
- *inductive interference* originates from the traction currents with their electrical and magnetic fields around substations and along the contact lines. Voltages and currents are induced in signalling and telecommunication cables, cable sheaths and industrial equipment
- *capacitive interference* results in voltages in insulated unearthed conductors close to AC overhead contact lines and could lead to dangerous voltages for any persons touching these conductors
- *high-frequency emissions* commonly occur from arcing between the contact line and pantographs or loose contacts at insulators. They mainly produce interference at radio transmission frequencies up to 1 MHz

The consequences can be categorised into disturbances, damage and hazards with limits that are not strictly defined:

- disturbances can affect signals in railway-owned installations for train control causing distorted audio in telecommunication lines and radio transmissions mainly in the AM band
- damage can be caused by stray currents, causing over-voltages and currents in communication circuits, digital equipment, metallic equipment and infrastructure
- hazards to animals and people from touch voltages and exceedingly high induced currents

The configuration of the traction power supply and the contact line as well as the return circuit determine the degree of interference. Design and operation of electrified railways must ensure that the electromagnetic properties do not exceed limits for electrical safety as given in EN 50 122-1, for stray current corrosion in EN 50 122-2 and for the high-frequency interference in EN 50 121 with respect to telecommunication and signalling installations.

8.2   Interferences due to electric traction systems

AC-three-phase lines for power transmission and distribution, as opposed to traction power supply networks, have conductors, which are arranged almost symmetrically for both directions of current flow, being three-phase overhead power lines suspended in the air at structures or multi-phase cables. The distance to the associated return conductors is small and the current distribution between the three conductors is almost symmetrical as long as no faults

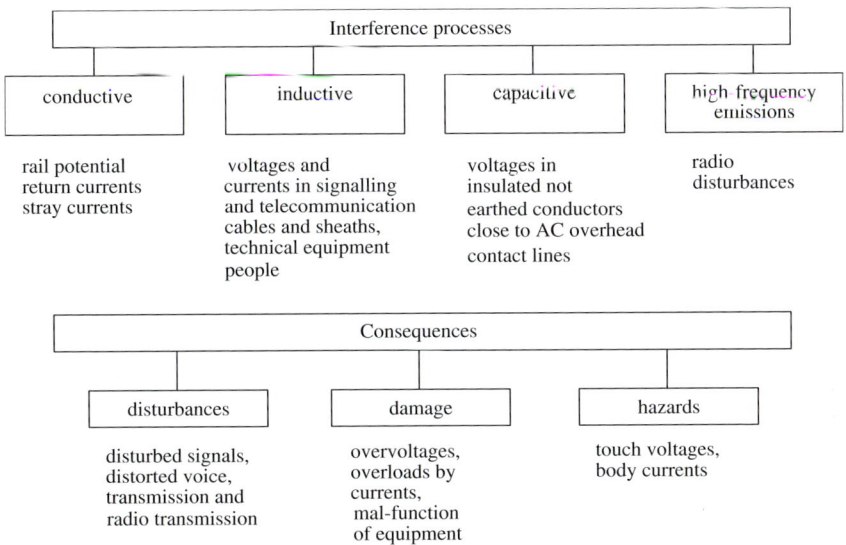

Figure 8.1: Interference processes and their consequences.

occur. The overhead contact lines for trolley buses are also operated symmetrically relative to earth.

In contrast to that, in electric traction systems with track return circuits, the traction current flows from the substation through the contact line system to the train and from there back via the tracks. This asymmetrical structure versus earth causes comparatively higher inductive, capacitive and galvanic interference because of return currents flowing through the running rails, the earth and any other parallel conductors. In the case of AC traction systems, the substation earthing equipment contributes to collecting the return currents from earth. Analogous considerations apply in the case of short circuits. The range of influence to be considered in the design depends on the traction voltage and currents as well as on the sensitivity of the influenced installations. In case of single-phase AC 16,7 Hz and 50 Hz electric railways influencing telecommunications networks DIN VDE 0228-3 indicates a range of influence of 500 m in urban areas and 2 000 m in other areas. Unless protective and stray current prevention provisions are made, the range of galvanic influence of DC railway traction power systems on any metallic conductors can extend over several kilometers apart from the tracks because of the resulting stray currents.

8.3 Coupling mechanisms

8.3.1 General

The interference due to voltages and currents in the contact line system is caused by various *coupling mechanisms*. If the wave length of the electromagnetic interference is considerably longer than the length of the installation, which generally applies to traction contact line installations, then the relations depicted in Figure 8.2, based on stationary conditions, describe the coupling mechanisms. These coupling mechanisms apply, in principle, to the contact lines

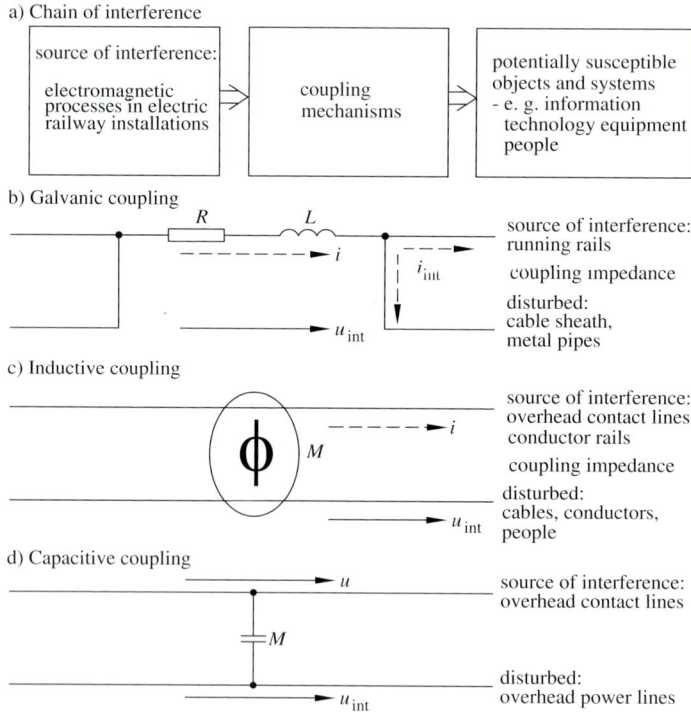

a) Chain of interference

b) Galvanic coupling

c) Inductive coupling

d) Capacitive coupling

Figure 8.2: Main coupling mechanisms determining interference caused by electric railways.

of both DC and AC railways. If the wave length is shorter and the length of the installation or the surge front is extremely short, e. g. because of a lightning impulse, the interference can be described by the *wave model* as explained in detail in [8.1].

In DC railways, it's mainly the *galvanic coupling* that must be considered. This causes touch voltages and *stray currents*, that are described in detail in Clause 6.5.

In traction contact line networks, the following *interference parameters* are to be considered:

– the traction *power network voltage*, described in terms of its nominal value and toler-ances, as well as the associated electric field it generates
– the *operating current* and the associated magnetic field
– the *short-circuit current* as well as the effective duration of any short circuit that can occur
– *harmonics* of the operating currents
– any *higher-frequency electromagnetic interference fields* caused by arcing between the collector strips and the contact lines or rails as well as by switching transients in the traction power supply network or traction vehicles

The contact line interconnections determine the current and voltage distribution. The geo-metric position of the interference source, that is the location of the individual conductors of the overhead contact line relative to the line or system subjected to interference, is another relevant factor.

The operating current flowing through a supply section is the fundamental quantity determin-ing the interference on other systems. For conventional railway traffic it is possible to deduce

a) Principle of arrangement

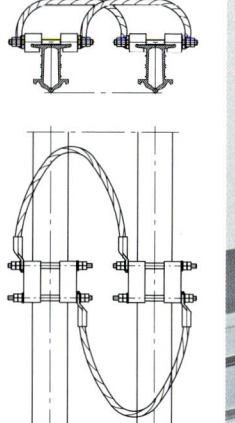

b) Connection at the upper face of the compound profile

Figure 11.83: Current connection between conductor rail sections using the conductor Sicat 8WL7075-0 and the feeder clamp Sicat 8WL7235-0A.

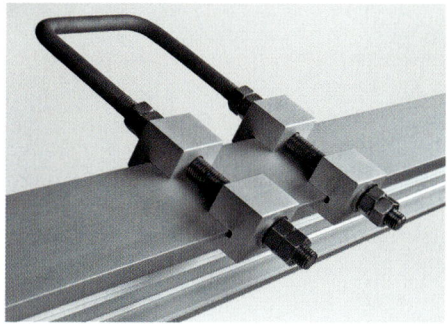

Figure 11.84: Earthing clamp Sicat 8WL7234-0A.

the design of a transition using conductor rail ramps. Up to 0,50 m variations in conductor rail length can be accommodated by using 1,40 m long jumpers. With such length variations and a temperature range between $-10\,°C$ and $+70\,°C$, the maximum section length can be 275 m. In the case of longer half sections, the current-connector conductors are suspended like feeding points (Clause 11.3.4.2). One feeder clamp is sufficient to transmit the total load current of the conductor rail overhead line. However, two feeder clamps are recommended at each connection point for redundancy.

11.3.4.4 Connection for earthing and short circuiting

For the *connection of earthing and short circuit devices*, earthing clamps are arranged along the conductor rail (Figure 11.84). The connection clamp U bolt has 16 mm diameter (M16). This clamp can withstand a 45 kA short-circuit for 100 ms. Three earthing clamps should be arranged at each section – one clamp immediately in front of the overlapings of the conductor rail at the section transition and one at the midpoint of the section.

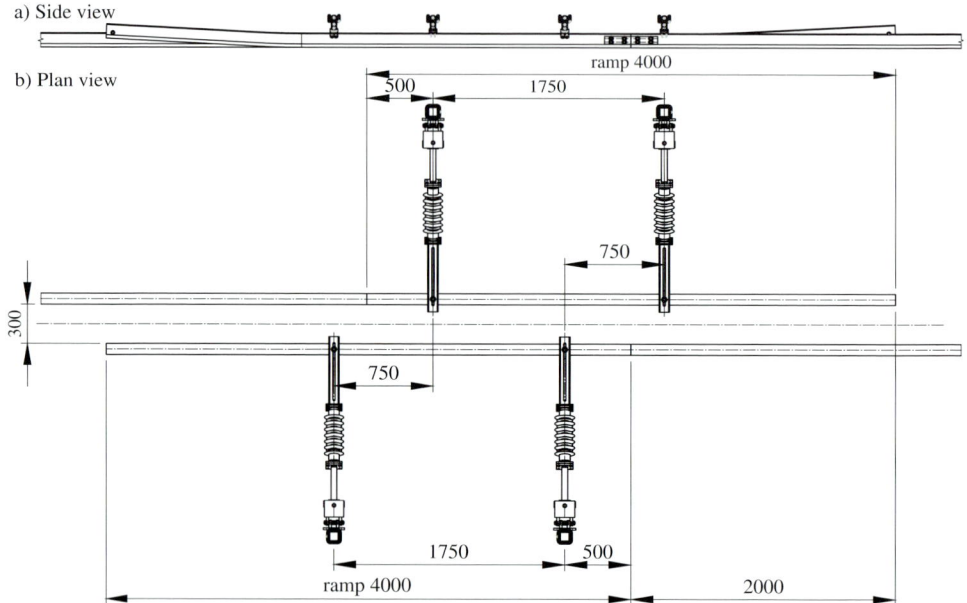

a) Side view

b) Plan view

Figure 11.85: Insulating section transition installed as an overlap.

11.3.5   Separating devices

11.3.5.1   Insulating section transition

Section transitions, as described in Clause 11.3.3.2, can be used as *electrical separations* provided that the electrical clearances prescribed in EN 50 119 are met between the overlapping conductor rails and supports over the full temperature range. Figure 11.85 shows an *insulating section transition* with supports arranged on both sides. This design is used for running speeds up to 140 km/h.

11.3.5.2   Section insulators

For running speeds above 140 km/h, a *section insulator* is inserted in the conductor rail (Figure 11.86), It is fixed at both ends to the conductor rails and insulated in between by a composite insulator. The pantograph is guided through this range by two copper runners. Section insulators and their application are described in Clause 11.2.4.

11.3.6   Transitions to flexible overhead contact lines

Transitions to flexible overhead contact lines are assemblies that accommodate the transition from a flexible overhead contact line to a rigid conductor rail contact line. The transitions smoothly transfer the differing elasticities of both designs to each other and guarantee good quality current transfer between the elastic overhead contact line and the relatively rigid conductor rail. Their function is to maintain the uniformity of contact line elasticity within the requirements of EN 50 119 as well as the minimum and maximum contact forces between

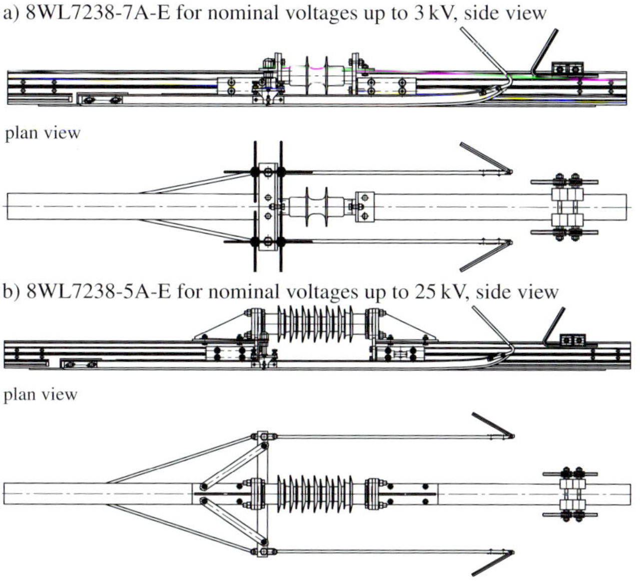

a) 8WL7238-7A-E for nominal voltages up to 3 kV, side view

plan view

b) 8WL7238-5A-E for nominal voltages up to 25 kV, side view

plan view

Figure 11.86: Section insulator for the Siemens overhead conductor rail.

structure portal

catenary wire

h

dropper
approx. 3,5 m 8WL7035-5

$5 \times h$

500 | 500 | 500 | 500 | 50

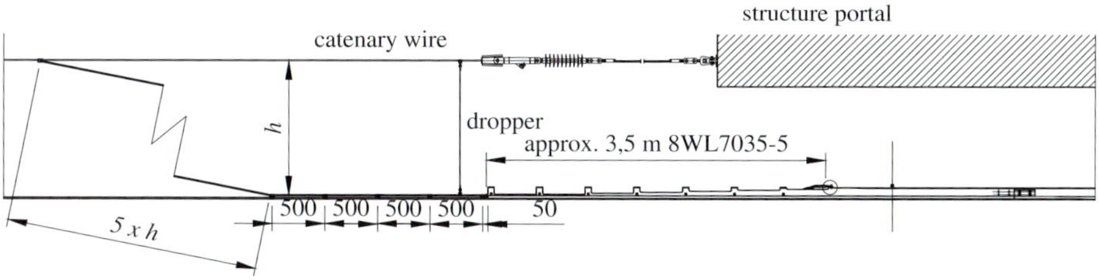

Figure 11.87: Arrangement of an overhead contact line transition between an elastic contact line and a rigid conductor rail using the transition element Sicat 8WL7230-2A.

overhead contact lines and pantograph. The 5 m long *transition element* 8WL7230-2A (Figures 11.87 and 11.88) reduces the moment of inertia in the direction of the flexible overhead contact line by increasing the size of cut-outs in the contact conductor rail profile and reaches at its end, an elasticity of 0,34 mm/N. With the constant contact force of the pantograph, the uplift increases continuously with distance from the rigid conductor rail. A current-connector conductor attached between the catenary wire and the conductor rail and then continued as a length of approximately 2,00 m in parallel to the contact wire, also assists in equalising the elasticity. By increasing the contact wire tensile force in the tensioning sections adjacent

Figure 11.88: Connection of the contact wire at the transition component Sicat 8WL7230-2A to the overhead conductor rail.

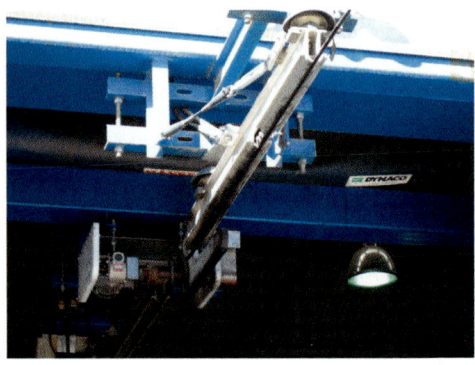

Figure 11.89: Transition from a contact wire to a overhead conductor rail in front of a roller door at a maintenance workshop at Madrid, Spain.

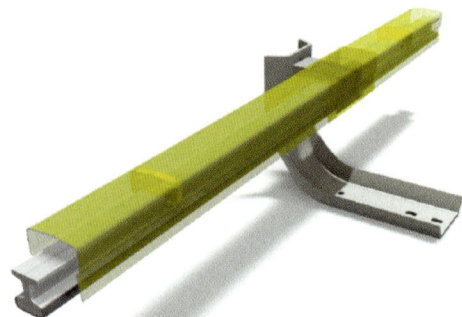

Figure 11.90: Third rail type SPL A][45-5330 with on top contact (see Figure 11.96 e)).

Figure 11.91: Third rail type SPL A][45-5330 with bottom contact (see Figure 11.96 e)).

to the conductor rail by approximately 2 kN, compared with the other tension sections, the difference in elasticity can be reduced even further.

The contact wire tensile force is transferred to the aluminum conductor rail and is guided to the building via a termination with a clamp 8WL7234-3. The connection is designed for contact wire tensile forces up to 15 kN. One meter of conductor rail is able to hold 1 kN of contact wire tensile force. Using a 12 kN operational contact wire tensile force and 2,5 design safety factor approximately 30 m of overhead conductor rail would be required to hold the contact wire tensile force. To not unfavourably affect the position of the conductor rail and its running performance, the vertical deviation of the anchor ropes should be limited to 6°. A support in front of the tension clamp equalises the low height displacement of the conductor rail. The lateral bending is limited to 8°.

In short sections such as under bridges or other underpasses, the contact wire can be installed uncut through the conductor rail and then re-integrated into the overhead contact line system as the contact wire tensile force does not affect the conductor rail. Another type of transition between the contact line and the conductor rail, in connection with a separator for roller doors, is shown in Figure 11.89. This device terminates the overhead contact line and also enables the roller door to be shut.

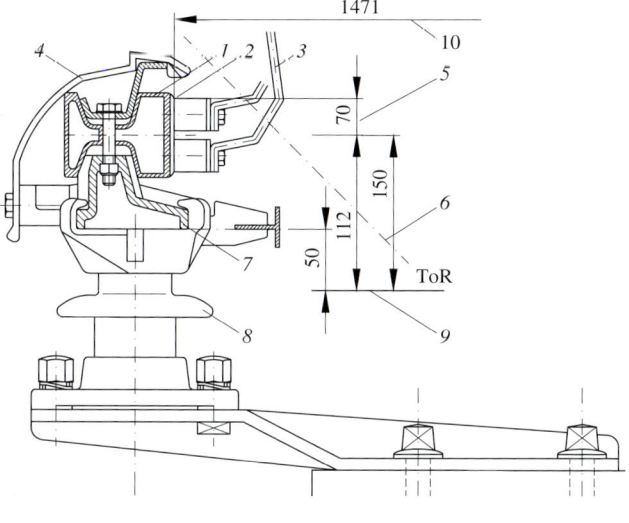

Figure 11.92: Third rail support of the Hamburg Metropolitan railway with side contact.
1 aluminum/stainless steel extruded composite third rail, *2* stainless steel contact surface, *3* contact shoes with E-Cu collectors, *4* insulated cover board, *5* maximum vertical working range of the current collectors, *6* standard vehicle gauge, *7* cast aluminum support, *8* porcelain support insulator, *9* top of running rails, *10* distance between contact surface and track

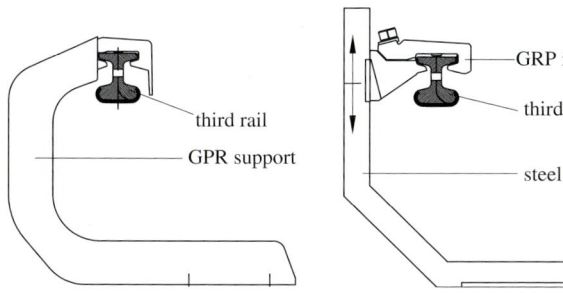

Figure 11.93: Third rail support made of GRP.

Figure 11.94: Third rail support made of steel.

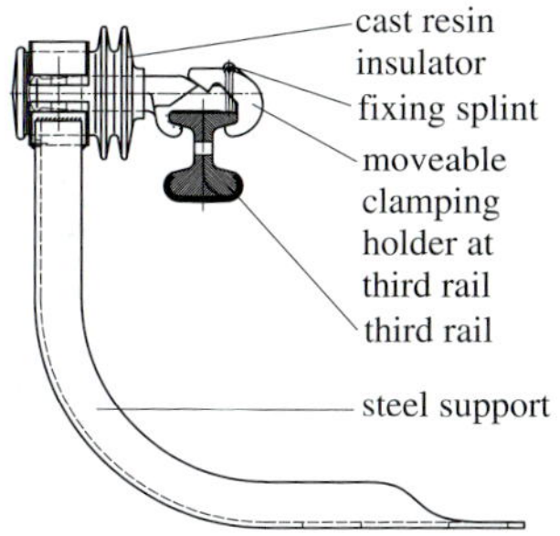

Figure 11.95: Third rail support made of steel.

11.4 Third rail installations

11.4.1 Supports

The *supports* carry the third rail and align it vertically and horizontally relative to the track. The supports are installed at 5 to 6 m spacings, with special screws, on the sleepers. The geometric position of the supports and the third rails affect the interaction with the current collectors. The service life of the contact materials depends on the lateral angle, height and height variations, the inclination in longitudinal direction and the change in inclination of the third rail. *Typical supports* are:

- made of GRP, steel or aluminum
- equipped with insulators made of GRP, cast resin or porcelain
- equipped with and without height adjustment
- used with on top contact (Figure 11.90), bottom contact (Figure 11.91) and on the side contact (Figure 11.92) of third rail

These types are also used in combination.

Table 11.16: Characteristics of third rails.

Material	Type	m' kg/m	A mm²	R' Ω/km	Used in	References
soft steel		40	5 100	0,0225	Berlin	Figure 11.96 a), Clauses 16.5.2.1
		60	7 600	0,0154	Vienna	–
		75	9 200	0,0128	New York	–
aluminum composite		6,4	2 100	0,0168	Barcelona	–
	A 5100	15,7	5 100	0,0069	Berlin	Figure 11.96 b), Clause 16.5.2.1
	A][Rail 37	12,0	4 520	0,0078	Vienna	Figure 11.96 d) [11.20]
	A][Rail 45	18,2	5 330	0,0066	Oslo	Figure 11.96 e), Clause 16.5.2.3
	A][Rail 47	18,8	5 490	0,0064	Vienna	Figure 11.96 f) [11.20]

GRP support for third rails with bottom contact

The *support* shown in Figure 11.93 consists completely *of glass-fibre reinforced, environmentally-proven plastic*, which attains a high strength due to production by injection molding. The support carries and guides the conductor rail on sliding elements made of polyoxymethylene which enable longitudinal travel of the third rail in case of temperature-related expansion. A compression plate reinforces the attachment of the support base at the sleeper. The height is fixed.

Support made of steel with a GRP insulator for third rails with bottom contact

The steel support welded from U-sections (see Figure 11.94) carries at its inner face, GRP-insulators that can be adjusted in height using an oblong hole. A clamping plate on top of the GRP insulator fixes the third rail. Guiding elements enable longitudinal movement of the conductor rail. The design of the steel support can be modified to different geometrical requirements of the individual installation.

Steel support with cast resin insulator for third rails with bottom contact

The steel support made of bent U sections carries a cast resin insulator fixed by a clamping clip. The cast resin insulator supports the third rail holder which secures the third rail with its clamping holder (see Figure 11.95). The moveable clamping holder is fixed by a split pin during installation. The design of the steel support can also be adjusted to the requirements of the installation. The height is fixed.

Aluminum support with porcelain insulators for side contact third rails

The supporting porcelain insulator is mounted on an aluminum third rail support via a base plate. An aluminum third rail holder, as shown in Figure 11.92 is on top of the insulator. The third rail holder supports the side contact *hollow third rail made of extruded aluminum* [11.19]. The height is fixed.

11.4.2 Third rail cross-sections and protecting covers

The *soft-steel conductor rails* used in the past (Figure 11.96 a), [11.21]) had a resistance $R' \leq 0,118\,\Omega\text{mm}^2\,\text{m}^{-1}$ according to Table 11.16, corresponding to a specific conductance of $8,5\,\Omega^{-1}\text{mm}^{-2}\text{m}$ according to DIN 17 122: 1978. To increase the capacity of DC railways of

a) Soft-iron third rail A5100 in
accordance with to DIN 43156

b) ACSR composite third rail
A5100 with metallurgical bond

c) ACSR composite third rail
A5300 with electric bond beween
aluminum and steel parts

d) ACSR composite third rail SPL
A][Rail 37-4520 with mechanical
bond between aluminum parts and
electric bond between aluminum
and steel parts

e) ACSR composite third rail SPL
A][Rail 45-5330 with mechanical
bond between aluminum parts and
electric bond between aluminum
and steel part

f) ACSR composite third rail SPL
A][Rail 47-5490 with mechanical
bond between aluminum parts and
electric bond between aluminum
and steel part

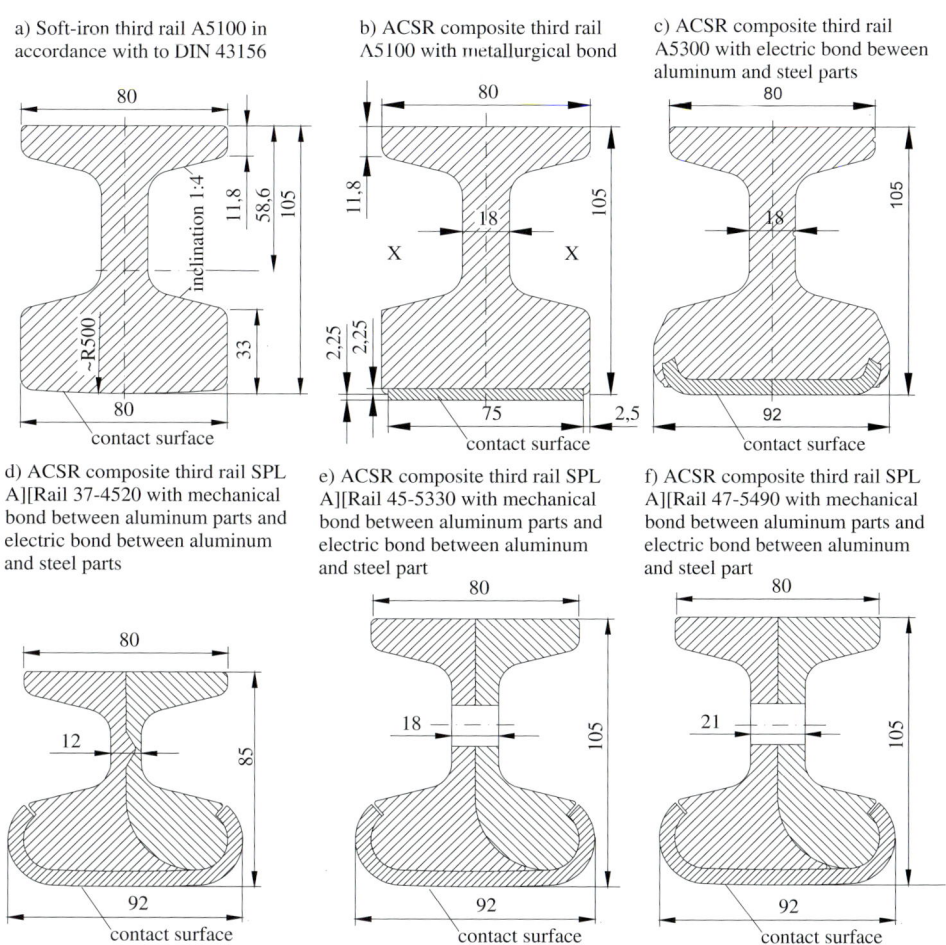

Figure 11.96: Third rail cross sections for third rail systems.

public mass transit systems and for upgrading existing installations, *aluminum-steel composite rails* are being used more frequently (Figure 11.96 b) to f)). The conductivity of AlMgSi alloy is approximately eight times higher than that of soft-steel. Its mass per unit length however, is less than half (Table 11.16).

Composite third rails enable large support spacings because of their mechanical properties [11.21] and are easier to install. While hollow composite third rails with $2\,100\,mm^2$ cross section are produced in a continuous compression-drawn process, composite third rails are milled as a solid section.

Composite third rails manufactured by extruding (Figure 11.96 b)) have a metallurgical bond between the steel and aluminum equivalent to a welded connection. With other types of composite third rails, the steel and aluminum are bonded mechanically (Figure 11.96 c)). The third rail types according to Figure 11.96 d) to f) are bonded mechanically between the two aluminum parts and bonded mechanically between the aluminum parts and the stainless steel part. The *stainless steel contact surface* has a tensile strength of at least $500\,N/mm^2$ [11.22] provides a high-wear resistance and thus a long service life.

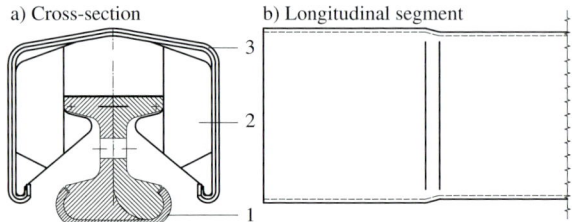

a) Cross-section   b) Longitudinal segment

Figure 11.97: Protective cover of an ACSR composite third rail.
1 ACSR composite third rail
2 spacer
3 third rail protective cover

To aid assembly, the delivery lengths of soft-steel rails are limited to 15 m and that of composite rails to 18 m. Considering the possible temperature range between 80 °C and -30 °C, is 110 K, a 15 m long soft-steel rail expands by 23,3 mm and an 18 m long composite rail by 46 mm.

Third rails are prone to *wear* during operation. For soft-steel third rails the wear should be limited

- on suburban lines to 10 % and
- on lines within the inner city to 15 %

The wear of the third rail can be evaluated by measuring the height h of the rail (Figure 11.96 a)). Figure 17.12 depicts the relationship between height measurements and the cross-section of the third rail A5100. The wear-resistant stainless steel contact surfaces increase the service life of the third rails. Reports on operational experience and further developments are described in [11.23].

Third rails with bottom or side contact have extruded plastic covers. In tunnels, halogen-free GRP is used and on sections in the open, weather-resistant PVC. They are provided in a brilliant yellow colour, e. g. RAL1018, making the cover clearly visible. The protective covers, as in Figure 11.97 a), have a U-shaped profile and are fixed to the third rail by spacers. They completely cover the top and side faces of bottom contact third rails. A protective cover for side contact third rails is included in Figure 11.92.

Each segment of the U-type protective cover has an expanded section on one end. During assembly, the individual segments are bedded into one another and then joined. At the third rail support, the protective cover is slotted on one side. For special components of the third rail system like the expansion joints, end ramps and cable connections, special covers made of GRP are available.

11.4.3   Mechanical connections and components

11.4.3.1   Third rail joint

Third rails are supplied in maximum lengths of 18 m and joined by plates (similar to fish plates) (Figure 11.98). Composite third rail splice plates are made of aluminum. They are fitted into the third rail segment and compressed by four connecting elements durably fitted into reamed holes. The *third rail joint* is mechanically solid and current-carrying. It has the same electrical conducting cross-section as the connected third rail. To minimise arcing, differences in height caused by tolerances between the third rail segments are equalised by grinding.

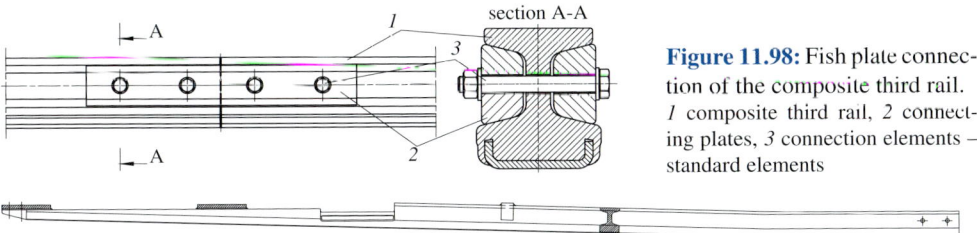

Figure 11.98: Fish plate connection of the composite third rail. *1* composite third rail, *2* connecting plates, *3* connection elements – standard elements

Figure 11.99: Third rail ramp for a bottom contact rail.

11.4.3.2 Third rail ramps, transitions between sections and points

Because of track geometry, the third rails for points and crossings on the same level cannot be installed continuously along the railway line. The gaps created need to be shorter than the minimum distance between electrically connected current collectors of a traction unit, to guarantee an uninterrupted power supply. E. g., in the case of Berlin City rail the current collector spacing is 25 m and the gap in the third rail may be 22 m long. At each end of the rails, *third rail ramps* (Figure 11.99) are provided. The required end sections enable the smooth passage of the current collectors, especially vertically. The current collector is guided into its upper end position when running into the third rail gap by the third rail end piece. The end piece inclined upwards takes the current collector at this position approximately one meter after the start of the following third rail sections and guides it back to the working position. The third rail intake which is supplied pre-bent by the manufacturer usually has a length of 5 700 mm.

On site, the *third rail intake* can be bent from a segment of the third rail. Third rail intakes in standard line sections are lifted by 60 mm and inclined by 1 : 50. In workshops, the inclination can be increased to 1 : 30 because of the low running speeds. The third rail intake can be lifted additionally by 30 mm in the last 600 mm in front of the gap. The bent third rail intake requires a space in the structure gauge for the current collector and under the car body.

The current collectors used for third rail installations are different to pantographs used for overhead contact lines because they are unsuitable to run across overlaps, which are arranged at the side of the track. For *non insulated transitions* between individual sections, extension joints or gaps are usually used because they are necessary for other reasons. Gaps are always used for *insulated transitions* between sections, between switching groups or supply ranges of substations (see Clause 11.4.4).

Intakes are also required at changes to the other side of the track. This can be caused by the track layout, location of points and before and after platforms. The third rail is usually installed on the left side in the running direction but needs to always be arranged on the side opposite to platforms.

11.4.3.3 Expansion joint

Because of the *thermal length variation* of third rails, expansion joints are required approximately every 90 m on lines in the open and every 120 m in tunnels. These are also called *dilatations* (Figure 11.100). Expansion joints enable temperature-dependent movement in the longitudinal direction across a connection consisting of oblong holes and gliding straps that are contacted by the current collector. A double expansion joint is characterised by two

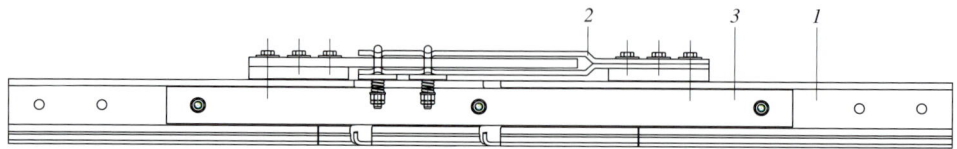

Figure 11.100: Expansion joint of an ACSR composite third rail.
1 ACSR composite third rail, *2* electrical connection, *3* mechanical connection

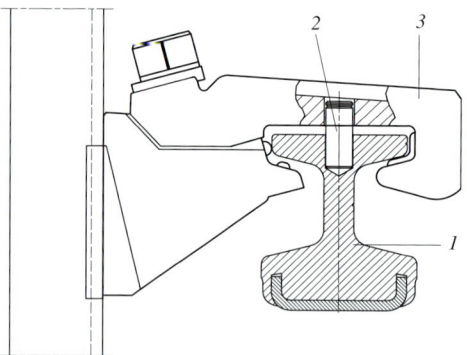

Figure 11.101: Third rail support with midpoint arrangement.
1 ACSR composite third rail
2 midpoint stainless steel bolt
3 GRP-insulator

oblique cuts giving an expansion distance of 200 mm. The current is transferred in this range by a copper contact plate, which glides into a spring contact, made from copper, parallel to the movement of the third rail (Figure 11.100). The ampacity of the connection is the same as that of the third rail.

11.4.3.4 Midpoint

The *midpoint of the third rail* is arranged approximately in the middle of a section, either between two third rail ramps, two expansion joints or between a ramp and an expansion joint. The midpoint is meant to fix the third rail and permit temperature-caused longitudinal movements in both directions away from the midpoint. In the case of short third rail lengths, the midpoint can also be arranged at one end. The midpoint shown in Figure 11.101 consists of stainless steel bolts, inserted into the insulator of a support and into the third rail.

11.4.4 Electrical connections

11.4.4.1 Ampacity

The *current-carrying connections* and the specific electrical connections as described in Clause 11.4.4.3 together with the feeder cables need to transfer the operating and short-circuit currents as specified for the individual system.

11.4.4.2 Feeding

At connection points the *feeder cables* coming from the substations are fixed to the third rail. The feeder is also used to electrically connect the individual third rail sections. It consists of aluminum straps (*4*) at the side of the third rail (*1*) opposite to the track at which a connection bolt (*5*) is fixed by a bimetallic sheet (*3*) for the cable lug (*6*) and feeder cable (*7*)

a) Side view

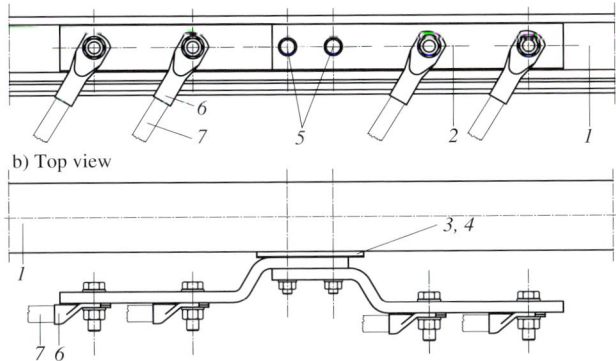

b) Top view

Figure 11.102: Third rail cable connection with four feeder cables (see Figure 11.103).

Figure 11.103: Third rail cable connection with removed feeder cables (see Figure 11.102).

(Figure 11.102). At third rails contacted on the bottom, the feeder cables are arranged at a distance of at least 50 mm from the gliding plane to ensure sufficient space for the current collector. Contact grease between the aluminum strap (*4*) and third rail (*1*) improves the electrical conductivity of the connection. The copper cables can be bolted to the cranked copper straps using crimp-type cable lugs. The number of cables is determined by considering the operational and short-circuit currents at the feed point. Because of their higher conductivity composite third rails require less feeding points than soft-steel rails.

11.4.4.3 Electrical connectors between individual sections of third rails

The joints of the soft-steel third rail are bridged by *electrical connectors*. The electrical resistance of this connection should be less than that of a five meter length of third rail section. At the site of the splice plates, the third rail is spray galvanized, improving the connection resistance to the equivalent of two meters of the third rail.

The joints of composite third rails carried out with aluminum plates are sufficiently conductive and do not require any cable connections (see Clause 11.4.3.1).

The *expansion joints* of the soft-steel rails are bridged by laminated connectors made of copper with a 600 mm^2 cross section or by flexible copper conductors. Contact plates, that slide into contact springs, are used for the electrical connection for composite rails (see Clause 11.4.3.3). Gaps within the third rail and non insulating transitions between sections are bridged directly by cables or by a disconnector (Figure 11.104) to separate the individual supply groups.

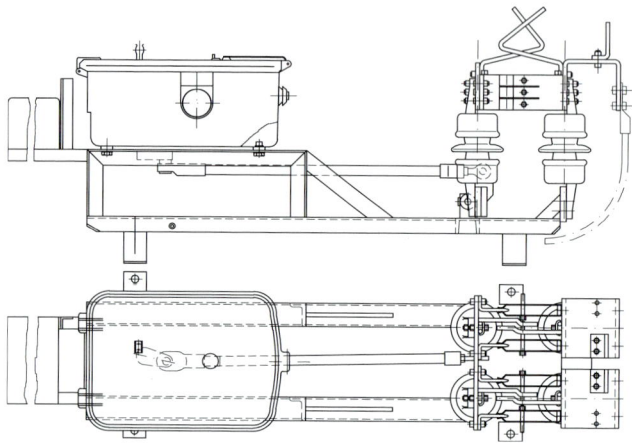

Figure 11.104: Conductor rail disconnector Siemens SBH 4000 [11.24].

11.4.5   Separations

The separations within third rail installations consist of gaps in the third rail that can be bridged electrically by disconnectors or by *separations in the third rail* where the rail is interrupted for a longer distance than in normal rail gaps. They also establish an insulating transition between rail sections and are arranged between the switching groups and between the supply sections of the substations. The distance between both third rail ramps at the rail gap in the longitudinal direction should be longer than the maximum distance between the electrically connected collectors of the traction units used to avoid bridging the separation during operation. At the Berlin city railway, the length of the separation is 36 m. Separations can be bridged, depending on the function, by substations or conductor rail disconnectors (see Clause 11.4.5). If the rail gaps cannot be designed to be long enough, the contact line protection, as in insulated overlaps in overhead lines, needs to be able to control the rapid variation in feeding currents when there is a change of supply sections.

11.4.6   Disconnectors and drives

Conductor rail disconnectors are required to enable power to be switched on and off in the individual feeding sections, tracks or groups of tracks in normal operation. These disconnectors enable the conductor rail installations to be separated longitudinally and laterally into switching sections. Figure 11.104 shows the design of a conductor rail disconnector of the series Siemens SHB 4000 [11.24]. This disconnector is designed to operate a rated voltage of DC 1 500 V and a permitted continuous current of 4 000 A. It is meant to be used for isolating conductor rail sections without load, however, it can break currents up to 400 A.

11.5   Testing of components and component assemblies

11.5.1   Introduction

The suitability of a component to meet the requirements of its envisaged application needs to be verified by tests according to EN 50 119, Chapter 8, and the product standards listed

Figure 11.105: Test laboratory at Siemens AG in Ludwigshafen.

there in. Component tests comprise three different types: *type tests*, *random sample tests* and *routine tests*.

Type tests are intended to establish design characteristics. Normally they are done only once and repeated only when design or material of the component is changed. The results of type tests are recorded as evidence of compliance with design requirements. If there is a group of similar components, it is sufficient for the type test to be performed on one kind of component of the group.

Random sample tests serve to monitor production. If statistical quality tests are performed for monitoring production, it is permissible for their results to be used as a substitute for acceptance.

Routine tests are intended to prove conformance of components with increased probability of failure to meet specific requirements, They are carried out on every component and the tests may not damage the component. After a positive test, the tested specimen may be used in an installation. Routine tests are used to detect latent defects that could affect the operating reliability of the component and take the form of load tests, magnetic crack tests, ultrasonic tests, X-rays, resistance and insulation tests or other non-destructive test methods.

The testing conditions, testing arrangements and testing parameters need to conform to the operating conditions and requirements or model as precisely as possible. Laboratories specialised in testing of contact line components are useful. Figure 11.105 showns such a laboratory.

Table 11.17: Tightening torques M_t in Nm for some bolts used in contact lines according to EN 50 119.

Thread dimension	Material of bolt	Unalloyed and alloyed steels according to EN ISO 898 Part 1 hot dip galvanized (tZn)			Rust and acid resistant steels steel groups A2 and A4		Copper-nickel alloy CU5
	Strength class	4.6	5.6	8.8	70	80	–
	Yield strength N/mm^2	240	300	640	450	600	540
M 8		–	–	23	16	22	20
M 10		–	–	46	32	43	39
M 12		25	38	80	56	75	68
M 16		60	90	195	135	180	165
M 20		120	180	390	280	370	330

11.5.2 Clamps, fittings and connecting components

11.5.2.1 Type test

At least four test specimens should be selected. All components, clamps or fittings selected for type testing should not be used for other tests or in service applications. All test conditions, test arrangements and test results should be recorded in a test report.

Material verification

Type tests should include *verification of materials* to ensure they are in accordance with the contract documents. Normally this verification should be carried out by inspecting documentation relating to material purchasing specifications, certificates of conformity or other quality documentation.

Verification of dimensions and visual test

Functional dimensions given in the drawing should be measured on the samples. A *visual examination* should include a check of the general condition and the surface of the sample for cracks, sink marks, unacceptable burrs or flashes, etc.

Functional test

Functional testing samples should be assembled according to the supplier's instructions, using all necessary connecting parts. Bolts should be tightened to 1,1 times the *tightening torque* stated in EN 50 119 (Table 11.17) or provided by the manufacturer. The dimensions relevant to the *function* should be checked. No permanent deformation should be permitted unless it is intended to do so by design.

Load testing on components

For *load testing*, specimens are tensioned or compressed in a tensile testing machine to achieve loadings as close as possible to those experienced in operation. These tensioning elements may not influence the test result. In the case of components with several load points per load side, it needs to be ensured that the forces are distributed accurately in every condition to represent the real load conditions.

The test specimens should be loaded with a constantly and smoothly increasing load up to the maximum force. The rate of change should be 5 to 10 N/mm^2 per second with respect to the conductor cross-section. A force-elongation or force-time diagram should be produced. From the force-elongation diagram, it should be possible to determine the force when transition from the elastic into the plastic range occurs. For components, the following characteristics need to be determined:

- the load at which a deformation occurs that infringes the function
- the force at which the test specimen fails
- the nature and location of failure on the test specimen

All values should be recorded in a test report together with the characteristic data for the test specimen, such as material, material condition, type of processing and surface condition. The test data should be used for determining the nominal force or permissible operating force. The maximum force on the test specimen must be equal to or greater than the required nominal force or the force specified in the appropriate standards or drawings.

Load testing of clamps

For *load testing* of clamps for conductors, anchor cables and wires are assembled with the intended conductors corresponding to the assembly instructions and inserted in the testing machine using suitable clamping parts. For compression clamps it should be ensured that conductor strands are not distorted from their intended alignment, resulting in an uneven stress distribution in the conductor. For bolted clamps, the bolts are tightened with a tightening torque corresponding to EN 50119 or with the data provided by the supplier.

The tensile force should be increased to 1,33 times the permissible operating force and held for one minute. Marks should be applied to the conductor where it exits the test specimen for observing possible slippage. For suspension clamps the maximum force in the direction of effective operational force should be determined.

For clamps on conductors, the following values need to be determined:

- force when the conductor starts to slip in the clamp
- maximum force of slipping through the clamp
- functional defects of the clamp

For stranded conductors the first wire break is considered as a breakage of the conductor.

Heat cycling tests

Heat cycling tests, as defined in EN 50119, are used to determine the long-term electrical behaviour of current-conducting connections.

11.5.2.2 Random sample test

Number of test specimens

Random sample tests should be performed only if a sufficient quantity of components are manufactured, typically more than 100 of the same component. The number p of the test specimens for the random sample test may be chosen depending on the batch size n:

$$100 < n \leq 500 \qquad p = 3$$
$$500 < n \leq 20\,000 \qquad p = 4 + n/1\,000$$
$$n > 20\,000 \qquad p = 15 + (0,5\,n)/1\,000$$

Selecting the test specimens and repeat tests

The test specimens are selected arbitrarily from the batch. If the specimens pass the random sample test, then the batch is accepted. If one specimen does not pass the random sample test, the test may be repeated but with twice the number of samples as previously selected from the batch. If all the new test specimens pass the tests, then the batch may be accepted. If any test specimen fails the random sample test, then the batch may be rejected. Specimens subjected to the random sample test may not be used for further testing or in service.

Material verification

Sample tests should include verification of materials, as performed for type testing, to ensure they accord with contract documents. This verification normally should be carried out by inspecting the documentation relative to material purchasing specifications, certificates of conformity or other quality documentation.

Verification of dimensions and visual test

To take account the influence of dimension and strength variations on the mechanical characteristics of the components, the dimensions influencing the characteristics are determined on all inspected specimens. All values should be recorded in the test report. The critical dimensions important for correct function should agree with the values determined in the appropriate standards and supplier's drawings. Sample tests should include visual examination, to ascertain conformity of the components, in all essential respects, with the design drawings.

Functional test

The *functional test* may be performed except for bolted clamps within the scope of the component assembly for the functional test as described in Clause 11.5.2.1. The components or clamps should be assembled according to the supplier's assembly instructions. Bolts should be tightened to 1,1 times the tightening torques stated in Table 11.17. The critical dimensions for correct function should be checked. All defects impairing the function of the components or clamps are unacceptable, except if the deformation is intended in the design.

Load testing

The requirements for *load testing* are given in Clause 11.5.2.1. The test specimens should be loaded initially with a load speed between 5 and 10 N/mm^2 per second, related to the conductor cross section, increasing constantly and smoothly up to the permissible operating force multiplied by the determined rating factor. Then, the permanent deformation should be determined either on a force-distance diagram or by checking the connection dimensions after relaxation and removal of the test specimen. The test specimen should then be reloaded with the same load speed up to the maximum force or up to failure of the component.

Evaluation of random sample tests

All *test values* should be recorded in the test report and statistically evaluated. In the case of fittings for insulators and support assemblies, the following characteristics should be determined:
- the component should not have a permanent deformation affecting its functionality after applying a load of 1,33 times the operating load
- the maximum force
- the nature and position of failure of the test specimen, e. g. rupture, deformation

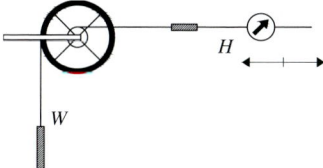

Figure 11.106: Example of a test on a tensioning device.

The force at which a deformation impairs the function should be greater than the permissible operating force required for the component. The maximum force of the test specimen should be equal to or greater than the nominal force or the force specified in the appropriate standards or drawings.

Evaluation of tests at suspension and dead-end clamps
The evaluation should be performed according to Clause 11.5.2.1. The force at which a permanent deformation occurs which affects the correct function has to be greater than 1,33 times the permissible operating force required for the component tested. The maximum force of the test specimen should be equal to or greater than the required nominal force or that force determined in the appropriate standards or drawings.

11.5.2.3 Routine tests

Where specified, every component of a production series has to be subjected to a routine test. Such routine test methods are carried out only for certain safety-critical components, e. g. wheels of tensioning devices. *Routine tests* may include a specified level of material verification, or testing where required by the contract quality plan. Load testing should be carried out in conditions similar to normal operational conditions and loaded by the routine test force. Initial cracks or fractures may not occur on the test specimens before the routine test force is reached.

11.5.3 Contact wires and other conductors

The requirements for *testing contact wires* are specified in EN 50 149. Testing of other wires or conductors is described in European, international or national standards. The following documents usually applied comprise:
- conductors made of aluminum, aluminum alloys and steel: EN 50 182, EN 50 183, EN 50 189, EN 50 326, EN 60 889, EN 61 232
- plastic ropes: EN 50 345
- copper and bronze stranded conductors should be tested in accordance with National Standards, e. g. DIN 48 201 and DIN 48 203

11.5.4 Tensioning devices

Tensioning devices with balance weights

Only type tests are required for *complete tensioning devices*. Where a fall-arresting device is installed, testing of a conductor break under the maximum operational load needs to be carried out under laboratory conditions.

Using the test in Figure 11.106, the *mechanical efficiency* of the tensioning device can be determined,

$$\eta = H / (W \cdot r) \quad ,$$

where

H mechanical tensile force of the conductor in N (Figure 11.106)
W load of the balance weights in N (Figure 11.106)
r reduction ratio of the tensioning device

The test should be carried out with the following considerations:
– wind and ice load need not to be taken into account
– at least four tests should be performed on four locations evenly distributed at least over one rotation of one of the wheels or a quarter of the complete range of movement, firstly in one direction of rotation, then in the opposite direction.

The tensile force of the conductor(s) should be recorded and comply with the data provided by the supplier. If a field test is carried out the results can be taken to describe the performance of the unit if the test lasted for at least one year. Additionally, it is necessary to verify, that fall-arresting devices function flawless at 1,3 times of the working load, see EN 50 119.

Tensioning devices without balance weights

The mechanical efficiency of the tensioning device should be determined with the following assumptions:
– wind and ice load need not be taken into account
– the efficiency should be measured at four locations evenly distributed over the complete range of movement
– if the tensioning device is intended to be used under ice conditions, the efficiency test should be performed under such conditions

The tensile force of the conductor or wire should remain within the specified range throughout the range of temperatures under consideration. If a field test is performed at an overhead contact line, the results may replace the efficiency test described above. The field test should run continuously over a minimum duration of one year with recording of the relevant data.

11.5.5 Droppers

Only type tests are required for droppers. The mechanical *fatigue test* consists of an alternate load and compression cycle, as shown in Figure 11.107. The *droppers* should be tested with their specific clamping devices. The test parameter should reflect the operational conditions. The compression amplitude should be specified between 20 mm and 200 mm and the internal force in the dropper should be specified between 100 N and 400 N. The frequency of the cycles should be between 0,5 and 10 Hz and a minimum of 2 000 000 cycles should be performed. Droppers for high-speed lines should be tested with a frequency of 5 Hz or more. The dropper may not break before the specified number of cycles is exceeded.

For special applications, such as low system height and high uplift locations, the total length and the dimension for compression should be reconsidered. The dropper clip should be installed on the corresponding wire according to the installation instructions. For contact and catenary wire clips, separate tests should be performed for each size of contact wire groove

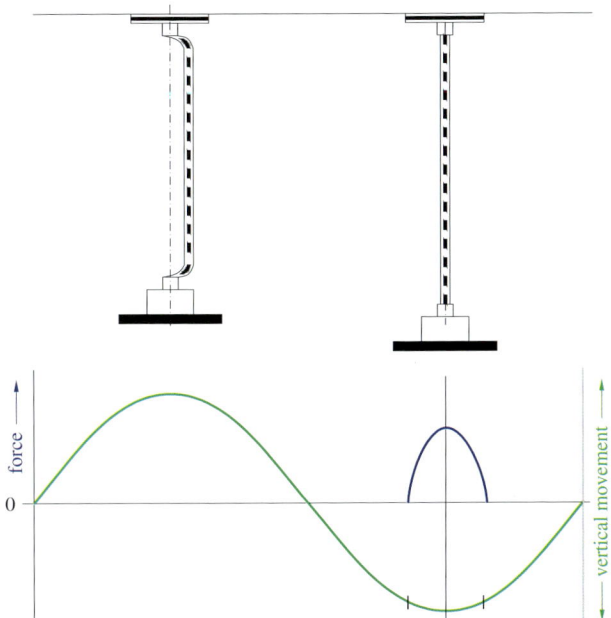

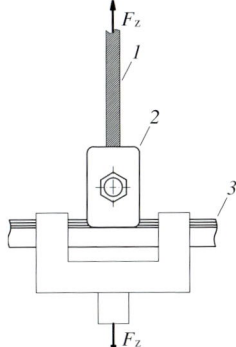

Figure 11.107: Example of a dropper test cycle.
green vertical movement of dropper wire,
blue force in dropper rope

Figure 11.108: Example of a dropper tension test assembly.
1 dropper wire, *2* dropper clamp, *3* contact wire

and catenary wire diameter. If the clip is designed for a variable range of contact wire grooves or catenary wire diameters, only the smallest and the largest sizes are required to be tested. An example of a dropper tensile test assembly is given in Figure 11.108. The clamp should not pull off the wire before a force of at least 3 kN is reached.

11.5.6 Electrical connections

Electrical connections include conductors and clamps. Only type tests are required for electrical connections. The clamps should be tested mechanically in accordance with the applicable requirements given in Clause 11.5.2. According to EN 50 119 the connections are to be tested electrically by means of a heat cycle test.

Mechanical *fatigue tests* should also be performed for electrical connections used on lines with a speed of 160 km/h and higher and if they are used on the contact wire. At least three complete electrical connections should be tested.

The following parameters should be applied, depending on contact line type and operational speed (Figure 11.109):

- the minimum specified length (l) of the connection
- the amplitude (a) of the movement should be up to 100 mm
- frequency should be 0,5 to 10 Hz
- minimum number of cycles should be 2 000 000

The electrical connection may not break before the specified number of cycles is exceeded. Electrical connections for high-speed lines should be tested with a frequency of 5 Hz or more.

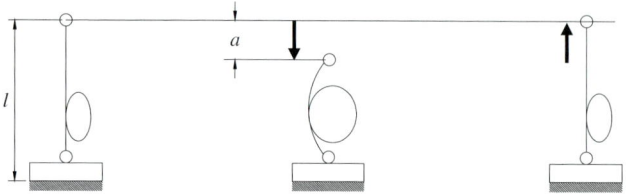

Figure 11.109: Example of a test cycle for an electrical connection.
l minimum specified length
a amplitude

11.5.7 Insulators

The *tests* should be carried out in accordance with the standards applicable to the type of insulator used:

- ceramic or glass insulator units: EN 60 383 (all parts), EN 60 672-2, IEC/TR 61 245 for insulators of AC installations, EN 61 325 for insulators in DC installations
- composite insulators: IEC 61 109 (tensile loaded insulators), EN 61 952 (insulators in bending)
- line post insulators: EN 60 168, EN 60 660

11.5.8 Section insulators

11.5.8.1 Type test

Test certification

A type test certificate should be provided for the *section insulator* verifying the electrical and mechanical parameters based on the standards used and manufacturer's information. Testing requirements and type tests should verify the suitability of the components. Specifications should be quoted for speed, weight, dimensions, clearances and creepage paths, nominal voltage, short-circuit current, permissible working load and breaking load.

The insulating body of the section insulator should be tested separately in accordance with Clause 11.5.7. Clamps and fittings of the section insulator device should be tested in accordance with Clause 11.5.2.

Load testing

Mechanical load tests should be carried out when assembled on to a contact wire length through which the load will be applied. The section insulator should be fixed in a tensile testing machine and loaded to the specified operational tensile load of the contact wire; the device should be adjusted so that the runners are in their operational condition. The load should be increased at a rate of 1 kN/s up to 1,33 times the operational tensile load of the contact wire and held for 1 minute. The load should then be reduced to the operational tensile load of the contact wire. No permanent deformation may occur. Then the tensile load can be increased at a rate of change of 1 kN/s until a break occurs. The break should occur at the contact wire.

Electrical testing

Electrical tests should be carried out for the section insulator and the suspension arrangement under operational conditions. The section insulator does not need to be loaded with the full mechanical tensile force. To establish the insulating capability of the sectioning device, a lightning impulse test and power frequency test in wet and dry conditions according

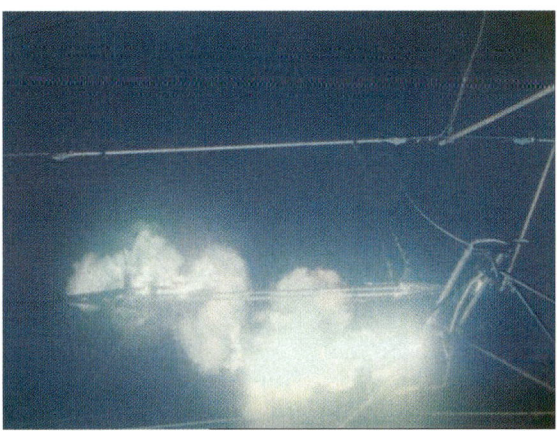

Figure 11.110: Testing the arc extinguishing performance at a short phase separation section consisting of neutral sections Sicat 8WL5545-4D.

to EN 60 383 should be carried out. Because of the shortened air gaps at the section insulator (see Clause 11.2.4.2) there is no verification of the testing parameters according to EN 50 124-1. The results are for information only. The test is designed to confirm that the flashover occurs at the planned position.

An *arcing test* should also be performed to ensure that the section insulator is capable of withstanding the expected levels of arcing without sustaining damage. The section insulator should be bridged by a wire at the same position where arcs are likely to be formed. The current should be limited to the maximum operational current of the section insulator as provided by the supplier. A voltage should be applied to the test specimen. Testing of a section insulator is shown in Figure 11.110. For this purpose, a traction unit with its pantograph raised was run across a neutral section without switching off its main circuit breaker. The flashover should occur at the planned position and be extinguished immediately. The flashover may not infringe the mechanical integrity of the section insulator.

A short-circuit test should also be performed to ensure that the sectioning device can withstand the flow of short-circuit current without damage. A short-circuit current of duration and magnitude as defined by the supplier should be applied. The effects of the short-circuit current may not affect the mechanical integrity of the section insulator.

Field testing

The electrical testing can be replaced by *field testing*. These tests should be carried out as agreed between the purchaser and the supplier. Additional testing in operation should be carried out if required by the purchaser. The operational testing should last at least one year.

11.5.8.2   Sample and routine tests

At least 2 % of specimens from each batch should be tested as follows:
  – visual inspection to verify identification marks
  – verification of dimensions and weight, as described in the supplier's drawing
All section insulators should undergo a visible inspection.

Figure 11.111: Testing the arc extinguishing performance of a disconnector.

11.5.9 Disconnectors and drives

Type and random sample tests for *disconnectors*, *drives* and *linkages* should consider the following specifications:

- EN 50 123-4 for DC disconnectors and
- EN 50 152-2 for AC disconnectors in conjunction with EN 62 271-102

EN 50 152-2 is applicable to single-phase AC single-pole disconnectors, earthing switches and switch disconnectors as designed for indoor or outdoor fixed installations for operation at frequencies of 16,7 Hz and 50 Hz on traction systems with a rated insulation level above 1 kV up to 52 kV.

EN 50 152-2 is also applicable to two-pole disconnectors, earthing switches and general purpose switches (interruption and universal switches) connected in the following manner, including:

- single pole supplying the connection to the contact line of the track, the other supplying the connection to the feeder cable which runs alongside the same track and used to boost the track voltage at regular intervals, in combination with auto-transformers or
- the two poles of the disconnector, earthing switch or general purpose switch connected in series to provide secure isolation with two breaks in series

EN 50 123-4 applies to DC switch-disconnectors, disconnectors and earthing switches for use in outdoor fixed installations of traction systems. The tests should be suitable for verifying rated current and voltage of the system. In Figure 11.111 testing of a disconnector is shown. Routine tests of disconnectors and drives should be carried out together on site.

11.6 Bibliography

11.1 *Siemens AG*: Contact line equipment for railway lines. Catalogue for materials. Siemens AG, Erlangen. 2012.

11.2 *Mier, G.*: Herstellung und Anwendung von Aluminium-Dritte Schienen (Production and application of third rails made of aluminium). In: Schweizer Aluminium-Rundschau (1984)3.

11.3 Hütte I: Des Ingenieurs Taschenbuch, Band I: Theoretische Grindlagen (The Engineer's hand book, Volume I: Theoretical basics). Wilhelm Ernst & Sohn Publishers, Berlin, 28th edition, 1955.

11.4 *Dubbel*: Taschenbuch Maschinenbau (Mechanical engineering handbook). Springer Publishers, Berlin – Heidelberg – New York, 11th edition, 1970.

11.5 *Tackmann, K. et al.*: Ermittlung des elektrischen Widerstandes je Kilometer für die Fahrschiene S49 (Determination of the electrical resistance per unit length for the rail S49). Measurements carried out by DR, Berlin, 1964.

11.6 *Markwardt, K. G.*: Power supply of electric railways (in Russian language). Publishers Transport, Moscow, 1984.

11.7 *Siemens AG*: Technische Tabellen, Größen, Formeln, Begriffe (Technical tables, quantities, formulae, terms). Siemens AG, Berlin–München, 2005.

11.8 VDV 551:2014-12: Overhead contact line poles, pole foundations and wall anchors. VDV, Association of German transport companies, 2014.

11.9 *Bausch, J.; Kießling, F.; Semrau, M.*: Hochfester Fahrdraht aus Kupfer-Magnesiumlegierung (High-strength contact wire made of copper magnesium alloy). In: Elektrische Bahnen 92(1994)11, pp. 295 to 300.

11.10 *Schmidt, H.; Schmieder, A.*: Stromabnahme im Hochgeschwindigkeitsverkehr (Power transmission to high-speed trains). In: Elektrische Bahnen 103(2005)4-5, pp. 231 to 236.

11.11 *Dölling, A.*: Tensioning devices based on wheel assemblies for overhead contact lines. In: Elektrische Bahnen 113(2015)INT2, pp. 59 to 66.

11.12 *Stephan, A.; Terfloth, S.*: Tensioning devices for overhead contact lines – Quo vadis? In: Elektrische Bahnen 113(2015)INT2, pp. 51 to 58.

11.13 *Hagedorn, P. J.; Schumacher, H.*: Die elektromechanische Nachspannvorrichtung für Oberleitungen (Electro-mechanical tensioning device for overhead contact lines). In: Elektrische Bahnen 82(1984)3, pp. 109 to 116.

11.14 *Schmieder, A.; Rankers, M.*: Disconnectors for AC overhead contact lines. In: Elektrische Bahnen 100(2002)6, pp. 207 to 214.

11.15 *Dölling, A.; Focks, M.; Gumberger, G.*: Fahrleitungserdung – automatisiert mit Sicat AES (Earthing of contact lines – automated by use of Sicat AES). In: Elektrische Bahnen 111(2013)3, pp. 172 to 184.

11.16 *Dölling, A*: Schalterstellungsmeldung Sicat DMS (Monitoring of disconnector positions in contact lines with Sicat DMS). In: Elektrische Bahnen 111(2013)12, pp. 770 to 776.

11.17 *Schmieder, A.; Rankers, M.*: Standardised composite insulators for AC overhead contact lines. In: Elektrische Bahnen 100(2002)6, pp. 215 to 219.

11.18 *Rosenke, D.; Uyanik, A.*: Neuentwicklung einer Stromschienenoberleitung für Tunnelstrecken (Development of an overhead conductor rail for tunnel sections). In: Verkehr und Technik (1985)5, pp. 136 to 138.

11.19 *Haupt, R.; Freidhofer, H.*: Elektrische Energieübertragung mit Aluminium-Verbundstromschienen bei der Berliner S-Bahn (Electrical energy transmission by means of aluminium-steel composite conductor rails for Berlin city railway). In: Elektrische Bahnen 50(1979)4, pp. 96 to 100.

11.20 *Jukl, T.; Miksche, M.*: Ausbau der U-Bahn in Wien (Extension of the Vienna underground network). In: Elektrische Bahnen 114(2016)1-2, pp. 90 to 96.

11.21 *Janetschke, K.; Freidhofer, H.; Mier, G.*: Einführung von neuen Stromschienenanlagen mit Aluminium-Verbundstromschienen bei der Berliner S-Bahn (Introduction of new third rail installations with aluminium-steel composite conductor rails at Berlin city railway). In: Elektrische Bahnen 80(1982)1, pp. 17 to 23.

11.22 *N. N.*: ALUSINGEN-Verbundstromschienen (ALUSINGEN composite conductor rails). Aluminium-Walzwerke Singen GmbH, Singen, 1979.

11.23 *Herrmann, S.*: Verbundstromschienen – Betriebserfahrungen und Weiterentwicklung (Compound conductor rails – operational experience and development). In: Elektrische Bahnen 98(2000)1-2, pp. 52 to 56.

11.24 *Siemens AG*: Stromschienen-Trennschalter SHB 4000 (Disconnector SHB 4000 for conductor rails). Siemens AG, Erlangen, 1980.

12 Project design

12.0 Symbols and abbreviations

Symbols	Definition	Unit
AC	Alternating Current	–
BS	Blade Start	–
C	length of overlapping area of contact lines in protective sections	m
CA	CAtenary wire	–
CW	Contact Wire	–
CW_k	crossing of contact wires within the points area	–
D	swaying area in straight line sections for the 1 600 mm pan head 0,200 m and for the 1 950 mm pan head 0,225 m	m
D_A	diameter of pole A at the height of the return conductor attachment	m
D_B	diameter of pole B at the height of the return conductor attachment	m
D_S	total length of neutral section of protective section, including overlap sections	m
D'_S	length of the neutral section of a protective section, excluding overlap sections	m
D_{CA-Bs}	vertical distance between upper surface of catenary wire at support and lower edge of bridge	m
D_{CA-Bx}	vertical distance between upper surface of catenary wire and lower edge of the building at verification site x	m
D_{CA-BM}	clearance between catenary wire and building at midspan	m
$D_{CA-BMmin}$	minimum clearance between the catenary wire and the building at midspan	m
D_{elf}	minimum clearance in air between conductor and earth components during fast-front lightning impulse overvoltage	m
D_{els}	minimum clearance in air between conductor and earth components during slow-front lightning impulse overvoltage	m
D_k	point of contact between the pantograph pan head and the contact wire	–
D_{ko}	swaying due to over-compensation	m
D_{ku}	swaying due to under-compensation	m
DC	Direct Current	–
E	inserting depth of a pole	m
ET, EB	end of contact wire run-up or run-off within the through-track or the branching-track respectively, between supports S_1 and S_2	–
FL	Feeder Line	–
F_R	radial force acting at the steady arm	N
F_{cc}	compensated centrifugal force	N
F_g	gravitational force	N
F_{ko}	centrifugal force due to over-compensation	N
F_{ku}	centrifugal force due to under-compensation	N
F_r	resulting force perpendicular to ToR	N
F_{ro}	resulting force due to over-compensation	N

Symbols	Definition	Unit
F_{ru}	resulting force due to under-compensation	N
G'	specific weight	N/m
G_C	conductor reaction coefficient considering the reaction of moving conductors to wind load. If the width of the gust effects the whole span, then $G_C = 1,0$ according to EN 50 341-2-4: 2016	–
G'_{CA}	specific weight of catenary wire	N/m
G'_{CW}	specific weight of contact wire	N/m
G'_{ice}	specific weight of ice and snow	N/m
G'_{OCL}	specific weight of the contact line consisting of contact wire, catenary wire, droppers, stitch wire and clamps	N/m
$G'_{OCL ice}$	specific weight of the contact line with ice	N/m
H_{CW}, H_{CA}	tensile force of contact wire or catenary wire, respectively	kN
H_{OD}	altitude of Top of Rail above sea level	m
H_S	height of rails	mm
H_y	tensile force of stitch wire	kN
KR	curvature of track	1/m
L	length of point	m
L_N	half tensioning section length	m
L_S	distance between the centre lines of the rails of a track	m
L_{SS}	minimum distance between the centre lines of adjacent pantographs	m
L'_{SS}	distance between the outer edges of adjacent pantographs	m
L''_{SS}	distance between the inner edges of adjacent pantographs	m
L_h	kinematic-mechanical pantograph gauge at height h	m
L_{hT}	kinematic-mechanical pantograph gauge at height h of through-track	m
L_{hB}	kinematic-mechanical pantograph gauge at height h of branching-track	m
L_o	kinematic-mechanical limit at verification height $h_o = 6,5\,m$	m
L_u	kinematic-mechanical limit at verification height $h_u = 5,0\,m$	m
L_W	length of point	m
P	provisional value for lateral position	m
P_M	variation of contact wire position caused by pole inclination under wind action	m
P_S	contact wire stagger tolerance at the support	m
P_T	displacement of the contact wire lateral position because of temperature dependent length variation of the contact wire	m
PC	Point Centre	–
PE	Point End	–
PS	Point Start	–
R	track radius	m
RC	return conductor	m
R_{RC-CA}	spacial distance between return conductor and catenary wire	m
RC	axis of return conductor	–
RC_A	return conductor swung out position	–
RC_R	return conductor in still air position	–
S_{exist}	existing electrical clearance in air. This distance includes the safety margin and, therefore, is more than the electrical clearance in air according to EN 50 119	m

Symbols	Definition	Unit
SB, ST	starting point of contact wire run-up for pantographs in branching-track or through-track respectively between supports S_1 and S_2	–
S'	projection as an exceedence of the gauge if the vehicle is situated in a curve and/or on a track with a gauge of more than 1,435 m	m
S_d	permanent minimum electrical clearance in air	m
S_k	short-term minimum electrical clearance in air	m
S_{min}	minimum clearance according to EN 50 341-2-4:2016	–
S_0	flexibility coefficient $S_0 = 0{,}225$ for vehicles with pantographs	–
$S_{1(2)}$	support 1 or 2 within the range of points	–
T	length of cross arm	m
$T_{CA\,height}$	tolerance of catenary wire height at the support	m
$T_{CW\,height}$	tolerance of contact wire height at the support in still air	m
T_{RC}	distance between pole centre and attachment of return conductor	m
T_{length}	longitudinal gradient which affects the contact wire height	‰
T_{cant}	track cant, which is effective on the overhead contact line	m
ToR	Top of Rail	–
U_e	pole extension above upper cantilever hinge	m
a	variable	m
a_{CAs-B}	distance between catenary support and edge of bridge or building in case of contact line lowering	m
a_{HS}	span width of head span	m
a_{PS-BS}	distance between start of points PS and start of blade BS	mm
a_{SO}	distance between signal and the first pole with twin cantilever of the overlapping in the main running direction	m
a_{Tr}	transverse span length	m
a_W	distance between point centre and radius end in branching-track	m
a_{min}	minimum spacing between conductors in still air	m
a_{som}	minimum flash-over distance at insulator sets	m
a_x	minimum distance between pole centre and verification point x	m
b	standard contact wire stagger at supports	m
b_{SO}	distance between the signal and the first pole with twin cantilevers within the overlap, opposite to the running direction	m
b_W	distance between start point and point centre	m
b_{1T}, b_{1B}	stagger of the contact wire within the through-track or branching-track at support 1 within the points area	m
b_{2T}, b_{2B}	stagger of the contact wire within the through-track or branching-track at support 2 within the points area	m
c	dimension to check the contact wire lateral position at midspan	m
c_W	distance between point centre and point end in through-track	m
d	conductor diameter	m
d_W	distance between radius end and point end in branching-track	m
d_p	width of run-up and run-off area in case of tangential wiring in point areas	m
d_k	creepage path of an insulating rod	m
$d_{s\,ins}$	length of section insulator	m
e	distance between Top of Rail and upper level of foundation or driven tube	m
e_W	horizontal distance between point ends in through- and branching-track	m

Symbols	Definition	Unit
e_g	lateral limit position of contact wire	m
e_{max}	maximum deflection of contact wire relative to track centre line (axis)	m
e_{po}	sway of pantograph being $0{,}170\,m$ at $h_o = 6{,}5\,m$	m
e_{pu}	sway of pantograph being $0{,}110\,m$ at $h_u = 5{,}0\,m$	m
e_{use}	useable contact wire lateral position	m
f	power frequency	Hz
f_{CA}	catenary wire sag without ice load and with a new contact wire	m
f_{CAM}	catenary wire sag in still air at midspan with a new contact wire	m
f_{CAMw}	catenary wire sag in still air in midspan due to contact wire worn by 20%	m
$f_{CAMw+dyn}$	catenary wire sag at midspan with 20% worn contact wire and lift after pantograph passes	m
f_{CAx}	catenary wire sag at position x distance from the catenary wire support	m
f_{CWMd-d}	additional contact wire sag at midspan between adjacent droppers without ice load	m
$f_{CWMd-dice}$	additional contact wire sag at midspan between adjacent droppers with ice load	m
$f_{CWMd-dice+dyn}$	additional contact wire sag at midspan between adjacent droppers with ice load and downwards oscillation after pantograph passes	m
$f_{CWMs-sice}$	additional contact wire sag at midspan between adjacent supports with ice load	m
$f_{CWMs-sice+dyn}$	additional contact wire sag at midspan between adjacent supports with ice load and downwards oscillation after pantograph passes	m
f_{RC}	return conductor sag	m
f_w	oblique distance between point ends in through- and branching-track	m
$f_{RCmax40}$	maximum return conductor sag at a conductor temperature of $40\,°C$	m
$f_{RCmax80}$	maximum return conductor sag at a conductor temperature of $80\,°C$	m
f_2	conductor sag at $40\,°C$ conductor temperature	m
g_w	distance between point ends rectangular to through-track	m
h	verification height above ToR	m
h_B	head room of bridge above ToR	m
h_{CAs}	height of the catenary wire at the support above ToR	m
h_{CAmax}	maximum catenary wire height at midspan above ToR	m
h_{CW}	contact wire height above ToR	m
h_{CWmin}	minimum contact wire height at midspan above ToR	m
h_{CWs}	contact wire height at the contact wire support above ToR	m
h_{CWmax}	maximum contact wire height above ToR	m
$h_{CWmax\,plan}$	maximum planned contact wire height above ToR	m
h_{CWmin}	minimum contact wire height above ToR occurring below bridges with simultaneous occurrence of ice load, possibly insulated catenary wire and the contact wire oscillation downwards after pantograph passage at midspan	m
$h_{CWmin\,plan}$	minimum planned contact wire height above ToR	m
h_{CWnom}	nominal contact wire height above ToR	m
h_M	pole length	m
h_{SH}	system height, vertical distance measured at the support between the lower surface of the contact wire and middle of the catenary wire	m
h_T	height of the overhead contact line above terrain	m

Symbols	Definition	Unit
h_{c0}	reference height of the rolling centre, $h_{c0} = 0,5\,\mathrm{m}$	m
h_f	versine as distance between the track axis in a curve and the secant between the supports within this curve	m
h_o	upper verification height above ToR	m
h_t	installation height of the lower pantograph joint above ToR	m
h_u	lower verification height above ToR	m
k	coefficient k according to EN 50 341-2-4:2016, Table 5.4.3/DE.2	–
k_R	distance between return conductor axis and contact line at support	m
l	span between two overhead contact line supports	m
l_{Al}	length of the cantilever	m
$l_{Al\,max}$	maximum length of standard cantilevers	m
l_A	working length of the pantograph pan head being – $l_A = 1,200\,\mathrm{m}$ for a 1 600 mm long pan head – $l_A = 1,550\,\mathrm{m}$ for a 1 950 mm long pan head	m
l_B	covering width of a building	m
l_G	limiting gauge as the distance between running edges of rails of a track according to stipulations of infrastructure manager: – secondary railways and secondary tracks $l_G = 1,470\,\mathrm{m}$ – tracks with $v \leq 160\,\mathrm{km/h}$ $l_G = 1,465\,\mathrm{m}$ – tracks with $v > 160\,\mathrm{km/h}$ $l_G = 1,463\,\mathrm{m}$	m
l_{M-s}	distance between the contact wire support and midspan	m
l_S	minimum collector strip length	m
l_{SE}	length of chord	m
l_W	length of the pan head being 1 600 mm or 1 950 mm	m
l_{SA}	pantograph width	m
l_{SK}	length of the secant in a track curve	m
l_{d-d}	distance between droppers	m
$l_{d\,min}$	minimum dropper length	m
l_{dy1}	dropper length of the stitch wire dropper 1	m
l_{dy2}	dropper length of the stitch wire dropper 2	m
$l_{d1}\ldots l_{d6}$	length of dropper 1 to 6 in the span	m
l_f	distance passed by the pantograph to the zero crossing of voltage	m
l_k	length of insulator set, minimum creepage path	m
l_p	projection length of pantograph horn	m
l_{total}	distance between signal and beginning of points	m
l_y	projection length of stitch wire	m
$l_{1(2)}$	length of span 1 (2)	m
n	number of supports	–
n	inverse of embankment inclination	–
n_{PS-M}	inclination of points	–
s_{sight}	visibility of signal	m
s'_0	flexibility of the vehicle with pantograph on the roof	-
t	distance between beginning of points and centre of points	m
t_r	transverse displacement of pantograph under action of a force of 300 N	m
u	cant of the track in curves	m
u_f	cant deficiency	m

Symbols	Definition	Unit
u_{f0}	reference value of the cant deficiency, $u_{f0} = 0{,}066\,\mathrm{m}$	m
u_0	reference value of the cant, $u_0 = 0{,}066\,\mathrm{m}$	m
$u_{1\,min}$	minimum distance between start of uplift for branching-track contact wire at position 1 and support S_1 for tangential point wiring	m
v	commercial speed of train	km/h
v_b	basic wind velocity at 10 m above terrain, averaged over 10 min and with a return period as required for the verification	m/s
$v_b(p)$	basic wind velocity with a probability of occurrence p, which deviates from a return period of 50	m/s
$v_{b,0,02}$	basic wind velocity with a probability of occurrence $p = 0{,}02$, this being once in 50 years	m/s
v_g	gust wind speed	m/s
v_{max}	maximum operational train speed	km/h
w_r	play in bogie of the reference vehicle	m
x	distance between contact wire crossing and point start in case of intersecting point wiring	m
x_i	distance between contact wire crossing and spot i for intersecting point wiring	m
x_B	distance between the catenary support and the edge of the building in case of contact line lowering	m
x_W	distance between the crossover droppers in points	m
y	uplift of the contact wire in the branching-track above points in addition to the nominal height between the contact wire crossing and the support S_1	m
y_{RC-CA}	horizontal distance between axis of return conductor and catenary wire	m
y_{RC-RC_A}	horizontal distance between axis of return conductor and swung return conductor	m
z	contact wire uplift of branching-track at points between contact wire crossing and support S_1	m
z_{RC-RC_A}	vertical distance between axis of return conductor and swung return conductor	m
z_{RC-ToR}	vertical distance between axis of return conductor and ToR	m
z_{RC-CA}	vertical distance between axis of return conductor and catenary wire	m
Δb_m	minimum contact wire displacement, conductor sweep	mm/m
Δf_z	minimum branching-track contact wire uplift at support S_1 without pantograph touching in through-track	m
Δf_{CAMw}	reduced catenary wire sag at midspan considering 20% contact wire wear	m
Δf_{CAMdyn}	additional catenary wire lift at midspan due to pantograph passage	m
$\Delta f_{CWMd-dice}$	additional contact wire sag at midspan between the droppers caused by ice or snow load	m
$\Delta f_{CWMs-sice}$	additional contact wire sag at midspan caused by ice load between the adjacent supports	m
$\Delta f_{CWMs-sdyn}$	oscillation amplitude of the contact wire downwards at midspan generated by the lift of the contact wire, whereby the contact wire is fixed at the supports and oscillates below its position in still air	m
Δf_{zi}	increase of contact wire height between contact wire crossing and support	m
Δh_{CA}	vertical difference of the suspension height of catenary wire at the two poles of a span	m

Symbols	Definition	Unit
$\Delta h_{\mathrm{Contact\,CW}}$	maximum difference in height between the highest and lowest position of the dynamic contact point within a span	m
Δh_{RC}	vertical difference of the suspension height of return conductor at the two poles of a span	m
$\sum j$	sum of horizontal supplements to consider random phenomena because of – asymmetry of loading – lateral transverse displacement of the track – cant tolerance and oscillations of the pantograph duen to track unevenness	m
$\sum j_{\mathrm{o}}$	sum of the horizontal supplements to take care of random phenomena at the upper verification point $h_{\mathrm{o}} = 6{,}5\,\mathrm{m}$	m
$\sum j_{\mathrm{u}}$	sum of the horizontal supplements to take care of random phenomena at the lower verification point $h_{\mathrm{u}} = 5{,}0\,\mathrm{m}$	m
α_{Dr}	turn angle of point triangle angle	degree
α_{CW}	branching-track contact wire angle at support S_1	degree
α_{W}	angle between the tangents of through- and branching-tracks at points	degree
α_{T}	points angle for point definition (see Table 12.12)	degree
δ	angle between the perpendicular to the line touching the rail heads and the centre line of the inclined vehicle	degree
η	angle betwcen the line touching the head of rails in curves to the horizontal	degree
θ_{r}	angle due to vehicle asymmetry	degree
ϑ	conductor temperature	°C
ρ	air density 1,225 at 15 °C and at 0 m sea level	kg/m^3
τ_{r}	manufacturing and installation tolerance of pantograph	m
φ	swing angle of a traction power line	degree
φ_{RC}	swing angle of return conductor	degree

12.1 Objective and process

The objective of *planning of contact lines* is to establish design documentation for a specific project, based on specific conditions, such as technical stipulations, line parameters and operator requirements that permit the installation and operation of a specific overhead contact line system to defined operational specifications and its efficient operation. The *planning* consists of the following steps:

- *preliminary design study*
- *design planning*
- *project implementation planning*
- *review* and preparation of review documents

The *preliminary design study* (Figure 12.1) forms part of the planning work for new lines as well as the reconstruction or electrification of existing lines. The design study examines various options for the overhead contact line system and identifies any additional infrastructure works required on track layout, tunnels, bridges etc. The compatibility of the electrification with other technical line equipment such as signalling is also examined during this phase. The *preliminary design study* produces technical solutions, including the design of the overhead contact line system, adaptation of the track layout, structural alterations to tunnels and bridges and an estimate of the implementation period and required investment. A summary report of the design study is then provided to railway entities for incorporation into their overall planning.

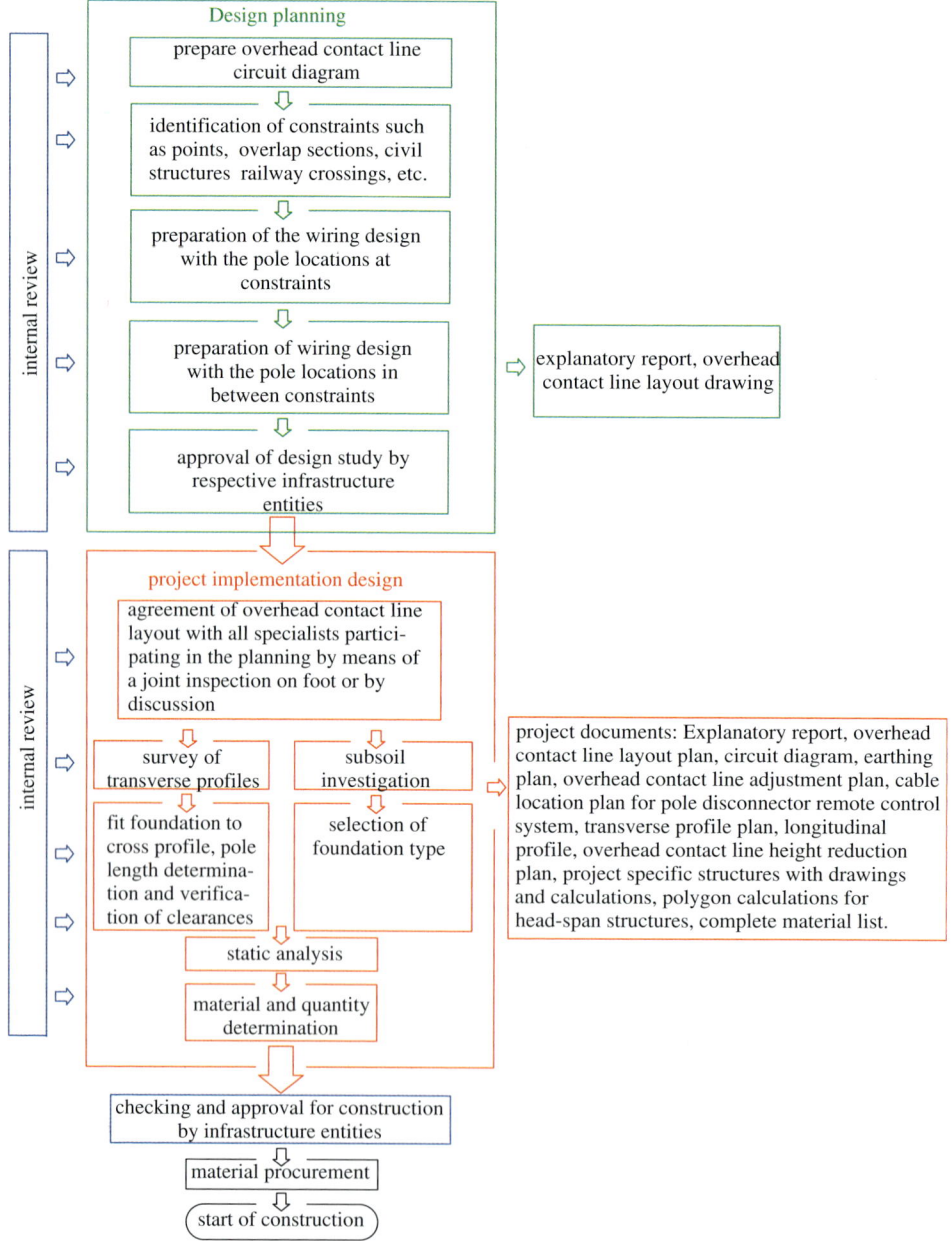

Figure 12.1: Flow chart for design planning, project implementation planning and reviews.

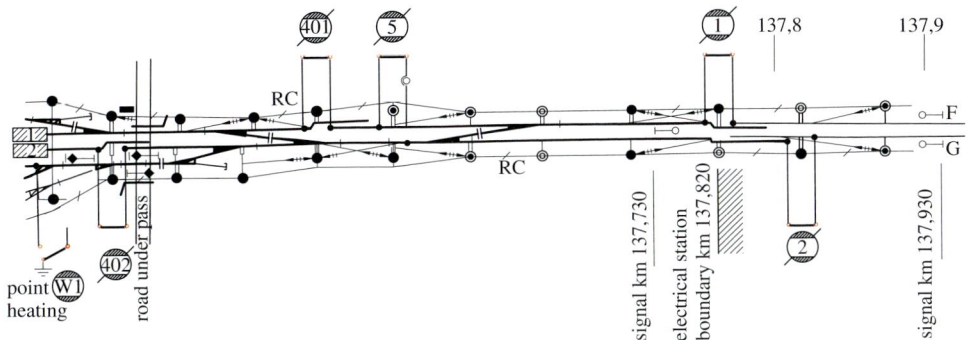

Figure 12.2: Master line diagram extract for a station (German Railway).

Design planning (Figure 12.1) commences with the preparation of an overhead contact line system circuit diagram (Figures 12.2 and 12.3) based on data provided in the preliminary design study. It contains feeders, circuit groups, subsections, disconnectors and traction power lines. The next stage is the identification of route constraints that impact technical aspects of the overhead contact wiring based on track layout and documentation on bridge structures, tunnels, signalling and telecommunication systems. The *wiring* at constraints such as buildings, line crossings, insulated sections before stations, points and cross overs is also defined during this early phase (see Clause 12.4). The wiring of the intermediate sections between the constraints is defined during the next phase (see Clauses 12.5 to 12.7). Supporting equipment such as single track cantilevers, multi-track cantilevers, head-span structures or portals are then assigned to the overhead contact line support locations whereby all relevant information contained in the planning documentation will be considered. The design study, including all planning documents, pole locations and *overhead contact line layout* diagram, is then distributed with an explanatory report to all interested parties for comment.

A site inspection of the line by all project participants is the first phase of *project implementation planning* (Figure 12.1) for the electrification of existing lines. This inspection assists with the identification of all installations that may clash with planned pole locations. Items to be considered include underground installations, culverts, drains, permanent way profile, signal visibility and neighbouring buildings. A report containing all information gathered during the inspection and conclusions drawn for planning the overhead contact line system is then compiled.

For new lines, a site inspection is not feasible. Consequently, design plans are agreed and confirmed by all project participants at a joint meeting. The departments responsible for planning track layout, civil engineering, bridge works, tunneling, signalling, telecommunications, general power supply, point heating equipment, substations and disconnector remote control systems need to consider all proposed pole locations during subsequent work. After fixing the pole locations along and across the track, a transverse profile survey is carried out at each pole location to establish foundation requirements and pole lengths. The distances between live components of the overhead contact line and other installations need to be checked. Pole sites and heights are then selected, taking into account the electrical clearances. The material requirements may be determined after concluding the planning phase. The combination of standardized components and assemblies simplifies the procurement of material and computerized production of material lists.

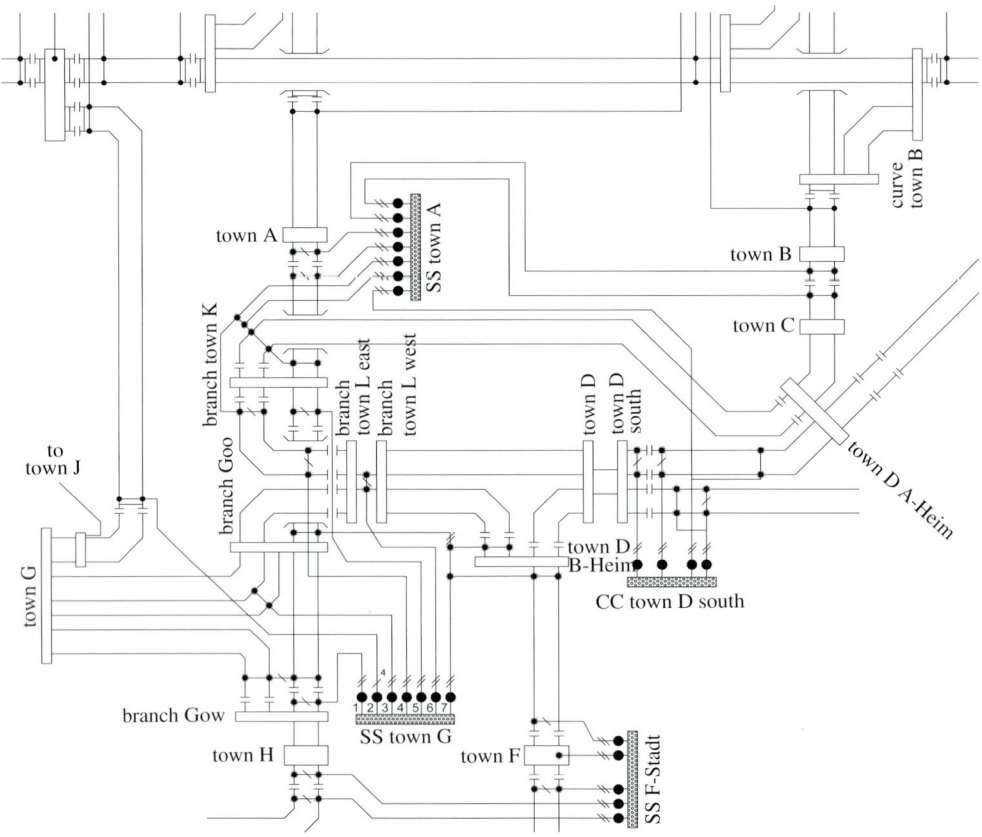

Figure 12.3: Overhead contact circuit diagram (German Railway).

The result of *project implementation planing* (Figure 12.1) includes:

  (1) an *explanatory report* covering the overhead contact line equipment and civil engineer-
      ing aspects of the project
  (2) *overhead contact line layout plans* to a standard scale 1 : 1 000 or 1 : 500
  (3) *transverse profiles* for interstation lines and for stations
  (4) *longitudinal profiles* for non-obvious routing of traction power lines
  (5) *longitudinal profiles* for *overhead contact line height reductions*
  (6) *project-specific structures* with drawings and calculations
  (7) *cantilever* and *dropper length calculations*
  (8) *polygon calculations* for head-span structures
  (9) *earthing plans* for stations
 (10) *cable layout plans* for the cables of disconnector remote control
 (11) complete *component list* consisting of *pole and foundation tables, overhead contact
      line tables* and lists of other material

An internal check of obtained results then follows. The process of project implementation
planning including the required checks is presented in Figure 12.1. The modifications to
planned design implemented during installation need to be recorded continuously in the
design documents. After commencing the installation, the *review documents* (Figure 12.1),

which represent the installed contact line, become the basis for operations and maintenance of the overhead contact line for the infrastructure manager.

12.2 Fundamentals, initial data and drawings

12.2.1 General

Overhead contact line system planning and construction work are based either on the standard specifications for a specific *design of overhead contact line* established by the infrastructure manager or on *functional specifications*. The latter provide the contractor with the flexibility to develop an overhead contact line system tailored to the project based on the contractor's experience. However, his responsibility is higher than when erecting an installation using existing and qualified infrastructure manager's directives. Both variants require the compilation of planning data that represents the *technical requirements*. This data set (see Clause 12.2.2) serves as an initial guide for all involved in the project and ensures a rapid and correct planning process.

In addition, before start of planning further documents are necessary (Clause 12.2.3), to establish the relationship of the line to existing installations and topography. These are classified as new, existing or already electrified reconstructed lines. In the following Clauses, these variants are dealt with separately.

12.2.2 Initial data

The *technical specifications and characteristics* are listed in a useful data summary. As an example the data summary presented in Table 12.1 includes the planning parameters of Siemens standard overhead contact line design Sicat S1.0 [12.4] for interoperable overhead contact lines.

12.2.3 Initial drawings

12.2.3.1 Electrification of new lines

The *planning documents* represent the line, whether recently installed or existing. They form the basis for the design of the overhead contact line system and include information relating to existing installations and the topography. Document formats and contents vary for new, existing and pre-electrified lines to be modified. These variants are treated separately.

After compiling the technical requirements for the overhead contact line, ongoing planning for a *new line* requires information regarding the *track layout*, *topography*, *soil conditions* and *system constraints*. Detailed documentation of this information is provided by:

- the *surveyor's layout plan*, which shows the track layout to a scale of 1 : 1 000 or 1 : 500, separately for interstation lines and stations. These drawings are provided either as paper drawings or in digital formats, where the latter is to be preferred. The documents need to include start and end of the track elements: straight lines, transition curves, curves with radii and cant and cant deficiency. The presentation of track axes (Figure 12.4 a)) suffices. A twin track presentation as shown in Figure 12.4 b) is less useful
- a *list of coordinates* for the track layout and gradients along the right-of-way

Table 12.1: Characteristics of Sicat S1.0 standard design.

General data	
rated voltage in kV / frequency in Hz	25 / 50 and 15 / 16,7
traction power supply system in kV	1×25 or 2×25, 15
speed on main lines / secondary lines in km/h	230 / 100
design for main lines / secondary lines in km/h	Sicat S1.0 / Sicat S1.0
length of pantograph pan head in mm	1 600 and 1 950
static contact pressure minimum / maximum in N	60/90
dynamic contact pressure minimum / maximum in N	> 0 / < 350
continuous current carrying capacity with parallel	
feeder 243-AL1 at °C / in A	80 / 1270
short-circuit capacity in kA/duration in s	17/1

Line information	
location of the line (e – exclusively, i – inclusively)	A-town (i) – B-town (e)
gauge in mm	1 435
cant u in mm	according to track layout
cant deficiency u_f in mm	according to track layout
sum of random lateral displacements $\sum j_o$ in mm	according to track layout
stipulation for structure gauge	GC according to EN 15 273
line length in km	75
number of tracks	2
minimum track curve radius in m	500
longitudinal profile of line is available	yes

Climatic information	
ambient temperature (average of annual extremes) in °C	-30 / $+40$
temperature working range in °C	-30 / $+70$
temperature for central position of the cantilevers in °C	$+20$
altitude H of the line above sea level in m	600
height of contact line h_C above terrain	according to longitudinal profile
wind zone (Germany)	2
Environmental pollution in industrial areas yes / no	yes

Technical information for overhead contact line	
contact wire	
contact wire type / tensile force in kN / maximum wear in %	AC-100 – CuAg / 12 / 20
fixed termination or automatically tensioned	automatically tensioned
catenary wire	
catenary wire type / tensile force in kN /	Bz II 50 / 10
fixed termination or automatically tensioned	/ automatically tensioned
stitch wire provided yes/no	yes
type / tensile force in kN / length in m	BzII 25/1,8-2,3/14-18
adjustable tensile force yes/no	no
dropper type	Bz 10
tensioning of contact and catenary wire, separately or commonly	separately
transmission ratio of tensioning device	1 : 3 or 1 : 1,5
maximum half tensioning length in m	880 m
reduction of tensioning length in curves	yes
maximum span length in m (1 950 mm long pan head)	80 m
determination of span length depending on the radius or basis wind speed	$i = f(v_b)$
determination of e_{use} in m	TSI EN CR
determination of e_{max} in m	$e_{use} - v$
contact wire pre-sag in % of span length	0,05 % of span length
standard / minimum / maximum contact wire height in m	5,50 / 4,95 / 6,50
system height on interstation lines / in stations in m	1,6 / 1,6

Table 12.1: Characteristics for Sicat S1.0 standard design (continued).

minimum clearance catenary to contact wire in mid span in m	0,5
maximum contact wire lateral displacement (stagger) at supports	
on straight tracks in m	$\pm 0,3$
in curves in m	according to analysis
useable contact wire stagger in straight line sections for 1 600 and 1 950 mm pan heads in m	$\leq 0,4\,\text{m} / \leq 0,5\,\text{m}$
catenary wire lateral displacement (stagger) in m	0
use of windstays in case of	$R > 1\,200\,\text{m}$
creepage path for insulators for 15/25 kV in mm	according to EN 50 124-1
separation of contact lines in insulated overlaps for 15/25 kV in m	0,45 / 0,50
separation of Contact lines in non-insulated overlaps for 15/25 kV in m	0,20 / 0,20
number of overlapping spans	3 or 5
tensioning weights concrete / malleable iron	concrete
point wiring: crossing or tangential	tangential
limitation of resetting forces	
under buildings for contact wire height yes/no	no
when calculating the span length yes/no	no
at railway crossings for contact wire height yes/no	no
Construction tolerances	
distance between top of rail and top of foundation or pile in mm	± 30
distance between track centre and pole front face in mm	$0 / - 50$
system height in mm	± 100
contact wire height at support in mm	± 20
maximum contact wire gradient	1 /1000
maximum contact wire gradient change	1 /2000
span length in m	± 1
provision v in m for – variation in contact wire position due to pole inclination under wind v_M in m	0,025
– tolerance of contact wire stagger at support v_S in m	0,030
– displacement of contact wire lateral position v_T due to temperature-dependent length variation	0,007
Lines	
feeder line, aerial	243-AL1, EN 50 182
feeder line, underground	N2XS2Y
bypass line	–
parallel feeder line	243-AL1, EN 50 182
disconnector line	Cu 95
return conductor type / aerial or underground conductor / insulated or not insulated	243-AL1, EN 50 182 / aerial conductor / not insulated
Insulators	
insulator for anchoring, switch lines and intermediate insulation type / material	Clause 11.2.6
cantilever insulator type / material	Clause 11.2.6
bird protection according to planning approval yes/no	yes
Cantilevers	
material: aluminum / steel	aluminum
Poles	
standard spacing between track axis and front face of pole in m	3,70
minimum distance between track axis and front face of pole in m	2,55
material: steel / concrete	steel
mounted or inserted steel poles	mounted
single poles, head-span structures or portals	single poles
termination poles with / without pole anchors(guys)	with pole anchors(guys)

Table 12.1: Characteristics for Sicat S1.0 standard design (continued).

Foundations	
standard spacing between track axis and track side of foundation in m	3,70 (concrete poles)
type of standard foundation	piling
type of foundations for difficult soil conditions	block, rock foundation
material for anchor bolts and nuts	steel, galvanized
reinforcement: yes / no / if necessary	if necessary
rating of foundations in accordance with	EN 50 119: 2009+A1: 2013
rating for earthquake risk yes/no	no
subsoil report available yes/no	yes
Railway earthing measures	
return conductor: yes / no	no
connection of pole directly to rail: yes / no	yes
type of connection	cable NYY-O or composite cable
Safety clearances	
minimum distance between railway energised and earthed parts - short duration in accordance with EN 50 119: 2009+A1: 2013 for 15/25 kV in mm	100/150
minimum distance between railway energised and earthed parts - long duration in accordance with EN 50 119: 2009+A1: 2013 in mm	150/270
minimum height of contact wire at level railway crossings in m	5,50
Electrical disconnector remote control	
cable type	NYY-J
location of control equipment	signal box Lh
use of trough channels yes / no	yes, where possible
Headroom	
overhead contact line under structures in m	5,90
overlapping sections under structures in m	6,20

- the permanent way *transverse profile (cross-section)* at each pole location provides the future transverse profile
- *signal position layouts* provide the locations of signals. The signal designs to be used follow from the technical specification of the signalling system. The dimensions of the signal, of the step treads and access ladders form the basis for verification of clearances between signals and energised parts of the overhead contact line system
- the *track insulation layout* and information on the short-circuit current provide the basis for the specification of the railway earthing
- *cable layout plans* and information related to underground installations form the basis for approving the pole foundation locations
- a list of *railway crossings* with chainage and crossing angle permits checking necessary clearances for the infrastructure gauge
- *bridge drawings* showing chainage, headroom, bridging width and crossing angle and any constraints form the basis for pole locations, the contact wire and system heights under the structure. Any necessary alterations to the bridge can be specified in this information
- information on *subsoil condition* is the basis for foundation type selecting and dimensions. The soil conditions also give an indication of the earthing resistance. If no soil documentation is available, then probing or other site investigations need to be carried out at selected locations during project implementation planning (see Clause 13.6).

a) With indications of track elements: straight line, transition curve, curve as presentation of track axis (Excerpt from the track layout plan DB line Nuremberg to Ingolstadt), lengths in m, cant in mm

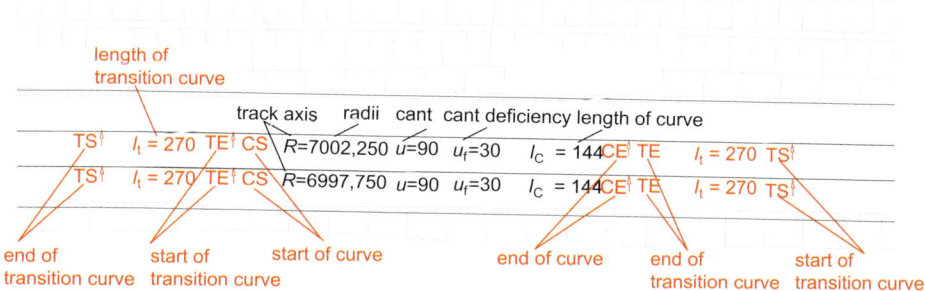

b) Without indication of track elements straight line, transition curve, curve as a twin track layout presentation (Excerpt from the track layout plan London to Bristol of the Great Western Electrification)

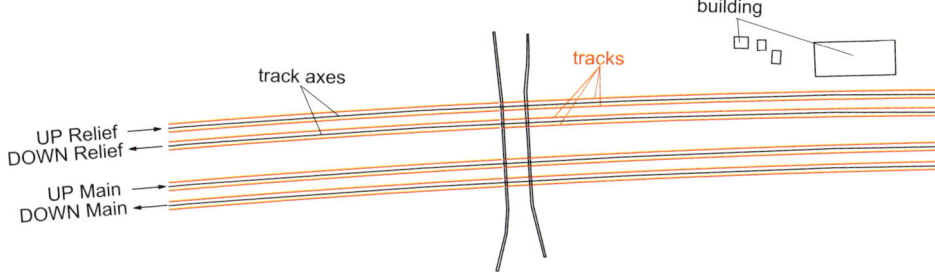

Figure 12.4: Track layout plan.

- all parties involved in *planning*, e. g. for future construction of buildings after electrification, conversion of tracks and extension of station platforms, should be consulted before the overhead contact line system is designed. This reduces subsequent alterations and planning and construction expenditures
- information regarding *tracks to be wired* together with loading gauges and specific route details and permitted out-of-gauge loads, if any, that form a further basis for the contact line layout plan
- *electrical sectioning plan* results in the arrangement of disconnectors, insulated subsections and section insulators
- information related to *traction power supply lines*, including bypass lines, parallel feeders, feeder lines and return conductor cables shown on the overhead system layout and longitudinal profile
- planning of the overhead contact line disconnectors for local or remote control requires information on the *control location*, the route and cable type. Endeavors should be made to coordinate cable laying
- an agreement between the infrastructure manager and the contractor on the *scope of the project* and its structure should avoid duplication of work and misunderstandings
- by using a *project master plan* project engineering, construction commencement and milestones can be controlled

A joint review of the documentation is carried out by planning engineers from both the infrastructure manager and the contractor. Agreement should be reached on the necessary provision of any missing documents.

12.2.3.2 Electrification of existing lines

The electrification of *existing lines* is often accompanied by some track layout modifications with overhead contact line planning for such sections carried out based on the documentation listed in Clause 12.2.3.1. Scaled layout drawings are required for line sections without track layout changes. These drawings need to show signal locations, cable positions, supply and drainage pipes, railway crossings, bridge overpasses and underpasses and information related to tracks to be equipped with overhead contact lines.

12.2.3.3 Modification of electrified lines

Conversions to overhead contact line systems are often preceded by track alterations similar to Clause 12.2.3.2. This work usually needs to be carried out in a number of stages. Staged construction conditions require additional information:
- track layout designed for each conversion stage
- as modifications in stations often require several *intermediate track stages* each of these requires track layouts and information related to the construction program
- an *inventory or revised plan* for the existing *overhead contact line* reveals the reference to the new layout. Invalid and outdated drawings may require that the overhead contact line system needs to be resurveyed

Meetings between parties involved in the project before and during planning assists co-ordination with other affected projects.

12.2.3.4 Construction and modification of contact lines on TEN lines

The design of infrastructure and contact line installations of lines of the trans-European railway system TEN is governed by the Interoperability Specifications (see Clause 2.1.4). Therefore, when planning contact line installations, the following requirements need to be observed:
- conventional lines < 250 km/h according to TSI ENE [12.5]
- high-speed lines ≥ 250 km/h according to TSI ENE [12.5]
- for national lines up to 200 km/h the specifications of the infrastructure managers

The classification of the line category, conventional lines, high-speed lines and national lines can be found in the infrastructure register of the operator, e. g. for German Railway DB in *infrastructure register* (ISR) [12.6]. According to directives 96/48/EC, 2004/250/EC and 2007/32/EC, each member state is obliged to establish an infrastructure register and to re-vise it annually. The regulation on the interoperability of the trans-European Railway system (TEIV) [12.7] implements the regulation into national legislation and commits the infrastruc-ture managers to publish an ISR with the most important parameters for the operation of their lines. For the contact line system at least the following data need to be included:
- voltage and frequency
- maximum train current
- for DC systems maximum current at standstill
- conditions for energy regeneration
- nominal contact wire height
- pantograph profile
- maximum line speed depending on the pantographs
- parameters of the contact line design

- minimum distance between adjacent pantographs
- maximum number of pantographs per train consist
- collector strip material
- arrangement of phase separation sections
- arrangement of system separation sections
- special cases and deviations from the TSI ENE

The following sections describe possibilities for the implementation of the interoperability specifications when planning overhead contact lines e. g. in view of selection of staggers and longitudinal spans.

12.2.3.5 Urban mass transit installations

Planning of contact lines for urban mass transit systems is carried out based on the the stipulations of the mass transit entity in charge and follow the relevant regional or national standards. In Germany, the document VDV 550 of the Association of German Mass Transit Operators (VDV) for overhead contact line systems establishes the design principles which implements the technical targets for mass transit of the European Standardisation Board CENELEC SC9XC 'Electric and electronic equipment of railways – Fixed installations' [12.8]. Further instructions for planning contact lines for mass transit installations are provided by the regulation on the construction and operation of tramways [12.9]. The contents and extent of documents for contact line planning is the same as for main lines.

12.2.3.6 Tracks and topography

Track layout and *topography* form the basis for overhead contact line planning with track layout shown in the layout plan. If an up-to-date layout plan is not available, then the track layout and terrain profile needs to be surveyed prior to commencing planning. The most common form of track and terrain surveying is the *terrestrial survey*, during which track layout and track profile are recorded with the aid of theodolites. Track layouts and transverse profiles at the pole locations are then created. However, *photogrammetric recordings* can survey tracks and railway profiles more rapidly. Stereo-infrared cameras record the line from a moving railway vehicle and digitalisation of the three-dimensional recordings is performed with the aid of a projector.

Aerial photos are suitable for simultaneous recording of track layout and transverse profiles. The flight with a stereo camera [12.10] and subsequent digitalisation provide three-dimensional documents from which longitudinal and transverse profiles can be produced. The accuracy depends on the experience of the analyser and vegetation on the ground but an accuracy of ±50 mm is achievable.

Terrain surveying with the aid of a *Global Positioning System* (*GPS*) is also an established method employed on new railway lines for surveying the track layout and the pole locations prior to track construction.

Correction programs [12.11] calculate the co-ordinates for the track layout and interesting terrain points based on the *world coordinate system WGS 84* using the recorded data. By conversion, co-ordinates based on the Gauss-Krüger co-ordinate system can achieve an accuracy better than ±10 mm. The planning documentation can be transferred into the information system used by the infrastructure manager, e. g. the DB-own coordinate system and associated data bank. The topography, together with track, structures and crossings, are also shown

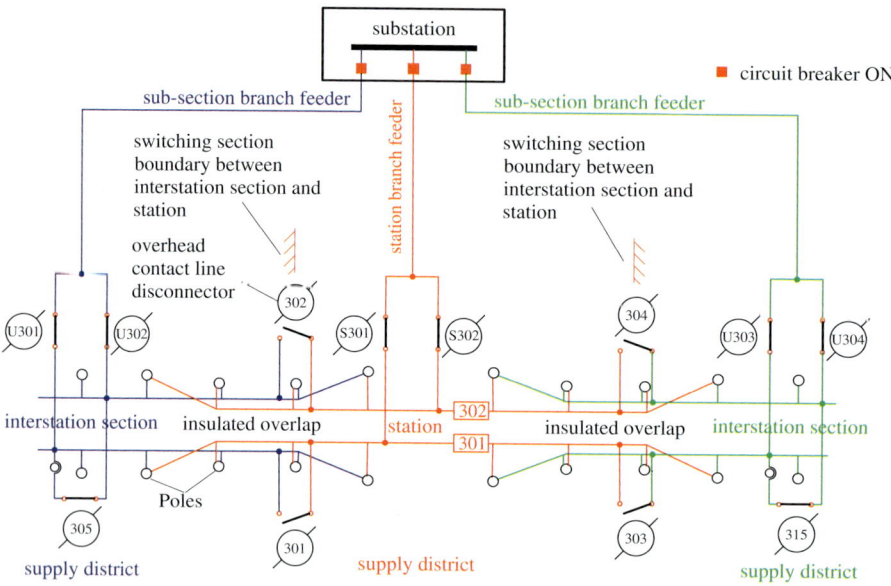

Figure 12.5: Supply districts (German Railway).

in the planning documents as these features affect the type and dimensions of supporting equipment, poles and foundations.

12.2.3.7   Circuit diagrams of overhead contact lines

The planning work also includes the production of the *overhead circuit diagram*, which is designed to suit the requirements of network and railway operations as well as protection and the overhead contact line design. A schematic track layout plan with signals, point connections and important buildings is the basis for this planning.

The substation supplies branch feeders with traction power using circuit breakers. In the case of double track lines, mostly three feeder branches are provided (see Figure 12.5). These feeders supply the station at the substation location and the adjacent interstation lines. Where a single-track line runs parallel to a double track line, all overhead contact lines are fed from a single branch feeder. When two double track lines run in parallel, each line is dealt with as a separate supply section. The supply sections are each subdivided longitudinally into switching sections. Overhead contact lines in stations and on the interstation sections form separate switching sections. Electrical boundaries between switching sections coincide with operational boundaries between interstation sections and stations (see Clause 5.4).

Switching sections of stations are subdivided into electrically separable *switching groups*. Main lines and secondary tracks each form separate switching groups. Sub-sectioning of switching groups on main tracks provides benefits for maintenance and the removal of disturbances in long stations.

The design of *electrical sectioning* of the overhead system should always be carried out in consultation with the relevant train operations department. Figure 12.5 shows an extract from a single-line diagram with switching instructions for a typical station. The normal disconnector position must be defined in this *overhead line switching diagram*.

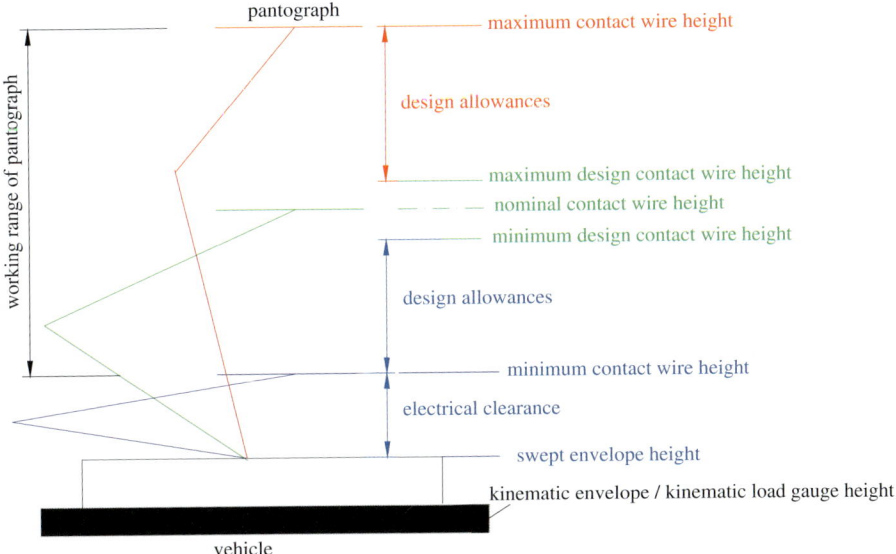

Figure 12.6: Definitions of contact wire heights in dependence on EN 50 119: 2009+A1: 2013.

Switching section boundaries are designed as insulating overlaps in through-tracks and are arranged so that no traction vehicles are forced to stop with a raised pantograph within the insulating overlap when the signal is in the stop position. Electrical disconnectors can connect the overhead lines at the overlaps. Switching sections and switching groups can be interconnected with disconnectors. Auxiliary loads are fed from the overhead line via disconnectors. Clause 5.4 lists the types of overhead line switching and the identification of disconnectors. When determining longitudinal sectioning of contact lines, switching section boundaries need to be protected by signals. Protection by a signal is achieved if no traction unit will stop in an insulated overlap with a raised pantograph in front of a signal that announces a stop. Disconnectors can connect the contact lines at supply and switching section boundaries. Longitudinal separations at stations should be carried out as switchable units. Then, the non-energized contact line sections can be limited in case of disturbances or maintenance. Supply and switching sections can be connected to each other by disconnectors (see also Clause 5.4. The *line supply diagrams* (Figure 12.3) are based on the overhead circuit diagrams (Figures 12.2 and 12.5). The *line supply diagram* shows the supply of one or more lines (Figure 12.3).

12.2.3.8   Placing of signals for electric traction

Signals for electric traction are required and their positions are in contact layout plans using symbols shown in Figures 17.47 and 17.48. Operator specifications form the basis for selection of these positions. An agreement with the operator is necessary (see Clause 17.5).

Table 12.2: Contact wire height and stagger for interoperable lines according to TSI ENE [12.5].

Description	≥ 250 km/h	< 250 km/h
nominal contact wire height mm	5 080 to 5 300	5 000 to 5 750
minimum design contact wire height contact wire height in mm	5 080	is obtained by adding: height of gauge G1 or G2, tolerance of track height, tolerance of contact wire height downwards, dynamic contact wire movement downwards, effects of ice loads and temperature variations on conductors.
maximum design contact wire height in mm	5 300	6 200 [1]
contact wire gradient	no gradient	see Table 12.3
useable contact wire lateral position in m	for 1 600 mm long pan head in straight line sections 0,40 for 1 950 mm long pan head in straight line sections 0,55	
maximum stagger at support in m	for 1 600 mm long pan head 0,30 for 1 950 mm long pan head 0,40	

[1] tolerances and uplift included the maximum contact wire height may not exceed 6 500 mm.

12.3   Technical basis

12.3.1   Selection of contact wire height

The kinematic vehicle gauge accounts for the vertical and horizontal movement of vehicles. Vehicles must not infringe this gauge even under dynamic impacts. By including the allowances to the kinematic vehicle gauge, the kinematic infrastructure gauge is obtained for the line in question. TSI ENE [12.5] refers to EN 50 367 and EN 50 367 refers to EN 15 273-1: 2013 to EN 15 273-3: 2013 regarding calculation of the kinematic pantograph gauge.

Therefore, the kinematic infrastructure gauge represents the interface between the space to be kept free for the unhindered passage of pantographs and the infrastructure. The infrastructure gauge GC (Figure 2.8) is specified for new high-speed railway lines. This gauge is also designated as the *mechanical pantograph gauge* and needs to be extended in its upper range by the electrical clearance. From this gauge, the *electrical pantograph gauge* can be determined and needs to be calculated as specified in TSI ENE [12.5] (see also Clauses 2.1.4, 4.4 and [12.12, 12.13]). The electrical gauge can be touched by earthed components, however, only the contact wire, steady arm, drop bracket of steady arm and components at the same voltage may approach the mechanical structure gauge. A distinction should be made between:

  – *minimum contact wire height*: According to Figure 12.6 the minimum value of contact wire height between top of rail (ToR) and lower edge of the contact wire results from the upper edge of the kinematic vehicle gauge and the lower edge of the contact wire [12.14]. This height enables a flawless interaction between pantograph and contact line. The contact wire height is defined by the distance between the connecting line of top of rails or, in the case of trolley bus lines, of road surface and the lower edge of the contact wire

  – *nominal contact wire height*: nominal value of the contact wire height at a support in still air

  – *minimum design contact wire height*: theoretical contact wire height including tolerances, designed to ensure that the minimum contact wire height is always achieved,

Table 12.3: Contact wire gradients according to EN 50 119: 2009+A1: 2013.

Commercial speed up to km/h	maximum gradient	‰	maximum gradient change	‰
50	1/40	25	1/40	25
60	1/50	20	1/100	10
100	1/167	6	1/333	3
120	1/250	4	1/500	2
160	1/300	3,3	1/600	1,7
200	1/500	2	1/1 000	1
250	1/1 000	1	1/2 000	0,5
> 250	0	0	0	0

after the following has been considered:
  – tolerances of track level, so far as not included in the kinematic gauge
  – tolerances of contact wire height downwards
  – downwards directed contact wire movements and
  – effects of ice loads and temperature changes on the conductor
– *maximum design contact wire height*: theoretical contact wire height including tolerances, movements etc., which may not result in exceeding the maximum contact wire height, including:
  – tolerances in the track level so far as not included in the kinematic gauge
  – uplift of the contact wire because of a passing pantograph
  – upwards directed movements of the contact wire when a pantograph passes
  – tolerances of contact wire height directed upwards
  – lift of contact wire due to wear
  – uplift of contact wire caused by temperature variations
– *maximum contact wire height*: maximum value of contact wire height which may not be exceeded in any possible case during operation and which the pantograph must reach
– *contact wire uplift*: vertical movement of the contact wire caused by the pantograph
– *related longitudinal gradient of the contact wire*: relation of the height difference of the contact wire above the track or road surface in case of trolley buses at two adjacent supports of the span

The contact wire should be installed at a uniform height and parallel to the track because variations in the contact wire height increase the contact wire wear. On high-speed lines > 250 km/h , a uniform contact wire height is required to avoid excessive wear (Table 12.3). The contact wire height, the gradient of the contact wire in relation to the track and the lateral displacement of the contact wire under the action of a cross-wind all govern the compatibility of the trans-European rail network (Table 12.2).

The vertical movement of the contact point between pantograph and contact wire within a span should be low and uniform, however, it is not limited according to TSI ENE [12.5]. The design engineer is able to control the contact wire uplift at the support by selecting the stagger and the resultant radial force. The difference in the radial forces at adjacent supports should be low, especially in the case of high-speed lines where the lateral movement of the contact wire relative to the pantograph should be at least 1,5 mm/m to avoid grooves on the collector strips (Clause 12.4). The terms concerning the contact wire height are presented in Figure 12.6.

Table 12.4: Specifications for contact wire height at DB, SNCF, ADIF and FS for new high-speed lines in mm.

Description	DB[1]	DB[2]	SNCF[1]		ADIF[1]	FS[3]
type of vehicle reference gauge G2	kin.	kin.	stat.	kin.	kin.	kin.
height of reference gauge	4 700[3]	4 680[2]	4 650[3]	4 700[3]	4 310	4 700[4]
extension in curves	107[5]	0	–	–	107	124
track gradient change	25[6]	0[6]	30	30	25	25
reserve for lifting at track maintenance	50[6]	0[6]	50	50	50	50
allowance for oscillations	18	0	0	0	408	1
Gauge height	4 900	4 680	4 730	4 780	4 900	4 900
Clearance	220[7]	150	270	170	270	220
Minimum contact wire height	5 120	4 830	5 000	4 950	5 170	5 120
tolerance for contact wire height	30	30		0	30	30
downwards movement of contact wire	50	0[8]		40	50	30
sag due to ice at maximum span	100[9]	100[9]		0	50	0
pre-sag of contact wire	0	0	30	30	0	30
allowance for air pollution	0	0	50	50	0	50
Sum	5 300	4 960	5 080	5 070	5 300	5260
minimum contact wire height	5 300	5 000	5 080		5 300	5 300
references	[12.15]	[12.14]	[12.16, 12.17]		[12.18]	[12.19]

kin. kinematic vehicle gauge
stat. static vehicle gauge
[1] acc. to UIC meeting, Munich, Juni,16 1994
[2] topical calculation of contact wire height acc. to [12.14]
[3] acc. to UIC meeting, Rome, September, 19 1995
[4] acc. to Figure 2.8
[5] acc. to [12.15]
[6] for ballasted tracks only; in case of slab track no allowances for track lifting and for gradient change is considered [12.15]
[7] for 25 kV acc. to [12.15], the former DIN VDE 0115:1982, Table 6
[8] the movement of the contact wire is considered as a short-term approach
[9] ice load acc. to $2,5 + 0,05 \cdot d$ in N/m, maximum span 65 m, contact wire tensile force 15 kN

On interoperable lines with speeds $\geq$ 250 km/h, the contact wire height can be designed between 5,08 m and 5,30 m and on lines with speeds < 250 km/h between 5,00 m and 5,50 m (Table 12.4).

Table 12.4 illustrates the nominal contact wire heights used in Germany at DB, in France at SNCF, in Spain at ADIF and in Italy at FS for high-speed lines. According to [12.14] and if no level crossings existed (see Clause 12.4.6), 5,0 m contact wire height would be continuously possible in Germany.

12.3.2 Selection of contact wire lateral position

12.3.2.1 Contact wire lateral position and radial forces

The length of pantograph types used on a certain line results in the kinematic-mechanical pantograph gauge L_h and the useable contact wire lateral position e_{use} (Figure 4.19), (see also Clause 4.9.11). The larger the sway of the pantograph due to over-compensation D_{ko} or under-compensation D_{ku} (Figure 12.7) the smaller the useable contact wire lateral position will be.

a) Over-compensated centrifugal force   b) Under-compensated centrifugal force

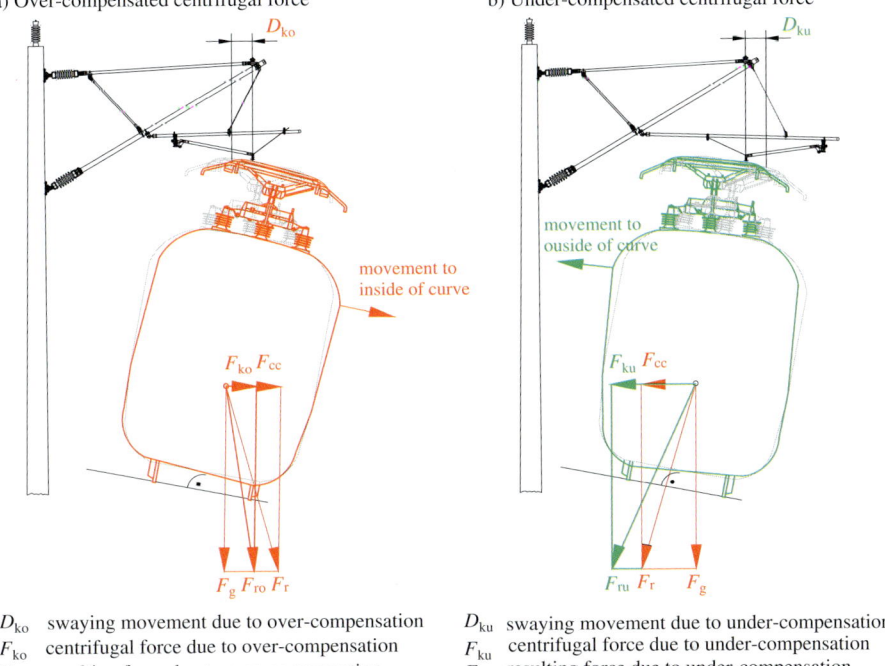

D_{ko} swaying movement due to over-compensation D_{ku} swaying movement due to under-compensation
F_{ko} centrifugal force due to over-compensation F_{ku} centrifugal force due to under-compensation
F_{ro} resulting force due to over-compensation F_{ru} resulting force due to under-compensation
F_g gravitational force F_{cc} compensated centrifugal force
F_r resulting force vertical to perpendicular to ToR

Figure 12.7: Displacement of the contact point D_k due to transverse movement of the vehicle and of the pantograph in relation to the contact wire (contact wire position in case of wind action and swaying of the vehicle (see Figure 12.10).

The maximum contact wire lateral position e_{max} is obtained by deducting the provisional value P from the useable contact wire lateral position e_{use}:

- under wind action the maximum contact wire lateral position e_{max}
- with the provisional value P the useable contact wire lateral position e_{use}
- with swaying movement D the working length l_A of the pantograph head

When designing the overhead contact line, the stagger b at the support shall comply with the following stipulations:

- utilising the maximum contact wire lateral position e_{max}(Clause 4.3.3)
- complying with the useable contact wire lateral position e_{use} (Clause 4.9.11)
- sweep of contact wire lateral position $\Delta b_m \geq 1,5$ mm/m (Clause 4.10.1, Table 4.14)
- radial force at support $80\,\mathrm{N} \geq F_R \leq 2\,000\,\mathrm{N}$ (Clause 4.10.1)
- difference of radial forces at adjacent supports as low as possible (Clause 4.1.5)
- spans as long as possible (Clause 4.8)

To wear the collector strips of the pantographs uniformly and guarantee continuous contact in curves and under wind action, the contact wire – as viewed in the plane of contact – is not installed parallel to the track axis but with an *alternating lateral displacement*, also called *zig-zag*, above the track (Figure 12.8 a), b) and c)).

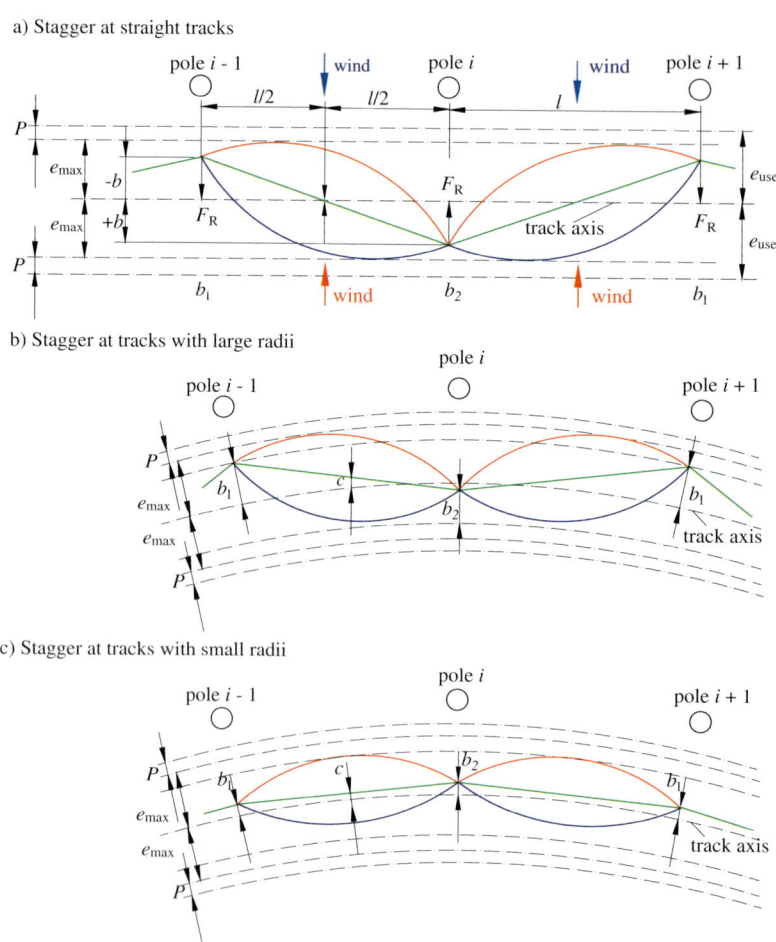

a) Stagger at straight tracks

b) Stagger at tracks with large radii

c) Stagger at tracks with small radii

Figure 12.8: Contact wire lateral position on straight tracks and curves.
l span length, e_{use} usable contact wire lateral position, b stagger at
support, P provisional value, e_{max} maximum contact wire lateral position, F_R radial force

The contact wire needs to be fixed at the supports such that the maximum contact wire lateral position e_{max} will not be exceeded within the span. To avoid grooving on collector strips, the contact wire should sweep laterally by at least 1,5 mm/m. In Figures 12.9 a) and b) a blue band marks the amount of lateral movement of the contact wire. Experience demonstrates that grooves on collector strips will not be formed if Δb_m is more than 1,5 mm/m on a spans less than 40 m. The calculation of the useable contact wire lateral position according to Clause 4.9.11 [12.5] taking into account the provisional value P results in a permissible maximum contact wire lateral position, which depends on the contact wire height (Figure 12.10). A limitation of the contact wire lateral position at the supports e. g. to $\pm 0,4$ m, $\pm 0,3$ m or $\pm 0,2$ m is not necessary. Especially, in long transition curves, staggers of more than e. g. 0,3 m but less than the useable contact wire lateral position enable longer spans.

In curves with large radii, alternating lateral positions can be used to advantage (Figure 12.8 a) and b)), in curves with small radii the contact wire at the supports is always on the outside of

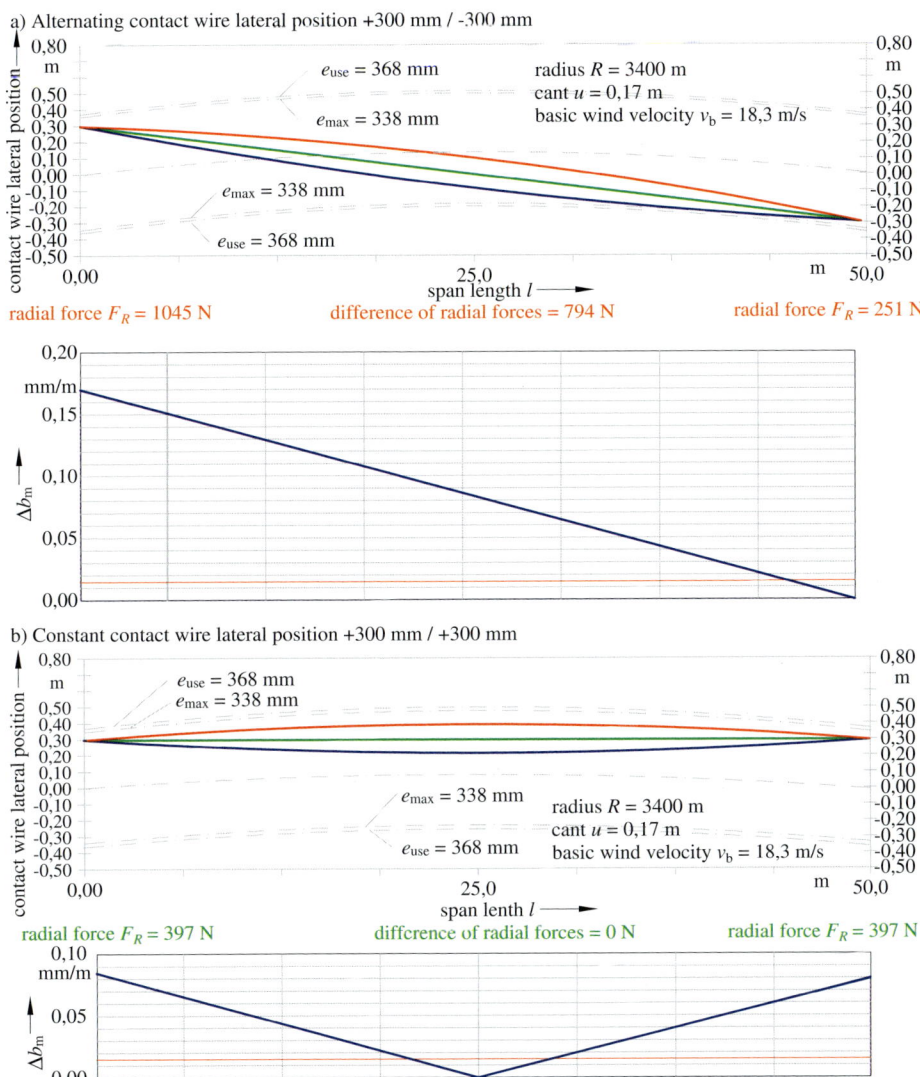

Figure 12.9: Difference in radial forces F_R as a result of contact wire lateral position at supports and of calculation of usable contact wire lateral position according to example 4.29 for the overhead contact line system Re300 with 27 kN contact wire tensile force, 21 kN catenary wire tensile force, 5,3 m contact wire height.

a) The utilization of usable contact wire lateral position e_{use} leads to higher radial force differences at the supports, but leads to more even wear of the pantograph carbon strips and a large change in contact wire lateral displacement Δb_m.

b) Even contact wire lateral position at supports b leads to even radial contact wire forces at the supports, but leads to slightly uneven wear of carbon strips because of only a small change in contact wire lateral position Δb_m within the span.

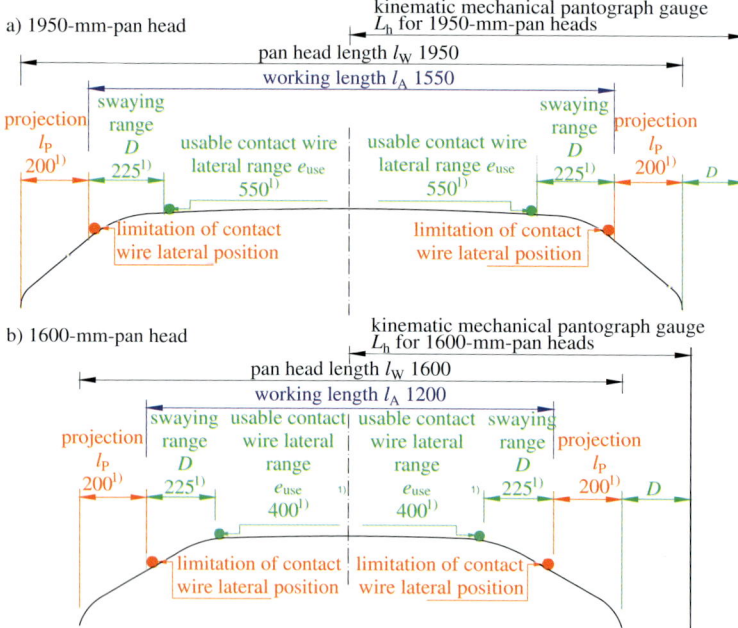

Figure 12.10: Contact wire lateral position at TSI - pan-heads.

the curve under consideration of possible stipulations for the contact wire variation transverse to the track Δb_m (Figure 12.8 c)).

By selecting the contact wire lateral position, the design engineer affects

- the contact wire radial force at the support
- the interaction between contact line and pantograph
- the contact wire wear
- the wear of the collector strips

The contact wire *radial force at the support* will be increased by reducing the span or increasing the stagger. Therefore, it is important to check the radial force as it may be possible to exceed the permissible range $80\,\mathrm{N} \leq F_R \leq 2\,000\,\mathrm{N}$ in tight curves under wind action. Falling below 80 N will exacerbate the wear of the steady arm joint, which could result in a failure and a contact line fault (see also Clause 17.2.3. The *difference between the radial forces* at supports bordering the span affect the interaction between contact line and pantograph. That difference should be as low as possible. High radial forces at the supports create inelastic spots whereas low forces result in elastic spots in the contact wire with the effect of low or high contact wire uplifts, respectively. Undesirable height variations of the pantograph and irregular contact wire wear will be the result (Figure 12.11). For a 50 m span in a 3 400 m radius curve and 0,17 m cant, the difference of radial forces is 794 N at 27 kN contact wire tensile force for the design according to Figure 12.9 a) while equal radial forces are obtained for the design according to Figure 12.9 b). By selecting the contact wire stagger, the design engineer also effects the *wear of collector strips*. Grooving and break-outs of the collector strip result in premature replacement of the collector strips (Figure 12.12) and contact line damage [12.20, 12.21].

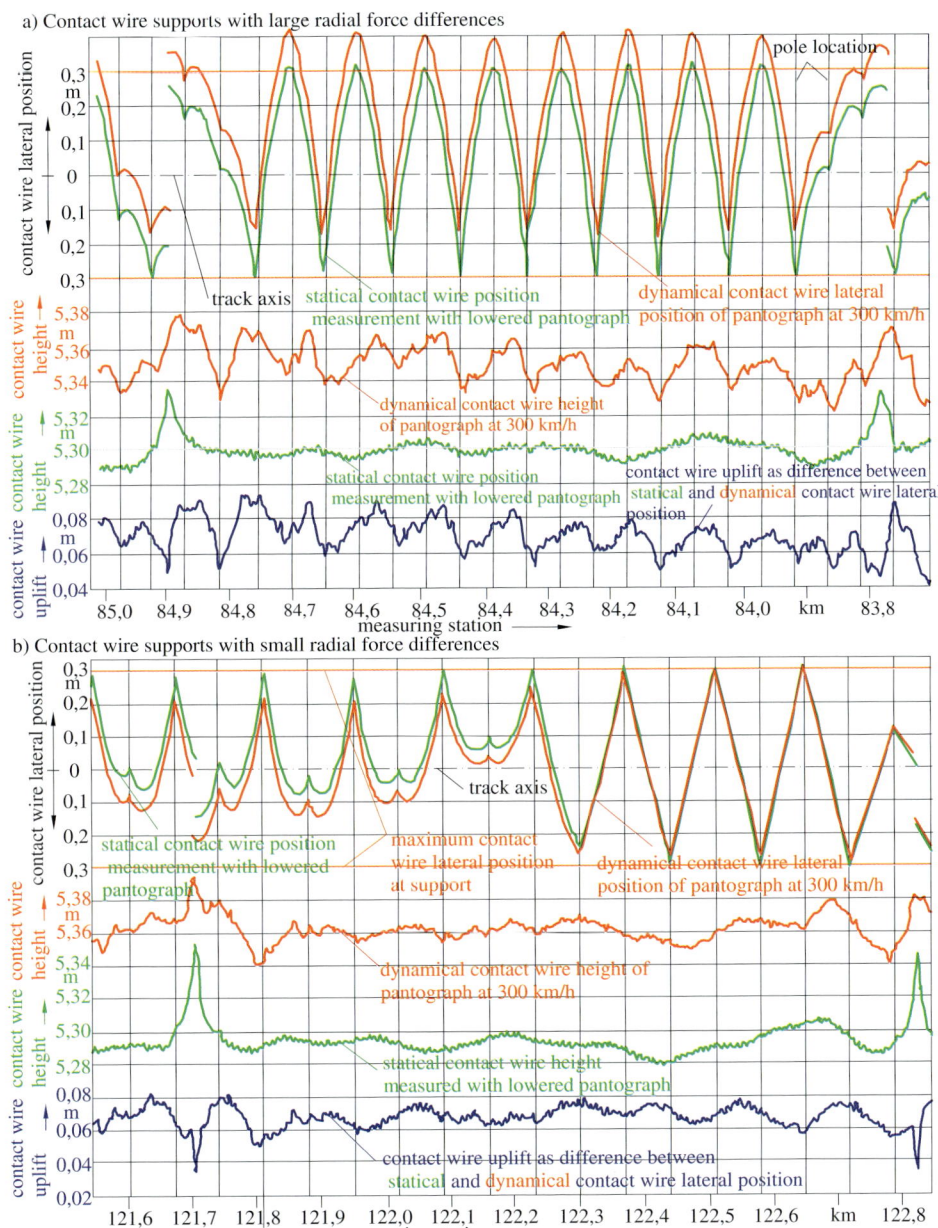

Figure 12.11: Measurement record for the interaction of an overhead contact line and a pantograph for a German high-speed line.

Static contact wire lateral position: measured at 300 km/h, not including the sway movements of test train.
Dynamic contact wire lateral position: measured at 300 km/h, including the sway movements of test train.

Figure 12.12: Roping and blow-outs at carbon strips.

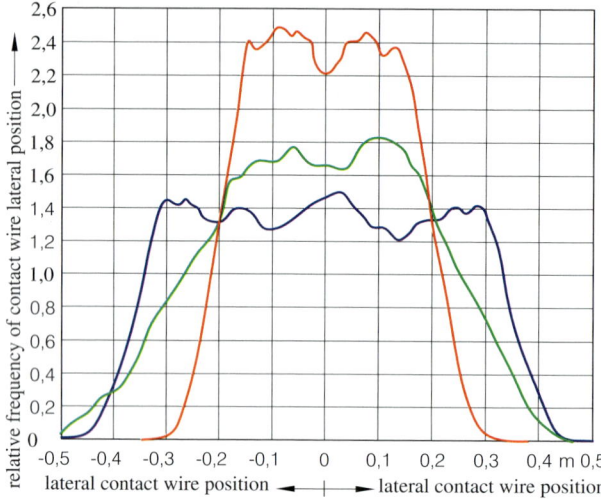

Figure 12.13: Relative frequency of contact wire lateral position at pan-head for predominant tangent high-speed lines for 300 km/h in Germany, a high-speed line with radii for 300 km/h in Germany and for the high-speed line Madrid–Lerida for 350 km/h with only 0,2 m contact wire lateral position.

Figure 12.13 shows the distribution of the contact wire lateral positions for two high-speed lines in Germany and the Madrid–Lèrida line in Spain. The related frequency of contact wire lateral position demonstrates the consequences of a predominately straight line (blue), a line with many curves (red) and the stagger of 0,2 m on the Madrid–Lèrida line (green). The data were gathered with DB's contact force measuring system (see Clause 10.4.3.5). The exceedings of the statically useable lateral positions are a consequence of sway at high speeds, at which the lateral displacements were evaluated.

The radial forces also serve as a basis for checking the torsional load on poles, refer to Clause 13.5.3, which can be created by two cantilevers at the same pole.

The radial force at the contact wire support can be determined as shown in Clause 4.1.5.

In mass transit systems, a varying contact wire lateral position aims at an equal wear of the collector strips. The same principles apply as for main line railways. Because of the transmission of higher currents the managers of mass transit systems stipulate higher requirements for the minimum lateral contact wire sweep Δb_m than for main line railways (see [12.8] and Clause 4.14).

An alternating contact wire lateral position to achieve a uniform wear of the collector strips is also adopted in commuter railway systems. Therefore, the same planning principles apply to these railways as in the case of long distance lines. The operators of such lines require higher values for the transverse sweep of the contact wire Δb_m (see [12.8] and Table 4.14) as compared with long distance lines.

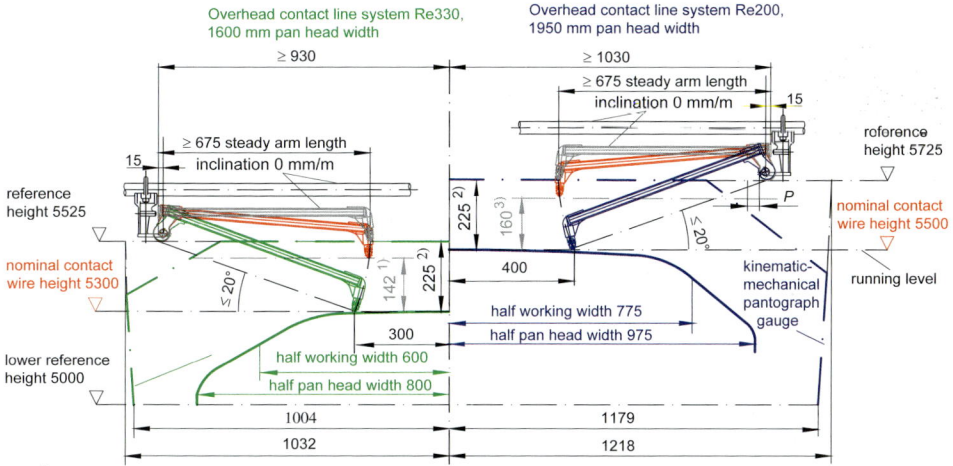

1) 82 mm maximum contact wire uplift at support according to technical dossier of interoperability constituent Re330 with an aditional 60 mm for the inclination of pan head and carbon strip wear according to EN 50206

2) 225 mm design clearance for steady uplift according to Deutsche Bahn drawing Ebs02.05.45 page 2 and page 3

3) 100 mm maximum contact wire uplift at support according to technical dossier of interoperability constituent Re200 and an additional 60 mm for the inclination of pan head and carbon strip wear according to EN 50206

contact wire stagger mm	steady arm length mm	installation site	pan-head length mm
400	675	cantilever and head span on straight tracks	1 950
300	675	cantilever and head span on straight tracks	1 600

Figure 12.14: Steady arm length dependent on the contact wire stagger and the mechanical pantograph gauge according to TSI ENE:2014 [12.5] (for parameters used for calculation of pantograph gauge see Table 4.8)

It can be advantageous to start with the selection of contact wire staggers in the curves since the staggers are determined by the track radii. In transition curves, a change of stagger from the inside of the curve to the outside of the curve may be required.

Past practice used the dimension c (Figure 12.8) as a check of the correct contact wire lateral position at the supports during construction activities. The distance between the contact wire in still air and the perpendicular to the line touching the rail heads at midspan is called dimension c. If the measured dimension c were less than the value listed in the table, the maximum contact wire lateral position would not be exceeded, under even wind action as proven.

Distortion of the contact line layout plans in the transverse direction, e. g. by a distortion scale of 10, simplifies checking the contact wire lateral position in curves, transition curves, overlaps and above points.

Table 4.15 and Clause 2.4.1 show staggers at supports for conventional and high-speed lines as stipulated by some selected railway line managers.

12.3.2.2   Contact wire stagger and length of steady arms

The contact wire lateral position at the support b and the length of the pan-head l_W determine the *steady arm length* . The *drop bracket of the contact wire steady arm* should stay outside the mechanical pantograph gauge and at least the provisional value P from the pantograph gauge (Figure 12.14). Small staggers result in long steady arms and larger staggers in short

a) Tight Z-type anchor rope with loose dropper

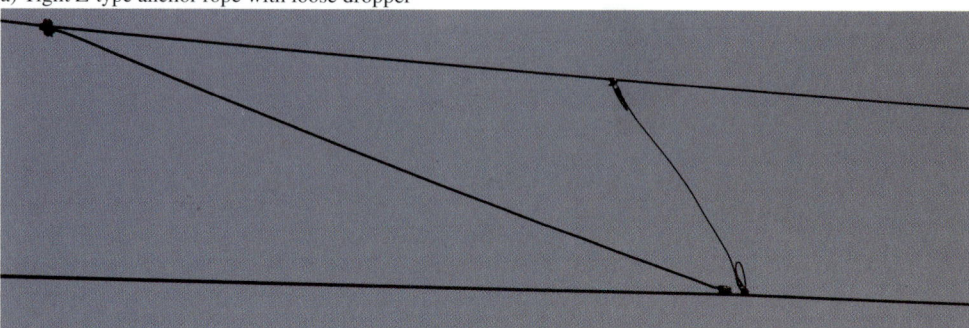

b) Loose Z-type anchor with tight dropper

Figure 12.15: Z-type anchor ropes due to the disturbed equilibrium of forces at the midpoint.

ones. In the case of dual operation with differently long pan-heads, the short governs the stagger and the long one the pantograph gauge and, therefore, the distance between track axis and drop brackets which lead to the *steady arm length*. The inclination of the steady arm being 20° at maximum relative to the plane of the contact, may result in a longer steady arm than necessary for the drop bracket being positioned completely outside the pantograph gauge (Figure 12.14 for Re330 and 1 600 mm pan-head).

Light weight steady arms need to be loaded to at least 80 N radial force, to minimise *wear at the steady arm joint*, in case of contact wire oscillations. The maximum radial force of the steady arm may not exceed 2 000 N. Compression loading of the steady arm is not permitted. No special provisions are required for passing of two high-speed trains. Wind stays might be required to stabilise the contact wire supports of contact lines with relatively low tensile forces. Such wind stays are only necessary on straight line sections and in curves with large radii. There are no transverse loads because of wind action in tunnels, therefore, no wind stays are required there.

Clause 4.1.5 contains the standard contact wire lateral positions at the supports for conventional and high-speed lines of some selected infrastructure managers.

12.3.3   Height and lateral position of catenary wire

Three alternatives are available for the lateral position of the catenary wire [12.22]:
  – arranged vertically above contact wire with the same lateral position
  – arranged vertically above track axis in case of alternating stagger of contact wire
  – inclined arrangement of catenary wire relative to contact wire such that catenary and
    contact wire assume opposite positions at supports

Table 12.5: Span length of SNCF contact lines depending on the curve radius.

Track radius in m	Span length in m
$200 \geq R \leq 300$	27,0
$300 \geq R \leq 400$	31,5
$400 \geq R \leq 500$	36,0
$500 \geq R \leq 650$	40,5
$650 \geq R \leq 850$	45,0
$850 \geq R \leq 1\,050$	49,5
$1\,050 \geq R \leq 1\,350$	54,0
$1\,350 \geq R \leq 1\,800$	58,5
$1\,800 \geq R$	63,0

The catenary wires of the contact line types Re100, Re200 and Sicat S1.0 are arranged vertically above the track axis on straight line sections, such that a semi-inclined contact line is obtained. In curves with radii less than 1 200 m, it is arranged vertically above the contact wire. The catenary wire is arranged on straight and curved line sections vertically above the contact wire in case of contact line types Re250, Re330 and Sicat H1.0.

Table 4.15 presents an overview of the catenary wire arrangement at some European railway operators. The arrangement vertically above the contact wire should be given priority. With the same lateral displacement of contact and catenary wire the same resetting forces result in the contact and catenary wire that reduce the tensile force between tensioning device and midpoint. In the case of differing displacements, the resetting force of contact and catenary wire also differs. Differing tensile forces within the contact and catenary wire, as a consequence of temperature-caused changes in length, the difference of which reaches its maximum at the midpoint causes differing movements of contact and catenary wires loading the anchor rope at the midpoint. The contact wire is lifted at the anchor rope (see Figure 12.15 a)) or will assume a sag there (see Figure 12.15 b)). Both effects will affect the interaction between contact line and pantograph.

The catenary wire supports the contact wire by droppers. The contact wire height, the shortest length of a flexible dropper at midspan and the sag of the catenary wire determine the height of the catenary wire at the supports and, therefore, the system height, which represents the distance between contact and catenary wire at the supports. Railway infrastructure managers adopt different system heights and, therefore, different catenary wire heights (Table 4.15). A shorter system height requires a dropper material which is sufficiently flexible i. e. resistant to alternating bending strain (Clause 11.2.2.7).

12.3.4   Span length

When *defining pole locations*, for reasons of economy, the maximum permissible span lengths should be exploited as far as possible (see Clause 4.9 and 4.11). Calculation of span lengths is described in Clause 4.9. The span length depends on:
  – the length of the working range of the pantograph head
  – the design wind velocity v_h
  – the tensile force of contact and catenary wire
  – the track radius
  – the required contact wire lateral movement Δb_m

Figure 4.44 illustrates the relationship between pantograph length and span length as specified by various railway entities. The Portuguese Railway (CP) and French Railway (SNCF)

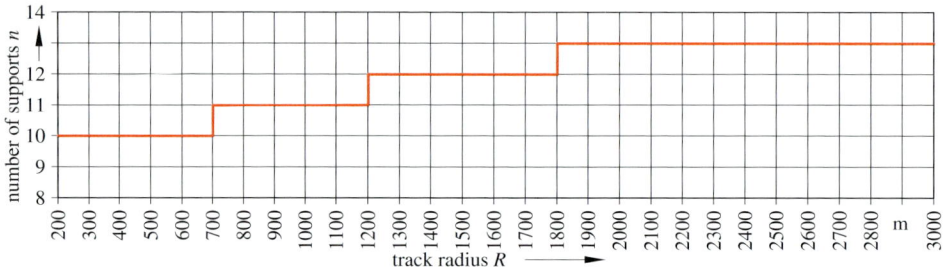

Figure 12.16: Relationship of the number of supports n in a half tensioning length for a track radius R for the DB contact line type Re200.

(Table 12.5) do not specify the span length as a function of the design wind velocity v_z. Their designs are based on track radius with consequently shorter span lengths. The calculation of span lengths of DB's contact line designs Re100 and Re200 follows the approach described in Clause 4.9. The contact line types Re250 and Re330 are adopted for high-speed lines on sections with radii above 3 000 m with spans up to 65 m. The *Siemens* type Sicat H1.0 employs spans up to 70 m. Due to dynamic performance, spans not more than 70 m are recommended. Shorter span lengths occur in crossovers, below structures and in overlaps.

12.3.5 Tensioning section length

In EN 50 119: 2009+A1: 2013, *the tensioning section length* is defined as the distance between the tensioning devices at the ends of the tensioning section. A contact wire or a catenary is strung starting at the midpoint, arranged approximately in the middle of the tensioning section in both directions. Therefore, the section between tensioning device and midpoint is designated as a half tensioning length L_N.

The lengths of the tensioning sections and the overlapping sections are relevant to the investment required. In the case of longer tensioning sections, the proportion of the overlaps is less important and investment decreases. Therefore, planning aims at long tensioning sections and short overlaps.

The length L_N depends on:
- the working range of tensioning devices with
    - travel of the weight stake
    - space for rope windings on the tensioning wheel
- variation of tensile forces if the tensioned conductors due to resetting forces and, therefore, on the number of spans in half a tensioning section
- the stagger and the distance between pole and track axis
- the operational tensile force depending on the conductor materials
- variation of the contact wire lateral position at the supports caused by thermal expansion of conductors, affecting the cantilever length
- the radii of track in curves
- the specified or expected wind velocities
- the temperature range of the contact line

The overhead contact line including steady arms and cantilevers moves in the direction of the tensioning device with increased temperatures. A component of the longitudinal contact

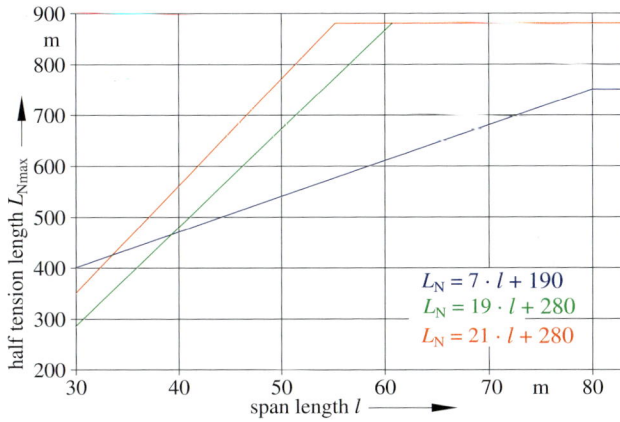

$$L_N = 7 \cdot l + 190$$
$$L_N = 19 \cdot l + 280$$
$$L_N = 21 \cdot l + 280$$

Figure 12.17: Relation between maximum tensioning section length and length of spans at DB [12.24, 12.25].

wire force acts through the cantilever in the direction of the pole because of swiveling of the cantilevers (Clause 4.1.5.6). This force is also known as the *resetting force*. The tensile force differences in individual spans are added to each other in curves and lead to the greatest differences between specified and actual value in the span at the midpoint anchor. To limit the tensile force differences near the midpoint anchor, the tensioning sections on curved tracks are reduced as the curve radii decrease. In total, a horizontal force reduction in each of the half tensioning sections of approximately 11 % can be permitted in the catenary and contact wires, which distributes itself as 8 % on the overhead contact line and 3 % on the tensioning device [12.23]. The maximum number of supports n depends on the track radius R and can be determined for a standard contact line Re200, with $H_{CA} = 10\,\text{kN}$ and $H_{CW} = 10\,\text{kN}$, wind velocity 26 m/s and a cantilever lengths of 2,5 m and 3,5 m in accordance with Figure 12.16. The half tensioning section length in a curve is determined from the span length as a function of the track radius and the wind velocity. The half tensioning section length L_N, therefore, depends on the achievable span length l as illustrated in Figure 12.17.
The relationships (12.1) and (12.2) can be used to determine the permissible half tensioning section length L_N depending on the cantilever length l_{Al}:

$$\text{Cantilever length } l_{Al} \text{ upto } 2,5\,\text{m}: \quad L_N = 19 \cdot l - 280 \quad \text{and for the} \tag{12.1}$$

$$\text{Cantilever length } l_{Al} \; 2,5\,\text{m to } 3,5\,\text{m}: \quad L_N = 21 \cdot l - 280 \quad, \tag{12.2}$$

where l is the span length. Equations (12.1) and (12.2) are valid for overhead contact lines with rated forces of 10 kN in the catenary wire and 10 kN in contact wire, a basic wind pressure of 390 N/m² as shown in Figure 12.17 as green and red lines, respectively. The blue line applies to the DB contact line type Re200 with 10 kN tensile force in the catenary and contact wire, 26 m/s design wind speed and 0,4 m stagger at the support, with the half tensioning section length limited to 750 m. The half tensioning section length for the contact line type Sicat S1.0 is 880 m [12.26].
Half tensioning section lengths longer than 1 000 m cannot be achieved for straight line sections in the open. In tunnels, however, half tensioning section lengths of more than 1 000 m can be achieved because of the lower ambient temperature variations.

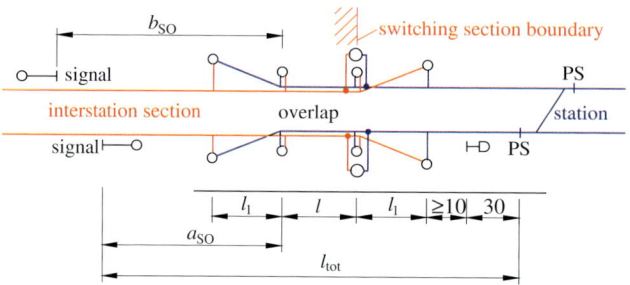

Figure 12.18: Distance between insulating overlaps and signals and points.
PS: Beginning of points
a_{SO}: in running direction
b_{SO}: in against running direction

12.3.6 Overlapping sections

12.3.6.1 Arrangement

Overlapping sections are arranged at the end of tensioning sections mechanically separating the adjacent tensioning sections. In the case of insulating overlaps, also called insulating sections, both electrical and mechanical separation of the adjacent sections is possible. These separations can also form the limits of switching sections (Clause 5.4.2). The insulated *overlapping sections* between station and the interstation sections need to be designed so that they are protected by a signal, i. e. a traction vehicle stopping at a stop signal with its pantograph raised may not be located within the insulated overlapping spans between the entry signal and the entry point. The distance a_{SO} according to Figure 12.18 defines the positioning of the *overlapping spans* relative to the signal and represents the minimum distance a required between the signal and the first pole of the overlap with two cantilevers. Experience shows that the traction current of starting trains decays within the distance a_{SO} to such a value that no arcs will occur between the contact wires because of potential differences and cause burnouts [12.27]. The distances a_{SO} and b_{SO} in Figure 12.18 depend upon:
  – between station and interstation line as standard design
    – disconnector normally open $a_{SO} \geq 100\,\text{m}$
    – disconnector normally closed $a_{SO} \geq 50\,\text{m}$
  – for lines with locomotive hauled trains and electric motor unit train sets
    – for regional railways $a_{SO}, b_{SO} \geq 200\,\text{m}$
    – for high-speed railways $a_{SO}, b_{SO} \geq 500\,\text{m}$
  – lines equipped with high-performance blocks $b_{SO} \geq 410\,\text{m}$
For standard lines operating at speeds up to 250 km/h, the distance l_{tot} between the signal and start of the first point of the station is determined as shown in Figure 12.18. Consequently the distance l_{tot} between the signal and point start should be at least 275 m for contact line type Sicat S1.0 with $l_1 = 70\,\text{m}$ and $l = 65\,\text{m}$ for a three-span insulating overlap. This ensures that an approaching traction vehicle with a raised pantograph has already reached an adequate speed when it passes the overlapping section, guaranteeing that spot heating of the contact wire due to current flowing between the switching sections via the pantograph does not lead to a contact wire burnout. Details on overlaps are described in Clause 3.4.8.5. Overhead line poles should be located at least 10 m from signals [12.28] Annex 2, Clause 2, Paragraph 12.

12.3.6.2 Contact wire lateral position and height

By symmetrically arranging the contact wires in overlaps, it is possible to avoid oscillations of the pantograph at the transition from one contact wire to the other. Figure 12.19 depicts

a) Uplifted contact wire within the maximum contact wire lateral position e_{max}

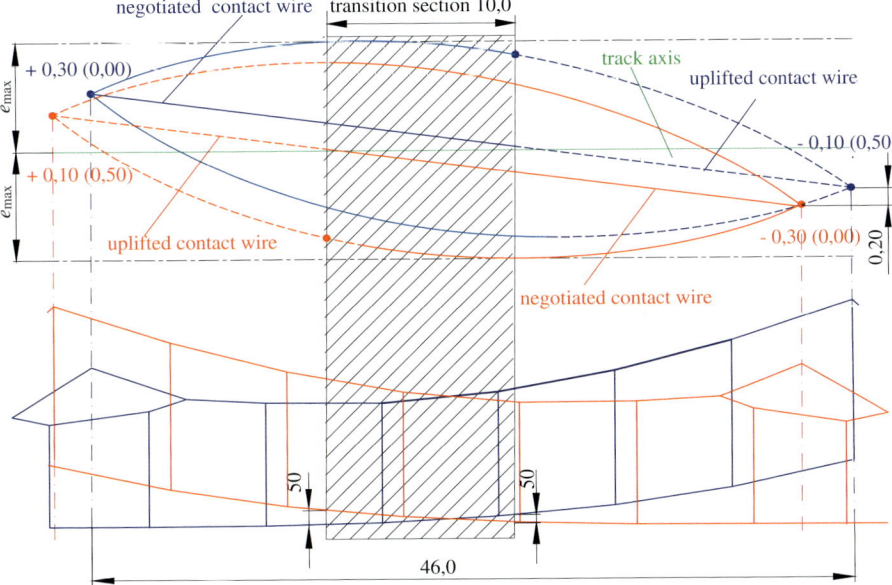

b) Uplifted contact wire outside the maximum contact wire lateral position

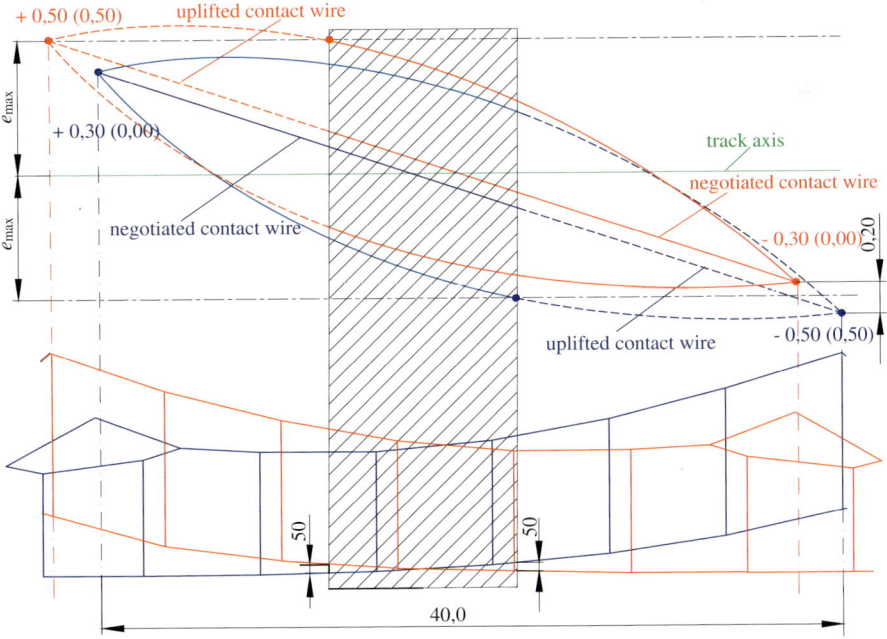

Figure 12.19: Arrangement of contact wires in overlaps for overhead contact line system Re200 with 0,2 m contact line spacing, 0.03 m provisional value, 1 600 mm pantograph length, wind zone W2 at 25 m/s, calculation of wind deflection according to TSI ENE and straight track. All dimensions in m.

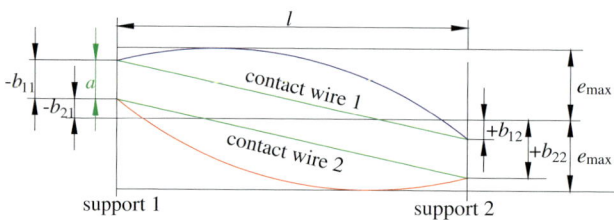

Figure 12.20: Contact wire lateral position for overlap sections.

two possibilities whereby in Figure 12.19 a) the lifted contact wire is placed between the track axis and the traversed contact wire. In Figure 12.19 b) the *lifted support* is placed on the opposite side of the traversed contact wire. In still air without wind action, a symmetrical arrangement exists in both cases. While one contact wire runs from above down to the contact wire nominal height, the other leaves the nominal height and rises out of running. For a short period both contact wires contact the collector strips of the pantograph. In the case of wind action, at least one contact wire needs to touch the collector strips. The contact wire transition range is hatched in Figure 12.19. The transition starts at a minimum contact wire uplift of 50 mm above the contact wire nominal height. Pantograph uplifts of less than 50 mm above nominal contact wire height at midspan need not be considered.

Longer spans other than as shown in Figure 12.19) can be designed according to Figure 12.19 a). The advantage is that it supports the lifting of the not negotiated contact wire in overlaps on interoperable lines operated with 1 600 mm long pantographs.

For contact line designs Re100 to Re330, Sicat S1.0 and Sicat H1.0 the crossing of the contact wires may be arranged before or after the transition span relative to the main running direction. In older contact line designs, the crossing was arranged before the transition span as viewed in the running direction.

12.3.6.3 Planning of overlaps

The distance between the parallel contact lines is fixed in overlaps, depending on whether the type of overlap is an *insulating overlap* or *non-insulating overlap* (Figure 12.20). Considering the maximum contact wire lateral position e_{max}, the design wind speed v_h and the contact wire height, span length and staggers at the supports of the overlap can be determined (see also Clauses 4.10 and 4.11). The arrangement of the contact wires in an overlap can be mirrored at the track axis, ensuring that a connection with the incoming contact line equipment will be possible. If the given situation does not permit a direct continuation, the contact wire lateral position needs to be changed before the overlap such that a contact wire support is placed close to the track axis. The radial force also needs to stay within the range $80 N \geq F_R \leq 2\,000 N$ at this support.

In insulating overlaps, the contact lines are installed with a spacing of 450 mm for 15 kV installations at German Railway DB. In *non-insulating overlaps* the spacing is 200 mm. The length of spans and the contact wire staggers at supports (Figure 12.19) results from the maximum wind velocity for the verification of serviceability and the useable contact wire lateral position e_{use}.

Swiss Railway (SBB) use a spacing of 300 mm for 15 kV insulating overlaps and 200 mm in non-insulating overlaps [12.29]. In 25 kV installations SNCF use a 500 mm spacing for insulating overlaps and 200 mm in *non-insulating overlaps*.

12.4   Constraints to planning

12.4.1   Introduction

From the point of view of *overhead contact line design*, several along track constraints are be noted:
- points
- signals with the required view
- crossings
- civil engineering structures, which effect the pole selection
- crossings of overhead power lines
- electrical isolations

Design of the overhead contact lines starts at the constraints and continues to the intermediate sections formed by the constraints.

12.4.2   Wiring of points

12.4.2.1   Introduction

Wiring of track points, crossings and double-slip crossovers can create higher contact forces between the contact wire and pantographs than normally occur, especially at speeds above 200 km/h because of:
- concentration of masses at contact wire crossings with crossing bars
- the pantograph touching several contact wires at once at the supports
- insufficient space for pantograph passage
- incorrect height and lateral position for the running-up contact wires
- high radial forces in combination with contact wire sections with low elasticity

Wiring of points should avoid these shortcomings and foster a favourable interaction between the contact line and pantograph in both running directions at points and crossings. Consequently, there are limited options for pole positions and staggers at points and crossings. Design of a single set of points provides the basis for the wiring, however, only consideration of the context between points and crossings as part of a cross-over connection at stations and local conditions like track spacing and position of the points relative to each other will result in appropriate wiring.

NB: The term *points* is used in this context as a generic term for *simple points*, *crossings* and *diamond crossings with points*.

Clause 12.4.3.6 discusses the selection of supports at points for new electrifications. When a set of points is relocated or exchanged during upgrade works for speed enhancement, the pole positions will also need to be modified. The wiring can be adjusted to the new situation by modifying the adjacent spans. If this is not possible, the poles can be relocated to a small extent within the points area, however, re-design of the contact line will be necessary.

Compliance with electrical clearances between the supports at points and other supports for other switching groups needs to be checked. In most cases, the required clearances can be attained by minor displacements of the supports. In cross-over connections between the through-tracks, the assigned supports must not be arranged directly opposite.

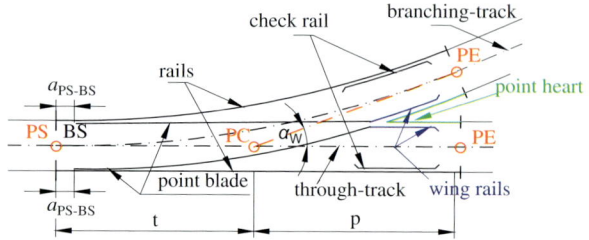

Figure 12.21: Layout of track points.
PS points start, PC points centre, PE points end and BS start of switch blade

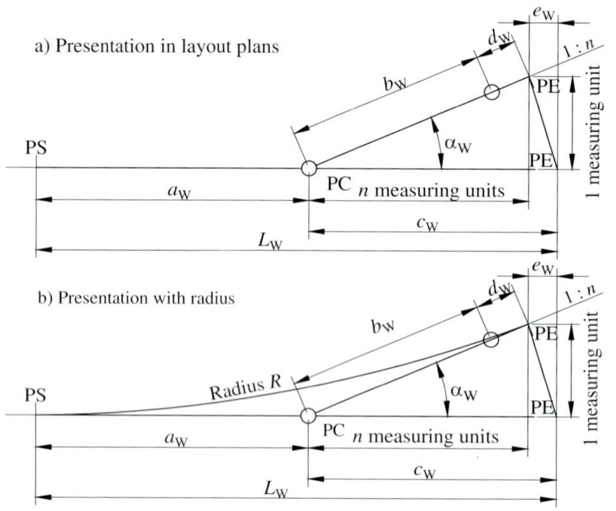

a) Presentation in layout plans

b) Presentation with radius

Figure 12.22: Illustration of simple points with point diamond apex in track layout plans.

12.4.2.2 Designation and presentation of track points in drawings

At points, a differentiation is made between through-tracks and *branching-tracks* [12.30]. A *through-track* is the designation allocated to the straight track at single points. At curved points, it is the straight track in the associated basic form, or the track with the higher operational priority or heavier load.

The *points start* PS serves as a reference point for wiring of points, however, it does not correspond to the start of the switch blade BS. Between the points start and the beginning of the radius there can be a straight section. If a circle is drawn from the beginning of the curve, with the radius of the *branching-track*, then the end of the point in the branching-track is located where the circle touches the tangent line 1 : n from the points centre (at the intersection of that tangent with the centre line of the through-track) as shown in Figure 12.21. Between the tangents for the through-track and the branching-track, the point angle α_W is obtained.

The tangent representation as shown in Figure 12.22 a) is used to illustrate points in layout diagrams in a simplified form. This however, is still inadequate for design of *points wiring*. Wiring design requires the track radius information to be provided as paper document diagrams or digital diagrams (see Figure 12.22 b)).

The parameters, *branching-track radius R* and *points branching inclination* 1 : n determine the type of points and points passage speed for the branching-track. Simple points can be traversed as follows [12.31]:

Table 12.6: Points types with dimensions, dimensions in mm [12.33].

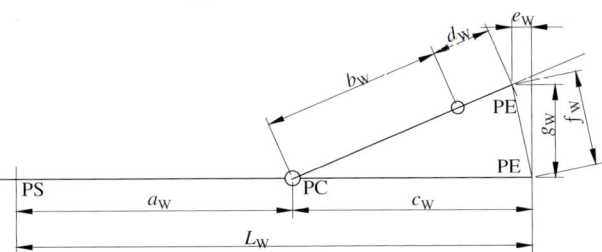

Point type	L_W	a_W	b_W	c_W	d_W	e_W	f_W	g_W	$v^{1)}$
Simple points with fixed diamond									
60-190-1:7,5	25 862	12 611	12 611	13 251	640	116	1 755	1 751	40
60-190-1:9	27 138	10 523	10 523	16 615	6 092	102	1 838	1 835	50
60-300-1:9	33 230	16 615	16 615	16 615	0	102	1 838	1 835	50
60-300-1:9/1:9,4	33 230	15 912	15 912	17 318	1 406	97	1 835	1 832	50
60-300-1:14	37 809	10 701	10 701	27 108	16 407	69	1 933	1 931	50
60-500-1:12	41 594	20 797	20 797	20 797	0	72	1 729	1 727	60
60-500-1:14	44 942	17 834	17 834	27 108	9 274	69	1 933	1 931	60
60-760-1:14	54 216	27 108	27 108	27 108	0	69	1 933	1 931	80
60-760-1:14/1:15	54 216	25 305	25 305	28 911	3 606	64	1 924	1 923	80
60-760-1:18,5	52 935	20 526	20 526	32 409	11 883	47	1 750	1 749	80
60-1200-1:18,5	64 818	32 409	32 409	32 409	0	47	1 750	1 749	100
60-1200-1:18,5/1:19,28	64 818	31 105	31 105	33 713	2 608	45	1 747	1 747	100
Simple points with movable diamond									
60-500-1:12	45 361	20 797	20 797	24 564	3 767	85	2 042	2 040	60
60-760-1:14	54 216	27 108	27 108	27 108	0	69	1 933	1 931	80
60-760-1:14/1:15	54 216	25 305	25 305	28 911	3 606	64	1 924	1 923	80
60-760-1:18,5	54 801	20 526	20 526	34 275	13 749	50	1 851	1 850	80
60-1200-1:18,5	66 615	32 409	32 409	34 206	1 797	50	1 847	1 846	100
60-1200-1:18,5/1:19,28	66 615	31 104	31 104	35 511	4 407	48	1 841	1 840	100
60-2500-1:26,5	94 306	47 153	47 153	47 153	0	34	1 778	1 778	130
60-2500-1:26,5/1:27,85	94 306	44 869	44 869	49 437	4 568	32	1 774	1 774	130
Clothoide type points with movable diamond									
60-2500/1000/∞-1:19,16	78 275	33 141	45 112	45 112	0	61	2 352	2 351	80
60-2500/1000-1:14,104	78 016	42 160	35 855	35 855	0	90	2 537	2 536	80
60-3000/1500/∞-1:23,73	89 485	38 410	51 075	51 075	0	45	2 150	2 150	100
60-3000/1500-1:18,132	89 416	47 624	41 792	41 792	0	63	2 302	2 301	100
60-4800/2450-1:24,257	111 016	57 672	51 343	51 344	0	1	2 115	2 115	130
60-6000/3700-1:32,5	122 253	64 569	57 684	57 684	0	27	1 774	1 774	160
60-7000/6000-1:42	154 266	80 104	74 162	74 162	0	21	1 765	1 765	200
60-10000/4000/∞-1:33,5	136 945	62 746	78 252	74 199	0	21	2 000	2 000	160
60-16000/6100/∞-1:41,5	168 823	80 899	95 329	87 924	0	7 724	2 000	2 000	220

$^{1)}$ maximum operational speed in branching-track

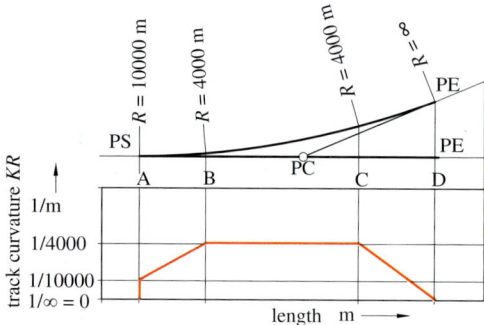

Figure 12.23: Development of curvature of clothoidal points.

Knowledge of the *track points designation* and track points type identifier is an essential prerequisite for wiring of points in practice. The track layout diagram contains the standardised *points designation*, from which the structure and geometry of the tracks follow. The conventions of UIC 711:1981 [12.32] are:

60 - 2 500 - 1 : 26,5 fb

└ supplementary designation: point diamond movable
└ inclination of tangent 1 : n
└ constant radius of branching-track 2 500 m
└ rail type UIC 60.

On high-speed lines, railway entities also use *clothoidal points* with variable radii in the branching-track. Such points enable running speeds up to 220 km/h in the branching-track. The designation is (Figure 12.22).

60 - 10 000/4 000/∞ -1 : 39,1131

└ inclination of tangent 1 : 39,1131
└ radii $R = 10 000/4 000/\infty$ m
└ rail UIC 60.

Figure 12.23 shows the development of the curvature of the discussed *clothoidal points*. Between the spots A and B as well as between C and D there are transition curves whereas between points B and C, the radius is constant and equal to 4 000 m. Table 12.6 contains the maximum operational speeds for *clothoidal points* in the branching-track [12.31].

Curved points provide track changeover operations in curves. They are formed by turning the points triangle about the points centre by the angle α_{Dr} while retaining the tangent length t and the points angle α_W as shown in Figure 12.24 a). If the curve centres of the through- and branching-tracks with radii R_1 and R_2, respectively, lie on opposing sides of the points after performing the rotation, then one refers to the points as a *contrary flexure turnout* as shown in Figure 12.24 b). The curve centres of *similar flexure points* lie on the same side of the points (Figure 12.24 c)).

If the points data are not known, they can be determined on site. The objective is to find the points start PS, which serves as the reference point for wiring and the location of the poles. The points start can be recognised by a welded rail joint (Figure 12.25).

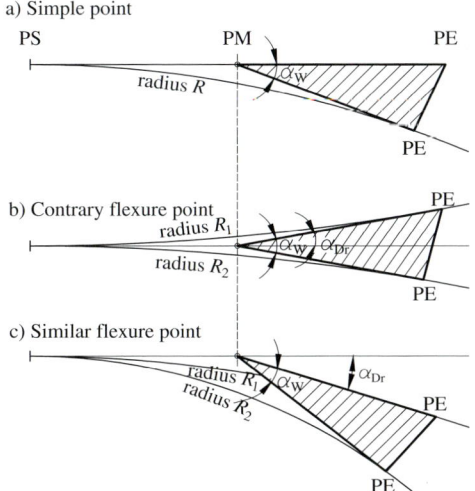

a) Simple point

b) Contrary flexure point

c) Similar flexure point

Figure 12.24: Design of curved points.

Figure 12.25: Welded joint at distance $a_{\text{PS}-\text{BS}}$ from the start of switch blade BS marking the points start PS.

Table 12.7: Distance $a_{\text{PS}-\text{BS}}$ between start of points PS and and start of switch blade BS (see Figure 12.11).

Point type rail shape UIC 60	Distance $a_{\text{PS}-\text{BS}}$ in mm
60 - 300 - 1 : 9	805
60 - 300 - 1 : 14	805
60 - 500 - 1 : 12	805
60 - 500 - 1 : 14	805
60 - 760 - 1 : 14	805
60 - 760 - 1 : 15	805
60 - 1200 - 1 : 18,5	805
60 - 2500 - 1 : 26,5	2 005
60 - 4800 / 2450 - 1 : 24,257	2 402
60 - 6000 / 3700 - 1 : 32,5	3 102
60 - 7000 / 6000 - 1 : 42	4 723

If there is no rail joint available, the points start can be obtained from the distance between the points and blade start according to Table 12.7. If otherwise unknown, the rail type can be obtained by measuring the rail height H_{S} and using the information shown in Table 12.8 and using the versine according to Figure12.26, the radius R of the branching-track is

$$R = l_{\text{SE}}^2/(8 \cdot h_{\text{f}}). \tag{12.3}$$

A check of the points position and longitudinal placing of the poles into stations or out onto interstations line is performed after determining the point start PS or the point end PE and making a PS mark on the rail web. Regularly used point types are presented in Table 12.5.

12.4.2.3 Principles of track points wiring

Both intersecting and tangential wiring is possible for track points, depending on the type of points and the pantograph length. The negotiable contact wires cross each other above the

Table 12.8: Height of rail types.

rail type	S 41	S 45	S 49	S 50	S 54	S 64	UIC 60	R 65
rail height H_S in mm	138	142	149	152	154	172	172	180

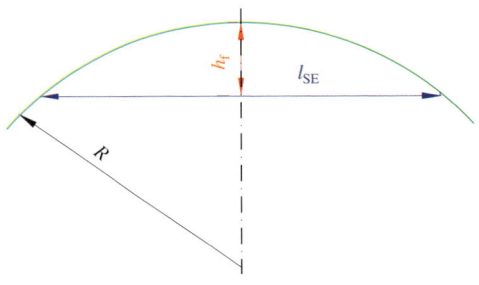

Figure 12.26: Determination of the branching-track radius using the *versine* h_f and the length of the chord l_{SE}.

points when *intersecting points wiring* is installed. The contact wires are connected vertically and horizontally to each other at the contact wire crossing with a *crossing bar*. The contact wire that is not being negotiated is also lifted by the crossing bar and *crossover droppers*, so that locally fixed positions are created for the *pantograph transition* from one contact wire to the next. The concentration of masses at the contact wire crossing consisting of two contact wires and the crossing bar affects the interaction between pantograph and contact line, especially for speeds above 120 km/h.

Tangential point wiring guides the contact lines in parallel to each other, similarly to an overlap. The pantograph running on the through-track does not touch the contact line of the branching-track. Therefore, crossover droppers between the catenaries are not necessary, since there is no interaction between the catenaries. Tangential point wiring is predominately designed for one pantograph length only. This wiring type, however, can be operated with differently long pantographs, such as the 1 600 mm and 1 950 mm pan-head lengths, whereby the longer pan-head determines the mechanical-kinematic pantograph gauge in the through- and branching-track L_{hT} or L_{hB}, respectively. Therefore, some railway operators use intersecting point wiring for lower speeds and tangential point wiring for higher speeds.

12.4.3 Intersecting points wiring

12.4.3.1 Requirements

To achieve the required quality of performance of the wiring above track points, the following requirements need to be observed:

(1) A *fitting-free area* ensures a safe transition between the contact lines and needs to be enforced for different lengths of pan-heads. For 1 950 mm pan-heads and 1 600 mm pan-heads (Figure 12.27) the distances are 1,05 m or 0,875 m, respectively, between the centre line of the negotiated track and the approaching contact wire, corresponding to the length of the pantograph. The contact wire run-up to the pan-head ends at distances of 0,60 m and 0,45 m, respectively, between the axis of the negotiated track and the running-up contact wire for the 1 950 mm pan-heads and 1 600 mm pan-heads. No clamps, except dropper clips, should be placed inside the fitting-free area and in the run-up area of the contact wire, i.e. in between the start and end of the run-up area. The run-up area is defined as a fitting-free area (Figures 12.30, 12.31 and 12.32).

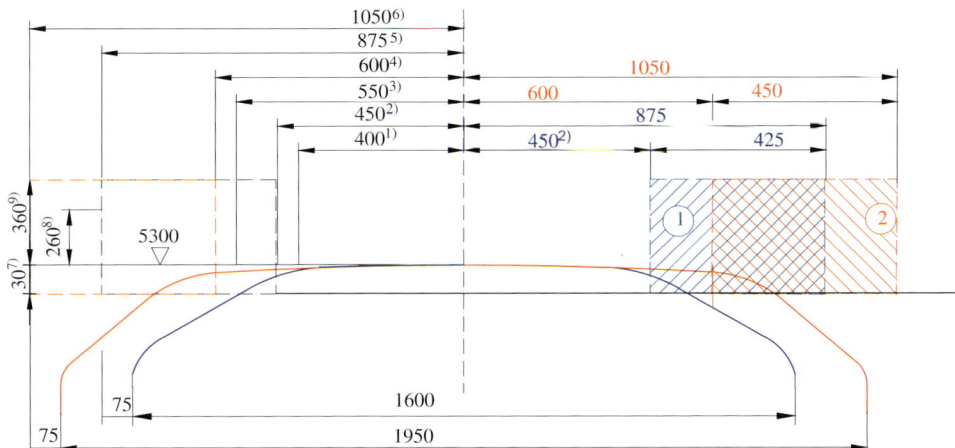

Figure 12.27: Design of the fitting free area for 1 950 mm and 1 600 mm long pan-heads.
[1] usable contact wire lateral position e_{use} for the 1 600 mm pan-head
[2] start of the fitting-free area for the 1 600 mm pan-head
[3] usable contact wire lateral position e_{use} for the 1 950 mm pan-head
[4] start of the fitting-free area for the 1 950 mm pan-head
[5] end of the fitting-free area for the 1 600 mm pan-head
[6] end of the fitting-free area for the 1 950 mm pan-head
[7] minimum contact wire height at support as a result of contact wire tolerance $T_{CW\,height} = 0,030$ m
[8] maximum height of the contact point at the support (see Table 4.7)
[9] maximum height of contact point at midspan (see Table 4.7)
1: fitting-free area for the 1 600 mm pan-head
2: fitting-free area for the 1 950 mm pan-head

(2) The spacing between the crossing of contact wires and the centre perpendicular to the plane of top of rails on the through-track should be 1/3 of the nominal stagger b at the support. The spacing between the crossing of contact wires and the centre perpendicular to the plane of top of rails on the branching-track should be 2/3 of the nominal stagger b. Therefore, the most favorable contact wire crossing is fixed with a spacing of the nominal stagger b between the track axis of the through-track and branching-track (Figure 12.33).

(3) From the point of contact wire run-up onwards, both contact wires – that of the through-track and of the branching-track need to be located between the two track centre lines. The leading contact wire prepares for the contact wire run-up by inclining the pan-head, as shown in Figure 12.28 a). A *zero position of a contact wire* is permissible, i.e. the contact wire is located at the canted centreline of the track (Figure 12.28 b)). The leading contact wire can pass beyond the track axis towards the opposite half of the pan-head if the running-up contact wire is located at least in the sway area of the pan-head.

If the traversed contact wire were to be located on the opposite half of the pan-head as viewed from the approaching contact wire (Figure 12.28 c)), the running-up contact wire could strike against the pan-head projection length end l_p and exert radial forces on the pantograph. Such an approach is known as a *pantograph trap*. This could lead

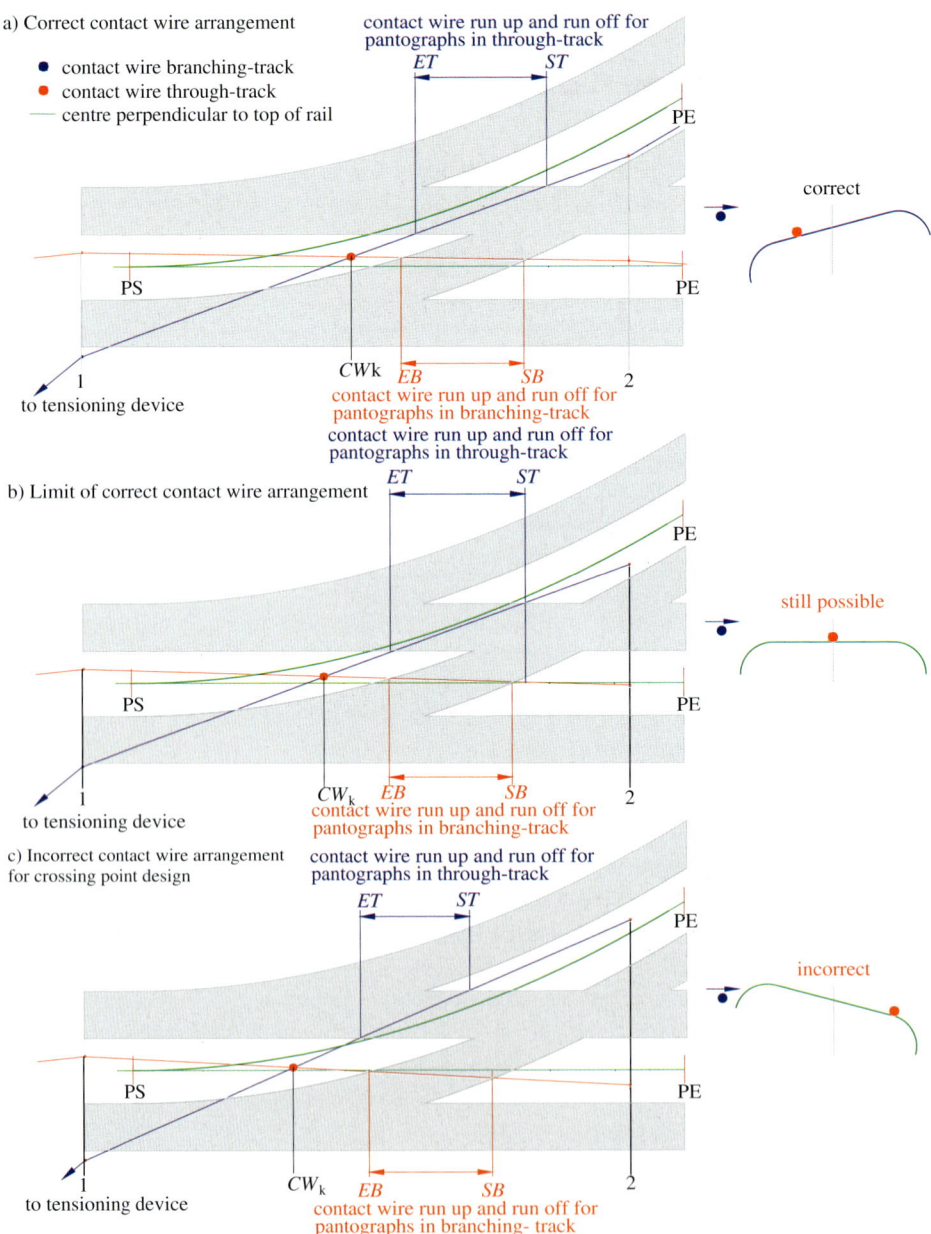

a) Correct contact wire arrangement

- contact wire branching-track
- contact wire through-track
- centre perpendicular to top of rail

Figure 12.28: Approach conditions for contact wires of contact lines above points with crossing design.

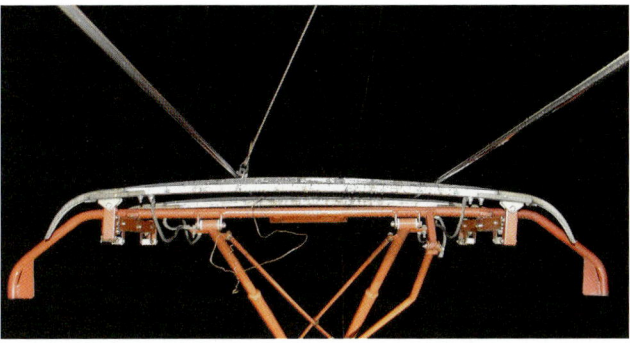

Figure 12.29: Unsuitable contact wire arrangement at an intersecting points wiring (see Figure 12.19 c)).

to high wear and eventually to damage of the contact wire and collector strip (Figure 12.29).

In the case of rigid (fixed anchored) catenary wires it is absolutely necessary to comply with this requirement, however, it is less important for automatically tensioned contact and catenary wires, positioned at the same height when being traversed.

Therefore, the span lengths between the points supports 1 and 2 need to be selected to ensure that the contact wires remain within the track axes for the through-track and the branching-track, including any adverse wind action. The radius of the branching-track consequently determines the achievable span length (Table 12.6), where no cant is used for simple points.

(4) To avoid hard spots, only one contact wire should be contacted by the pantograph at the supports. Maximum uplift of the contact wire, at supports, should not exceed 100 mm for conventional lines with $v_{max} \leq 200$ km/h and for high-speed lines where $v_{max} > 200$ km/h. When complying with the requirement to lift the branching contact wire by more than 150 mm at support S_1 (Figure 12.35), the pantographs should not be able to touch the lifted contact wire at support S_1 (see also Figure 12.27).

(5) The *contact wire crossing* should be arranged as far as possible from support S_1 so the branching-track contact wire can be raised at support S_1 by $\Delta f_z \geq 150$ mm relative to the nominal contact wire height (Figure 12.35), a height which the pantograph can not reach even under dynamic uplift on the through-track (see requirement (4)).

(6) The bending angle of the contact wire at support S_1 in the branching-track creates a contact wire radial force at the support. The radial force must be limited to 2 000 N for light-weight steady arms or to 3 000 N when the contact wire is fixed at the registration arm. This is achievable, if the branching-track contact wire can be raised by more than 200 mm at support S_1 (see requirement (4)).

(7) Irrespective of other criteria, such as displacement by wind and the restriction of span lengths, the pole spacing should not exceed 65 m between supports S_1 and S_2.

(8) To achieve uniform lift of both contact wires at the run-up/run-off position, crossover droppers are installed in overhead contact lines for speeds up to 200 km/h. However, these crossover droppers are not effective for speeds above 200 km/h and, therefore, can be dispensed with at higher speeds.

(9) The contact lines for both the through-track and branching-track should move with temperature changes in the same direction with approximately the same distance at the crossing. Therefore, the midpoints of both contact lines should be situated on the same side as seen from the crossing.

(10) Running from the branching-track into the through-track lifts the contact wire. The crossing bars (Figure 12.37) or *crossing clamps* (Figure 12.38) should transfer this lift to the through-track. The use of crossover droppers creates a smooth run-up of the contact wire on the through-track. Crossover droppers are well suited for speeds up to 200 km/h, since the mass concentration affects the interaction of the pantograph and the overhead contact line to a low extent but only at speeds up to 200 km/h. Crossing clamps are better suited at operational speeds above 200 km/h, since they possess a lower mass only (Figure 12.38).

The catenary wires should cross without touching, including in the swung condition under wind action. Therefore, sufficient clearance needs to be provided between the catenary wires. This can be achieved by reducing the system height of the branching-track catenary. The minimum system height is obtained with the minimum dropper length at midspan. If sufficient distance between the catenary wires is not achievable, protective sleeves can be applied to the catenary wires.

(11) The maximum contact wire lateral position e_{max} of the contact wire for the through- and branching-tracks should not exceed the track axes under wind action because in this situation the run-up conditions of the contact wire to the pantograph would not comply with the stipulations according to Figure 12.28.

(12) The radial forces at the negotiated support should be in the range $80\,\text{N} \leq F_R \leq 2\,000\,\text{N}$. Then, no extraordinary wear will be produced at the *steady arm articulation* and *light weight steady arms* can be used. The difference of the radial forces at the negotiated supports of the through-track should be as low as possible to achieve an elasticity as uniform as possible.

Planning should comply with these criteria at speeds above 160 km/h. Compromises are possible concerning the requirements (3), (4) and (5) for speeds < 160 km/h.

12.4.3.2   Fitting-free area

The pantograph contacts both contact wires above the track points over a short section in the case of intersecting wiring of points. The contact wire of either the branching-track or that of the through-track runs up the horn of the pantograph from one side. A risk of collision between the contact strip and any inclined fittings is present due to dynamic uplift or assembling flaws. Investigations carried out at Deutsche Reichsbahn localised this reason as a source of faults and lead to the definition of a *fitting-free area* at a distance between 500 mm to 1 050 mm on both sides of the track axis [12.34]. At that time the conditions were:

– fixed catenary wire with temperature related variations of height of the contact wire
– low contact wire tensile force
– heavy steady arm and clamp material
– flat collector strips without a radius

German Railway DB shifted the start of the fitting-free area from 500 mm as was stipulated in 1942 [12.34] to 600 mm and, therefore, reduced the fitting-free area from originally 550 mm to 450 mm for the 1 950 mm pan-head (marked in Figure 12.27 by ②). The fitting-free area starts at a distance of 50 mm from the maximum lateral contact wire position e_{max} in straight lines at 600 mm from the track axis. The fitting-free area exceeds the pantograph length by 75 mm (Figure 12.27). This definition can also be adopted to establish the fitting-free area for other pantograph types, such as the 1 600 mm long pantograph pan-head marked in Figure 12.27 by ① [12.35].

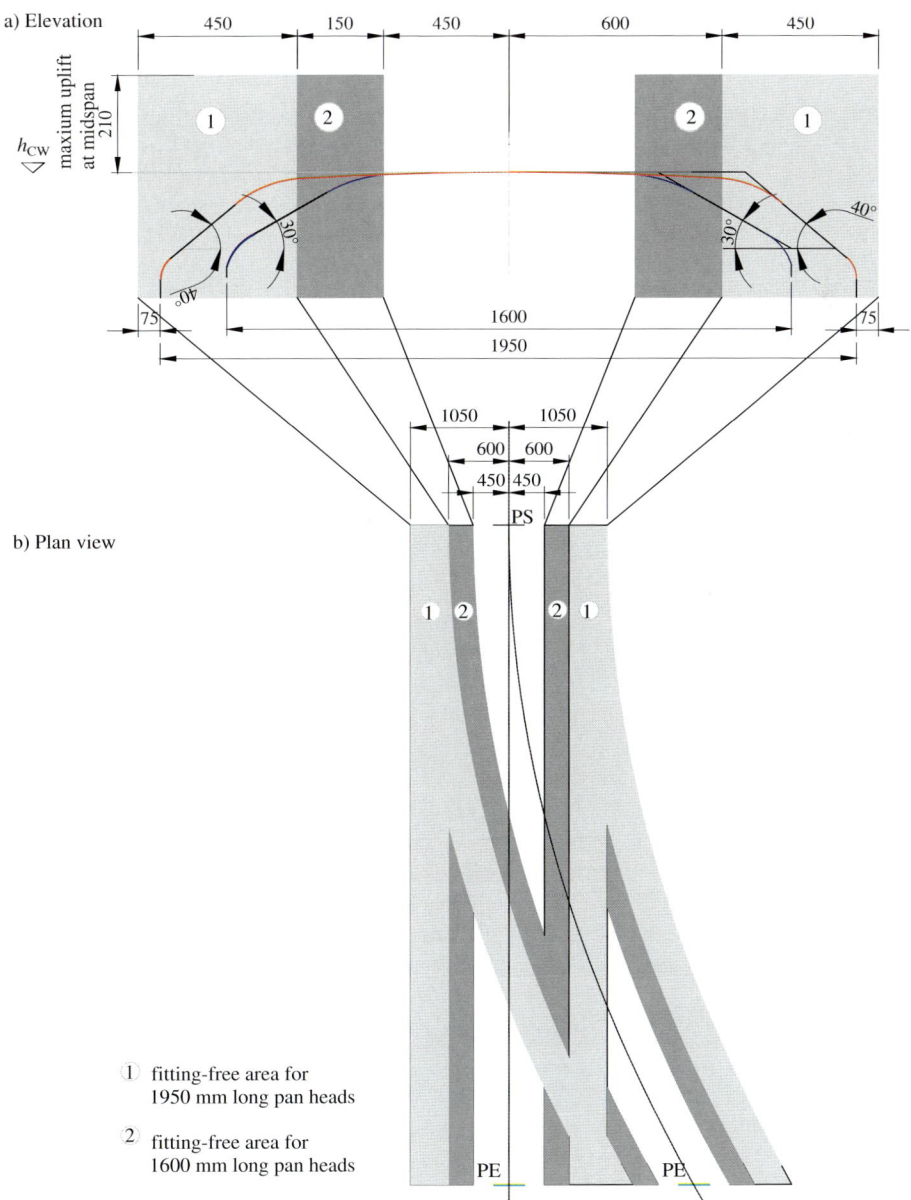

a) Elevation

b) Plan view

① fitting-free area for
1950 mm long pan heads

② fitting-free area for
1600 mm long pan heads

Figure 12.30: Fitting-free area for dual-mode operation with 1 950 mm and 1 600 mm long pan-heads. Dimensions in mm

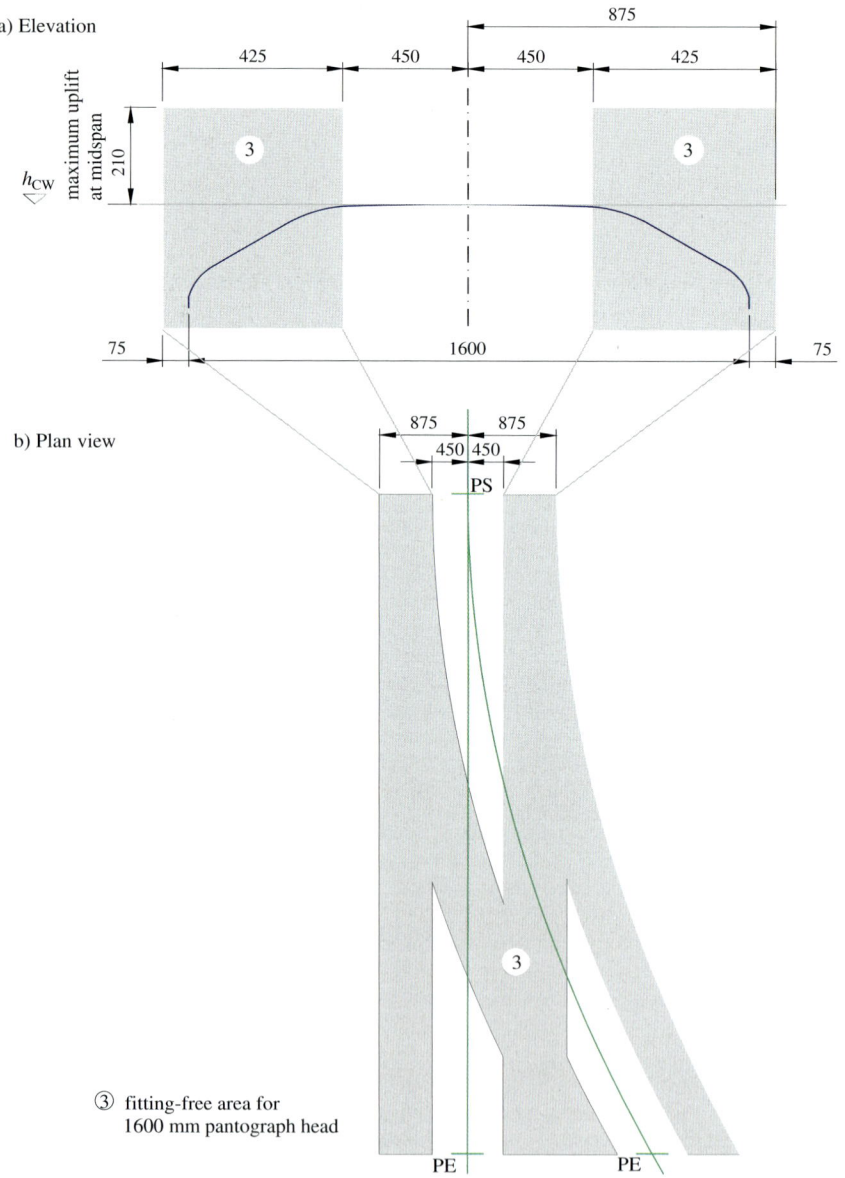

Figure 12.31: Fitting-free area for operation with the 1 600 mm long pan-head.
Dimensions in mm

Figure 12.32: Contact wire run-up with contact wire splice in the fitting-free area for the 1 600 mm pan-head. Note: Contact wire splices are not permitted in the fitting-free area (left in the picture), however dropper clamps are permitted in the fitting-free area (centre of picture).

The fitting-free area to the left and right of the track centre line, measured from the centre perpendicular from top of rails is kept clear of:

- feeder fittings, contact wire fittings, stitch wire fittings and anchor rope fittings
- wedge and cone type dead-end fittings and contact wire splices and
- dropper clips

as far as possible. Railway entities in Austria, Germany, Switzerland, Norway, Spain and Russia define different *fitting-free areas* depending on the geometry of their pantograph pan-heads.

Following the introduction of the 1 600 mm long pan-head (Figure 12.27) it became necessary to ensure that flawless dual mode operation with the contact wires within the range of track points was possible with both the 1 600 mm long and 1 950 mm long pan-head. The comparison of the geometry of a 1 950 mm long pan-head with that of the 1 600 mm long pan-head (Figure 12.30 a)) reveals that the flanks of the 1 600 mm long pan-head are designed with approximately the same slope as the 1 950 mm pan-head. Consequently, a fitting on the contact wire could collide with the pantograph horn during the lateral take-in of the contact wire to the 1 600 mm long pantograph. A fitting-free area is also necessary for this pantograph. In Figure 12.32 a contact wire splice in the fitting-free area is present with potential to create damage to the pantograph if there was a slight inclination of the splice. This situation exists for similar pan-heads in Germany, Austria, Norway, Sweden and Russia. The dimension of the fitting-free area depends on the length of the pan-head l_W, the swaying movements D of the pantograph and the height of the contact point above the track. On lines with dual-mode operation, e. g. with the 1 600 mm long and the 1 950 mm long pan-heads, inner and outer limits result, according to the shorter or longer pan-head, respectively (Figures 12.27, 12.30 and 12.31). For contact lines that are negotiated by one pantograph type only, its geometry alone determines the fitting-free area.

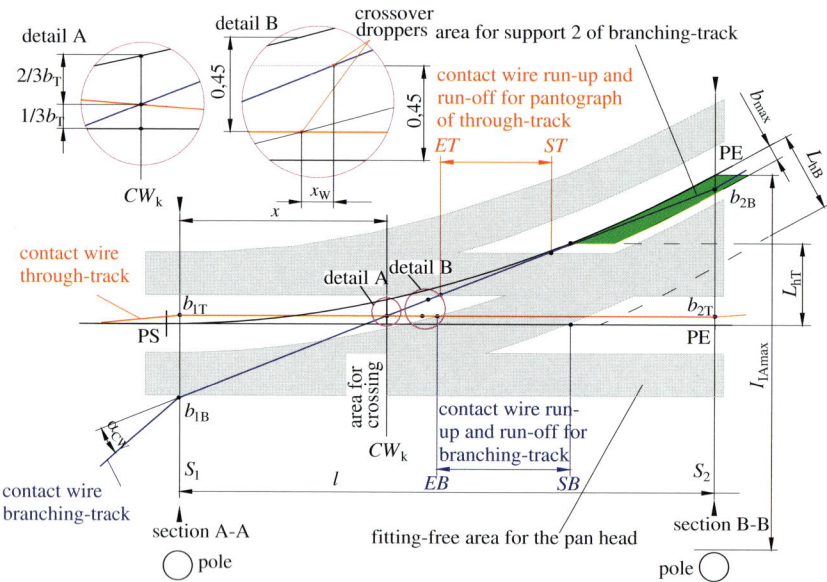

Figure 12.33: Markings in distorted scale track layout diagrams of points.

PS   points start
PE   points end
S_1 support 1
S_2 support 2
L_{hT} kinematic limit of the mechanical pantograph gauge at the height h for the through-track
L_{hB} kinematic limit of the mechanical pantograph gauge at the height h for the branching-track
ST   start of the contact wire run-up or run-out in the through-track
ET   end of the contact wire run-up or run-out in the through-track
SB   start of the contact wire run-up or run-out in the branching-track
EB   end of the contact wire run-up or run-out in the branching-track
b_{1B} contact wire stagger of branching-track at support 1
b_{1T} contact wire stagger of through-track at support 1
b_{2B} contact wire stagger of branching-track at support 2
b_{2T} contact wire stagger of through-track at support 2
CW_K crossing point of contact wires
b_B standard stagger at the support on straight track sections in the branching-track
b_T standard stagger at the support on straight track sections in the through-track
b_{max} maximum stagger at the support
l_{Almax} maximum standard length of a cantilever

12.4.3.3   Contact wire height at track points

In the area of track points, vertical movement of pan-heads and concentrated masses should
be avoided because they degrade the interaction between pantograph and contact line. Correct
contact wire heights h_{CW} and avoidance of mass concentrations are important.

The contact wire for the through-track is arranged below the branching-track contact wire
(Figure 12.35). At the crossing point CW_k, the through-track contact wire is set 20 mm higher
than the *nominal contact wire height* h_{CWnom} [12.36] and the increased contact wire height
starts and ends at the droppers adjacent to the crossing bar. The contact wire on branching-
track is adjusted at crossing to be 40 mm higher than the nominal contact wire height h_{CWnom}.

a) Section A-A at support S_1 b) Section B-B at support S_2

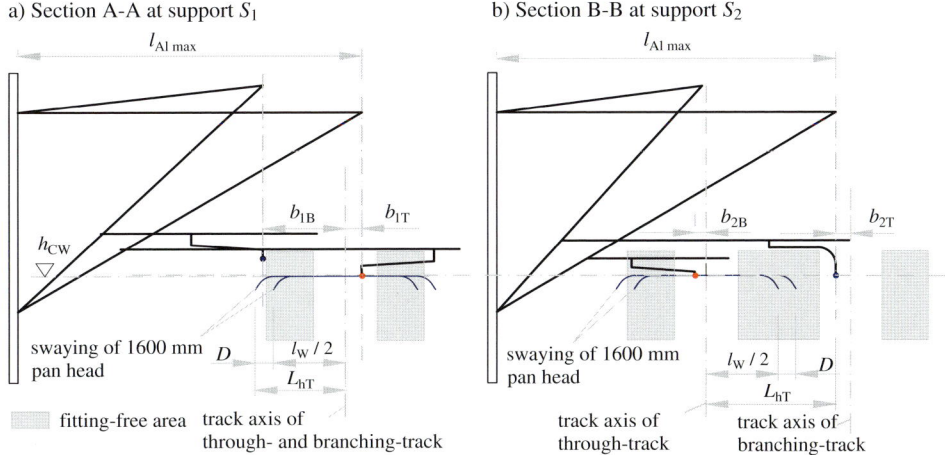

Figure 12.34: Transverse profile at supports S_1 and S_2 (see also Figure 12.33).

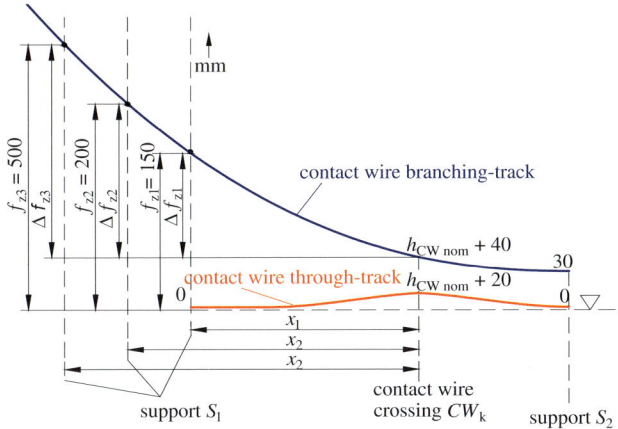

Figure 12.35: Contact wire heights in the points area. Dimensions in mm

The contact wire should remain at this height at following droppers in direction towards support S_2 on branching-track contact line. This means the contact wire in the overlapping area is approximately 30 mm higher on the branching-track than on the through-track.

Starting from the contact wire crossing, the contact wire on the branching-track, in the direction of the points start, is lifted towards support S_1 with the shape of a quadratic parabola by a further 110 mm. At support S_1 this results in a height difference of 150 mm between the contact wire on the branching-track and the nominal contact wire height of the traversed contact wire on the through-track contact line. With a dynamic uplift less than 100 mm, the pantograph cannot reach the raised contact wire at support S_1 (Figures 12.34 and 12.35). In addition to an exact contact wire height, the *dynamic uplift* of the negotiated contact wire is transferred to the over-crossing contact wire by *crossover droppers* before it is traversed.

If the pantograph should only touch the contact wire of the through-track at support S_1 the distance x between the contact wire crossing and support S_1 should be sufficiently long to permit raising the branching-track contact wire to ≥ 150 mm, or better to 200 mm. At the supports in the five-span overlaps of the contact line Re250, experience reveals that a contact

Table 12.9: Distances x between the contact wire crossing and the support S_1 (see also Figure 12.10).

Type of contact wire	Tensile force of contact wire	Specific weight of contact wire	Distance x_1 at $f_{z1} = 150\,\mathrm{mm}$ $\Delta f_{z1} = 110\,\mathrm{mm}$ at support S_1	Distance x_2 at $f_{z2} = 200\,\mathrm{mm}$ $\Delta f_{z2} = 160\,\mathrm{mm}$ at support S_1	Distance x_3 at $f_{z3} = 500\,\mathrm{mm}$ $\Delta f_{z3} = 460\,\mathrm{mm}$ for installation of an insulator
	H_{CW} in N	G'_{CW} in N/m	x_1 in m	x_2 in m	x_3 in m
AC-80 – Cu	10 000	6,77	18,03[1]	21,74	36,86
AC-100 – Cu	10 000	8,46	16,13	19,45	32,98
AC-120 – CuAg	15 000	10,15	18,03	21,75	36,87
AC-120 – CuMg	27 000	10,15	24,19	29,18	49,47

[1] $x_1 = \sqrt{2 \cdot z \cdot H_{CW}/G'_{CW}} = \sqrt{2 \cdot 0,110 \cdot 10000/6,77} = 18,03\,\mathrm{m}$

wire raised by 150 mm will not be touched by pantographs. The clearance at the support is limited to 225 mm according to Ebs 02.05.45 P. 1 and P. 2 for DB's standard designs [12.37, 12.38] (see also Figure 12.14). A higher contact wire uplift at the support is not possible with DB's or Siemens' overhead contact line designs.

A contact wire of the branching-track that is raised less than 150 mm above the contact wire of the through-track is considered as *negotiated in the raised state* by DB. The support S_1 for the branching-track is to be configured with a maximum lateral contact wire stagger that does not infringe the fitting-free area (Figure 12.33). If the contact wire of the branching-track is 200 mm above the running contact wire of the through-track (see Table 12.9), then the support S_1 can be located inside the fitting-free area and the contact wire may be directly fixed at the registration arm. This is the ideal design method and results in the position of the branching-track contact wire at support S_1 (Figures 12.33 and 12.35).

Verification of the *contact wire height increase* Δf_z between the contact wire crossing and support S_1 can be confirmed using (12.4) with

$$\Delta f_z = G'_{CW} \cdot x^2 / (2 \cdot H_{CW}) \tag{12.4}$$

where

Δf_z contact wire height increase between contact wire crossing and support S_1 in m,
G'_{CW} specific weight of the contact wire in N/m,
x distance between contact wire crossing and support S_1 in m (see Table 12.9) and
H_{CW} tensile force of the contact wire in N.

The minimum distances x_1 and x_2 for the contact wire uplift Δf_z to 150 mm or 200 mm respectively, are listed in Table 12.9 for several contact wires and their tensile forces. If the geometry of the points does not enable the contact wire lift, then the stagger in the branching-track at support S_1 can be placed outside the outer limit of the fitting-free area (Figure 12.23). Checking the *radial contact wire forces* at the supports follows in the next step. They should fall within the range $80\,\mathrm{N} < F_R < 2\,000\,\mathrm{N}$ for *light-weight steady arms* and should not exceed $3\,000\,\mathrm{N}$ for a direct fix to the registration arm. Eventually, the *opposing turning of twin cantilevers* and their mutual distance needs to be examined.

12.4.3.4   Arrangement of crossover droppers in point wirings

Crossover droppers facilitate the run-up of the laterally incoming contact wire on the pan-head of the pantograph within the points area. These droppers are arranged between the catenary wire of the through-track contact line and the contact wire of the branching-track contact

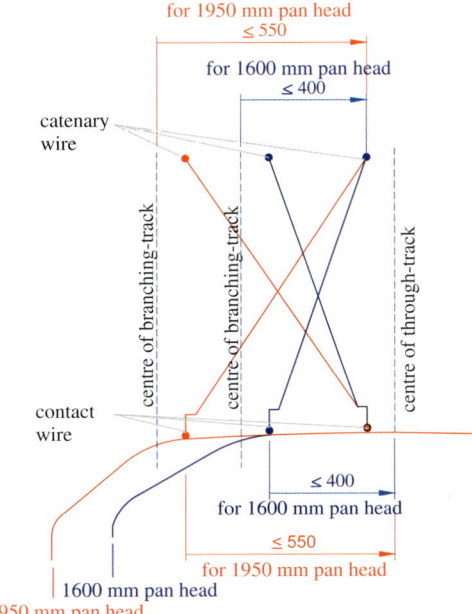

for 1950 mm pan head
≤ 550

for 1600 mm pan head
≤ 400

catenary wire

centre of branching-track

centre of branching-track

centre of through-track

contact wire

≤ 400
for 1600 mm pan head

≤ 550
for 1950 mm pan head

1600 mm pan head
1950 mm pan head

Figure 12.36: Arrangement of crossover droppers (see also Figure 12.33, detail B).
Note: crossover droppers for the 1 600 mm pan-head
crossover droppers for the 1 950 mm pan-head

line as well as between the catenary wire of the branching-track contact line and the contact wire of the through-track (Figures 12.33 and 12.36). The uplift of the contact wire, created by a pantograph running on the through-track, unloads the catenary wire in the through-track via the droppers. The crossover droppers fixed to this catenary wire lift the contact wire of the branching-track preparing for the run-up of the contact wire of the branching-track onto the pantograph running on the through-track. When running on the branching-track, the pantograph lifts the contact wire and unloads the catenary wire of the branching-track via the droppers and the catenary wire is lifted. The crossover droppers fixed to this catenary wire lift the contact wire above the through-track and facilitate the run-up of the pantograph running on the branching-track to the contact wire of the through-track. The crossover droppers are arranged 50 mm to the inner edge of the fitting-free area, at a distance of 1,0 m from each other. On railway lines operated with 1 600 mm long pan-heads a pair of crossover droppers arranged in the run-up area is adequate. In the case of dual mode operation with 1 950 mm and 1 600 mm long pan-heads two pairs of crossover droppers improve the run-up behaviour (Figure 12.36). On lines operated at speeds ≥ 250 km/h, experience confirms that crossover droppers are not required.

12.4.3.5 Connection of crossing contact lines above track points

When running from the branching-track into the through-track, the contact wire is lifted. The contact wire of the through-track is also lifted via the crossing bar. As a consequence, the crossing bar (Figure 12.37), in conjunction with the crossover droppers, ensures a smooth run-up of the pantograph to the contact line of the through-track.

Unfavourable running characteristics, especially at high speeds, can occur close to the contact wire crossing because of the mass concentration of two contact wires and up to 2,68 m long crossing bars [12.39] with the potential to cause high wear of the contact wire. Crossing

a) Fastening of crossing bar   b) Crossing with crossing bar

Figure 12.37: Crossing bar above points.

a) Fastening of crossing clamp   b) Crossing clamp at crossing

Figure 12.38: Crossing clamp for high-speed points in Norway.

clamps [12.40], as shown in Figure 12.38, are more suited for high speeds than crossing bars because of their lower weight. The contact wires of high-speed contact lines cross each other at an angle of less than 10°. Considering requirements (8) and (9) of Clause 12.4.3.1, only small differences in longitudinal movements and lateral displacements occur. Using a dropper, the crossing clamp can be lifted by 20 mm compared with the contact wire nominal height and the required flow of the contact wire height can be complied with in the area of the crossing points (see Clause 12.4.3.1).

Only rigidly terminated catenary wires are connected by a fixed crossing clamp, however, connection clamps are not required for automatically tensioned designs. The crossing catenary wires are arranged so that they will not touch each other during pantograph passage and induced movements. The crossing contact lines are connected electrically by current-carrying connections.

12.4.3.6   Lateral position of contact wire at track points

The definition of the *contact wire layout at track points* and the supports at points can be performed either on a track layout diagram or locally on site. The first procedure using a layout diagram will be explained. A plan with a distorted scale transverse to the track centre line, e. g. with a ratio 1 : 10, is helpful for designing the wiring on a drawing or by using CAD software (Figure 12.33).

At first, the fitting-free area should be identified in the distorted points layout diagram, as shown in Figure 12.33 and Clause 12.4.2, requirement (1)).

Table 12.10: Kinematic-mechanical limit of pantograph gauge in branching-track L_{hB} and in through-track L_{hT} according to TSI ENE:2014 [12.5]. Dimensions in mm

Point type	Kinematic-mechanic pantograph gauge for				Commercial speed
	ballasted superstructure		slab track		
horizontal allowances [12.3]	$\sum j_u = 79, \sum j_o = 99$		$\sum j_u = 25, \sum j_o = 32$		in
pan head length	1 600	1 950	1 600	1 950	branching-track
verificaction height h	5 525	5 725	5 525	5 725	km/h
60-190-1:7,5	1 045	1 231	987	1 171	40
60-190-1:9	1 045	1 231	987	1 171	40
60-300-1:9	1 040	1 226	982	1 166	50
60-300-1:9/1:9,4	1 040	1 226	982	1 166	50
60-300-1:14	1 040	1 226	982	1 166	50
60-500-1:12	1 037	1 223	978	1 162	60
60-500-1:14	1 037	1 223	978	1 162	60
60-760-1:14	1 035	1 221	977	1 161	80
60-760-1:14/1:15	1 035	1 221	977	1 161	80
60-760-1:18,5	1 035	1 221	977	1 161	80
60-1200-1:18,5	1 034	1 220	976	1 159	100
60-1200-1:18,5	-	-	-	1 151[1]	100
60-1200-1:18,5/1:19,277	1 034	1 220	976	1 159	100
60-2500-1:26,5	1 033	1 219	974	1 158	130
60-2500-1:26,5/1:27,85	1 033	1 219	974	1 158	130
60-4800/2450-1:24,257	1 033[2]	1 219[2]	974[2]	1 158[2]	130
60-6000/3700-1:32,5	1 033[3]	1 218[3]	974[3]	1 158[3]	160
60-7000/6000-1:42	1 032[4]	1 218[4]	974[4]	1 158[4]	200
radius of through-track ∞	1 032	1 218	973	1 157	5)
Rounded gauge dimension	**1 050**	**1 250**	**1 000**	**1 200**	

[1] 1 950 mm pan head and verification height 5,525 m, [2] radius 2 450 m, [3] radius 3 700 m, [4] radius 6 000 m, [5] depending on radius of branching-track, $u = 0,0$ m, $u_0 = u_{f0} = 0,066$ m

(1) Design of intersecting points wiring starts with the production of a 10 to 1 distorted scale track layout diagram (Figure 12.33).

(2) Marking the fitting-free area for the relevant length pantograph pan-head ensures collision free passage of the pantograph with inclined clamps. For dual-mode operation the fitting-free areas of both pantographs need to be marked (Figure 12.33).

(3) The next step involves marking the kinematic limits of the mechanical pantograph gauge at height h for the through-track L_{hT} and branching-track L_{hB} (Figure 12.33), ensuring collision free passage of the pantograph for both the through- and branching-tracks.

(4) The crossing point CW_k (see Figure 12.33) should be located at a distance, of one third of the stagger, $1/3 b_T$ that is $1/3 \cdot 0,3$ m $= 0,1$ m from the axis of the through-track (Figure 12.33 and Clause 12.4.2). Therefore the distance of CW_k from the branching-track is $2/3 b_B$ that is $2/3 \cdot 0,3$ m $= 0,2$ m creating a favourable contact performance.

(5) Support S_1 near the points start should be located at a distance of at least x from the contact wire crossing (Figure 12.33). Under this condition, the branching-track contact wire can be lifted sufficiently before reaching support S_1. The minimum distances x_1 or x_2 can be taken from Table 12.9.

(6) The next step concerns the arrangement of the contact wire for the through-track, through the crossing point CW_k, which should be located $1/3 b_T$ from the track axis

of the through track. Since both contact wires are connected at the crossing point, wind displacement will not result in the same displacement as in other similar spans. Experience shows that the span-dependent displacement of the contact wire will be limited by one third. The contact wire displaced by wind should not cross the centre line of the through-track.

(7) Selection of the location for support S_2, for the branching-track, determines the span length between the supports, S_1 and S_2, (Figure 12.33). The support S_2 for branching-track must be placed within the shaded area in Figure 12.33 defined by:

- the maximum lateral displacement b_{2B} of the contact wire of the branching-track at support S_2
- the position of the contact wire support b_{2B} outside the kinematic limit of the mechanical pantograph gauge L_{hT} of the pantograph running on the through-track
- the maximum standard cantilever length $l_{Al\,max}$

In the case of dual-mode operation, the longer of both pantograph pan-heads determines the kinematic limit of the mechanical pantograph gauge L_{hT}, e. g. the 1 950 mm long pantograph pan-head. L_{hT} depends on the contact wire height and the track radius (Clause 4.9). High contact wire heights and small track radii increase the value L_{hT}. Table 12.10 specifies the values L_{hT} for the 1 600 mm and 1 950 mm long pan-heads. For standardisation, a contact wire height of 6,0 m forms the basis for Table 12.10. The values L_{hT} include the tolerances for the lateral contact wire displacement. The support S_2 should be placed where the distance between the through-track axis and the start of the shaded area is more than L_{hT}.

The span length between S_1 and S_2 is determined by the radius of the branching-track and the specification that the contact wire displaced by wind should also remain between both track axes (see Figure 12.41 and Clause 12.4.2, requirement (3)).

(8) With the support S_2 of the branching-track located in the shaded area, the position of the branching-track contact wire towards support S_1 can be determined. Commencing from within the marked area, at the branching-track support S_2, the contact wire continues across the contact wire crossing CW_k to support S_1. At this location, select an intersecting point with the position of S_1. The position of S_1 should be equal or greater than x to the contact wire crossing. The intersecting point should be outside of the fitting-free area (Figure 12.33). For the type of points shown in Figure 12.33 a contact wire position outside the fitting-free area is possible. If this contact wire lateral position is placed outside the fitting-free area (Figure 12.33) the contact wire of the branching-track can be fixed directly to the registration arm and then guided directly to the tensioning device. An additional cantilever to guide the branching-track contact line should not be necessary.

(9) Then the contact wire lateral position b_{2T} at the support S_2 of the through-track can be determined using the selected arrangement of the contact wire above the through-track according to step (6). Since both supports S_{2B} and S_{2T} are arranged at the same pole, however, on different cantilevers, both supports need to be arranged approximately at the same position. Starting with the selected stagger b_{2B} at support S_2 of the branching-track, the position of support S_2 for the through-track is 1,2 m towards the end of the points. The dimension 1,2 m corresponds to the distance between the cantilevers at support S_2 (Figure 12.33). For the contact wire position b_{2T} at support S_2 a smaller value needs to be selected. Thereby, the incoming contact-wire of the through-track can

be arranged from both spans at that support (Figure 12.33). The contact wire support b_{2T} needs to be placed outside the kinematic limit of the mechanical gauge L_{hT} of the pantograph running on the branching-track.

(10) Compliance with the requirements is checked as the last step in Clause 12.4.2.

(10.1) Is the crossing placed between the track axes and positioned $1/3 \cdot b_T$ from the through-track axis (see requirement (1), Clause 12.4.2)?

(10.2) Can the contact wire of the branching-track be lifted by at least 150 mm above the contact wire nominal height at support S_1 (see requirement (2), Clause 12.4.2)?

(10.3) Does the contact wire of the through-track run through the crossing point CW_k and are the supports of the through-track contact wire b_{1T} and b_{2T} placed outside the fitting-free area (see requirement (3), Clause 12.4.2)?

(10.4) Is the marked area in Figure 12.33 correctly placed and is the support S_2 of the branching-track arranged within this area (see requirement (4), Clause 12.4.2)?

(10.5) Is the support S_1 of the branching-track contact wire placed outside the fitting-free area (see requirement (5), Clause 12.4.2)? If the contact wire support of the branching-track is placed between the track axis and the outer rim of the fitting-free area, Δf_z needs to be more than 150 mm. If the support S_1 of the branching-track contact wire is situated between 150 mm and 200 mm above the contact wire nominal height a steady arm must register the contact wire. If it is placed more than 200 mm above the contact wire nominal height, it can be directly fixed to the registration arm. In this case, the contact line can be terminated at the following pole. If the stagger b_{1B} of the support S_1 for the branching-track contact wire is placed outside the outer limit of the fitting-free area, the length of the cantilever needs to be checked. This length should not be less than 2,5 m.

(10.6) Is the support S_2 of the through-track correctly placed (Clause 12.4.2, requirement (6))? Both contact wires need to be placed at the markings ST and SB in between the track axis. The contact wire stagger b_{2T} should be close to zero to enable an arbitrary connection of the incoming contact wire of the through-track.

(10.7) Do the selected contact wire lateral positions b_{1T}, b_{1B}, b_{2T} and b_{2B} comply with the maximum stagger e_{max}? If the maximum permissible contact wire lateral position e_{max} is exceeded, it is necessary to reduce the span length between the supports S_1 and S_2 or to modify the *lateral position of the contact wire* at the supports. The possibilities of modifications are limited because of the fitting-free area. The span length between the supports S_1 and S_2 should not be more than 65 m to limit the dynamic effects on the contact line and to obtain an adequate interaction of pantograph and contact line in the points area (Clause 12.4.2, requirement (7)).

The examples 12.1 and 12.2 illustrate the process of planning *intersecting points wiring*.

12.4.3.7 Examples

Example 12.1: The point EW 60-1200-1:18,5 on a slab track is to be wired intersecting with the overhead contact line type Re330, consisting of a contact wire AC-120−CuMg tensioned at 27 kN for operation with an 1 950 mm long pan-head considering the fitting-free area (Figure 12.40). The wind speed is 20,3 m/s with 10 years return period. The steps for the points wiring are marked in Figure 12.40 and include:

(1) *Distortion of the layout plan of the points*: According to Table 12.6 this set of points, including the branching-track radius, can be designed with a 1 to 10 distorted scale factor transverse to the track axis as explained in Clause 12.4.3.3.

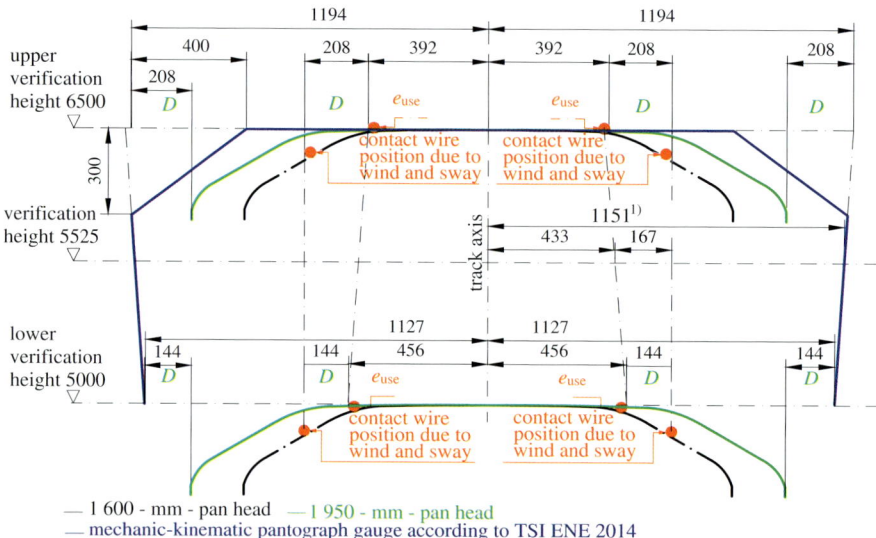

Figure 12.39: Evaluation of the kinematic-mechanical pantograph gauge L_{hT} and L_{hB} for through-track and branching-track, respectively, in the case of dual-mode operation with 1 600 mm and 1 950 mm long pan-heads for the high-speed contact line Sicat H1.0 with slab tracks above EW60-1200-1:18,5 points, whereby the kinematic-mechanical pantograph gauge and the useable contact wire lateral position follow from the radius of the branching-track. Dimensions in mm, [1] 1 950 mm pan-head for 5 525 mm verification height (see also red marked value in Table 12.10)

(2) *Marking of the fitting-free area*: Marking the fitting-free area at 0,60 m parallel to the track axis guarantees collision-free passage of the pantograph relative to the clamps within the points area.

(3) *Designation of the position for L_{hT}*: The kinematic limit of the mechanical pantograph gauge L_{hT} will be 1,20 m for the 1 950 mm long pan-head on the slab track according to Table 12.10.

(4) *Designation of the position for CW_k*: The contact wire crossing CW_k is placed at 0,4 m track spacing (i.e. distance between the track axes) and 130 mm from the axis of the through-track.

(5) *Designation of position for S_1*: The support S_1 is situated at least 24,2 m from the contact wire crossing towards the point start (Table 12.9).

(6) *Determination of contact wire position on the through-track*: A line parallel to the through-track at a distance of 0,1 m on the side of the branching-track locates the contact wire of the through-track. The distance of 0,10 m corresponds to the value $1/3 \cdot b_T$ (see Clause 12.4.3.6, requirement (6)).

(7) *Determination of support S_2 location for the branching-track*: The range of this support results from the maximum contact wire stagger at the support $b_T = 0,40$ m, $L_{hT} = 1,20$ m and the maximum length of the standard cantilever $l_{Almax} = 5,40$ m for the contact line design Re330 [12.41] at a distance of 3,40 m between the pole and the axis of the through-track (Figure 12.40). In Figure 12.31, the support S_2 of the branching-track is situated at a distance of 1,20 m to the axis of the through-track.

(8) *Determination of the support S_1 location for the branching-track*: A straight line starts at the contact wire support S_2 of the branching-track, runs through the crossing point CW_k and ends either before the fitting-free area or outside the fitting-free area. The contact wire support S_1 of the branching-track is situated in Figure 12.40 at 1,1 m laterally to the through-track axis outside the outer limit of the fitting-free area. Thereby, the contact wire can be fixed at the registration arm and guided directly from the support S_1 to the flexible tensioning device.

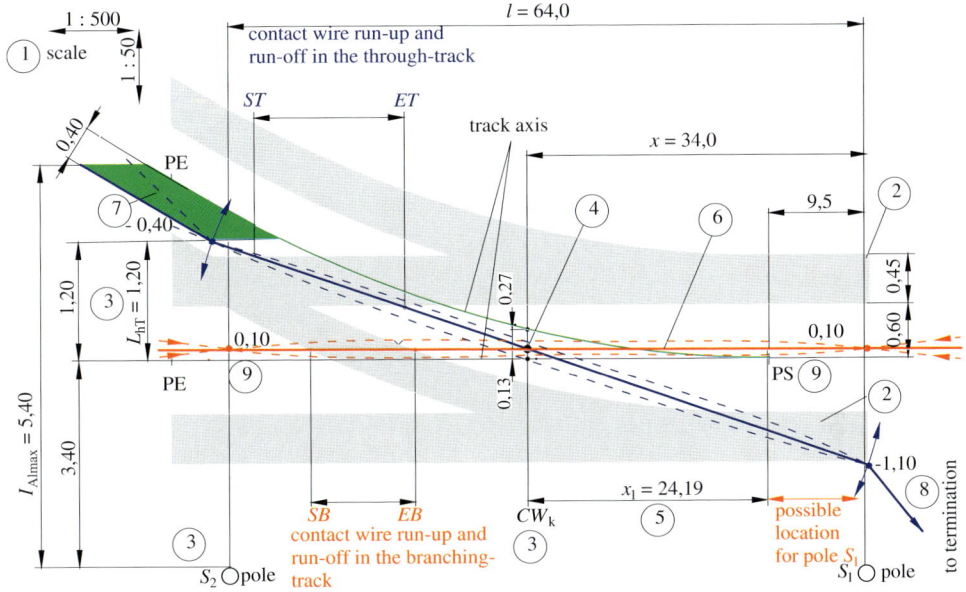

track	radial forces at support in N	
	support S_1	support S_2
through track	231[1]	231[2]
branching track	83[3]	83[4]

[1] preceding span length 70 m, preceding contact wire stagger $-0,4$ m
[2] subsequent span length 70 m, subsequent contact wire stagger $-0,4$ m
[3] preceding span length 70 m, preceding contact wire stagger at the tensioning device $-3,6$ m
[4] subsequent span length 70 m, subsequent contact wire stagger $-0,4$ m

Figure 12.40: Wiring of points EW 60-1200-1:18,5 with the contact line design Sicat H1.0 and a fitting-free area valid for the 1 950 mm long pan-head, according to example 12.1 for exclusive operation with a 1 950 mm long pan-head including radial forces at the supports. Dimensions in m.

(9) *Determination of supports S_1 and S_2 for the contact wire on the through-track*: With a span length of 64 m the supports S_1 and S_2 of the contact wire on the through-track are situated at the same positions as the supports S_1 and S_2 for the branching-track.

(10) *Checking the selected wiring*:

    – the supports are placed outside the fitting-free area (requirement (1) of Clause 12.4.3.1)

    – the distance of the contact wire crossing from the through-track axis is $b_T/3$ (requirement (2) of Clause 12.4.3.1)

    – both contact wires are located between the track axes in the run up/run off area (requirement (3) of Clause 12.4.3.1)

    – only one contact wire will be touched by the pantograph at supports S_1 and S_2 (requirement (4) of Clause 12.4.3.1)

    – the distance x between the contact wire crossing and the support S_1 is 34 m and corresponds to the specification in Table 12.9 (requirement (5) of Clause 12.4.3.1). At support S_1, the contact wire of the branching-track is situated 217 mm above the nominal contact wire height

    – the span length $l = 64$ m is less than the specified span length 65 m (requirement (7) of Clause 12.4.3.1)

- cross-over droppers are not necessary for high-speed overhead contact lines (requirement (8) of Clause 12.4.3.1)
- requirement (9) of Clause 12.4.3.1 will be complied with in the following design procedure
- requirement (10) of Clause 12.4.3.1 will be followed during the installation
- the contact wire lateral position e_{max}, marked as a broken line in Figure 12.40 and reduced around 1/3, does not exceed, at any position, the track axis (see requirement (11) of Clause 12.4.3.1)
- compliance of the radial forces of the contact wire at the supports S_1 and S_2 can be checked using Equation (4.7). According to Table 12.11 the radial forces at the supports stay within the permissible range $80\,N \leq F_R \leq 2\,000\,N$ for the selected wiring. Unwanted high radial contact wire forces at the supports S_1 and S_2 must not be expected (requirement (6) of Clause 12.4.3.1). The difference of the radial forces at the supports S_1 and S_2 is less than 200 N. Therefore, uniform elasticity exists within the range of the points (see requirement (12) of Clause 12.4.3.1).

Figure 12.41 shows the wiring for points EW 60-1200-1:18,5.

Example 12.2: Points EW 60-1200-1:18,5 located on a slab track are to be wired using contact line type Re330, consisting of a contact wire AC-120 – CuMg tensioned at 27 kN tensile force for dual operation with 1 600 mm and 1 950 mm long pan-heads, under consideration of the fitting-free area applicable for this condition with an intersecting design (Figures 12.39 and 12.41). The basic wind speed with 10 years return period is 20,3 m/s. The steps for the design procedure follow example 12.1:

(1) *Distortion of the points track layout*: According to Table 12.6 the points, including the branching track, can be designed and distorted with a factor of 10 transverse to the through-track, as explained in Clause 12.4.3.1.

(2) *Marking the fitting-free area*: Marking the fitting-free area parallel to the track axes at a distance of 0,45 m and a width of 0,60 m takes care of operations with 1 600 mm and 1 950 mm long pan-heads and the collision-free passage of the pantographs in relation to clamps in the points area.

(3) *Identification of the position for L_{hT}*: The limit of the mechanical pantograph gauge L_{hT} for the 1 950 mm pan-head is around 1,20 m according to Table 12.10.

(4) *Marking the contact wire crossing CW_k*: The contact wire crossing is situated at a 0,30 m track spacing (between the track axes) and a distance of 0,10 m from the axis of the through-track.

(5) *Marking the position of support S_1*: The support S_1 is situated at least $x = 24,2\,m$ from the contact wire crossing CW_k towards the start of point (Table 12.9).

(6) *Determination of the contact wire position within the through-track*: A line parallel to the through-track axis at a distance of 0,1 m to the side of the branching-track defines the contact wire on the through-track. The distance 0,10 m corresponds to the value $2 \cdot e_{max}/3$.

(7) *Determination of the support S_2 location for the branching-track*: The range results from the maximum contact wire lateral position at the support $b_T = 0,30\,m$, $L_{hT} = 1,20\,m$ and the maximum length of the standard cantilever $l_{Al\,max} = 5,40\,m$ for the contact line design Re330 [12.41] with 3,40 m clearance between the pole and the axis of the through-track (see Figure 12.41). The stagger $-0,30\,m$ at the position of support S_2 for the branching-track contact wire complies with the specification that this support should be situated at a distance of more than L_{hT} to the axis of the through-track. For the example of Figure 12.41 the support S_2 for the branching-track is situated at a distance of 1,15 m from the through-track axis .

(8) *Determination of the location for support S_1 for the through-track*: A straight line starts at the contact wire support S_2 of the branching-track and runs through the crossing point CW_k in the direction of the points start and ends either before the fitting-free area or outside the outer limit of the fitting-free area. The contact wire support S_1 is situated at a distance of 1,1 m to the through-track axis and therefore, outside the fitting-free area (Figure 12.41).

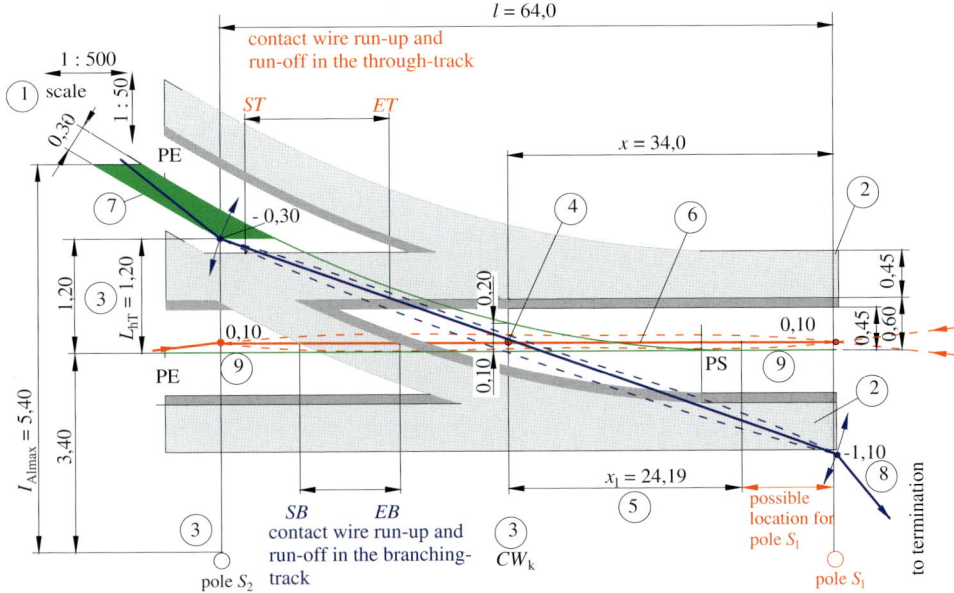

Figure 12.41: Wiring of point EW 60-1200-1:18,5 on a slab track with fitting-free area valid for 1 600 mm and 1 950 mm long pan-heads of overhead contact line type Re330 according to example 12.2 for dual-mode operation with 1 600 mm and 1 950 mm long pan-heads. All dimensions in m.

track	radial forces at support in N	
	support S_1	support S_2
through-track	154[1]	154[2]
branching-track	148[3]	148[4]

[1] preceding span length 70 m, preceding contact wire stagger $-0,3$ m
[2] subsequent span length 70 m, subsequent contact wire stagger $-0,3$ m
[3] preceding span length 70 m, preceding contact wire stagger $-3,4$ m at the tensioning device
[4] subsequent span length 70 m, subsequent contact wire stagger $-0,3$ m, 4,7 m track distance

(9) *Determination of the supports S_1 and S_2 for the through-track contact wire*: With a span-length of 60 m the supports S_1 and S_2 of the through-track contact wire are situated at the same positions as the supports S_1 and S_2 for the branching-track.

(10) *Checking the selected solution*:

  – the supports are placed outside the fitting-free area (Requirement (1) of Clause 12.4.2.1)
  – the contact wire crossing is situated at $b_T/3$ from the axis of the through-track (Requirement (2) of Clause 12.4.2.1)
  – both contact wires are located between the track axes in the run-up/run-off area (Requirement (3) of Clause 12.4.2)
  – only one contact wire will be touched by the pantograph at S_1 and S_2 (Requirement (4) of Clause 12.4.2.1)
  – the distance x being 30 m between the contact wire crossing and the support S_1 corresponds to the specification in Table 12.9 (Requirement (5) of Clause 12.4.2.1)
  – the span length $l = 60$ m is less than the specified maximum span of 65 m (Requirement (7) of Clause 12.4.2.1)
  – crossover droppers are not necessary for high-speed overhead contact lines (Requirement (8) of Clause 12.4.2.1)

        – requirement (9) of Clause 12.4.2.1 will be complied with in the following design steps
        – requirement (10) of Clause 12.4.2.1 will be followed during the installation
        – requirements (11) of Clause 12.4.2.1: The contact wire stagger e_{max}, shown as dashed
          lines in Figure 12.41 and reduced by $1/3$, never exceeds the track axis
        – requirements (6) and (12) of Clause 12.4.2.1: The observance of horizontal forces of the
          contact wire at the supports S_1 and S_2 can be checked with the aid of equation (4.6).
          The following horizontal forces (Table 12.14) at the supports for the selected wiring lie
          within the permitted range $80\,\mathrm{N} \leq F_R \leq 2\,000\,\mathrm{N}$. The difference of the radial forces at the
          supports S_1 and S_2 is less than $200\,\mathrm{N}$. Therefore, a uniform elasticity within the points
          area exists.

Figure 12.41 shows the wiring of points EW 60-1200-1:18,5 for dual pantograph operation.

12.4.4   Tangential track points wiring

12.4.4.1   Requirements

Tangential points wiring is a parallel routing of the contact lines in the overlap area, where
the pantograph running on the through-track only touches the contact wire of this track and
not the contact wire of the branching-track. The contact wire of the branching-track is situated outside the kinematic-mechanical pantograph gauge of the through-track or has been
lifted already. Favourable conditions result for the interaction between pantographs and overhead contact lines because of the missing contact wire intersection including a crossing bar,
especially for high-speed contact lines.

Requirements for design of tangential points wiring:

   (1) *Fitting-free areas* only needs to be observed when running from and into the branching-
      track. Running on the through-track should not be a concern. This wiring provides a
      flawless transition between the contact wires of the through- and branching-track and
      into the branching-track but needs to be adjusted for different lengths of pan-heads.
      The *contact wire run-up* to the pan-head starts, for the $1\,950\,\mathrm{mm}$ and $1\,600\,\mathrm{mm}$ long
      pan-heads (Figure 12.27) at $1{,}05\,\mathrm{m}$ and $0{,}875\,\mathrm{m}$, respectively, distance between the
      axis of the traversed track and the contact wire run-up. The contact wire run-up to
      the pantograph pan-head ends at $0{,}60\,\mathrm{m}$ and $0{,}45\,\mathrm{m}$ distance, respectively, between the
      axis of the traversed track to the running-up contact wire. Only dropper clips should be
      located within the run-up section of the contact wire between the start and end of the
      run-up procedure because that area is defined as a fitting-free area, see Clause 12.4.4.2.

   (2) The *stagger of the branching contact wire* needs to be selected to ensure the contact
      wire of the branching-track is situated outside the kinematic-mechanical pantograph
      gauge of the through-track and so that the pantograph running on the through-track does
      not touch that contact wire (Figure 12.42). Should the contact wire of the branching-
      track be guided between the supports S_1 and S_2, into the kinematic-mechanical pan-
      tograph gauge of the through-track, then the contact wire of the branching-track is
      situated above the high level of uplift of the pantograph running on the through-track.
      At least $100\,\mathrm{mm}$ clearance is required so that the pantograph is not able to touch the
      contact wire of the branching-track in this situation (Figure 12.43). Tangential points
      wiring for a certain type of pantograph enables parallel guiding of the contact wire of
      the branching-track to the axis of the through-track outside the kinematic-mechanical

pantograph gauge (Figure 12.42). For dual operation the contact wire of the branching-track may not be guided in parallel (Figure 12.43).

(3) The *contact wire height of the branching-track* is installed 50 mm lower than the contact wire height of the through-track (Figure 12.42). Consequently, the contact wire of the through-track runs within the range of the useable contact wire lateral position, however, at least within the working range of the pantograph pan-head (see Clause 12.4.4.3). After both contact wires run within the working range of the pan-head, the contact wire of the branching-track can be lifted. This contact wire should be lifted to at least 200 mm above the nominal contact wire height at support S_1.

(4) The *lateral position of the contact wire of the through-track* at support S_2, should always be on the same side as the contact wire of the branching-track, having the maximum contact wire lateral position. Starting at support S_1 the contact wire of the through-track should run parallel to the track axis (Figure 12.42) or with a changing lateral position to support S_2 (Figure 12.43).

(5) The *contact wire lateral position of the branching-track* at support S_2 should always be on the through-track side with the maximum contact wire lateral position (Figure 12.42). The lateral position of the contact wire of the branching-track at support S_1 is determined by requirement (6), meaning that within the working range or within the useable contact wire lateral position both contact wires should be situated at least within the working range of the pantograph. For the 1 950 mm long pan-head, this requirement can be complied with by the parallel position of the contact wire of the branching-track to the axis of the through-track and the changing contact wire position of the contact wire within the through-track (Figure 12.42). For dual operation with 1 950 mm and 1 600 mm long pan-heads a non-parallel position of the contact wire is required for the contact wire of the through-track and non-parallel guiding of the contact wire of the branching-track (Figure 12.43). Both alternatives are presented in Figures 12.42 and 12.43.

(6) The *run-up condition*, when running from and to the branching-track, involves the contact wires being, at least for a short distance, within the working range of the pantograph pan-head before the contact wire of the through-track or the branching-track leaves the pantograph to the side or upwards. This run-up condition avoids de-wiring. In Figure 12.43, these conditions are presented by position 1 for the 1 950 mm long and by position 2 for the 1 600 mm long pan-head types.

(7) The *contact wire radial forces* should be between $80\,\text{N} \leq F_R \leq 2\,000\,\text{N}$ at the active supports. Contact wire lifted by more than 200 mm can be directly fixed to the registration arm.

(8) The *difference of the radial forces* at the contacted supports of the through-track should be as low as possible to achieve uniform elasticity (see Clause 12.3.2.1).

(9) The *displacement due to wind* needs to be considered, including for tangential wiring and determines the distance between the supports S_1 and S_2. The tolerances for spacing between the poles needs to be considered when designing the span length.

(10) *Crossover droppers*, as in the case of intersecting point wiring, are not required for tangential points wiring (see Clause 12.4.4.5).

(11) *Current connectors* between the contact lines of the branching- and through-track are required. These connectors should be placed, if possible, not in the contacted section of the overhead contact line, but as close as possible to the tangential arrangement (see Clause 12.4.4.6).

Table 12.11: Minimum distances $u_{1\,\text{min}}$ and $u_{2\,\text{min}}$ between the start of uplift for branching-track at position 1 and position 2 respectively between position 1 and support S_1 (Figures 12.42 and 12.43).

Contact wire type	Tensile force of contact wire H_{CW} in N	Specific weight of contact wire G'_{CW} in N/m	Distance u_1 for $f_{z1} = 100\,\text{mm}$ $\Delta f_{z1} = 150\,\text{mm}$ at support S_1 u_1 in m	Distance u_2 for $f_{z2} = 200\,\text{mm}$ $\Delta f_{z2} = 250\,\text{mm}$ at support S_1 u_2 in m
AC-80–Cu	10 000	6,77	21,05	27,18
AC-100–Cu	10 000	8,46	18,83	24,31
AC-120–CuAg	15 000	10,15	21,06	27,18
AC-120–CuMg	27 000	10,15	28,25[1]	36,47

[1] $u_2 = \sqrt{2 \cdot \Delta f_{z2} \cdot H_{\text{CW}}/G'_{\text{CW}}} = \sqrt{2 \cdot 0,150 \cdot 27\,000/10,15} = 28,25$ m (Figure 12.43)

(12) The track catenaries of the through- and the branching-track should move in the case of *change in temperature* in the same direction and with the same amount of length variation. Therefore, the midpoints of the through- and branching-track catenaries should be arranged approximately at the same location.

International experience shows that for tangential points wiring, both contact wires do not need to be situated on the same half of the pan-head at the moment of running up. A precondition, however, is that the catenary and contact wires are automatically tensioned. Both contact wires should, at least, be situated within the working range of the pan-head.

12.4.4.2 Fitting-free area

Experience proves the *fitting-free area* is a necessity for intersecting wiring of points to avoid disturbances. In the case of tangential wiring the pantograph should not touch the contact wire of the branching-track. In this case, consideration of a fitting-free area is not required. When running from the branching-track into the through-track, the contact wire runs up the pantograph from the side. Therefore, consideration of a fitting-free area is necessary and should also be observed for tangential wiring of points.

12.4.4.3 Contact wire height at track points

For fixed catenary wire, as specified in the past, the contact wire was lower in summer than in winter. For fixed catenary wires in branching-tracks interfacing with tensioned catenary wires on through-tracks, during winter the run-up condition for the contact wire was more favourable than it was in summer because the contact wire was 200 mm lower in the summer. Because of the application of separate tensioning devices for contact wire and catenary wire, the contact wire is situated at approximately the same height in winter as in summer such that the run-up of the contact wire no longer depends on the temperature and is approximately always the same. Therefore, stable run-up conditions, independent of the temperature can be assumed.

For tangential points wiring, the contact wire should be situated at the supports of the through-track, at the nominal height, e. g. at 5 300 mm. The contact wire of the branching-track should be situated at supports 1 and 2 approximately 50 mm lower than that of the through-track, that is at 5 250 mm. The reduction of the contact wire height by e. g. 50 mm corresponds to a

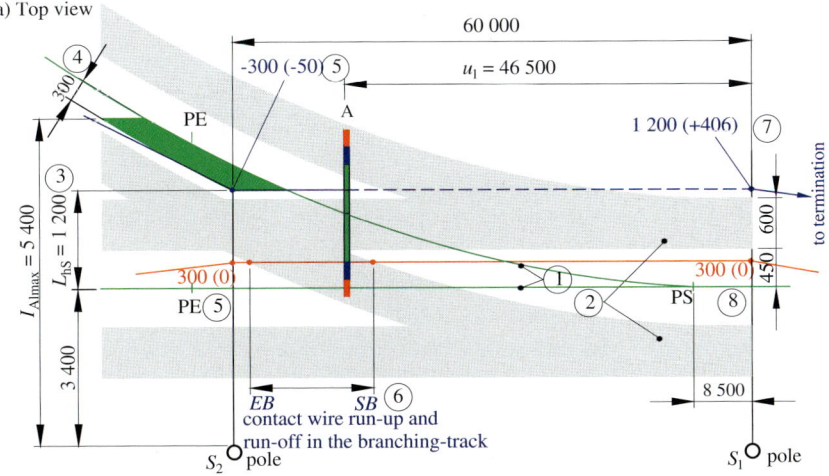

a) Top view

b) Cross section at position 1 for pantograph type 1950 mm

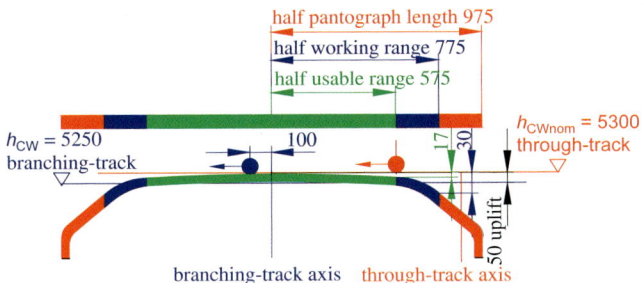

Track	Radial force at support in N	
	Support S_2	Support S_1
through-track	$270^{2)}$	$270^{1)}$
branching-track	$1\,946^{4)}$	$2\,337^{3)}$

[1] preceding span 60 000, preceding stagger −300, [2] subsequent span 60 000, subsequent stagger −300, [3] preceding span 47 100, preceding stagger +300, points in the opposite track, [4] subsequent span 60 000, subsequent span stagger at the tensioning equipment 3 600

Figure 12.42: Tangential wiring of the points EW 60-1200-1:18,5 with the contact line type Re330 and the fitting-free area valid for the 1 950 mm long pan-head corresponding to example 12.3 for operation with the 1 950 mm long pan-head, all dimensions in mm. blue: contact wire of the branching-track, red: contact wire of the through-track, green: track axis

50 mm contact wire uplift at maximum running speed in the branching-track. If in the case of long points for higher running speeds the contact wire uplift is higher, then the contact wire height of the branching-track should be reduced by this amount.

This allows the contact wire to run, for movements to and from the branching-track, laterally from above into the range of the useable contact wire sides of pantograph types with 1 950 mm and 1 600 mm heads.

12.4.4.4   Contact wire lateral position at track points

The contact wire lateral position at the supports needs to be selected to ensure that the pantograph running on the through-track will not touch the contact wire of the branching-track. The contact wire of the through-track should be positioned starting at support S_2 towards the branching-track. When running from the branching track into the through-track the contact wire of the through-track runs at first on the working range of the pan-head and then into the range of the useable contact wire lateral position (Figures 12.42 and 12.43). After the contact wire of the through-track is situated within the useable pan-head range of the useable contact wire lateral position, the contact wire of the branching-track can leave the working range on the opposite side of the pan-head.

As a result of this procedure the final run-up and run-off of the contact wires for the through- and branching-tracks are on different sides of the pantograph. Because of aspects mentioned in Clause 12.4.4.2 this situation has been considered as problematic for intersecting points wiring and undesirable until now.

Experience from tangential points wiring of high-speed lines shows that the advantages of the tangential points wiring prevails and running-up from both sides of the pantograph does not result in disadvantages regarding the interaction of overhead contact lines and pantographs (Figure 12.43), if the contact wire height and lateral position are designed in accordance with Clause 12.4.4.3. When running from and into the branching-track, the contact wires of the through- and branching-track, during the run-up, should both be situated for a short distance in the working range of the pantograph before they leave this range on their respective sides (Figures 12.42 and 12.43).

12.4.4.5   Arrangement of crossover droppers at track points

In the case of intersecting points wiring, *crossover droppers* improve the run-up of the contact wires to the pantograph for a rigid catenary wire. The positive effect of the crossover droppers cannot be verified for automatically tensioned catenary wires with correct contact wire heights, up to 200 km/h. Within the high-speed range, the use of crossover droppers is unnecessary for intersecting points wiring because the cross droppers react too slowly for the running speed. Tangential points wiring can be compared to standard overlaps, so crossover droppers are not required.

12.4.4.6   Connection of contact lines at track points

Both of the overhead contact lines in tangential points wiring need to be connected together by current connectors, as with intersecting points wiring. The contact lines can assume different potentials which could lead to arcing with running pantographs and possible contact wire burn-out and contact line disturbances. Current connectors contribute to equalising the potential between contact lines and avoiding potential differences.

12.4.4.7   Examples

Tangential points wiring can be observed in France, Spain, Turkey and Sweden for all operational speeds and in China, there only for high-speed overhead contact lines.

a) Top view

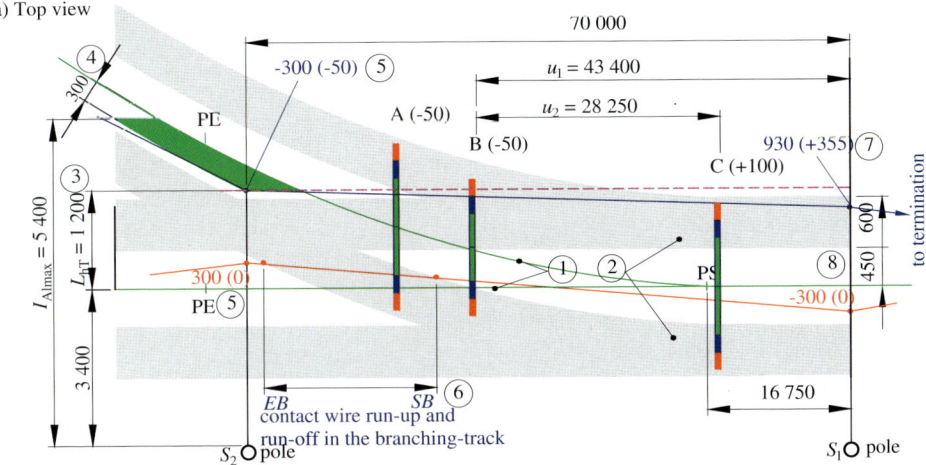

b) Cross section at position 1 for pantograph type 1950 mm   c) Cross section at position 2 for pantograph type 1600 mm

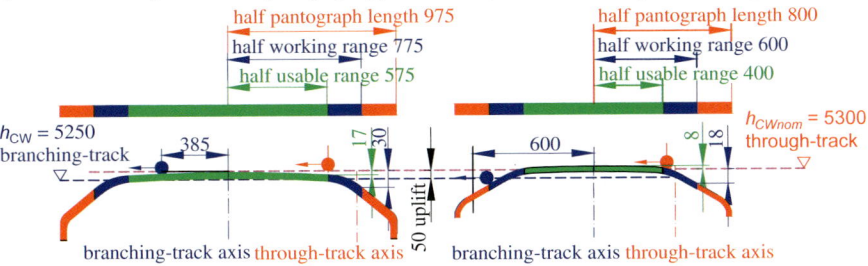

Track	Radial force at support in N	
	Support S_2	Support S_1
through-track	$463^{2)}$	$463^{1)}$
branching-track	$1\,921^{4)}$	$2\,477^{3)}$

[1] preceding span 70 000, preceding stagger -300
[2] subsequent span 70 000, subsequent stagger 300
[3] preceding span 47 100, preceding stagger $+300$, points in the opposite track
[4] subsequent span 70 000, subsequent stagger at the tensioning equipment 3 600

Figure 12.43: Tangential wiring of points EW 60-1200-1:18,5 on slab track with the overhead contact line type Re330 and for dual-mode with the fitting-free area valid for the 1 600 mm and the 1 950 mm long pan-heads corresponding to the example 12.4, all dimensions in mm. blue: contact wire of the branching-track, red: contact wire of the through-track, green: track axis

Example 12.3: The points EW 60-1200-1:18,5 on a concrete slab should be wired with the overhead contact line type Re330, consisting of an AC AC-120–CuMg contact wire tensioned at 27 kN tensile force for operation with a 1 950 mm long pan-head under consideration of the fitting-free area applicable for this condition with tangential wiring (Figure 12.42). The basic wind speed will be 20,3 m/s corresponding to a return period of ten years.

The following design steps are required for tangential wiring:

(1) *Distortion of the points layout plan*: The geometry of the points can be designed using Table 12.6, based on the radius of the branching-track and can be distorted by a factor of 10 transversely to the track as explained in Clause 12.4.4.3.

(2) *Marking the fitting-free area*: The fitting-free area should suitable for dual-mode operation with the 1 600 mm and the 1 950 mm long pan-heads. Initially, the line will be operated with 1 950 mm long pantographs. In the layout plan of the points the fitting-free area can be designed parallel to the track axis, offset at a distance of 0,45 m and with a width of 0,60 m.

(3) *Marking the position for L_{hT}*: The width of the kinematic-mechanical pantograph gauge L_{hT} is 1,20 m for the 1 950 mm long pantograph pan-head on a slab track according to Table 12.10.

(4) *Marking support S_2 for the branching-track*: The support S_2 of the branching-track contact wire is situated at the intersection of $b_T = 0,30$ m and $L_{hT} = 1,20$ m and 50 mm below the contact wire nominal height, which is presented in parentheses in Figure 12.42.

(5) *Marking the location of the S_2 through-track support*: The through-track support S_2 is situated at the same position as the branching-track support S_2 at $b_{2B} = 0,30$ m.

(6) *Determining the starting point of the through-track contact wire run-up and run-off*: The starting point (ST) and end point (ET) of the contact wire run-up for the through-track contact wire, when running from or into the branching-track happens at the intersection of the through-track contact wire with the fitting-free area of the branching pantograph (Figure 12.42).

(7) *Determining where to start lifting the contact wire of the branching-track*: At position 1, the through track contact wire is within the useable contact wire lateral position of the pantograph pan-head and the contact wire of the branching-track can be lifted from here towards support S_1. (Figure 12.42). Between position 1 and the support S_1 the contact wire is lifted in the form of a parabola (see (12.4)). In a distance of $u_{1\min}$ the contact wire can be lifted a maximum of 200 mm above the nominal contact wire height at the support S_1 (Table 12.11).

(8) *Determining the position of support S_1 for the branching-track*: At a distance of $u_1 = 46,5$ m from position 1, the branching-track contact wire can be lifted to 406 mm. The support S_1 of the branching-track contact wire can be located between this position and the start of the points and it will be clear of pantographs running on the through-track (Figure 12.42).

(9) *Determining the location of support S_1 for the through-track*: The support S_1 of the through-track contact wire should be placed at the same position as support S_1 for the branching-track contact wire. At this position the through track contact wire lateral position is $b = 0,30$ m and the branching track contact wire lateral position is $b = 1,200$ m . Figure 12.42 shows the span between the supports S_1 and S_2 as 60 m, which can be accepted for the specified wind velocity. Within the contacted area of the branching-track contact wire starting at the support S_2 up to the position 1 the branching-track contact wire is situated within the useable contact wire lateral position. This requirement is also complied with for the contact wire of the through-track.

(10) *Checking the selected wiring*:

  – The supports are placed outside the fitting-free area (requirement (1) of Clause 12.4.4.1).
  – The branching-track contact wire is situated outside the mechanical pantograph gauge (requirement (2) of the Clause 12.4.4.1).
  – The branching-track contact wire height is lower by 50 mm than that of the through-track (requirement (3) of the Clause 12.4.4.1).
  – The lateral position of the through-track contact wire at support S_2 is situated to the side of the branching-track (requirement (4) of the Clause 12.4.4.1).
  – The lateral position of the branching-track contact wire at support S_2 is situated on the side of the through-track (requirement (5) of the Clause 12.4.4.1).
  – When running from the branching-track into the through-track both contact wires are situated at least within the working range of the pantograph at the moment of contact wire run-up (requirement (6) of the Clause 12.4.4.1).
  – According to the table within the Figure Bild 12.42 the radial forces of the contact wire at supports S_1 and S_2 are within the permissible range 80 N $< F_R < 2\,000$ N for the selected wiring (requirement (7) of the Clause 12.4.4.1).

– The difference in radial forces at supports S_1 and S_2 is less than $200\,N$ within the through-track. Thereby, approximately equal elasticity is achieved within the range of the points (requirement (8) of the Clause 12.4.4.1). The speed in the branching-track is limited to $100\,km/h$. The radial force at support S_2 of the branching-track contact wire can be accepted, therefore. The support S_1 of the branching-track contact wire is lifted by more than $200\,mm$ and will not be traversed by the pantograph.

– The contact wire lateral position in case of wind action does not exceed at any position e_{max} (requirement (9) of the Clause 12.4.4.1).

– Crossover droppers are not planned for tangential wiring (requirement (10) of Clause 12.4.4.1).

– The requirements (11) and (12) of Clause 12.4.4.1 will be observed during planning.

Example 12.4: The point EW 60-1200-1:18,5 on slab track is to be wired with the overhead contact line type Re330, consisting of a contact wire AC-120 – CuMg tensioned at $27\,kN$ tensile force for dual operation with the $1\,600\,mm$ and $1\,950\,mm$ long pan-head under consideration of the valid fitting-free area to be wired tangentially (Figure 12.43). A basic wind speed of $20,3\,m/s$ with a ten year return period is to be used. The following steps are required for tangential wiring:

(1) Steps (1) to (6) are the same as for example 12.3.

(7) *Determine the starting point for lifting the branching-track contact wire*: For dual operation of both types of pan-heads, the branching-track contact wire should diverge from the axis of the through-track, so it does not run in parallel, however, it should approach the axis of the through-track in the direction of the support S_1. The through-track contact wire is guided from support S_2 to support S_1 changing contact wire lateral position. The branching-track contact wire lateral position at support S_1 should be selected to ensure a sufficiently large distance between position 2 of the $1\,600\,mm$ pan-head and position 3 of the $1\,950\,mm$ pan-head that will allow the through-track contact wire to be lifted between positions 1 and 2 by at least $100\,mm$. Furthermore, the span should be long enough to be able to lift the branching-track contact wire at support S_1 by at least $200\,mm$ above nominal contact wire height.

At the position where both contact wires are situated within the working range of the pan-head, or better, where the through-track contact wire is within the useable contact wire lateral position 2 for the shorter $1\,600\,mm$ pan-head, the branching-track contact wire can be lifted (Figure 12.43).

(8) *Determine the location of support S_1 for the branching-track*: Using the length $u_1 = 43,4\,m$ between position 2 and support S_1 the contact wire of the branching-track can be lifted with the shape of a parabola and will reach more than $200\,mm$ uplift at support S_1. At support S_1 the branching-track contact wire can be directly fixed to the registration tube and then guided directly to the tensioning equipment.

(9) *Determine the support S_1 location of through-track*: The support S_1 for the through-track contact wire should be at the same location as support S_1 for the branching-track contact wire (lateral position $b_B = 0,93\,m$). The distance between the supports S_1 and S_2 is $70\,m$ (Figure 12.43), and between the positions 2 and 3, $u_2 = 28,25\,m$. Therefore, according to equation (12.4) the branching-track contact wire situated at position 3, is $0,10\,m$ above nominal contact wire height and the pantograph running on the through-track will not touch this contact wire.

(10) *Checking the selected solution*:

– the requirements (1) to (6) of Clause 12.4.4.1 are met

– the radial forces of the contact wire at supports S_1 and S_2 are within the permissible range (requirement (7) of Clause 12.4.4.1)

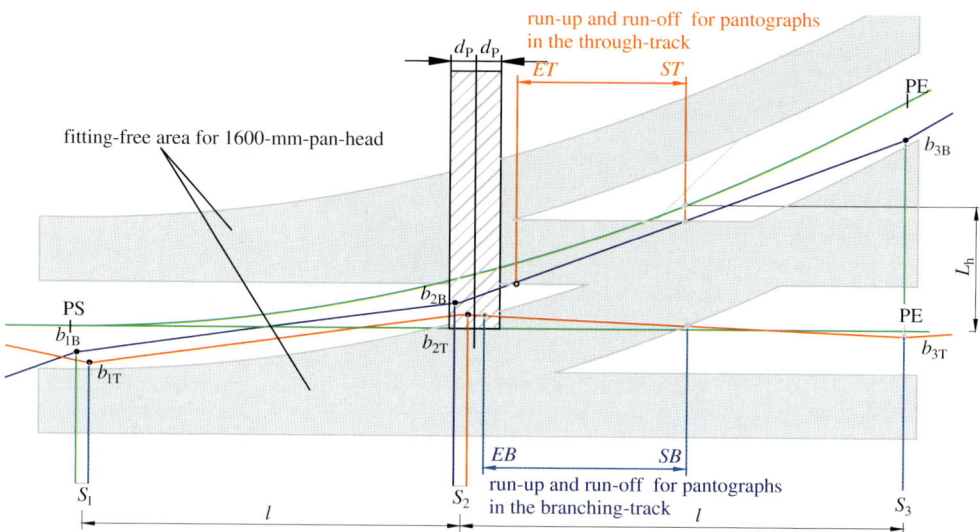

Figure 12.44: Tangential point design of point type EW60-1200-1:18,5 at SNCF for secondary tracks (Table 12.12).

– the difference in the radial forces at the supports S_1 and S_2 is less than 200 N. therefore, an approximately constant elasticity is available within the range of the points (requirement (8) of Clause 12.4.4.1)
– the contact wire lateral position does not exceed e_{max} (requirement (9) of Clause 12.4.4.1)
– requirements (10) to (12) of Clause 12.4.4.1 can be considered during further planning and the subsequent construction

From the examples it can be concluded that for tangential points wiring:
– the tangential points wiring and the intersecting points wiring comply with the specifications in Clause 12.4.3.1
– the advantage over intersecting points wiring is that the pantograph on the through-track does not touch the contact wire of the branching-track, resulting in lower contact wire wear, especially because of the absence of a contact wire crossing
– when running from or into the branching-track, both contact wires are always situated on the same half of the pan-head. As demonstrated by international experience, no disadvantages result from this arrangement concerning the interaction between overhead contact line and pantograph

12.4.4.8 Tangential wiring at SNCF

The French Railway (SNCF) uses, for secondary tracks up to 120 km/h, simple tangential wiring consisting of two contact lines. Corresponding with this type of points, the distance d_P and therefore, the position of the pole S_2 can be determined from Table 12.12 and Figure 12.44.

By touching both contact wires at support S_2, a higher contact wire wear is produced in this area. More favourable run-up conditions can be accomplished by points wiring corresponding to the Figures 12.42 and 12.43.

Table 12.12: The distance d_P determines the position of support S_2 (Figure 12.44).

Point type	Point angle $\tan \alpha_T$	Distance d_P m
EW 60-1200-1:18,5	0,054	±4,00
EW 60- 760-1:14	0,067	±3,30
EW 60- 500-1:12	0,083	±2,30
EW 60- 300-1: 9,4	0,106	±2,00
EW 60- 300-1: 9	0,111	±1,80
EW 60- 190-1: 7,5	0,133	±1,50

12.4.4.9 Crossover connections with tangential points wiring

Instead of section insulators, SNCF uses *tangential wiring* for crossover connections and overlaps in crossover connections. These connections are wired with three contact lines. On high-speed lines with diverging moves at 200 km/h, an additional contact line, also designated as an auxiliary contact line, guides the pantograph according to Figures 12.45 a), 12.45 b), 12.46 a) and 12.47 similar to guiding a pantograph in an overlap.

Figure 12.45 a) shows the tangential wiring of high-speed points EW60-17 000/ 7 300-1: 500-CC-TC, which complies with the requirements on wiring of points. The wiring within Figures 12.45 b) and 12.46 a) only partly complies with these conditions. At the support, the pantograph touches two contact wires simultaneously. Contact force measurements reveal high contact force peaks for such arrangements which result in high wear of the contact wire. The arrangement of tangential points wiring with auxiliary contact lines at portals (Figure 12.47).

12.4.5 Signals and signal visibility

Signals require a minimum spacing between the signal pole and the overhead contact line pole, e. g. this is 10 m at DB. Minimum electrical clearances according to EN 50 122-1 between earthed parts of the signals and the energised components of the overhead contact line and earthed signal components are also to be observed (Figure 12.93).

Unhindered line of sight for the cab driver to the signal is a safety requirement for railway operation. The overhead line poles should be located so that they permit uninterrupted signal visibility so the cab driver is able to fully recognise and comprehend complete signal information for at least 6,75 s from a specified distance s_{sight} onwards. Short-term obscuring of the signal by overhead contact line components such as poles, cantilevers, steady arms and insulators is permitted and does not require verification of visibility. The required signal visibility s_{sight} depends on the maximum running speed of the line according to [12.42] (see Table 12.13).

If the day-light main and advance signals are arranged in straight line sections at a clearance of 3,15 m to the track axis and the poles at a distance of 3,00 m, then special information on the signal sighting is not required [12.46]. If the distances of the signals to the track are larger, then *sighting verification* needs to be established.

Table 12.13: Signal visibility distance s_{sight} according to German DB regulation 819.0202 [12.42].

Line speed in km/h	Minimum signal visibility s_{sight} in m to	
	advance signal	main signal
≤100	200	300
≤120	250	400
>120	300	500

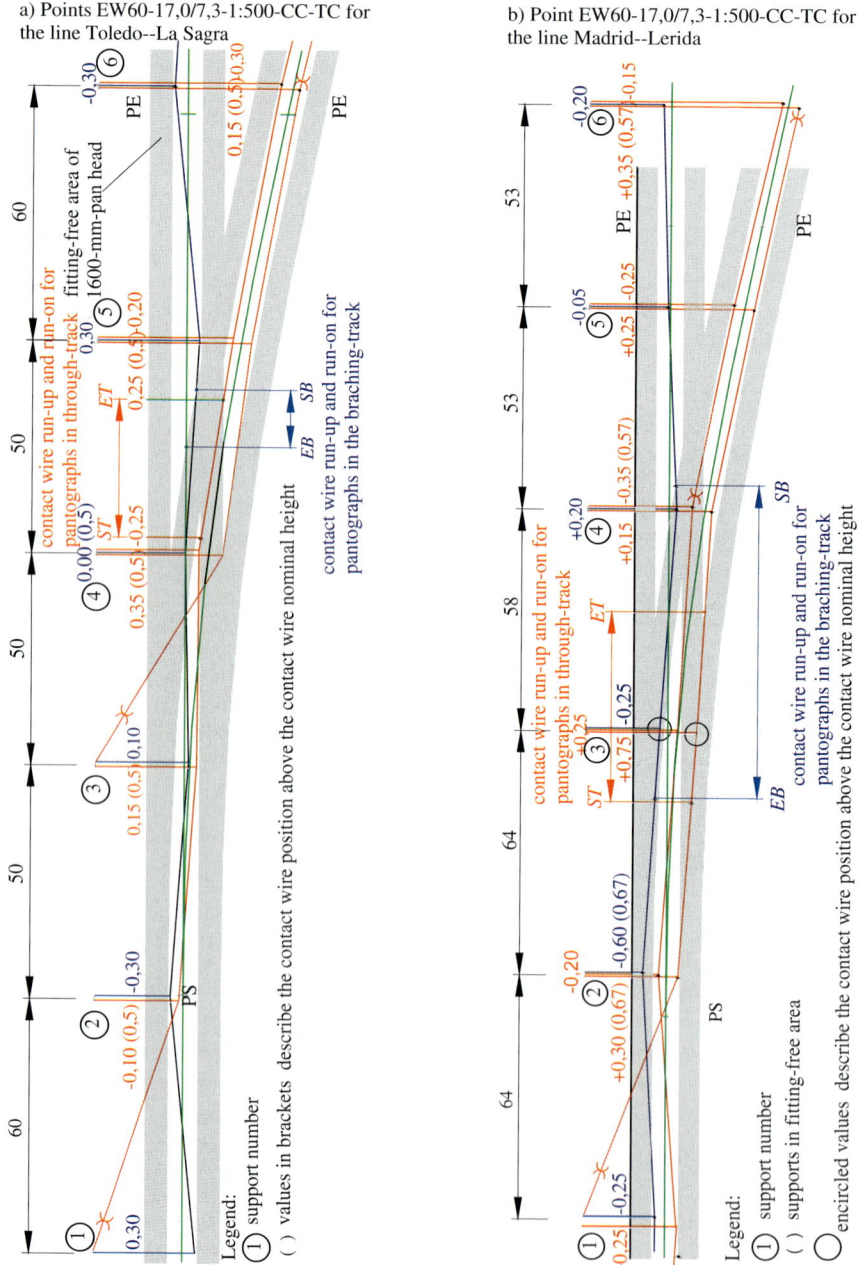

Figure 12.45: Tangential points wiring for high-speed lines. track axis, through-track contact wire, branching-track contact wire

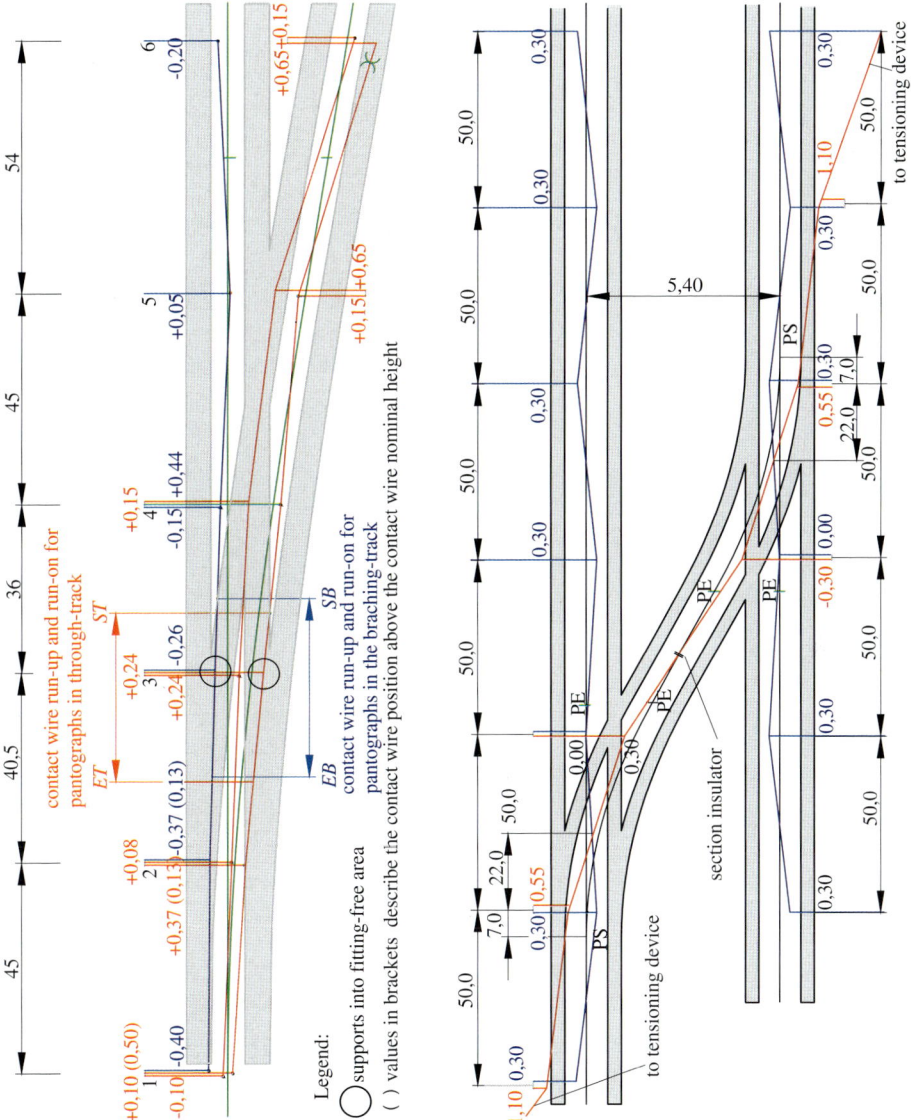

Figure 12.46: Points wiring of crossover connections. a) Tangential wiring with insulating overlap in the points connection for high-speed lines in France, b) intersection wiring with section insulators and points 60-1200-1:18.5 with fitting-free area for the Norwegian Railway, through-track contact wire, branching-track contact wire, track axis

Figure 12.47: Tangential point design with insulating overlap for the Atlantic high-speed line, France.

Table 12.14: Contact wire heights at railway level crossings according to EN 50 122-1: 2011.

Contact wire height h_{CW} above upper surface of road	Vehicle height [12.43]	Height limiting sign No. 265 [12.44]	Minimum height of profile gate lower edge
$h_{CW} \geq 5,50\,m$	4,0 m	no	–
$5,50\,m > h_{CW} \geq 5,00\,m$	4,0 m	yes	–
$5,00\,m > h_{CW} \geq 4,80\,m$ [12.14]	4,0 m	yes	$4,10\,m^{1)}$

[1)] 0,1 m between lower edge of the profile gate and the highest point of road vehicle

12.4.6 Overhead contact lines at level crossings

Level crossings between main line railways and roads are permitted only for limited speeds, in Germany up to 160 km/h [12.3]. For higher running speeds, level crossings need to be replaced by road-over or road-under crossings because safe use of railway crossings by road traffic is to be guaranteed during the planning of the overhead contact line. In continental Europe the road vehicle height is limited at 4,0 m, which is stipulated in Germany by the Directive on registration of road traffic (StVZO) [12.43]. Clause 14.7 describes temporary overhead contact line arrangements for oversize transports higher than 4,0 m.

The minimum contact wire height at level crossings is at least 5,50 m according to EN 50 122-1 including *contact line movements*, *ice loads* and *tolerances*. In case of contact lines with voltages exceeding AC 1 000 V or DC 1 500 V, a minimum distance of 1,5 m is required be-

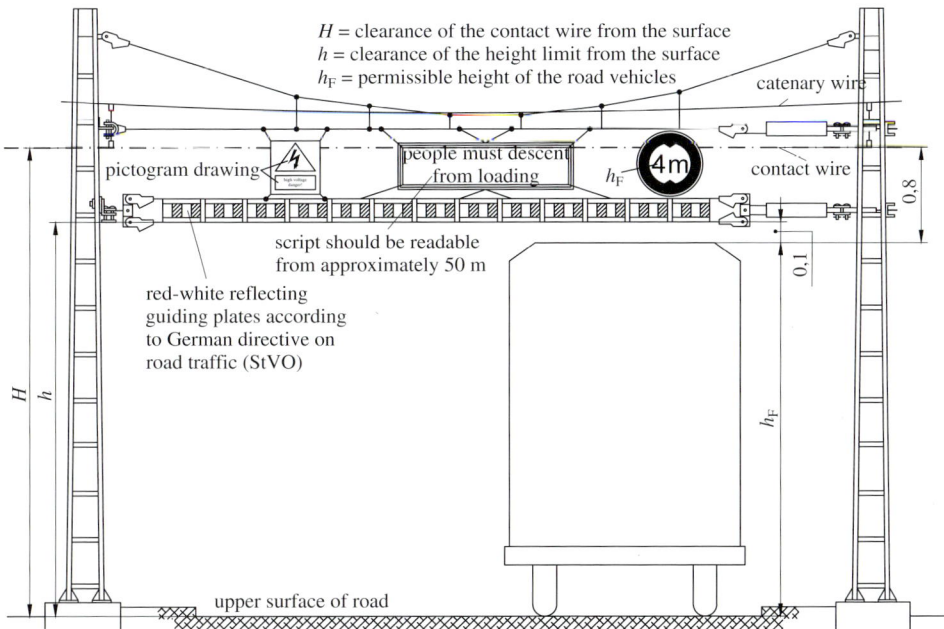

Figure 12.48: Height restriction arrangement with profile gate [12.45]. Dimensions in m

tween the 4,0 m height of a vehicle roof and contact wire lower surface. Then, *traffic warning signs* to limit the height of vehicles are unnecessary (see Table 12.14).

According to EN 50 122-1:2011 (Table 12.14) there is at least 1,0 m *safety margin* required between a vehicle top side and the contact wire lower surface. Then, *traffic warning signs* to limit the height of vehicles are necessary.

If it is necessary to install the contact wire at a height below 5,0 m [12.14] then *profile gates* (Figure 12.48) need to be installed on both sides of the railway crossing. By using *profile gates* the safety margin between the contact wire and vehicles can be reduced to a minimum of 0,5 m according to EN 50 122-1 (Table 12.14).

Figure 12.48 shows a height limitation installation with a *load barrier*, the lower edge of which is arranged 0,8 m below the contact wire height of 4,8 m. A clearance between vehicle and lower edge of profile gate needs to be at least 0,1 m.

After determination of minimum contact wire height in midspan the contact wire height is determined at the supports next to the level crossing depending on span length and contact line features such as tensile force and type of contact wire and catenary wire. The example 12.5 demonstrates the determination of the contact wire height at supports at railway level crossing. Accordingly, the contact wire height at the support amounts to at least 5,67 m in still air. German Railway DB installs contact wires at a height of 5,75 m, in still air, at the supports for the automatically tensioned contact lines Re100 and Re200.

Figure 12.49: Sign No. 265 height restriction to 4,0 m at level crossing [12.44].

Example 12.5: To prevent infringing on the minimum clearance of 5,50 m the contact line must be installed with an *increased contact wire height* of a 45 m span of accounting for the factors:

– minimum clearance between overhead contact line and road surface	5,50 m
– provision for lifting of tracks	0,05 m
– sag at specified ice load	0,07 m
– tolerance for installation	0,03 m
– contact wire movement downwards	0,05 m
– sag between two droppers	0,02 m
The contact wire height (h_{CW}) in still air is	5,67 m

The tensile forces of the contact and catenary wires vary because of resetting forces as described in Clause 5.9 of EN 50 119: 2009+A1: 2013. The maximum permissible variation of tensile forces in overhead contact lines should be taken into consideration (Clause 4.1.5.6 and [12.35]). The contact force variation needs to be considered when discussing the extension of the service period of the contact wires [12.47].

In the case of contact lines with fixed catenary wire, the contact wire height at supports needs to be designed to a height of more than 5,75 m considering span length and the contact wire sag at maximum contact and catenary wire temperature. At railway crossings with a minimum contact wire height of 5,5 m, a contact line lift to at least 5,75 m can be planned in accordance with permissible contact wire gradients and their changes as prescribed in EN 50 119: 2009+A1: 2013. In the case of a reduced contact wire height, the entity in charge of the road maintenance needs to plan the required traffic signs and if any, the profile gates. According to EN 50 122-1: 2011 and the German Tramway Design and Operational Regulation (BOStrab) [12.9] a minimum head room of 4,7 m above the road surface to live overhead components of the contact line, with voltages of DC 1,5 kV and AC 1,0 kV, needs to be planned. This height may be reduced to 4,5 m if a traffic sign No. 265 (Figure 12.49), according to the German highway code [12.44], announces the height limit. On the traffic signs, 4,0 m needs to be declared as the permissible vehicle height, i. e. the minimum contact wire height minus 0,5 m security margin according to EN 50 122-1. If the contact wire height needs to be lowered to 4,3 m and 4,0 m vehicle height maintained, profile gates as for main lines can be provided to limit the vehicle height. In this case the head room between vehicles and the contact wire lower surface may be reduced to 0,3 m according to EN 50 122-1.

12.4.7   Traction power lines at railway crossings

In contrast to automatically tensioned overhead contact lines, traction power lines are prone to temperature-dependent sag variations that cannot be compensated. The span length and temperature determine the sag of power lines. Their minimum clearance to the road surface is 6,0 m. The *minimum clearance of the power line* to the road surface needs to be calculated for maximum conductor temperature and/or ice load, where the more unfavourable value determines the support height of the power line.

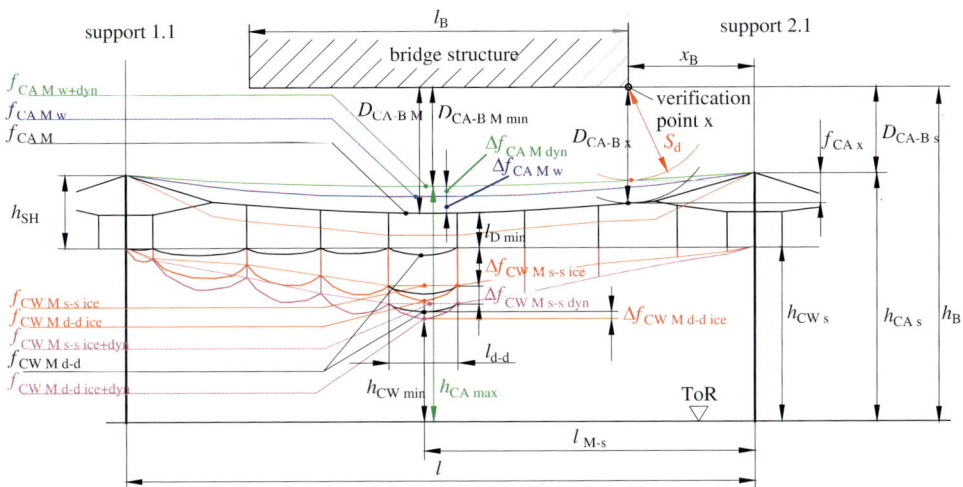

Figure 12.50: Clearances and sags of a contact line below a structure.
catenary and contact wire position in still air without wear and snow and ice, contact wire position with ice, catenary wire position at 20 % contact wire wear, catenary wire position with contact wire wear and pantograph passage, contact wire position with ice and pantograph passage

12.4.8   Traction power lines above road bridges

When a *road bridge* crosses railway tracks and traction power lines running parallel to tracks that cannot be routed below the road bridge together with the contact line, or installed as an underground cable, the power line can be routed above the bridge. A clearance of 7 m between the lowest point of the traction power line and road surface on the bridge needs to be provided in accordance with EN 50 341-1:2013. There is no spatial relation between the traction power lines above the road and the overhead contact line below a road bridge [12.48]. The minimum clearance required is 7 m in accordance with EN 50 341-1:2013.

12.4.9   Overhead contact lines below structures

Bridges and other buildings form *constraints for wiring* because they limit the free selection of pole positions and the wiring needs to be tailored to suit the individual structures (Figure 12.50). Space is often lacking below bridges and other structures to allow the arrangement of contact lines without limitations or special modifications at the structure. The arrangement of *overlaps* or *points wiring under bridges* is difficult. Compliance with minimum clearances between energised parts of the contact line and the structure and between the the contact wire and the top of rails needs to be verified during the design of the contact line. The existing clearance S_{exist} to the closest energised part of the overhead contact line can be determined from the bridge data including headroom, width, crossing angle, inclination of the bridge in parallel and transversely to the tracks and the profile of the underside of the bridge. The minimum clearances S according to EN 50 119: 2009+A1: 2013 are listed in Table 2.2. Railway operators often do not fully utilise these values but increase these clearances or insulate the catenary wires. At DB 0,6 m minimum clearance S_{d} is observed below and beside structures thus avoiding arcs caused by birds, icicles in winter and water streams. For distances smaller than 0,6 m between structure and catenary wire, DB adopts insulated catenary wires.

a) After installation   b) Six months later

 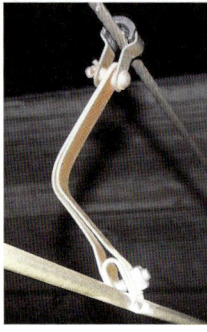   

Figure 12.51: Sliding dropper, unsuitable design.

Figure 12.52: Dropper triangle below buildings with sliding clamp on catenary wire and fixed distance between catenary and contact wire (Photo: Kruch, Austria) [12.40].

In this case, the dropper clamps at the catenary wire need to be insulated (Figure 12.53) and the contact wire should be strung without pre-sag below buildings. The weight of the insulated catenary wire is taken into account when determining the length of the droppers.

For *verification of clearances*, the wear of the contact wire needs to be considered. A wear of 20 % unloads the catenary wire resulting in a more elevated position of the catenary wire. Under this condition, the minimum clearance must not be infringed.

If the minimum clearance S_d, according to Figure 12.50 and Table 2.2, for the nominal contact wire and system height are not achieved, the distance between the building and catenary wire could be increased by extending the span length. The parabolic shape of the catenary wire sag results in a greater distance between the catenary wire and the building at the edges of the buildings. If this is not sufficient, the catenary can be lowered by reducing the system height h_{SH}. A limit is provided by minimum length of droppers at midspan (Figure 12.50). Below buildings, flexible droppers, that do not disturb the interaction between pantograph and contact line should be adopted. The commercial speed v effects the length and material of flexible droppers. For droppers made of bronze BzII the minimum lengths l_{Dmin} are:

– 300 mm for $v \leq 120$ km/h,
– 500 mm for $120 \, \text{km/h} < v \leq 230$ km/h
– 600 mm for $v \geq 230$ km/h

Shorter *minimum lengths of droppers* are possible through the use of fatigue-resistant dropper conductors or oblique droppers (Figures 12.53 and 11.40). If the catenary wire height needs to be reduced further, *sliding droppers* approximately 70 mm high can be installed instead of conventional droppers (Figure 11.41 a)). However, the consequences are extremely variable elasticity, unequal wear and bowing of the contact wire in the upward direction. Sliding droppers are only suitable for speeds up to 80 km/h. Sliding droppers according to Figure 12.51 are not suitable and need to be constantly monitored. Clause 11.2.2.7 gives further information on sliding droppers.

If the reduction of the catenary height is not sufficient, the contact wire can be reduced to the *minimum contact wire height* of 5,0 m (Table 12.5 and Figure 12.50).

The use of double or triple contact wires without a catenary wire offers further possibilities to minimise the space for installation of overhead contact lines. Due to the low elasticity of

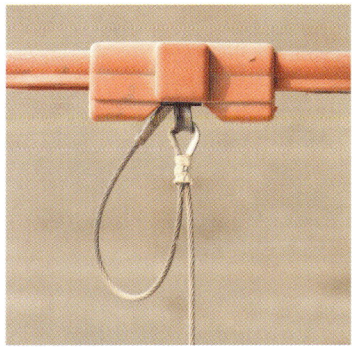

Figure 12.53: Catenary wire protection below a bridge for the 25 kV high-speed contact line.

Figure 12.54: Insulating plate and insulated catenary wire at Oslo S, Norway (Foto: Bane Nor, T. Pedersen).

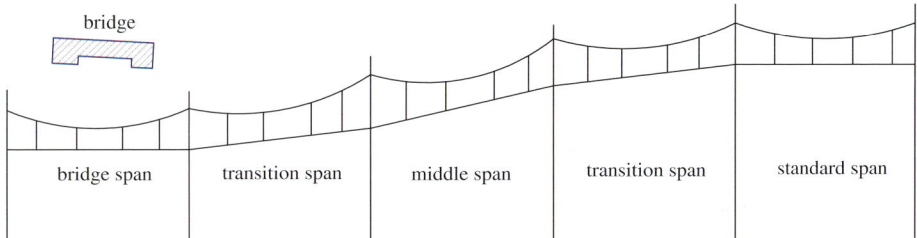

Figure 12.55: Arrangement of transition and centre spans at overhead contact line height reductions.

such contact lines, only low contact wire lifts occur. Insulating plates at the lower surface of a bridge allow reduced clearance without risk of the contact wire or pantograph touching the insulating plates (Figure 12.54).

For low bridges, the catenary wire can be anchored at the bridge face and guided under the bridge or lowered to the contact wire level and guided through as a second contact wire below the building. For restricted space conditions, protective sections are installed before and after the bridging structure. Then the contact wire can be earthed at the bridge enabling *reduced clearance* between contact wire and building to only mechanical requirements with consideration for dynamic uplift. However, switching off the vehicle circuit breaker will be necessary at each passage under the building.

Rigid conductor rails do not provide a big advantage below low bridges when compared with flexible contact lines. The design height of conductor rail is similar or even higher than that of twin or triple contact wires. The transition from flexible contact line to the rigid conductor rail affects the interaction between pantograph and contact line at speeds above 100 km/h. If the above mentioned measures are insufficient, lowering the rail level or lifting or reconstructing the bridge will be required.

Once the contact wire height, system heights and *span length in the span under the bridge* have been determined, the layout of the *contact wire gradient* in the neighbouring spans can proceed. Distinction is made between transition and centre spans (Figure 12.55). The maximum gradients in the transition and centre spans depend on the maximum operational speed and should be based on the running speeds as shown in Table 12.3. The lowest point of

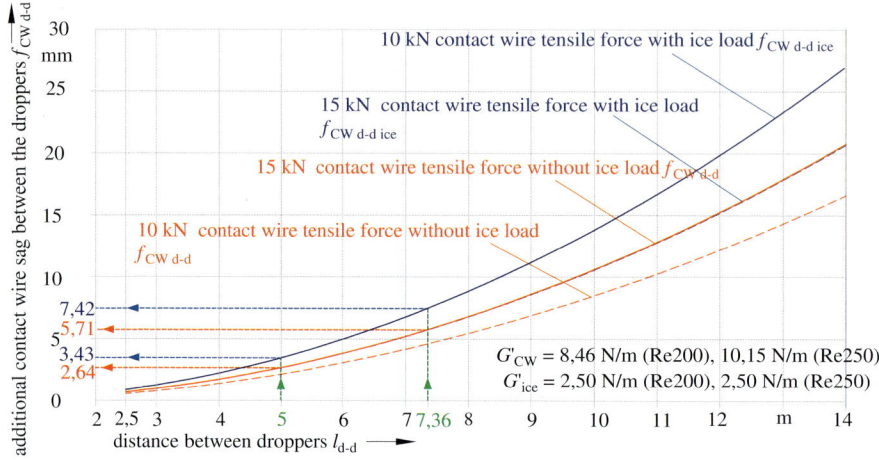

Figure 12.56: Maximum contact wire sag between droppers with and without ice loading, ice load zone I1 for a contact line with AC-100 or AC-120 contact wire.

the contact wire in any of the transition or centre spans may not be lower than in bridge span. The design of the *contact wire lowering* starts with the span under the obstacle by evaluating the minimum contact wire height $h_{CW\,min}$ (Figure 12.50), the correct determination of which is important as it affects the investment of the project. The minimum contact wire height $h_{CW\,min}$ is determined by:

– Height of the static vehicle gauge

   Determination of the minimum contact wire height $h_{CW\,min}$ is based on the static and kinematic gauge GC of the vehicle according to DIN EN 15 273-2: 2014, Figure B.1 and B.2, respectively [12.14]. The heights of the gauges GC amount to 4 650 mm and 4 700 mm, respectively (Figure 2.9 a) and b)). The static vehicle gauge already includes a 50 mm lifting provision so that provision for track lifting only needs to be considered for greater amounts. For concrete slab permanent way no lifting provision is required, see also Table 12.2.

– Minimum electrical clearance

   The minimum electrical clearance in air according to Table 2.2 is related to the upper vehicle gauge limit and depends on the nominal voltage of the contact line. This clearance is different for static and dynamic conditions during short-period exposures. To calculate the minimum contact wire height $h_{CW\,min}$ for 15 kV overhead contact lines, 150 mm clearance is observed for permanent conditions and 100 mm for short-term clearance [12.14].

– Lateral track inclination

   In the case of a cant u below the building a lateral inclination $T_{cant} = 2/3\,u$ at contact line height of the contact line needs to be considered. The cant u is obtained from the layout plan.

– Longitudinal track gradient

   In the case of tracks with a longitudinal gradient or change of gradient under structures, a track longitudinal gradient T_{length} at the contact line height needs to be considered. Depending on the structural degree of coverage, this will result in a reduction of the minimum contact wire height.

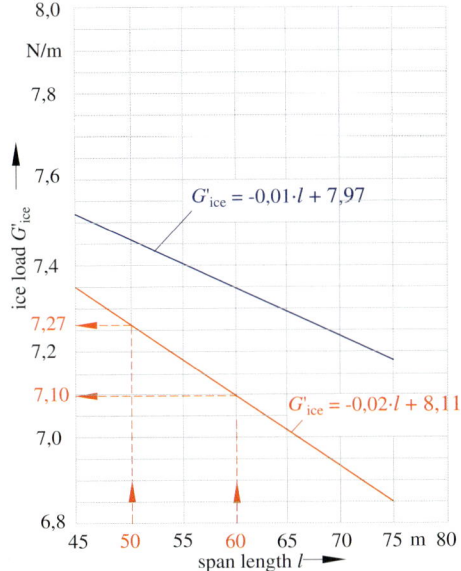

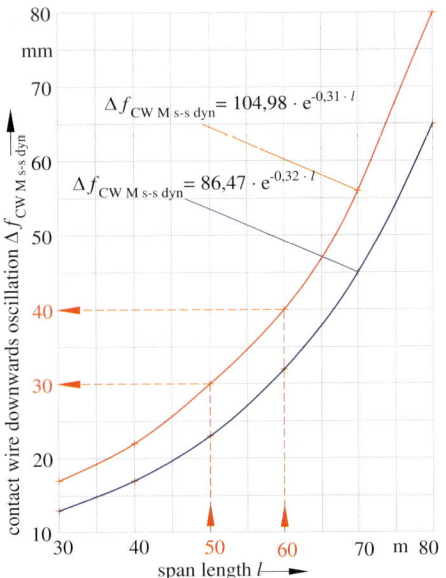

Figure 12.57: Ice load G'_{ice} per meter contact line, ice load zone I1, new contact wires AC-100 for Re200 and AC-120 for Re250 depending on the span.

Figure 12.58: Contact wire downwards oscillation at midspan of Re200 and Re250 after pantograph passage at 200 km/h and 250 km/h, respectively [12.51].

– Contact wire sag between droppers due to ice accretion
  Between the droppers, a climatic-dependent additional sag of the contact wire occurs. Figure 12.56 shows the contact wire sag with and without ice load for tensile stresses of 10 kN and 15 kN, whereby the ice load depends on the local ice load area (Clause 2.4.3) and the conductor diameter. The contact wire sag due to ice accretion $\Delta f_{CW\,d-d\,ice}$ between adjacent droppers depends on the dropper spacing and the contact wire tensile force.

– Contact wire sag between supports due to ice accretion
  Ice accretion on catenary wire, contact wire, stitch wire and droppers increases the load G'_{OCL} to $G'_{OCL\,ice}$ related to the span length by the ice load G'_{ice}. There is an additional contact wire sag at midspan between the adjacent supports $\Delta f_{CW\,Ms-s\,ice}$, which needs to be considered when calculating the minimum contact wire height. In Figure 12.57 the length-related ice load G'_{ice} is shown according to EN 50 341-1-3:2012 (see Figure 2.29) for ice load area I1 for contact lines Re200 and Re250 taking into consideration the effective conductor lengths, the droppers included. For short spans, the relation between conductor length and span length is greater than for long spans, so a non-linear ratio between ice load and span length results. In Figure 12.59, the contact wire sag $\Delta f_{CW\,Ms-s\,ice}$ is presented according to the related ice load as shown in Figure 12.57.

– Contact wire oscillations downwards after passing of a pantograph
  After the passing of a pantograph, downward oscillations of the contact wire with amplitude $\Delta f_{CW\,Mdyn}$ are created by the contact wire lift, so the contact wire oscillates below its position in still air. This dynamic movement lasts for a short period only,

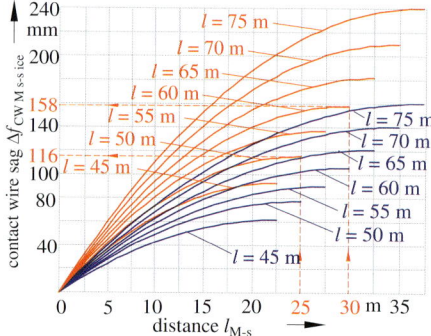

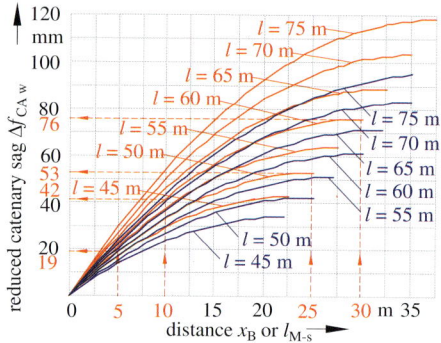

Figure 12.59: Additional contact wire sag at midspan between supports $\Delta f_{CWMs-sice}$ due to ice load in zone I1 for the contact line types Re200 and Re250, ice load per meter contact line according to Figure 12.57.

Figure 12.60: Reduced catenary wire sag Δf_{CAMw} in a distance x_B from the catenary wire support to the verification point x at 20% contact wire wear for the contact line types Re200 and Re250.

where events lasting for less than 5 s are considered as short term and ones lasting more than 5 s as long term. In this case, 100 mm vertical electrical clearance applies in air. In Figure 12.58 the contact wire downwards oscillation Δf_{CWMdyn} is shown depending on the span length.

– Tolerance of contact wire height

The tolerance $T_{CWheight}$ depends on the contact wire type and the contract agreements. For contact lines Re200 and Re250 these are +/-100 mm [12.49, 12.50]. German Rail DB reduces these tolerances to +/10 mm for contact line types Re200 and Re250 [12.50, 12.15] at railway crossings and under structures.

The lowest possible contact wire height h_{CWmin} is determined by adding the parameters:
– limit height of the vehicle gauge GC
– permanent or short-term electrical clearance in air S_d or S_k
– track lateral inclination T_{cant} and track longitudinal gradient T_{length}
– contact wire sag between supports due to ice load $\Delta f_{CWMs-sice}$
– contact wire sag between droppers due to ice load $\Delta f_{CWMd-dice}$
– contact wire downwards oscillation at midspan $\Delta f_{CWMs-sdyn}$ by pantograph passage
– contact wire height tolerance $T_{CWheight}$.

After evaluating the lowest contact wire height, the calculation of maximum catenary wire height follows, considering the electrical minimum clearances in air between catenary wire and structures (Figure 12.50). The electrical minimum clearance in air may not be infringed, especially for the catenary wire under both static and dynamic conditions. The maximum catenary wire height consists of:

– Catenary wire sag between supports

The catenary wire sag f_{CA} depends on the tensile force and the weight per unit length of the catenary wire. (Table 11.7). In Figure 12.61, the catenary wire sag f_{CA} between the catenary wire support and bridge edge x_B or between the catenary wire support and midspan l_{M-s} is shown for different spans l of the overhead contact lines Re200 and Re250.

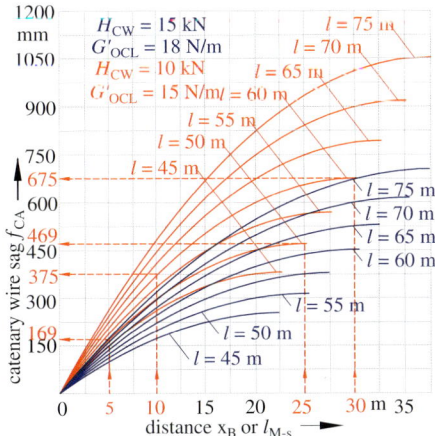

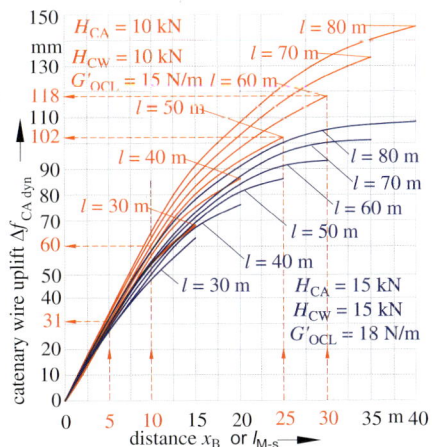

Figure 12.61: Catenary wire sag f_{CA} without ice load for the contact line types Re200 and Re250 with new contact wire in a distance x_B between catenary wire support and reference point x.

Figure 12.62: Catenary wire uplift $\Delta f_{CA\,dyn}$ after pantograph passage for contact line types Re200 and Re250 at 200 km/h and 250 km/h, respectively, at a distance x_B between catenary wire support and the reference point x [12.52].

- Reduced catenary wire sag in the case of 20 % contact wire wear
  The catenary wire sag is reduced with a decrease in the weight per unit length of the contact line, reaching its minimum at a contact wire wear of 20 %. The catenary wire sag is reduced by the amount Δf_{CAMw}. Figure 12.60 shows the catenary wire uplift Δf_{CAMw} depending on the parameter x_B starting from the support in the direction to bridge edge and from the support in the direction to midspan l_{M-s} for differing span lengths l for the contact lines Re200 and Re250.
- Catenary wire uplift when a pantograph passes
  The catenary wire sag is decreased for the short period of time, during which the pantograph passes under the building. Because of the static and dynamic contact force, the pantograph lifts the contact wire. The contact wire load, the dropper load and the catenary wire load are reduced with the result that the catenary wire is lifted for a short period and approaches the building. In Figure 12.62, the catenary wire lift Δf_{CAMdyn} is shown for the contact line types Re200 and Re250 if a pantograph passes at 200 km/h and 250 km/h under the building. The catenary wire lift in Figure 12.62 is based on the mean contact forces according to EN 50 367 for 200 km/h and 250 km/h, respectively.
- Tolerance of catenary wire height
  The tolerance $T_{CA\,height}$ of the catenary wire height is a maximum of ± 100 mm for contact lines Re100, Re200 and Re250.

The electrical minimum clearance $S_d = 150$ mm in air needs always to be observed between the catenary wire and the bridge for:

- catenary wire sag f_{CA}
- catenary wire uplift Δf_{CAMw} considering a 20 % contact wire wear
- catenary wire tolerance $T_{CA\,height}$

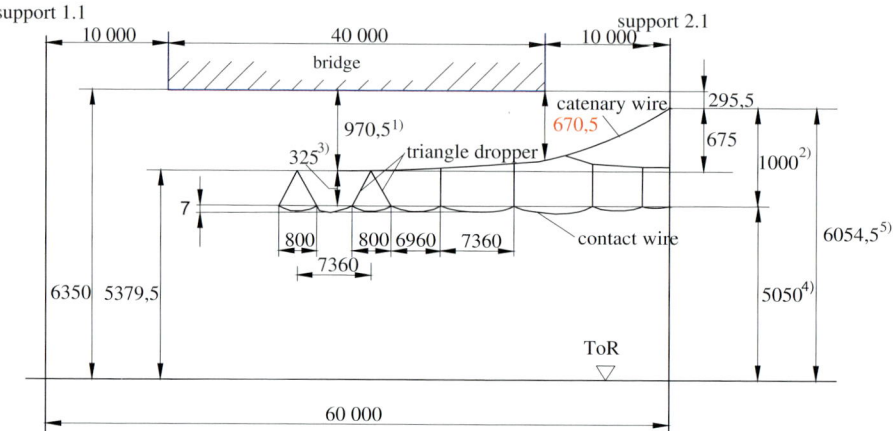

Figure 12.63: Arrangement of span under the bridge according to Example 12.6. Dimensions in mm

1) distance between lower surface of the bridge and upper edge of catenary wire
2) height between lower edge of contact wire and centre of catenary wire
3) distance between lower edge of contact wire and centre of catenary wire (catenary wire diameter 9 mm, contact wire diameter 12 mm)
4) distance between ToR and lower edge of contact wire
5) distance between ToR and upper edge of catenary wire

The minimum dynamic electrical clearance $S_k = 100$ mm in air between catenary wire and bridge must not be infringed for:

- catenary wire sag f_{CA} with the corresponding contact line weight
- catenary wire uplift Δf_{CAMw} considering a 20 % contact wire wear
- tolerance of catenary wire height $T_{CA\,height}$
- catenary wire uplift $\Delta f_{CA\,dyn}$ in the case of a passing pantograph

The clearance between catenary wire and structure needs to at least correspond to the permanent electrical clearance S_d or the short-term clearance S_k. Some European rail infrastructure managers, e g. German Rail DB, also require plastic-insulated catenary wires or insulating plates as bird protection below structures, if the clearance is below 0,6 m [12.53]. The protection of the catenary wire needs to be rated for the nominal voltage of the contact line (Figures 12.53 and 12.54, respectively). After the minimum contact wire height and the selected maximum catenary wire height have been determined, the space for the contact line can be determined and the shortest possible dropper length at midspan can be calculated. The shortest dropper length should not be less than the values given in Clause 12.4.9.

Example 12.6: The contact line lowering is to be designed for a line with ballasted track and dimensions of a bridge according to Figure 12.64. The aim is to design the minimum contact wire height, the maximum catenary wire height and the system height for the minimum contact wire height and the maximum catenary wire height for the span under the bridge for two options. In option 1, the span under the bridge is 60 m long with a dropper spacing of 7,36 m. In option 2 these data are 50 m and 5,00 m respectively. After determining the minimum contact wire height and the maximum catenary wire height, the contact wire heights, catenary wire heights and system heights can be selected for implementation considering the minimum dropper length $l_{D\,min}$. For the range of the contact line lowering adjacent to the bridge, the contact wire and the maximum catenary wire heights need to be determined. The reference for minimum contact wire height will be 4,65 m for the static vehicle gauge

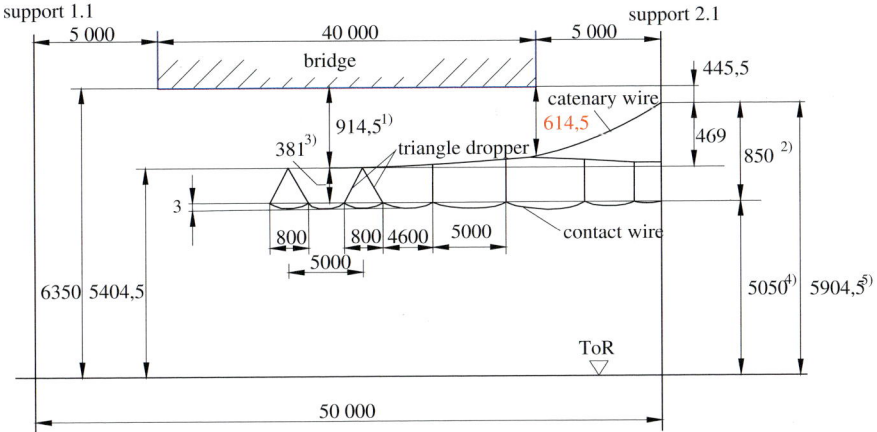

Figure 12.64: Arrangement of a 50 m span and 5,0 m dropper spacing under the bridge according to example presented in Example 12.6. Dimensions in mm, legend see Figure 12.63

according to EBO [12.3] and gauge GC [12.14]. The contact wire displacement, caused by wind below the bridge can be neglected. At the site, ice load zone I1 applies. Sliding droppers according to Figure 12.51 can not be applied but, if required, triangle droppers according to Figure 12.52 can be used. The data to design the contact line lowering in the Figures 12.58 and 12.59 for the contact line Re200 at $v = 200$ km/h are:

- maximum commercial speed   200 km/h
- nominal voltage   15 kV
- overhead contact line type   Re200
- tensile force of contact/catenary wire   10 kN/10 kN
- length l of span under bridge, option 1   60,0 m
- length l of span under bridge, option 2   50,0 m
- headroom under bridge h_B 6,35 m
- width of bridge   40,0 m
- crossing angle between bridge and tracks   100,0 gon
- distance x_B between overhead line support and bridge, option 1   10,0 m
- distance x_B between overhead line support and bridge, option 2   5,0 m
- cant and longitudinal gradient   0 m / 0 %
- specific load of overhead contact line G'_{OCL} 15,0 N/m
- nominal system height h_{SH} 1,8 m
- nominal contact wire height $h_{CW\,nom}$ 5,5 m
- contact force of pantograph at 200 km/h according to EN 50 367: 2017   100 N
- minimum short-term electrical clearance S_k for 15 kV   0,10 m
- minimum permanent electrical clearance S_d for 15 kV   0,15 m
- contact wire height tolerance $T_{CW\,height}$ according to [12.50]   ± 0,010 m
- limited catenary wire height tolerance $T_{CA\,height}$ according to [12.50] ± 0,030 m

– Calculation of minimum contact wire height

The minimum contact wire height was determined in Table 12.15 and may not fall below 5 016 mm and 4 960 mm for options 1 and 2, respectively.

– Calculation of maximum catenary wire height at the support. Dimensions in mm

The maximum catenary wire height at the support is calculated in Table 12.16. The values marked in blue and red at the last table line are the maximum electrical clearances required and need to be

Table 12.15: Calculation of the minimum contact wire height. Dimensions in mm

Option	1		2	
span length l	60 000		50 000	
distance of droppers at midspan l_{d-d}	7 360		5 000	
reference gauge	static and kinematic vehicle gauge GC according to EN 15 273-2: 2013, Bild B.1 and Bild B.2 respectively			
reference height	4 650	4 700	4 650	4 700
extensions at track cant changes in curves	0			
extensions at track gradient changes	0			
track lifting reserve	already included in the vehicle gauge			
electrical clearance S_{el}	permanent	short-term	permanent	short-term
	150	100	150	100
tolerance of contact wire height $T_{CW\,height}$	10	10	10	10
contact wire downwards oscillation $\Delta f_{CW\,Ms-sdyn}$ at midspan (Figure 12.58)	0	40	0	30
additional contact wire sag between droppers with ice load $\Delta f_{CW\,Md-dice}$ (Figure 12.56)	8	8	4	4
additional contact wire sag between supports $\Delta f_{CW\,Ms-sice}$ at midspan due to ice load (Figure 12.59)	158	158	116	116
minimum contact wire height $h_{CW\,min}$ at support	4976	5 016	4 930	4 960

Note: permanent: contact line is in a static condition with no trains running under the structure,
short-term: a train with pantograph passes under the structure and creates downwards oscillations

adopted for the following steps. Some infrastructure managers, including German Rail DB, adopt 0,6 m as the minimum electrical clearance between bare catenary wires and structures.

– Calculation of the system height

Table 12.17 shows the calculation for the maximum system height.

– Selection of catenary wire, contact wire and system height at the support.

Table 12.18 shows the selection of catenary height, contact wire height and system height at the support, midspan and bridge edge. The distance between contact and catenary wire at midspan is less than 0,5 m resulting in the adoption of triangle droppers. Consequently, a plastic-clad catenary wire is not required.

– Calculation of the existing clearance between structure and catenary wire

The selected arrangements permit a minimum contact wire height of 5,05 m. In the case of ice load, the contact wire height does not fall below 4,80 m [12.14]. For options 1 and 2, the length of the inclined droppers does not fall below 0,5 m, which is the minimum dropper length according to Clause 12.4.9 (Figures 12.63 and 12.64).

The clearances between the catenary wire upper edge and bridge edge are 670,5 mm and 614,5 mm, respectively, and therefore, more than 0,6 m as is the minimum clearance requirement according to [12.54]. Insulated catenary wires or insulting protection under the bridge will not be necessary. Figures 12.63 and 12.64, show the arrangements for 60 m and 50 m spans.

– Selection of contact wire gradients in the adjacent spans

The contact wire heights at adjacent supports can be obtained from Table 12.3, up to 250 km/h under consideration of the allowable gradients. Above 250 km/h, contact wire gradients are not possible.

The contact wire heights at the adjacent supports can be determined by using the allowable gradients

Table 12.16: Calculation of maximum catenary wire height at the support. Dimensions in mm

Option	1		2	
span length	60 000		50 000	
horizontal distance between catenary wire support and bridge edge x_B	10 000		5 000	
electrical clearance in air	permanent	short-term	permanent	short-term
minimum clearance between bridge and catenary wire $S_{d\,min}$	600[1]	600[1]	600[1]	600[1]
catenary wire lift Δf_{CAMdyn} at bridge edge due to pantograph passage[1] (Figure 12.62)	0	60	0	31
catenary wire lift Δf_{CAMw} at bridge edge due to 20 % contact wire wear (Figure 12.60)	42	42	19	19
catenary wire tolerance $T_{CA\,height}$	30	30	30	30
minimum clearance S_{min} between catenary wire and bridge[1)2)] after installation	672[2]	732[2]	649[2]	680[2]
catenary wire sag at bridge edge f_{CAx} without ice load (Figure 12.61)	375	375	169	169
vertical difference between bridge lower face and catenary wire at support	−297	−357	−480	−511
catenary wire sag at midspan without ice load f_{CAM} according to (Figure 12.61)	675	675	469	469
distance between catenary wire upper face and bridge lower face at midspan	$675 - (-357) = 1\,032$		$469 - (-511) = 949$	
headroom of bridge h_B	6 350			
maximum height of catenary wire upper face at support above ToR h_{CAs}	$6\,350 + (-357) = 5\,993$		$6\,350 + (-511) = 5\,839$	
maximum height of catenary wire upper face at midspan above ToR	$5\,993 - 675 = 5\,318$		$5\,839 - 469 = 5\,370$	

[1] for the catenary wire lift, a maximum contact force according to EN 50 367: 2017

[2] for clearances below 0,6 m German Rail DB adopts plastic-clad catenary wires

according to Table 12.3. Within the range of the contact line lowering on both sides of the bridge the contact wire heights should develop with the shape of a sinusoidal graph resulting in an approximately constant vertical acceleration of the pantograph between the transition and standard spans.

Figure 12.65 shows the required data for the design of the contact wire lowering according to Example 12.6 for option 1.

The system height h_{SH} should increase as soon as possible to enable longer droppers and consequent improvement in dynamic behaviour. It follows that the system height will be increased by 0,80 m, from 1,00 m at support 2.1 to 1,8 m standard system height at support 2.2.

If low system heights are increased to standard system height at the adjacent support, the catenary wire supports may not be unloaded at low temperatures. In such cases, it is recommended, that the vertical load components at the catenary wire support are verified for lowest height at the lowest temperature (Clause 4.3.3).

At speeds > 250 km/h, the contact forces distort the dropper clamps in contact line lowerings, with reduced system height, which can result in disturbances. According to EN 50 119: 2009+ A1: 2013 this is one reason for avoiding contact line lowering in lines for speeds > 250 km/h (see Figure 12.51 b)).

Table 12.17: Maximum system height based on minimum contact wire height and maximum catenary wire height. Dimensions in mm

Option	1	2
maximum catenary wire height at support catenary wire upper edge h_{CAs}	5 993,0	5 839,0
radius of the catenary wire	4,5	4,5
minimum contact wire height at support	5 016	4 960
system height[1] SH	$5 993 - 4,5 - 5 016 = 972,5$	$5 839 - 4,5 - 4 960 = 874,5$

[1] system height according to DB definition: distance from catenary wire center to lower face of contact wire

Table 12.18: Selection of catenary wire, contact wire and system height. Dimensions in mm

Option	1	2
selected contact wire height at the support h_{CWs}	5 050	5 050
minimum distance between catenary and contact wire at the triangle dropper at midspan with 200 mm at support additionally 100 mm at midspan of the line category II [12.13] (minimum dropper length l_{Dmin})	$200 + 100 = 300$	
catenary wire sag at midspan without ice load f_{CAM} according to Figure 12.61	675	469
minimum system height with triangle droppers SH_{min} according to Figure 12.61	$300 + 675 = 975$	$300 + 469 = 769$
selected system height SH	1 000	850
available distance between catenary wire and contact wire at midspan as the dropper length l_D	$1 000 - 675$ $= 325$	$850 - 469$ $= 381$
available catenary wire height upper edge at support above ToR h_{CAs}	$5 050 + 1 000 + 4,5$ $= 6 054,5$	$5 050 + 850 + 4,5$ $= 5 904,5$
available catenary wire height upper edge at midspan above ToR h_{CAM} after installation	$6 054,5 - 675$ $= 5 379,5$	$5 904,5 - 469$ $= 5 435,5$
head room of bridge h_B	6 350	
available distance between catenary wire and bridge at midspan D_{CA-BM} after installation	$6 350 - 5 379,5$ $= 970,5$	$6 350 - 5 435,5$ $= 914,5$

Table 12.19: Available distance between bridge and catenary wire. Dimensions in mm

Option	1	2
catenary wire upper edge height at support h_{CAs}	6 054,5	5 904,5
catenary wire upper edge sag at the bridge edge	375	169
catenary wire height at the bridge edge	$6 054,5 - 375$ $= 5 679,5$	$5 904,5 - 169$ $= 5 735,5$
head room of bridge h_B	6 350	
distance between catenary wire and bridge edge S_d	$6 350 - 5 679,5$ $= 670,5$	$6 350 - 5 735,5$ $= 614,5$

support 1.1   support 2.1 support 2.2   support 2.3   support 2.4   support 2.5 support 2.6 support 2.7

bridge

span length	60	65	70	75	75	75	75
span type	bridge 0	transition 1	middle 2	middle 3	transition 4	transition 5	standard 6
gradient in ‰	0	1,00	2,00	2,00	1,00	0,27	0,00
gradient change in ‰ 1,00	1,00	1,00	0,00	-1,00	-0,73	0,00	0,00
system height	1,00	1,00	1,80	1,80	1,80	1,80	1,80	1,80
contact wire height 5,050	5,050	5,115	5,255	5,405	5,480	5,500	5,500

Figure 12.65: Arrangement of spans adjacent to the bridge as per Example 12.6 for option 1. Dimensions in m

a) In Germany

b) In Norway

Figure 12.66: Pole foundation on a bridge.

12.4.10   Railway bridges

Railway bridges form *constraints* to the selection of pole sites, since they can usually only be placed at the columns of the bridge. The longitudinal girders on new lines can accommodate pole foundations (Figure 12.66) and sites for pole foundations need to be selected through close cooperation between planning engineers for the bridge and overhead contact line.

12.4.11   Crossings of overhead power lines above contact lines

For crossings of contact lines by overhead power lines specific agreements are required between the entity in charge of the power line and the railway infrastructure manager. In Germany, agreements according to [12.55] need to be made between the power line operator and German Railway DB. These agreements contain specifications for the planning, construction and maintenance of power lines crossing the railway line.

The agreements distinguish between new crossings and modification or dismantling of existing installations. In the case of new crossings, the costs-by-cause principle applies, i. e. the entity constructing the new installation bears the expenditures for the crossing. When both installations are constructed at the same time, the expenditures will be shared.

EN 50 119: 2009+A1: 2013 refers to EN 50 341-1: 2012 concerning clearances to overhead power lines. According to EN 50341-1 an analytical verification of the lowest point of the

Table 12.20: Minimum distances L_{SS} for raised pantographs according to EN 50 367: 2012.

Commercial speed	AC			DC 3,0 kV			DC 1,5 kV		
$v \leq$ 80 km/h	8	8	8	8	8	8	20	8	8
80 < v ≤ 120 km/h	20	15	15	20	15	15	35	20	15
120 < v ≤ 160 km/h	85	85	35	20	20	20	85	35	20
160 < v ≤ 250 km/h	200	85	35	200	115	35	200	85	35
v ≥ 250 km/h	200	200	200	200	200	200	200	200	35
type of line [1]	A	B	C	A	B	C	A	B	C

[1] the adopted shortest pantograph spacings L_{SS} result in the line types A to C which the infrastructure manager determines and lists in the infrastructure register

power line is required for the crossing of electrified railways. The clearance between power line and overhead contact line can be determined using the suspension heights of the conductors at the poles and the sag. The longitudinal plan enables a graphical demonstration of compliance with the minimum clearance S_{min} between the crossing power line and the contact line. For power lines with voltages above 45 kV, at least 2,6 m vertical clearance is required between crossing conductors and the component of the contact line installation at maximum conductor temperature, ice load, wind load or 11,0 m to the top of rail where electrification does not exist, but is intended. According to EN 50 341-1: 2012, Table 5.12, the clearance needs to comply with the special case related to the swinging of the over-crossing conductor due to differing wind loads at +5 °C and the under-crossing conductor of the traction power line having its lowest sag.

According to EN 50 341-1: 2012, Table 5.12, a minimum clearance $S_{min} = 2m + D_{elf}$, however, at least 2,6 m needs to be maintained between, for example, a 400 kV power line and components of the contact line installation. D_{el} is the minimum clearance in air and is required between live and earthed components to avoid arcing during fast-front or slow-front over-voltages. For a crossing with a 400 kV power line $D_{elf} = 2,92$ m and $D_{els} = 3,20$ m result according to the example in Annex E.4.3 of EN 50 341-1: 2012 at altitudes up to 1 000 m above sea level. Therefore, the minimum clearance S_{min} will be for fast-front over-voltages

$$S_{min} = 2 + D_{elf} = 2 + 2,92 = 4,92 \, m \tag{12.5}$$

and for slow-front over-voltages

$$S_{min} = 2 + D_{els} = 2 + 3,20 = 5,20 \, m \quad . \tag{12.6}$$

The greater of both values S_{min} is selected after consideration of lightning and switching over-voltages to be complied with. No flashovers due to over-voltages should occur between overcrossing and undercrossing lines. Therefore, the minimum clearance between live and earthed components at the supports needs to be at least 1,1 times the minimum flash-over distance at the insulator sets at three poles in front and three poles behind the crossing. Therefore, it applies

$$a_{som} \cdot 1,1 > 2 + D_{el} \quad . \tag{12.7}$$

The larger of both components in Equation (12.6) is to be used as minimum clearance S_{min}. In case of the example of an overcrossing 400 kV line a_{som} is 4,9 m. According to (12.6) it results $a_{\text{som}} \cdot 1,1 = 4,9 \cdot 1,1 > 2 + D_{\text{els}} = 2 + 3,2 = 5,20$ m. In this case a minimum clearance $S_{\text{min}} = 5,39$ m should be utilised between the overcrossing power line and the uppermost earthed component of the existing overhead contact line. In case of an envisaged electrification according to EN 50 341-1: 2012, a minimum clearance of $S_{\text{min}} + 11,5$ m needs to be obeyed between conductor and top of rail. As a result, deducing 2 m, the highest point of the contact line planned in future will be less than 8,5 m. The height of poles will be 8,0 m above top of rail. If a parallel feeder line was arranged at the top of pole the minimum of 5,39 m to the overcrossing 400 kV line could be fallen below.

12.4.12 Arrangement of electrical separations

12.4.12.1 Basics

Electric separations like *section insulators, phase and system separations* form *constraints for planning* of an overhead contact line.

The local position of an electric separation is defined in the circuit diagram and is defined in an early phase of contact line planning (Clause 12.2.3.7), whereby signal sites, station entry points and switching circuits need also to be considered.

This applies especially to *insulating overlaps* which are arranged between stations and interstation sections as well as *phase and system separations*.

When traversing overlaps and section insulators voltage differences up to 1 200 V are permitted at DB [12.35], however, severe arcing could occur as a consequence. Because of the resulting sparking, insulating overlaps and section insulators may not be installed at platforms or in tunnels. Experience gained in Norway reveals that the power supply with autotransformers used there results in severe arcing when traversed by a pantograph already at low line loadings [12.56, 12.57]. If this type of power supply is applied, then insulating line separations cannot be arranged at platforms or in tunnels.

12.4.12.2 Arrangement of pantographs

To negotiate the specified types of separation sections, the maximum spacing L''_{SS} of pantographs on a train needs to be less than 400 m. On lines traversed by trains with three consecutive pantographs, the span covering three consecutive pantographs needs to be $L''_{\text{SS}} \geq D_{\text{S}}$ and $D_{\text{S}} \geq 8$ m (Figures 12.67 and 12.72). The pantograph in the middle may be arranged at any position within this distance, whereby the minimum spacing L''_{SS} for protective section type 1 should be 143 m or more than 79 m for type 2. Depending on the shortest distance L_{SS} between two adjacent lifted pantographs and on the width l_{SA} of pantographs, the infrastructure manager needs to declare the type of line A, B and C and the maximum operational speed for each one (Table 12.28). The type of line determines the design of protective sections with the characteristic parameters *total length* D_{S}, length of neutral section D'_{S} and when section insulators are used in protective sections, the length of the section insulator $d_{\text{s ins}}$ (Figure 12.67). The infrastructure manager supervises the arrangement of pantographs using suitable diagnostic devices [12.58].

According to EN 50 367:2012 + AC:2013 + A1:2016, there may not be any electrical connections between lifted pantographs of AC trains. In Figure 12.72, the possible arrangements of pantographs are shown for interoperable lines. In the case of DC railways, electrical connections between the pantographs may exist but they need to be interruptible by disconnectors.

Table 12.21: Electrical clearances between live and earthed contact lines in electrical separations according to EN 50 119: 2009+A1: 2013.

[1] only for existing systems.

Nominal voltage kV	Recommended clearances	
	permanent S_d mm	short-term S_k mm
DC 0,6[1]	100	50
DC 0,75	100	50
DC 1,5	100	50
DC 3,0	150	50
AC 15	150	100
AC 25	270	150

Protective sections form constraints and also influence the wiring especially in tunnels, points connections and in stations.

12.4.12.3 Electrical clearances

During maintenance activities in protective sections, one contact line can be live and the adjacent contact line can be switched off and earthed. I this situation the air clearances between earthed and live parts of the overhead contact line are defined in Table 12.21. The clearances stipulated in Table 12.21 should also be applied to clearances between adjacent live parts of contact lines of different electrical sections with the same voltage and phase.

Different clearances for static and dynamic cases are justified by different probabilities of these conditions. For example, it is improbable that an over-voltage surge will occur at the same moment that a pantograph passes a bottleneck in a tunnel. In such a dynamic case, the use of clearances for short-term events is justified. The values in Table 12.21 do not apply to section insulators where the dynamic clearances with reduced values are applied permanently. Reference is made to EN 50 122-1 and prIEC 60 913: 2013 for reduced electrical clearances for section insulators.

For 2 AC 15/30 kV and 2 AC 25/50 kV auto-transformer systems, there is a phase difference of 180° between live parts connected to the feeder line and live parts connected to the overhead contact line. Similarly, a phase difference between 120° and 180° results at phase separations of single phase AC systems.

For an overhead contact line system with electrical sections supplied by different voltage phases, a phase-to-phase voltage higher than the nominal voltage occurs. Table 12.22 provides requirements for th desired clearance in still air between live parts under all conditions.

12.4.12.4 Arrangement of phase separation sections and protective sections

Phase separation sections and *protective sections*, henceforth referred to as protective sections, separate line sections which are supplied by different phases of the three-phase power

Table 12.22: Clearance between different phases according to EN 50 119: 2009+A1: 2013.

Nominal voltage kV	Phase difference degree	Relative voltage kV	Recommended clearance	
			permanent mm	short-term mm
15	120	26,0	260	175
15	180	30,0	300	200
25	120	43,3	400	230
25	180	50,0	540	300

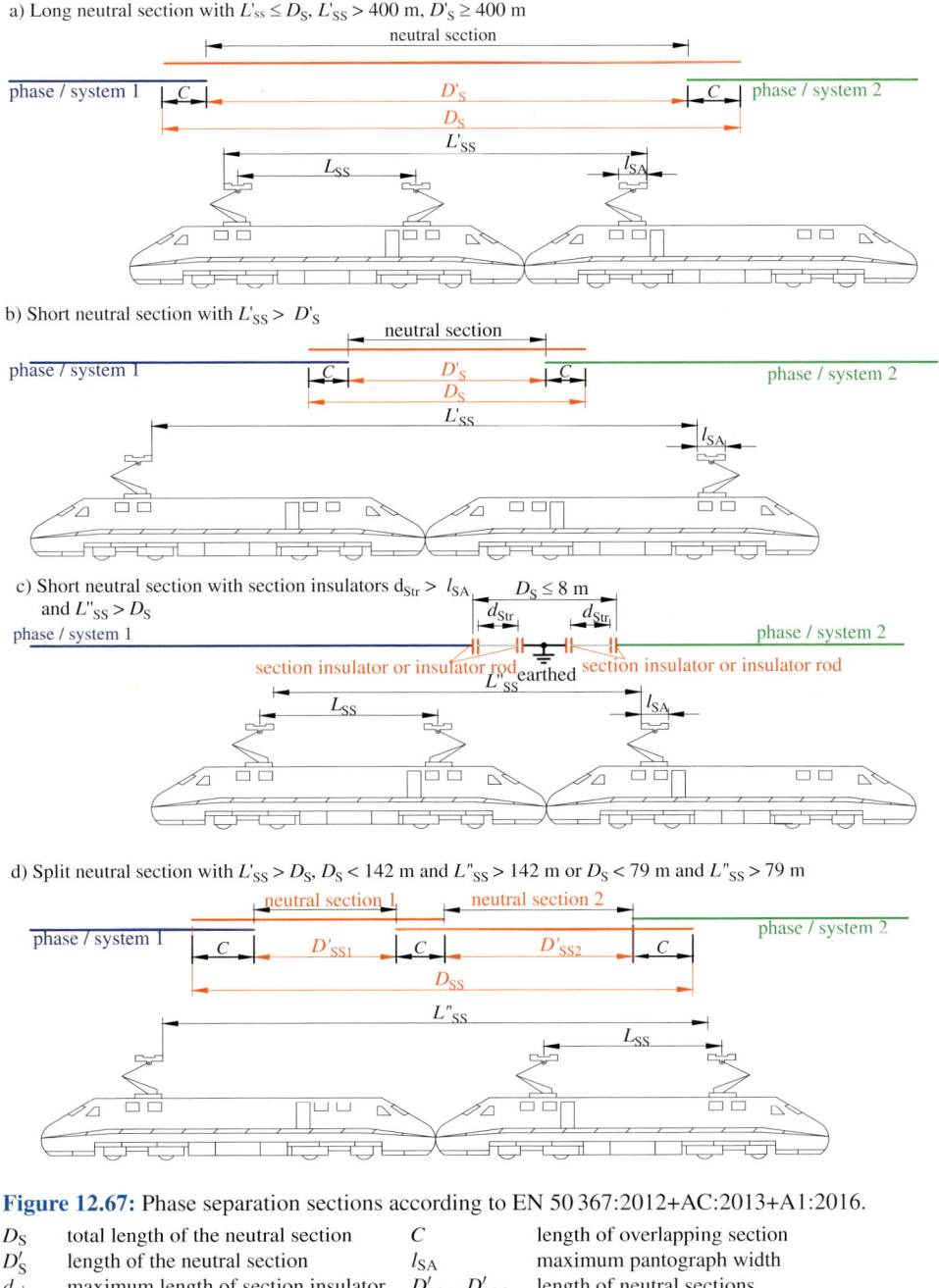

Figure 12.67: Phase separation sections according to EN 50 367:2012+AC:2013+A1:2016.

D_S	total length of the neutral section	C	length of overlapping section
D'_S	length of the neutral section	l_{SA}	maximum pantograph width
$d_{s\,ins}$	maximum length of section insulator	D'_{SS1}, D'_{SS2}	length of neutral sections
L_{SS}	minimum spacing of adjacent pantographs (see Table 12.28)		

a) Long neutral section with $L'_{ss} \le D_S$, $L'_{SS} > 400$ m, $D'_S \ge 400$ m

b) Short neutral section with $L'_{SS} > D'_S$

c) Short neutral section with section insulators $d_{Str} > l_{SA}$ and $L''_{SS} > D_S$

d) Split neutral section with $L'_{SS} > D_S$, $D_S < 142$ m and $L''_{SS} > 142$ m or $D_S < 79$ m and $L''_{SS} > 79$ m

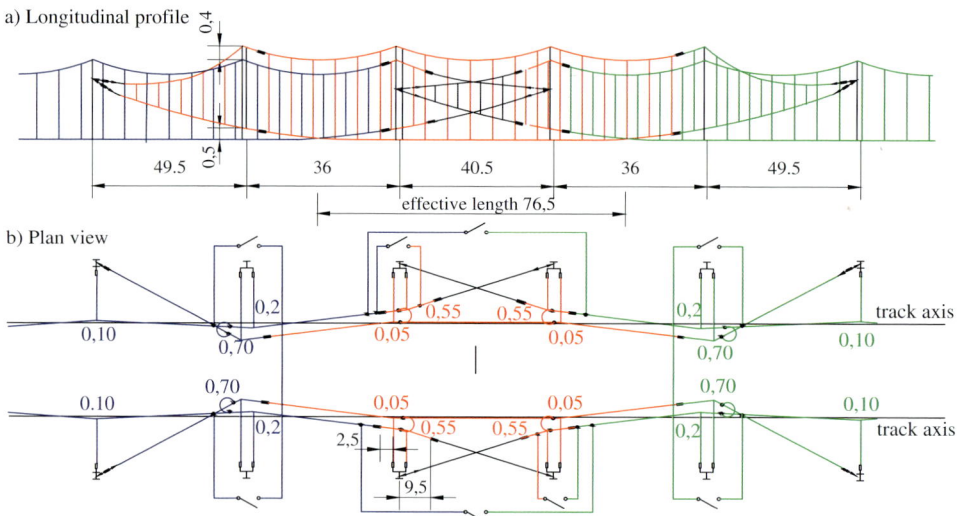

Figure 12.68: Short protective section of the conventional line Bucharest North, Romania (Graphic: Balfour Beatty Rail). Dimensions in m

grid. Trains may not electrically bridge such sections (Figure 12.67). Suitable procedures also need to be established to restart trains that come to a standstill within the phase separation section. The neutral section needs to be connected with the next supply section in front of the separation by a contact line disconnector and to be designed as short as possible minimising the probability of trains coming to a standstill. In Figure 12.67 the principles of phase separation sections are shown according to EN 50 367:2012 + AC:2013 + A1:2016:

- long neutral sections where all the pantographs of the longest trains with the maximum spacing L'_{SS} can be accommodated within the neutral section D'_S. The neutral section D_S needs to be at least 400 m long in this case. Depending on the operational speed, they will be equipped with insulated overlaps or with section insulators (Figure 12.67 a)). The maximum spacing L'_{SS} between the outermost pantographs is limited to 400 m or
- short neutral sections stretching over a total length D_S being less than the minimum spacing L''_{SS} between the pantographs. In EN 50 367:2012 + AC:2013 + A1:2016 a length of 200 m is specified as the minimum interoperable spacing L''_{SS} between the pantographs (Figure 12.67 b)) or
- short neutral sections formed by neutral section insulators and stretching over a total length $D_S \leq 8$ m (Figure 12.67 c)). The calculation of creepage length of insulator rod see also 12.4.12.5 or
- split neutral sections with three insulating overlaps as shown in Figure 12.67 d). The overall length D_S of this design is limited to 142 m including clearances and tolerances or 79 m. Such a design should be avoided as far as possible.

Experience demonstrates that the long neutral sections do not impose restrictions on operations. For high-speed lines designed for speeds above 200 km/h, long, short or subdivided protective sections can be applied as shown in Figure 12.67 c) and d). In the case of short protective sections with section insulators, the central part needs to be connected to the railway earth. The neutral sections D_S according to Figure 12.67 c) can be equipped with insulating rods or twin section insulators and need to be longer than ≥ 8 m. The length d_{sins} needs to be chosen in accordance with the system voltage, the maximum line speed and the maximum pantograph width. As a stipulation for the described designs, the simultaneously active pantographs can not be connected electrically. The protective sections shown in Figure 12.67 need to be agreed upon with the infrastructure manager before the start of detailed overhead contact line planning. The effective length of a neutral zone of a protective section starts and ends at the position where the contact wire is lifted by at least 210 mm above the nominal contact wire height. At that position, the pantograph can only touch the live contact wire, that runs upwards, in the improbable case of a simultaneous occurrence of maximum tolerances, maximum uplift at the support and maximum difference between the lowest and highest up-lift position. Especially, in the case of high-speed lines, a sufficiently long neutral section is necessary in order to avoid flashovers between the phases, e. g. L1 and L2, also in the case of switching over-voltages. Figure 12.68 shows a short phase separation of a conventional line at Bucharest North in Romania. Clause 14.5.5 contains further information on protective sections.

12.4.12.5   Short protective sections

Short-protective sections (SPS) consist of two section insulators with an earthed section in between that should be at least three metres long and supported by a cantilever connected to the return circuit (Figure 12.69). During normal operation, the main circuit breaker of the traction vehicle is switched-off automatically before traversing the SPS. If the control of the main circuit breaker fails and the traction unit continuously draws energy from the contact line, an arc between the live overhead contact line section and the earthed intermediate section will occur. The circuit breaker in the substation would switch off this fault to earth avoiding a short circuit between the two supply sections with different voltages or phases and electrically separated by the SPS. The section insulators consist of insulating rods and runners that do not overlap. Because of arcing risks, these section insulators are equipped with arcing horns (Figure 12.69 and 12.71) so this type of SPS cannot be installed at platforms and in tunnels. SPS extend over a shorter effective length (Figure 12.70) than insulated overlaps enabling them to be used in limited spatial conditions such as curves with radii of 180 m for example as electric separation of supply sections. SPS need only an effective length of 8 m, however, the operational speed is limited to 140 km/h. At the earthed cantilever, the contact and catenary wire should be arranged at the track axis. SPSs need protection against twisting (Figure 12.69 and 12.71) to guarantee a parallel position to the plane of the top of rails.

Short protective sections are used in connection with a coupling post to supply the adjacent line sections. The signalling of a SPS is presented in Figure 12.69. Signals EL1 and EL2 are arranged over a distance of three metres in front of the earthed cantilever. When the vehicle passes the SPS, the main circuit breaker of the traction vehicle, which was switched off in front of the SPS, can again be switched on.

By running carbon collector strips over the insulating rods of the section insulators, a pollution layer of carbon is formed on the lower side of the insulating rods that does not depend on the humidity and may not result in arcing between the live and earthed overhead contact line section.

The distance between the live and earthed runners of the separation device is designated as the *insulating length* but not the length of the insulating rod. The calculation of the insulating length in short protective sections is carried out according to EN 50 367. There are two

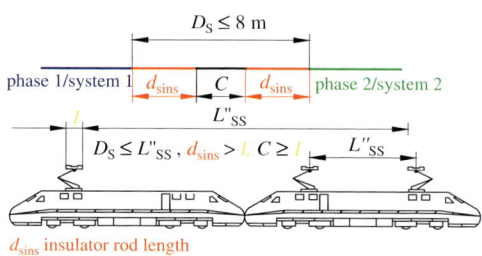

d_{sins} insulator rod length

Figure 12.70: Short protective section according to EN 50 367:2012+AC:2013+A1:2016.

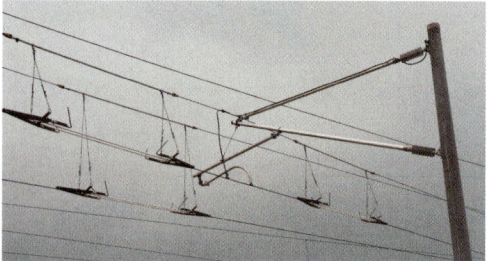

Figure 12.69: Signalling at a short protective section.

Figure 12.71: Implemented short protective section with protection against twisting.

methods, the first approach considers the air gap and the second considers the creepage path between the live and earthed overhead contact line section. Longer creepage paths need to be used for selection of the length of the insulating rods.

The calculation of the creepage path d_k for the insulating rod considers:
- the width l_{SA} of the pan-head of the pantograph
- the permanent electrical clearance S_d and the short-term electrical clearance S_k
- the commercial speed v of the trains at the SPS
- the power frequency f
- the distance the pantograph runs until the zero crossing of voltage is reached

Three cases can be distinguished:
1) the insulating rod separates the earthed section and the adjacent live sections and pantographs do not run along the insulating rod. The insulating rod should comply with requirements concerning the creepage path
2) the insulating rod is traversed by the pantograph and should comply with the conditions concerning the required air gap
3) the insulating rod is traversed by the pantograph and should comply with the requirements regarding the creepage path

The longest of the calculated creepage paths forms the basis for selection of the length of the insulating rod (Figure 12.69).

In the first case, the static consideration, the specific creepage path is assumed as 45 mm/kV according to Table 2.3 between the earthed section and the live overhead contact line for 15 kV 16,7 Hz to be

$$L_{S\,min\,stat\,15} = l_k \cdot U_{max} - 54 \cdot 17,25 - 931 \, mm \tag{12.8}$$

and for the 25 kV 50 Hz

$$L_{S\,min\,stat\,25} = 54 \, mm/kV \cdot 27,5 \, kV = 1485 \, mm \quad . \tag{12.9}$$

In the second case, dynamic consideration, the creepage path between the earthed section and the live overhead contact line is calculated by

$$L_{S\,min\,dyn} = l_W + S_k + l_f \quad , \tag{12.10}$$

whereby the length, after which the operating voltage has reached the zero crossing, is calculated by

$$l_f = v/3,6 \cdot 1000 \cdot 1/(f \cdot 2) \tag{12.11}$$

where:
$L_{S\,min\,dyn}$ minimum length of insulating rods at the SPS in mm
l_W width of the pantograph pan-head in mm
S_k short-term electric clearance in air, live part to earth, in mm
l_f way passed by the pantograph up to the zero crossing of the operating voltage, in mm
l_k minimum creepage path in mm/kV
v operating speed of the train in km/h
f frequency of the operating current in 1/s
For a 15 kV 16,7 Hz power supply, 140 km/h operating speed, 650 mm pantograph width and 100 mm short-term electrical clearance, $L_{S\,min\,dyn}$ is obtained from

$$L_{S\,min\,dyn\,15} = 650 + 100 + 1164 = 1914 \, mm$$

and for a 25 kV 50 Hz power supply from

$$L_{S\,min\,dyn\,25} = 650 + 150 + 389 = 1189 \, mm \quad .$$

The length of insulating rods can be rounded to 2000 mm or 1200 mm respectively.
The third case is based on dynamic considerations. The pantograph traverses the insulating rod which has to comply with requirements on creepage strength. The minimum length of the insulating rod results from the required creepage path between the pan-head and the live end of the insulating rod or between the pan-head and the earthed end of the insulating rod. Because of the layers of carbon on the insulating rods, a specific creepage path of 54 mm/kV, according to condition PD 7 is valid and assumed for severely polluted surfaces. The minimum length of the insulating rod for 15 kV 16,7 Hz power supply will be

$$L_{S\,min\,dyn} = l_W + U_{max} \cdot l_k = 650 + 17,25 \cdot 54 = 1582 \, mm \tag{12.12}$$

and for 25 kV 50 Hz power supply

$$L_{S\,min\,dyn\,25} = 650 + 27,5 \cdot 54 = 2135 \, mm \quad . \tag{12.13}$$

The length of the insulating rod can be rounded to 1 600 mm and 2 000 mm respectively.
After comparing the minimum length for the insulating rods of the SPS, as calculated for the three different cases, a 2 m long rod for a 15 kV 16,7 Hz power supply and a 2,2 m long insulating rod for a 25 kV 50 Hz power supply was used.

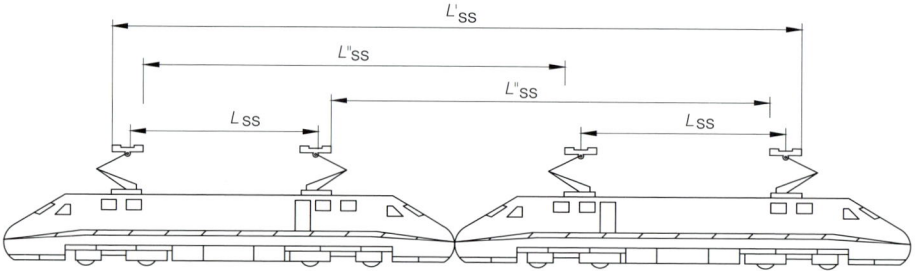

Figure 12.72: Pantograph arrangements for protective sections according to Figure 12.67.
$L'_{SS} < 400\,\text{m}$, $L''_{SS} > D_S$ for three consecutive pantographs $L''_{SS} > 143\,\text{m}$

12.4.12.6 Arrangement of system separation sections

The separation sections between different power supply systems must guarantee that trains can run from one power supply system to the adjacent different power supply system without bridging both supply systems. Design and operation of system separation sections is explained in Clause 14.4.

There are two possibilities for a train to run through such separation sections:
- with the pantograph raised and touching the contact wire
- with the pantograph lowered and not touching the contact wire

– Raised pantographs contacting the contact wire

If system separation sections are traversed with pantographs raised to the contact wire, the following conditions apply according to EN 50 367:2012 + AC:2013 + A1:2016:
- the geometry of the components of the overhead contact line needs to prevent pantographs from short-circuiting or bridging both power systems
- provisions need to be made in supply sections to avoid bridging the line sections adjacent to neutral sections, should the opening of the on-board circuit breaker(s) fail
- for lines with speeds above 250 km/h, the height of contact wires in both sections adjacent to the neutral section need to be the same

An example of the arrangement of system separation sections is given in Figure 12.67.

– Pantographs lowered without touching the contact wire

This option needs to be chosen if the conditions of operation with pantographs raised cannot be met. If a system separation section is traversed with pantographs lowered, it needs to be designed to avoid bridging by an unintentionally raised pantograph. This can be achieved by devices that detect short circuits and switch-off the power supply systems behind the train, as viewed in running direction.

Further options of separation sections are described in Clause 14.4.2.

12.5 Layout plan

12.5.1 Objectives and information

A *track layout plan*, containing the information in Table 12.1 and described in Clause 12.2, is used for the preparation of the overhead contact line circuit diagram (Figure 12.2 and 12.3) and the *overhead contact line layout plan*. A track layout plan contains:

- track elements such as curves and transition sections with lengths, radii, cant and cant deficiency (Figure 12.4 a))
- contact wire routing with overlaps, anchors and midpoints
- electrical connections, electrical isolations and disconnectors
- cross-span equipment, such as single cantilevers and multi-track cantilevers, horizontal registrations, flexible cross-spans and portals
- poles with numbering, anchors and information related to the span lengths between adjacent poles
- distances between poles and structures, start, end and type of points
- information on reference tracks and rails for surveying of poles
- information on railway earthing
- traction power supply lines and cables
- disconnectors with numbering and type of operating mechanism
- areas with contact line height reductions, contact line lifting at level crossings
- signals for electrical operations
- comments and legends, e. g. magnitude of short-circuit current

The *adjustment diagram* shows the overhead contact line routing to a distorted scale. The *stagger of the contact wire* and the system height can be seen in addition to the type of support, either pull-off or push-off. The layout plan and adjustment diagram assist with the installation, the re-installation and re-adjustment of the contact line after fault situations.

The *earthing diagram* shows all items of equipment for the return current circuit, such as:
- longitudinal and transverse rail bonds
- transverse track bonds
- reactance coil joint bonds
- diagonal bonds
- Z-bonds at points
- return current cables and conductors
- connections between running rails and poles as well as connections to conducting components within the overhead contact line range, but do not belong to the contact line system (see Figure 12.76)

The *contact line layout plan* is used to show the earthing system in stations and on the interstation line. If the wiring of the interstation line and the station is distributed over several sheets, then the overlaps of plans are to be arranged to ensure no repetitions occur.

12.5.2   Overhead contact line symbols

All overhead contact line components are represented in the layout plan using *overhead contact line symbols*. As an example, Tables 12.23 to 12.26 illustrate the symbols employed by DB in layout plans.

12.5.3   Contact line supports and pole locations

The *contact line support* defines the attachment location of the contact and catenary wire allowing determination of the contact wire run.

The *local track layout*, the conditions on the superstructure and the requirements for mechanical separation of overhead contact lines define the type of transverse supporting element such as *single cantilever*, *portal* or *pull-off*. The pole locations are selected considering the

Table 12.23: Symbols for the overhead contact lines.

Designation	Symbol	Example	Designation	Symbol	Example
track with overhead contact line system (OLS)			midpoint anchor in tunnel with anchoring to the ceiling		
track without OLS			midpoint anchor in tunnel with anchoring to the wall		
track with planned OLS			contact wire crossing provided with a cross-contact bar (negotiated crossing)		
OLS without catenary wire automatic-tensioned			contact wire crossing without touching (without cross-contact bar)		
fixed termination			contact wire crossing at double point with clamped contact wires		
OLS with fixed catenary wire automatic-tensioned			electrical connection of two lines		
fixed termination			electrical connection between catenary wire and contact wire		
OLS with tensioned catenary and contact wires automatic-tensioned			section insulators		
fixed termination			insulation in contact wire and/or catenary wire or cross-span wire		
OLS with tensioned catenary and twin contact wires automatic-tensioned			bracket between two contact wires (compression and tension)		
fixed termination			contact wire-conductor connection without insulation		
planned termination			contact wire-conductor connection with intermediate insulation		
midpoint			overlap		
midpoint anchor at head-span structure			contact wire stagger		

possible span length, the type of superstructure, location of underground cables and pipes, drainage channels, traction power lines strung on the overhead contact line poles, minimum clearances to objects and the subsoil conditions. For double track lines, facing pole locations are preferred. The supports associated with points can lead to *staggered pole locations* at point connections. In curves on single-track lines the poles are located, if possible, on the outside of the curve. The *span lengths*, which designate the distance between the adjacent cantilevers at their central position, are to be recorded on the layout plan. The layout of the overhead contact line, the contact line supports and the poles should be represented on the layout plans using the symbols presented in Tables 12.23 to 12.26.

Table 12.24: Symbols for transverse supports and for poles and soffit posts.

Designation	Symbol	Example	Designation	Symbol	Example
cantilever for one track on pole			double-channel pole		
two cantilever on pole			lattice steel pole		
cantilever across two one tracks			double-channel pole on bracket		
pull-off for one track			lattice steel pole on bracket		
pull-off for two or more tracks			concrete support pole		
head span suspension			concrete tension pole		
tensioning portal structure with intermediate pole			H-beam pole		
portal			anchor		
support in tunnel on wall without steady arm			pole with foundation protection		
double support in tunnel on wall without steady arm			soffit post in tunnel on a wall without steady arm		
support in tunnel on wall with one steady arm			soffit post on ceiling in tunnel with one steady arm		
support in tunnel on wall with two steady arms			soffit post on ceiling in tunnel with two steady arms		

12.5.4 Single poles

The distance between *single poles* along the track is determined by the required location of the contact line supports. Standard lateral clearances between the track axis and the track-side face of the poles (dimension *TP*) should be used where ever possible to allow the use of standard cantilevers with advantages for installation and maintenance. The standard distance results from summing the half-width of the track sub-ballast footing, the width of the duct channel to be laid in front of the pole (if required) and the construction tolerances to be considered. For example, the standard dimension *TP* is 3,7 m for DB high-speed lines.

The maximum structural dimensions of cantilevers and the length of working platforms of maintenance vehicles limit the distance of poles from the track. The *vehicle gauge envelope*, with extensions in curves, limits the proximity of poles to the track. The minimum clearance between pole foundations and the track is determined by summing the half-width of the vehicle gauge envelope GC being 2,5 m, the construction tolerance of 0,05 m and the margin for curve effect on the inside of the curves. The vehicle gauge also needs to be considered when arranging poles between tracks. In stations where overhead contact lines are to be separated mechanically, pole aisles between the tracks may be necessary but they will also affect planning of the track layout.

Table 12.25: Symbols for traction power lines and for disconnectors.

Designation	Symbol	Example	Designation	Symbol	Example
traction power line (e.g. 2 reinforcing line feeders E-AL-240)		E-AL-240	disconnector: open		
planned traction power line (considered in design)	----	E-AL-240	disconnector: closed		
traction power cable, e.g. 15 kV cable (e.g. Cu 95mm²)	-----	Cu 95²	disconnector: with earth contact		
traction power line cross arm	|		disconnector: with hand operated mechanism, triangular key	No.	16
single suspension with strap	EH	EH	disconnector: with hand operated mechanism, square key	No.	6
double suspension	DH	DH	disconnector: motor-driven, for 1000 A, locally controlled	No.	601
V-suspension	V	V	disconnector: motor-driven, for 1000 A, remotely controlled	No. xxxxA	W1 1000A
traction power line termination at pole	T	T	disconnector: motor-driven, for 1700 A, locally controlled	No. xxxxA	15 1700A
termination at traction power line cross arm	T	T	disconnector: motor-driven, for 1700 A, remotely controlled	No.	412
termination at cross arm for switching lines	T	T	disconnector: motor-driven, for 2000 A, locally controlled	No.	2
double termination at traction power line cross arm	DT	DT	disconnector[1]: motor-driven, for 2000 A, remotely controlled	No.	4
double termination at cross arm for switching lines [1]	DT	DT	disconnector[1]: motor-driven with short-circuit indicator	No.	1
intermediate anchoring at traction power line cross arm	T T	T T	switching transverse line cross-arm[1]	No. xxxxA	11 1000A
[1] designation and symbol of former Deutsche Reichsbahn			control cable (e.g. 3 core 1,5 mm²)	----	1,5 /3
			circuit breaker		

[1] designation and symbol of former Deutsche Reichsbahn

Poles in station areas and in front of platform approaches should not hinder passenger movement thus pole anchors and flexible tensioning devices are to be avoided in these areas. If necessary, protective baskets need to be installed at the weight stack of tensioning devices. Overhead lines on spur tracks needs to be continued beyond the spur track end and, if possible, anchored at the next overhead contact line pole. It should be noted that poles can only be provided 20 m beyond the end of the track or outside the standard gauge.

Symbols for supports on layout plans can be selected from Table 12.24. The poles can be assigned temporary numbers during the construction phase and also need to be provided with *equipment identifiers* that indicate location and pole number at least by commissioning.

Table 12.26: Symbols for traction return circuit, protective earthing and miscellaneous symbols.

Designation	Symbol	Example	Designation	Symbol	Example
rail longitudinal bond			earthing plate		
rail bond			transformer		
track transverse bond			electrical complementary signal with / without directional arrow		
support earthing over cross wire			central switching section boundary		
connection line between rail and equipment			substation boundary[1]		SS1 SS2
right rail insulated, left rail earthed	E	$0 \longrightarrow E \longrightarrow x$	maintenance boundary		
left rail insulated, right rail earthed	E	$0 \longrightarrow E \longrightarrow x$	sealing end		
connection line connected to right rail	/E	$0 \longrightarrow /E \longrightarrow x$	current transformer		
connection line connected to left rail	/	$0 \longrightarrow / \longrightarrow x$	location of railway earthing fixture		
insulated rail joint in left single rail, insulation in direction of datum		$0 \longrightarrow F \longrightarrow x$	location of live-line tester		
insulated rail joint in right single rail, insulation in direction of datum		$0 \longrightarrow L \longrightarrow x$	location of earthing fixture (only for rescue train)		
insulated rail joint in left single rail, insulation in opposite direction of datum		$0 \longrightarrow 7 \longrightarrow x$			
insulated rail joint in right single rail, insulation in opposite direction of datum		$0 \longrightarrow \dashv \longrightarrow x$			
voltage limiter					

[1] designation and symbol of former Deutsche Reichsbahn

12.5.5 Flexible head spans

A *flexible head span* supports the contact lines of several tracks and fixes them laterally. *Head span wires* with a deep sag are arranged between two *head span poles*, to support the vertical loads of the overhead contact line supports. The upper and lower *cross-span wires*, strung without sag support the horizontal forces of the catenary and contact wire respectively (Figure 12.82). Normally the upper cross-span wires are earthed. If the distances between the earthed upper cross-span wire and the live catenary wire are insufficient because of the inclination of the insulators then a live upper cross-span wire needs to be installed. The distances between head-span poles and the track are unrestricted. The use of *head-spans* requires that all overhead contact line wire supports attached to the head-span are located at the same longitudinal track co-ordinate. Pole aisles, such as required for single poles, are not necessary. From a technical viewpoint, *head-span lengths* are not limited but should be restricted to approximately 80 m for practical reasons. Crossing loading platforms with head-span structures should be avoided. Head-span poles enable either fixed or flexibly terminated overhead contact lines. However, it can be expedient to provide separate termination poles to carry the longitudinal loads to limit the overhead contact line length or to avoid sharp bends and intersections (see also Clauses 11.2.1.3 and 13.2.4).

12.5.6 Multiple-track cantilevers

Multiple-track cantilevers span a maximum of three tracks. The accommodated contact wire supports need to be arranged at the same co-ordinate. The overhead contact lines, supported by multiple-track cantilevers, are considered as mechanically separate from each other. As with head-span structures, multiple-track cantilevers can be positioned more flexibly and perpendicularly to the tracks after considering the kinematic pantograph gauge (see also Clauses 11.2.1.2 and 13.2.3).

12.5.7 Portals

Portals serve purposes similar to head spans or multiple-track cantilevers. They enable greater numbers of tracks and widths to be spanned. Portals can also be arranged obliquely to the track centre-line, in the same way as head-span structures. Portals also offer advantages at points and crossovers of high-speed lines because they can accommodate overhead contact line supports that cannot be carried by single poles. They also allow mechanical separation of overhead contact lines and do not transfer oscillations of one contact line to another (see also Clauses 11.2.1.4 and 13.2.5).

12.5.8 Tunnel supports

The selection of *tunnel supports* is determined by the tunnel cross-section and the ceiling height. The type of support or head-span equipment, such as *elastic supports* or cantilevers is selected to suit the available headroom and the requirements specified for the overhead contact line. The arrangement of supports between tracks or on the tunnel wall influences their design (see also Clause 11.2.1.7).

12.5.9 Electrical connections

Electrical connections provide a current-carrying bond between the contact wire and the catenary wire, two overhead contact lines or between an overhead contact line and *parallel feeders*. They provide defined connections between crossing overhead lines at track points or between negotiated and non-negotiated overhead contact lines such as in overlaps or neutral sections. However, if the catenary wire is interrupted under bridges, the provision of an electrical connection before and after the structure distributes the current load. The same applies to the installation of additional electrical connections near earthed structures, such as signals and interlockings. The installation of further electrical connectors is expedient in sections

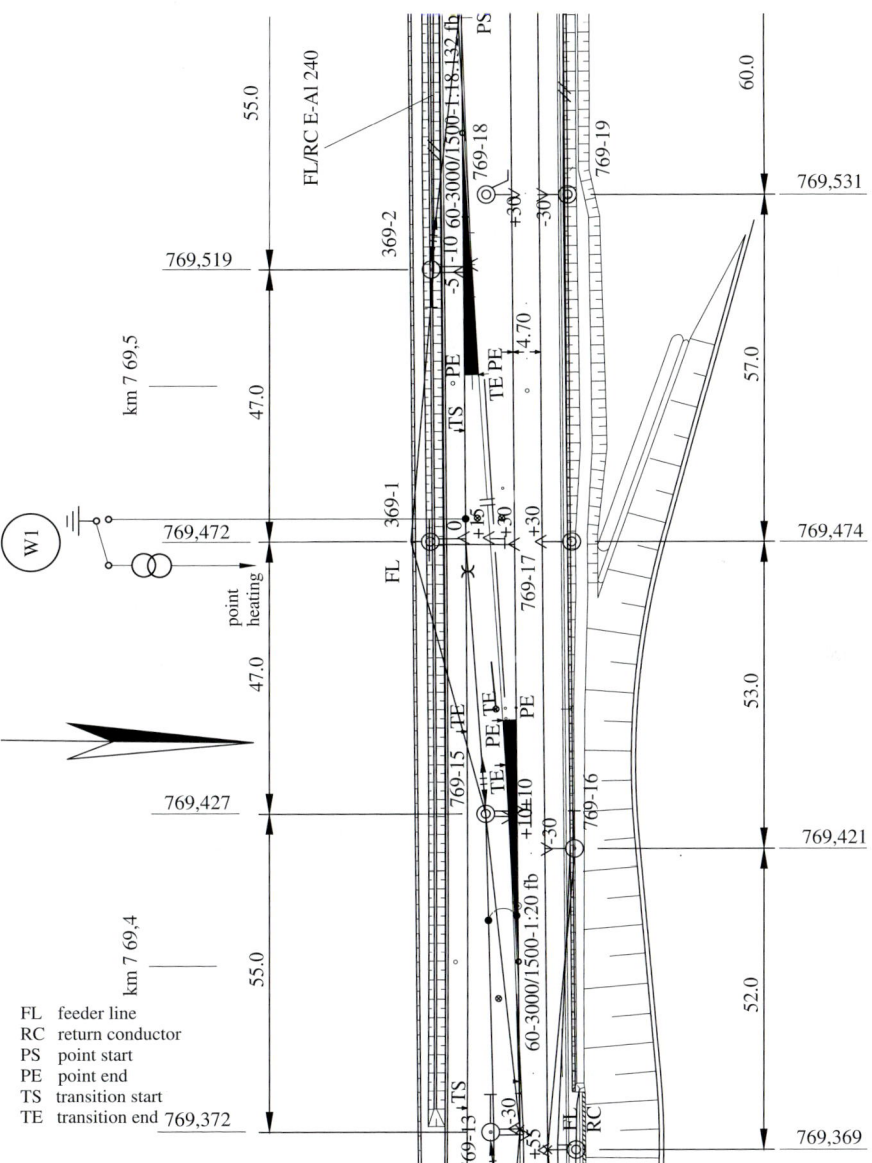

Figure 12.73: Overhead contact line layout plan (partial view).

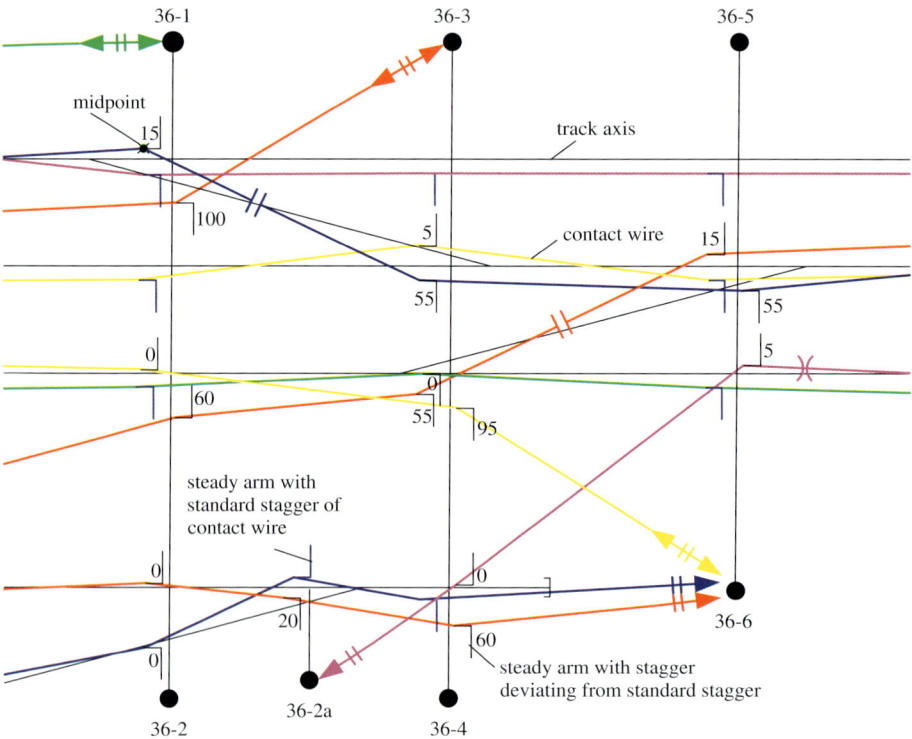

Figure 12.74: Adjustment plan (partial view).

with high electrical loads drawn by electric traction vehicles, on steep gradients for example. The Z-type anchor between the catenary wire and the contact wire fulfills the function of an electrical connector, making the provision of electrical connections at the midpoint anchor unnecessary. Droppers can also be designed to serve as electrical connections (see Clauses 6.6.4 and 11.2.2.7).

12.5.10 Return current circuits and protective earthing

Provisions for the traction *return current circuits* and the *protective earthing* of structures need to be defined in detail during the planning of the overhead contact line and recorded in the planning documents, especially on the *earthing diagram*. The basic *design rules* are dealt with in Clauses 6.5.4, 6.6.3 and 6.6.4, In Figures 6.39 to 6.44 some actual examples are shown. The basic design is dealt with in Chapter 6.

The individual designs for return circuits need to be agreed with signal planning and design. This process is instrumental in creating an earthing diagram, which can be established as a separate document and can supplement the contact line layout plan or can be directly integrated in the layout plan.

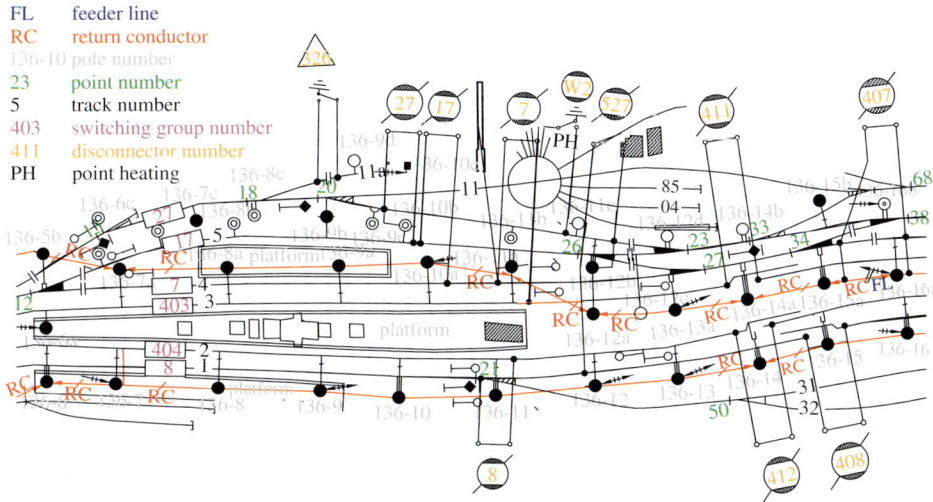

FL   feeder line
RC   return conductor
136-10 pole number
23   point number
5   track number
403   switching group number
411   disconnector number
PH   point heating

Figure 12.75: Layout drawing (partial view).

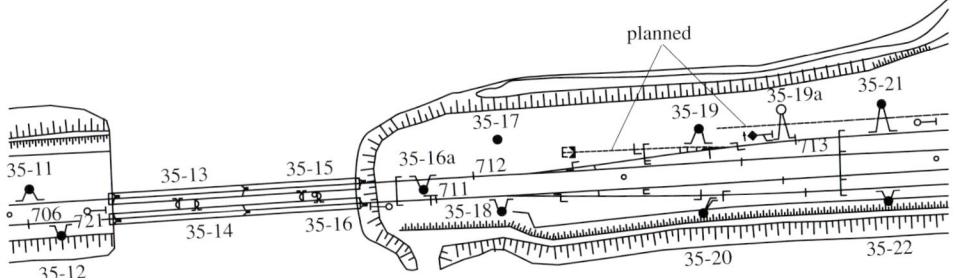

Figure 12.76: Earthing diagram (partial view).

12.5.11 Establishing layout plans

The *overhead contact line layout* should contain all important information within a 15 m right-of-way from the centre line of the outer tracks. The interaction between the overhead contact line and other equipment can be seen from the agreed symbols (Figure 12.73).

The *adjustment plan* is used in stations to show complex contact line routing (Figure 12.74). It is prepared to a scale of 1 : 500 longitudinally and 1 : 50 transversely to the track.

In stations, the *circuit diagram*, also known as the electrical section diagram, shows the circuit of the overhead contact lines. It is used for *switching operations* in the overhead line network and represents the normal state of disconnectors (Figure 12.75).

Many designs do not include a special *earthing diagram* for stations because the contact layout plans should contain all the components of the current return circuit and protective earthing (Figure 12.76).

The layout diagram of the *disconnector remote control* enables the identification of their routing, type and connection points of the control cables for the remote control of the overhead line disconnectors.

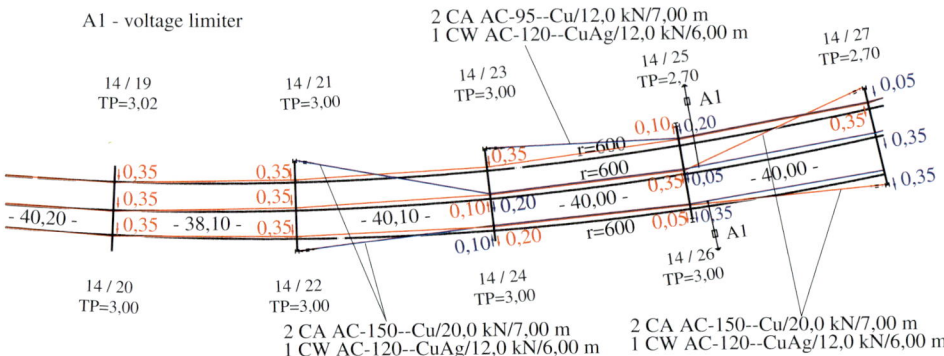

Figure 12.77: Overhead contact line layout for a mass transit system (partial view).

12.5.12   Layouts for mass transit overhead contact lines

The *layouts for mass transit overhead contact lines* identify the necessary details of adjacent buildings, installations and land boundaries, that the planning engineer has to consider. The planning engineer uses the track and road layouts, as well as the adjacent buildings, to establish the overhead line layout plans, including where adjacent buildings may carry the overhead contact line supports. To ensure that noise from pantographs is not transmitted to supporting buildings, design engineers adopt Parafil ropes as damping elements.

The symbols that characterise overhead contact line components vary between mass transit entities. Consequently, software for overhead contact line design can only be used for mass transit systems with extensive customisation. In most cases, overhead contact line layouts are sufficient for mass transit systems (Figure 12.77) and transverse and longitudinal plans are rarely needed.

12.6   Transverse profile diagram

12.6.1   Objective and information

The *transverse profile diagram*, also known as an overhead cross-section, shows the arrangement and type of supports, head-span equipment, poles with the traction power supply lines and geometric dimensions on a section through the railway *permanent way*. This assists in determining material, installation and maintenance requirements.

12.6.2   Types of poles and their classification

The *selection* of *poles* for individual locations is also part of designing overhead contact lines. Distinctions between several types of poles in the overhead contact line, such as a *suspension pole*, *midpoint anchor pole* or *termination pole*, are made to correspond with their function in accordance with EN 50 119: 2009+A1: 2013 (see Clause 13.3.1). The project engineer prepares an overview of the pole types to be used, based on the characteristic data of the overhead contact line to be designed and the static analysis, from which the pole will be selected. Figure 12.78 illustrates the pole types for the Re330 contact line.

Figure 12.78: Types of poles for overhead contact line type Re330.
1 suspension pole with single cantilever, 3 and 4 suspension pole with twin cantilevers in overlaps,
5 midpoint pole, 6, 7 midpoint anchor pole, 8 tensioning pole.

a) Bolt-mounted pole b) Inserted pole c) Pull-over pole

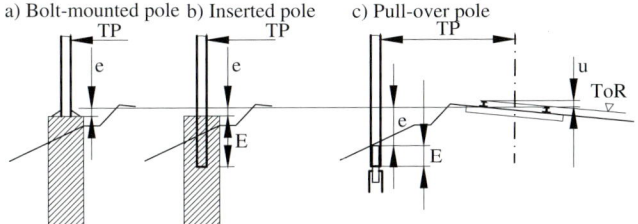

Figure 12.79: Interface between pole and foundation.
ToR: top of rail, TP: distance between the track centre line and the track-side face of the pole, e: distance between top of rail and top of foundation, E: insert depth of pole into the foundation, u: track cant

Poles can be mounted with anchor bolts on top of the foundation (Figure 12.79 a)), inserted into the foundation (Figure 12.79 b)) or pulled over a tube (Figure 12.79 c)). The *pole length* depends on the design of the connection between pole and foundation, to be determined during the design work. The distance between the track centre line and the track-side face of the pole is designated as dimension TP. The difference in height between top of rail and the top surface of the foundation is dimension e and the pull-over or *insertion length* is the dimension E (see Figure 12.79). In Figure 12.83 a typical pole geometry is shown for the contact line Re330 and in Table 12.27 Sicat H1.0 respectively with their identifiers.

12.6.3 Transverse switching lines, disconnectors on poles

Transverse switching lines connect disconnectors to the overhead contact lines. These feeders run directly to the contact line overlaps where the transverse disconnector is located adjacent to the overhead line to be connected (Figure 12.80 a)). Overhead lines on tracks, that are not

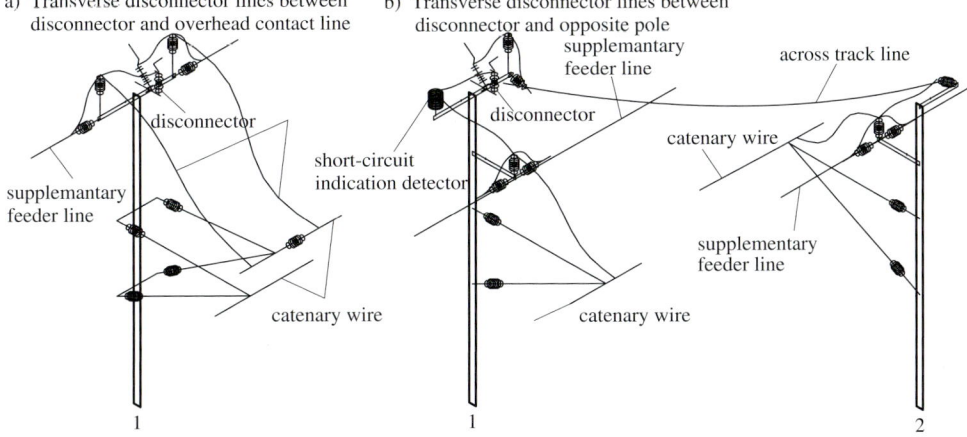

a) Transverse disconnector lines between disconnector and overhead contact line

b) Transverse disconnector lines between disconnector and opposite pole

Figure 12.80: Schematic diagram of transverse disconnector lines.

a) Concrete pole on a driven pile foundation   b) H-type pole on a on-site cast concrete foundation

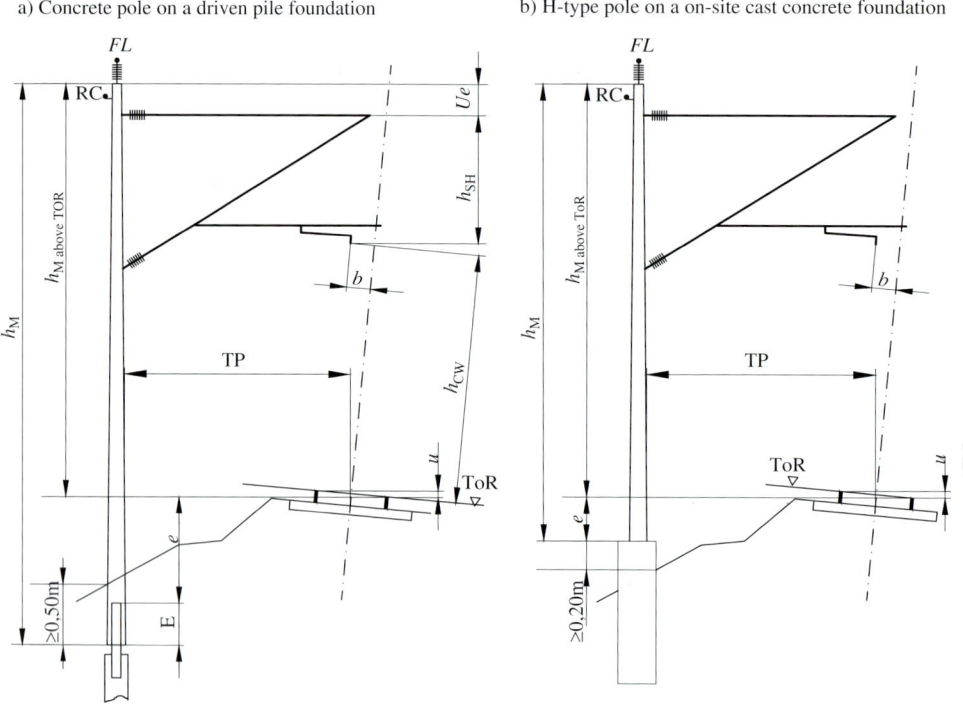

Figure 12.81: Arrangement of poles in embankment location and curves.
Symbols see Figure 12.79

located close to the pole carrying the disconnector, are connected by across track feeders and *drop feeds*. (Figure 12.80 b)). Across track feeders are often found in stations with flexible head-span structures.

The arrangement of the drop feeds needs to take into account the effect of wind and contact line movement. These feeders across the track need to be allowed for when selecting poles. Disconnectors with their feeders are fixed by the location of switchable overlaps or section insulators. The positioning of the overhead contact line disconnectors on the poles for the *connection of circuit groups* in stations can be chosen within certain limits. Short cable routes between the disconnector and the control location, as well as the common use of cable ducts should be considered.

12.6.4   Determination of pole lengths

The dimension TP between track axis and pole face should be defined before fixing the *pole lengths*, i. e. the pole location needs to be defined in the transverse profile. Pull-on, inserted and *mounted poles* result in different dimensions e (Figure 12.79) being the vertical distance between the top surface of the driven pile, tube or concrete foundation and the top of the lowest rail of the reference track. Concrete poles, which are pulled over a steel tube welded onto a driven steel pile require an earth covering 0,5 m thick. The driven tube should reach into the pole by at least 0,50 m when measured from the pole lower end. The foundation for bolt-mounted and inserted poles should project at least 0,2 m above the terrain level to avoid water gathering on the foundation surface and freezing during winter.

The pole length follows from these conditions. The railway transverse profile existing at the location, the contact wire and system heights, a pole extension Ue above the upper cantilever swivel hinge and any planned traction power lines should also be evaluated. The pole length is illustrated in Figure 12.81 a) for concrete poles with driven pile foundations and in Figure 12.81 b) for bolt-mounted poles with block foundations, each in a curve. The pole length for *pull-over concrete poles* is the sum of:

- pull-over dimension E
- distance e between the top foundation surface and the top of rail of reference track
- supplement for cant $2/3\,u$
- contact wire height h_{CW}
- system height h_{SH} and
- pole over-length Ue

Example 12.8: Find the pole length h_M of a concrete pole for contact line Re330. The dimension $e = 1,15$ has already been determined from the track profile. The cant is 0,15 m.

over-length Ue	0,30 m
system height h_{SH}	1,80 m
contact wire height h_{CW}	5,30 m
supplement for cant $2/3\,u$ for $u = 0,15$ m	0,10 m
Sum: pole length $h_{M\,above\,ToR}\,h_M$	7,50 m
difference e between ToR and top surface of foundation	1,15 m
pull-over dimension E	0,50 m
Sum: Total pole length h_M	9,15 m

The pole length should be increased to comply with the increment step of 0,25 m, therefore, to 9,25 m. The e-dimension is to be corrected correspondingly and implemented at 1,25 m.

12.6.5 Cantilevers

Determination of the *cantilever type* and the corresponding lengths of main components enable materials procurement. *Cantilever calculation programs* are used to determine the *cantilever type* with respect to configuration and lengths of components. The contact line data such as specific weights, support data and track data are necessary for this purpose. A dimensional check is performed after erecting the pole and the cantilever calculation and manufacturing are carried out after considering any deviations that might arise during the construction phase.

The overhead contact line data, specific weights and support track geometry data are required to determine the type of cantilever and lengths of components. The contact line data consists of the type of insulation, temperature range, tensile forces in catenary and contact wires and other wires and conductor types. The following support data is also necessary: pole number, pole type, span lengths, contact wire height, contact wire and catenary wire stagger, distance between track side face of pole and track centre line, pole inclination, track radius and line gradient. The results are presented in a graphic or tabular format with dimensions for *manufacturing the cantilever* (Figure 12.82).

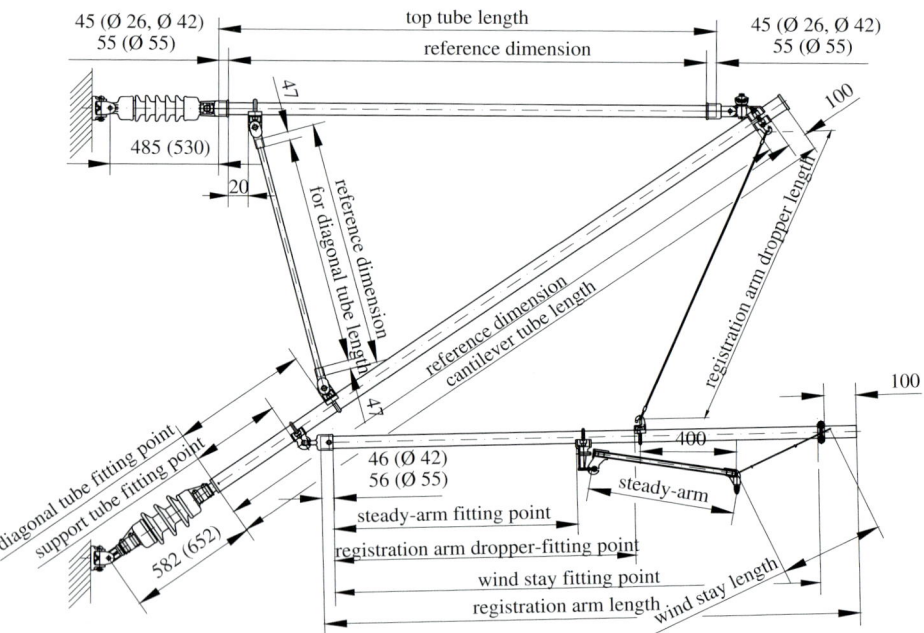

Figure 12.82: Dimensions for manufacturing a cantilever with pull-off contact wire support.

12.6.6   Pole and foundation selection

The *selection of pole design*, its configuration and mechanical strength is carried out for each individual pole site. Selection tables provide assistance for typical applications. Where required, because of non-standard conditions, static ratings for individual pole locations (see Clause 13.3 and 13.7) are carried out. Table 12.27, shows available pole types and their range of application for contact line type Sicat H1.0. The pole type NB 3 refers to the application of a *concrete pole* of type 3 as shown in Figure 12.78 and Figure 13.5 for new lines.

Example 12.9:   A concrete pole is to be found for carrying twin cantilevers in an overlap with the contact line Sicat H1.0 on a curve with a track radius $R = 4000$ m, without cant, e, g. at a platform. The pole carries a parallel line feeder FL at the pole head in addition to the contact line. A driven pile, fitted with a welded tube, supports the concrete pole. Dimension $e = 0,7$ m was determined from the track transverse profile. The pole type is extracted from Figure 12.78 as type 3 or 4 and Figure 12.83 indicates pole design 21 matches the example criteria. Pole type NB 4 is selected with a pull-over dimension $E = 0,5$ m from Table 12.27 for a radius $R = 4000$ m.

Table 12.27: Concrete pole types for contact line Sicat H1.0 for 33 m/s wind velocity and radius ≥ 2000 m.

Pole type	Diameter of driven pile tube	Pole type 1	Pole types 3, 4 and 6	Pole type 5	Pole type 8
21	R 1	NB 2	NB 4	NB 3	NB 5
22	R 2	NB 2	NB 4	NB 3	NB 5
23	R 3	NB 2	NB 4	NB 3	NB 5
24	R 2	NB 2	NB 4	NB 3	NB 5

configuration 21 with
- catenary
- return conductor
- parallel feeder wire
  at pole head

configuration 22 with
- catenary
- return conductor
- parallel feeder wire
  on inner side of pole

configuration 23 with
- catenary
- return conductor
- parallel feeder wire
  on inner side of pole
- reinforcing line

configuration 24 with
- catenary
- return conductor
  on inner side of pole
- parallel feeder wire
  at pole head

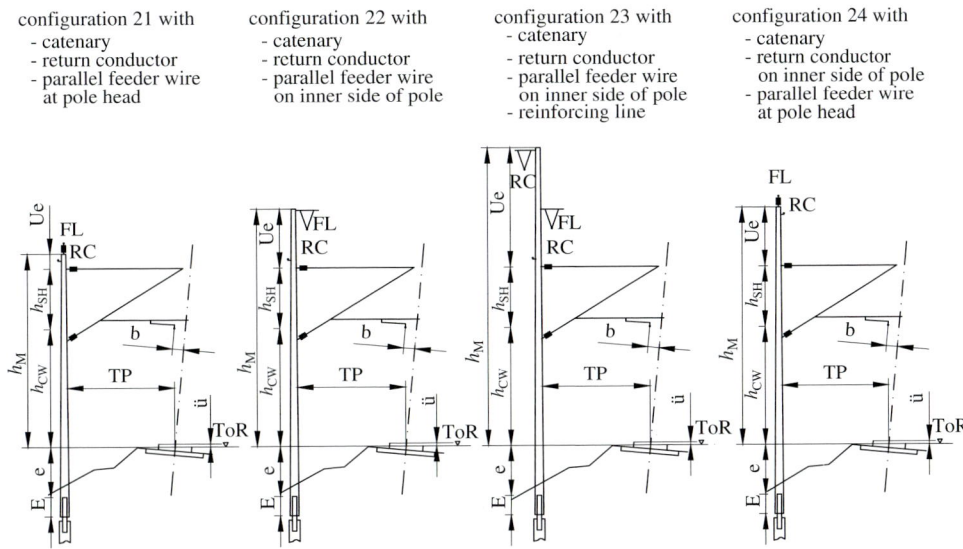

Figure 12.83: Pole designs for overhead contact line type Re330. Symbols see Figure 12.79

a) Concrete block
foundation for
bolt-mounted poles

b) Concrete block
foundation for
inserted poles

c) Driven pile
foundation

d) Driven tube
foundation

e) Directly embedded
concrete pole

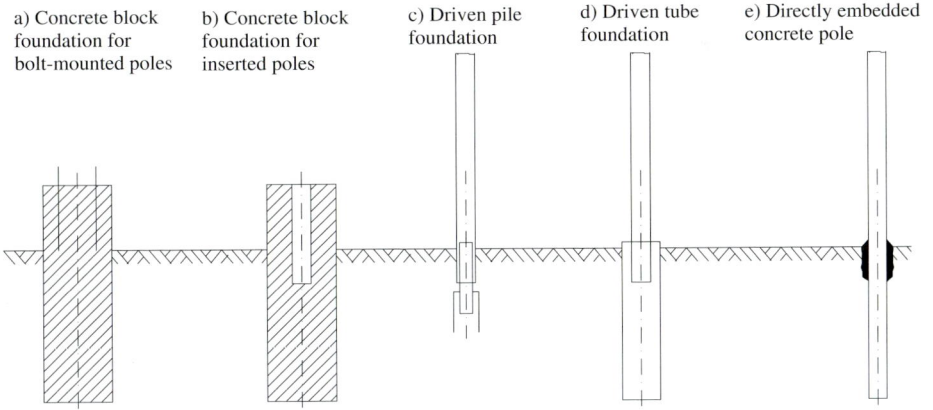

Figure 12.84: Standard types of foundations.

The foundation type is determined by the soil properties, construction resources and the pole design. Steel poles with a base plate require *concrete block foundations* with a square or round cross-section, or *driven pile foundations* with anchor bolts (Figure 12.84 a)). Poles may also be attached to structures such as bridge girders. *Concrete foundations* with bore holes accept *inserted poles* of steel or concrete, embedded into concrete or chippings after erection (Figure 12.84 b)). Concrete poles can also be embedded directly in the ground (Figure 12.84 e)). They can be mounted on steel or concrete driven piles and cast in mortar (Figure 12.84 c)). Steel or driven pile tubes can also be used to support concrete or steel poles (Figure 12.84 d)). Foundations are selected depending on the type of pole, its loadings, type of soil and the available installation equipment. The foundations can be selected either from prepared tables for a certain contact line type or they may be designed individually with the aid of static

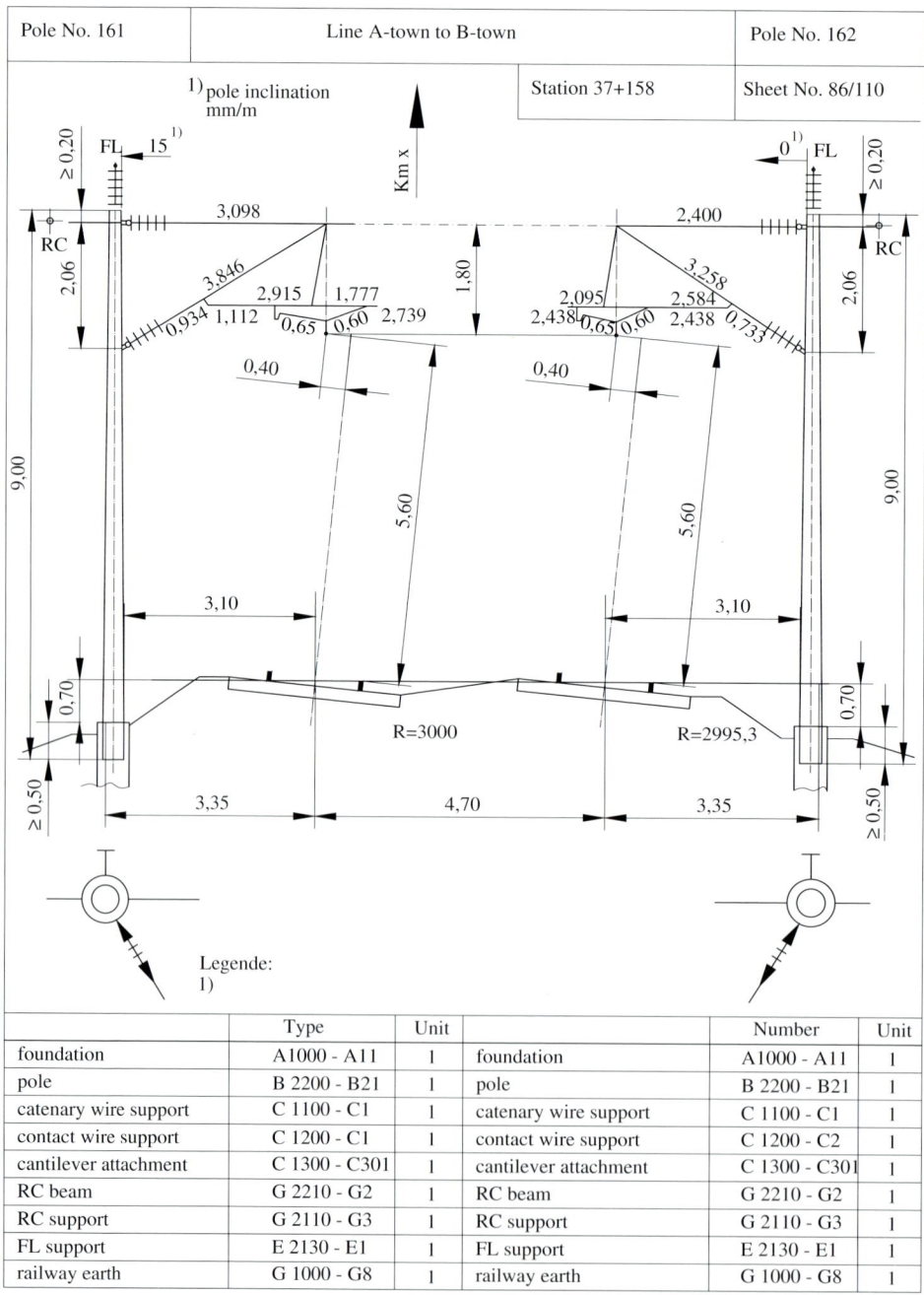

Figure 12.85: Transverse profile of a double track line with individual poles.
[1] inclination of pole in mm

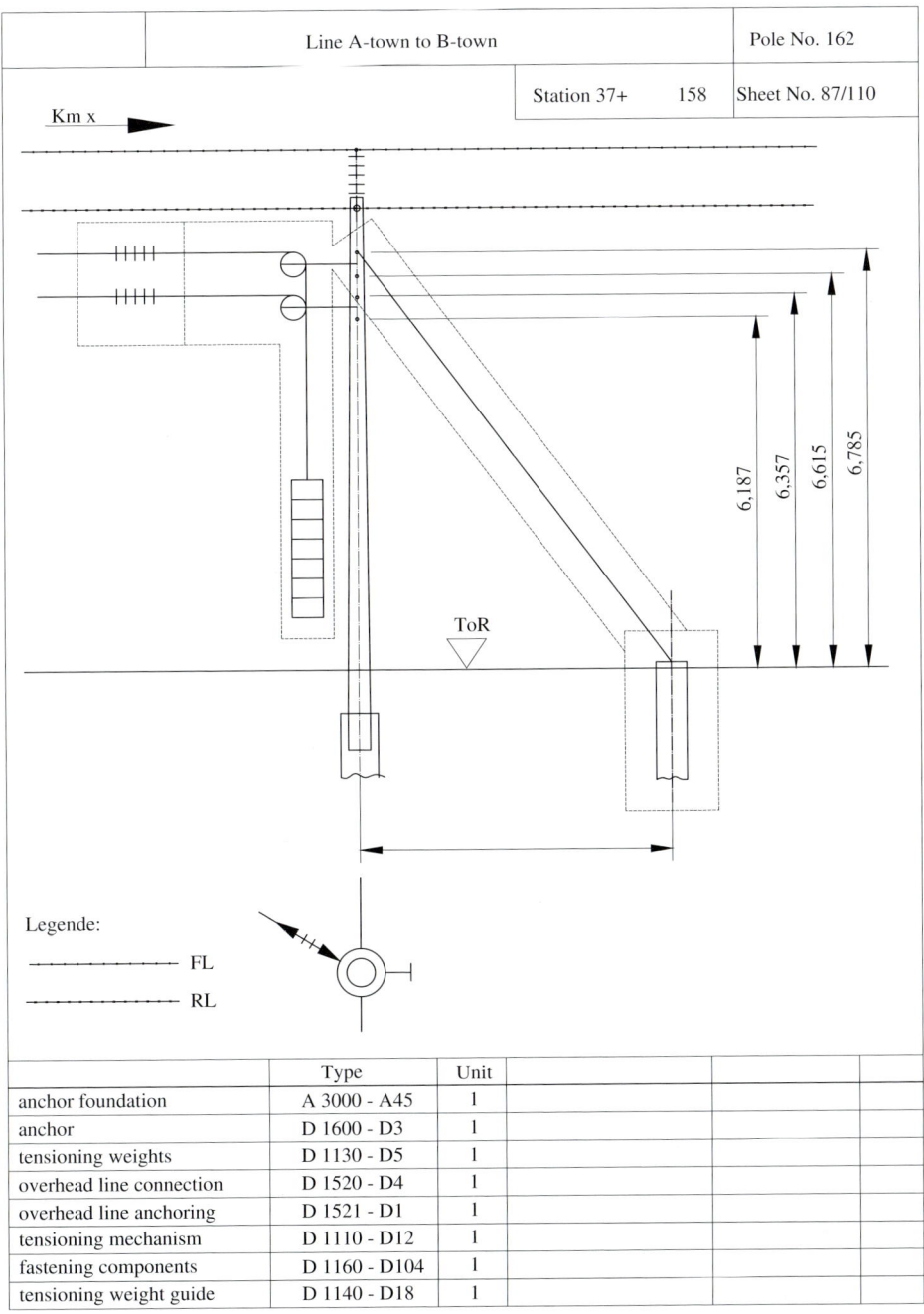

	Line A-town to B-town		Pole No. 162
		Station 37+ 158	Sheet No. 87/110

Km x

6,187
6,357
6,615
6,785

ToR

Legende:

———————— FL

———————— RL

	Type	Unit			
anchor foundation	A 3000 - A45	1			
anchor	D 1600 - D3	1			
tensioning weights	D 1130 - D5	1			
overhead line connection	D 1520 - D4	1			
overhead line anchoring	D 1521 - D1	1			
tensioning mechanism	D 1110 - D12	1			
fastening components	D 1160 - D104	1			
tensioning weight guide	D 1140 - D18	1			

Figure 12.86: Longitudinal profile of contact line termination.

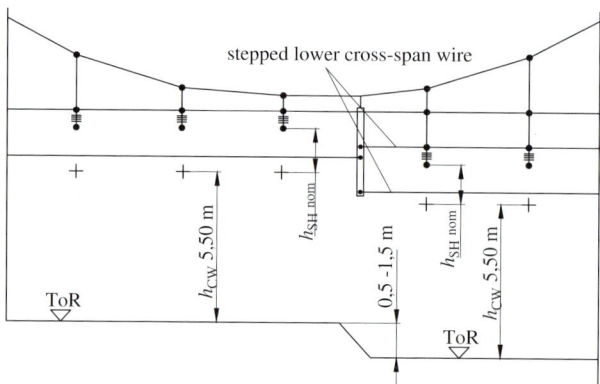

stepped lower cross-span wire

h_{CW} 5.50 m

$h_{SH\,nom}$

ToR

0.5 -1.5 m

$h_{SH\,nom}$

h_{CW} 5.50 m

ToR

Figure 12.87: Accommodation of track height differences in head spans using a stepped lower cross-span wire.

analysis (see Clause 13.7). Figure 12.87 shows some commonly used standard foundation types. Transverse profile diagrams in combination with with longitudinal profiles provide a clear picture of the type and geometric dimensions of foundations (Figures 12.88 and 12.89).

12.6.7   Head-span structures

When planning and analysing *head-span structures*, a distinction needs to be made between earthed and live upper cross-span wires as explained in Clause 12.5.5. The *upper cross-span wire* may be earthed for track radii of more than 800 m because of lower radial forces and the inclination of the insulators and should be energised for track radii less than around 800 m. The type of *catenary wire support* follows accordingly. The arrangement of the sectionalising groups determines the intermediate insulation in the *lower cross-span wire* and in the energised *upper cross-span wire*. *Cross-span tensioning springs* are used to compensate the temperature-dependent conductor length variations and the associated conductor stresses.

Track height differences up to 0,5 m can be accommodated by varying the system height. Stepped lower cross-span wires are required for larger track height differences (Figure 12.87). Generally, at least two *head-span wires* are provided due to reliability reasons. Four head-span wires are sometimes used, depending upon the number of supports and span of the head-span. The head-span wire sag depends on the span width of the head-span. The relationship between sag and span width a_{HS} of the head-span should be between $1/5$ and $1/10$ of the span width. The poles are then fitted into the transverse profile diagram and their lengths determined. The *cross-span polygon* shows the lengths of wires for the head-span structure and permits prefabrication in the workshop. Disconnectors, across-track feeders and jumpers, together with intermediate insulators in the head-span wires, are contained in the transverse profile diagram.

Configuration programs provide a representation of the geometry of the head-span structure with material requirements (Figure 12.88). This type of representation simplifies the installation and provides material data for reconstruction of a head-span structure after damage (see also Clause 11.2.1.3).

12.6.8   Portals

The modular design of *portals* with standardised end and intermediate lattice framework as used by the Norwegian Long Distance Railways (Bane Nor), (Figure 12.89), simplifies

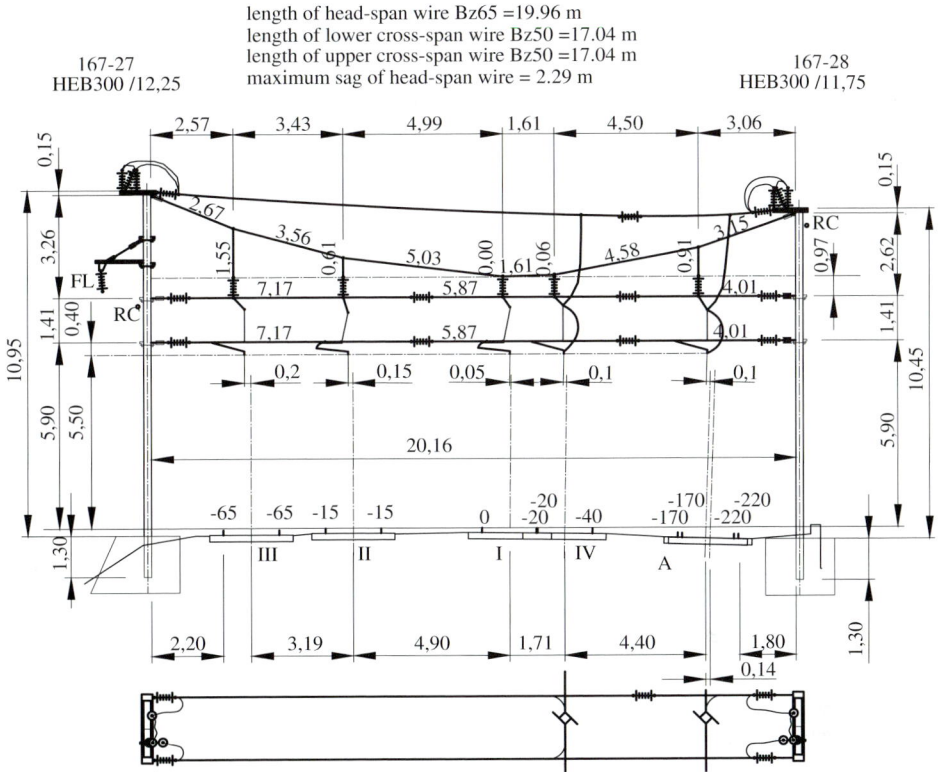

length of head-span wire Bz65 =19.96 m
length of lower cross-span wire Bz50 =17.04 m
length of upper cross-span wire Bz50 =17.04 m
maximum sag of head-span wire = 2.29 m

167-27
HEB300 /12,25

167-28
HEB300 /11,75

Figure 12.88: Transverse profile in stations with a flexible head-span and disconnectors, across-track feeders and jumpers at the Portuguese State Railway CP.

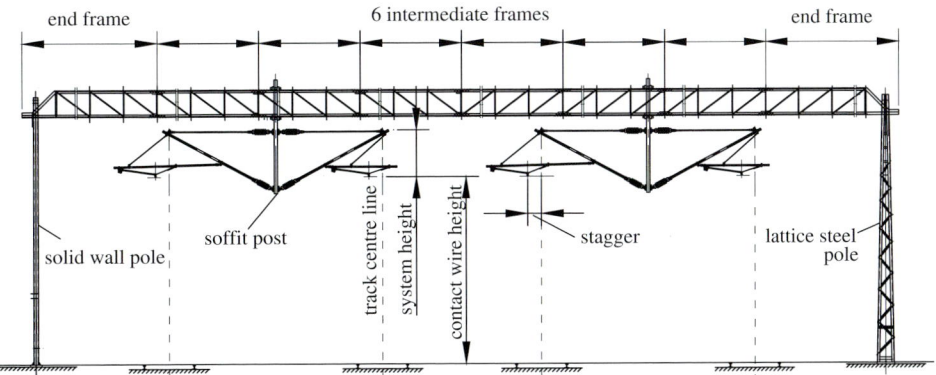

end frame   6 intermediate frames   end frame

solid wall pole   soffit post   track centre line   system height   contact wire height   stagger   lattice steel pole

Figure 12.89: Portal type adopted by Norwegian Railways Bane Nor.

Table 12.28: Portal types at Norwegian Railways Bane Nor.

Portal type	12	14
portal length in m	11–33	28–43
number of tracks	2	> 2

Table 12.29: Arrangement of stitch wires and droppers for contact line type Sicat S1.0.

Span length in m	Stitch wire length in m	Number of stitch wire droppers per support	Spacing stitch wire dropper to support in m	Spacing stitch wire dropper to stitch wire dropper in m	Number of droppers in remaining span length
support type A, pull-off support					
$l \geq 79$	18	4	2,5	3,5	6
$l \geq 71$	14	2	2,5	2)	6
$l \geq 65$	12	2	2,5	2)	4
$l \geq 42$	1)	1)	1)	1)	5
support type B, push-off support					
$1 \geq 79$	14	2	2,5	2)	6
$1 \geq 79$	18	4	2,5	3,5	6
$1 \geq 71$	14	2	2,5	2)	6
$1 \geq 65$	12	2	2,5	2)	4
$1 \geq 42$	1)	1)	1)	1)	5

1) no stitch wire, 2) only two stitch wire droppers arranged symmetrically about support

planning. These portals carry the *supports in stations* and on winding interstation lines, to avoid *push-off contact wire supports*. A maximum of nine tracks can be spanned.

Bane Nor has employed portal types 12 and 14 since 1997. These have different angle sections and can be applied as shown in Table 12.28. Portal type 12 is employed to span two tracks with cross-span widths up to 33 m, and portal type 14 for more than two tracks with cross-span widths between 28 m and 43 m.

The *portal bridge length* results from the spacing of the pole centres, determined from the track system and rounded up to the nearest whole metre.

Soffit posts are mounted on the portal to support standard cantilevers. Track height differences are accommodated by varying the system height at the cantilever or changing the height of the soffit posts on the portal.

The *radial forces in curves* and the portal length determine the pole types to be used to support the portal. Where the radial loads are less than 6 000 N and the portal lengths shorter than 30 m, a double-channel pole is used on one side of the portal and a lattice steel pole manufactured from angle sections on the other side. Lattice steel poles are used on both sides of the portal where the radial loads are greater than 6 000 N and the portals longer than 30 m.

12.7 Longitudinal profiles

12.7.1 Contents

Longitudinal profiles indicate contact wire heights in areas such as overlaps, contact wire height reductions and elevations above and below obstacles, traction power supply lines and dropper arrangements.

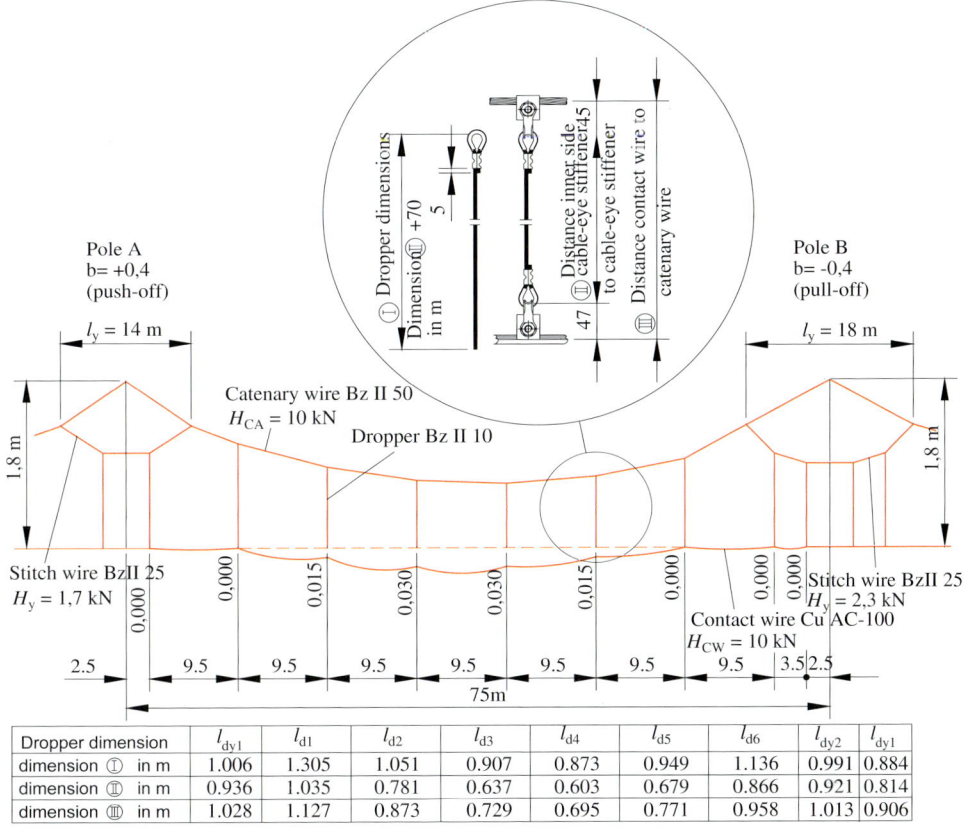

Dropper dimension	l_{dy1}	l_{d1}	l_{d2}	l_{d3}	l_{d4}	l_{d5}	l_{d6}	l_{dy2}	l_{dy1}
dimension ① in m	1.006	1.305	1.051	0.907	0.873	0.949	1.136	0.991	0.884
dimension ② in m	0.936	1.035	0.781	0.637	0.603	0.679	0.866	0.921	0.814
dimension ③ in m	1.028	1.127	0.873	0.729	0.695	0.771	0.958	1.013	0.906

Figure 12.90: Dropper spacing and lengths in a 75 m span of contact line type Re200 for non-current carrying droppers.

12.7.2 Dropper arrangement

Dropper lengths are calculated by computer software using the following parameters:
– *contact wire stagger* at the support
– *longitudinal line gradient*
– *track cant, track cant deficiency* and *track radius*
– *system height*
– *tensile force in catenary wire and contact wire*
– *employment of stitch wires and their data*
– *dead weight per unit length*

The *dropper spacing* in the spans depends on the overhead contact line type and the span lengths. For contact line type Re200, the arrangement of the stitch wire droppers can be obtained from Table 12.29. A spacing of 9,5 m between the droppers results for a 75 m span (Figure 12.90).

Dropper calculation during the design phase assists *material procurement* as in the case of cantilever calculation. The exact dropper lengths can be calculated and manufactured at the construction site after measuring the position of the contact line supports.

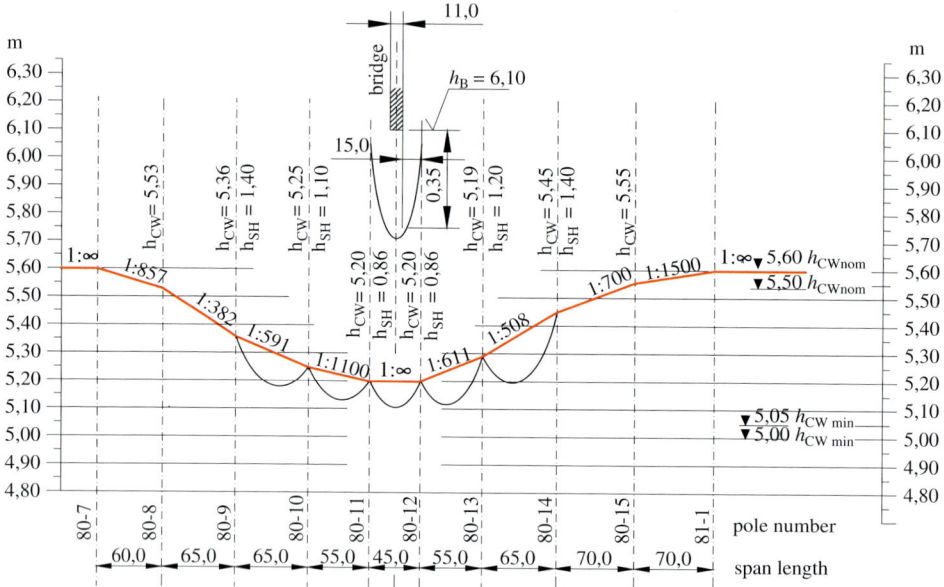

Figure 12.91: Extract from a contact wire height reduction for contact line type Re200. Dimensions in m.

12.7.3 Contact line height reductions

In longitudinal profile diagrams, the *contact line height reductions* are shown in accordance with the calculation described in Clause 12.4.9. They illustrate the contact wire gradients and system heights to a distorted scale: longitudinal scale 1 : 100 and vertical scale 1 : 10 (Figure 12.91). The exaggerated vertical scale highlights the effects of *sag*, *system height* and *clearances*.

The contact wire sag under ice load is depicted over four to five spans adjacent to a bridge structure, to verify that the minimum contact wire height has been met.

12.7.4 Traction power line longitudinal profile

Longitudinal profiles for *traction power lines* contain the clearances to structures and other equipment. The line *longitudinal profiles* are also drawn to distorted scales, usually with a scale of 1 : 500 along the line and 1 : 100 vertical to the line longitudinal axis. Figure 12.92 illustrates an extract from a longitudinal power line profile.

12.7.5 Minimum clearances to overhead lines and traction feeder lines

12.7.5.1 Introduction

The required *minimum clearances* between live parts of the overhead contact line system, such as contact lines, supports, disconnectors to third party objects and to the electrical infrastructure gauge depend on the operating voltage (see Figure 12.93). They include effects caused by temperature, ice loads and wind.

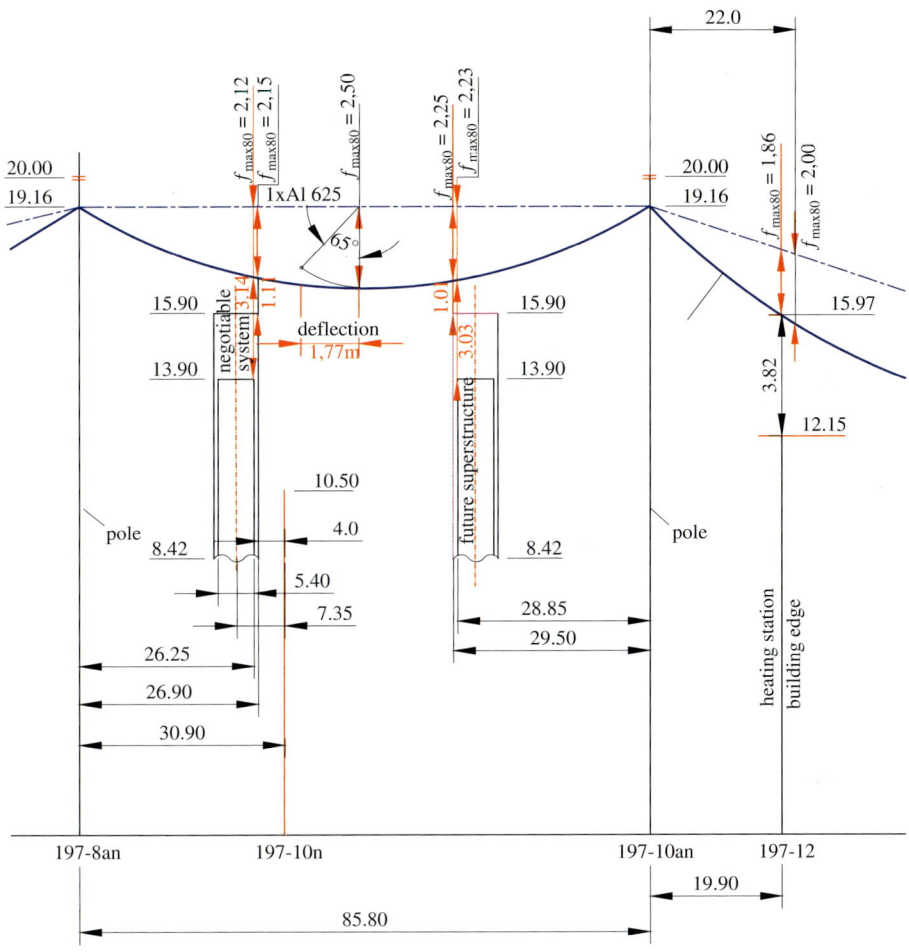

Figure 12.92: Extract from a power line longitudinal profile. All dimensions in m

f_{max80} : Maximum line sag at conductor temperature 80°C

If attached to the contact line structures, traction power lines form part of the contact line system. Then, minimum clearances apply according to EN 50 119 and EN 50 122-1 (Table 12.30) or to the operator's stipulations like German Railway directive 997.

Minimum clearances in open terrain detailed in EN 50 341-1 and EN 50 341-2-4 (see Clauses 12.4.3 and 12.4.4) apply to traction power lines attached to dedicated poles, rather than contact line poles. Table 12.30 summarises these minimum clearances. To ensure protection from live parts of the overhead contact line installation and overhead power lines, the practices of 'protection by clearance' and 'protection by obstacles' are applied. The less complicated method is protection by clearance, achieved by sufficiently large clearances in air (Figure 12.93). Where the available clearance is insufficient, obstacles or barriers are established to avoid direct touching of the live parts of the installations.

Table 12.30: Minimum clearances from various objects to energised components of the overhead contact line according to EN 50 122-1: 2011+A1: 2011+AC: 2012+A2: 2016 and EN 50 119: 2009+A1: 2013.

No.	Object	Direction from object	Clearance in m AC 1 kV DC 1,5 kV	Clearance in m DC 3 kV AC 15 kV AC 25 kV
1	standing surface for electrically skilled staff, electro-technically instructed persons and railway system instructed persons	downwards sideways upwards	1,35 1,35 2,60	1,50 1,50 3,50[1] 2,75[2]
2	standing surface for general public	downwards sideways upwards	2,50 1,45 3,00	3,00 2,25 3,50
3	platforms	upwards	4,50	4,50
4	obstacles with solid walls (mesh width less than 1 200 mm^2)	downwards sidewards	0,60 0,30	1,50[1] 0,60
5	structures such as platform roofs, superstructures, tunnels, buildings, bridges	all	0,10	0,15[3] 0,27[4]
6	signal or lighting poles, working platform parts of signals, that can be stepped on	all		1,50[5]
7	barriers open	all		1,00
8	unclimbable structures, such as signal vanes, with climbing barriers	all		0,60
9	windows in buildings for electrically skilled staff, electro-technically instructed persons and railway system instructed persons	sideways	1,35	1,50 3,57
10	windows in buildings for general public	sideways	1,45	2,25 4,07
11	road surface at crossings – minimum clearance to vehicles established by signs – minimum clearance to vehicles protected by gauge (profile) gates	upwards all all	4,70 0,50 0,30	5,50 1,00 0,50
12	contact lines of another circuit group	all	1,50	1,50
13	feeder line	downwards		2,00
14	return conductor	upwards sideways		0,50 1,25
15	across-track feeder	all		2,00
17	contact line installation – across-track feeders in same circuit group	all		0,10

^{1}) not in public area, ^{2}) 0,5 m if solid-wall obstacles are adopted, ^{3}) up to AC 15 kV, ^{4}) up to AC 25 kV, ^{5}) may be reduced to 0,6 m under certain circumstances

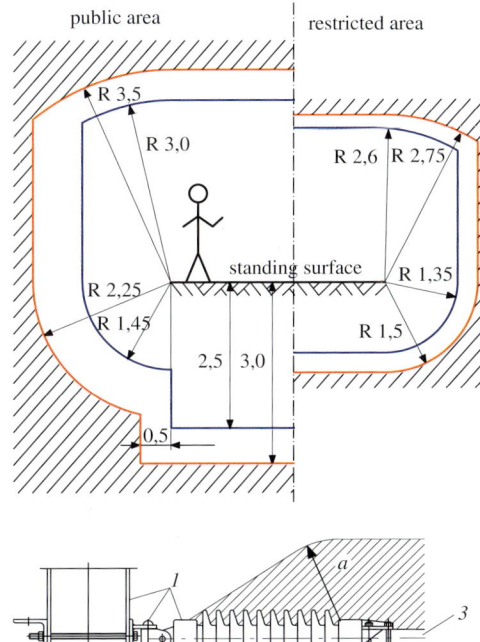

Figure 12.93: Minimum clearances between persons and live components according to EN 50 122-1:2011.
AC 15 kV and AC 25 kV
DC 1,5 kV and DC 3,0 kV

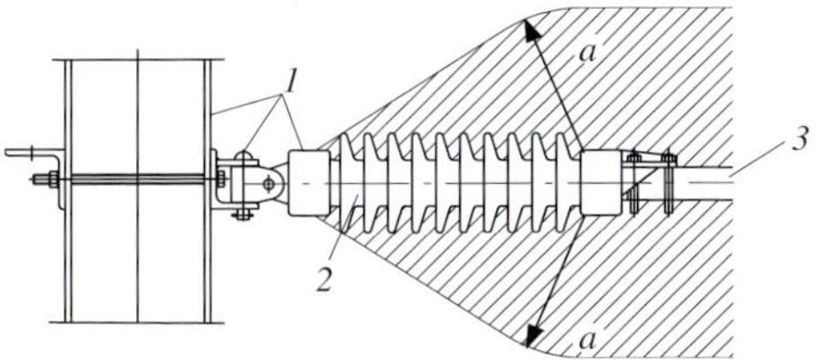

Figure 12.94: Electrical clearances in air (minimum air clearances) between insulators and structures (proposal for prEN 50 119:2017).
a 270 mm for AC 25 kV, 150 mm for AC 15 kV and for DC 3 kV, 100 mm up to DC 1,5 kV
1 earthed components, *2* insulator sheds,
3 live top anchor

12.7.5.2   Protection by clearance

People must not be able to reach a live component from any standing surface according to EN 50 122-1. All areas wider than 100 mm × 100 mm are considered to be possible standing surfaces. Components of the overhead contact line installation, the power line or parts of the vehicle outside surfaces such as pantographs, lines on vehicle roofs and resistors are considered as active parts of the installation. The clearance is measured in a straight line and without any remedies as shown in Figure 12.93. The clearances given in Figure 12.93

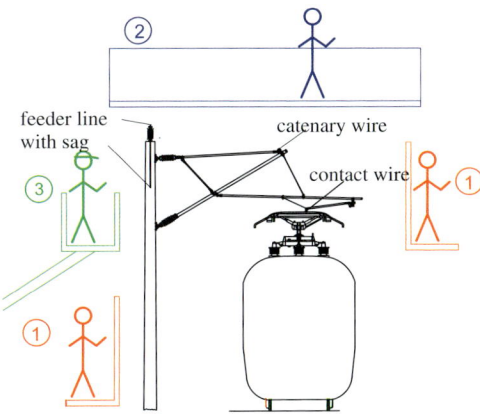

Figure 12.95: Standing surfaces close to live parts.
standing surface beside live components (examples see Figure 12.96 a) and b))
standing above live components (examples see Figure 12.97 a) and b))
working platform in workshops (example see Figure 12.98)

a) Solid wall obstacle and meshed wire
at Cologne–Frankfurt high-speed line

b) Glass solid wall design at HSL Zuid high-speed line

Figure 12.96: Examples of obstacles for standing surface adjacent to live parts.

are minimum values for the public and the reserved parts of the installation that needs to be complied with under all temperatures as well as additional and exceptional loads on the lines. The insulators that connect the live part of the installation with the inactive part are considered to be live as far as the clearances are concerned. The protective clearance to be complied with in the case of *working close to the voltage or at overhead contact line installations* is described in EN 50 119: 2009+A1: 2013 and starts at the insulator shed of the earthed cap of insulator. The diagram of electrical clearance is shown in Figure 12.94. The clearances are

– 1,25 m for DC 1,5 kV and DC 3,0 kV and
– 1,50 m for AC 15 kV, AC 25 kV or 2×25 kV.

A minimum clearance of 2,50 m is required in still air in accordance with to EN 50 122-1 between branches of trees or bushes and live components of the contact or traction power line. German Rail DB increase this clearance to 5,0 m where tree heights are above 4,0 m.

12.7.5.3 Standing surfaces

EN 50 122-1 specifies that a *standing surface* is any point on a surface where persons may stand or walk about without great effort. Standing surfaces can be situated in the public or private area where there is a minimum area of more than 10 cm × 10 cm.

12.7.5.4 Protection by obstacles or barriers

Where the clearances detailed in Clause 12.7.5.2 cannot be complied with, *protection by obstacles or barriers* needs to be adopted. The shape and the arrangement of the obstacles depends on the:

– position of the standing surface relative to the live component
– distance between the obstacle and the live part and
– arrangement of the standing surface in private or public areas

a) Combination of solid wall and meshed wire obstacle

b) Inclined concrete or
steel apron (canopy)

Figure 12.97: Examples of obstacles on bridge structures above live parts.

Figure 12.95 shows typical cases where the obstacle protects people on the standing surface from unintentional or random touching of live parts in a straight direction. The obstacles need to consist of:
- solid walls or solid wall doors without holes or gaps or
- lattice structures

Obstacles should consist of non-conductive material or plastic covered metal. However, if plastic covered or non-conductive obstacles are installed, they need to consist of a solid wall edged with a bare rail-bonded conductor. Obstacles:
- need to be mechanically stable and reliably fixed to avoid movement and reduced minimum clearances to live parts
- may not be able to be dismantled in public areas with standard tools without being destroyed

Standing surfaces beside live parts of the overhead contact line and active parts of vehicles (Figure 12.95, case 1)

The obstacles need to reach 1,8 m in height and be designed with a solid wall up to a height of 1,0 m. The upper 0,8 m section of the obstacle needs to consist of a screen with meshes not larger than $1\,200\,\text{mm}^2$. The clearance between the obstacle and the live part needs to be at least 1 m in the public area and at least 0,6 m in the private area. If this clearance cannot be achieved, then a solid wall is required. Figure 12.96 shows typical obstacles at standing surfaces close to active parts.

Standing surfaces above active parts of the overhead contact line or active parts of vehicles (Figure 12.95, case 2)

The obstacle at standing surfaces above active parts needs to:
- be installed as a solid wall
- cover the pantograph area by ±2,0 m as seen from the track axis
- project at least 0,5 m on both sides of the active parts
- be limited on both sides by obstacles

The obstacles need to be arranged vertically according to Figure 12.97 a) or horizontally according to Figure 12.97 b). The obstacle according to Figure 12.97 a) corresponds to Figure 12.96. Figure 12.97 b) shows a solid concrete obstacle or a steel element with an inclination of at least 20° away from structures. This design should preclude walking on the obstacle. The minimum clearance between railing or terrain surface and live component is 2,25 m.

Figure 12.98: Obstacle at a working platform in a depot.

Figure 12.99: Control panel with keys.

Standing surfaces for linesmen (Figure **12.95, case 3**)

In the case of working staff, the clearances according to Figure 12.95 need to be complied with. If it is possible to approach closer than these clearances, then permissible reductions need to be stipulated in the operational instructions. Roofs of platforms, working platforms and working bridges, signalling bridges, work places at signals, maintenance ladders, work platforms of lifting vehicles, work platforms of maintenance cars that can only be used to carry out work activities at overhead contact line installations are not considered as standing surfaces according to cases 1 to 3 in Figure 12.95. Working on such installations is considered in operational specifications for labour safety. Figure 12.98 shows an obstacle for the standing surface in a rolling stock maintenance depot.

The metallic door shown in Figure 12.98 at the top of the stairs can only be opened if the conditions for entering the work platform are met. This is the case if:

- the overhead contact line along the working platform has been de-energized
- the contact line is earthed on both sides by disconnecting earthing switches
- checking the contact line section to confirm that the earthing on both sides has been carried out and the isolating distances are visible

Only after these actions are completed can the cover to the control board be opened (Figure 12.99). The key to open the metal door at the top of the stairs, enabling entry to the work platform, can then be taken from the control board (see also Clause 14.2).

Warning signs (Figure **12.100**)

Warning signs need to be arranged at entries where there is a dangerous proximity to the contact line installation. The signs should be located at a clearly visible position close to the entrance and needs to comply with ISO 3864: 2011.

Figure 12.100: Warning sign.

Table 12.31: Minimum clearances between public installations and traction power lines and their components.

No	Object	Direction	Clearance in m DC 1,5 kV AC 1,0 kV	Clearance in m DC 3 kV AC 15 kV AC 25 kV	Standard
1	loading platforms	vertically		12,0	EN 50 122-1
2	ground surfaces negotiable	vertically	4,70	5,5	EN 50 122-1
3	climbable trees	vertically	2,50	2,5	EN 50 122-1
4	windows in buildings				
	– private area	horizontally		1,5	EN 50 341-1
	– public area	horizontally		2,25	EN 50 341-1
5.1	buildings with roof slope:				
	roof slope > 15°and fire-resistant roof	vertically		3,0	EN 50 341-1
	roof slope ≤ 15°and fire-resistant roof	vertically		5,0	EN 50 341-1
5.2	buildings with roof slope: > 15° or ≤ 15° without fire-resistant roofs and above inflammable equipment like gas stations	vertically		10,6	EN 50 341-1
5.3	buildings with roof slope: > 15°or ≤ 15°with or without fire-resistant roofs	horizontally		3,0	EN 50 341-1
6	Antennas, lightning protection poles flag poles, advertising posts etc.	horizontally		2,0 2,6	EN 50 341-1

12.7.6 Traction power lines

12.7.6.1 Definitions and requirements

Traction power lines such as feeder lines, bypass lines and return conductors are part of the electric traction system and are frequently installed on the overhead contact line poles. This arrangement should not affect safe operation and, as for as possible, not require switching off during inspections. The minimum clearances depend on the nominal voltage (see Table 12.30) and the climatic conditions.

12.7.6.2 Line arrangement at poles

Traction power lines (TPL) are attached to the top of contact line poles by line post or suspension insulators. If several conductors of traction power lines are arranged at the same contact line support, cross-arms are arranged at the pole. Cross-arms ensure that minimum clearance is maintained between the line and the pole and other objects.

Sectioning supports form rigid points in the line and are equipped with dead-end insulator sets. Dead-end supports are found at the beginning and end of a line section. Table 12.30

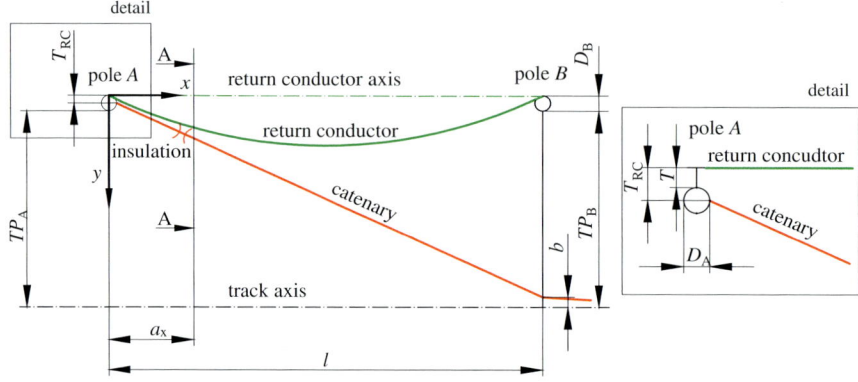

Figure 12.101: Plan view of terminating contact line and the deflected return current conductor.

TP_A distance between track-side face of pole A and the track centre line in m
TP_B as above for pole B
D_A diameter of pole A at the height of the return current conductor suspension
D_B as above for pole B
b contact wire stagger of the terminating contact line system in m
T_{RC} distance of return conductor attachment to axis of pole A in m
T cross arm length in m
l span length between poles A and B in m
a_x distance from the centre line of pole A to the verification point in m, whereby a measures to the end of insulator live side

shows minimum clearances between objects within the railway property and Table 12.31 the clearances applicable when traction power lines are routed through terrain not owned by the railway infrastructure manager.

12.7.6.3   Clearance verification

The routing of a traction power line and its arrangement at the supports determines the line geometry. Temperature variations, wind action and ice loads cause variations of conductor positions that may not violate the minimum clearances between the conductors themselves and between conductors and other objects. Acceptable clearances need to be verified for specified adverse cases of conductor temperatures and loading. The conditions are specified in EN 50341-1 for Europe and in national standards. To care for wind action the minimum spacing a_{min} between conductors in still air needs to be

$$a_{min} = k \cdot \sqrt{f_2 + l_k} + 0,75 D_{el}. \tag{12.14}$$

The coefficient k can be taken from EN 50341-1:2012, Annex F, Table F.1 or national and operator specifications which serve to aid the determination of required minimum distances at midspan. To determine the coefficient k, which depends on the swing angle φ and sag f_2, it is sufficient to consider a conductor temperature of 40 °C. The length l_k is the part of the insulator set swinging perpendicularly to the line.

Verification of clearances between adjacent conductors, as well as to the ground and any objects under and close to the line, is required with *sags* and conductor positions occurring under these temperatures and loadings. The clearances between return current conductor and contact line will be demonstrated by an example. The following conditions apply:

- acceptable clearance is to be verified at the conductor position resulting in the closest proximity, both in still air and in swung condition
- the arrangement of return current conductor supports at poles needs to consider the minimum dynamic clearances between attachments of return current conductors and live parts of feeder lines. An appropriate selection of the conductor type and its tensile stress can ensure compliance with this requirement. Tables 12.30, 12.31, 12.21 and EN 50 119: 2009+A1: 2013, Table 5.9, specify the clearances to earthed parts under static and dynamic conditions
- a conductor temperature of 60 °C is assumed in still air for return conductors 243-AL1 sagged with 20 N/mm^2 tensile stress and a temperature 40 °C in swung condition by an angle $\varphi_{RL} = 65°$ according to EN 50 341-1
- the return current conductor must not sag below the contact wire at the bottom of the sagging curve. The specified clearance between return conductor and terrain surface needs to be complied with

The *verification of minimum clearances* is carried out as follows:
- determination of the point a_x of the shortest clearance between the return current conductor and the terminated contact line
- determination of the position of the verification point, i. e. at position a_x
- determination of the position of the return current conductor centre line at position a_x
- calculation of the return current conductor sag f_{RC} at point a_x
- calculation of the spatial clearance R_{RC-CA} between the swung return current conductor and the catenary wire at point a_x

In Figure 12.101, the arrangement of the return current conductor and the terminated contact line is shown in the distorted plan view. In Figure 12.102, section A-A is shown.

The sag of the return current conductor f_{RC} at point a_x, viewed from support A, is obtained from (4.64) and (4.67):

$$f_{RC} = (G'/2H) \cdot a_x(l - a_x) = 4 f_{RC\,max\,40} \cdot a_x(l - a_x)l^2 = 4 \cdot f_{RC\,max\,40} \cdot (1 - a_x/l) \cdot (a_x/l)$$

with

$$f_{RC\,max\,40} = (G'/8H) \cdot l^2$$

where G' is the length-related conductor weight and H the conductor tensile force. Distances z_{RC-RC_A} and y_{RC-RC_A} at point a_x follow from (Figure 12.102)

$$z_{RC-RC_A} = 4 \cdot f_{RC\,max\,40} \cdot (1 - a_x/l)(a_x/l)\cos\varphi \qquad (12.15)$$

$$y_{RC-RC_A} = 4 \cdot f_{RC\,max\,40} \cdot \left(1 - \frac{a_x}{l}\right) \cdot \frac{a_x}{l} \cdot \sin\varphi \quad . \qquad (12.16)$$

The sag of the return current conductor f_{RC} at point a_x is calculated for different support point heights from (4.73)

$$f_{RC} = G' \cdot a_x \frac{l - a_x}{2H} + \Delta h_{RC} \cdot \frac{a_x}{l} = 4 f_{RC\,max\,40} \cdot \left(1 - \frac{a_x}{l} + \frac{\Delta h_{RC}}{4 \cdot f_{RC\,max\,40}}\right) \cdot \frac{a_x}{l} \quad ,$$

whereby the value Δh_{RC} is the height difference between the suspension points of the return current conductor at pole A and pole B. The quantities z_{RC-RC_A} and y_{RC-RC_A} follow from

$$z_{RC-RC_A} = 4 \cdot f_{RC\,max\,40} \cdot \left[1 - \frac{a_x}{l} + \frac{\Delta h_{RC}}{4 f_{RC\,max\,40}}\right] \cdot \frac{a_x}{l} \cdot \cos\varphi \qquad (12.17)$$

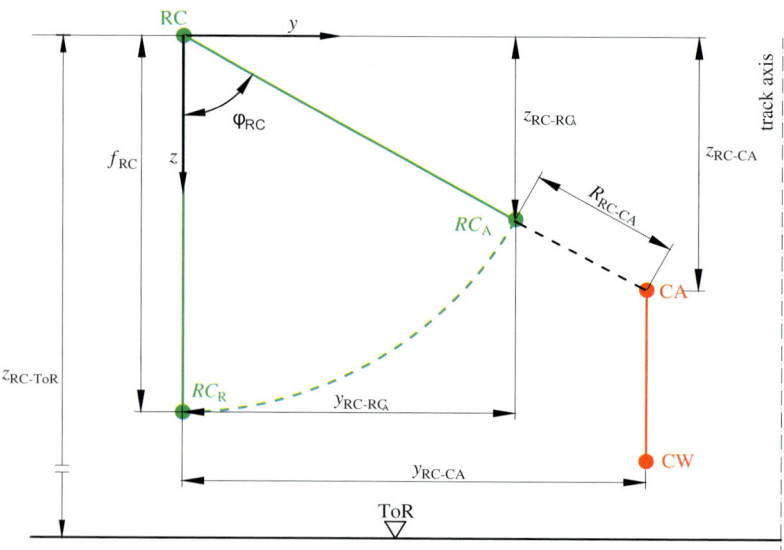

Figure 12.102: Section AA (see Figure 12.101) of the terminating contact line and return current conductor at point a_x.

RC_A return conductor position deflected under wind action
RC return conductor axis
RC_R position of return conductor in still air
CA catenary wire axis
CW contact wire axis
R_{RC-CA} spatial clearance between deflected return conductor and catenary wire
z_{RC-CA} vertical distance between return conductor support and catenary wire
z_{RC-RC_A} vertical distance between return conductor support and deflected return conductor line
y_{RC-CA} horizontal distance between return conductor and catenary wire
y_{RC-RC_A} horizontal distance between return conductor centre line and deflected return conductor line
z_{RC-ToR} vertical distance between return conductor support and top of rail)
f_{RC} sag of return conductor at $\vartheta = 40\,°C$ at point a
φ_{RC} deflection angle of return conductor in degrees

$$y_{RC-RC_A} = 4 \cdot f_{RC\,max\,40} \cdot [1 - a_x/l]\,(a_x/l)\sin\varphi \quad . \tag{12.18}$$

The spatial position of the catenary wire at the verification point a_x with reference to pole A is determined according to Figure 12.103 from

$$y_{RC-CA} = (D_A/2 + TP_A - b) \cdot a_x/l + T_{RC} \tag{12.19}$$

and

$$z_{RC-CA} = z_{RC_A-TR} - z_{CA_A-TR} - (z_{CA_B-TR} - z_{CA_A-TR}) \cdot a_x/l$$
$$+ 4f_{CA\,max\,40} \cdot [1 + \Delta h_{CA}/(4 \cdot f_{CA\,max\,40}) - a_x/l] \cdot a_x/l, \tag{12.20}$$

whereby $T_{RC} = D_A/2 + T$. It applies $T = 0$ and $T_{RC} = D_A/2$ to the attachment of the return current conductor directly to the pole. The spatial clearance R_{RC-CA} at point a_x is

$$R_{RC-CA} = \sqrt{(y_{RC-CA} - y_{RC-RA})^2 + (z_{RC-CA} - z_{RC-RA})^2} \quad . \tag{12.21}$$

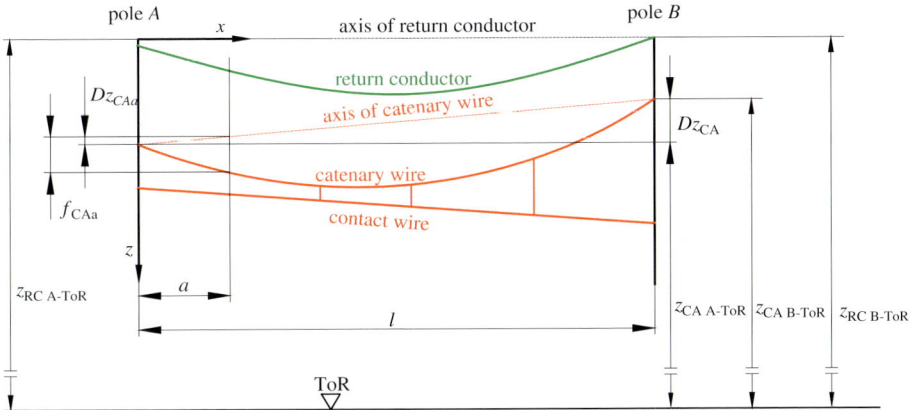

Figure 12.103: Side view of the terminating contact line system with return conductor line.

The distance R_{RC-CA} needs to be greater than the minimum clearance S_k:

$$R_{RC-CA} \geq S_k \quad . \tag{12.22}$$

Example 12.10: The clearance between the return conductor and the anchored contact line is to be found at point a_x (see Figure 12.103).

Initial data:

– return conductor	243-AL1
– tensile stress of return conductor σ_{RC}	20 N/mm²
– span length l between pole A and B	65 m
– position of verification point a_x	25 m from pole A
– suspension points of return current conductor attachment	equal height
– maximum sag $f_{RC\,max\,40}$	1,87 m
– length T of crossarm for the return current conductor	0,55 m
– maximum sag $f_{CA\,max\,40}$ of the catenary wire of contact line Re330 seen from the higher support point	0,739 m
– pole type	concrete pole NB 4

The geometrical layout corresponds to Figures 12.101 to 12.103 with $z_{CAA-ToR} = 6,70$ m, $z_{CAB-ToR} = 7,30$ m, $z_{RCA-ToR} = z_{RCB-ToR} = 7,10$ m, $D_A/2 = 0,15$ m, $TP_A = TP_B = 3,60$ m.

Calculation of the distances y_{RC-CA} and z_{RC-CA}

$$f_{RC} = 4 \cdot 1,87 \cdot \frac{25}{65} \left(1 - \frac{25}{65} \right) = 1,770\,\text{m}$$

$$T_{RC} = 0,15 + 0,55 = 0,70\,\text{m}$$

$$y_{RC-CA} = 0,70 + (0,15 + 0,70 + 3,60 - 0,40) \cdot 25/65 = 2,257\,\text{m}$$

$$\begin{aligned}
z_{RC-CA} &= 7,10 - 6,70 - (7,30 - 6,70)25/65 \\
&\quad + 4 \cdot 0,739 \cdot 25 \left[1 + 0,60/(4 \cdot 0,739) - (25/65) \right] / 65 \\
&= 1,099\,\text{m}
\end{aligned}$$

The minimum clearance R_{RC-CA} is reached if the return conductor and the catenary wire are on the connecting line between the return conductor support and the catenary wire (Figure 12.102).

$$
\begin{aligned}
\varphi_{RC} &= \arctan(2,257/1,099) = 64,1° \\
z_{RC-RC_A} &= 1,770 \cdot \cos\varphi_{RC} = 0,775\,\text{m} \\
y_{RC-RC_A} &= 1,770 \cdot \sin\varphi_{RC} = 1,591\,\text{m}
\end{aligned}
$$

Calculation of the minimum clearance between catenary wire and return conductor

$$
R_{RC-CA} = \sqrt{(2,257 - 1,591)^2 + (1,099 - 0,775)^2} - 1,770 = -1,03\,\text{m} \quad .
$$

With a cross-arm length of 0,55 m the return conductor touches the catenary conductor in a swung position. Therefore, the cross-arm needs to be extended to at least 1,5 m, whereby $R_{RC-CA} = 0,20\,\text{m}$ and $R_{RC-CA} > S_k = 0,10\,\text{m}$ result. No violation of the minimum clearances occurs if the construction tolerances are observed.

If the minimum clearance is violated, it is possible to increase the installation height of the return conductor or employ a longer cross arm. Normally the return conductor is installed at the height of the catenary wire. The return conductor is attached to a cross arm at poles with terminating contact lines and at midpoint anchor poles. The same procedure can be adopted to verify clearances between feeder and auxiliary feeder lines and other objects. If the local conditions are changed the distance R_{RC-CA} needs to be rechecked.

12.8 Project documentation

The *project documentation* includes all information necessary for approval, material procurement, implementation and maintenance. These are detailed in:
- list of contents and list of modifications
- *approvals for construction*
- *explanatory reports* with instructions for construction
- *overhead line layout plans*, adjustment diagrams, earthing plans, layout plans for cables and layout plans of disconnector remote control
- *transverse profile diagrams*, such as track transverse profile diagrams and polygons for cross-span structures
- longitudinal profiles, such as contact line height reductions and line height diagrams
- project related structures with drawings and calculations and if necessary the
- *materials list* including the pole and foundation table

The objective of the explanatory report is to establish the planning principles and assumptions made for the project, to inform the inspector and subsequent construction manager of the configuration constraints. This ensures that information is not lost during handover of the project for construction. The *explanatory report* contains the technical requirements, planning documents, technical explanations for equipment and approvals for project implementation. The *technical requirements* for the configuration of the overhead contact line can be found in Clause 12.2.2. If the configuration is based on a standard design, then the technical requirements are defined in advance and it is sufficient to state the type of overhead contact line. The *planning documents* corresponding to Clause 12.2.2, upon which the configuration is based, should also be cited. Reports, such as inspection reports and other meeting reports, containing information relevant to the system layout, should also be listed.

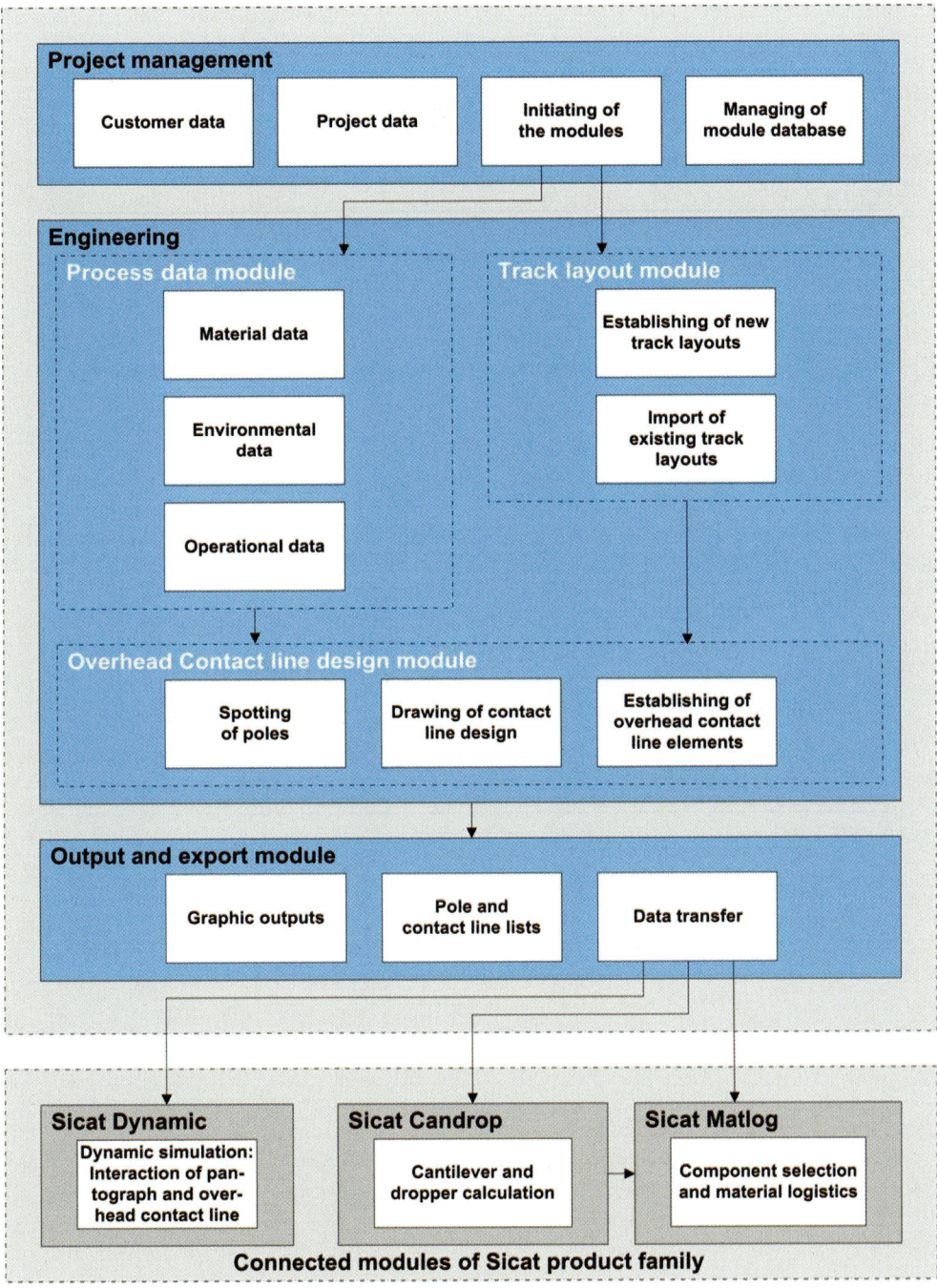

Figure 12.104: Modular structure of Sicat Master family.

The *technical explanations* relating to project equipment are subdivided into poles, cross-span elements, overhead contact line, traction power supply line, return current system, railway earthing and protective measures and profile clearance and special structures with calculation notes and drawings, as necessary. The description of the subsoil based on subsoil investigations and the type of foundations should be included in the technical explanations to the construction project part and any special foundations also listed.

The *overhead line layout plans* follow the explanatory report. Earthing diagrams, adjustment diagrams and transverse and longitudinal profiles are to be attached to the project as needed.

The *material list*, which includes the pole and foundation table, forms the basis for the construction of the overhead contact line and subsequently for the spare parts inventory. It is expedient to use data bases to administer all project data, that will assist the operator to perform maintenance and achieve quicker fault repairs.

12.9   Computer supported planning

12.9.1   Objectives

Currently, *computer-supported planning* of contact line installations is widely applied [12.59, 12.60, 12.61]. The programs used support individual engineering steps or integrated planning processes based on interactive modules such as the Sicat Master [12.59, 12.60] program system developed by Siemens, which enables the engineer to carry out the design steps interactively with the computer. The main objectives are:

- *processing the wiring* and related calculations
- *preparation of diagrams*, drawings and lists for procurement and construction
- *selection of materials*
- redundancy-free *implementation* and administration *of changes*

Design of contact lines both in stations and on interstation sections is possible.

12.9.2   Structure and modules

The Sicat Master program is structured into several modules (Figure 12.104) that systematically link to each other and are used interactively to carry out the design steps. Important modules are:

- project management module
- planning modules
    - process data module (Figure 12.105)
    - track layout module (Figure 12.106)
    - overhead contact line design module (Figure 12.107)
- output and export modules
    - graphical output
    - pole and catenary lists
    - data transfer

The tasks of the modules are described in separate Clauses. All data, such as general data relating to materials and type of contact line, customer data or project data, are stored without redundancy in central databases and can be updated with the aid of master data editors.

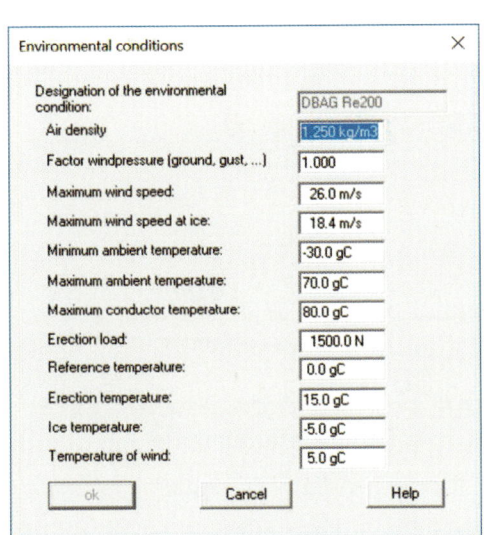

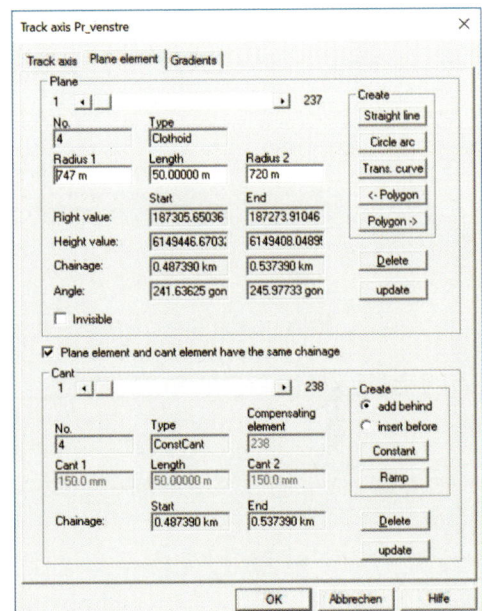

Figure 12.105: Process data module editor for environmental data.

Figure 12.106: Track layout module editor for track elements.

12.9.3 Project management module

The *project management module* is the core of the program in which the customer data, the general project data are organised and the project engineering functions are initiated. The project engineering itself is carried out in the modules for the process data, track layout and contact line design. The databases are also managed with the project management module.

12.9.4 Process data module

In the *process data module* values are specified for the project engineering and are stored on the project level. The process data are separated into material, environmental and operating data. All the parameters required for design, such as characteristics, materials, components, ambient conditions etc. are assigned to the overhead contact line type (Figure 12.105). During project engineering, these values are then used to design the overhead contact line installation. The characteristics and parameters of overhead contact lines that conform to European interoperability requirements such as Sicat H and Sicat S, as well as the standard types used by German Railway, such as Re100, Re200, Re250 and Re330, are stored together with their respective characteristics in the process data module as standard contact line designs. Any other types of contact lines required for a specific application can be input here as well.

12.9.5 Track layout module

The track layout data – a three-dimensional description of the track design – is an important starting point for the engineering of contact lines. The track layout description plays a crucial

a) Undistorted scale

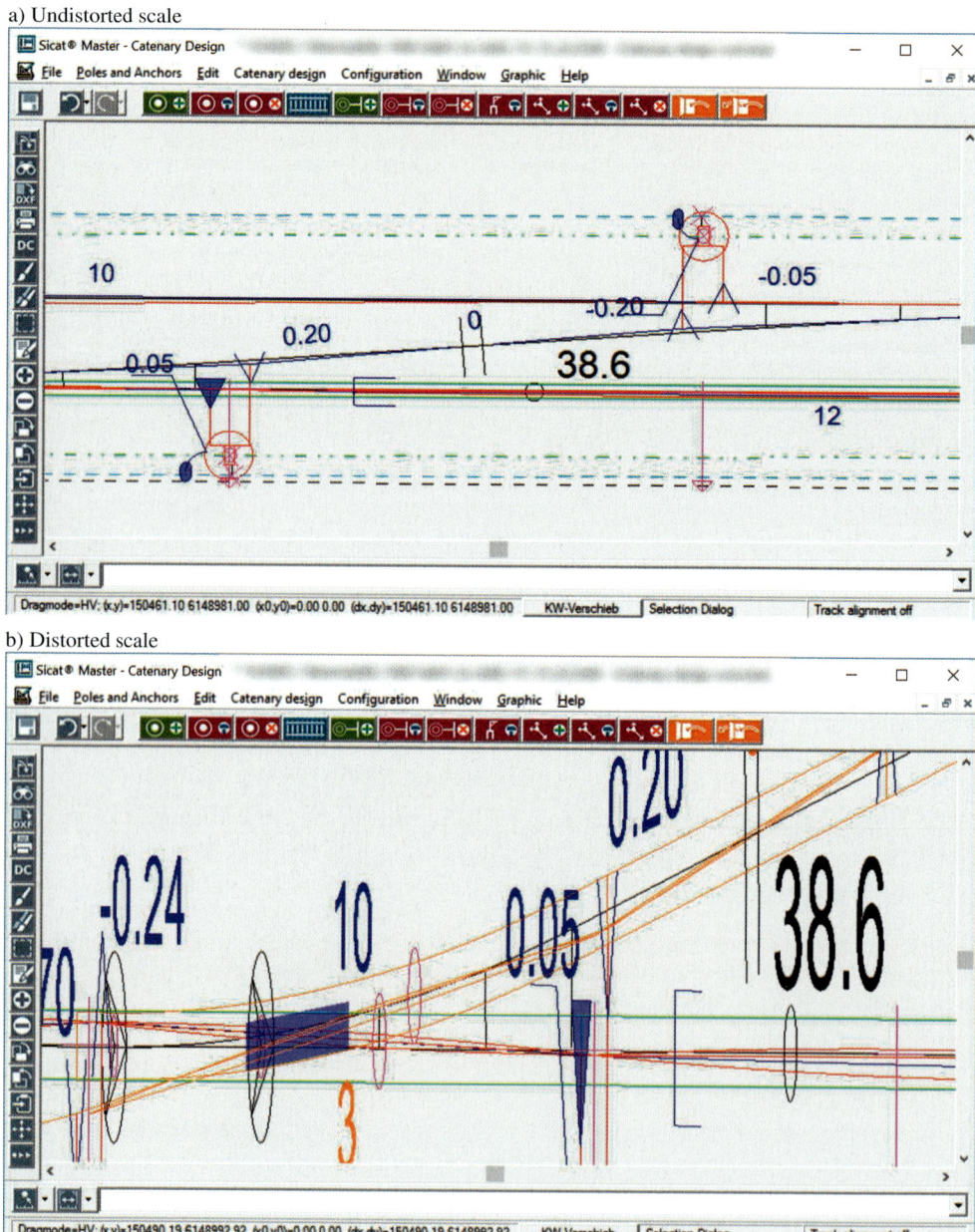

b) Distorted scale

Figure 12.107: Entry of contact line elements into the contact line design module.

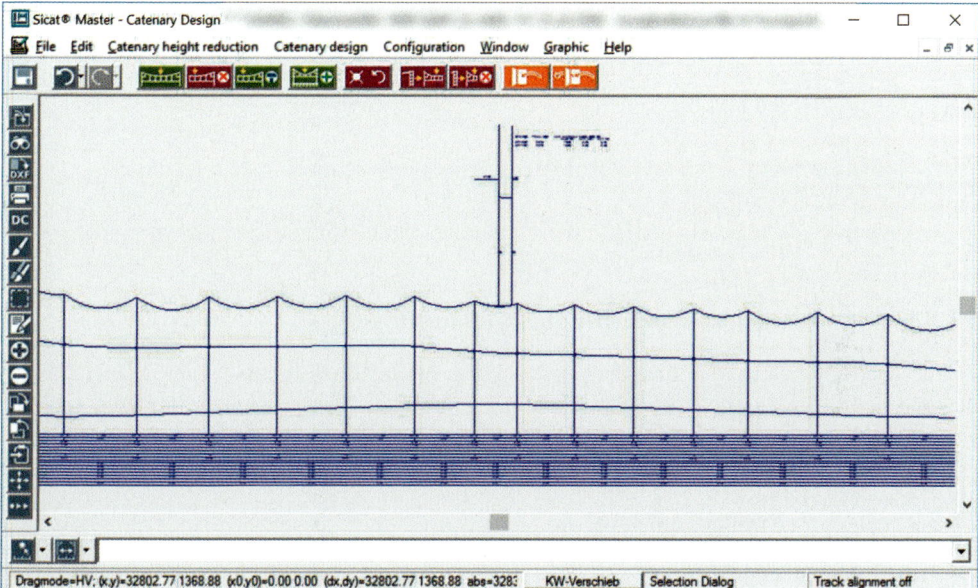

Figure 12.108: Interactive window for contact line system height reduction under structures.

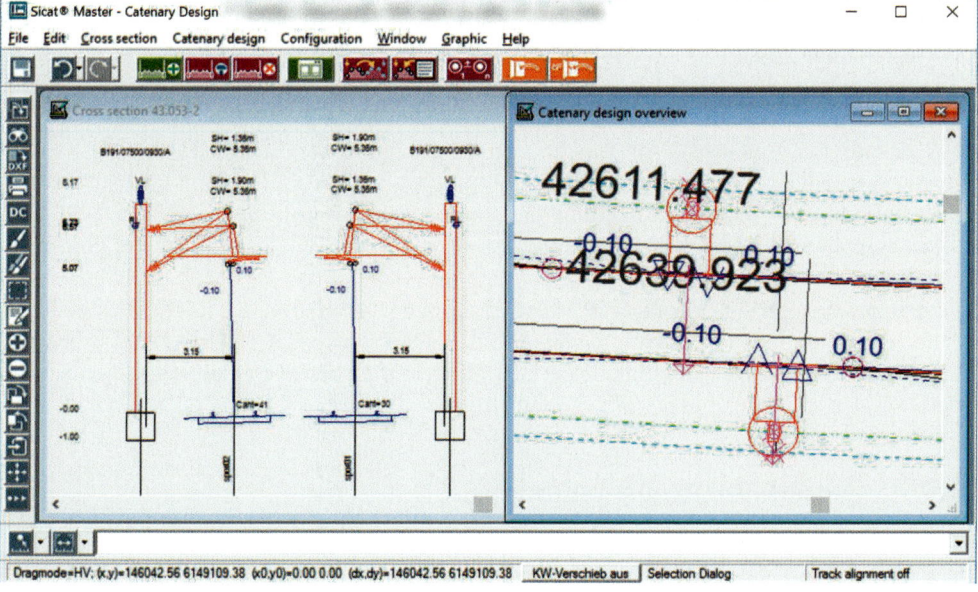

Figure 12.109: Contact line cross-section.

role in the quality of project engineering and of the installation to be designed. With this module, a new track layout can be established and existing track layouts can be imported (Figure 12.106). The track layout data can be recorded by:

(1) entry based on Gauss-Krueger coordinates using
   – structured text file
   – graphic dialog box
   – data import from the DB AG geometrical information system (DBGIS)
(2) copying from dxf-files

A comprehensive library of standard track point designs facilitates the input of point data.

12.9.6 Overhead contact line design module

Using the track layout data and the process data, the overhead contact line is interactively designed with the overhead contact line design module. The engineer is provided with all required functions and various display formats that support the engineer in the design process:

(1) functions for establishing, processing and deleting overhead contact line components
   – contact lines
   – support points and poles
   – cantilevers and multiple-track cantilevers
   – tunnel supports
   – head spans and portals
   – disconnectors, insulators, connectors and section insulators
   – traction power lines
(2) display of line cross-sections (Figure 12.109)
(3) calculation and display of contact line height reductions (Figure 12.108)
(4) designing and processing of traction power lines (reinforcing feeders, feeder lines, return conductors)
(5) insertion of section insulators and tensioning devices based on a comprehensive library
(6) functions for automatic generation of dimensions and
(7) description of planning

Advanced calculation tools, e. g. for longitudinal span length optimisation and for the definition of standard contact line sections that can be used several times and ensure efficient project engineering. These tools also permit analysis of the mechanical loading of pole components.

The precise engineering of contact lines under structures is important. The routines included in the contact line design module enable the calculation of exact clearances to structures, including where the track is inclined, has curvatures in the horizontal and vertical planes and is canted.

12.9.7 Output and export module

The *output and export module* contains various options for output of the data:
   – layout plans, cross- sections and
   – longitudinal profiles

The output module allows the export of planning data such as dxf or dwg files and the data can be provided for other CAD programs. The design data can be provided for

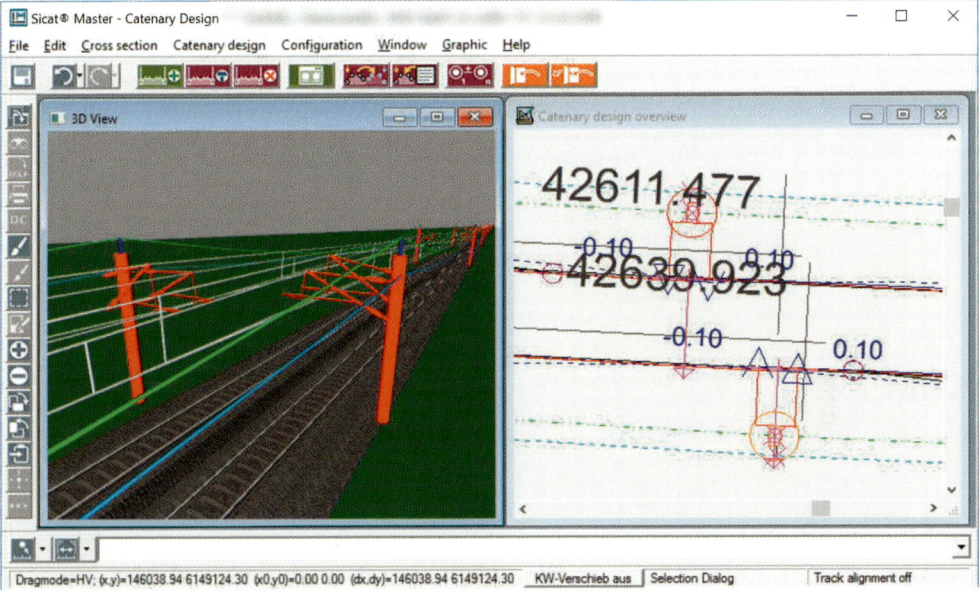

Figure 12.110: 3D presentation of overhead contact line into the overhead contact line design module.

- pole lists
- contact line lists
- material lists

All of which support the material procurement and project logistics.

The three-dimensional presentation of the overhead contact line system (Bild 12.110) offers, at the computer screen as well as in DWG format, an additional overview of complicated design situations that also aid the identification of conflicts that are difficult to recognise.

Interfaces to other IT-tools complete the integration of the project engineering process carried out by the Sicat program family.

Using the DXF/DWG format, background pictures with information on the terrain can be inserted to arrange pole sites in the terrain. The same applies to the ECW raster format with reference to the GEO system.

In both cases, the graphic module integrated in Sicat Master aligns the graphic data files with the world coordinate system. A simulation of the interaction between overhead contact line and pantograph (Figure 12.111) is enabled because the program Sicat Dynamic is incorporated in the Sicat Master software [12.62]. By evaluating the contact forces and the contact wire uplift, a quantitative assessment of the overhead contact line can be accomplished as well as a specific analysis of critical sections of the catenary.

Sicat Candrop is a program for calculating cantilevers and droppers using the track geometry and pole sites as provided by the Sicat Master program (Figure 12.112). This module is designed to establish the cutting lengths of cantilever tubes, attachment points for fittings and the lengths of droppers. Consequently, construction is partially shifted into the workshop where precision is increased and the track occupations can be used more effectively.

The program Sicat Matlog supports the selection of components and material logistics. Its interface to the Sicat Master software (Figure 12.113) refers to the most recent design revision

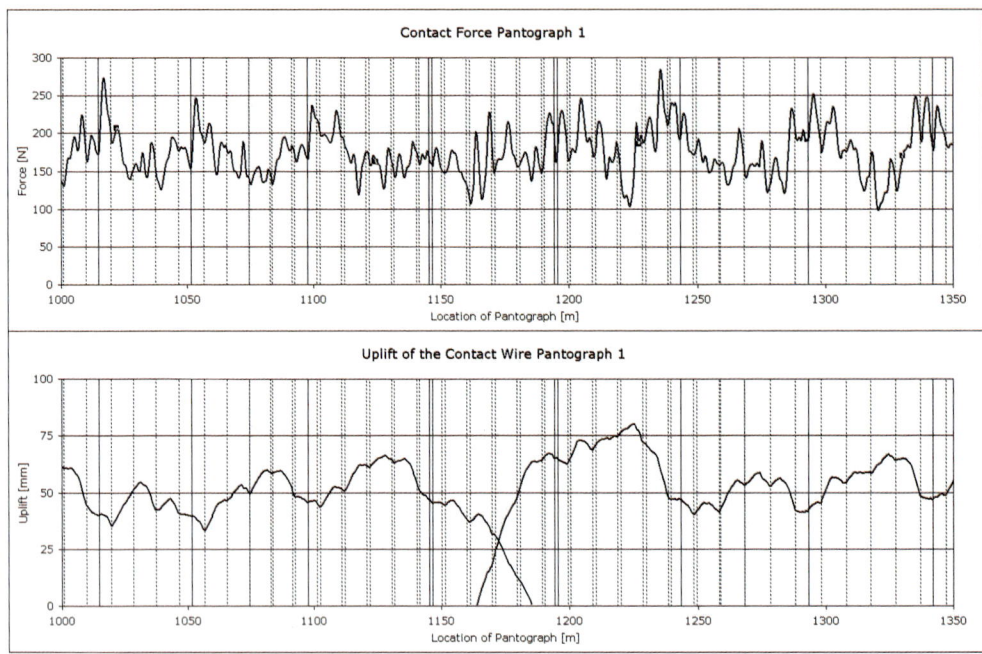

Figure 12.111: Contact wire uplift simulation for Sicat H1.0 by Sicat Dynamic.

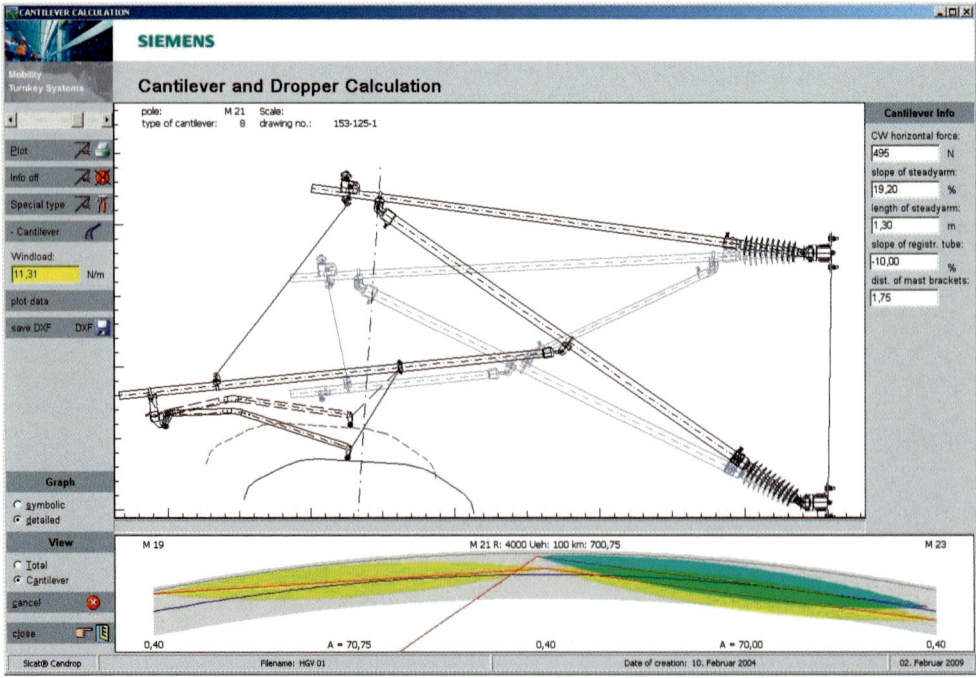

Figure 12.112: Cantilever and dropper calculation by Sicat Candrop.

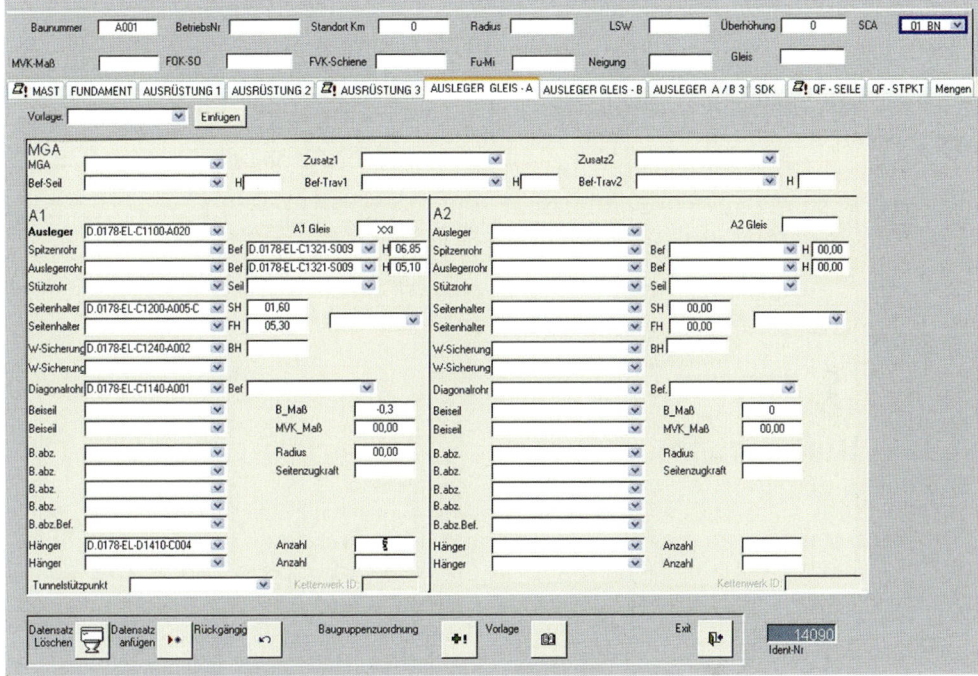

Figure 12.113: Menu for material logistics by Sicat Matlog.

making continuous supervision of the currently planned materials regarding inquiries, offers, orders and supplies possible. A simple comparison is also possible between the originally assessed and ordered materials with the detailed demand according to Sicat Master planning. The consequence is on demand knowledge of materials for individual line sections and all additional materials required. Sicat Matlog improves documentation, for example during the As-Built documentation process and offers advantages for the maintenance of the contact line installation.

12.9.8 Hardware and software

The Sicat Master program runs on a PC platform so a standard high-performance computer for graphical applications is sufficient. CAD programs such as AutoCAD are recommended for graphic processing and for plotter outputs. The object-oriented programming languages C++, C# and .net were chosen for the development and application of the system because of certain benefits provided.

12.9.9 Application

The structure of the program system corresponds to the steps for project processing as shown in Figure 12.104. The coordination of individual modules compels the project engineer to use standard processes, which ensure uniform quality of the design. The flexibility and ease of use of the system results in considerable time and cost benefits, especially during alterations to the project specifications such as modifications of track layout and signal locations. All planning documents and a complete database for an installation are available at the completion of design work and can be used for line management. The Sicat Master contact line planning program is not only used internally by Siemens but also by several other railway entities. Examples for previous important applications of Sicat Master are:

- high-speed line Cologne–Frankfurt Sicat H1.0
- high-speed line Amsterdam–Rotterdam–Antwerpen Sicat H1.0
- high-speed line Leipzig/Halle–Erfurt Re330
- high-speed line Erfurt–Ebensfeld Re330
- upgrade line Ebensfeld–Nuremberg Re200 und Re250
- Siemens depot Dortmund-Eving Re100
- contact lines Denmark Sicat SX

12.10   Bibliography

12.1   *Decision 2011/291/EU*: Technical specification for interoperability relating to the rolling stock subsystem – 'Locomotives and passenger rolling stock' of the trans-European conventional rail system (TSI LOC&PAS CR). In: Journal of the European Union No. L139 (2011), pp. 1 to 151.

12.2   *Decision 2008/232/EC*: Technical specification for interoperability relating to the rolling stock subsystem of the trans-European high-speed rail system. In: Official Journal of the European Communities, No. L84 (2008) pp. 1 to 105.

12.3   *Federal Republic of Germany*: Eisenbahn-Bau- und Betriebsordnung EBO (Railway installation and operation directive). Bundesrepublik Deutschland, May 12, 1967. In: Deutsches Bundesgestzblatt BGBl. II p. 1563, last amendment March 19, 2008 BGBl. I p. 467.

12.4   *Grimrath, H.; Reuen, H.*: Elektrifizierung der Strecke Elmshorn–Itzehoe mit der Oberleitung Sicat S 1.0. (Electrification of Elmshorn–Itzehoe line section with catenary type Sicat S1.0). In: Elektrische Bahnen 96(1998)10, pp. 320 to 325.

12.5   *Regulation 1301/2014/EU*: Technical specification for interoperability of the energy subsystem of the rail system in the Union. In: Official Journal of the European Union Nr. L 356 (2014), pp. 179 to 227.

12.6   *Deutsche Bahn*: Register of Infrastructures. Internet: http://fahrweg.dbnetze.com/fahrweg-de, 2013.

12.7   *TEIV*: Transeuropaen railway interoperability regulation. In: Official Journal of the European Union Nr. L 356 (2014), pp. 179 to 227.

12.8   *VDV document 550*: Oberleitungsanlagen für Straßen- und Stadtbahnen (Overhead contact line systems for tramways and City railways). Association of German Traffic Entities, Cologne, 2003.

12.9   *Federal Republic of Germany BOStrab*: Verordnung über den Bau und Betrieb der Straßenbahnen (Straßenbahn-Bau- und Betriebsordnung - BOStrab) (Regulation on the installation and operation of tramways). Bundesrepublik Deutschland, 1987 BGBl. I p. 2648, last amendment 2007 BGBl. I p. 2569.

12.10   *Grunder, H.; Kocher, M.; Waeckerlig, W.*: Rechnergestützte Fahrleitungsprojektierung beim Umbau des Bahnhofs Spiez (Computer-based planning of overhead contact lines for conversion of the Spiez railway station). In: Elektrische Bahnen 91(1993)4, pp. 125 to 130.

12.11   *Geissler, G.*: Einführung in die Vermessung mit GPS-Systemen. (Introduction into surveying using GPS). Information brochure Engineering company for geodetic systems, Munich, 1994.

12.12   *Decision 2011/274/EU*: Technical specification for interoperability of the energy subsystem of the trans-European conventional rail system. In: Official Journal of the European Union No. L126 (2011), pp. 1 to 52.

12.13   *Decision 2008/284/EU*: Technical specification for interoperability of the energy subsystem of the trans-European high-speed rail system. In: Official Journal of the European Union No. L104 (2008), pp. 1 to 79.

12.14   *Berthold, G.*: Mindestfahrdrahthöhe bei der Deutschen Bahn (Minimum contact wire height at German Railway). In: Elektrische Bahnen 100(2002)10, pp. 404 to 408.

12.15   *Deutsche Bahn AG Dirctive 997.0101*: Oberleitungsanlagen, Allgemeine Grundsätze, Freizuhaltende Räume bei Oberleitungen (Overhead contact lines, general basics, spaces to be kept free for overhead contact lines). Deutsche Bahn AG, Frankfurt, 2001.

12.16   *Cabirol, M.*: Pantograph/OHL interaction. SNCF engineering studies and concepts developed. In: Rail International, 31(2000)9, pp. 23 to 30.

12.17   *Dupont, C.*: Instandhaltung der Fahrleitungen bei der SNCF (Maintenance of overhead contact lines at the SNCF). In: Elektrische Bahnen, 90(1992)2, pp. 68 to 73.

12.18   *Payan Cuevas, F.; Puschmann, R.; Vega, T.*: Instandhaltung der Oberleitungsanlage für die Hochgeschwindigkeitsstrecke Madrid–Lérida (Maintenance of overhead contact line of the high-speed line Madrid to Lérida). In: Elektrische Bahnen, 106(2008)5, pp. 211 to 221.

12.19   *Cattaneo, C.; Klinge, R.; Fasciolo, S.*: Hochgeschwindigkeitsverkehr in Italien am Beispiel Rom–Neapel (High-speed operation in Italy at the example Rome–Naples). In: Elektrische Bahnen, 106(2005)4-5, pp. 253 to 261.

12.20   *Puschmann, R.*: Maximale Fahrdrahtseitenlage und Spannweiten für interoperable Strecken (Maximum lateral deflection of contact wire and span lengths of interoperable lines). In: Elektrische Bahnen 110(2012)7, pp. 336 to 348.

12.21   *Puschmann, R.*: Contact wire lateral position and span lengths of interoperable lines. In: Elektrische Bahnen 110(2012)11, pp. 612 to 632.

12.22   *Hoerning, D. O.*: Elektrische Bahnen (Electric railways). Walter de Gruyter & Co. Publishers, Berlin-Leipzig, 1926.

12.23   *Süberkrüb, M.*: Technik der Bahnstromleitungen (Technology of overhead contact lines). Wilhelm Ernst & Sohn Publishers, Berlin-München-Düsseldorf, 1971.

12.24   *German Rail Document Ebs 02.05.70*: Ermittlung der maximal zulässigen Nachspannlänge (Determination of permissible tensioning length). Deutsche Bahn AG, Frankfurt, 2001.

12.25   *German Rail Document Ebs 07.04.01*: Anordnung des Kettenwerkes (Arrangement of the catenary). Deutsche Bahn AG, Frankfurt, 1986.

12.26 *Schwab, H.-J.; Ungvari, S.*: Entwicklung und Ausführung neuer Oberleitungssysteme (Development and design of new overhead contact line systems). In: Elektrische Bahnen 104(2006)5, pp. 238 to 248.

12.27 *German Rail Directive 997.0301: 2005*: Oberleitungsanlagen; Speisung und Schaltung der Oberleitungsanlage planen (Overhead contact lines; Planning of supply and circuits of overhead contact lines). Deutsche Bahn AG, Frankfurt, 2005.

12.28 *German Rail Directive 800.01: 1993*: Bahnanlagen entwerfen - Allgemeine Entwurfsgrundlagen (Planning of railway systms - General design basics). Deutsche Bahn AG, Frankfurt, 1993.

12.29 *Drawing SBB 0162.2010.0003*: Anordnung der Parallelführung mit Auslegern – Grundlagen (Arrangement of overlappings with cantilevers – Basic principles). SBB, Bern, 1980.

12.30 *Berg, G.; Henker, H.*: Weichen (Points). VEB, Publishers for traffic systems, Berlin, 1976.

12.31 *Lichtenberger, B., et al.*: Handbuch Gleis (Track manual). Tetzlaff Publishers, Hamburg, 2004.

12.32 *UIC 711*: Geometry of points and crossings with UIC rails permitting speeds of 100 km/h or more on the diverging track. In: UIC-codex, Paris, 1981.

12.33 *BWG*: Datenblätter (data sheets on points). BWG Gesellschaft mbH & Co. KG, Butzbach, 2011.

12.34 *German Railway*: Vorträge bei den Unterrichtskursen mit Erfahrungsaustausch über Konstruktion, Bau und Betrieb von Fahrleitungen (Lessons at the information and experience exchange on design, installation and operation of overhead contact lines). German Railway Central Office, Munich, 1942.

12.35 *German Rail Directive 997.0102*: Oberleitungsanlagen, Oberleitungsanlagen planen und errichten (Overhead contact lines, Planning and installations of overhead contact lines). Deutsche Bahn AG, Frankfurt, 2001.

12.36 *Kießling, F.*: Studie zur Entwicklung einer Oberleitung für hohe Geschwindigkeiten (Study on development of an overhead contact line for high speeds). Siemens AG, Erlangen, 1992.

12.37 *German Rail Document Ebs 02.05.45 Bl.1*: Seitenhalter-Stellung (Abnahme), Gs-OL, Re100, Re160-S-Bahn, Re160 und Re200 (Steady arm position at approval for contact lines Re100, Re160-S-Bahn, Re160 and Re200). Deutsche Bahn AG, München, 1992.

12.38 *German Rail Document Ebs 02.05.45 Bl.2*: Seitenhalter-Stellung (Abnahme), Re250 und Re330 (Steady arm position at approval for contact lines Re250 and Re330). Deutsche Bahn AG, Frankfurt am Main, 2007.

12.39 *German Rail Document Ebs 07.41.10 Bl.1*: Oberleitungskreuzung (Overhead contact line crossing). Deutsche Bahn AG, Frankfurt, 2000.

12.40 *Kruch*: Hänger, Produktinformation (Droppers, information on product). Ing. Karl und Albert Kruch Gesellschaft m.b.H. & Co KG, Wien, 2008.

12.41 *German Rail Document Ebs 05.08.52,*: Schwenkausleger, angelenkt. (Pull-off swivelling cantilever), Deutsche Bahn AG, Frankfurt, 1995.

12.42 *German Rail Directive 819.0202*: Bautechnik, Leit-, Signal- und Telekommunikationstechnik (Civil engineering, controlling signalling and telecommunication technology). Deutsche Bahn AG, Frankfurt, 2008.

12.43 *Federal Republic of Germany*: Straßenverkehrszulassungsordnung [Directive on registration of road traffic). In: BGBl. I p. 1793, September 1988, new edition BGBl. I, p. 679. April 2012, latest amendment BGBl. I, p. 2083, July 2013.

12.44 *Federal Republic of Germany*: Straßenverkehrsordnung (Directive on road traffic) In: BGBl. I p. 1565, November 1970, new edition BGBl. I p. 367, March 2013.

12.45 *German Rail Document Ebs 19.01.01*: Höhenbegrenzung mit Profiltor vor Bahnübergängen (Height restriction at level railway crossings with a profile gate). Deutsche Bahn AG, Frankfurt am Main, 2002.

12.46 *German Rail Document Ebs 02.05.11*: Sehkeile vor Signalen (Visibility in front of signals). Deutsche Bahn AG, Frankfurt, 1962.

12.47 *German Rail Directive 997.9115*: Maschinen-, Energie- und Elektrotechnik, Werkstattwesen – Oberleitungsanlagen – RZ-Maßnahmen– Maßnahmenkatalog für Oberleitungen (Mechanical, energy and electric engineering technology, workshops – Overhead contact lines – RZ-measures – Catalog of measures at overhead contact lines). Deutsche Bahn AG, Frankfurt, 2004.

12.48 *German Rail Ltd.*: Entscheidung zur Führung von Bahnenergieleitungen über Straßenbrücken (Decision on the guidance of traction powerline below road bridges). Deutsche Bahn AG, Frankfurt, 2002.

12.49 *German Rail Document Ebs 02.05.29, Sheet 1*: Toleranzen für Oberleitungen (Tolernances of overhead contact lines). Deutsche Bahn AG, München, 1982.

12.50 *German Rail Directive 997.9113*: Oberleitungsanlagen, Regeln und Kriterien für Planung, Lieferung, Errichtung und Instandhaltung (Overhead contact lines, specifications and criteria for planning, supply, installation and maintenance). Deutsche Bahn AG, Frankfurt, 2001.

12.51 *German Rail*: Simulation of dynamic interaction Pantograph SSS400 and overhead lines S25 and S20 (Norway) – Movement of contact wire and messenger wire. Report, Munich, 2017.

12.52 *German Rail Document Ebs 02.05.17, Sheet 2.2*: Oberleitungsanordnung bei Überbauten, Ermittlung von Einzelwerten (Arrangement of overhead contact lines in case overbrigding structures, determination of singular data). Deutsche Bahn AG, Frankfurt, 2007.

12.53 *German Rail Document Ebs 19.11.12*: Vogelschutz unter Bauwerken (Bird protection under structures). Deutsche Bahn AG, Frankfurt, 1997.

12.54 *German Rail Document Ebs 19.01.24*: Vogelschutz unter Bauwerken (Bird protection under structures). Deutsche Bahn AG, Frankfurt, 1997.

12.55 *Federal Republic of Germany EKRG*: Eisenbahnkreuzungsgesetz (Law for railway crossings). Bundesrepublik Deutschland, 1963, BGBl. I, p. 631, last amendment 1971, BGBl. I, p. 337.

12.56 *Johnsen, F.; Nyebak, M.*: Schaltungsaufbau im Oberleitungsnetz der norwegischen Eisenbahn Jernbaneverket (Circuitry of overhead contact line network at norwegian railway Jernbaneverket). In: Elektrische Bahnen 102(2004)4, pp. 195 to 200.

12.57 *Martinsen, F.; Nordgård, M.; Schütte, T.:* A new type of autotransformer system for the railway in Norway. In: Elektrische Bahnen 114(2010)7, pp. 334 to 343.

12.58 *Peter Eichenberger, P.; Hayoz; P.; Schweller, M.:* Fahrleitungs-Parallelführungen im Gotthard-Basistunnel – Schutz vor Fahrdrahtbrüchen (Overlaps in the overhead contact line of the Gotthard base tunnel – protection against contact wire failures). In: Elektrische Bahnen 102(2016)4, pp. 370 to 375.

12.59 *Burkert, W.; Puschmann, R.:* System zur interaktiven Projektierung von Oberleitungsanlagen (System for interactive planning of overhead contact lines). In: Elektrische Bahnen 93(1995)3, pp. 104 to 109.

12.60 *Burkert, W.:* Oberleitungsplanung mit der erweiterten Software Sicat MASTER (Overhead contact line planning using the extended software Sicat MASTER). In: Elektrische Bahnen 108(2010)8-9, pp. 377 to 384.

12.61 *Hofbauer, G.; Hofbauer, W.:* Oberleitungsplanung und Simulation des Stromabnehmerlaufs (Planning of overhead contact lines and simulation of pantograph running). In: Elektrische Bahnen 107(2009)1-2, pp. 104 to 109.

12.62 *Reichmann, T.:* Simulation des Systems Oberleitungskettenwerk und Stromabnehmer mit der Finite-Elemente-Methode (Simulation of the system overhead contact line and pantograph by the Finite-Element-Method). In: Elektrische Bahnen 103(2005)1-2, pp. 69 to 75.

13 Cross-span structures, poles and foundations

13.0 Symbols and abbreviations

Symbol	Definition	Unit
A	cross-sectional area	mm^2, cm^2
A	accidental action	n/kN
A_K	characteristic accidental action	N, kN
A_S	shear area of a bolt	mm^2, cm^2
A_eff	effective cross-sectional area	mm^2, cm^2
A_ins	insulator area exposed to wind	m^2
A_net	net cross-section of member	mm^2
A_o	area at the surface of a foundation	m^2
A_r	area exposed to friction	m^2
A_str	pole area exposed to wind	m^2
A_t1, A_t2	lattice tower area of wall 1, 2 exposed to wind	m^2
A_u	area of a foundation base	m^2
C_C	aerodynamic drag factor for conductors	–
C_ins	aerodynamic drag factor for insulators	–
C_str	aerodynamic drag factor for poles	–
C_t	subgrade modulus	MN/m^3
C_t1, C_t2	aerodynamic drag factor for lattice tower face 1, 2	–
D_I	diameter of ice covered conductor	m
$D_\text{x,yd}$	design load of the bracing in face x, y	kN
E	modulus of elasticity	kN/mm^2
E_d	design value of an action	N
E_p	earth pressure	N, kN
E_t	inserting depth of a pole	m
F	acting force	N, kN
F_A	longitudinal force of a cantilever tube	N
$F_\text{A(B)x}$	horizontal force at a head-span pole A(B)	N, kN
F_AHd	design value of the horizontal force of a anchor foundation	N, kN
F_AVd	design value of the vertical force of a anchor foundation	N, kN
F_D	longitudinal force of diagonal strut or registration arm	N
F_Dr	dropper force	N
$F_\text{Hä}$	longitudinal force of the dropper	N
F_K	characteristic load	N
F_Kr	characteristic strength of an anchor foundation	kN
F_Ro	resistance of the block foundation above the pivot	kN
F_Ru	resistance of the block foundation below the pivot	kN
F_W'	wind force on conductor related to length	N/m

Symbol	Definition	Unit
F_d	design load	N
$F_{lcs\,max}$	maximum force of the lower cross-span wire	kN
$F_{lcs\,min}$	minimum force of the lower cross-span wire	kN
F_{pre}	pre-stressing force	N, kN
F_{st}	force of the registration arm	N, kN
F_{top}	longitudinal force of the top anchor	N
F_{ucs}	force of the upper cross-span wire	kN
$F_{ucs\,max}$	maximum force in the upper cross-span wire	kN
$F_{ucs\,min}$	minimum force in the upper cross-span wire	kN
G	shear modulus	N/mm^2
G	permanent actions	N, kN
G'	length-related weight force of a conductor	N/m
G_C	reaction coefficient of conductors	–
G_K	characteristic permanent action	N, kN
G_{OCLi}	weight of a catenary at support i	N, kN
G'_{ice}	length-related ice load of a conductor	N/m
G_{ins}	reaction coefficient of insulators	–
G_{str}	reaction coefficient of poles	–
H	horizontal tensile force of a conductor	N, kN
H_{Qy}	horizontal tensile force of head-span wire	N, kN
H_R	horizontal tensile force of return conductor	N, kN
H_{ax}	horizontal component of head-span wire	N, kN
H_{ik}	horizontal tensile force of a span wire i,k	N, kN
I	moment of inertia	mm^4, cm^4
I_T	torsional modulus	mm^4, cm^4
K_1	shape factor	–
M	acting moment	Nm, kNm
M_{Ad}	design value of an acting moment	Nm, kNm
M_{B2}	bending moment of the top tube	Nm, kNm
M_{B4}	bending moment of the cantilever tube	Nm, kNm
M_{B6}	bending moment of the registration arm	Nm, kNm
M_{Kr}	characteristic moment of friction	kNm
M_R	moment of resistance	kNm
M_{max}	maximum equivalent moment of the head-span	Nm, kNm
$M_{max\,z}$	maximum bending moment of pole	Nm, kNm
M_x	bending moment at the position x	kNm
$M_{x(y,z),d}$	design moment at the pole basis in x(y,z) axis	Nm, kNm
$M_{x(y),ToF}$	moment at upper foundation face	Nm, kNm
$M_{x(y)pl,Rd}$	plastic moment of inertia in x(y) direction	Nm, kNm
NN_i	height of a support i above sea level	m
N_{10}, N_{30}	number of blows at probing	–
N_{bd}	design value of bearing	N, kN
$N_{bpl,Rd}$	plastic bearing strength	N, kN
N_{dd}	design value at compression	N, kN
$N_{dpl,Rd}$	plastic compression strength	N, kN

Symbol	Definition	Unit
N_{sd}	design value at shearing	N, kN
$N_{spl,Rd}$	plastic shearing strength	N, kN
$N_{tpl,Rd}$	plastic tensile strength	N, kN
N_{td}	design value at tension	N, kN
Q	variable action	N, kN
Q_{CA}	horizontal load on a catenary wire	N, kN
Q_{CAHk}	permanent horizontal loads on a catenary wire k	N, kN
Q_{CAWk}	wind load at catenary wire k	N, kN
Q_{CK}	characteristic conductor tensile force	N, kN
Q_{CW}	horizontal loads on a contact wire	N, kN
Q_{CWHk}	permanent horizontal load on a contact wire k	N, kN
Q_{CWWk}	wind load on a contact wire k	N, kN
Q_{EH}	horizontal force of a traction power line	N, kN
Q_{EW}	wind load on a traction power line	N, kN
$Q_{Hi(k)x(y)}$	horizontal action i, (k) in x(y) direction	N, kN
Q_{IK}	characteristic ice action	N, kN
Q_K	characteristic value of a variable action	N, kN
Q_P	construction and maintenance action	N, kN
Q_{PK}	characteristic action due to construction and maintenance	N, kN
Q_{RW}	wind load on a return conductor	N, kN
Q_{WC}	wind load on conductor	N, kN
Q_{WK}	characteristic wind action	N, kN
Q_{Wins}	wind load on insulator	N, kN
$Q_{Wi(k)x(y)}$	wind load on conductor or pole parallel or rectangular to track	N, kN
Q_{Wt}	wind load on lattice tower	N, kN
Q_{strW}	wind load on a pole	N, kN
Q_{td}	total equivalent force (force at top of pole)	N, kN
Q_{xd}	transverse force at pole base, horizontal force in y direction	N, kN
$Q_{x(y)}$	transverse force	N, kN
Q_{yd}	transverse force at pole base, horizontal force in z direction	N, kN
Q_z	vertical load of foundation	N, kN
Q_{zC}	weight force of concrete	N, kN
Q_{zE}	weight force of soil	N, kN
R	track radius	m
R_K	characteristic value of resistance	kN, kNm
R_d	total design value of resistance	kN, kNm
$R_{o,u}$	components of reaction forces of block foundation	kN
S, S_0, S_y, S_z	static cross-section modulus	mm^3, cm^3
S_d	design load for leg member	kN
S_i	projection	m
$S_{x(y),d}$	design load for leg member due to moment related to the x,(y)direction	kN
T	absolute temperature	K
$T_{x(y,z)d}$	design load in x(y,z)direction	N, kN
T_r	skin friction value	N/m^2
ToF	Top of Foundation	–

Symbol	Definition	Unit
ToR	Top of Rail	–
$V_{A(B)z}$	vertical load A (B) of a head-span	N, kN
V_{CA}	vertical load at the catenary wire support	N, kN
V_{CAN}	vertical load of a cantilever	N, kN
V_{CW}	vertical load at the contact wire support	N, kN
V_E	vertical load of a power line	N, kN
V_G	vertical load of pole head equipment	N, kN
V_{OCL}	vertical load at a contact line support	N, kN
V_{RC}	vertical load of a return conductor	N, kN
$V_{i,k}$	vertical load at a new support i, k	N, kN
$W_{x(y)el}$	elastic section modulus	mm^3, cm^3
$W_{x(y)pl}$	plastic section modulus	mm^3, cm^3
X_K	characteristic value of a resistance	N, kN
X_d	design value of a resistance	N, kN
X_{id}	design value of a resistance i	N, kN
a	transverse span length of a head span	m
a_{01}	distance of points 0 and 1 at cross-span	m
a_{mi}	subsection of the length of a head span i	m
a_o	distance between force application and point of deflection	m
$b_{1,2}$	width of legs of a section	m
b_{Ey}	distance to embankment	m
b_F	width of foundation	m
$b_{Fx,y}$	width of foundation in x and y direction	m
b_R	distance between foundation and closest rail	m
b_T	width of an H-beam	mm
b_i	stagger at support i	m
b_{mik}	sub-width of horizontal registration	m
$b_{x,yk}$	width of tower at height of load actions k	m
$b_{x,(y),o,(u)}$	width of tower at the upper end of a panel in the face x, (y), o, (u)	m
c	effective cohesion	kN/m^2
c_u	undrained shear strength	kN/m^2
d	conductor diameter	m
$d_{e(i)}$	external (internal)tube diameter	mm
d_l	diameter of hole	mm
e	distance between ToR and ToF	mm
e_1	edge distance in direction of the force	mm
e_2	distance of holes in direction of the force	mm
e_3	edge distance rectangular to the force	mm
$e_{A(B)}$	difference in height between rail and foundation upper face at pole A (B) of the head span	m
e_m	difference in height between rail and ToF	m
e_o	earth resistance	kN/m^3
e_p	soil resistance	N/m^2
$e_{x(y)}$	eccentricity of the vertical load of a block foundation	m
f	deflection	mm

Symbol	Definition	Unit
f_{CAH}	deflection under permanent loads at height of catenary wire	mm
$f_{CA(H+W)}$	deflection under maximum loads at height of catenary wire	mm
f_{CAN}	deflection of cantilever tube	m
f_{CWW}	deflection under wind in height of contact wire	mm
f_a	deflection at position a	mm
f_m	deflection of cantilever tube	mm
f_{pile}	displacement at the pile head	mm
g	gravitation on Earth	m/s^2
h	height above ground or fixing	m
h_A	height of cantilever	m
$h_{A(B)}$	height at head-span pole	m
h_{AL}	length between traction power line and fixing of cantilever	m
h_{CA}	catenary wire height	m
h_{CAN}	distance between traction power line and top of pole	m
$h_{CS(B)}$	height of head-span wires at pole A (B)	m
h_{CW}	contact wire height	m
h_E	height of power line	m
h_{EH}	distance between upper swivel hinge and power line	m
$h_{FA(B)}$	system height of horizontal registration	m
$h_{HSA(B)}$	system height of horizontal registration arrangements at pole A (B)	m
h_{LSW}	height of lower cross-span wire	m
h_M	height or length of pole	m
h_{SCA}	height of catenary support within a head-span	m
h_{SDi}	fixing height of span wire i at pole	m
h_{SH}	system height	m
h_{SP}	structural support height	m
h_T	height of an H-beam	mm
h_{TWi}	height of tension wire i of a horizontal registration	m
h_b	distance between contact wire and registration arm fixing at cantilever tube	m
h_m	height of head-span	m
h_{uCR}	height of upper cross-span wire in a head-span	m
h_{uR}	height of lower cross-span wire in a head-span	m
$h_{x(y)}$	effective impact height of actions in x,y direction	m
h_z	impact height of transverse force	m
i	inertia radius	cm
i_ξ	minimum inertia radius	cm
k_d	ice load factor	N/(m·mm)
$k_{x(y)}$	factor for increasing the bending moments	–
l_{1-2}, l_{2-3}	sub-lengths of top tube	m
l_{3-4}	sub-length of cantilever tube	m
l_{3-5}	length of cantilever tube	m
l_{4-5}	sub-length of cantilever tube	m
$l_{4-6(7,7')}, l_{6-7}$	sub-lengths of registration arm	m
l_A	length of cantilever	m

Symbol	Definition	Unit
l_E	eccentricity of the traction power line	m
l_d	decisive length of bracing	m
l_i	length of a span i	m
l_{mik}	sub-length horizontal registrations	m
$l_{x,y}$	tower width between central axes of leg members	m
m_1	number of forces	–
m_D	number of diagonals in a panel	–
m'_{OCL}	contact line mass related to length	kg/m
n_{FKi-k}	inclination of span wire in a horizontal registration	–
n	number of catenaries in a head-span	–
n_l	number of holes in the cross-section of a member	–
n_s	number of parallel wires in a head-span	–
n_1	number of forces	–
n_{s1}	number of shearing areas	–
n_2	number of bolts	–
p	probability of exceeding in general	–
p_t	soil resistance parameter	$kN/(m^2 \cdot m)$
q'	length-related load	N/m
q_b	basic wind pressure	N/m^2
q_z	wind pressure at the height z	N/m^2
s	distance of centre of gravity at block foundations	m
s_T	thickness of web of an H-beam section	mm
s_k	buckling length	m
$s_{o, u}$	position of centre of forces at block foundations	m
t	depth for reference of soil pressure	m
t_0	foundation depth, depth of embedment, embedding depth	m
t_E	depth of insertion, depth of embedment	m
t_T	thickness of flange of an H-beam	mm
t_e	depth of embankment	m
t_l	thickness of section	mm
t_{pile}	pile length	m
t_x	distance between top of foundation and soil surface	m
t_z	non-bearing section of a pile	m
t_{z1}	top of foundation above soil surface	m
t_{z2}	distance between intersection of foundation and soil surface	m
x_1	thickness of a non-bearing soil layer	m
y_E	coordinate of power line	m
y_K	coordinate of cantilever	m
y_i	distance of support i in a head-span to pole	m
z_i	sag in a head-span at support i	m
z_{ki}	effective lever of vertical force $(k)i$	m
z_m	position of maximum bending moment of a pile	m
z_{max}	maximum sag in a head-span i	m
Δh	difference of height in a head-span	m
$\Delta x(y)$	increase of latitude in panel $x(y)$	mm/m

Symbol	Definition	Unit
$\sum j$	random-related lateral displacements	m
ϕ_R	angle of internal friction	degree
ϕ_k	auxiliary parameter for member buckling	–
α	imperfection factor	–
α_{mi}	inclination angle of a span wire	degree
α_s	factor for shearing	–
β_{CB}	concrete strength for calculations	N/mm^2
β_{CS}	yield strength of concrete steel	N/mm^2
β_{CWN}	concrete nominal strength	N/mm^2
β_{CWS}	concrete series strength	N/mm^2
β_E	angle of earth frustum	degree
β_W	embankment gradient	degree
β_P	shape factor	–
γ_A	partial factor for accidental loads	–
γ_C	partial factor for conductor tensile forces	–
γ_{CC}	partial factor for concrete stress	–
γ_{CS}	partial factor for concrete steel	–
γ_E	unit weight force of soil without buoyancy	kN/m^3
γ_{Eb}	unit weight force of soil with buoyancy	kN/m^3
γ_F	partial factor for loadings	–
γ_G	partial factor for permanent actions	–
γ_I	partial factor for ice action	–
γ_M	partial factor for materials	–
γ_W	partial factor for wind action	–
γ_P	partial factor for loads due to construction and maintenance	–
γ_g	unit weight force of foundation	kN/m^3
γ_v	partial factor for pre-stressing	–
ε_b	strain of concrete	–
ε_s	strain of concrete steel	–
ϑ	torsional angle around pole longitudinal axis	degree
ι	angle between crucial wind direction and conductor	degree
κ	factor to adapt the ultimate soil pressure to depth	–
κ_n	factor for consideration of embankment gradient	–
λ	slenderness	–
$\overline{\lambda}$	slenderness ratio	–
λ_P	earth pressure coefficient	–
μ	factor for consideration of the cross-section shape at buckling	–
ρ	air density 1,225 at 15 °C and 0 m altitude H above sea level	kg/m^3
ρ_I	density of ice	kg/m^3
σ	available soil pressure	kN/m^2
$\sigma_{1,5}$	limit of soil pressure at 1,5 m depth	kN/m^2
σ_c	concrete compression stress	N/mm^2
σ_{cb}	soil compression stress	N/mm^2
σ_f	yield strength of material	N/mm^2
σ_{per}	permissible soil pressure	kN/m^2

Symbol	Definition	Unit
σ_{st}	available steel stress	kN/m^2
σ_t	soil pressure at the depth t	kN/m^2
σ_u	tensile strength of the material	N/mm^2
τ_r	value of skin friction	kN/m^2
φ	angle between wind direction and the perpendicular to the conductor or to the tower face	degree
χ	buckling factor for compression loads	–

13.1   Loads and their actions on contact lines

13.1.1   Introduction

Contact lines are subject to different external loads and actions. *Dead loads* from conductors, fittings, insulators and supports act permanently and can be determined accurately from technical data and dimensions. The *conductor tensile forces* also act permanently. However, non-automatically tensioned conductor tensions also depend on the conductor temperature. Unless installed in tunnels or other protected areas, contact lines are exposed to the weather and occasionally experience heavy additional loads due to *wind action* on conductors and structural components and loads caused by *ice accretion* on conductors. These loads can be determined by statistically evaluating records of long-term weather observations. Wind and ice are randomly distributed variables; their frequency of occurrence can be described by the laws of probability, refer to IEC 60 826 and [13.1]. During construction and maintenance, contact lines can be subjected to additional loads that need to be withstood by the structures and must be considered to ensure *personnel safety*. Design loads should also include adequate provisions for construction and maintenance loads. Loading requirements are dealt with by EN 50 119 and its national annexes [13.2].

13.1.2   Classification of actions

Loads and actions on contact lines can be classified by their application, by their nature and structural response.

 – **Permanent actions G** are dead weight of supports including foundations, fittings and fixed equipment, dead weight of conductors and the effects of applicable conductor tensile forces. Normally the characteristic value of permanent actions can be determined as one value G_K because the variability of permanent actions is very small.
 – **Variable actions Q** comprise wind loads, ice loads and other imposed loads. Wind loads, ice loads and applicable temperatures are climatic conditions which can be assessed using probability methods or on a deterministic basis or taken from applicable standards. Conductor tensile force effects caused by wind and ice and temperature deviations can be variable actions as well, depending on the contact line design. For variable actions, the characteristic value Q_K corresponds to a nominal value used for deterministic based actions or an upper value of an intended probability of not being exceeded or a lower value with an intended probability of not being lower during a reference period.
 – **Accidental actions A** are failure containment loads etc. relating to security aspects. The representative value is generally a characteristic value A_K corresponding to a specified

condition. The dynamic actions after the seizure of automatic tensioning equipment or breaking of contact or catenary wire may be modelled by an equivalent static action.

– **Construction and maintenance loads** Q_P take into account working procedures, temporary guying, lifting arrangements etc. Their characteristic values Q_{PK} are deterministic values provided to assure the safety of personnel.

13.1.3 Permanent actions

The *dead loads* of structures, fittings and conductors act vertically and the loads resulting from *conductor tensile forces* act horizontally in the case of level attachment points. Dead loads are independent of the conductor temperature and result from the installation dimensions. During the life cycle of an installation, they vary only because of the contact wire wear. The conductor tensile forces are approximately constant with automatically tensioned contact and catenary wires. However, fixed terminated wires and conductors depend on the conductor temperature and vary as a function of ambient temperature, current loading and ice loads. Structural design must consider the imposed maximum tensile forces.

The *vertical load* at support i results from (see Clause 4.1.3)

$$V_i = G'(l_i + l_{i+1})/2 + H\left[(NN_i - NN_{i-1})/l_i + (NN_i - NN_{i+1})/l_{i+1}\right] \quad , \qquad (13.1)$$

where l_i and l_{i+1} are the span lengths adjacent to the support i. NN_i is the reference height of the support i and H the horizontal conductor tensile force.

The *horizontal conductor tensile force* results from conductor tensile stress and cross-section and determines also the horizontal (radial) loads at the supports associated with the deviation angle of the conductor. The determination of these load components is dealt with in Clause 4.1.5. The radial force at the support i with the track radius is

$$Q_{Hi} = H\left[(l_i + l_{i+1})/(2R) + (b_i - b_{i-1})/l_i + (b_i - b_{i+1})/l_{i+1}\right] \quad . \qquad (13.2)$$

The value b_i is the horizontal *stagger* at support i.

13.1.4 Variable actions

13.1.4.1 General

Variable actions due to climatic conditions are added to the permanent loads. All contact lines are exposed to wind in the open. *Ice accretion* on conductors occurs in many regions in addition to wind action. Extreme climatic conditions result in the most adverse maximum mechanical loads on contact lines.

13.1.4.2 Wind loads

In Clause 2.5.2.2 the basic wind load q_z is determined, where z is the height above ground. *Wind force on conductors* causes forces transverse to the line direction. From two adjacent spans the wind force on a support from conductors is:

$$Q_{WC} = q_z \cdot G_C \cdot C_C \cdot d \cdot (l_1 + l_2)/2 \cdot \cos^2 \varphi \quad , \qquad (13.3)$$

where

q_z characteristic dynamic wind pressure, see Clause 2.5.2.6,
G_C structural response factor for conductors taking into account the response of move-
         able conductors to wind load. The factor G_C can be determined according to national
         experience. Typically $G_C = 0,75$,
C_C drag factor of the conductor being 1,0 specified in EN 50 119.

Where twin conductors are run in parallel, a reduction in the wind load may be taken on the
leeward conductor to 80% of the windward conductor, if the spacing between the conductor
axes is less than five times the diameter.

The *wind force on an insulator* and line fittings acts at the attachment point to the support in
wind direction and is equal to

$$Q_{W\,ins} = q_z \cdot G_{ins} \cdot C_{ins} \cdot A_{ins} \cdot \cos^2 \varphi \quad , \tag{13.4}$$

where

G_{ins} structural resonance factor for insulator sets. It can be taken as 1,05,
C_{ins} drag factor for insulators, to be taken as 1,2.

Wind forces on solid poles or structures can be calculated from

$$Q_{W\,str} = q_z \cdot G_{str} \cdot C_{str} \cdot A_{str} \quad , \tag{13.5}$$

where

G_{str} structural resonance factor for a pole. For self-supporting steel and concrete structures
         G_{str} is typically 1,00 to 1,15,
C_{str} the drag factor depending on the shape and surface roughness of the structure. The
         values according Table 13.2 apply:
         – tubular steel and concrete structures,
         – with circular cross-section $C_{str} = 0,7$,
         – with dodecagonal cross-section $C_{str} = 0,85$,
         – with hexagonal or octagonal cross-section $C_{str} = 1,0$,
         – steel and concrete structures with square or rectangular cross-section $C_{str} = 1,4$,
         – poles made of H-beams $C_{str} = 1,6$.

Wind forces on lattice structures of rectangular cross-sections can be calculated from (Figure 13.1)

$$Q_{Wt} = q_z \cdot G_{str} \left(1 + 0,2 \sin^2 2\varphi \right) \left(C_{t1} \cdot A_{t1} \cos^2 \varphi + C_{t2} \cdot A_{t2} \sin^2 \varphi \right) \quad . \tag{13.6}$$

Alternatively, the wind forces may be calculated in accordance with Eurocodes.

The drag factors $C_{t1,2}$ depend on the solidity ratio as described in EN 1991-1-4. An appropriate value is $C_{t1,2} = 2,8$ for lattice structures with square or rectangular cross-section made of
angle sections.

The *wind forces acting on cross-beams*, head spans and cross-spans as well as on cantilevers
may be determined by summation of wind actions on the individual elements.

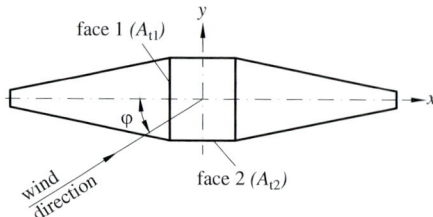

Figure 13.1: Wind action on lattice steel structures.

Figure 13.2: Ice load at an overhead contact line (Photograph: G. Hahn).

13.1.4.3 Ice loads

Ice loads with different characteristics occur on conductors and poles. Ice accretion on conductors may lead to operationally restraining conditions and additional loads, especially those ice accretions from precipitation characterised by high density and strong adhesion. Rain freezes on conductors at temperatures around 0 °C. The ice is clear or opaque, has an approximately circular cross-section and adheres to the conductor with high density. Such ice accretion can prohibit the pantograph from contacting the contact wire. In the case of automatically tensioned installations, the sags also increase considerably and impair operations. In Figure 13.2 the ice accretion is shown at an overhead contact line.

As with wind loads, ice loads also follow the statistical rules discussed in IEC 60 826 and EN 50 341-1. For contact line design, ice loads are often taken from the design for overhead transmission lines standards, such as EN 50 341-2-4:

$$G'_{\text{ice}} = (G'_{0\text{ice}} + k_\text{d} \cdot d) \qquad \text{in N/m} \quad , \tag{13.7}$$

where d is the conductor diameter in millimetres, $G'_{0\text{ice}}$ is 5 to 20 N/m and $k_\text{d} = 0{,}1, 0{,}2, 0{,}3$ and $0{,}4$ N/(m $\cdot$ mm) depending on the ice load zone.

EN 50 125-2 indicates four classes of ice loads specified in Table 13.1. These values apply to conductor diameters between 10 and 20 mm.

13.1.4.4 Combined action of wind and ice

Wind action on ice-covered conductors increases the mechanical loading on conductors and supports and should be considered when ice formation is reinforced by wind action, see

Table 13.1: Ice loads on conductors according to EN 50 125-2.

Class	Ice load N/m
I0 (without ice)	0
I1 (low)	3,5
I2 (normal)	7,0
I3 (heavy)	15,0

EN 50 341-1. When *combined actions of wind and ice* load need to be considered for the design of overhead contact line installations and structures, it may be assumed that 50 % of the wind load according to Clause 13.1.4.2 acts on structures and equipment without ice and on conductors covered with ice according to Clause 13.1.4.3. The unit weight force ρ_I of the ice may by taken as $7\,500\,\text{N/m}^3$, the aerodynamic drag factor as 1,0. The equivalent diameter D_I in metres of the ice accretion may be calculated from

$$D_I = \sqrt{d^2 + 4\,G'_{\text{ice}} / (\pi \cdot \rho_I)} \quad , \tag{13.8}$$

where d conductor diameter in metres and G'_{ice} characteristic ice load in N/m.

13.1.4.5 Temperature effects

The temperature effects are related other climatic actions. The following cases should be considered according to EN 50 119:
  – minimum temperature $-20\,°\text{C}$ to be considered with no other climatic actions
  – ambient temperature $+5\,°\text{C}$ in case of extreme wind loads
  – temperature $-5\,°\text{C}$ with ice and combined wind loads

13.1.5 Loads due to construction, maintenance and other conditions

During *construction and maintenance* of a contact line, additional loads occur because of linesmen, temporary anchoring, fastening of tools and conductor stringing operations. These additional loads should also be considered when designing contact line structures. Vertically acting *construction loads* of at least 1,0 kN should be assumed for horizontal cross-beams of suspension structures and at least 2,0 kN for other types of structures (see EN 50 119, Clause 6.2.8). These forces should be assumed as acting at the individually most unfavourable nodes of the cross-beams or at the attachment points of conductors to the structures. Cantilevers need not to be designed for such loads where appropriate working practices are adopted.

For all members of lattice structures which can be climbed with inclined bracing at an angle less than 30° to horizontal a construction load of 1,0 kN acting vertically in the centre of an element should be assumed, however, without any other loads. This load considers the weight of linesmen. Step-bolts and stirrups should be rated for a concentrated load of 1,0 kN acting vertically at a statically adverse position.

Accidental loads are specified to take care of failure containment and emergency situations. In general, at any conductor attachment point to a structure the relevant residual static load resulting from the release of the tension of a contact wire, catenary wire or feeder line should be applied. In general, it is sufficient to consider accidental loads for structures at the end of tensioning sections or for anchor structures of mid-span anchors. Details should be specified for each project.

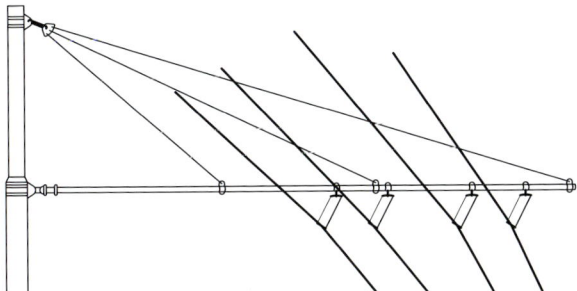

Figure 13.3: Cantilever for two trolley bus contact lines.

When overhead contact lines are to be constructed in seismically active regions, consideration should be given to forces due to earthquakes and/or seismic tremors. Refer to EN 50 341-1, Annex C, for information. Actions on contact lines caused by short circuits should be considered if they could result in an essential loading.

13.2 Transverse supports and registrations

13.2.1 Types of transverse support equipment

Grouped under this heading are *cantilevers* of varying designs including cantilevers across several tracks, *flexible cross-spans* and *rigid portals*. Clauses 3.3 and 11.2.1 illustrate relevant designs of these components and describe their functions within the overhead contact line. In general, they are used to support the contact line equipment, register the lateral position of contact and catenary wires and reliably transfer the loads acting on the overhead contact line to the poles.

13.2.2 Swivel cantilevers

Swivel cantilevers, also known as hinged cantilevers, are made of tubular elements and are the most frequently used design for individual supports in an overhead contact line installation. They are fixed to poles, buildings or other structures and swivel around a vertical axis. Cantilevers rigidly fixed to poles are no longer used for new installations but can still be found in service on older routes. Figure 11.4 depicts a cantilever design using a *catenary wire swivel clamp* which is fixed to the cantilever tube. In addition to supporting the catenary wires, the catenary wire clamp connects the cantilever tube with the top tube or top anchor rope. With this design, mainly tensile forces occur in the top anchor, permitting the use of ropes instead of tubes in many cases. Only axial forces act on the structural cantilever elements. The cantilever shown in Figure 11.5 utilises a *catenary wire clamp* which is moveable along the top tube. Consequently, bending moments occur in the top anchoring element requiring a bending-resistant rating.

Mechanically effective cantilever elements can consist of steel tubes, steel or copper alloy wires, aluminum tubes and bars or tubes made of glass fibre reinforced plastics, that can also serve as insulators.

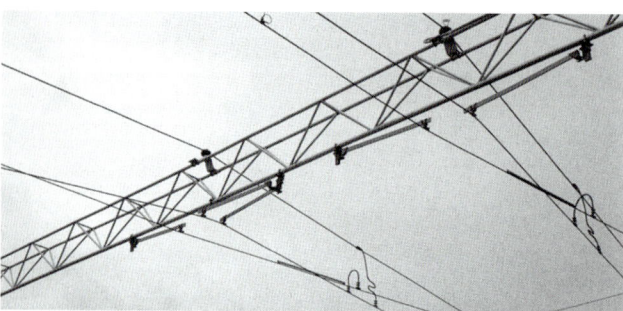

Figure 13.4: Lattice-type cross-beam made of aluminum (Photograph: Müller-Gerlach).

13.2.3 Cantilever across several tracks

If supports can only be installed on one side of a multi-track line, *cantilevers across several tracks* are a viable alternative. The design shown in Figure 11.17 consists of a cantilever arm made of two U-channels connected face to face or of a square tube. The cantilever arm is fixed to the pole by a hinge and carries posts with *swivel cantilevers* attached. The vertical loads are supported by rope-type guys arranged obliquely between cantilever arm and pole. Cantilevers across several tracks, made of glass fibre reinforced plastic (GRP) tubular sections, arranged in parallel, are often adopted for mass transit installations. These cantilevers are used to support the bridles of trolley contact lines or *trolley bus contact lines* (Figure 13.3). Rope-type guys, arranged obliquely, are also used in these cases to carry the vertical loads. In contrast to main line railways, the plastic components are used to provide insulation, making separate insulators superfluous. Generally, these designs are adopted for installations supplied by DC 600 V or DC 750 V. Their lengths should be limited to approximately 10 m because of construction and maintenance difficulties.

13.2.4 Flexible transverse support equipment

Head-span supports for contact lines offer benefits in stations with more than two parallel tracks because space to install poles between tracks is not available. The individual contact lines can be arranged within the cross-span as desired. This is advantageous, especially for wiring station ends with many turnouts. Flexible cross-spans have reached lengths up to 80 m. In practice, the length should be limited to 40 m because of operational and maintenance practicalities. Figures 3.19 and 11.20 show the principle arrangement. The vertical loads, resulting from the loads of the individual contact lines are carried by head span wires. The *horizontal registration* is provided by cross-span wires. The upper *cross-span wire* carries the horizontal loads resulting from registration of the catenary wires and the lower those from the contact wires.

The head-span and cross-span wires mechanically connect the contact lines and movements are transmitted between the lines. This is considered undesirable for installations of high-speed overhead contact lines. Consequently, DB's directive 997.0101 [13.3] recommends separate supports for each track carrying high-speed lines.

The *head-span wires* have a sag between one eighth and one tenth of the cross-span length. Design of head span equipment involves calculation of the head-span wire loads and selection of wire dimensions.

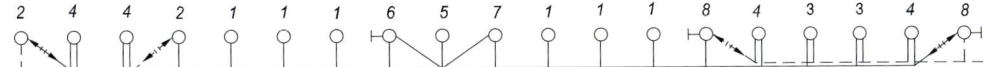

Figure 13.5: Tensioning section of an overhead contact line. *1* suspension pole, *2* and *8* tensioning poles, *3* and *4* intermediate suspension poles, *5* midpoint pole, *6* and *7* midpoint anchor poles

13.2.5 Portal structures

In the case of *portal structures*, a rigid horizontal cross-beam carries the individual contact lines. The beam is supported by poles on both sides of the line. Some designs use posts with swivel cantilevers fixed to the cross-beam, while with others the cross-beam carries the vertical loads and is used for the lateral guidance of the catenary wires. A lower *cross-span wire* registers the contact wires. Figure 11.23 shows a cross-beam adopting a lattice girder. Portal structures are used for lengths up to 40 m. For long portals, lattice steel design proves to be advantageous. For spans up to 25 m H-beams or hollow steel sections may also be used. To reduce maintenance, aluminum portals were adopted for some mass transit installations (Figure 13.4). Cross-beams are loaded by vertical forces and by moments from posts for cantilevers in their vertical plane. Additional bending in the horizontal plane occurs if midpoints or tensioning equipment is arranged at a cross-beam.

13.3 Poles

13.3.1 Types of poles

Poles used to support contact lines perform various functions. Figure 13.5 depicts a *tensioning section* of a contact line outside stations (interstation line). The *contact line* starts at a *tensioning pole* (type 2 or 8). A tensioning pole supports both the load from the cantilever and the forces exerted by termination tensions of contact and catenary wires. In some cases, stays are arranged to counteract the tensioning force acting in line direction (type 8).

Intermediate poles (type 3 and 4), within the *overlap section*, carry two cantilevers arranged on brackets. These poles are loaded in bending and by torsional moments due to differing radial forces of the contact lines. These poles require torsional stiffness.

Suspension poles (type 1) are equipped with just one cantilever and withstand loads created by contact wire and catenary wire stagger, radial forces on curves and wind loads. A *midpoint pole* (type 5) is arranged approximately in the middle of the tensioning section. It is loaded by the contact line and the midpoint anchors. There are *midpoint anchor poles* for termination of midpoint anchors (type 6 and 7), loaded by forces in line direction and generally anchored by stays.

Poles for *head-spans* carry loads from the head-span wires, cross-span wires as well as those from tensioning equipment acting in line direction, if any. The height of these poles must consider the sag of the head-span wire.

Poles carrying cross-beams are loaded by vertical and transverse loads only, since the cross-beams are fixed to the poles by hinged joints, to avoid bending-resistant joints. Loads from cross-span wires, cantilevers and terminated contact lines act in addition to the loads from cross-beams. Some poles in contact line installations are only used for *radial contact line registration* or terminations without cantilevers.

Table 13.2: Loads for design of contact line structures.

Designation of pole[1]	Type of pole	Permanent loads	Variable loads
1	suspension pole with one cantilever	– dead loads of conductors, cantilevers and poles – forces due to radial action and stagger	– wind loads
3, 4	intermediate suspension pole with twin cantilever	– as 1 – torsional moment due to radial forces and stagger	– as 1 – torsional moment due to wind action
5	midpoint pole	– as 1 – loads due to anchoring of contact line	– as 1
6, 7	midpoint anchor pole	– as 1 – loads due to anchoring of contact line	– as 1
2, 8	tensioning pole	– as 1 – loads due to anchoring of contact line	– as 1

[1] see Figure 13.5

Traction power lines are often installed on the overhead contact line poles resulting in different conductor configurations and additional loads. The traction power lines are usually supported with suspension or strain insulators. Table 12.27 depicts some pole configurations as adopted for Sicat H1.0 high-speed lines. If the power lines are *parallel feeder* lines having the same electric potential as the contact line, they only require a reduced clearance to the contact line. *Supply feeders* or *by-pass feeders* can be switched separately from the contact line. It is standard practice to provide a clearance of at least 2 m to the contact line. Pole types for power lines may be classified as *suspension poles* equipped with suspension insulator sets, intermediate strain poles and dead end poles equipped with tension insulator sets.

The dimensions of the pole top need to be determined in correlation with the traction power line arrangement. This ensures compliance with minimum clearances to earth potential as specified by relevant standards such as EN 50 341-1. The required clearances are 0,15 m for 15 kV and 0,27 m for 25 kV.

13.3.2   Loading assumptions

Various types of external loads act simultaneously on the supports of overhead contact lines (see Table 13.2). Depending on the utilization of the poles, these loads should be combined to cover all possible *loading combinations* during operation and minimize the risk of failure. Relevant standards such as EN 50 119 stipulate these *loading assumptions* and combine them to six load cases:

Load case A: Loads at minimum temperature
Minimum temperature loads and components of permanent conductor tensile forces acting in the relevant direction at the minimum ambient and design temperatures are considered, refer to Clause 13.1.4.5.

Load case B: Maximum wind loads
Permanent loads, conductor tensile forces increased by the action of wind and wind loads on each element according to Clause 13.1.4.2 acting in the most unfavourable direction. The ambient temperature under this condition should be as specified in Clause 13.1.4.5.

Table 13.3: Load cases for the design of structures according to EN 50 119.

Type of structures	Load cases
cantilevers hinged	A, B, C, D, E
cantilevers rigid	A, B, C, D, E, F
head-spans	A, B, C, D, E, F except pulley supports
portals/cross-beams	A, B, C, D, E, F except pulley/hinged supports
suspension structures	A, B, C, D, E
pull-off structures	A, B, C ,D ,E
midpoint anchors	A, B, C, D, E, F
midpoint structures	A, B, C, D, E
structures for flexible and rigid cross supports	A, B, C, D, E, F[1]
structures for horizontal catenary arrangements	A, B, C, D; F
tensioning structures	A, B, D, E
structures with feeder and parallel reinforcing lines	A, B, C, D, E, F[2]
overhead line structures carrying additional power lines	A, B, C, D, E, F
anchor supports dependent on type of anchor	A, B, C, D, E, F

[1] if midpoint, [2] except hinged supports

Load case C: Ice loads

Permanent loads, ice loads on conductors and structures and conductor tensile forces increased by ice loads according to Clause 13.1.4.3, if applicable.

Load case D: Combined action of wind and ice loads

Permanent loads and ice loads, conductor tensile forces increased by ice loads and wind loads, a combined wind and ice load according to Clause 13.1.4.4 if applicable. The wind load acts in the most unfavourable direction.

Load case E: Construction and maintenance loads

Permanent loads, ice loads, conductor tensile forces and construction and maintenance loads according to Clause 13.1.5 together with a reduced wind load and reduced ice load as specified in EN 50 119 or a project specification.

Load case F: Accidental loads

Permanent loads and permanent conductor tensile forces together with the unintentional reduction of one or several conductor forces. In this case, the tensile forces of the contact and catenary wires act particularly on the midpoint anchor.

The load cases for the design of structures are summarised in Table 13.3. For rating poles, the load case resulting in the maximum stress has to be selected.

13.3.3   Partial factors for actions

European standards for steel structures (Eurocode EN 1993: Design of steel structures) and concrete structures (Eurocode EN 1992: Design of concrete structures), use *partial factors*. They are separated into partial factors for actions and partial factors for material. The relevant partial factors for actions are specified in Table 13.4.

The partial factor for permanent actions due to dead weight γ_G is 1,3. Where the dead weight acts favourably on an element, i. e. reducing the loading, the partial factor γ_G is to be assumed as 1,0.

Table 13.4: Partial factors for actions according to EN 50 119.

Type of loads	Load cases					
	A	B	C	D	E	F
permanent γ_G	1,3	1,3	1,3	1,3	1,3	1,0
variable action γ_{cV}	1,3	1,3	1,3	1,3	1,3	1,0
relieving γ_G, γ_{cP}, γ_{cV}	1,0	1,0	1,0	1,0	1,0	1,0
permanent loads						
– load increasing γ_G, γ_C	1,3	1,3	1,3	1,3	1,3	1,0
– relieving γ_G, γ_C	1,0	1,0	1,0	1,0	1,0	1,0
wind γ_W, γ_C	–	1,3	-	1,3	–	1,0
ice γ_I, γ_C	–	–	1,3	1,3	–	–
accidental γ_A	–	–	–	–	-	1,0
construction γ_P	–	–	–	–	1,5	–

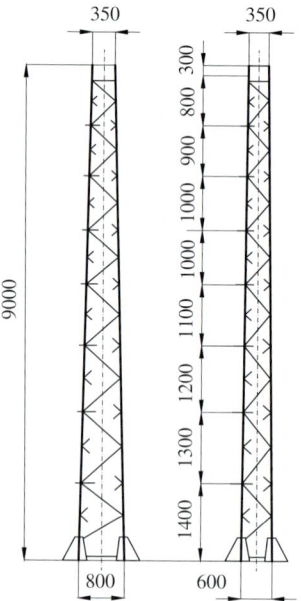

Figure 13.6: Lattice steel pole for overhead contact lines.

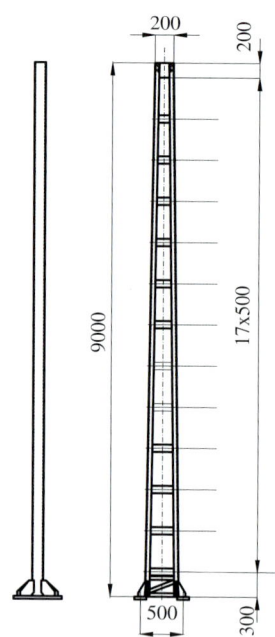

Figure 13.7: Double channel pole.

The partial factors γ_W for wind loads and γ_I for ice loads should be equal to 1,3. The partial factor γ_C applicable to conductor tensile forces should be 1,3.

In the case of accidental load cases, the partial factors γ_G for permanent loads, γ_C for conductor tensile forces and γ_A for accidental loads can be assumed as 1,0.

The partial factor for construction and maintenance loads γ_P is 1,5 combined with partial factors γ_G and γ_C equal 1,3 for permanent actions.

13.3.4   Structural design and materials

A large variety of pole types are used for contact line supports. *Lattice steel poles* as shown in Figure 13.6 consist of four leg members made from angle sections and diagonal bracing.

Figure 13.8: H-beam steel pole.

Figure 13.9: Tubular steel pole for contact line of the high-speed line HSL Zuid in the Netherlands.

Their strength can be adapted to the required loading condition. DB AG uses a pole family starting with dimensions of 800 mm × 600 mm at the bottom and angle sections L80 × 8 up to dimensions 1 600 mm × 2 000 mm with angle sections L150 × 14. These are used for tensioning poles and head span supports.

Double channel poles (Figure 13.7) consist of two U-channels (also known as parallel flange channels – PFCs) connected by flat steel plates at a spacing of approximately 500 mm. They are tapered in the vertical direction. Channels [100, [120, [140 and [160 are generally used. Double channel poles are characterised by differing bending resistances around either axes and are used primarily for support locations without tensioning equipment.

Many overhead line installations use *steel poles made of H-beams*, which are readily available ex-stock. However, their relatively high weight/strength ratio, low resistance to *deflection* and low resistance to torsion are disadvantages especially when used with twin cantilevers. Figure 13.8, shows an H-beam pole.

Tapered, *thin walled hollow steel poles* are an interesting alternative as their dimensions can be matched to the loading requirements. Generally, these poles are used for mass transit installations in urban areas. They are produced by rolling, drawing or welding and enable the manufacture of poles with tailored cross-sectional dimensions, strength and torsional rigidity. Figure 13.9 shows a pole made of a tapered steel tube as used for the high-speed line HSL Zuid in the Netherlands.

Spun concrete poles are also used for contact line installations (Figure 13.10). They are characterised by circular cross-sections, a hollow core and produced with a conical increase in

Figure 13.10: Spun concrete pole of an overhead contact line.

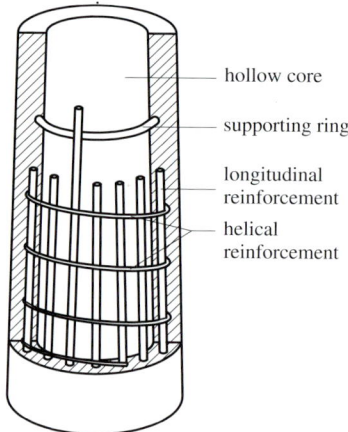

hollow core

supporting ring

longitudinal reinforcement

helical reinforcement

Figure 13.11: Structure of a spun concrete pole.

diameter from the top to the bottom by at least 15 mm/m. They are spun in two-part horizontal casings that are rotated around their longitudinal axis. The spinning process achieves a high *concrete compressive strength* of 70 N/mm^2 in accordance with C70/80 and recently up to 100 N/mm^2 in accordance with C95/105. The high density of the concrete protects the reinforcement from corrosion and prevents cracking.

Concrete pole reinforcement can be either slack reinforcement bars or pre-stressed reinforcement using high-tensile steel wires. *Pre-stressed poles* have become increasingly popular for railway applications. Pre-stressing of the steel wires is carried out before spinning. After the concrete sets, the pre-stressing strands are cut at the mould ends, producing a compressive force in the concrete pole. Under bending, this pre-stressing must be exceeded before the concrete experiences tensile stresses and subsequent cracking. The structure of a spun concrete pole is shown in Figure 13.11.

In the past, different kinds of defects were reported in spun concrete poles. These defects included cracks along the separating joints of the moulds, longitudinal cracks of differing lengths and widths, transverse cracks and torsional cracks. Results of various investigations suggest that the cracks were due to lacks in structural design including inadequate concrete thickness and reinforcement cover, insufficient helical reinforcement and mishandling during manufacturing. Modifications to the standards and to quality assurance measures taken by manufacturers, indicate that similar defects will not re-occur in the future and spun concrete poles may be considered as long-lasting components.

Some railway operators, e. g. the ÖBB, use *concrete poles with a solid core*, produced on vibration tables with the reinforcement arranged into rectangular casings. Concrete is poured in and compacted by external vibrators. As a consequence of the lower concrete strength and the solid cross-section, these types of poles are considerably heavier than spun concrete poles. When produced to best practice, concrete poles achieve a long life free of maintenance. For overhead contact line installations, they have proven benefits, especially when associated with

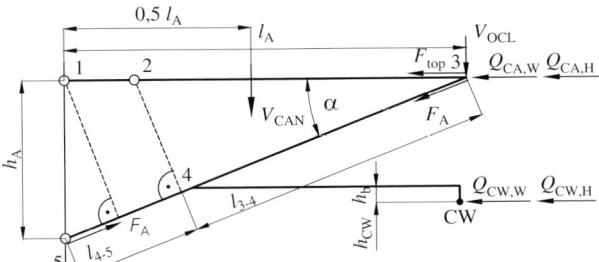

Figure 13.12: Loading of a cantilever.

direct-planted foundations [13.3]. However, for tensioning poles relatively massive cross-sections are required if stays cannot be used. They may also look clumsy.

13.4 Rating of cross-span supports

13.4.1 Introduction

Rating the *cross-span support* includes determining the internal forces and moments and designing the components in accordance with the relevant standards. The following sections will deal with typical designs of *cantilevers* and *flexible head-spans*. Other designs may be rated by adopting the forces detailed here and using standard civil engineering design methods.

13.4.2 Cantilevers

13.4.2.1 Loading and internal forces and moments

Top anchor, cantilever tube, registration arm and diagonal strut, if any (Figure 11.4), of a *tube-type swivel cantilever* carry the loads and register the contact and catenary wire.
Various load combinations act on the cantilevers and poles, depending on:
  - *pole position*, whether outside or inside the curve
  - *type of support*, pull-off or push-off
  - *wind action* and
  - *ice effects*, if any
The individual load components result from:
  - *vertical load due to the contact line* according to Equation (13.1)
  - *vertical load due to the ice-covered contact line* Equation (13.7)
  - *radial load on wires* refer to Equations (13.2),
  - *wind loads* on the catenary and contact wire refer to Equation (13.3)
  - *dead load* of the cantilever, acting approximately at half of its length
Figure 13.12 shows the loads which act on a cantilever. Wind on conductors can act in either direction. To simplify the analysis, the individual load inputs are summed to vertical and horizontal components.
  - Vertical load

$$V_{CA} = V_{OCL} + V_{CAN} \cdot 0,50 \quad , \tag{13.9}$$

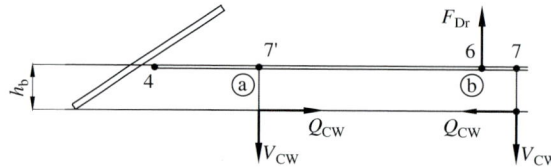

Figure 13.13: Loading of a registration arm.
a pull-off support
b push-off support

- Horizontal load on the catenary wire

$$Q_{CA} = Q_{CAH} \pm Q_{CAW} \quad \text{and} \tag{13.10}$$

- Horizontal load on the contact wire

$$Q_{CW} = Q_{CWH} \pm Q_{CWW} \quad . \tag{13.11}$$

The force F_{top} acting on the top anchor is,

$$F_{top} = (V_{CA} \cdot l_A - Q_{CW} \cdot h_b - Q_{CA} \cdot h_A)/h_A \quad . \tag{13.12}$$

The force F_A in the cantilever tube between points 3 and 4 is

$$F_A = -V_{OCL} \cdot \sqrt{1 + (l_A/h_A)^2} \tag{13.13}$$

and between points 4 and 5

$$F_A = -V_{OCL} \sqrt{1 + (l_A/h_A)^2} - Q_{CW} \Big/ \sqrt{1 + (h_A/l_A)^2} \quad . \tag{13.14}$$

The forces Q_{CA} and Q_{CW} are positive if acting in the direction indicated in Figure 13.12. Without a diagonal strut, the maximum bending moment M_{B4} in the cantilever tube occurs at point 4, where the registration arm is fixed

$$M_{B4} = Q_{CW} \cdot h_A \cdot l_{3-4} \cdot l_{4-5} \Big/ (l_A^2 + h_A^2) \quad . \tag{13.15}$$

The *diagonal strut* modifies the system to a statically indeterminate one. However, since the diagonal strut is applied as closely as possible to point 4 (Figure 13.12) it is reasonable to assume that the strut carries all the forces acting perpendicularly to the cantilever tube and transfers them to the top tube. The subsequent force F_D acting in the strut is:

$$F_D \approx Q_{CW} \Big/ \sqrt{1 + (l_A/h_A)^2} \quad . \tag{13.16}$$

The bending moment acting in the top tube at point 2 is determined by

$$M_{B2} = Q_{CW} \cdot h_A \cdot l_{2-3} \cdot l_{1-2} \Big/ (l_A^2 + h_A^2) \quad . \tag{13.17}$$

Figure 13.13 shows the loading of the *registration arm* caused by the loads from the contact wire. Thus, V_{CW} is the weight force of the contact wire length supported by the *steady arm* and includes the dead weight of the steady arm, the fittings and the registration arm. In the case of a push-off cantilever (position 7), the *dropper* force F_{Dr} is

$$F_{Dr} = (V_{CW} \cdot l_{4-7} + Q_{CW} \cdot h_S)/l_{4-6} \tag{13.18}$$

and in the case of a pull-off support (position 7')

$$F_{Dr} = (V_{CW} \cdot l_{4-7'} - Q_{CW} \cdot h_b)/l_{4-6} \quad .$$ (13.19)

If $Q_{CW} \cdot h_b \geq V_{CW} \cdot l_{4-7'}$ a compression force would occur in the dropper leading to uplift. Such a situation should be avoided. Where necessary, a strut has to be provided instead of a slack dropper. The bending moment at the attachment of the dropper is

$$M_{B6} \approx V_{CW} \cdot l_{6-7} + Q_{CW} \cdot h_b \quad .$$ (13.20)

13.4.2.2 Rating based on EN 50 119

The design is based on the *partial factor approach* which is described in EN 50 119 and also in Eurocodes, e. g. EN 1993 for steel structures. The partial factor approach verifies that the effects of design actions do not exceed the resistance or the serviceability limits of the structure. The design load F_d of an action is expressed by

$$F_d = \gamma_F \cdot F_K \quad ,$$ (13.21)

where γ_F is the partial factor and F_K the characteristic value of the considered action. The partial factor γ_F depends on the type of action and its variation and takes care of the uncertainties of the design model. The partial factors for contact lines are given in Table 13.4. Generally, the design value X_d of a material property is defined as:

$$X_d = X_K/\gamma_M \quad .$$ (13.22)

The partial factor for the material property γ_M covers unfavourable deviations from the characteristic value X_K, inaccuracies in applied conversion factors and uncertainties in the geometric properties and the resistance model. When designing any component or its connections, the following condition must be satisfied:

$$E_d \leq R_d \quad ,$$ (13.23)

where E_d is the total design value of the effect of actions such as internal forces or moments or their combination. E_d follows from

$$E_d = f\left(\gamma_G \, G_K; \gamma_W \, Q_{WK}; \gamma_I \, Q_{IK}; \gamma_P \, Q_{PK}; \gamma_C \, Q_{CK}\right) \quad ,$$ (13.24)

where $\gamma_G, \gamma_W, \gamma_I, \gamma_P, \gamma_C$ associated partial factors according to Table 13.4.
R_d is the corresponding structural design resistance combining all structural properties with the respective property X_{nd} in accordance with

$$R_d = f\{X_{1d}; X_{2d}; \ldots\} \quad .$$ (13.25)

For further details refer to EN 50 119, Clause 6.1.

The *internal forces and moments* within cross-span supports and structures are determined using the principles of structural analysis for rigid and flexible structures for either statically determinate or indeterminate structures or flexible rope systems. Example models and procedures can be found in EN 1993-1-1 for steel structures, EN 12 843 for concrete poles, EN 1992-1-1 for concrete structures, EN 50 341-1 and in recognized publications on structural analysis.

The elements of cross-span supports are loaded in compression, tension, bending or torsion. The calculation of the resistance of elements needs to consider the type of loads and the buckling stability if required and the analysis of connections.

For the analysis of resistance of steel structures, reference is made to EN 1993-1-1 members of lattice steel structures and their connections, for solid-wall and double channel steel structures and to EN 12 843 for reinforced concrete structures.

The resistances of solid wires, stranded conductors of metallic material loaded by tensile forces follow from the relevant standards, e. g. EN 50 182. EN 50 345 applies to ropes made of polymeric materials.

The partial factors γ_M for steel material in accordance with EN 50 119 are as follows:
  – cross-sections under simultaneous tensile forces and bending $\gamma_M = 1,1$
  – members to buckling $\gamma_M = 1,1$
  – connections under shearing and bearing $\gamma_M = 1,25$
  – net cross-sections based on ultimate tensile stress under tensile load $\gamma_M = 1,25$
  – welded connections $\gamma_M = 1,25$
  – bolts in tension $\gamma_M = 1,25$
  – metallic ropes under tensile force $\gamma_M = 1,5$

The partial factors for concrete structures are as follows:
  – prestressing force $\gamma_M = 0,9$ or $1,2$ depending whether the action is adverse or not for the calculated effect
  – concrete $\gamma_M = 1,5$
  – reinforcing steel (convential or prestressed) $\gamma_M = 1,15$

The partial factors for polymeric ropes according to EN 50 345 should be $\gamma_M = 3,0$.

The *design values* may be moments as well.

The verification of capacity must be carried out using the interaction relation where components are loaded in a combination of tension or compression and bending:

$$N_{zd}/N_{zpl,Rd} + M_{x,d}/W_{xplRd} + M_{yd}/W_{yplRd} \leq 1 \quad . \tag{13.26}$$

In (13.26) N_{zd}, $M_{x,d}$ and $M_{y,d}$ represent the design values of actions in tension and bending and $N_{zpl,Rd}$, W_{xplRd} and W_{yplRd}, respectively, the plastic strength according to

$$N_{zpl,Rd} = A \cdot \sigma_f/\gamma_M \tag{13.27}$$

and

$$M_{pl,Rd} = W_{pl} \cdot \sigma_f/\gamma_M = 2 \cdot S \cdot \sigma_f/\gamma_M \quad . \tag{13.28}$$

Here, σ_f is the *yield strength* of the material, A the net cross-section, W_{pl} the *plastic section modulus* and S the static modulus. In case of tubes the static modulus is

$$S = (1/12)\left(d_e^3 - d_i^3\right) \quad , \tag{13.29}$$

where d_e is the external and d_i the internal tube diameter.

In the case of compression and bending, the calculation must demonstrate them

$$\frac{N_{zd}}{\chi \cdot A \cdot \sigma_f/\gamma_M} + \frac{k_x \cdot M_{x,d}}{W_{xpl,Rd} \cdot \sigma_f/\gamma_M} + \frac{k_y \cdot M_{y,d}}{W_{ypl,Rd} \cdot \sigma_f/\gamma_M} \leq 1,0 \quad . \tag{13.30}$$

Table 13.5: Characteristics of aluminum and steel tubes used for overhead contact lines.

Diameter, thickness	Cross-section A	Moment of inertia I	Radius of inertia i	Section modulus, elastic W_{el}	Section modulus, plastic W_{pl}
mm	mm^2	10^4 mm^4	mm	10^3 mm^3	10^3 mm^3
$26 \times 3,5$	247,40	1,603	8,13	1,233	1,786
$32 \times 3,5$	313,37	3,230	10,15	2,019	3,433
42×4	477,52	8,715	13,51	4,150	5,798
55×4	640,88	20,960	18,08	7,624	10,425
55×6	923,63	28,136	17,45	10,231	14,778
70×5	1 020,50	54,210	23,05	15,490	21,167
70×6	1 206,37	62,309	22,73	17,803	24,648
80×6	1 394,87	96,106	26,25	24,027	32,928

Here, χ is determined from

$$\chi = 1 \Big/ \left[\phi_k + \left(\phi_k^2 - \overline{\lambda}^{-2} \right)^{0,5} \right]$$

(13.31)

and $k_{x,y}$ from

$$k_{x,y} = 1 - \mu \cdot N_{sd} / (\chi \cdot A \cdot \sigma_f) \quad .$$

(13.32)

If (13.32) yielded a higher value, $k_{max} = 1,50$ would apply. The non-dimensional slenderness $\overline{\lambda}$ applying to tubes is given by:

$$\overline{\lambda} = \lambda \Big/ \left(\pi \cdot \sqrt{E / \sigma_f} \right) \quad ,$$

(13.33)

where E is the modulus of elasticity, λ the slenderness ratio and σ_f the material yield strength. The parameter ϕ_k follows from

$$\phi_k = 0,5 \left[1 + \alpha \left(\overline{\lambda} - 0,2 \right) + \overline{\lambda}^2 \right] \quad .$$

(13.34)

The *imperfection factor* α may be taken as 0,21 in case of tubes. The value μ in (13.32) is obtained from

$$\mu = \overline{\lambda} \left(2 \beta_P - 4 \right) + (W_{pl} / W_{el} - 1) \le 0,9 \quad .$$

Since $\beta_P = 1,3$ applies to tubes in cantilevers, there is

$$\mu = -1,4 \overline{\lambda} + (W_{pl} / W_{el} - 1) \quad .$$

(13.35)

The data for W_{pl} and W_{el} for tubes may be taken from Table 13.5. The example in Clause 13.8 demonstrates the verification process.

To assure *serviceability* for use, the deflection of components in cantilevers should be limited to $1/100$ of the member length. This limitation may govern the rating in cases of cantilevers without diagonal struts. In case of a proof based on the first order theory, the component of the force F_D in the registration arm perpendicular to the cantilever tube according to (13.16) causes the deflection. Using nomenclature according to Figure 13.12 and the condition that l_{5-4} is less than l_{4-3} the *maximum deflection* follows from [13.4] to

$$f_m = F_D \cdot (l_{5-4} / l_{5-3}) \cdot \left(l_{5-3}^2 - l_{4-5}^2 \right)^{1,5} \Big/ \left(9 \sqrt{3} \cdot EI \right) \quad .$$

(13.36)

In the case of cantilevers with a complicated design, it is recommended that deformation be calculated by means of commercially available computer software.

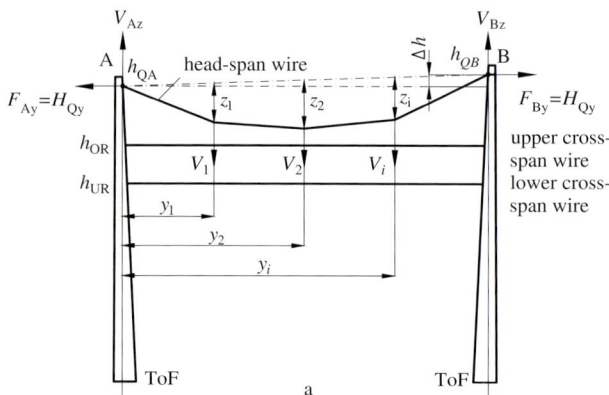

Figure 13.14: Forces and sags within a head-span.

13.4.3 Flexible cross-supporting structures

13.4.3.1 Introduction

Flexible cross-supporting structures, called *head spans*, carry the vertical loads, caused by the self weight forces of contact lines and their supports, by tensile forces in the head span wires. The headspan wire geometry is determined by the position and vertical load imposed from each supported contact line. The head span rating designs the head span wires and determines the loads acting on the supporting poles.

13.4.3.2 Loading, internal forces and sag of head-span wires

The vertical forces caused by the contact lines are obtained from the spans l_1 and l_2 on both sides of the head span

$$G_{\mathrm{OCL}i} = m'_{\mathrm{OCL}} \cdot g\,(l_1 + l_2)\,/2 \quad . \tag{13.37}$$

The dead weight of each individual support can be assumed as equivalent to a vertical load of 220 to 250 N. Therefore, the total vertical load is

$$V_{\mathrm{OCL}i} = G_{\mathrm{OCL}i} + 250 \quad . \tag{13.38}$$

In addition to the support loads, the dead load of head-span and cross-span wires, of section insulators and a construction load of conventionally 1 kN acting at the most adverse position are considered. These loads should be distributed to the individual supports to simplify the analysis.

Head-span wires are strung with a sag $y_{\max}$ of 10 % to 15 % of the head-span length a (Figure 13.14). Thereby, the effect of temperature-dependent variation of length on the *sag of head-span wires* may be neglected. The calculation of forces and sags of head-span wires may be determined graphically or by means of equivalent moments. The graphical method to determine forces and sags is considered in detail in [13.5].

In Figure 13.14, a head span is shown supporting three contact lines. The analytical calculation of head-span wires is explained below. A and B, known as top reaction forces, have the same value:

$$F_{\mathrm{Ay}} = F_{\mathrm{By}} = H_{\mathrm{Qy}} \quad . \tag{13.39}$$

Since the *sag* z_{max} is given, the horizontal component H_{Qy} of total force in the head-span wire is obtained from the maximum equivalent moment M_{max}:

$$H_{Qy} = M_{max}/z_{max} \quad . \tag{13.40}$$

The vertical head-span wire components at the poles are obtained from

$$V_{Az} + V_{Bz} = \sum_{i=1}^{n} V_i \quad , \tag{13.41}$$

where n is the number of contact line supports. The equilibrium around A gives

$$V_{Bz} \cdot a = \sum_{i=1}^{n} V_i \cdot y_i + H_{Qy} \cdot \Delta h \quad ,$$

the *vertical component of support force* at pole B is obtained from

$$V_{Bz} = \left(\sum_{i=1}^{n} V_i \cdot y_i + H_{Qy} \cdot \Delta h \right) \Big/ a \quad . \tag{13.42}$$

From (13.41) and (13.42) the vertical force at pole A can be derived:

$$V_{Az} = \sum_{i=1}^{n} V_{OCLi} - \left(\sum_{i=1}^{n} V_{OCLi} \cdot y_i + H_{Qy} \cdot \Delta h \right) \Big/ a \quad . \tag{13.43}$$

As can be seen from Figure 13.14, the equilibrium at the contact line support k yields to

$$V_{Az} \cdot y_k = H_{ay} \cdot z_k + \sum_{i=1}^{k-1} (y_k - y_i) \cdot V_i \quad . \tag{13.44}$$

At the position of the maximum sag the value z_k is equal to z_{max}. Then, from (13.44) the *horizontal component of the head-span wire* force can be obtained as

$$H_{Qy} = \left(V_{Az} \cdot y_k - \sum_{i=1}^{k-1} V_i(y_k - y_i) \right) \Big/ z_{max} \quad . \tag{13.45}$$

The value z_{max} follows from the assumed sag and the selected lengths of poles.

13.4.3.3 Determination of head-span pole lengths

The *contact wire height* h_{CW} above ToR (see Figure 13.15) is the reference for determining *pole lengths*:

- h_{LSW}, the lower cross-span support wire height, is installed 0,40 m to 0,50 m above the contact line
- h_{SH}, the system height, of the contact line is between 1,4 m and 2,0 m
- h_{SP}, the structural support height, is likely to be between 0,8 m and 1,2 m depending on the desirable length of the shortest head-span dropper or insulator depending on whether the upper cross-span wire is live or earthed

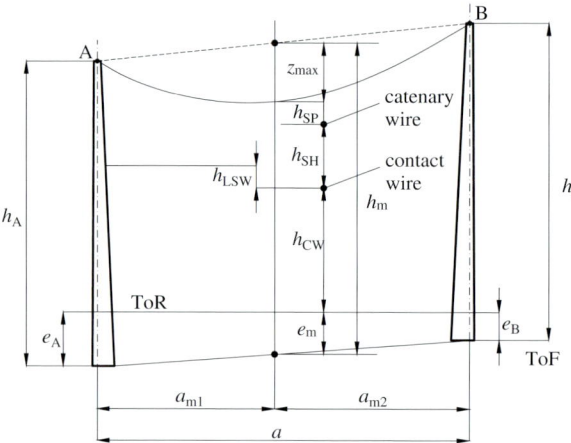

Figure 13.15: Determination of pole lengths within a head-span.

h_{CW} contact wire height above ToR
ToR top of rail
h_{SH} system height
h_{SP} structural support height
h_{LSW} height of lower cross-span
      wire above contact wire
ToF top of foundation

With the dimensions shown in Figure 13.15, the median height h_m of the straight line, which connects the fixing points A and B of the cross-span wires at both poles measured at the position, where the maximum cross-span wire moment occurs, should be at least:

$$h_m = h_{CW} + h_{SH} + h_{SP} + e_m + z_{max} \tag{13.46}$$

The sag of the head-span wire z_{max} forming the last term in (13.46) is chosen between 10 and 15 % of the span. The dimension e_m results from the difference in height of the top of both foundations and the top of rail of the reference rail being

$$e_m = e_A - (e_A - e_B) \cdot a_{m1}/a \quad . \tag{13.47}$$

The relation between the heights of head-span wire fixing and the value h_m according to (13.46) is then (Figure 13.15)

$$h_m = h_A - (h_A - h_B)a_{m1}/a \quad . \tag{13.48}$$

From a stock list of available pole lengths, h_A and h_B are selected and the sag y_{max} can be determined from Equation (13.46):

$$z_{max} = h_m - (h_{CW} + h_{SH} + h_{SP} + e_m) \quad . \tag{13.49}$$

The value z_{max} obtained from (13.49) is used to calculate H_{Qy} from (13.45).

13.4.3.4 Loading and internal forces of cross-span wires

The cross-span wires are loaded by the pre-stressing force F_{pre}, the radial forces of the contact line and the wind loads. The radial forces act in geometric directions set by the support positions, while the wind loads change their direction. The *maximum force F_{ucs} in the upper cross-span* wire follows from

$$F_{ucs\,max} = F_{pre} + \sum_{k=1}^{n} Q_{CAHk} + \sum_{k=1}^{n} Q_{CAWk} \tag{13.50}$$

and the minimum from

$$F_{\text{ucs min}} = F_{\text{pre}} - \sum_{k=1}^{n} Q_{\text{CAH}k} - \sum_{k=1}^{n} Q_{\text{CAW}k} \quad . \tag{13.51}$$

The pre-stressing force F_{pre} should be selected such that $F_{\text{ucs min}}$ is positive.
The forces $Q_{\text{CAH}k}$ and $Q_{\text{CAW}k}$ result from (13.2) and (13.3) respectively for the catenary wires. Analogously, it applies to the force F_{lcs} in the lower cross-span wire

$$F_{\text{lcs max}} = F_{\text{pre}} + \sum_{k=1}^{n} Q_{\text{CWH}k} + \sum_{k=1}^{n} Q_{\text{CWW}k} \tag{13.52}$$

and

$$F_{\text{lcs min}} = F_{\text{pre}} - \sum_{k=1}^{n} Q_{\text{CWH}k} - \sum_{k=1}^{n} Q_{\text{CWW}k} \quad . \tag{13.53}$$

The forces $Q_{\text{CWH}k}$ and $Q_{\text{CWW}k}$ result from (13.2) and (13.3), respectively. *Cross-span wire springs* are installed at the pole that experiences the lower load resulting from the radial forces. The prestressing forces F_{pre} need to be calculated for both poles. When determining the cross-span wire forces from the loads of the individual contact lines, due consideration should be given to a situation where the radial forces and wind forces balance each other. E. g., the wind loads need not be considered for pole B if the radial forces acting on pole A are greater than the wind forces.

13.4.3.5 Rating of head-span wires and cross-span wires

The *head-span* and *cross-span wires* are rated for the maximum forces determined:

$$H_{\text{ay}} \gamma_{\text{F}} \leq n \cdot A \cdot \sigma_{\text{u}} / \gamma_{\text{M}} \quad , \tag{13.54}$$

where γ_{F} and γ_{M} are partial factors, n is the number of parallel head-span wires and A their cross-section. Generally two or four copper alloy CuMg0,5 (BZII) wires with cross sections between 50 and 120 mm^2 and $\sigma_{\text{u}} = 580\,\text{N/mm}^2$ are required. Following EN 50 119, γ_{F} can be taken as 1,3 and γ_{M} as 1,5.
The poles are loaded by the head spans with the forces H_{ax}, F_{ucs} and F_{lcs}. Loads from traction feeder lines, from termination of contact lines and from cantilevers fixed directly to the poles must be considered. The rating of the poles is considered in Clause 13.5 and this methodology can also be applied for rating cross-span poles. The foundations can be designed using the methods described in Clause 13.7.

13.4.4 Horizontal registration arrangements

Horizontal registration arrangements and *cross-spans* also called *side-bridle arrangements*, can be found in central urban areas of mass transit installations. They allow the support of trolley wire systems from buildings or poles relatively far from the tracks. They also accommodate the arrangement of contact lines above crossing and branching tracks at large squares. The poles can be installed at locations where they do not interfere with the road traffic.
Figure 13.16 shows a cross-span for a double-track line. The loads V_1 and V_2 in accordance with (13.1) follow from the contact wires, the radial forces Q_{CWH1} and Q_{CWH2} in accordance

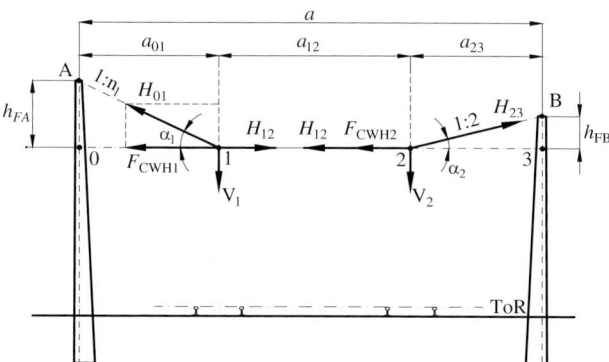

Figure 13.16: Cross-span arrangement of a trolley contact line.

with Equation (13.2). Since the radial forces act in the direction of pole A the wire 0–1 is designed to carry the load V_1. The gradient of the wire can be chosen between 1 : 10 and 1 : 15. The following applies to the tensile force of the wire 0–1:

$$H_{01} = V_1/\sin\alpha_{m1} \sim V_1/\tan\alpha_{m1} = V_1 \cdot a_{01}/h_{FA} \quad . \tag{13.55}$$

The transverse wire in between the supports carries the load

$$H_{12} = Q_{CWH1} + H_{01}\cos\alpha_{m1} \approx Q_{CWH1} + V_1 a_{01}/h_{FA} \quad . \tag{13.56}$$

The wire 2–3 acting at support 2 carries the resultant force from H_{12}, Q_{CWH_2} and V_2 and has to be arranged in the direction of their action. Since the gradient of the tensile wires is low, it applies

$$H_{23} \sim H_{12} + Q_{CWH2} = Q_{CWH1} + Q_{CWH2} + V_1 a_{01}/h_{FA} \quad . \tag{13.57}$$

The gradient of the wire 2–3 follows from

$$\tan\alpha_{m2} = V_2/\left(Q_{CWH1} + Q_{CWH2} + V_1 a_{01}/h_{FA}\right) \quad . \tag{13.58}$$

From the distance a_{23} to the pole the difference between the height of support 2 and the attachment at the pole follows:

$$h_{FB} = a_{23}\tan\alpha_{m2} \quad . \tag{13.59}$$

The *rating of the support wires* and ropes can be determined using equation (13.54). The poles A and B will be rated for the forces H_{01} and H_{23} respectively, which act at a height corresponding to the sum of contact wire height, the design height of the supports and the values h_{FA} and h_{FB}.

For horizontal registration of contact wires, arrangements in accordance with Figure 13.17 are used. Where the distances between the individual supports do not exceed 20 m the stagger is distributed over several spans. Therefore, only vertical loads can occur at some supports, the wind load excepted. Tensile forces and design heights can be approximated from the following:

$$H_{53} \approx H_{56} \approx H_{64} = V_5 \cdot n_{FK5-3} \quad , \tag{13.60}$$

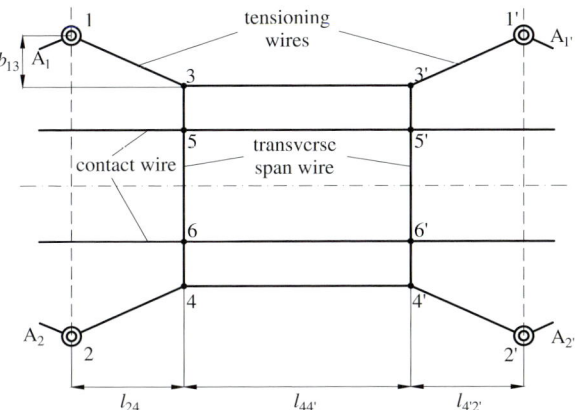

taking advantage of the symmetries, n_1 should again be selected between 10 and 15. The height of the span wire at point 3 will be

$$h_{\mathrm{SD3}} = h_{\mathrm{FH}} + h_{\mathrm{SH}} + l_{\mathrm{m53}}/n_{\mathrm{FK5-3}} \quad . \tag{13.61}$$

The tensile force between points 1 and 3 is

$$H_{13} = H_{53} \cdot \sqrt{l_{\mathrm{m13}}^2 + b_{m13}^2} \;/\; b_{m13} \tag{13.62}$$

and between points 3 and $3'$ the force

$$H_{33'} = H_{13} \cdot l_{\mathrm{m13}}/\sqrt{l_{\mathrm{m13}}^2 + b_{\mathrm{m13}}^2} = H_{53} \cdot l_{\mathrm{m13}}/b_{\mathrm{m13}} \tag{13.63}$$

acts. The *height* h_{SD1} *of fixing* at a pole is finally

$$h_{\mathrm{SD1}} = h_{\mathrm{SD3}} + V_5 \cdot \sqrt{l_{\mathrm{m13}}^2 + b_{\mathrm{m13}}^2} \;/\; H_{13} \quad . \tag{13.64}$$

The supports will be rated with forces H_{13} and their heights of application. It is likely that several tensioning forces act at the same pole and can be combined geometrically. Details on calculating horizontal registrations can be found in [13.6] and [13.7].

13.5 Rating of poles

13.5.1 Introduction

The poles used for contact line installations can be classified as single pole types including *lattice steel designs*. The loading of the poles is characterised by the applied bending moments which must be adequately restrained. The *rating of poles* includes the determination of lengths, internal forces or moments and the selection or design of appropriate pole types.

13.5.2 Determination of pole length

Pole length determination includes the allowances for support of overhead contact lines, the arrangement of traction power lines and the top of foundation level in relation to top of rail. On interstation lines the length of poles is determined by:

- distance e between top of rail (ToR) and top of foundation (ToF)
- contact wire height h_{CW}
- system height h_{SH}
- distance between upper swivel bracket and suspension or termination of traction power lines h_{EH}
- required space for insulation of traction power lines
- additional length h_{AL} at the top of pole above the traction power line or the fixing of cantilever. Generally, adopted values of h_{AL} are 0,10 to 0,15 m in case of steel poles and 0,20 to 0,30 m in case of concrete poles
- insertion depth E for insertion poles

Example 13.1: For $e = 0{,}70$ m, $h_{CW} = 5{,}50$ m; $h_{SH} = 1{,}80$ m and $h_{AL} = 0{,}10$ m a pole length of 8,10 m would result. Since pole lengths are generally standardized in steps of 0,25 m or 0,50 m, a length of 8,00 m is selected. Adjustment can be achieved by selecting the value $e = 0{,}60$ m for the construction of the top of foundation.

13.5.3 Loadings and internal forces and moments

Forces from various sources act on contact line supports.
Loads acting through the cantilevers include:

- loads from contact lines, including dead loads and wind loads and
- dead loads of the cantilevers themselves (see Clause 13.4.2.1)

Loads acting at terminations include:

- tensile forces resulting from the terminated contact lines and
- loads due to the dead weights of the tensioning equipment

At midpoints the loads include:

- forces caused by the midpoint anchors and
- forces caused by the termination of the midpoint

Traction power line loads include:

- wind loads
- radial loads and
- loads from intermediate or dead end terminations

At cross-span supports there are:

- loads from head-span and cross-span wires

Loads from disconnectors, transformers and lighting equipment must be considered when applicable. Dead loads and wind on structures act on all types of poles.

Dead weights and tensile forces act permanently; wind and ice loads are variable. In accordance with the rules stipulated by EN 50 119 and European standards, the *design loads* E_d of the actions follow from *permanent actions* G_K and *variable actions* Q_K

$$E_d = f\left(G_K \cdot \gamma_G;\ Q_{WK} \cdot \gamma_W;\ Q_{IK} \cdot \gamma_I;\ Q_{CK} \cdot \gamma_C\right) \quad , \tag{13.65}$$

where $\gamma_G, \gamma_W, \gamma_I$ and γ_C are partial factors as to Table 13.4 or selected specifically.

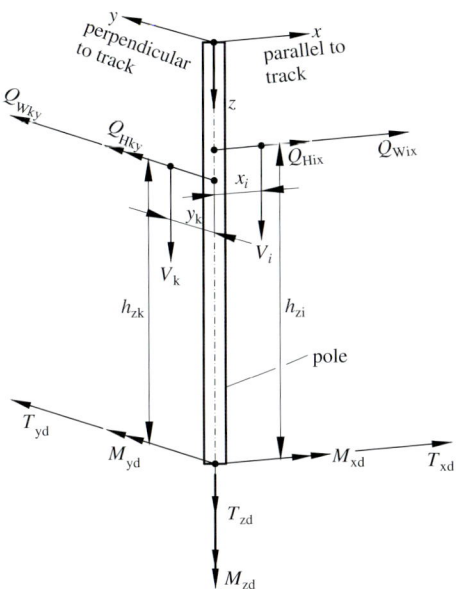

Figure 13.18: Loads and structural capacity at an overhead contact line pole.

The bending moments and transverse forces at the top of the foundation can be calculated in accordance with Figure 13.18. Loads perpendicular to the track result in the moment M_x and the transverse forces T_y, while loads parallel to the track produce the moment M_y and the forces T_x. The force T_z represents the sum of the vertical loads. The moment M_z which represents a torque around the pole axis, may result from asymmetrical action of loads in relation to the pole axis. At the top of the foundation the vertical forces are

$$T_{zd} = \sum_{k=1}^{n_1} \gamma_G V_k + \sum_{i=1}^{m_1} \gamma_G V_i \quad , \tag{13.66}$$

where γ_G represents the individual partial factor and V_i, V_k the vertical loads, which result from Equation (13.1) for conductors. The loads perpendicular to the track result from

$$T_{yd} = \sum_{k=1}^{n_1} (\gamma_G Q_{Hiy} \pm \gamma_W Q_{Wiy}) \quad . \tag{13.67}$$

The forces Q_{Hky} result from the contact lines and traction power lines in accordance with Equations (13.2), (4.15) to (4.20). The forces Q_{Wky} are wind loads from Equation (13.3) for contact lines. They may act in any direction.
In most cases, the forces parallel to the track

$$T_{xd} = \sum_{i=1}^{m_1} (\gamma_G Q_{Hix} \pm \gamma_W Q_{Wix}) \tag{13.68}$$

are derived from terminations of contact and traction power lines. Wind action needs to be considered only if there are no cantilevers attached to the pole.
The design bending moments M_{yd} and M_{xd}, caused by the transverse forces and vertical forces acting at distances x_i and y_k from the poles central axis are

$$M_{yd} = \sum_{i=1}^{m_1} (\gamma_G Q_{Hxi} \pm \gamma_W Q_{Wxi}) \cdot h_{zi} + \sum_{i=1}^{m_1} \gamma_G \cdot V_i \cdot x_i \tag{13.69}$$

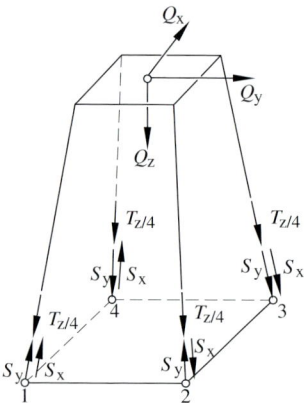
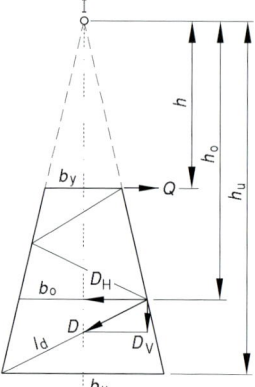

Figure 13.19: Leg member forces of a lattice pole.

Figure 13.20: Bracing forces of a lattice steel pole.

and

$$M_{xd} = \sum_{k=1}^{n_1} \left(\gamma_G Q_{Hyk} \pm \gamma_w Q_{Wyk} \right) \cdot h_{zk} + \sum_{k=1}^{n_1} \gamma_G \cdot V_k \cdot y_k \quad . \tag{13.70}$$

Loads from terminating contact lines and other loads may act eccentrically to the pole central axis and create torsional moments M_z around the central axis. This applies to poles equipped with twin cantilevers if the loads from the individual contact lines differ. If y_k and x_i effect an eccentric load action, it applies

$$M_{zd} = \sum_{k=1}^{n_1} \left(\gamma_G \cdot Q_{Hxk} \pm \gamma_w Q_{Wxk} \right) \cdot y_k + \sum_{i=1}^{m_1} \left(\gamma_G Q_{Hiy} \pm \gamma_w Q_{Wiy} \right) \cdot x_i \quad . \tag{13.71}$$

The internal forces and moments according to Equations (13.66) to (13.71) are used for *rating of* poles and *foundations* or for selecting them from available tables of resistances. In the case of slender structures, e. g. H-beams, the deflections should be limited at contact wire height, to ensure railway operations are not impaired (see Clause 13.5.4.6).

13.5.4   Rating of structural components

13.5.4.1   Introduction

The *rating of components* is based on the standards for design of steel structures and steel-reinforced concrete structures. The rating includes the selection of materials, e. g. the steel grades. Since the pole dead weights and wind loads on poles affect the internal forces and moments, the determination of required dimensions is an iterative process.

13.5.4.2   Lattice steel poles

Lattice steel poles are three-dimensional trusses. The members are formed predominantly by angle sections loaded by tensile and compression forces. Design analysis of contact line poles with rectangular cross-section consider the pole faces as two-dimensioned truss-structures. Figure 13.19 shows the *leg member forces* are

$$\pm S_d = \pm S_{xd} \pm S_{yd} \pm T_{zd}/4 = \pm M_{xd}/(2\,l_y) + M_{yd}/(2\,l_x) \pm T_{zd}/4 \quad , \tag{13.72}$$

where l_x and l_y denote the distances between centroidal axis of the leg members. Compression forces receive a negative sign and tensile forces a positive.

The *bracing forces* of faces y and x can be calculated (Figure 13.20 based on [13.1]) from

$$D_{yd} = \left[\sum_{k=1}^{m_1} \left(\gamma_G Q_{Hyk} \pm \gamma_W Q_{Wyk} \right) b_{yk} + \Delta y \sum_{k=1}^{n_1} \gamma_G V_k \cdot y_k \right] \frac{l_d}{m_D \cdot b_{yo} \cdot b_{yu}} \tag{13.73}$$

and

$$D_{xd} = \left[\sum_{i=1}^{m_1} \left(\gamma_G Q_{Hxi} \pm \gamma_W Q_{Wxi} \right) b_{xi} + \Delta x \sum_{i=1}^{n_1} \gamma_G V_i \cdot x_i \right] \frac{l_d}{m_D \cdot b_{xo} \cdot b_{xu}} \quad . \tag{13.74}$$

In Equations (13.73) and (13.74) b_{yk} and b_{xi} denote the pole width at the load application points and l_d is the system length of the bracing. Data b_{xo}, b_{xu} as well as b_{yo}, b_{yu} are the widths of the pole above and below the bracing being analysed. Δy and Δx denote the increase of width in pole faces y and x, respectively. The factor m has the value 2 for single warren and 4 for double warren truss.

Each structural member must be analysed to ensure that cross-sections are able to withstand applied tensile and compression forces and that joints have adequate capacity.

European standardisation requires adoption of EN 1993-1-1. For compression loaded members force N_{dd} it should be verified that

$$N_{dpl,Rd} = \chi \cdot A_{eff} \cdot \sigma_f / \gamma_M \quad , \tag{13.75}$$

where χ_{min} follows from Equation (13.31), A_{eff} is the cross section, σ_f the yield stress and γ_M the partial factor for material being 1,1 in accordance with EN 50 119.

For members loaded by tensile forces it should be verified that

$$N_{tpl,Rd} = A_{net} \sigma_u / \gamma_M \quad , \tag{13.76}$$

where A_{net} is the net cross-section of the member, σ_u the ultimate tensile strength and γ_M the partial factor, being 1,25 in this case. If both angle legs are connected A_{net} is

$$A_{net} = 0,9 \cdot (A - n_2 \cdot d_1 \cdot t_1) \quad , \tag{13.77}$$

where n is the number of holes in the relevant cross-section, d the hole diameter and t the thickness. If only one leg of the angle section is connected by one bolt then

$$A_{net} = (b_1 - d_1) \cdot t_1 \quad , \tag{13.78}$$

where b_1 is the width of the connected angle leg. Where there are two or more bolts, the net cross-section A_{net} is found from

$$A_{net} = (b_1 - d_1 + b_2/2) \cdot t_1 \quad , \tag{13.79}$$

where b_2 is the width of the leg without holes. The strength of a bolted connection having n_1 shearing cross-sections A_S follows from

$$N_{\mathrm{spl,Rd}} = n_{\mathrm{s1}} \cdot 0,6\,\sigma_{\mathrm{u}} \cdot A_{\mathrm{s}}/\gamma_{\mathrm{M}} \quad . \tag{13.80}$$

The verification of bearing capacity for joints with n_2 bolts is carried out by

$$N_{\mathrm{bpl,Rd}} = n_2 \cdot \alpha_{\mathrm{s}} \cdot \sigma_{\mathrm{f}} \cdot d_1 \cdot t_1/\gamma_{\mathrm{M}} \quad , \tag{13.81}$$

where α_{s} is the lowest of the following values

$$\alpha_{\mathrm{s}} = \left\{ \begin{array}{l} 1,20 \cdot (e_1/d_1) \\ 1,85 \cdot (e_1/d_1 - 0,5) \\ 0,96 \cdot (e_2/d_1 - 0,5) \\ 2,30 \cdot (e_3/d_1 - 0,5) \end{array} \right.$$

There, e_1 is the edge distance in direction of the force, e_2 the spacing of holes in direction of the force and e_3 the edge distance perpendicularly to the direction of force.

Example 13.2: The strength is to be determined for a leg member L100 · 10, S235, $A = 1920\,\mathrm{mm}^2$, buckling length 1,95 m, inertia radius $i_\xi = 1,95\,\mathrm{cm}$, connected by 4 bolts M20 5.6 in both legs following the EN 1993-1-1 approach. $\sigma_{\mathrm{f}} = 235\,\mathrm{N/mm}^2$; $E = 210000\,\mathrm{N/mm}^2$; bolthole diameter $d = 22\,\mathrm{mm}$. $\alpha = 0,49$ (buckling line c)

$$\lambda = s_k/i_\xi = 195/1,95 = 100$$

$$\overline{\lambda} = \lambda / \left(\pi \sqrt{E/\sigma_{\mathrm{f}}} \right) = 100/ \left(\pi \sqrt{210000/235} \right) = 1,065$$

$$\phi = 0,5 \left[1 + \alpha \left(\overline{\lambda} - 0,2 \right) + \overline{\lambda}^2 \right] = 0,5 \left[1 + 0,49\,(1,065 - 0,2) + 1,065^2 \right] = 1,279$$

$$\chi = 1/ \left(1,279 + \sqrt{1,279^2 - 1,065^2} \right) = 0,503$$

$$N_{\mathrm{dd}} = 0,503 \cdot 1920 \cdot 235/1,1 \cdot 10^{-3} = 206,4\,\mathrm{kN} \quad .$$

$$N_{\mathrm{td}} = 0,9\,(1920 - 2 \cdot 22 \cdot 10)\,355/1,25 \cdot 10^{-3} = 378,3\,\mathrm{kN} \quad .$$

For bolts 5.6 the ultimate tensile strength is $\sigma_{\mathrm{u}} = 500\,\mathrm{N/mm}^2$ and $A_{\mathrm{s}} = 20^2 \cdot \pi/4 = 314\,\mathrm{mm}^2$

$$N_{\mathrm{spl,Rd}} = 8 \cdot 0,6 \cdot 500 \cdot 314/1,25 \cdot 10^{-3} = 603\,\mathrm{kN} \quad .$$

For $e_1 = 40$; $e_2 = 60$; $e_3 = 40\,\mathrm{mm}$ it follows $\alpha_{\mathrm{s}} = 2,09$;

$$N_{\mathrm{bpl,Rd}} = 4 \cdot 2,09 \cdot 20 \cdot 10 \cdot 355/1,25 \cdot 10^{-3} = 475\,\mathrm{kN} \quad .$$

The compression loading limits the strength capacity of the leg member and, therefore, of the pole. The permissible bending moment is given by multiplying the compression force by twice the pole width (see Equation (13.72)). Assuming the width of the poles as $b_{\mathrm{x}} = (800 - 2 \cdot 32) = 736\,\mathrm{mm}$, the design moment M_{dx} is $2 \cdot 206,4 \cdot 0,736 = 304\,\mathrm{kN \cdot m}$.

In practice, railway companies use pre-designed *pole families* where the poles are characterised by their strength capacities in both directions. Figure 13.21 shows the characteristic moments for a family of lattice steel poles with base widths of $600 \times 800\,\mathrm{mm}$ and $800 \times 1000\,\mathrm{mm}$ with different angle sections and pole lengths. To verify a certain pole the interaction relationship applies with the partial factor $\gamma_{\mathrm{M}} = 1,1$:

$$M_{\mathrm{dx}}/(M_{\mathrm{kxd}}/\gamma_{\mathrm{M}}) + M_{\mathrm{dy}}/(M_{\mathrm{kyd}}/\gamma_{\mathrm{M}}) \le 1 \quad . \tag{13.82}$$

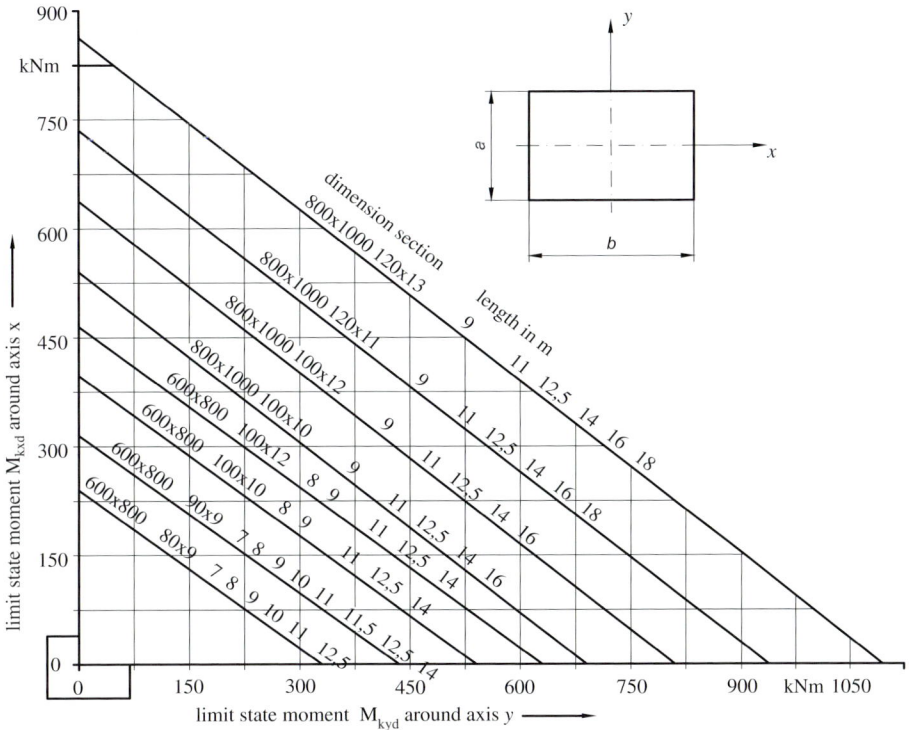

Figure 13.21: Characteristic limit state moments M_k for lattice steel poles.

13.5.4.3 Double channel poles

Double channel poles consist of two channels connected by stay plates with the channel spacing decreasing towards the top of the pole. The spacing of the stay plates is approximately 500 mm (Figure 13.7). These poles possess high strength in the direction of their transverse axis, while perpendicular to this direction, only the bending strength of the individual channels is effective. Consequently, they are used where the loads act predominantly in one direction, such as in the case of suspension poles on interstation lines.

The strength of such poles can be evaluated by treating them as Vierendeel girders, refer to [13.8]. DB use poles with profiles U100, U120, U140 and U160. Their characteristic ultimate strengths are shown in Figure 13.22. The following equation must be satisfied with $\gamma_M = 1,1$ according to the EN 50 119

$$M_{xd} \leq M_{kx}/\gamma_M \quad . \tag{13.83}$$

Double channel poles are connected to their foundations by M30 anchor bolts for U100 types and M36 for U120, U140 and U160 types.

13.5.4.4 H-beam poles

In comparison to lattice steel poles and double channel poles, *poles made of H-beams*, also known as universal columns (UC), require minimum fabrication resource. They are especially

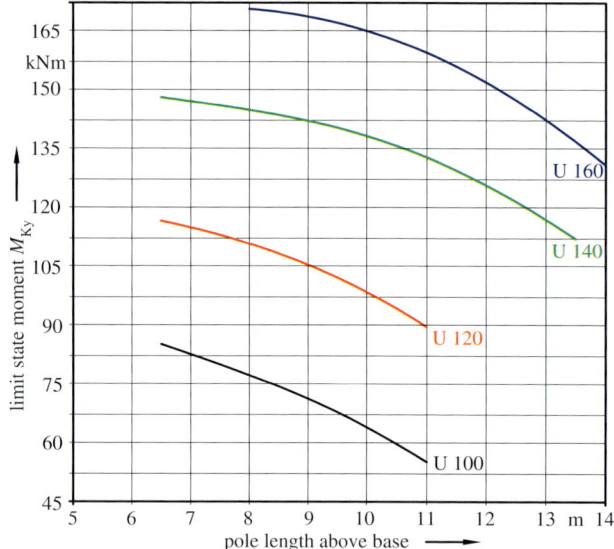

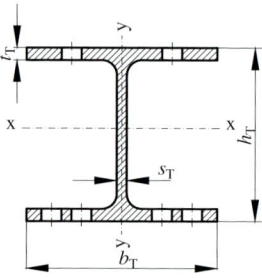

Figure 13.22: Characteristic limit state moments M_{Ky} for double channel poles.

Figure 13.23: H-beam cross-section designations.

suited for sites with limited space, e. g. in between tracks. However, their strength to weight ratio is low. They are weak in bending and suffer higher *deflections* at the same static loading capacity as lattice steel or double channel poles. H-beams have disproportionate strengths between their major and minor axes (Figure 13.23). To reinforce the weaker axis, two beams can be arranged in parallel and directly welded together or connected by batten plates. Single H-beams possess a low *torsional stiffness* around their longitudinal axis. When used with twin cantilevers, e. g. pole type 4 according to Figure 13.5, the torsion of the poles must be verified. The torsional rotation should not exceed 6 grad. The limited deflection at the height of the contact wire is significant on selection cross-section. H-poles are cheaper than any other form of structural supports.

Design standards require that

$$N_{zd}/N_{zdpl,Rd} + M_{xd}/\left(M_{xpl,Rd}/\gamma_M\right) + M_{yd}/\left(M_{ypl,Rd}/\gamma_M\right) \leq 1 \quad , \tag{13.84}$$

If σ_f is the yield stress,

$$N_{zdpl,Rd} = A_{net} \cdot \sigma_t/\gamma_M \quad , \tag{13.85}$$

$$M_{xpl,Rd} = 2\,S_x \cdot \sigma_f \quad \text{and} \tag{13.86}$$

$$M_{ypl,Rd} = 2\,S_y \cdot \sigma_f \tag{13.87}$$

are the characteristic strengths, where there is a complete plastic condition of cross-sections. For H-beams the first order moments S_x and S_y may be obtained from

$$S_x = b_T \cdot t_T \cdot (h_T/2 - t_T/2) + (h_T/s_T - t_T)^2 \cdot s_T/2 \tag{13.88}$$

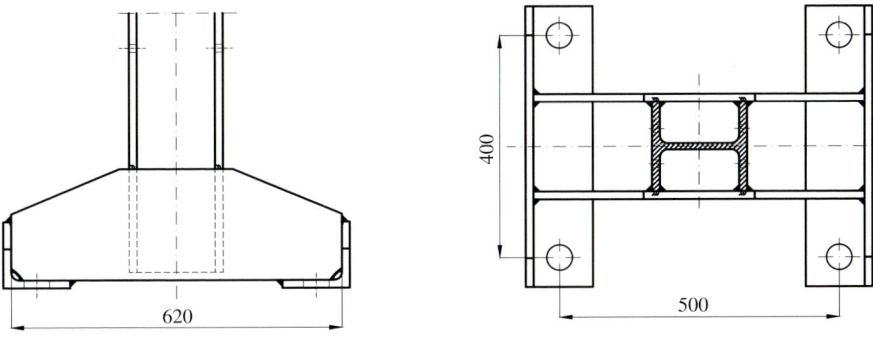

Figure 13.24: Typical design of base for poles made of H-beam sections.

and

$$S_y = t_T \cdot b_T{}^2/4 + (h_T - 2t_T) \cdot s_T{}^2/8 \quad . \tag{13.89}$$

The dimensions b_T, h_T, t_T and s_T are shown in Figure 13.23 and properties for [13.9] are published in the standards and industry section. Figure 13.24 shows a typical design of the base for poles made from H-beams.

Because H-poles are weak in torsion both bending and torsion need to be verified, when being loaded by a torsional moment. H-beams are weak in torsion because they are open sections. The torsional rotation measured in degrees can be obtained in accordance with [13.10] for a H-beam with a length h_M loaded by the moment M_{zd} from

$$\vartheta = h_M \cdot M_{zd}/(1,3 I_T \cdot G)(180/\pi) \quad , \tag{13.90}$$

where the torsional modulus I_T is

$$I_T = \left(2 \cdot b_T \cdot t_T{}^3 + (h_T - 2t_T) \cdot s_T{}^3 \right)/3 \quad . \tag{13.91}$$

The shear modulus G is $8 \cdot 10^4$ N/mm^2 in case of steel.

13.5.4.5   Steel reinforced concrete poles

The external loads of *steel reinforced concrete poles* result from Clause 13.5.3, and can be used to determine the required equivalent design load. This is the horizontally acting total force at the top of the pole without wind load on the pole in accordance with

$$Q_{td} = \left(\sqrt{M_{xd}^2 + M_{yd}^2} \right) / h_M \quad , \tag{13.92}$$

where M_{xd} and M_{yd} can be calculated from (13.69) and (13.70) respectively and h_M is the pole length. With the value Q_{td}, the required pole can be selected from manufacturer's catalogues or selection tables for pre-stressed or slack-reinforced poles. The cross sections can be designed in accordance with the relevant standards. The acting moments are determined from the *equivalent design loads* and the wind load or from individual design forces according to the second order theory. The standard EN 12 843 applies to concrete poles manufactured in a workshop. In accordance with EN 12 843, concrete of class C35/45 is the minimum that

Table 13.6: Concrete for steel-reinforced concrete poles.

Designation	Nominal strength β_{CWN} N/mm²	Series strength β_{CWS} N/mm²	Value for calculation β_{CB} N/mm²
C35/45	45	50	21
C45/55	55	60	26
C55/65	65	70	30
C75/95	95	100	44

Table 13.7: Partial factors related to the limit state of strength.

			Loading condition 2 normal load	Loading condition 3 transportation and construction loads
For strength analysis				
1	prestressing acting favourably	γ_{vf}	0,80	1,00
2	acting unfavourably	γ_{vf}	1,20	1,00
3	concrete	$\gamma_{CC}^{1)}$	1,50	1,30
4	spun concrete	$\gamma_{CC}^{1)}$	1,40	1,25
5	concrete and prestressing steel	γ_{CS}	1,25	1,10
For deflection analysis according to the second order theory				
6	concrete and spun concrete	$\gamma_{CC}^{2)}$	1,20	—
7	concrete and prestressing steel	γ_{CS}	1,15	—

[1] related to $0,7\,\beta_{CWN}$, [2] related to 0,85 times the maximum internal moments

may be used for concrete poles. For pre-stressed spun concrete poles high-strength concrete classes C55/65 and C75/95 are preferable.

To determine the most adverse stresses, the following loading conditions have to be assessed according to EN 12 843:

– loading condition 1: *permanent loads*
– loading condition 2: *normal loading* and
– loading condition 3: *loadings due to transport and construction*

The vertical loads and the tensile forces of the conductors act as *permanent loads*. In the case of *normal loading*, the wind load on conductors and on the pole are also taken into account. The internal design load stresses for loading conditions 2 and 3 are determined for the ultimate limit state of resistance using the theoretical values of the strength, which can be obtained by dividing the *nominal strength β_{CWN} of concrete* according to Table 13.6 by the partial factor γ_{CC} in accordance with Table 13.7 lines 3, 4 and 6. The design strength of reinforcement steel is obtained by dividing the yield strength β_{CS} by the partial factor γ_{CS} in accordance with able 13.7 lines 5 and 7.

A stress-strain curve in accordance with Figure 13.25 can be assumed for concrete and a curve to Figure 13.26 for slack and pre-stressed steel reinforcement. For pre-stressed poles, concrete tensile stresses are not permissible under the action of permanent loads or under action of 40 % of the moment caused by normal loads.

The internal load stresses in the limit state of resistance are determined using γ_F times the normal load, taking into account the pole deformation (second order theory). An unintentional

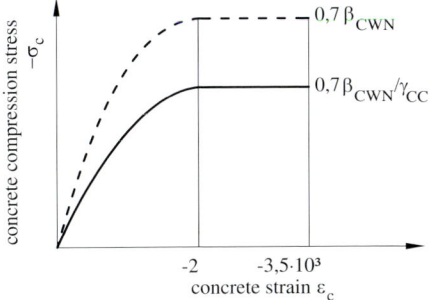

Figure 13.25: Stress-strain curve for concrete for calculation of the permissible internal stresses and the limit condition of resistance (parabola-rectangle-diagram).

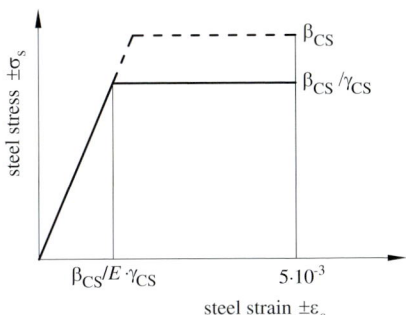

Figure 13.26: Stress-strain curve for concrete steel and pre-stressing steel for calculation of internal stresses and deformations at limit state of resistance.

tilting of the unloaded pole of 5 mm/m should be assumed. The tilting is assumed to include effects caused by curvature because of unequal heating.

The effects of pole deformation may be neglected if the additional moment due to deformation and tilting is less than 5 % in the cross-section at the top of foundation or is less than 10 % in case of the most adverse section, respectively. For this proof, estimation on the safe side suffices.

To eliminate damage during operations, the following items should be considered:

- adequate helical reinforcement to withstand peripheral tensile stresses at the surface. The helical reinforcement should consist of ribbed concrete steel and should be provided independently of the static demands. The reinforcement should be as follows:
    - 5 mm diameter steel and a maximum pitch of 60 mm
    - 4 mm diameter steel and a pitch of less than 40 mm
    - up to 3 mm diameter and a pitch of less than 30 mm
- the concrete should be at least 40 mm thick
- the spacing between unidirectional reinforcement rods needs to be only half of the rod diameter, with exception of overlapping sections. It should be at least as wide as the diameter of the maximum aggregate size
- the *concrete coverage* should be at least 15 mm above the helical reinforcement or 20 mm above the pre-stressed steel
- the concrete poles need ventilation to avoid internal moisture and condensation

The *water to cement ratio* is reduced to below 0,4 during the *spinning process*. Spun concrete poles are produced using casings that can be split longitudinally. The pre-formed helical reinforcement is arranged first in the casing and then the slack or pre-tensioned longitudinal reinforcement is fixed. In the case of pre-tensioning, a head is put on each rod. Sockets and other elements used to connect contact line components are arranged along the pole rein-

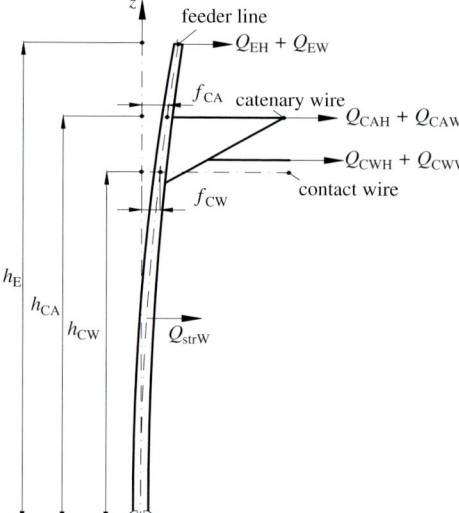

Figure 13.27: Calculation of deformations.

forcing. After preparing the reinforcement, the concrete is poured in and the casing is bolted together. The reinforcement rods are pre-tensioned with stresses up to $800 \, N/mm^2$.

The number of revolutions during spinning depends on the diameter of the pole and the casing. Centrifugal accelerations between 10 and 50 g are normal. After 12 to 15 minutes, the spinning process is complete and the casing with the spun pole is stored in a heating chamber where, for a short period, steam is passed along the outside of the casing. The temperature in the curing chamber should not exceed $50 \, °C$ and the poles should remain in the casings for 24 hours. Heat treatment at higher temperature was used in the past, however, this was one cause of longitudinal cracking. Poles manufactured to the above specifications attain 70 % of their nominal strength during storage in the casings. To avoid premature damage, reliable production and good engineering are prerequisites for long service life of poles.

13.5.4.6   Deflection

All supports deflect under load because they are made of elastic materials. Lattice steel poles with wide base dimensions are relatively rigid structures and only tall head-span poles will show visible *deflection*. Usually, these poles are raked opposite to the load direction, so that the pole stands vertically after application of the loads. Double channel poles are also relatively stiff about their major axis. Generally, verification of the deflection is not necessary with fabricated poles. At relatively weak poles such as H-beams the limitation of the deflection is likely to govern the design of the section properties. The deflection at a height h above the base of a pole, with a variable section modulus along the vertical axis loaded by a bending moment $M_{x,y}(z)$ will be:

$$f = \frac{1}{E} \int_0^h M_{x,y}(z) / I(z) \cdot z \, dx \quad .$$

(13.93)

As shown in Figures 13.18 and 13.27, the coordinate x is from the point where the deflection is to be determined. Since the integral in (13.93) cannot always be solved analytically, numerical methods, especially computer software, were developed to determine the deflection for poles in accordance with (13.93). Refer to [13.1].

For poles of H-beam sections, the moment of inertia I is constant along the beam and (13.93) can be solved analytically. In accordance with [13.4] the *deflection* at a distance a_o from the load position caused by the application of force F at height h above the foundation surface is:

$$f_a = \frac{Fh^3}{6EI_y}\left[2 - 3\frac{a_o}{h} + \left(\frac{a_o}{h}\right)^3\right] \quad , \tag{13.94}$$

and due to a moment M

$$f_a = \frac{Mh^2}{2EI_y}\left[1 - 2\frac{a_o}{h} + \left(\frac{a_o}{h}\right)^2\right] \tag{13.95}$$

and due to the uniformly distributed load q for a beam with a total length h

$$f_a = \frac{q' \cdot h^4}{24 \cdot E \cdot I_y}\left[3 - 4\cdot\frac{a_o}{h} + \left(\frac{a_o}{h}\right)^4\right] \quad . \tag{13.96}$$

For contact lines the following cases are of special interest:
- deflection under wind load only, at contact wire height h_{CW}. It is recommended that this deflection is limited to 25 mm
- deflection under permanent loads at the catenary wire height h_{CA}. It is recommended that this value be limited to 1 % of the catenary wire height
- deflection under maximum loads at the catenary wire height h_{CA}. It is recommended that this value be limited to 1,5 % of the catenary wire height.

Using the designation according to Figure 13.27, the following formulae can be used for the calculation of the deflection:
- At the height h_{CW} of the contact wire under wind load only:

$$
\begin{aligned}
f_{CWW} = \Big[& Q_{CAW}h_{CW}^2(h_{CA} - h_{CW}/3) + Q_{CWW}\cdot 2h_{CW}^3/3 \\
& + Q_{EW}h_{CW}^2(h_E - h_{CW}/3) \\
& + Q_{strw}h_{CW}^2\left(h_E/2 - h_{CW}/3 + h_{CW}^2/(12h_E)\right)\Big] \cdot 238/I_y \;\text{(mm)}.
\end{aligned}
\tag{13.97}
$$

- At the height h_{CA} of the catenary wire under permanent loads

$$
\begin{aligned}
f_{CAH} = \Big[& Q_{CAH}h_{CA}^3 2/3 + Q_{CWH}\cdot h_{CW}^2(h_{CA} - h_{CW}/3) + Q_{EH}h_{CA}^2(h_E - h_{CA}/3) \\
& + (V_{OCL}\cdot y_K + V_E\cdot y_E)h_{CA}^2\Big] \cdot 238/I_y \;\text{(mm)} \quad .
\end{aligned}
\tag{13.98}
$$

- At the height h_{CW} of the catenary wire under maximum loads

$$
\begin{aligned}
f_{CA(H+W)} = \Big\{ & (Q_{CAH} + Q_{CAW})\cdot h_{CA}^3\cdot 2/3 \\
& + (Q_{CWH} + Q_{CWW})\cdot h_{CW}^2(h_{CA} - h_{CW}/3) \\
& + (Q_{EH} + Q_{EW})h_{CA}^2(h_E - h_{CA}/3) \\
& + Q_{strw}h_{CA}^2\left[h/2 - h_{CA}/3 + h_{CA}^2/(12h)\right] \\
& + (V_{OCL}\cdot y_K + V_E\cdot y_E)h_{CA}^2\Big\} \cdot 238/I_y \;\text{mm} \quad .
\end{aligned}
\tag{13.99}
$$

In Equations (13.97) to (13.99) the forces Q_{CWH}, Q_{CAH} and Q_{EH} can be obtained from (13.2) to (13.8), Q_{CWW}, Q_{CAW} and Q_{EW} from Equation (13.3) as well as V_{OCL} and V_E in accordance with Equation (13.1). Q_{EH}, Q_{EW} and V_E are related to the feeder lines supported by the pole. Q_{PW} is the total wind force on the pole according to Equation (13.5). In these equations the modulus of elasticity used for steel was $2,1 \cdot 10^5$ N/mm^2. For I the unit is cm^4, as commonly used in standard tables. The forces have to be inserted in kN, the lengths in m. The results will have the unit mm. The example given in Clause 13.8 demonstrates the use of these formulae.

13.6 Subsoil

13.6.1 Introduction

Foundations of overhead contact line supports reliably transfer the various structural loads resulting from different loading scenarios, into the subsoil without unacceptable movement of the foundation bodies. Since the *subsoil conditions* at the support sites are crucial to the selection and the design of foundations, they should be known to the necessary extent before rating the foundations. *Subsoil investigations* provide this information by classifying the encountered soil according to the standards ISO 14 688 – Part 1 and Part 2 or to national standards and supplying the soil characteristics needed for selection and design of foundations.

Soil mechanical engineering classifies the subsoil forming the crust of the Earth, as *undisturbed soil* (loose rock), *rock* (solid rock) and *soil fill*. Loose rock is a natural heap of mineral particles. Without applying any force, it can be separated into the existing particle sizes. In the case of rock, the application of force is necessary for separation. This classification is characteristic of civil engineering and differs from terms currently used in geology.

13.6.2 Undisturbed soil

13.6.2.1 Classification

Undisturbed soil was formed by an ancient geological process on Earth, by chemical and physical weathering and decomposition of rock or it may have an organic origin. For the purposes of civil engineering, undisturbed soil is classified as inorganic or organic. Inorganic subsoils consist of two main types, non-cohesive, friable soils and cohesive soils. They are distinguished by particle sizes. Most of the subsoils encountered in the field are mixtures of different particle sizes and will be classified depending on the principal fraction in terms of mass.

13.6.2.2 Non-cohesive, granular

Non-cohesive subsoils are characterised by particle sizes above 0,063 mm. They are subdivided into boulders, cobbles, gravel and sand (Table 13.8).

13.6.2.3 Cohesive soils

Cohesive soils are characterised by particle sizes less than 0,063 mm which cannot be distinguished by the unassisted eye. Cohesive soils are also subdivided by their particle sizes (Table 13.8).

Table 13.8: Particle size fractions of non-cohesive and cohesive soils according to ISO 14 688-1.

Soil fraction	Subfraction	Symbol	Particle sizes in mm	
non-cohesive soil				
very coarse soil	Large boulder	LBo	above 630	
	Boulder	Bo	above 200	to 630
coarse soil	Cobbles	Co	63	to 200
	Gravel	Gr	2	to 63
	Coarse gravel	CGr	20	to 63
	Medium gravel	MGr	6,3	to 20
	Fine gravel	FGr	2,0	to 6,3
	Sand	Sa	0,063	to 2,0
	Coarse sand	CSa	0,63	to 2,0
	Medium sand	MSa	0,2	to 0,63
	Fine sand	FSa	0,063	to 0,2
cohesive soil				
fine soil	Silt	Si	0,002	to 0,063
	Coarse silt	CSi	0,02	to 0,063
	Medium silt	MSi	0,0063	to 0,02
	Fine silt	FSo	0,002	to 0,0063
	Clay (finest)	Cl	below 0,002	

13.6.2.4   Composite soils

Most soils are *composite soils* consisting of principal and secondary fractions. They are designated by a noun (main term) describing the principal fraction and by one or more adjectives (qualifying terms) describing the secondary fractions, e. g. sandy gravel saGr, gravely clay grCl. The principal fraction in terms of mass determines the engineering properties of the soil. Secondary and further fractions do not determine but will affect the engineering properties of the soil. The secondary fractions are placed as adjectives in front of the principal fraction in order of their relevance, as shown by the following examples:
  – sandy gravel (saGr)
  – medium sandy silt (msaSi)
  – fine gravelly coarse sand (fgrCSa)
  – silty fine sand (sifSa)
  – medium sandy clay (msaCl)
If secondary fractions are present in particularly small or large portions, the term "slightly" or "very" should precede the qualifying term (see ISO 14 688-1). A composite soil is considered as non-cohesive if it contains less than 15 % by mass of particle sizes less than 0,063 mm. The non-cohesive components determine the engineering properties of the composite soil. Otherwise the soil is classified as a cohesive soil with coarse-particle fractions.

13.6.2.5   Organic soils

Organic soils contain residues of decomposed plants and animal organisms. Besides purely organic soils, composite soils having characteristics close to clay and silt with substantial organic fraction are called *mud*. Since the compressibility of these soil types is high, they are not suitable as a subsoil to carry loads. Because of the determining effect of the organic content on the soil properties, non-cohesive and cohesive soils with more than 5 % in by mass of organic contents are considered as organic soils. For identification and description of organic soil see ISO 14 688-1.

13.6.3 Rock

The term *rock* includes all solid subsoils which form the hard and solid part of the Earth's crust. Their degree of weathering is important for loading capacity.

13.6.4 Soil fill

Soil fill is commonly encountered in artificially constructed railway embankments. For these engineered embankments, the soil material is selected carefully to achieve a compact and dense structure. The material extracted in tunnel boring is often used in embankments. The design of foundations to be installed in compacted railway embankments should be based on soil density, measured by probes.

13.6.5 Soil investigation

The supports of an overhead contact line extend over long line sections where varying soil conditions may be encountered. *Soil investigations* aim at providing information to enable a decision on the most suitable type of foundations and subsequent design. The soil investigation forms the basis for static analysis of foundations. For this reason, the extent of soil investigations and the depth of individual studies should be adapted to the needs of the design. The depth of soil investigation depends on the type of foundation and applied loads. If good bearing soil is found up to the ground surface, in most cases it might be sufficient to carry out soil tests to 1 m below the expected sub-base of the foundation.

Because soil investigations are rarely carried out at each individual support site, the foundation design is based on available documentation and on general assumptions. The adequacy of the assumptions is then verified during foundation installation. Soil investigations may not be waived when obviously low-bearing soils, such as mud, will be encountered. In such cases, the position of high-bearing soil layers must be investigated.

Line section inspections may serve to decide on the type of soil investigation. Preliminary information on soil may be obtained from the installation of the permanent way or any other structures that the infrastructure manager may have. Soil investigation methods should be adequate to the type of soil, to obtain soil samples and to determine the density of soil layers.

13.6.6 Methods of obtaining soil samples

13.6.6.1 Introduction

The methods of obtaining soil samples are standardized in EN ISO 22 475-1 and national standards. For contact lines, it is sufficient to gain disturbed samples, as an investigation of soil characteristics in laboratory tests is not required. The disturbed soil samples are used to study the sequence of soil layers, their boundaries, the type of layers, the distribution of particle sizes, the consistency, the ground water table and organic contents. Low quality level investigations are sufficient and they do not need to be performed at each site.

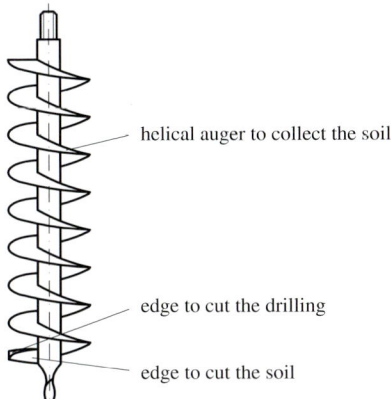

helical auger to collect the soil

edge to cut the drilling

edge to cut the soil

Figure 13.28: Helical auger.

13.6.6.2 Investigation by boring

Non-lined *investigation bores* of 300 to 500 mm in diameter will yield suitable results for the definition of soil types, water table, soil stiffness and density of the stratification. For investigation boring, low-duty boring rigs with auger diameters of 100 to 150 mm are also used. Helical augers as shown in Figure 13.28 are used to obtain soil samples. These may also be used to investigate less firm and water-containing layers. Depths down to 12 m can be reached easily and the profile of layers can be recognized where the soil adheres to the flighting of the auger.

13.6.6.3 Investigation by probes

Probe boring using a penetrometer to take disturbed samples of soil is suited for soil investigation of sites along a contact line. The *penetrometer* consists of a grooved probe rod, with a longitudinal groove, 1,0 m long at its end. After driving the probe, a soil sample is gathered in the longitudinal groove and recovered when hauled from the bore hole. Sands below the ground water table cannot be recovered since they are washed out. For such soil conditions, probes are equipped with a tube sampler for soil recovery. Investigations using probes deliver continuous soil profiles in soils that are not too firm.

13.6.7 Probing

13.6.7.1 Introduction

Probes are either driven or pressed into the soil and the soil resistance is recorded during probe penetration. Probing complements soil profiles by quantitative findings on the *stratification density*. For many overhead contact line installations, probing alone may be sufficient to obtain the necessary data for foundation design. Many types of probes are in use for soil investigations in general; those that are used for overhead contact line construction are discussed here after.

Table 13.9: Data for driven probes according to EN ISO 22 476-2.

Device	Weight of hammer in kg	Falling height in m
light probe (DPL)	10	0,50
medium probe (DPM)	30	0,50
heavy probe (DPH)	50	0,50

13.6.7.2 Driven probes in accordance with EN ISO 22 476-2

Dimensions of *driven probes* and instructions for their use are given in EN ISO 22 476-2 and national standard such as DIN 4094 for Germany. Under specified conditions, a rod with a cone-type end is driven into the soil, the number of blows necessary for a penetration of 100 mm (N_{10}) being counted. This number represents a reference for the density of stratification and consistency of layers explored. Table 13.9 reveals some characteristics of various driven probes. For contact lines foundations, the light probing apparatus (DPL) is used in most cases.

The results of probing depend on various parameters and conditions (see EN ISO 22 476-2). So it is not possible to draw conclusions on the type of soil from the indentation resistance of the probe. According to [13.11] the number of blows and the density of layers are correlated as shown in Table 13.10.

13.6.7.3 Standard Penetration Test

The *Standard Penetration Test* (SPT) according to EN ISO 22 476-3 was developed in the USA and is now used worldwide. Firstly, a lined borehole is drilled. Then a cylinder with 35 mm internal diameter installed at the end of a rod is inserted into the hole and driven to a depth of 150 mm by a hammer weighing 63,2 kg and falling from a height of 760 mm. The number of blows necessary to indent the probe by 300 mm is counted (N_{30}). The soil sample, pressed into the cylinder during indentation, can be recovered and studied. According to [13.12], a correlation between condition of *stratification of non-cohesive soils* as well as *consistency of cohesive soils* is proposed as shown in Table 13.11.

13.6.8 Evaluation of soil investigation

Information on the assessment and *classification of rock* and its weathering conditions can be found in [13.13]. The range between loose and solid rock is divided into six classes and distinguished by the degree of weathering of the rock and the particle compound. They are named as follows:

Table 13.10: Guide data for correlation between number of blows N_{10} for a penetration of 100 mm and stratification density according to EN ISO 22 476-2.

Type of soil, Stratification	Number of blows N_{10}	
	Light probe	Heavy probe
Sand,		
medium density	≥ 15	≥ 5
dense	≥ 30	≥ 10
Sand-gravel mixture		
medium density	≥ 15	≥ 5
dense	$-$ [1]	≥ 18

[1] light probe not suited

Table 13.11: SPT probing in non-cohesive and cohesive soil (Standard Penetration Test).

Standard Penetration Test in			
Non-cohesive soil		Cohesive soil	
Number of blows $N_{30}^{1)}$	Stratification	Number of blows $N_{30}^{1)}$	Consistence
0 to 4	very loose	0 to 2	very soft
4 to 10	loose	2 to 4	soft
10 to 30	medium dense	4 to 8	medium
30 to 50	dense	8 to 15	stiff
> 50	very dense	15 to 30	very stiff
		> 30	hard

[1)] number of blows for 300 mm penetration

Table 13.12: Assessment of the degree of aggressiveness of ground water and soils of predominantly natural composition.

	Degree of aggressiveness		
	slightly aggressive pH-value 6,5 to 5,5	heavily aggressive pH-value 5,5 to 4,5	very heavily aggressive pH-value below 4,5
Types of water: chalk soluble carbon acid (CO_2) in mg/l determined by the marble test	15 to 40	40 to 100	above 100
ammonium (NH_4^+) in mg/l	15 to 30	30 to 60	above 60
magnesium (Mg^{2+}) in mg/l	300 to 1 000	1 000 to 3 000	above 3 000
sulfat (SO_4^{2-}) in mg/l	200 to 600	600 to 3000	above 3 000
Soils: sulfat (SO_4^{2-}) in mg per kg air-dry soil	2 000 to 5 000	above 5 000	

- w_0 non-weathered rock
- w_1 minor weathering
- w_2 moderately weathered
- w_3 highly weathered
- w_4 completely weathered
- w_5 soil

Rock foundations, e. g. rock anchors, can be used in weathering conditions of w_0 to w_2.

Ground water and soils can attack concrete if they contain free acids, sulphides, sulphates, magnesium salts, ammonium salts or grease and oil. Ground water that could attack concrete can be recognized by a dark colouring, a muddy smell and by emerging gas bubbles. If it is suspected that concrete could be damaged by ground water, samples must be taken and investigated in the laboratory in accordance with DIN 4030, Clause 5. Table 13.12 summarizes information on *hazards of ground water to concrete*.

13.6.9 Soil characteristics

Soil investigations are evaluated according to ISO 14 688-1 and ISO 14 688-2 and recorded in *bore hole logs*. The *soil characteristics* specified in EN 50 119 can be used as a basis for the design of contact line foundations. These characteristics are presented in Table 13.13. For

Table 13.13: Geotechnical soil characteristics of some standard soils according to EN 50 119, Annex C.

1	2	3	4	5	6	7	8	9
Type of soil	γ_E kN/m³	γ_{Eb} kN/m³	ϕ_R degree	c kN/m²	c_u kN/m²	C_t MN/m³	$\sigma_{1,5}$ MN/m²	κ
Marl, compact	20±2	11±2	25±5	30±5	60±20	>200	0,64	8,0
Marl, altered	19±2	11±2	20±5	10±5	30±10	50±10	0,48	6,5
Gravel, graded	19±2	10±2	38±5	–	–	150±10	0,64	8,0
Sand, loose	18±2	10±2	30±5	–	–	60±10	0,32	5,5
Sand, semi-dense	19±2	11±2	33±5	–	–	80±10	0,48	6,5
Sand, dense	20±2	12±2	35±5	–	–	100±10	0,64	8,0
Sandy silt	18±2	10±2	25±5	10±5	30±10	60±10	0,32	5,5
Clayey silt	19±2	11±2	20±5	20±10	40±10	50±10	0,32	5,5
Loam, silt, malleable	17±2	7±2	20±5	–	20±10	35±5	0,48	6,5
Clay, soft	17±2	7±2	12±5	25±5	60±20	25±5	0,16	4,0
Clay, stiff	19±2	9±2	15±5	25±5	60±20	30±5	0,32	5,5
Clay, very stiff	20±2	10±2	20±5	25±5	60±20	40±5	0,48	6,5
Clay till	20±2	10±2	30±5	12±7	450±350	–	0,48	6,5
Clay, with organic addition	15±2	5±2	15±5	–	–	–	–	–
Peat, marsh	12±2	2±2	–	–	–	–	–	–
Backfill, embankment medium compaction	19±2	10±2	25±5	–	15±5	20±5	0,48	6,5

2 γ_E specific weight dry
3 γ_{Eb} specific weight under buoyancy
4 ϕ_R angle of internal friction
5 c cohesion (effective)
6 c_u undrained shear strength
7 C_t subgrade modulus at 2 m depth
8 $\sigma_{1,5}$ ultimate soil pressure at 1,5 m depth
9 κ factor to adapt the ultimate soil pressure to depth (see Equation (13.100))

individual types of soil, the *unit weight*, the *angle of internal friction* and the *characteristic soil pressure* are listed. The use of these values requires compaction of soil after placing fill material. The characteristic soil pressure depends on the unit weight. The ultimate pressure according to Table 13.14 applies to a depth of 1,5 m. The ultimate pressure σ_t at the depth t rises by the additional loading according to the increase in depth multiplied by the factor κ (Table 13.14).

$$\sigma_t = \sigma_{1,5} + (t - 1,5)\gamma_E \cdot \kappa \quad . \tag{13.100}$$

In the case of ground water, the unit weight reduced by buoyancy is considered according to Table 13.14, column 3. Alternatively, soil characteristics are given in Table 13.14 depending on typical soil conditions usually provided as a result of soil investigations.

13.6.10 Practical application

With respect to the most frequently adopted foundation types for contact line installations, namely concrete block and driven pile foundations, the soil investigations are aimed predominantly at

Table 13.14: Soil characteristic for design of foundations acc. to EN 50 341-3-4.

1	2	3	4	5	6	7
Type of soil	Unit weight force γ_E (Values for design)		Angle of internal friction ϕ_R degree	Ultimate soil pressure at a depth of 1,5 m kN/m^2	Coefficent κ	Earth frustum angle β_E degree
	naturally humid kN/m^3	with buoyancy kN/m^3			–	
Undisturbed soils						
Non-cohesive soils						
1 sand, loose	17	9	30	320	5,5	8
2 sand, semi loose	18	10	32,5	480	6,5	8
3 sand, dense	19	11	35	640	8,0	10
4 gravel, bolder, uniform	17	9	35	640	8,0	12
5 gravel-sand, uniform	18	10	35	640	8,0	12
6 bolder, stones, macadam, graded	18	10	35	640	8,0	12
Cohesive soils						
7 very soft	16	8	0	0	1,6	—
8 soft (easy to knead), purely cohesive	18	9	15	64	3,2	4
9 soft, with non-cohesive additions	19	10	17,5	64	4,0	4
10 firm (difficult to knead), purely cohesive	18	9	17,5	160	4,0	6
11 firm, with non-cohesive additions	19	10	22,5	160	4,8	6
12 stiff, purely cohesive	18	10	22,5	320	4,8	8
13 stiff, with non-cohesive additions	19	11	25	320	5,5	8
14 hard, purely cohesive	18	—	27,5	640	5,5	10
15 hard, with non-cohesive additions	19	—	30	640	6,5	10
Organic soils, and soils with organic additions	5 to 16	0 to 7	15		1,6	
Rock 16 with considerable fissuring or unfavourable stratification in sound not weathered condition,	20			1600		
17 with minor fissuring or favourable stratification	25			4800		
Made up ground and fill uncompacted embankment	12 to 16	6 to 10	10 to 25	48 to 160	3,2	4 to 10
compacted embankment	classification according to type of soil, density of stratification and consistency					

- *bearing capacity* in case of compression
- *depth of good bearing soil strata* and density of stratification and
- *suitability* of soil *for pile driving* to the depth of expected pile length

Accordingly, the low-duty driven probe in accordance with EN ISO 22 476-2 and the penetrometer with grooved probe (see Clause 13.6.6.3) are adopted.

An initial indication of the soil conditions to be expected can be obtained from the railway infrastructure manager and, in the case of new lines, from the companies installing the permanent way or by bridge management. Further knowledge can be gained by inspecting the line, observing height and slope of embankments, sections with surface rock, wetlands, drainage

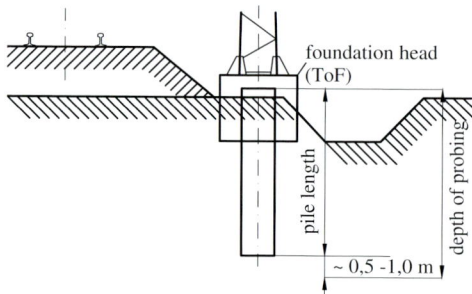

Figure 13.29: Soil investigation for a pile foundation.

installations etc. Based on the line inspection and on the type of foundation envisaged, the *extent of soil investigations* can be determined. Soil investigations should aim at reliable and continuous information about the soil conditions along the line. Investigations at each individual pole site would be an optimum from the technological point of view but expensive and time-consuming. With respect to the continuous character of railway lines, the soil investigation may be limited to areas of varying soil conditions and on sites for dead-end or mid-point poles.

Tests using the low-duty driven probe (DPL), with more than 8 blows for 100 mm of penetration depth, indicate bearing soil. Probing should then be continued for another 3,5 m in the case of suspension poles and 4,5 m in the case of dead-end poles and stopped when a depth of 0,5 to 1,0 m below the point of pile is reached.

In Figure 13.29, the provisions for probing at a dead-end pole site are shown and in Table 13.15, the number of blows achieved with the low-duty probe is an example. More than 8 blows were required to reach 0,5 m below the surface. The probing was continued to a depth of 5,0 m. Below 1,5 m, the soil stratification is medium to dense and below 3,0 m it is dense to very dense with characteristic soil pressures of 480 and 640 N/mm^2, respectively. For the design of a pile foundation, the strata 0,5 m below surface may be considered as presenting good lateral-bearing capacity.

13.7 Foundations

13.7.1 Basis of design

The *type of foundation* depends on the pole type, the loading, the soil conditions and the available technology for foundation installation. Since there is a close correlation between pole and foundation design, the selection of poles should consider foundation requirements. The foundations for contact line poles may be classified as *compact foundations* characterised by supporting the pole by a single foundation body. In this case, the main loads are moments

Table 13.15: Number of blows for the low-duty driven probe DPL according to Table 13.9 for the example according to Figure 13.29.

Depth	Number of blows N_{10}									
0 to 1,0	4	4	2	2	6	8	10	10	12	13
1,0 to 2,0	12	16	14	20	20	18	22	24	20	20
2,0 to 3,0	19	24	24	22	26	23	23	24	28	30
3,0 to 4,0	30	32	26	30	32	38	40	42	38	40
4,0 to 5,0	36	40	40	46	50	53	52	55	54	56

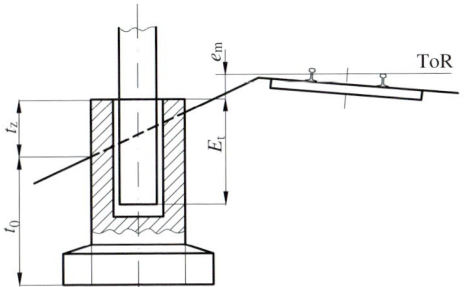

Figure 13.30: Designations for foundations for contact line poles.

e_m difference in height between rail head and top of foundation

E_t insertion depth of poles inserted into the foundation

t_z distance between top of foundation and lowest soil surface level

t_0 embedding depth of foundation

ToR   top of rail (rail head)

combined with horizontal and vertical forces. The loads are transmitted to the subsoil by *soil pressure* in the foundation sub-face or by *lateral earth resistance*, depending on the type of compact foundation.

The transition to new approaches to foundation designs is under way where the verification is no longer carried out for working loads but for limit loads and limit strength. This design approach also forms the basis of new European Civil Engineering standards. When the *limit strength of a foundation* is exceeded, the foundation will no longer fulfill its task and fail.

The requirements and basis of design are presented in the current standard EN 50 119. The designations given in Figure 13.30 were introduced by German Railway DB for overhead contact line foundations and will be used in the following sections.

The loads resulting from poles and structures as well as the dead load of the foundation itself, act on the foundation. They are expressed with their ultimate design values. For geotechnical design by calculation, the general design formula is

$$E_d \leq R_K / \gamma_M \quad , \tag{13.101}$$

where E_d is the design value of the structural load, R_K is the characteristic value of the foundation resistance and γ_M is the partial factor of resistance. This partial factor depends on the type of foundation and should be at least 1,20.

13.7.2 Concrete block foundations with side-bearing faces

Concrete block foundations are usually constructed with vertical faces or with one or more changes of section. With this foundation type, the soil pressure in the sub-face as well as the lateral earth resistance add to the limit strength. The lateral earth resistance may only be considered where the soil remains undisturbed. The earth resistance depends on the strata density and the soil characteristics. Where the foundation height is generally larger than the width, the loading is mainly transmitted by lateral pressure (EN 50 341-1). As a first approximation, the contribution of the sub-face may be neglected. Using the following methods concrete foundations with circular cross sections may also be designed.

Several assumptions are made to determine the resistance of block foundations:

– the external loads are predominantly transmitted to the subsoil by the pressure between the foundation face perpendicular to the loading direction and the soil

– the foundation rotates under a load like a rigid body around a pivot assumed at two thirds of the foundation depth (see Figure 13.31)

– a linear increase of the soil resistance with the depth is assumed:

$$\sigma_{per} = p_t \cdot z, \tag{13.102}$$

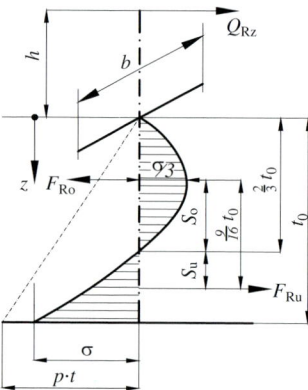

Figure 13.31: Load carrying performance of a block foundation without steps.

where p_t is a soil resistance parameter and t the depth. The parameter p_t can be taken as $200 \, \text{kN}/(\text{m}^2 \cdot \text{m})$ for hard and stiff or dense soils and as $160 \, \text{kN}/(\text{m}^2 \cdot \text{m})$ for firm and semi-dense soils. With this assumption, the inclination is less than one degree.

The assumption of a pivot at two thirds of the foundation depth yields a parabolic pattern of the pressure between foundation and soil

$$\sigma = -2 \, p_t z + 3 \, p_t/t_0 \cdot z^2 \quad . \tag{13.103}$$

With this distribution of the pressure, the *total reaction forces* above and below the pivot can be obtained. Above the pivot

$$F_{\text{Ro}} = b_F \int_0^{2/3 \cdot t_0} \sigma \, \mathrm{d}z = b_F \int_0^{2/3 \cdot t_0} (-2 \, p_t z + 3 \, p_t/t_0 \cdot z^2) \mathrm{d}z = -4/27 t_0^2 \cdot b_F \cdot p_t \tag{13.104}$$

and below the pivot

$$F_{\text{Ru}} = b_F \int_{2/3 \cdot t_0}^{t_0} \sigma \, \mathrm{d}z = b_F \int_{2/3 \cdot t_0}^{t_0} (-2 \, p_t z + 3 \, p_t/t_0 \cdot z^2) \mathrm{d}z = 4/27 t_0^2 \cdot b_F \cdot p_t \quad . \tag{13.105}$$

In (13.104) and (13.105) b_F is the dimension of the foundation perpendicular to the force. The distance between both forces F_{Ro} and F_{Ru} results from the centres of pressure s_o and s_u, which are obtained from

$$
\begin{aligned}
s_o &= \frac{b_F}{F_{\text{Ro}}} \int_0^{2t_0/3} \sigma(2/3 t_0 - z) \mathrm{d}z = \frac{p_t b_F}{F_{\text{Ro}}} \int_0^{2t_0/3} (4 z^2 - 4/3 t_0 \cdot z - 3 z^3/t_0) \mathrm{d}z \\
&= t_0/3
\end{aligned}
\tag{13.106}
$$

and

$$
\begin{aligned}
s_u &= \frac{b_F}{F_{\text{Ru}}} \int_{2t_0/3}^{t_0} \sigma(z - 2/3 t_0) \mathrm{d}z = \frac{p_t b_F}{F_{\text{Ru}}} \int_{2 \cdot t_0/3}^{t_0} (4/3 t_0 \cdot z - 4 z^2 + 3 z^3/t_0) \mathrm{d}z \\
&= 11/48 t_0 \quad .
\end{aligned}
\tag{13.107}
$$

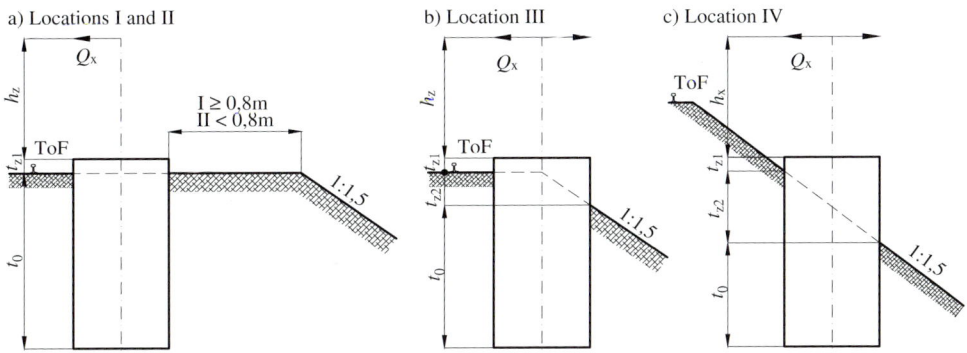

Figure 13.32: Arrangement of block foundations.

The spacing of the centres of soil pressure s is then

$$s = (1/3 + 11/48) \cdot t_0 = 9/16 t_0 \quad . \tag{13.108}$$

Since $F_{Ro} = F_{Ru}$ there is no resulting horizontal force left to counteract the horizontal loads. As a hypothesis, it is assumed that the friction in the foundation faces in parallel to the loading direction counteracts the horizontal loads. The foundation is loaded by moment $M_{x(y)ToF}$ acting at the top of the foundation. This moment can be taken as being generated by a horizontal force $Q_{y(x)}$ acting at the height h_z above the foundation, which is assumed to be 8,0 m.

$$M_{x(y)} = Q_{y(x)} \cdot h_z$$

The balance of forces of resistance around the assumed pivot results in

$$M_R = F_{Ro} \cdot s_o - F_{Ru} \cdot s_u = F_{Ro} \cdot s = 9/16 \cdot 4/27 \, p_t t_0^3 \cdot b_F = 1/12 \, p_t t_0^3 \cdot b_F \quad . \tag{13.109}$$

According to Figure 13.32, four conditions can be distinguished concerning the location and arrangement of a block foundation:

- in a plane, where the foundation edge is more than 0,8 m away from the edge of embankment (location I)
- in a plane, where the foundation edge is less than 0,8 m away from the edge of embankment (location II)
- in the edge of embankment (location III) and
- in the embankment (location IV).

When determining the characteristic resistances (moments), the situation on site has to be taken into account as shown in Figure 13.32.

For location I: $h = h_z + t_{z1}$, $t = t_0$ and $p_t = 200 \, \text{kN}/(\text{m}^2 \cdot \text{m})$, $h_z = 8,0 \, \text{m}$, $t_{z1} = 0,2 \, \text{m}$ are assumed.

The acting moment is $M_y = Q_x h_z$, hence

$$Q_x \left(h_z + t_{z1} + 2/3 \, t_0 \right) = p_t t_0^3 b_F / 12$$

and

$$M_R = p_t \cdot t_0^3 \cdot b / \left(12 + 12 t_{z1}/h_z + 8 t_0/h_z \right) = p_t t_0^3 \cdot b / \left(12, 3 \, t + t_0 \right) \quad .$$

Table 13.16: Characteristic moments M_R for block foundations in kN·m.

$M_R = p_t \cdot 10^3 \cdot b_F \cdot t_0 \cdot K_1/K_2$

		Location I	Location II	Location III		Location IV	
				towards track	away from track	towards track	away from track
	$p_t^{1)}$	200	160	160	160	160	160
	K_1	t_0^2	t_0^2	$(t_0+0,3)^2$	t_0^2	$(t_0+0,3)^2$	t_0^2
	K_2	$12,3+t_0$	$12,3+t_0$	$12,6+t_0$	$12,75+t_0$	$13,05+t_0$	$13,2+t_0$
Width b (m)	Depth t_0 (m)						
1,00	1,60	59	47	65	46	63	44
1,00	1,80	83	66	88	64	86	62
1,00	2,00	112	90	116	86	112	84
1,20	2,30	200	160	200	155	194	150
1,20	2,50	254	203	250	197	242	191

$^{1)}$ p_t in kN/(m^2·m)

In general, the moment can be expressed by

$$M_R = p_t \cdot t_0 \cdot b \cdot K_1(t_0)/K_2(t_0) \quad . \tag{13.110}$$

The assumptions for location I also apply to location II, however p_t is reduced to $160\,\mathrm{kN/(m^2 \cdot m)}$.

In locations III and IV, the action of the moments towards the track and away from track need to be distinguished. With $t_{z2} = 0,3$ m it is obtained:

In location III it applies towards the track

$$K_1 = (t_0 + 0,3)^2 \quad \text{and} \quad K_2 = 12,6 + t_0$$

and away from the track

$$K_1 = t_0^2 \quad \text{and} \quad K_2 = 12,75 + t_0 \quad .$$

In location IV $t_{z2} = 0,6$ m and

$$K_1(t_0) = (t_0 + t_{z2}/2)^2 = (t_0 + 0,3)^2 \quad ,$$

$$K_2(t_0) = 12 + 1,5 \cdot (t_{z1} + t_{z2}/2) + (t_0 + t_{z2}/2) = (13,05 + t_0)$$

are obtained towards the track as well as

$$K_1(t_0) = t_0^2 \quad ,$$

$$K_2(t_0) = 12 + 1,5 \cdot (t_{z1} + t_{z2}) + t_0 = 13,2 + t_0 \quad .$$

away from track. In Table 13.16, examples are presented for block foundations with $1,0 \times 0,9$ m and $1,2 \times 0,9$ m cross section. It should be verified that

$$M_{x(y)d} < M_R/\gamma_M = M_R/1,2.$$

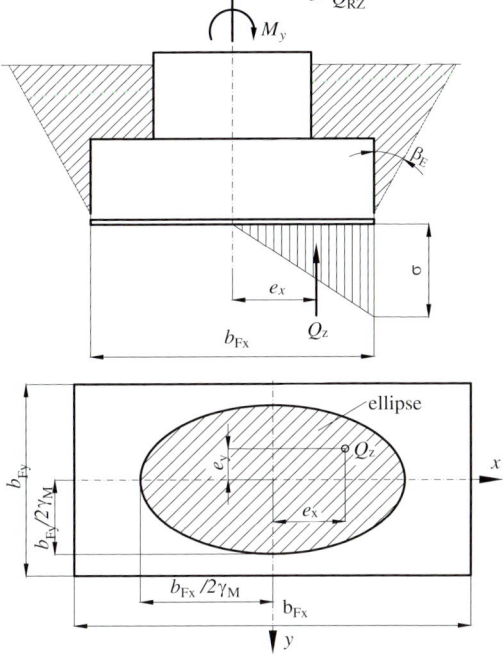

Figure 13.33: Load carrying performance of a block foundation with step.

Figure 13.34: Permissible range of eccentricities e_y and e_x of the total load Q_z in a rectangular foundation sub-face.

13.7.3 Block foundations with steps

Block foundations with steps, also called gravity or slab foundations, are selected if the soil is not suitable for lateral-bearing foundations. High external loads may require stepped block foundations as well. Therefore, this foundation type is often adopted for head-span poles.

For block foundations with steps, the loading is mainly transmitted to the soil by the sub-face area of the foundation. In most cases, the vertical faces add very little to the resistance because a lateral constraint is no longer effective in the backfill.

Design methods for *momentum-loaded block foundations* are readily available. In contact line design practice, it is often assumed that the loading is predominantly transmitted through the sub-face area (Figure 13.33). The lateral constraint is considered by an equivalent weight force of a soil body starting at the foundation sub-face with an angle β_E towards the vertical and ending at the surface. The angle β_E can be obtained from Table 13.14 The soil body has the shape of a truncated pyramid; the volume of the foundation body itself being subtracted. If a foundation with a rectangular sub-face is loaded by the overturning moments M_{xd} and M_{yd} acting on the sub-face, the stability is given by Figures 13.33 and 13.34

$$(e_x/b_{Fx})^2 + (e_y/b_{Fy})^2 \leq 1/\left(4\,\gamma_M^2\right) \quad , \tag{13.111}$$

where $e_x = M_{xd}/Q_z$ and $e_x = M_{xd}/Q_z$. The moments M_{xd} and M_{yd} correspond to the limit state with partial factors included. The resulting force Q_z is the sum of all vertical loads, including the dead weight of the pole, the contact lines, the pole equipment, the dead weight of the foundation and the effective weight of the overburden soil. Q_z is a characteristic load without partial factors. The overburden soil can be assumed as having the shape of a truncated pyramid with an angle β_E starting at the foundation sub-face. Values for the angle β_E can be found in Table 13.14. The partial factor γ_M should be at least 1,2.

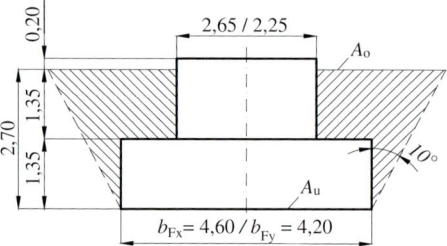

Figure 13.35: Example of a stepped block foundation.

In addition to the proof of stability using Equation (13.111), the soil pressure in the sub-face is to be verified using the equation (see Figure 13.33)

$$\gamma_F \cdot Q_z \leq \left(b_{Fy} - 2/e_y\right) \cdot \left(b_{Fx} - 2/e_x\right) \cdot \sigma_{ult}/\gamma_M \quad . \tag{13.112}$$

Equation (13.112) assumes that the ultimate vertical load $\gamma_F \cdot Q_z$ is equally distributed on a rectangular area with the dimensions $(b_{Fy} - 2/e_y)$ times $(b_{Fx} - 2/e_x)$. The partial factor γ_M should be at least 1,2.
An alternative approach to design block foundations is described in [13.14] and known as Sulzberger's method.

Example 13.3: The permissible design bending moment is to be determined for a stepped block foundation according to Figure 13.35. Soil: Sand, dense. Unit weight of concrete 22 kN/m³ and of soil 19 kN/m³. Ultimate soil pressure 640 kN/m², earth frustum angle 10°. Weight of pole, conductors and equipment 20 kN.
Weight force of the concrete body

$$Q_{zC} = (4,60 \cdot 4,20 \cdot 1,35 + 2,65 \cdot 2,25 \cdot 1,55)\, 22,0 = 35,3 \cdot 22,0 = 777\, \text{kN} \quad .$$

Weight force of the soil body:
Sub-face area

$$A_u = 4,60 \cdot 4,20 = 19,3\, \text{m}^2 \quad .$$

Area at ground level

$$A_o = (4,60 + 2 \cdot 2,7 \tan 10)(4,20 + 2 \cdot 2,7 \tan 10) = 28,6\, \text{m}^2 \quad .$$

$$Q_{zE} = \left[(2,7/3)(19,3 + 28,6 + \sqrt{28,6 \cdot 19,3}\,) - 34,1 \right] \cdot 19,0 = 30,2 \cdot 19,0 = 573\, \text{kN} \quad .$$

The total characteristic weight force is $Q_z = 20 + 777 + 573 = 1\,370\, \text{kN}$.
To determine the ultimate design moments, the conditions (13.111) and (13.112) need to be verified.
With $e_y = 0$, the maximum permissible e_x is obtained from condition (13.112) for $\gamma_F = 1,30$, $b_{Fx} = 4,60$ m and $b_{Fy} = 4,20$ m

$$1,30 \cdot 1370 = (4,6 - 2 \cdot e_x)\, 4,2 \cdot 640/1,2$$

or

$$e_x = 1/2\,[4,6 - (1,30 \cdot 1370 \cdot 1,2)/(4,2 \cdot 640)] = 1,903\, \text{m}$$

as well as, assuming $e_x = 0$ and $\gamma_M = 1,2$

$$1,30 \cdot 1370 = (4,2 - 2 \cdot e_y) \cdot 4,6 \cdot 640/1,2$$

Table 13.17: Factor κ_n for the effect of embankment gradient on the strength of stepped block foundations. (For symbols used see Figure 13.36).

Distance b_{Ey}	Ratio of height of embankment/ to depth of foundation t_e/t_0		Factor κ_n		
m	from	to	n=1,00	1,50	2,00
0	0,25	0,50	1,10	1,07	1,04
	0,50	0,75	1,20	1,15	1,09
	1,0	$\geq 1,0$	1,30	1,22	1,13
0,20	0,25	0,50	1,08	1,06	1,04
	0,50	0,75	1,17	1,12	1,07
	0,75	$\geq 1,0$	1,25	1,18	1,11
0,40	0,25	0,50	1,06	1,05	1,03
	0,50	0,75	1,13	1,09	1,05
	0,75	$\geq 1,0$	1,19	1,14	1,08
0,60	0,25	0,50	1,05	1,03	1,02
	0,50	0,75	1,09	1,07	1,04
	0,75	$\geq 1,0$	1,14	1,10	1,06
0,80	0,25	0,50	1,03	1,02	1,01
	0,50	0,75	1,05	1,04	1,03
	0,75	$\geq 1,0$	1,08	1,06	1,04
1,00	0,25	0,50	1,01	1,01	1,00
	0,50	0,75	1,02	1,01	1,01
	0,75	$\geq 1,0$	1,03	1,02	1,01
1,10	0,25	$\geq 1,0$	1,00	1,00	1,00

or

$$e_y = 1/2(4,2 - (1,30 \cdot 1370 \cdot 1,2))/(4,6 \cdot 640) = 1,737\,\text{m} \quad .$$

From condition (13.111) it is obtained with $e_y = 0$

$$e_x = b_{Fx}/(2 \cdot 1,2) = 4,6/2,4 = 1,917\,\text{m}$$

and for $e_x = 0$

$$e_y = b_{Fy}/(2 \cdot 1,2) = 4,2/2,4 = 1,750\,\text{m} \quad .$$

Condition (13.112) (ultimate soil pressure) limits the permissible design moments.

$$M_{xd} = Q_z \cdot e_x = 1370 \cdot 1,903 = 2600\,\text{kN} \cdot \text{m}$$

and

$$M_{yd} = Q_z \cdot e_y = 1370 \cdot 1,737 = 2380\,\text{kN} \cdot \text{m} \quad .$$

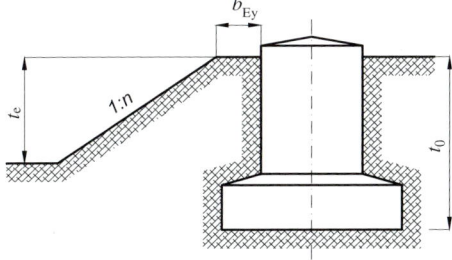

Figure 13.36: Stepped concrete block foundation installed in an embankment.

a) Lateral view   b) Plan view

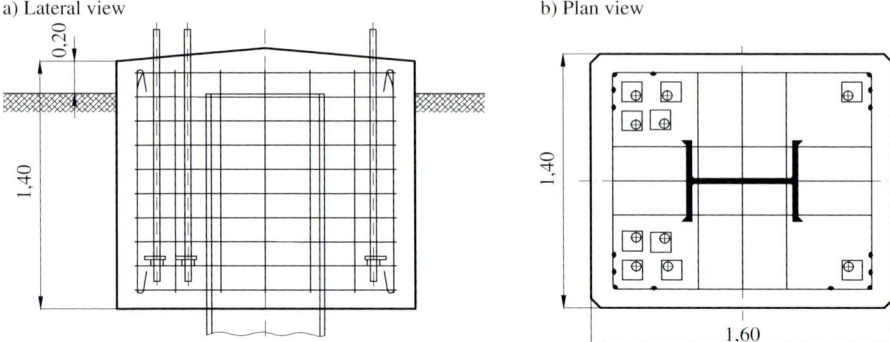

Figure 13.37: Driven steel pile with a concrete header.

The stability margin can be obtained from Equation (13.111)

$$\gamma_M = e_x/2\,b_{Fx} = 4,60/(2 \cdot 1,903) = 1,209$$

or

$$\gamma_M = e_y/2\,b_y = 4,20/(2 \cdot 1,737) = 1,209 \quad .$$

The permissible design moments would be $M_{xd} = 2\,280\,\text{kN·m}$ and $M_{yd} = 2\,082\,\text{kN· m}$ for the ultimate soil pressure of $400\,\text{kN/m}^2$. While the ultimate soil pressure is reduced to 63 %, the ultimate moments are reduced to 88 %.

If stepped block foundations are installed close to, or directly in, embankments, their loading capacity is reduced depending on the installation conditions and the gradient and height of the embankment. The ultimate moments determined by the approach described above have to be reduced by dividing them with the relevant factors κ_n according to Table 13.17.

13.7.4   Driven pile foundations

Driven pile foundations are an economic alternative to cast in-situ concrete foundations and are especially suited to locations with good lateral-bearing soil only occurring at greater depth or with a high *ground water table*, which otherwise would require a costly shuttering of the excavation and drainage of ground water. A variety of steel piles selected according to the pole type gained increasing importance. Piles are also suitable at locations with limited space. They require very limited excavation and minimize the disturbance of soil and stratification in critical embankments.

Steel piles with sheet-wall profiles, as shown in Figure 13.37, can be suitable for lattice steel, double-channel and bolt-mounted H-beam poles. A concrete header on the piles accommodates the pole base using standard anchor bolts and with the pile cross sections designed for the loading. Two individual sections can be welded together to form a twin section pile to transfer high loads acting in two orthogonal directions.

For spun concrete poles, *driven tubes* are used, onto which the pole is fixed [13.3] (Figure 13.38). Since the external diameters of the tubes have to be less than the inside diameter of the *spun concrete poles*, relatively thick-walled, heavy and expensive tubes are necessary. Sheet-wall profiles with a tube welded at the top can be adopted. This design forms

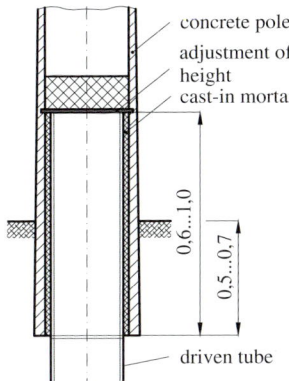

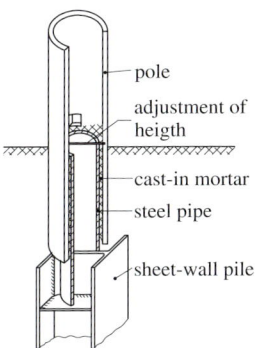

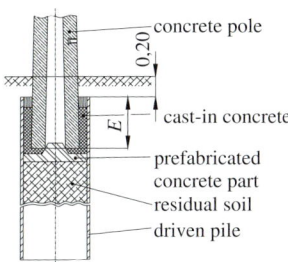

Figure 13.38: Spun concrete pole set on a driven tube foundation.

Figure 13.39: Spun concrete pole on a sheet-wall pile with a tube welded on pile top.

Figure 13.40: Spun concrete pole inserted into a driven steel tube.

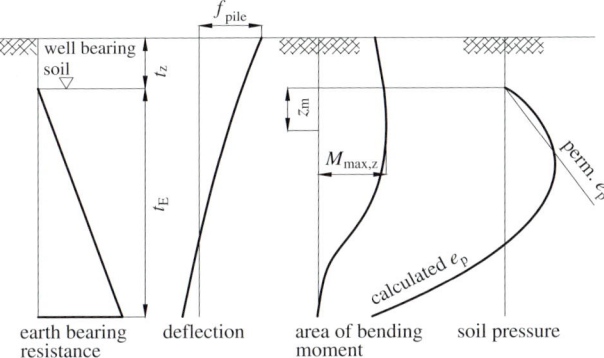

Figure 13.41: Basis of pile design according to [13.15].

a favourable combination of spun concrete poles and high-capacity driven pile foundations (Figure 13.39). After aligning the pole, the space between the concrete pole and steel tube is grouted with mortar. The concrete pole protects the steel pile against corrosion in the air/soil transition area. Other corrosion protection provisions are unnecessary.

As an alternative design, concrete and H-beam poles can be inserted into large diameter steel tubes (Figure 13.40). The space is filled with concrete or grit. However, corrosion protection is necessary if steel tubes encase steel poles.

For verification of the *geotechnical stability* of a pile foundation, the method described in [13.15] can be used. The method was developed for design of large-size piles in harbours and does not require high accuracy in soil investigations, but still results in a reliable design. This approach is based on the assumptions indicated in Figure 13.41. The *earth bearing resistance* is assumed to increase linearly with the depth starting at the top of the good-bearing soil stratum. To create a *reaction moment* the pile must undergo deflections. The adequacy of these assumptions was confirmed by many applications under test conditions without failure of the pile foundations. The approach uses the earth resistance $e_o = \gamma_E \cdot \lambda_p$, where the earth pressure coefficient λ_p in flat terrain follows from

$$\lambda_p = \tan^2(45° + \phi_R/2) \quad . \tag{13.113}$$

Table 13.18: Earth pressure coefficient λ_p for piles in embankments according to [13.10] depending on the angle of internal friction ϕ_R as a soil caracteristic (data for internal friction see Table 13.14) and the embankment gradient β_w. β_w is positive if the loading acts towards the embankment and negative if it acts away from the embankment.

Embankment	Angle of internal friction ϕ_R								
gradient β_w	40°	35°	30°	25°	20°	15°	10°	5°	0°
+40°	70,923								
+35°	34,051	18,817							
+30°	21,592	13,226	8,743						
+25°	14,929	9,951	6,982	5,075					
+20°	11,062	7,822	5,737	4,319	3,312				
+15°	8,570	6,331	4,807	3,723	2,926	2,321			
+10°	6,840	5,228	4,080	3,235	2,595	2,099	1,704		
+ 5°	5,572	4,375	3,492	2,823	2,304	1,894	1,564	1,291	
0°	4,599	3,690	3,000	2,464	2,039	1,698	1,420	1,191	1,000
− 5°	3,826	3,124	2,577	2,143	1,792	1,504	1,262	0,992	
−10°	3,193	2,643	2,204	1,848	1,552	1,295	0,970		
−15°	2,660	2,224	1,866	1,566	1,299	0,933			
−20°	2,201	1,848	1,548	1,277	0,883				
−25°	1,796	1,502	1,231	0,821					
−30°	1,428	1,163	0,750						
−35°	1,076	0,671							
−40°	0,587								

In (13.113) ϕ_R represents the angle of internal friction (see Table 13.14). For foundations in an embankment with the angle β_W the *earth pressure coefficient* λ_p can be taken from Table 13.18. The unit weight γ_E of the soil should be taken as $10\,\mathrm{kN/m^3}$ in view of the ground water table. Using this approach and the information from Figure 13.41, the location z_m of the maximum bending moment can be obtained from

$$z_m^3 + 3 \cdot b \cdot z_m^2 = 6 \cdot Q_{yd}/f_W \quad . \tag{13.114}$$

The transverse ultimate force Q_{xd} represents the sum of horizontal forces and correlates with the design moment at top of the piles by

$$Q_{yd} = M_{xd}/h_z \quad .$$

For pile foundations with loads in the direction of both pile axis, the corresponding resultant actions must be used for M_{yd} and Q_{xd}. The maximum moment is

$$M_{x\mathrm{max}} = Q_{yd}(h_z + z_z + z_m) - e_p(b_F \cdot z_m^3/6 + z_m^4/24) \quad . \tag{13.115}$$

Often the second, load decreasing term in (13.115) is neglected and, in a simplified manner, the maximum bending moment is obtained from

$$M_{x\mathrm{max}} = Q_{yd}(h_z + z_z + z_m) \quad , \tag{13.116}$$

where z_z is the thickness of the non-bearing soil stratum.

The *embedding depth* into the bearing soil is obtained from $t_0 = 1{,}2 \cdot t_E$ in accordance with [13.15], where t_E is received from

$$t_E^3(t_E + 4 \cdot b_F)/(t_E + h_z) - 4 z_m^2(z_m + 3 \cdot b_F) = 0 \quad . \tag{13.117}$$

Equation (13.117) can be solved numerically. The total pile length should be

$$t_{pile} = t_0 + z_z + t_z - 0,20 \text{ m} \quad , \tag{13.118}$$

where it is assumed that the pile ends 0,2 m below the top of the foundation. The definition of t_z can be seen in Figure 13.30. In practice, the pile length may be rounded to 0,5 m steps where up to 0,15 m, rounding down is acceptable. For cantilever poles, the minimum embedding depth t_0 should be 3,0 m, for intermediate poles of overlaps it should be 3,5 m and 4,5 m for dead-end poles.

If the poles situated in an embankment are loaded in parallel to the track, the design may be carried out by adding an ideal non-bearing soil stratum with the thickness $z_z = 0,945/n$ where $1/n$ is the gradient of the embankment. In embankments with a height greater than the piling depth in a level terrain minus 1,0 m the embedding depth should be taken as $t_0 = 1,7 \cdot t_E$, if the direction of loading is at right angles to the embankment.

In addition to the *geotechnical stability*, the strength under bending of the steel pile must be verified. For a steel grade of S235 in accordance with EN 10025, a partial factor $\gamma_M = 1,25$ should be adopted.

Furthermore, the *displacement at the pile head* can be obtained from

$$f_{pile} = \frac{Q_y}{E \cdot I} \left[\frac{(h_z + z_z + 0,65 \cdot t_0)^3}{3} - (h_z + z_z + 0,65 \cdot t_0)^2 \cdot \frac{h_z}{2} + \frac{h_z^3}{6} \right] \quad , \tag{13.119}$$

where h_x is the equivalent height of application of the resulting transverse force. The verification of displacement is carried out for characteristic loads Q_z without partial factors on the load site. The horizontal pile displacement should be limited to 30 mm or 0,005 times the pile length, where the lower value should apply. An example for pile design following this approach is shown in Clause 13.8.

13.7.5 Anchor foundations

For overhead contact line poles, anchors are used to react to permanent longitudinal loads present in a given direction, e. g. loads from terminations of contact or catenary wires which otherwise would lead to high loads on poles and their foundations. As shown in Figure 13.42, *anchor foundations* are loaded by anchor forces in the vertical direction by the component F_{AV} and horizontally by the component F_{AH}. The characteristic resistance against being pulled out is created by the dead weight of the foundation and the skin friction against the surrounding soil:

$$F_{Kr} = \gamma_g V_g + A_r \tau_r \quad , \tag{13.120}$$

with V_g volume of foundation, γ_g unit weight of foundation, e. g. of concrete, A_r friction area, τ_r skin friction value.

The friction value τ_r depends on the material used for foundation and on the type of soil. Table 13.19 lists commonly adopted data for friction between concrete and soil.

The anchor foundation is sufficiently designed in view of uplift if

$$F_{AVd} \leq F_{Kr}/\gamma_M \quad . \tag{13.121}$$

The partial factor γ_M should be at least 1,2, if F_{AVd} is the ultimate design load.

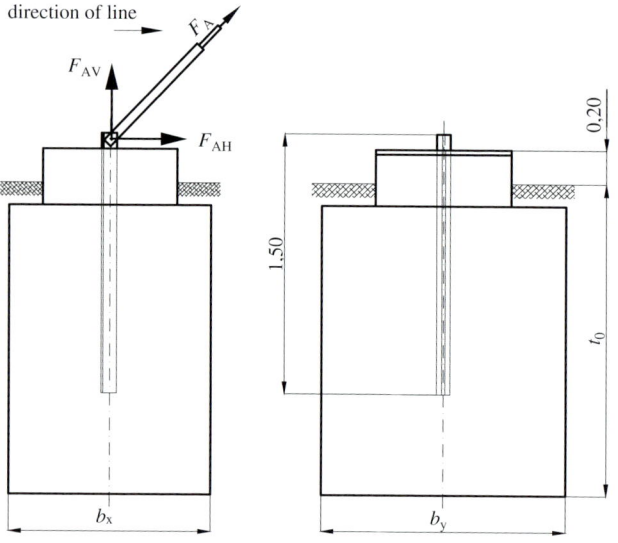

Figure 13.42: Foundation for guy anchors.

Table 13.19: Friction values τ_r between soil and concrete.

Type of soil	Friction value τ_r in kN/m^2
sand, very dense	20
sand, dense	15
sand, medium dense	10
sand, loose	5
clay, hard	12
clay, semihard	6

The horizontal loading is counteracted by the *passive earth pressure*, which may be assumed as increasing linearly with the depth. In total, the earth pressure of a body with the depth t_0 and the width b_F is then

$$E_P = 0{,}5 \cdot \gamma_E \cdot t_0^2 \cdot b_F \cdot (\tan^2(45 + \phi_R/2) - \tan^2(45 - \phi_R/2)) \quad , \tag{13.122}$$

where γ_E is the unit weight of the soil and ϕ_R is the *angle of internal friction* (see Table 13.12). Stability is ensured if the resistance modified by the partial factor γ_M is higher than the horizontal load. Therefore,

$$F_{AHd} \leq E_P/\gamma_M \quad . \tag{13.123}$$

Overturning of the anchor foundation can not occur if the acting moment is less than the characteristic resistant moment divided by the partial factor γ_M

$$M_{Ad} \leq M_{Kr}/\gamma_M \quad , \tag{13.124}$$

where (see Figure 13.42)

$$M_{Kr} = F_{Kr} \cdot b_{Fy}/2 + E_P \cdot t_0/3 \quad . \tag{13.125}$$

Table 13.20: Data of overhead contact line type Sicat H1.0.

		Contact wire AC-120 – CuMg	Catenary wire Bz II 120	Return feeder 243 - AL1	Parallel feeder 243 - AL1
mass per unit length	kg/m	1,07	1,06	0,67	0,67
diameter	mm	13,2	14,0	20,3	20,3
tensile force	kN	27	21	4,8	8,5
wind load[1] per unit length	N/m	9,8	10,4	11,9	11,9

[1] clamps and droppers considered accordingly

Example 13.4: An anchor ultimate force F_{Ad} of 50 kN acts at an angle of 50°. In this case $F_{AVd} = F_{Ad} \cdot \sin 50° = 38,3$ kN and $F_{AHd} = F_{Ad} \cdot \cos 50° = 32,1$ kN. The soil is assumed to be medium densely stratified sand characterised by $\gamma_E = 16$ kN/m³, $\varphi_R = 30°$ and $\tau_r = 10$ kN/m². The anchor foundation has the dimensions $t_0 = 2,00$ m, $b_{Fy} = 1,20$ m, $b_{Fx} = 1,00$ m (Figure 13.42). Therefore, with $V_g = 1,7 \cdot 1,0 \cdot 1,2 = 2,04$ m³ and if the upper 0,5 m is considered as non-bearing, $A_r = 1,5 \cdot 2 \cdot (1,0 + 1,2) = 6,60$ m², Equation (13.120) yields

$$F_{Kr} = 22 \cdot 2,04 + 10 \cdot 6,60 = 110,9 \text{ kN} \quad .$$

According to (13.122) it is obtained

$$E_p = 0,5 \cdot 16 \cdot 2,0^2 \cdot 1,2 \cdot (\tan^2 60 - \tan^2 30) = 102,4 \text{ kN} \quad .$$

From Equation (13.121) it can be seen that

$$38,3 \leq 110,9/1,2 = 92,4 \text{ kN} \quad .$$

The stability against overturning is verified using Equation (13.124) applied to the sub-face

$$M_{Kr} = 32,1 \cdot 2,0 + 38,3 \cdot 0,5 = 83,3 \leq (110,9 \cdot 0,5 + 102,5 \cdot 2,0/3)/1,2 = 103,2 \text{ kN} \cdot \text{m} \quad .$$

The selected anchor foundation complies with the requirements.

13.8 Example

13.8.1 Contact line data

This example of the *design of cantilevers*, poles and foundations is for a pole for a high-speed line equipped with the Sicat H1.0 type contact line and swivelling cantilevers using European standards where applicable. A wind velocity of 27,5 m/s with a basic wind pressure of 390 N/m² according to class W2 (normal) is assumed for the example (see Table 2.11). The line runs in open terrain (category II), where ice load class I2 (normal according Table 13.2) with 7,0 N/m ice load applies.

Relevant data for the contact line is given in Table 13.20. The following additional information also applies

contact wire height	5,30 m
system height	1,80 m
stagger	0,30 m
track radius	10 000 m
span length	65 m

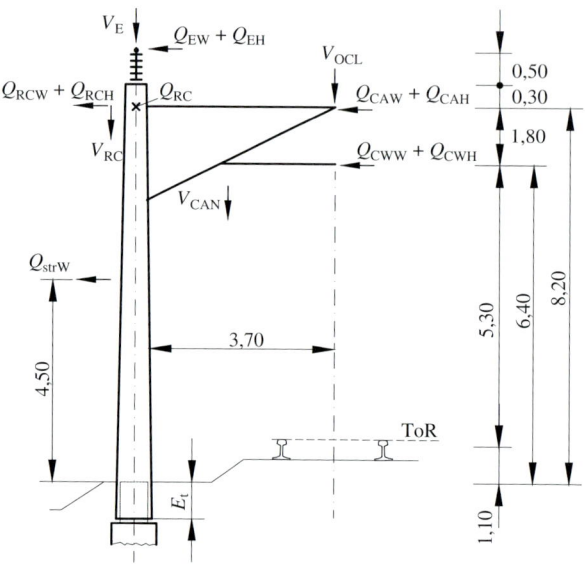

Figure 13.43: Dimensions of and actions on an overhead contact line pole, contact line Sicat H1.0.

A termination for the return feeder is arranged at the pole. Dimensions and designation of forces are shown in Figure 13.43.

In this example, only the numerical data is given for the intermediate steps of the calculation. Units are used as noted above.

The relevant static wind pressure is obtained from Equation (2.5) for heights up to 7 m $q_z = 1,5 \cdot 390 = 585\,\mathrm{N/m^2}$. The wind force on conductors is therefore, with $G_C = 1,0$ and C_C and according to Clause 2.5.2.7:

Contact wire with clamps and droppers[1]: $Q_{CWW} = 585 \cdot 0,0132 \cdot 1,1 \cdot 1,15 = 9,8\,\mathrm{N/m}$

Catenary wire with clamps and droppers[1]: $Q_{CAW} = 585 \cdot 0,014 \cdot 1,1 \cdot 1,15 = 10,4\,\mathrm{N/m}$

Parallel feeder and return conductor: $Q_{EW} = Q_{RW} = 585 \cdot 0,0203 \cdot 1,0 = 11,88\,\mathrm{N/m}$

[1] A value of 1,15 is assumed for wind action on droppers and clamps.

13.8.2   Loads

Vertical loads (see (13.1))

contact line:	V_{OCL}	$= 22,5 \cdot 65$	$= 1\,465\,\mathrm{N}$
cantilever:	V_{CAN}		$1\,500\,\mathrm{N}$
parallel feeder line:	V_E	$= 6,7 \cdot 65$	$= 440\,\mathrm{N}$
return conductor:	V_{RC}	$= 6,7 \cdot 65/2$	$= 220\,\mathrm{N}$
equipment at pole head:	V_G	$=$	$= 1\,000\,\mathrm{N}$

Wind loads (see (13.3))

catenary wire:	Q_{CAW}	$= 10,40 \cdot 65$	$= 676\,\mathrm{N}$
contact wire:	Q_{CWW}	$= 9,80 \cdot 65$	$= 637\,\mathrm{N}$
parallel feeder:	Q_{EW}	$= 11,88 \cdot 65$	$= 772\,\mathrm{N}$
return feeder:	Q_{RW}	$= 11,88 \cdot 65/2$	$= 386\,\mathrm{N}$
pole (HE-B260):	Q_{strW}	$= 1,6 \cdot 585 \cdot 0,26 \cdot 8,5 = 2\,070\,\mathrm{N}$	

Horizontal components of conductor tensile forces (see (13.3)), $R = 10\,000\,\mathrm{m}$, $b = \pm0,3\,\mathrm{m}$

catenary wire:
$$Q_{CAH} = 21\,000 \cdot 65/10\,000 = 137\,N$$
$$+ \quad 4 \cdot 0,30 \cdot 21\,000/65 = 388\,N$$
$$= 525\,N$$

contact wire:
$$Q_{CWH} = 27\,000 \cdot 65/10\,000 = 175\,N$$
$$+ \quad 4 \cdot 0,30 \cdot 27\,000/65 = 500\,N$$
$$= 675\,N$$

parallel feeder: $\quad Q_{EH} = 8\,500 \cdot 65/10\,000 = 55\,N$

return feeder: $\quad Q_{RH} = 4\,800 \cdot 65/(10\,000 \cdot 2) = 15\,N$

Termination of return feeder: $\quad H_R = 20 \cdot 240 = 4\,800\,N$

In EN 50 119 the following partial factors and factors for actions are specified:

Permanent actions $\quad \gamma_G \quad = 1,30$ (if increasing the stress)

$\quad\quad\quad\quad\quad\quad\quad\quad\quad \gamma_G \quad = 1,00$ (if decreasing the stress)

Variable actions $\quad \gamma_W, \gamma_I \quad = 1,30$

The dead weight of conductors, cantilevers and poles as well as the radial forces of conductors are considered as permanent actions while wind and ice are variable actions.

The loads calculated above must be multiplied by *partial factors* when calculating the internal forces and moments.

13.8.3 Design of pole

Design of shear forces and bending moments at base of pole

$$Q_{yd} = 1,30\,(525 + 675 + 55 + 15) +$$
$$1,30\,(637 + 676 + 772 + 386 + 2070) = 7554 \approx 7,6\,kN$$
$$M_{xd} = 1,30\,(1465 \cdot 3,7 + 1500 \cdot 0,50 \cdot 3,70 + 525 \cdot 8,2 + 675 \cdot 6,4 + 55 \cdot 9,0$$
$$+ 15 \cdot 8,20) + 1,30\,(676 \cdot 8,2 + 637 \cdot 6,4 + 772 \cdot 9,0 + 386 \cdot 8,2$$
$$+ 2070 \cdot 8,5/2) = 59\,760\,Nm \approx 60\,kNm$$
$$Q_{xd} = 1,30 \cdot 4800 = 6240\,N \approx 6,3\,kN$$
$$M_{yd} = 1,30 \cdot 4800 \cdot 8,2 = 51\,170\,N \approx 51,2\,kN$$

Vertical forces

– increasing the internal forces

$$Q_z = 1,30\,(1465 + 1500 + 440 + 220 + 1000) = 6013\,N \approx 6,0\,kN \quad ,$$

– decreasing the internal forces

$$Q_z = 1,00\,(1465 + 1500 + 440 + 220 + 1000) = 4625\,N \approx 4,6\,kN \quad .$$

Analysis of strength

Yield stress $\sigma_f = 235\,N/mm^2$; $\gamma_M = 1,1$

$\quad\quad W_{ypl} = 1283 \cdot 10^3\,mm^3$ (see [13.9], section HE–B 260)

$\quad\quad W_{xpl} = 602 \cdot 10^3\,mm^3$; $A = 11\,800\,mm^2$

$\quad\quad I_z \quad = 14\,920\,cm^4$

Plastic strength, axial force (see (13.27))

$$N_{\mathrm{xpl,Rd}} = 235 \cdot 11\,800/1,1 = 2\,520\,\mathrm{kN}.$$

Plastic strength, bending moment (see (13.28))

$$
\begin{aligned}
M_{\mathrm{ypl,Rd}} &= 235 \cdot 1\,283 \cdot 10^3/1,1 = 274\,\mathrm{kN \cdot m}, \\
M_{\mathrm{xpl,Rd}} &= 235 \cdot 602 \cdot 10^3/1,1 = 129\,\mathrm{kN \cdot m}.
\end{aligned}
$$

From (13.26) it follows

$$6,0/2520 + 60/274 + 51,2/129 = 0,62 < 1,00 \quad .$$

The design strength is higher than the design loads formed by internal forces and moments, including partial factors.

Verification of deflection

The serviceability for use, namely the deflection, has to be verified without any partial factors. The deflection perpendicularly to track is critical. The deflection under wind action at the height of contact wire is, according to (13.97)

$$
\begin{aligned}
f_{\mathrm{CWW}} &= [0,676 \cdot 6,4^2(8,2 - 6,4/3) + 0,637 \cdot 2 \cdot 6,4^3/3 \\
&\quad + 0,772 \cdot 6,4^2(9,0 - 6,4/3) + 0,386 \cdot 6,4^2(8,2 - 6,4/3) \\
&\quad + 2,07 \cdot 6,4^2(8,5/2 - 6,4/3 + 6,4^2/(12 \cdot 8,5))] \cdot 238/14\,920 \\
&= 806 \cdot 238/14\,920 = 13,0\,\mathrm{mm} < 25\,\mathrm{mm} \quad .
\end{aligned}
$$

The deflection at the height of the catenary wire under action of permanent loads is calculated according to (13.98)

$$
\begin{aligned}
f_{\mathrm{CAH}} &= [0,525 \cdot 2 \cdot 8,2^3/3 + 0,675 \cdot 6,4^2(8,2 - 6,4/3) \\
&\quad + 0,055 \cdot 8,2^2(9,0 - 8,2/3) + 0,015 \cdot 8,2^3 \cdot 2/3 \\
&\quad + (1,465 \cdot 3,7 + 1,500 \cdot 0,5 \cdot 3,7)8,2^2] \cdot 238/14\,920 \\
&= 940 \cdot 238/14\,920 = 15,0\,\mathrm{mm} < 0,01 \cdot 8\,200 = 82\,\mathrm{mm} \quad .
\end{aligned}
$$

The deflection at the height of the catenary wire under the action of maximum loads is calculated according to (13.99)

$$
\begin{aligned}
f_{\mathrm{CA(H+W)}} &= [1,201 \cdot 2 \cdot 8,2^3/3 + 1,312 \cdot 6,4^2(8,2 - 6,4/3) \\
&\quad + 0,827 \cdot 8,2^2(9,0 - 8,2/3) + 0,401 \cdot 2 \cdot 8,2^3/3 \\
&\quad + 2,070 \cdot 8,2^2(8,5/2 - 8,2/3 + 8,2^2/(12 \cdot 8,5)) \\
&\quad + (1,465 \cdot 3,7 + 1,500 \cdot 0,50 \cdot 3,7)8,2^2] \cdot 238/14\,920 \\
&= 2\,117 \cdot 238/14\,920 = 33,7\,\mathrm{mm} < 0,015 \cdot 8\,200 = 123\,\mathrm{mm} \quad .
\end{aligned}
$$

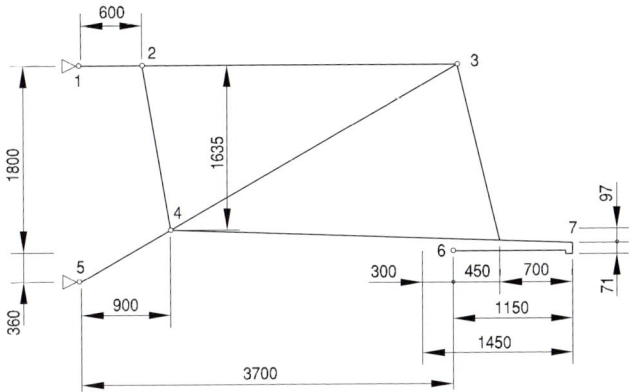

Figure 13.44: Geometry of a push-off cantilever.

13.8.4 Cantilever

As an example, the design of a cantilever made of aluminum alloy is verified for a push-off support. Figure 13.44 shows the cantilever geometry.

Lengths and dimensions of tubes are

	length	dimension
top tube	$l_{2-3} = 3\,700$ mm	42×4
cantilever tube	$l_{5-3} = 4\,285$ mm	70×6
diagonal strut	$l_{4-2} = 1\,657$ mm	$26 \times 3,5$
registration tube	$l_{4-7} = 3\,951$ mm	55×6

Two combinations of loads are verified

- wind load, no ice load (load case B)
- half of design wind load, ice accretion on catenary wire and cantilever (load case D), ice load 7,0 N/m

Vertical loads

- contact line equipment without ice: $V_{OCL} = 1\,465$ N
- contact wire: $V_{CW} = 10,7 \cdot 65 = 700$ N
- catenary wire with ice accretion: $V_{CA} = (1,06 \cdot 9,81 + 7,0) \cdot 65 = 1\,455$ N
- proportionate cantilever dead weight: $V_{CAN} = 750$ N
- with ice: $V_{CAN} = 850$ N

Horizontal loads

- without ice (load case B): see Clause 13.8.3
- with ice and wind (load case D):
- catenary wire:

  diameter with ice: $D_I = [7,0 \cdot 4/(\pi \cdot 7500) + 0,014^2]^{0,50} = 0,0372$ m $\approx 0,037$ m
  $$Q_{CAW} = 585/2 \cdot 65 \cdot 0,037 \cdot 1,15 = 809 \text{ N}$$
- contact wire: $Q_{CWW} = 585/2 \cdot 0,014 \cdot 65 \cdot 1,1 \cdot 1,15 = 337$ N.

The loads have to be multiplied with partial factors.

Verification of top tube 42×4, AlMgSi1, F31

Internal forces at top tube (see (13.12))

Load case B:

$$
\begin{aligned}
F_{\text{top}} &= \big[(1\,465+750)\cdot 1,30\cdot 3,70+(675+676)\cdot 1,30\cdot 0,36 \\
&\quad + (525+637)\cdot 1,30\cdot 2,16\big]/2,16 = 6\,736\,\text{N} \approx 6,75\,\text{kN} \quad .
\end{aligned}
$$

Load case D:

$$
\begin{aligned}
F_{\text{top}} &= \big[(1\,465+850+455)\cdot 1,30\cdot 3,70+(675+337)\cdot 1,30\cdot 0,36 \\
&\quad + (525+809)\cdot 1,30\cdot 2,16\big]/2,16 = 8\,122\,\text{N} \approx 8,1\,\text{kN} \quad .
\end{aligned}
$$

Internal forces at diagonal strut (see (13.16))
Load case B:

$$
F_D = \pm(675\cdot 1,30+637\cdot 1,30)/\sqrt{1+(3,7/2,16)^2} = \pm860\,\text{N} \quad .
$$

Load case D:

$$
F_D = \pm(675\cdot 1,30+319\cdot 1,30)/\sqrt{1+(3,7/2,16)^2} = \pm651\,\text{N} \quad .
$$

Internal moment at the top tube at attachment of the diagonal strut (see Equation (13.17) and Figure 13.44)
Load case B:

$$
\begin{aligned}
M_{B2} &= (675\cdot 1,30+637\cdot 1,30)\cdot 2,16\cdot 0,6\cdot 3,1/(3,7^2+2,16^2) = 373\,\text{Nm} \\
&\approx 0,38\,\text{kNm} \quad .
\end{aligned}
$$

Load case D:

$$
\begin{aligned}
M_{B2} &= (675\cdot 1,30+319\cdot 1,30)2,16\cdot 0,6\cdot 3,1/(3,7^2+2,16^2) = 283\,\text{Nm} \\
&\approx 0,28\,\text{kNm} \quad .
\end{aligned}
$$

Equation (13.26) applies to the strength in case of loading by axial forces and moments if the shear force is low. $N_{\text{pl,Rd}}$ can be obtained from (13.27) using $\sigma_f = 260\,\text{N/mm}^2$, $A = 477,5\,\text{mm}^2$ (Table 13.1) and $\gamma_{\text{MO}} = 1,1$

$$
N_{\text{ypl,Rd}} = 477,5\cdot 260/1,1 = 113\,\text{kN} \quad .
$$

The value $M_{\text{xpl,Rd}}$ results from (13.28) and Table 13.1

$$
M_{\text{xpl,Rd}} = 5,798\cdot 260/1,1 = 1,37\,\text{kNm} \quad .
$$

Then, it is obtained for the maximum loading (case B)

$$
6,7/113+0,38/1,37 = 0,34 < 1,0 \quad .
$$

Therefore, the strength of the top tube is verified.

Verification of the cantilever tube 70×6, AlMgSi1, F31

Internal forces at the cantilever tube (see Equation (13.14))

Load case B (compression loading)

$$
\begin{aligned}
F_A \;=\; & -(1\,465+750)\cdot 1,30\sqrt{1+(3,7/2,16)^2} \\
& -(637+675)\cdot 1,30 \,/\, \sqrt{1+(2,16/3,70)^2} = -7\,184\,\text{N} \approx -7,2\,\text{kN} \quad .
\end{aligned}
$$

Load case D

$$
\begin{aligned}
F_A \;=\; & -(1\,465+850+455)\cdot 1,30\sqrt{1+(3,7/2,16)^2} \\
& -(316+675)\cdot 1,30 \,/\, \sqrt{1+(2,16/3,70)^2} = -8\,255\,\text{N} \approx -8,3\,\text{kN} \quad .
\end{aligned}
$$

Load case D is decisive.

Verification according to (13.30) without bending moments.

$s_k = 4\,280\,\text{mm}$; $i = 22,73\,\text{mm}$ (Table 13.5); $\lambda = 4\,280/22,73 = 188$;

$\overline{\lambda} = 188/(\pi \cdot \sqrt{70\,000/260}) = 3,65$ according to (13.33)

$\phi = 0,5\,(1+0,21\,(3,65-0,2)+3,65^2) = 7,5$ according to (13.34)

$\chi = 1/(7,5+\sqrt{(7,5^2-3,65^2)}) = 0,071 < 1$

Without moments the following is obtained from (13.30)

$$
8\,255/(0,071 \cdot 1\,206,4 \cdot 260/1,1) = 0,41 < 1,0 \quad .
$$

Therefore, the strength of the cantilever tube is verified.

Verification of registration tube 55×6, AlMgSi1, F31

Use as a push-off support

Internal force at the registration tube, load case B

$$
F_{St} = -(637+675) = -1\,312\,\text{N} \quad ,
$$

$$
V_{FD} = (65 \cdot 10,7) + 100 \sim 800\,\text{N} \quad .
$$

Internal moment (see (13.20))

$$
M_{B6} = 1,30 \cdot [1\,312 \cdot 0,071 + 800 \cdot 0,7] = 850\,\text{Nm} \quad ,
$$

$s_k = 3\,950\,\text{mm}$; $i = 17,45\,\text{mm}$

$\lambda = 3\,950/17,45 = 226$

$\overline{\lambda} = 226/(\pi \cdot \sqrt{70\,000/260}) = 4,39$ according to (13.33)

$\phi_k = 0,5\,(1+0,21\,(4,39-0,2)+4,39^2) = 10,6$ according to (13.34).

$\chi = 1/(10,6+\sqrt{(10,6^2-4,39^2)}) = 0,049$.

$k_y = 1,50$ (13.30) yields: $1,3 \cdot 1\,312/(0,049 \cdot 923,6 \cdot 260/1,1) +$
$1,5 \cdot 850 \cdot 1,3 \cdot 10^3/(14\,778 \cdot 260/1,1) = 0,16+0,47 = 0,63 < 1,0$.

Therefore, the strength of the cantilever tube is verified.

Verification of the diagonal strut $26 \times 3,5$, AlMgSi1, F31

Internal force in the diagonal strut: $-860\,\text{N}$.

$s_k = 1\,662\,\text{mm}$; $i = 8,13\,\text{mm}$; $\lambda = 1\,662/8,13 = 205$;

$\bar{\lambda} = 205/(\pi \cdot \sqrt{70\,000/260}) = 3,97$

$\phi_k = 0,5\,(1 + 0,21\,(3,97 - 0,2) + 3,97^2) = 8,78$

$\chi = 1/(8,78 + \sqrt{(8,78^2 - 3,97^2)}) = 0,060$

$N_{\text{dd}}/N_{\text{pl,Rd}} = 860/(0,060 \cdot 247,4 \cdot 260/1,1) = 0,25 < 1,0$.

Therefore, the strength of the diagonal strut is verified.

13.8.5   Foundation

A driven pile foundation with a H-beam steel wall pile Psp370 (steel grade S235) is adopted for this example. The data of the Psp370 pile are:

$$
\begin{array}{llll}
W_{\text{xel}} = 2\,290 \cdot 10^3 & \text{mm}^3 & I_y = 42\,350 \cdot 10^4 & \text{mm}^4 \\
W_{\text{yel}} = 804 \cdot 10^3 & \text{mm}^3 & I_z = 15\,280 \cdot 10^4 & \text{mm}^4
\end{array}
$$

Effective width $b = 0,38\,\text{m}$, pile length selected $5,0\,\text{m}$.

Site: Plain terrain, bearing soil: sand, $1,0\,\text{m}$ below surface. Surface of soil $0,5\,\text{m}$ below top of foundation, high water table.

Loading with partial factors γ_F included:

$$
\begin{array}{lll}
Q_{\text{yd}} = 8,5\,\text{kN} & M_{\text{xd}} = 63,5\,\text{kN} \cdot \text{m} & h_z = 63,7/8,5 = 7,47\,\text{m} \\
Q_{\text{xd}} = 6,3\,\text{kN} & M_{\text{yd}} = 51,2\,\text{kN} \cdot \text{m} & h_z = 51,2/6,3 = 8,13\,\text{m}
\end{array}
$$

Resulting moment: $M_{\text{Rd}} = \sqrt{60^2 + 51,2^2} = 78,9\,\text{kN} \cdot \text{m}$,

Resulting horizontal force: $Q_{\text{Rd}} = \sqrt{7,6^2 + 6,3^2} = 9,9\,\text{kN}$.

Effective height for action of resulting force above bearing soil

$$h_{\text{zd}} = 78,9/9,9 + 0,5 + 0,5 = 9,0\,\text{m} \quad .$$

Unit weight of soil $\gamma_E = 10\,\text{kN/m}^3$; angle of internal friction $\phi_R = 30°$

Specific soil pressure according to Table 13.18: $\lambda_p = 3,0$; depth of maximum internal moment (see Equation (13.114))

$$z_m^3 + 3 \cdot 0,38 \cdot z_m^2 = 6 \cdot 9,9/(10 \cdot 3,0) = 1,98 \quad .$$

This yields $z_m = 0,97\,\text{m} \sim 1,00\,\text{m}$.

Embedding length t_E to be obtained from (see Equation (13.117))

$$t_E^3(t_E + 4 \cdot 0,38)/(t_E + 9,00) - 4z_m^2(z_m + 3 \cdot 0,38) = 0 \quad .$$

Therefore,	$t_E \approx 2,80\,\text{m}$
Embedding length	$t_0 = 1,2 \cdot 2,80 = 3,40\,\text{m}$
Total length of pile (13.118)	$t_{\text{pile}} = 3,40 + 0,5 + 0,5 - 0,2 = 4,20\,\text{m}$
Permissible steel stress of pile	$235/1,25 = 188\,\text{N/mm}^2$

$$
\begin{aligned}
M_x &= 60 + 8,5(0,5 + 1,0 + 1,0) \sim 82\,\text{kN} \cdot \text{m} \\
M_y &= 51,2 + 6,3(0,5 + 1,0 + 1,0) \sim 67,0\,\text{kN} \cdot \text{m} \\
\sigma_{\text{st}} &= 82 \cdot 10^6/2\,290 \cdot 10^3 + 67,0 \cdot 10^6/804 \cdot 10^3 = 120\,\text{N/mm}^2 < 188\,\text{N/mm}^2.
\end{aligned}
$$

Displacement of pile top under characteristic loads without partial factors (see (13.119))

$$
\begin{aligned}
f_{\mathrm{pile}x} = {}& (8,5/1,3) \cdot 10^3/(210000 \cdot 42350 \cdot 10^4) \cdot [(7,90 + 1,0 + 0,65 \cdot 3,4)^3/3 \\
& - (7,90 + 1,0 + 0,65 \cdot 3,4)^2 \cdot 7,90/2 + 7,90^3/6] \cdot 10^9 = 3,8\,\mathrm{mm}
\end{aligned}
$$

and

$$
\begin{aligned}
f_{\mathrm{pile}y} = {}& (6,3/1,30) \cdot 10^3/(210000 \cdot 15280 \cdot 10^4) \cdot [(8,13 + 1,0 + 0,65 \cdot 3,4)^3/3 \\
& - (8,13 + 0,5 + 0,65 \cdot 3,4)^2 \cdot 8,13/2 + 8,13^3/6] \cdot 10^9 = 8,0\,\mathrm{mm} \quad .
\end{aligned}
$$

The foundation complies with the requirements.

13.9   Bibliography

13.1   *Kiessling, F.; Nefzger, P.; Nolasco, J. F.; Kaintzyk, U.*: Overhead power lines – Planning, design, construction. Springer Publishing, Berlin – Heidelberg – New York, 2003.

13.2   *Wiesner, W.*: Oberleitungen–Statische Bemessung nach DIN EN 50 119 (Overhead contact lines – Static analysis according to EN 50 119). In: Elektrische Bahnen 108(2010)10, pp. 446 to 452.

13.3   *Bauer, K.-H.; Stotz, W.*: Rammrohrgründungen für Betonmaste (Driven tube foundations for concrete poles). In: Elektrische Bahnen 78(1980)10, pp. 260 to 264.

13.4   *Dubbel*: Taschenbuch Maschinenbau (Mechanical engineering hand book). Springer Publishing, Berlin – Heidelberg – New York, 11th edition, 1970.

13.5   *Altmann, S.*: Die grafische Bestimmung der Quer- und Richtseillängen bei Fahrleitungen für 15 kV und 16,7 Hz (Graphical determination of head-span and cross-span wire lengths for AC 15 kV 16,7 Hz contact lines). In: Signal und Schiene 6(1962)11, 12, pp. 410 to 415, pp. 455 to 458 and 7(1963)1, pp. 25 to 32.

13.6   *Sachs, K.*: Die ortsfesten Anlagen elektrischer Bahnen (The fixed installations of electric railways). Orell-Füssli Publishing, Zürich – Leipzig, 1938.

13.7   *Süberkrüb, M.*: Technik der Bahnstrom-Leitungen (Technology of overhead contact lines). Wilhelm Ernst & Sohn Publishing, Berlin, 1971.

13.8   *Petersen, C.*: Stahlbau (Steel structures). Vieweg Publishing, Braunschweig, 3rd edition, 1993.

13.9   *Kindmann, R.*: Stahlbau kompakt (Steel structures). Stahleisen Publishing, Düsseldorf, 3rd edition, 2014.

13.10 *Hütte I*: Des Ingenieurs Taschenbuch, Theoretische Grundlagen (The engineer's hand book, Volume I, basic theories). Wilhelm Ernst & Sohn Publishing, Berlin, 28th edition, 1955.

13.11 Grundbautaschenbuch (Soil mechanics hand book). Wilhelm Ernst & Sohn Publishing, Berlin, 3rd edition, 1980.

13.12 *Terzaghi, K.; Pech, R.*: Bodenmechanik in der Baupraxis (Soil mechanics in civil engineering practice). Springer Publishing, Berlin – Heidelberg – New York, 1961.

13.13 *Heitfeld, K. H.*: Ingenieurgeologische Probleme im Grenzbereich zwischen Locker- und Fest-
       gestein (Geological engineering problems within the transition between loose soil and rock).
       Springer Publishing, Berlin – Heidelberg – New York, 1985.

13.14 *Sulzberger, G.*: Die Fundamante der Freileitungstragwerke und ihre Berechnung (The founda-
       tions for overhead line supports and their calculation). Bull. des Schweizerischen Elekrotech-
       nischen Vereins 36(1940), pp. 240 to 243.

13.15 *Blum, H.*: Wirtschaftliche Dalbenformen und deren Berechnung (Economic design of piers
       and their calculation). Bautechnik 9(1932)2, pp. 50 to 55.

14 Designs for special applications

14.0 Symbols and abbreviations

Symbol	Definition	Unit
H	clearance of contact wire from road surface	m
h	clearance of height limits from road surface	m
h_{veh}	permissible height of road vehicles	m

14.1 Introduction

Contact lines for special applications are systems or components, which differ by their mechanical and electrotechnical specialties from the standard solutions specified in catalogues for contact line types. Contact lines in *maintenance workshops*, *loading facilities*, *crossings of differing installations* and *special railway systems* belong to this category.

Such designs, often only traversed at lower speeds than the standard contact lines, form provisional or permanent alternatives on upgraded lines to expensive constructions or modifications of bridges, which would jeopardize the productivity of electrification. This chapter introduces selected, historically and technically interesting systems and advices on planning and designing of such special systems.

14.2 Maintenance installations

As a rule, the *maintenance of electric traction vehicles* is generally carried out at indoor depots. The overhead contact line is extended into the depot and designed so that it may be disconnected on each maintenance track. Within indoor depots, *overhead conductor rails*, as described in Clause 11.3, are used as an alternative to conventional elastic contact lines. Examples are presented in [14.1, 14.2]. Figure 14.1 shows the transition between the conductor rail in the depot and overhead contact line in front of the depot.

Doors of depots are designed to accommodate the contact wire with openings lined with insulating material to provide electrical clearances according to EN 50 124-1 whilst keeping birds out. The gap of the contact line needs to meet the insulation coordination requirements according to EN 50 124-1. *Separators for rolling doors* permit closing of such doors (Figure 14.2).

Contact line disconnectors are used to isolate and earth overhead contact lines in depots and maintenance facilities. They are equipped with an earth contact. In the disconnected condition they can be barred and locked. The key, which can only be removed in the disconnected and earthed condition, enables opening of the entrance to the work platforms. In the case of installations with more than two tracks, the overhead contact lines of all tracks can be switched-off by a *master switch* arranged outside the building. Figure 14.3 shows a typical example of such an arrangement used in DB's Munich depot.

Using local control panels, the disconnectors supplying the individual maintenance tracks, can be controlled and their position monitored. The earthing of the contact line is carried out

Figure 14.1: Guiding a Siemens conductor rail through the door of a maintenance facility at DB Munich.

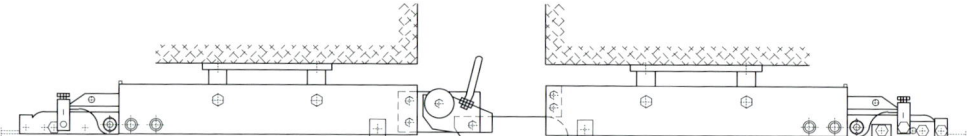

Figure 14.2: Separator for rolling doors.

by means of a manually operated *earthing switch*. Its switching position is also displayed at the local control console. A gate arranged above the control console, equipped with six captive keys, locks the earthing switch in its closed position and protects maintenance personnel from unauthorized and premature release of the earthing. Only when all authorized maintenance personnel insert their keys into the gate, the earthing disconnector can be operated and the overhead contact line re-energized.

Figure 14.4 shows a working platform in DMRC's depot Mukundpur. Another example for disconnecting and earthing overhead contact lines in workshops is shown in Figure 14.5, installed in DB's Berlin-Rummelsburg workshop. Here, disconnecting and earthing is carried out automatically. Initiated by the control command: *contact line to be earthed* the automatic control opens the feeding disconnectors section by section then connects the overhead conductor rail to earth in front of and within the workshop, activates the traction signal EL6 (see Figure 17.48) and triggers the closing of the visible motor-operated earthing switch at site. After locking of the earthing switch the automatic control releases the safety keys and the power supply of the lifting platforms.

Removal of the earthing and reclosing of the overhead contact line power following work activities can only be carried out if the lifting platforms returned to their parking position and all keys were returned to the key lock at the control console.

All connecting and transition conditions are signalled at the control console. In emergency cases the simultaneous earthing of all overhead contact lines within the workshop can be initiated by several emergency buttons which are installed at various positions within the workshop together with the automatic control and acoustic and optical alarm signals. This easy to handle and reliable automated switching-off and earthing installation has been well proven under frequent operational conditions since 2002. The application of the automated earthing system Sicat AES for workshops is described in Chapter 17.1.7.2.

Figure 14.3: Device to disconnect and earth the overhead contact lines in a railway depot.

Figure 14.4: Working platform in DMRC's depot Mukundpur.

a) Cover closed

b) Cover open

Figure 14.5: Control console
with key cover closed and open
in DB's Berlin-Rummelsburg
workshop (Photo: DB).

Figure 14.6 shows the *turntable equipment* in front of a circular engine shed in DB Schenker's depot in Nuremberg. The radially arranged contact wires supply the electric locomotives on all running tracks via one of both pantographs. Turning of the turntable is only permitted when the pantographs are lowered. The return current flows to the main tracks by cable connections. Many newly constructed locomotive depots have a rectangular layout and the distribution of the locomotives to the individual maintenance tracks is done with a travelling platform, with the pantographs of the locomotives lowered. The tracks are equipped with conventional overhead contact lines.

Specifically designed contact line installations are used in *workshops for washing and de-icing* of electric traction units. The contact line is isolated to avoid any contact between the water and the live contact line when washing the vehicle roofs. In such a section the vehicles are moved by winches or with the second pantograph which is outside and isolated from the washing area. The latter is the case with the German high-speed train ICE 1. To be able to clean other high-speed trains like the DB's ICE 2 and ICT the washing facility in the Munich depot has been re-fitted with a winch in the centre section and an overhead contact rail in the adjacent section. An overhead contact rail supplies the trains in the adjacent sections.

a) Depot Nuremberg   b) Depot Cologne

Figure 14.6: Contact line installation over a turntable (Photo 14.6 a): Nitzinger).

a) Operational condition   b) Cutting procedure   c) Contact wire cutted

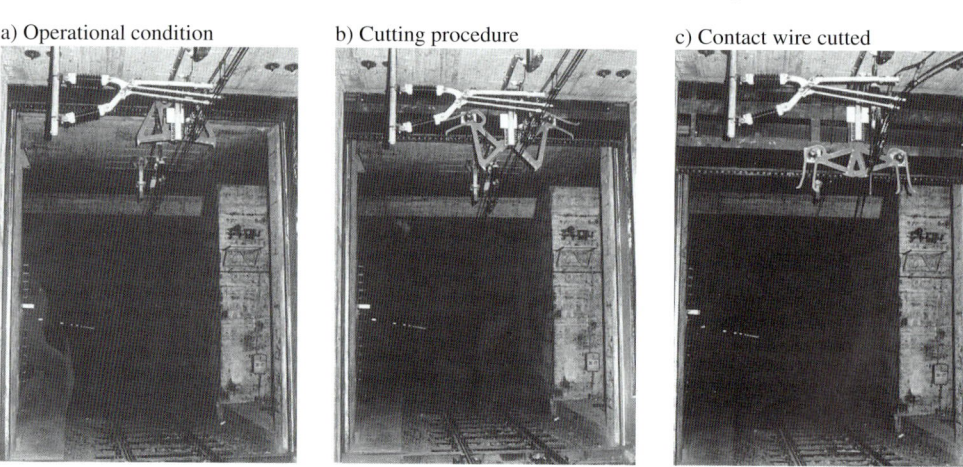

Figure 14.7: Overhead contact line with clipping devices at a flood gate door in Frankfurt (Photo: Liebig).

14.3   Tunnel seals

Flood gate doors are used to seal tunnel sections against flood waters. They are designed as drop gates which remain functional even when the power supply fails. The report [14.4] describes two different designs for removing the overhead contact lines from the operating area of flood gate doors.

In Munich, the railway passes below the river Isar. In this tunnel, the overhead contact line on both sides of the flood gate doors are terminated and the gaps between the two contact lines are closed by 6 m long *conductor rails*. The falling gate unlocks the conductor rails at one end using rollers. The conductor rail is released and pivots around a hinge at the far end like a pendulum. The overhead conductor rail then falls out of the clearance gauge of the flood gate door. These contact line sections are arranged in the starting and braking section of the line and are negotiated at 40 km/h. There is a stepwise transition in elasticity from the elastic contact line to the rigid conductor rail with the latter showing high wear and needing regular replacement.

Another design was adopted for the urban mass transit railway which crosses the river Main in Frankfurt. There, *clipping devices* are installed to cut the overhead contact line equipment during the lowering of the flood gate doors after automatic disconnection and earthing of the contact lines (Figure 14.7).

Latching in of the tension wheel arrangements limits the damage of the overhead contact line to the sections cut. As an additional security measure, auxiliary midpoint anchors before and after the flood gate doors help to keep the contact lines in position. A special device for unloading the overhead contact line enables the installation of prepared contact wires and conductor sections after clearing the disturbance. This special device is also used to dismantle the contact line between splices during the annual functional checks of the flood gate doors. These installations have worked satisfactorily during train operations at 80 km/h and during the annual checks.

14.4 Interface between electrification systems

14.4.1 Introduction and requirements

In Chapter 1 various railway energy supply systems were described. Their geographical use is shown in Figure 1.1. The nature of international railway traffic means that electric trains must be able to operate beyond the limits of individual supply systems.

This is of special significance for the interoperability of the trans-European rail system. Therefore, the TSI Energy [14.5] requires that trains should be able to move from one energy supply system to an adjacent one that uses a different type of energy supply without bridging the two systems.

The operations and design of the interface section depend on the type of both systems as well as on the arrangement of pantographs on trains and on the running speed. There are two possibilities for the installation of interfaces:

– installation of special interfaces on open lines. This method requires the locomotives to be designed for operation with at least two power supply systems. The locomotives can traverse the interface without stopping
– equipment of a railway station with the two adjacent power supply systems. This method requires some tracks of the station to be operated with either of the two power supply systems. The locomotives do not require dual equipment. The trains stop in the station and the locomotives are changed.

The first transition section in a mainline railway were traversed in Modane in France in 1930 [14.6]. The paper [14.7] refers to the type and frequency of traffic and the type of adjacent supply systems to be connected as crucial to the design of the electrical installation.

14.4.2 System separation sections on interstation lines

System separation sections on interstation lines are traversed by *multiple system traction units*. With reference to [14.5], there are two methods for the train to run through system separation sections on interstation lines:

– with pantograph raised and touching the contact wire, e. g between AC 15 kV and AC 25 kV
– with pantograph lowered and not touching the contact wire, e. g between DC 3 kV and AC 25 kV

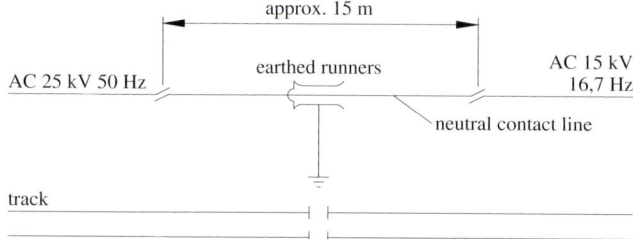

Figure 14.8: System separation between AC 25 kV 50 Hz and AC 15 kV 16,7 Hz, France to Germany (Photo: Behmann).

Figure 14.9: System separation between AC 25 kV and AC 15 kV with earthed cantilevers, France to Germany (Photo: Behmann).

If system separation sections are negotiated with pantographs raised to the contact wire,

- devices on locomotives should automatically open the circuit breaker before reaching the separation section and automatically recognize the voltage of the new power supply system at the pantograph in order to switch in the corresponding circuits,
- the geometry of different elements of the overhead contact line should prevent pantographs from short-circuiting or bridging both power systems and
- provisions should be taken in the energy subsystem to avoid bridging of both adjacent power supply systems when the opening of the on-board breaker fails.

If the *system separation sections* are traversed with pantograph lowered,

- at least 200 m before running through separation sections the traction units' main circuit breaker should be opened triggered by control signals without the driver's intervention,
- the pantograph should be lowered without the driver's intervention, triggered by control signals and
- the design of the interface section between differing power supply systems should ensure that bridging, by a pantograph, of the two power supply systems is avoided and isolating both supply sections is triggered immediately. Triggering of a short-circuit is ensured by operation within insulated sections with an earthed cantilever.

At system separation sections between DC and AC systems, the pantographs are changed usually because DC systems require higher currents at the same power. Pantographs for DC systems may have increased head masses and contact forces compared to AC pantographs.

A system separation section between SNCF's AC 25 kV 50 Hz network and DB's AC 15 kV 16,7 Hz network is installed on the line Metz (France) to Saarbrücken (Germany) (Figures

Figure 14.10: Earthed cantilevers of system separation between AC 25 kV and AC 15 kV in detail, France to Germany (Photo: Behmann).

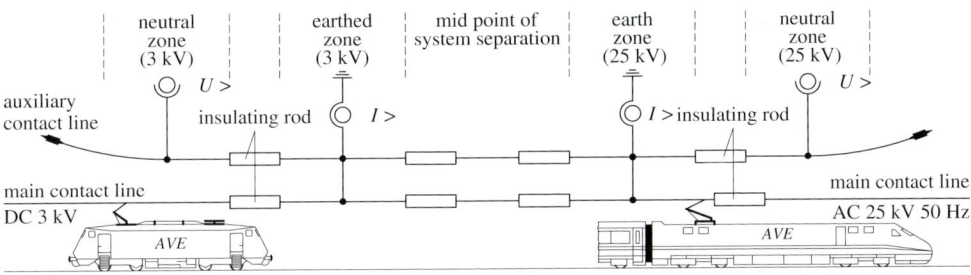

Figure 14.11: System separation section on the Madrid–Seville line. In operation from 1992 to 2002 [14.9].

14.8 to 14.10). There is an approximately 15 m long neutral section between two section insulators. Since changing the pantographs is required, the system separation is passed with dropped pantographs, the train coasting through the section. If a pantograph was unintentionally raised voltage could be transmitted to the neutral section by an arc at the section insulator. In this case the pantograph would trigger a short-circuit at the earthed cantilever and the feeder lines would be switched-off to prevent short-circuiting of both supply systems. The design has proved to be satisfactory since 1974 [14.8].

The separations between AC 25 kV 50 Hz and DC 3 kV which were in operation on the Madrid–Seville line from 1992 to 2002 (Figure 14.11) [14.9] before the DC sections were converted to AC 25 kV [14.10]. Voltage transformers are arranged at the neutral section of auxiliary contact lines to trigger switching-off of the supplying substation circuit breaker should pantographs be inadvertently raised. If this device failed, running into the earthed section would result in a short-circuit and switching-off of both adjacent power supply systems by the contact line protection. Separate earthed line sections were arranged for the DC 3 kV and the AC 25 kV sides. Section insulators made of synthetic material are used to insulate the different contact wire sections. They enable the passage of pantographs in emergency cases up to speeds of 280 km/h.

A system separation section of the SNCF between AC 25 kV and DC 1,5 kV is shown in Figure 14.12. It is formed by *protective sections* with earthed parts designed as overhead contact

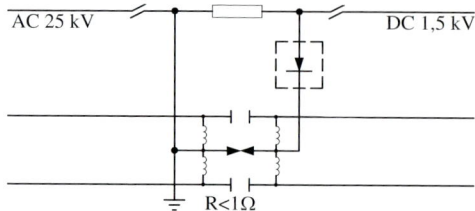

Figure 14.12: Scheme of a system separation section between AC 25 kV and DC 1,5 kV in the SNCF network, France.

lines equipped with a section insulator. It is equipped with a power electronic diode, insulated track sections and an impedance bond for the separation of the return current systems. It is operated with lowered pantographs.

Where system separations are between closely related power supply systems with differing voltage levels, for example DC 1,5 kV to DC 3 kV or 15 kV AC 16,7 Hz to 25 kV 50 Hz, neither pantograph changes nor dropping are required. However, as in cases of phase separation sections, the main circuit breaker on-board must be switched-off. If the switching operation is not carried out, the *unintentional energizing of the earthed section* will result in a short-circuit and switching-off of the feeding substation circuit breaker, triggered by the contact line protection.

14.4.3   Stations with two power supply systems

Multi-system traction units need a higher investment than single-system traction units. Where heavily trafficked electrified lines extend on both sides of power system borders, a large number of multi-system traction units would be necessary to provide unrestricted operation on both systems. The alternative is to use stations equipped with both adjacent traction power supply systems to allow more economic single system traction units. These *system separation stations* enable the arrival of trains hauled by traction units with one system and its departure hauled by traction units of the adjacent system from the same track. The stations also provide shunting facilities for changing the traction unit. The layout designs and circuit diagrams for stations provided with two types of traction power supply vary widely, because stations with existing track layouts needed to be equipped with two energy supply systems. There are stations with longitudinal and transverse separation of the contact lines (Figure 14.37) without switching the contact line, as well as stations where the contact lines of several tracks can be supplied with both power supply systems.

Where contact lines are for one supply system only, the single system traction units negotiate the system separation sections coasting with dropped pantographs. Shunting locomotives with independent drives, e. g. diesel engines are used to return them to their home system. If traction unit pantographs were not dropped in front of the system separation section, running into the earthed section would trigger a short-circuit and cause the opening of the circuit breakers in the supplying substation.

The number of tracks to be equipped with *switchable overhead contact lines* depends on the track layout and the operational usage. The section insulators arranged above track insulating sections divide the overhead contact line into individual switching sections that are fed by specific switching posts. The simplified diagrams, Figures 14.13 and 14.14 show switching diagrams for stations equipped with two power supply systems.

When locking-in the route for the trains, automatic switching of the power supply system is carried out. Because track points, disconnectors and signals are interlocked, running from

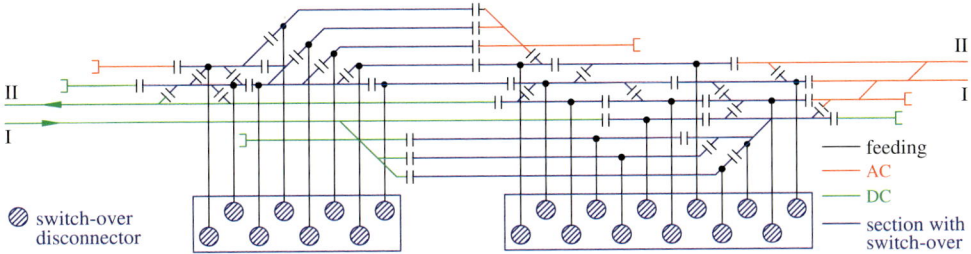

Figure 14.13: Overview of a system change between AC 25 kV 50 Hz and DC 3 kV in a railway station in Russia [14.11].

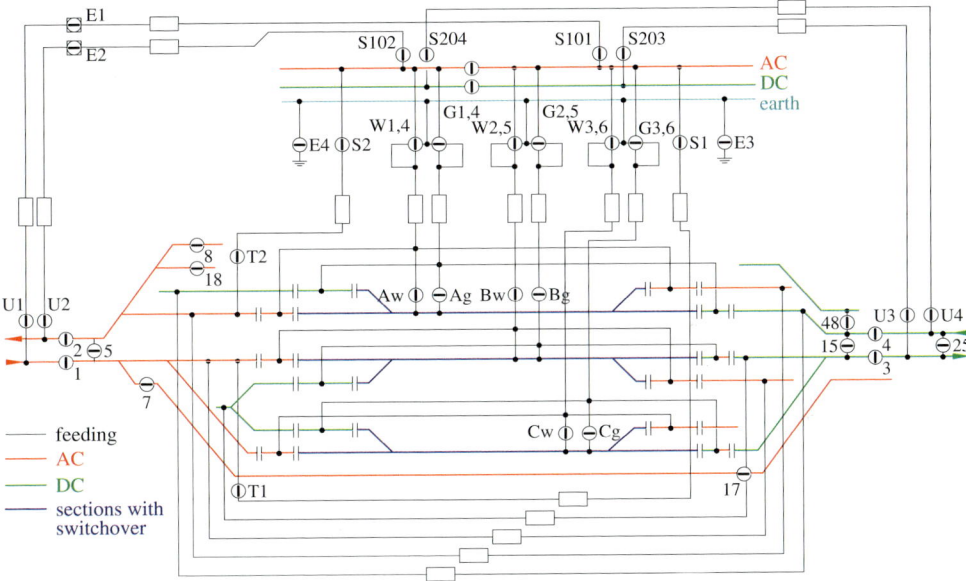

Figure 14.14: Circuit diagram of a system change railway station Emmerich between AC 15 kV 16,7 Hz and DC 1,5 kV, Germany to Netherlands.

disconnectors 1, 2, 3, 4 separation to interstation line; 5, 15, 25 cross coupling; 7, 17, 8, 18 supply of sidings; E1, E2, E3, E4 earthing; G1,4, G2,5, G3,6 supply of switchover tracks; W1,4, W2,5, W3,6 supply of AC tracks; S101, S102 supply of AC system; S203, S204 supply of DC system; S1, S2, T1, T2 supply of AC end section; U1, U2, U3, U4 disconnectors for feeder lines, Aw, Bw, Cw AC supply of switchover tracks; Ag, Bg, Cg DC supply of switchover tracks

one switching section into an adjacent one is only possible if both are fed by the same power supply system [14.11].

The insulation level of the overhead contact line corresponds to the system with the higher nominal voltage. Protection against stray current corrosion is necessary for all poles at the station.

Stations provided with two types of traction power supply permit mixed operation of single-system and multi-system traction units. For slow trains, single-system traction units with engine changes are generally used while rapid trains adopt multi-system traction units which do not require time consuming locomotive changes with associated shunting operations.

14.4.4   Operating AC and DC trains on the same track

Similarly to stations with dual power supply systems where several track sections are used by trains supplied by differing electrification systems *common track sections* are operated alternatively by AC or DC systems. The tracks used for both systems also can serve common platforms. This may ease changing of trains at the same platform. Hohenneuendorf north of Berlin is an example with a common platform track for the mainline and commuter system with an AC 15 kV 16,7 Hz overhead contact line and a DC 0,6 kV conductor rail aside the tracks. This track section is isolated from the mainline track system and the traction current for the mainline is returned via an isolating transformer.

Close to Zürich in Switzerland the Ütliberg railway supplied by DC 1,5 kV and the Sihltal railway supplied by AC 15 kV 16,7 Hz run on the same tracks between Giesshübel and Selnau [14.12]. Each system uses its own overhead contact line which is isolated from the other. The DC overhead contact line is arranged 1,30 m aside the track axis and is negotiated by a current collector arranged laterally at the traction unit. Each current collector device respects the clearance gauge of the other system. The overhead contact lines arrangement and tracks as well as track magnets and point lanterns avoid short-circuiting of the two systems due to fault propagation. At the crossing of both overhead contact lines a protective section is arranged which can be supplied by AC 15 kV 16,7 Hz as required. The connection condition of this section is signalled to the cab driver via signal lights. To avoid stray current problems the common line section was constructed adopting design rules for DC railways. The poles are insulated from the foundations and bonded to the track. The steel reinforcement of tunnels of crossing AC 15 kV 16,7 Hz lines is equipped with cathodic protection against stray currents. EN 50 122-3 stipulates the requirements for electric safety for earthing and return conductor in case of parallel operation of AC and DC.

14.5   Movable bridges

14.5.1   General

Movable bridges enable crossings between railway lines and shipping lanes without limiting the clearance height for shipping or constructing bridges with high clearances. When equipping such bridges with contact lines, the constraints of both traffic systems must be met.

In the Netherlands, some movable parts of bridges were not equipped with contact lines but are operated in a coasting condition with the pantographs dropped. *Overhead conductor rails* similar to those described in Clause 14.7.2 on both sides of the bridges would gradually guide the pantographs to their upper position and push them down to their normal level after passing the bridge, if the driver failed to drop the pantograph. A pre-condition for such a procedure is that the development of pantographs exceeds the maximum contact wire height only by a small amount. Where bridges are situated close to stations or signal positions, stopping of coasting trains is likely and an uninterrupted power supply to the traction units is necessary.

Turning, folding and lifting railway bridges were constructed many decades ago and many are monuments of the art of engineering. The electrification of such bridges, must be preceded by a thorough check of the bridge's loading capability. Some bridges cannot carry the loads from dead-end poles and from tensile forces of elastic overhead contact lines that provide optimum current collection. These bridges are often equipped with overhead conductor rails, the limitations of which, together with the vibrations from the bridges impose limits on the

Figure 14.15: Folding bridge close to the City of Papenburg, North-West Germany.

running speed. Trains usually coast over these bridges with pantographs dropped. The overhead contact line serves as an emergency running surface for pantographs unintentionally left in the raised position or for feeding starting traction units with pantographs on the contact wire. Paper [14.13] describes installations of *overhead conductor rails* on movable bridges when electrifying the line New Haven–Boston with AC 25 kV.

14.5.2   Folding bridges

Folding bridges with balance beams offer sufficient space for conductor rails. In the case of folding bridges equipped with counter balance weights moving on the opposite side of the pivot into the gauge of the railway line, runners or conductor rails are used which are moved out of the operational range of the balance weights before the folding process of the bridge is started. An example is given by the two Papenburg folding bridges in North-West Germany [14.14], where the balance beam was extended and used as a support for the conductor rails. The counter balance weight of both bridges had to be enlarged correspondingly. A pair of runners similar to those used for section insulators (Figure 14.15) provides the connection to the elastic contact lines terminated on both banks of the river. An elastic cantilever above the runners and above the conductor rail is used to adjust the running level for the pantograph and damps the oscillations. Contact wire pieces clamped to the contact line in front of the bridge reduce the elasticity continuously in direction to the bridge.

DB's railway line Bremen to Emden crosses two *folding bridges* close to the city of Oldenburg [14.15]. There, portals on both banks are used to terminate the contact lines. They also support the swivelling parts of the conductor rails and the corresponding drives and operating linkage on the side of the bridge where the balance weights are arranged. The four short overhead contact line sections arranged on the flaps are terminated rigidly at the dead-end portal in the middle of the bridge and flexibly at the weight casings by means of spring-type tensioning devices.

The folding bridge across the *River Peene close to Anklam* in East Germany and that across the stream Ziegelgraben close to Stralsund are equipped with *conductor rail overhead lines* on their movable parts and the adjacent bridge sections. Their swivelling cantilevers are provided with a drive to turn the overhead contact line out of the reach of the balance weights (Figure 14.16 a) and b)) before opening the bridges. In view of the adverse dynamic charac-

a) Swivelling cantilevers are provided with a drive to turn the overhead contact line out

b) Bridge in position *open*

c) New half of the bridge with a compound conductor rail

Figure 14.16: Folding bridge across the river Peene close to Anklam, Germany (Photo: K. Schatkowski) .

Figure 14.17: Bascule bridge with overhead conductor rails type Furrer+Frey, on New Haven–Boston line, USA (Photo: Furrer+Frey).

teristics of the overhead contact line, which is affected by the vibrations of the bridges and as a result of running tests, the maximum running speed with raised pantograph was limited to 20 km/h for the Ziegelgraben bridge and to 10 km/h for the Peene bridge. However, usually the latter bridge was traversed with dropped pantographs and the train in coasting condition. During 2012/2013 the folding bridge across the Peene River was renovated. The new bridge is equipped with a hydraulic drive and does not need counter weights [14.16]. Therefore, the design of the conductor rail contact line was less sophisticated and the line on the bridge can be traversed at 120 km/h with pantographs lifted (Figure 14.16 c)).

The New Haven to Boston line electrified by Amtrak in 1999 crosses the Connecticut River, Niantic River and Thames River on bascule bridges which were equipped with overhead conductor rails (Figure 14.17) designed for 145 km/h maximum speed [14.13]. To enable bridging of the gap between movable and fixed parts of the bridge a movable conductor rail unit is used. It comprises of a portal structure which moves on running rails mounted on girders arranged between the pins. The movable conductor rail unit is mounted on the portal structure and provided with a mechanism that enables to move the contact line out of the line gauge (Figure 14.17). Interlocks in the control system prevent the bascule bridge span to be opened before the movable conductor rail unit has been fully retracted.

14.5.3 Swivelling bridges

Swivelling bridges, also known as swing bridges, rotate by 90° around a vertical pivot supported on a pillar arranged in the middle of the bridge and open shipping lanes on both sides. The swivelling bridge across the river Hunte close to Elsfleth, North Germany, (Figure 14.18)

Figure 14.18: Swivelling bridge across the river Hunte close to the City of Elsfleth, North Germany, turning in a lifted position.

Figure 14.19: Swivelling bridge with swivelling overlaps made of overhead conductor rails type Furrer+Frey on New Haven–Boston-line, USA (Photo: Furrer+Frey).

Figure 14.20: Lifting bridge crossing the Kattwyk waterway within the Hamburg seaport, Germany.

is an example of such an installation which was electrified at the time of DB's Bremen–Hude–Nordenham line electrification in 1980 [14.15]. Before starting the swivelling operation, the track locks are released and the bridge is lifted by 0,36 m at its centre and by 0,20 m at the free ends. On the swivelling part of the bridge there is a 200 mm^2 overhead conductor rail supported by rectangular hollow galvanised steel sections. The overhead conductor rail can accommodate temperature-induced length variations by means of movable supporting clamps. Suspension insulators in a triangular arrangement provide the connection to the supporting beams. Laterally latching runners, arranged at the moving part of the bridge, form overlaps with the counterparts of the overhead contact line arranged at the adjacent rigid parts of the bridges or on the river banks.

The Amtrak line referred to in Clause 14.5.2 also crosses Shaw's Cove River and Mystic River on *swivelling bridges*. The movable parts are equipped with conductor rails. The transition from the overhead contact line to the conductor rail is arranged on the fixed bridge heads. Swivelling components bridge the gap between fixed and movable parts at both ends and motor-operated allow for enough space for bridge operation (Figure 14.19) [14.13].

a) Operation position   b) Lifted position

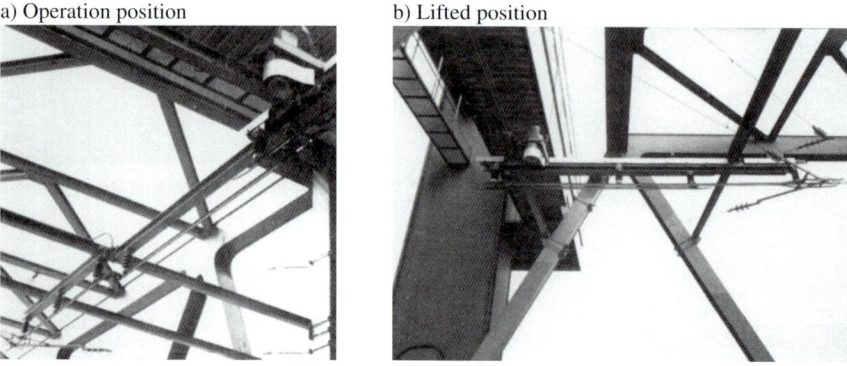

Figure 14.21: Movable conductor rails of the Kattwyk lifting bridge.

14.5.4 Lifting bridges

The movable section of a *lifting bridge* glides vertically along pillars arranged at both sides of the shipping lane. In the position "opened for the navigation", the raised section of the bridge limits the clearance for ships, the tallest vessels operating on the shipping lane and the maximum water level determine the required lifting height. The lifting height of bridges varies from a few metres up to 63 m as with the bridge crossing the river Maas in Belgium.

The lifting bridge across the Kattwyk waterway within the Hamburg seaport, Germany, depicted in Figure 14.20 was commissioned in 1973 and is used for rail and road traffic. Its lifted section is 106 m long and raised by around 46 m. In 1983, the lifting section of the bridge was equipped with an AC 15 kV 16,7 Hz *overhead conductor rail* supported by the structural steel work of the bridge [14.17]. The elastic contact line equipment is terminated on both sides 11 m before the end of the rigid parts of the bridge and is continued by an 8 m long rigid overhead conductor rail. This conductor rail ends 3 m before the transition from the fixed parts of the bridge to the movable part, since the balance weights of the lifting bridge reach into the clearance of the railway and road in the raised position of the bridge. The remaining gaps are bridged by approximately 8 m long movable overhead contact line sections, which are moved by steel ropes from the fixed part and can be locked with the contact line on the lifting part of the bridge (Figure 14.21). These devices are equipped with 5,5 m long conductor rails, the contact blades of which act as the electrical connection with the contact line on the lifting section. Three meter long conductor rails serve as transitions to the rigidly terminated contact line equipment on the movable bridge section.

14.5.5 Electrical connections and signalling

Electrical connections and signalling must co-ordinate and interlock the moving of the bridge with railway operation and energizing of the contact line. The contact line is de-energized before opening the movable part of the bridge and can only be re-energized after closing and locking the movable parts. On either side of the bridges there are interfaces or neutral sections which are earthed after de-energizing the bridge section. The arrangement depicted in Figure 14.22 accommodates switching operations and the continuation of supply to successive feeding sections by by-pass lines, for which cables are often installed in the river bed or in conduits. When constructing overhead power lines for this purpose the clearance for naviga-

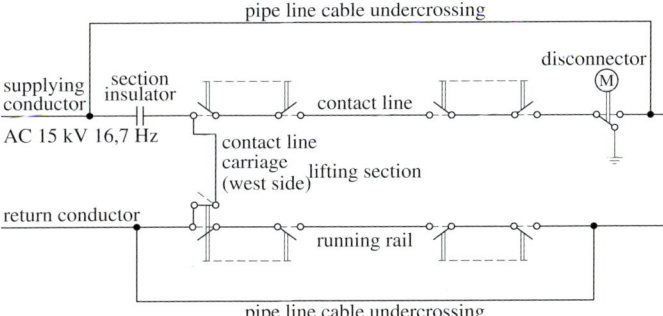

Figure 14.22: Schematic connection diagram of the AC 15 kV 16,7 Hz overhead contact line at the Kattwyk bridge in Hamburg, Germany.

tion being 65 m must be observed. When planning the installation, the security of the *traction return current* via overhead power lines or cable should also be considered. To ensure security of the return current during operation of a traction unit on the bridge, the movable parts of the Kattwyk bridge were equipped with *contact blades for the return current*, to give an example. Cables connect the running rails of the individual parts of the bridge.

A stop signal is displayed to trains approaching movable bridges in the open position; a separate signal indication of the switching condition of the contact line is not necessary. Procedures and operation manual provide instructions and speed restrictions for bridges where trains traverse with the pantograph raised to the overhead contact line at low speeds and with pantographs dropped at high speeds.

14.6 Level crossings of lines with differing supply systems

14.6.1 Crossings between mainline railways and tramways

Level crossings between mainline railways and tramways are uncommon. Some examples are known in Germany:
- DB railway Schalke-Wanne (AC 15 kV 16,7 Hz) with the tramway Bismarckstraße (DC 0,6 kV) in Gelsenkirchen
- DB railway Huckarde Süd branch to Deusen in Dortmund (AC 15 kV 16,7 Hz) and Dortmund city lightrail system (DC 0,6 kV)
- DB railway Leipzig to Altenburg (AC 15 kV 16,7 Hz) and local tramway in Markkleeberg (DC 0,6 kV) close to Leipzig

Because only pantographs are used on the tramway vehicles, it was feasible to wire the individual lines crossing approximately at a right angle with standard overhead contact line equipment (Figure 14.23).

The contact wire of the mainline railway is arranged underneath the contact wire for the tramway system thus providing more favourable running conditions for the mainline. The difference in height between the crossing contact wires is reduced using additional short contact wire runners clamped to the side of the contact wires and the tramway pantograph is guided to the level of the mainline contact wire. To avoid any longitudinal movement of the contact lines because of temperature changes, the crossing is designed similar to a midspan

Figure 14.23: Level crossing between a mainline railway and a light-rail tramway in Markkleeberg close to Leipzig, Germany.

anchor. The running speeds are limited in the crossing area to avoid arcing and increased wear. They reach 50 km/h for mainlines and 30 km/h for the tramways.

The design of level crossings between mainline railways and local light-rail systems must ensure that connecting the mainline system to the power supply of the light-rail system is not possible. Consequently, the crossing overhead contact lines are equipped with neutral sections or section insulators in all four directions (Figure 14.24).

On the Markleeberg system, when the light-rail vehicles are operating, a disconnector feeds the 0,6 kV potential into the common overhead contact line section. The short neutral sections of the DB contact line are equipped with a permanently earthed control section. End position contacts arranged in the barrier beams control the opening of the disconnectors when the barriers are closed. A neutral potential is achieved at the common overhead contact line section when the crossing is open for mainline operation. The traction units of the main line pass the crossing in a coasting condition with their on-board circuit breakers open. If a traction unit driver forgets to open the on-board circuit breaker, arcing will be initiated at the neutral section with the short-circuit leading to switching-off the feeding circuit breaker in the AC 15 kV substation. The line gradients enable the vehicles to coast out of the crossing section without drive.

An alternative solution is shown in Figure 14.25 that enables the feeding of AC 15 kV into the neutral sections of the mainline railways as well as the central sections. In addition to the usual signalling of the level crossings, the coasting sections of the mainline are protected by El 1 "main circuit breaker open" signals before the crossing and El 2 "main circuit breaker closed" signals after the crossing (see Clause 17.5).

In Figure 14.26 a) a level crossing between a double-track tramway and a double-track commuter railway in Melbourne is shown and in Figure 14.26 b) the adopted crossing component. The trains and tramways run at 20 km/h at the crossing. The trolley wire contact line crosses the catenary of the train. The crossing section is isolated by eight section insulators in the

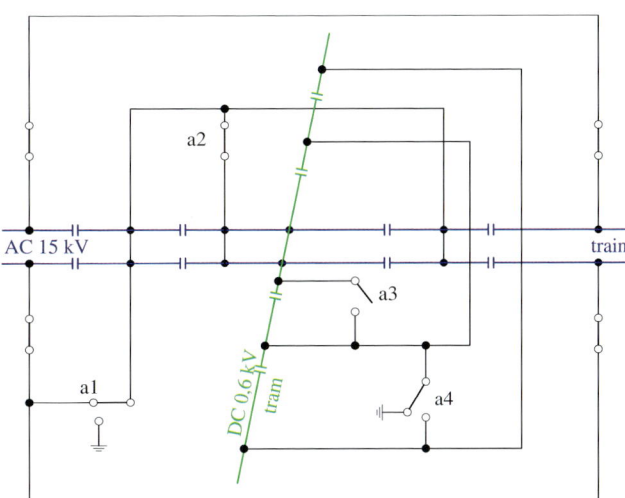

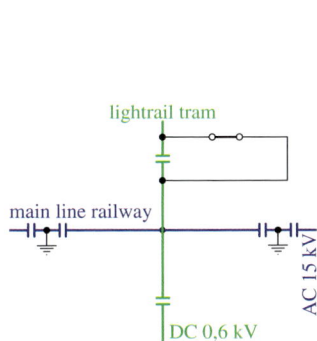

Figure 14.24: Schematic connection diagram of the crossing between mainline railway (AC 15 kV 16,7 Hz) and lightrail tramway at Markkleeberg (DC 0,6 kV) close to Leipzig, Germany.

Figure 14.25: Schematic connection diagram of the crossing between mainline railway and light-rail tramway system at Gelsenkirchen in Germany showing the AC 15 kV 16,7 Hz operational condition.

a) Total view

b) Contact wire crossing element

Figure 14.26: Level crossing between commuter railway and tramway in Melbourne, Australia.

contact wires and four insulators at the messenger wires and is supplied with voltage via a changeover switch and a switching line. In case of open barrier the voltage of the crossing section is DC 0,6 kV and at closed barriers DC 1,5 kV [14.18]. In Figure 14.27 alternatives for contact wire crossing components are shown.

14.6.2 Crossings between tramway and trolley bus lines

The following information will concentrate on *crossings between tramway* systems *and trolley bus lines* [14.19]. The crossing components are designed taking into consideration the pantographs and current collectors used. The majority of tramway systems use pantographs that run under the contact wire. However, the trolley collectors of trolley buses are equipped

a) Melbourne   b) Budapest

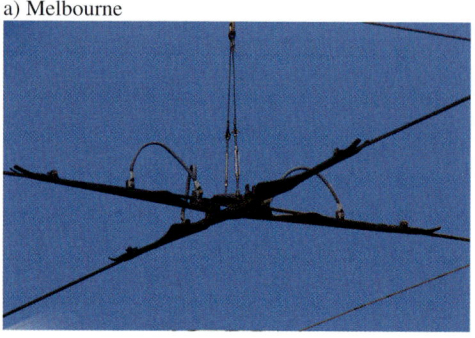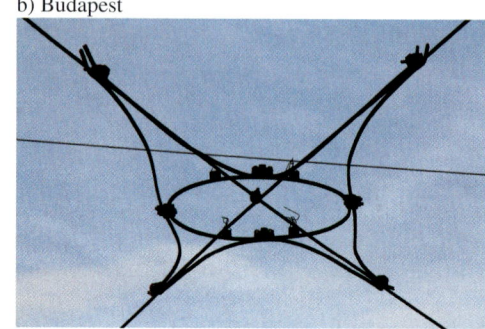

Figure 14.27: Crossing components in tramway contact lines.

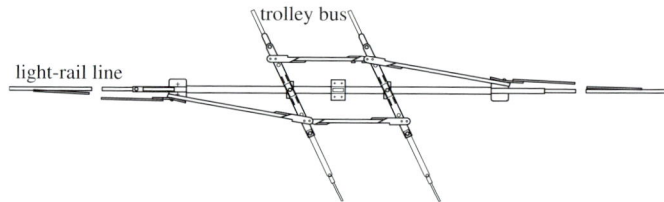

Figure 14.28: Gliding shoe   **Figure 14.29:** Contact line arrangement for an oblique crossing be-
of a trolley current collector.   tween a tramway and a trolley bus.

with *gliding shoes* which are held to the contact wire laterally by the sides of the shoe (Figure 14.28). The trolley shoe must have lateral guidance at all times requiring more complex crossing of the contact wires with pantograph systems. The trolley gliding shoes should be able to cross the contacting level of the tramway pantographs without any obstacles at any time. Two configurations are used depending on the angle of the line crossing. An alternative solution for oblique crossings is shown in Figure 14.29.

For crossings (Figure 14.29) with angles between 15° and 75° a crossing component is used where the light-rail pantograph is guided across the gaps for the trolley contact wires by *insulated gliding runners*.

In the case of perpendicular crossings, often encountered at tramway and trolley bus crossings within city areas, the gaps are more difficult to traverse. With the configuration shown in Figure 14.30 the pantographs of the tramway pass the crossing section utilising the width of the collector strips and the inertia of the pantograph head to bridge the gap. The gap and insulation lengths are too short to damage the collector strips. Because no switching is used with crossings between tramway systems and trolley bus systems, the crossing components have to be insulated with respect to the different potentials of the two trolley bus overhead contact wires.

The design shown in Figure 14.31 is suited for crossings with angles between 15° and 60° [14.19]. Running skids push the pantograph of the tramway approximately 0,10 m below the crossing trolley bus line contact wires. Where the tramway contact lines are automatic-tensioned, ducts above the crossing level enable longitudinal movements of the contact wires. Headspan wires support the contact line crossing in position.

a) Elevation

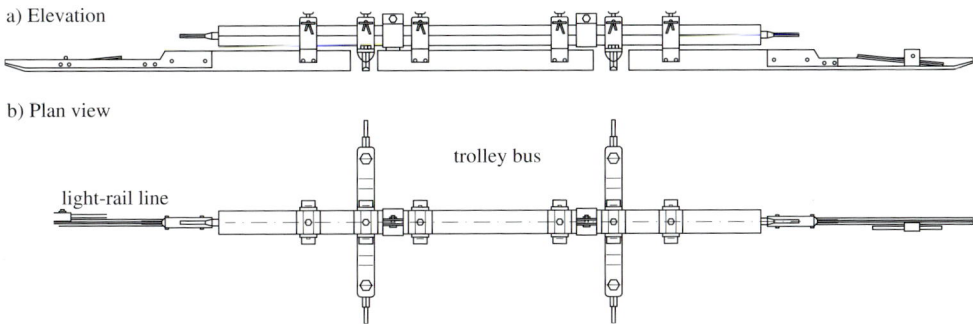

b) Plan view

light-rail line

trolley bus

Figure 14.30: Contact line design for perpendicular crossings between tramway and trolley bus.

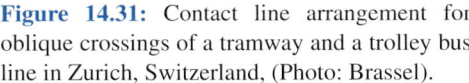

Figure 14.31: Contact line arrangement for oblique crossings of a tramway and a trolley bus line in Zurich, Switzerland, (Photo: Brassel).

Figure 14.32: Rectangular crossing between tramway and trolley bus in Zurich with adjustable crossing elements.

In case of rectangular crossings often encounted between tramways and trolley bus lines within cities it is more difficult to traverse the gaps. The tramway pantographs slip above the crossing utilising the width of the contact strips and the gravity of the pan head as shown in Figure 14.30. The gap and the insulation lengths are too short to damage the contact strips. Since there are no switching operations when traversing such crossings the crossing elements need to be insulated for the differing voltages of the two trolley bus contact wires. A rectangular crossing between tramway and trolley bus in Zurich is shown in Figure 14.32.

Figure 14.33: Cantilevers pivoting in a vertical plane at a level crossing in Biesenthal, Germany.

14.7 Contact line design above road level crossings

14.7.1 Standard and oversize transports

The *road vehicle height* is limited in continental Europe to 4,00 m. In Germany, the Directive on registration of road traffic [14.20] sets out this height. There, level crossings with mainline railway are permitted only up to a maximum speed of 160 km/h. The design of overhead contact lines at level crossings may not passing of vehicles which obey the height requirement. In Clause 12.4.5 the contact line design and additional measures are dealt with which are taken not to violate the clearance for passing of road vehicles. Exceeding the limit of 4,00 m is sometimes necessary by *oversize transports* which have to be declared to and approved by the traffic police. The special arrangements at the overhead contact lines to enable such oversize transports are described hereafter.

If there is no restriction on the contact wire height caused by other structures, the permissible height for passing road vehicles can be increased by lifting the contact line height to suit. For example, for regular oversize transports the *maximum possible contact wire height* at level road crossings is approximately 6,00 m on the DB rail system. This results from the *maximum pantograph development* reaching 6,50 m. In accordance with UIC Code 608 [14.21] this is reduced by contact wire uplift when pantographs pass by and the additional effects like ice and tolerances are considered .

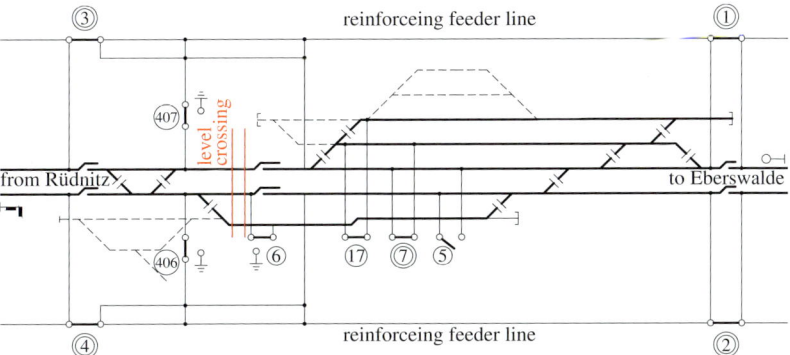

Figure 14.34: Schematic circuit diagram for the station and oversize level crossing at Biesenthal, Germany. 1 to 4 disconnectors for insulating overlap between station and interstation section, 5 cross coupling between main tracks, 6, 7, 17 coupling of sidings, 406, 407 earthing switch of contact line above oversize crossing

14.7.2   Temporary lifting of contact lines by movable cantilevers

If a level crossing is frequently traversed by oversize transports requiring contact wire modification, it may be economic to install contact lines with *adjustable heights* to increase the clearance at level crossings.

DB's Berlin to Stralsund railway line crosses an international route for heavy-weight and oversize vehicle transportation at Biesenthal station. When electrifying the line in 1988 and in view of proposed 300 transports per year with a height of 7,00 m it was decided, as an alternative to the installation of a bridge, to install an d test a *contact line lifting system* will [14.23]. A total of six cantilevers were equipped and fixed to the poles by means of parallelogram-type linkages and corresponding supports which enable the cantilevers to pivot in a vertical plane (Figure 14.33). Each of the three parallel contact lines is lifted by a separate steel wire rope mechanism that is electrically driven and controlled synchronously. Should a failure occur in the electric drives, crank-operated gears enable manual lifting and lowering the cantilevers. The operation takes 20 s electrically and 1 min manually for each direction. Therefore, the crossing of an oversize transport can be carried out during a break between trains. In the lowered position, the overhead contact line permits an unrestricted running speed. The almost doubled tensile force at the stitch wires in the raised position requires increased conductor cross-sections. The weights of the tensioning device do not move during the raising process. Guided by the parallelogram linkage, the cantilever provides the advantage of secure return of the overhead contact line to the initial position, even in case of ice accretion on the lines.

An isolated power supply section with bypass feeder lines (Figure 14.34) enables switching-off the contact lines at the railway crossing and by-passing the power. A similarly designed installation for a double track railway crossing was installed at Jacobsdorf station when electrify the railway line from Berlin to Frankfurt/Oder.

a) Movement upwards   b) Movement downwards

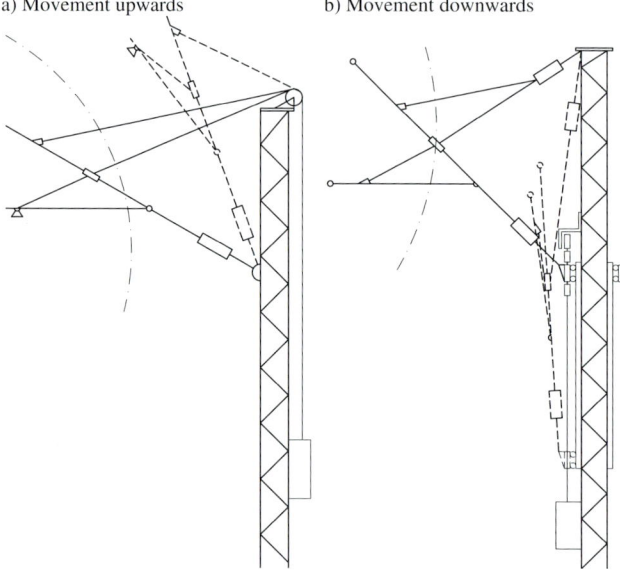

Figure 14.35: Vertically swivelling cantilevers for container stations.

14.7.3 Temporary lifting or removing of contact lines

If passing of over-size transports is a rare event, one of the following alternatives can be adopted:

- *raising the contact wire by means of installation tools*: At crossings where this alternative is likely to be used then its application can be eased by increasing the system height in the vicinity of the level crossing and by considering the resulting loading when designing the supports
- *partial or complete dismantling*: for the passage of oversize machines used in open pit mines with heights well above 10 m, the contact lines are dismantled and temporarily deposited between the tracks

Both alternatives require extended track occupations, de-energizing and earthing of the contact line as well as an increased commitment of personnel, construction tools and vehicles.

14.8 Container terminals, loading and checking tracks, railway lines in mines

14.8.1 Swivelling contact lines

Container terminals will often not be electrified to avoid disturbing loading operations by portal cranes. This results in a change of engines after loading operations and requires the availability of diesel or hybrid engines. The *swivelling of overhead contact line* design described below and applied on parts of or the total length of the loading track, permits secure loading and unloading of container trains exclusively using electric traction.

Installations as described in [14.24] enable swivelling in vertical planes of all cantilevers of an overhead contact line section around their pivots at the poles. Swivelling upwards as per Figure 14.35 a) or downwards as per Figure 14.35 b) of the cantilevers form viable alter-

a) Arrangement of the contact line equipment

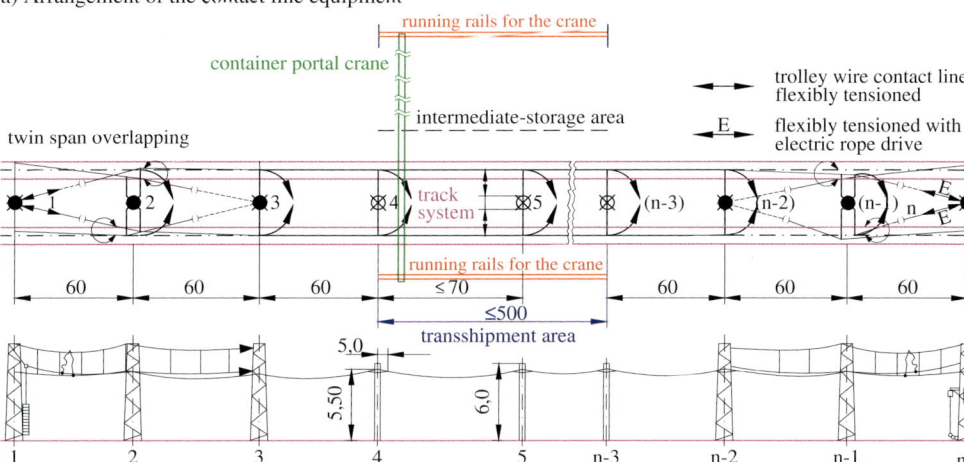

b) Arrangement in relation to the gauge

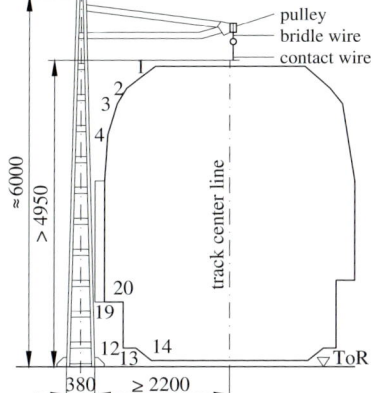

Figure 14.36: Horizontally swivelling overhead contact line for container terminals.

natives. Because of the lateral displacement, the contact line is moved out of the clearance gauge of the track allowing unhindered loading operations. A separate operating mechanism is required for each pole.

Figure 14.36 shows a *horizontally swivelling contact line* consisting of a trolley wire overhead contact line with bridle wires at swivelling cantilevers [14.25]. The contact line can be negotiated at 75 km/h. The tensioning equipment at one end of the overhead contact line section permits a lateral movement of the overhead contact line initiated by an operating mechanism at the other end. This movement leads to a turning of around 85° and moves the contact line out of the loading gauge within a period of 2 min.

Other designs with cantilever rotating in vertical direction are described in [14.26]. Utilizing the maximum working development of the pantographs of 6,50 m is an alternative to swivelling the overhead contact line in container terminals. In this case, the contact wire is installed at a height of 6,30 m allowing for operation also at *maximum permissible contact wire uplift*. The containers are then loaded or unloaded by hydraulically operated swivelling arms underneath the contact line.

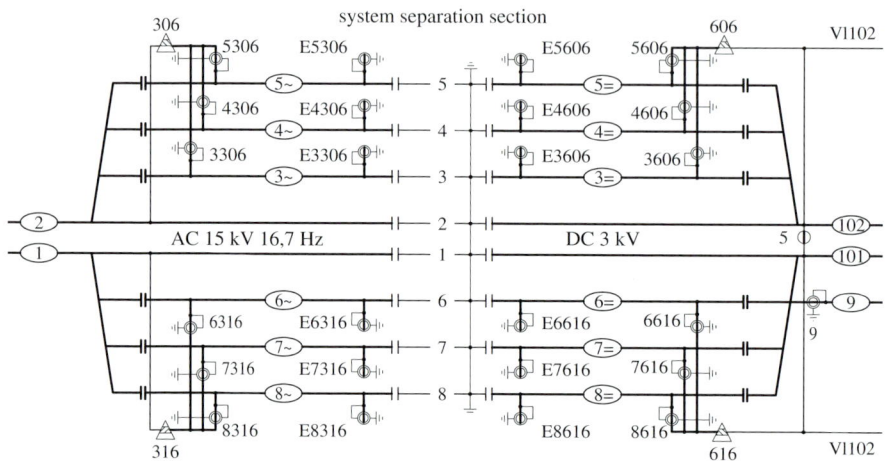

Figure 14.37: Schematic diagram for customs management track combined with a system separation between AC 15 kV and DC 3 kV in the station Oderbrücke near Frankfurt/Oder, Germany. disconnectors 3306, 4306, 5306, 6316, 7316, 8316 AC 15 kV power supply; disconnectors 3506, 4606, 5606, 6616, 7616, 8616 DC 3 kV power supply; earthing switches E3306, E4306, E5306, E6316, E7316, E8316, E3606, E4606, E5606, E6616, E7616, E8616; power supply disconnectors 306, 316, 606, 616

Figure 14.38: Contact line supported by poles clamped to the tracks and moved with the track.

Figure 14.39: Laterally arranged current collector on a mine locomotive (Photo: M. Hoffmann).

14.8.2   Circuit diagrams for loading and checking tracks

Before swivelling the contact line or carrying out loading operations within its reach, the *contact line above loading or checking tracks* must be de-energized and earthed. Frequently required switching and earthing operations, e. g. for overhead contact line over loading and customs management tracks, can be eased by switching devices and operating equipment similar to those shown in Figure 14.3. Figure 14.37 depicts an arrangement known to railway operators as a *customs or safety connection* combined with a system separation section in the railway station Oderbrücke near Frankfurt/Oder, Germany. The overhead contact line is

a) Total view   b) Cantilever at support

Figure 14.40: Laterally arranged trolley wires (Photo: M. Hoffmann).

separated by section insulators and fed or earthed through individual disconnectors track by track. As there is a separation section in the station, switching devices exist twice for each track, one for each power supply system. At each end of a track, a control panel is arranged where authorized persons can carry out switching operations. Using special keys and push buttons, they can request permission to approach to the contact line closer than the minimum clearances and initiate interlocking against re-energizing of the overhead contact line. There, the local area train operations manager then de-energizes and earths the line from the local area control panel without consultation with the master control centre. The de-energized conditions of the contact line are then signalled to the control panel. Re-energisation is dependent on completion of a captive key authorisation procedure.

14.8.3   Movable stopes and laterally arranged overhead contact lines

Trolley-type overhead contact lines, which are supported by poles mounted directly onto movable tracks of railways in mines are known as *stope-type contact lines* (Figure 14.38).
A drum arranged at the end of the track, called stope end blocking, compensates the change of length occurring in mine railway service. Before moving the track the overhead contact line is de-energized by opening a circuit breaker and earthed, then. Electric mine locomotives are equipped with additional *current collectors* (Figure 14.39) arranged laterally to secure power supply also within track sections, where trolley wire overhead contact lines arranged laterally provide the required space for loading and unloading manoeuvres (Figure 14.40).

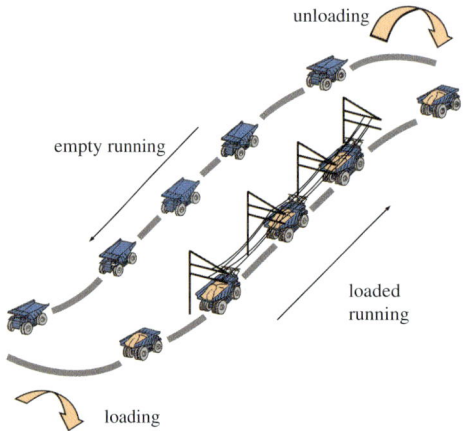

Figure 14.41: Simplified presentation of running a dump truck with a contact line.

Figure 14.42: Electrically driven lorry equipped with contact line type Sicat TT in the Lumwana Mine in Zambia.

14.9   Contact lines for haul trucks

14.9.1   Contact lines for trucks in open-pit mining

To transport ores and excavation material from mining in open-pit copper and gold mines heavy trucks with hybrid diesel-electric drives having 2 MW power are used. They reach 450 t total load with with 250 t payload and the dimensions of 14 m in length, 8 m in width and 7 m in height. On the six to ten percent steep ramps of mines a supply by electric energy is commercial and environmental friendly (Figure 14.41). The electric power supply system called *Truck Trolley* (TT) supply consists of several mobile container substations and a *twin-pole contact line* Sicat TT operated at DC 1,2 to 2,0 kV voltage. Along the route steel pipes are placed into drilled foundations by concrete. On their top bolts are installed to accommodate the contact line poles. These poles carry cantilevers with the two contact lines consisting of both staggered arranged contact wires (Figure 14.42). Tensioning wheel sets load the conductors at 10 to 12 kN each. When changing the route the metal pole can be dismantled and erected again at the new site together with the cantilevers.

For current collection the pantographs are equipped with a twin pole pantograph and sensors which register the contact wire position on the carbon contact strips and display it in the driver's cabin. Therefore, the driver is able to lift the pantograph at the correct position and to steer the truck such that the contact wires remain within the working range of the contact strips. If the contact wires leave the provided range of contact strips the pantograph will be lowered and the drive is switched to the diesel-electric propulsion.

Table 14.1 presents data of a payoff calculation for *electrical operation of trucks* in a South African copper mine. The cycle shown in Figure 14.41 is 15 km long, whereby 6 km are equipped with contact line. On these data the return of investment for the electrification of the trucks was determined as two years, depending on the price of electric and diesel energy. The shorter travelling times enhance the productivity of the mine. The number of trucks can be reduced and the life cycle period of drives increased.

Table 14.1: Data for electrical operation of trucks in a copper mine.

		Without contact line	With contact line at ramps
data of whole line			
– mean speed	km/h	10	12,45
– time for 15 km cycle	min	90	72
data within the 6 km long ramp			
– running speed	km/h	11	24
– running time	min	33	15
– diesel consumption per cycle	l	191	8
– electric energy per cycle	kWh	–	895
– energy cost savings per cycle	US $	–	10
– fuel and energy cost savings at 280 000 cycles with 50 trucks per year	US $	–	2 800 000

14.9.2   Contact lines for lorries and busses on public roads

Overhead contact lines supply busses with energy in cities. They are as well suited for *supplying electric energy to road vehicles* on larger distances outside cities.

Already in the year of founding the Siemens & Halske entity some ideas arose for electric vehicles. In 1847 Werner von Siemens wrote: "When I will have leisure and money I will design an electro-magnetic taxi which will not keep me stuck in mud …".

Some first electrified busses were operated at the beginning of the 20th century. Without track, however, with contact line they are bound to a track but not guided by track. The current is supplied by two contact wires and collected by two approximately 6 m long arms mounted on the roof of the vehicle which enable passing other vehicles and approaching to stations 4,5 m right and left of the contact line. The voltage is DC 0,6 or 0,75 kV in most cases. Trolley busses are characterized by a smooth noiseless running, high potential of acceleration and smooth braking.

Trolley busses are operated in approximately 300 cities, most of then in Central Europe, Asia and North America. In Moscow the largest trolley bus network is operated. In some cities also lorries are operated electrically on the trolley bus network for local freight transport and for repairs. Trolley bus lines for connections between cities existed in Russia and in Italy.

Already in the 30th proposes were made to electrify the motorways. The investment and operation costs were assessed as relatively low already in case of the not dense traffic at that time [14.27, 14.28].

This idea was newly considered by Siemens in 2009 and thoroughly studied assuming a doubling of freight transport until 2050. Alternative solutions to hybrid drives such as energy storage, cells and inductive solutions will not be suited as the sole drive for heavy lorries. Therefore, lorries and buses for long distances should be equipped with classic hybrid drives and current collectors and supplied on main lines from overhead contact lines. On sections without contact lines they could run by diesel-electric drives. This design would combine the electric power supply as proven for 100 years for railways and the flexibility of the road.

These considerations formed the starting point of the project *electro-mobility of heavy lorries in view of exoneration of environment of densely populated areas* whereby Siemens state-aided by the German Federal Ministry for Environment, Protection of Nature and Reactor Safety from July 2010 to September 2011 prepared a study for the use of electrical energy for freight transport on roads and tested the prototypes of central components and sub-systems

Figure 14.43: Siemens test track for electric utility vehicles.

Figure 14.44: Siemens pantograph.

[14.29]. The technical feasibility was practically demonstrated on a test line close to Berlin. A DC substation which is designed for regeneration supplies the line by DC 650 V as required by the vehicle.

The 1 500 m long twin-pole overhead contact line (Figure 14.43) consists of two overhead catenaries having 1,35 m distance with a messenger wire BzII 120 mm^2 tensioned by 10 kN and a contact wire Cu 150 tensioned by 20 kN. The contact wire height is 5,15 m above the motorway.

Without new technologies for rail transport the CO_2 reduction by 80 % in Europe and within the transport sector by 60 % in comparison to 1990 could not be achieved according to studies carried out by the European Commission. When supplied electrically through contact lines the vehicles do not exhaust any harmful substances. The fighting against hazardous substances when procuring electric energy on the basis of fossil fuels is moved to more efficient stationary big plants.

14.10   Contact lines in historical city centres

14.10.1   Architecturally designed overhead contact lines

At the beginning of use of electrically driven railways for urban transit the current was transmitted to the vehicles through the running rails as feeder and return conductors or arranged in a duct below a slotted rail (Clause 3.1.1). Such arrangements frequently caused electric accidents due to touching of both rails or to short circuits and disturbantions because rain water flew into the duct or it was heavily polluted. Therefore, since more than hundred years conductor rails and overhead contact lines have been used to feed electrically driven vehicles. Since approximately 100 years conductor rails and overhead contact lines have been adopted for supply of vehicles. Conductor rails designed as third rails within the track range are used mainly for underground railways, in order to keep the tunnel cross section and, therefore, the investments low. Third rails supply as well commuter trains running on reserved permanent right-of-ways without crossings. For main line railways they are used rarely, for example in Southern UK. Overhead contact lines have proven their qualification for main lines, tramways and trolley busses for many decades.

As soon as introducing overhead contact lines discussions started on their *architectural appearance* in city centres and an alternative solution. Continuously there were trials to conceive

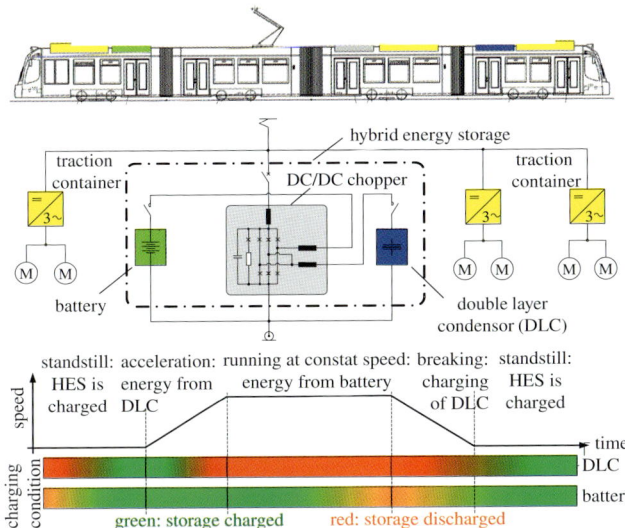

Figure 14.45 labels:

hybrid energy storage

traction container   DC/DC chopper   traction container

battery

double layer condensor (DLC)

standstill: acceleration: running at constat speed: breaking: standstill:
HES is   energy from   energy from battery   charging   HES is
charged   DLC   of DLC   charged

speed

time

DLC

battery

charging condition

green: storage charged   red: storage discharged

Figure 14.45: Electric equipment of a tramway with hybrid energy storages HES by Siemens.

technical solutions without contact lines. The alternatives described in Clauses 14.10.2 and 14.10.3 increase investments and operational costs in total.

The architectural pleasant design of contact lines forms such an alternative:

- use of less visible single contact lines and horizontal arrangements
- fixing of cross span wire at buildings by shapely rosettes
- slim poles adjusted to the city sights similar to old street lightnings or a combination with them
- short, inconspicuous and pleasant cantilevers and steady arms made of colourful glass fibre reinforced plastics
- use of plastic ropes for damping purposes and of loop insulators instead of large conspicuous components
- use of high tensile forces for contact wires to reduce the number of cross spans
- tensioning equipment within the poles or their sections
- avoiding of tramway crossings on historic places or at least no sections insulators at that areas

14.10.2   Loading stations for vehicles with energy storages

For some years storages have been used for the braking energy and the supply of hybrid busses with diesel-electric drives and of tramways. Siemens developed hybrid energy storages which are re-loaded at the stations via pantographs and DC choppers and supply the motors by inverters during running (Figure 14.45). The storage consists of double layer condensers for high starting currents and the battery unit for steady running (Figure 14.45) The unit can also resorb the braking energy. Tramways equipped with this technology are able to run approximately 2,5 km without overhead contact line to the following loading station or the next overhead contact line stations.

The *loading units* are equipped with overhead conductor rails, which the pantographs contact after the vehicle having entered the *station* (Figure 14.46). The charging process takes 20 to 30 s and, therefore, does not exceed the usual stopping periods of tramways. The contact rail

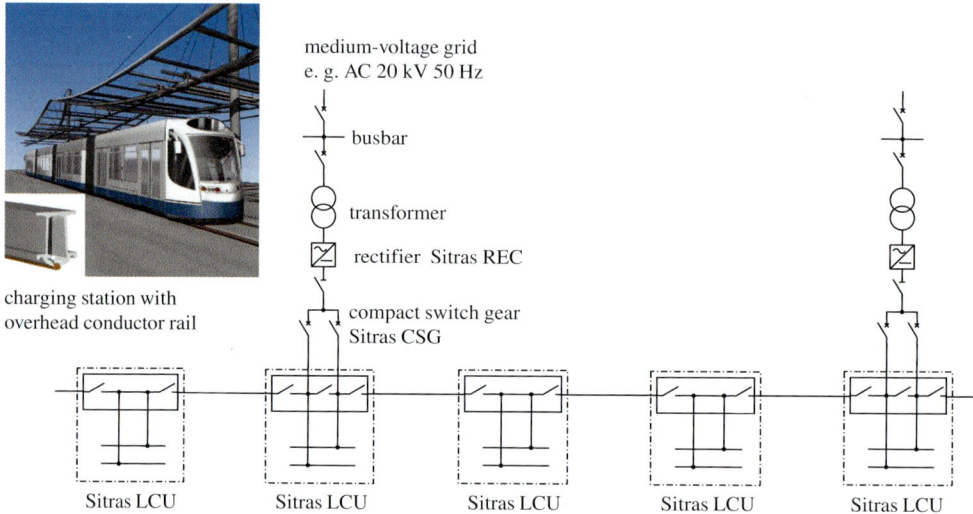

charging station with
overhead conductor rail

medium-voltage grid
e. g. AC 20 kV 50 Hz

busbar

transformer

rectifier Sitras REC

compact switch gear
Sitras CSG

Sitras LCU   Sitras LCU   Sitras LCU   Sitras LCU   Sitras LCU

Figure 14.46: Loading stations for tramways power supply from the DC substation.
LCU Local Charging Unit

Figure 14.47: Power supply of
the APS system: Underground
conductor rail between tracks;
control box down left.

as well as the pantograph are designed for the short-turn charging currents. After leaving the
station the charging current is interrupted and the pantograph lowers. The charging units are
supplied by conventional DC substations via cables, which are designed for high charging
currents.

14.10.3   Conductor rail under the ground surface

Below-surface lines, also called *under-ground conductor rails*, are contact lines under the
surface which supply the vehicle with current from underneath. As described in Clauses 3.1.1
and 14.10.1 they were used in the first half of the 20th century to supply electric tramways.
Due to missing safety – electrical accidents due to inadmissible touching voltages – as well
as technical and operational problems by rainwater, icing, pollution and many defects they
were replaced by overhead contact lines.
Since the begin of the 21st century there are new technical developments and utilizations of
underground conductor rails. The APS system of the Alstom Company for power supply of
tramways within historical cities is one of these developments. APS stands for the French

Figure 14.48: System separation section between the APS system and a contact line. The APS systems starts infront of the vehicle.

designation Alimentation Par Sol (power supply from the ground). The visible part consists of a conductor rail which is arranged on surface level between the rails. To avoid inadmissible touch voltages the conductor rail is put under voltage only under neath the vehicles. It is separated into 11 m long sections consisting of an 8 m long conductive and a 3 m long insulating segment (Figure 14.47).

The conductive segment will be put under voltage by a coded signal only for a short period, when it will be completely covered by the tramway and is connected galvanically with the rails. In operation two segments are live at maximum. Each tramway vehicle commands on two underfloor pantographs, whereby one of them contacts always a live segment. Before leaving this segment the other pantograph is already in contact with the segment following in running direction. Therefore, a permanent power supply of the vehicles and the electrical protection within the sections is guaranteed. However, at least 33 m long vehicles are required as well as a steady on- and off-switching. For this purpose each 22 m control boxes are installed in the subsoil which control two sections each. If in case of a failure a section is not switched off, the whole line section will be switched off triggered by a control unit arranged in the train station and the failing segment is discovered, de-activated and the line will be re-activated. The tramway is able to pass 150 m at 3 km/h over failed segments supplied by a battery.

The APS system was first adopted in the historic city of Bordeaux on a length of ten kilometres and often failed during the first years of operation. Meanwhile this system has been also installed in other cities, e. g. Reims (2 km) and Angers (1,5 km). Outside the centres the tramways are supplied by overhead contact lines. The system separation sections (Figure 14.48) between the APS system and the overhead contact line are situated at stations. The switching between current collection from the underground to the overhead contact line is carried out by the cab driver. If he forgets switching an alert will be started.

14.11   Bibliography

14.1 *Matthes, U.*: Stromschienenoberleitung in der Fahrzeughalle Erfurt (Overhead conductor rail in the rail vehicle maintenance shop at Erfurt). In: Elektrische Bahnen 98(2000)1-2, pp. 58 to 62.

14.2   *Tessun, H.*: Deckenstromschiene – konstruktive Gestaltung (Overhead conductor rail – structural design). In: Elektrische Bahnen 104(2006)4, pp. 177 to 181.

14.3   *Dilger, R.; Zielinski, I.; Fröde, D.*: Abschalt- und Erdungsautomatik in der großen Wagenhalle in Berlin-Rummelsburg (Automated switching and earthing in the workshop Berlin-Rummelsburg). In: Elektrische Bahnen 101(2003)11, pp. 493 to 499.

14.4   *Liebig, A.*: Oberleitungen an Wehrkammertoren (Overhead contact lines at flood gate doors). In: Elektrische Bahnen 95(1997)1-2, pp. 42 to 46.

14.5   *Decision 2008/284/EC*: Technical specification for interoperability relating to the energy subsystem of the trans-European high-speed rail system. In: Official Journal of European Communities, No. L104 (2008), pp. 1 to 79.

14.6   *Schwach, G.*: Oberleitungen für hochgespannten Einphasenwechselstrom in Deutschland, Österreich und der Schweiz (Overhead contact lines for high-voltage single phase currents in Germany, Austria and Switzerland). Villingen-Schwenningen, Wetzel Publishing, 1989.

14.7   *Kahler, P.*: Technische und wirtschaftliche Probleme an den Stoßstellen zwischen verschiedenen Bahnstromsystemen. (Technical and economical problems at the interface between differing traction power supply systems). HfV "Friedrich List" Dresden, doctoral thesis, 1962.

14.8   *Behmann, U.*: Schutz gegen Schäden durch Lichtbögen an Oberleitungstrennstellen (Protection against arcs at overhead contact line separation sections). In: Elektrische Bahnen 109(2011)6, pp. 304 to 305.

14.9   *Vega, T.*: Schnellfahrstrecke Madrid–Sevilla durchgehend mit AC 25 kV 50 Hz betrieben (High-speed line Madrid–Seville operated completely with AC 25 kV 50 Hz). In: Elektrische Bahnen 101(2003)3, pp. 134 to 135.

14.10 *Braun, E.; Kistner, H.*: Systemtrennstellen auf der Schnellfahrstrecke Madrid–Sevilla (System separation sections on the high-speed line Madrid–Seville). In: Elektrische Bahnen 92(1994)8, pp. 229 to 233.

14.11 *Freifeld, A. W.*: Planning of overhead contact line installations (in Russian language). Transport Publishing, Moskau, 1984.

14.12 *Wili, U.*: Parallelbetrieb mit Gleichstrom und Wechselstrom auf einem Gleis (Operating AC and DC trains on the same track). In: Elektrische Bahnen 10(2003)11, pp. 507 to 513.

14.13 *Cox, S. G.; Nünlist, F.; Marti, R.*: Deckenstromschienen für Dreh- und Klappbrücken (Overhead conductor rails on moveable bridges). In: Elektrische Bahnen 99(2001)1-2, pp. 90 to 93.

14.14 *Schäfer, H.-D.*: Elektrifizierung der Strecke Salzbergen–Emden–Norddeich (Mole) (Electrification of the Salzbergen–Emden–Norddeich (Mole) line in Germany). In: Elektrische Bahnen 78(1980)10, pp. 265 to 269.

14.15 *Koswig, J.; Freidhofer, H.*: Oberleitungsanlagen bei beweglichen Brücken im Raum Bremen/Oldenburg (Overhead contact line installations on moveable bridges in the Bremen/Oldenburg area). In: Elektrische Bahnen 78(1980)10, pp. 278 to 282.

14.16 *Schatkowski, K.*: Oberleitung auf der Klappbrücke über die Peene in Anklam (Overhead contact line on the folding bridge across the Peene river near Anklam). In: Elektrische Bahnen 111(2013)4, pp. 267 to 272.

14.17 *Hofer, R.*: Die Ausrüstung Europas größter Hubbrücke mit einer 15-kV-Oberleitung (Equipment of Europe's largest lifting bridge with an AC 15 kV overhead contact line). In: Elektrische Bahnen 85(1987)3, pp. 80 to 85.

14.18 *Jozsa, B.; Puschmann, R.*: Elektrischer Betrieb bei Bayside- und Hillside-Trains in Melbourne/Australien (Network and development at Melbourne's Bayside and Hillside trains). In: Elektrische Bahnen 99(2001)4, pp. 162 to 166.

14.19 *Brassel, W.-U.*: Parallelbetrieb und Kreuzungen von Trolleybus- und Straßenbahnoberleitungen (Parallel operation and crossings between trolley bus and tramway contact lines). In: Elektrische Bahnen 102(2004)7, pp. 296 to 301.

14.20 *Federal Republic of Germany*: Straßenverkehrszulassungsordnung [Directive on registration of road traffic). In: BGBl. I p. 1793, September 1988, new edition BGBl. I, p. 679. April 2012, latest amendment BGBl. I, p. 2083, July 2013.

14.21 *UIC Code 608*: Conditions to be complied with for the pantographs of tractive units used in international services. UIC, Paris, 3rd edition 2003.

14.22 *Deutsche Bahn AG, 4Ebs 19.01.01*: Höhenbegrenzung mit Profiltor vor Bahnübergängen (Height restriction at level railway crossings with a profile gate). Deutsche Bahn AG, Fankfurt am Main, 2002.

14.23 *Schmieder, A.*: Fahrleitungshebeeinrichtung für die Durchfahrt von Großraumtransporten an niveaugleichen Bahnübergängen (Contact line lifting equipment for the passage of oversize transports at level railway crossings). In: Elektrische Bahnen 103(2005)9, pp. 432 to 434.

14.24 Patent Germany 180 37 62, Class 20 k. 9/01: Anordnung von Fahrleitungen in Verladezonen (Arrangement of overhead contact lines in loading areas).

14.25 *Schmidt, P.*: Energieversorgung für den elektrischen Zugbetrieb auf Containerbahnhöfen (Power supply for the electrical railway operation on container terminals). In: Die Eisenbahntechnik 21(1973)4, pp. 157 to 159.

14.26 *Solka, M.*: Kippfahrleitung als Wettbewerbselement im Schienenverkehr (Tilting catenary as a competition factor in railway freight traffic). In: Elektrische Bahnen 103(2005)9, pp. 432 to 434.

14.27 *Dönges, F.*: Die Verstromung der Reichsautobahnen. Ein Beitrag zur Synthese zwischen Schiene und Straße (Electrification of German state motorways. A contribution to the synthesis between road and rails). In: VTW – Verkehrstechnische Woche 30(1936)3, pp. 29 ff.

14.28 *Oertel, W.*: Elektrischer Schnellverkehr mit Oberleitungs-Omnibussen auf der Reichsautobahn (Electric fast traffic by means of trolley busses on the state motorway). In: VTW – Verkehrstechnische Woche 30(1936)8, pp. 93 ff.

14.29 *Gerstenberg, F.; Lehmann, M.; Zanner, F.*: Elektromobilität bei schweren Nutzfahrzeugen (Electrical mobility of heavy duty trucks). In: Elektrische Bahnen 110(2012)8-9, pp. 452 to 460.

15 Construction and acceptance

15.0 Symbols and abbreviations

Symbol	Definition	Unit
BETRA	application on track blocking	–
EBA	German Federal Railway Authority	–
EBC	German Federal Railway-CERT	–
EdA	declaration of acceptability	–
EVU	railway operation entity	–
FMW	contact line installation vehicle	–
GUV	mandatory accident insurance	–
HIOB	maintenance vehicle with raising platform for overhead contact lines	–
IFO	maintenance vehicle for overhead contact lines	–
INDUSI	inductive train protection	–
IOC	interoperability testing report	–
MTW	installation and maintenance vehicle	–
OMF	overhead contact line installation vehicle	–
OSE	local control panel	–
PA	distance track axis to pole face	m
PZB	train control at spots	–
Sifa	safety control connection	–
TEIV	transeuropean interoperability regulation	–
ToR	Top of Rail	–
TPL	Traction Power Lines	–
WBL	pantograph for high-speed lines	–
Zes	DB's power supply control centre	–
h_{CW}	contact wire height	m
h_H	distance between measuring equipment and contact wire	m
h_{SH}	system height	m
v	running speed	km/h

15.1 Basics and principles

The *construction* of an overhead contact line system includes all installation and assembly work using components designed, produced and tested to assure their quality and *acceptance* (see Chapter 11). The *testing operation* is part of the acceptance process and begins the commissioning of the overhead contact line (see Chapter 17). The *construction and assembly activities* are summarised in Table 15.1: They start with the preparation of the pole *foundations* and are followed by the *setting of poles*, *assembly of cantilevers* and *cross- spans*, *tensioning devices* and *midpoints* in preparation for the assembly of the *contact line installation*. The *earthing* of components and parts completes the construction activities. Installing documents such as contact line plans, pole and foundation schedules, material lists and cross-span draw-

Table 15.1: Working steps for construction of overhead contact lines.

Tasks	Working steps
installation of foundations	survey of locations, search for cables and pipelines, securing of ballast, excavation, scaffolding, pile drilling, driving, fixing of anchor bolts, installation of foundation earthing, pouring and compacting of concrete, remove of scaffolding, refilling of soil
setting of poles	attachment of brackets, cleaning of excavation or hole, mounting or inserting of poles, alignment of poles, pouring of mortar and underfilling of poles
pre-assembling of cantilevers	measuring of dimensions at poles, calculation of cantilever dimensions, cutting of tubes and brackets, mounting of fittings and insulators
cantilever assembling	assembling of cantilevers at poles and securing against rotation
mounting of tensioning devices	inserting and secure tensioning devices, install weight guidance, complete weight stack as well as messenger and contact wire termination
midpoint anchor installation	installing of midpoint anchor, attaching ropes for messenger wire midpoint anchor and tensioning to the specified force
preparation of droppers and stitch wires	measuring of contact wire support clamps, calculating and manufacturing droppers, cutting of stitch wires to length
installing contact line	stringing messenger and contact wire indivually or together, connecting with the tensioning device at the start of the tensioning section, stringing under tension; clamping into the cantilever and terminating at the midpoint anchor and the tensioning device at the end with the specified tensile force, releasing the tensioning device, installing the stitch wires and droppers during the stringing work, installing the contact wire midpoint anchor, tensioning the stitch wires
adjustment of contact line	checking the contact wire height and stagger and adjusting if necessary by modifying the cantilevers and droppers
installing electrical connectors	installing electrical connectors at connected overlaps, at insulated overlaps, at midpoints, at bridges, at contact line disconnectors, between contact line and parallel feeder line, between contact lines above points
installing overhead contact line disconnectors	installing and adjusting the drive mechanism, the linkage and disconnector parts, connecting remote control cable and power feeder or switching lines
installing insulators	installing insulators into the overhead contact line
installing traction power line	stringing of traction power lines, stringing to the final tensile force and fixing at clamps
mounting of plates and labels	attaching pole, disconnector and traction power line number plates, attention and warning plates at the described positions
installation of railway earthing	connecting metal parts in the contact line range and poles with the selected rails, laying and connecting earthing connections, installing voltage limiters in DC systems
revision of documents	revision of the documents to comply with the installation (As-Built Drawings)

ings with reference to the *drawings for the individual overhead contact line type* (see Chapter 12) form the basis for these activities. The planning engineer divides the documents into the plans for stations and lines in between. The amount of work depends on the length and number of the tracks to be wired. Installation procedures and *equipment* used are determined by local site conditions, such as the construction of a new line without train operations or refurbishment between trains without electrical operations or installation or conversion during breaks of electrical train operation.

Figure 15.1: Constructing an auger-bored foundation at a high-speed line with slab track.

Infrastructure managers may establish their own *specification for construction* of overhead contact line systems on the basis of international and national standards (Annex 1). The planning described in Chapter 12 and approval by experts form a precondition for installation, acceptance and the subsequent operation.

15.2   Construction of foundations

The type of foundations for poles is selected depending on type of poles, static loadings, soil conditions and involved expenditures. They can be classified with respect to:
- type of installation in auger-bored, pile and excavated foundations
- static function in flat and deep foundations
- foundation shape as block, stepped block, tube-pile and H-beam pile foundations

Pile foundations have been used successfully in Germany since 1990. Various types of steel profiles are installed into the soil by pile driving and then filled with mortar with the concrete pole placed on top of the pile (Clause 13.7.4). Driven pile foundations are characterized by

a) Earthing insert on the shuttering before concreting   b) Earthing cable at the earth sleeve after concreting

Figure 15.2: Rail earthing at a foundation on a viaduct.

rapid installation and avoidance of extensive concrete works during rail operations. *Drilled pile foundations* are favoured in rock and on high-speed lines with rigid permanent way (Figure 15.1).

Rail wagons transport the required driving or drilling machines as well as the concrete mixing vehicles to the foundation site. In the case of foundation construction before track laying, as is often the case for new lines, the foundations are installed from a construction road or from the track permanent way. The contractors for extensive concrete works along the lines use *concrete mixer trains*. Where there are no breaks in train operations, the foundations can be installed by access from outside the tracks.

For new civil engineering structures, the foundations or the attachments of the supports are integrated into the structures and are available before starting the installation of the overhead contact line.

Rail earthing needs to be considered during foundation installation (see Chapter 6). Where foundations are cast in-situ without reinforcement (see Clause 13.6.9) no special provisions for earthing are necessary. In the case of reinforced foundations prepared on site, the reinforcement needs to be connected to an earthing insert. The earthing is arranged within the concrete body of the foundation and has a female thread (Figure 15.2 a)), that can accommodate a bolt for the earthing connection to the track (Figure 15.2 b)).

For driven tube or pile foundations, the earthing conductor is welded to the driven tube or pile and is encased in the cast in-situ foundation between the earthing busbar (Figure 15.3 a)) and the earthing insert (Figure 15.3 b)). In the case of concrete poles which are placed on driven piles or within tubes, the earthing conductor welded to the driven tube or pile is brought out and after setting of the pole, bolted to an earthing insert arranged at the lower part of the concrete pole (Figure 15.4 a) and b)). If several earthing connections, e. g. noise protecting walls and handrails at viaducts need to be connected to the earthing insert of a pole an earthing busbar is used (Figure 15.4 c)).

a) Driven pile with earthing bar for earthing rail

b) Driven pile with earthing bar for earthing cable

Figure 15.3: Railway earthing at a foundation with driven piles.

a) Earthing bushing for one cable

b) Earthing bushing for 2 cables

c) Earthing bushing for 3 cables

Figure 15.4: Railway earthing at a concrete pole.

15.3 Pole setting

When using an installation train on the track, a mobile crane placed on the train *installs the poles* onto pre-prepared foundations. When working from outside the track, a mobile crane (Figure 15.5 a)) or a dual-mode vehicle installs the poles on or into the foundation, e. g. on new lines from the permanent way, before placing the tracks (Figure 15.5 b)). The crane holds the concrete poles during pouring and hardening of the rapid-setting mortar. The lifting equipment of the dual-mode vehicle places the steel poles on the foundation where they can be fixed temporarily and subsequently aligned according to the specification. The setting of poles with helicopters accelerates installation and is also possible between trains running on the line. The poles are pre-assembled with the necessary fixing brackets, tensioning devices and all components for the cantilevers at the assembly site (Figure 15.5 c)).

a) Installation by mobile crane from outside track   b) Installation by crane from slab track

c) Installation by a helicopter on an upgraded line

Figure 15.5: Pole setting.

a) Pre-fabricated cantilevers   b) Installation of cantilevers

Figure 15.6: Manufacturing and installation of cantilevers.

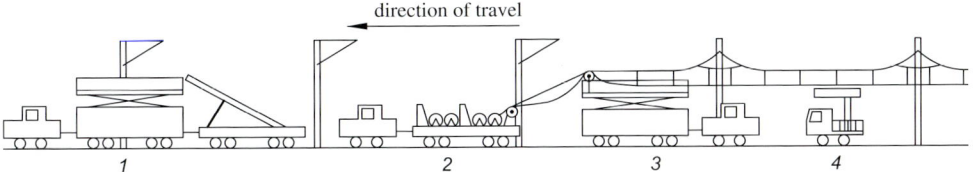

Figure 15.7: Installation train adopted for integrated overhead contact line installation.
1 first work unit with high-rail platform car and flat car, installation of cantilevers
2 second work unit with reel stand car and pull wheel tensioner, stringing of catenary wire and contact wire
3 third work unit with high-rail platform car, installation of contact line
4 fourth work unit with high-rail platform for adjusting the overhead contact line

15.4   Installation of cantilevers and cross-span supports

After setting the poles, a site survey is carried out as a basis for determining the cutting lengths and dimensions of fitting the cantilever components. Special software is used to calculate cantilever dimensions and cutting lengths in advance. The cantilevers are pre-assembled in the assembly workshop and transported by an installation train to the installation site. The cantilevers are assembled at their corresponding pole in accordance with the allocated pole numbers (Figure 15.6).

Before assembling the *cross-span arrangements*, the tracks crossed need to be blocked for train operations for a short period. Portals and cantilevers across several tracks can be fixed to the poles by cranes. For the installation of pre-assembled cross-spans and cantilevers across *several tracks*, *pulley blocks* and *helicopters* are also used.

15.5   Installation of contact lines

15.5.1   Contact line equipment

Completion of foundations, support setting and track laying are pre-requisites for the installation of cantilevers and contact lines. The position of the tracks must correspond to the design position. However, the method of contact line installation depends on the availability of the construction plant and the time required for the planned activity. Figure 15.7 shows the *mechanised installation of an overhead contact line* with individual supports. This method can also be used on sections between stations [15.1]. The breaks of train operation need to be planned so that at least the catenary wire or contact wire or both can be installed within a tensioning section in a continuous working period.

The tensioning devices and midpoint anchors are installed prior to the overhead contact line installation. The cantilevers, pre-assembled in the workshop are placed on a flat wagon of the installation train and a conveyer belt transports them to the working platform of the first work unit (Figure15.7). The linesmen are able to install the cantilevers on the poles from a laterally turnable working platform.

After fixing the catenary wire and the contact wire at the fixed tensioning device of the terminating pole, work unit 2 with reel stand car, follows work unit 1. The pull-wheel tensioner on the reel stand car strings the catenary wire with the specified catenary wire tensile force and the contact wire with approximately 3 kN. Observing these values, especially during acceleration and breaking of the work unit, avoids bends and kinks in the contact wire and minimizes waviness. Laterally swiveling and site-adjustable rollers guide the catenary wire and the contact wire into position and enable their attachment to the cantilevers using the platform of work unit 3. After fixing the catenary wire to the catenary wire suspension clamps the contact wires and stitch wire are installed using temporary droppers. Work unit 3 completes installation of the catenary, midpoint and the termination of the contact line at the opposite terminating pole. The described procedure also allows for simultaneous installation of twin contact wires.

If, after stringing the overhead contact line, immediate use of the track by electric traction units is not necessary, the height of the catenary wire suspension clamp at the cantilevers can be measured for use in calculations for *dropper production*. This intermediate step has proven to be advantageous because droppers can be produced in the workshop and a high precision contact wire position is achieved.

If the overhead contact line is to be used for electric traction vehicles immediately after contact line stringing, work unit 3 installs the pre-fabricated droppers to support the incompletely strung contact wire. The data marked on poles provides the calculation basis for dropper fabrication before installation of the contact line. In this case, a multi-purpose platform car commences the adjustment work after work unit 3 has reached the midpoint and the tensile force of the contact wire has been adjusted to the design value at the released tensioning device. Work unit 4 starts with adjusting the contact line (see Table 15.1 and Figure 15.7). If work unit 3 has reached the terminating pole the tensile force of the catenary and contact wire is increased to the designed value. Then, the installation train runs back to the midpoint and adjusts the contact line in the second half of the tensioning section.

The *adjustment work* includes such tasks as clamping the droppers, tensioning the stitch wires and checking the overhead contact line with respect to its designed geometry and the contact wire uplift when loaded. Deviations from the set points that are outside the permissible tolerances need to be corrected.

Appropriate modified installation technologies are employed on densely trafficked existing electrified tracks, on which the existing contact line or the contact wire only needs to be dismantled first. Such dismantling can have several reasons. Often, the existing pole spacing is unsuitable for the intended speed increase or components such as poles or contact lines have reached the end of their serviceable life because of wear or aging. In these circumstances, an additional installation train precedes the the first installation train shown in Figure 15.7 to dismantle the existing contact line.

Increasing mechanisation and reduced work periods influences the development of installation methods. The installation vehicles and equipment described in Clause 17.3.5 permit the dismantling, complete installation and partial adjustment of a tensioning section in approximately six hours. An integrated *complete mechanical installation* is not possible in shorter break periods or at stations with contact wire crossings or cross-spans. In this case, the catenary wire and contact wire are strung separately and consecutively.

Integrated *installation units* can also be employed instead of individual installation trains. The duration of track closures and the type and scope of the work significantly influence the use of heavy-duty construction equipment. The expenditures to owners and, the effort required from the necessary operating staff are high. The closure duration of use and, therefore, the efficiency is mostly low due to short closure periods for working on the line. Mobile ladders are still employed for a part of the work including dismantling or adjustment of the contact line. The ladders can be lifted from the operating track manually to clear the track at short notice.

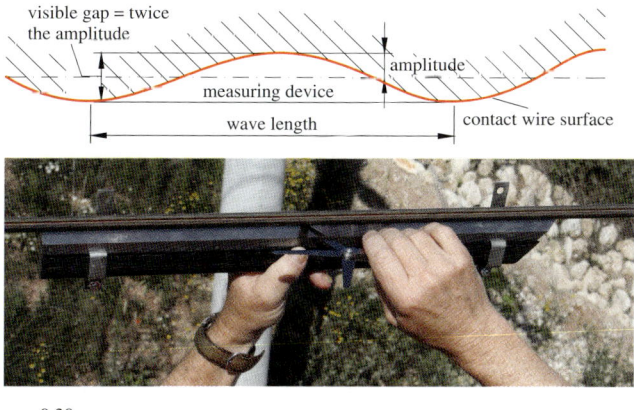

Figure 15.8: Determining unevenness (ripples) in contact wires.

Figure 15.9: Testing unevenness in contact wires.

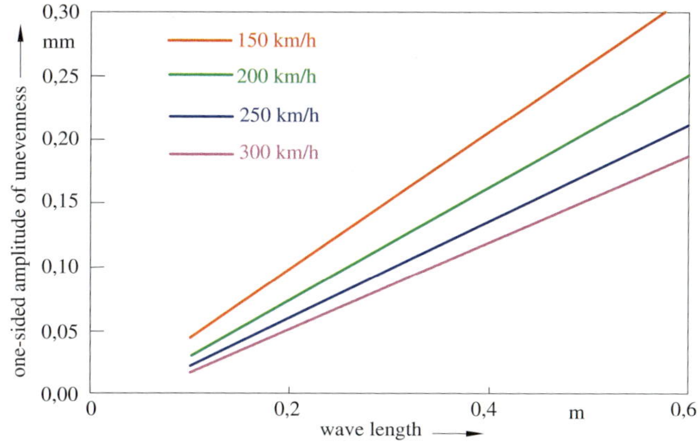

Figure 15.10: Recommended limits for slide unevenness in contact wires according to [15.2].

Installation-friendly designs such as aluminum cantilevers and brackets or GRP tubes combined with copper-aluminum alloy fittings, compression connections or plug-in clamps simplify and accelerate the work.

Two to four fitters are employed on each installation unit. Traction vehicle drivers, installation train drivers, equipment operators, look-outs and supervisors are also necessary. The experience of the personnel, the efficiency of the work planning and the available installation equipment determines the length of the track closures. The track closure needs 1,5 to 8 hours. The wiring of cross-overs requires short-term closure of several tracks to regular traffic. Work on upgraded lines can become very complicated because of operational constraints, track closure delays caused by train delays, switching, earthing and release procedures that are necessary before work can commence. Additional time must be allowed for travel into blocked sites and the instigation of the necessary protection measures for the construction site. In exceptional cases, with especially dense traffic loading, the overhead contact line can also be installed at night only. The prerequisite for this is the provision of satisfactory lighting on the installation trains and additional protection measures against the danger caused by night-time train operations. A sufficiently accurate adjustment of the overhead contact line is difficult under such conditions. Longer track closure periods are beneficial for work efficiency. Working in a completely blocked track, twelve tensioning sections on one track can be completed between Friday evening and Monday morning.

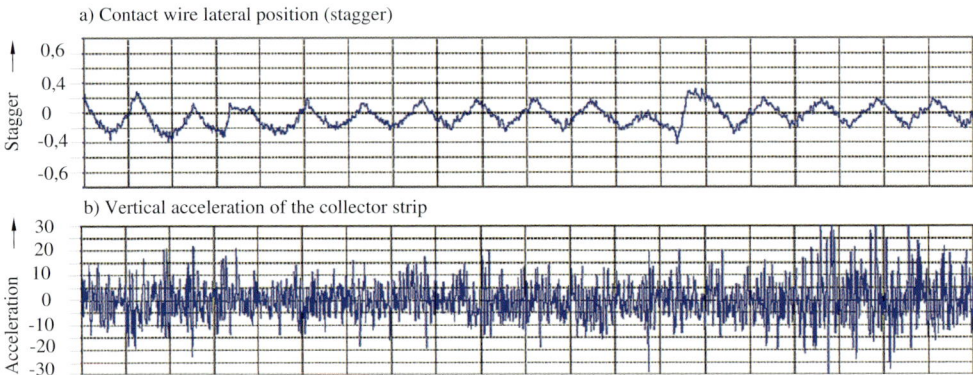

Figure 15.11: Acceleration of the leading collector strip of the pantograph DSA 380E at 300 km/h, with low pass filter frequency of 100 Hz.

Construction work restrictions can be caused by weather, for example during heavy frost and with wind speed above 10 m/s, depending upon the deployment regulations for the work platforms and cranes. Work on the overhead contact line should be interrupted during approaching thunderstorms to ensure personnel safety.

15.5.2 High-speed contact lines

The development of low-maintenance *high-speed overhead contact lines* since 1990 has placed increasing demands upon the quality of installation, *reliability* and the *life cycle* of components which can be effected by:
  – careful transport of materials to the construction site
  – correct installation of the fittings and insulators
  – greasing of current-carrying clamps in accordance with manufacturer's specifications
  – observance of the specified torque for bolt connections

Use of special high-tensile materials such as copper-tin or copper-magnesium alloys for contact wires as a pre-condition for higher contact wire tensile stresses and, therefore, higher running speeds require special tools and corresponding knowledge by fitters.

International experience has shown that, particularly during manufacturing and installation of contact wires with very high tensile strength unevenness also called rippling may occur. Depending on the type of pantographs used, this may increase the occurrence of arcing in some cases and impair the quality of the contact between the contact wire and the collector strips [15.2, 15.3, 15.4]. Considerable experience and suitable measuring devices are required to recognize slide unevenness in the contact wire. Rippling can be determined, e. g. using a one metre long steel ruler and check gauges with 0,05 mm increments (Figures 15.8 and 15.9). The gap when measured in this way, represents the double amplitude of the ripple in the contact wire.

Experience gained from high-speed lines [15.4] has shown that arcing can be reduced if the double amplitude of the ripples is less than 0,15 mm. Limits recommended in Japan are shown in Figure 15.10 [15.2].

Unevenness in the contact wire can also be determined by measuring the interaction of the overhead contact line with the pantograph and by evaluating the vertical acceleration of the

Figure 15.12: Contact wire straightening tool for conventional contact lines.

Figure 15.13: Contact wire straightening tool to smooth out micro waviness of high-tensile contact wires developed by Siemens and nkt cables.

Figure 15.14: Stringing unit used by Siemens for high-speed overhead contact lines.

Figure 15.15: Twisting lever to eliminate contact wire irregularities by twisting.

collector strips of the pantograph (Figure 15.11). In the right hand part of Figure 15.11 increased vertical acceleration values of the leading collector strip can be seen; caused by ripples in the contact wire. Using contact wire straightening tools as shown in Figure 15.12 unevenness and kinks can be smoothed in a copper contact wire for conventional lines. However, attempts to remove, by these means, the micro waviness caused by production and installation of copper alloy contact wires with a tensile strength of more than $400 \, \mathrm{N/mm^2}$ failed. Only the use of the newly developed straightening tool (Figure 15.13) succeeded in smoothing out the high-strength materials. During the stringing procedure, the contact wire is guided from the reel through a pull wheel to increase the tensile force and then through the pressing rollers of the straightening tool. Eventually, guiding wheels with large diameters lift the contact wire to the final installation level (Figure 15.14). Twisted contact wires frequently result in irregularities but can be corrected by the twisting lever shown in Figure 15.15.

Well trained staff is a prerequisite for the installation of high-quality overhead contact lines. The quality of interaction between contact wire and pantograph is directly affected by the geometrical accuracy of the contact line [15.5]. Table 15.2 contains important *tolerances* for a high-speed overhead contact line.

Table 15.2: Permissible tolerances for a high-speed contact line.

Parameter	Tolerance
distance between the rail and foundation top surface or driven tubes	± 50 mm
distance between track centre line and pole front face	± 50 mm
pole inclination	$\pm 0,3°$
pole twisting	$\pm 5°$
span length	± 500 mm
system height h_{SH}	± 30 mm
contact wire stagger at steady arm	± 30 mm
contact wire height at support	± 10 mm
difference of contact wire height between two consecutive droppers	± 10 mm

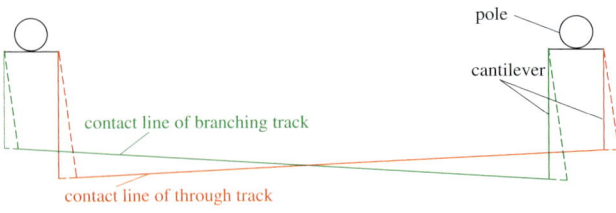

Figure 15.16: Arrangement of cantilevers for contact lines above track points.

15.5.3   Section insulator

The *installation of a section insulator* begins by marking its installation position, which for cross-over connections should be midway between both mainline tracks. They need 1,65 m horizontal clearance between the runners of the section insulators and the corresponding track centre lines. The centreline of the section insulator should be, as far as possible, at a maximum of 150 mm from the track centre line [15.6]. The section insulator is installed on the tensioned contact wire after attaching the contact wire dead end clamps. The contact wire between the clamps can be cut and removed. The section insulator and runners should be arranged parallel to the track both longitudinal and transversely.

An insulator or insulating rod and the suspension of the section insulator needs to be installed above the section insulator. The distance between the uppermost component of the section insulator and the catenary wire should be more than 0,46 m to avoid melting the catenary wire in case of arcs [15.7, 15.8].

Because of their mass, section insulators cause variations in elasticity within the overhead contact line. Correct height adjustment of the run-in and and run-out of the runners effects the running of the pantograph and helps avoiding premature wear. Section insulators should be set approximately 20 mm above the contact wire nominal height. As experience has shown, section insulators with runners are unsuitable in conductor rail overhead lines. The oscillating runners are worn to their permissible limits of wear within a few months.

15.5.4   Contact lines above points

When *installing contact lines above points* it is important to check whether both contact wires in the pantograph entry area are located on same half of the collector head since collector head traps could otherwise arise (Clause 12.4.2 and Figure 12.19 c)).

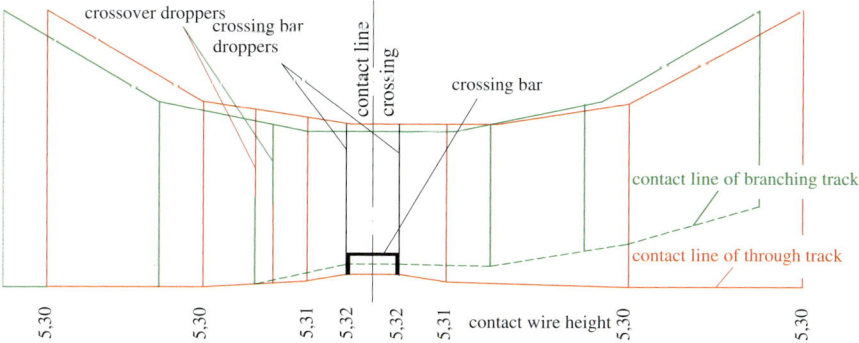

Figure 15.17: Contact wire height and arrangement of droppers above track points.

The droppers in contact lines above points are installed as follows:
- check of correct cantilever arrangement for contact lines above points (Figure 15.16)
- check the correct stagger for the contact lines above the points at the supports
- measure the distance between the support centres
- calculate the spacing between the droppers
- when installing droppers at a contact wire crossing, the temperature dependent positions of the cantilevers and the distances between the contact wire crossing and the corresponding midpoint have to be considered (see Figure 15.17)
- install the droppers on the catenary wire and temporarily on the contact wire by not crimping the sleeves of the dropper clamps at the contact wire – the droppers should not be folded
- check the contact wire height at the contact wire crossing, which should be 20 mm above the contact wire nominal height on the main track. At the droppers in front of the contact wire crossing, the contact wire height should be 10 mm above the nominal contact wire height. The crossing bar should be installed horizontally (Figure 15.17)
- adjust the contact wire height on the main track and crimp dropper clips
- adjust the contact wire height on the branching track and crimp the sleeves
- test if there is pre-sag within the range of the points, which should not be the case
- install the cross-over droppers (see Clause 12.4.3.4 and Figure 12.36)
- check the distance between the cross-over droppers. The distance should allow for the temperature dependent movement of the contact lines above the points. When calculating the distances between the cross-over droppers, the existing distance between the contact wire crossing and the midpoint needs to be considered
- a final measurement is required to confirm the correct contact wire vertical and horizontal position and the correct installation of the contact line

15.5.5   Earthing connections

In the case of AC systems *earthing connections* connect the components to be earthed with the return circuit (see also Chapter 6). At the earthing inserts connected to the pole reinforcement the existing return and earthing conductors are connected to the upper part of the pole (Figure 15.2 and 15.4). Metallic installations within the overhead contact line zone such as noise protection walls, handrails on viaducts and civil engineering structures need to be directly

Figure 15.18: Fixing an earthing connection to the rigid permanent way of a high-speed line.

connected to the connection points at the track or to the earthing inserts in the lower part of the poles. Since these activities do not need track blocking, they can be carried out before and after closure periods to spread the work load of staff. Tested bolted connections may be used to attach railway earthing connectors to the rail. Correct earthing installation, such as sufficient ballast cover is required to provide protection against damage by track laying machines. On high-speed lines, the earthing connections are stressed by suction effects of passing high-speed trains. Consequently, the earthing connections are connected to the rigid permanent way by cable clips around the conductors (Figure 15.18).

15.5.6   Traction power lines

Traction power lines (TPL) can be installed using traditional conductor stringing methods [15.9] or with adopting helicopters. The co-pilot of the helicopter operates the brake device on the conductor reel and ensures a constant tensile force in the conductor. V-shaped catching aids on the post insulators simplify the insertion of the conductor. After dismantling the catching aids the conductor can be sagged according to the sagging table and then clipped in. The final measuring of dimensions forms the basis for the revision documents (as-built drawings) for the operator and verifies that the standard clearances to adjacent objects have been complied with.

15.6   Equipment for installation and maintenance

15.6.1   General

To comply with the requirements for qualified installation and maintenance of overhead contact lines [15.10, 15.11] *specialized equipment* and *special purpose vehicles* are necessary in addition to the usual tools and vehicles which permit correct, rapid and safe working. It is normal to use the same equipment for preventive and corrective maintenance. Simple handling and continuous operational readiness during rough railway operations under all weather conditions is essential for high availability.

15.6.2   Helicopters

Helicopters can be employed to set *poles* and *portals*, string railway traction power lines or even string *head span structures* in large railway stations without hindering train operations.

a) Driving of a pile from the track bed

b) Setting a concrete pile with the pile driver

Figure 15.19: Pile driving.

Weighing the expenditures of helicopter use (Figure 15.5 c)) against those for track closure for traditional installation methods determines the use of the most advantageous method for the respective project.

15.6.3   Road vehicles

The foundation type, pole type and foundation construction methods explained in Clause 15.2 determine the equipment to be used. *Excavators* with special claws prepare the foundation pit for block foundations. *Drilling machines* are normally used to excavate earth for circular foundations. Pollution of the ballast bed by drilling debris should be avoided. *Explosive pile drivers* are used to install the piles (Figure 15.19). *Vibration pile drivers* can only be used conditionally because of possible danger to superstructure. Road mixers transport the concrete to the foundation site using public roads. However, if foundation sites are not accessible from outside the railway tracks, then concrete mixers, arranged on flat wagons, prepare the concrete and pour it from the track into the foundation excavations.

Motorised road cranes (Figure 15.20) or railway cranes are used for setting poles.

Specialised installation vehicles with work platforms are used for the installation of overhead contact lines of tramways in inner city areas. The insulated platforms also allow live line working (Figure 15.21 a)). When rail vehicles approach, it is possible to clear the work site quickly. The contact and catenary wire reels are carried on trailers (Figure 15.22).

Figure 15.20: Pole installation using a road-going crane.

a) Dual-mode railroad vehicle

b) Road installation vehicle with working platform

Figure 15.21: Installation vehicles.

a) Trailer for a dual-mode vehicle

b) Trailer for a road-going vehicle

Figure 15.22: Trailer for contact and catenary wire reels.

a) Gauge-free rotating crane Multitasker 100

b) Non gauge-free rotating crane Multitasker 250

Figure 15.23: Railway rotating crane.

15.6.4 Rail-bound vehicles

Railway rotating cranes can be adopted for *pole setting* from the tracks (Figure 15.23). The rotating crane shown in Figure 15.23 a) is able to work within the gauge without disrupting train operations on the neighbouring track; so blocking of the neighbouring track is not required. However, a speed restriction is established at the crane work site.

Contracting companies predominantly use *rail-bound work vehicles* with or without their own drive for installation of overhead contact lines. Work train vehicles with their own propulsion do not need locomotives and are, therefore, shorter than working trains with a separate locomotive. The *self-propelled assembly vehicles* with work platforms shown in Figures 15.24 a) and 15.27 include the following typical equipment for installation and maintenance vehicles:

– enlarged cabin with a workshop and racks for fittings and components
– hydraulically driven work platform
– loading crane with work basket
– equipment for installation of contact and catenary wire
– measuring pantograph
– inspection facilities

Modern multifunctional self-propelled vehicles enable completely mechanized overhead contact line assembly as described in Clause 15.5. Figure 15.24 b) shows a *contact line installation vehicle* FMW of DB that is transported to the installation site and while there, by a work train locomotive. The vehicle FMW includes a 16,2 m long lifting platform and a swiveling lateral platform that permits work on components 5 m from the track axis.

a) MTW 100 with own drive and reel wagon

b) Installation vehicle FMW without own drive

Figure 15.24: Contact line installation vehicle.

a) Large maintenance vehicle HIOB type 711 of DB

b) Small maintenance vehicle IFO type 703 of DB

Figure 15.25: Overhead contact line installation vehicles.

The *raised working platform maintenance vehicle* type HIOB 711 (Figure 15.25 a)) and the overhead contact line *maintenance vehicle* type IFO 703 (Figure 15.25 b)) comply with the requirements of DB and other railway companies. Both vehicles belong to a new vehicle family and satisfy requirements for short response times with self-propelled maximum speeds of 120 km/h and 90 km/h, respectively. Both vehicles are equipped with swiveling raised work platforms 1,5 m long and 1,6 m wide on the HIOB and 2,0 m long and 1,4 m wide on the IFO vehicle. Figure 15.26 shows the working radius of the raised work platforms of both vehicles. Pantographs of the WBL 85 design with individually sprung contact strips are installed for testing purposes. The viewing cockpit permits observation of the contact line and monitoring with a video camera. The *overhead contact line installation vehicle* type OMF1 of DB (Figure 15.27) is another vehicle used for installation, maintenance and repair of overhead contact lines. This vehicle reaches a maximum speed of 160 km/h. The cabin is designed on ergonomic principles and has a 5 m wide loading platform on both sides, permitting pre-assembled components to be loaded. The accessible roof is 8,9 m long and provides storage space for material and tools during the installation of the overhead contact line. The work platform has a maximum raised height of 14,3 m and a lateral radius of 12,9 m. A work basket, with a reach of 21 m above the top of rail and a maximum lateral extension of 18 m, allows work to be performed on contact line components that are difficult to reach. It also has a crane arm with a load moment of 240 MNm that can also be used to set poles and

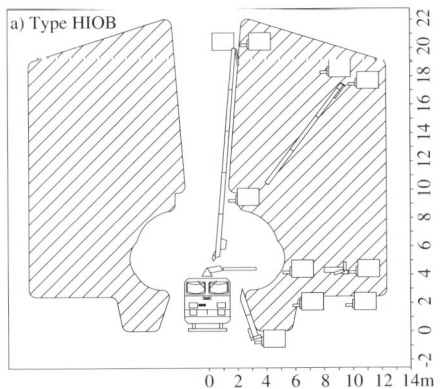
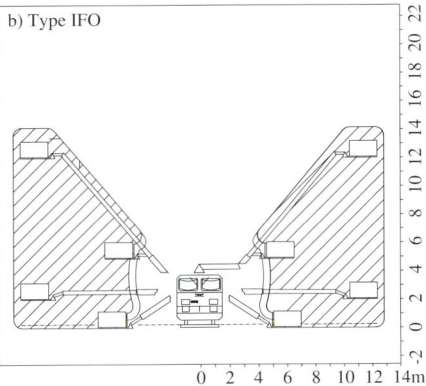

Figure 15.26: Working radius of raised platform for maintenance cars.

Figure 15.27: Overhead contact line installation vehicle type OMF1.

a pantograph for checking the contact line geometry. The pantograph insulated for 25 kV and suited for earthing to the railway earth, if necessary, completes the equipment on this universal vehicle.

Overhead contact line installation cars – also called *reel cars* [15.12], permit completely mechanized overhead contact line installation as described in Clause 15.5. The equipment and tools installed on this car can be used to string contact and catenary wire with relative ease. The installation vehicle type 575 (Figures 15.28 and 15.29), with conductor raising equipment installed on both ends of the car, permit stringing of contact wire and catenary wire at maximum working height of 7,5 m. This car has bogies that permit a 120 km/h maximum running speed.

The reel stands for winding conductors and wires have an independent drive and brake that permits stringing of contact and catenary wires with constant pre-tension. Located between the winding reel stands and the lifting gantry, additional reel stand are available on both ends for storage of additional reels. A crane with hydraulic drive facilitates the swift loading and changing of reels and the setting of *provisional poles*. On-board lighting provides sufficient

Figure 15.28: Overhead contact line installation car of type 575 of DB for stringing of contact and catenary wire.

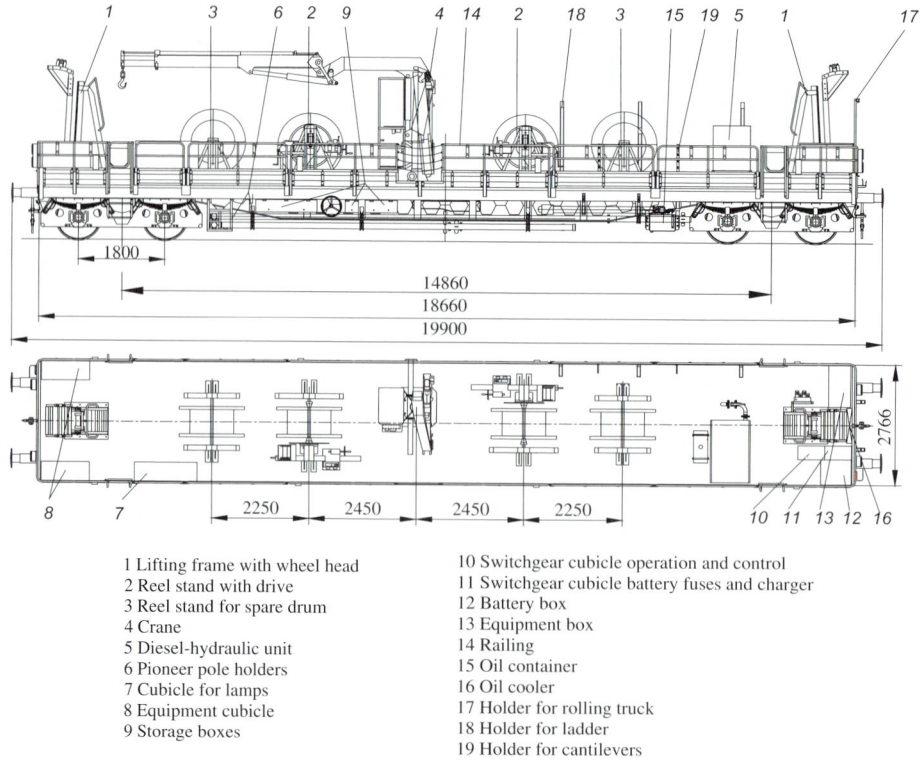

1 Lifting frame with wheel head	10 Switchgear cubicle operation and control
2 Reel stand with drive	11 Switchgear cubicle battery fuses and charger
3 Reel stand for spare drum	12 Battery box
4 Crane	13 Equipment box
5 Diesel-hydraulic unit	14 Railing
6 Pioneer pole holders	15 Oil container
7 Cubicle for lamps	16 Oil cooler
8 Equipment cubicle	17 Holder for rolling truck
9 Storage boxes	18 Holder for ladder
	19 Holder for cantilevers

Figure 15.29: Equipment layout for the overhead contact line installation car type 575.

Figure 15.30: Putting on tracks of a dual-mode vehicle on a new railway line.

Figure 15.31: Dual-mode vehicle during overhead contact line installation.

illumination for night works. For operations at contact line breakdowns with considerable damage, all required wires, conductors and 8 m and 11 m long provisional poles, also called pioneer poles, cantilevers, insulators and fittings are available on the reel stand car.

Catenary wire and contact wire replacements on railway lines with high traffic frequency is only possible with the aid of mechanised processes. The technology for the renewal of contact line as described in Clause 15.5 requires the deployment of the *overhead contact line reconstruction unit* type MTW100 (Figure 15.24 a)). Track closure times of only 5,5 hours per contact line section are sufficient if the work is well organised. The installation train consists of two independent units, each with its own drive, one for the removal and the other for the installation of the overhead contact line. The first installation unit removes the droppers and cantilevers using horizontally mobile work platforms located on the roof of the *installation wagon*. Reel wagons wind in the dismantled catenary wire and contact wire. The pre-assembled cantilevers are installed from the second unit. After the contact wire and catenary wire are strung with pre-tensioning that can be adjusted between 5 and 12 kN, the new droppers are installed from the work platform with the position of the train and the platform height displayed. When installation activities are completed; the overhead contact line can be used at maximum line speed [15.13]. Railway flat cars with raising platforms can also be used for overhead contact line installation.

15.6.5 Dual-mode vehicles

Dual-mode vehicles are able to operate on roads as well as on the tracks. They can be used for the installation of overhead contact lines for local and main line railways (Figure 15.21 a)) and allow working at positions that are not possible from running ladders. The on-track drive of the vehicles permits running at 80 km/h. Overhead contact line works frequently need to be carried out at night, so the work platform is lit depending on the running direction. The on-tracking of the vehicles can be carried out at crossings with an angle up to 90° or at specially designed access positions (Figure 15.30). While the vehicles operate on the track the linesmen can operate the vehicle from the work platform (Figure 15.31). The vehicle has a working and a moving mode making inspection or fast running possible. The vehicles are provided with safety requirement and railway radio, in compliance with the requirements of the railway operator. The track running equipment has driven track wheels.

The manufacturers of these vehicles use the following types of drives depending on the weight of vehicle [15.14]:

- friction wheel device for running on tracks with a wheel set and rubber tires (Figure 15.32 a))
- articulated wheels with track guidance on a wheel set and rubber tires on tracks used as drive (Figure 15.32 b))
- hydrostatic driven track propulsion device consisting of two brackets for each wheel set (Figure 15.32 c))
- hydrostatic driven track propulsion device consisting of a bracket with a wheel set and a bogie with two wheel sets (Figure 15.32 d))
- hydrostatic driven track propulsion device with two bogies with two wheel sets each (Figure 15.32 e))
- hydrostatic driven track propulsion device with two wheels sets and an insulated working platform (Figure 15.32 f))

There are two designs of the friction wheel track propulsion device. With the first, the tires of the vehicle run on a friction drum to drive the track propulsion device (Figure 15.32 a)). The second has an articulated roller guidance system with the vehicle tires running directly on the rails (Figure 15.32 b)). The driving and braking forces to be transferred are smaller in the first case than in the second design [15.15].

Track propulsion devices with wheel sets that lift the vehicle and tires off the rails (Figure 15.32 c) to f)) use hydrostatic drives to propel the track drive devices and guidance devices keep the vehicle on the track. This type of propulsion gear may consist of brackets (Figure 15.32 c)), bogies (Figure 15.32 d)) or a combination of both (Figure 15.32 e)).

Using dual-mode vehicles (Figure 15.32 f)) with insulated lifting platforms live-line working is possible. Live-line working is not permitted on installations for main lines because there is no *double insulation* of the overhead contact line. On installations for local traffic with voltages up to DC 750 V and double insulation live-line working is advantageous because expensive breaks in train operations and safety protection measures can be avoided.

Dual-mode vehicles with single wheel sets are able to run on tracks with a radius down to 25 m and those equipped with bogies can run on tracks with a 40 m radius. Single wheel sets are able to carry 18 t and those with bogies 36 t loads for each wheel set [15.14]. Although dual-mode vehicles are able to start and brake on gradients up to 80 ‰ [15.17]. DB only permits the operation of dual-mode vehicles on track gradients up to 40 ‰ [15.19]. Basic vehicles with a track running device can be equipped, according to the requirements of the assembly or maintenance works, with:

- large cabin with workshop (Figure 15.33 a) by SRS Sjölanders AB [15.18])
- scissor-type work platform (Figure 15.33 b) by ZWEIWEG International Ltd.)
- articulated work platforms (Figure 15.33 c) by ZWEIWEG International Ltd. [15.17])
- telescopic crane platform (Figure 15.33 d) by SRT Schörling Rail Tech Ltd. [15.17])

Vehicles with a large cabin (Figure 15.33 a)) can be equipped with a workshop and an office for diagnostic measurements and evaluations. Vehicles equipped with air conditioning and thermal insulation are suited for the use in regions with temperature extremes. Scissor-type work platforms (Figure 15.33 b)) can only move in the vertical direction. Lateral movements

a) Friction wheel track propulsion device and rubber tire [15.15]

b) Articulated roller guidance system, tires run directly on the rails [15.16]

c) Hydrostatic bracket device for running on tracks [15.15]

d) Hydrostatic bracket/bogie rail-running device [15.17]

e) Hydrostatic propulsion for running on rails for two bogies [15.17]

f) Dual-mode vehicle for working under live-line working [15.15]

Figure 15.32: Dual-mode vehicles.

a) Vehicle with large cabin

b) Scissors working platform

c) Articulated working platform

d) Telescopic working platform

Figure 15.33: Dual-mode vehicles.

are not possible, however, the platform can be turned around. The scissor-type work platform can be raised up to 11 m above top of rail (ToR) and reach the highest components such as the disconnectors on individual poles. This type of work platform can be used for inspection of overhead contact lines in cities and main lines on inter-station sections. Articulated work platforms (Figure 15.33 c)) enable vertical and horizontal movements of the platform.

Articulated and telescopic work platforms have a wider working area (Figure 15.33 c) and d)) than e. g. scissor-type platforms (Figure 15.33 b)). Articulated and telescopic work platforms enable movements of the platforms in and against the running direction without moving the vehicle forwards or backwards. With telescopic work platforms (Figure 15.33 d)) working at large heights is possible within stations. The work baskets are equipped with a display of the height. Working on track cants up to 170 mm is possible without additional stabilizers. Movements of the work platforms up to the limits of stability is controlled thus avoiding dangerous situations. Movements of the work platforms into the gauge of the parallel track can also be excluded.

The type of assembly and maintenance works determines the size of the work platform. In the case of assembly of overhead contact lines, long work platforms are necessary; for inspections during maintenance, relatively short platforms are sufficient. Also in the case of large track cants the work platforms can be stabilized horizontally and the rolling of material or tools can be avoided. Dual-mode vehicles can also be equipped with measuring pantographs, requiring sway compensation.

Figure 15.34: Moveable ladder.

Figure 15.35: Testing for voltage absence on the line and installation of earthing and short-circuiting device (Photo: Walter, DB Ulm).

15.6.6   Ladders

Simple leaning, step, extension and *track-moveable ladders* with lengths between 4 and 12 m are frequently used for assembling overhead contact lines (Figure 15.34). Preparations for working with ladders, including their assembly, can be carried out clear of the track gauge. The ladders can be set on the tracks quickly when the operational track possession starts. For works at concrete poles, scaling ladders, that can be fixed to the poles can be used.

15.6.7   Communications

When working on railway lines continuous communication between the persons involved is necessary, especially during breaks in train operations. Using state of the art radio communication such as portable walkie talkies is important. The applicant may ask for power to be switched off by the central switching command (Zes), operational blocking of the track from the traffic supervisor and order the installation of earthing equipment. The person in charge of the whole activity keeps in contact with the leading fitters, the applicant for switching, the leader of the work train and the safety supervisor. The people in charge of the safe working should also be provided with a walkie-talkie or a mobile phone.

a) Manual pressing pliers   b) Hydraulic high-pressure pliers with pump

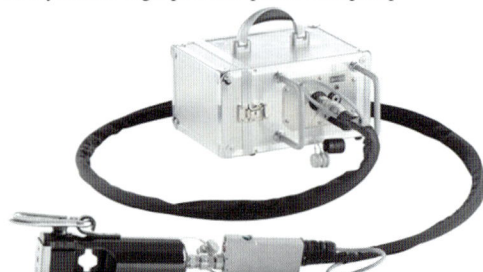

Figure 15.36: Compression tool.

15.6.8 Testing and earthing equipment for overhead contact lines

The earthing of overhead contact lines is carried out in compliance with the *five safety rules* according to EN 50 110-1:

- switching-off the overhead contact line
- securing against reenergising
- testing of absence of voltage
- earthing and short-circuiting of the overhead contact line (Figure 15.35)
- informing personnel about the limits of the working area and adjacent live overhead contact lines.

The switching applicant is responsible for establishing all preconditions for working close to and at overhead contact line installations. For earthing, at least one voltage tester and two sets of earthing equipment are necessary per switching group. The earthing equipment needs to be installed so that the earthing clamp does not kink the contact wire. If the earthing equipment should not infringe the gauge, it is necessary to fix it close to the pole at the cantilever.

15.6.9 Signaling and safety equipment

Signals are used to protect the construction site. They block the tracks where work is being performed with safety signs lit at night. For an electric blocking of the tracks, signals for electric traction are necessary (Figures 17.22 and 17.23). When preparing for the work activities and applying for track blocking, the operational and electrical constraints need to be considered. The signaling and safety equipment needs to be checked before commencing activities. Only fault-free equipment can be used. Sufficient surplus materials also need to be available. *Voltage testers* display the voltage of the overhead contact line. After registering the voltage-free status of the overhead contact line *earthing* and a *short-circuiting device* secures the work area on both sites (Figure 15.35). Close to the substations two sets of devices are installed.

15.6.10 Lighting devices

When working at night or in tunnels, lighting of the working area is required. If the lighting installations on the vehicles are insufficient, additional lighting equipment will be installed along the track before blocking the line. The lighting equipment needs to be tested regularly in accordance with EN 12 464-1.

a) Gall's flat link with come-along

b) Conductor stringing clamp type Siemens

c) Conductor span clamp type Pfisterer

d) Conductor stringing clamp type Pfisterer

Figure 15.37: Conductor stringing clamps.

a) Stringing wheel

b) Conductor pulley block

c) Chain come-along type

Figure 15.38: Conductor stringing wheels and pulley blocks.

15.6.11 Compression tools

Compression techniques, also known as crimping, can produce long lasting connections of conductors and wires. This is especially important for mechanically robust connections within the main current path. *Special tools* like mechanical hydraulic pliers (Figure 15.36 a)) or *hydraulic high-pressure pliers* with either electric or mechanical drive are required for the installation of compression clamps such as C or E clamps (Figure 15.36 b)) by *pressing*.
Manual pressing pliers are suitable for 35 mm^2 copper and 50 mm^2 aluminum cross-section overhead contact line materials. Hydraulic high-pressure pliers need to be used for compressing larger cross-sections. All types of compression connections can be carried out with these tools, equipped with run-in and back, oil-proof pipe coupling and controlled by valves.

15.6.12 Stringing and load-carrying equipment

Come-alongs with Gall's flat links (Figure 15.37 a)) or ropes enable conductor or wire ends to be strung together. For example, a come-along takes the contact wire tension when installing insulators or section insulators. Short conductor sections with thimbles on both ends, also called *strops*, are used when fixing wheel tensioning devices. Hooks and *conductor tensioning clamps* (Figure 15.37 b) and c)) are used to anchor wires and conductors when changing insulators or supports. *Contact wire* and *conductor tensioning clamps* are used to attach the come-along, for example, to the contact wire.

Traction power lines can be strung over pulley wheels (Figure 15.38 a)), fixed to a cross-arm at poles. After the adjustment of the conductor, it is fixed to the insulator.

15.6.13 Workshop equipment

Pre-assembly of cantilevers, droppers, earthing connections, disconnector linkages etc. is carried out in the construction site workshop. Tools and equipment required for pre-assembly include work benches with vices, anvils, metal saws, cut-off grinders, pair of snips, drill presses and bench grinders. In addition, a portable power generator set of at least 3 000 W, a motor chain saw, a rail drilling machine, a hand drill, an electric die stock, a contact wire twisting lever and a contact wire straightening tool should be continuously available.

15.6.14 Personal safety equipment

The personal safety equipment of the fitters should at least consist of:
– protective boots with steel cap, ankle and sprain protection
– protective clothing such as reflective vest and gloves
– thermoplastic helmet, five years old at maximum with date of production
– rebound strap as protection against falling when climbing poles
– breathing masks with date of production in case of coating works
– eye protection for grinding and welding activities and
– ear protection during grinding activities

This equipment is a precondition for safe and accident free work on overhead contact line installations.

15.7 Acceptance

15.7.1 General

The basis of technical acceptance procedures for overhead contact lines are different for conventional lines with speeds up to 230 km/h and high-speed lines with speeds above 230 km/h. The acceptance procedure of both line categories stems from the Technical Specification for the Interoperability of the Energy Subsystem (TSI ENE) [15.20], European standards (EN), national standards as well as the directives and specifications of the railway operators. The technical specifications for the Energy Subsystem and the corresponding EN standards that were implemented into the national laws of the EU members are mandatory for conventional and high-speed lines.

15.7.2 Tasks of the contractor

The contractor of a system, designated as the producer within [15.20], has to provide information on the operational limits of all components of the overhead contact line to the infrastructure manager. Information should be provided on all items that can change during operations, such as permissible contact wire wear and the permissible tolerances. The contractor hands over to the infrastructure manager the necessary documents for the following operation of the overhead contact line installation. Included with these documents are the revised construction documents (As-built drawings), remarks on operation, recommendations concerning maintenance as well as the provision of spare parts (EN 50 119:2013, Clause 9.5).

15.7.3 Tasks of the Acceptance Engineer

The *Acceptance Engineer* for railway installations must be approved by the National Notified Body as an *expert* for the acceptance of railway installations. He can also be accepted as an expert for the railway CERT (EBC). The expert can work in the whole European Union carrying out this function.

The expert works independently, impartially, free of instructions, conscientiously and personally. Conditions for approval as an acceptance engineer of EBC are [15.21]:
  – graduation as an engineer
  – at least three years relevant occupational experience
  – at least 30 years of age
  – personal aptitude, e. g. more than average knowledge regarding
      – railway operations
      – planning of railways
      – construction and operation of installations
      – technology of the installations as well as
      – legal specifications and the generally accepted codes of practice
  – practical experience and the ability to establish expertise and testing reports
  – knowledge and experience in using the tools for carrying out acceptances
  – orderly economical conditions
  – impartiality and independence
  – previous approval as an expert for checking documents

The acceptance engineer is in charge of and responsible for:
  – fully carrying out the acceptance procedure and supervision of defects rectification
  – summarising the results of acceptance testing in a written acceptance report and personally signing off
  – the presentation of any departures from the standard, of unusual facts or of cases in doubts
  – cooperation with other approved experts, also called co-experts, for interdisciplinary acceptances
  – keeping the acceptance documents for ten years
  – his own further education as well as the necessary exchange of experience gained from the required verification

15.7.4 Preparation of the acceptance procedure

Pre-conditions for acceptance are:
  – completed assembly of the overhead contact line system
  – the contractor having completed and documented an internal quality review

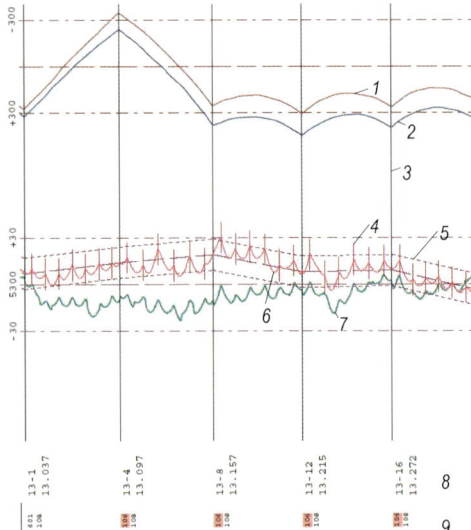

Figure 15.39: Recording of contact wire position measurement.
1 contact wire lateral position,
    measured on Dec. 2., 2003
2 contact wire lateral position,
    measured on Oct. 13., 2004
3 pole position
4 dropper position
5 tolerance range of contact wire height
6 contact wire height,
    measured on Oct. 13., 2004
7 contact wire height,
    measured on Dec. 2., 2003
8 pole number and position in km, position of
    irregularities *9* code of irregularities

— the contractor's announcement by the *declaration of readiness* for acceptance that the installation was installed correctly and completely. This includes the contractor's declaration that the system was correctly assembled
— release of the construction documents by the *authorized engineer for building documentations*
— all released construction documents are in the hands of the acceptance engineer together with the *report on testing the planning documents*
— the construction documents were at least revised by hand.

With the declaration of readiness for acceptance the constructor confirms termination of installation of the system and compliance with laws, standards, specifications and correct execution corresponding to the requirements of the project. In addition, the engineer supervising the construction confirms in this document the correct construction of those parts of the system no longer accessible. These comprise the foundations, earthing installations, equipotential bonding of buildings, protection against touching and laying of cables.

Prior to acceptance, the acceptance engineer reviews the technical section of the contract for the erection of the overhead contact line system. The technical specification agreed upon such as tolerances, type of pantographs and bird protection form the basis for acceptance together with the normative specifications and directives of the railway operator. The contractor of the overhead contact line system provides the necessary drawings, assembly plans and contact line layout plans. The acceptance engineer agrees with the contractor:

— the program and technical plan for the acceptance
— the persons to be involved
— the required vehicles and measurement tools
— the required documents prior to the acceptance
— the necessary documents after the end of the acceptance

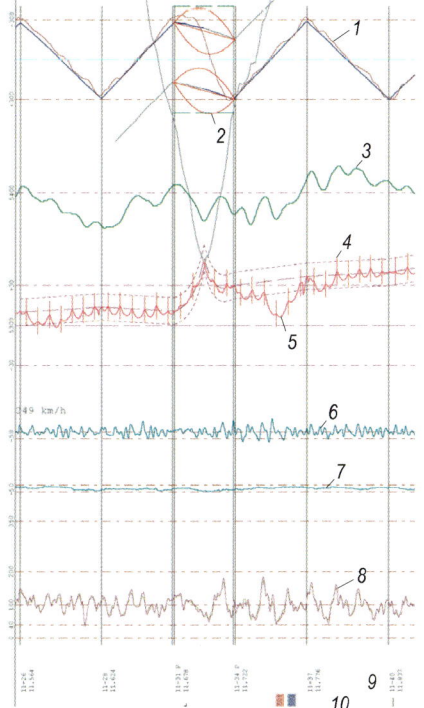

Figure 15.40: Recording of contact wire lateral position, sensor forces with aerodynamic correction and contact forces at 249 km/h.

1 statically and dynamically recorded contact wire lateral position
2 exceeding the permissible contact wire lateral position at wind speeds of 33 m/s
3 dynamic contact wire height
4 dropper position
5 contact wire position in still air
6 vertical acceleration of collector strip
7 acceleration of the vehicle roof
8 sensor forces with aerodynamic correction and contact forces according to EN 50 317
9 pole numbers and pole location in km position
10 position of irregularities

15.7.5 Acceptance procedure

The requirements for acceptance are set out in EN 50 119:2013, Clause 8.15. The acceptance refers to technical and functional aspects as well as safety aspects and starts with a visual inspection of components close to the ground. Measurement of contact wire height and lateral position follows (Figure 15.39). Then, the overhead contact line, cantilevers, disconnectors and feeder lines are checked by using an inspection vehicle. The acceptance engineer drafts a report on the results, listing all deficiencies. After correction of the faults, the acceptance engineer confirms that the inspected section is free of deficiencies or at least he confirms the correction of the safety relevant deficiencies.

After the additional conditions are complied with, the following activities are carried out:

- confirmation of the correction of deficiencies after the acceptance
- handing over of the revised design documents together with the official permissions of the Railway Authorities to the operator:
    - contact line layout plans
    - earthing plans
    - pole and foundation lists
    - drawings especially established for the project
    - cross-sections
    - verification of crossings
    - underground cable layout plans
    - circuit plans of the contact line
    - measuring and test reports for cables

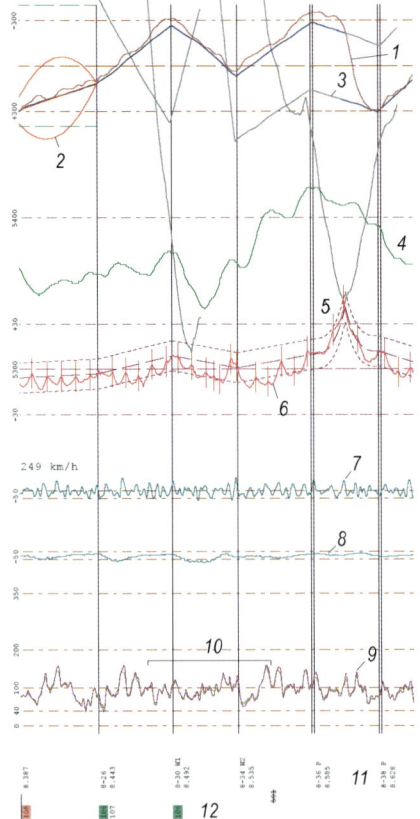

Figure 15.41: Recording of contact wire lateral position, sensor forces with aerodynamic correction and contact forces according to EN 50 317 at 249 km/h above track points.

1 dynamic contact wire lateral position
2 exceeding permissible contact wire lateral position at wind velocity of 33 m/s
3 contact wire lateral position in still air
4 dynamic contact wire height
5 dropper position
6 contact wire position in still air
7 vertical acceleration of collector strip
8 acceleration of the vehicle
9 sensor forces with aerodynamic correction and contact forces according to EN 50 317
10 position of point
11 pole number and pole location in km position
12 position of irregularities

- supply of the tools and equipment necessary for operations, such as earthing equipment, voltage testers, keys to the operator
- handing over of the operating instructions for the local control panels and the remote control devices to the operator
- handing over of the quality testing reports of the system components
- confirmation of the adjustment of the overhead contact line protection within the substations
- confirmation of the publication of the commissioning in the business reports, within the public press and by warning signs
- confirmation of the instructions concerning the energising of the overhead contact line
- confirmation of the distribution of the switching circuit plans for the operation of the overhead contact line

After this procedure, the acceptance engineer recommends energisation of the overhead contact line.

Then the measuring runs follow, these test the interaction between the overhead contact line and pantographs. To assess the results and localise the deficiencies, a digital comparison of the measurement of the contact wire position with the contact force measurements has proven advantageous. Figure 15.40 demonstrates measuring records for contact wire position and contact forces for a section between stations and Figure 15.41 for an overhead contact line above a set of points. By comparing the contact wire position and contact force, the reasons for contact forces outside the accepted range can be determined. Excessive contact forces can be caused by:

(1) contact wire position outside the accepted range
(2) kinks and waves in the contact wire
(3) differences of elasticity within the overhead contact line
(4) reflection of running waves in the overhead contact line
(5) unfavourable aerodynamic features of structures on the line, e. g. tunnel entrances

The reason for (1) is closely related to changes in contact wire height. The necessary corrections can be established and carried out. Follow up additional measurements are advisable, to check that the deficiencies were corrected. The reason for (2) requires a new inspection of the overhead contact line in the area of the observed deficiency. One reason for this may also be a wind stay at the cantilever without the necessary play. The reason for (3) can be assumed to be in the planning and design of the overhead contact line and can only be partly corrected after installation of the contact line. Crossing point wiring, overlaps and section insulators cause differences in elasticity which are in the nature of the system and can not be completely corrected. The reason for (4) stems from the design and can only be corrected by changing the design of the overhead contact line. The reason for (5) results from civil engineering structures not designed to specifications for high-speed lines. Such irregularities cannot be corrected after the construction of the structures. The acceptance engineer classifies the observed deficiencies according to reasons (1) to (5) and supervises their correction.

The assessment of the energy subsystem corresponding to the specifications of the TSI ENE [15.20] supplements the functional-technical and safety-relevant acceptance and forms the basis for the acceptance of overhead contact lines of interoperable lines. The TSI ENE [15.20] refers to the European standards that also need to be considered.

Since [15.20] only concerns those components of an installation that are relevant for the interoperability, other additional standards and specifications need to be considered, e. g. for foundations and poles. The acceptance procedure, therefore, is divided into interoperable and technical aspects that are covered by the TSI Energy and other specifications, respectively. The acceptance report summarises the results of the investigation records and consists of the interoperable and general aspects of the acceptance procedure. The provisional acceptance report can be used by the authorities for providing permission for provisional use (Figure 15.42) on the basis of which the system can be energised for test operations. After finalising the acceptance runs, the final acceptance report is drafted for the authority in charge to use as a basis for the testing report on interoperability (IOC). The EBC can grant the EC testing certificate for the energy subsystem on the basis of the final testing report (Figure 15.42). The final acceptance report is the last step of the acceptance procedure and the basis for the contractual transfer of the overhead contact line system by the operator of the installation. The papers [15.22, 15.23] describe examples of acceptance procedures for overhead contact lines.

15.8   Commissioning

15.8.1   Procedure

After finishing his/her activities, the acceptance engineer hands over his/her provisional re-port to the *person in charge of the commissioning* [15.24]. The person in charge checks for completeness and correctness of the contents and files the application on the approval for op-eration or in case of interoperable overhead contact lines the application on approval of com-missioning with his/her own approval for commissioning at the Railway Authority in charge. The Railway Authority checks the safety-related aspects of the project and approves for use or for commissioning, possibly with specifications for the commissioning. This approval is handed to the person in charge of the commissioning; he/she implements the specifications for use or for commissioning prior to commissioning of the overhead contact line installation. In the case of comprehensive commissioning of overhead contact line installations the per-son responsible for the commissioning drafts a commissioning program, which can include a switching program.

Without an *approval for use* or for *commissioning* provided by the Railway Authority an overhead contact line installation, as a structural subsystem may not be set into operation. The person in charge of the commissioning reports to the Railway Authority within two days after commissioning. In the notice of commissioning, the specifications for the operation need to be presented which, if any, may result from deficiencies still existing and parts of the installation not yet completed. He/she supervises the correction of those deficiencies. The acceptance procedure is presented in Figure 15.42.

15.8.2   Person in charge of commissioning

The *person in charge for the commissioning* is an employee of the operator and represents the infrastructure manager; he/she is authorised by nomination for this position. An employee can act as the person in charge for the acceptance as long as they have sufficient knowledge about line equipment including civil engineering structures, railway superstructure, buildings and railway operations and comply with the following conditions [15.24]:

- degree in a relevant subject or
- member of the higher technical services section at the infrastructure manager and at least two years of relevant experience or
- education as an operating engineer at the infrastructure manager or
- approval as an especially competent employee

He/she has the following tasks:

- signing the installation declaration to the Railway Authority
- providing the technical safety report to the Railway Authority
- assessing the readiness of the installation for commissioning
- deciding on the commissioning of the installation
- reporting on the commissioning
- reporting on the commissioning to the Railway Authority by a relevant announcement
- supervising the correction of deficiencies

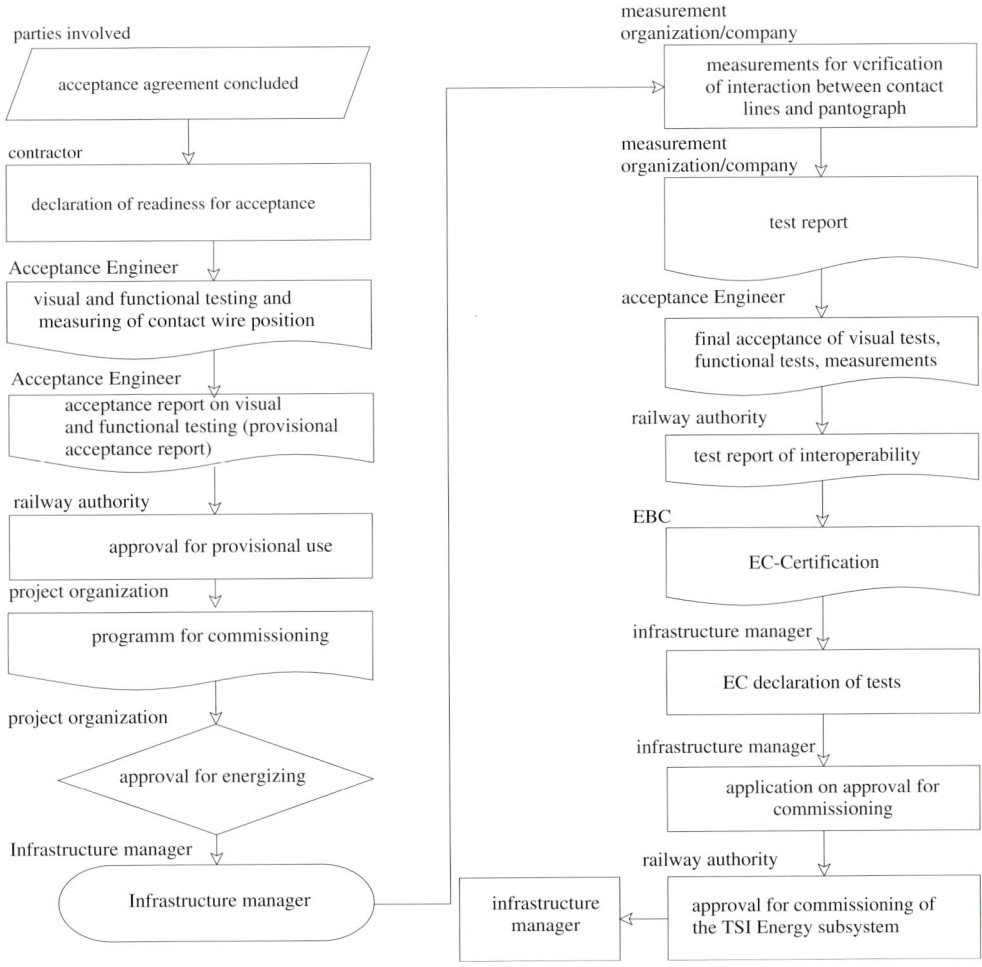

Figure 15.42: Acceptance procedure for an interoperable high-speed line using the rules in Germany as an example.

15.8.3   Tasks of the Notified Body

A Notified Body checks the *conformity* and *serviceability* of *interoperability components* and conducts the *EC testing procedure* for subsystems. After successful testing, the Notified Body establishes the EC declaration for components and for the EC declaration of the subsystem confirming compliance with the specifications from TSI ENE. The Directive 96/48/EC defines interoperability components as assemblies, sub-assemblies and complete material assemblies that are installed in a subsystem or are planned to be installed in such a system and on which the interoperability of railway systems depends directly or indirectly. These are the overhead contact line and the pantographs. The component overhead contact line consists, according to the TSI ENE, of the contact line equipment with catenary wire, contact wire, droppers, stitch wires, Z anchors and anchors at midpoints with the corresponding connecting elements, insulators and other connecting components including feeders and current connectors. The supports like cantilevers, poles and foundations as well as return conductors, auto-transformer feeder lines, disconnectors and other insulators are not part of the interoperability component *overhead contact line*. The component *pantograph* is assessed according to the TSI rolling stock and not part of the TSI ENE [15.20].

The infrastructure manager can select a suitable Notified Body from those Notified Bodies which are accepted in Europe. E.g. in Germany the Railway CERT (EBC) works as a Notified Body according to the Directive 2004/50/EC for European infrastructure managers and carries out testing of components and subsystems.

Prior to commissioning, the interoperability components require an EC conformity declaration and/or an EC declaration for serviceability, where the manufacturer declares that a product placed on the market complies with the basic health and safety requirements of the relevant European Directives, confirming conformity with those Directives. The EC declaration of serviceability of individual interoperability components according to the EC Directive 96/48/EC, Annex IV, is a special form of the EC conformity declaration, if these components are used within railway engineering surroundings or if interfaces are given.

The assessment of the component overhead contact line or of the subsystem energy, including maintenance, is carried out according to the modules of the TSI ENE [15.20] and ends with the EC declaration for the components and the EC testing declaration for the energy subsystem.

15.8.4   Tasks of the Railway Authority

The Railway Authority serves as the authority responsible for the safety of the railway in the member state. The authority is responsible for:

- the supervision of the operational installations of the railways
- the control of risks which originate in the operation of the railway or of their operational installations
- the exertion of sovereign power, survey and cooperation rights
- the approval of commissioning for structural subsystems of the trans-European high-speed and conventional rail systems and
- the supervision of construction of operational installations of government owned railways

Before commencing installation of overhead contact lines, the Railway Authority needs to be informed by advanced notices and a notice of start of installation activities. The Railway Authority can then be in a position to carry out their supervisory responsibilities for the installation activities. The Railway Authority must be informed about the acceptance so that they may participate as required. Following completion of the acceptance, the authority receives information on the commissioning. The procedure and the cooperation between authorities for commissioning of interoperable overhead contact lines is presented in Figure 15.42.

15.9 Bibliography

15.1 *Irsigler, M.; Kohel, J.*: Oberleitungen – Neubau, Umbau und Instandhaltung (Overhead contact lines – installation, reconstruction and maintenance). In: Elektrische Bahnen, 104(2006)1-2, pp. 59 to 69.

15.2 *Nagasaka, S.; Aboshi, M.*: Measurement and estimation of contact wire uneveness. In: Quarterly Report of Railway Technical Research Institute Japan 4(2004)2, pp. 86 to 91.

15.3 *Bauer, K. H.; Gerichten, F.; Kiessling, F.; Lerner, F.*: Einsatz von Aluminium für die Oberleitung der Neubaustrecken der Deutschen Bundesbahn (Utilization of aluminium for the overhead line of the new high-speed lines of the German Federal Railway). In: Elektrische Bahnen, 84(1986)10, pp. 298 to 306.

15.4 *Schmidt, H.; Schmieder, A.*: Stromabnahme im Hochgeschwindigkeitsverkehr (Current collection at high-speed traffic). In: Elektrische Bahnen 103(2005)4-5, pp. 231 to 236.

15.5 *Rux, M.; Schmieder, A.; Zweig, B.-W.*: Qualitätsgerechte Fertigung und Montage hochfester Fahrdrähte (Quality-oriented production and installation of high-strength contact wires). In: Elektrische Bahnen, 105(2007)4-5, pp. 269 to 270 and 272 to 275.

15.6 *Deutsche Bahn AG Ebs 25.04.034*: Richtlinie zum Streckentrennereinbau (Directive for installation of section insulators). Deutsche Bahn AG, Munich, 1978.

15.7 *Deutsche Bahn AG Ebs 02.15.12*: Erforderliche Systemhöhen für Streckentrenner (Required system height for section insulators). Deutsche Bahn AG, Munich, 1964.

15.8 *Deutsche Bahn AG Ebs 02.05.27*: Erforderliche Systemhöhen für Streckentrenner in Gleichstromoberleitungen der DB (Bahnhöfe) (Required system height for section insulators at DB's DC contact lines (stations)). Deutsche Bahn AG, München, 1987.

15.9 *Kiessling, F.; Nefzger, P.; Nolasco, J. F.; Kaintzyk, U.*: Overhead power lines – Planning, design, construction. Springer Publishing, Berlin – Heidelberg – New York, 2003.

15.10 *Borgwardt, H.*: Instandhaltungskonzeption für Oberleitungsanlagen (Organisation of maintenance for overhead contact line installations). In: Construction and maintenance of railway installations, Edition ETR, Hestra Publishing, 1993.

15.11 *Borgwardt, H.*: Druckschrift DS 462 – Grundlage einer sicheren Betriebsführung im Oberleitungsnetz der Deutsche Bundesbahn (Document DS 462 – Basis of a reliable operation of German Railway's overhead contact line network). In: Elektrische Bahnen, 89(1991)4, pp. 106 to 113.

15.12 *Borgwardt, H.*: Schienenfahrzeuge zur Oberleitungsentstörung und -instandhaltung für die Deutsche Bahn (Catenary emergency repair and maintenance railcars for Deutsche Bahn). In: Elektrische Bahnen 94(1996)11, pp. 337 to 340 and 12, pp. 349 to 356.

15.13 *Schneider, B.; Wagner, E.*: Mechanisierte Oberleitungsmontage bei den SBB (Mechanised overhead contact line installation at SBB). In: Der Eisenbahningenieur 49(1998)2, pp. 27 to 30.

15.14 *Hilton Kommunal GmbH*: Spezialist für Zweiwegefahrzeuge (Specialist for dual mode vehicles). Product information, Gehrden, 2013.

15.15 *Lübke, D.*: Einsatzmöglichkeiten von Zweiwegefahrzeugen (Applications for dual mode vehicles). In: ETR (2006)6, pp. 334 to 338.

15.16 *Mercedes-Benz*: Wartung der Gleisinfrastruktur (Maintenance of the rail-track infrastructure). Product information, Stuttgart, 2013.

15.17 *ZWEIWEG Schneider GmbH & Co. KG*: Oberleitungsmontagefahrzeuge (Overhead contact line installation vehicles). Product information, Leichlingen, 2013.

15.18 *SRS Sjölanders AB*: Road to rail, road- rail-road vehicles, product overview. Product information. Osby (Sweden), 2013.

15.19 *Deutsche Bahn AG, Directive 931.0103*: Maschinen-, Energie und Elektrotechnik, Werkstättenwesen: Nebenfahrzeuge"=Bauart und Instandhaltung-Bauanforderungen für Zweiwegefahrzeuge (Department for Machines, Electric engineering, Workshops: Requirements on types and maintenance of auxiliary vehicles, requirements on dual mode vehicles). Deutsche Bahn AG, Berlin, 2004.

15.20 *Regulation 1301/2014/EU*: Technical specification on the interoperability relating to the Energy subsystem of the rail system in the Union. In: Official Journal of European Union. No. L356 (2014), pp. 179 to 227.

15.21 *Eisenbahn-Bundesamt PRÜF-STE*: Richtlinie über die fachtechnischen Voraussetzungen und die Anerkennung von Gutachtern und Prüfern für Signal-, Telekommunikations- und Elektrotechnische Anlagen (Directive on the technical conditions and acceptance of experts and examiners for signalling, telecommunication and electrical engineering installations). Eisenbahn-Bundesamt, Bonn, 2002.

15.22 *Behrends, D.; Vega Vega, T.*: Assessment of interoperable overhead contact line system EAC 350. In: Elektrische Bahnen, 104(2006)1-2, pp. 237 to 241.

15.23 *Grimm, R.; Puschmann, R.; Rux, M.*: Oberleitungsabnahme am Beispiel der Neubaustrecke Köln-Rhein/Main (Acceptance and approval of overhead contact lines for new Cologne Rhine/Main high-speed line). In: Elektrische Bahnen, 101(2003)4-5, pp. 200 to 207.

15.24 *Eisenbahn-Bundesamt VV BAU-STE*: Verwaltungsvorschrift für die Bauaufsicht über Signal-, Telekommunikations- und Elektrotechnische Anlagen (Administrative directive on the supervision of signalling, telecommunication and electrical engineering installations). Eisenbahn-Bundesamt, Bonn, 2010.

16 Implemented contact line installations

16.0 Symbols and abbreviations

Symbol	Definition	Unit
BN	BaneNor (Norwegian Railway)	–
CP	Comboios de Portugal (Portuguese Railway)	–
H_{CW}	contact wire tensile force	kN
H_{CA}	messenger wire tensile force	kN
H_y	stitch wire tensile force	kN
HSB	Hamburger S-Bahn	–
JR	Japan Railways	–
ÖBB	Austrian Federal Railways	–
RER	Réseau Express Régional d'Île-de-France	–
SBB	Swiss Federal Railways	–
SNCB	Société Nationale des Chemins de Fer Belges	–
SNCF	Société Nationale des Chemins de Fer Français	–
ToR	Top of Rail	–
h_{CW}	contact wire height above top of rail	m
h_{SH}	system height	m

16.1 Railway types and their contact lines

16.1.1 Main line traffic

Conventional railways transport people and goods at speeds up to 200 km/h over distances of more than 50 km. The infrastructure of main lines must comply with the Technical Specification for Interoperability (TSI) [16.1], with national standards and directives for construction and operation of railways, for example EBO [16.2] in Germany. Main line railways are equipped with various contact line types depending on the type of power supply and running speed as described in Clause 3.3.3. On high-speed lines, trains run with speeds above 200 km/h. The components of the railway line, especially the overhead contact line, need to be adjusted to the speed requirements and the TSI in Europe.

16.1.2 Mass transit traffic

16.1.2.1 Types of traffic

Mass transit systems comprise according to [16.3]
- regional railways,
- rapid-transit railways,
- tramways,
- city railways,

a) Rapid-transit railway in Berlin   b) Tramway in Dresden

Figure 16.1: Rapid-transit railways and tramways.

- metros and
- trolley buses.

These railways are used mainly to transport people and are predominantly driven by electric energy. The contact line types are adjusted to the voltage applied and the right-of-way used.

16.1.2.2   Regional railways

Regional railways use frequently their own right-of-way and often the power supply installations of main line railways. They are also governed by directives for construction and operation of railways like EBO [16.2] as well as the TSI ENE [16.4]. Signaling systems are those used for main line railways. Regional railways transport people between the centres of larger cities and suburban areas. They close the gap between main line railways connecting bigger cities and mass-transit railways which mainly operate within the cities. Electrically operated regional railways are mainly equipped with overhead contact lines (see Clause 3.3.3 and 3.4).

16.1.2.3   Rapid-transit railways

Rapid-transit railways comprise railways running on regional and main lines utilising the main line infrastructure and lines in city centres, running on their own right-of-way, and lines energised by DC systems. Depending on the type of operation, rapid-transit railways use either the main line signaling system or their own system. Overhead contact lines (see Clause 3.3.3) or third rails systems (see Clause 3.5) are used to energise these railways. Figure 16.1 a) shows the Berlin rapid-transit railway system supplied by third rail.

16.1.2.4   Tramways

Tramways (Figure 16.1 b)) are part of road traffic and predominantly do not run on own right-of-ways but depend on the roads provided for other traffic. The distance between stations are short, so they run at a 20 to 25 km/h speed. They are governed in Germany by BOStrab [16.5] directive on tramways which differs from the directives for main line railways EBO [16.2].

a) Tramway in Karlsruhe

b) Metro line in Nuremberg

Figure 16.2: Tramways and metros.

Increasingly, tramways are on their own right-of-way and also utilise the infrastructure of main lines, including the AC power supply. Depending on requirements, overhead contact lines for tramways vary considerably (see Clause 3.1.3).

The transition from tramways sharing their right-of-way with other traffic partners and city transit railways running on their own right-of-way is indistinct. For the tramway used in Karlsruhe, trams run on roads through the inner city (Figure 16.2 a)) and outside the city centres on main line railway tracks, running as rapid-transit railways combining the advantages of tramways and rapid-transit railways. Dual mode vehicles enable the connection of traffic on main lines supplied by AC with DC operation in the city centre, avoiding the need for passengers, who wish to travel from suburban areas into the city centres, to change vehicles.

16.1.2.5 City railways

City railways predominantly run on their own right-of-way, independently of other road traffic. They can reach considerably higher average speeds than tramways because of the crossing-free right-of-way without interruptions from other traffic. City railways need to comply with the BO-Strab [16.5] directives for tramways in Germany and consequently are considered as part of the road traffic. City railways frequently run underground in centres and on the surface outside of city centres. City railways often use typical tramway vehicles, with types of overhead contact lines depending on individual conditions, e. g. tunnel cross-sections and conditions in the open (see Clauses 3.3.2, 3.3.3, 3.3.4 and 3.4).

16.1.2.6 Metros

Metros form the most efficient mass transit systems, They can be found mainly in major cities, usually supplied by energy from third rail and are referred to as a metro, subway, U-Bahn etc. They run on their own infrastructure, predominantly underground and are covered by the German BO-Strab [16.5] specification (Figure 16.1 a)).

a) Trolleybus type AG 300 T in Solingen   b) Trolleybus type Solaris in Eberswalde

Figure 16.3: Trolley buses.

16.1.2.7   Trolley buses

Trolley buses are propelled by one or more electric motors, powered by traction energy from twin-pole overhead contact lines arranged above the running lane like tramways. Trolley buses are fixed to a lane but not guided by running tracks. However, they must comply with the directives for the operation of traffic companies transporting people (in Germany BO-Kraft) whilst their infrastructure complies with the directive for construction and operation of tramways (BO-Strab [16.5]).

Trolley bus installations commenced operation at the beginning of the 20th Century and are now mainly found in Central and Eastern Europe and in China. In Germany, trolley buses are operated in the cities of Esslingen, Eberswalde and Solingen (Figures 16.3 a) and b)). Their overhead contact lines are discussed in Clause 3.1.4.

16.1.3   Industrial railway lines

16.1.3.1   Railways running on tracks

Industrial railways with standard or narrow-gauge tracks transport goods on company-owned premises to the main line network or to company-owned power plants. Transport companies for ports, coal mining, chemical and steel production operate electrified industrial railways.

In Germany, the directive for installation and operation of connecting railways BOA/EBOA [16.6] specifies the requirements for installation and operation of industrial railways. Since industrial railways are supervised by the provinces, the BOA/EBOA is part of provincial law. Electrically operated industrial railways use different types of power supply. The 350 km long industrial railway is operated by Vattenfall-European Mining Ltd. in Spremberg with DC 2,4 kV (Figure 16.4 a)). The network of RWE Ltd. is 300 km long [16.7]. The overhead contact lines are adjusted to the requirements of the power supply and running speeds.

a) Track-bound industrial railway of Vattenfall European Mining Ltd. in Spremberg

b) Road-bound industrial vehicle in an uranium mine in Lumwana/Zambia

Figure 16.4: Industrial railway and electrically supplied lorry.

a) Essential features

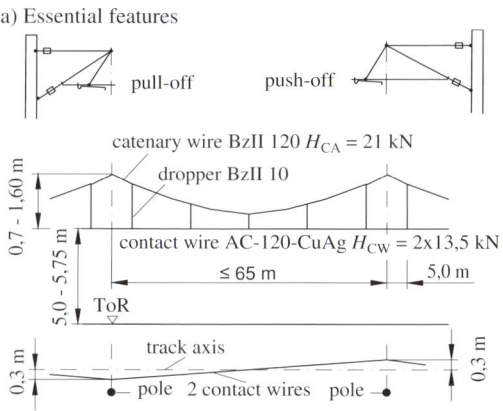

b) Pole with cantilever

c) Twin contact wire clamp

d) Twin steady arm

e) Twin steady arm drop bracket

Figure 16.5: Overhead contact line design Sicat SD for conventional DC lines in the Netherlands.

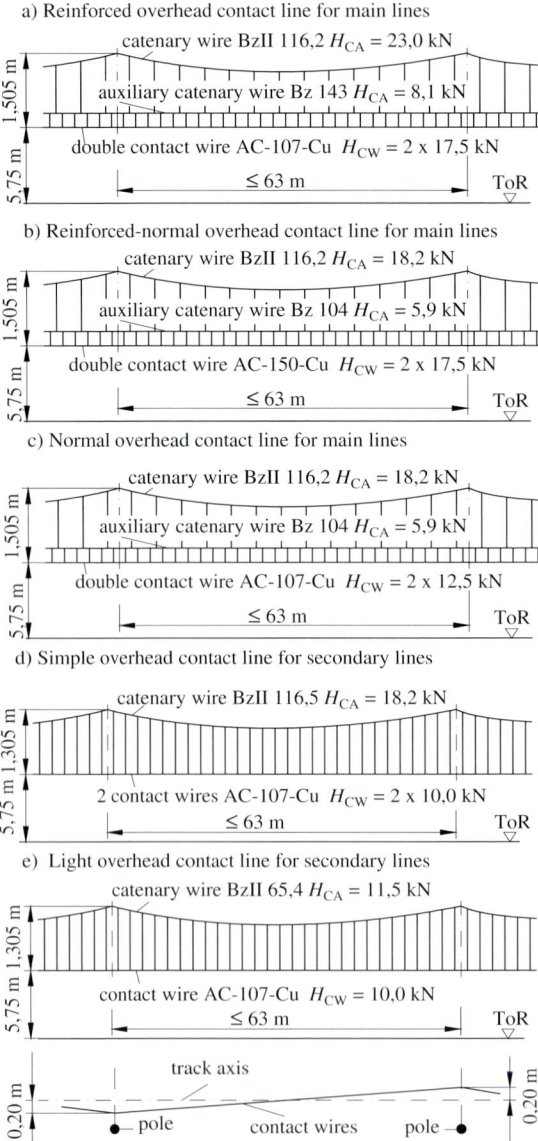

a) Reinforced overhead contact line for main lines

catenary wire BzII 116,2 H_{CA} = 23,0 kN

auxiliary catenary wire Bz 143 H_{CA} = 8,1 kN

double contact wire AC-107-Cu H_{CW} = 2 x 17,5 kN

≤ 63 m   ToR

1.505 m   5.75 m

b) Reinforced-normal overhead contact line for main lines

catenary wire BzII 116,2 H_{CA} = 18,2 kN

auxiliary catenary wire Bz 104 H_{CA} = 5,9 kN

double contact wire AC-150-Cu H_{CW} = 2 x 17,5 kN

≤ 63 m   ToR

1.505 m   5.75 m

c) Normal overhead contact line for main lines

catenary wire BzII 116,2 H_{CA} = 18,2 kN

auxiliary catenary wire Bz 104 H_{CA} = 5,9 kN

double contact wire AC-107-Cu H_{CW} = 2 x 12,5 kN

≤ 63 m   ToR

1.505 m   5.75 m

d) Simple overhead contact line for secondary lines

catenary wire BzII 116,5 H_{CA} = 18,2 kN

2 contact wires AC-107-Cu H_{CW} = 2 x 10,0 kN

≤ 63 m   ToR

5.75 m 1,305 m

e) Light overhead contact line for secondary lines

catenary wire BzII 65,4 H_{CA} = 11,5 kN

contact wire AC-107-Cu H_{CW} = 10,0 kN

≤ 63 m   ToR

5.75 m 1,305 m

track axis

0.20 m   0.20 m

pole   contact wires   pole

Figure 16.6: DC 1,5 kV overhead contact lines of SNCF.

16.1.3.2 Electrically supplied lorries

Electrically supplied lorries are used mainly in open pit mining to transport heavy loads on steep gradients. They are supplied by pantographs (Figure 16.4 b)) from a twin-pole DC contact line arranged above the running lane. This is an economic solution in pits with steep gradients. Dump trucks equipped with a hybrid diesel electric drive are supplied in mines from DC 0,75 kV or higher contact line. Figure 16.4 b) shows a hybrid dump truck in a pit.

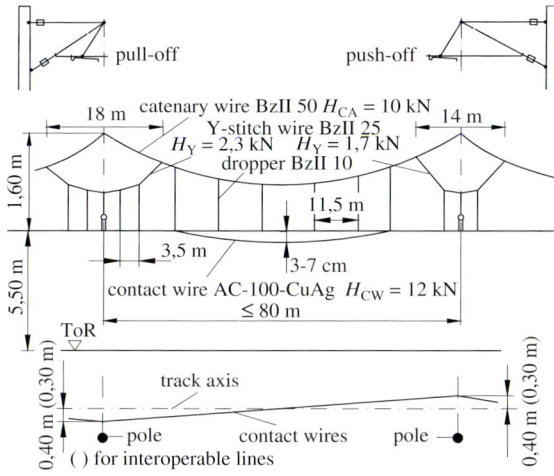

Figure 16.7: Design of Siemens overhead contact line type Sicat SA for the airport line Cologne-Bonn .

16.2   Installations of conventional main lines

16.2.1   DC 1,5 kV overhead contact lines of ProRail in the Netherlands

The construction of the AC 25 kV 50 Hz high-speed line *Hogesnelheidslijn Zuid* (HSL Zuid) in 2009 required connections between the HSL and the conventional DC 1,5 kV network. In connecting curves between the HSL and conventional lines, the overhead contact line type Sicat SD for DC 1,5 kV (Figure 16.5 a)) provided the transition between the overhead contact line type Sicat HA and the existing DC overhead contact lines [16.8].

Galvanised steel poles were erected on pre-fabricated concrete foundations (Figure 16.5 b)). The anchor bolts with leveling nuts beneath the base plate fix the overhead contact line pole approximately 100 mm above the surface of the foundation. This allows the humidity to evaporate from the upper surface of the foundation avoiding damage during winter.

Droppers support both contact wires by pendulum clamps (Figure 16.5 c)), allowing vertical movement of the contact wires and promoting uniform wear when traversed by pantographs. Two light-weight aluminum steady arms, fixed by twin drop brackets, control the contact wires (Figure 16.5 d) and Figure 16.5 e)). An electrical current connector is installed between each of the steady arms and the twin drop bracket.

16.2.2   Overhead contact lines for DC 1,5 kV of SNCF in France

In December 2012, the *French State Railway* (SNCF) was operating a 5 863 km long DC 1,5 kV network. On main lines, the *compound contact line equipment* shown in Figure 16.6 a) is mostly used. On secondary lines and tracks an overhead contact line with a simple and light-weight design is used, as shown in Figure 16.6 b).

The maximum span on straight tracks is 63 m and pulley-wheel tensioning devices with a pulley ratio of 1 : 5 are mainly used. Single steel poles made of H-beams, also called universal columns, support the cantilevers with the contact line.

Table 16.1: Characteristic data of the contact line design Sicat SA.

	Sections in the open	Sections in tunnels
contact wire	AC-100 – CuAg	AC-100 – CuAg
tensile force in kN	12	12
messenger wire	Bz II 50	Bz II 50
tensile force in kN	10	10
stitch wire	Bz II 25	Bz II 25
tensile force in kN	1,8/2,3	1,8/2,3
length in m	14/18	14/18
span lengths in m	75	50
system height in m	1,60	1,10
stagger in m	±0,30	±0,30
contact wire height in m	5,0–5,5	5,0–5,5
maximum tensioning length in m	1 760	1 760
overlapping	over three spans	over three spans
span lengths in m	70 + 70 + 70	50 + 50 + 50
insulated overlap	over three spans	over three spans
span lengths in m	70 + 70 + 70	50 + 50 + 50
brackets	steel, galvanized	steel, galvanized
cantilever tubes	aluminum cast alloy F31	aluminum cast alloy F31
poles	spun concrete poles	steel posts, galvanized
foundations	driven steel tubes	–

a) Essential features

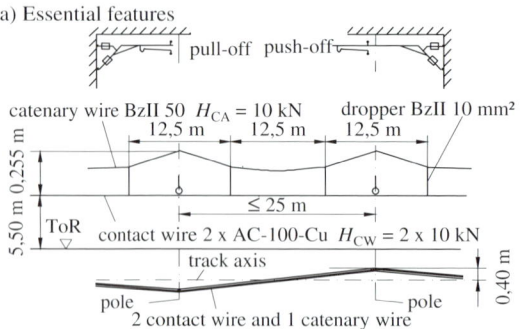

b) Pull-off support

Figure 16.8: DB overhead contact line design Re200 for tunnels.

16.2.3   AC 15 kV 16,7 Hz overhead contact line Sicat SA, Germany

Siemens designed the *overhead contact line design Sicat SA* for speeds up to 230 km/h (Figure 16.7). This contact line consists of a catenary wire BzII 50 and a contact wire AC 100 – CuAg tensioned to 10 kN and 12 kN, respectively (see Table 16.1). Stitch wires 14 m long, are used for push-off supports and 18 m long for pull-off supports. The system height at standard supports is 1,6 m. This overhead contact line design has been used successfully on the airport line Cologne–Bonn, the line Stendal–Uelzen and Itzehoe–Elmshorn in Germany [16.9].

16.2.4   AC 15 kV 16,7 Hz contact line Re200 of DB for tunnels, Germany

In small tunnel profiles and under structures, DB uses type Re200 contact line with a small system height adjusted to narrow tunnel profiles. This type uses catenary wire and contact

a) Sicat SX design features

b) Aluminum cantilever

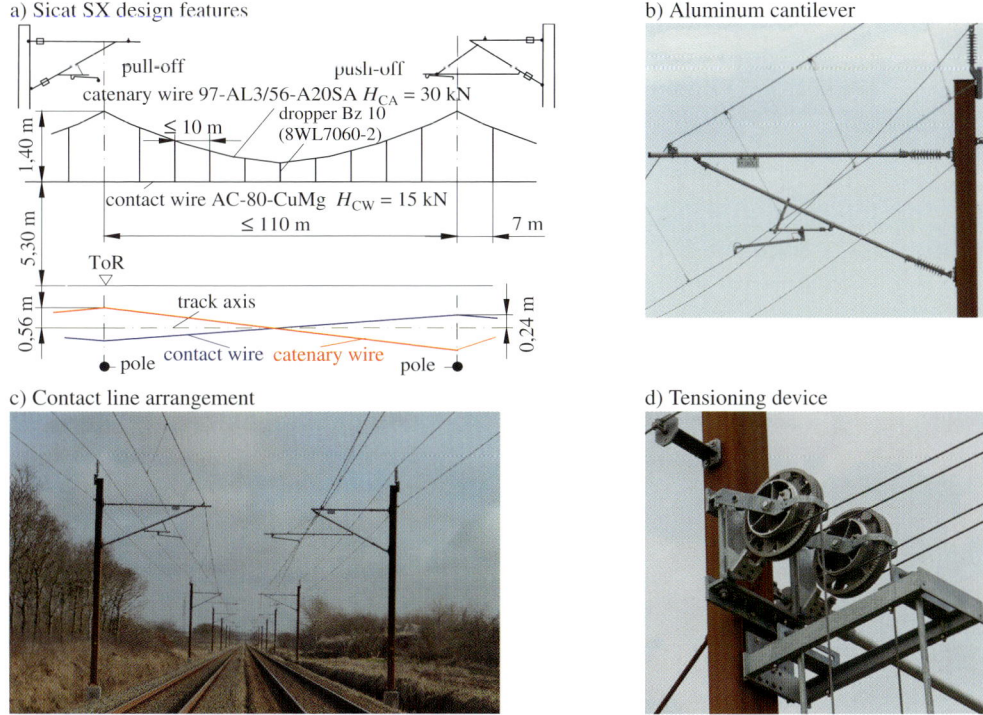

c) Contact line arrangement

d) Tensioning device

Figure 16.9: Overhead contact line Sicat SX in Denmark (Photos: Siemens AG).

wire with features as shown in Figure 16.8 a). Figure 16.8 b) shows a pull-off-support and in Figure 16.8 c) a push-off support below a road bridge. The necessary space is similar to that for an overhead conductor rail.

16.2.5 AC 25 kV 50 Hz overhead contact line Sicat SX, Denmark

The *Danish railway infrastructure owner Banedanmark* (BDK) electrifies approximately 1 300 km and potentially another 300 km of existing and newly built railway lines until 2026. The main tracks for line speeds of up to 250 km/h will be equipped with the inclined overhead catenary system *Sicat SX* designed by Siemens AG [16.10] with the help of *Sicat Dynamic* (see Chapter 12). For siding tracks and depot areas the *Sicat SX light* overhead catenary system will be used which is characterised by lower tensile forces. The project commenced in June 2015 and will last until the year 2026 to electrify up to 15 railway lines in Denmark, 6 of them as an option. The intention of Banedanmark is to connect major cities with traveling durations below one hour. The first double track section Esbjerg – Lunderskov was handed over to Banedanmark for operational use in April 2017. This efficient contact line design (Figure 16.9 a)) features span lengths of more than 100 m, tension sections up to 2 000 m, wheel tensioning device with 1 : 1,5 gear ratio, no stitch wires, a temperature range of 110 K and for operation with a 1 600 mm and 1 950 mm pantographs. Figures 16.9 b), 16.9 c) and 16.9 d) show the installed overhead contact line with cantilevers made of aluminum tubes and the tensioning device. The cantilevers and droppers for the Sicat SX overhead contact line were calculated with *Sicat Candrop*.

a) Transition at the tunnel portal   b) Support

c) Midpoint   d) Conductor rail splice

Figure 16.10: Overhead conductor rail Sicat SR in the Zimmerberg tunnel in Switzerland (Photos: Siemens AG, F. Jung).

16.2.6   AC 15 kV 16,7 Hz overhead conductor rail system, Zimmerberg tunnel, Switzerland

The Zimmerberg tunnel in the Swiss Zurich Canton has a single track and is 1 984 m long. The tunnel, operated by the *Swiss Federal Railways (SBB),* was opened in 1897 and electrified in 1923. It is situated on the line from Zurich to Zug forming part of the line from Zurich to the Gotthard tunnel. Between Zurich and Horgen-Oberdorf and from Baar to Zug, the line has already been supplemented by a second track, however, between Thalwil and Litti/Baar including the Zimmerberg tunnel there is one track only.

After 90 years of operation, the SBB decided to renew the entire contact line using a new design. Due to the high short-circuit currents resulting from close proximity to the Sihlbrugg substation, SBB also decided to install an overhead conductor rail that had to be approved by the Swiss Authority for Traffic (BAV).

Figure 16.10 a) depicts the transition used between the flexible overhead contact line on the open sections and the *Sicat SR conductor rail* in the tunnel. Figure 16.10 b) shows the corresponding support adjusted to the structure gauge and to the small tunnel cross-section. Figures 16.10 c) and 16.10 d) show the midpoint between expansion joints and the conductor rail joint using connection straps.

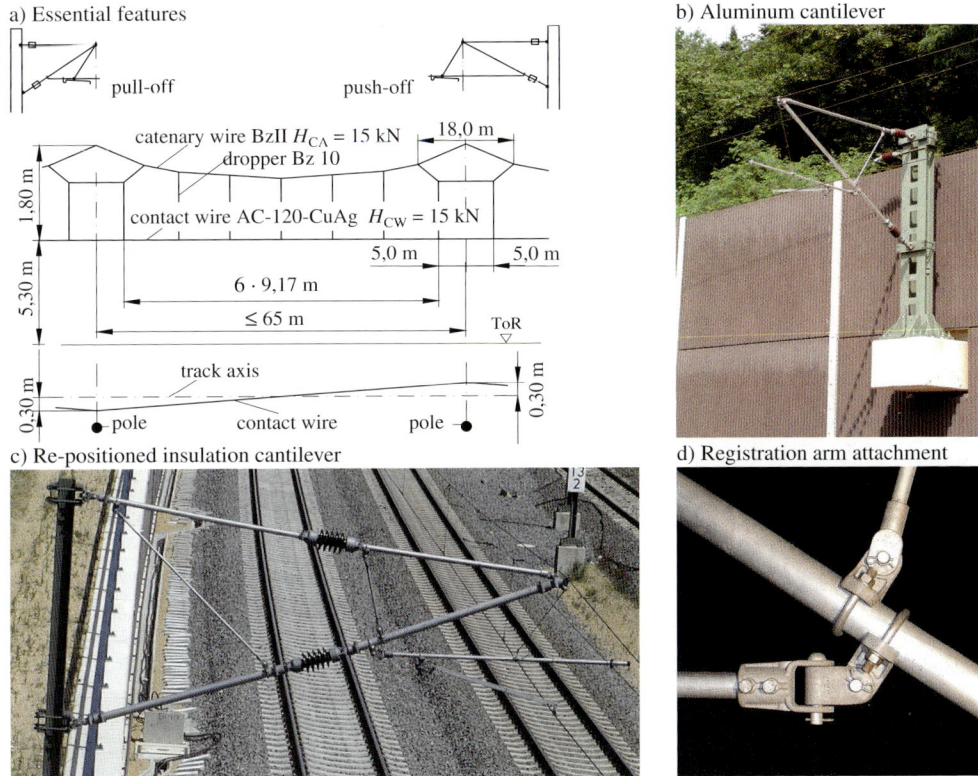

a) Essential features

pull-off   push-off

b) Aluminum cantilever

catenary wire BzII H_{CA} = 15 kN
dropper Bz 10

18,0 m

contact wire AC-120-CuAg H_{CW} = 15 kN

5,0 m   5,0 m

1,80 m

5,30 m

$6 \cdot 9{,}17$ m

≤ 65 m   ToR

track axis

0,30 m   0,30 m

pole   contact wire   pole

c) Re-positioned insulation cantilever

d) Registration arm attachment

Figure 16.11: Overhead contact line Re250 on the Cologne–Düren high-speed line.

16.3 Installations for high-speed main lines

16.3.1 AC 15 kV 16,7 Hz overhead contact line Re250, Germany

The *high-speed line Cologne–Düren* forms a subsection of the trans- European corridor from Cologne to Brussels for 250 km/h. The line originally planned for 160 km/h commercial speed was transformed to a high-speed line and equipped with *overhead contact line type Re250*. Figure 16.11 a) shows the essential features of this design which was also applied to the first high-speed lines in Germany between Würzburg and Hanover. Figure 16.11 b) shows a standard cantilever. At retaining walls, accessible to the public, cantilevers with re-positioned insulators serve as a protective measure using distance (Figure 16.11 c)). Re250 uses hinges for connections between registration arm and cantilever tube (Figure 16.11 d)).

16.3.2 AC 15 kV 16,7 Hz overhead contact line S25, Norway

Norge Jernbaneverket (JBV) operate an AC 15 kV 16,7 Hz network, where the *overhead contact line design S20* is used for lines with speeds up to 200 km/h and the design *S25* for lines with speeds up to 250 km/h. The *high-speed line between Oslo and the Gardermoen airport* uses the design S25 which corresponds to the German Re250. Typical spans of this overhead contact line with stitch wires are shown in Figure 16.12 a) and described in [16.11].

a) Features   b) Glass insulator cantilever

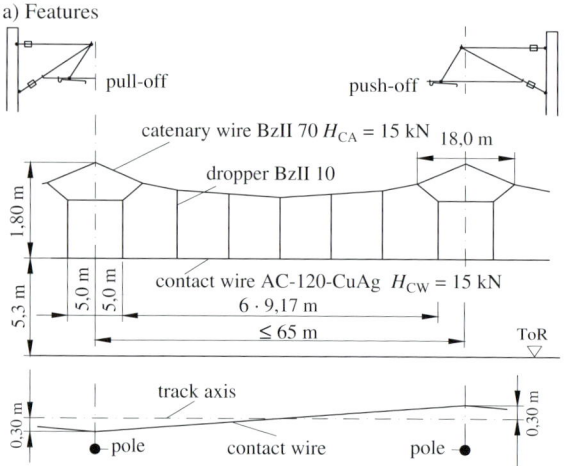

c) Portal-type cross-span   d) Tensioning device

Figure 16.12: Overhead contact line design S25 of JBV in Norway (Photos: JBV, T. Pedersen).

The cantilevers, which only require *low maintenance*, are designed with a horizontal top tube and adjustable catenary wire suspension clamp and adjustable hook end fittings on the cantilever tube, enabling adjustment of the contact wire lateral position (Figure 16.12 b)). Portals mounted on solid steel poles in stations as shown in Figure 16.12 c) form a modular building kit, that enables adjustment for different cross-spans or portal lengths.

On sections between stations, a design with individual poles predominates. Tensioning devices with a pulley ratio of 3 : 1 as in Figure 16.12 d) individually tension the contact and the catenary wire. Return currents are directed into the return conductors by booster transformers. Figure 16.12 d) shows the insulated fixing of the booster lines at the poles. JBV use five-span overlaps in the contact line equipment on line sections in the open and in tunnels.

16.3.3 AC 15 kV 16,7 Hz overhead contact line Re250, Switzerland

The 35 km long *Lötschberg Basis Tunnel (LBT)* connects the cities of Thun in the Bern Alps, in the North and Brig in the Rhone valley, south of the tunnel and has been in operation since December 9, 2007. The tunnel establishes an essential part of the new railway connection

a) Features Re250 LBT-T II

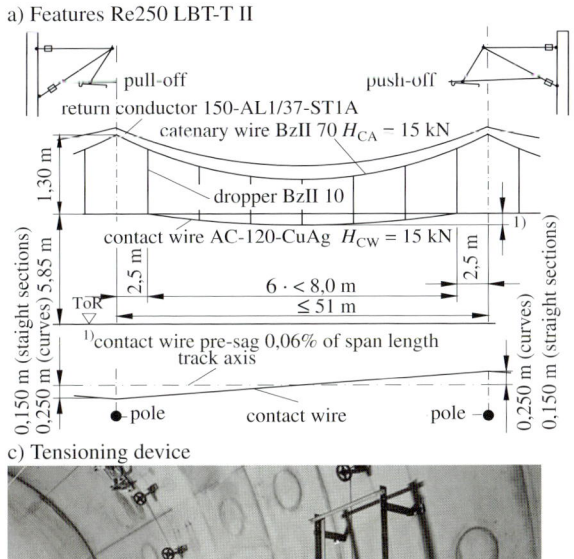

b) Tunnel cantilever

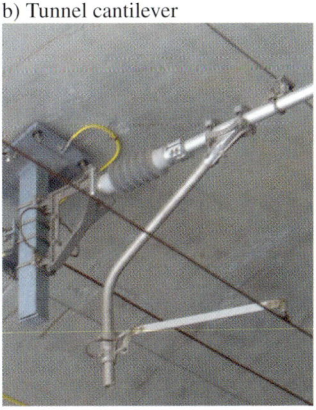

c) Tensioning device

Figure 16.13: Overhead contact line design Re250 LBT-T II for the Lötschberg Base Tunnel.

across the Alps (NEAT) and belongs to the trans-European high-speed rail system. High-speed trains traverse the tunnel at 250 km/h and freight trains with 4 000 t total weight with speeds between 80 and 160 km/h. The upgraded rail link with track gradients up to 30 ‰ [16.12] relieves road traffic.

Essential features of the *overhead contact line design Re250 LBT-T II* are presented in Figure 16.13 a). The design enables the use of 1 320 mm and 1 450 mm long Swiss standard pantographs and the 1 600 mm Euro pan heads. Simulation of the interaction between the overhead contact line and pantographs yielded favorable results [16.13] with:

- two pantographs with 185 m spacing at 275 km/h
- two pantographs with 18,5 m spacing at 200 km/h and
- three pairs of two pantographs spaced at 18,5 m and at least 185 m spacing between the individual pairs of pantographs.

The intended traffic with passenger cars and trucks on shuttle trains required provisions to adjust the overhead contact line to the enlarged structure gauge. The contact wire height is 5,85 m and the system height 1,30 m. The contact wire wear is limited to 30 %. The span length was limited to 51 m so that the dropper lengths were not less than 0,5 m. The cantilevers consist of corrosion resistant materials and compound insulators. Short-circuit-proof connectors bridge the joints within the cantilevers (Figure 16.13 b)). The tensioning devices are shown in Figure 16.13 c).

a) Essential features type 2.1

b) Open line cantilever

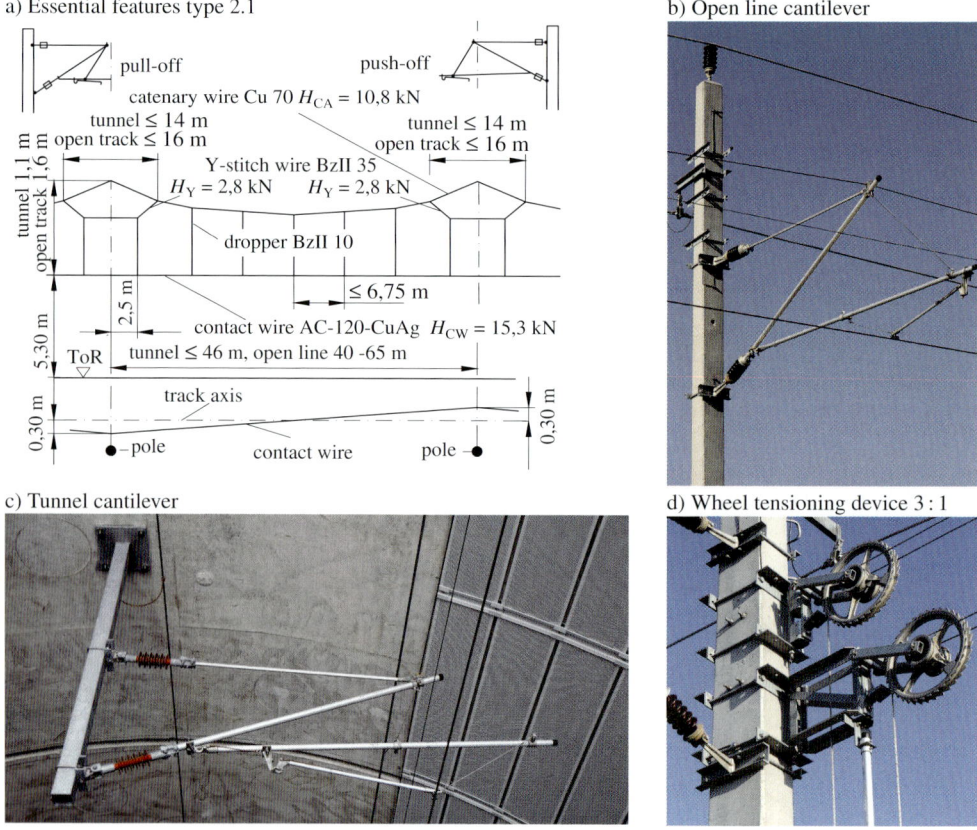

c) Tunnel cantilever

d) Wheel tensioning device 3 : 1

Figure 16.14: Overhead contact line design 2.1 for the high-speed line Vienna–St. Pölten (Photos: SPL Powerlines Austria GmbH & Co KG, R. Herrmann).

16.3.4   AC 15 kV 16,7 Hz overhead contact line type 2.1, Austria

The 60 km long new high-speed line Vienna–St. Pölten is part of the trans-European rail system (TEN) and has been in operation since 2012. The line, equipped with four tracks on some sections, has increased the capacity for freight transport and also accommodates shorter passenger journeys. The Austrian Federal Railways (ÖBB) [16.14] equipped the 44 km long high-speed section *Vienna–St. Pölten* with the contact line design 2.1 which was designed for a commercial speed of 250 km/h (Figure 16.14 a)).

Figure 16.14 b) shows a cantilever on an interstation section with a top anchor tube inclined into the direction of the track centre line. The wheel tensioning devices shown in Figure 16.14 d) take care of tensioning the catenary wire and contact wire separately on the interstation line, with the overlaps stretching over five spans. The ÖBB uses rectangular concrete poles both on this high-speed line and their conventional lines (Figure 16.14 b), d) and e)).

16.3.5   AC 15 kV 16,7 Hz overhead contact line Sicat H, Germany

The *high-speed line Frankfurt–Cologne* is 180 km long with gradients up to 40 ‰. It was designed mainly for passenger trains and is operated at a commercial speed of 300 km/h. The

a) Features Sicat H1.0

b) Tensioning device

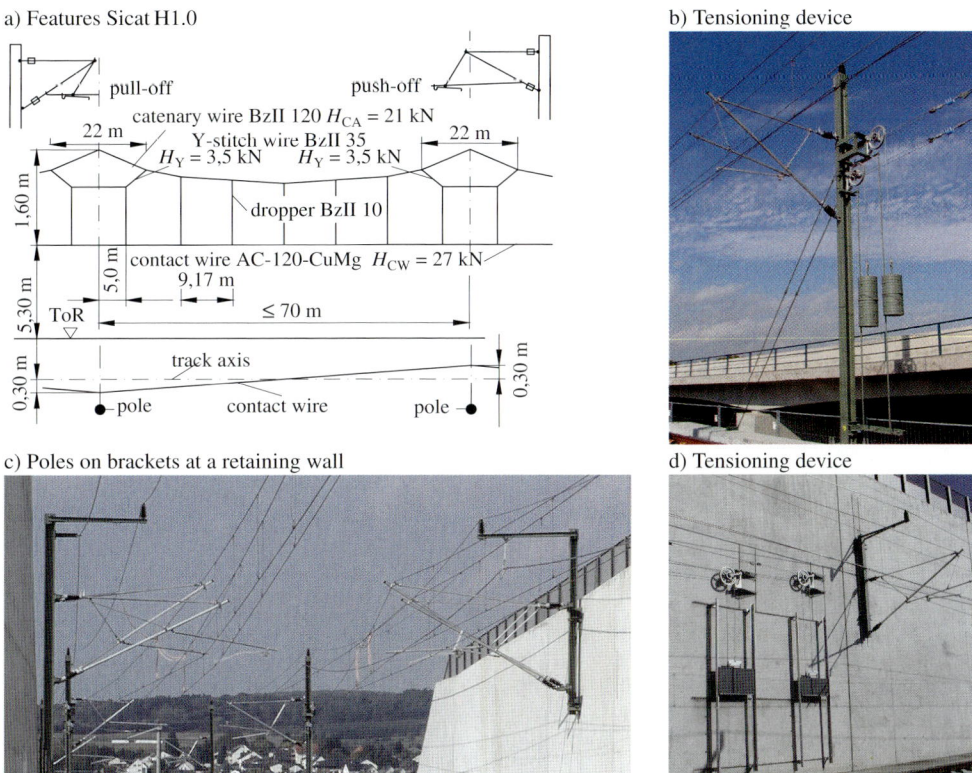

c) Poles on brackets at a retaining wall

d) Tensioning device

Figure 16.15: Overhead contact line Sicat H1.0 for the *high-speed line Frankfurt–Cologne* (Photos: Siemens AG, H. Schmidt).

line was commissioned in 2002. The *overhead contact line design Sicat H1.0 was designed by Siemens* for speeds up to 400 km/h (Figure 16.15 a)) and was installed for the first time on this high-speed line [16.15, 16.16, 16.17]. In Table 16.2 technical details of this contact line design are presented.

16.3.6 AC 15 kV 16,7 Hz overhead contact line Re330, Germany

In December 2015 the newly erected line Leipzig/Halle–Erfurt was commisioned, which is part of the high-speed connection between Berlin and Munich as well as of the railway line Berlin–Munich–Palermo of the transEuropean railway network. The design speed of the new 123 km long line section is 300 km/h; it runs through three tunnels with a total length of 5,4 km as well as over six viaducts with a total length of 14,4 km, wherby the 6 465 m long Saale-Elster viaduct forms the longest bridge in Germany and the longest main line railway viaduct in Europe. The adjacent sections between Erfurt and Ebensfeld as well as between Ebensfeld and Nuremberg will be commissioned end of 2017 and after 2020, respectively. For the line sections between Halle and Nuremberg DB selected the overhead contact line design Re 330 [16.18] which proved its qualification on several other lines in Germany.

Table 16.2: Characteristic data of the Siemens contact line type Sicat H1.0.

features	Lines in the open	Lines in tunnels
contact wire	AC-120 – CuMg	AC-120 – CuMg
tensile force in kN	27	27
catenary wire	Bz II 120	Bz II 120
tensile force in kN	21	21
stitch wire	Bz II 35	without
tensile force in kN	3,5	–
length in m	22	–
span length in m	70	50
system height in m	1,60	1,10
stagger in m	$\pm 0,30$	$\pm 0,30$
contact wire height in m	5,30	5,30
maximum tensioning length in m	1 400	1 400
overlaps	over three spans	over five spans
span lengths in m	70 + 70 + 70	50 + 50 + 50 + 50 + 50
insulating overlaps	over five spans	over five spans
Spans in m	70 + 70 + 62 + 70 + 70	50 + 50 + 50 + 50 + 50
brackets	steel, galvanized	steel, galvanized
cantilever tubes	aluminum alloy	aluminum alloy
poles	spun concrete poles	—
foundations	bored piers, driven steel piles	—

The *overhead contact line design Re330* (Figure 16.16 a)) is suited for 330 km/h speed and was first installed on the high-speed lines Berlin–Hanover and Nuremberg–Ingolstadt. The Re330 design differs from the overhead contact line design Re250 (Clause 16.3.1) using higher tensile forces for the catenary and contact wire and a larger catenary wire cross-section.

In Figure 16.16 a) the essential features of the contact line Re330 are shown. Swiveling cantilevers made of corrosion-resistant aluminum alloys are adopted for carrying the contact line. The cantilevers are fixed to the concrete poles by sockets which are connected to the earthing system of the poles. Depending on the loading, the cantilevers are designed with diagonal struts between cantilever and top tube (Figure 16.16 c)). In Figure 16.16 c) the contact line in a tunnel is shown. The interaction between this contact line consisting of a AC-120–CuMg contact wire, a 70 mm^2 bronze catenary wire, 10 mm^2 bronze droppers and stitch wires and pantographs is favourable even at speeds of 300 km/h and above. The new wheel tensioner (Figure 16.16 d)) avoids the distortion of the contact line in case of a contact or catenary wire failure. New overhead contact lines with similar technical performances as Re250 were developed in Spain, Norway, Turkey and Switzerland. Similarly, the contact line Re330 formed the basis for some contact line designs with slightly modified features such as the Siemens contact line type Sicat H1.0 (see Clauses 16.3.5 and 16.3.7), the Spanish design EAC 350 (Clause 16.3.8) and the Chinese type SiFCAT 350 (Clause 16.3.11).

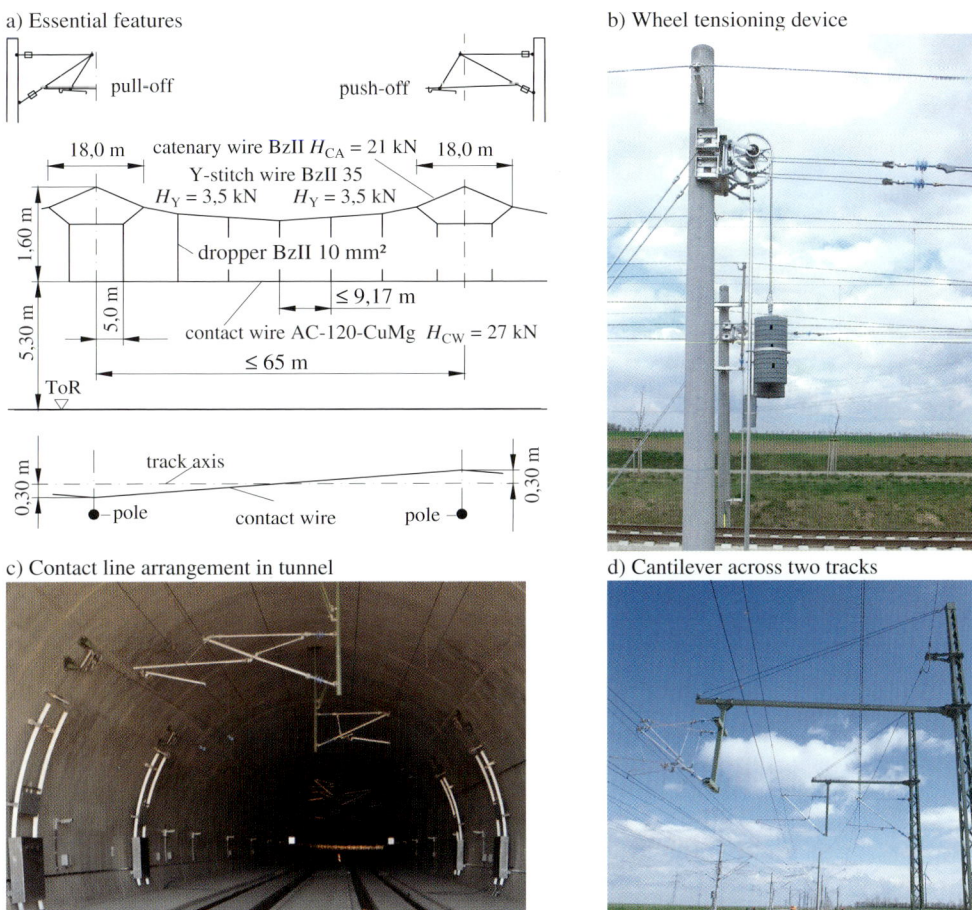

a) Essential features

pull-off   push-off

18,0 m   catenary wire BzII H_{CA} = 21 kN   18,0 m
Y-stitch wire BzII 35
H_Y = 3,5 kN H_Y = 3,5 kN

1,60 m

5,0 m

dropper BzII 10 mm²

≤ 9,17 m

5,30 m

contact wire AC-120-CuMg H_{CW} = 27 kN

≤ 65 m

ToR

0,30 m   track axis   0,30 m

pole   contact wire   pole

b) Wheel tensioning device

c) Contact line arrangement in tunnel

d) Cantilever across two tracks

Figure 16.16: Overhead contact line Re330 for the high-speed line Leipzig/Halle–Erfurt (Photos: SPL Powerlines Germany GmbH, R. Hickethier).

16.3.7   AC 25 kV 50 Hz overhead contact line Sicat H, Netherlands

The *high-speed line HSL Zuid* connects Amsterdam with the railway systems in Belgium, France and Germany [16.8, 16.17]. The line supplied by 2 AC 25 kV 50 Hz is suitable for speeds of 300 km/h and equipped with overhead contact line Sicat H1.0 (Figure 16.17 a)). Figure 16.17 b) shows the adopted cantilevers and circular steel poles. The tensioning device equipped with dry bearings, especially developed for this installation, are maintenance-free and show an improved latch-in performance in case of a contact wire or catenary wire un-loading (Figure 16.17 d)). On long viaducts, changes in the length of the bridge girders occur because of temperature variations. This can lead to variations in the length and sags of return conductors or feeder lines. The suspension of the negative feeder and the return conductor at poles with pulleys permits longitudinal displacements of poles on a viaduct without impacts on the sags of the conductors (Figure 16.17 e)).

The overhead contact line design Sicat H also proved successful on the Segovia–Valladolid and Toledo–La Sagra high-speed lines in Spain which are supplied at AC 25 kV 50 Hz.

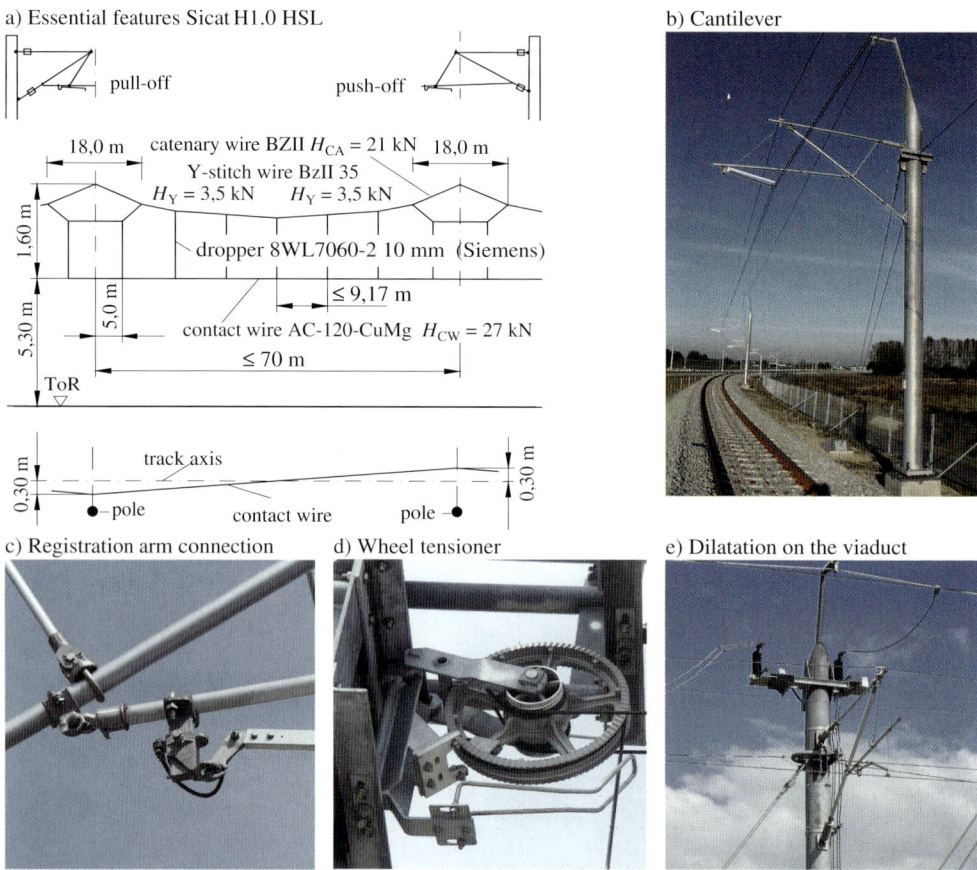

Figure 16.17: Overhead contact line Sicat H for the high-speed line Amsterdam–Rotterdam–Antwerp.

16.3.8 AC 25 kV 50 Hz overhead contact line EAC 350, Spain

The Spanish infrastructure manager *Administrador de Infraestructuras Ferroviarias (ADIF)* installed another high-speed line designed for 350 km/h between Madrid and Valencia, commissioned in December 2010. This line is equipped with the overhead contact line design EAC 350 and forms a part of the Spanish high-speed network which in 2015 reached a length of 1 600 km.

Figure 16.18 a) shows the features of the *overhead contact line EAC 350*. The tensioning sections are at maximum 1 400 m long and the longitudinal spans 63 m. The minimum length of the 25 m^2 copper droppers is 250 mm.

The distance between the two parallel contact lines in non-insulating overlaps is 0,2 m and 0,45 m in insulated overlaps. Figure 16.18 c) shows an aluminum cantilever in tunnels. The wheel tensioning device of catenary wire has a pulley ratio of 3 : 1, and contact wire is 5 : 1 (Figure 16.18 d)). The overhead contact line is designed for a temperature range of 80 K.

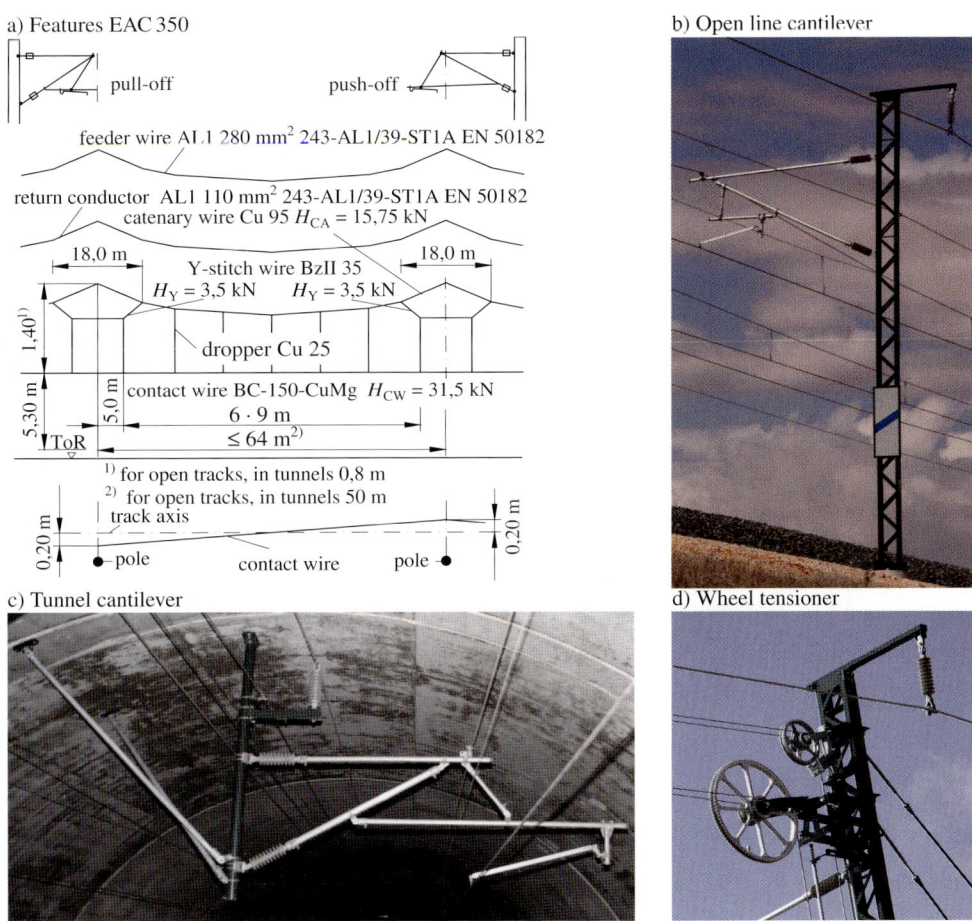

a) Features EAC 350

pull-off   push-off

feeder wire Al 1 280 mm² 243-AL1/39-ST1A EN 50182

return conductor AL1 110 mm² 243-AL1/39-ST1A EN 50182
catenary wire Cu 95 H_{CA} = 15,75 kN

18,0 m   18,0 m

Y-stitch wire BzII 35
H_Y = 3,5 kN H_Y = 3,5 kN

1,40 m

dropper Cu 25

5,30 m

5,0 m

contact wire BC-150-CuMg H_{CW} = 31,5 kN
6 · 9 m
≤ 64 m[2]

ToR

[1] for open tracks, in tunnels 0,8 m
[2] for open tracks, in tunnels 50 m
track axis

0,20 m

pole   contact wire   pole

0,20 m

b) Open line cantilever

c) Tunnel cantilever

d) Wheel tensioner

Figure 16.18: Overhead contact line EAC 350 for the high-speed line Madrid–Motilla–Valencia.

Table 16.3: Important TGV lines in France and their types of overhead contact lines.

Features	Unit	Paris–Lyon	Paris–Le Mans/Tours	Paris–Lille/Calais	Valence–Marseille	Paris–Strasbourg
length of line	km	820	560	660	600	600
commercial speed	km/h	260[1]	300	300	300/320	320/350
contact wire – type – tensile force	– kN	15	AC-120–Cu 20	20	AC-150–CuMg 25	26
catenary wire – type – tensile force	– kN	BzII 65 14	BzII 65 14	BzII 65 14	BzII 116 20	BzII 116 20
stitch wire	–	yes	no	no	no	no
pre-sag	%[2]	0,1	0,1	0,05	0,05	0,05

[1] in sections 300 km/h in 2009, [2] related to the span length

a) Features SNCF high-speed overhead contact line

b) Pole on Paris–Strasbourg line

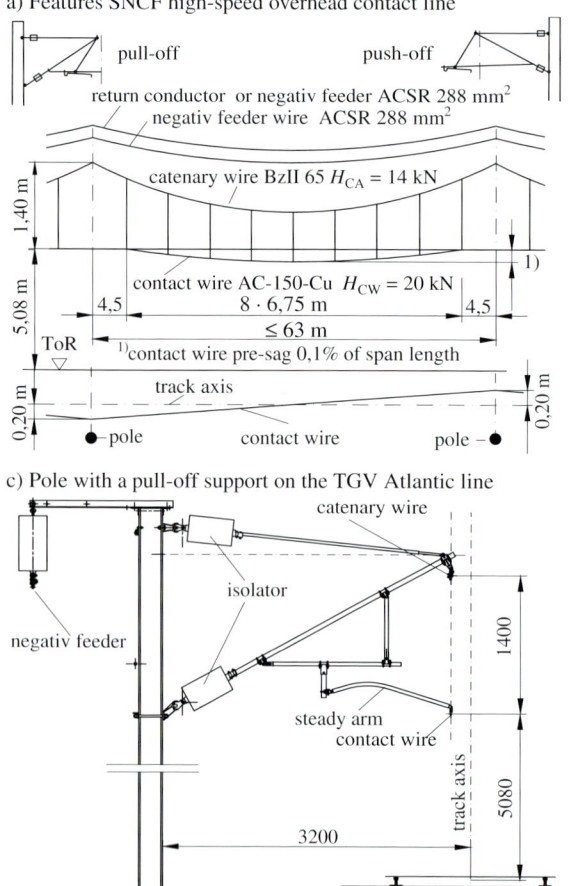

c) Pole with a pull-off support on the TGV Atlantic line

Figure 16.19: Overhead contact lines of SNCF in France.

.

16.3.9   AC 25 kV 50 Hz overhead contact line of SNCF, France

French State Railway SNCF operates 1 840 km high-speed lines with operational speeds up to 300 km/h, supplied by 1 or 2 AC 25 kV 50 Hz, listed in Table 16.3. The adopted contact lines can be subdivided into four generations. The main components like clamps, steady arms etc. are common for all these contact line types.

The features of the SNCF high-speed contact line, e. g. Paris–Tours, are shown in Figure 16.19 a). On the line *Paris–Lyon*, an overhead contact line with stitch wires was adopted. The tensile force of the copper contact wire AC-120 – Cu is 15 kN. The line originally designed for 260 km/h has been operated in sections at 270 km/h or 300 km/h since 2009.

For the *Paris–Le Mans/Tours* line, SNCF raised the commercial speed to 300 km/h and adopted a contact wire AC-150 – Cu (Figure 16.19 a)), tensioned to 20 kN. Figure 16.19 c) shows an individual pole with a cantilever and a support with a negative feeder for 2 AC 25 kV. The design of the contact wire support enables staggers up to 400 mm [16.19]. SNCF adopts *steel poles made of H-beams, also called universal columns*. The *tensioning device* with a pulley ratio of 5:1 compensates variations in length of the contact and catenary wires due to temperature variations and current loading.

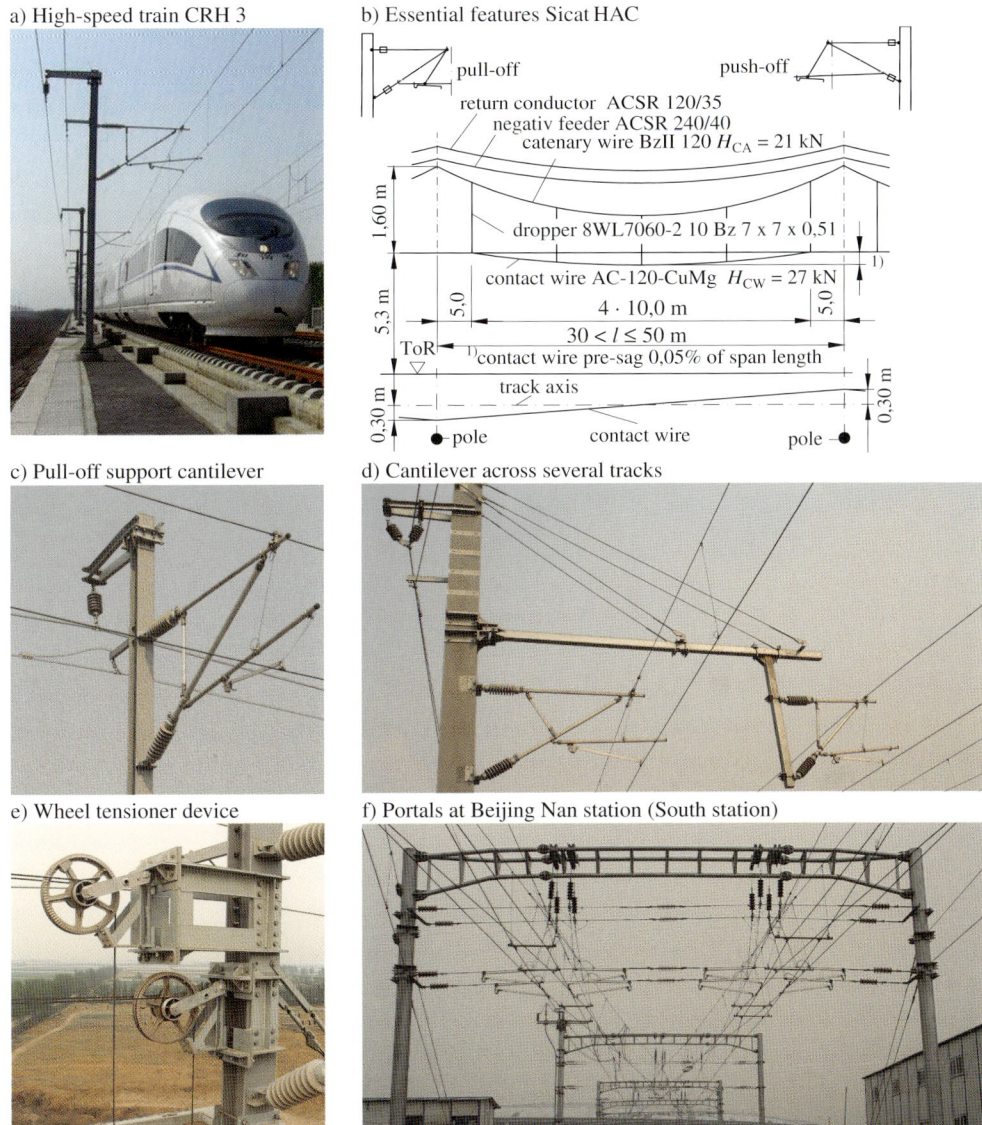

a) High-speed train CRH 3

b) Essential features Sicat HAC

pull-off   push-off

return conductor ACSR 120/35
negativ feeder ACSR 240/40
catenary wire BzII 120 H_{CA} = 21 kN

dropper 8WL7060-2 10 Bz 7 x 7 x 0,51

1,60 m

contact wire AC-120-CuMg H_{CW} = 27 kN

5,3 m

5,0

4 · 10,0 m

5,0

1)

30 < l ≤ 50 m

ToR   1) contact wire pre-sag 0,05% of span length

track axis

0,30 m

pole   contact wire   pole

0,30 m

c) Pull-off support cantilever

d) Cantilever across several tracks

e) Wheel tensioner device

f) Portals at Beijing Nan station (South station)

Figure 16.20: High-speed line Beijing–Tianjin in the Peoples Republic of China.

On 5 May 1990, a TGV train achieved a speed of 515 km/h on the line Paris–Tours, which was a *world record on rails* at that time. The overhead contact line design and a cantilever for this line are shown in Figures 16.19 a), 16.19 b), respectively [16.20] (see Clause 10.9).

16.3.10   AC 25 kV 50 Hz overhead contact line Sicat HAC, China

The *high-speed line Beijing–Tianjin* is the first subsection of the line Beijing to Shanghai and was designed for a commercial speed of 300 km/h as specified by the *Chinese Ministry of*

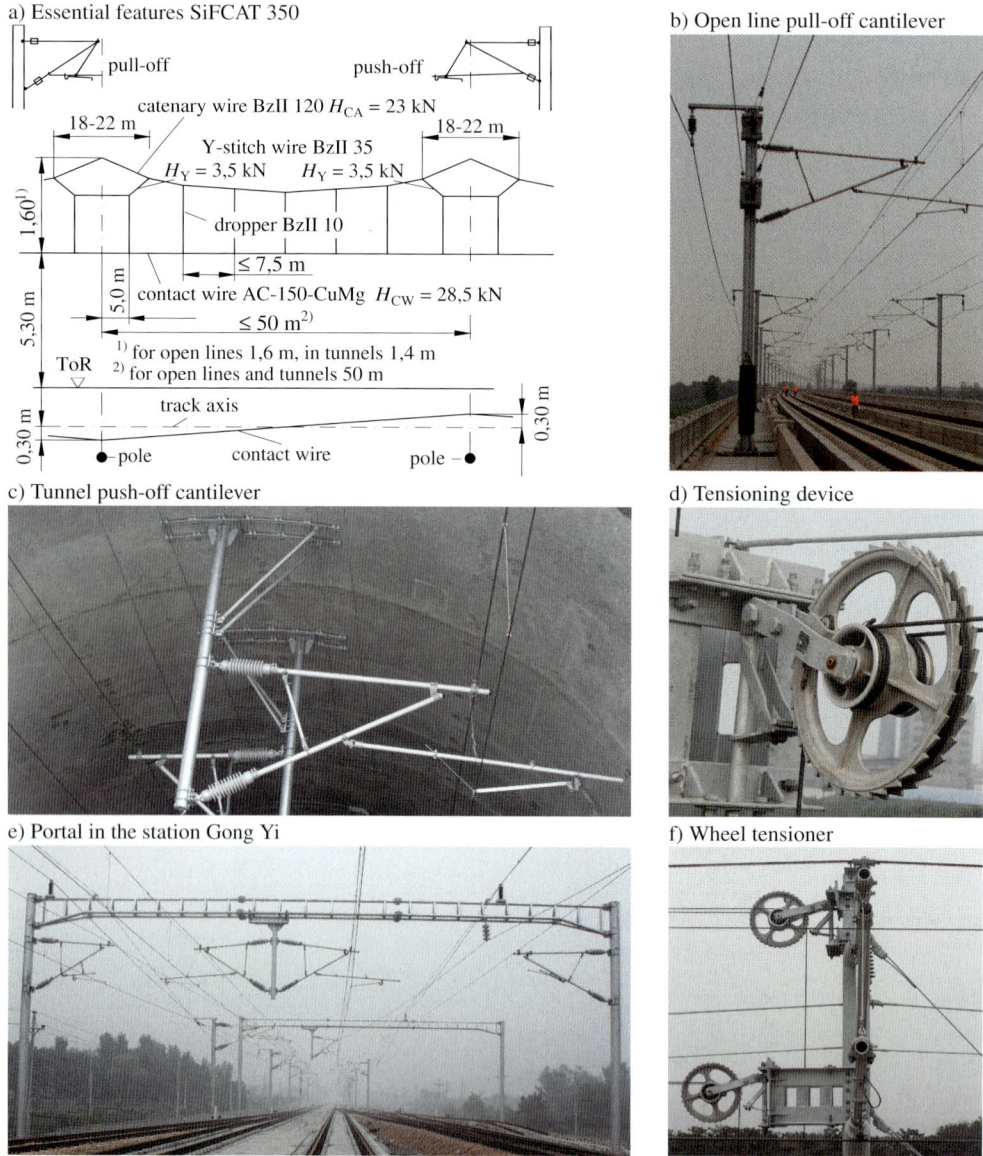

Figure 16.21: High-speed overhead contact line SiFCAT 350 Zhengzhou–Xi'an in China.

Railways (MOR). This first high-speed line in China was inaugurated in July 2008, before the Olympic Games. The *overhead contact line design Sicat HAC* (Figure 16.20 b)) without a stitch wire, with a pre-sag and with spans limited to 50 m ran successfully at speeds up to 350 km/h. Figure 16.20 b) shows the essential features of this overhead contact line design. The contact wire height is 5 300 mm and the pre-sag equates to 0,05 % of the span length or 30 mm (Figure 16.20 b)). The system height is 1 600 mm. Figures 16.20 c) and d) show a pull-off cantilever made of aluminum and a cantilever across several tracks, each with negative feeder and return conductor.

The overlaps between the individual contact line sections stretch over five spans. Wheel tensioning devices with a gear ratio of 3 : 1 compensate the changes in length of contact and catenary wire caused by thermal effects (Figure 16.20 e)). The contact lines above switches are designed with crossing contact wires. Within railway stations, portals carry the overhead contact line supports (Figure 16.20 f)) and steel poles made of H-beam sections support the cantilevers. The foundations of the poles are integral with the 115 km long railway viaduct. The individual supply sections are isolated by phase separation sections with 475 m long neutral zones.

16.3.11 AC 25 kV 50 Hz overhead contact line SiFCAT 350, China

The 485 km long *high-speed line from Zhengzhou to Xi'an* connects the metropolitan areas of the provincial capitals of Henan and Shaanxi with the northern lines to Beijing, with Lanzhai to the west, with Hong Kong in the south and Lianyungang on the East Chinese Ocean to the east. The 460 km long high-speed section operates with 2 AC 50/25 kV 50 Hz and was designed for a commercial speed of 350 km/h.

The *overhead contact line design SiFCAT 350* is named after the designers Siemens AG and FSDI (China Railway Fourth Survey and Design Institute Group Ltd.) and the design speed of 350 km/h, version 1 – the result of a German-Chinese cooperation. This interoperable contact line was designed in Europe for 1 600 mm long pantographs and is traversed by CRH 2 and CRH 3 high-speed trains in China with 1 950 mm long pantographs, the Ministry of Railways (MOR) stipulated this as the standard type for China.

The contact line consists of a AC-150 – CuMg contact wire and a BzII-120 catenary wire with BzII-10 droppers. The system height is 1 600 mm on open line sections and 1 400 mm in tunnels. The maximum span length never exceeds 50 m in the open nor in tunnels. Essential features of the overhead contact line type SiFCAT 350.1 are shown in Figure 16.21 a). In Figure 16.21 b), a typical pull-off cantilever is shown with a top tube inclined to the track centreline and porcelain insulators in the top and cantilever tube. The catenary wire support is separate from the cantilever and top tube connection. The cantilevers consist of aluminum tubes and fittings made of aluminum alloys and stainless steel bolts and split pins. Figure 16.21 c) shows a pull-off cantilever in a tunnel.

Only the push-off cantilevers are equipped with wind stays. The overlaps stretch over five spans and the overhead contact line is flexibly terminated by wheel tensioners with a gear ratio of 3 : 1 for contact and catenary wires (Figure 16.21 d) and f)).

Steel poles made of appropriately sized H-sections for their use as simple suspension poles, poles in overlaps, poles at midpoints or strain poles carrying cantilevers. Depending on their use, these poles may also carry the return conductors and feeder lines. Portals support the drop brackets and cantilevers in stations (Figure 16.21 e)). The contact lines above switches are designed tangentially on the eastern line section and with crossings on the western part of the line. The cross-overs are equipped with two section insulators and can be negotiated at 120 km/h. The neutral sections are mostly ≥ 450 m long. In insulating overlaps, both contact lines run in parallel, spaced at 500 mm. Where space is limited between two tunnel portals, there are neutral sections with an effective length of 58 m (see Clause 12.4.4.9).

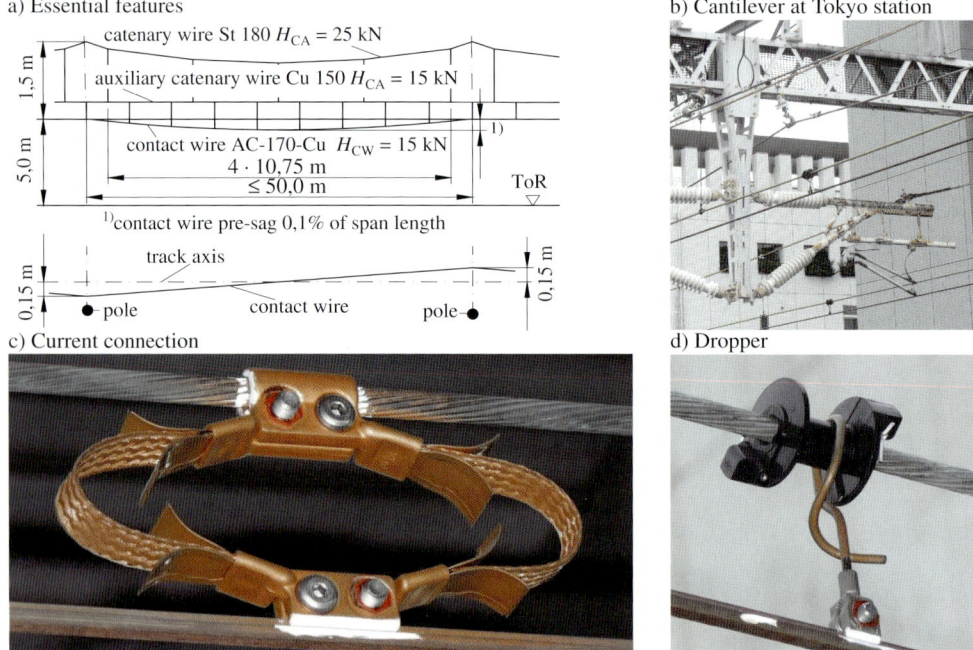

a) Essential features

b) Cantilever at Tokyo station

c) Current connection

d) Dropper

Figure 16.22: Overhead contact line of the Tokaido high-speed line in Japan (Photos: Siemens AG, M. Hoffmann).

16.3.12   AC 25 kV 60 Hz overhead contact line Tokaido, Japan

The *Japanese Railways* (JR) used the 1 435 mm standard gauge to construct the 515 km *Shinkansen railway* from Tokyo to Osaka. More than 400 000 passengers use the trains on the Tokaido high-speed line, daily. This railway line is considered to have the highest passenger load on a high-speed line worldwide.

The *Tokaido high-speed line* was supplied by a 1 AC 25 kV 60 Hz system enabling a commercial speed of 210 km/h at commencement of operations in 1964. The *compound overhead contact line* with auxiliary catenary wire provides relatively constant elasticity over the individual spans (Figure 16.22 a)). The 180 mm^2 steel catenary wire is tensioned to 25 kN, the 50 mm^2 copper cadmium auxiliary catenary wire and the hard-drawn 170 mm^2 copper contact wire are tensioned to 15 kN each [16.21].

Separate steady arms register the stagger of the contact wire and *auxiliary catenary wire* at 150 mm (Figure 16.22 b)). Both steady arms are fixed to the registration arm. The contact wire design height is 5,0 m. The catenary wire can be adjusted along the top tube corresponding to the track lateral positions. At the push-off support, the catenary wire can be adjusted between the pole and the end of the cantilever. The pull-off support permits the movement of the catenary wire along the projecting top tube. At the top tube there is a plate with holes to fix the catenary wire clamp and to connect the cantilever tube with the top tube.

Damping elements between the contact wire and the auxiliary catenary wire limit the oscillations in the contact line (Figure 16.22 c)). The contact wire is carried by bending-resistant droppers without a fixed connection at the auxiliary catenary wire (Figure 16.22 d)).

Figure 16.23: Portal with cantilevers arranged at drop posts on the Direttissima Rome–Florence, in the northern line section near Valdarno, Italy (Photo: B. Puschmann).

Figure 16.24: Connection of the portals with the foundations on the DC 3 kV Rome–Florence line, (Photo: B. Puschmann).

Individual supports are the preferred solution on sections with large track radii, however, portals are preferred in stations. The tensioning sections are 1 500 m long and overlap sections stretch over five spans to accommodate the transition between individual tensioning sections.

16.3.13 DC 3,0 kV overhead contact line Rome–Florence, Italy

Commissioned in 1991, the *Italian State Railway* (FS) operates the 238 km long high-speed line Rome–Florence with DC 3 kV up to 250 km/h commercial speed [16.22]. In the southern section of the line, the contact line consists of a catenary wire, two contact wires and no stitch wires. The catenary wire is tensioned to 27,5 kN and the contact wires to 15 kN each. In the northern region, the contact line consists of two AC-150 – Cu contact wires tensioned to 15 kN each and two 160 mm^2 cadmium copper catenary wires both tensioned to 15 kN. At the supports, two stitch wires help ensure uniform elasticity within the span. *Pulley type tensioning devices* tension the catenary and contact wire in both parts. Overlaps, extending over three spans, are adopted for the transition between the individual tensioning sections. The tensioning devices are arranged above the contact lines at *portals* where deviation pulleys guide the tensioning ropes to the tensioning weights arranged within the poles.

Swiveling cantilevers carry the overhead contact line with ropes used as top anchors. Lattice type galvanised drop posts installed on portals carry the cantilevers. The use of portals predominates (Figure 16.23) with the poles attached to the foundations by pivots that are able to transfer vertical and horizontal forces but not bending moments to the foundation (Figure 16.24).

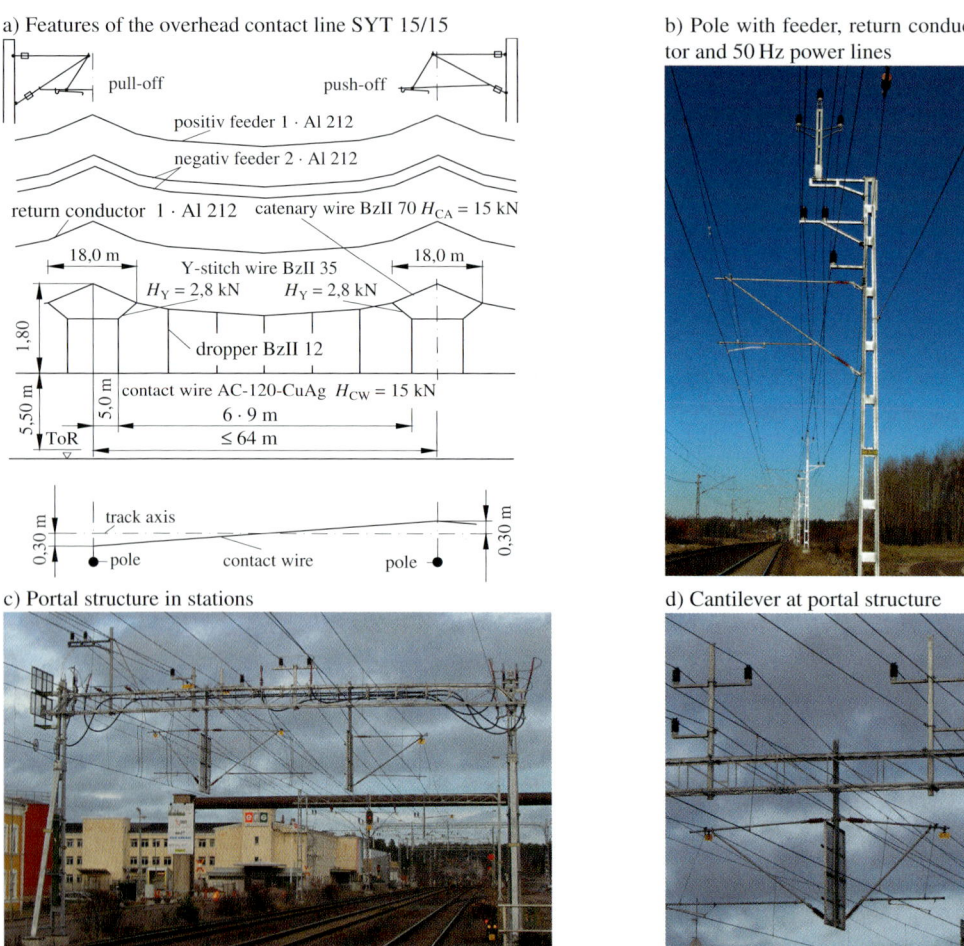

a) Features of the overhead contact line SYT 15/15

pull-off   push-off

positiv feeder 1 · Al 212

negativ feeder 2 · Al 212

return conductor 1 · Al 212   catenary wire BzII 70 H_{CA} = 15 kN

18,0 m   Y-stitch wire BzII 35   18,0 m
                 H_Y = 2,8 kN H_Y = 2,8 kN

1,80   dropper BzII 12

5,0 m   contact wire AC-120-CuAg H_{CW} = 15 kN
5,50 m   6 · 9 m
ToR   ≤ 64 m

0,30 m   track axis
                                              0,30 m
pole   contact wire   pole

b) Pole with feeder, return conductor and 50 Hz power lines

c) Portal structure in stations

d) Cantilever at portal structure

Figure 16.25: 2AC 25 kV 50 Hz overhead contact line SYT 15/15 Mjölby–Nässjö, Sweden, (Photos: J.-P. Marquass, SPL Powerlines Sweden).

16.4   Regional railways

16.4.1   2AC 25 kV 50 Hz overhead contact line type SYT 15/15, Sweden

The Swedish national Railways operate a network with approximately 14 000 km kilometers of tracks, of which 12 000 km are wired with overhead contact lines. The Swedish infrastructure manager Trafikverket was created in 2011 by merging of the former railway operator Banverket, Vägverket and partly of the responsibilities of Luftfartsverket and Sjöfartsverket as well. Trafikverket generate traction power distributed by converting of power from the 50 Hz public network into the 16,7 Hz railway frequency by rotating machines or static converters. At present 48 rotating or static converter stations supply into the overhead contact line network. The overlaying high-voltage network which currently exists only between the Boden and Mälaren region, will be extended step by step. Trafikverket introduces gradually

Table 16.4: Characteristic data of contact line designs STY 15/15 and SYT 9,8/9,8.

Feature	Unit	SYT 15/15	SYT 9,8/9,8
contact wire	–	AC-120 – CuAg	AC-100 – Cu
– tensile force kN	kN	15	9,8
messenger wire	–	Bz II 70	Cu 70
– tensile force	kN	15	9,8
stitch wire	–	Bz II 35	Bz II 35
–tensile force	kN	2,8	2,0
–length	m	18	14
droppers	–	Bz II 12	Bz II 12
system height	m	1,80	1,55
contact wire stagger	m	±0,30	±0,30
contact wire height	m	5,5	5,5
maximum tensioning length	m	1 200	1 320
uninsulated overlaps	–	five spans	three spans
insulated overlaps	–	five spans	three spans
negative feeder	–	2 x 212 AL1	2 x 212 AL1
– tensile stress	N/mm^2	40	40
parallel feeder line	–	1 x 212 AL1	1 x 212 AL1
– tensile stress	N/mm^2	40	40
return conductor	–	1 x 212 AL1	1 x 212 AL1
– tensile stress	N/mm^2	20	20
cantilever tubes steel	–	dia 42 steel pipe	dia 42 steel pipe
poles steel	–	U160 8,1 m	U160 8,1 m
	–	U200 9,0 m	U200 9,0 m

the power supply by auto-transformers (AT) in addition to the power supply by booster transformers (BT) which was the standard supply so far. The Swedish National Railways have already installed auto-transformers supply by 2AC 15 kV 16,7 Hz between on the branch line from Ratsi to the ore mining station Svappavaara close to Kiruna in 1998 [16.23, 16.24]. Since 1998 the newly electrified lines have been equipped consequently with an AT power supply. The line sections Mjölby–Nässjö and Astorp – Hässleholm of the Södra Stambanan were transformed in February 2014 and September 2013, respectively.

The overhead contact line on the line section Mjölby–Nässjö is designed for 200 km/h commercial speed. The contact line type SYT 15/15 is equipped with 15 m long stitch wires (Figure 16.25 a)), their contact and catenary wires are tensioned at 15 kN each. The tensioning and insulating overlapping sections stretch over five spans. On lines with many curves and radii below 1 200 m the contact line type SYT 9,8/9,8 is adopted whereby contact and catenary wire are commonly tensioned by a double-lever type tensioning device which results in a force of 9,8 kN each on both components. This design utilizes 14 m long stitch wires; the overlaps stretch over three spans. Together with the contact line the parallel feeder line forms the positive feeder of the AT supply. The parallel feeder is connected with the contact line at a spacing of 300 m by 70 mm^2 flexible copper conductors. These current connectors are fixed at the upper cantilever tube and connected with the catenary wire from there. The return conductors are strung at 20 N/mm^2 tensile stress and terminated at their ends. Two bare 50 mm^2 copper wires connect the poles with the earthing rail of the track in a spacing of 300 m. At these positions transverse track inter-meshing exists. Cable lugs and bolted joints connect the

a) arrangement on single poles

b) cantilevers across two tracks

Figure 16.26: Overhead contact line design Re100 of DB Netz in between Borna and Geithain (Photo: SPL Powerlines Germany GmbH, R. Hickethier).

earthing cable to the poles and soldered joints to the earthing rail. Important features of the installed overhead contact line are presented in Figure 16.25 a) and Table 16.4.

The negative feeder of the AT system consists of two overhead power line conductors (Figure 16.25 a)). An additional return conductor is installed to reduce the rail potentials and induced interferences as well as to guarantee the return circuit. All conductors are arranged at the poles (Figure 16.25 b)). Trafikverket predominantly adopts double channel poles with U160 sections and 8,1 m in length as well as U200 sections with 9,0 m in length. The taller U200 poles are used to string additional return conductors. The conductors are formed by 212 mm^2 aluminum conductors. In Figure 16.25 b) the standard arrangement of the lines at the contact line poles is shown. Below bridges the negative feeder is guided underground.

The parallel feeder line ends in front of bridges and starts again after it. In both cases a connection to the contact line is provided. In stations with more than two tracks brackets on the portals support the conductors (Figure 16.25 c)). The feeders supplying the contact line of the track in question are arranges above that track on the portal (Figure 16.25 d)). Twin-pole disconnectors connect the contact lines of both tracks within stations of double-track lines.

16.4.2   AC 15 kV 16,7 Hz overhead contact line Re100, Germany

The line Borna–Geithain branches off in Neukieritzsch from the main line, Leipzig to Hof and runs via Borna and Geithain to Chemnitz. Since the inauguration of the Leipzig City Tunnel in 2013, electrically operated city railways and regional railways run from Leipzig via Borna to Geithain. To prepare for the start of operations the *Deutsche Bahn Netz AG (DB Netz)* electrified the 18 km long subsection from Borna to Geithain using the overhead contact line type Re100 in 2010.

The AC 15 kV 16,7 Hz overhead contact line design Re100 corresponds to the requirements of the city and regional railway. On the interstation sections, individual poles predominate (Figure 16.26 a)). In the short sections of double track line cantilevers across several tracks support the overhead contact lines of both tracks (Figure 16.26 b)).

a) Essential features Re200

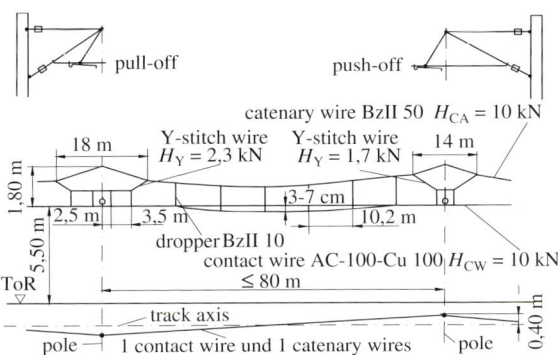

b) 110 kV traction power line at concrete pole

c) Contact wire stringing with a dual-mode vehicle

d) Support on the bridge

Figure 16.27: Design Re200 used for the sections in between stations of the rapid-transit railway line Nuremberg–Lauf–Hartmannshof (Photos: SPL Powerlines Germany GmbH, R. Hickethier).

The foundations at the station are constructed with cast in-situ concrete, they carry lattice steel poles and on the section in between the stations driven piles support the concrete poles (Figure 16.26 b)). Aluminum cantilevers support the contact line designed without stitch wires. On the overhead contact line poles, a 243-AL1 feeder line is arranged to supply the line sections that are distant from the switching post Neukieritzsch.

16.5   Rapid-transit railways

16.5.1   Rapid-transit railways with overhead contact lines

16.5.1.1   AC 15 kV 16,7 Hz overhead contact line Re200, Germany

The rapid-transit railway within the Nuremberg–Fürth–Erlangen agglomeration started in September 1987 with the first line from Nuremberg to Lauf. Today it runs four lines with total length of 224 km. The operator DB Regio Franconia, transports 105 000 passengers a day. A two track line was created in 2011 by the extension of the first rapid-transit line from Lauf to Hartmannshof. Figure 16.27 a) shows the essential features of the *overhead contact line Re200* for the section between the stations. On the Lauf–Hartmannshof section a

a) Essential features type ÖBB 1.2

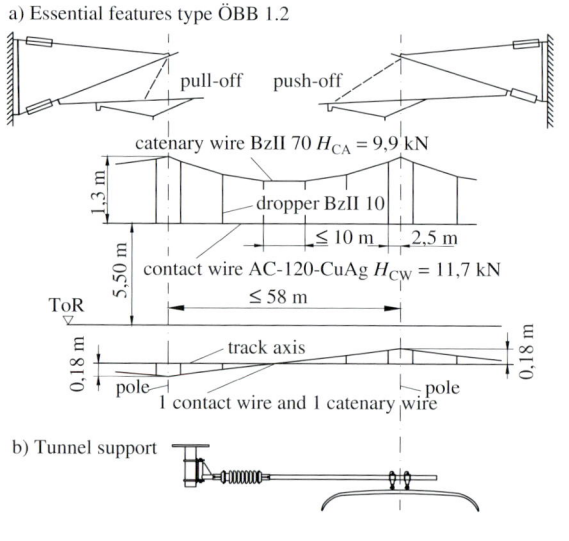

c) AC 110 kV 16,7 Hz traction
power line on concrete poles

b) Tunnel support

d) Support in tunnel

e) Tunnel portal with a tensioning
device in the metropolitan area

Figure 16.28: Features of the overhead contact line design 1.2 for the line to the airport Vienna–
Schwechat (Photos: SPL Powerlines Austria GmbH & Co KG, R. Herrmann).

110 kV 16,7 Hz traction power line was installed on the concrete contact line poles (Figure
16.27 b)), which avoided additional power lines and the need of additional land.

Figure 16.27 c) shows contact wire stringing with two dual-mode vehicles on a bridge across
the A9 motorway. The cantilevers are attached to the bridge by drop posts (Figure 16.27 d)).

16.5.1.2   AC 15 kV 16,7 Hz overhead contact line type 1.2, Austria

The Vienna rapid-transit railway was opened in 1962 and since then has served the transport
needs of the Vienna agglomeration. The 13 lines are used daily by approximately 300 000
passengers. The Austrian Railways (ÖBB) operate the Vienna rapid-transit railway with the
same AC 15 kV 16,7 Hz power supply as used for the main line railways.

The rebuilding of rapid-transit railway line 7 was completed in 2003 ushering in an half hour
service between Vienna centre and the airport Schwechat. Using the rapid-transit airport train
(CAT), passengers travel between the Vienna centre station and the airport within 16 minutes.
The overhead contact line design type ÖBB 1.2 of the Vienna rapid-transit railway (Figure

a) Features of the city railway contact line for rectangular tunnels

b) Pull-off support in the city railway tunnel in Munich

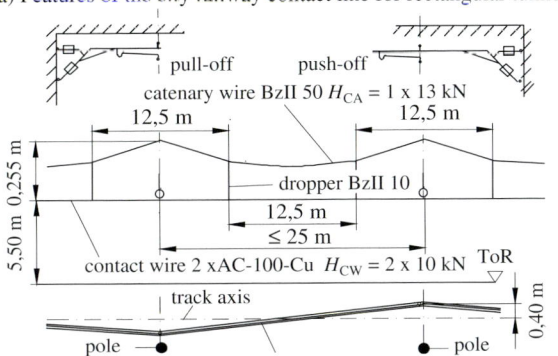

Figure 16.29: City railway overhead contact line of Deutsche Bahn in tunnels (Photos: A. Bauer).

16.28 a)), uses aluminium cantilevers supported by rectangular concrete poles. In special areas, the concrete poles also carry an AC 110 kV 16,7 Hz traction power line (Figure 16.28 b)). A return conductor 257-AL1/60-ST1A is also installed on the poles.

The right-of-way passes through several Vienna city districts where many especial designs were tailored to local requirements (Figure 16.28 c)). The Vienna Centre Station is reached through a tunnel with 5,33 m clearance (Figure 16.28 d)). Figure 16.28 a) presents the essential features of overhead contact line design 1.2 and a tunnel support (Figure 16.28 b)).

16.5.1.3 AC 15 kV 16,7 Hz rapid-transit railway DB overhead contact line, Germany

The DB's *rapid-transit railway overhead contact line* is designed for tunnels with rectangular cross-sections, installed in open cut or in tunnels with circular cross-sections, driven by underground means and for running speeds of 100 km/h. Figure 16.29 a) shows the essential features of this contact line type. A catenary wire tensioned to 13 kN tensile force and two contact wires AC-100 – Cu tensioned to 10 kN each are typical for this contact line type. In Figure 16.29 b) a pull-off support is shown with two steady arms. In overlap sections, the paired cantilevers support the contact wires of both tensioning sections.

16.5.1.4 DC 1,5 kV overhead contact line of RER, Paris

The 587 km long network Réseau Express Régional d'Île-de-France (RER) is the rapid-transit railway operated in the agglomeration of Paris. The RER lines serve the connections between suburbs and city centre, transporting one million passengers per day (Table 16.5). The independent Paris transport administration (*Régie autonome des transports Parisiens* – RATP) and the French National Railway (*Société nationale des chemins de fer français* – SNCF) operate the rapid-transit lines A and B. The lines C, D and E are operated by SNCF alone.

Dual-mode vehicles for DC 1,5 kV and AC 25 kV 50 Hz run on lines A, B, C, and D. They are supplied on the RATP and the SNCF lines south-west, south and south-east of Paris with DC and in the north-western, northern, north-eastern regions of Paris with AC.

Figure 16.30 a) shows the design of a simple DC 1,5 kV overhead contact line while Figure Bild 16.30 b) shows the reinforced version of that contact line used at RER. The simple overhead contact line consists of a catenary wire and two contact wires. The reinforced overhead

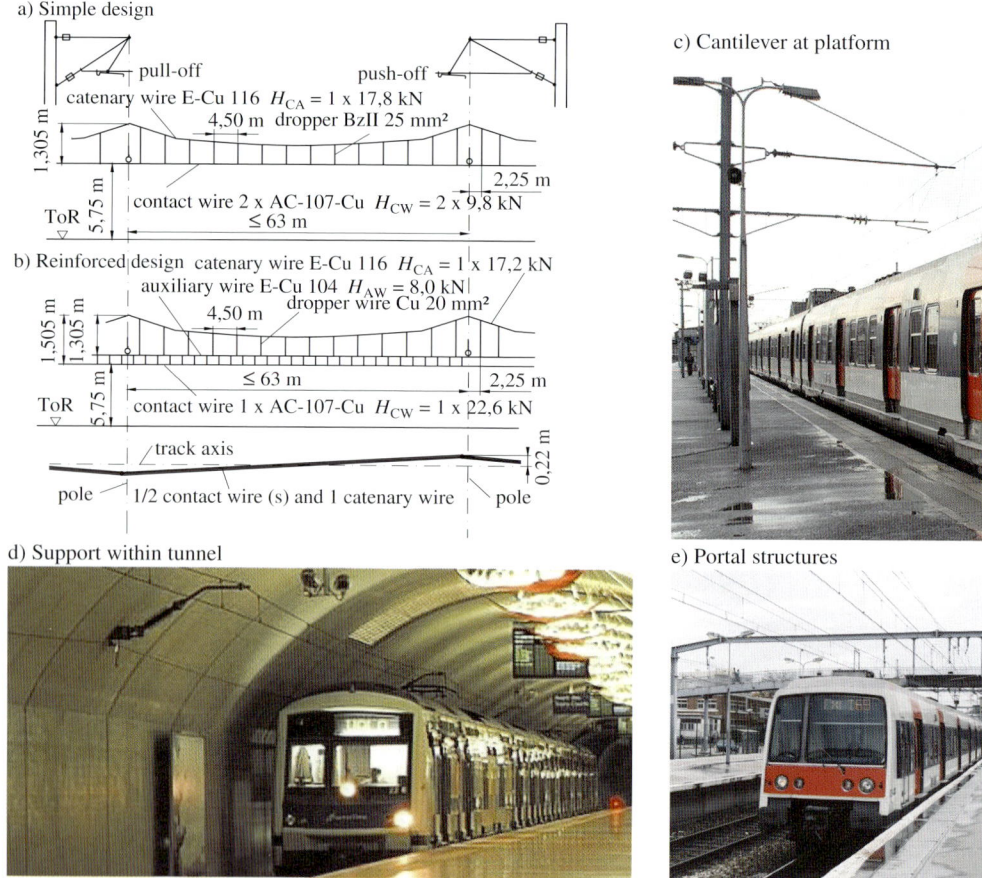

Figure 16.30: Features of the overhead contact line of RER Paris in 2013.

contact line version uses one contact wire and one catenary wire. The droppers for the simple system are made of a flexible copper alloy rope (Figure 16.30 a)). For the reinforced version the droppers between the catenary wire and the auxiliary catenary wire are made of copper conductors. Between the auxiliary catenary wire and the contact wire loop-type plastic profiles are used as droppers to avoid any current exchange between contact wire and auxiliary catenary wire. Consequently, current connectors are installed at regular spacings. At the midpoint, the cantilever is fixed using anchors.

16.5.1.5   DC 1,5 kV overhead contact line Santo Domingo, Dominican Republic

A rapid increase in road traffic led to the decision to use a rapid-transit railway network to relieve traffic congestion in Santo Domingo. The first line began operating one 29 January 2009. *Oficina Para el Reordenamiento del Transporte* (OPRET) operates this and other rapid-transit railways in Caribbean countries, including the Tren Urbano in San Juan, Puerto Rico. The line runs from the city centre through a tunnel to the terminal at Mamá Tingó in the north. In open sections, vehicles are supplied from an overhead contact line type Sicat LD (Figures 16.31 a) and 16.31 b)) by DC 1,5 kV.

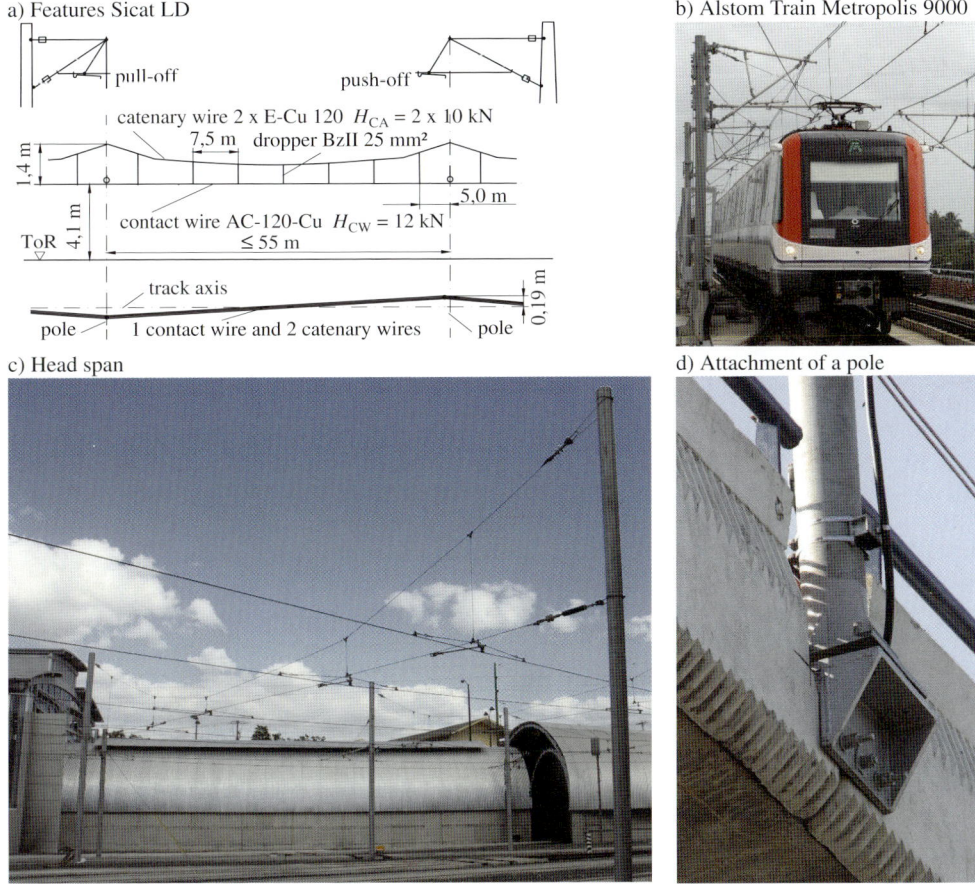

a) Features Sicat LD

pull-off   push-off

catenary wire 2 x E-Cu 120 H_{CA} = 2 x 10 kN
7,5 m   dropper BzII 25 mm²

1,4 m

5,0 m

contact wire AC-120-Cu H_{CW} = 12 kN
≤ 55 m

ToR 4,1 m

track axis

pole   1 contact wire and 2 catenary wires   pole   0,19 m

b) Alstom Train Metropolis 9000

c) Head span

d) Attachment of a pole

Figure 16.31: Overhead contact line Sicat LD in Santo Domingo.

In Santo Domingo, the overhead contact line type Sicat LD uses steel tube poles. Through stations in the open, the cantilevers are supported by drop posts which are integrated with the structural system. The contact line type Sicat LD [16.25] uses cantilevers made of aluminum alloy. The contact line consists of a contact wire AC-120 – Cu and two catenary wires 120-Cu-ETP and droppers made of flexible copper conductors with a 25 mm² cross-section. The contact wire and both catenary wires are tensioned by wheel tensioning devices where one device tensions both catenary wires. The individual supply sections are separated by light-weight section insulators, type Sicat 8WL5545-7A [16.27]. Above points, the contact line

Table 16.5: Proportion of travellers on RATP in Paris.

Type of transportation	Traffic volume in %
metro	48
bus	34
Regional express railway (RER)	15
tramway	2
metro Orlyval	0,5
funicular in Montmartre	0,5

a) Features of the conductor rail type rapid-transit railway Berlin

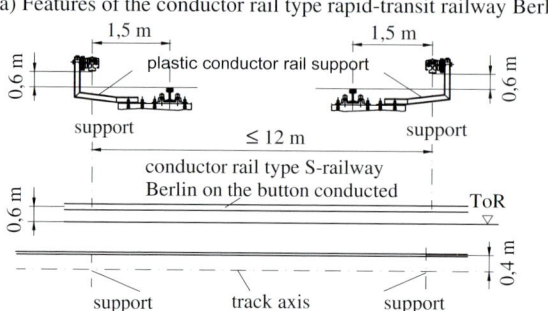

b) Conductor rail support

Figure 16.32: Conductor rail of rapid-transit railway, Berlin (Photo: Rehau AG).

is designed with crossing contact wires. In the workshops, approximately 10 km of contact lines are suspended from head spans with 25 m long cross-spans because of the very narrow space available (Figure 16.31 c)). On open lines cantilevers support the trolley-type contact line with a contact wire AC-120 – Cu. At the workshop entrances, section insulators separate the contact line in the workshop premises from the system outside the depot.

16.5.2 Rapid-transit railways with conductor third rails

16.5.2.1 DC 0,75 kV rapid-transit railway conductor third rail, Berlin, Germany

The network of the *rapid-transit railway Berlin* comprises 332 km with 15 lines and 166 stations. The network is operated by the S-Bahn Berlin Ltd. as a subsidiary of Deutsche Bahn and a partner of the transport association Berlin–Brandenburg.

In 1900, Siemens & Halske inaugurated electrical suburban operations on the 11,9 km long line Wannsee–Zehlendorf with a motor train unit consisting of ten compartment and traction units with three wheel sets operated at DC 0,75 kV at both ends. The conductor rail was contacted on the top. There were 15 train journeys per day in each direction. In 1903, electrical operations commenced on the 9,3 km long line from Potsdam suburban station to Groß-Lichterfelde-East on DC 0,55 kV using a conductor rail designed by Union Electrizitäts-Gesellschaft Berlin (UEG) and later with an AEG conductor rail contacted on the top. In 1924, operations started from the Stettin suburban station, today known as North Station, to Bernau. Conversion of electrical operations to a DC 0,75 kV conductor rail contacted on the bottom of the Potsdam suburban station to Groß-Lichterfelde-East line resulted in the formation of the 220 km rapid-transit railway, abbreviated to *S-Bahn*.

Figure 16.32 a) shows the essential features of conductor rail of Berlin rapid-transit railway. The cross-section of the aluminum compound conductor rail used today is described in Clause 11.3.2. Plastic supports carry the aluminum compound conductor rail contacted from the bottom. These supports also fix the plastic protection covers provided as touch protection (Figure 16.32 b)). The conductor rail support can be used to adjust the height and clearance from track of the conductor rail, varying the attachment of the support at the sleepers.

16.5.2.2 DC 1,2 kV rapid-transit railway conductor third rail, Hamburg, Germany

The Hamburg-Altona rapid-transit railway, originally commissioned in 1906 by Siemens & Halske as an overhead contact line, was redesigned in 1937 by Deutsche Reichsbahn into

a) Laterally contacted conductor rail

b) Current collector at the conductor rail

c) Cross-sections of types of conductor rails

conductor rail cross-section 1939   conductor rail cross-section bulge   conductor rail cross-section HSB

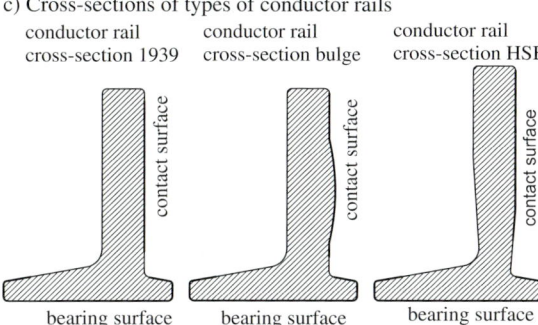

Figure 16.33: Conductor rail type Hamburger S-Bahn (HSB) for rapid-transit railway, Hamburg, 2012 (Photos: SPL Powerlines Germany GmbH, R. Hickethier)

an installation operated at DC 1,2 kV and supplied by side contact conductor rails (Figure 16.33 a) and b)). The conductor rail supports, made of steel and fixed to the sleepers at 5 m spacings were insulated from the rail by ceramic insulators.

The *Hamburg rapid-transit railway* uses a soft-iron conductor rail with a cross-sectional area of 5 100 mm^2 (Figure 16.33 c) [16.28]). The conductor rail support guides the foot of the rail (Figure 16.33 a)) and accommodates variations in length of the conductor rail caused by temperatures changes in the range of $-30\,°C$ to $+70\,°C$. The conductor rails supplied in lengths of 18 m are welded together into 72 m lengths that are transported to site by special trains. The expansion joints installed at 72 m spacings caused considerable wear on vehicle pantographs. The solution, discovered as part of a development project, enabled the length of the individual rail sections without gaps, to be increased to 1 000 m [16.28].

PVC protection covers replaced the timber protection covers (Figure 16.33 a)) used before. Figure 16.33 c) shows the cross-sections of various types of conductor rails used.

16.5.2.3 DC 0,75 kV conductor third rail rapid-transit railway, Oslo, Norway

The *rapid-transit railway in Oslo* was created from several suburban and tramway lines. Since 1966, six lines have been connected to the rapid-transit railway network, totaling 84 km and 104 stations by 2012. Within the city centre the railway runs through tunnels and is known

a) Features of the A][Rail conductor rail type of SPL Powerlines

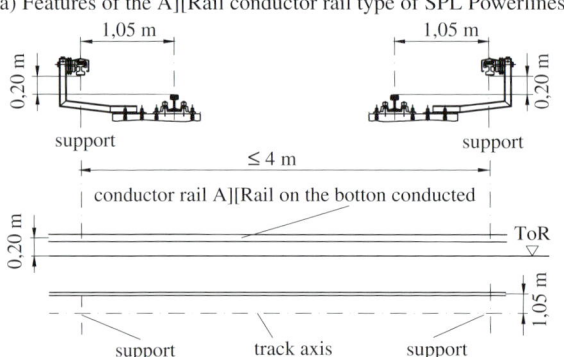

b) City railway vehicle MX3000

c) Conductor rail support with third rail type SPL

d) A][Rail conductor rail profile

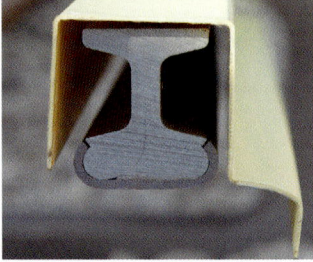

e) Ramp of conductor rail

f) Cross-over conductor rails

Figure 16.34: City railway (T-Bane) Oslo with A][Rail conductor rail type of SPL Powerlines (Photos: SPL Powerlines Norway, A. Rød, 2013).

as the T-Bane (Tunnelbahn). The rapid-transit railway is the second most popular means of travel in Oslo after buses, transporting approximately 165 000 passengers per day.

In 2012 the modernisation of line from Kolsåsbahn to Gjønnes and in 2014 to Kolsås, provided an additional high-performance rapid-transit railway line into the Oslo suburban area.

The Kolsås city railway line, which is operated at DC 0,75 kV, was modernised with a conductor rail with essential features presented in Figure 16.34 a). Figure 16.34 b) depicts the Siemens rapid-transit railway vehicle MX3000. The conductor rail support type SPL (Figure 16.34 c)) carries the conductor rail type SPL A at 4 m spacings. The conductor rail has

a) Station with a third rail

b) Arrangement of the third rails on a viaduct

Figure 16.35: Third rail contact line BTS Bangkok, Thailand.

a $5\,332\,\mathrm{mm}^2$ cross-section and consists of two aluminum parts clamped by a stainless steel section at the lower section of the conductor rail (Figure 16.34 d)). A special welding process connects both aluminum parts at the upper part of the conductor rail and can be separated to exchange individual conductor rail sections. Individual parts of the conductor rail type can be replaced during maintenance. The current collector contacts the stainless steel plate of the conductor rail.

The conductor rails are interrupted at cross-over connections and level crossings. At these locations, conductor rail ramps ensure smooth running of the collector onto the conductor rail (Figure 16.34 e)). The conductor rail ramp drops the collector down to working height and can be traversed at 80 km per hour. The arrangement of the conductor rails at points in stations is shown in Figure 16.34 f).

16.5.2.4 DC 0,75 kV conductor third rail BTS, Bangkok, Thailand

In Bangkok, a city with approximately ten million inhabitants, the *Bangkok Transit System (BTS)* operates as part of an urban commuter railway with a total length of 200 km. A 23 km long section, called the green line was commissioned in 1999 and operates with DC 0,75 kV and is equipped with a third rail [16.29]. The system uses aluminum steel compound rails (Figures 16.35 a) and b)). The conductor rail system cannot be divided electrically to achieve a straight forward switching diagram. Only within the depot the third rail can be disconnected by a disconnector. During normal operations, all sections of the third rails are connected to the DC 0,75 kV substations but if there is a substation failure, the supply sections can be interconnected by the track circuit breakers and DC bus.

Insulating plastic support pads between the rails and the sleeper reduce stray currents that can leak from the rails into the earth and building structures. To increase the conductivity of the rails along the tracks, the rail joints are connected longitudinally. Bonding is also installed between the rails and the individual tracks. Within the stations, remote controlled short-circuiters connect the running rails with the building earth when the permissible touch voltage of the rails is exceeded.

a) Poles with single cantilevers

b) Cantilever across two tracks

c) Flexible head span

d) Flexible head span at the structure of a bridge

Figure 16.36: City railway overhead contact line RNV (Photos: SPL Powerlines Germany GmbH, R. Hickethier).

16.6   City railways equipped with overhead contact lines

16.6.1   DC 0,75 kV overhead contact line, Mannheim, Germany

The *Rhein-Neckar-Verkehr Ltd.* (RNV) operates local passenger traffic in the Rhine-Neckar area with Heidelberg, Mannheim and Ludwigshafen on the Rhine over a network length of 307 km. In 2005 RNV extended the suburban network by the Mannheim-East subsection of the city line, connecting the SAP event centre and the May Market Mannheim into the RNV city railway network.

The contact line used corresponds to the RNV standard design with single cantilevers (Figure 16.36 a)), cantilevers across two tracks (Figure 16.36 b)) and flexible head spans (Figure 16.36 c)). The cantilevers consists of aluminum or glass fibre reinforced plastic. The contact line consists either of a twin catenary wire, a vertically arranged contact line with a catenary wire or a simple trolley wire design without catenary wire. The designer of the overhead contact line considered the architectural requirements and selected an inconspicuous head span design (Figure 16.36 d)).

16.6.2   DC 0,75 kV overhead contact line, Houston, USA

The *MetroRail Houston (METRO)*, forms the second light-rail system in Texas after the *Dallas Area Rapid Transit*. In 2004 operations commenced on the 12 km long line, in the open with 16 stations. The operating company *Metropolitan Transit Authority of Harris County (METRO)* transports passengers with low floor Siemens Avanto type vehicles.

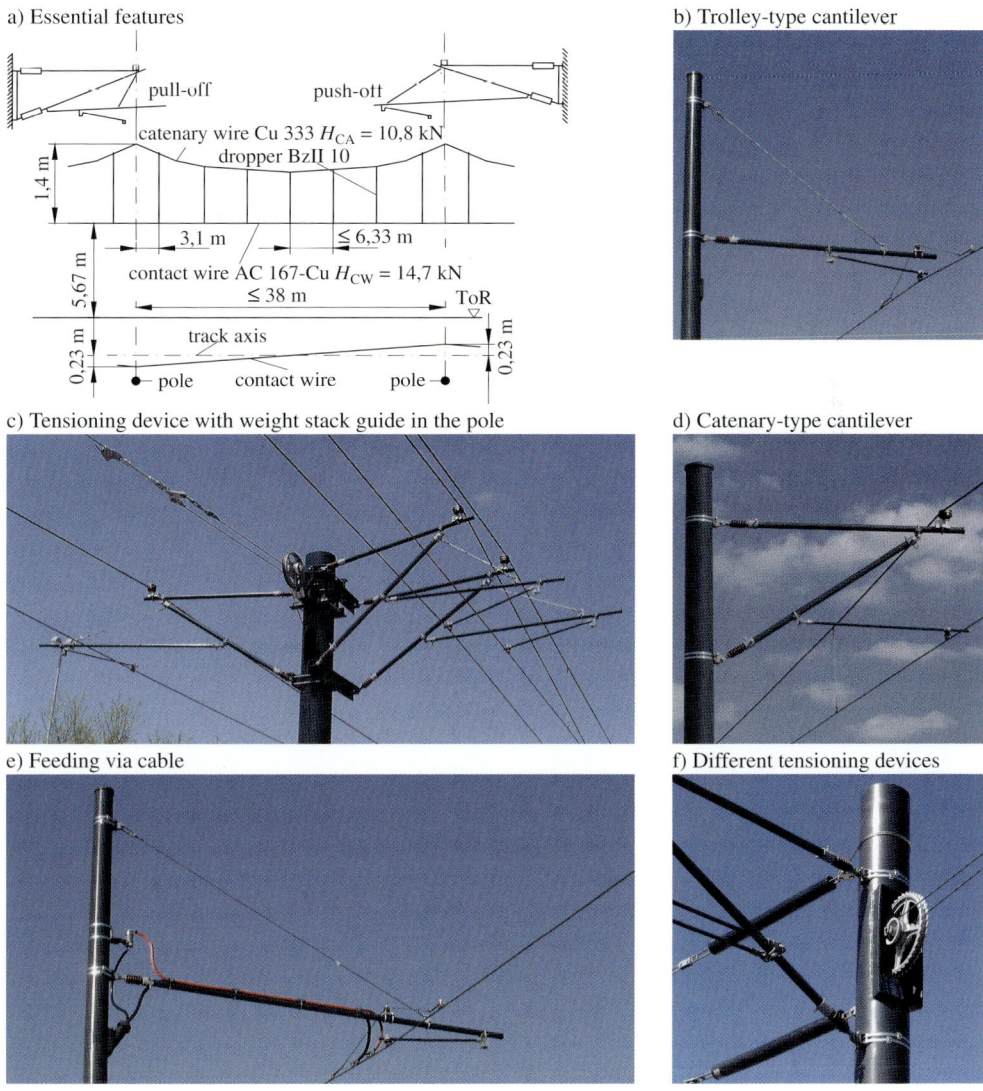

a) Essential features

pull-off   push-off

catenary wire Cu 333 $H_{CA} = 10{,}8$ kN
dropper BzII 10

1,4 m

3,1 m   ≤ 6,33 m

5,67 m

contact wire AC 167-Cu $H_{CW} = 14{,}7$ kN
≤ 38 m   ToR

0,23 m

track axis

0,23 m

pole   contact wire   pole

b) Trolley-type cantilever

c) Tensioning device with weight stack guide in the pole

d) Catenary-type cantilever

e) Feeding via cable

f) Different tensioning devices

Figure 16.37: City railway overhead contact line in Houston.

Nine substations supply the overhead contact line with DC 0,75 kV from the 34,5 kV or 12,5 kV medium voltage network. The contact line is designed as a single or catenary supported system. Figure 16.37 a) shows the essential design features of the catenary supported contact line, which considers the special climatic conditions and temperature range between −30 °C and 50 °C. The cantilevers, equipped with double insulation, enable live-line working at the overhead contact line. The weights of the tensioning devices run within the poles so there is no need to provide for protection of weight stacks accessible to the public. At crossovers, a contact line without a catenary wire and with a 10 kN spring tensioning device, is used for the contact wire.

a) Cross-section of the copper conductor rail Sicat 8WL7006-0A

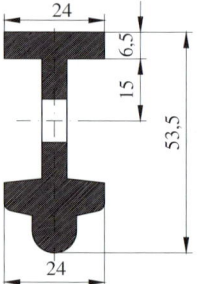

b) Insulated overlaps with a conductor rail support Sicat 8WL 3584-6
with a disconnector

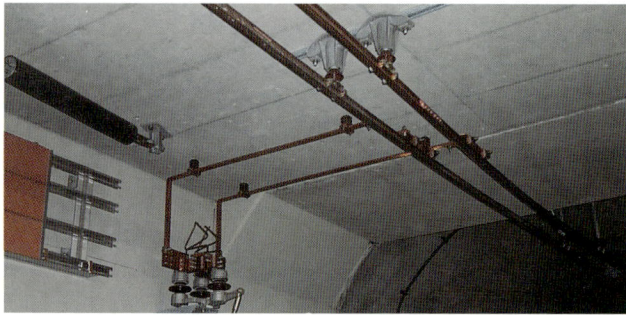

c) Arrangement of insulated overlaps at the tunnel ceiling

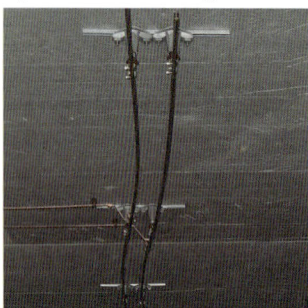

d) Transition of the flexible overhead contact line to the conductor rail
overhead contact line

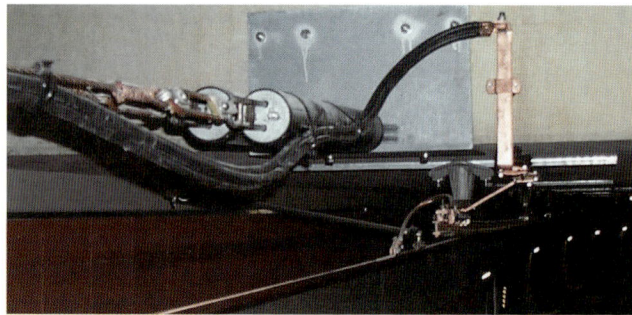

Figure 16.38: City railway overhead conductor rail Sicat 8WL7006-0A in Bielefeld (Photos: SPL Powerlines Germany Ltd., K. Schemmel).

16.6.3   DC 0,75 kV overhead contact line, Bielefeld, Germany

The *Bielefeld city railway* has served public passengers since 1900 and emerged from the original Bielefeld tramway. Between 1978 and 1991, a tunnel was constructed through the city centre and the suburban lines were granted new rights-of-way, allowing the railway to run independently of road traffic. The operator of the city railway *moBiel Ltd.* as a subsidiary of *Stadtwerke Bielefeld*, purchased new M type vehicles replacing the previous Düwag vehicles. Since the installation of a new line towards the university in 2000, these new vehicles have serviced the four 1 m gauge lines of the 71 km network. The city tunnel forms the central part of the network and begins north of the main station, from *Jahnplatz* station south to the town hall. In the city tunnel, an overhead conductor rail supplies vehicles with DC 0,75 kV. Outside the tunnel, trolley type and catenary suspended overhead contact lines are used.

The conductor rail overhead contact line has a low system height and consists of a 600 mm^2 Cu ETP section as shown in Figure 16.38 a). The conductor rail insulated supports allow adjustment of the conductor rail from the tunnel ceiling (Figure 16.38 b)). The individual supply sections are separated by insulated overlaps that can be bridged by contact line disconnectors (Figure 16.38 b) and c)). The transition between the flexible overhead contact line and the conductor rail is shown in Figure 16.38 d).

a) Conductor rail support

b) Overlap with feeding

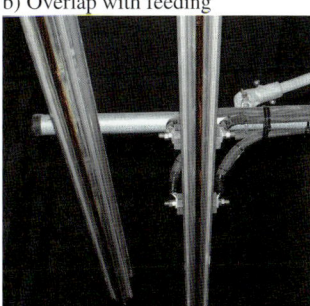

c) Guiding the conductor rails above points

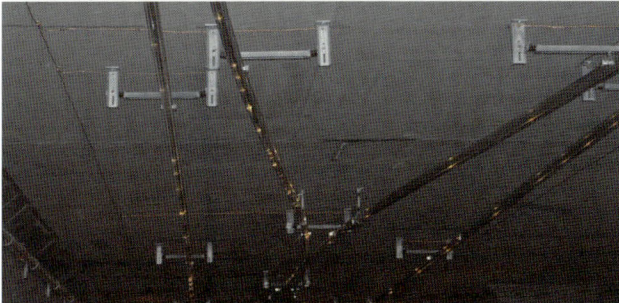

d) Support in a station area

Figure 16.39: City railway overhead conductor rail Sicat SR in Calgary/Canada (Photos: Siemens AG, D. Pfeffermann).

16.6.4 DC 0,75 kV overhead conductor rail system Calgary, Canada

The *city railway West LRT in Calgary* in the Canadian province of Alberta expanded their existing network by 8 km from the city centre to *69. Street SW*. The new line serves six stations and is supplied by seven substations. DC 0,75 kV overhead conductor rail lines were installed in tunnels 1024 m and 270 m in length. The *Calgary Transit* operates this new line which opened in 2012. The supports of the Sicat SR conductor rail, adjustable at the tunnel ceiling, are spaced at a maximum of 12,5 m and supported by galvanized channel section posts (Figure 16.39 a)). The insulating overlaps electrically separate the supply sections of the substations which supply the overhead conductor rail at these overlaps with DC 0,75 kV (Figure 16.39 b)). At points the conductor rails of the mainline supply the pantographs while the branching conductor rail is lifted slightly to prevent pantographs on the mainline from touching them (Figure 16.39 c)). The nominal height of the conductor rail is 4,2 m. and within station areas, the conductor rail drop posts are integrated with the architecture of the ceiling design (Figure 16.39 d)).

16.6.5 DC 3,0 kV overhead conductor rail system Fortaleza, Brazil

The Brazilian City Railway operator *Metrofor* modernised the regional commuter network, In 2012, with the aim of offering reliable and comfortable passenger transport during the football World Cup in 2014 for the neighbouring cities of Fortaleza, Caucaia, Marazion and Pacatuba *Metrofor* expanded and electrified the main lines with a second track. Since 2014,

a) Section insulator within the conductor rail   b) Overlap above points

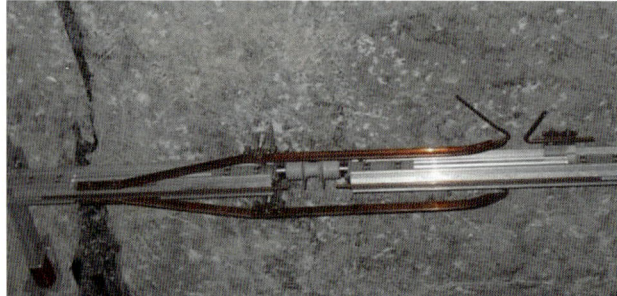 

Figure 16.40: City railway overhead conductor rail in Fortalezza/Brazil (Photos: Siemens AG, T. Schmidt).

this modernisation has enabled the network to transport 700 000 passengers a day.

The *city railway in Fortaleza* uses a DC 3,0 kV conductor rail overhead line on the 24 km long metric gauge section of the line south from their workshop at Vila Flores to the city centre within the tunnels. The purchaser *Cia. Cearense de Transportes Metropolitanos* operates this line, which was installed in 2013, at a commercial speed of 120 km/h. Between Benfica and João Felipe, 4 km of the line runs in a tunnel with conductor rail supports fixed to the tunnel ceiling by galvanized box-type sections at spacings of 8 to 12 m. The height of the 2300 mm^2 cross-section conductor rail is 4,35 m from the lower edge of the contact wire to the top of rail. Section insulators (Figure 16.40 a)) or insulating overlaps separate the line into individual sections, substations supply the rails at the insulating overlaps (Figure 16.40 b)). Mid way between the overlaps, a midpoint fixes the conductor rail overhead contact line.

At points, the conductor rail of the branching track is lifted slightly relative to the height of the conductor rail of the main track (Figure 16.40 d)), so the pantograph does not touch the conductor rail of the branching track.

16.7 Tramways

16.7.1 DC 0,75 kV overhead contact line, Nuremberg, Germany

Nuremberg has approximately 500 000 inhabitants and has been serviced by an electrically operated tramway since 1896. It began with approximately 13 km of 1 435 mm gauge track, from Maxfeld to Main Station to Lorenzkirche–Plärrer–Fürth. By 1939, the network had grown to a length of 73 km, however, because of damage during the Second World War and line removal in 2013, the tramway network was reduced to five lines totaling 37 km with approximately half on a separate right-of-way. The *VAG Verkehrs-Aktiengesellschaft Nuremberg* transports 125 000 passengers per day. The DC 0,75 kV overhead contact line utilises three designs (Figure 3.16) because of the different periods of development:

– fixed terminated contact wire with single suspension and support distances up to 15 m
– tensioned contact wire with stitch wire suspension and 30 m support distances
– a contact line with catenary suspension and support distances up to 60 m

The essential features of the contact line with catenary suspension can be seen in Figure 16.41 a). The trolley wire contact line with bridle-type suspension and single support sus-

a) Features of the contact line suspended by catenary

pull-off   push-off

catenary wire BzII 50 $H_{CA} = 10$ kN

1,2 m

dropper BzII 10

≤ 60 m   2,5 m

5,60 m

ToR   contact wire AC-100-Cu 100 $H_{CW} = 10$ kN
or AC-120-CuAg $H_{CW} = 12$ kN

0,3 m   track axis   0,3 m

pole   contact wire   pole

b) Fixed terminated contact wire

c) Automatically tensioned contact wire with stitch wire

d) Cantilever for catenary

e) Conductor rail overhead contact line in a depot

f) Two track GRP cantilever

Figure 16.41: Tramway overhead contact line in Nuremberg (Photos: R. Knode, VAG Nürnberg).

pension consists of automatically tensioned or fixed terminated contact wire AC-100 – Cu or AC-120 – Cu (Figure 16.41 b) and Figure 16.41 c)), respectively. The contact line with catenary suspension has contact wires AC-100 – Cu tensioned to 10 kN and AC-120 – Cu tensioned to 12 kN. The catenary wire, BzII 50 mm² suspends the contact wire by BzII10 droppers. Cantilevers made of GRP and flexible head spans support the contact line with catenary suspension (Figure 16.41 d) and Figure 16.41 f)), respectively. The German regulation BO-Strab [16.5] permits contact wire wear up to 40 %, while VAG only permits 30 %. Separate tensioning devices are used for the contact and catenary wires and the weights of the tensioning devices are arranged within the poles. For short tensioning sections, *VAG* uses spring-type tensioning devices. Within the depot overhead conductor rails supply the vehicles (Figure 16.41 e)).

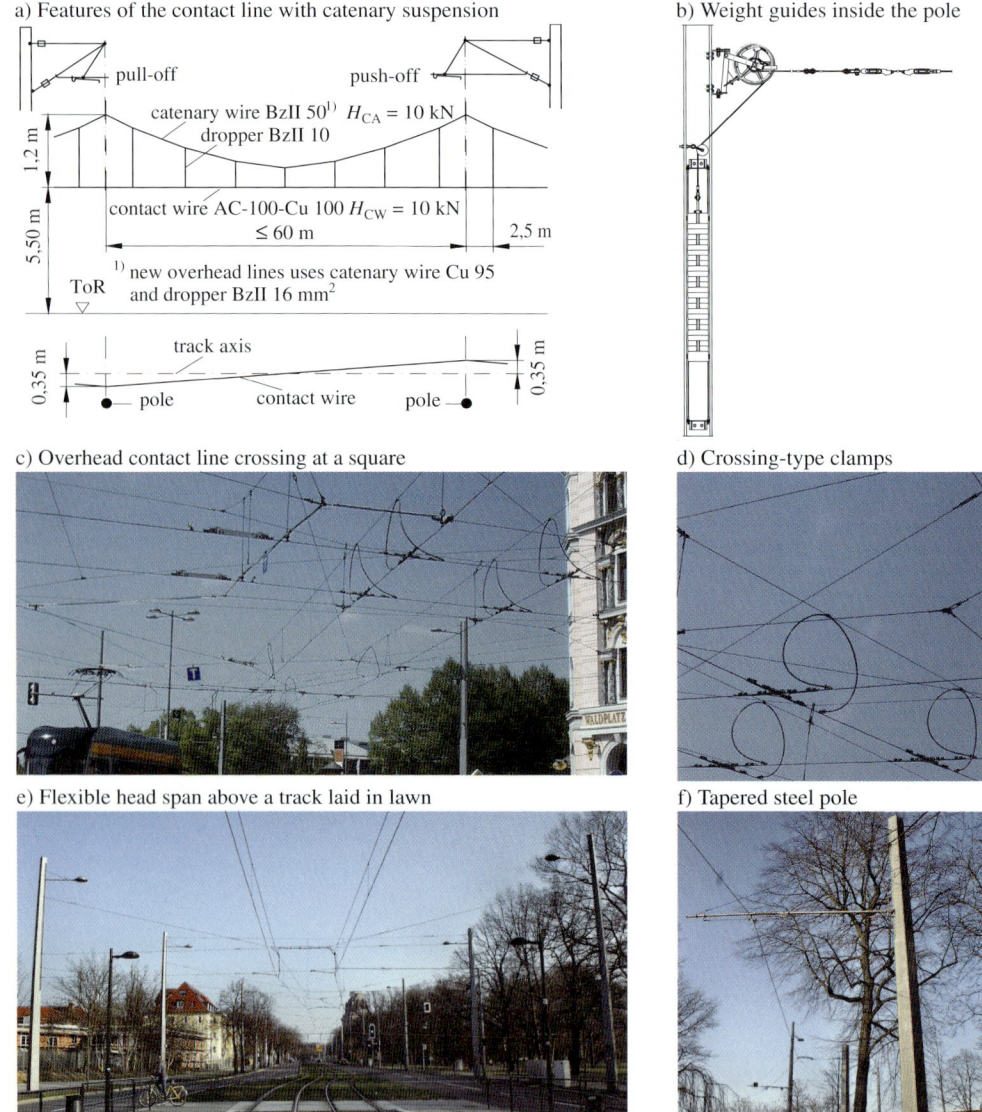

Figure 16.42: Tramway overhead contact line in Leipzig (Photos: SPL Powerlines Germany GmbH, S. Mühl, R. Voigt).

16.7.2   DC 0,6 kV overhead contact line of tramway, Leipzig, Germany

The *Leipziger Verkehrsbetriebe (LVB)* operate the second largest tramway system within Germany and it services 530 000 inhabitants and transports 140 million passengers per year on the 143 km long network. The tramway commenced operations in 1896.

The DC 0,6 kV overhead contact line consists of a contact wire AC-100 – Cu and a catenary wire BzII-95, automatically tensioned by wheel-type tensioners in common or separately. On new lines, *LVB* adopts catenary wires made of Cu-95 mm^2 and BzII-16 mm^2 droppers.

Table 16.6: Types and dimensions of pantographs at LVB Leipzig.

Manufacturer	Type	Length of pantograph mm	Length of collector strip mm
Lekov	EPDE 01-2600	1 680	1 050
Lekov	EPDE 02-2600	1 680	1 050
CKD	–	1 628	1 050
Stemmann	FB50	1 680	1 050
Stemmann	FB80	1 680	1 050
Stemmann	FB500	1 700	1 050
Stemmann	FB800	1 680	1 050

The essential features of the contact line are shown in Figure 16.42 a). The weights of the new tensioning devices run within the poles (Figure 16.42 b) but for short tensioning lengths *LVB* also uses *Tensorex* spring tensioning devices. Crossing terminal clamps fix the contact wires at 90°crossings (Figures 16.42 c) and d)). Flexible head spans (Figure 16.42 e)) or GRP cantilevers (Figure 16.42 f)) support the overhead contact line. Tapered steel poles (Figure 16.42 e)) with octagonal or hexagonal cross-sections and conical reinforced concrete poles support the cross-span equipment. *LVB* uses conventional pantographs and single arm pantographs (Table 16.6).

16.8 Underground railways

16.8.1 AC 25 kV 50 Hz overhead conductor rail, New Delhi, India

On December 24, 2002, *Delhi Metro Rail Corporation* (DMRC) opened the first *underground line* in New Delhi, capital of *India*. Delhi Metro is the world's 12th longest metro system and 10th largest in average daily ridership with 2,76 Million and a total length of 218 kilometers serving 164 stations including six on Airport Express line. The system is a mix of underground, at-grade and elevated sections using both broad-gauge and standard-gauge tracks. The development of network can be divided into three phases. Phase I containing three lines was completed by 2006, and Phase II in 2011. Phase III is scheduled for completion by 2018. The complete of Delhi metro Line 6 extension from Mandi House to Kashmere Gate has been opened for passenger service since May 28, 2017.

The conductor rail types used for the tunnel sections were supplied by different manufactures. Siemens installed the Sicat SR type on Line 6 extension from Central Secretariat to Kashmere Gate and Line 7. The essential features of which are shown in Figures 16.43 a). The Sicat SR overhead conductor rail system consists up to 11 m long aluminum profiles with a box-type cross-sectional shape into which the contact wire is clamped (Figures 16.43 b)). Conductor rail supports are fixed at 7 to 11 m spacings to the tunnel ceiling or to the tunnel wall (Figure 16.43 d)). Overlaps arranged at approximately 500 m spacings compensate temperature-depenent length variations with midpoints installed midway between the overlaps (Figure 16.43 c)). Separations are arranged between the individual feeding sections which are supplied by the substations.

In case power supply with AC 25 kV 50 Hz, neutral sections are arranged approximately in the middle between the substations. In Figure 16.41e) the schematic structure of a neutral section is shown. This design is called a split neutral section according to EN 50 367 (Figure 16.43 e)) and consists of a permanently earthed conductor rail section in the center and two

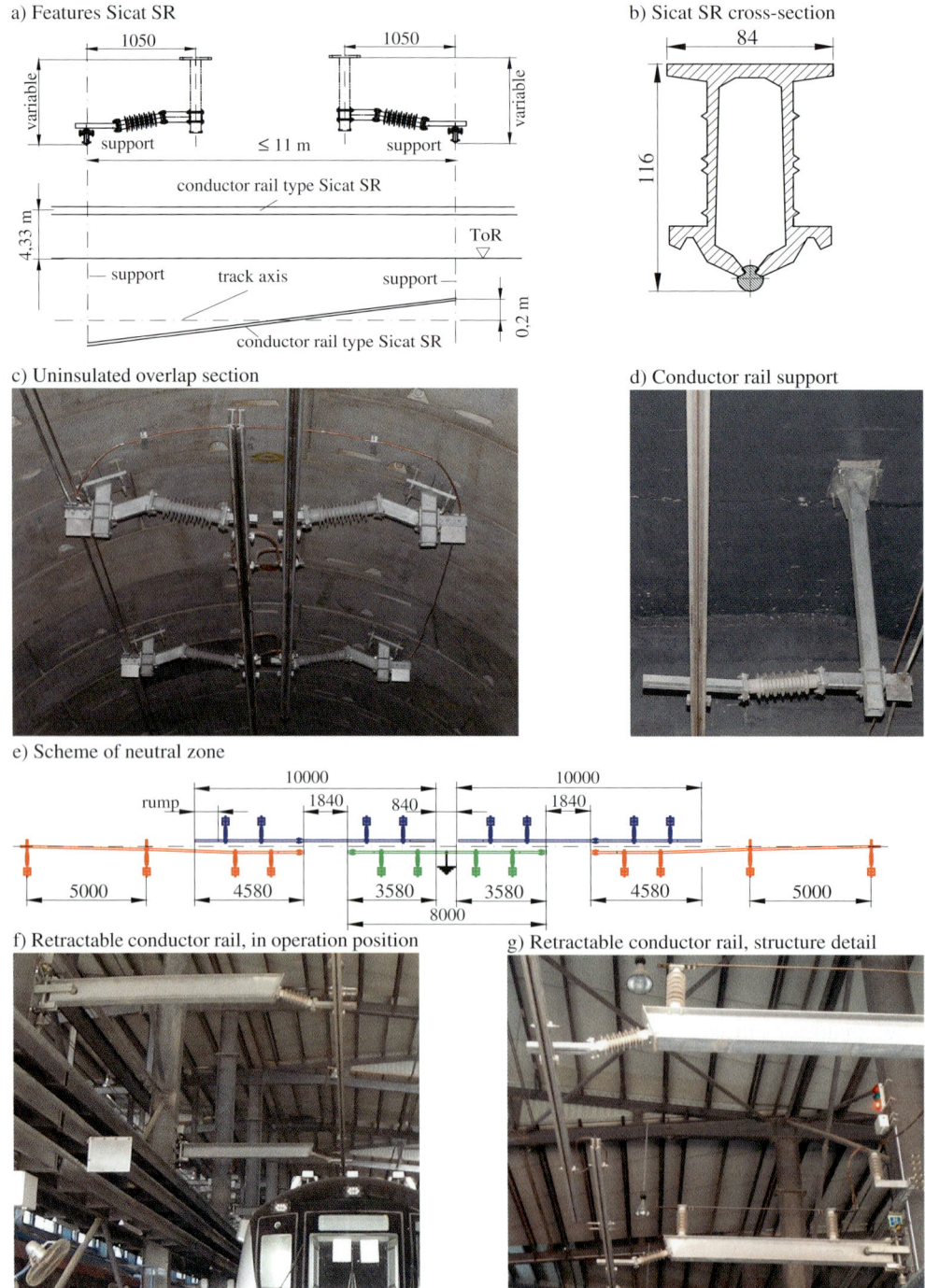

a) Features Sicat SR

b) Sicat SR cross-section

c) Uninsulated overlap section

d) Conductor rail support

e) Scheme of neutral zone

f) Retractable conductor rail, in operation position

g) Retractable conductor rail, structure detail

Figure 16.43: AC 25 kV conductor rail overhead contact line Sicat SR in New Delhi, India.

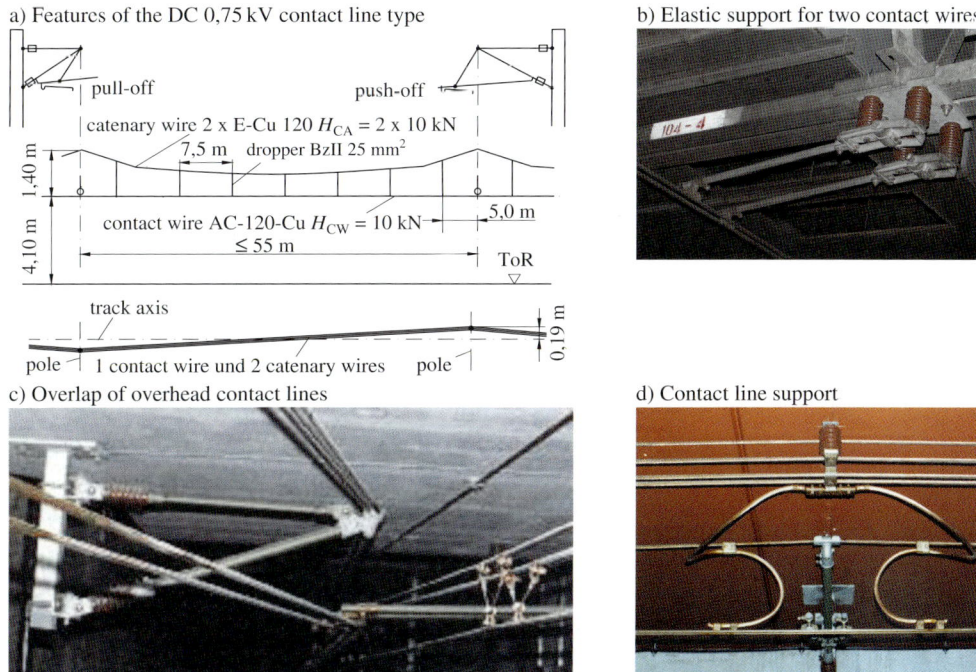

a) Features of the DC 0,75 kV contact line type

pull-off   push-off

catenary wire 2 x E-Cu 120 H_{CA} = 2 x 10 kN

7,5 m   dropper BzII 25 mm²

1,40 m

4,10 m

contact wire AC-120-Cu H_{CW} = 10 kN

≤ 55 m

5,0 m

ToR

track axis

0,19 m

pole   1 contact wire und 2 catenary wires   pole

b) Elastic support for two contact wires

c) Overlap of overhead contact lines

d) Contact line support

Figure 16.44: Underground railway Hong Kong with overhead contact line.

neutral sections, which guide the pantographs. The earthed conductor rail section in the center prevents an electrical bridging of the adjacent live sections and, therefore, a short circuiting between the substations. Figure 16.43 e) shows the lengths of insulating overlaps and the earthed section.

A retractable conductor rail overhead system called RCR was installed in the depots Mukund-pur and Vinodnagar both of the New Delhi Metro Line 7 for inspection of vehicles (Figure 16.43 f)). These depot tracks are equipped with cranes, roof accesses, working platforms, lifting jacks and other maintenance equipment. For inspecting and maintaining vehicles, some clearance areas without any overhead contact lines are necessary above the vehicles. The maintenance staff is able to enter the vehicle roof and use movable cranes, when the conductor rail has been retracted (Figure 16.43 g)).

The conductor rail is designed to provide a reliable power supply by an overhead contact line and, once it is swiveled aside, enough space for inspection and maintenance work on top of the train. The approximately 160 m long retractable conductor rail sections consists of a 2 300 mm² aluminum profile according to Siemens type 8WL7230-0A with a BC 150 contact wire. The contact wire type BF 150 is used in the depot area outside the buildings. The conductor rail is suspended every 8 m at cantilevers provided with insulators. The conductor rail overhead system Sicat SR is described in Clause 11.3.

16.8.2   DC 1,5 kV overhead contact line MTR, Hong Kong

The *Mass Transit Railway (MTR)* is an extensive underground network and is the most important means of commuter transport in Hong Kong. The railway began operations in 1979 and now includes 212 km with seven lines, 155 stations and transports approximately 2 500 000 passengers per day. As well as operating the underground network in Hong Kong, the *MTR Corporation Limited* operates networks in Beijing, Melbourne, Shenzhen, Hangzhou and Stockholm. The *(MTR)* in Hong Kong uses a DC 1,5 kV overhead contact line with essential features shown in Figure 16.44 a).

The lines run predominantly in tunnels. Because of the low ceiling height in the rectangular and round tunnels of the Kwun Tong line, catenary suspended contact line could not be installed. Therefore, *elastic supports* carry twin contact wires AC-120 – CuAg (Figure 16.44 b)), tensioned to 24 kN using weights. Figure 16.44 c) shows the overlap of the contact lines, consisting of twin catenary wires and twin contact wires.

Four 150 mm^2 copper *line feeders* transfer energy in the tunnels to the vehicles and the contact wires. On the more recent Tsuen Wan line, swivelling cantilevers with tubes as shown in Figure 16.44 d) and e) support overhead contact lines with twin contact wires AC-120 – CuAg and twin catenary wires Cu 150. Separate steady arms for each contact wire fix the stagger at $\pm 0,20$ m. Both contact wires and catenary wires are commonly terminated at a tension of 24 kN.

16.8.3 DC 0,75 kV underground railway, Nuremberg, Germany

The *Nuremberg underground railway* opened in 1972 on the Langwasser Süd–Bauernfeind-straße line. There are now three lines with a total length of 35 km, of which 30,5 km run in tunnel, 4,5 km on open terrain and 1 km on a viaduct. Up to 400 000 passengers are transported daily. The Verkehrs-Aktiengesellschaft (VAG) Nuremberg operates the lines in the cities of Nuremberg and Fürth. The VAG is also responsible for maintaining the infrastructure. The underground and the Nuremberg tramway form part of the joint passenger transport services in the metropolitan area of Nuremberg.

After Siemens commissioned the first DT3 type motor unit for the driverless underground for the VAG on 15 January, 2004, the first automatic underground line in Germany, U3, started on 15 March, 2008. The motor units, equipped with eight driving motors, reach a commercial speed of 80 km/h. The VAG has also operated the U2 fully automated the line since 2009. The conversion of a third line, U1 to automatic driverless operation is not yet planned.

The line sections in the open and in the depot are constructed with ballasted track. The line section in tunnels is constructed as a rigid permanent way where both track installations use rail types S41 and S49, with a gauge of 1 435 mm. The traction power line is adjusted to suit these three types of permanent way.

The third rails are contacted from the bottom and supply the underground vehicles with DC 0,75 kV traction power. The essential features of the third rail installation in Nuremberg can be seen in Figure 16.45 a). The third rail supports are fixed at every eighth sleeper in straight line sections corresponding to a 5,36 m spacing. In curves, the spacing between the supports is reduced to 2,07 m, i.e. at each third sleeper. Until 1998, the VAG used supports made of steel with insulators but these are now Balfour Beatty (BBRail) type GRP plastic supports. From 1998 onwards, the VAG adopted composite third rails with a 5 100 mm^2 cross-section instead of steel conductor rails with the same cross-section. Fish-plate joints with bolts and self-locking nuts connect the individual rail sections. To compensate for length variation caused by temperature changes, expansion joints are arranged on the section in the open every 90 m. In tunnels, however, the spacing is 180 m because of the smaller temperature range. Between the expansion joints and for short conductor rail segments, fixed point clamps lock the conductor rail at three supports to avoid unplanned movement at these locations.

a) Features of the DC 0,75 kV underground railway third rail

b) Conductor rails in a station

c) Conductor rails on an interstation line

d) Conductor rails at points

Figure 16.45: Underground rail with third rail of subway in Nuremberg (Photos: R. Knode).

The conductor rail has plastic covers to prevent accidental touching by persons. These covers are fixed by spacers clamped to the conductor rail and they also protect feeding points, end ramps and expansion joints.

The third rail is arranged on the outer side of the tracks but at platforms, the conductor rail is located on the opposite side to the platform as shown in Figure 16.45 b). On sections with small structure gauges on open lines, in tunnels and within depots, the conductor rails are located between the tracks away from inspection paths (Figure 16.45 c)). Gaps in the conductor rail in stations and at points separate the individual supply sections from the substations.

In point areas the conductor rail is interrupted because of the branching track and arranged on the opposite side of the track (Figure 16.45 d)). Conductor rail end ramps with gradients of 1 : 50 on main lines and 1 : 30 in storage tracks and workshops enable the smooth intake of collector shoes and transition to conductor rail running height of 0,19 m above the top of rail.

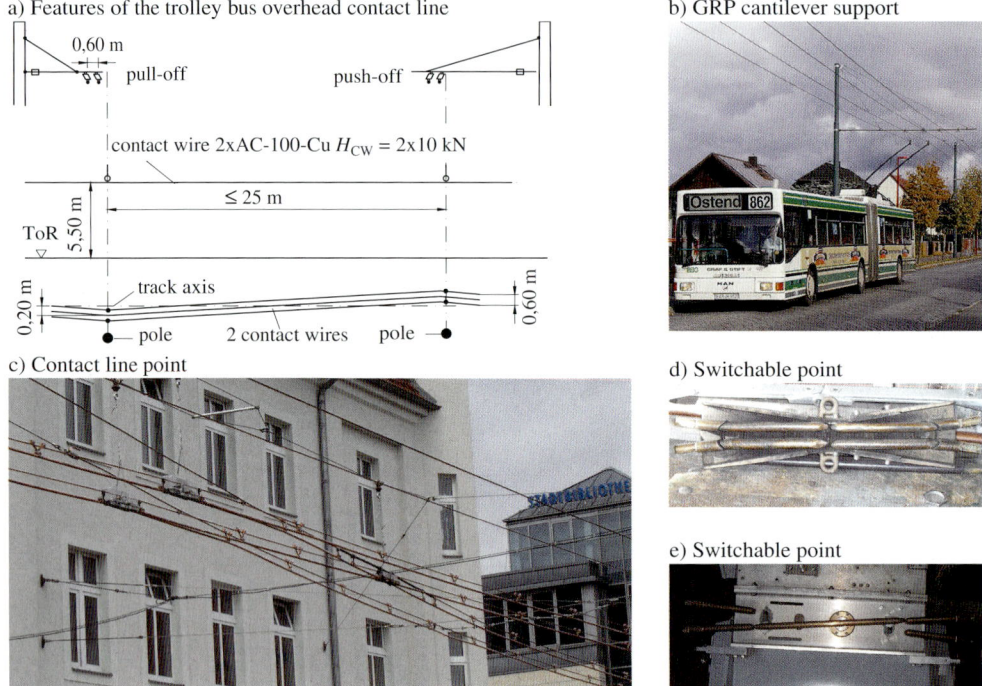

Figure 16.46: Trolley bus overhead contact line Eberswalde, Brandenburg (Photos: H. Bülow).

The 6 mm thick layer of stainless steel covering the foot and both halves of the conductor rail are ultrasonically tested every 6 months to check the thickness of the stainless steel layer. The wear data gained over a number of years provides valuable information regarding wear over the total underground railway network.

16.9 Trolley bus installations

16.9.1 DC 0,75 kV trolley bus overhead contact line, Germany

The *trolley bus in Eberswalde* was commissioned in 1940 and remains the oldest system in Germany. The operator transports approximately 11 500 passengers a day on two lines over a 37 km long network. Three DC 0,66 kV rectifier substations supply the overhead contact line via cable connections every 2,5 to 3,0 km.

Figure 16.46 a) shows the essential features of the Eberswalde contact line. The contact line consists of two contact wires, cantilever supports (Figure 16.46 b)), flexible head spans, points (Figure 16.46 c), d) and e)), feeders (Figure 16.47 a)), crossings (Figure 16.47 b)), section insulators (Figure 16.47 c)) and cross bonding of the contact lines (Figure 16.47 d)). The points crossings in Figure 16.46 c) lead the trolley collector at branch-offs of the contact lines in the desired direction. The collectors control the points. Overhead contact line crossings as shown in Figure 16.46 d) are installed above road crossings and do not permit a change of running direction as in the case of points.

a) Feeding point of the contact line   b) Crossing

c) Section insulator   d) Cross-bonding with a cable connection

Figure 16.47: Trolley bus overhead contact line Eberwalde, Brandenburg (Photos: H. Bülow).

Separations between two adjacent supply sections (Figure 16.47 c)) are indicated by electrical signals and need to be traversed without drawing current. Cross bonding with cable connections between the contact lines of different directions as shown in Figure 16.47 a) and d) connect the negative and positive poles of the overhead contact line for both directions, respectively.

Viewed in the running direction, the positive pole contact wire is on the left hand side and the negative pole contact wire is on the right. The trolley bus collects the traction energy with collector heads that glide along the contact wire and consist of sliding shoes (Figure 16.48 a)). Trolley collectors lead the energy from the collector heads to the vehicle (Figure 16.48 b)). The nominal contact wire height above the road is 5,5 m and must not fall below 4,5 m or infringe the gauge.

Trolley bus line contact wires have a flat profile and are spaced at 0,6 m in all the designs up to 1989. More recently they are spaced at 0,7 m, with pendulum-like suspensions. The

a) Head of trolley collector

b) Trolley collector

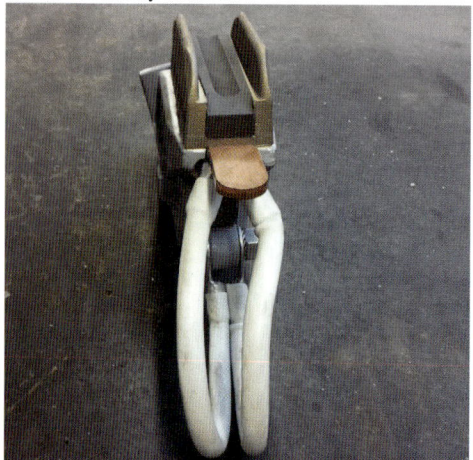

Figure 16.48: Trolley bus current collector parts, Eberswalde, Brandenburg (Photos: H. Bülow).

contact wire clips at the steady arms, which are shorter than railway types, must not impede the sliding heads of the trolley collector.

On lines with pendulum-type suspension and rigidly terminated contact wires, the spans between the concrete poles are 27 m, on lines without contact wire stagger. With automatically tensioned contact wires, they are 50 m. The relatively short span avoids extreme contact wire sags.

In Eberswalde, steel tube or lattice steel towers and concrete poles support the cross spans, the contact wire supports and the cross-span wires (Figure 16.47 a)). Cross-spans are often attached to wall anchors of adjacent buildings. Cantilevers for the new contact line types are manufactured from GRP rods.

With pendulum suspensions, the changing stagger of the contact wire compensates the thermal expansion of the endlessly strung contact wire. Spacers installed on the contact wires avoid clashing of both contact wires.

On lines erected after 1989, the contact wire has alternating stagger for pendulum type supports, such that the change of angle of the contact wire at the supports is 1° to 2°. This change of angle does not lead to dewirement of the trolley collectors and the contact wire is fixed terminated on these lines.

Vertical and horizontal oscillations of the contact wires are caused by the pressure of the trolley collector, by the lateral movements of the trolley bus and also by the unevenness of the road. The oscillations are more severe than with vehicles running on tracks and affect the interaction of the trolley collector and contact line. Due to the oscillations, the carbon collector strips of the sliding shoes wear more severely than with completely elastic pendulum type suspensions (Figure 16.47 c)). The non-uniformity of the elasticity is reduced by elastic contact wire suspensions and contact wire wear is also reduced.

16.9.2   DC 0,75 kV trolley bus overhead contact line, Beijing, China

Preparations for the operation of the first tramway in Beijing started with the founding of the *Beijing Electric Trolley Joint Stock Company* in 1921. Operations on the first nine kilome-

a) Trolley bus in Beijing

b) Pendulum-type suspension

c) Sliding suspension with insulator

d) Crossing at 90°

Figure 16.49: Trolley bus overhead contact line, Beijing (Photos: B. Puschmann).

tre line between Qianmen and Xizhimen started in 1924, with a further seven lines brought into service in 1949. The tramway network expanded to 43 km and was an important transport system. However, the tramway vehicles imported from France proved to be obsolete, of poor quality and were noisy. Because of their low speed, further expansion of the town was hindered. Consequently, the *Beijing City Trolley Bus Company* commenced *trolley bus operations in Beijing* in 1957 and by 1959, the trolley buses had completely replaced all tram lines and were the most important service vehicle in Beijing. Its advantages of low noise, being environmentally friendly and energy efficient, rapidly led to a further extension of the network to the current 15 lines totalling approximately 190 km (Figure 16.49 a)).

The contact wire supports consist of pendulum-type suspensions. In the vicinity of the contact wire support, a 10 mm diameter galvanised steel rope is clamped in parallel to the contact wire using parallel groove clamps. At this pendulum the droppers are fixed, guaranteeing the vertical position of the parallel groove clamps. In the dropper wires two insulators are

a) Trolley bus starts lifting the trolley collector

b) Trolley collector heads at the entry funnel

c) Trolley collector heads at the conductor rail

d) Trolley collector runs on the contact wire

Figure 16.50: Automatic entry of trolley bus trolley collector in Beijing (Photos: B. Puschmann).

installed to achieve double insulation (Figure 16.49 b)) and galvanised cantilever tubes carry the contact wire supports.

In Beijing, hybrid trolley buses operate, using batteries of super-capacitors to bridge line sections without overhead contact lines. At the beginning of the contact line intake, funnels are arranged below which, the trolley busses lift the trolley collector (Figure 16.50). At the inclined plane of the intake funnel, the trolley collector head slides onto a conductor rail. In Figure 16.49 c) the trolley collector head can be seen sliding out of the intake funnel. After approximately 3 m, the conductor rail ends and the contact wire, clamped into the conductor rail, takes on the interaction with the trolley collector.

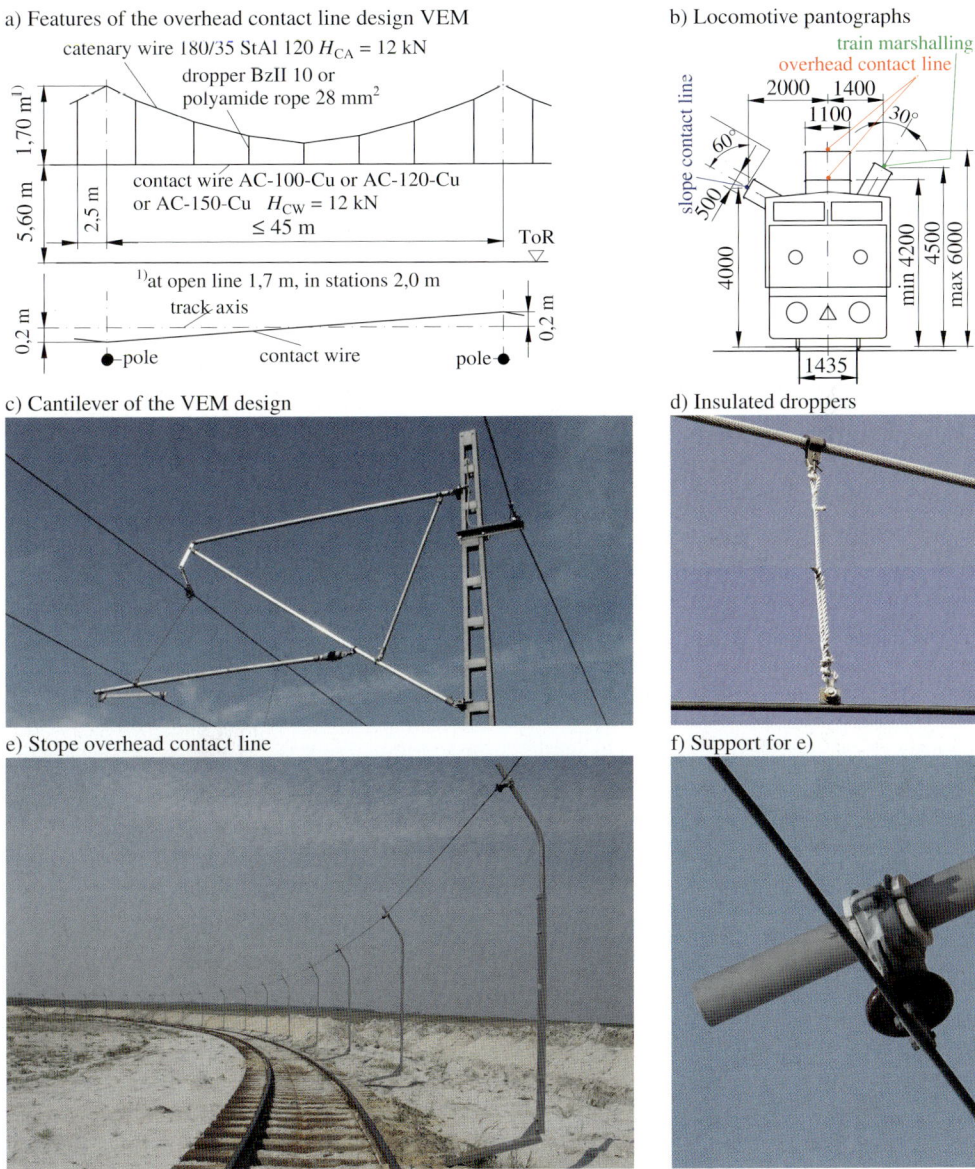

a) Features of the overhead contact line design VEM

catenary wire 180/35 StAl 120 H_{CA} = 12 kN

dropper BzII 10 or
polyamide rope 28 mm²

contact wire AC-100-Cu or AC-120-Cu
or AC-150-Cu H_{CW} = 12 kN
≤ 45 m

¹⁾at open line 1,7 m, in stations 2,0 m
track axis

pole   contact wire   pole

b) Locomotive pantographs

train marshalling
overhead contact line

c) Cantilever of the VEM design

d) Insulated droppers

e) Stope overhead contact line

f) Support for e)

Figure 16.51: Industrial railway overhead contact line at VEM (Photos: Siemens AG, R. Henniges).

16.10   Industrial railways

16.10.1   DC 2,4 kV overhead contact line VEM, Brandenburg, Germany

In 1990, the *Lausitzer Braunkohle AG Laubag (Lausitz Lignite Ltd.)* was formed from the lignite public company, Senftenberg. In 2002, this company merged with the Hamburgischen Electricitäts–Werke (HEW), the Vereinigte Energiewerke AG (VEAG) and the *Berlin Städtische Elektrizitätswerke AG (BEWAG)* to form the new *Vattenfall Europe AG (VEM)*. The open pit mining and refining plants at *Jänschwalde, Cottbus-North, Welzow-South, Nochten and Reichwalde* belong to *Vattenfall Europe Mining AG* as well as the lignite refining plant in *Schwarze Pumpe*, that supplies the lignite to the power plants of *Vattenfall Europe Generation Ltd.* in Jänschwalde, Spremberg and Boxberg with raw lignite using electrified railways. The overhead contact line installation, supplied at DC 2,4 kV comprises 380 km of tracks. The voltage tolerance is between −33 % and +20 %, the polarity of the contact line is negative with a positive return system. Figure 16.51 a) demonstrates the essential features of the *VEM* overhead contact line.

VEM Railway uses a special structure gauge, smaller than that used for main lines in Germany. The overhead contact line, the stope overhead contact line and the train marshalling installation are designed to this reduced structure gauge (Figure 16.51 b)).

The VEM uses aluminum cantilevers with cast aluminum components (Figure 16.51 c)). A multi-strand polyamide conductor is used for the droppers (Figure 16.51 d)).

Stope overhead contact lines have their poles attached to the sleepers, enabling the contact line to be re-positioned with the tracks. In the lignite area and at waist dumps the contact line is arranged centrally above the tracks while in the loading area it is 4,0 m high and beside the track (Figure 16.51 e)). Figure 16.51 f) shows a stope contact wire support. The maximum spacing between poles is 15 m while in curves the span length depends on the radius. The sections with the stope contact lines are electrically separated from the other overhead contact line sections and straps bridge the rail joints longitudinally.

16.10.2   AC 6,6 kV 50 Hz overhead contact line, Hambach, Germany

RWE Power Line Ltd. operates an industrial railway to transport coal from the lignite open pit mine at Hambach to the lignite power plants. Since 2013 the line has run on a new 15 km long right-of-way. The line, equipped with two tracks and completely electrified, is designed for the special wide gauge coal wagons with a vehicle gauge 3,80 m wide and 6,0 m track spacings on the interstation line. The Hambachbahn is operated 24 hours a day, three days per week and must have high availability for coal transport between the open pit mine and the power stations.

The overhead contact line [16.30] essentially uses the Re100 standard overhead contact line design of DB. It mainly uses DB components (Figure 16.52 a)).

The Hambachbahn is supplied by an overlying 3 AC 380/110 kV 50 Hz grid. A 380/30 kV transformer supplies the Hambach substation and from this main substation a 2 AC 30 kV 50 Hz traction line (Figure 16.52 b)) supplies the substations along the line. Figure 16.52 c) shows a switch disconnector arranged on an overhead contact line pole which transfers the energy from the 2 AC 30 kV 50 Hz line to the substation along the railway. In this substation, transformers transfer the energy from the 2 AC 30 kV 50 Hz network into the 1 AC 6,6 kV 50 Hz system and supply the overhead contact lines through four circuit breakers with each

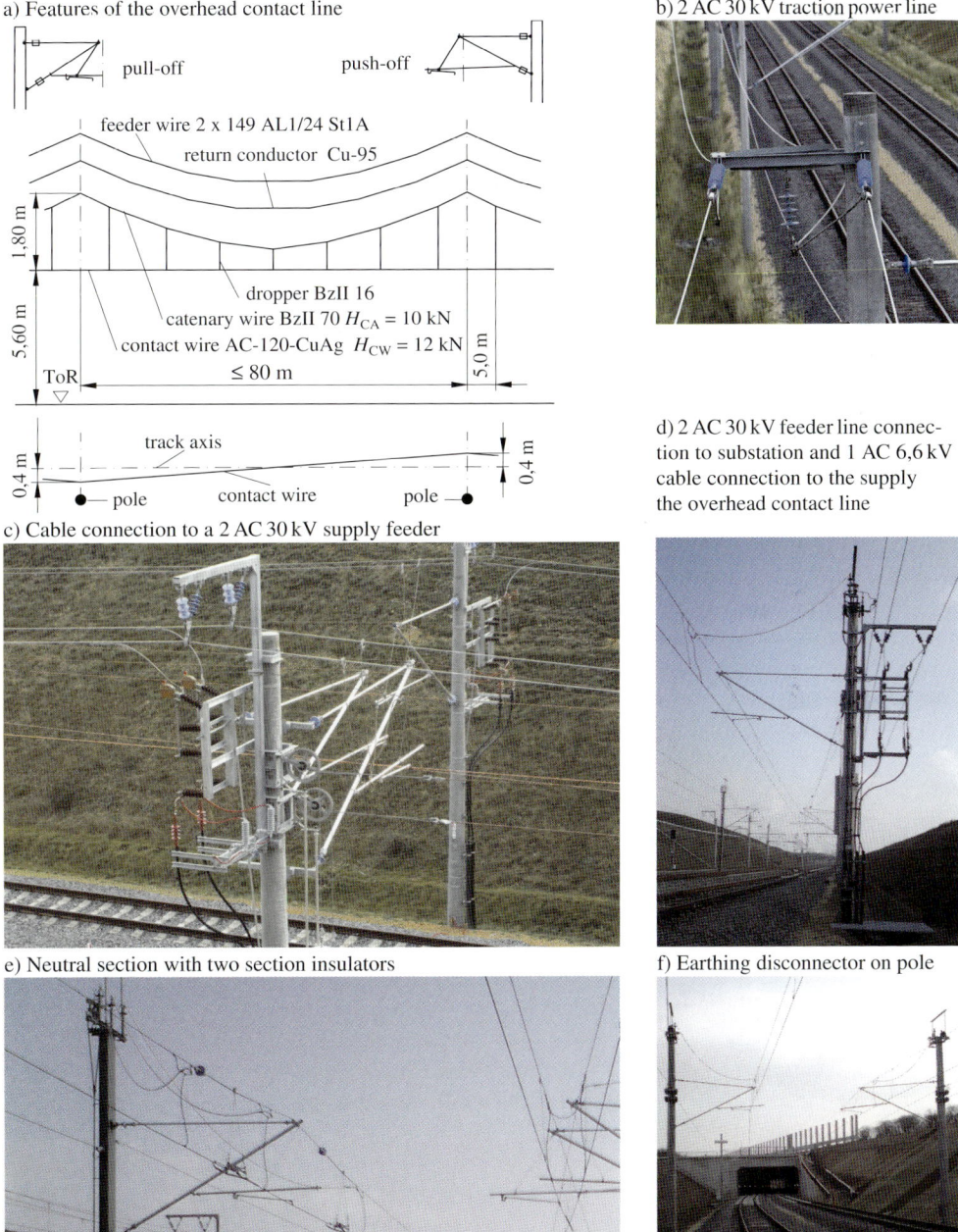

a) Features of the overhead contact line

pull-off   push-off

feeder wire 2 x 149 AL1/24 St1A
return conductor Cu-95

1,80 m

5,60 m

dropper BzII 16
catenary wire BzII 70 H_{CA} = 10 kN
contact wire AC-120-CuAg H_{CW} = 12 kN

ToR

≤ 80 m

5,0 m

track axis

0,4 m

pole   contact wire   pole

0,4 m

b) 2 AC 30 kV traction power line

c) Cable connection to a 2 AC 30 kV supply feeder

d) 2 AC 30 kV feeder line connection to substation and 1 AC 6,6 kV cable connection to the supply the overhead contact line

e) Neutral section with two section insulators

f) Earthing disconnector on pole

Figure 16.52: Industrial railway overhead contact line, Hambach (Photos: SPL Powerlines Germany GmbH, M. Vousten).

a) Loading station with laterally arranged overhead contact line

b) Lateral overhead contact line

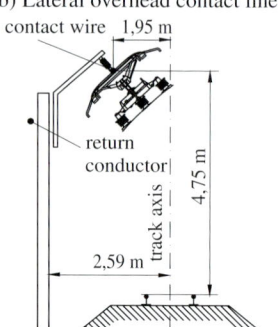

Figure 16.53: Industrial railway overhead contact line, Hambach (Photos: SPL Powerlines Germany GmbH, M. Vousten).

circuit breaker supplying approximately 4 km of overhead contact line. Figure 16.52 d) shows the 2 AC 30 kV 50 Hz cable connection to the substation and the 1 AC 6,6 kV 50 Hz cable connection that supplies the overhead contact lines.

One of the tracks is insulated in track lengths up to 200 m; in tracks with lengths greater than 200 m, both rails are insulated. A 95-mm^2 Cu cable, designated as a Z–connector, connects the rails conducting the return current in case the insulated rail is changed. For tracks with two insulated rails, 95 mm^2 cross-section copper cables conduct the return current from the rails to the track reactor and a central tap.

A 95 mm^2 copper overhead line, in parallel to the rails (Figure 16.53 a)), conducts the return current to the substation and is connected to components to be earthed, that could become live in case of a failure. To increase impedance, reactors are arranged in the connection between the return conductor and the rails at pole sites. The overhead contact line is separated into individual supply sections by protective sections consisting of two section insulators, each with a neutral section in between. The design of such protective sections (Figure 16.52 e)) separates live and temporarily switched-off supply sections that could be bridged by traction units with two pantographs in the case of a failure. The motor-operated disconnectors at the protective sections supply the protective section in the basic configuration and separate them in case of a failure.

The overhead contact lines of both tracks in the existing tunnel can be isolated in case of emergency by disconnectors and earthed by four earthing disconnectors (Figure 16.52 f)) to permit access by rescue teams and can be protected against resetting.

The loading stations at the lignite open pit Hambach are equipped with laterally arranged overhead contact lines that supply energy to laterally arranged pantographs (Figure 16.53 a)). The laterally arranged overhead contact line permits the loading of the wagons from above with conveyor belts and the simultaneous moving of the wagons (Figure 16.53 b)). Outside of the loading station the traction vehicle uses the pantographs arranged centrally on the roof.

16.11   Bibliography

16.1 *Regulation 1299/2014/EU*: Technical specification for the interoperability of the infrastructure subsystem of the railway system in the European Union. In: Official Journal of the European Union No. L 356 (2014), pp. 1 to 109.

16.2 *EBO*: Eisenbahn-Bau- und Betriebsordnung (Ordinance on construction and operation of rail-ways). Bundesrepublik Deutschland, BGBl. 1967 II pp. 1563, with last modification by article I of the ordinance of July 25, 2012 (BGBl.I pp. 173).

16.3 *Verband Deutscher Verkehrsunternehmen VDV*: Stadtbahnen in Deutschland (City railways in Germany). Alba Publishing GmbH & Co. KG, Düsseldorf, 2000.

16.4 *Regulation 1301/2014/EU*: Technical specification for the interoperability of the energy sub-system of the railway system in the European Union. In: Official Journal of the European Union No. L 356 (2014), pp. 179 to 227.

16.5 *BOStrab*: Verordnung über den Bau und Betrieb der Straßenbahnen (Straßenbahn-Bau- und Betriebsordnung) (Regulation on the installation and operation of tramways (Tramway installation and operation regulation)). Bundesrepublik Deutschland, BGBl. I No. 58, 1987, pp. 2648, last edition in: BGBl. I No. 57, 2007, pp. 2569.

16.6 *BOA/EBOA*: Verordnung über den Bau und Betrieb von Anschlussbahnen (Regulation on the construction and operation of secondary railways). German provincial laws.

16.7 *RWE Power*: Ihr zuverlässiger Partner für Schienenfahrzeuge, Instandhaltung Bahn im Technikzentrum (Your reliable partner for railbound vehicles, maintenance of railways in the technology centre). In: RWE Power, product information, Essen, 2013.

16.8 *Abst, S.; et al.*: Elektrifizierung der Hochgeschwindigkeitstrecke HSL Zuid in den Niederlanden (Electrification of the high-speed line HSL Zuid in the Netherlands). In: EI Der Eisenbahningenieur 58(2007)11, pp. 46 to 57.

16.9 *Grimrath, H.; Reuen, H.*: Elektrifizierung der Strecke Elmshorn–Itzehoe mit der Oberleitung Sicat S1.0 (Electrification of the line Elmshorn–Itzehoe with the overhead contact line Sicat S1.0). In: Elektrische Bahnen 96(1998)10, pp. 320 to 326.

16.10 *Kökényesi, M.; Kunz, D.* Overhead contact line Sicat SX – Approval and operational experience in Hungary. In: Elektrische Bahnen 112(2014)INT, pp. 104 to 107.

16.11 *Thoresen, Th. E.; Gjertsen, E.*: Neue Oberleitungen der Norges Statsbaner (New overhead contact line of Norges Statsbaner). In: Elektrische Bahnen 94(1996)4, pp. 115 to 119.

16.12 *Lörtscher, M.; Aeberhard, M.; Schär, R.*: Bahnenergiebedarf für die Lötschberg-Strecken (Traction energy demand for the Lötschberg lines. In: Elektrische Bahnen 105(2007)11, pp. 54 to 554.

16.13 *Hahn, G.*: Oberleitungstechnische Ausrüstung des Lötschberg-Basistunnels (Overhead contact line equipment of the Lötschberg basis tunnel). In: Elektrische Bahnen 105(2007)4-5, pp. 284 to 289.

16.14 *Kurzweil, F.*: Neubaustrecke Wien–St. Pölten – TSI-konforme Oberleitungen (New railway line Vienna–St. Pölten – TSI conform overhead contact lines). In Elektrische Bahnen 111(2013)6-7, pp. 418 to 424.

16.15 *Altmann, M.; Matthes, R.; Rister, S.*: Die Elektrifizierung der Hochgeschwindigkeitsstrecke HSL Zuid (Electrification of the high-speed line HSL Zuid). In: Elektrische Bahnen 104(2005)4-5, pp. 248 to 252.

16.16 *Kohlhaas, J.; et al.*: Interoperable Oberleitung Sicat H1.0 der Schnellfahrstecke Köln–Rhein/ Main (Interoperable overhead contact line Sicat H1.0 of the high-speed line Cologne–Rhine/ Main). In: Elektrische Bahnen 100(2002)7, pp. 249 to 257.

16.17 *Grimm, R.; Puschmann, R.; Rux, M.*: Oberleitungsabnahme am Beispiel der Neubaustrecke Köln–Rhein/Main (Overhead contact line approval at the example of the new line Cologne– Rhine/Main). In: Elektrische Bahnen 101(2003)4-5, pp. 200 to 207.

16.18 *Grimrath, H.*: Elektrifizierung der Strecke Lehrte–Oebisfelde mit der Oberleitung Bauart Re330 (Electrification of Railway line Lehrte–Oebisfelde with the overhead contact line type Re330). In: Elektrische Bahnen 96(1998)1-2, pp. 24 to 28.

16.19 *Luppi, J.; Lamon, J.-P.*: Histoire de la caténaire 25 kV (History of the 25 kV overhead contact line). In: Revue Générale des Chemins de Fer (1992)3, pp. 35 to 52.

16.20 *Chambron, E.*: La conduite du projet TGV Atlantique et les travaux de génie civil (Construction of the TGV Atlantic line and the civil engineering structures). In: Revue Générale des Chemins de Fer (1986)12, pp. 567.

16.21 *Watanabe, K.*: Review and perspective of SHINKANSEN. In: Elektrische Bahnen 83(1985)5, pp. 145 to 152.

16.22 *Hardmeier, W.; Schneider, A.*: Direttissima Italien, die Schnellfahrstrecken Bologna–Florenz und Florenz–Rom (Direttissima Italy, the high-speed lines Bologna–Florence and Florence– Roma). Orell Füssli Publishing, Zurich and Wiesbaden, 1989.

16.23 *Deutschmann, P.; Nilsson, A.*: Kraftförsörjningsanläggningar / Autotransformatorensystem – Systembeskrivning (Traction power supply plants/autotransformator system system description). BANVERKET, Verksamhetssystemet Standart – BVS 1542.11601 (Vorschrift), 18.12.2009.

16.24 *Schütte, T.; Thiede, J.*: Kombinierte Streckenspeisung mit Auto- und Saugtransformatoren. In: Elektrische Bahnen 98 (2000), H. 7, S. 249–253.

16.25 *Kiessling F.; et al.*: Líneas de Contacto para Ferrocarriles Electrificados (Contact lines for electrified railways). Siemens AG, Erlangen, 2008.

16.26 *Burkert, W.*: Oberleitungsplanung mit der erweiterten Software Sicat MASTER (Planning of overhead contact lines with the extended software Sicat MASTER). In: Elektrische Bahnen 108(2010)8-9, pp. 377 to 385.

16.27 *Siemens AG*: Product catalog 2012. Siemens.com/rail-electrification.

16.28 *Hickethier, R.*: Die Stromschienenanlage und deren Planung mit Unterstützung von rechnergestützten Zeichenprogrammen (Third rail installation and its planning with the aid of computer-supported drawing programmes). Ellert & Richter Publishing, Hamburg, 1996.

16.29 *Weitlaner, E.; Schneider, E.*: Bahnstromversorgung für die Stadtbahn BTS Bangkok (Traction energy supply for the city railway BTS Bangkok). In: Glasers Analen 123(1999)6, pp. 253 to 260.

16.30 *Hübner, F.-K.; Kießling, F.; Meyer, H.-H.*: Projektierung einer Oberleitung im Rheinischen Braunkohlenrevier (Hambachbahn) (Planning of an overhead contact line in Rhenish lignite area). In: Elektrische Bahnen 82(1984)11, pp. 359 to 366.

17 Management and maintenance

17.0 Symbols and abbreviations

Symbol	Definition	Unit
$A(t)$	probability for the functionality of an installation	–
A_D	constant availability	–
A_n	nominal cross-section	mm^2
A_{res}	residual cross-section	mm^2
ACLT	Automatic overhead Contact Line Testing	–
ADIF	Admistrador De Infrastructuras Ferroviarias	–
CMS	Catenary Monitoring System	–
$\overline{D}_{100}$	mean maintenance duration	h
H	distance between measurement device and contact wire	m
JBV	Norwegian railway infrastructure manager (Jernbaneverket)	–
LCC	life cycle cost	EUR
MTBF	mean time between failure	–
N_0	starting set of units	–
N_t	starting set of units at the begin of a time interval	–
S_i	fault number i	–
ÖBB	Austrian Federal Railway	–
$P(t)$	probability that the installation is able to function	–
PMS	Pantograph Monitoring System	–
pH-value	magnitude to determine the acid or basic character of a solution in water	–
$R(t)$	probability of the disturbance-free working	–
RFID	Radio-Frequency Identification	–
RMS	Rail poential Monitoring System	–
$\overline{T}$	mean duration of function	h
$\overline{T}_L$	service life	h
Z	content of cement	kg/m^3
$Z_{0(1)}$	condition of an installation at the time $t_{0(1)}$, for which the probability $P_{0(1)}$ applies	–
Zes	central network control centre at DB	–
$f(t)$	failure intensity function	–
h	height of conductor rail	m
h_{CW}	contact wire height	m
t	period of time	h
t_T	residual period of use	h
t_a	time of failure	h
t_{ai}	time of occurrence of the failure i	h
t_{wi}	time of restart of operation after a failure	h
t_0	start of operation	h
t_W	time of resuming operation	h

Symbol	Definition	Unit
w/z	water cement value	–
Δt	time step	–
α	angle between perpendicular and distance line between Mephisto device and contact wire	degree
λ_0	constant failure rate	–
$\lambda(t)$	failure rate	$1/(100\,\mathrm{km\,a})$
$\mu(t)$	repair rate	$1/(100\,\mathrm{km\,a})$

17.1 Management

17.1.1 Definitions

The *management* of an overhead contact line system includes the *operation of electrical installations* and *working in the electrotechnical system* with the aim of maintaining high availability. Figure 17.1 shows the operational activities involved.

17.1.2 Education and training of staff

The *operation of overhead contact line systems* requires suitably trained and experienced staff. Infrastructure managers direct the sequence of operations and behaviour, that enables new personnel to acquire requisite knowledge of operations and employment in overhead contact line installations.

The aim of *staff education* is the transfer of knowledge regarding safe and correct work methods, including the recognition of possible dangers of irregularities and incorrect behaviour. For working on or close to overhead contact line installations, the following responsibilities need to be distinguished depending on the complexity:

- *nominated persons in control* are responsible for the operation of an *electrical installation* in accordance with EN 50 110-1. If necessary, some of these responsibilities can be transferred to other people.
- *nominated persons in control of the work activity* are responsible for carrying out the work. If necessary, some of these responsibilities can be transferred to other people (EN 50 110-1).
- *responsible electrically skilled persons* are electrically educated persons. They have the relevant education, knowledge and experience to assume the technical and supervisory responsibility and are nominated by the contractor. In Germany their responsibilities are defined in DIN VDE 1000-10.
- *electrically skilled persons* are able, because of their technical education, knowledge and experience including a knowledge of relevant standards, to assess the activities assigned to them and to recognize possible danger (DIN VDE 1000-10.).
- *electrically instructed persons* (EuP) are electrically skilled persons familiar with their assigned activities and possible dangers resulting from incorrect actions. They are also informed about necessary protection equipment, personal protection equipment and protection methods (DIN VDE 1 000-10).
- *instructed persons on railway technology* carry out activities in the non public area in accordance with EN 50 122-1. They learn about the work activities assigned to them

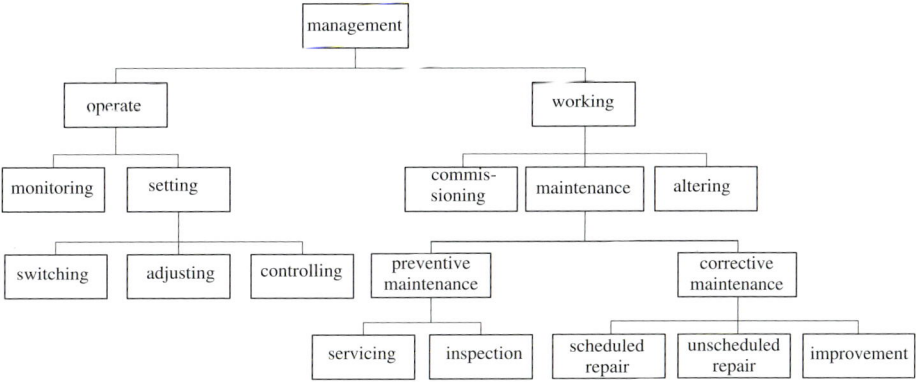

Figure 17.1: Structure of management activities.

and possible dangers of incorrect actions, including the rules of conduct. The instruction on railway technology is carried out by persons who because of their education, knowledge and experience recognise and evaluate possible dangers especially in electric railway operations. Persons instructed on railway technology may not independently carry out activities on electrical installations of vehicles and overhead contact lines, however, they may make use of minimum clearances for the non-public domain in accordance with Clauses 4.1.2.1 and 5.1.2.1 of EN 50 122-1 and guards in accordance with Clauses 4.1.3 and 5.1.3 of that standard.

– *ordinary person* is any person who is neither an electrically skilled person nor an electrically instructed person (EN 50 110-1).

Depending on the type of activities in the overhead contact line installation electro-technical knowledge and operating experience form the preconditions to prudent execution of management activities. The electrically skilled person must according to EN 50 110-1 have completed their electro-technical education as an electrical engineer, electrical master craftsman or journeyman. Specific profiles of requirements for electrically skilled persons establish minimum knowledge and experience requirements for work and conduct in operational railway environments. Electrically skilled personnel instruct electrically instructed persons on the type and amount of activities and supervises them [17.1]. Railway technically instructed people carry out activities on electrified lines, though, not on overhead contact line installations. They, by instruction of an electrically skilled person, may recognise danger due to *electric railway operations* and can conduct themselves appropriately.

Educational training includes:

– design of the overhead contact line installation
– master plans with switching instructions
– danger due to electric railway operations
– control and operation of overhead contact line disconnectors
– recording of switching calls
– switching under own responsibility and
– behaviour in case of danger

Staff demonstrate their knowledge through regular tests that record, renew and reinforce knowledge. Over two years, DB instruct their staff in all important electrotechnical standards of conduct, the company-owned directives and the analysis of disturbances and accidents.

17.1.3   Electro-technical standards on conduct and directives

Technical stipulations support users and operators in recognising the danger, that can arise from incorrect actions on electrotechnical repairs and installations. International standards serve as the generally agreed rules of technology.

Product and installation standards such as EN 50 119 and EN 50 122-1 contain technical requirements and specifications. *Standards on operations* such as EN 50 110 stipulate behaviour and processes for management and users. Company directives such as DB Directive 462, *Operation of the overhead contact line installation* supplement the general standards for railway-specific conditions [17.2]. The structure of the modularly designed DB Directive 462 [17.2] into *principles*, *management*, and *working on and close to overhead contact line installations* echoes the definitions in EN 50 110-1 and VDE 0105-103 supplementing the separation of operations into operations and works.

The Austrian Federal Railways summarise the *in-house requirements* in the Directive EL 52 [17.3] for electrical operations, subdivided into *general stipulation*, *provisions for safety for working* and *specific specifications*, concerning the operation itself.

The Swiss Federal Railways specify the requirements on working at and close to overhead contact line installations in Directive RTE 20600 *Safety in case of working in the area of railway power supply installations*. This directive is part of the basic rules of *technology railway* published by the Swiss Association of Public Transport [17.4].

17.1.4   Five basic rules of safety

According to VDE 0105-100 five rules of safety need to be considered before starting work on overhead contact line installations:

- isolating
- securing against re-closing
- verifying the isolation from all points of supply
- earthing and short-circuiting
- covering neighbouring live parts or providing a barrier

Under *isolating* it is understood that all live components of an electrical installation are separated from other live components. For isolation, differently long air gaps need to be established between live and dead parts of the installation, depending on the operational voltage. When operating on electrical equipment with less than $\geq$ AC 50 V or $\geq$ DC 120 V isolating is always necessary in Germany, Austria and Switzerland. Otherwise special provisions for working under live-line conditions should be made.

Securing against re-closing should ensure that whilst working activities are carried out, an installation is not accidentally re-closed. Re-closing can be avoided by:

- dismantling fuses that can be replaced by lockable isolating elements and locking out breakers, switching cubicles or switching boxes in low-voltage networks.
- employing special tools so that ordinary persons can not re-close in public areas
- warning plates against re-closing at the entrances to closed electrical compartments
- lock-out devices in remote-controlled high-voltage installations that inform the operator in the control centre of the reclosing prohibition
- warning plate against re-closing

The *absence of voltage needs to be checked* by persons on site with voltage testing equipment capable of demonstrating all-pole disconnections. Voltage testing devices for installations above 1 kV nominal voltage are designed for one pole with an electrically insulated lance having a length up to several meters, manually placed on to the overhead contact line. Using a capacitance potential divider, the existence of a high voltage is indicated optically or acoustically by a testing circuit on the lance. The testing device functionality needs to be checked at a known live overhead contact line before and after use. Before use, because the testing device could be faulty and after testing because it may have failed during testing. Single-pole high-voltage testers for nominal voltages above 1 kV include integrated test circuitry simplifying the functional test.

Earthing and short-circuiting is established after testing for isolation from supply by connecting the overhead contact line and the return connector using a short-circuit proven earthing and short-circuiting device. This should ensure that the protection device of the supplying circuit breaker will trip in the event of erroneous re-closing and switch-off the circuit breaker. The earthing and short-circuiting devices should be situated in line of sight of the work place. This is followed by the *provision of a barrier* between neighbouring live components and the work site, using either an insulating cover or physical barrier. This can only be carried out in substations and not on overhead contact lines.

In the case of overhead contact lines, safe guarding is replaced by special instructions where the nominated person in control of the work activity instructs staff on site before starting activities including:

– parts of the installations still under voltage
– working limits identified by earthing and short-circuiting devices
– neighbouring tracks still in operation
– specific dangers at the work place

17.1.5   Switching and earthing

The active parts of the overhead contact line system are normally energised. Operational management, maintenance work and disturbance events necessitate *switching operations*. The circuit diagram (see Clause 12.2.3.7) shows the designation and normal position of the disconnectors, their assignment to *switching groups*, conduct during hazards and the location of earthing and short-circuiting devices and voltage testers. Only appropriately trained persons may perform switching operations - the switching command controller, the switching applicant and the disconnector operator.

The *switching command controller* and the *switching service manager* in the area power supply control centres have the highest qualifications in switching services. They must gain and demonstrate their knowledge in a training course with a subsequent examination that indicates they can perform switching operations independently under their own authority, or issue switching instructions for the execution of switching operations by other persons. The *switching command controllers* operate remotely-controlled disconnectors. Switching operators who received training as *switching applicants*, can switch locally controlled or manually operated switches after training as a switching applicant. These persons include:

– traffic superintendent on electrified lines
– staff of a technical department and
– staff carrying out construction and supervision work

The switching operation itself is performed on the basis of a switching dialog, the process being formally recorded. The switching request contains:

– name of the system parts to be switched, e. g. X-town station, switching group I
– type of switching operation, e. g. open

- permission from the traffic superintendent to operational closure of the track
- identification of the switching applicant with code number.

After approval of the switching request the central control centre can issue a switching instruction to open disconnectors or perform a *switching operation* itself. After opening the disconnector and protecting it against unintentional re-closing the switching command controller at the central control centre confirms the execution of the switching operation to the switching applicant.

The master control centre can transfer the responsibility for a disconnector switching request to the switching applicant. Within the German Railway DB, this process is called switching within ones own authority. The overhead contact line disconnector is identified by the switching command controller in the master switching control centre. If other disconnectors supply the switching group or groups concerned by the switching application, the switching command controller in the master control centre also safeguards these disconnectors against unintentional re-closing. Only after the operational closure of all relevant tracks of the switching groups to be disconnected by the traffic superintendent, the switching applicant can open the disconnector and safeguard it against unintentional re-closing.

Maintenance work can commence after checking that the line is de-energised, that *earthing devices are applied in front of and behind the work location* and the supervisor has verbally instructed the maintenance team about the working limits and special hazard situations. The switching applicant must be constantly available during the disconnection period of the switching group.

After completion of the work, the supervisor reports the safe operational state of the overhead contact line system to the switching applicant. The switching applicant reports to the switching command controller that the contact line is ready for reclosing and reinstates the normal disconnector position after receipt of an instruction, or independently if it is under his own authority and cancels the operational closure. If several switching permits had been issued for a switching group, the reclosing may be performed only after all applicants reported readiness for reclosing to the switching command controller.

The *switching dialogue* requires approval of the traffic superintendent for the operational closure of the tracks and all subsequent information relating to the switching operation must be documented in the telephone log for switching operations or recorded on the *voice recorder* in the central control centre.

The equipment needed to check the voltage on the overhead contact line and for earthing is located at the stations and should be inspected at intervals of two to five years. Repeated instruction on service regulations relating to switching serves to avoid errors and reinforce safe activity sequences. Regular *accident prevention training* assists the review and consolidation of knowledge.

17.1.6   Tasks of the infrastructure manager for operations

The *infrastructure manager* of an installation is obliged to maintain the stipulated performance of the energy supply during service life. The infrastructure manager must establish a *maintenance plan* in accordance with TSI ENE [17.5] to guarantee that the performance of the energy supply remains within defined limits. The maintenance plan references all documents describing the maintenance of the infrastructure. In particular, the maintenance plan needs to include:

- the number of staff and their professional qualifications within the plan identifying further training
- the personal protective equipment requirements
- the periodic inspection program
- the equipment held at the maintenance centres and in vehicles
- the proactive and reactive maintenance spares
- a fault clearance program

Maintenance procedures include safety instructions such as not interrupting return current, protection against over-voltages and the recognition of short circuits. The infrastructure manager is responsible for notifications on safety relevant incidents, shortcomings and frequent disturbances to the railway supervising authority - in Germany the Federal Railway Authority (EBA) to the Provincial Railway Authority (LEA) or equated national officials.

17.1.7 Earthing installations for contact lines

17.1.7.1 General

The range of applications for emergency earthing and short-circuiting of contact lines includes emergency situations in tunnels, in public areas of local and main line traffic as well as in maintenance works for electrically driven vehicles. Emergency earthing installations enable immediate earthing and short-circuiting of contact lines and commencement of rescue operations.

The Technical Specification, *safety in railway tunnels* (TSI SRT) [17.6] applies to both new and rehabilitated tunnels more than 1,0 km long on interoperable lines.

Additionally, railway operating and infrastructure entities need to apply on their safety-critical installations the Regulation 402/2013/EC [17.7] also called Common Safety Methods (CSM) regarding a common safety method for evaluation and assessment of risks according to Article 6, Paragraph 3a) of the Directive 49/2004/EC [17.8].

In the event of accidents or breakdowns in tunnels, unimpeded access via the tunnel entrances and emergency exits is essential for evacuation purposes. In such an event, the overhead contact line needs to be de-energised and earthed and connected to the return circuit before rescue personnel enter the tunnel and approach the accident location.

The required equipment is provided at the tunnel portals or emergency exits on the tunnel wall. Railway operators in the EU use different types of switching devices for *emergency earthing* in tunnels. Clauses 17.1.7.2 and 17.1.7.3 include examples from Germany and the Netherlands.

17.1.7.2 Emergency overhead contact line earthing in Germany

For any accidents in tunnels, the Technical Specification *Safety in Railway Tunnels* (TSI SRT) [17.6] requires that the infrastructure manager provides a means of de-energising the overhead contact line or conductor rails, before rescue personnel enter the tunnel. Since civil rescue personnel are likely to reach the tunnel entrance before the contact line staff of railway [17.9], the overhead contact line must be switched-off and connected to the return circuit by the rescue personnel using the automatic overhead contact line testing (ACLT) [17.10].

When there is a tunnel accident, the de-energising of overhead contact lines and traction energy lines is carried out by the central network control centre (Zes), independently of the ACLT by opening the relevant contact line disconnectors. In the event of a disconnector

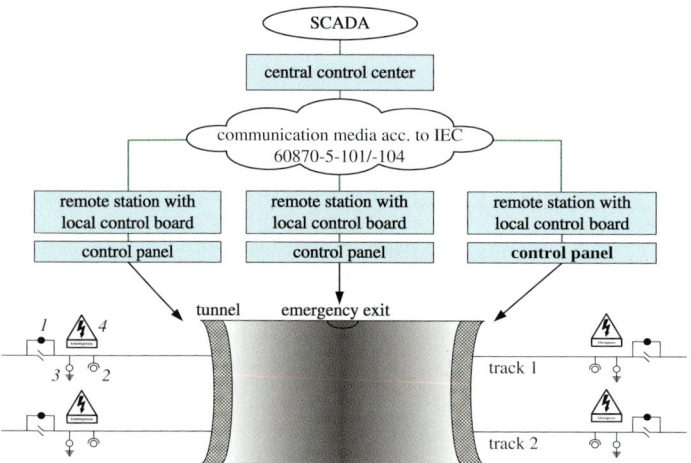

Figure 17.2: Schematic design of the overhead contact line voltage testing system for emergency earthing of contact lines in tunnels. *1* disconnector or load switch to disconnect the contact line *2* voltage transformer or isolation amplifier. *3* earthing disconnector or earthing load switch. *4* sign to define the limits of the work area.

failure, the central control centre switches off the circuit breakers in the supplying substations of the supply region. The absence of voltage is displayed on the local control board of the ACLT and within the Zes. When the absence of voltage is confirmed, the ACLT releases control of the overhead contact line disconnectors (with earth contact) and, therefore, the contact line earthing. Earthing and short-circuiting, using earth contact earthing switches, can be carried out by remote-control from the Zes in charge or from one of the tunnel entrances using the local control board of the ACLT remote station [17.9].

The overhead contact line voltage testing system displays the status of de-energised, earthed and connected to the return circuit of contact lines and traction power lines locally to rescue personnel, complying with the stipulations of EN 50 110 that establish the rules for working close to contact line systems (Clauses 15.6.8 and 17.1.4) encompassing the five safety rules. The earthed condition of the overhead contact line is signalised at the limits of the work area and at the entrances to tunnels. Figure 17.2 depicts the schematic structure of the ACLT for emergency earthing of the contact line.

The equipment of the ACLT comprises disconnectors to de-energise and earth the switching groups, voltage testing equipment, earthing switches or in-rush resistant voltage starting disconnectors and warning signs to display the working limits (Figure 17.3). The earthing switches are equipped with rotation angle sensors to monitor the intake and the end position of the moveable switching contacts (Clause 11.2.5). ACLT systems are suited to AC and DC railways of main line and local area traffic.

In maintenance depots (Clause 14.2) the contact line power above individual tracks can be switched off by the ACLT emergency earthing. The absence of voltage can be tested, the contact line can be earthed and subsequently the access gate to the work platform at roof level, e. g. close to the conductor rail overhead contact line, can be released (Clause 3.5 and 14.3). This condition is blocked until released by key switch safety interlocks.

17.1.7.3   Emergency earthing in tunnels in the Netherlands

On the first high-speed line in the Netherlands, the HSL Zuid (Clause 16.3.7), the infrastructure manager, Infraspeed, uses manually operated contact line earthing switches for emergency earthing of overhead contact lines in tunnels (Figure 17.4). As in the German network

a) Overhead contact line is live, the work
   boundary area is not visible

b) Overhead contact line is de-energised, earthed
   and the work boundary area sign can be seen

Figure 17.3: Moveable warning sign at the boundary of the area in which rescue services are permitted to work – example of the airport connection Cologne/Bonn.

(Figure 17.2), these earthing switches are installed at the tunnel portals. Trains should as far as possible be able to leave the tunnel even in the case of emergency. If a train must stop within the tunnel as a result of an accident, the overhead contact line voltage is switched off by the control centre in Rotterdam (Figure 17.5). The earthing switches are activated locally by the rescue services. Voltage testing has not yet been considered. If the contact line voltage is not switched-off in time, the closing of the make-proof earthing switches leads to a short circuit and switching-off of the supplying circuit breaker in the substation so that the rescue can continue.

Figure 17.4: Earthing switch on the HSL Zuid line, Netherlands.

Figure 17.5: Control centre in Rotterdam, Netherlands.
The screens show measured data and the state of substations and overhead contact lines

17.1.8   Irregularities and their recognition

The operator provides a current reporting and action plan for *irregularities* and *disturbances* in the overhead contact line network to aid recognition of necessary actions and information flow. Reports are conveyed by telephone to the switching command controller in the central control centre, for action.

In case of circuit breaker trips, the ACLT performs an automatic check for the absence of short circuits (Clause 1.5.3.7). The supply section will be re-closed if the result is positive. If a sustained *short circuit* is present, the fault position should be localised as accurately as possible to assist the repair team to target and locate the fault quickly. The disturbed line section is immediately blocked by the traffic superintendent, isolated and transferred to the switching applicant in the repair team with a notice of isolation. An immediate operational blocking and earthing of dc-energised overhead contact lines is necessary to avoid vehicle travel into neutral sections with consequent arcing and damage to overlapping or section insulators. After clarification of the required repair time, the leader of the repair team (see Clause 15.6.7) provides his time estimate to railway operations management and then attends to the rapid removal of the damage. Provisional solutions with *dropped pantograph sections* can help to restart train operations and reduce the build up of train traffic. It is then possible to reinstate the condition of the overhead contact lines during track closures arranged on a more long-term basis.

Irregularities and deployments for *overhead contact line system fault clearance* are recorded on prepared forms. These are circulated to a pre-defined distribution list within certain time limits, assisting statistical analysis and future incident prevention (Clause 17.3.6).

17.2   Wear and aging

17.2.1   Classification of components

The *components* in an overhead contact line system can be differentiated by the type of loads they carry:
- predominantly mechanical loads, such as poles and support equipment
- electrical and mechanical loads, such as catenary components, energy supply lines
- predominantly electrical loads, such as electrical connectors and clamps

Components of contact line systems are subjected to *aging* and electrical and mechanical *wear*, depending on the period of use and magnitude and duration of the load. Knowledge of wear and aging processes are essential for maintenance management.

17.2.2   Concrete poles and foundations

Corrosion is the destruction of material resulting from chemical and electrochemical processes. During electro-corrosion, metals oxidise as a result of a chemical reaction and usually accompanied by current flow, including *stray current corrosion*.

Electro-corrosion, however, also occurs without an electrical current source, e. g. by metals with different positions within the *electro-chemical series* or as a result of inhomogeneity of different surface sections of metal, that provoke potential differences. Air humidity with soil as an electrolyte can start the reaction. In accordance with Faraday's First Law of electrolysis as in Equation (6.31), the anode components are corroded and the metal erosion is proportional to the amount of current flowing [17.11]. Concrete consists of solid, liquid and gaseous components. The protective effect for reinforcement is caused by the alkalinity of the interstitial water with ph values between 12,5 and 13,5 forming a protective layer but not from hermetic inclusion. This helps slow the corrosion by forming a protective layer.

This process reverses after a drop in the concentration of calcium oxide hydrate. The cause of this drop can be cracks and carbonation or the presence of activators in the concrete. Possible activators are calcium chloride to accelerate the setting time and sodium chloride as frost protection during pole manufacture in winter or as de-icing material on roads. *Continuous protection* for reinforcement ceases when the cracks exceed 1 mm or concrete cover layers are less than 20 mm. The following factors affect concrete:

- mechanical loads and compression
- water that washes out calcium chloride, precipitating the formation of calcium carbonate, recognisable by white stains
- carbon dioxide in water and in the atmosphere, that leads to chemical decomposition
- stresses in the capillaries in the concrete and inside the poles caused by repeated freezing, with a volume increase of 9 %, and water thawed by solar radiation and cooling air currents

Damage to poles that do not have electrical causes, such as crack formation, the separation of the concrete from the reinforcement, the effects of aggressive materials etc. are less dangerous than stray current corrosion but occur more frequently. The effective prevention of this damage is possible by following specified concrete production quality control measures, including follow-up treatment. The concrete strength is selected from the viewpoints of stability and durability. Experience [17.11] shows that satisfactory protection against weathering, acids and carbonation can be achieved with a *water w to cement z ratio* $w/z < 0,45$ and a cement component $Z > 300 \, \text{kg/m}^3$. The *concrete poles* employed in Central Europe during the last 15 years have values $w/z \approx 0,35$ and $Z \approx 400 \, \text{kg/m}^3$ and, therefore, posses good qualities for railway applications. Ventilation of hollow poles has proven helpful to avoid cracks.

Stray current corrosion occurs on DC railways, when return currents flow through underground sections of foundations and poles. To avoid this phenomenon, the electrical interconnection of poles, foundations and rails either individually or in the form of a collective earth is essential for protective tripping during insulation faults, normally only executed by *voltage limiters* or spark gaps. If they are defective or if a direct contact is made, then a continuous current flow can occur depending on the potential differences and resistances present in the ground. Even a current density of $0,06 \, \text{A/m}^2$ can cause the start of electrical corrosion of underground equipment. The ground resistance of a concrete pole can be between 3 and $3\,000 \, \Omega$ but does not normally exceed $30 \, \Omega$. Damage to concrete foundations has been observed on DC railways mainly at depths of 0,4 to 1,0 m and over a length of 0,5 to 1,0 m. The destruction of the protective effect of the concrete cannot be reversed. This means that the reinforcement continues to corrode even after removal of the cause of the stray current corrosion. Therefore, the prevention of current flow through concrete poles and foundations gains special significance. Experience demonstrates that the *service life of concrete poles* under normal operational conditions can be 60 years and more. Stray current corrosion is dealt with in Clauses 6.3.1, 6.3.2 and 6.3.7.

17.2.3   Steel poles, cantilevers and other support structures

Damage to metal structures can be classified as follows:

- corrosion
- deformation due to external influences, such as train derailments
- brittleness at low temperatures and deformation at high temperatures
- mechanical overload due to errors during planning or installation and
- electrical corrosion

Limited durability and delayed refurbishment of the *corrosion protection of steel components* is the primary reason for corrosion other than aggressive environments, especially steel poles that corrode at the interface with concrete foundations or foundation caps. These parts can only be durably protected by elastic coating systems. Corrosion under continental climate conditions is assisted by sulfurous gases, mainly at temperatures between 0 °C and 15 °C and under coastal conditions by salts with their chloride ions – a fluid electrolyte layer being a pre-requisite. The rate of corrosion is up to six times higher in industrial regions than in rural areas because of the increased air pollution. Ventilation is required to avoid corrosion inside hollow poles.

Since 1860 *Hot-dip galvanizing* has been employed as corrosion protection for steel components. The zinc forms an active coating as well as a passive corrosion protection layer. The passive protection is due to the barrier effect of the zinc coating. The active corrosion protection is formed by the cathodic effects of the zinc covering. Iron is more positive on the electro-chemical voltage series than zinc, therefore, zinc is less precious than iron. As a galvanic element on the surface of a component, zinc forms the anode and iron the cathode. Iron, therefore, is cathodically protected as the more precious metal until the zinc has completely corroded. In comparison with more precious metals like iron, zinc serves as a sacrificial anode.

In air, zinc in combination with carbon dioxide forms a protective layer with zinc oxide and carbonate, also called patina or zinc carbonate and looks darker and less brilliant than a surface with a new layer of zinc. If there is a shortage of carbon dioxide and in conditions with chloride or sulfate conditions close to the coast or during winter with a higher content of SO_2 in the air, white rust can be formed as a loose and voluminous cover. White rust decomposes the zinc layer at increasing rates with the air humidity.

The protective zinc carbonate layer, however, is eroded by chloride or sulfate containing conditions in humid air. The original, approximately 85 µm thick zinc layer on a cantilever or a steel pole can be reduced each year by 2 µm in rural areas, 3 µm in urban areas and up to 20 µm in industrial or coastal areas [17.12]. Supplementary coatings are necessary when the residual thickness has reached 40 µm. Corrosion also depends on the design and arrangement of the components since these offer varying conditions for the accumulation of dust and moisture.

Since approximately 1985, *aluminum cantilevers* have become popular in Germany as an alternative to *hot-dip galvanised steel cantilevers with hot-dip galvanised malleable cast iron fittings* [17.13]. Aluminum has proven to have a relatively high resistance to corrosion, as it forms a dense surface oxide layer. The protection effect is not lost after mechanical damage, because the protective layer renews itself. Aluminum posses favourable properties in the case of short circuits because of its conductivity, which is ten times higher than that of steel and its specific heat which is twice that of steel. The service life of hot-dip galvanised components, maintained by timely renewal of the coatings, is estimated to be greater than 70 years. Experience with aluminum components already shows a service life of over 80 years without corrosion protection measures.

The attachment of *steady arms* to the drop bracket with a loose fit can lead to mechanical wear. Therefore, a steady arm minimum tensile force of 80 N is recommended.

Electrical erosion occurs on DC railways when partial currents flow through moveable, non-insulated connections (Figure 17.11). Voltage differences of only 15 to 20 V can lead to the erosion of metallic parts because of electric arcs. These phenomena can be avoided by the provision of electrically conducting bypass cables or electrical insulation at these points.

17.2.4   Traction power supply lines, catenary wires, droppers and connectors

Traction power supply lines, catenary wires, droppers and connectors are subjected to a high electrical loading, mechanical stresses from tensile loads and, climatic factors and vibrations that can lead to *fatigue phenomena*, wear, corrosion and glow-out. Vibrations, especially near mass concentrations such as clamps and insulators, are the cause of fatigue.

The high degree of corrosion resistance of *aluminum components* can be explained by the formation of a protective oxide layer. When polluted by alkaline and salt-laden substances, aluminum corrodes faster than copper. *Bimetallic copper-clad steel conductors*, which are employed in several countries for contact wires and catenary wires, are subject to corrosion in sulfurous air.

Glow phenomena arise in the mentioned elements because of overloading by electric current in the event of incidents and locally under conducting clamps. Clamp heating is caused by an increase in the *transition resistance*, e. g. due to oxide layers on the contact surfaces, reduction of bolt tightening torque or distortion after severe temperature variations. If only a few external conductor strands are in contact with the contact surfaces of the clamp, they become overloaded and glow-out. The current flow shifts to other, internal strands and must now overcome a higher transition resistance, that leads to a rapid increase in heating. The rate of *aging of connectors* is directly dependent on the current loading and is correspondingly higher on DC railways. A satisfactorily applied contact grease layer or crimped connections counteract these phenomena. The *failure of connectors* can lead to the glow-out of droppers that do not have sufficient current-carrying capacity. Potential differences of 15 to 20 V can form electric arcs and result in electrical erosion at the connection points. Droppers are also subject to mechanical loads from friction and bending during the passage of pantographs. The degree of wear increases with increasing stiffness of the dropper, e. g. with current conducting droppers, thicker wire cross-sections and reduced dropper lengths.

17.2.5   Contact wires

17.2.5.1   General

The operational and environmental loads typical for bare electrical conductors strung in the open also act on *contact wires*, together with the added demands caused by the passage of pantographs and the current collection. These cause mechanical and electrical wear. The rubbing of the pantograph on the contact wire causes mechanical wear reducing the cross-section. Because of the energy transmission from the substation to the running vehicles and the sliding of the pantograph against the contact, wire joules and friction heat are produced. Since the contact wire exhibits microwaves and cannot be strung exactly in parallel to the track, arcs can occur. The joules and friction heat and the arcs result in electrical wear which reduces the contact wire strength. The correlation between mechanical and electrical wear cannot be determined from operating contact lines. Consequently, test stands with an acceptable replication of field conditions [17.14] are used to simulate suitable possibilities, to investigate these correlations. Contact wire thickness measurements can confirm the results of test stand simulations.

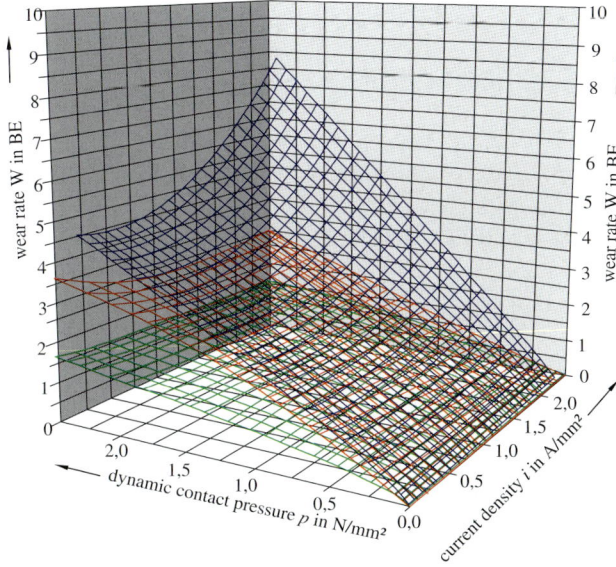

Figure 17.6: Contact wire wear for contact wires AC-100–Cu (blue), AC-120–CuAg0,1 (red) and AC-120–CuMg0,4 (green) with hard carbon collector strips on ICE pantographs at 100 km/h running speed on a test stand [17.14].

17.2.5.2 Mechanical wear

The collector strips of the pantograph act with a vertical contact force on the contact wire. Increased running speed causes the contact force, also called contact pressure p (Figure 10.26), to rise and depending on the number of pantograph passages also the rate of wear W as a related magnitude, approximately proportionally (Figure 17.6) [17.14]. The *rate of wear of the contact wire* also depends on the material used for the contact wire and the collector strip. The layer of pollution at the contact wire and collector strip interface affects their contact behaviour [17.15]. The contact wire, consisting of pure or a low-alloyed copper, is covered by a polluted layer within a short period of time. With collector strips made of carbon or impregnated carbon, this effect is not observed. The polluted layer of copper caused by oxidation, also known as patina, consists predominantly of copper I oxide (Cu_2O), copper II oxide (CuO) and to a lesser extent copper sulfate (Cu_2SO_4). Atmospheric emmissions affect the creation of these oxides.

The friction process of the collector strips causes the abrasion products of carbon and dust to be stored in the pollution layer. Because of the lower friction value of the copper patina, compared with pure copper, the sliding features improve and reduce wear of the collector strip. Since the patina on the lower surface of the contact wire is harder than the copper itself, the wear of the contact wire is reduced. However, the patina is characterized by a higher electric resistivity as copper oxides behave like semi-conductors.

During rain or hoar frost, a water or ice layer forms on the contacts. A layer of rain water acts like a lubricant, while a hoar frost layer increases the wear rate and results in arcing. The arcing may burn holes into the collector strip holder and affect the collector strips. Therefore, a sufficient lateral displacement of the contact wire Δb_m should be considered when planning the contact line (Table 4.14).

Planning, installation and maintenance of overhead contact lines should aim for low and uniform contact wire wear. Unwanted locally increased *contact wire wear* is often caused by masses in the overhead contact line such as section insulators and crossing contact wires with

a) Contact wire with ripples and kinks

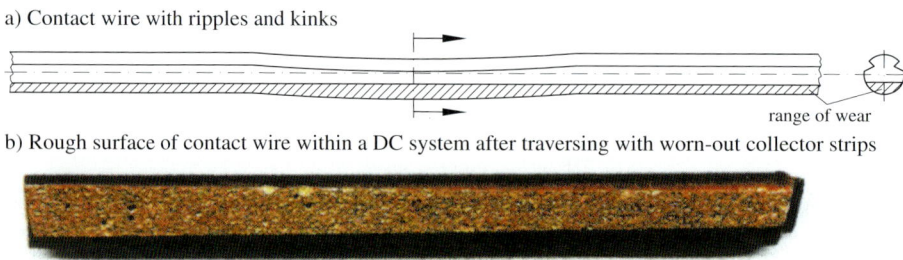

range of wear

b) Rough surface of contact wire within a DC system after traversing with worn-out collector strips

Figure 17.7: Reduction of contact wire cross-section.

crossing bars. It can also be caused by high radial forces at contact wire supports that lead to increased wear from abrasion due to higher contact forces and arcing. Similar phenomena occur from the superposition of oscillations in the overhead contact line. Running pantographs under *ripples* and *kinks* resulting from incorrect installation or during traffic operations, e. g. lose tarpaulins on freight wagons can cause local cross-section reductions (Figure 17.7 a)). Aerodynamically unfavourably designed tunnel portals cause strong and abrupt changes of the air stream acting on the pantograph. The pulsed forces directed upwards or downwards act on the pantograph and consequently on the contact wire causing increased wear.

17.2.5.3   Electrical wear

Electrical wear is created by high temperatures from melting, evaporating and/or burning of material at current bridges formed between the contact wire and collector strip. Less than acceptable minimum contact forces or non-conductive water and ice layers can cause arcing, leading to high temperatures.

With arc-free operation, heat is created at the current bridges by current and friction and can lead to plastic contact wire elongations if the limit temperatures are exceeded (Table 2.8), leading to reduced tensile strength and increased wear (Clause 7.2). The electrical wear, caused as a consequence, depends on the duration of the action and the running speed of the vehicle. At standstill, the vehicle experiences only the static uplift contact pressure, while the dynamic contact pressure is zero (Figure 17.6) [17.14]. With increasing speed and consequentially rising contact pressure p, the electrical wear also rises and is designated as related wear rate in Figure 17.6. It is approximately linear, but then becomes parabolic with a minimum of the wear at average current densities. For speeds above 100 km/h for the parameters given in Figure 17.6 [17.14] no wear minimum is formed. In this situation an approximately linear pattern can be observed.

Clauses 11.2.2.2 and 7.2 also describe the processes of current transition from the contact wire to the collector strips.

Investigations show that electrical wear is a maximum with carbon collector strips. The ratio between electrical and mechanical wear was quoted as 80 % to 20 % [17.16].

17.2.5.4   Permissible wear

The contact wire section with the greatest local wear determines the *service life of the contact wire*. If 20 % of the cross-section in the case of DB systems or 30 % in tramway systems has been worn away, contact wire splices and new contact wire sections are installed at the

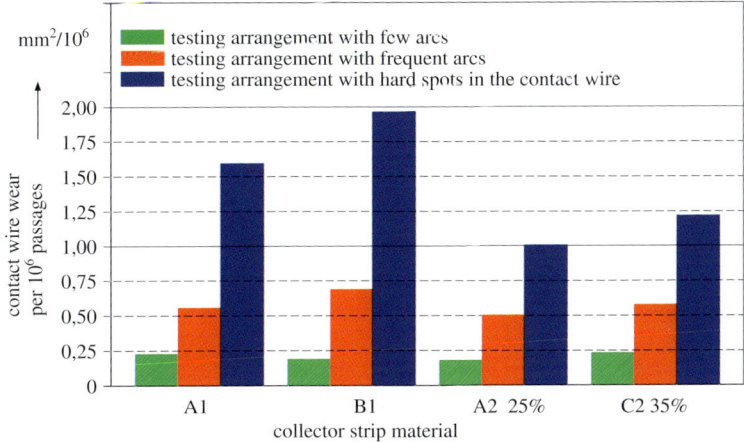

Figure 17.8: Contact wire wear with carbon and metalized collector strips from different manufacturers [17.17].
A to C Manufacturer of collector strips, A1 and B1 pure carbon collector strips of the manufacturers A and B, A2 25% and C2 35 with 25% and 35%, respectively, copper impregnated carbon of the manufacturers A and C, respectively

affected positions of the contact lines. The maximum number of *contact wire splices* per tensioning section is four for DB overhead contact line type Re 100 and one for contact lines Re 200 to Re 330. When these numbers are exceeded, the contact wire over the complete tensile section must be replaced, as stipulated in the DB Directive 997.01.149, Table 2. The permissible wear differs for individual railway operators.

With twin contact wires on DC railways, differing stresses can be observed in the individual contact wires. Local heavy wear points alternate between the two contact wires. This is caused by non-uniform pantograph contact force distribution over both contact wires, resulting in different transition resistances between the contact wires and the collector strips. Consequently, the current in some sections is only conducted by one of the two contact strips.

Heavy arcing can be caused by unfavourable combinations of overhead contact lines and pantographs or by rough unserviceable collector strips leading to a rough and partially annealed contact wire surface (Figure 17.7 b)). This surface can only be made smooth again by considerable loss of cross-section of the contact wire and collector strips.

Experience by the Spanish infrastructure manager, Administrador de Infraestructuras Ferroviarias (ADIF) shows that a running distance of 40 000 to 100 000 km and approximately two million pantograph passes are the wear limits of collector strips and contact wires, respectively, for AC railways using *carbon collector strips*. Correspondingly, the values for highly loaded DC railways with more than 2 000 A per pantograph are only 20 000 to 30 000 km and less than 100 000 pantograph passages.

Until 2010, the DB only used pantograph with pure carbon collector strips because of the low wear record of the contact wire and collector strips combined. The DB expected higher contact wire wear with metal impregnated collector strips. In the frame of a UIC study on test stands, the contact wire wear of pure and metalized carbon collector strips was measured [17.17]. In Figure 17.8, the most important results of these tests are shown and conclude that the use of pure and metalized carbon collector strips for AC and DC lines leads to the same results up to a mean current drain as far the service life of the contact wires and the collector strips are concerned. The effects on wear are virtually the same. Therefore, a suitable pantograph may be equipped with pure carbon or metalized collector strips.

Because of statistically proven different operational conditions, precise information on the *expected absolute service life* is not possible. Absolute wear values were determined in tests described in the publications [17.18] to [17.19]. The following inferences can be drawn for the given operational conditions:

– contact wires made of electrolytic copper wear faster than silver alloys and these wear faster than magnesium alloys (Figure 17.6)
– the wear rates show minima dependent upon current, tending in the direction of higher currents with increasing speed of travel (*current lubricating effect*, see also Clause 10.7)
– the wear rate increases with increasing contact force (Figure 17.6)
– at a lower contact force with arcing and increasing current density, electrical wear also increases (Figure 17.6)
– under experimental conditions, the total wear decreases with increasing running speed and use of carbon collector strips

The following conclusions can be drawn regarding the expected *contact line service life* with a permitted wear of 20 %:

– the average value for the service life of a contact wire with a nominal cross-section of $100 \, mm^2$ with a rate of wear as experienced on the Russian State Railway is approximately one million pantograph passes. With an average train headway of 10 minutes, the calculated service life is approximately 20 years.
– under test conditions, a wear rate of $8 \, mm^2/10^6$ pantograph passes can be assumed for a silver alloy contact wire as used on high-speed lines. Consequently, a service life of three million pantograph passages results. At 6 minutes head way for such a contact wire, 34 years of service life result.

Deviations from the designed height and stagger of the contact wire relative to the track are possible, resulting from changes in the overhead contact line system or the track geometry during operations. Low-bearing capacity soils, that were inadequately considered during foundation installation may lead to pole inclination. External impacts and the described wear phenomena may lead to changing contact force distributions and to geometrical changes within the overhead contact line. Experience gained in Germany with the installation of *high-duty overhead contact lines* shows that such effects can be reduced to a minimum.

Furthermore, changes to the height and alignment of the rails occur during train operations and as a result of permanent way maintenance. Regular inspections are necessary to ensure that these processes do not lead to *pantograph dewirement* or to increased wear. Figure 17.9 shows the calculation of the contact wire service life for an AC-100–Cu and AC-120–Cu contact wire. Uniformly applicable, metalized collector strips for 1 600 mm and 1 950 mm long pantographs according to TSI LOC&PAS [17.20] should reduce the number of different pantographs used on interoperable lines.

17.2.6 Insulators

The *behaviour and aging of insulators over time* is determined by their mechanical and electrical stresses. It also depends on the type of design and the materials employed. If an insulator in a cantilever failed under tensile, compression or bending loads, this could lead to damage to pantographs resulting in the tearing down of the overhead contact line along the full brak-

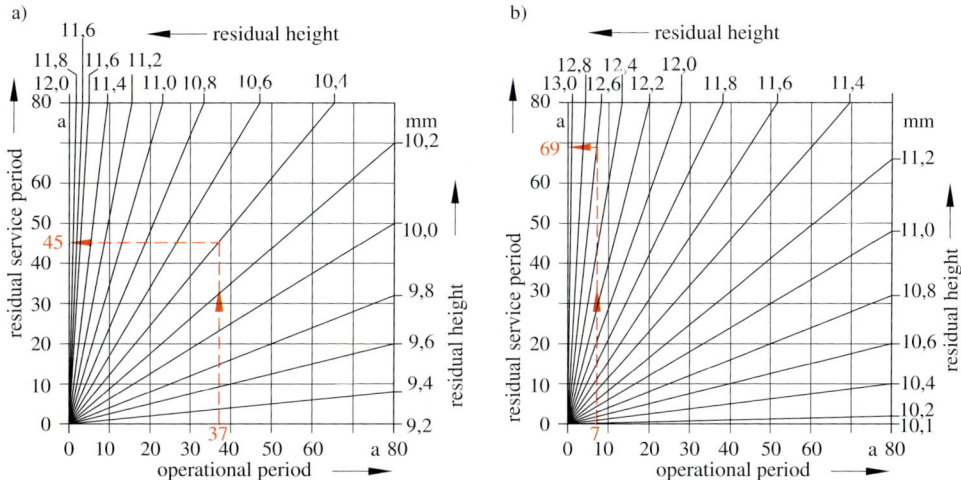

Figure 17.9: Evaluation of the residual life of contact wires.

Note: The calculation is based on the same number of pantograph passes per year and the same quality of the interaction of pantograph and contact line.

Example a): The result of a thickness measurement of the contact wire AC-100 was 10,4 mm. The previous contact wire life was 37 years. Figure 17.9 a) also shows that the remaining life to contact wire renewal will be 45 years.

Example b): The result of a thickness measurement of the contact wire AC-120 was 12,6 mm. The previous contact wire life was 7 years. From Figure 17.9 b), the remaining life to renewal of contact wire will be 69 years.

ing distance of the train. Similar effects could ensue from failures of dead-end insulators, stressed by tensile forces and oscillations.

Arcs occur at the insulators as a result of *flashovers*, e. g. caused by birds, lightning overvoltages or severe pollution. They damage glazes and polymer surfaces by forming burn tracks and partially destroy the insulator sheds leading to possible insulator fractures (Figure 17.10). Partially damaged insulators should be located with the aid of short-circuit location techniques and replaced, as they can become unstable because of the defects and moisture penetration. Erosion effects and early aging can occur on moist and polluted *plastic insulators* due to electrostatic partial discharges. The degree of pollution on contact line systems is more severe than with overhead power lines because of mixed traffic with diesel traction, the swirling up of dust and the transport of raw materials that react aggressively in the atmosphere. The pollution particles contain ion-forming materials that combine with the moisture in the atmosphere to form electrolytes. A moisture layer, composed of small droplets of dew or drizzle, is especially dangerous. The resulting *creepage currents* heat the surface and lead to increased conductivity of the electrolyte. It possesses a positive temperature coefficient and is accompanied by simultaneous surface drying. The creation of the aforementioned partial discharges and flashovers depends on these processes.

While *long-rod insulators* are puncture-proof, punctures occur more easily in *porcelain or glass cap-and-pin insulators* because of their shape. Because of the ball and socket connection between the sheds of the cap-and-pin insulators (Figure 11.63 b)), the damage caused by glass or porcelain fractures does not always lead to a collapse of the contact line and consequential damage is minimised. Creepage currents of up to 150 μA created on insulators of DC

Figure 17.10: Damaged insulator after a flash-over.

Figure 17.11: Electrical erosion caused by DC currents.

railways occurring as a result of climatic conditions and pollution leads to corrosion damage at the cap connection fittings. The reduction of the insulator pin diameter is between 0,15 and 0,6 mm per annum and requires the replacement of tunnel insulators every few years.

Porcelain insulators are widely used because of their high mechanical strength, chemical and heat resistance and favourable electrical properties. The *aluminum oxide porcelain* used today avoids the disadvantages of quartz-porcelain and achieves higher strength [17.21]. The advantages include:

- high strength and reliability
- temperature range $-50\,°C$ to $+550\,°C$
- durability under sudden temperature changes, e. g. at short circuits
- little aging effects

Temperature changes also affect the aging of the cement that connects the porcelain body to the end fittings. Manufactured from malleable cast iron, it must compensate the differing expansion properties of these materials. Sealants using Portland and sulphur cement are more severely affected than those manufactured from *lead-antimony alloy*, but they possess a greater resistance to higher temperatures than lead, e. g. during short circuits.

The *service life of modern porcelain insulators* without flashovers is estimated to be 50 to 60 years. Failure rates of glass insulators are determined by their greater sensitivity to arcing and temperature changes compared to porcelain.

Increasingly popular *plastic insulators* are especially resistant to external influences such as vandalism. The amount of wear caused by weathering and UV radiation depends on the surface material. *Silicone materials* have proven especially robust and long-living. They simultaneously display hydrophobic properties and have been in use since 1980. The aging of composite insulators particularly depends on the connection between the GRP rod, fittings and plastic covering. Unstable bonding materials and compressions used for the attachment of the sheds to glass fibre reinforced plastic rods and fittings cannot prevent the penetration of moisture into the intermediate spaces, and thus corrosion and internal flashovers occur. Tests in accordance with EN 62 217 can help to avoid such flaws.

Since 1980, the use of *glass fibre reinforced plastic cantilevers* in urban mass transit installations and in AC 25 kV 50 Hz overhead contact lines in Denmark, close to the sea, has been a positive experience. The resins used to bond the glass fibres are subject to aging caused by weathering in the form of alternating moisture and surface drying combined with UV radi-

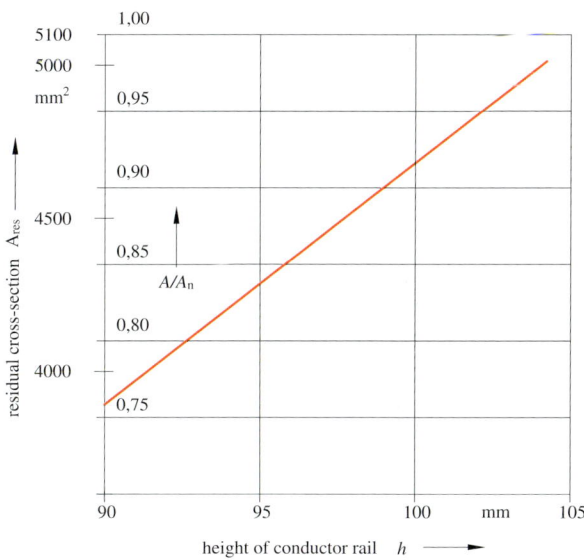

Figure 17.12: Residual cross-section of a conductor rail A-5100 made of steel, after wear. A_n nominal cross-section

ation. As a result, the resin layers erode and glass fibres become exposed. The penetration depth over a period of 50 years is estimated to be only a few tenths of a millimetre [17.22] with minimum influence on the strength. This process can be retarded, for example, by applying a synthetic fabric close to the surface with a thicker resin layer.

17.2.7 Disconnectors and section insulators

The *disconnectors on poles* and drives employed on electric railways must withstand several thousand switching operations over decades of outdoor operation. Wear and aging affect the linkages, contact surfaces, arcing horns and the lubricating and contact greases in the case of former designs. The adjustment and replacement of contact elements after frequent switching operations under load and renewal of the greases at regular intervals ensures a long service life for the equipment. The disconnectors of the Siemens type 8WL6144-1 for 15 kV and 25 kV are free of servicing, since the contact surfaces are protected by a special silver-graphite covering (see Clause 11.2.5.2).

The passage of pantographs *over section insulators* causes wear of the runners because of increased contact forces, loosening of the bolt connections and to commutation processes. The latter cause arcing, which deflects upwards by the arcing horns and then extinguishes. The material erosion occurring at the arcing entry and exit points requires regular inspections and component replacement, especially on DC railways. Section insulator runners in overhead conductor rail installations suffer high wear rates as shown by experience from Oslo and Copenhagen. Consequently, installation of section insulators in conductor rail contact lines should be avoided.

17.2.8 Conductor rail installations

Conductor rails experience wear during normal operation. For the iron conductor rails of the Berlin City railway (S-Bahn) the following *rates of wear* have been established, after which

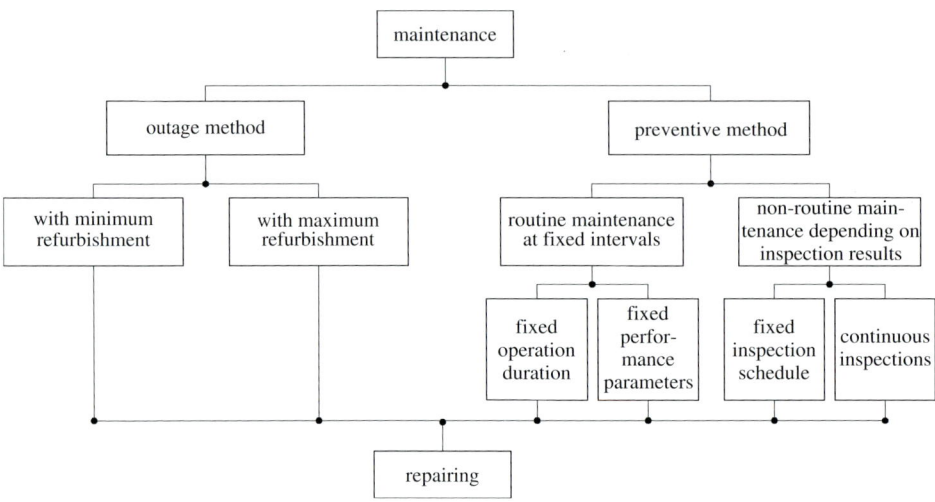

Figure 17.13: Maintenance methods.

the conductor rail should be replaced [17.23]:
 – on suburban lines 10 %
 – on city centre lines 15 %
 – on lines in depots 20 %
The wear of conductor rails can be determined and recorded by measuring the height h of the conductor rail (see Figures 11.75 and 11.76). Figure 17.12 shows the relationship between these measurements and the cross-section of the conductor rail A-5100.

17.3   Maintenance

17.3.1   Scope of maintenance

According to EN 13 306 *maintenance* includes all measures to *preserve the planned status*, to *determine and evaluate the actual state* and to *restore the planned state* of operating equipment and installations. As shown in Figure 17.13 the terms *servicing, inspection, refurbishment* and *improvement* are assigned to these steps. Servicing is not necessary on contact lines of modern designs. Maintenance, therefore, consists of inspection, repair and improvement. According to [17.24] *maintenance methods* can be classified as shown in Figure 17.13. The *outage method* by which the components are replaced only after damage occurs is unsuitable for overhead contact lines since they lack redundancy and damaged installations would have negative effects on train operations. *Preventive maintenance* based on fixed cycles ensures high availability and effective planning of staff, machinery and track closures but at high cost. Consequently numerous European railway companies have adopted *non-routine maintenance depending on inspection results*. The overhead contact line diagnoses is performed to a programed schedule that considers experience, the importance of lines and pre-condition of systems.

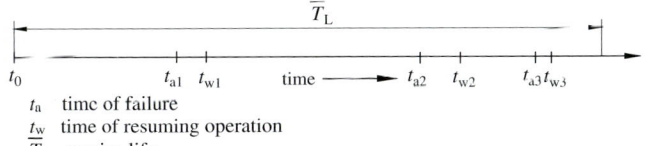

t_0 t_{a1} t_{w1} time t_{a2} t_{w2} $t_{a3}t_{w3}$

t_a time of failure
t_w time of resuming operation
$\overline{T}_L$ service life

Figure 17.14: Probability of faultless availability of overhead contact lines.

17.3.2 Third rails

Conductor rails arranged as a *third rail close to the running tracks at ground level* offer advantages for maintenance in comparison with overhead contact lines. Because of easier access, inspections are conducted more easily and with less expense. Working on the ground eliminates fall hazards for linemen. Because of the simpler design of ground level conductor rails, without foundations, poles and tensioning devices, repairs and maintenance are easier to carry out. Conductor rails can endure a higher number of pantograph passages than overhead contact lines before renewal is necessary. Third rails are less prone to damage from pantograph/shoe derailments which can lead to severe damage to overhead contact lines.

From the maintenance perspective, third rails also have disadvantages because of their close proximity to the ground and dangers posed to persons through violation of safety clearances. Because of the gap in conductor rails close to points and crossings, cost-effective designs with cables and vehicles with pickup shoes on both sides are necessary. Two or four sliding shoes on each side of vehicles are necessary to bridge the gaps within third rails. The arrangement of several sliding shoes ensures power supply even if a sliding shoe is destroyed at the designed weak point, in the case of a collision with an obstacle.

The arrangement of conductor rails close to the ground also results in pollution of the insulation. The rigid running surface of the sliding shoes on the rigid conductor rail necessitates a running speed limited to approximately 120 km/h. Snow banks and snow clearing can hinder the operation of third rail systems.

For maintenance activities within the track, reliable de-energising of the conductor rail is necessary. With *current testing boxes* and short-circuiters, the conductor rails can be tested and earthed after de-energising.

17.3.3 Reliability

An overhead contact line installation is a complex system from the aspect of *reliability* and possesses no redundancy for technical reasons. As with other operating equipment in the railway energy supply system, e. g. transformers and circuit breakers, it represents a *regenerative object*, i. e. its use does not cease at the instant of failure, but is only interrupted. It is repaired and continues its functions. This situation is illustrated graphically in Figure 17.14. The overhead contact line installation is functional during the time period t_{wi} to t_{ai+1}, but not between t_{ai} and t_{wi}.

The behaviour of the service life of individual components and the overhead contact line installation can be described as a random variable, with the exception of the contact wire. Significant parameters for the characterisation of the behaviour of the service life are the probability of failure-free work $R(t)$ and the failure rate $\lambda(t)$. The following applies:

- $R(t)$ is the probability that the time of a failure of the studied unit, e. g. 100 km of contact line, does not occur within a studied period of service t. $R(t)$ is also known as

Table 17.1: Mean time between failures (MTBF) in years between two failures of selected elements in the overhead contact lines related to 100 km contact lines.

Element	DR[1]	RZD[2]
Poles	17,6	10,5
Supports	2,8	2,4
Contact lines	1,5	1,5
Section insulators	21,5	30,4

[1] Deutsche Reichsbahn, German Railways
[2] Russian Railways

Table 17.2: Average fault duration values for the overhead contact line system at DR and RZD (in brackets) for the years 1975 to 1977.

Element	$\overline{D}$ h	$\overline{D}_{S100}$ h/100 km
Contact wire	5,3 (5,4)	5,4 (2,4)
Steady arm	3,3 (8,5)	2,1 (1,7)
Insulator	4,4 (3,2)	5,1 (1,6)
Miscellaneous	4,2 (11,0)	18,1 (5,5)
Contact line	4,2 (6,8)	30,7 (11,2)

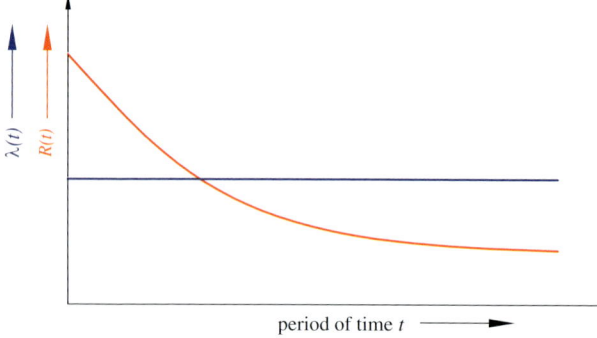

Figure 17.15: Development of probability of disturbance-free working $R(t)$ for overhead contact lines with time for a constant failure rate $\lambda(t)$.

the *survival probability*. It is calculated in practice as follows

$$R(t) \approx 1 - \frac{1}{N_0} \sum_{t=1}^{n} S_i \quad . \tag{17.1}$$

where N_0 is the starting set of studied units and S_i is the number of failures in the period i, e. g. per one year.

– $\lambda(t)$ is the probability of the failure of a studied unit within a time period of $(t, t + \Delta t)$ or failure rate (Figure 17.15), when the studied unit has already had a service life t. It is, therefore, the number of studied units that failed during a service period related to the number of studied units at the start of the period.

With the number N_t of units at the start of the time interval Δt, the *failure rate* is calculated as

$$\lambda(t) \approx (N_t - N_{t+\Delta t}) / (N_t \cdot \Delta t) \quad . \tag{17.2}$$

– since contact line installations are repaired immediately after their failure, the failure rate per 100 km of contact line or electrified tracks is determined in practice by the number of faults per year related to this contact line length.
– statistically founded statements can be made for components and various designs of overhead contact lines, on the basis of numerous evaluations of contact line failures that the respective failure rates are constant values [17.25, 17.26]. This statement does not apply to the contact wire in overhead contact line installations, which shows an increasing failure rate with increasing wear [17.25, 17.27].

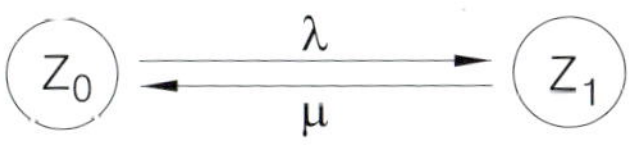

Figure 17.16: State diagram for the contact line. λ failure rate, μ correction rate

– a constant failure rate $\lambda(t) = \lambda_0$ is important for the practice. Then the relationship between probability of disturbance-free working $R(t)$ and failure rate is [17.28]

$$R(t) = e^{-\lambda_0 \cdot t} \quad , \tag{17.3}$$

whose trend can be seen in Figure 17.15.

The *causes of failure for contact lines* and their components are manifold. One detailed analysis showed that for overhead contact line installations, failure due to *design short-comings* of the overhead contact line under real operating conditions lies between three and five percent of the evaluated total failure rate for the overhead contact lines. *External impacts* on the overhead contact line predominate in the failure rates specified in Figure 17.46. These include affects from train operations such as defective pantographs, loading gauge violations, civil engineering activities, climatic influences and railway crime. The stated failure rates should, therefore, be described as *site-related failure rates*. They are significantly higher than the failure rates for the overhead contact line hardware itself and depend on location. There, the network of a railway company can be viewed as a location, for example. If the failure rate is constant, then failure behaviour does not depend upon the preceding loading history. Then, the *average functional life* $\overline{T}$, also known as the *mean time between failures* (MTBF), will be

$$\overline{T} = 1/\lambda_0 \quad . \tag{17.4}$$

The expected *remaining service life* t_T also equals $1/\lambda_0$. An expected value for the mean time between failures and the remaining service life of 100 km of contact line of 83 days is calculated for DB using the data in Figure 17.46. Observed values lie between a few hours and three years. The average mean time between failures does not permit a prediction of the next failure, however, it allows for sound planning of maintenance work.

Figure 17.16 demonstrates the transition from the normal, undisturbed condition Z_0 to the failed condition Z_1 described by the failure rate λ and back by the correction rate μ.

The *downtime* caused by the repair work includes the period between the instant of occurrence of the failure and the restart of train operations. This variable $\overline{D}$, also known as *mean time to repair*, can be calculated starting from Figure 17.14

$$\overline{D} = \frac{1}{n} \sum_{i=1}^{n} (t_{ai} - t_{wi}) \quad . \tag{17.5}$$

The mean time to repair includes the following significant time components
   – mean duration from occurrence of the non-functionality until the start of measures to repair
   – mean travel time for the repair vehicles from the depot location to the fault location and
   – mean working time for the removal of the non-functionality of the contact line.

It follows from the fault analysis that one half to two thirds of the total mean disturbance time is taken up by work time to correct the fault.

The mean time to repair is a random variable that can be described by the *normal* or *Erlang k-distribution* [17.29]. It can, as shown by the fault statistics of railway companies, be considered as a constant parameter for a location in the defined context. Measured values of the mean time to repair for components and contact lines at two railway companies are contained in Table 17.2. $\overline{D}_{100}$ is the mean time to repair per 100 km of contact line. The real duration of a fault varies between approximately five minutes and more than fifty hours. Function and organisation-dependent downtimes are not included in the mean time to repair.

The inverted value of the mean maintenance downtime is the *correction rate* $\mu(t)$, also known as the *intensity of maintenance*.

$$\mu(t) = 1/\overline{D} \tag{17.6}$$

The *reliability models* cited by Markov [17.30, 17.31] are suitable for the description of the properties or renewable systems. The system states are defined by nodes and the relationships by directional graphs with corresponding transition rates (Figure 17.16). Two states are applicable for a contact line system:

Z_0 contact line is functional

Z_1 contact line has failed

The transition between the states is given by the failure rate λ and the above mentioned correction rate μ. The state diagram can be drawn as shown in Figure 17.16. The probability that the state Z_0 is given is P_0 and for state Z_1 and P_1. The probabilities of states can be determined by the differential equations

$$\begin{pmatrix} P_0'(t) \\ P_1'(t) \end{pmatrix} = \begin{pmatrix} -\lambda & \mu \\ \lambda & -\mu \end{pmatrix} \begin{pmatrix} P_0(t) \\ P_1(t) \end{pmatrix} \quad . \tag{17.7}$$

The solution of (17.7) yields the probability that the system is in working order

$$P_0(t) = \frac{\mu}{\mu + \lambda} + \frac{\lambda}{(\mu + \lambda) \cdot e^{(\lambda + \mu)t}} = A(t) \quad . \tag{17.8}$$

In (17.8) the *availability* $A(t)$ expresses the probability that the contact line can completely fulfil its tasks under defined conditions at a certain point in time t.

The *constant availability* A_D is sufficiently accurate to characterise the failure behaviour for railway energy supply systems. The constant *long-term availability* of contact lines in electric railways is achieved approximately 24 h after commissioning. After that, it applies

$$A_D = \mu/(\mu + \lambda) = \frac{\overline{T}}{\overline{T} + \overline{D}} \quad . \tag{17.9}$$

as results from Equations (17.4), (17.5) or (17.8).

Example 17.1:   Calculate the availability A_D under the following conditions:

Two failures per 100 km and year, failure rate $\lambda = 2 \ (100 \ \mathrm{km \cdot a})^{-1}$

Mean time to repair 10 h

Correction rate $\mu = (1/\overline{D}) \cdot (8\,760) = (8\,760/10) = 876 \cdot (100\,\mathrm{km \cdot a})^{-1}$.

The availability follows from (17.9)

$$A_D = 876/(876 + 2) = 876/878 = 0{,}99772 \quad .$$

This would mean that the 100 km long overhead line is not available for the $1 - 0{,}99772 = 0{,}00228$ part of one year or 20 hours per year.

The *failure rates* for overhead contact line systems on main line railways vary between 1 to 5 per 100 km and year [17.26] and reach values higher than 50/(100 km·a) for tramways [17.32]. At DB, the average failure rate of contact lines was observed as 4,55/(100 km·a) and the correction rate as 2738/(100 km·a). This value corresponds to a mean time to repair of 3,3 h. Using the failure and correction rates unavailability of 0,9983 is obtained for contact lines. An availability of 0,9983 means that the 100 km long contact line will not be available on 14,9 h per year.

The number of mainline tracks and the distances between stations or crossing points influences the *effects of the non-availability* of the contact line *on train operations*. The availability of the overhead contact line quoted in the example for a 100 km single track line directly affects the *availability of train operations* on the whole line. On multi-track lines, electric train operations can continue operating with restrictions by travelling around the obstacle in the affected section on other line sections without restrictions. Availability of overhead contact lines over 100 km in length only affects the availability of electric train operations under consideration of the line routing. Grave effects follow the failure of cross-span structures crossing all parallel tracks.

The availability can be increased by:

- *high-quality overhead contact lines* that operate for long periods without faults and maintenance
- *maintenance-friendly components* whose state can be easily diagnosed and allow rapid repair or simple replacement
- *reduction of repair times* is achieved by fast fault localisation and swift arrival of repair staff with suitable repair material and
- properly trained and experienced linesmen who also contribute to a high overhead contact line availability

For more information on RAMS requirements see also EN 50 126 which deals with requirements on reliability, availability, maintainability and safety of railway energy supply installations and as well CLC/TS 562 which specifies measures and verification recording for the reliability of traction energy supply systems.

17.3.4 Diagnostics

17.3.4.1 Basics

Contact line diagnostics according to [17.24] is understood to determine and analyse the state of a contact line system on the basis of measurable or externally recognisable properties, as far as possible without significantly influencing train operations. Its objective is to reduce the cost of necessary maintenance work and to perform this at the correct time, while making full use of the remaining service life of the equipment and with minimum impact on train operations. Diagnostics form the basis for the transition from rigid maintenance cycles to *condition-related maintenance*. The subject of overhead contact line diagnostics is not a locally restricted, compact system, but a wide-spread energised installation that is inaccessible without aids.

17.3.4.2 Inspection plan at DB

As an example, DB carries out inspections and tests corresponding to the time periods stated in Table 17.3 for the listed categories of lines during *preventive maintenance* on a fixed schedule. The first order category contains the overhead contact lines on long-distance main lines,

Table 17.3: Schedule for the inspection periods in months of overhead contact lines at DB according to Directive 997.0140.

Type of inspection	Abbreviation	Contact lines of 1st order	Contact lines of 2nd order
Check of conditions	Z1	6	24
	Z2	24	24
	Z3	12	12
	Z4	72	72
Functional test	F1	6	12
	F2	12	12
	F3	24	24
	F4	24	24
	F5	84–12	84–12
	F6[1]	24[2]	–
		6[3]	–
	F7	when required	when required

[1] only for main tracks, [2] for $v \leq 160$ km/h, for [3] $v > 160$ km/h

long-distance branch lines and city lines, intersecting contact lines and head-span structures and specially defined overhead contact lines, such as older types of design and especially endangered systems. All other overhead contact lines belong to the second order category.

Condition checks serve to determine and assess the actual state of various overhead contact line components with the help of binoculars or simple measuring devices during line inspections on foot, without the necessity for switching measures or operational closures.

Condition check Z1 includes overhead contact lines, supports and tensioning devices. Condition check Z2 includes all other contact line elements, such as feeder and other lines, cable termination seals, disconnectors, foundations, poles, head-span structures, railway earths, local control devices, EL signals, warning signs and gauge infringement. Damaged components and connectors, pollution, corrosion, the temperature dependent position of the contact lines etc. are recorded.

Condition check Z3 includes inspection of catenaries, supports, tensioning devices, head spans, foundations, poles and earthing devices of out-of-service installations.

Functional tests include the function of the overhead contact line/pantograph system with the aid of inspection or measurement vehicles. They include:

- Test F1: Determination of the *position of intersecting contact wires* at high-speed, applying a contact force $F_{stat} = 150$ N
- Test F2: Determination of the *contact wire stagger* position, the inclination of the registration arms and steady arms and the position of clamps to avoid pantograph strikes against clamps with $v \leq 40$ km/h and $F_{stat} = 150$ N
- Test F3: Determination of the *contact wire height* at critical positions, e. g. at contact wire height reductions below 5,10 m and at railway crossings with contact wire heights below 5,75 m in still air
- Test F4: Review of *minimum electrical clearances* between the overhead contact line and bridges and tunnels after disconnection and earthing with $F_{stat} = 250$ N
- Test F5: Visual *examination of the contact wire* over its entire length and measurement of the contact wire thickness at the locations along a tensioning section suspected of having the greatest wear. Determination of the sequence for check measurements depending upon the number of pantograph passes and wear

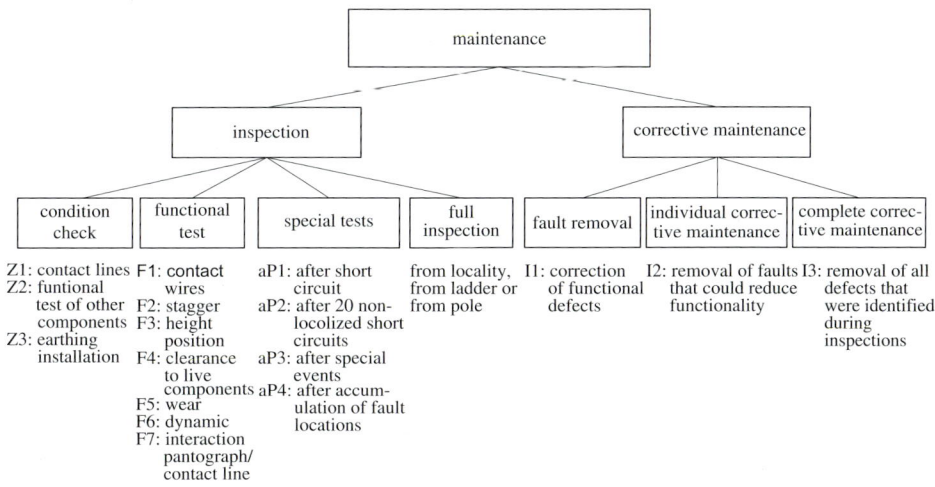

Figure 17.17: Organisation of the overhead contact line maintenance at DB.

 - Test F6: *Testing of* the *dynamic behaviour* of the overhead contact line system at the line speed and energised overhead line using the measurement car
 - Test F7: Observe the *passage of a pantograph* after reconstruction or repair of the overhead contact line

Special condition and functional checks additional to the planned diagnostic measurements are known as extraordinary checks:

 - Test aP1: Determination of the *position of a short circuit* and examination of the overhead contact line within a narrow range of neighbouring supports and railway earths in the short-circuit path by inspection on foot
 - Test aP2: *Examination of the de-energised overhead contact line* with vehicles or ladders after 20 non-localised short-circuits
 - Test aP3: Inspection of the line on foot or by vehicle after special events, such as storms, extreme temperatures, icing, etc.
 - Test aP4: Determination of the *contact wire position in still air*, if accumulated contact force peaks and higher vertical accelerations of the pantograph were experienced during functional test F6

The complete check includes a comprehensive visual inspection and measurement of contact wire wear from vehicles or ladders. It should be carried out after special events or depending on train frequency, at intervals staggered over 48 months for a very high frequency or up to 10 years for low frequency, including checks of condition Z1 and Z2. An overview of all inspections and maintenance performed at DB is shown in Figure 17.17.

The revised planning documents and results of the inspections, such as *contact wire wear*, contact wire stagger and contact forces, are elements of the operational handbook.

The handbook also contains *master cards* with all characteristic data and their modifications, as operating sheets with inspection results and damage record, defects and repairs. *Maintenance overviews* serve to review inspection data and plan rectification work. They also include fault positions and short-circuit locations. The compilation of data considers the urgency of corrective maintenance works. For instance, if the indication device for contact wire stagger indicates in the range ≥ 750 mm during functional test F1, then the adjustment errors need to

Table 17.4: Assignment of contact wire thickness to categories of wear.

Category	%	Contact wire thickness in mm		
		AC 100	AC 120	AC 150
I	100–94	12,0–11,0	13,2–12,0	14,5–13,3
II	95–77	10,9–10,2	11,9–11,0	13,2–12,3
III	78–80	10,1– 9,2	10,9–10,0	12,2–11,1

a) Measurement of contact wire position

b) Measurement device on rails

Figure 17.18: Laser measurement.

be corrected immediately and as soon as possible, if in the range ≥ 550 mm. The measured contact wire thickness is classified into three categories in accordance with Table 17.4.

The correction of contact wire position is considered urgent at spots with low contact wire wear because maximum advantage can be gained by correcting the contact wire position, there. For category III wire, which is no longer economical to repair, short-interval inspections are preferred. e. g. within three months to ensure that the contact wire thickness does not fall below the limit at damaged locations. After the limit is reached DB replaces the contact wire over the whole tensioning section.

Thermovision is another diagnostic tool for current-carrying parts of energy supply systems. It uses an infrared camera to visualise the increased temperature of faulty components, such as poor electrical connections and reduced cross-sections of wires and cables carrying electrical current, compared to intact components [17.32]. The precondition for use of this technology is the availability of defined constant currents, which can only be achieved by feeding the line in absence of train operations (see Clause 17.3.5.12).

17.3.5   Measuring and diagnostic equipment

17.3.5.1   Contact wire geometrical position measurements

The position of the contact wire determines the quality of the interaction between pantographs and the contact line and, as a consequence, the wear of the contact wire and pantograph. Measurements of the contact wire position should be carried out at regular intervals after instal-

a) Measurement principle

b) Application of device

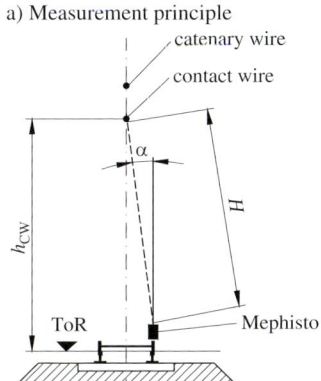

Figure 17.19: Geometrical position measurement of contact wire using *Mephisto*.
h_{CW} contact wire height, H distance between measurement device and contact wire, α angle between perpendicular and distance line between measurement device and contact wire

lation and during the maintenance period (see Table 17.3). The inspection intervals should be separately specified for inspection of overhead contact lines above rigid superstructures (fixed track) and for overhead contact lines above ballasted superstructures.

DB inspects the stagger twice a year and the contact wire height every second year. There is no differentiation between conventional lines, high-speed lines or rigid and ballasted superstructures.

For supervisory measurements, contactless procedures, that do not affect the contact wire position during the measurement are preferable. When using a laser as in Figure 17.18, an insulated metal bar is laid across the rail heads. The laser measuring device records the contact wire height between the beam on the rail heads and the lower edge of the contact wire [17.33, 17.34] with the measuring device arranged vertically on the bar. The position of the measuring device on the bar relative to the track centre line represents the horizontal position of the contact wire. The lateral position of the contact wire at the support corresponds to the stagger. The measurement procedure described above can only be used over short sections such as in areas of points or overlaps, because of the slow measuring progress.

Using the mobile measuring device *Mephisto*, up to 2 km long sections can be traversed and measured in one hour. A laser measuring device is arranged on a mobile frame and records the contact wire position by triangulation (Figures 17.19 a) and 17.19 b)) [17.35].

The use of ultrasonic measuring principles is an alternative to laser technology. The ultrasonic measuring device is arranged on a frame mounted on a mobile track trolley (Figure 17.20). With the mobile measuring device OHL Wizard [17.36] an individual record resolution of 50 mm can be achieved at a measuring speed up to 10 km/h. However, a device to compensate for vehicle sway during measurement is required in these cases.

Laser measuring devices [17.37, 17.38] installed on an inspection train are suitable for speeds up to 400 km/h. In Figure 17.21, infrared laser measurement equipment is shown for measuring the *contact wire position* using avalanche photo diodes. Because of the 400 Hz scanning frequency and the use of a polygon mirror, the effect of the train speed on the measurement results can be disregarded. The measuring device takes a set of measurements at a distance of 0,45 m at 350 km/h running speed. At a range of 10 m, the contact wire height and contact wire stagger is determined from the measured angle and the distance to the contact wire.

Figure 17.20: Measurement of contact wire position using ultrasonics.
Note: OHL Wizard is the yellow equipment above the trolley frame.

a) IPM measurement device

b) DB measurement device

Figure 17.21: Contact wire geometrical position measurement devices [17.37, 17.38].

The sway of the vehicle body is considered when evaluating the contact wire position. The measuring device is harmless to human eyes at distances of more than 2,5 m. The laser is automatically switched off if the running speed falls below a given limit or if the polygon mirror comes to a standstill (Figure 17.21 a)). The measuring device is able to determine the position of up to four contact wires relative to the track centreline and inspection runs can be integrated into the normal schedule without delaying other trains.

Figure 17.21 b) shows a contact wire position measuring device according to [17.38]. This laser measuring device records the *contact wire height*, the *contact wire lateral position* and the *contact wire thickness* without touching the contact wire. The contact wire position and thickness are recorded by triangulation using four high-resolution line scanning cameras with spotlights for lighting mounted on a vibration-proof beam. During daylight, the contact wire is recognized as a dark object in front of a clear sky and in tunnels, using spot lights to light the contact wire, as a light object against a dark background. The measuring device uses spotlights and lasers as light sources. The measuring interval, defined as the spacing between two measurements, depends on the required processing time and, therefore, on the

Figure 17.22: Equipment on the diagnostic vehicle [17.39].

running speed. The interval is approximately 100 mm at 120 km/h. There is no system-related limit speed for the utilization of the measuring and processing system. The measuring device was tested up to 350 km/h. Measurements are possible, irrespective of the environmental conditions, with the exception of heavy snow fall.

17.3.5.2 Optical contactless inspection

Increased utilization of the capacity and headway of railway lines reduces the period available for carrying out conventional visual inspections of overhead contact lines. Consequently, contactless inspections at speeds up to 160 km/h are required with subsequent records processing. This enables digital inspections during short breaks between trains. The inspection device [17.39] records the contact wire height, the contact wire lateral position, the contact wire thickness and the geometry of the contact line and its transverse supporting structures (Figure 17.22). By processing the recorded data, shortcomings can be identified, their tendency concerning further development can be forecast and repair activities planned.

Recording of contact wire position

Laser systems record without contact the height and lateral position of up to four contact wires. The swaying of the measuring car on straight line sections and in curves is eliminated from the records by a compensation unit. The recording of the contact wire position can be carried out either with a dropped pantograph or with a pantograph lifted to the contact line. The static uplift force of the lifted pantograph is 100 to 120 N. In Figure 17.23 a record of a contact wire position measurement is shown together with error codes.

Measurement of the contact wire thickness

Contact wire wear can be determined from the width of the contact surface and additionally from the residual thickness of the contact wire (Figure 17.22). Two sensors record the contact wire profile and the contact surface of the contact wire. The evaluation results in the residual diameter of the contact wire at spacings of 20 mm with an accuracy of 0,25 mm at 120 km/h. Sections with increased wear can be determined online (Figure 17.23).

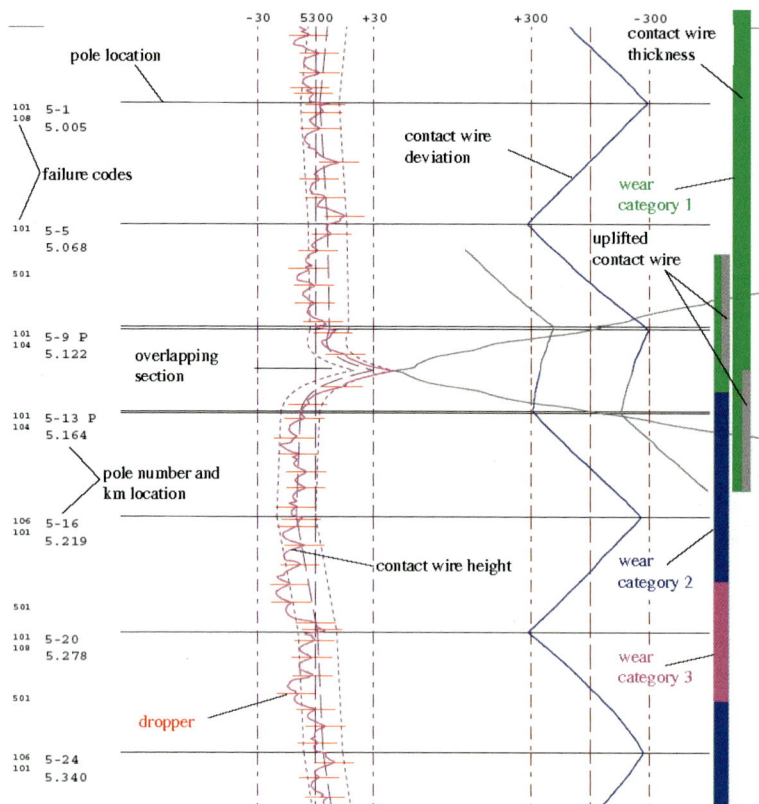

Figure 17.23: Record of contact wire position measurement and contact wire thickness measurement [17.39].

Determination of the pole position

To check the longitudinal spans of existing overhead contact line installations the measuring car uses a laser-based recording system which records the pole positions (Figure 17.22). The pole position recording system consists of two locating units installed on the left and right sides of the roof of the measurement car in the vicinity of the bogies. Each locating unit consists of two laser distance sensors. The measuring ray of the distance sensors is directed vertically upwards. Where both laser rays of a locating unit are reflected simultaneously, it is assumed that the object is a steady arm and, therefore, a contact line support.

Inspection of the contact line

Two high-speed diode line cameras (Figure 17.24) take continuous pictures of the contact line from underneath. Infrared laser lighting provides sufficient lighting to enable inspection runs during daylight and at night with diminished visibility. An infinitely long recording of the contact line from both sides is produced. The recording is referenced to the longitudinal track code. These high-resolution pictures relieve the need for further processing of the pictures since a visual inspection at the computer is possible (Figure 17.24 a)). In the next processing step, picture screening software carries out a comparison with the first recording of the contact line during the acceptance procedure and marks geometric deviations with a red frame (Figure 17.24 b)).

a) Without evaluation   b) With evaluation

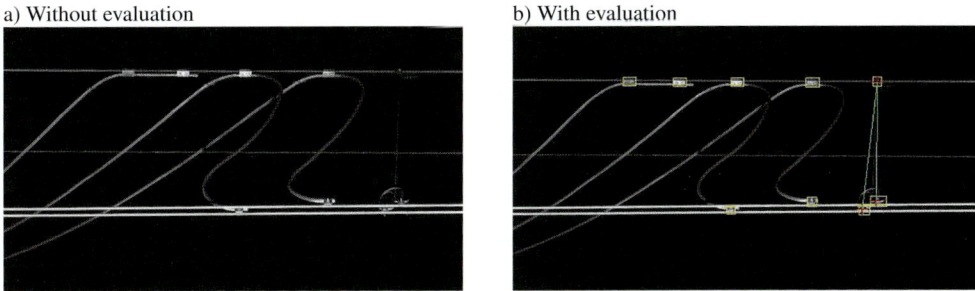

Figure 17.24: High-definition picture of overhead contact line. green frame: geometry corresponds to the geometry after acceptance test, red frame: geometry does not corresponds to the geometry after acceptance test

a) Without evaluation   b) With evaluation

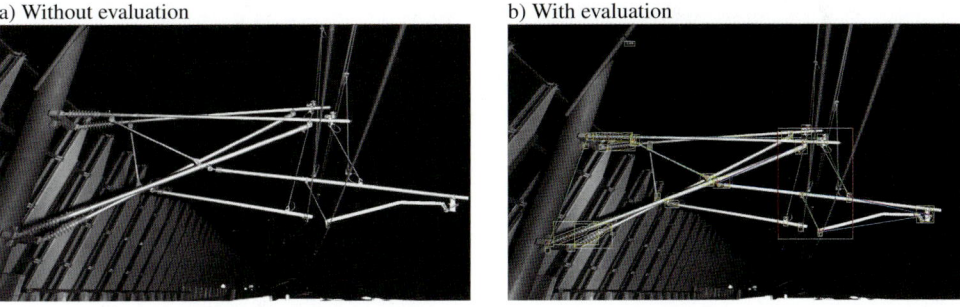

Figure 17.25: High-definition picture of a cantilever. green frame: geometry corresponds to the geometry after acceptance test, red frame: geometry does not corresponds to the geometry after acceptance test

Inspection of transverse line supports

Two high-speed line scanning cameras arranged at the heads of the vehicle record the transverse support installations in front and behind (Figure 17.25). The high-resolution photos enable a preliminary evaluation directly after the inspection run (Figure 17.25 a)). During the following processing step, photo recognition software carries out a comparison with the first records taken after the acceptance of the overhead contact line and marks geometric differences with a red frame (Figure 17.25 b)).

17.3.5.3   Measurement of pole and steady arm inclination

During installation and maintenance inclination testing is required to check the correct position of contact line poles, registration arms and steady arms at the cantilevers. The inclination is recorded in relation to the horizontal or the vertical line. A measuring instrument as shown in Figure 17.26 can be used for this purpose. With this measurement, for example, the inclination of the steady arm can be determined as shown in Figure 17.26.

17.3.5.4   Measurement of the layer thickness of corrosion protection coatings

Corrosion protection systems avoid the loss of metal through rust as in the case of steel poles and DB requires that corrosion protection is applied. After the application of corrosion protection, verification is required that sufficient thickness of cover was applied. To measure the layer thickness, measuring devices are available to determine the thickness of:

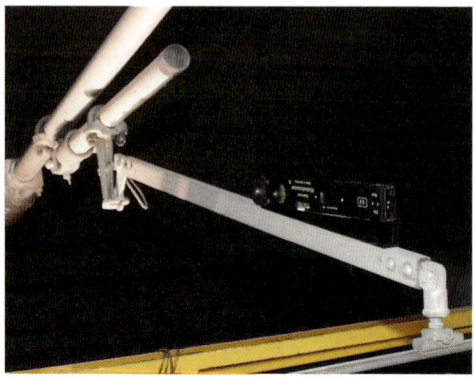

Figure 17.26: Measurement device of inclination of the steady arm.

a) Tool for testing galvanic layers

b) Tool for testing paint layers

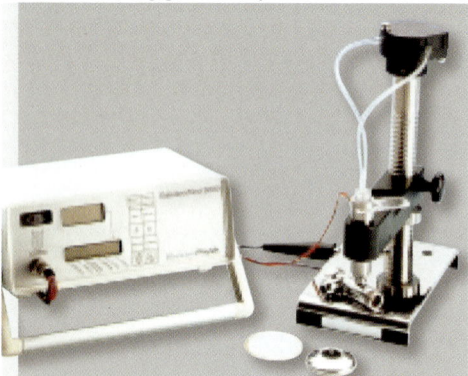 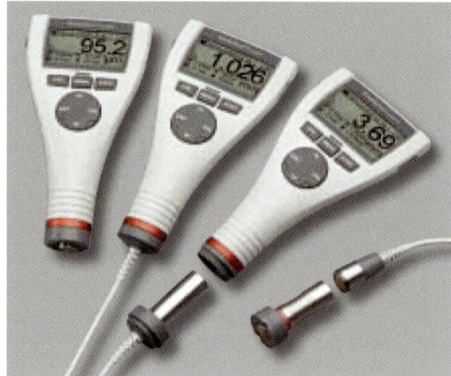

Figure 17.27: Devices to measure the thickness of corrosion protection layers [17.40], [17.41].

- – paints, finishes and other non-ferromagnetic layers on steel
- – insulating layers on non-ferrous metals
- – galvanic layers on conducting ground and
- – multilayers on insulating ground

Figure 17.27 a) shows a device for measuring the thicknesses of galvanically applied layers with a minimum thickness of 0,05 μ or measuring the thickness of galvanic single- or multilayers [17.40]. Figure 17.27 b) shows a device for measuring the thickness of paints applied on galvanized zinc layers [17.41].

17.3.5.5 Measurement of conductor tensile forces

The tensile forces of conductors can be checked using tension measuring gauges, (Figures 17.28 a) to d)) without damaging the conductor. Gauges are useful during installation, acceptance and maintenance of overhead contact line installations. The device shown in Figure 17.28 a) [17.42] is designed to measure forces between 0 N and 10 kN, e. g. at the installed stitch wire. Figures 17.28 b) [17.43], 17.28 c) [17.44] and 17.28 d) [17.45] depict devices similar to that shown in Figure 17.28 a). With the device shown in Figure 17.28 d) [17.45] the type of conductor and the range of tensile forces are pre-selected on the display. After the device has been applied to the conductor and the adjustable screw in the middle of the device has been tightened, the tensile force of the conductor can be read on the display.

a) Device type Honigmann   b) Device type PEWA   c) Device type FK Tensio Sauter

d) Device type Pfeifer

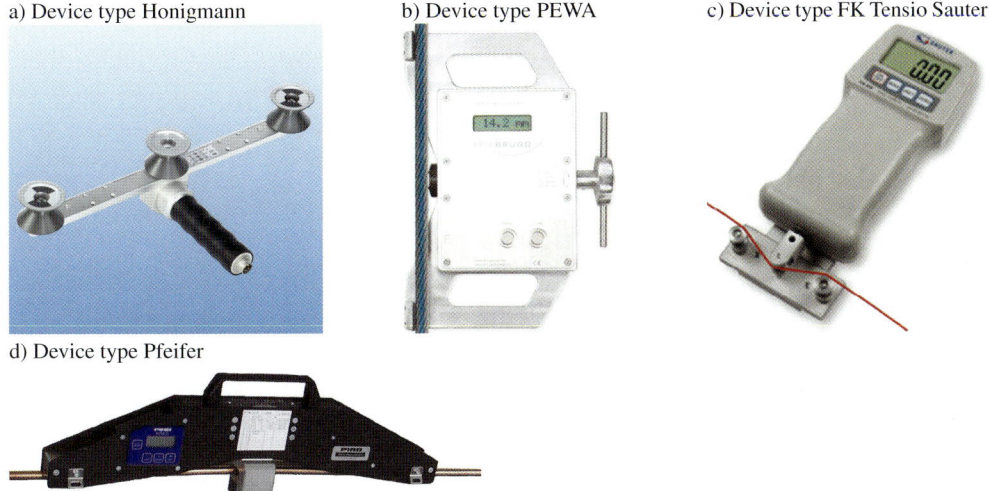

Figure 17.28: Devices for measuring conductor tensile forces [17.42, 17.43, 17.44, 17.45].

a) Measuring device type Vetter   b) Measuring device type Grube

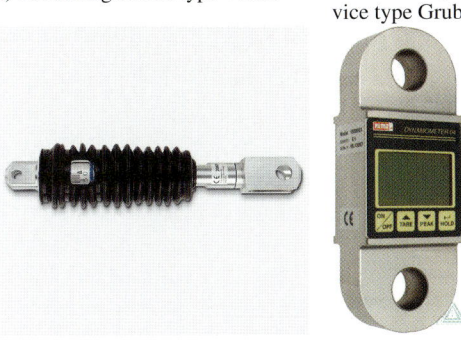

Figure 17.29: Conductor dynamometer with mechanical and digital display [17.47, 17.48].

Contact force measuring devices, also called dynamometers, cannot be placed onto the conductors like the devices shown in Figures 17.28 a) to d). They need to be inserted in series with conductors to be measured e. g. into the stitch wire. One end of the stitch wire is fixed to the catenary wire by a double U-clamp and the other end, at which the tensile force is to be measured, with a dynamometer, is fixed to a come-along (Figure 17.29 a) and b) [17.47, 17.48]). By using the come-along, the tensile force in the stitch wire is increased until the planned tensile force can be read at the dynamometer. Then the U-clamp is tightened and the stitch wire is fixed on both ends to the catenary wire with double U-clamps.

17.3.5.6   Uplift measurement

Supervision of contact wire uplift during operations is important for timely recognition of irregularities in the interaction between the pantograph and overhead contact line. Jernbaneverket in Norway records the uplift by permanently installed measuring equipment as shown in Figure 17.30 a). Figure17.30 b) shows the normal pattern of uplift during the passing of a pantograph [17.49].

a) Uplift measuring device (Photo: T. Sørensen)   b) Measured uplift

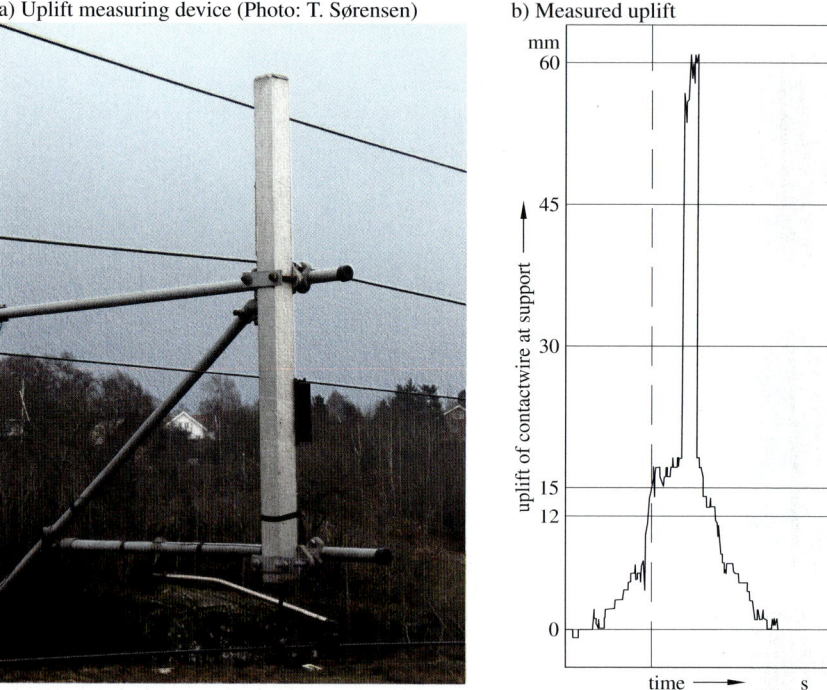

Figure 17.30: Measuring equipment for contact wire uplift.

The *pantograph monitoring system Sicat PMS* enables the measurement of contact wire uplift at supports (Bild 17.31). An insulated electric measuring sensor is placed between the contact wire and catenary wire and records the vertical variation in the contact wire height. An optical measuring sensor is fixed to the registration arm and records the lift of the steady arm. These measurements are used to prepare reports on the contact force of the pantograph, the quality of interactions at the contact line and the condition of pantographs. Unmaintained or defective pantographs on traction units are recognised by the control centre who direct defective traction units to the workshops.

The uplift measuring equipment of DB is installed temporarily at cantilevers. An insulated conductor fixed to the contact wire is carried over pulleys to a potentiometer. The rotation angle of the potentiometer is translated into electrical signals that correspond to the magnitude of the uplift (Clause 10.4.6.1 [17.50]).

During trial runs of the overhead contact line, measurements taken of the uplift simultaneously with measurements of the contact forces should confirm simulation predictions. The measured uplift may not exceed the simulated values (Clause 10.4.6.2).

Measurements of the uplift at the support are possible at the first and last dropper and in the middle of the span according to the requirements of the TSI Energy by using mobile measuring equipment. The measurements require four cameras - two of them record the contact wire uplift close to the support. The third camera records the contact wire uplift at midspan and the fourth camera is directed at the approaching train and measures the train's running speed (Figure 17.32) [17.51]. With this measuring equipment, which is easy to install outside the danger zone close to the tracks, testing of the contact points is possible during acceptance runs for overhead contact lines and during conformity tests according to TSI.

Figure 17.31: Siemens measuring device PMS for contact wire uplift.

a) Camcorder for uplift measurement

b) Uplift of the contact wire as a split picture

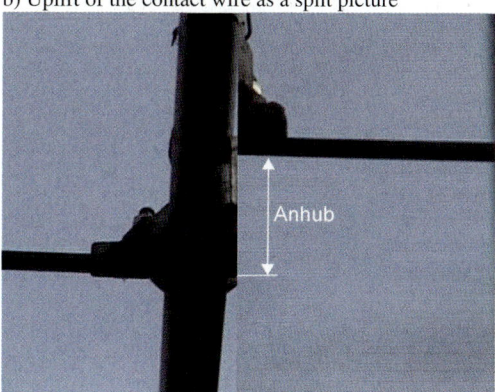

Figure 17.32: Mobile measurement equipment for contact wire uplift [17.51].

17.3.5.7   Contact line monitoring

External impacts can cause damage to overhead contact lines and interruptions to operations, making reliable and fast detection of faults advantageous. The contact line monitoring device (Sicat CMS) [17.52] aims at permanent monitoring of contact lines and immediate triggering of alarms when there are irregularities.

In the past, such events were only detected by the protection technology within the substations if the contact line protection detected that the established threshold values of current and voltage or the derived parameters had been exceeded or fallen below [17.53].

The Sicat CMS device monitors the inclination of the tensioning wheel lever, relative to the horizontal, which is typically 18° to 20° in the initial position at commissioning. The change of force in a tensioned conductor results in a change of inclination of the lever (Figure 17.33). The change of position of the lever allows operators to draw conclusions on disturbing events

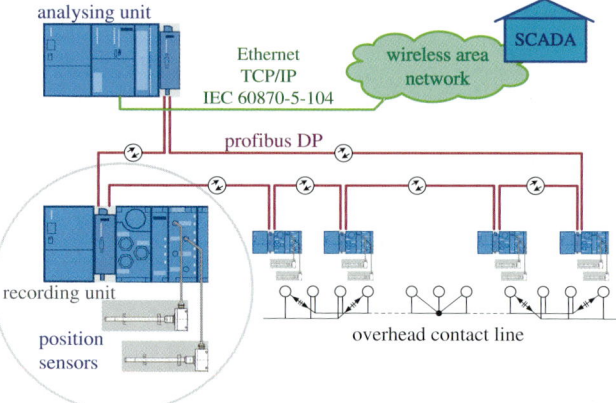

Figure 17.33: Arrangement of the   **Figure 17.34:** Structure of a Sicat CMS installation.
sensor at a tensioning wheel lever.

Figure 17.35: Contact line monitoring device at
the high-speed line HSL Zuid, Netherlands.

within a tensioning section, e. g. a fallen tree in the overhead contact line or variations of
the tensile forces after theft of conductors [17.52]. The degree of efficiency of the tensioning
device can be determined from the detected inclination of the lever and the horizontal tensile
force deviations.

In the case of the CMS device, the position of the tensioning wheel lever is monitored, as
shown in Figure 17.33, by a magneto-strictive displacement sensor. This sensor functions
without contact and, therefore, without wear. The evaluation unit analyses the recorded values of all sensors which are connected to the remote terminal unit (RTU). The RTUs are
connected to the evaluation units by profibus, profinet or mobile radio. Mistakes, warnings
and information on maintenance are transmitted from the evaluation unit by the SCADA
communication network of the operator to the operations control centre (Figure 17.34).

Uses for Sicat CMS system include:

- obtaining information on the strain that components are subjected to during operation
- to observe component aging and to establish models for aging and tolerance limits for
  condition monitoring

a) Measuring equipment   b) Laser and anemometer

Figure 17.36: Measuring the contact wire deflection due to wind in Norway (Photo: T. Petersen).

– to investigate the effects of climatic variables on the position of the overhead contact
   line
– to check stipulated tolerances of ratings

The Sicat CMS device and additional measuring sensors can be used to gather and evaluate:

– tensile forces in the catenary and contact wire using contact force sensors and, consequently, the decrease of tensile forces in certain tensioning sections
– efficiency of the tensioning device
– mean temperature and wear depending on the change of length of the conductors using
   the height movement of the weight stack of the tensioning device
– change of the lateral position of the last cantilever and supervision of the lateral movement of contact and catenary wire
– current loading and simultaneous measurement of climatic environmental conditions
– climatic environmental conditions like air temperature, humidity, wind velocity and
   direction and solar radiation

CMS devices are installed in the tunnels of the high-speed line Segovia–Valladolid in Spain
and on lines in the open of the HSL Zuid line in the Netherlands (Figure 17.35).

17.3.5.8   Contact wire wind deflection measurements

The contact lines of the railway operator Jernbaneverket in Norway are prone to high wind
velocities caused by their exposed position. Until 2012, the determination of the longitudinal
span and the stagger at the supports was based only on the past assumptions for calculations
as shown in Chapters 2 and 4. To check the accuracy of the calculation in context with the

a) Recorder   b) Badly worn rear pantograph carbon collector strip

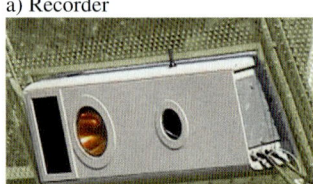
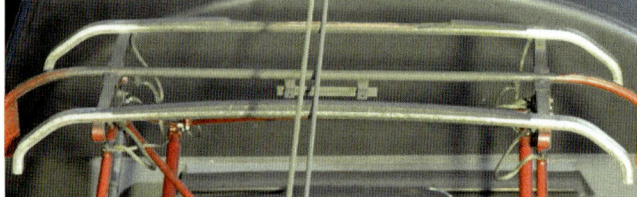

Figure 17.37: Pantograph monitoring system.

extension of the Norwegian railway network within the next few years, the contact wire deviation caused by wind was recorded. The measuring device shown in Figures 17.36 a) and
b) records the deviation of the contact wire caused by the wind on the line Kristiansand–
Stavanger, depending on the recorded wind velocity, wind direction and temperature. This
line runs in open terrain and is prone to extremely high ocean winds.

The measurement of the contact wire movement is carried out by laser sensors (see Clause
4.8 and Figure 17.36 a) and b). The frequency of measurements is determined by the wind
velocity. A high wind velocity results in a high measuring rate. The measured signals are
transferred by the GSM network (GPRS) to a data bank. The results of these measurements
over five years confirm agreement with the calculations according to Chapters 4.8 and 4.9.
The measuring installation to record wind velocities and wind direction is also used to determine if running speeds should be reduced on lines equipped with overhead contact lines,
if extreme wind velocities occur, or even to stop traffic during extraordinary wind velocities.
The aim is to avoid damage to the overhead contact line caused be de-wiring of the pantograph from the contact wire.

17.3.5.9 Pantograph monitoring system

Defective pantographs may create unusual wear and even contact line damage. Pantograph
monitoring systems should detect defects in the pantographs or unusual wear on the collector
strips. Monitoring pantograph carbon strips using the method described in [17.54] enables
detection without interrupting traffic. The radar sensor first detects the radar echo from the
current collector when the approaching train is approximately 100 m away. As the train moves
towards the sensor, the echo signal is traced until the pantograph is seven meters in front of the
camera. At this point, the image quality is optimal and a picture is taken. The flash is equipped
with a red filter to protect the train driver from being blinded by the flash, (Figure 17.37 a)).
After the camera captures the image of the pantograph, the digital image is transferred to
the master controller and then to the image processing unit which separates and checks the
images of the carbon strips. If the check shows that the carbon strip has been damaged or
worn (Figure 17.37 b)), an alarm is transmitted via Ethernet or a GSMR/GPRS modem.

17.3.5.10 Monitoring the rail potential in DC traction systems

With Radio Frequency Identification technology (RFID) rail potentials can be measured without any physical contact (Figure 17.38 a) and b)). The potential can be checked during normal
train operations and provides the first step towards an integral monitoring system for contact
line systems. Sicat Rail Potential Measurement system (RMS) [17.55] is based on the prin-

a) Monitoring principle   b) Sensor tag on the track

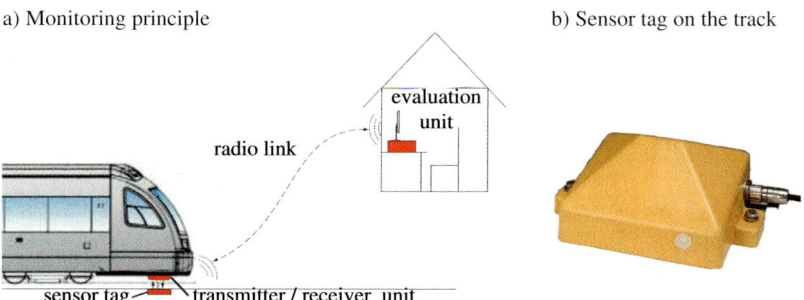

Figure 17.38: Rail potential monitoring by Sicat RMS.

ciple of the Siemens surface acoustic wave identification system *Sofis*, which has already proven its reliability in railway operations. Sicat RMS consists of
- a sensor tag, permanently installed at the track,
- a transmitter and receiver unit on a vehicle and
- a stationary evaluation unit.

These three units communicate wirelessly. The kind of identification technology used ensures that sensor tag and transmitter and receiver unit exchange data for only a short period. This enables reliable transmission of measured values even at high running speeds.

Sensor tag

In DC traction systems there is always a potential difference between earth and rail because of train operations and associated voltage differences along the line. Because of this and on account of the simplified communication between transmitter and receiver unit, the sensor tag is mounted in the track area, preferably on the sleeper. When the train approaches, the rail to earth potential is measured and converted to a specific frequency. This gives rise to an additional frequency-dependent reflection, which is scanned and evaluated by the transmitter and receiver unit. The sensor tag does not need a separate energy supply because the voltage being measured is also the supply voltage. Should the sensor tag report no rail potential, this means, there is a rail to earth connection in the affected track area, which in turn means there is a risk of stray-currents in a DC railway installation.

Transmitter and receiver unit

The transmitter and receiver unit consists of a scanner, interface converter, power supply, data logger with integrated GSM radio link and an antenna. The transmitter/receiver unit is mounted in or on a rail vehicle and transmits a radio signal in the gigahertz frequency range, through an antenna, which points towards the tag, to receive and evaluate signal reflections from the tag.

Evaluation unit

A PC, acting as a stationary evaluation unit, communicates at set intervals with the transmitter and receiver unit on the vehicle with a customized evaluation software package. The identification of an individual sensor tag in conjunction with associated measurements enables a rail-earth connection to be easily located from a central control centre. The evaluation software can also be linked to an installed maintenance program.

Typical application

Voltage limiting devices are installed at stationary railway installations along the track. They protect the systems by continuously connecting affected parts to the return circuit and by

Figure 17.39: Temperature measurement device using a laser.

triggering the protection system in the substation when necessary. After being punctured, the voltage limiting devices need to be replaced as soon as possible. If they are not replaced quickly there is a risk of conducting DC currents and damaging earthed structures by stray-current corrosion. The use of RFID technology in train operations enables punctured voltage limiters to be identified rapidly, by scanning from rail vehicles with prompt evaluation. No rail potential measured over a defined period indicates a punctured voltage limiting device. In the past, tracks had to be walked by maintenance staff at regular intervals to check for failed voltage limiting devices. Regular inspections for checking voltage limiters which were carried during night-time without train services are superfluous. Individual devices are only replaced as required [17.55].

The device FCS 601 according to [17.56] is another system used to supervise voltage limiting devices and therefore, the rail potential.

A simplified tag containing only an ID address can also be attached to cables within the permanent way. A missing tag indicates the absence of the cable, e. g. if stolen.

17.3.5.11   Temperature measurement

To monitor the temperature-dependent position of the cantilevers and the weight stack at the tension wheel, measurements of the contact and the catenary wire temperatures are required. For these measurements, laser temperature measuring units with a laser pointer can be used (Figure 17.39). During installation of the contact line, this measurement is carried out in a de-energised condition. For later adjustment of cantilevers and tensioning devices following commencement of operations, the temperature measurement can be carried out at live overhead contact lines from a safe distance using the device shown in Figure 17.39.

17.3.5.12   Thermovision measuring

Thermovision diagnostics are another measuring principle for power supply installations, able to visualise extreme heating in damaged components. These components include badly conducting electrical connections and reduced cross-sections of wires and conductors with current flow compared with intact components [17.32]. Defined currents form a pre-condition that cannot always be produced during electric railway operations.

Thermovision is increasingly used at maintenance of overhead contact line installations. Jernbanverket in Norway and German Railway DB use thermovision to locate hot spots within

a) Measuring device with evaluation unit

b) Hot spots at earthing conductor connection

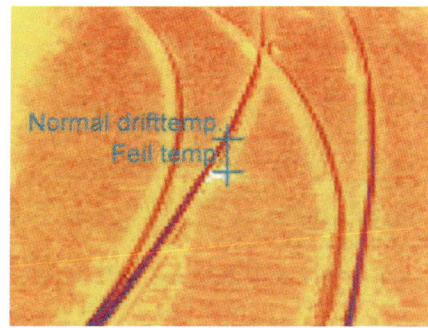

c) Hot connections at insulator

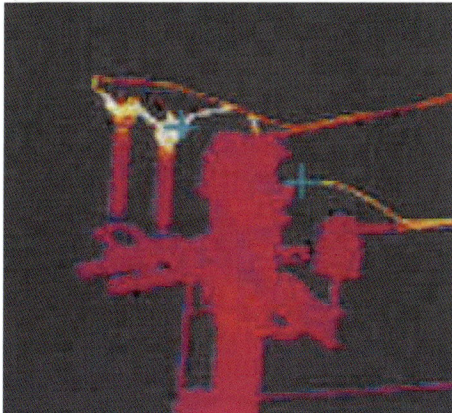

d) Current connection between two catenaries

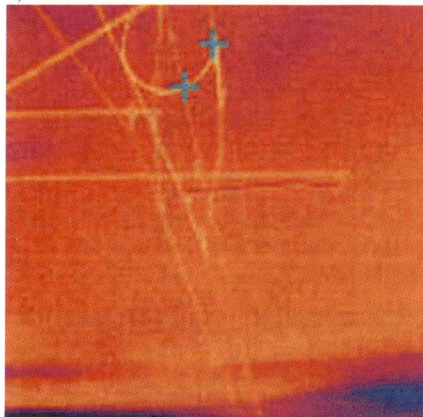

Figure 17.40: Thermovision at Jernbaneverket, Norway.

a) Current connector as a thermovision photo

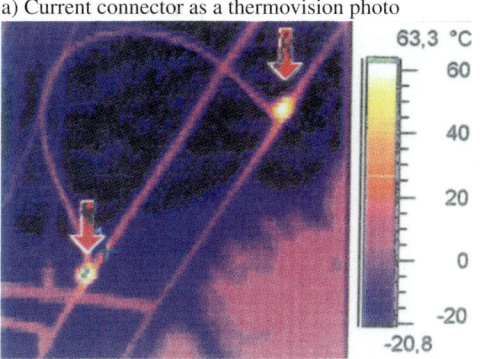

b) Standard photo

Figure 17.41: Diagnostics by thermovision.

Figure 17.42: DB's inspection (measuring) train on the high-speed line HSL Zuid, Netherlands.

their overhead contact line installations, as shown in Figure 17.40 a). By comparing such spots with the immediate surroundings, loose bolts at current connectors, defective contacts at disconnectors or defective return conductor connections are identified. The observed temperature is associated with a certain colour, dark colours correspond to low temperatures and light colours to high temperatures. The measurement enables determination of differences in temperature with surroundings. A report on the diagnosis summarises the critical items and an evaluation determines the urgency of inspections and repair of shortcomings. After the repairs, a report summarises the results of the inspection and removal of defects.

Figures 17.40 b) to d) show hot spots at connections of earthing conductors and feeder lines. Figure 17.41 shows a thermovision image of a current connector conducting 350 A measured current. The clamp on the contact wire was heated to 63 °C and attained a temperature 45 K higher than the contact wire and connector. The inspection and repair team are directed to remove these detected shortcomings. The availability of the overhead contact line installation is enhanced and the inspection costs decreased.

17.3.5.13 Contact force measurement

Contact force measurements reveal the quality of the interaction between overhead contact lines and pantographs, and are used to verify correct installation after completing contact line installations. During the maintenance period some infrastructure managers also use contact force measurements as a functional check. For example, at DB contact force measurements are carried out at six-month intervals on high-speed lines and annually for conventional lines. By using increasingly more precise contact-less measuring devices to record contact wire heights, lateral positions and contact wire thicknesses, some infrastructure managers are able to dispense with the more expensive contact force measurements. Some infrastructure managers operate inspection trains equipped for contact force measurements, that also record track condition, enabling the monitoring of the interaction between track, vehicles and overhead contact line. If the contact force measurement is only used for commissioning or after replacement of the contact wire, a mobile measuring unit can be used. A pantograph, equipped with recording sensors is installed on an operational traction unit for this purpose. Such a unit can be a locomotive, as shown in Figure 17.42, or a complete measuring train. The companies Bombardier in Henningsdorf, Euraltest in France, ADIF in Spain, Schunk in

a) Diagnostics from upper surface

b) Diagnostics from bottom

c) Steady increase of current collector height at a conductor rail ramp (black line)

d) Unsteady increase of current collector height at a conductor rail ramp (black line)

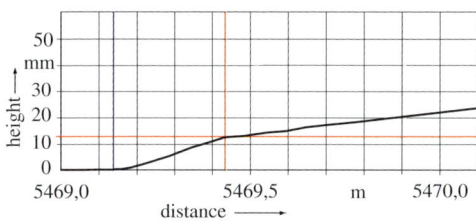

e) Diagnostics from the upper surface

f) Vehicle with diagnostic equipment

Figure 17.43: Diagnostics of conductor rails close to the ground and collector shoes of city railway and metro vehicles.

Table 17.5: Classification of disturbance causes at DB.

Overhead contact line shortcomings	Internal railway impacts	External impacts

Manufacturing
– insulators
– material defects
Power supply management
– switch gear disturbances
– protection tripping
– voltage differences
– incorrect switching operations
Installation and maintenance
– installation
– maintenance
	Operation
	– adverse management
	– railway operational accidents
	– other reasons
	Traction vehicles
	– defects of electric
	  traction vehicles
	– non observance of EL-signals
	– railway operation accidents
	– stopping in insulated overlaps
	– pantograph operation
	– defective wagons, incorrect
	  freight loadings
	Works at installations
	– civil engineering work
	– work on superstructures
	– work on signalling systems
		Third party impacts
		– flashovers caused by animals
		– climatic influences
		– road vehicles, construction
		  machinery, tanks, etc.
		– third party railways
		– trees, branches
		– unauthorized persons on and
		  near overhead contact lines
		– fires at houses, forests and
		  embankment
		– objects in overhead
		  contact lines
		– third party working
		– railway crime, stone throwing,
		  and projectiles, vandalism

Stockholm and German Railway DB in Germany currently offer their inspection trains for contact force measurements to other infrastructure managers (Clause 10.4.3.5).

17.3.5.14 Diagnostics of third rail installations

The interaction of pantographs/collector shoes with third rails can be monitored similarly to the interaction between pantographs and flexible overhead contact lines. When measuring the interaction of conductor rails close to the ground with pantographs, also known as contact shoes, the sensors are fixed to the contact shoes (Figure 17.43 a)). The measurements, recorded at commercial speed include:

- the height of the collector shoe under or on the contact rail and including the height of the conductor rail
- the contact line ramps
- the contact forces between the collector shoe and the conductor rail
- the current flowing across the collector shoe
- the voltage progress at the conductor rail

A camera records the movements of the current collector shoe. The video record is used in conjunction with dedicated software to evaluate the measurements, particularly for significant observations such as when the conductor rail height varies from designed values: the lateral position of the conductor rail can also be determined.

The conductor shoe can be calibrated, before measurements are recorded, by mounting it on a test rig at a defined position and applying downward forces with various speed to determine the transfer function of the contact forces.

Following measurements and evaluation, conclusions can be drawn on the position and the condition of the conductor rail and the collector shoe. Images of incorrect conductor rail positions can be used to program corrective maintenance (Figure 17.43 b)) [17.57, 17.58].

Table 17.6: Selected data related to operational statistics during 1995 for DB and RZD.

Parameter	Unit	DB	RZD
– electrified line	km	17 125[1]	39 100
– electrified track	km	44 809[1]	91 250
– proportion of electrified lines in network	%	42,9	44,6
– proportion of electric traction related to	%	83,7	74,3
– transport volume			
– electrical energy consumption	MWh/km	188	299
– contact line disturbances with	1/100 track km	1,93	1,12
– train delays, total %			
– portion caused by internal effect	%	0,42	0,62
– of electric traction			
– portion due to external impacts	%	1,51	0,50
– number of damage cases to overhead line	$1/10^6$ tv-km	1,09	1,02
– per $1/10^6$ traction vehicle per km			
– number of short circuits	1/100 track km	34	55[2]
– portion sustained short circuits	%	1,07	3,2
– self-breakage of insulators per 1 000 track km	1/1 000 track km	0,98	1,46
– failure of fittings per 1 000 track km	1/1 000 track km	0,20	1,72
– labour force for maintenance of overhead			
contact line, staff members per 100 track km	manpower/100 track km	4,4	10,8

[1] without urban transportation systems in Hamburg and Berlin, [2] only AC

17.3.6   Statistical recording and analysis of faults

The *statistical recording* and *analysis* of *material and operating data, inspections* and *corrective maintenance*, as well as irregularities and disturbances form an important basis for planning maintenance and further development of overhead contact lines. Forms specified by the infrastructure manager are used for recording of all important data relating to a fault, including location, affected system components, assigned staff, time of the occurrence, the arrival of the maintenance team and the restoration of electric train operations. The *fault reports* are distributed to specified departments depending upon the type of fault within defined time limits and are included in the operational statistics. DB specifies the *classification of disturbances* shown in Table 17.5 for the recording of fault causes.

Comparisons between the overhead contact line system statistics and their analysis for individual railway companies is only conditionally possible because of the different methods applied to maintenance and statistic recording. It is, therefore, only possible to compare a limited selection of data on the operational statistics of DB and RZD in Table 17.6.

Based on DB statistics Figure 17.44 shows clearly that the overhead contact line system plays only a minor role in the total number of faults, but a significantly larger role in the causes of time delays. The reason for this is the lack of redundancy. This allows the conclusion that high-quality, reliable overhead contact lines are an essential prerequisite for punctual train operations. The subdivision into the classification of all faults, and the classification into parties responsible for causing damage to the overhead contact line system at DB and delays longer than 10 minutes during 1995, is contained in Figure 17.45. The development of *overhead contact line disturbances* at DB between 1976 and 1995 can be seen in Figure 17.46. The number of disturbances with damage and delays was continuously reduced over the period. The defects caused by manufacturing, installation, maintenance and operational management are few in comparison to other railway companies.

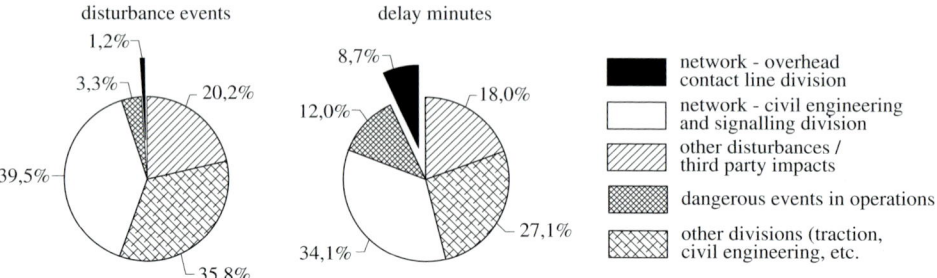

Figure 17.44: Classification of technical train operational disturbances at DB during 1995.

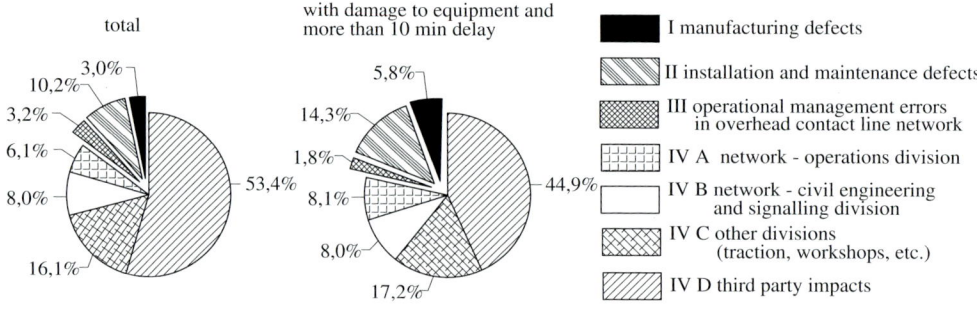

Figure 17.45: Classification of disturbances in overhead contact line installations at DB with respect to the parties causing the disturbances.

17.3.7   Corrective maintenance

DB subdivides *corrective maintenance*, depending on the main objective, into fault repairs, individual repair and full repair. *Fault repair* comprises the immediate restoration of the overhead contact line to function and the removal of safety related defects detected during inspections either completely, or as far as absolutely necessary, to avoid long delays to trains. The *individual repairs* correct defects that could lead to impaired functionality, e. g. replacement of clamps after glow-out. They are performed during pre-planned work deployments and extend the interval until the next full inspection.

Complete corrective maintenance includes the removal of all defects observed during preceding inspections and examinations. It requires long-term planning, co-ordination with other activities on the line and additional staff deployment. It should be combined with a full inspection for economic efficiency.

Overhead contact line disturbances can be avoided by timely removal of branches and bushes within a distance of 2,5 m from poles and lines.

The *partial renewal* of overhead contact line components requires special techniques, taking into account available track closure times and installation equipment. *Contact wire replacement* commences with the release of the contact wire from the termination and its coiling onto an empty contact wire reel. The new contact wire is then strung to the tensioning device in place of the old one and fastened. The release from and clamping onto the droppers and steady arms and coiling and unwinding old and new contact wires, respectively, under pretensioning, can be performed simultaneously with the help of a common contact wire reel

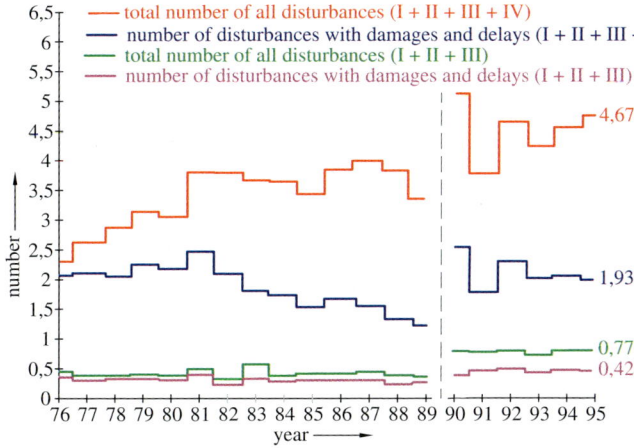

Figure 17.46: Total number of all disturbances with damage per 100 track kilometers by faults in manufacturing, installation or maintenance, of the operational management in the contact line net, up to 1989 in West Germany, from 1990 West and East Germany .

wagon. Wheel tensioning devices and steady arms are provisionally secured using installation aids. The replacement of a contact wire requires approximately 1,5 h for a tensioning section.

During the replacement of a *catenary wire*, the new catenary wire is drawn, without or with little pre-tensioning and attached provisionally to the old catenary wire. The release of the old catenary wire and the connection of the new one to the supports, droppers and stitch wires follow. The old catenary wire is finally coiled onto a reel and the contact line adjusted.

The replacement of mechanically stressed components such as clamps, insulators, cantilevers and bolt-mounted poles requires the alleviation of the load by installation equipment or aids, first. The replacement and loading of new components can follow.

Normally new head-span structures and embedded poles can be installed adjacent to the existing units, and the loads then transferred. Subsequent adjustment and the removal of the old parts complete the work.

Partial renewals are always costly and impair train operations because of the necessary track closures. They can be avoided or reduced to a minimum by employing *long-life components* and high-quality overhead contact lines.

17.4 Recycling and disposal

17.4.1 Dismantling

The most common case of partial *dismantling* occurs during the replacement of the contact wire described in Clause 17.3.7. Complete renewal of the system, or the termination of electric train operations on individual lines, requires the complete *removal of the overhead contact line*. For contact lines, the contact wire is first released and coiled up, then the droppers and stitch wires are removed and then, finally, the catenary wire coiled onto a reel. The same equipment as used for the installation can be employed for the individual work stages. If only short track closures are available, then the contact line can be cut down in pieces and loaded onto flat wagons. Cantilevers and other cross-span structures are unbolted from the poles. Steel poles can be dismantled ready for re-use. Concrete poles are normally destroyed at the base by the removal process. Tube foundations are an exception and allow the pole to

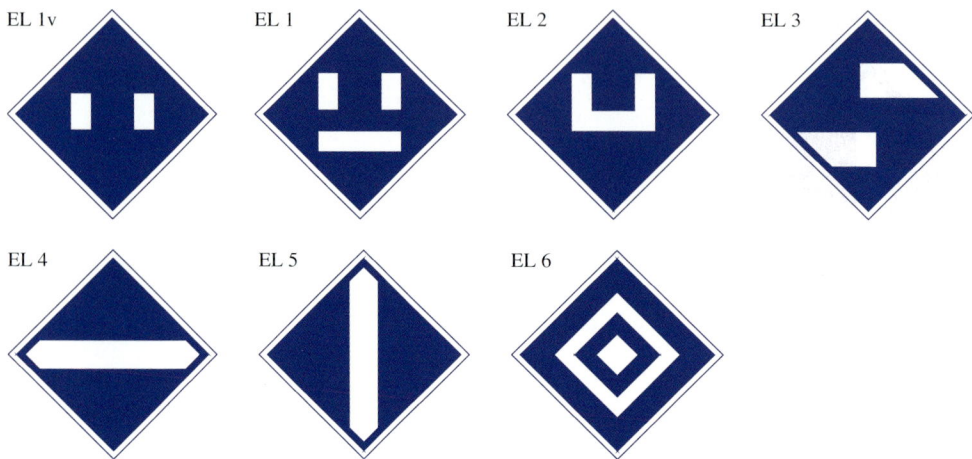

Figure 17.47: Signals for electric traction according to DB Directive 301.

be lifted after removal of the concrete cover and stone chippings. The *removal of foundations* is very expensive, the section below ground is normally left in place.

17.4.2   Suitable preparation and disposal of materials for recycling

The dismantled components are processed in workshops, depending on the *recycling method*. This includes suitable *disposal and separation* of different *materials* and cutting metals to suitable lengths by the metal mills. The disposal of the individual components can be divided into the following categories:

Re-use at the same level: Steel from poles and cantilevers, non-alloyed copper from contact wires and clamps and the alloyed contact wire CuAg, are all materials that can be reused for the manufacture of the same components, after melting down and appropriate preparation.

Re-use at a lower level:
- CuMg contact wires and aluminum parts
- concrete poles
- plastics

Disposal: Worn porcelain and glass insulators are mostly disposed of on a *waste dump*.

17.5   Signals for electric traction

Requirements arise from the operation of overhead installations with respect to *signalisation* of de-energised, earthed and unserviceable sections, of neutral sections and of unwired tracks. The significance of the signals for electric operations is explained using DB's practice described in DB Directive 301 as an example (Figure 17.47).

The traction vehicle driver recognises the actions required by *EL-signals*, installed along side or above the track. The EL-signals consist of square blue *panels* with black and white borders mounted on one corner and displaying white switch signals. Figure 17.47 illustrates the EL-signals used by DB.

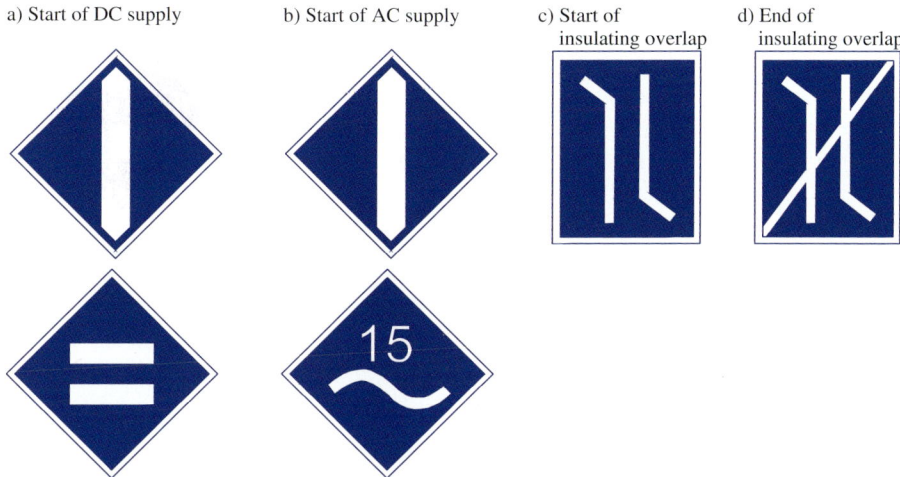

a) Start of DC supply b) Start of AC supply c) Start of insulating overlap d) End of insulating overlap

Figure 17.48: Signals for electric traction.

The signal El 1v announces the signal El 1 and is arranged at a distance equal to half of the braking distance in front of the signal El 1. Signal El 1 identifies the latest location at which the main circuit breaker of the electric traction vehicle has to be switched off and the signal El 2, the earliest location at which the main circuit breaker may be switched on again. Both signals are permanent installations and are illuminated at night.

They are located before or after neutral sections and coupling posts. The signals at neutral sections cannot be changed and continuously show EL 1 at the start of the neutral section, on the rear side EL 2 and at the end of the neutral section EL 2 and on the rear side EL 1.

For switchable neutral sections, e. g. at coupling posts, switchable signals are also installed. If the coupling post is closed, the signal EL 2 indicates to the driver at the beginning of the neutral section that he should not switch off the circuit breaker.

In the case of an open coupling post the signal EL 1 indicates that the neutral section is effective and the circuit breaker must be opened.

EL 1 is arranged directly under EL 2 for short neutral sections. The driver of the traction vehicle recognises from this signal arrangement that the main circuit breaker must be switched off at the location of the signal and may be switched on again after passing the signal.

Signals El 3 to El 5 mark overhead contact line sections that may not be passed with the pantograph raised. Signal EL 3 is located as an announcing signal at least 250 m before the following signal EL 4, which shows the signal EL 5 on its rear side. The signal EL 4 is located 30 m in front of a section to be traversed with a lowered pantograph. At the Signal EL 4 the pantographs must be lowered.

The pantographs may be raised again after passing the signal EL5, which is located 30 m beyond the section to be passed with dropped pantographs. The signals EL 3 to EL 5 are installed during construction work and in emergencies, however, not permanently.

The signal EL 6 is permanently installed and announces the *end of the overhead contact line* and, therefore, *stop for electric traction vehicles* with raised pantographs. It is located 10 m before the end of a serviceable overhead contact line section.

At system separation sections the signal according to Figure 17.48 c) announces to the vehicle driver that after passing the system separation section, an AC overhead contact line starts. The

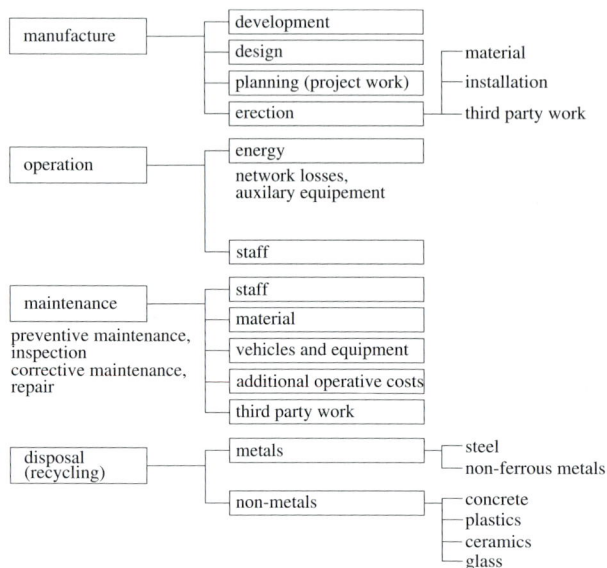

Figure 17.49: Elements of life cycle costs for overhead contact line systems.

signal according to Figure 17.48 d) announces the begining of a DC overhead contact line. Above the sine curve the voltage can also be indicated. The vehicle driver may close the circuit breaker of the vehicle only after passing the system separation section.

The signal Figures 17.48 c) and d) are arranged on the poles within the isolated overlap which forms a border for the overhead contact line supply network. The signal shown in Figure 17.48 c) announces the beginning and the signal shown in Figure 17.48 d) the end of a section, in which the vehicle may not stop with a raised pantograph. If an electrically driven vehicle comes to a stop between the beginning and end of the overlap marked by signals shown at Figures 17.48 c) and d), the pantographs must be lowered.

Closing the contact line disconnector raising the pantograph and continuing of travel are possible only after consulting the central control center (Zes).

17.6   Life cycle considerations

Increasingly, decisions regarding the design and use of installations are made by operators not only on the basis of the initial investment but also after considering the total expenditure and costs expected during the entire lifetime of the installation. The expected costs over the lifetime of a component are known as *life cycle costs* (LCC). They permit an integrated economic viability evaluation.

The physical, i. e. real expected *service life of overhead contact lines* for electric railways is high compared to other equipment. The physical service life of overhead contact lines has been estimated to be between 20 and 70 years (Clause 17.3). This long service life is an essential reason why LCC examinations were not considered in the past. The service life of contact lines is furthermore dependent upon the development of electrically hauled transport. It is influenced by long-term line and speed development. These reasons also support the view that overhead contact lines should be analysed more strictly in accordance with LCC.

The life cycle costs include:

  – *manufacturing expenditures*
  – *operating and operator costs*
  – *maintenance costs*
  – *disposal costs* (dismantling and recycling)

The individual costs can be seen in Figure 17.49. Objective comparisons of overhead contact lines are possible on the basis of life cycle costs. High-quality overhead contact lines that are more expensive to manufacture display often, the lowest life cycle costs in system comparisons.

The contact wire is the wear-intensive element of an overhead contact line, whose service life has a decisive influence upon the life cycle costs. High costs are associated with the *replacement of the contact wire* under operational conditions. The contact wire wear, therefore, has great significance for the life cycle costs. The rating of the conductor cross-sections influences the energy losses and the quality of the supplied voltage for electric train operations.

High reliability and simple repair of the overhead contact line system after a disturbance is of fundamental importance for maintenance costs. Overhead lines with durable components are less sensitive to vandalism, electrical flashovers and atmospheric affects and entail lower maintenance costs. Durable systems will remain operational for a longer period. Failures cause not only repair costs but also a series of *consequential costs* such as train delays, loss of public image and passengers, etc., which cannot always be evaluated monetarily.

Examination shows that the *disposal costs* of overhead contact lines are substantial. Considerable costs can be saved by dismantling and reusing overhead contact line components and by recouping the remaining material value.

17.7   Bibliography

17.1 *Borgwardt, H.*: Instandhaltungskonzeption für Oberleitungsanlagen (Organization of maintenance for overhead contact line installations). In: Construction and maintenance of railway installations, Edition ETR – Hestra-Publishing, Darmstadt, 1993.

17.2 *Borgwardt, H.*: Druckschrift DS 462 – Grundlage einer sicheren Betriebsführung im Oberleitungsnetz der Deutsche Bundesbahn (Leaflet DS 462 – Basis of a reliable operation of German Railways overhead contact line network). In: Elektrische Bahnen, 89(1991)4, pp. 106 to 113.

17.3 *Directive DV EL52*: Dienstvorschrift für den Betrieb der Leitungsanlagen der elektrisch betriebenen Haupt-, Neben- und Anschlussbahnen (Official instructions for the operation electrical installations of electric traction main lines, local lines and connecting lines). Austrian Railways, Vienna, 2017.

17.4 *Directive R RTE 20600*: Sicherheit bei Arbeiten im Bereich von Bahnstromanlagen (Safety when working close to railway traction energy installations). In: Regelwerk Technik Eisenbahn (RTE), Editor Verband öffentlicher Verkehr, 2012.

17.5 *Regulation 1301/2014/EC*: Technical specification of the interoperability of the subsystem energy of the railway system in the European Union (TSI ENE). In: Journal of the European Union No. L 356 (2014), pp. 179 to 227.

17.6 *Regulation 1303/2014/EC*: Technical specification of the interoperability for safety in railway tunnels of the railway system in the European Union (TSI SRT). In: Journal of the European Union No. L 356 (2014) pp. 394 to 420.

17.7   *Regulation 402/2013/EC*: Common safety methods for the evaluation and assessment of risks. In: Journal of the European Union No. L 121 (2014), pp. 8 to 25.

17.8   *Directive 49/2004/EC*: Directive on the licensing of railway undertakings and the allocation of railway infrastructure capacity. In: Official Journey of the European Union, No. L164(2004) pp. 44 to 113.

17.9   *Directive 123*: Emergency management, fire protection. Deutsche Bahn AG, Frankfurt, 2015.

17.10 *Directive 997.9117*: Oberleitungsspannungsprüfeinrichtung (OLSP) (Overhead contact line voltage detector (ACLT)). Deutsche Bahn AG, Berlin, 2014.

17.11 *Weigler, H.*: Widerstand des Betons und der Bewehrung (Corrosion resistance of concrete and reinforcement). In: Concrete poles under the impact of weather. Conference of the German concrete and pre-fabricated concrete industry e.V., Darmstadt, 1994, pp. 59 to 80.

17.12 *Kleingarn, J.-P.*: Feuerverzinken von Einzelteilen aus Stahl, Stückverzinken. (Hot-dip galvanizing of steel parts, galvanizing of individual parts). In: Information brochure of the consulting organisation for use of steel and application of galvanizing, part 293.

17.13 *Bauer, K. H.; Gerichten, F.; Kießling, F.; Lerner, F.*: Einsatz von Aluminium für die Oberleitung der Neubaustrecken der Deutschen Bundesbahn (Utilization of aluminium for the overhead line of the new high-speed lines of the German Federal Railway). In: Elektrische Bahnen, 84(1986)10, pp. 298 to 306.

17.14 *Becker, K.; Resch, U.; Wilde, J.; Zweig, B.-W.*: Werkstofftechnische Optimierung des Systems Fahrdraht/Stromabnehmer (Material technological optimization of the system contact wire/pantograph). In: ETG-Fachbericht 74, 2. Internationale Konferenz Elektrische Bahnsysteme, pp. 255 to 260, Berlin 1999.

17.15 *Borgwardt, H.*: Verschleißverhalten des Fahrdrahtes der Regeloberleitungen der Deutschen Bahn (Wear performance of the contact wire of the standard overhead contact lines of German Railways). In: Elektrische Bahnen, 87(1989)10, pp. 287 to 295,

17.16 *Wiessler, U.*: Qualität und Ausführung von Kohleschleifstücken – 100 Jahre Stromabnahme vom Fahrdraht mit Kohlenstoff – Rückschau und Ausblick (Quality and design of carbon collector strips – 100 years of current collection from the contact wire by carbon – Historical view and outlook). In: Schunk-Bericht aus Forschung und Entwicklung, No. 39, Schunk Kohlenstofftechnik, Giessen 1985.

17.17 *Auditeau, G.; Avronsart, S.; Courtois, C.; Krötz, W.*: Carbon contact strip material – Testing of wear. In: Elektrische Bahnen, 11(2013)3, pp. 186 to 195.

17.18 *Becker, K.; Resch, U.; Rukwied, A.; Zweig, B.-W.*: Lebensdauermodellierung von Oberleitungen (Modelling of life cycle of overhead contact lines). In: Elektrische Bahnen, 94(1996)11, pp. 329 to 336.

17.19 *Becker, K.; Resch, U.; Rukwied, A.; Zweig, B.-W.*: Das Verschleißverhalten der Regeloberleitung Re 250 unter den Bedingungen des Hochgeschwindigkeitsschienenverkehrs (Wear performance of the standard overhead contact line Re 250 under the conditions of a high-speed rail traffic). In: ZEV Glasers Annalen, 120(1996)6, pp. 244 to 251.

17.20 *Regulation 1302/2014/EC*: Technical specification of the interoperability relating to the Rolling Stock subsystem – Locomotives and passenger rolling stock of the European rail system (TSI LOC&PAS). In: Journal of the European Union No. L 356 (2014), pp. 228 to 393.

17.21 *Liebermann, H.*: Technische Vorzüge von Tonerdeporzellan für die Zuverlässigkeit von Hochspannungsisolatoren (Technical advantages of aluminium oxide porcelain on the reliability of high-voltage insulators). In: Keramische Zeitschrift 47(1995)6, pp. 461 to 464.

17.22 *Wolf, S.*: Untersuchung zur Entwicklung eines Oberleitungsstützpunktes ohne Isolatoren (Development of an overhead contact line support without insulators). Fachhochschule Wiesbaden, thesis for diploma, 1996.

17.23 *DR-M 24.71.010*: Abnutzung von Stromschienen (Wear of conductor rails). Deutsche Reichsbahn, Berlin, 1980.

17.24 *Zweig, B.-W.*: Ein Beitrag zur optimalen Gestaltung der Fahrleitungsinstandhaltung bei der DR unter besonderer Berücksichtigung der Einführung diagnostischer Methoden und Geräte (A contribution to the optimum organization of overhead contact line maintenance with specific consideration of new diagnostic methods and devices). HfV Dresden, doctorate thesis, 1984.

17.25 *Puschmann, R.*: Ermittlung der Ausfallrate von Fahrleitungen und dUfw (Determination of the failure rate of overhead contact lines and substations). HfV Dresden, thesis for diploma, 1974.

17.26 *Schmidt, P.*: Energieversorgung elektrischer Bahnen (Power supply of electrical railways). Publishing transpress, Berlin, 1988.

17.27 *Wüsthoff, W.*: Beitrag zum Zusammenhang von Kurzschlußbeanspruchung und mittleren Ausfallabstand für Elemente des Hauptstromkreises eines Bahnenergieversorgungssystems mit 16,7 Hz (Contribution on the context of short-circuit stresses and the mean time between failure for elements of the main current circuit of a 16,7 Hz traction power supply system). HfV Dresden, doctorate thesis, 1977.

17.28 *Fischer, K.*: Zuverlässigkeits- und Instandhaltungstheorie (Theory of reliability and maintenance). Publishing transpress, Berlin, 1984.

17.29 *Häse, P.*: Ein Beitrag zur Bestimmung optimaler Instandhaltungsmethoden für Baugruppen von Kettenwerksfahrleitungen unter besonderer Berücksichtigung von Elementen der Zuverlässigkeits- und Erneuerungstheorie (Contribution on the determination of optimum maintenance methods for components of overhead contact lines with specific application of reliability and refurbishment theory). HfV Dresden, doctorate thesis, 1979.

17.30 *Koslow, B. A.; Uschakow, I. A.*: Handbuch zur Berechnung der Zuverlässigkeit für Ingenieure (Manual on calculation of reliability for engineers). Hanser-Publishing, München, 1979.

17.31 *Kochs, H.-D.*: Zuverlässigkeit elektrotechnischer Anlagen (Reliability of electrotechnical installations). Springer-Verlag, Berlin–Heidelberg–New York, 1984.

17.32 *Petrausch, D.*: Thermische Modellierung und Thermovision bei Fahrleitungsanlagen (Thermal modelling and thermovision at overhead contact line installations). In: Elektrische Bahnen, 88(1990)2, pp. 80 to 84.

17.33 *Pfisterer:* Contact wire geometrical position measurement equipment. In: Product information, Winterbach, Germany, 2008.

17.34 *Feinmess:* Catenary measuring instrument FM1. In: Product information, Dresden, Germany,
2008.

17.35 *CEMAFER:* Movable catenary laser measuring instrument type Mephisto. In: Product information, Breisach, Germany, 2008.

17.36 *Puschmann, R.; Wehrhahn, D.:* Fahrdrahtlagemessung mit Ultraschall (Contact wire geometrical position measurement by means of ultra sonic). In: Elektrische Bahnen, 109(2011)7,
pp. 323 to 330.

17.37 *Frauenhofer-Institut für Physikalische Messtechnik:* Determining the position of contact
wires at 250 km/h. Product information, www.ipm.fraunhofer.de, Freiburg, 2013.

17.38 *Sarnes, B.:* Inspektion von Fahrdrahtlage und -stärke bei beliebiger Geschwindigkeit (Inspection of contact wire position and thickness at any speed). In: Elektrische Bahnen, 99(2001)12,
pp. 490 to 495.

17.39 *Richter, U.; Schneider, R.:* Automatische optische Inspektion von Oberleitungen (Automati-
cal optical inspection of overhead contact line systems). In: Elektrische Bahnen, 99(2001)1-2,
pp. 94 to 100.

17.40 *ElektroPhysik:* Schichtdickenmessung – GalvanoTest (Layer thickness measurement – GalvanoTest). In: ElektroPhysik product information, Cologne, 2012.

17.41 *ElektroPhysik:* Schichtdickenmessung – MiniTest Serie 700 (Layer thickness measurement –
MiniTest series 700). In: Product information ElektroPhysik , Cologne, 2012.

17.42 *Honigmann:* Zugkraftsensor Tritens 136.3 S30 zur mobilen und stationären Zugkraftmessung
(Tensile force sensor Tritens 136.3. S30 for mobile and stationary tensile force measurement).
In: Product information Honigmann, Cologne, 2012.

17.43 *PEWA:* Seilspannungsmessgerät (Conductor tension testing device). In: Product information,
PEWA Messtechnik GmbH, 2013.

17.44 *SAUTER:* Kraftmessgerät FK Tensio (Contact force measurement device FK Tensio).
In: Product information SAUTER GmbH Messtechnik, 2013.

17.45 *Pfeifer:* Seilvorspannung (Conductor pre-tensioning). In: Product information Pfeifer Seil-
und Hebetechnik GmbH, 2011.

17.46 *PIAB:* Dynamometer. In: Product information PIAB, Witten, Germany, 2008.

17.47 *Vetter:* Technik für Kabelverlegung und Freileitungsbau (Technology for cable laying and
overhead line construction). In: Product information Vetter GmbH, 2013.

17.48 *Grube:* Dynamometer Zugkraftmessgerät (Dynamometer tensile force measurement device).
In: Grube Catalogue No. 54, 2013.

17.49 *Thoresen, Th. E.:* System for measuring the uplift from all pantographs passing the installed
system. In: Product information, Oslo, Norway, 2008.

17.50 *Möller, H.; Grebner, L.; Hofmann, D.*: Stromabnehmerdiagnose im laufenden Betrieb durch stationäre Anhubmessung (Pantograph diagnosis under operation by stationary uplift measurement). In: Elektrische Bahnen, 100(2002)6, pp. 198 to 203.

17.51 *Hietzge, J.; Stephan, A.*: Berührungslose Messung des Oberleitungsanhubs (Non-contact uplift measurement at overhead contact lines). In: Elektrische Bahnen, 105(2007)4-5, pp. 276 to 279.

17.52 *Bechmann, J. et al.*: Überwachungseinrichtung für Oberleitungskettenwerke (Contact line monitoring system). In: Elektrische Bahnen, 106(2008)8-9, pp. 400 to 407.

17.53 *Wili, U.; Zenglein, S.*: Auswirkung der RAMS-Normen auf elektrotechnische Anlagen (The application of the RAMS standards to fixed installations). In: Elektrische Bahnen, 105(2007)4-5, pp. 246 to 253.

17.54 *Sensys Traffic*: Pantograph monitoring system. In: Product information Sensys Traffic AB, Jönköping, Sweden, 2008.

17.55 Sicat RMS: Überwachung des Schienenpotenzials mit Sicat RMS (Monitoring of the rail potential by Sicat RMS). In: Product information Siemens, Erlangen, 2008.

17.56 *Voigt, B.; Liebert, R.*: FCS 601 – ein System zur Überwachung von Spannungsdurchschlagssicherungen (FCS 601 – a system to monitor overvoltage limiters). In: Signal + Draht, 92(2000)10, pp. 47 to 48.

17.57 *Deutzer, M.; Engelmann, R.; Wilmes, T.*: Messsystem zur Prüfung einer Metro-Stromschienenanlage und eines Dritte-Schiene-Stromabnehmers (Measuring system for testing a metro conductor rail installation and of a conductor rail collector). In: Verkehr+Technik, 57(2004)5, pp. 1 to 7.

17.58 *Deutzer, M.*: Prüfung und Verbesserung der Stromentnahme in Dritte-Schiene-Anlagen (Testing and improvement of current collection in third rail installations). In: Elektrische Bahnen, 108(2010)8-9, pp. 368 to 376.

Appendix: Standards

A.1 IEC Publications

IEC	Year	Title
IEC 60 050-811	1991	International electrotechnical vocabulary; chapter 811: electric traction
IEC 60 112	2003	Method for determining the comparative and the proof tracking indices of solid insulating materials under moist conditions
IEC 60 273	1990	Characteristics of indoor and outdoor post insulators for systems with nominal voltages greater than 1000 V
IEC 60 305	1995	Insulators for overhead lines with a nominal voltageabove 1000 V – Ceramic or glass insulator units for a.c. systems – Characteristics of insulator units of the cap and pin type
IEC 60 364-1	2005	Low-voltage electrical installations – Part 1: Fundamental principles, assessment of general characteristics, definitions
IEC 60 364-4-41	2017	Low-voltage electrical installations – Part 4-41: Protection for safety – Protection against electric shock
IEC 60 383	Series	Insulators for overhead lines with nominal voltage above 1000 V
IEC 60 433	1998	Insulators for overhead lines with a nominal voltage above 1000 V – Ceramic insulators for a.c. systems – Characteristics of insulator units of the long rod type
TSE IEC/TS 60 479-1	2010	Effects of current on human beings and livestock – Part 1: General aspects
IEC 60 672-2	1999	Ceramic and glass insulating materials - Part 2: Methods of test
IEC/TS 60 815	Series	Guide for the selection of insulators in respect of polluted conditions
IEC 60 826	2017	Loading and strength of overhead transmission lines
IEC 60 865-1	2011	Short-circuit currents – Calculation of effects – Part 1: Definitions and calculation methods
IEC 60 870-5-101	2003	Telecontrol equipment and systems – Part 5-101: Transmission protocols; Companion standard for basic telecontrol tasks
IEC 60 870-5-104	2006	Telecontrol equipment and systems – Part 5-104: Transmission protocols – Network access for IEC 60870-5-101 using standard transport profiles
IEC 60 913	2013	Railway applications – Fixed installations – Electric traction overhead contact lines
IEC 61 000-5	2014	Electromagnetic compatibility (EMC) – Part 4-5: Testing and measurement techniques – Surge immunity test
IEC 61 089	1991	Round wire concentric lay overhead electrical stranded conductors
IEC 61 109	2008	Insulators for overhead lines - Composite suspension and tension insulators for a.c. systems with a nominal voltage greater than 1000 V – Definitions, test methods and acceptance criteria
IEC/TS 61 245	2015	Artificial pollution tests on high-voltage insulators to be used on d.c systems
TSE IEC/TR 3 61 597	2002	Overhead electrical conductors – Calculation methods for stranded bare conductors
IEC 61 936-1	2010	Power installations exceeding 1 kV a.c. – Part 1: Common rules
IEC 61 952	2008	Insulators for overhead lines – Composite line post insulators for a.c. systems with a nominal voltage greater than 1000 V – Definitions, test methods and acceptance criteria
IEC 62 128-1	2013	Railway applications – Fixed installations – Part 1: Protective provisions relating to electrical safety and earthing
IEC 62 128-2	2013	Railway applications – Fixed installations – Part 2: Protective provisions against the effects of stray currents caused by d.c. traction systems
IEC 62 128-3	2013	Railway applications – Fixed installations – Electrical safety, earthing and the return circuit – Part 3: Mutual interaction of a.c. and d.c. traction systems
IEC 62 305-2	2010	Protection against lightning – Part 2: Risk management

IEC 62 305-3	2010	Protection against lightning – Part 3: Physical damage to structures and life hazard
IEC 62 917	2016	Railway applications – Fixed installations – Electric traction – Copper and copper alloy grooved contact wires

A.2 European Standards

EN	Year	Title
EN 1706	2010	Aluminium and aluminium alloys – Castings – Chemical composition and mechanical properties
EN 1991-1-4	2005	Eurocode 1: Actions on structures – Part 1-4: General actions – Wind actions
EN 1992	Series	Eurocode 2 – Design of concrete structures
EN 1993	Series	Eurocode 3 – Design of steel structures
EN 1993-1-1	2005	Eurocode 3: Design of steel structures – Part 1-1: General rules and rules for buildings
EN 1997	Series	Eurocode 7: Geotechnical design
EN 10 025	Series	Hot rolled products of non-alloy structural steels; technical delivery conditions
EN 10 204	2004	Metallic products: Types of inspection documents
EN 12 464-1	2011	Light and lighting – Lighting of work places – Part 1: Indoor work places
EN 12 843	2004	Precast concrete products – Masts and poles
EN 13 306	2010	Maintenance. Maintenance terminology
EN ISO 14 688-1	2002	Geotechnical investigation and testing – Identification and classification of soil – Part 1: Identification and description
EN 15 273-1+A1	2016	Railway applications – Gauges – Part 1: General – Common rules for infrastructure and rolling stock
EN 15 273-2+A1	2016	Railway applications – Gauges – Part 2: Rolling stock gauge
EN 15 273-3+A3	2016	Railway applications – Gauges – Part 3: Structure gauges
EN ISO 22 475-1	2006	Geotechnical investigation and testing – Sampling methods and groundwater measurements – Part 1: Technical principles for execution
EN ISO 22 476-2	2005	Geotechnical investigation and testing – Field testing – Part 2: Dynamic probing
EN ISO 22 476-3	2005	Geotechnical investigation and testing – Field testing – Part 3: Standard penetration test
EN 50 110-1	2013	Operation of electrical installations – Part 1: General requirements
EN 50 119	2009	Railway applications – Fixed installations – Electric traction overhead contact lines
DIN EN 50 119 Beiblatt 1	2011	Railway applications – Fixed installations – Electric traction overhead contact lines – Supplement 1: National Annex
EN 50 119/A1	2013	Railway applications – Fixed installations – Electric traction overhead contact lines
EN 50 121-1	2017	Railway applications – Electromagnetic compatibility – Part 1: General
EN 50 121-2	2017	Railway applications – Electromagnetic compatibility – Part 2: Emission of the whole railway system to the outside world
EN 50 121-3-1	2017	Railway applications – Electromagnetic compatibility – Part 3-1: Rolling stock – Train and complete vehicle
EN 50 121-3-2	2016	Railway applications – Electromagnetic compatibility – Part 3-2: Rolling stock – Apparatus
EN 50 121-4	2016	Railway applications – Electromagnetic compatibility – Part 4: Emission and immunity of the signalling and telecommunications apparatus
EN 50 121-5	2017	Railway applications – Electromagnetic compatibility – Part 5: Emission and immunity of fixed power supply installations and apparatus
EN 50 122-1	2011	Railway applications – Fixed installations – Electical safety, earthing and bonding – Part 1: Protective provisions against electric shock

EN 50 122-2	2010	Railway applications – Fixed installations – Electrical safety, earthing and the return circuit – Part 2: Provisions against the effects of stray currents caused by d.c. traction systems
EN 50 122-3	2010	Railway applications – Fixed installations – Electrical safety, earthing and the return circuit – Part 3: Mutual interaction of a.c. and d.c. traction systems
EN 50 123-1	2003	Railway applications – Fixed installations – D.C. switchgear – Part 1: General
EN 50 123-4	2003	Railway applications – Fixed installations – D.C. switchgear – Part 4: Outdoor d.c. in-line switch disconnectors, disconnectors and d.c. earthing switches
EN 50 124-1	2017	Railway applications – Insulation coordination Part 1: Basic requirements; Clearances and creepage distances for all electrical and electronic equipment
EN 50 125-2	2002	Railway applications – Fixed equipment – Environmental conditions for equipment – Part 2: Fixed installations
EN 50 126	Series	Railway applications – The specification and demonstration of reliability, availability, maintainability and safety (RAMS)
EN 50 149	2012	Railway applications – Fixed installations – Electric traction – Copper and copper alloy grooved contact wires
EN 50 152-2	2012	Railway applications – Fixed installations – Particular requirement for a.c. switchgear – Part 2: Single-phase disconnectors, earthing switches and switches with Um above 1 kV
EN 50 162	2004	Protection against corrosion by stray current from DC systems
EN 50 163	2004	Railway applications – Supply voltages of traction systems
EN 50 182	2001	Conductors for overhead lines – Round wires concentric lay stranded conductors
EN 50 183	2000	Conductors for overhead lines – Aluminium-magnesium-silicon alloy wires
EN 50 189	2000	Conductors for overhead lines – Zinc coated steel wires
EN 50 206-1	2010	Railway applications – Rolling stock – Pantographs: Characteristics and tests – Part 1: Pantographs for main line vehicles
EN 50 206-2	2010	Railway applications – Rolling stock – Pantographs: Characteristics and tests – Part 2: Pantographs for metros and light rail vehicles
EN 50 317	2012	Railway applications – Current collection systems – Requirements for and validation of measurements of the dynamic interaction between pantograph and overhead contact line
EN 50 318	2002	Railway applications – Current collection systems – Validation of simulation of the dynamic interaction between pantograph and overhead contact line
EN 50 326	2002	Conductors for overhead lines – Characteristics of greases
EN 50 341-1	2012	Overhead electrical lines exceeding AC 1 kV – Part 1: General requirements – Common specifications
EN 50 341-2	2001	Overhead electrical lines exceeding AC 45 kV – Part 2: Index of National Normative Aspects
EN 50 341-2-4	2016	Overhead electrical lines exceeding AC 1 kV – Part 2-4: National Normative Aspects (NNA) for Germany
EN 50 341-3	2001	Overhead electrical lines exceeding AC 45 kV – Part 3: Set of National Normative Aspects
EN 50 345	2009	Railway applications – Fixed installations – Electric traction – Insulating synthetic rope assemblies for support of overhead contact lines
EN 50 367	2012	Railway applications – Current collection systems – Technical criteria for the interaction between pantograph and overhead line
EN 50 388	2012	Railway Applications. Power supply and rolling stock. Technical criteria for the coordination between power supply (substation) and rolling stock to achieve interoperability
CLC/TR 50 488	2006	Railway applications – Safety measures for the personnel working on or near overhead contact lines
EN 50 633	2016	Railway applications. Fixed installations. Protection principles for AC and DC electric traction systems

EN 60 034-1	2010	Rotating electrical machines – Part 1: Rating and performance
EN 60 060	Series	High-voltage test techniques
EN 60 071	Series	Insulation coordination
EN 60 112	2003	Method for the determination of the proof and the comparative tracking indices of solid insulating materials
EN 60 146-1-3	1993	Semiconductor convertors; general requirements and line commutated convertors; part 1-3: transformers and reactors
EN 60 168	1994	Tests on indoor and outdoor post insulators of ceramic material or glass for systems with nominal voltages greater than 1000 V
EN 60 305	1996	Insulators for overhead lines with a nominal voltage above 1 kV – Ceramic or glass insulator units for a.c. systems – Characteristics of insulator units of the cap and pin type
EN 60 383-1	1996	Insulators for overhead lines with a nominal voltage above 1 kV – Part 1: Ceramic or glass insulator units for a.c. systems – Definitions, test methods and acceptance criteria
EN 60 383-2	1995	Insulators for overhead lines with a nominal voltage above 1000 V – Part 2: Insulator strings and insulator sets for a.c. systems – Definitions, test methods and acceptance criteria (IEC 60383-2:1993)
EN 60 433	1998	Insulators for overhead lines with a nominal voltage above 1 kV. Ceramic insulators with a.c. systems. Characteristics of insulator units of the long rod type
EN 60 437	1997	Radio interference test on high-voltage insulators
EN 60 529	1991	Degrees of protection provided by enclosures (IP code)
EN 60 660	1999	Insulators – Tests on indoor post insulators of organic material for systems with nominal voltages greater than 1 kV up to but not including 300 kV
EN 60 664-1	2007	Isolation coordination for equipment within low-voltage systems – Part 1: Principles, requirements and tests
EN 60 672	Series	Ceramic and glass insulating materials
EN 60 865-1	2012	Short-circuit currents – Calculation of effects – Part 1: Definitions and calculation methods
EN 60 870-5-103	1998	Telecontrol equipment and systems – Part 5-103: Transmission protocols – Companion standard for the informative interface of protection equipment
EN 60 889	1997	Hard-drawn aluminium wire for overhead line conductors
EN 61 109	2008	Insulators for overhead lines – Composite suspension and tension insulators for a.c. systems with a nominal voltage greater than 1000 V – Definitions, test methods and acceptance criteria
EN 61 232	1995	Aluminium-clad steel wires for electrical purposes
EN 61 302	1996	Electrical insulating materials – Method to evaluate the resistance to tracking and erosion – Rotating wheel dip test
EN 61 325	1995	Insulators for overhead lines with a nominal voltage above 1000 V – Ceramic or glass insulator units for d.c. systems – Definitions, test methods and acceptance criteria
EN 61 773	1996	Overhead lines – Testing of foundations for structures
EN 61 952	2008	Insulators for overhead lines – Composite line post insulators for a.c. systems with a nominal voltage greater than 1000 V – Definitions, test methods and acceptance criteria
EN 62 217	2013	Polymeric HV insulators for indoor and outdoor use – General definitions, test methods and acceptance criteria
EN 62 271-102	2002	High-voltage switchgear and controlgear – Part 102: Alternating current disconnectors and earthing switches
EN 62 271-103	2011	High-voltage switchgear and controlgear – Part 103: Switches for rated voltages above 1 kV up to and including 52 kV
EN 62 271-200	2012	High-voltage switchgear and controlgear – Part 200: AC metal-enclosed switchgear and controlgear for rated voltages above 1 kV and up to and including 52 kV
EN 62 621	2016	Railway applications – Fixed installations – Electric traction – Specific requirements for composite insulators used for overhead contact line systems

Index